San A................,

(512) 633-9746

-16

CASES AND MATERIALS

THE LAW OF OIL AND GAS

NINTH EDITION

by

PATRICK H. MARTIN
Campanile Professor of Mineral Law, Emeritus
Louisiana State University

BRUCE M. KRAMER
Maddox Professor of Law, Emeritus
Texas Tech University

FOUNDATION PRESS
2011

THOMSON REUTERS™

© 1956, 1964, 1974, 1979, 1987, 1992, 2002 FOUNDATION PRESS
© 2007 THOMSON REUTERS/FOUNDATION PRESS
© 2011 By THOMSON REUTERS/FOUNDATION PRESS
      1 New York Plaza, 34th Floor
      New York, NY 10004
      Phone Toll Free 1–877–888–1330
      Fax 646–424–5201
      foundation–press.com

Printed in the United States of America

ISBN 978–1–60930–050–0

Mat #41180582

Dedications

To Ann, my loving wife, and my seven children, Patrick, Shannon, Michelle, Andrew, Luke, Thomas, and Helen. Without their patience and support, this work could not have been done.

Pat Martin

To Marilynn, Cassie, Marissa, Hannah & Danielle, my wife and daughters who have taught me more than my years of research and whose support and kindness have been immeasurable over the years.

Bruce Kramer

PREFACE TO NINTH EDITION

Although fundamental principles of contract and property law do not change in so short a period as a decade, there are enough variations on existing themes in the new cases to warrant a new edition. This edition builds upon the foundation of the eight prior editions. It retains most of the organizational structure of the prior works, but updates the material to reflect recent developments and trends in oil and gas case law.

The first chapter introduces the student to oil and gas geology and development. It has now incorporated discussion of new exploration, drilling and production techniques, including 3-D seismic, horizontal drilling, coal methane gas production, hydraulic fracturing, the locations of new shale gas plays, and Outer Continental Shelf developments, together with several new diagrams.

Chapter 2 in several earlier editions was on energy policy. The topic of energy policy remains very important to United States law, but the demise of price controls of earlier decades has rendered much of the materials dated, though the ghosts of natural gas regulation haunt the royalty materials of Chapter 3. Chapter 2 provides the foundation for understanding the role that basic property law, concepts and policy play in oil and gas law. It notes the different approaches taken to some fundamental questions about the ownership of oil and gas and how those approaches have led to different results to common questions among the oil and gas jurisdictions. New cases have been substituted or added on subsurface storage of natural gas (p. 30), on subsurface trespass in relation to hydraulic fracturing (p. 75), and on concurrent ownership after exercise of a pooling clause (p. 129). To reflect the current emphasis on the relations of parties involving split estates, the materials from earlier editions on surface easements has been moved to Chapter 2 (p. 159) and updated.

Chapter 3, on oil and gas leases, has been modified with many of the cases contained in the earlier editions being retained. Throughout the chapter emphasis may be found on how courts characterize the property interests created by an oil and gas lease. These characterizations have important ramifications in dealing with such issues as payment of delay rentals, interpretation of leasehold savings provision, the availability of equitable defenses in actions seeking to terminate the lease and the application of the temporary cessation of production doctrine. Recent developments pertaining to lease termination for failure to produce in paying quantities (p. 270), royalty calculation on post-production costs, and class actions, as well as cases focusing on calculation of overriding royalties, are reflected in this edition.

The chapter on implied covenants, Chapter 4, incorporates newer discussion of the implication of covenants and the application of this analysis to oil and gas leases and other contracts, together with recent common lessee developments. The section added to Chapter 4 in the eighth edition to discuss the obligation to restore the surface that is found in some leases and in some cases as an implied covenant has been updated to reflect the most recent developments.

Chapters 5 and 6 on conveyancing and transfers subsequent to a lease have incorporated new cases to reflect developments in the interpretation of the word "minerals", the problems of fractional interests, the Mother Hubbard clause (579), the "two-grant" theory, and an effort to "wash-out" an overriding royalty (p. 765). The chapter on pooling and unitization, Chapter 7, has been updated primarily in the notes following the cases.

The final chapter, on public lands, has been revised. The description of the leasing process after 1987 has been rewritten, two older cases have been replaced by a newer case on environmental aspects of leasing (p. 997) and another on the relationship of state and federal authority (p. 1048).

The book is a bit longer than the ninth edition, and we believe it can still be conveniently used in a one semester, three hour course. The cases and other materials have been pruned and footnotes, where retained, have been renumbered. Editing of opinions has not always been indicated.

ACKNOWLEDGEMENTS

We wish to express our appreciation to the following authors, periodicals and publishers for permission to reproduce material from their publications:

Allen, *Horizontal Drilling—A Key to Enhanced Recovery, The* Interstate Oil and Gas Compact & Committee Bulletin, June, 1988. Reprinted by permission.

John Wheaton and Teresa Donato, *Coalbed-Methane Basics: Powder River Basin, Montana*, Montana Bureau of Mines and Geology, Information Pamphlet 5 (2004), p. 2, from http://www.mbmg.mtech.edu/pdf/ip_5.pdf. (Fig 10 p. 13 *infra*). Reprinted by permission.

Bruce M. Kramer and Patrick H. Martin, *The Law of Pooling and Unitization*. Copyright 2010 by Matthew Bender & Co., Inc., publishers. Excerpts are printed with the permission of the publisher.

Patrick H. Martin & Bruce M. Kramer, *Williams & Meyers Oil and Gas Law*. Copyright 2010 by Matthew Bender & Co., Inc., publishers. Excerpts are printed with the permission of the publisher.

Meyers and Crafton, *The Covenant of Further Exploration—Thirty Years Later*, 32 Rocky Mt. Min. L. Inst. 1—1 (1986). Copyright © 1986 by Matthew Bender & Co., Inc., publishers. Excerpts are printed with the permission of the publisher.

Shell Oil Company, *The Story of Petroleum*. Reprinted by permission.

Other illustrations are from public domain sources.

SUMMARY OF CONTENTS

APPENDICES

TABLE OF CONTENTS

Table of Contents

Table of Contents

Table of Contents

Table of Contents

Table of Contents

APPENDICES

TABLE OF CASES

Principal cases are in ALL CAPS type. Non-principal
cases are in small caps and roman type. References are to Pages.

xxi

Chapter 1

A BRIEF INTRODUCTION TO THE
SCIENTIFIC AND ENGINEERING BACKGROUND

Petroleum is a generic name for certain combustible hydrocarbon compounds found in the earth. The molecular structure of these hydrogen and carbon compounds varies from the very simple structure of *methane* (CH_4), a component of the fuel natural gas, to more complex structures, such as that of *octane* (C_8H_{18}), a component of crude oil. In addition, impurities are often associated with petroleum (the sulphur compound that contaminates *sour gas* and oil is one), and these should be removed prior to marketing the product.

Of the many physical properties of petroleum studied by scientists, three are fundamental for an understanding of oil and gas production. First, petroleum occurs in nature in the *gaseous, liquid* and *solid states,* usually as gas or a liquid. Wherever it occurs as a liquid there is almost always some gas also present in solution. Since gas expands when pressure is reduced, there is energy available for the propulsion of the oil to the surface. How much energy is available depends on the amount of gas present in the reservoir, at least in part. We say "in part" because other sources of energy may also be present in the reservoir, as we shall see in a later paragraph.

Another important property of petroleum is its *specific gravity* or *density.* In the case of solids and liquids, specific gravity expresses the ratio between the weights of equal volumes of water and another substance measured at a standard temperature. The weight of water is assigned a value of 1. Liquid petroleum normally being lighter than water, its specific gravity is a fraction. For example, the specific gravity of octane is .7064. In the oil industry, however, the specific gravity of oil is commonly expressed in *A.P.I. degrees.* On this scale the ratio is inverted so that oil with the least specific gravity has the highest A.P.I. gravity. Most crude oils range from 27° to 35° A.P.I. gravity. Other things being equal, the higher the A.P.I. gravity the better the price for the oil.

The third property to be noticed is *viscosity,* which is an inverse measure of the ability of a fluid to flow. The less viscous the fluid the greater its mobility. There is a relationship between specific gravity and viscosity, for usually the less dense a petroleum compound is the less viscous it is. The viscosity of oil in a reservoir is also affected by the amount of gas present in solution, for gas is the less viscous of the two fluids. Production methods which permit gas to escape from solution before the oil has reached the well bore decrease ultimate recovery from the reservoir by increasing the viscosity of the oil as well as by dissipating the reservoir energy.

To understand how petroleum is found and produced, we need to know something about petroleum geology. This is a big subject and what follows is only rudimentary. All rocks are divided into three basic classifications: *igneous* (granite is an example), *metamorphic* (slate and marble are examples), and *sedimentary,* of which three kinds are especially important in petroleum geology: sandstone, limestone and shale. The crust of the earth. is composed of layers of these rocks overlain in some places with a thin coating of top soil, and any single layer (or *stratum*) will normally contain only one kind of rock. These strata having been deposited at different periods of time, the deepest layer will ordinarily be the oldest. Igneous and metamorphic rocks are also called *basement rock,* since, being older, they ordinarily occur beneath sedimentary deposits. Nearly all commercial oil and gas production is from some form of sedimentary rock. This is accounted for in one theory by the absence of source material for the manufacture of petroleum prior to the time of sedimentary deposits. According to this theory, oil and gas was formed from animal and vegetable life in the sea, and it was the sea that deposited sedimentary strata.

Under another theory some natural gas is of non-biological origin. Proponents believe that the interior of the earth is rich in carbon and that deep-earth carbon is held in various hydrocarbons beneath the crust, in the mantle, under extraordinarily high pressures and temperatures. If this theory is correct, many deep targets and areas that have been written off for lack of biologically rich sediments may prove richly productive through wells drilled to deeper horizons.

Regardless of the correct theory of the origin of petroleum, there is another reason for its presence being confined to sedimentary rocks. Unlike igneous and metamorphic rocks, many sandstones and limestones and some shales possess two physical properties necessary for the accumulation of petroleum in commercial quantities, viz., *porosity* and *permeability.* Porosity is a measure of the void spaces in a material, the volume within rock that can contain fluids. Permeability of rock is its capacity for transmitting a fluid. It is not enough that reservoir rock be capable of holding petroleum; it must also allow the petroleum to move through it. Usually porosity and permeability conjoin, but this is not invariably true. In summary, a commercial oil deposit requires the presence of a porous, permeable rock formation containing oil of marketable A.P.I. gravity and of producible viscosity.

It is the business of petroleum geologists, aided by geophysicists and other scientists, to search for these deposits. At present, however, there is no way of finding oil and gas short of drilling wells. What the geologists look for are *reservoir traps* -- underground formations favorable to the accumulation of oil and gas. A reservoir trap is better illustrated than defined, but it can be generally described as a tilted layer of sedimentary rock overlain by an impervious substance and folded, broken, or otherwise so formed as to stop the natural migration of the petroleum upward. Some of the important types of traps are shown in the drawings that follow.

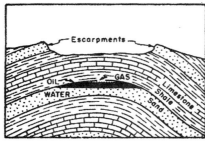

Fig. 1.—An anticline containing an oil and gas reservoir.[2]

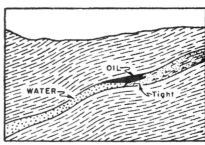

Fig. 2.—A monocline. The upper boundary of the oil trap is non-porous impermeable rock. (After LeRoy)

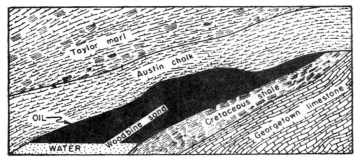

Fig. 3.—East-west cross section of East Texas reservoir.

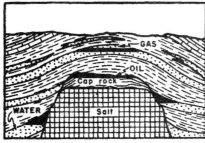

Fig. 4.—Reservoir in a piercement type salt dome.

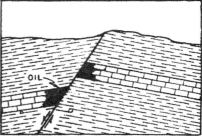

Fig. 5.—Fault traps. (After LeRoy)
[A8854]

[Figures 1–5 and 7–8 in this chapter are reprinted from "Oil and Gas Production," compiled by the Interstate Oil Compact Commission. Copyright by University of Oklahoma Press. Used by Permission.]

A few observations about these drawings may be useful before we turn to petroleum exploration methods. Geologists recognize two fundamental types of reservoir traps, *structural* and *stratigraphic,* and a third type which combines characteristics of the other two. Figures 1 and 5 represent simple structural traps. In Figure 1 the trap could have resulted from a gentle uplift which produced an arch in a stratum that was probably level when laid down. In Figure 5 the trap is caused by faulting, which is the cracking and breaking of a rock plane. These breaks may be small and the displacement slight or they may extend for many miles and involve hundreds of feet of displacement. A variety of causes accounts for faulting, from violent upheavals that thrust great masses of rock upward to the gradual and gentle work of gravity. The effect of the latter is found along the Gulf Coast, where the weight of the sea and the silt washed into it caused the rock planes of the region to *dip* (*i.e.,* slope) to the south and then in places to crack and break. Figure 5 shows how a limestone monocline broke and slipped down, leaving the limestone to face onto impervious shale, thus forming the trap. A stratigraphic trap results from a change in the character of reservoir rock (*e.g.,* from permeable to impermeable) or from a break in its continuity. A simple stratigraphic trap is not illustrated above, but one example is a lens of porous, permeable sandstone surrounded by impermeable shale. Oil and gas deposits have been found in such traps despite the absence of any structure. More common, however, are the combination traps like those in Figures 2 and 3. In Figure 2 the upper part of a monoclinal structure "tightens up," that is, loses its porosity and permeability. The incline is the structural feature and the tightening up is the stratigraphic feature of this combination trap. Figure 3 shows another combination trap in which the sandstone "pinches out" between two planes of impervious rock. The East Texas pool (shown there in cross-section) is the largest in the lower 48 states with a capacity of about five billion barrels of recoverable oil. Salt domes (Fig. 4) are also frequently associated with petroleum reservoirs, for the intrusion of the salt plug forms traps of all three types.

Oil and gas exploration is the search for reservoir traps. To explain how that search is carried on perhaps we may best resort to an imaginary situation. The management of the Flush Production Company has allocated $25 million for new exploration during the year. It is the job of the Exploration Department to turn up prospects worth spending this money on. From study of generally available geological data -- regional surveys, surface geology reports, *etc.* -- the department has become interested in an area that lies over a buried inland sea. Geologists know that these great basins, like the Williston Basin of the Northern Great Plains, frequently contain valuable oil and gas deposits. But the area over the whole basin is too large for the company to explore in full -- the Williston Basin, for example, covers parts of three states and Canada -- so the geologists set about to reduce the area of interest to a size that can be accommodated in the company's exploration budget. One way to do this, and a very common exploration method in *wildcat territory* (that is, unproven and largely unexplored territory) is the *geophysical survey.* This survey might employ several devices, *gravity-meter* or *magnetometer* for instance, but *seismograph* is selected as the instrument best suited to the territory.

The "shooting" of an area by seismograph is intended to furnish the company's exploration department with a contour map of the subsurface. A seismograph records a shock wave, set off in the ground by an explosion, as the wave is reflected and refracted by the various layers of rock encountered. With the geological and geophysical information now available, the geologists are in a position to recommend what land to lease and where to locate the exploratory (*wildcat*) well. The Flush Oil Co. land department sends out its landmen to secure oil and gas leases from the landowners. On an operation of this kind, the operator usually seeks a large block of leases before drilling. The title to the test area is then examined by the legal department, and if any title defect is observed, it would be cured before a well is drilled. The site selected for the initial test well appears on the geophysical contour map as the high point on a faulted anticline, but when the well is drilled it comes in a dryhole. The search for oil does not end here.

Data from the well can be immensely valuable in adding to the geologists' knowledge of the subsurface. *Core tests* from the well indicate the strata penetrated and the depth at which each was encountered. The *electrical well log* (sometimes called by a trade-name, Schlumberger) gives about the same information and becomes especially useful when logs from other wells are available for correlation. With the new information from the dry hole supplementing the old, a second test well is located and drilled. So the process of drilling, coring, and electrical logging goes on until the budget for the project is exhausted or petroleum is found. If luck is with the company and a discovery is made, the remaining exploration problem is to establish the boundaries of the field.

Of course there are almost an infinite number of variations on this story. Certainly not all wildcatting is done by the major companies; in fact a significant share of it is done by the independents, whose exploration department may amount to one geologist and whose whole force may easily come to less than ten persons, excluding the drilling crew, who will usually be the employees of an independent drilling contractor. Nor does everyone in search of petroleum have the good fortune to start out with $25 million. Many test wells are one-shot propositions, on limited acreage, financed by a syndicate organized for the purpose and disbanded if the well is dry. Also in contrast to the description above, land is often leased before any geological study has been made of it, and wells are sometimes drilled that way too, although the days of divining rod and divine inspiration are about gone. With exploratory wells often costing several million dollars and with only one wildcat out of nine a producer, more substantial guidance is usually sought. It is also true that a great deal of exploratory drilling is done, not by the original lessee, but by an assignee under a deal known as a "farm-out." We will study the legal aspects of these agreements in Chapter 6.

Suffice it now to know that many operators, especially the major companies, have more land under lease than they can explore. The farm-out is an arrangement by which another operator drills the test well in return for a percentage interest in the lease. Other financial arrangements include the *dry hole* and *bottom hole letter,* by which a person with leases in the area contributes towards the cost of a test well on nearby land in return for access to information from the well. Many other variations could be mentioned, but the fundamental elements in the description of petroleum exploration given here are repeated every day: leasing the land, careful geological study of it, making a location for a test well, clearing the legal title to the land, and drilling the well.

Two methods of oil well drilling have been used. *Cable tool drilling,* the older method, operates on a hammer principle. A heavy, sharp-pointed bit is raised and dropped continuously so that it chips and pounds a hole in the rock. The bottom of the hole is kept full of mud and water, and the dropping motion is so regulated that at the moment the bit hits bottom it starts up again, adding a vacuum effect to the pounding. The pulverized stone left in the hole, called *cuttings,* is removed by a bucket-like device called a *bailer,* which is lowered into the hole periodically. The fact that all the contents of the hole can be examined at will is one of the advantages of the cable tool rig.

In wider use, however, is the *rotary drilling* rig. (Fig. 6). As its name implies, rotary drilling operates on the principle of boring a hole by continuous turning of a *bit.* The bit is the key tool; the rest of the rig is designed to make it effective. While bits vary in design and purpose, one common type consists of a housing and three interlocking, movable wheels with sharp teeth, looking something like a cluster of gears. The bit, which is hollow and very heavy, is attached to the *drill stem,* composed of hollow lengths of pipe leading to the surface. As the hole gets deeper, more lengths of pipe can be added at the top. Almost as important as the bit is the *drilling fluid.* Although known in the industry as *mud,* it is actually a prepared chemical compound, sold at a price belying its humble name. Mud expense will run to $10-35,000 on relatively shallow wells (*e.g.,* 4500 feet) and $50-200,000 on wells 10,000 feet in depth. Mud costs on offshore wells are much higher. The drilling mud circulates continuously down the drill pipe, through the bit, into the hole and upwards between the hole and the pipe to a surface pit, where it is purified and begins the trip all over again. The flow of mud removes the cuttings from the hole without removal of the bit (unlike cable tool drilling), lubricates and cools the bit in the hole, and prevents blowouts in high pressure strata.

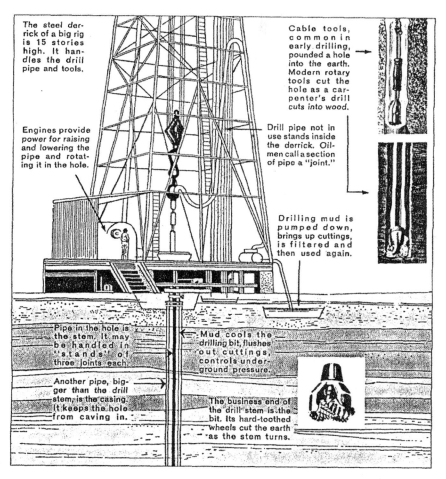

The steel derrick of a big rig is 15 stories high. It handles the drill pipe and tools.

Cable tools, common in early drilling, pounded a hole into the earth. Modern rotary tools cut the hole as a carpenter's drill cuts into wood.

Engines provide power for raising and lowering the pipe and rotating it in the hole.

Drill pipe not in use stands inside the derrick. Oilmen call a section of pipe a "joint."

Drilling mud is pumped down, brings up cuttings, is filtered and then used again.

Pipe in the hole is the stem. It may be handled in "stands" of three joints each.

Mud cools the drilling bit, flushes out cuttings, controls underground pressure.

Another pipe, bigger than the drill stem, is the casing. It keeps the hole from caving in.

The business end of the drill stem is the bit. Its hard-toothed wheels cut the earth as the stem turns.

[**Fig. 6**. Oil and gas drilling rig. (Shell Oil Company, *The Story of Petroleum*, p. 13. Reprinted by permission).]

We have noticed already that during the drilling it is customary to take cores -- samples of the strata which the bit passes through -- and electrical logs. Another standard procedure is the *drill stem test,* which is run whenever the well reaches potential oil or gas bearing formations. A special tool is lowered in the hole and placed next to its wall; the drilling mud is removed from the vicinity; and the fluids of the formation are allowed to flow into the tool while an instrument measures the pressure. If no petroleum flows from the formation, the well will be drilled to the next potential producing stratum for a similar test, or it will be abandoned as a dry hole and *plugged* to prevent harm to shallow fresh water formations. If the test shows the presence of petroleum but deeper production is also anticipated, the higher formation may be sealed off and drilling continued. With discovery of more petroleum in the lower formation, the well can be put into production by *dual completion,* which permits it to produce from two formations at once.

Most wells, however, are single completion wells. After a successful drill stem test, a new string of pipe, called *casing,* is run into the hole. The casing is then *set* by forcing oil well cement between it and the wall of the hole. After the cement hardens, the well is *perforated* by a special device that blows holes through the casing and cement, allowing the oil and gas to enter the well bore. Sometimes it is necessary to increase the flow of petroleum into the hole by *acidizing* or *hydraulic fracturing* the formation. Each process seeks to increase the permeability of the reservoir rock, the former by chemical treatment and the latter by pressure which cracks the rock. The surface activities for completing an oil well consist of installing the *Christmas tree* (a complex set of gauges and valves controlling the flow of oil and gas from a well-head) and *stock tanks* and connecting the well with a pipeline, if one exists in the area. A variety of special equipment may also be installed to treat the oil prior to its delivery to the pipeline. If there is no pipeline the oil is shipped by truck or railroad to its destination, usually a refinery. A natural gas well requires no measuring tanks but must have a pipeline connection for the gas to reach the market and a meter to measure gas produced. In some cases, where no market is available, the liquid components are removed from the gas, which may then be returned to the formation. This process ranges from the inexpensive but wasteful *stripping plant* process to the multi-million dollar, efficient *recycling* operation. At one time, gas was flared after the gasoline was removed, but this enormously wasteful practice has been greatly reduced by state regulation.

A consideration of the mechanics of oil and gas production closes this brief discussion of the scientific background of oil and gas law.

Three fluids may be found singly or in combination in a reservoir trap: oil, gas and water, usually salt water (Fig. 7). If each is present in its natural state, the water will be at the bottom, the oil next, and free gas on top. (You will recall that water has the greatest density, oil next, and gas the least.) The lines separating these fluids (called *oil-water* and *gas-oil* contact lines) are not sharply defined; at the gas-oil contact line, for example, there is likely to be a zone of very high A.P.I. gravity oil heavily saturated with gas. Also present in the typical reservoir will be *connate water,* a thin film of water around each grain of the stone, but very little of this is produced by the well. Free gas does not always occur in a reservoir, but some gas is almost always present in solution in the oil, most of which becomes free gas when the oil reaches the reduced pressure of the surface. Such gas, known as *casinghead gas,* was customarily flared in the 1930s, but now it is common (though by no means uniform) practice to remove its liquid components and sell it.

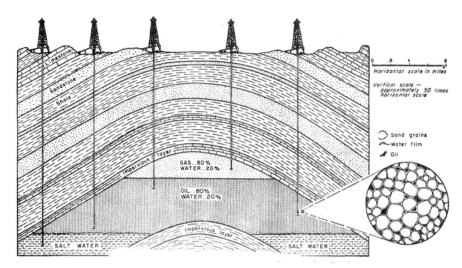

[**Fig. 7**. Distribution of fluids within a reservoir. At right is a magnified view of sand grains, with film of water around each grain and oil in remainder of pore space.]

Both natural and artificial means are used to produce oil. During primary production natural energy propels the petroleum to the well bore, where artificial energy can then be used to lift it to the surface, if necessary. The natural sources of reservoir energy are: (1) gas expansion, (2) water encroachment, and (3) gravity. One of these forces is always present in a commercial oil field, and often a combination of all three. Gas expansion reservoirs are the most common. A reduction of pressure from opening the well allows the gas to expand, forcing the oil to the well bore and lifting it to the surface. If some of the gas is free, the field is known as a *gas-cap* field; if not, as a *solution-gas* field. In either event, maximum ultimate recovery depends on conserving the gas pressure. Hence, it is as improper to produce gas from the gas cap as it is to produce oil from wells with high *gas-oil ratios*. In many states the gas-oil ratio is carefully regulated, and inefficient wells dissipating the reservoir pressure are shut in or put on limited production.

A *water-drive* field derives its energy mainly from edge or bottom-water in the formation, though gas expansion may give an assist. Water is only slightly compressible but when tremendous volumes of it are present, as is frequently true in reservoir traps, the effect of the slight compression is greatly magnified. With a reduction of pressure, the water expands, pushing the oil ahead of it. Recovery of a very high percentage of the oil in place can be achieved in water-drive fields because the water has the effect of flushing out the recalcitrant oil and washing it free toward the well bore, but such recoveries depend on use of proper production methods. Water pressure should be maintained and the water table should rise uniformly. Accordingly, wells with high *water-oil ratios* should not be allowed to produce. Nor should the rate of oil production be so high that channels form between the water table and the well bore, bypassing oil in the less permeable parts of the formation. Fig. 8 illustrates the contrast in recoveries between good and bad production methods in a combination gas-cap, water-drive field.

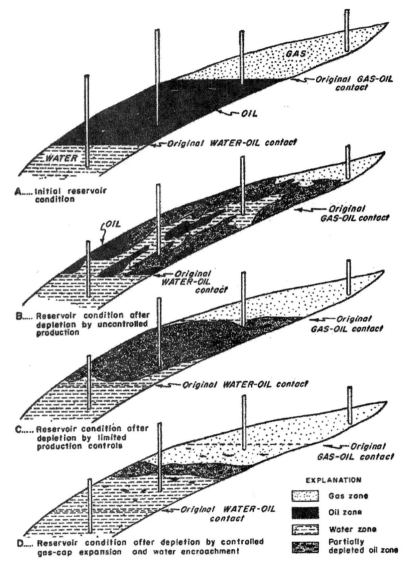

[Fig. 8. Sketches that show how oil recoveries increase in a water-drive reservoir if production rates are controlled.]

Rate of production, gas-oil and water-oil ratios are the primary factors affecting recoveries. Of less importance is *well spacing*. The dense drilling of the '20s and early '30s was wasteful certainly, but not so much in harm done to reservoirs as in the expense of useless wells. It is now recognized that one oil well can efficiently drain twenty to eighty acres; a gas well may efficiently drain as much as eight hundred acres (depending on the reservoir); most states have some sort of regulation requiring uniform spacing of wells within specified limits. Irregular spacing replaced overcrowding as a spacing problem. Some states have granted liberal exceptions to the uniform drilling pattern. Even this causes no serious problem by itself, but such exceptions have been usually accompanied by a disproportionately large production allowable for the small tract. For example, in a field drilled on a 40–acre pattern, the wells drilled on 5–acre sites as exceptions may receive ninety percent of the allowable for a standard site. Not only is this unfair to the larger site owners, it can result in such damage to the reservoir as the channelling described above. The solution to the problem is integration of small tracts into drilling units of the proper size, a process called *pooling*. Legal aspects of well spacing and pooling are considered in Chapter 7.

New drilling and production techniques may be used to increase recovery of oil or gas. The late 1980s saw a substantial increase in the use of *horizontal drilling*. (*See* Moore, *Horizontal Drilling—New Technology Bringing New Legal and Regulatory Challenges*, 36 Rocky Mt. Min. L. Inst. 15–1 (1990)). Petroleum engineers have for many years employed directional drilling (or *whipstocking*) successfully to control the bottom hole location of a well (such as beneath a river bed) so that the surface location can be at a more favorable spot (such as the river's bank). Horizontal drilling goes well beyond directional drilling by significantly increasing the amount of reservoir open to the well bore. Three major methods of horizontal drilling are shown in Figure 9.

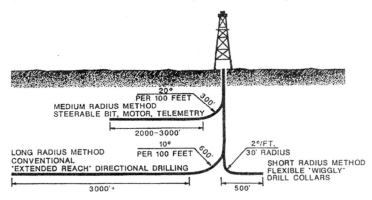

[**Fig. 9**. Horizontal drilling – 3 Major Methods (Source: Allen, "Horizontal Drilling – A Key to Enhanced Recovery," *The Interstate Oil and Gas Compact & Committee Bulletin*, June 1988. Reprinted by Permission).]

Production of methane gas from coalbeds (*coalbed methane*) is another promising technology that is being used to increase the production of gas in the United States. This has developed since the late 1970s in six regions—the San Juan Basin (New Mexico and Colorado), the Black Warrior Basin (Alabama), the Northern and Central Appalachian basins, the Piceance Basin (Colorado), and the Powder River Basin (Wyoming). A process called *hydrofracturing* is used. After a well is drilled into a coal seam, a water and sand fluid mixture is injected that causes fractures to open in the seam. This increases the available gas flow by as much as five-fold. Such gas production could be as much as 11% of the total U.S. gas in the near future. (Levy and Rosenlieb, *Producing Natural Gas From Coal Seams*, 127 Pub. Util. Fort. 53 (June 15, 1991); Cohen, *Leasing of Coalbed Methane Gas Rights—Are Oil and Gas Lease Clauses Analogous?*, 15 Cum.L.Rev. 703 (1985)). *See also* the papers from the *Wyoming Law Review* 2004 Coalbed Methane Symposium, 4 Wyo. L. Rev. 541 *et seq.* An indication of the rapid development of this resource is a recent report that from 1997 to 2007, the number of producing CBM wells in the Powder River Basin went from under 500 to more than 22,000.

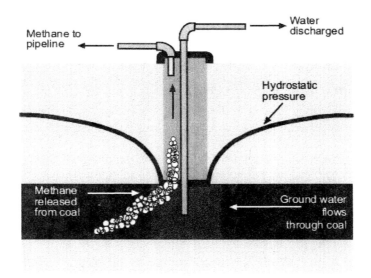

[Fig.10. Coalbed Methane Production]
Figure reproduced with permission from the Montana
Bureau of Mines and Geology (Wheaton and Donato, 2004)

When primary production of oil declines or ceases from loss of reservoir energy, it may often be restored by artificial reservoir *repressuring operations* or by other enhanced recovery methods. These production techniques may be classified into several categories: (1) *pressure maintenance* involves the injection of a fluid into a reservoir just beginning to show production and pressure decline. Its object is to maintain primary production by keeping pressure up. An excellent field example is the salt-water injection program in the East Texas oil field, which is a water-drive reservoir. A serious decline in pressure was halted in 1942 by injecting, in the lower part of the stratum, the salt water produced by the wells near the oil-water contact line. There was also disposed of thereby the salt water, which posed a serious pollution problem on the surface. Gas injection operations are also common, especially in gas-cap fields. (2) The term *secondary recovery* is often applied to worn-out fields, where the pressure is about gone and the wells are on the pump. A usual method employed in these fields is water flooding. A five-spot water-flood program works like this. Four wells, one in each corner of a square, are designated input wells. A well in the center of the square is utilized as a producing well. Water is then circulated through the input wells into the reservoir, washing the remaining oil before it toward the production well. Other patterns may be used.

Some further production methods are referred to as (3) *tertiary recovery* or, more simply, *enhanced recovery.* Thermal methods of enhanced recovery introduce heat into a reservoir to improve recovery, including cyclic steam injection (huff-and-puff), steam flooding, and in-situ combustion. Miscible fluid injection involves pumping a gas or liquid substance into a reservoir that has the effect of dissolving the oil so that recovery will be increased. Carbon dioxide and hydrocarbon material are often used. Injection of an immiscible fluid simply pushes oil before it, sweeping the oil to a production well. Chemical flooding may use miscible or immiscible fluids to improve sweep efficiency of a water flood and enhance recovery of oil.

The producing states and the United States give special tax or other incentives to encourage use of these special methods of increasing oil production. Most pressure maintenance, secondary recovery, cycling and recycling, and other enhanced recovery operations require the cooperation of the operators and the landowners in the field for financial and engineering reasons, so that operations can be conducted without regard for property lines. New techniques of drilling and producing oil and gas wells present important and often novel legal questions. Administrative agencies may have to adapt spacing and allowable regulations to the new methods in order to promote maximum efficient recovery and to protect correlative rights. Questions may be raised by lessors whether their lessees have a right to undertake new production techniques or whether they have a duty to employ such production enhancing methods. Adjacent owners of property rights may assert claims of trespass or injury from subsurface operations. The special legal problems that arise from some of the enhanced recovery methods are reflected in a number of issues addressed in the readings in this book. The complexity and novelty of these issues contribute to the great interest that this fascinating business and field of law holds for so many people.

Several technical developments have stimulated oil and gas drilling and development activity in the United States in the last few years. Despite the fact that the United States is a high-cost area in which to produce oil and natural gas, the technical advances have allowed American producers to get more production for their dollars than would have otherwise been possible. Horizontal drilling has already been described in this casebook, and its use continues to increase. The other technological contributions are from three dimensional seismic imaging and the use of tension leg platforms for deep offshore production.

Three dimensional seismology or "*3-D imaging*" uses sound waves that are projected underground. The waves are bounced off structures thousands of feet below the surface and are correlated across grids. The technology allows geologists to see for the first time through layers of sediment that cover many unexplored or partially explored oil and gas fields. While the use of the technology is expensive, it allows much more accurate depiction of subsurface structures and greatly increases the success rate for drilling of wells. The success ratio for wildcat wells using the technology has jumped from 20 percent to 50 percent, and in some fields development drilling is successful as much as 100% of the time. A consequence for leasing activity is that companies interested in drilling are probably leasing in more precisely defined blocks. Another geological tool that may prove useful is *four-dimensional seismic*. This is the name given to time-lapse 3-D seismic. The technique employs comparison of a series of 3-D seismic surveys of the same producing area over time. Geologists or geophysicists can then detect changes in the flow of hydrocarbons and other materials in the reservoir and thereby seek to predict future production.

Above we mentioned hydrofracturing (or "fracking") in connection with coalbed methane production. Fracking has also made possible recovery of natural gas (and oil) from shales with low porosity and permeability. This "fracking" process of creating fissures in tight underground formations allows natural gas and oil to flow. The two diagrams below depict the fracking technique and the geographic areas in which shale gas is being developed.

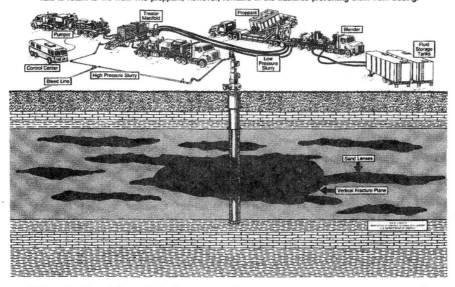

[Fig. 11. Fracking shale for natural gas – Source – NETL, public domain**].**

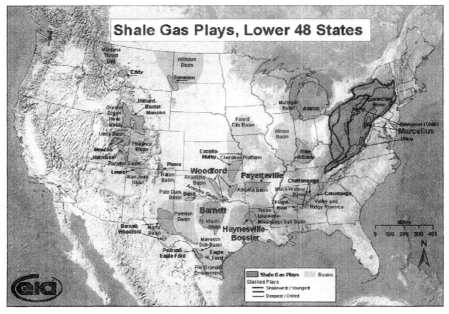

[Fig. 12. Shale gas plays]

The development of natural gas in these shale plays has dramatically increased assessments of the long-term supplies available for domestic consumption.

New drilling techniques have made possible the drilling of wells in deep waters. Only a few years ago, most wells offshore were at depths of several hundred feet. Deepwater drilling and production (in depths greater that 1000 feet) has increased dramatically in the last twenty years; about 4,000 wells have been drilled in Gulf water depths in excess of 1,000 feet and 700 wells in 5,000 feet depth or greater. By 2010, 80 percent of offshore oil production and 45 percent of natural gas production occurred in water depths in excess of 1,000 feet. As of April 1, 2011 the Gulf of Mexico has 6,323 active leases, 64 percent of which are in deepwater. The approximately 43 million leased OCS acres account for about 15 percent of America's domestic natural gas production and about 27 percent of America's domestic oil production. Since 2000, there have been dozens of industry-announced discoveries in water depths greater than 7,000 feet. The diagram below will illustrate the state-of-the-art technology used in deepwater oil and gas activity.

[Fig.13. Offshore Drilling and Production]

Left to Right - 1, 2) conventional fixed platforms; 3) compliant tower; 4, 5) vertically moored tension leg and mini-tension leg platform; 6) Spar ; 7, 8) Semi-submersibles ; 9) Floating production, storage, and offloading facility; 10) sub-sea completion and tie-back .

Development of oil and gas on the OCS suffered a setback with the April 20, 2010 blowout of the Macondo well drilled by the Deepwater Horizon. Eleven lives were lost and as much as 4.9 million barrels of oil were released into the Gulf of Mexico.

Chapter 2

THE NATURE AND PROTECTION OF INTERESTS
IN OIL AND GAS

SECTION 1. NATURE OF OWNERSHIP IN OIL AND GAS

(See Martin & Kramer, *Williams and Meyers Oil and Gas Law* §§ 203-204.9)

A. THE RULE OF CAPTURE

ELLIFF v. TEXON DRILLING CO.

Supreme Court of Texas, 1948.
146 Tex. 575, 210 S.W.2d 558, 4 A.L.R.2d 191.

FOLLEY, JUSTICE. This is a suit by the petitioners, Mrs. Mabel Elliff, Frank Elliff, and Charles C. Elliff, against the respondents, Texon Drilling Company, a Texas corporation, Texon Royalty Company, a Texas corporation, Texon Royalty Company, a Delaware corporation, and John L. Sullivan, for damages resulting from a "blow-out" gas well drilled by respondents in the Agua Dulce Field in Nueces County.

The petitioners owned the surface and certain royalty interests in 3054.9 acres of land in Nueces County, upon which there was a producing well known as Elliff No. 1. They owned all the mineral estate underlying the west 1500 acres of the tract, and an undivided one-half interest in the mineral estate underlying the east 1554.9 acres. Both tracts were subject to oil and gas leases, and therefore their royalty interest in the west 1500 acres was one-eighth of the oil or gas, and in the east 1554.9 acres was one-sixteenth of the oil and gas.

It was alleged that these lands overlaid approximately fifty per cent of a huge reservoir of gas and distillate and that the remainder of the reservoir was under the lands owned by Mrs. Clara Driscoll, adjoining the lands of petitioners on the east. Prior to November 1936, respondents were engaged in the drilling of Driscoll–Sevier No. 2 as an offset well at a location 466 feet east of petitioners' east line. On the date stated, when respondents had reached a depth of approximately 6838 feet, the well blew out, caught fire and cratered. Attempts to control it were unsuccessful, and huge quantities of gas, distillate and some oil were blown into the air, dissipating large quantities from the reservoir into which the offset well was drilled. When the Driscoll–Sevier No. 2 well blew out, the fissure or opening in the ground around the well gradually increased until it enveloped and destroyed Elliff No. 1. The latter well also blew out, cratered, caught fire and burned for several years. Two water wells on petitioners' land became involved in the cratering and each of them blew out. Certain damages also resulted to the surface of petitioners' lands and to their cattle thereon. The cratering process and the eruption continued until large quantities of gas and distillate were drained from under petitioners' land and escaped into the air, all of which was alleged to be the direct and proximate result of the negligence of respondents in permitting their well to blow out. The extent of the emissions from the Driscoll–Sevier No. 2 and Elliff No. 1, and the two water wells on petitioners' lands, was shown at various times during the several years between the blowout in November 1936, and the time of the trial in June 1946. There was also expert testimony from petroleum engineers showing the extent of the losses from the underground reservoir, which computations extended from the date of the blowout only up to June 1938. It was indicated that it was not feasible to calculate the losses subsequent thereto, although lesser emissions of gas continued even up to the time of the trial. All the evidence with reference to the damages included all losses from the reservoir beneath petitioners' land without regard to whether they were wasted and dissipated from above the Driscoll land or from petitioners' land.

The jury found that respondents were negligent in failing to use drilling mud of sufficient weight in drilling their well, and that such negligence was the proximate cause of the well blowing out. It also found that petitioners had suffered $4620 damage to sixty acres of the surface, and $1350 for the loss of 27 head of cattle. The damages for the gas and distillate wasted "from and under" the lands of petitioners, due to respondents' negligence, was fixed by the jury at $78,580.46 for the gas, and $69,967.73 for the distillate. These figures were based upon the respective fractional royalty interest of petitioners in the whole amount wasted under their two tracts of land, and at a value, fixed by the court without objection by the parties, of two cents per 1000 cubic feet for the gas and $1.25 per barrel for the distillate.

The findings as to the amount of drainage of gas and distillate from beneath petitioners' lands were based primarily upon the testimony of petitioners' expert witness, C.J. Jennings, a petroleum engineer. He obtained his information from drilling records and electric logs from the high pressure Agua Dulce Field. He was thereby enabled to fairly estimate the amount of gas and distillate. He had definite information as to porosity and bottom-hole pressure both before and after the blowout. He was able to estimate the amount of gas wasted under the Elliff tract by calculating the volume of the strata of sands and the voids which were occupied by gas. Under his method of calculation the determining factor was the decrease in bottom-hole pressures of the sands caused by the blowout. He estimated that 13,096,717,000 cubic feet of gas had been drained from the west 1500 acres of the Elliff land, and that 57,625,728,000 cubic feet had been drained from the east 1554.9 acres as a result of the blowout. The distillate loss was calculated by taking the gas and distillate ratio from the records of the Railroad Commission. Jennings estimated that 195,713 barrels had been drained from the west 1500 acres and 802,690 barrels from the east 1554.9 acres, as a result of the blowout.

On the findings of the jury the trial court rendered judgment for petitioners for

$154,518.19, which included $148,548.19 for the gas and distillate, and $5970 for damages to the land and cattle. The Court of Civil Appeals reversed the judgment and remanded the cause. 210 S.W.2d 553.

The reversal by the Court of Civil Appeals rests upon two grounds. The first was that since substantially all of the gas and distillate which was drained from under petitioners' lands was lost through respondents' blowout well, petitioners could not recover because under the law of capture they had lost all property rights in the gas or distillate which had migrated from their lands. The second theory was that the recovery cannot stand because the trial court had submitted the wrong measure of damages in that petitioners' claim "is for trespass in and to a freehold estate in land and the proper measure of damage is the reasonable cash market value before and after the occurrence complained of."

In our opinion the Court of Civil Appeals was without authority to pass upon the propriety of the measure of damages adopted by the trial court for the simple reason that no such assignment was presented to that court. Although such an objection was raised in the trial court, we do not find an intimation of it brought forward to the Court of Civil Appeals. The question is therefore not before us, and our subsequent conclusions as to the rights of the parties are without reference to the correctness of the measure of damages, and we express no opinion on that question.

Consequently, our attention will be confined to the sole question as to whether the law of capture absolves respondents of any liability for the negligent waste or destruction of petitioners' gas and distillate, though substantially all of such waste or destruction occurred after the minerals had been drained from beneath petitioners' lands.

We do not regard as authoritative the three decisions by the Supreme Court of Louisiana to the effect that an adjoining owner is without right of action for gas wasted from the common pool by his neighbor, because in that state only qualified ownership of oil and gas is recognized, no absolute ownership of minerals in place exists, and the unqualified rule is that under the law of capture the minerals belong exclusively to the one that produces them. . . . [Those] decisions rested in part on the theory that "the loss complained of was, manifestly, more a matter of uncertainty and speculation than of fact or estimate." In the more recent trend of the decisions of our state, with the growth and development of scientific knowledge of oil and gas, it is now recognized "that when an oil field has been fairly tested and developed, experts can determine approximately the amount of oil and gas in place in a common pool, and can also equitably determine the amount of oil and gas recoverable by the owner of each tract of land under certain operating conditions." Brown v. Humble Oil & Refining Co., 126 Tex. 296, 83 S.W.2d 935, 940, 87 S.W.2d 1069, 99 A.L.R. 1107, 101 A.L.R. 1393.

In Texas, and in other jurisdictions, a different rule exists as to ownership. In our state the landowner is regarded as having absolute title in severalty to the oil and gas in place beneath his land. The only qualification of that rule of ownership is that it must be considered in connection with the law of capture and is subject to police regulations. Brown v. Humble Oil & Refining Co., supra. The oil and gas beneath the soil are considered a part of the realty. Each owner of land owns separately, distinctly and exclusively all the oil and gas under his land and is accorded the usual remedies against trespassers who appropriate the minerals or destroy their market value.

The conflict in the decisions of the various states with reference to the character of ownership is traceable to some extent to the divergent views entertained by the courts, particularly in the earlier cases, as to the nature and migratory character of oil and gas in the soil. 31A Tex.Jur. 24, Sec. 5. In the absence of common law precedent, and owing to the lack of scientific information as to the movement of these minerals, some of the courts have sought by analogy to compare oil and gas to other types of property such as wild animals, birds, subterranean waters and other migratory things, with reference to which the common law had established rules denying any character of ownership prior to capture. However, as was said by Professor A.W. Walker, Jr., of the School of Law of the University of Texas: "There is no oil or gas producing state today which follows the wild-animal analogy to its logical conclusion that the landowner has no property interest in the oil and gas in place." 16 T.L.R. 370, 371. In the light of modern scientific knowledge these early analogies have been disproven, and courts generally have come to recognize that oil and gas, as commonly found in underground reservoirs, are securely entrapped in a static condition in the original pool, and, ordinarily, so remain until disturbed by penetrations from the surface. It is further established, nevertheless, that these minerals will migrate across property lines towards any low pressure area created by production from the common pool. This migratory character of oil and gas has given rise to the so-called rule or law of capture. That rule simply is that the owner of a tract of land acquires title to the oil or gas which he produces from wells on his land, though part of the oil or gas may have migrated from adjoining lands. He may thus appropriate the oil and gas that have flowed from adjacent lands without the consent of the owner of those lands, and without incurring liability to him for drainage. The non-liability is based upon the theory that after the drainage the title or property interest of the former owner is gone. This rule, at first blush, would seem to conflict with the view of absolute ownership of the minerals in place, but it was otherwise decided in the early case of Stephens County v. Mid–Kansas Oil & Gas Co., 1923, 113 Tex. 160, 254 S.W. 290, 29 A.L.R. 566. Mr. Justice Greenwood there stated, 113 Tex. 167, 254 S.W. 292, 29 A.L.R. 566:

> The objection lacks substantial foundation that gas or oil in a certain tract of land cannot be owned in place, because subject to appropriation, without the consent of the owner of the tract, through drainage from wells on adjacent lands. If the owners of adjacent lands have the right to appropriate, without liability, the gas and oil underlying their neighbor's land, then their neighbor has the correlative right to appropriate, through like methods of drainage, the gas and oil underlying the tracts adjacent to his own.

Thus it is seen that, notwithstanding the fact that oil and gas beneath the surface are subject both to capture and administrative regulation, the fundamental rule of absolute ownership of the minerals in place is not affected in our state. In recognition of such ownership, our courts, in decisions involving well-spacing regulations of our Railroad Commission, have frequently announced the sound view that each landowner should be afforded the opportunity to produce his fair share of the recoverable oil and gas beneath his land, which is but another way of recognizing the existence of correlative rights between the various landowners over a common reservoir of oil or gas.

It must be conceded that under the law of capture there is no liability for reasonable and legitimate drainage from the common pool. The landowner is privileged to sink as many wells as he desires upon his tract of land and extract therefrom and appropriate all the oil and gas that he may produce, so long as he operates within the spirit and purpose of conservation statutes and orders of the Railroad Commission. These laws and regulations are designed to afford each owner a reasonable opportunity to produce his proportionate part of the oil and gas from the entire pool and to prevent operating practices injurious to the common reservoir. In this manner, if all operators exercise the same degree of skill and diligence, each owner will recover in most instances his fair share of the oil and gas. This reasonable opportunity to produce his fair share of the oil and gas is the landowner's common law right under our theory of absolute ownership of the minerals in place. But from the very nature of this theory the right of each land holder is qualified, and is limited to legitimate operations. Each owner whose land overlies the basin has a like interest, and each must of necessity exercise his right with some regard to the rights of others. No owner should be permitted to carry on his operations in reckless or lawless irresponsibility, but must submit to such limitations as are necessary to enable each to get his own. Hague v. Wheeler, 157 Pa. 324, 27 A. 714, 717, 22 L.R.A. 141, 37 Am.St.Rep. 736.

While we are cognizant of the fact that there is a certain amount of reasonable and necessary waste incident to the production of oil and gas to which the non-liability rule must also apply, we do not think this immunity should be extended so as to include the negligent waste or destruction of the oil and gas.

In 1 Summers, Oil and Gas, Perm.Ed., § 63 correlative rights of owners of land in a common source of supply of oil and gas are discussed and described in the following language:

These existing property relations, called the correlative rights of the owners of land in the common source of supply, were not created by statute, but held to exist because of the peculiar physical facts of oil and gas. The term 'correlative rights' is merely a convenient method of indicating that each owner of land in a common source of supply of oil and gas has legal privileges as against other owners of land therein to take oil or gas therefrom by lawful operations conducted on his own land; that each such owner has duties to the other owners not to exercise his privileges of taking so as to injure the common source of supply; and that each such owner has rights that other owners not exercise their privileges of taking so as to injure the common source of supply.

In 85 A.L.R. 1156, in discussing the case of Hague v. Wheeler, *supra,* the annotator states:

> . . . The fact that the owner of the land has a right to take and to use gas and oil even to the diminution or exhaustion of the supply under his neighbor's land, does not give him the right to waste the gas. His property in the gas underlying his land consists of the right to appropriate the same, and permitting the gas to escape into the air is not an appropriation thereof in the proper sense of the term.

In like manner, the negligent waste and destruction of petitioners' gas and distillate was neither a legitimate drainage of the minerals from beneath their lands nor a lawful or reasonable appropriation of them. Consequently, the petitioners did not lose their right, title and interest in them under the law of capture. At the time of their removal they belonged to petitioners, and their wrongful dissipation deprived these owners of the right and opportunity to produce them. That right is forever lost, the same cannot be restored, and petitioners are without an adequate legal remedy unless we allow a recovery under the same common law which governs other actions for damages and under which the property rights in oil and gas are vested. This remedy should not be denied.

In common with others who are familiar with the nature of oil and gas and the risks involved in their production, the respondents had knowledge that a failure to use due care in drilling their well might result in a blowout with the consequent waste and dissipation of the oil, gas and distillate from the common reservoir. In the conduct of one's business or in the use and exploitation of one's property, the law imposes upon all persons the duty to exercise ordinary care to avoid injury or damage to the property of others. Thus under the common law, and independent of the conservation statutes, the respondents were legally bound to use due care to avoid the negligent waste or destruction of the minerals imbedded in petitioners' oil and gas-bearing strata. This common-law duty the respondents failed to discharge. For that omission they should be required to respond in such damages as will reasonably compensate the injured parties for the loss sustained as the proximate result of the negligent conduct. The fact that the major portion of the gas and distillate escaped from the well on respondents' premises is immaterial. Irrespective of the opening from which the minerals escaped, they belonged to the petitioners and the loss was the same. They would not have been dissipated at any opening except for the wrongful conduct of the respondents. Being responsible for the loss they are in no position to deny liability because the gas and distillate did not escape through the surface of petitioners' lands.

We are therefore of the opinion the Court of Civil Appeals erred in holding that under the law of capture the petitioners cannot recover for the damages resulting from the wrongful drainage of the gas and distillate from beneath their lands. However, we cannot affirm the judgment of the trial court because there is an assignment of error in the Court of Civil Appeals challenging the sufficiency of the evidence to support the findings of the jury on the amount of the damages, and another charging that the verdict was excessive. We have no jurisdiction of those assignments, and, since they have not been passed upon, the judgment of the Court of Civil Appeals is reversed and the cause remanded to that court for consideration of all assignments except those herein decided.

NOTES

1. What should the measure of damages be in *Elliff*? On remand, the Court of Civil Appeals rejected defendant's contention that the proper measure of damages was the diminution in market value of plaintiff's property interest, saying:

> The measure suggested is the one applicable to suits for injury to real property. We have no doubt that in cases of injuries, such as that sustained by appellees here, an action for damages to real property would lie, and we so held in our former opinion. Damages recoverable in such action would be measured by the difference in the values of the property before and after the injury. We also held in our former opinion that if appellees were to recover at all under the facts of this case, it must be as and for damages to real estate. This holding was reversed by the Supreme Court. That Court decided appellees were entitled to relief under another and additional theory, that is, that appellees could recover as and for the taking and destruction of gas and distillate which appellees owned or in which they had a property interest. The difference of opinion between this Court and the Supreme Court is neither narrow nor elusive, but is broad and fundamental. It goes to the concept of the nature of the action. The theory of the action necessarily indicates to some extent the factors which must be considered in ascertaining damages. A proper standard for the measure of damages to real property is inappropriate for measuring damages for the taking or destruction of personal property. It would be anomalous to say that appellees could elect to recover as and for gas taken and destroyed rather than for injury to real property, but that having made such election, their damages would nevertheless be governed by the same measure as that relating to damages to real property. For all practical purposes, it would restrict the appellees to an action for injury to real property and nullify the Supreme Court's holding. Texon Drilling Co. v. Elliff, 216 S.W.2d 824, 830–31 (Tex.Civ.App.1948, error ref'd n.r.e.).

2. The Louisiana jurisprudence on the liability of an adjoining landowner for drainage of oil or gas was summarized in Breaux v. Pan American Petroleum Corp., 163 So.2d 406, 20 O.&G.R. 476 (La.App.1964), *writ ref'd*, 246 La. 581, 165 So.2d 481, 20 O.&G.R. 501 (1964), as follows:

> We think a correct interpretation of the jurisprudence hereinabove cited is that a landowner is not entitled to recover *damages* from the *owner or lessee of adjoining lands* on the ground that oil and gas have been drained from beneath

his property by wells located on the adjacent tract, unless it is established that the oil or gas withdrawn from the common reservoir has been wasted, that the waste was caused by the negligence of defendant or by his willful intent to injure plaintiff, and that the quantity of oil and gas withdrawn from beneath plaintiff's property can be measured or that the damages sustained by plaintiff can be ascertained with some degree of accuracy. In the instant suit there are no allegations to the effect that the oil or gas produced from the adjoining tract has been wasted, and there are no allegations that defendants were negligent or that they willfully intended to injure plaintiffs. According to the established jurisprudence, therefore, the petition in this suit fails to state a cause of action for damages against defendants, *insofar as it is directed against them as lessees of the adjoining tract of land or as operators of the draining well located on adjacent lands.* 163 So.2d at 412, 20 O.&G.R. at 482–483.

If this is an accurate statement of the Louisiana law of the time, how does it differ from *Elliff*?

3. The Louisiana Mineral Code, Title 31 of the Louisiana Revised Statutes, became effective on January 1, 1975. Consider the following Articles in relation to the *Elliff* problem. Has the existing law as summarized been modified?

Art. 6. Ownership of land does not include ownership of oil, gas, and other minerals occurring naturally in liquid or gaseous form, or of any elements or compounds in solution, emulsion, or association with such minerals. The landowner has the exclusive right to explore and develop his property for the production of such minerals and to reduce them to possession and ownership.

Art. 8. A landowner may use and enjoy his property in the most unlimited manner for the purpose of discovering and producing minerals, provided it is not prohibited by law. He may reduce to possession and ownership all of the minerals occurring naturally in a liquid or gaseous state that can be obtained by operations on or beneath his land even though his operations may cause their migration from beneath the land of another.

Art. 9. Landowners and others with rights in a common reservoir or deposit of minerals have correlative rights and duties with respect to one another in the development and production of the common source of minerals.

Art. 10. A person with rights in a common reservoir or deposit of minerals may not make works, operate, or otherwise use his rights so as to deprive another intentionally or negligently of the liberty of enjoying his rights, or that may intentionally or negligently cause damage to him. This article and Article 9 shall not affect the right of a landowner to extract liquid or gaseous minerals in accordance with the principle of Article 8.

Art. 14. A landowner has no right against another who causes drainage of liquid or gaseous minerals from beneath his property if the drainage results from drilling or mining operations on other lands. This does not affect his right to relief for negligent or intentional waste under Articles 9 and 10, or against another who may be contractually obligated to protect his property from drainage.

4. The Comments to Article 10 of the Louisiana Mineral Code cite with approval the *Elliff* decision. In Coon v. Placid Oil Co., 493 So.2d 1236, 92 O.&G.R. 501 (La. App.1986), *writ denied* 497 So.2d 1002 (La. 1986), a Lousiana Court of Appeal held that where a neighboring well was damaged by a blowout of a well on adjacent property, the owner of the blowout well was held not to be liable for damages that were merely speculative. The court said that: "When damages are claimed their amount must be proven with certainty ... Proof which establishes only possibility, speculation or unsupported probability does not establish a damage claim." Despite extensive expert testimony for the plaintiffs, the appeals court said the evidence was insufficient to prove subterranean damages. Does a rule against speculative damages effectively overcome the Mineral Code protection of correlative rights? A later court decision has anwered this question in the negative, indicating that Louisiana has adopted the approach of *Elliff v. Texon* in Article 10 of the Mineral Code. Mobil Exploration & Producing, U.S., Inc. v. Certain Underwriters Subscribing to Cover Note 95-3317(A), La. App. 2001-2219, 837 So. 2d 11, 34, 159 O.&G.R. 271 (La. Ct. App. 2002), *writs denied*, 841 So. 2d 805, 843 So. 2d 1129, 1130 (La. 2003) ("we affirm the finding as to the damage sustained by the State in the impairment of its right to produce hydrocarbons from the reservoir in question"). The measure of damages is now the loss of hydrocarbons that could have been recovered.

5. In *Elliff* the plaintiffs "owned the surface and certain royalty interests" in the drained land. In discussing the *Elliff* case Professor A.W. Walker noted that "the court assumed without expressly deciding that the lease executed by the plaintiffs had not divested them of title in place to their royalty one-eighth of the gas. Where the royalty on gas is a fractional share of the market value of the gas it is usually held that title to all of the gas is vested in the lessee." *See* Walker, *Nature of the Landowner's Interest in Oil and Gas*, 17 Mont.L.Rev. 22 at 30 (1955). If such is the nature of plaintiffs' interest in *Elliff*, should it affect any aspect of the case?

6. Does the fact that the *Elliff* court allows the plaintiffs to bring their own suit mean that the plaintiffs' lessee does not have the authority to bring a suit against defendants on behalf of their lessors? For a ruling that only the lessor can bring a claim for its loss in an *Eliff* situation, *see* HECI Exploration Co. v. Neel, 982 S.W.2d 881, 143 O.&G.R. 142 (Tex. 1998), discussed *infra*, p. 519. If the lessee has the power or even the duty to market the royalty owner's share of production, why should the lessee be precluded from representing the lessor in a claim for damages?

7. Suppose that the negligence of *A,* as in *Elliff*, caused minerals to migrate from *C's* land to *B's*, which latter parcel is located between *A's* and *C's* land, the result being net loss of minerals beneath *C's* land but not beneath *B's*. Would *A* be liable to *C* ? To *B* ? Assuming that some of the minerals originally in place under *B's* land blew out of *A's* negligently operated well, could the *Elliff* measure of

damages result in a double recovery? *See* Nix, *What is the Proper Remedy in the Slant–Hole Suit?*, 18 Southwestern L.J. 486 at 490 (1964).

8. In Texas and other states, production may be restricted by state regulatory agencies. Assuming that a landowner may recover damages for negligent drainage, should the present and anticipated allowable production from the plaintiff's land affect the recovery?

9. Does the theory adopted in the state as to the nature of the landowner's interest in oil and gas affect:

(a) determination of whether the landowner has the exclusive right or privilege to drill a well upon his land for the purpose of producing oil and gas?

(b) determining the classification of the interest of a mineral or royalty owner or lessee as corporeal or incorporeal? *See infra* Sec. 2.

(c) the determination of whether there is liability under the facts of *Elliff*?

(d) the measure of damages to be employed in such a case as *Elliff*, where liability is found?

(e) the determination of whether or not a tort action lies in favor of the owner of Blackacre against the owner of Whiteacre when ordinary, prudent production on Whiteacre causes drainage from Blackacre? *See* Billeaud Planters, Inc. v. Union Oil Co. of California, 245 F.2d 14, 7 O.&G.R. 798 (5th Cir.1957).

(f) the right of the royalty owner to demand an accounting from the lessee where the lessee is allegedly draining oil from the owner's leasehold to an adjacent leasehold?: Williams v. Humble Oil & Refining Co., 432 F.2d 165 (5th Cir.), on motion for rehearing, 435 F.2d 772 (5th Cir.1970) (p. 461 *infra*).

(g) the determination of whether or not wasteful conduct by a landowner may be enjoined by an adjoining landowner? *See* Louisville Gas Co. v. Kentucky Heating Co., 117 Ky. 71, 77 S.W. 368 (1903) and Hague v. Wheeler, 157 Pa. 324, 27 A. 714 (1893).

(h) the scope of governmental power to regulate operations and production? *See* U. V. Industries, Inc. v. Danielson, 184 Mont. 203, 602 P.2d 571, 64 O. & G.R. 454 (1979); Strong, *Application of the Doctrine of Correlative Rights by the State Conservation Agency in the Absence of Express Statutory Authorization*, 28 Mont.L.Rev. 205 (1967).

(i) allocation of the benefits of ownership when "gas" rights have been separated from "oil" rights? *See* Note, *Phase Severance of Gas Rights from Oil Rights*, 63 Tex.L.Rev. 133, 156 (n. 100) (1984).

10. Economists have been interested in the problems relating to the allocation of scarce resources that exist in a common pool for many years. Do you agree with the following statement that appeared in a recent article that explored the economic ramifications of a rule of first possession: "Economically, crude oil has more in common with grizzly bears than it does with silver and

gold. By nature, oil and bears are not anchored to a single, well defined tract of land. Accordingly, oil (and gas) law is remarkably similar to wildlife law and, at the same time, remarkably dissimilar to the law of hard rock minerals." Dean Lueck, *The Rule of First Possession and the Design of the Law*, 38 J.L. & Econ. 393, 425-26 (1995). For a vigorous defense of the rule of capture and the ferae naturae analogy for oil and gas law, see Rance L. Craft, *Of Reservoir Hogs and Pelt Fiction: Defending the Ferae Naturae Analogy Between Petroleum and Wildlife*, 44 Emory L.J. 697 (1995).

11. The historical development of the rule of capture, especially reliance on the analogy to *ferae naturae* or wild animals is explored in Bruce M. Kramer & Owen L. Anderson, *The Rule of Capture – An Oil and Gas Perspective*, 35 Env't L. 899 (2005). Should an ownership doctrine that is based on an inapt analogy continue to serve as the basis for the ownership theory for oil and gas? Is the rule of capture a rule of non-liability? An early definition of the rule of capture quoted almost verbatim by the Elliff court says: "The owner of a tract of land acquires title to the oil and gas which he produces from wells drilled thereon, though it may be proved that part of such oil or gas migrated from adjoining lands." Robert E. Hardwicke, *The Rule of Capture and Its Implications as Applied to Oil and Gas*, 13 Tex. L.Rev. 393, 393 (1935). Does that definition deal with what happens to the oil and gas once produced or to the potential damage to the common source of supply by over-drilling?

12. Does the Hardwicke definition create an ownership rule or a rule of non-liability? Is there a functiional difference between the two rules. In Petro Pro, Ltd. v. Upland Resources, Inc., 279 S.W.3d 743, 753, 169 O.&G.R. 607 (Tex.App.— Amarillo 2007, rev. denied), the court described one's ownership interest in oil and gas as follows:

> The rule of capture is a rule of non-liability that determines what remedies, if any, that an adjacent landowner has against his neighbor for production of migratory subsurface minerals belonging to the owner-in-place. Essentially, the rule provides that, absent malice or willful waste, landowners have the right to produce migratory subsurface minerals that they can capture without being liable to their neighbor, even if in doing so they deprive their neighbor of their ownership interest in the actual minerals.

Is there anything in the *Elliff* opinion that would restrict one's ownership interest in captured oil and gas where the capturer engaged in malice, willful waste, or as in *Elliff*, negligent production techniques? *Compare* Hague v. Wheeler, 157 Pa. 324, 27 A. 714 (1893) *with* Ohio Oil Corp v. Indiana, 177 U.S. 190, 20 S. Ct. 576, 44 L. Ed. 729 (1900) *and* Manufacturers Gas and Oil Co. v. Indiana Natural Gas Co., 155 Ind. 461, 57 N.E. 912 (1900).

———

B. STATE REGULATION AND THE MODIFICATION OF THE RULE OF CAPTURE

Kramer & Martin, *The Law of Pooling and Unitization* § 2.02

(A more complete treatment of conservation regulation and the effects of pooling and unitization is given in Chapter 7 *infra*).

Judicial recognition of the changes made to the common law rule of capture by state conservation statutes has been reasonably slow, notwithstanding the judiciary's recognition of the problems caused by the rule's [of capture] application. This recognition is most often developed through an agency's attempt to protect the correlative rights of the owners over a common source of supply. . .

The ability of a statutory conservation scheme to modify the common law rule of capture is exemplified in Schrimsher Oil & Gas Exploration v. Stoll. [19 Ohio App.3d 274, 484 n.E.2d 166, 88 O.&G.R. 158 (1984). In *Schrimsher*, the lessee had received a state well-drilling permit without the written pooling agreements of each proeprty owner in the unit required by the agency. The well was located near the landowner's property line and was draining oil and gas from beneath his property. Under classic rule of capture principles the landowner had one option: drill his own offest well to intercept the flow of hydrocarbons across his property line. In addition, he could have attacked the issuance of the well-drilling permit through direct administrative and judicial procedures. When the producer sued for declaratory judgment as to the extent of royalty owed to the landowner, the landowner counterclaimed for damages for drainage because the well has been placed too close to the property line. The state permit could not be collaterally attacked, so the only avenue of redress was a direct assault on *Kelly v. Ohio Oil Co.*, [57 Ohio St. 317, 49 N.E. 399 (1897)] which had clearly adopted the rule of capture and the offset well rule to deal with the issue of drainage of oil and gas. The *Schrimsher* court concluded that the rule of capture, as adopted in *Kelly*, had been superseded by the Ohio conservation statutes relating to well spacing requirements. *Schrimsher* was thus essentially a drainage case, albeit one colored by the fact that the landowner, after initially protesting the well's illegal location withdrew his protest and acquiesced in the new drilling site when the producer promised to pay him a percentage of the well's royalty. The landowner based his counterclaim on the fact that the well, as located, was draining hydrocarbons from under his property. The court recognized that the landowner was entitled to some compensation for the diminution of the resource, and held that the company was liable for the amount of lease royalty that they would have had under a compulsory pooling order. The court upheld the use of the straight acreage formula found in the compulsory pooling statute to determine the amount of compensation due to the landowner. This had been the basis for the parol agreement, so the producer was in the same position as if it had entered into a written pooling agreement with the landowner.

The statutory modification of the rule of capture was achieved indirectly through the creation of a new tort. This new tort recognized a damage recovery for violation of state statutory and regulatory rules relating to the location and operation of oil and gas wells. It is related to the traditional common law remedies of trespass and conversion that would normally not be applicable in the situation where the neighboring well was merely draining hydrocarbons that had been originally located under the adjacent tract. While the state statutory and regulatory mechanisms provided a panoply of administrative and judicial remedies to deal with the regulatory scheme, the Schrimsher decision goes beyond those remedies to interpret the conservation statutes as modifying the rule of capture that had been announced in *Kelly*.

NOTES

1. A number of states have statutes that both preserve common law remedies and create a new statutory cause of action for statutory or administrative violations. See e.g., Tex.Nat.Res.Code § 85.321. In Exxon Corp. v. Emerald Oil & Gas Co., L.C., 331 S.W.3d 419, 54 Tex. Sup. Ct. J. 342 (2010), the Texas Supreme Court concluded that § 85.321 allowed a private party to sue where the defendant allegedly violated provisions prohibiting waste and prescribing the plugging of wells.

2. A Tenth Circuit opinion, applying Oklahoma law, made the following comments relating to the impact of Oklahoma's oil and gas conservation laws and regulations on the rule of capture: " Everything the plaintiffs claim the defendants here should have done required the approval of the OCC.[Oklahoma Corporation Commission, eds.] The OCC is required to consider the correlative rights of other interest holders in the common source of supply. . . . Where the state, through the OCC, undertakes to protect the correlative rights of owners in a common source of supply, the law of capture—the theoretical underpinning for the plaintiffs' proposed rule—does not apply. . . Fransen v. Conoco, Inc., 64 F.3d 1481, 1491, 131 O.&G.R. 331 (10th Cir. 1995), *cert. denied*, 516 U.S. 1166 (1996).

3. The mere enactment of a conservation statute will not necessarily change the rule of capture. For example, in Trent v. Energy Development Corp., 902 F.2d 1143, 110 O.&G.R. 572 (4th Cir. 1990), the court found that in the absence of a compulsory pooling order issued by the appropriate state conservation agency the rule of capture applied so as to deny a neighboring tract owner a share of the production where the well was located off of the plaintiff's tract.

4. There is a judicial canon that requires courts to narrowly construe statutes that purport to modify the common law. In Calfisch v. Crotty, 2 Misc.3d 786, 774 N.Y.S.2d 653, 161 O.&G.R. 779 (Sup.Ct. 2003) the court used that canon to reaffirm New York's commitment to the rule of capture even in light of the compulsory pooling statute that had been enacted.

5. The concept that there are limitations on the ability of a mineral owner to capture oil and gas is often described by the use of the term "correlative rights." In many state oil and gas conservation statutes, the overall purposes of the statutes are stated to be the prevention of waste, the conservation of natural resources and the

protection of correlative rights. In Kingwood Oil Co. v. Corporation Commission, 1964 OK 241, 396 P.2d 1008, 1010, 21 O.&G.R. 620, the court defined "correlative rights" as:

> . . . a convenient method of indicating that each owner of land in a common source of supply of oil and gas has legal privileges as against each other owners of land therein to take oil and gas therefrom by lawful operations conducted on their own land, limited, however, by duties to other owners not to take an undue proportion of oil and gas.

6. Is there anything in the *Elliff* decision that supports the limitation on the rule of capture relating to not taking an "undue proportion of oil and gas"? If the well in *Elliff* had not blown out but had merely produced substantial amounts of natural gas that had been originally located under the neighboring tract would the neighbor have had a cause of action for taking an "undue proportion" of the natural gas?

C. SUBSURFACE STORAGE OF GAS

Because of the difficulty of storing natural gas above ground, many natural gas utilities and industrial users use depleted underground formations to store gas. Public utilities were the first to use this approach to deal with the seasonal fluctuations in demand that could not be met because the pipeline delivery systems that were available could not provide adequate supplies during maximum demand periods. Because of the importance of this issue, many states have enacted statutes which regulate various aspects of gas storage. See e.g., La. Rev. Stat. tit. 30, § 22; Okla. Stat. Ann. tit. 52, §§ 36.1 et seq.; Tex. Nat. Res. Code §§91.171 et seq.. Under many of these statutes the gas storer can condemn the right to store the gas underground. By condemning lands that entirely cover the storage reservoir the gas injector avoids any problem. But in many situations, accurate information is lacking and the cost may be prohibitive to achieve total ownership of the underground storage area. That leads to disputes regarding ownership of the stored gas under a rule of capture regime.

NORTHERN NATURAL GAS CO. v. L.D. DRILLING, INC.

United States District Court, District of Kansas, 2010
--- F.Supp.2d ---

WESLEY E. BROWN, Senior District Judge. This matter is before the court on Northern's Motion for Preliminary Injunction (Doc. 341), which seeks an order prohibiting the defendants from operating their gas wells in "the Expansion Area" pending a ruling on the merits of the complaint. The request for injunction is based on a claim that the continued operation of the defendants' wells constitutes a nuisance that unreasonably interferes with Northern's gas storage facility. For the reasons stated herein, the motion for preliminary injunction will be granted, subject to certain conditions.

Although the case involves a number of difficult issues, two points became clear at the injunction hearing. First, Northern presented strong evidence that the defendants' wells are in fact producing storage gas that has migrated out of the Cunningham Storage Field. Defendants more or less deny that allegation, although one of their own experts confirmed that storage gas is likely migrating to the expansion area, and after two years of litigation the defendants have cited no substantial evidence to the contrary. L.D. Davis (owner of L.D. Drilling) and his consulting geologist Kim Shoemaker, as well as Nash Oil & Gas owner Jerry Nash, all testified they do not believe they are producing storage gas. But those denials do not appear to be based on an objective consideration of all the evidence. Their skepticism is due in part to the significant distance from their wells to the (previous) border of the storage field. But Northern has produced gas composition analysis, seismic data, and historical pressure and production data, all of which tend to show that the defendants' production consists mainly of storage gas. FERC previously considered much of the same evidence and concluded that storage gas was in fact migrating to the Expansion Area and was being produced by the defendants' wells. Defendants' disbelief may also spring in part from indignation at being accused of conspiring to draw gas out of the storage field, or at being blamed by Northern for a situation that was apparently caused in part by Northern's own erroneous assessment that an underground fault in the storage field would contain the injected storage gas. Be that as it may, if blame is set aside and one looks only at the evidence, it is hard to escape a conclusion that the defendants' wells in the Expansion Area are producing or have produced storage gas.

The second point now clear is that the landscape has been altered significantly by Northern's filing of a condemnation action. Northern seeks in that action to condemn the property rights necessary to allow it to include the Expansion Area in the Cunningham Storage Field. Like the differences in pressure causing gas to migrate to the expansion area, the arguments in this case are moving inexorably toward the condemnation case, where the ultimate issue is how much Northern will have to pay to acquire the defendants' property rights. Counsel for all of the parties have invoked the condemnation to some degree in this case, and the pendency of that action is a fact the court cannot ignore in assessing the request for an injunction.

I. *Facts.*

A. *Background.*

Natural gas in the Viola formation of the Cunningham Field was discovered in 1932, and primary production began in 1934. Over the next several decades, about 79 billion cubic feet of natural gas was produced from the Viola formation. By 1974, primary depletion had reduced the Cunningham Field reservoir pressure down to 76 psi (absolute) from an original reservoir pressure of 1695 psi. The field was depleted by gas production prior to 1977.

Northern is a natural gas company engaged in the interstate transportation of natural gas within the meaning of the Natural Gas Act (NGA), 15 U.S.C. §

717f(a). As such, it is subject to the jurisdiction of FERC, which is charged with implementation of the NGA.

If FERC finds the construction or extension of an interstate gas transportation facility is in the public interest, it may grant a natural gas company a Certificate of Public Convenience and Necessity. Such a certificate grants the holder the right to use eminent domain to acquire the property rights necessary to construct or extend the facility, if the holder cannot obtain the rights through negotiation. 15 U.S.C. § 717f. Like federal law, Kansas law recognizes that the underground storage of natural gas promotes the public interest by building reserves for orderly withdrawal in periods of peak demand. K.S.A. § 55-1202.

In 1977 and 1978, Northern obtained certificates from the Kansas Corporation Commission and from FERC to develop and operate the Cunningham Storage Field in Pratt and Kingman Counties, Kansas. The storage field initially included the Viola formation under about 23,000 acres of property.

When the Cunningham Field was first certified, it was believed that two underground faults, including one running toward the northeast side of the field, would contain and trap the injected gas, making the Viola an isolated reservoir within the field. Northern began injecting storage gas into the field by 1979. A "fill-up" period of several years followed in which Northern continued to inject storage gas, thereby re-pressuring the field. Northern's evidence, including hysteresis curves showing the relationship between gas inventory and storage field pressure, shows that beginning in 1985 and continuing for about the next ten years or so, the storage field was essentially stable, with no evident migration of storage gas.

In 2002, Northern filed suit against the Trans Pacific Oil Corporation, claiming the latter was producing storage gas from two wells just outside the northern boundary of the Cunningham Storage Field. In 2004, Northern sued Nash Oil & Gas, Inc., claiming Nash wells located about four miles north of the field were producing storage gas. A jury in the *Trans Pacific* case found Northern had not shown that storage gas migrated to the two Trans Pac wells after July 1, 1993, the effective date of K.S.A. § 55-1210. In March 2007, Northern's claims against Nash in the 2004 case were dismissed on summary judgment. Both of these judgments were affirmed on appeal to the Tenth Circuit.

[The court then discusses the various judicial and administrative actions filed following the 2004 cases which include the FERC's issuance of a certificate of public convenience and necessity to expand the boundaries of the storage facility, litigation against the gas purchasers from the wells allegedly producing storage gas asserting conversion of Northern's personal property in the stored natural gas and an eminent domain case seeking to condemn the mineral estates within the expanded storage area. Eds.]

In the FERC proceeding, Northern submitted evidence that an aquifer had originally served as a hydrostatic seal for the Cunningham reservoir rather than the underground fault that was thought to be the containment mechanism.

According to Northern's evidence, during primary production of the field the aquifer acted as a limited water drive, pushing native gas out of pore spaces and allowing water to move in, with a corresponding reduction in pressure of the aquifer north of the fault. Later, during the Cunningham storage field "fill up" period, storage gas injected by Northern was pushed past the northern fault and into the aquifer, thereby increasing pressure in the aquifer north of the field. The field stabilized in 1985 and remained stable for about ten years thereafter. Around 1994, hysteresis curves began to indicate gas loss from the Cunningham reservoir, which Northern argued to FERC was caused by third-party production from Nash and others that began about that time. Even though such production was located several miles north of the storage field, Northern asserted that the production of increasing volumes of gas and water by third-party producers north of the fault was creating pressure sinks that caused increased gas migration from the storage field. Northern presented evidence that about 13 Bcf of storage gas has since migrated from the field.

Northern's current motion seeks to enjoin production from the following wells, which are located exclusively within the Authorized Expansion Area: [26 wells are listed, Eds.]

Northern's motion seeks to enjoin "the continued interference by the Defendants, through operation of such wells, from interfering with Northern's use and enjoyment of the Cunningham Storage Field; and enjoin Defendants from otherwise interfering with Northern's right and ability to protect the integrity of the Cunningham Storage Field." The motion alleges that without the injunction, Northern will be irreparably harmed in its ability to meet FERC's deadlines and to restore the containment mechanisms and stability of the Cunningham Storage Field."

B. *Summary of Evidence & Findings.*

[The court provides an extensive summary of the competing evidence relating to whether or not the defendants' wells are producing storage gas that is migrating from the existing underground storage facility. The court also finds that the FERC findings relating to migration that support Northern's position are not binding on the court because the underlying legal issues are different. Eds.]

II. *Preliminary Injunction Standards.*

When seeking a preliminary injunction, the moving party must demonstrate: (1) a likelihood of success on the merits; (2) a likelihood that the movant will suffer irreparable harm in the absence of preliminary relief; (3) that the balance of equities tips in the movant's favor; and (4) that the injunction is in the public interest. In addition, the movant must establish "a relationship between the injury claimed in the party's motion and the conduct asserted in the complaint." A mandatory preliminary injunction-i.e., one that requires the nonmoving party to take affirmative action-is an extraordinary remedy and is generally disfavored. Before a court may grant such relief, the movant must "make a heightened showing of the [ ] four factors." The requested injunction in this case would alter the status quo and would require the defendants to take affirmative action to cease production and shut in their wells. As such, it qualifies as a disfavored mandatory injunction.

III. Discussion.

A. *Likelihood of Success on the Merits.* Northern claims the defendants' operation of their wells in the expansion area constitutes a nuisance because it interferes with Northern's use and enjoyment of the Cunningham Storage Field. Under Kansas law, " 'A nuisance is an annoyance, and any use of property by one which gives offense to or endangers the life or health, violates the laws of decency, unreasonably pollutes the air with foul, noxious odors or smoke, or obstructs the reasonable and comfortable use and enjoyment of the property of another may be said to be a nuisance." *Smith v. Kansas Gas Service Co.*, 285 Kan. 33, 169 P.3d 1052 (2007). In the absence of an actual trespass, the interference with the use of the land must be substantial in order to constitute nuisance. *See Williams v. Amoco Production Co.*, 241 Kan. 102, 734 P.2d 1113, 1124 (1987). Moreover, the interference must be of such a nature, duration or amount as to constitute an unreasonable interference with the use and enjoyment of land. *Id.* at 1125.

The evidence presented clearly establishes that continued production from the defendants' wells in the expansion area will substantially interfere with Northern's use and enjoyment of the Cunningham Storage Field. There is strong and clear evidence that storage gas from the previously-certified areas of the field (i.e. from the pre-2010 borders) is migrating out to the expansion area and has been doing so for some time, with wells even in the northern portion of the expansion area producing primarily storage gas, even though some of those wells are more than 6 miles from the underground fault. The defendants' production of substantial amounts of storage gas and water will likely continue to draw storage gas beyond the underground fault and out of the storage field as long as such production continues, threatening the continued viability of the storage facility.

The gravity of the harm from the continued operation of defendants' wells in the expansion area weighs in favor of a finding of nuisance. Federal and state law recognize the importance of underground gas storage facilities and the public interests they further. The continued operation of these wells will prevent containment of a large and important storage field from being reestablished. Moreover, continued production will likely cause increased migration to occur, according to the evidence, as the migration pathway grows larger and becomes more saturated.

The evidence presented at the hearing shows that the Cunningham Storage Field and the Expansion Area are likely in communication with one another, and that migration of storage gas results from a pressure differential between the storage field and the expansion area that came about as a result of defendants' production of gas and water beginning in the mid 1990's. This is the most reasonable interpretation of the evidence, including evidence showing a substantial increase in pressure in Expansion Area or nearby wells after 1979, and evidence of an extended period of storage field stability that ended about the same time production by one or more defendants commenced in the Expansion Area. There is also some evidence that the defendants' wells are not experiencing a normal decline in production as would be expected from production of native gas. The defendants' production creates "pressure sinks" that draw gas to the wells. As defendants note, oil and gas production almost by definition involves a lowering of pressure to draw in hydrocarbons from the surrounding area, and the mere fact that a differential exists is not proof of unreasonable interference. The defendants have a right to produce native gas from the Expansion Area. There is a conflict in the evidence as to whether the pumps being used by the defendants and the amount of water being produced are unusually large, and whether the defendants should have known that their production would draw in gas from the storage field. Some of the pumps used by the defendants appear to be of significant size, although the evidence fails to show clearly that the use of such pumps was unprecedented for this area. Similarly, there was some evidence that production of over 200 barrels of water for gas wells was high, but the evidence failed to show clearly that it was so unusual as to show an intent to affect the storage field. At the same time, the defendants have failed to address Northern's evidence that drilling numerous productive wells "down dip" of the formation should have suggested a possible connection to the storage field. The evidence also indicates what might be described as a lack of curiosity by the defendants as to whether or not they were producing storage gas. At any rate, what cannot be disputed is that after the June 2, 2010 FERC order-which the defendants chose not to appeal-the defendants were clearly on notice that their wells were producing primarily if not entirely storage gas, and that their production of significant amounts of water was likely influencing the migration of storage gas from the Cunningham field. Whatever the defendants' intentions up to the point of the FERC order, their continued production thereafter with knowledge that they were causing storage gas to migrate can now be viewed as an intentional and substantial interference with Northern's use of the Cunningham Storage Field.

The utility of the defendants' conduct is relevant to the nuisance claim. *See Restatement (Second) of Torts,* § 826 ("An intentional invasion of another's interest in the use and enjoyment of land is unreasonable if ... the gravity of harm outweighs the utility of the actor's conduct...."). The defendants have a right to produce natural gas to which they or their lessors hold title in the expansion area. Moreover, Northern's erroneous assessment of the underground fault appears to have contributed at least in part to the situation it now claims to be a nuisance, and Northern could also bear potential liability for what amounts to an unauthorized use of the Viola formation in the expansion area. But several factors unique to this case nevertheless support Northern's claim that the defendants' production currently constitutes an unreasonable interference. First, Northern has obtained a certificate from FERC to include the expansion area in the Cunningham Storage Field. Whatever the parties' rights may be as to storage gas that migrated before July 1, 1993, or with respect to storage gas that migrated after 1993 but before the FERC certificate, Kansas law appears clear that the certificate means Northern retains title at least to any storage gas that has migrated or will migrate to the Expansion Area from June 2, 2010 and thereafter. *See Union Gas System, Inc. v. Carnahan,* 245 Kan. 80, 774 P.2d 962 (1989) ("As soon as Union's storage gas operation became authorized and its gas identifiable, the gas was no longer *ferae naturae* and subject to the rule of capture. The title to Union's gas remained in Union."). Persons other than the injector have no right "to produce, take, reduce to possession, either by means of the law of capture or otherwise, waste, or otherwise interfere with or exercise any control over such gas." K.S.A. § 55-1210(b). In these circumstances, even though the defendants have received a permit from the Kansas Corporation Commission to operate their wells, Kansas law does not grant the defendants an unfettered right to continue producing storage gas. Second, there is strong evidence that all of the wells at issue would, if allowed to continue operating during this litigation, produce primarily storage gas. According to Northern's expert, the production from numerous wells consists entirely of storage gas while several others contain more than 85% storage gas. Four wells in the northernmost portion of the expansion area contain about 65% storage gas-meaning up to 35% of that production may be native gas-but there is also evidence that the percentage of storage gas in those wells is increasing over time. Defendants have cited no gas composition evidence to contradict Dr. Boehm's opinions. Third, the defendants' property rights are now the subject of a condemnation action, and, notwithstanding pending challenges to Northern's authority to condemn, it is likely that the rights will be acquired by Northern sometime in the future. Of course, Northern will have to pay just compensation to acquire the rights-which include all interests relating to natural gas in the Viola and Simpson formations, including royalty interests, overriding royalty interests, working interests, and "any and all rights of any type or nature incident or appurtenant thereto." The court recognizes that Northern has not yet paid for any such rights, and the defendants thus retain the right to produce native gas unless and until Northern pays compensation pursuant to a lawful condemnation. But at the same time, the defendants do not possess a right to produce storage gas to which Northern retains title, nor do they have a right to unreasonably interfere with a certified storage field. *See* K.S.A. § 55-1210.

The right of condemnation granted natural gas companies under federal or state law is the primary means of resolving claims of interference with a gas storage facility. Northern has ostensibly been granted that authority and is seeking to condemn the property involved in the defendants' production. There will inevitably be a delay before any condemnation can be completed, however, and in the meantime the defendants' continued production-the vast majority of which appears to constitute storage gas-would work a substantial and unreasonable interference with the existing storage field. Mr. Davis himself candidly admitted that he did not think a reasonable operator should be drawing gas out of a storage field and producing it. Under the unique circumstances of the case, the defendants' claim of a right to continue producing native gas does not justify the substantial interference such production would cause with Northern's right to the use and enjoyment of the storage field while the merits of the nuisance claim-or alternatively the merits of the condemnation complaint-are decided. The court concludes that Northern has satisfied the requirement of showing a likelihood of success on the merits.

[The court discusses the other factors in granting a mandaotry preliminary injunction, including the likelihood that the movant will suffer irreparable harm in the absence of preliminary relief, the balance of equities and whether or not the injunction is in the public interest. The court concludes that all of the factors favor the issuance of the injunction. Eds.]

IV. *Conclusion.*

Northern Natural Gas Company's Motion for Preliminary Injunction is GRANTED subject to the conditions set forth herein. It is ordered that by February 21, 2011, defendants L.D. Drilling, Inc., Val Energy, Inc., and Nash Oil & Gas, Inc., shall cease and refrain from further production of natural gas from the following wells in the Viola and/or Simpson formations of the Expansion Area certified by FERC in its June 2, 2010 Order. . .

NOTES

1. This common law nuisance action, as all such actions, is predicated on an unreasonable interference with the use and enjoyment of one's property? Does the court identify Northern's property interest that is being interfered with? It clearly states that Kansas, at least since 1993, recognizes that the injector of non-native natural gas into a licensed underground storage facility does not lose its ownership interest in its personal property?

2. Northern Natural Gas has been fighting and for the most part losing in its attempt to use the judicial system to protect its property interests in the injected gas. See Northern Natural Gas Co. v. Trans Pacific Oil Corp., 529 F.3d 1248, 170 O.&G.R. 643 (10th Cir. 2008); Northern Natural Gas Co. v. Trans Pacific Oil Corp., 248 Fed.Appx. 882 (10th Cir. 2007); Northern Natural Gas Co. v. Nash Oil & Gas, Inc., 526 F.3d 626, 168 O.&G.R. 258 (10th Cir. 2008), *aff'g*, 506 F.Supp.2d 520, 168 O.&G.R. 249 (D.Kan. 2007).

3. Similar issues were raised in Pacific Gas & Electric Co. v. Zuckerman, 189 Cal.App.3d 1113, 234 Cal.Rptr. 630, 93 O.&G.R. 9 (1987) which involved similar facts relating to alleged drainage of injected, non-native gas into a FERC-certificated underground storage facility to wells located near the boundary of the facility. Unlike the *Northern* case, however, the wells producing the stored gas were also operated by the storage facility owner. The California Court of Appeals also held that the injected, non-native gas was owned by the injector in the context of a condemnation case involving the valuation of the condemned mineral estate.

4. Does it matter whether a jurisdiction adopts the ownership-in-place or

non-ownership theory in deciding that the injector of gas does not lose title to that gas upon its re-injection into an underground natural gas storage facility. *See* Ellis v. Arkansas Louisiana Gas Co., 609 F.2d 436, 64 O.&G.R. 503 (10th Cir.1979), *cert. denied*, 445 U.S. 964 (1980) (applying Oklahoma law); White v. New York State Natural Gas Corp., 190 F.Supp. 342, 14 O. & G.R. 253 (W.D.Pa. 1960) (applying Pennsylvania law); Humble Oil & Refining Co. v. West, 508 S.W.2d 812, 48 O.&G.R. 516 (Tex. 1974); Lone Star Gas Co. v. Murchison, 353 S.W.2d 870, 16 O. & G.R. 816, 94 A.L.R.2d 529 (Tex.Civ.App. 1962, writ ref'd n.r.e.). For cases finding that the injector could not prevent an owner from producing the injected gas *see* Hammonds v. Central Kentucky Natural Gas Co., 255 Ky. 685, 75 S.W.2d 204 (1934), *overruled in part,* Texas American Energy Corp. v. Citizens Fidelity Bank & Trust Co., 736 S.W.2d 25, 99 O.&G.R. 258 (Ky. 1987); Anderson v. Beech Aircraft Corp., 237 Kan. 336, 699 P.2d 1023, 85 O.&G.R. 83 (1985); Union Gas Systems, Inc. v. Carnahan, 245 Kan. 80, 774 P.2d 962 (1989).

5. Would the issue of ownership be affected if the injected gas migrated across property lines so that the producing well was located outside of the storage owner's boundaries? *See* Oklahoma Natural Gas Co. v. Mahan & Rowsey, Inc., 786 F.2d 1004, 89 O.&G.R. 152 (10th Cir. 1986), *cert. denied*, 479 U.S. 853 (1986). *See also* Reese Exploration, Inc. v. Williams Natural Gas Co., 983 F.2d 1514, 123 O.&G.R. 17 (10th Cir. 1993) where the issue concerned gas storage injection interfering with oil production.

6. What is the impact of a state regulatory system on the rights of the surface owner and the injector? If the statute only allows public utilities to inject gas, or requires that the injector receive a state permit or certificate of convenience, will the ownership of the injected gas depend on whether the party has received that permit or certificate? In Anderson v. Beech Aircraft, note 4 *supra,* the court said:

> we hold ... that ... where a natural gas public utility was not involved, where no certificate authorizing an underground storage facility had been issued ... and where the defendant had used the property of an adjoining landowner for gas storage without authorization or consent, the defendant, as the owner of non-native natural gas, lost title thereto when it injected non-native gas into the underground area and the gas was then produced from the common reservoir located under the adjacent property. 699 P.2d at 1032.

7. In 1993 Kansas enacted a statute which preserves to the injector title to injected gas even if it should escape the storage horizon, where the injector can establish by a preponderance of the evidence that such gas was originally injected into the underground formation. K.S.A. § 55-1210. Williams Natural Gas Co. v. Supra Energy, Inc., 931 P.2d 7, 136 O.&G.R. 26 (Kan. 1997) upheld the constitutionality of this statute insofar as the statute authorizes the injector to perform tests, at its own expense, on other mineral owner's wells in order to protect against the migratory loss of injected gas. Query: Is the Rule of Capture a property right of an adjacent owner that is abrogated by the Kansas statute?

8. Who is entitled to compensation for the use of the depleted producing

formation, the surface owner, the royalty owner, the mineral owner or the lessee? *See* Central Kentucky Natural Gas Co. v. Smallwood, 252 S.W.2d 866, 2 O.&G.R. 19 (Ky. 1952), *overruled in part,* Texas American Energy Corp. v. Citizens Fidelity Bank & Trust Co., 736 S.W.2d 25, 99 O.&G.R. 258 (Ky. 1987); Ellis v. Arkansas Louisiana Gas Co., 609 F.2d 436, 64 O.&G.R. 503 (10th Cir.1979), *cert. denied,* 445 U.S. 964 (1980) (applying Oklahoma law); Miles v. Home Gas Co., 35 A.D.2d 1042, 316 N.Y.S.2d 908, 37 O.&G.R. 445 (1970); Tate v. United Fuel Gas Co., 137 W.Va. 272, 71 S.E.2d 65, 1 O.&G.R. 1459 (1952); Emeny v. United States, 188 Ct.Cl. 1024, 412 F.2d 1319, 34 O.&G.R. 53 (1969); United States v. 43.42 Acres of Land, 520 F.Supp. 1042, 71 O.&G.R. 31 (W.D.La.1981); Cassinos v. Union Oil Co. of Calif., 18 Cal.App.4th 1770, 18 Cal.Rptr.2d 574, 125 O.&G.R. 472 (1993); Mapco, Inc. v. Carter, 808 S.W.2d 262 (Tex.App. 1991), *rev'd on other grounds,* 817 S.W.2d 686 (Tex. 1992).

Perhaps the use of the term "surface estate" (and its status as "servient") in connection with land when a "mineral estate" has been created may suggest to some that a mineral owner owns a "subsurface estate" that gives the mineral owner control of all subsurface activity. Would it not be more accurate to say that the landowner continues to own both the surface and the subsurface, subject to certain rights to the minerals in the mineral owner? If so, the landowner (i.e. surface owner) is the proper party to deal with subsurface storage or disposal of saltwater, carbon dioxide, or other substances, so long as such activity does not impede or diminish the recovery of minerals by the mineral owner. In Dick Properties, LLC v. Paul H. Bowman Trust, 221 P.3d 618 (Kan.App., 2010) the court held that the landowner was the party with whom a saltwater disposal agreement was to be made, not the mineral owner, particularly as the agreement was to allow saltwater disposal from off the land in question. The mineral owner, the court said, had a "court-created implied covenant" to dispose of saltwater but this was "limited to saltwater produced on the lease and does not extend to saltwater produced on other leases." Put another way, the landowner continues to own the pore-space beneath the surface, absent agreement to the contrary. See Michael B. Donald, *Carbon Sequestration,* 56th Ann. Inst. on Min. Law 437 (2009); Phillip Marston & Patricia Moore, *From EOR to CCS: The Evolving Legal and Regulatory Framework for Carbon Capture and Storage,* 29 Energy L.J. 421 (2008); Jeffrey Moore, *The Potential Law of On-Shore Geologic Sequestration of CO2 Captured From Coal-Fired Power Plants,* 28 Energy L.J. 443 (2007); Interstate Oil & Gas Compact Commission, Task Force on Carbon Capture & Geologic Storage, *Storage of Carbon Dioxide in Geologic Structures: A Legal and Regulatory Guide for States and Provinces* (2007).

9. Can a gas storage lessor produce native oil and gas from the leased storage strata? Storck v. Cities Service Gas Co., 575 P.2d 1364, 61 O. & G.R. 205 (Okl. 1977); Union Gas Systems, Inc. v. Carnahan, 245 Kan. 80, 774 P.2d 962 (1989).

10. Able is a mineral estate owner who has mined salt from underneath Baker's surface estate. Able wants to stop mining salt and use the excavated salt

caverns for storage purposes. Does Able have the right to the use and enjoyment of the chambers left vacant by the extraction of the salt? *See* International Salt Co. v. Geostow, Inc., 878 F.2d 570, 104 O.&G.R. 276 (2d Cir.1989).

11. Able executes an oil and gas lease to Baker. Able is also the owner of the surface estate. Baker has drilled a producing gas well but now seeks to convert the producing well into an injection well for the purposes of storing gas. Does the oil and gas lease give Baker the right to store gas in the reservoir which no longer contains sufficient quantities of native gas to make it economically feasible to continue production? *See* Pomposini v. T.W. Phillips Gas & Oil Co., 397 Pa.Super. 564, 580 A.2d 776 (1990).

12. Right Refinery's operations allow certain hydrocarbons which are being processed at their facility to escape into the ground. These hydrocarbons migrate into a shallow reservoir or formation. Do the owners of the adjacent parcels have the right to drill wells and produce the escaped hydrocarbons? *Compare* Champlin Exploration, Inc. v. Western Bridge & Steel Co., Inc., 1979 OK 108, 597 P.2d 1215, 64 O.&G.R. 160 *with* Frost v. Ponca City, 1975 OK 141, 541 P.2d 1321, 53 O.&G.R. 370.

13. With the injection of hydrocarbons or other fluids into underground formations, problems may arise when the injected fluids migrate across property lines and affect neighboring interests? That issue is discussed in Chapter 7, Section 7, *infra.*

─────────

SECTION 2. CLASSIFICATIONS AND CONSEQUENCES

A. INTRODUCTION

Martin & Kramer, *Williams and Meyers Oil and Gas Law* § 202

§ 202. Interests in Oil and Gas Which Landowner May Create by Grant or Reservation

The owner of land in fee simple absolute was said at common law to own the land to an indefinite extent, upwards as well as downwards; "cujus est solum, ejus est usque ad coelum et ad inferos." The landowner alone was entitled to prospect for, sever and remove from the land anything found on or beneath the surface.

The owner may, of course, grant to others certain interests in the land. In general, the types of interests which the landowner may create by grant or reservation in oil, gas and other minerals are leasehold interests, mineral interests, and royalty interests.

§ 202.1 Leasehold interest

The landowner may execute the type of instrument generally described as an oil and gas lease. The lessee under this instrument is given the exclusive authorization to go upon the land for the purpose of prospecting for oil and gas (and usually other minerals also), severing and removing the same. Typically the duration of the lessee's authorization is for a stated term of years "and so long thereafter" as minerals are produced in paying quantities. Frequently the term of years may be extended by operations other than production from the premises, *e.g.,* by designated drilling operations. The consideration for the authorization given the lessee is most frequently an agreement by the lessee to deliver to the lessor a share of the product of the land, or the proceeds thereof, free of cost, described as a royalty. The landowner may also receive a cash consideration on the execution of the lease and be given the benefit of certain covenants or promises by the lessee relating to his operations on the premises.

The very name "lease" is unfortunate inasmuch as it tends to give the impression to the uninformed that the relationship arising between the parties to an oil and gas lease is the same as that of landlord and tenant under a common law lease of land, whereas, except in Louisiana, the dissimilarities are more important than the similarities.

The most common oil and gas lease in the mid-continent area is the so-called "Producers 88" lease, or the "Producers 88 lease—revised." These terms do not describe a specific instrument or indicate the interests created thereby. Apparently every landowner leasing his land for oil and gas desires a "Producers 88" lease, and the "lease hound", landman, or other person seeking to obtain a lease obligingly provides a "Producers 88" lease for his signature. Thus in a leading Texas case, specific performance of a contract in writing to execute "an 88 form lease" on certain land was denied, the court declaring: "As we see it the reference to 'an 88 form lease' is as incapable of definite application as if the term 'oil and gas lease form' had been used instead."

§ 202.2 Mineral interest

A mineral interest may be created by grant or by reservation or exception. The mineral interest thus created may be the entire mineral interest in the premises or it may be an undivided interest in the minerals; it may be in fee, fee simple defeasible, for life or for years.

The owner of the full mineral interest in particular premises normally has the right to go upon the premises for the purpose of prospecting for, severing and removing therefrom all minerals (or specifically described minerals). A severed mineral interest normally includes development and executive rights, *i.e.,* the right to drill or to execute an oil and gas lease. With increasing frequency, however, instruments creating interests in oil and gas grant or reserve the executive rights for the entire tract of land, *e.g.,* A "to B, excepting and reserving ½ of the minerals and the exclusive right to drill upon and/or lease the above described land."

§ 202.3 Royalty interest

A royalty interest may be created by grant or by reservation or exception, just as a mineral interest. This interest differs from a mineral interest in that the owner is not authorized to go upon the premises in which the royalty interest exists for the purpose of prospecting for, severing and removing minerals, although he may be given certain rights of ingress and egress for the purpose of informing himself as to the details of operations on the land with respect to exploration and development. The owner is, however, entitled to share in such minerals as are severed, or the proceeds thereof.

A royalty interest is a normal incident of an oil and gas lease; the lessor reserves such interest in production by the lessee. A royalty interest may, however, be created apart from a lease. A landowner before lease may convey a royalty interest in the land (or convey the land or minerals, reserving such royalty interest); after leasing the land he may convey all or some portion of his reserved royalty under the lease or he may convey a royalty interest for a duration not related to the life of the particular lease, *e.g.,* in fee, for life, for a term of years, or for a term of years "and so long thereafter" as oil and gas is produced.

More complex interests in oil and gas may be created. The oil and gas lease is considered in Chapter 4. Mineral, royalty and other landowner interests are considered in Chapter 6.

B. THE CORPOREAL—INCORPOREAL DISTINCTION
(*See* Martin & Kramer, *Williams and Meyers Oil and Gas Law* §§ 208–215).

GERHARD v. STEPHENS

Supreme Court of California, 1968.
68 Cal.2d 864, 69 Cal.Rptr. 612, 442 P.2d 692, 31 O.&G.R. 28.

TOBRINER, JUSTICE. Plaintiffs brought these four actions to quiet title to undivided mineral interests underlying section 31, township 16 south, range 11 east, Mt. Diablo base and meridian, in San Benito County (section 31). They claimed these interests as successors of stockholders in two now defunct corporations, Ashurst Oil, Land and Development Company (Ashurst) and California Oil Products Company (COP), which obtained specified mineral rights in section 31 in 1910. [The conveyances transferred "all petroleum, coal oil, naptha, asphalt, maltha, brea, bitumen, natural gas and other kindred or similar substances and deposits and rocks, gravels or other formations containing or yielding any of said substances" lying underneath described premises and the right to enter on the property with the necessary equipment and extract the described substances. The rights were transferred to the corporation, its successors and assigns forever. (Eds.)]

. . .

In 1956 Shell began drilling operations [under a lease from the surface owners. Eds.] and struck oil on section 31. Plaintiff Gerhard, who had no connection with any of the transactions outlined above, read of the discovery and began investigating the title to section 31. Learning of the interests transferred to

COP and Ashurst he approached the successors of the stockholders, obtained some of the claims, and instituted legal action. Other heirs of the stockholders brought separate actions. The trial court entered judgment in favor of defendants [the surface owners and their lessees. Eds.]

. . .

1. Abandonment

A. *Plaintiffs' interests are subject to abandonment.*

The trial court found that "the interests in section 31 claimed by plaintiffs . . . to the oil, gas and other fugacious minerals have been abandoned." Plaintiffs challenge the propriety of this finding. They point to the general rule that legal title to a fee simple can never be abandoned. They now claim ownership of such an estate in the fugacious minerals underlying section 31 and accordingly conclude that such an estate could not be abandoned. Defendants, on the other hand, rely on the general rule that property interests in the nature of *incorporeal hereditaments* can be abandoned. They so categorize the interests obtained by plaintiffs' predecessors in the fugacious minerals and therefore contend that these rights are subject to abandonment.

Our courts have not spoken on this precise question. Since, however, the question involves title to interests in real property we must, if possible, find our solution within the rich but rigid context of the historical background of the common law and the California cases. As we explain, however, not only the precedents but also the considerations underlying these precedents lead to the identical conclusion: these interests may be abandoned.

We shall point out that . . . the corporations [had] the exclusive and perpetual privilege of drilling for oil and gas. Under California law we classify such an interest as a profit *a prendre,* an incorporeal hereditament. We then explain that profits *a prendre* are essentially indistinguishable from easements and, thus, like easements and other incorporeal hereditaments they can be abandoned. In so holding we reject plaintiffs' argument that since California cases describe a perpetual profit *a prendre* as an "estate in fee" such a perpetual interest is not subject to abandonment under the general rule that "fee interests" in real property cannot be abandoned; the cases have used the term "fee interest" in two different contexts: (a) to indicate the perpetual nature of the interest; (b) to indicate the possessory, or corporeal, nature of the interest. The cases in which the courts have declared that a "fee interest" cannot be abandoned use the term in the latter sense; the cases that hold that a perpetual profit is a "fee interest" use the term in the former sense.

In support of our conclusion we note that the historical reason for the refusal of courts to recognize abandonment of certain interests in real property—the possibility of voids in titles—does not apply here. To the contrary the recognition that holders of ancient titles and their successors may abandon such interests tends to improve the marketability of titles. It removes encrusted impediments to titles and thus permits the present-day sale and disposition of land so that this prime resource may be utilized to the maximum.

1. *The deeds gave to plaintiffs the exclusive privilege of drilling for oil and gas, which constitutes a profit a prendre, an incorporeal hereditament.*

We begin with consideration of the leading case of Callahan v. Martin (1935) 3 Cal.2d 110, 43 P.2d 788, 101 A.L.R. 871. In that case this court definitively resolved the conflict in the early California cases as to the nature of ownership of oil and gas interests in this state. We there rejected the view held by the majority of concerned jurisdictions that the assignee of an oil and gas interest possesses a defeasible fee in definite corporeal real property. (*See, e.g.,* Texas Co. v. Daugherty (1915) 107 Tex. 226, 176 S.W. 717, L.R.A.1917F, 989; *see also* 1 Williams & Meyers, Oil and Gas Law, § 203.3; 1 Kuntz, Oil & Gas, § 2.4, p. 66.) The concept of the defeasible fee, we held, did not sufficiently recognize the practical difference between oil and gas and solid minerals: the latter remain in place beneath the surface of land, but the former, fugacious and vagrant may be drawn from beneath the surface of other lands. (3 Cal.2d at pp. 116–117, 43 P.2d 788.)

We held that "an operating lessee under a lease for a term of years . . . has an interest or estate in real property in the nature of a profit *a prendre,* which is an incorporeal hereditament, and that the assignee of a royalty interest in oil rights under an assignment by the landowner[1] also has an interest or estate in real property in the nature of an incorporeal hereditament." (P. 118, 43 P.2d p. 792.)[2] We further stated, "The holders of oil interests under a landowner's assignments have rights of profit *a prendre,* which are estates in real property, and are properly described as real property, or real estate, where, as in the instant case, they are to endure in perpetuity." (Pp. 127–128, 43 P.2d p. 797.)

In Dabney–Johnston Oil Corp. v. Walden (1935) 4 Cal.2d 637, 52 P.2d 237, we further elaborated the California position. "The owner of land has the exclusive right on his land to drill for and produce oil. This right inhering in the owner by virtue of his title to the land is a valuable right which he may transfer, *The right when granted is a profit a prendre, a right to remove a part of the substance of the land. A profit a prendre is an interest in real property in the nature of an*

---

[1] Although the court in *Callahan* terms the assignee's interest a "royalty interest," it was similar to those created in the instant case. We construed the conveyance in *Callahan* as giving the assignee the right, subject to any outstanding leases, to enter on the property, drill for, and produce oil.

[2] The term "corporeal hereditament" generally refers to tangible real property. An incorporeal hereditament is a right stemming from a corporeal thing. It may relate to land, in which case it constitutes "a limited interest or estate in land . . .. Where this limited interest is to endure in perpetuity or for life, it is a freehold interest, and real property or real estate, as well as an estate in real property." (Callahan v. Martin, *supra,* 3 Cal.2d at p. 120, 43 P.2d at p. 793.)

A profit *a prendre* is an incorporeal hereditament; it is a right to take something from the land of another.

incorporeal hereditament. . . . The profit *a prendre,* whether it is unlimited as to duration or limited to a term of years, is an estate in real property. If it is for a term of years, it is a chattel real, which is nevertheless an estate in real property, although not real property or real estate. Where it is unlimited in duration, it is a freehold interest, an estate in fee, and real property or real estate. Thus, although the oil and gas in place doctrine is rejected, interests in oil rights which are estates in real property may be granted separate and apart from a grant of surface title." (Italics added.) (P. 649, 52 P.2d p. 243.)

With this background we proceed to analyze the grants to the two corporations. We have no doubt that Abrams and Brandt purported to convey to the corporations the entire interest of the partnership in the oil, gas, and other hydrocarbonic substances underlying section 31. As we pointed out in La Laguna Ranch Co. v. Dodge (1941) 18 Cal.2d 132 at page 135, 114 P.2d 351 at page 353, 135 A.L.R. 546, however, "[T]he owner of land does not have an absolute title to the oil and gas in place as corporeal real property, but rather has the 'exclusive right' to drill for oil and gas upon his premises." Since an owner may not effectively transfer rights in property greater than those he himself is able to enjoy (23 Am.Jur.2d, Deeds, § 289), the corporations received no title to the oil and gas as corporeal real property but rather obtained the exclusive privilege of drilling for these substances. This right, as explained by *Callahan* and *Dabney–Johnston,* is a profit *a prendre,* an incorporeal hereditament. Since the right is to endure in perpetuity, it is a "freehold interest, an estate in fee, and real property or real estate."

2. Plaintiffs' perpetual profits a prendre, like easements, are subject to abandonment.

Following the common law this court has classified profits *a prendre,* together with easements, as incorporeal hereditaments. As we shall point out, for the purposes of the issue now before us, easements and profits *a prendre* are indistinguishable. "[T]he term 'easement' is so used [in the Restatement of Property] as to include within its meaning the special meaning commonly expressed by the term 'profit.' . . . In phrasing the rules applicable to each of these interests it has been found . . . that in no case was there a rule applicable to one of these interests which was not also applicable to the other. . . . " (Rest., Property, § 450, Note; Costa v. Fawcett (1962), 202 Cal.App.2d 695, 21 Cal.Rptr. 143; Civ.Code, § 802.)

. . .

. . . Although no California authority precisely states whether the perpetual profit may be abandoned, we submit that the cases discussing abandonment of perpetual easements created by grant in real property disclose the appropriate rule.

The cases have held that perpetual easements created by grant in real property *can* be abandoned.

We conclude that profits, like easements, can be abandoned. In so holding, we follow the general rule that all incorporeal hereditaments can be abandoned.

As the leading authorities in the field of oil and gas law, Professor Williams, Dean Maxwell, and Professor Meyers, have declared, such perpetual profits *a prendre,* constituting incorporeal hereditaments, were subject to abandonment. "At common law, incorporeal estates but not corporeal estates were subject to extinguishment by abandonment. If the mineral grantee or lessee has merely an incorporeal estate, then the courts may be able to clear the title of the interest of such grantee or lessee who makes no effort for a long period of time to produce minerals from the land by holding that his nonaction is sufficient evidence of intent to abandon his incorporeal interest." (Williams, Maxwell, Meyers, Cases and Materials on the Law of Oil and Gas (1st ed. 1956); *see also* 1 Williams & Meyers, Oil & Gas Law, § 210.1; Smith, Methods for Facilitating the Development of Oil and Gas Interests (1964) 43 Tex.L.Rev. 129, 161; *cf.* Woodward, Ownership of Interests in Oil and Gas (1965) 26 Ohio St.L.J. 353, 365–366.)

The property law of this state thus compels the conclusion that the perpetual profits *a prendre* here involved are subject to abandonment.

3. The cases which hold that a "fee interest" cannot be abandoned use the term in its possessory rather than durational sense; cases using "fee interest" in describing an incorporeal hereditament refer only to its duration and do not hold that it cannot be abandoned.

We next demonstrate the fallacy in plaintiffs' argument that, since *Callahan* and *Dabney–Johnston* describe a *perpetual* profit as a "fee interest," the interest is because of its "fee" nature not subject to abandonment. We shall explain that in *Callahan* and *Dabney–Johnston* the court used the term "fee" to designate merely the duration of the estate; this designation in no way alters our conclusion that a perpetual incorporeal hereditament can be abandoned.

It is true, as plaintiffs point out, that many California cases have declared that a *fee simple* interest in *real property* may not be abandoned. . . . [T]he courts have also held, however, that perpetual easements in real property are also "fee interests" yet that such interests *can* be abandoned.

The explanation of the superficial disparity in the consequences of ownership of these interests in real property becomes clear when we consider that our courts have employed the term "fee" to describe two different concepts.[3] In one sense the courts use "fee" to describe the totality of the *possessory and corporeal* rights of ownership in real property. Hence in some cases they have spoken of a *fee* interest as distinguished from an *easement.*

On the other hand, the courts describe fee ownership as any estate of inheritance. . . .

The holdings in *Callahan* and *Dabney–Johnston* clearly negate the possibility

---

[3] We rarely find occasion to employ the term "fee" in a third sense, its most traditional meaning: indicating possessory rights in land held of someone else as contrasted with allodial ownership. (*See* 1 Co.Litt. *1.a.–1.b; 2 Blackstone, Commentaries *104–105.)

that this court used the description "fee" as an indication of anything but the *duration* of the interests. Thus *Dabney–Johnston* speaks of a profit held in fee in contradistinction to one held for a term of years. (4 Cal.2d 637, 649, 52 P.2d 237.) And the thrust of *Callahan* is that the holder of a profit to remove oil and gas enjoys no definitive possessory ownership of these substances until they are severed from the realty and become personal property.

We conclude that the rulings . . . that a "fee" interest in real property cannot be abandoned are explicable only upon an analysis of the facts involved in the particular cases; in each case the court concerned itself with title to corporeal real property. Thus these cases are entirely consistent with the cases we follow here, holding that easements "in fee" may be abandoned.

To summarize, we look to the *nature* of the interest as well as its *duration* to define plaintiffs' rights. Incorporeal interests, as distinguished from corporeal ones, may be abandoned, whatever their life, whether limited or unlimited in time, whether "fee" or a term, whether perpetual or restricted. In short, the temporal life of the hereditament does not tell us for this purpose what *kind* of a legal interest it is;   we must classify it according to its genus, not merely its durational characteristic.

4. Principle and policy support the conclusion that plaintiffs' interests are subject to abandonment.

Although the historical rationale for the rule that certain interests in real property cannot be abandoned finds little articulation in the cases, the reason appears to be that society cannot tolerate voids in the ownership of land. (2 American Law of Property, Easements, § 8.98, p. 304.) As stated by the court in Kern County Land Co. v. Nighbert, *supra,* 75 Cal.App. 103 at 108, 241 P. 915 at 917, "The title to land is not a nebulous thing; it must rest somewhere."[4] No such abeyance in ownership arises, however, if the owner of a perpetual profit chooses to abandon his interest.   Comment (a) to section 504 of the Restatement of Property states: "That there is an ownership ready to take the benefit resulting from an abandonment of an easement is the probable explanation of the tolerance of the law toward the abandonment of such interests.   In many cases, of which the ownership

---

[4] This rationale probably stems from feudal society's disinclination to tolerate an abeyance of seisin. "But a state of abeyance was always odious, and never admitted but from necessity, because in that interval, there could not be any seisin of the land, nor any tenant to the *praecipe,* nor any one of the ability to protect the inheritance from wrong, or to answer for its burdens and services. . . . The title if attacked, could not be completely defended, because there was no one in being whom the tenant could pray *in aid* to support his right. . . . The particular tenant could not be punishable for waste, for the writ of waste could only be brought by him who was entitled to the inheritance.   So many operations of law were suspended by this sad theory of an estate in abeyance that great impediments were thrown in the way of it, and no acts of the parties were allowed to put the immediate freehold in abeyance by limiting it to commence *in futuro;* and we have seen, that one ground on which the rule in Shelley's case is placed, was to prevent an abeyance of the estate." (4 Kent, Commentaries *259.)

of land in fee is an example, an abandonment, if permitted, would result in a void in the ownership of the affected thing, the filling of which would be largely a question of chance and would probably produce grave uncertainty of title." (*Cf.* 2 Pollock & Maitland, History of English Law (1895) 80.)

If interests in real property can be and are abandoned, they do not become, as in the case of personal property, the property of the first appropriator (*see* 1 Cal.Jur.2d, Abandonment, § 10), but instead return to the estate out of which they were carved. The abandonment of a profit *a prendre,* therefore, because the profit in essence is an easement, does not become subject to the void in ownership that the common law of land title sought to avoid. If a perpetual right of way or other easement is abandoned, the property interest reverts to the servient estate. Similarly, a perpetual right to remove oil and gas (see Civ.Code, §§ 801, 802) would ordinarily revert to the surface estate,[5] thereby freeing that estate of its burden and permitting its owner more complete utilization and enjoyment of his property.

Commentators have noted that "The abandonment concept, when applied, frequently serves the very useful purpose of clearing title to land of mineral interest of long standing, the existence of which may impede exploration or development of the premises by reason of difficulty of ascertainment of present owners or of difficulty of obtaining the joinder of such owners."[6] (1 Williams & Meyers, Oil and Gas Law, § 210.1, pp. 104–105; *see also* Smith, Methods for Facilitating the Development of Oil and Gas Lands Burdened with Outstanding Mineral Interests, supra, 43 Tex.L.Rev. 129, 129–136; Street, The Need for Legislation to Eliminate Dormant Royalty Interests (March 1963) 42 Mich.State B.J. 49; *cf.* Audrain, Title Insurance of Oil and Gas Leases (1956) 3 U.C.L.A.L.Rev. 523, 530.)[7]

---

[5] We state no absolute rule on this point because under certain circumstances not present here it may not apply. Thus, if the grantor carves out a perpetual profit to remove oil and gas and later conveys to another the fee ownership in an underground corporeal stratum (*see, e.g.,* Nevada Irr. Dist. v. Keystone Copper Corp., supra, 224 Cal.App.2d 523, 36 Cal.Rptr. 775), upon abandonment by the profit holder the ownership of the oil and gas removed from the stratum might well rest in the subsurface owner.

We further recognize the possibility, not discussed by the parties, that when an incorporeal hereditament is held in cotenancy, the interest abandoned by one or more cotenants might revert to the remaining cotenants instead of to the servient estate. In view, however, of plaintiffs' pleadings we decline to comment on this possibility at this time.

[6] Several jurisdictions, confronted perhaps more frequently with this problem have enacted legislation designed to terminate dormant oil and gas interests, whether limited or unlimited in duration, following a specified period of nonuse or failure to rerecord the interests. (La.Stat.Ann.Civil Code, arts. 789, 3546 (10 years); Mich.Stats.Ann. § 26.1163(1), C.L.Mich.1948, § 554.291[P.A.1963, No. 42] (20 years); Tenn.Code Ann. § 64–704 (10 years); Va.Code Ann., §§ 55–154, 55–155 (35 years).) [The Louisiana legislation is now a part of the Louisiana Mineral Code, Art. 27, La.Rev.Stat., Title 31. Eds.]

[7] Reviewing the cases in those jurisdictions whose courts have had occasion to consider this question, we find no universally established rule contrary to the one we adopt. The Wyoming

. . . By describing rights identical to those granted to the corporations as incorporeal hereditaments our court fore ordained the conclusion we now reach. Moreover, a ruling that incorporeal hereditaments of the type involved here may be abandoned tends to promote the marketability of title by facilitating the clearing of titles. To that extent it better fulfills the demands of a modern economic order. Further, it reduces the possibility of the resurrection of the ghosts of abandoned claims by which title searchers and forgotten owners collect the windfalls of accidental profit.

In recognizing the freedom of an owner of an unlimited profit *a prendre* to abandon his interest, we realize, of course, that factual issues may determine the ownership of such interests. The establishment of title by the resolution of factual, as opposed to legal, questions, however, does not create undesirable uncertainty as to the ownership of real property, since trial courts and juries must frequently make such resolutions determinative of title. In the instant case we protect the required stability of title by adhering to the *rules of law* governing ownership of real property.

Moreover, our holding leaves open no possibility that owners of valuable property rights of the type herein involved will, contrary to their desire, lose these rights.[8] As we shall point out, abandonment hinges upon the intent of the owner to forego all future conforming uses of his property, and the trier of fact must find the conduct demonstrating the intent "so decisive and conclusive as to indicate a clear

---

Supreme Court holds that an oil and gas lease is a profit *a prendre;* it reasons that since a profit is an incorporeal hereditament, it can be abandoned. (Boatman v. Andre (1932) 44 Wyo. 352, 362, 12 P.2d 370.) Illinois, which the commentators declare follows a theory of ownership similar to that enunciated in *Callahan,* holds that an unlimited interest in oil and gas may not be lost by abandonment. (Jilek v. Chicago W. & F. Coal Co. (1943) 382 Ill. 241, 253–254, 47 N.E.2d 96, 146 A.L.R. 871; *see* 1 Williams & Meyers, op. cit. *supra,* § 203, p. 31.) Other states, which may also be classified as adopting the "nonownership" rule, have held that an oil and gas interest may not be lost by mere nonuser. (Piney Oil & Gas Co. v. Scott (1934) 258 Ky. 51, 63, 79 S.W.2d 394; Noble v. Kahn (1952) 206 Okl. 13, 16, 240 P.2d 757, 35 A.L.R.2d 119; Gill v. Fletcher (1906) 74 Ohio St. 295, 78 N.E. 433 (gypsum and plaster); *see* 1 Williams & Meyers, op.cit. *supra,* § 203, p. 31). Since both nonuser and intent to abandon must concur in order to produce abandonment of an incorporeal hereditament (Smith v. Worn, supra, 93 Cal. 206, 212, 28 P. 944), these cases are consistent with our holding. Three states, which are generally recognized as adherents of the "absolute ownership" theory, refuse to permit the owner of an oil and gas interest to abandon it or to lose it by nonuser. (Hillegust v. Amerada Petroleum Corp. (Tex.Civ.App.1955) 282 S.W.2d 892, 897 (cannot abandon fee simple determinable interest); Sanford v. Alabama Power Co. (1951) 256 Ala. 280, 288, 54 So.2d 562 (following severance a mineral estate is a *incorporeal* hereditament and cannot be lost by mere nonuser); Bodcaw Lumber Co. v. Goode (1923) 160 Ark. 48, 61, 254 S.W. 345, 29 A.L.R. 578 (mineral interest cannot be lost by nonuser); *see* 1 Williams & Meyers, op. cit. *supra,* § 203, p. 31.) In none of the cases above cited did the courts elaborate on the problem herein involved; we find no discussion or analysis to assist us.

[8] A conscientious owner of such a right can, of course, fully protect himself, e.g., by applying for separate assessment (Rev. & Tax.Code, § 2803, *cf.* Smith v. Anderson (1967) 67 Cal.2d 635, 652, 63 Cal.Rptr. 391, 433 P.2d 183).

intent to abandon . . .." (Smith v. Worn, supra, 93 Cal. 206, 213, 28 P. 944, 945.)

B. *Evidence of abandonment: general principles.*

We now consider the propriety of the trial court's finding that all plaintiffs or their predecessors had abandoned their interests in the fugacious minerals underlying section 31. . . .

We have pointed out, supra, that we cannot differentiate profits *a prendre* and easements in terms of the legal consequences stemming from ownership. Accordingly, in determining whether defendants have established an abandonment, we may with propriety consider the cases that discuss abandonment of perpetual[9] easements created by grant.[10]

" 'As a general rule, in order to constitute an abandonment of an easement . . . there must be a nonuser accompanied by unequivocal and decisive acts on the part of the [dominant tenant], clearly showing an intention to abandon." ' (People v. Southern Pacific Co., *supra,* 172 Cal. 692, 700, 158 P. 177, 180.) As other cases point out, however, the owner's nonuser itself may under some circumstances constitute such an act. "Where non-user is evidence of an abandonment of a right, the question is one of intention, depending on the circumstances . . .." (Smith v. Worn, *supra,* 93 Cal. 206, 213, 28 P. 944, 945.)[11] . . .

. . .

Several California cases which have upheld a trial court's determination that a dominant tenant has abandoned an easement created by grant have found sufficient evidence of the requisite intent by evaluating the implications of nonuser in terms of "economic exigencies" and "external realities." (Ocean Shore R.R. Co. v. Doelger, *supra,* 179 Cal.App.2d 222, 232, 3 Cal.Rptr. 706; People v. Ocean Shore Railroad, *supra,* 32 Cal.2d 406, 419, 196 P.2d 570.) Thus, in the *Ocean Shore* cases, the courts sustained findings of abandonment when the evidence showed that the restoration of service would be prohibitively expensive and that effective

---

[9] The California courts have held that "Abandonment will be more readily found in the case of oil and gas *leases* than in most other cases." (Hall v. Augur (1927) 82 Cal.App. 594, 599, 256 P. 232; Romero v. Brewer, *supra,* 58 Cal.App.2d 759, 763, 137 P.2d 872.) (Italics added.) Such a leasehold interest is an incorporeal hereditament (Callahan v. Martin, *supra,* 3 Cal.2d 110, 118, 43 P.2d 788), but leasehold interests in oil and gas are, of course, of limited duration. Moreover, a grantor typically gives the lessee the exclusive right to develop and take the oil and limits himself to a compensation based on the production of the lessee. Thus if a lessee fails to drill, it would be unfair to preclude the owner of the servient estate from making other arrangements to gain the hoped-for benefits.

[10] Civil Code, section 811, subdivision 4, provides that nonuse for the prescriptive period extinguishes an easement acquired by *adverse user.*

[11] The Restatement points out that "Conduct from which an intention to abandon an easement may be inferred may consist in a failure to make the use authorized. Non-use does not of itself produce an abandonment no matter how long continued. It but evidences the necessary intention. Its effectiveness as evidence is dependent upon the circumstances." (Rest., Property, § 504, com. d . . ..)

competition with other carriers was realistically impossible. . . .

Other California cases point out that the dominant tenant's nonuse does not tend to demonstrate the requisite intent merely because the activation of the easement would produce no *immediate* benefit sufficient to justify the necessary expense. As we pointed out in Faus v. City of Los Angeles, *supra,* 67 Cal.2d 350 at 364, 62 Cal.Rptr. 193, at 202, 431 P.2d 849, at 858, "the holder must . . . manifest an intent to forswear all *future* conforming uses of the easement." (Italics added.) Thus in People v. Southern Pacific Co., *supra,* 172 Cal. 692, 158 P. 177, we rejected the trial court's finding that the railroad had abandoned its easement: "There is no proof that the needs of commerce during this period were such that . . . the existing or probable immediate future traffic would have justified the cost of building such tracks . . .." (P. 699, 158 P. p. 179.) In Parker v. Swett (1919) 40 Cal.App. 68, 180 P. 351, the court reversed the judgment for defendants: "The evidence of plaintiff's predecessor in interest is to the effect that he never exercised his right to lay the pipe for the reason that he did not require the water during his ownership of the ranch and the installation of a pipe line was quite expensive." (P. 74, 180 P. p. 353.)

We have pointed out *supra* that "While nonuse alone does not establish intent, the court may consider the length of the nonuse in ascertaining the existence of such intent." (Ocean Shore R.R. Co. v. Doelger, *supra,* 179 Cal.App. 222, 232, 3 Cal.Rptr. 706, 713.) We must determine whether the trial court could reasonably find the necessary intent on the basis of plaintiffs' 47 years of nonuser,[12] their apparent lack of concern with their interests, and their failure to give any visible indication of intent to make future use of the property.

Abrams and Brandt purchased the Syncline Ranch in 1905, during an "oil boom" in the area. They organized Ashurst in order to obtain financial backing for the contemplated exploration for, and production of, oil and gas. After the corporation's dissolution in 1915, however, the oil interests were no longer owned and managed by a unified corporate structure; instead these interests embraced 148 owners, many of whom, as evidenced by the judgment roll in action 1870, could not be located.

Any one of these owners individually could have entered upon section 31 and explored for oil and gas. In doing so, however, he would, if unsuccessful, have been compelled to bear the entire expense and risk of the operations. If he had succeeded in his endeavor, he could not have excluded from section 31 other cotenants who had desired to drill[13] , and, as to the remaining cotenants, he would

---

[12]The trial court found that Abrams and Brandt conveyed a portion of the oil and gas rights in section 31 to COP in October 1910 and that COP never drilled on the property. The trial court found that Abrams and Brandt conveyed the oil and gas rights to Ashurst in March *1910* and that Ashurst drilled three unproductive wells between *1900* and *1903* (*sic*). Plaintiffs filed their complaints in 1957 and 1958.

[13] In *Callahan* we pointed out that "If numerous holders of oil rights in a single parcel of land are

have been required to render an accounting for the profits.[14]

An extended nonuser, when considered in light of these "economic exigencies" and "external realities" (Ocean Shore R.R. Co. v. Doelger, *supra*, 179 Cal.App.2d 222, 232, 3 Cal.Rptr. 706) cannot *by itself* support the trial court's finding of intent to abandon. Although the acquisition of such an interest in oil and gas may be entirely speculative and the owner may well intend to foresake his interest if the "oil boom" which induced his purchase collapses, plaintiffs and their predecessors, incurring no significant detriment by retaining their interests, might well have contemplated that the property might again become valuable because of the efforts of others.

In order to protect the owner of an unlimited profit *a prendre* or other incorporeal hereditament against "involuntary" abandonment under circumstances in which conflicting inferences may be drawn from his nonuser we hold that the trial court must find either that the owner's future use of the right could result only from a palpably unsound business judgment[15] or that the owner has given a further indication of his intent to abandon. Applying this principle, derived from the cases cited supra, to the instant situation we find no substantial evidence to support the finding that plaintiffs Gerhard, Solomon, and Mettler abandoned their interests in section 31.

[The court's discussion of the abandonment by other plaintiffs of their interest is omitted. A finding of abandonment was sustained on the basis of the rejection of stock in Ashurst and COP combined with the extended period of nonuser. Eds.]

Our holding that the trial court properly found that the Weber plaintiffs abandoned their interests in the oil and gas and other fugacious minerals underlying section 31 does not completely dispose of that appeal. The grants to Ashurst and COP included Abrams' and Brandt's interest in certain specified hydrocarbon substances other than oil and gas. Although most of these substances are liquid or semiliquid in their natural state, one such substance possesses nonfugacious qualities.[16] We have pointed out, supra, that the partnership's deed purports to

---

unable to agree upon an operating lessee or upon the terms of an oil lease, we are inclined to think that the powers of a court of equity may be invoked to formulate a just and reasonable plan for the development and production of oil upon the land, and to settle the controversy in accordance therewith. But this can be determined as the question may arise in future litigation." (3 Cal.2d at p. 126, 43 P.2d at p. 796.)

[14] Of course a coowner could attempt to benefit from his interests yet avoid the risk of nondiscovery if he could obtain a lessee. A potential lessee, however, might well be unwilling to undertake operations in view of the small size of a single owner's interest and the possibility that other cotenants would not ratify the lease.

[15] We would be reluctant to assume that an *individual* abandons his right merely because sound business judgment would call for this action. The intent to abandon is subjective, and we recognize the possibility that personal pride in, and attachment to, property rights may cause an owner to adopt a less functional approach in the management of his possessions.

[16] Asphalt, for instance, consists "chiefly of hydrocarbons, ranging from hard and brittle to plastic

convey a complete possessory estate in the specified minerals, but, because of the fugacious nature of oil and gas, our courts could not give effect to a transfer of the latter substances as corporeal real property but would instead treat the deed as conveying ownership of an unlimited profit *a prendre* in the fugacious minerals. As to any nonfugacious minerals underlying section 31, however, Abrams and Brandt had title to them in place and therefore could convey them as corporeal real property.[17]

We have concluded, however, that the grantors, in conveying their rights in the enumerated minerals in a single deed to each corporation, did not intend to convey two different types of estate in the same instrument. The grantors had the power and right to convey a profit *a prendre* in the nonfugacious minerals. In making the conveyance they were primarily concerned with transferring the *oil and gas* rights to the corporations; as is evidenced by the then extant oil boom in the area and by the names of the corporations (Ashurst Oil, Land and Development Company and California Oil Products Company). In treating the incidental conveyance of the asphalt as a conveyance of a profit *a prendre* we are in accord with Callahan v. Martin, *supra,* 3 Cal.2d 110, 43 P.2d 788, and Dabney–Johnston Oil Corp. v. Walden, *supra,* 4 Cal.2d 637, 52 P.2d 237, in which the interests disputed were rights in oil, gas, and other hydrocarbons; yet this court did not distinguish, or even mention, nonfugacious hydrocarbons in its analysis of the nature of the interests in question.

[The judgment in favor of the defendants was reversed as to those plaintiffs found not to have abandoned their interests.]

TRAYNOR, C.J., and McCOMB, PETERS, MOSK, BURKE, and SCHAUER, JJ., concur.

Rehearing denied; McCOMB, J., dissenting. SCHAUER, J., sitting in place of SULLIVAN, J., who deemed himself disqualified.

NOTES

1. The *Gerhard* opinion makes reference to the usefulness of the abandonment doctrine in clearing the way for surface development. What are its drawbacks in promoting either surface or mineral development? Legislation was proposed in California to allow the surface owner to designate specific drillsites with the approval of the State Oil and Gas Supervisor. Although not adopted, legislation of this kind appears to be a good solution to the difficult problem of encouraging the most beneficial use of the surface despite the burden on the surface of easements of a mineral owner. For leases more than twenty years old California has provided an

---

in form." (Webster's New Internat. Dict. (3d ed.).)

[17] "The owner of real property may divide his lands horizontally as well as vertically, and when he conveys the subsurface mineral deposits separately from the surface rights, or reserves them from a conveyance of such surface rights, he creates two separate fee simple estates in the land, each of which has the same status and rank." (Nevada Irr. Dist. v. Keystone Copper Corp., *supra,* 224 Cal.App.2d 523, 527, 36 Cal.Rptr. 775, 778 (quartz, gold and silver deposits).)

action under Code of Civil Procedure § 772.030 to free the surface when it is "not reasonably needed for oil and gas operations" by a lessee. *See* Donlan v. Weaver, 118 Cal.App.3d 675, 173 Cal.Rptr. 566, 69 O.&G.R. 230 (1981).

2. Consider the court's suggestion that the owners of severed mineral interests can fully protect their interests by "applying for a separate assessment" for property tax purposes. Are there practical problems with the court's suggestion? *See* Dodge, *The Gerhard Doctrine of Abandonment—Outlook for California's Oil and Gas Industry*, 6 Land and Water L.Rev. 511, 520 (1971).

3. MOORER V. BETHLEHEM BAPTIST CHURCH, 272 Ala. 259, 130 So.2d 367, 16 O. & G.R. 491 (1961). Landowner executed an oil and gas lease and then brought suit against others to quiet title to the land. The complaint did not allege the terms of the lease and it was not attached to the papers filed in court. *Held:* Action dismissed because, without a construction of the lease, it does not affirmatively appear that plaintiff has the necessary possessory right to maintain a quiet title action. Would outstanding leases limited to oil and gas create such a problem in a jurisdiction adhering to the qualified or nonownership theory? Consider the effect of classification of oil and gas interests on jurisdiction and venue. *See* Kelly Oil Co. v. Svetlik, 975 S.W.2d 762, 140 O.&G.R. 605 (Tex. App.--Corpus Christi, 1998) (overriding royalty under leases of Mississippi lands; no jurisdiction for Texas courts to resolve title); CLK Company, L.L.C. v. CXY Energy Inc., 98-0802 (La.App. 4 Cir. 9/16/98), 719 So.2d 1098, 140 O.&G.R. 464 (overriding royalty classified as an incorporeal immovable; accordingly, the only proper venue was the parish where the land was located, not where the contract was made). What other consequences follow the classification of an interest as corporeal or incorporeal and what is the basis on which the classification should be made? *See* Martin & Kramer, *Williams and Meyers Oil and Gas Law* §§ 209–211.

Is there any good reason for the interests in oil and gas obtained by leasing from a landowner in an ownership "in place" state to be considered "inchoate" until the minerals are produced and reduced to possession? *See* Budd v. Ethyl Corp., 251 Ark. 639, 474 S.W.2d 411, 40 O. & G.R. 335 (1971).

4. Because an interest is classified as incorporeal does it follow that it is not an interest in real property? What answer do you find in *Gerhard? Compare* Phillips v. Springfield Crude Oil Co., 76 Kan. 783, 92 P. 1119 (1907), where a mechanics lien claimed for labor performed in drilling oil and gas wells was denied on the ground that an oil and gas lease "is a mere license to explore, an incorporeal hereditament, a profit a prendre" and thus not an interest in the real estate sufficient to support a mechanics lien. The legislative response to *Phillips* was the enactment in 1909 of an Act providing for a laborer's and materialman's lien on oil and gas leaseholds and providing that the lien shall be enforced against the leasehold in the same manner as is provided by law for enforcing mechanic's or other liens against real estate. Kan. Stat. Ann. §§ 55–207 to 55–210.

5. Can a state treat an oil and gas lease as personalty in some contexts and

realty in other contexts? In Utica National Bank & Trust Co. v. Marney, 233 Kan. 432, 661 P.2d 1246, 77 O. & G.R. 61 (1983) the Kansas Supreme Court found that oil and gas leases can be either personalty or realty depending on the circumstances. The court did note that leases "constitute personal property except in those specific instances when that classification is changed by statute for a specific purpose." An oil and gas lease was held not to be realty subject to the provisions of the statute providing for the automatic attachment of a judgment lien. *In accord*: High Oil, Ltd. v. National City Bank of Cleveland, Ohio, 22 Kan.App.2d 968, 925 P.2d 846 (1997)(judgment liens that attach to real property do not attach to oil and gas leases which are personal property). For other important legal consequences that may flow from the "realty" or "personalty" label, *see* Woodward, *Ownership of Interests in Oil and Gas*, 26 Ohio St.L.J. 353, 366 (1965).

6. The *Gerhard* opinion cites case authority for the proposition that "[a]bandonment will be more readily found in the case of oil and gas *leases* than in most other cases." Why should this be so?

7. The *Gerhard* opinion suggests that "when an incorporeal hereditament is held in cotenancy, the interest abandoned by one or more cotenants might revert to the remaining cotenants instead of to the servient estate." Under what conditions of ownership should this occur? *See* Discussion Notes, 56 O.&G.R. 53.

8. Is long-time non-user sufficient to constitute an abandonment of an oil and gas lease? What evidence is required to show the necessary intent to abandon in addition to non-user? *See* Olson v. Schwartz, 345 N.W.2d 33, 81 O.&G.R. 125 (N.D.1984); Shannon v. Stookey, 59 Ill.App.3d 573, 16 Ill.Dec. 774, 375 N.E.2d 881, 61 O.&G.R. 43 (1978).

9. Can a breach of the implied covenant to further develop the leasehold be used as a basis for finding an abandonment of the lease? If so, would the lessor have to provide notice to the lessee of the putative breach if the lease provided that the lessee must be notified of any breach of an implied covenant and given an opportunity to cure the breach? *See* Rook v. James E. Russell Petroleum, Inc., 235 Kan. 6, 679 P.2d 158, 80 O. & G.R. 471 (1984).

10. Would the breach of an express provision requiring the payment of a minimum annual royalty constitute an abandonment of the lease or justify a court declaring a forfeiture? M & C Oil, Inc. v. Geffert, 212 Kan.App.2d 267, 897 P.2d 191, 132 O.&G.R. 263 (1995).

11. Can a coal lease in a state which treats ownership of coal as a corporeal hereditament be abandoned? Citing several oil and gas cases, the West Virginia Supreme Court of Appeals found that such an interest can be abandoned after a review of several factors, including length of time of non-use, cause and reasonable diligence of the lessee. Christian Land Corp. v. C. & C. Co., 188 W.Va. 26, 422 S.E.2d 503, 123 O.&G.R. 358 (1992). See also Jolynne Corp. v. Michels, 191 W.Va. 406, 446 S.E.2d 494 (1994) where the court found that the oil and gas lease expired on its own terms where there was a failure to produce for over 10 years, although the court did use abandonment language to discuss the

nature of the lessor's claims.

12. Many oil and gas leases contain surrender clauses whereby the lessee may surrender all or a portion of the leased premises. Would a unilateral surrender by the lessee be effective without the consent of the lessor if the jursidiction treated the lease as a corporeal hereditament? An incorporeal hereditament? See e.g., Kennecott Corp. v. Union Oil Co. of California, 196 Cal.App.3d 1179, 242 Cal.Rptr. 403 (1987); Superior Oil Co v. Dabney, 147 Tex. 51, 211 S.W.2d 563 (1948).

13. West Virginia provides, by statute, for a "rebuttable legal presumption" of intent to abandon where a lessee fails to produce and market oil and/or gas from the leased premises. W.Va. Code § 36-4-9a. In Howell v. Appalachian Energy, Inc., 205 W.Va. 508, 519 S.E.2d 423 (1999), the court in applying the statute construed it in favor of finding abandonment by limiting the lessee to the four listed exceptions for rebutting the presumption.

TEXACO, INC. v. SHORT

Supreme Court of the United States, 1982.
454 U.S. 516, 102 S.Ct. 781, 70 L.Ed.2d 738, 72 O.&G.R. 217.

JUSTICE STEVENS delivered the opinion of the Court.

In 1971 the Indiana Legislature enacted a statute providing that a severed mineral interest that is not used for a period of 20 years automatically lapses and reverts to the current surface owner of the property, unless the mineral owner files a statement of claim in the local county recorder's office.[18] The Indiana Supreme Court rejected a challenge to the constitutionality of the statute. Ind., 406 N.E.2d 625 (1980). We noted probable jurisdiction, 450 U.S. 993, 101 S.Ct. 1693, 68 L.Ed.2d 192, and now affirm.

As the Indiana Supreme Court explained, the Mineral Lapse Act "puts an end to interests in coal, oil, gas or other minerals which have not been used for twenty years." The statute provides that the unused interest shall be "extinguished" and that its "ownership shall revert to the then owner of the interest out of which it was carved."[19] The statute, which became effective on September 2, 1971, contained a 2-year grace period in which owners of mineral interests that were then unused and subject to lapse could preserve those interests by filing a claim in the recorder's office.

----

[18] The statute is entitled the Dormant Mineral Interests Act, and is more commonly known as the Mineral Lapse Act. Ind.Code §§ 32–5–11–1 through 32–5–11–8 (1976), as added by 1971 Ind.Acts, Pub.L. 423, § 1.

[19] "Any interest in coal, oil and gas, and other minerals, shall, if unused for a period of 20 years, be extinguished, unless a statement of claim is filed in accordance with section five hereof [sic], and the ownership shall revert to the then owner of the interest out of which it was carved." Ind.Code § 32–5–11–1 (1976).

The "use" of a mineral interest[20] that is sufficient to preclude its extinction includes the actual or attempted production of minerals, the payment of rents or royalties, and any payment of taxes;[21] a mineral owner may also protect his interest by filing a statement of claim with the local recorder of deeds.[22] The statute contains one exception to this general rule: if an owner of 10 or more interests in the same county files a statement of claim that inadvertently omits some of those interests, the omitted interests may be preserved by a supplemental filing made within 60 days of receiving actual notice of the lapse.[23]

---

[20] As defined by the Act: "A mineral interest shall be taken to mean the interest which is created by an instrument transferring, either by grant, assignment, or reservation, or otherwise an interest, of any kind, in coal, oil and gas, and other minerals." Ind.Code § 32–5–11–2 (1976). The Indiana Supreme Court described the nature of this interest as follows: "Interests or estates in oil, gas, coal and other minerals lying beneath the surface of the land are interests in real estate for our purposes here, and as such are entitled beyond question to the firmest protection of the Constitution from irrational state action. They are vested property interests separate and distinct from the surface ownership. The State has no power to deprive an owner of such an interest without due process of law. They are entitled to the same protection as are fee simple titles. They are themselves of great utility and benefit to the society as a means of facilitating the development of natural resources." Ind., 406 N.E.2d, at 627.

[21] "A mineral interest shall be deemed to be used when there are any minerals produced thereunder or when operations are being conducted thereon for injection, withdrawal, storage or disposal of water, gas or other fluid substances, or when rentals or royalties are being paid by the owner thereof for the purpose of delaying or enjoying the use or exercise of such rights or when any such use is being carried out on any tract with which such mineral interest may be unitized or pooled for production purposes, or when, in the case of coal or other solid minerals, there is production from a common vein or seam by the owners of such mineral interests, or when taxes are paid on such mineral interest by the owner thereof. Any use pursuant to or authorized by the instrument creating such mineral interest shall be effective to continue in force all rights granted by such instrument." Ind.Code § 32–5–11–3 (1976).

[22] "The statement of claim provided in section one above shall be filed by the owner of the mineral interest prior to the end of the twenty year period set forth in section two [sic] or within two years after the effective date of this act, whichever is later, and shall contain the name and address of the owner of such interest, and description of the land, on or under which such mineral interest is located. Such statement of claim shall be filed in the office of the Recorder of Deeds in the county in which such land is located. Upon the filing of the statement of claim within the time provided, it shall be deemed that such mineral interest was being used on the date the statement of claim was filed." Ind.Code § 32–511–4 (1976).

[23] "Failure to file a statement of claim within the time provided in section 4 shall not cause a mineral interest to be extinguished if the owner of such mineral interest:

"1) was at the time of the expiration of the period provided in section four, the owner of ten or more mineral interests, as above defined, in the county in which such mineral interest is located, and;

"2) made diligent effort to preserve all of such interests as were not being used, and did within a period of ten years prior to the expiration of the period provided in section 4 preserve other mineral interests, in said county, by the filing of statements of claim as herein required, and;

The statute does not require that any specific notice be given to a mineral owner prior to a statutory lapse of a mineral estate. The Act does set forth a procedure, however, by which a surface owner who has succeeded to the ownership of a mineral estate pursuant to the statute may give notice that the mineral interest has lapsed.[24]

[Two cases were consolidated on appeal. In both it was stipulated that the mineral interests had not been used for 20 years, nor had a statement of claim been filed within the two year grace period. Eds.]

At all stages of the proceedings, appellants challenged the constitutionality of the Dormant Mineral Interests Act. Appellants claimed that the lack of prior notice of the lapse of their mineral rights deprived them of property without due process of law, that the statute effected a taking of private property for public use without just compensation, and that the exception contained in the Act for owners of 10 or more mineral interests denied them the equal protection of the law; appellants based these arguments on the Fourteenth Amendment of the United States Constitution. Appellants also contended that the statute constituted an impairment of contracts in violation of Art. 1, § 10, of the Constitution. The state trial court held that the statute deprived appellants of property without due process of law, and effected a taking of property without just compensation.

...

I

Appellants raise several specific challenges to the constitutionality of the Mineral Lapse Act. Before addressing these arguments, however, it is appropriate to consider whether the State has the power to provide that property rights of this character shall be extinguished if their owners do not take the affirmative action required by the State.

In Board of Regents v. Roth, 408 U.S. 564, 577, 92 S.Ct. 2701, 2709, 33

"3) failed to preserve such interest through inadvertence, and;

"4) filed the statement of claim herein required, within sixty (60) days after publication of notice as provided in section seven herein [sic], if such notice is published, and if no such notice is published, within sixty (60) days after receiving actual knowledge that such mineral interest had lapsed." Ind.Code § 32–5–11–5 (1976).

[24] "Any person who will succeed to the ownership of any mineral interest, upon the lapse thereof, may give notice of the lapse of such mineral interest by publishing the same in a newspaper of general circulation in the county in which such mineral interest is located, and, if the address of such mineral interest owner is shown of record or can be determined upon reasonable inquiry, by mailing within ten days after such publication a copy of such notice to the owner of such mineral interest. The notice shall state the name of the owner of such mineral interest, as shown of record, a description of the land, and the name of the person giving such notice. If a copy of such notice, together with an affidavit of service thereof, shall be promptly filed in the office of the Recorder of Deeds in the county wherein such land is located, the record thereof shall be prima facie evidence, in any legal proceedings, that such notice was given." Ind.Code § 32–5–11–6 (1976).

L.Ed.2d 548, the Court stated:

> Property interests, of course, are not created by the Constitution. Rather, they are created and their dimensions are defined by existing rules or understandings that stem from an independent source such as state law—rules or understandings that secure certain benefits and that support claims of entitlement to those benefits.

The State of Indiana has defined a severed mineral estate as a "vested property interest," entitled to "the same protection as are fee simple titles." Through its Dormant Mineral Interests Act, however, the State has declared that this property interest is of less than absolute duration; retention is conditioned on the performance of at least one of the actions required by the Act. We have no doubt that, just as a State may create a property interest that is entitled to constitutional protection, the State has the power to condition the permanent retention of that property right on the performance of reasonable conditions that indicate a present intention to retain the interest.

...

The Indiana statute is similar in operation to a typical recording statute. Such statutes provide that a valid transfer of property may be defeated by a subsequent purported transfer if the earlier transfer is not properly recorded. In Jackson v. Lamphire, 3 Pet. 280, 7 L.Ed. 679, the Court upheld such a statute, even as retroactively applied to a deed that need not have been recorded at the time delivered. The Court stated:

> It is within the undoubted power of state legislatures to pass recording acts, by which the elder grantee shall be postponed to a younger, if the prior deed is not recorded within the limited time; and the power is the same whether the deed is dated before or after the passage of the recording act. Though the effect of such a law is to render the prior deed fraudulent and void against a subsequent purchaser, it is not a law impairing the obligation of contracts; such too is the power to pass acts of limitations, and their effect. Reasons of sound policy have led to the general adoption of laws of both descriptions, and their validity cannot be questioned. The time and manner of their operation, the exceptions to them, and the acts from which the time limited shall begin to run, will generally depend on the sound discretion of the legislature, according to the nature of the titles, the situation of the country, and the emergency which leads to their enactment. *Id.*, at 290.

These decisions clearly establish that the State of Indiana has the power to enact the kind of legislation at issue. In each case, the Court upheld the power of the State to condition the retention of a property right upon the performance of an act within a limited period of time. In each instance, as a result of the failure of the property owner to perform the statutory condition, an interest in fee was deemed as a matter of law to be abandoned and to lapse.

It is also clear that the State has not exercised this power in an arbitrary manner. The Indiana statute provides that a severed mineral interest shall not terminate if its owner takes any one of three steps to establish his continuing interest in the property. If the owner engages in actual production, or collects rents or royalties from another person who does or proposes to do so, his interest is protected. If the owner pays taxes, no matter how small, the interest is secure. If the owner files a written statement of claim in the county recorder's office, the interest remains viable. Only if none of these actions is taken for a period of 20 years does a mineral interest lapse and revert to the surface owner.

Each of the actions required by the State to avoid an abandonment of a mineral estate furthers a legitimate state goal. Certainly the State may encourage owners of mineral interests to develop the potential of those interests; similarly, the fiscal interest in collecting property taxes is manifest. The requirement that a mineral owner file a public statement of claim furthers both of these goals by facilitating the identification and location of mineral owners, from whom developers may acquire operating rights and from whom the county may collect taxes. The State surely has the power to condition the ownership of property on compliance with conditions that impose such a slight burden on the owner while providing such clear benefits to the State.

II

Two of appellants' arguments may be answered quickly. Appellants contend that the Mineral Lapse Act takes private property without just compensation in violation of the Fourteenth Amendment; they also argue that the statute constitutes an impermissible impairment of contracts in violation of the Contract Clause. The authorities already discussed mandate rejection of each of these arguments.

In ruling that private property may be deemed to be abandoned and to lapse upon the failure of its owner to take reasonable actions imposed by law, this Court has never required the State to compensate the owner for the consequences of his own neglect. We have concluded that the State may treat a mineral interest that has not been used for 20 years and for which no statement of claim has been filed as abandoned; it follows that, after abandonment, the former owner retains no interest for which he may claim compensation. It is the owner's failure to make any use of the property—and not the action of the State—that causes the lapse of the property right; there is no "taking" that requires compensation. The requirement that an owner of a property interest that has not been used for 20 years must come forward and file a current statement of claim is not itself a "taking."

Nor does the Mineral Lapse Act unconstitutionally impair the obligation of contracts. In the specific cases under review, the mineral owners did not execute the coal and oil leases in question until after the statutory lapse of their mineral rights. The statute cannot be said to impair a contract that did not exist at the time of its enactment. Appellants' right to enter such an agreement of course has been impaired by the statute; this right, however, is a property right and not a contract right. In any event, a mineral owner may safeguard any contractual obligations or rights by filing a statement of claim in the county recorder's office. Such a minimal "burden" on contractual obligations is not beyond the scope of permissible state action.

III

Appellants' primary attack on the Dormant Mineral Interests Act is that it extinguished their property rights without adequate notice. In advancing this argument, appellants actually assert two quite different claims. First, appellants argue that the State of Indiana did not adequately notify them of the legal requirements of the new statute. Second, appellants argue that a mineral interest may not be extinguished unless the surface owner gives the mineral owner advance notice that the 20–year period of nonuse is about to expire. When these two arguments are considered separately, it is clear that neither has merit.

A

The first question raised is simply how a legislature must go about advising its citizens of actions that must be taken to avoid a valid rule of law that a mineral interest that has not been used for 20 years will be deemed to be abandoned. The answer to this question is no different from that posed for any legislative enactment affecting substantial rights. Generally, a legislature need do nothing more than enact and publish the law, and afford the citizenry a reasonable opportunity to familiarize itself with its terms and to comply. In this case, the 2–year grace period included in the Indiana statute forecloses any argument that the statute is invalid because mineral owners may not have had an opportunity to become familiar with its terms. It is well established that persons owning property within a State are charged with knowledge of relevant statutory provisions affecting the control or disposition of such property.

It is also settled that the question whether a statutory grace period provides an adequate opportunity for citizens to become familiar with a new law is a matter on which the Court shows the greatest deference to the judgment of state legislatures. . A legislative body is in a far better position than a court to form a correct judgment concerning the number of persons affected by a change in the law, the means by which information concerning the law is disseminated in the community, and the likelihood that innocent persons may be harmed by the failure to receive adequate notice.

In short, both the Indiana Legislature and the Indiana Supreme Court have concluded that a 2–year period was sufficient to allow property owners in the State to familiarize themselves with the terms of the statute and to take any action deemed appropriate to protect existing interests. On the basis of the records in these two proceedings, we cannot conclude that the statute was so unprecedented and so unlikely to come to the attention of citizens reasonably attentive to the enactment of laws affecting their rights that this 2–year period was constitutionally inadequate. We refuse to displace hastily the judgment of the legislature and to conclude that a legitimate exercise of state legislative power is invalid because citizens might not have been aware of the requirements of the law.

B

We have concluded that appellants may be presumed to have had knowledge of the terms of the Dormant Mineral Interests Act. Specifically, they are presumed to have known that an unused mineral interest would lapse unless they filed a statement of claim. The question then presented is whether, given that knowledge, appellants had a constitutional right to be advised—presumably by the surface owner—that their 20–year period of nonuse was about to expire.

In answering this question, it is essential to recognize the difference between the self-executing feature of the statute and a subsequent judicial determination that a particular lapse did in fact occur. As noted by appellants, no specific notice need be given of an impending lapse. If there has been a statutory use of the interest during the preceding 20–year period, however, by definition there is no lapse—whether or not the surface owner, or any other party, is aware of that use. Thus, no mineral estate that has been protected by any of the means set forth in the statute may be lost through lack of notice. It is undisputed that, before judgment could be entered in a quiet title action that would determine conclusively that a mineral interest has reverted to the surface owner, the full procedural protections of the Due Process Clause—including notice reasonably calculated to reach all interested parties and a prior opportunity to be heard—must be provided.

...

We have held that the State may impose on an owner of a mineral interest the burden of using that interest or filing a current statement of claim. We think it follows inexorably that the State may impose on him the lesser burden of keeping informed of the use or nonuse of his own property. We discern no procedural defect in this statute.

[The court rejected the equal protection claims based on the statutory exception for multiple interests and affirmed the Indiana Supreme Court's decision upholding the validity of the Act. Eds.]

NOTES

1. Would a conveyance of a mineral be a "use" within the meaning of the Indiana Dormant Minerals Act? Would a lease be a "use?" *See* McCoy v. Richards, 581 F.Supp. 143, 82 O.&G.R. 262 (S.D.Ind.1983), 623 F.Supp. 1300, 87 O.&G.R. 495 (S.D.Ind.1984), *aff'd*, 771 F.2d 1108, 87 O.&G.R. 514 (7th Cir.1985). Would the payment of advance royalties be a "use" under the Indiana Dormant Minerals Act? See Miller v. Weber, 839 N.E.2d 204 (Ind.Ct.App. 2005).

2. In addition to the civil law doctrine of prescription liberandi causa applicable to certain mineral interests in Louisiana, several other states have adopted variants of a dormant mineral act which terminate mineral interests which do not meet certain requirements. The supreme courts of Illinois, Nebraska and Wisconsin invalidated their dormant mineral acts but valid acts are in effect in California, Florida, Georgia, Indiana, Kansas, Kentucky, Michigan, North Carolina, North Dakota, Oklahoma and Virginia. Alabama has a "rule of repose" which bars

claim to mineral interests after a lapse of twenty years of non-user. *See* Boshell v. Keith, 418 So.2d 89 (Ala.1982). *See generally* Martin & Kramer, *Williams & Meyers Oil and Gas Law*, §§ 216.1 to 216.8.

3. Can the United States require mining claimants on public lands to register their claims through a federal recording system or lose their otherwise valid claim? *See* United States v. Locke, 471 U.S. 84, 84 O.&G.R. 299 (1985).

4. Would it matter if the state treats mineral interests as corporeal or incorporeal interests? Is there an intent requirement as in *Gerhard?*

5. The language of Dormant Minerals Acts must be carefully examined to determine what type of actions will prevent the mineral estates from lapsing. For example, where an Act provides that the interest will lapse if it is not "sold, leased, mortgaged or transferred by [recorded] instrument" does the mere running, rather than the execution, of a lease comply with any of those requirements so as to toll the running of the 20 year lapse period. In Energetics, Ltd. v. Whitmill, 442 Mich. 38, 497 N.W.2d 497, 128 O.&G.R. 245 (1993) the Michigan Supreme Court found that the language of the Act was not the same as if it contained the words "subject to a lease" and found that the lapse period would not be tolled merely while the lease was being kept alive by the payment of delay rentals. But the court did conclude that upon the termination of the lease there was a transfer of an interest that would trigger the beginning of a new 20 year period. As to that finding would it matter whether the reversionary interest was a possibility of reverter or a power of termination? Which one becomes a possessory estate immediately upon the termination of the prior possessory estate? For other factual and interpretational problems caused by such acts see Fisch v. Randall Mill Corp., 262 Ga. 861, 426 S.E.2d 883, 122 O.&G.R. 462 (1993).

6. Is the Indiana Dormant Mineral Act self-executing? Must the surface owner take affirmative steps to achieve the merger of title of the severed mineral estate? For an example of a non-self executing dormant mineral statute, see Kan.Stat.Ann. §§ 55-1601, *et seq.* analyzed in Scully v. Overall, 840 P.2d 1211 (Kan.App. 1992).

7. Another solution to the problems caused by having dormant minerals cloud titles may be to provide for the creation of trusts for unlocatable mineral, leasehold or working interest owners, giving to the trustee the power to execute leases, division orders or other instruments as may be appropriate. Mont.Code Ann. § 82-1-301 et seq.

8. The rationale of the principal case can also be applied to Marketable Record Title Acts which purport to extinguish claims to title after a period of time has elapsed. The Marketable Record Title Act is self-executing, much like the Indiana Dormant Mineral Act. Relying on the constitutional argument made in the principal case, the Oklahoma Marketable Record Title Act withstood a due process challenge in Bennett v. Whitehouse, 690 F.Supp. 955, 99 O.&G.R. 552 (W.D.Okla. 1988).

9. What type of notice, if any is required to be sent to the mineral owner whose severed mineral estate has allegedly lapsed under the applicable Dormant Minerals Act? A series of cases applying the North Dakota Dormant Minerals Act explore the impact of record notice of ownership of the severed mineral estate on the notice requirements imposed upon the surface owner seeking to unify title. See Johnson v. Taliaferro. 2011 ND 34, 793 N.W.2d 804; Sorenson v. Felton, 2011 ND 33, 793 N.W.2d 799, and; Sorenson v. Alinder, 2011 ND 36, 793 N.W.2d 797.

NOTES ON DEED-LEASE DISTINCTION

1. Several cases early in the development of oil and gas jurisprudence struggled with determining whether a particular written instrument was a mineral deed or a mineral lease. In LOOMIS V. GULF OIL CORP., 123 S.W.2d 501 (Tex.Civ.App. 1938), after concluding that the instrument was a deed and not a lease, the court listed the following factors in dealing with this interpretational issue: "On principle and reason this interpretation is believed to be correct: (1) The instrument contains all the essential elements of a valid deed of conveyance such as competent parties, proper subject matter, valuable consideration, apt words of conveyance, proper words of execution, *etc.*. . . (2) It conveys "all" said minerals without limitation or qualification. (3) Its language manifests an intention to convey the fee simple title to the minerals in and under the land. (4) The consideration of $ 250 recited in the instrument taken alone, or in connection with the whole consideration, is a valuable one. . . (5) In the event of production, the grantor is to have none of the minerals, as such, but only 'one equal 1/8 of all the net proceeds realized from the sale of the minerals.' (6) It further provides 'and if said land is now under valid and binding mineral lease to any person or corporation other than grantee, that since this instrument is intended to and does pass all royalty reserved in such lease to the grantee herein, that such grantee, its successors or assigns may be entitled to under such lease.' (7) The grantee is to 'have and to hold' the minerals 'forever.'" 123 S.W.2d at 503.

2. What other factors are relevant in considering whether an instrument is a deed or a lease? Does the instrument contain a royalty provision? *Compare Loomis with* Dougherty v. Greene, 218 Miss. 250, 67 So.2d 297, 2 O.&G.R. 1402 (1953). Did a substantial cash payment accompany the transfer of the interest? *Compare* Ramsey v. Yunker, 311 Ky. 820, 226 S.W.2d 14 (1950) *with* Commissioner v. Felix Oil Co., 144 F.2d 276 (9th Cir.1944).

3. In Kansas, Kan. Stat. Ann. § 79–420 requires recordation of an instrument severing minerals or a separate return for tax purposes of the value of the severed mineral interest. This statute has been held inapplicable to a lease, and hence it is necessary to determine whether a particular instrument is a lease or mineral deed. *See* Hushaw v. Kansas Farmers' Union Royalty Co., 149 Kan. 64, 86 P.2d 559 (1939), *appeal dism'd*, 307 U.S. 615, 59 S.Ct. 1046, 83 L.Ed. 1496 (1939), rehearing denied, 308 U.S. 634, 60 S.Ct. 70, 84 L.Ed. 527 (1939).

4. The owner of lands in Texas subject to the provisions of the Texas

Relinquishment Act is authorized to execute an oil and gas lease but is not authorized to convey the minerals. Under this Act it becomes necessary to classify a particular instrument as a permissible "lease" or a non-permissible mineral "deed." *See* Texas Co. v. State of Texas, 154 Tex. 494, 281 S.W.2d 83, 4 O.&G.R. 727, 1571 (1955). Such a classification process is also necessary under a Tennessee statute setting a maximum term of ten years during which an oil and gas lease but not "a fee simple estate for a particular deposit or stratum" can be held without production. *See* Layne v. Baggenstoss, 640 S.W.2d 1, 76 O.&G.R. 634 (Tenn.App.1982).

5. Consider whether the classification of an instrument as a deed or as a lease should be viewed as significant in the adjudication of controversy over the following matters:

(a) The implication of exploration or development covenants in the instrument. *See* Dallas Power & Light Co. v. Cleghorn, 623 S.W.2d 310, 72 O.&G.R. 153 (Tex.1981); Danciger Oil & Refining Co. of Texas v. Powell, 137 Tex. 484, 154 S.W.2d 632, 137 A.L.R. 408 (1941).

(b) Termination of the granted interest by abandonment or by forfeiture for nondevelopment. *See* Crain v. Pure Oil Co., 25 F.2d 824 (8th Cir.1928); Clark v. Wilson, 316 S.W.2d 693, 9 O.&G.R. 646 (Ky.1958).

6. The Louisiana Mineral Code [La.R.S. Title 31] distinguishes mineral servitudes and mineral leases:

Art. 21. A mineral servitude is the right of enjoyment of land belonging to another for the purpose of exploring for and producing minerals and reducing them to possession and ownership.

Art. 22. The owner of a mineral servitude is under no obligation to exercise it. If he does, he is entitled to use only so much of the land as is reasonably necessary to conduct his operations. He is obligated, insofar as practicable, to restore the surface to its original condition at the earliest reasonable time.

Art. 23. The owner of a mineral servitude may conduct his operations with the freedom and subject to the restrictions that apply to a landowner. He may protect his right against interference or damage by all of the means available to a landowner.

. . .

Art. 114. A mineral lease is a contract by which the lessee is granted the right to explore for and produce minerals. . . .

Art. 122. A mineral lessee is not under a fiduciary obligation to his lessor, but he is bound to perform the contract in good faith and to develop and operate the property leased as a reasonably prudent operator for the mutual benefit of himself and his lessor. Parties may stipulate what shall constitute reasonably prudent conduct on the part of the lessee.

SECTION 3. SURFACE AND SUBSURFACE TRESPASS
(*See* Martin & Kramer, *Williams and Meyers Oil and Gas Law* §§ 225–232.)

Some of the cases that follow bear on the following hypothetical case:

X has a lease on a section of land executed by *A*. The lease is in the last year of the primary term and will expire unless a producing well is drilled. *X* has just begun drilling operations when he receives a letter from *B* advising him that *B* claims fee simple title to the land, demanding that *X* surrender possession, and warning *X* that he will be held strictly liable for any damages that his continued possession may cause. The territory is quite active: lease bonus has been as high as $100 per acre. Royalty has been sold in the area for $1000 per royalty acre. *X* consults his lawyer (who, incidentally, approved title to the land) about his next move. What would you advise?

———

A. TRESPASS WITHOUT PRODUCTION

HUMBLE OIL & REFINING CO. v. KISHI

Commission of Appeals of Texas, 1925, 1927.
276 S.W. 190, 291 S.W. 538.

BISHOP, J. In this case both the Humble Oil & Refining Company and K. Kishi have filed applications for writs of error, and there is here presented for review the holding of the Court of Civil Appeals on all questions discussed in its opinion. 261 S.W. 228. We agree with the conclusion reached by the Court of Civil Appeals on all questions except its holding that proof of the market value of the leasehold interest in the land involved did not in law furnish the measure of damages which should be awarded, and only such statement of the case is here made as is deemed necessary to a discussion of this holding.

K. Kishi, the owner of all the surface and three-fourths undivided interest in the oil and mineral rights, and Isaac Lang, the owner of the remaining one-fourth interest in the oil and mineral rights of 50 acres of land in Orange County, Tex., executed to the Humble Oil & Refining Company a lease granting to it the exclusive right to enter upon said land and drill oil wells and take therefrom the oil. This lease was of date December 23, 1919, but was not signed and acknowledged by Lang until January 29, 1920, and was thereafter delivered to said Humble Oil & Refining Company. By its provisions it was to remain in force for no longer period of time than three years from its date, unless within said three years drilling for oil was commenced. No drilling was begun within the time provided, and the lease expired. After the expiration of this lease, in January, 1923, oil was found on an adjoining tract of land in a well drilled near this 50 acres. On January 23, 1923, the Humble Oil & Refining Company entered upon this 50 acres of land, and began drilling an oil well thereon, claiming the exclusive right to the leasehold interest therein. It claimed that the lease had not expired, and that under its terms it did not expire until three years after it was signed and acknowledged by Lang and delivered. Kishi protested against this entry, and advised the Humble Oil & Refining Company that he would hold it responsible for any damages that might accrue to him. Lang, however, consented to the entry under the claim made.

The Humble Oil & Refining Company remained in possession under this entry until it completed the drilling of the well, which resulted in the failure to find oil, and it relinquished possession on May 10, 1923. As a result of the discovery of oil on the adjoining tract of land, the leasehold interest in the 50–acre tract was of the market value of $1,000 per acre. At the time the Humble Oil & Refining Company relinquished possession, and thereafter the leasehold interest had no value by reason of the failure to find oil on this tract.

In suit by Kishi against the Humble Oil & Refining Company for damages sustained by him, the district court awarded him nominal damages in the sum of $1 only, holding that the amount of damages sustained by him was uncertain and not susceptible of proof. On appeal from this judgment the Court of Civil Appeals held that he was under the facts entitled to recover the actual damages occasioned by reason of the wrongful entry and ouster which was the value to him of his three-fourths undivided leasehold interest, but that proof of the market value of the entire leasehold interest was not in law sufficient upon which to base the amount of his recovery.

Oil in place under the land is real estate. The exclusive right to enter upon the land, drill wells thereon, and remove therefrom the oil to exhaustion, paying therefor a portion of the oil when extracted or the equivalent of such portion, is a property right which the law protects. The Humble Oil & Refining Company, wrongfully claiming to own this right over the protest of Kishi, and excluding him therefrom, entered upon his 50 acres of land for the purpose of drilling the well. This was clearly a trespass and ouster. This right had a market value of $50,000, being $1,000 per acre. Had Lang not consented, and had he and Kishi joined in a suit to recover their damages for the wrongful entry, the measure of their damages would have been the market value of the leasehold interest which is here shown to be $1,000 per acre. Lang would have been entitled to one-fourth and Kishi to three-fourths of the amount recovered. We can conceive of no reason why Kishi should be permitted to recover either a larger or smaller amount, because it is shown that Lang consented to the entry.

The Humble Oil & Refining Company insists that it should not be required to pay as damages Kishi's proportionate share of the market value of this leasehold interest, because it entered upon the land in good faith, believing that its lease had not expired. Though it did so in good faith, without any intention to injure Kishi, it asserted a right it did not have. This right, at the time it was wrongfully asserted, had a market value. Had the oil company acquired this right by purchase before its entry, the presumption of law is that it would have been required to pay the market value therefor. Had it done so, Kishi would have been entitled to receive three-fourths of the market value of the leasehold interest. We think that in this case three-fourths of the market value of the leasehold interest was the measure of the damages which Kishi was in law entitled to recover, and that proof of the market value was in law sufficient upon which to determine the amount of judgment in his favor. Gulf, Colorado & Santa Fé Ry. Co. v. Cusenberry, 86 Tex. 529, 26 S.W. 43.

We therefore recommend that the judgment of the Court of Civil Appeals be reversed, and that the judgment of the district court be so reformed as to allow K. Kishi judgment against the Humble Oil & Refining Company for the sum of $37,500, with legal interest, and, as so reformed, that the judgment of the district court be here affirmed.

CURETON, C.J. Judgment of the Court of Civil Appeals reversed, and that of the district court reformed, and, as reformed, affirmed, as recommended by the Commission of Appeals.

On second motion for a Rehearing, 291 S.W. 538 (Tex.Com.App.1927):

BISHOP, J. The Humble Oil & Refining Company insists that the holding made in the original opinion (276 S.W. 190) denies to a cotenant the right to the use and enjoyment of the land owned in cotenancy. No such holding was intended, and we think the opinion makes it clear that no such holding is made. The entry upon this land as a cotenant of Kishi would have been lawful and would have occasioned no injury to Kishi's right to the value of his property. The entry made upon this land was unlawful, not because the company had no right to make entry, but because the entry made was in denial of Kishi's right. The character of the entry made was unlawful, and was the sole cause of the injury complained of. The company had no right to deprive Kishi of the value of his property by making the unwarranted claim that the lease theretofore executed by him gave it the right of entry, and its entry under this claim was wrongful. This wrongful act destroyed the value of Kishi's property, and the company should be required to respond in damages for the injury resulting from its unlawful act.

We have concluded that the recommendation made in our opinion that the judgment of the Court of Civil Appeals (261 S.W. 228) be reversed was erroneous. Both the trial court and the Court of Civil Appeals found that the leasehold value of the land was $1,000 per acre. A three-fourths undivided interest in the leasehold estate may or may not have been three-fourths of $1,000 per acre. The fact that it was an undivided interest may have affected its value. There is no proof of this record showing the value of this interest other than that the leasehold estate as a whole was of the value of $1,000 per acre. From the proof of this value, we think the trial court could have inferred that an undivided three-fourths interest was three-fourths of $1,000 per acre, and, had the trial court so found, we think this finding would have been binding on the appellate courts. However, we have concluded that an appellate court is not warranted in indulging in this inference in the absence of a specific finding to this effect by the trial court. The value of Kishi's three-fourths undivided interest in the leasehold was the amount of damages he was entitled to recover, and evidence as to whether he was offered any specific amount, or would have accepted such offer, is not necessary to the proof of the legal measure of damages. No such evidence was adduced on the trial, and we are not here deciding whether such evidence would be admissible as material to the establishment of the value of Kishi's interest. The amount of his damages is measured by the loss in the value of his leasehold interest caused by the wrongful conduct of which complaint is made.

We recommend that the judgment heretofore rendered be set aside and that the judgment of the Court of Civil Appeals reversing and remanding the cause be affirmed, with direction that the district court ascertain the amount of damages under the measure herein indicated and render judgment accordingly.

CURETON, C. J. Motion for rehearing granted. . . . We approve the holding of the Commission of Appeals on the questions discussed in its opinion.

NOTES

1. On a second trial it was objected that witnesses qualified to testify as to the value of the leasehold interest as a whole could not testify as to the value of an undivided interest in the leasehold "because such is the mere opinion of the witness, not based on knowledge of sales of such interest." On second appeal, 299 S.W. 687 (Tex.Civ.App.1927, error ref'd), the admission of this testimony over the objection was sustained, citing the language of the Commission in the principal case and other authority.

2. Did Kishi have to prove loss of a specific lease offer? *See* Kidd v. Hoggett, *infra.* Is there a problem with the damages being too speculative? Recall Elliff v. Texon Drilling Co., *supra. See also,* Veazey v. W.T. Burton Industries, Inc., 407 So.2d 59, 72 O.&G.R. 60 (La. App. 1981); Fite v. Miller, 192 La. 229, 187 So. 650 (1939). *Compare* Mobil Exploration & Producing, U.S., Inc. v. Certain Underwriters Subscribing to Cover Note 95-3317(A), 837 So. 2d 11, 159 O.&G.R. 271 (La. Ct. App. 2002),

For the critical discussion of the *Kishi* case by Leon Green, *see What Protection Has a Landowner Against a Trespass which Merely Destroys the Speculative Value of his Property*, 4 Tex.L.Rev. 215 (1926), with an addendum in 5 Tex.L.Rev. 323 (1927). A. W. Walker discusses the general problem in *Fee Simple Ownership of Oil and Gas in Texas*, 6 Tex.L.Rev. 71, 132 (1928), where (at p. 137) he makes the interesting suggestion that if the court in the *Kishi* case was awarding damages for decrease in the lessor's interest in a "leasehold," the plaintiff in that case did not get full compensation. "He should also have received compensation for the decrease in the market value of his royalty interest."

See also Qualls, *Damages for Dry Hole Trespass*, 40 N.D.L.Rev. 65 (1964).

3. Texas is listed as included among the states permitting a cotenant to remove minerals without liability for waste. Under what theory, then, did the court in *Kishi* hold that the plaintiff was entitled to recover damages for the loss of speculative values by reason of the drilling of a dry hole by the lessee of a cotenant? Can a cotenant challenge the validity of another cotenant's interest while at the same time producing the jointly held oil and gas? *See* Byrom v. Pendley, 717 S.W.2d 602, 93 O.&G.R. 419 (Tex.1986) and Shell Oil Co. v. Howth, 138 Tex. 357, 159 S.W.2d

483 (1942). Should it matter whether the entry occurred before or after the termination of the cotenant's right of entry? In Davis v. Byrd, 238 Mo.App. 581, 185 S.W.2d 866 (1945), it was suggested that a tenant in common commits no wrong in entering upon the common property for the purpose of carrying on mining operations in the usual way and therefore cannot be held to be a trespasser. The court declared that the removal of ore without willful injury to the common property or unnecessary destruction caused by negligence or unskillfulness does not constitute waste, and that no injunction will lie at the instance of one cotenant to restrain another cotenant or his lessee from conducting mining operations on the common property unless it appears that such tenant in common or his lessee, sought to be enjoined, has excluded or prevented the complaining cotenant from exercising the same rights and privileges or has been guilty of an act or acts denying his interest in the common property.

4. In MARTEL v. HALL OIL CO., 36 Wyo. 166, 253 P. 862, 52 A.L.R. 91 (1927), a judgment for the defendant in a situation similar to that in the *Kishi* case was affirmed. The main basis of the decision is stated as follows:

> . . . [P]laintiffs had . . . the right to lease or sell their interest in the land, and their main claim for damages is . . . based upon the theory that defendants destroyed the sale value of their interest by drilling on the land, and proving that the land contained no oil or gas. In other words, plaintiffs contend that defendants, by their acts, made the quality of the land as to oil and gas, known to the public, thereby destroying the sale value as aforesaid. They complain, in short, that the defendants, by their trespass, made the truth known, and that defendants thereby deprived the plaintiffs of the chance or opportunity to make a good bargain by selling or otherwise disposing of their interest. Now we should not lose sight of the nature, meaning and significance of that chance or opportunity, which is claimed to have been destroyed. It may be, though there is no showing to that effect, that plaintiffs might have sold their rights for a considerable sum of money. They would then have been the gainers and the purchaser would have been the loser. They would, upon their own theory, have pocketed a lot of money, but for what? What would they have given in return? Nothing. They would have sold something of no value whatever. They would, upon their own theory, have received something for nothing in return. This is the sort of opportunity which we are asked to protect, and that, too, in the absence of a contract or a bona fide offer for purchase from any one, showing a voluntary desire and consent to make such payment, and which, and which only, could furnish the plaintiffs with a moral basis for taking or exacting of sum of money for which nothing of value would, in fact, be given in return. That a wealthy corporation might have been the purchaser furnishes no answer; it might have been a poor widow instead.

5. There are very few reported decisions on the issues raised by *Kishi* and *Martel*, although those cases are about 75 years old. Does that suggest the problem had no lasting importance?

6. What is the legal position of a landowner who silently but knowingly suffers unauthorized drilling of his land until the operations result in a dry hole? Would the doctrines of waiver and estoppel apply in that situation? *See* Mullins v. Ward, 1985 OK 109, 712 P.2d 55, 87 O.&G.R. 340, and Brixey v. Union Oil Co., 283 F.Supp. 353, 28 O.&G.R. 541 (W.D.Ark.1968).

7. In AMERICAN SURETY CO. OF NEW YORK v. MARSH, 146 Okl. 261, 293 P. 1041 (1930), plaintiff brought suit on a supersedeas bond given by his adversary in an unsuccessful appeal from a judgment for plaintiff in a previous action to try title. During the appeal a dry hole was drilled on adjacent lands. The case made by the plaintiff and the court's decision are summed up in the following statement:

> Conceding that the defendant's claim to and possession of the premises was asserted in good faith, as said in the Kishi case, 'it asserted a right it did not have,' and its conduct operated as a barrier against the right of the plaintiff to enjoy his full and complete property right in the lease. There is little merit to the contention that his right was merely speculative. His proof was sufficient to show that he had an opportunity to sell, and that for a substantial sum and could have done so but for the asserted rights and conduct of the defendant, who was successful in holding it until plaintiff's bargain was gone and until adjacent developments had depreciated the value to a minimum. A party must be entitled to recover against another for causing him such a substantial loss. The conduct of the defendant was just as effective to damage plaintiff in his valuable right as if it had held the property by force of arms.

The relevant clause of the bond in *Marsh* simply bound the defendant to pay "all other damages the obligor may be legally liable for." Aside from the proof of a specific opportunity to sell, is there as sound a basis for giving a cause of action in *Marsh* as there is in *Kishi* or *Martel* ? What additional element might be required for a cause of action in *Marsh*?

KIDD v. HOGGETT

Court of Civil Appeals of Texas, San Antonio, 1959.
331 S.W.2d 515, 12 O.&G.R. 730, error refused, no reversible error.

POPE, JUSTICE. Pierce A. Hoggett and wife sued Barron Kidd, A. W. Cherry, and others, to remove the cloud of an unreleased but expired oil and gas lease, and for damages. There was no jury. Upon trial defendants disclaimed with respect to the oil and gas lease. The court rendered judgment removing the cloud but also awarded damages against defendants Kidd and Cherry in the amount of $8,400. Only Kidd and Cherry have appealed from the judgment. Kidd and Cherry urge that the judgment for damages was erroneous because (1) they were under no duty to release the oil and gas lease because it was still in force and effect when demand was made upon them, (2) this is an action for slander of title and the plaintiffs, prior to the motion for judgment, failed to prove malice, which is an essential element of such actions, (3) an implied finding of malice is against the great weight and preponderance of the evidence, and (4) the action for damages is barred by the two-year statute of limitations. We affirm the judgment.

On April 24, 1944, the then owners, executed an oil and gas lease to 2,871 acres of land. Plaintiffs are the present owners of 2,831 acres of the land, and Howell Wright owns the other forty acres. On January 25 and 26, 1954, by assignments, A. W. Cherry and Barron Kidd became the owners of the oil and gas lease. This was a short time before the primary term of the lease expired, so they commenced a well on the forty-acre Wright tract. Cherry and Kidd took the position that the well was a producing well but there was no market for the gas from February 23, 1954, the completion date. The lease contained a shut-in gas clause which provided:

> . . . where gas from a well producing gas only is not sold or used, Lessee may pay as royalty $50 per well per year, and upon such payment it will be considered that gas is being produced within the meaning of Paragraph 2 hereof. . . .

The well was shut in and lessees paid shut-in royalty for the year commencing in April, 1954, and again for the year commencing in April, 1955.

Plaintiffs became suspicious about the well, because the gas from the well was not sold at a time when the Junction Natural Gas Company needed gas. In November, 1955, after they had accepted the shut-in royalty, plaintiffs demanded a release of the oil and gas lease, but defendants refused upon assurances that the well was a producer. On January 9, 1956, plaintiffs entered into a specific lease with Ray Albaugh by which he agreed to pay $2.00 an acre bonus and $1.00 an acre for the first year's rental. This lease was subject to plaintiffs' obtaining a release from Kidd and Cherry of the oil and gas lease of record. When plaintiffs failed to obtain the release, Albaugh refused to enter into his lease and in January, 1957, released to plaintiffs. At the time of the trial, the tract had no value for oil and gas purposes. Plaintiffs, therefore, proved actual damages of at least $8,493. Shell Oil Co. v. Howth, 138 Tex. 357, 159 S.W.2d 483.

Plaintiffs on June 27, 1958, filed suit. When the case was tried in March of 1959, defendants disclaimed, and the court correctly removed the cloud of the oil and gas lease from the title. Kidd and Cherry do not complain of that part of the judgment which removes the cloud from the title. A greater portion of the briefs is devoted to the study of the nature of this action. Plaintiffs take the position that in a suit for removal of cloud and for damages, that malice is not a necessary element, but that in any event they both pleaded and proved malice. Defendants take the position that this is a suit for slander of title, that malice is a necessary element, and that it was not proved.

A lessee is under a duty in Texas to release of record an expired oil and gas lease. If the duty is imposed by the lease, an action rests on contract. Where, as here, there is no such contractual provision, there is nevertheless a duty. Speaking of a refusal to release an expired oil and gas lease, it was said in Witherspoon v. Green, Tex.Civ.App., 274 S.W. 170: "The law charged appellee with the duty of removing this cloud from appellant's title by the execution of a release of his apparent, though not actual, interest in the land."

There may be some confusion as to nature of the action for damages for breach of the duty to release an oil and gas lease. Is malice necessary in such an action as this, and if so, what kind of malice? When there is an actual trespass upon property, an ouster of and a repudiation of the title of the true owner, allegations of malice are not required. Also, by way of analogy, in the case of a wrongful levy of an attachment, malice was not essential to the recovery of actual damages.

Malice is an essential element to the recovery of actual damages in a suit for slander of title. To recover damages against one who, to protect himself, buys an outstanding title there must be proof of a bad faith purchase of said title for the purpose of maliciously asserting a claim against and a repudiation of the lessor's title. Shell Oil Co. v. Howth, 138 Tex. 357, 159 S.W.2d 483. Malice is an element in a suit for damages for the wrongful filing of an abstract of judgment.

One who sues to remove the cloud of an unreleased oil and gas lease and goes further and seeks actual damages occasioned by the loss of a specific sale of an oil and gas lease is, in our opinion, seeking recovery for the disparagement of his title, or, as it is termed, slander of title. The one clear case on the point is Wheelock v. Batte, 225 S.W.2d 591. In that case Mr. Justice Hughes reasons, and we agree, that a lessee should be permitted the right to assert his good faith claims to valuable properties without incurring great risk of damages just because he leaves disputed rights to the decision of courts. Hence, a plaintiff should prove malice on the part of the one refusing to give the release, and a defendant may defend against actual damages by proof of his good faith claim.

How malice is defined, however, is another matter. Malice as a basis for recovery of actual damages, as distinguished from punitive damages should mean that the act or refusal was deliberate conduct without reasonable cause. *See,* 8 Institute on Oil and Gas 357. Malice as a basis for recovery of punitive damages should mean actual malice, that is, ill will, bad or evil motive, or such gross indifference to or reckless disregard of the rights of others as will amount to a wilful or wanton act.

The decision in this case is not as difficult as the law on the subject may be indefinite, since whether malice, of whatever kind, is necessary or not, it was amply alleged and proved by the plaintiffs. The recovery was only for actual and not punitive damages. Since the case was tried without a jury, we presume all fact issues supported by the evidence were found in support of the judgment. In our opinion the evidence supports plaintiffs' contentions that the well was not a producing well when completed or afterwards. The evidence and the inferences from the evidence support a finding of malice and that finding is not against the great weight and preponderance of the evidence. . . .

. . .

There is no merit to defendants' contention that plaintiffs' action was barred by the two-year statute of limitations, Vernon's Ann.Civ.St. art. 5526. The sale to

Albaugh was lost in January, 1957. The suit was brought on June 27, 1958. In a slander of title suit, one necessary element of a plaintiff's proof is that he suffered special damages. Unless he can allege and prove that fact he has no cause of action. If frustration of a specific sale is necessary to prove a cause of action, the cause of action did not mature until the frustration occurred. Shell Oil Co. v. Howth, 138 Tex. 357, 159 S.W.2d 483. Any other rule would mean that limitation was running against a plaintiff before he had a cause of action. *See* 8 Institute on Oil and Gas 374.

The judgment is affirmed.

NOTES

1. Why must plaintiff show loss of an opportunity to sell in *Kidd* but not in *Kishi?* Why is good faith a defense in *Kidd* but not in *Kishi?* Was defendant a trespasser in *Kidd?* Does it matter?

2. Kidd v. Hoggett, *supra,* and Shell Oil Co. v. Howth, 138 Tex. 357, 159 S.W.2d 483 (1942), were relied on by the Texas Supreme Court in Ellis v. Waldrop, 656 S.W.2d 902 (Tex.1983) to "reaffirm the long-standing general rule ... [that] the plaintiff must allege the loss of a specific sale" in a slander of title action. To the same effect is A.H. Belo Corp. v. Sanders, 632 S.W.2d 145 (Tex.1982). These cases were also followed in Williams v. Jennings, 755 S.W.2d 874, 107 O.&G.R. 167 (Tex.App.1988, writ denied). Exemplary damages of $65,000 were upheld in *Williams* for not filing a release of a lease for more than a year; the finding of malice was based on a jury finding of "conduct with such gross indifference to the rights of plaintiff as to amount to a willful and wanton act." The loss of a specific sale was established by means of plaintiff's testimony as to a verbal offer. Should this suffice? *See* Judge Sears' dissent in *Williams* urging that it should not.

3. On slander of title, *see* Kuntz, *Liability for Clouding Title to Oil and Gas Interests*, 8 Sw. Legal Fdn. Oil & Gas Inst. 331 (1957), cited in the principal case, and Martin & Kramer, *Williams and Meyers Oil and Gas Law* § 232.

4. The stringent requirements of a slander of title claim are illustrated and discussed in Hill v. Heritage Resources, Inc., 964 S.W.2d 89, 140 O.&G.R. 78 (Tex.App 1997, writ denied). The court lists five elements for slander of title and rejects the claim that there was a slander of title as the party alleging the injury did not have title to the interest allegedly slandered. The court also notes that the same facts may give rise to either a tortious interference with contract claim or a slander of title claim.

5. For a successful assertion of a slander of title claim by an assignee of a lease against the assignor who had filed an affidavit of lease termination and reversion, see Duncan Land & Exploration, Inc. v. Littlepage, 984 S.W.2d 318, 141 O.&G.R. 366 (Tex.App.--Ft. Worth 1999, pet. denied). The elements of a slander of title suit were listed as 1) the uttering and publishing of disparaging words; 2) falsity; 3) malice; 4) special damages; 5) possession of an estate or

interest in the property disparaged; and 6) the loss of a specific sale. The court found malice even though plaintiff consulted with his attorney, and it distinguished the malice requirement in a slander of title action with the malice requirement for a punitive damage award.

6. The consequences of a successful slander of title claim can be drastic for the defendant. See TXO Production Corp. v. Alliance Resources Corp., 419 S.E.2d 870 (1992), 121 O.&G.R. 326, *aff'd*, 509 U.S. 443, 126 O.&G.R. 576 (1993). The West Virginia Supreme Court of Appeals, and the United States Supreme Court, upheld a ten million dollar punitive damage award where the actual damages were found to be $19,000. The West Virginia court reviewed the old English slander of title cases and the Restatement Second of Torts and confirmed findings of false statement, malice and special damages (though not loss of a specific sale) against TXO.

7. In addition to slander of title, a party may bring an action for tortious interference with a contract or a contractual relationship in situations such as what occurred in the principal case or in situations were parties execute top leases while the base lease is still in existence. It is generally perceived that a party asserting damages for toritous interference with contact need not allege the loss of a specific sale. See Voiles v. Santa Fe Minerals, Inc., 1996 OK 13, 911 P.2d 1205, 134 O.&G.R. 268; Marrs & Smith Partnership v. D.K. Boyd Oil & Gas Co., 223 S.W.3d 1, 164 O.&G.R. 605 (Tex.App. 2005); Santa Fe Energy Co. v. Carrillo, 948 S.W.2d 780, 136 O.&G.R. 199 (Tex.App. 1997).

———

B. DEFINING A SUBSURFACE TRESPASS

COASTAL OIL & GAS CORP. v. GARZA ENERGY TRUST

Supreme Court of Texas, 2008
268 S.W.3d 1

HECHT, JUSTICE. The primary issue in this appeal is whether subsurface hydraulic fracturing of a natural gas well that extends into another's property is a trespass for which the value of gas drained as a result may be recovered as damages. We hold that the rule of capture bars recovery of such damages. We also hold:

• mineral lessors with a reversionary interest have standing to bring an action for subsurface trespass causing actual injury;

[the court also resolved issues relating to the implied covenant to prevent drainage and the implied covenant to develop, bad faith pooling and evidentiary matters relating to the admission of a document containing a racial slur against Mexican-Americans that eventually let the court to remand the matter for a new trial, Eds.]

I

Respondents, to whom we shall refer collectively as Salinas, own the minerals in a 748-acre tract of land in Hidalgo County called Share 13, which they and their ancestors have occupied for over a century. At all times material to this case, petitioner Coastal Oil & Gas Corp. has been the lessee of the minerals in Share 13 and an adjacent tract, Share 15. Coastal was also the lessee of the minerals in Share 12 until it acquired the mineral estate in that 163-acre tract in 1995. A natural gas reservoir, the Vicksburg T formation, lies between 11,688 and 12,610 feet below these tracts.

From 1978 to 1983, Coastal drilled three wells on Share 13, two of which were productive, the M. Salinas No. 1 and No. 2V, though the other, the B. Salinas No. 1, was not. In 1994, Coastal drilled the M. Salinas No. 3, and it was an exceptional producer. The No. 3 well was about 1,700 feet from Share 12. The closest well on Share 12 was the Pennzoil Fee No. 1, but Coastal wanted one closer, so in 1996, Coastal drilled the Coastal Fee No. 1 in the northeast corner of Share 12, as close to Share 13 as Texas Railroad Commission's statewide spacing *Rule 37* permitted -- 467 feet from the boundaries to the north and east. That location was too close to the Pennzoil Fee No. 1, and the Commission refused Coastal an exception because both wells would drain from Share 13. So Coastal shut in the Pennzoil Fee No. 1, a producing well, in order that it could operate the Coastal Fee No. 1 well near Share 13. In February 1997, Coastal drilled the Coastal Fee No. 2, also near Share 13.

[In response to a lawsuit filed by the plaintiffs, Coastal drills several additional wells on both Share 12 and Share 13. Eds.]

The Vicksburg T is a "tight" sandstone formation, relatively imporous and impermeable, from which natural gas cannot be commercially produced without hydraulic fracturing stimulation, or "fracing", as the process is known in the industry. This is done by pumping fluid down a well at high pressure so that it is forced out into the formation. The pressure creates cracks in the rock that propagate along the azimuth of natural fault lines in an elongated elliptical pattern in opposite directions from the well. Behind the fluid comes a slurry containing small granules called proppants -- sand, ceramic beads, or bauxite are used -- that lodge themselves in the cracks, propping them open against the enormous subsurface pressure that would force them shut as soon as the fluid was gone. The fluid is then drained, leaving the cracks open for gas or oil to flow to the wellbore. Fracing in effect increases the well's exposure to the formation, allowing greater production. First used commercially in 1949, fracing is now essential to economic production of oil and gas and commonly used throughout Texas, the United States, and the world.

Engineers design a fracing operation for a particular well, selecting the injection pressure, volumes of material injected, and type of proppant to achieve a desired result based on data regarding the porosity, permeability, and modulus (elasticity) of the rock, and the pressure and other aspects of the reservoir. The design projects the length of the fractures from the well measured three ways: the hydraulic length, which is the distance the fracing fluid will travel, sometimes as far as 3,000 feet from the well; the propped length, which is the slightly shorter distance the proppant will reach; and the effective length, the still shorter distance within which the fracing operation will actually improve production. Estimates of these distances are dependent on available data and are at best imprecise. Clues about the direction in which fractures are likely to run horizontally from the well may be derived from seismic and other data, but virtually nothing can be done to control that direction; the fractures will follow Mother Nature's fault lines in the formation. The vertical dimension of the fracing pattern is confined by barriers -- in this case, shale -- or other lithological changes above and below the reservoir.

For the Coastal Fee No. 1, the fracing hydraulic length was designed to reach over 1,000 feet from the well. Salinas's expert, Dr. Michael J. Economides, testified he would have designed the operation to extend at least 1,100 to 1,500 feet from the well. The farthest distance from the well to the Share 13 lease line was 660 feet. The parties agree that the hydraulic and propped lengths exceeded this distance, but they disagree whether the effective length did. The lengths cannot be measured directly, and each side bases its assertion on the opinions of an eminent engineer long experienced in hydraulic fracturing: Economides for Salinas, and Dr. Stephen Allen Holditch for Coastal. Holditch believed that a shorter effective length was supported by post-fracing production data.

All the wells on Share 12 and Share 13 were fraced.

***

The jury found:

- Coastal's fracing of the Coastal Fee No. 1 well trespassed on Share 13, causing substantial drainage, which a reasonably prudent operator would have prevented, and $ 1 million damages in lost royalties;

- Coastal acted with malice and appropriated Salinas's property unlawfully, and should be assessed $ 10 million punitive damages;

- Salinas's reasonable attorney fees for trial were $ 1.4 million.

We begin with Salinas's contention that the incursion of hydraulic fracturing fluid and proppants into another's land two miles below the surface constitutes a trespass for which the minerals owner can recover damages equal to the value of the royalty on the gas thereby drained from the land. Coastal argues that Salinas has no standing to assert an action for trespass, and even if he did, hydraulic fracturing is not an actionable trespass. Because standing may be jurisdictional, we address it first.

A

As a mineral lessor, Salinas has only "a royalty interest and the possibility of reverter" should the leases terminate, but "no right to possess, explore for, or produce the minerals." [18] Texas courts have occasionally stated that "[t]he gist of an action of trespass to realty is the injury to the right of possession." [19] Since Salinas has no possessory right to the minerals in Share 13, Coastal argues he has no standing to sue for trespass.

But courts have stated the rule too broadly. At common law, trespass included several actions directed to different kinds of wrongs. Trespass *quare clausum fregit* was limited to physical invasions of plaintiff's possessory interest in land; trespass on the case was not and provided an action   for injury to a non-possessory interest, such as reversion. Professors Prosser and Keeton explain:

> Thus a landlord cannot sue for a mere trespass to land in the occupation of his tenant. He is not without legal remedy, in the form of an action on the case for the injury to the reversion; but in order to maintain it, he must show more than the trespass -- namely, actual permanent harm to the property of such sort as to affect the value of his interest. [W. Page Keeton, Dan B. Dobbs, Robert E. Keeton & David G. Owen, Prosser & Keaton on the Law of Torts § 13 at 78 (5th ed. 1984).]

Salinas's reversion interest in the minerals leased to Coastal is similar to a landlord's reversion interest in the surface estate. By his claim of trespass, Salinas seeks redress for a permanent injury to that interest -- a loss of value because of wrongful drainage. His claim is not speculative; he has alleged actual, concrete harm whether his leases continue or not, either in reduced royalty revenues or in loss of value to the reversion. This gives him standing to sue for a form of trespass, and under our liberal pleading rules, unlike the common law, he was not required to specify which form. At common law, choosing the wrong form of action was fatal to the case, but modern civil procedure has abandoned such rigid distinctions. It is important to note, however, that Salinas's claim of trespass does not entitle him to nominal damages (which he has not sought). He must prove actual injury.

B

Had Coastal caused something like proppants to be deposited on the surface of Share 13, it would be liable for trespass, and from the ancient common law maxim that land ownership extends to the sky above and the earth's center below, one might extrapolate that the same rule should apply two miles below the surface. But that maxim -- *cujus est solum ejus est usque ad coelum et ad inferos* -- "has no place in the modern world." Wheeling an airplane across the surface of one's property without permission is a trespass; flying the plane through the airspace two miles above the property is not. Lord Coke, who pronounced the maxim, did not consider the possibility of airplanes. But neither did he imagine oil wells. The law of trespass need no more be the same two miles below the surface than two miles above.

[The court analyzes a number of earlier Texas opinions dealing with the issue of hydraulic fracturing including Gregg v. Delhi-Taylor Oil Corp., 162 Tex. 26, 344 S.W.2d 411, 14 O.&G.R. 106 (1961); Geo-Viking, Inc. v. Tex-Lee Operating Co., 817 S.W.2d 357, 117 O.&G.R. 341 (Tex.App. 1991), *writ denied per curiam*, 839 S.W.2d 797, 117 O.&G.R. 357 (Tex. 1992) and Railroad Commission v. Manziel, 361 S.W.2d 560, 17 O.&G.R. 444 (Tex. 1962). While there is language in some of these cases suggesting that a hydraulic fracturing operation that crosses a boundary line constitutes a trespass, the majority opinion distinguishes those cases and finds that the issue had not been decided prior to this case. Eds.]

* * *

We need not decide the broader issue here. In this case, actionable trespass requires injury, and Salinas's only claim of injury -- that Coastal's fracing operation made it possible for gas to flow from beneath Share 13 to the Share 12 wells -- is precluded by the rule of capture. That rule gives a mineral rights owner title to the oil and gas produced from a lawful well bottomed on the property, even if the oil and gas flowed to the well from beneath another owner's tract. The rule of capture is a cornerstone of the oil and gas industry and is fundamental both to property rights and to state regulation. Salinas does not claim that the Coastal Fee No. 1 violates any statute or regulation. Thus, the gas he claims to have lost simply does not belong to him. He does not claim that the hydraulic fracturing operation damaged his wells or the Vicksburg T formation beneath his property. In sum, Salinas does not claim damages that are recoverable.

Salinas argues that the rule of capture does not apply because hydraulic fracturing is unnatural. The point of this argument is not clear. If by "unnatural" Salinas means due to human intervention, the simple answer is that such activity is the very basis for the rule, not a reason to suspend its application. Nothing is more unnatural in that sense than the drilling of wells, without which there would be no need for the rule at all. If by "unnatural" Salinas means unusual, the facts are that hydraulic fracturing has long been commonplace throughout the industry and is necessary for commercial production in the Vicksburg T and many other formations. And if by "unnatural" Salinas means unfair, the law affords him ample relief. He may use hydraulic fracturing to stimulate production from his own wells and drain the gas to his own property -- which his operator, Coastal, has successfully done already -- and he may sue Coastal for not doing so sooner -- which he has also done, in this case, though unsuccessfully, as it now turns out. [39]

Salinas argues that stimulating production through hydraulic fracturing that extends beyond one's property is no different from drilling a deviated or slant well -- a well that departs from the vertical significantly -- bottomed on another's property, which is unlawful. Both produce oil and gas situated beneath another's property. But the rule of capture determines title to gas that drains from property owned by one person onto property owned by another. It says nothing about the ownership of gas that has remained in place. The gas produced through a deviated well does not migrate to the wellbore from another's property; it is already on another's property. The rule of capture is justified because a landowner can protect himself from drainage by drilling his own well, thereby avoiding the uncertainties of determining how gas is migrating through a reservoir. It is a rule of expedience. One cannot protect against drainage from a deviated well by drilling his own well; the deviated well will continue to produce his gas. Nor is there any uncertainty that a deviated well is producing another owner's gas. The justifications for the rule of capture do not support applying the rule to a deviated well.

We are not persuaded by Salinas's arguments. Rather, we find four reasons not to change the rule of capture to allow one property owner to sue another for oil and gas drained by hydraulic fracturing that extends beyond lease lines.

First, the law already affords the owner who claims drainage full recourse. This is the justification for the rule of capture, and it applies regardless of whether the drainage is due to fracing. If the drained owner has no well, he can drill one to offset drainage from his property. If the minerals are leased and the lessee has not drilled a well, the owner can sue the lessee for violation of the implied covenant in the lease to protect against drainage. If an offset well will not adequately protect against drainage, the owner (or his operator) may offer to pool, and if the offer is rejected, he may apply to the Railroad Commission for forced pooling. The Commission may also regulate production to prevent drainage. No one suggests that these various remedies provide inadequate protection against drainage.

Second, allowing recovery for the value of gas drained by hydraulic fracturing usurps to courts and juries the lawful and preferable authority of the Railroad Commission to regulate oil and gas production. Such recovery assumes that the gas belongs to the owner of the minerals in the drained property, contrary to the rule of capture. While a mineral rights owner has a real interest in oil and gas in place, "this right does not extend to *specific* oil and gas beneath the property"; ownership must be "considered in connection with the law of capture, which is recognized as a property right" as well. The minerals owner is entitled, not to the molecules actually residing below the surface, but to "a fair chance to recover the oil and gas in or under his land, *or* their equivalents in kind." . . . The Commission's role should not be supplanted by the law of trespass.

Third, determining the value of oil and gas drained by hydraulic fracturing is the kind of issue the litigation process is least equipped to handle. One difficulty is that the material facts are hidden below miles of rock, making it difficult to ascertain what might have happened. Such difficulty in proof is one of the justifications for the rule of capture. But there is an even greater difficulty with litigating recovery for drainage resulting from fracing, and it is that trial judges and juries cannot take into account social policies, industry operations, and the greater good which are all tremendously important in deciding whether fracing should or should not be against the law. While this Court may consider such matters in fashioning the common law, we should not alter the rule of capture on which an industry and its regulation have relied for decades to create new and uncertain possibilities for liability with no more evidence of necessity and appropriateness than this case presents.

Fourth, the law of capture should not be changed to apply differently to hydraulic fracturing because no one in the industry appears to want or need the change. The Court has received amicus curiae briefs in this case from the

Railroad Commission, the General Land Office, the American Royalty Council, the Texas Oil & Gas Association, the Texas Independent Producers & Royalty Owners Association, the Texas Alliance of Energy Producers, Harding Co., BJ Services Co., Halliburton Energy Services, Inc., Schlumberger Technology Corp., Chesapeake Energy Corp., Devon Energy Corp., Dominion Exploration & Production, Inc., EOG Resources, Inc., Oxy Usa Inc., Questar Exploration and Production Co., XTO Energy, Inc., and Chief Oil & Gas LLC. These briefs from every corner of the industry -- regulators, landowners, royalty owners, operators, and hydraulic fracturing service providers -- all oppose liability for hydraulic fracturing, almost always warning of adverse consequences in the direst language. Though hydraulic fracturing has been commonplace in the oil and gas industry for over sixty years, neither the Legislature nor the Commission has ever seen fit to regulate it, though every other aspect of production has been thoroughly regulated. Into so settled a regime the common law need not thrust itself.

Accordingly, we hold that damages for drainage by hydraulic fracturing are precluded by the rule of capture. It should go without saying that the rule of capture cannot be used to shield misconduct that is illegal, malicious, reckless, or intended to harm another without commercial justification, should such a case ever arise. But that certainly did not occur in this case, and no instance of it has been cited to us.

* * *

We reverse the court of appeals' judgment, render judgment that Salinas take nothing on his claims for trespass and breach of the implied covenant to protect against drainage, and remand the remainder of the case for a new trial.

WILLETT, JUSTICE concurring. James Michener may well be right: "Water, not oil, is the lifeblood of Texas ... " But together, oil and gas are its muscle, which today fends off atrophy.

At a time of insatiable appetite for energy and harder-to-reach deposits--iron truths that contribute to $ 145 a barrel crude and $ 4 a gallon gasoline --Texas common law should not give traction to an action rooted in abstraction. Our fast-growing State confronts fast-growing energy needs, and Texas can ill afford its finite resources, or its law, to remain stuck in the ground. The Court today averts an improvident decision that, in terms of its real-world impact, would have been a legal dry hole, juris-imprudence that turned booms into busts and torrents into trickles. Scarcity exists, but *above-ground* supply obstacles also exist, and this Court shouldn't be one of them. . . .

Bottom line: We are more and more over a barrel as "our reserves of fossil fuels are becoming harder and more expensive to find." Given this supply-side slide, maximizing recovery via fracing is essential; enshrining trespass liability for fracing (a "tres-frac" claim) is not. I join today's no-liability result and suggest another reason for barring tres-frac suits: Open-ended liability threatens to inflict grave and unmitigable harm, ensuring that much of our State's undeveloped energy supplies would stay that way--undeveloped. Texas oil and gas law favors drilling wells, not drilling consumers. Amid soaring demand and sagging supply, Texas common law must accommodate cutting-edge technologies able to extract untold reserves from unconventional fields.

JOHNSON, JUSTICE, concurring in part and dissenting in part. I join the Court's opinion except for Part II-B. As to Part II-B, I would not address whether the rule of capture precludes damages when oil and gas is produced through hydraulic fractures that extend across lease lines until it is determined whether hydraulically fracturing across lease lines is a trespass.

I. Rule of Capture

The rule of capture precludes liability for capturing oil or gas drained from a neighboring property "whenever such flow occurs solely through the operation of natural agencies in a normal manner, as distinguished from artificial means applied to stimulate such a flow." The rationale for the rule of capture is the "fugitive nature" of hydrocarbons. They flow to places of lesser pressure and do not respect property lines. The gas at issue here, however, did not migrate to Coastal's well because of naturally occurring pressure changes in the reservoir. If it had, then I probably would agree that the rule of capture insulates Coastal from liability. But the jury found that Coastal trespassed by means of the hydraulic fracturing process, and Coastal does not contest that finding here. Rather, Coastal contends that a subsurface trespass by hydraulic fracturing is not actionable.   In the face of this record and an uncontested finding that Coastal trespassed on Share 13 by the manner in which it conducted operations on Share 12, I do not agree that the rule of capture applies. Coastal did not legally recover the gas it drained from Share 13 unless Coastal's hydraulic fracture into Share 13 was not illegal. Until the issue of trespass is addressed, Coastal's fracture into Share 13 must be considered an illegal trespass. I would not apply the rule to a situation such as this in which a party effectively enters another's lease without consent, drains minerals by means of an artificially created channel or device, and then "captures" the minerals on the trespasser's lease.

In considering the effects of the rule of capture, the underlying premise is that a landowner owns the minerals, including oil and gas, underneath his property. . . . Coastal concedes that gas must be legally produced in order to come within the rule of capture. The key word is "legally." Without it, the rule of capture becomes only a license to obtain minerals in any manner, including unauthorized deviated wells, and vacuum pumps and whatever other method oilfield operators can devise.

Today the Court says that because Salinas does not claim the Coastal Fee No. 1 well violates a statute or regulation, the gas that traveled through the artificially created and propped-open fractures from Share 13 to the well "simply does not belong to him." But that conclusion does not square with the underlying rationale for the rule of capture, and as seems only logical and just: an operator such as Coastal owns the oil and gas that is *legally* captured. And "legally" should not sanction all methods other than those specifically prohibited by statute or rule of the Railroad Commission. It simply cannot be a legal activity for one person to trespass on another's property. The question the Court does not answer, but which it logically must to decide this case, is whether it was legal for Coastal to hydraulically fracture into Share 13. The answer to the question requires us to address Coastal's primary issue: does hydraulic fracturing across lease lines constitute subsurface trespass.

We have held that a trespass occurs when a well begun on property where the operator has a right to drill is, without permission, deviated so the well crosses into another's lease. Coastal argues that there are differences between taking minerals from another's lease through fracturing and taking them by means of a deviated well. Maybe there are, even though both involve a lease operator's intentional actions which result in inserting foreign materials without permission into a second lease, draining minerals by means of the foreign materials, and "capturing" the minerals on the first lease. The question certainly is not foreclosed. To differentiate between a deviated well and a fractured well, the Court says that gas extracted from a neighboring lease through a deviated well is not subject to the rule of capture for two reasons: the neighbor cannot protect from such drainage by drilling a well, and there is no uncertainty that the deviated well is producing another owner's gas. I fail to follow the Court's logic. As to the first reason, the neighbor can protect from either a fracture extending into the neighbor's property or a deviated well. Both simply provide the means for gas to flow to an area of lower pressure and from there to the drilling operator's property where it is captured. The only difference is the degree of drainage that can be prevented by offset wells, and a fracture's exposure to the reservoir may be greater than that of the deviated well and thus drain more gas. As to the second reason, the purpose of both a deviated well and a hydraulic fracture is for gas to flow through them to be gathered at a distant surface. Coastal fractured its well so gas would flow through the fractures to the wellbore, and no one contends that gas did not do so. The evidence showed that the effective length of a fracture can be fairly closely determined after the fracture operation.

NOTES

1. In Stop the Beach Renourishment, Inc. v. Florida Department of Environmental Protection, --- U.S. ---, 130 S.Ct. 2592 (2010), Justice Scalia speaking for a plurality of 4 justices posits the view that court decisions can just as much violate the 5th Amendment's prohibition against the taking of private property without the payment of just compensation as can legislation or administrative agency decisions. Does the decision in *Coastal Oil & Gas* raise regulatory takings issues in that the earlier Texas judicial opinions appeared to recognize a trespassory cause of action where the injection of "frac" fluids crossed boundary lines? Was there a clearly recognized common law property right to be free of cross-boundary frac fluids prior to the *Coastal Oil & Gas* decision?

2. Are the 4 reasons cited by the Supreme Court relevant in determining the scope and/or breadth of a property interest? Do the reasons focus on private or

public matters?   Does the court mention the impact of its conclusion on the private property rights of the owners of the land receiving the "frac" fluids and/or proppants?

3. Other states that have dealt with the issue of cross-boundary migration of "frac" fluids have not directly resolved the trespass issue. See Zinke & Trumbo, Ltd. v. State Corporation, 749 P.2d 21, 103 O.&G.R. 201 (Kan. 1988) and ANR Production Co. v. Kerr-McGee Corp., 893 P.2d 698, 134 O.&G.R. 631 (Wyo. 1995).

4. The problem of disposing of oil field wastes, including brine, produced water or salt water creates both practical and legal difficulties. The basic problem has been stated as follows:

> The migration of injected fluids or gases into underground formations does not end at the property line.   Once injected, these fluids may cause extraterritorial effects that damage other individuals' mineral or surface rights.   The rule of capture does not deal with this situation because it emphasizes the withdrawal and not the injection of materials.   Because injection of fluids is the primary method by which secondary or tertiary recovery projects operate, the potential liability of the injectors for injuries caused to neighboring owners could act as an impediment to the implementation of these techniques and a subsequent loss of recoverable oil and gas reserves. [1 Kramer & Martin, *The Law of Pooling and Unitization* 2-29.]

Is there a subsurface trespass to the mineral estate if the mineral lessee injects wastewater that is produced from off-lease wells without the mineral lessor's permission? Can there be a trespass to the lessor's reversionary interest, which by definition must be non-possessory? In Cassinos v. Union Oil Co. of Calif., 14 Cal.App.4th 1770, 18 Cal.Rptr.2d 574 (1993) the court found that there was an "unauthorized entry" onto the land of another by the act of injecting off-leasehold wastewater which constituted a trespass.   This issue of subsurface trespass is taken up further *infra*, Chapter 7. *See* also *Corbello v. Iowa Production, infra* p. 531.

5. Disposal of oil field or other wastes that cross boundary lines are typically authorized by state agencies. Should the receipt of a permit insulate the injector or disposer of such wastes from a common law trespass action? *Compare* FPL Farming, Ltd. v. Environmental Processing Systems, 305 S.W.3d 739 (Tex.App. 2009) *with* Berkley v. Railroad Commission of Texas, 282 S.W.3d 240 (Tex.App. 2009).

NOTE ON GEOPHYSICAL TRESPASS

1. In Phillips Petroleum Co. v. Cowden, 241 F.2d 586, 7 O.&G.R. 1291 (5th Cir. 1957), the owner of a mineral estate sued a party who had engaged in geophysical operations on the surface above the mineral estate, as well as upon neighboring lands where the geophysical operator had the consent of both the

surface and mineral owners. After determining that it is the owner of the severed mineral estate who owns the property interest in geophysical operations, the Fifth Circuit concluded that because there had been a surface trespass by the geophysical operator on a small portion of the mineral estate, the geophysical operator had committed a trespass. Where there is no surface trespass, however, the "transmission" of sound waves from lands owned by third parties that cross the boundary line and thus would provide information regarding the potential for mineral development under those lands would not constitute a trespass. In this first opinion, the court reversed a judgment awarding damages based on the alleged value of a geophysical lease on the entire 2682 acre tract owned by the plainitff even though the surface encroachment was only on a 81.2-acre portion of the tract. The geophysical operator also alleged that the quality of the data was so poor that no determination could be made as to whether there were or were not minerals underlying the 2682 acre tract. On a second appeal, the redetermination by the trial court that 2682 acres had been invaded and that plaintiff was damaged in the same amount as found in the first trial, $ 53,640, or $ 20.00/acre was sustained. Phillips Petroleum Co. v. Cowden, 256 F.2d 408, 9 O.&G.R. 110 (5th Cir. 1958).

2. Does the owner of an unrecorded mineral lease have a cause of action against a surface owner who engages in good faith geophysical operations? Would the state's recording statute defeat the claim of a tortious invasion of the leasehold rights? *See* Grynberg v. City of Northglenn, 739 P.2d 230, 95 O.&G.R. 28 (Colo.1987). *See also* Picou v. Fohs Oil Co., 222 La. 1068, 64 So.2d 434, 2 O.&G.R. 525 (1953).

Does the oil and gas lessee have the right to engage in seismic operations? See Musser Davis Land Co. v. Union Pacific Resources, 201 F.3d 561, 145 O.&G.R. 282 (5th Cir. 2000) discussed page 173 *infra*. Does the lessee have the exclusive right to engage in such operations? Ready v. Texaco, Inc., 410 P.2d 983, 24 O.&G.R. 521, 28 A.L.R.3d 1419 (Wyo.1966), held that lessees under federal and state oil and gas leases did not have exclusive rights to make geophysical surveys. *See* Taylor, *Non–Exclusive Rights of Lessees to Conduct Geophysical Exploration—Federal and Wyoming State Oil and Gas Leases*, 3 Land & Water L.Rev. 103 (1968). The *Ready* case is followed in Mustang Production Co. v. Texaco, Inc., 549 F.Supp. 424, 74 O.&G.R. 462 (D.Kan.1982), aff'd, 754 F.2d 892 (10th Cir.1985), where a lease on private land specified an easement for geophysical exploration but did not expressly describe it as exclusive. See Lomenick, *The Oil and Gas Lessee's Right to Geophysical Exploration: Incidental or Exclusive?* 20 Tulsa L.J. 97 (1984).

3. Modern geophysical exploration involves the use of 3-D or 4-D seismic technology. Sound waves are passed through the rock to provide a multi-dimensional view of the underground formation. If a geophysical operation emits sound waves from Blackacre where it has permission to do so and the waves cross over the property line and provide information about the adjacent Whiteacre tract

has there been a trespass? In Villareal v. Grant Geophysical, Inc., 136 S.W.3d 265, 160 O.&G.R. 99 (Tex.App. 2004), the court follows the *Cowden* rationale and finds no trespass as long as the surface facilities are located on lands where the geophysical explorer has permission to locate. Should the good or bad faith of the geophysical explorer effect the issue of whether there is a trespass? In theory no, since trespass is a strict liability tort, but the court in *Villareal* appeared to be effected by the geophysical explorer's efforts to not discover information about neighboring lands. The newly developed geophysical exploration tools are discussed in Anderson & Pigott, *Seismic Technology and Law: Partners or Adversaries?*, 24 Energy & Min'l L. Inst. 11-1 (2003). The impact of the rule of capture on geophysical operations is analyzed in Bruce M. Kramer & Owen L. Anderson, *The Rule of Capture – An Oil and Gas Perspective*, 35 Env't. L. 899 (2005).

4. Should a geophysical trespasser be liable for the decline in market value of leasing rights and royalty interests caused by the dissemination by the trespasser of unfavorable information about plaintiff's land? Should this element of damage be added to that recovered by plaintiff in the principal case? *See* Angelloz v. Humble Oil & Refining Co., 196 La. 604, 199 So. 656 (1940).

5. Is the proper measure of damages the value of the information gained by the trespass to the trespasser, on the theory that liability should be based on principles of unjust enrichment and restitution? *See* Malone, *Ruminations on a New Tort*, 4 La.L.Rev. 309 (1942).

6. Does the owner of the right to engage in geophysical exploration bear the burden of proof on the damages issue? How much certainty is required in showing damages? *See* Tinsley v. Seismic Explorations, Inc., 229 La. 23, 117 So.2d 897, 12 O.&G.R. 76 (1960). Under the Louisiana Mineral Code, La.Rev.Stat. 31: 175, co-owners of at least 80% of the mineral servitude can conduct geophysical operations on the land. But another statute, La.Rev.Stat. 30:217 requires consent of the owner of the land before geological surveys by torsion balance, seismograph explosions, mechanical device or other method can be used. See Jeanes v. G.F.S.Co., 647 So.2d 533 (La.App. 1994, writ denied) for a discussion of those two statutes where the geophysical explorer did not get the consent of a mineral servitude owner. The latter statute has since been revised.

7. Suppose the geophysical operator does not enter plaintiff's land but vibrations from explosions of dynamite on nearby land penetrate plaintiff's land and the geophysicist uses data thus obtained from plaintiff's land in plotting contours for the entire area, including plaintiff's land. Can plaintiff recover? *See* Kennedy v. General Geophysical Co., 213 S.W.2d 707 (Tex.Civ.App.1948, error ref'd n.r.e.). If the vibrations result in physical injury to the plaintiff's land or structures quite different questions arise. Teledyne Exploration Co. v. Dickerson, 253 So.2d 817, 40 O.&G.R. 47 (Miss.1971), followed earlier cases in imposing absolute liability for damage resulting from geophysical explosions on neighboring land. *See* Keeton and Jones, *Tort Liability and the Oil and Gas Industry*, 35

Tex.L.Rev. 1 (1956). Technological advances are ending the necessity for shot holes and dynamite to create sound waves for seismic surveys and are making it easier to obtain subsurface information without physically entering land which is a subject of interest. Should these developments have an impact on the state of the law? The law reviews are searching for new theories of liability. See Mark D. Christiansen, *Oil and Gas: Improper Geophysical Exploration - Filling in the Remedial Gap*, 32 Okla L. Rev. 903, 907 (1979); Comment, *The Surreptitious Geophysical Survey: An Interference with Prospective Advantage*, 15 Pac.L.J. 381 (1984); Note, *Oil and Gas: Recovery for Wrongful Geophysical Exploration— Catching Up with Technology"* 23 Washburn L.J. 107 (1983); Harry L. Blomquist, III, *Geophysical Trespass? The Guessing Game Created by the Awkward Combination of Outmoded Laws and Soaring Technology*, 48 Baylor L. Rev. 21, 29-31 (1996). For cases involving aerial reconnaissance see Ratliff v. Beard, 416 So.2d 307, 73 O.&G.R. 532 (La.App.1982) and Gulf Coast Real Estate Auction Company, Inc. v. Chevron Industries, Inc., 665 F.2d 574, 73 O.&G.R. 98 (5th Cir.1982).

———

C. TRESPASS ACTIONS INVOLVING PRODUCTION

CHAMPLIN REFINING CO. v. ALADDIN PETROLEUM CORP.

Supreme Court of Oklahoma, 1951.
205 Okl. 524, 238 P.2d 827, 1 O.&G.R. 93.

JOHNSON, JUSTICE. . . . [In an earlier action it was determined that the state and its lessee, Champlin, did not have title to the land in question. Eds.]

These appeals are from judgments directing Champlin Refining Company to account to defendants in error for converted oil and gas at the highest market value between the conversion and trial. The judgments were rendered after the mandate heretofore mentioned was filed in the trial court.

Hereafter, we shall, for the sake of brevity, refer to plaintiff in error as Champlin, who is now the only real party in interest, and to the defendants in error as Aladdin defendants and Oldham defendants or as defendants.

There are two material issues presented. One is whether the trial court correctly directed Champlin to account to the mineral owners for the oil and gas taken from their land at the highest market price between the date of conversion and the judgment of the Court. The other is whether Champlin was properly denied credit for the expense incurred in drilling a directional dry hole to a point under the Oldham property.

There is no controversy as to the essential facts, which are briefly as follows:

The State of Oklahoma, through the Commissioners of the Land Office, sold an oil and gas lease to Champlin Refining Company, covering a portion of the South Half of the Arkansas River bed in Pawnee County, Oklahoma, on July 12, 1942, at public sale. The attorney for Champlin, the Chief Counsel for the Commissioners of the Land Office, and the Attorney General of the State of

Oklahoma were all of the opinion that the State of Oklahoma had title to said property and all so advised the Company. Champlin, relying on this advice and opinions of title, purchased the lease and drilled two producing wells thereon, costing $157,471.20. Thereafter, the trial court, complying with the mandate in the Aladdin case, *supra*, awarded one of the Champlin wells to Aladdin defendants, case No. 9640 and the other well to Phillips Petroleum Company, *et al.,* or Oldham defendants, case No. 9641.

Thereupon, Champlin delivered possession of the lease and producing wells thereon to Aladdin defendants and Oldham defendants and paid them the entire proceeds of the oil and gas produced by it from the two wells at the market value of oil and gas on date of production in the aggregate sum of $507,906.68 less expenses of development and operation in the sum of $197,676.39, making a total sum of $310,230.29, including the royalty, which Champlin paid to Aladdin and Oldham defendants.

However, after the mandate had been filed in the title litigation, Aladdin case, *supra*, the Aladdin and Oldham defendants filed pleadings in the trial court asking for monetary damages, and for the first time, elected to sue for the highest market value of the oil and gas in an accounting which they contend is a suit for conversion. Approximately five years had elapsed since the date of the alleged conversion in the Aladdin case, and a year and seven months had elapsed after conversion was first alleged in the Oldham case.

There is no dispute as to Champlin's production figures nor as to the expenses of development and operation or amount of royalties paid to the Aladdin and Oldham defendants.

It was established that the market price of top grade oil at the time the first well was completed was $1.25 per barrel; the highest market price thereafter was $2.65 per barrel. The trial court ordered Champlin to pay to defendants the difference between the highest market price of the oil and the actual market price at the time of production. The trial court further found that Champlin was not guilty of such wilful conduct as would deprive it of the right to recoup the expenses incurred by it in recovering the oil and gas and ordered recoupment accordingly. From this last phase of the order no exception or appeal was taken and it is now final.

The court further ordered that one of the directional branches of the second well was non-producing and that Champlin should not be allowed to recoup the cost of this well.

Briefly summarized this appeal is from the order which: (a) requires Champlin to pay to defendants the claim aggregating $268,723.32, the difference between the actual market price of the oil when produced or converted and the highest market value between the conversion and the verdict; (b) denies reimbursement to Champlin for the costs of drilling the non-producing branch of the directional producing well in the sum of $56,135.50.

It is contended by Champlin that defendants are not entitled to recover the highest market value between the time of conversion of the oil and trial or verdict, because they did not establish that they had commenced and prosecuted their action with reasonable diligence; and that the conversion was not wilfully wrongful but was in good faith by reason of its bona fide lease from the State of Oklahoma whom it believed to be vested with title. On the other hand, it is contended by defendants that they did commence and prosecute their actions with reasonable diligence; that Champlin knew of their claims of title and that under Title 23, Sec. 64, O.S.1941,[25] regardless of the wilful wrongfulness or good faith of Champlin, they were entitled to the highest market value between the time of conversion of the oil and verdict.

Conversion is any distinct act or dominion wrongfully exerted over another's personal property in denial of or inconsistent with his rights therein.

But, in construing Title 23, Sec. 64, O.S.1941, cited by defendants, we have said that in order to recover the highest market value between the time of conversion of personalty and verdict, owner must establish that he commenced and prosecuted his action with reasonable diligence; that 15 months unexplained delay in bringing action for wrongful conversion would preclude the owner from recovering highest market value of property between time of conversion and trial; that the measure of damages for oil produced by lessee taking possession in good faith under a void or invalid lease is value of oil less reasonable cost of production. Miller v. Tidal Oil Co., 161 Okl. 155, 17 P.2d 967, 87 A.L.R. 811.

The record shows that defendants failed to exercise reasonable diligence in the prosecution of their action and are not entitled to the highest market value of the oil between the time of conversion and trial. Especially is this true in this case since Champlin acted in good faith.

There is nothing in the record to indicate Champlin's lack of good faith in taking peaceable possession of the property leased to it by the State. It is evident however that it erred in judgment as to the true or rightful ownership of the land, as it was thereafter determined that its lease was void because the State of Oklahoma had no title, but this does not impeach its "good faith" which unquestionably existed when it purchased the lease and developed it. Therefore, we conclude that the court erred in not applying the measure of damages announced in the Miller v.

---

[25] This provision is in part as follows: "The detriment caused by the wrongful conversion of personal property is presumed to be:

"1. The value of the property at the time of the conversion with the interest from that time; or,

"2. Where the action has been prosecuted with reasonable diligence, the highest market value of the property at any time between the conversion and the verdict, without interest, at the option of the injured party. . . . "

For a general discussion of the rules governing the recovery of the highest market value of property of fluctuating value see McCormick, Damages § 48 (1935). Wronski v. Sun Oil Co., *supra* p. 226 rejected the application of the highest market value of oil converted over a period of substantial market fluctuations. [Eds.]

Tidal Oil Co. case, *supra.* There in syllabi 2 and 3, we said:

2. An unexplained delay of fifteen months after the cause of action accrued in bringing an action for wrongful conversion shows such want of reasonable diligence in commencing and prosecuting his action as will preclude plaintiff from recovering the highest market value of the property between the time of conversion and the trial.

3. Where a person in good faith enters into peaceable possession of land upon which he owns an oil and gas lease and produces oil and gas therefrom, and thereafter said lease is declared void or invalid, the measure of damages to the landlord in an action for an accounting for the oil and gas produced from said premises by the lessee is the value of the oil at the surface or in pipe line or tanks wherever the same may be, less the reasonable cost of producing the same.

It is next contended by Champlin that the court erred in denying it reimbursement for the costs expended in drilling the non-producing directional branch of the second well under the Oldham property.

We are of the opinion that the cost of drilling the unprofitable directional branch of a producing well is part of reasonable costs of development and production, and must be deducted in accounting. Bailey v. Texas Pacific Coal & Oil Co., 168 Okl. 275, 32 P.2d 709; Moody v. Wagner, 167 Okl. 99, 23 P.2d 633.

The record discloses that after Champlin had obtained production in the river bed No. 1 well, permission was granted by the Corporation Commission to use the same surface location for the second well, because the high banks at the location where the second well was to be bottomed made the cost of building an additional pier prohibitive, and for this reason permission was granted it to drill a directional well, and it was so drilled to a bottom location which was advised by its geologist but upon completion the well was dry and in a non-productive portion of the formation.

Champlin then plugged back to within 300 feet of the surface of the non-producing directional well and branched off from there with another directional hole which was bottomed only 250 feet west of the first branch and thereby obtained the second producing well. Thus it is seen that 300 feet of the non-producing directional well, the unprofitable directional branch, became a part of the directional producing well. The well was drilled in good faith and the cost thereof, being reasonable and necessary, is deductible in the action for an accounting. Moody v. Wagner, *supra,* and Miller v. Tidal Oil Company, *supra.*

The judgment is reversed with directions for further proceedings in accord with the views herein expressed.

WELCH, CORN, GIBSON and O'NEAL, JJ., concur.

LUTTRELL, V.C.J., concurs in part and dissents in part.

DAVISON, J., concurs in conclusion.

HALLEY, J., dissents.

LUTTRELL, VICE CHIEF JUSTICE (dissenting in part). I dissent to that part of the opinion allowing Champlin costs expended in drilling the nonproductive well. The cost of the drilling of such well conferred no benefit upon the property, and added no value thereto. I am unable to see how Champlin is entitled to offset the cost of drilling a dry hole against the value of the oil and gas for which it is liable. In allowing the actual cost of drilling productive wells such allowance is upon the theory that the property has been benefited to the extent of such drilling costs. Here the dry hole added nothing to the value, and, in my opinion, the cost thereof should not be allowed as an offset.

NOTES

1. Did the court in *Champlin* allow recovery of the costs associated with the dry hole because it added value to the owner's interest or because it was money expended in good faith? What if an owner drilled a dry hole on a spacing unit prior to the entry of a pooling order affecting the working interest owners? Could that owner recover if the work was done in good faith? *See* Wilcox Oil Co. v. Corporation Commission, 393 P.2d 242, 21 O.&G.R. 67 (Okl.1964).

2. In HOUSTON PRODUCTION CO. v. MECOM OIL CO., 62 S.W.2d 75 (Tex.Com.App.1933), undisputed facts showed that "defendant in error drilled the well for which it has been allowed compensation, after plaintiff in error had served notice in writing that it held a valid lease upon this land. In addition to this, the defendant in error had full knowledge of the pendency of the cross-action by plaintiff in error to recover the land at the time it drilled the well for which compensation was allowed." The court overturned a jury finding of good faith stating:

> We are inclined to adhere to the well established rule that, where one enters into possession of land and makes improvements thereon with full knowledge of the pendency of an action to enforce an adverse claim to the premises, he cannot be considered a trespasser in good faith so as to entitle him to recover the cost of his improvements.

> We are not impressed with the force of the reasoning that a distinction should be made between an ordinary trespasser and an oil and gas trespasser in regard to recovery for improvements. The basis for the claimed distinction is that, where oil and gas land is involved, it is often necessary to drill the same in order to prevent its being drained by adjoining wells. We think a complete answer to this proposition is that, when a court of equity is confronted with such a situation, it has ample authority to take such action as will prevent the property's being drained of its oil and gas pending the final adjudication of title.

Compare the following statement from Gulf Production Co. v. Spear, 125 Tex. 530, 84 S.W.2d 452 (1935):

> Good faith under one state of facts would not be good faith under a slightly

different state of facts. In general, it may be said that to act in good faith in developing a tract of land for oil and gas one must have both an honest and a reasonable belief in the superiority of his title.

Houston Production Co. v. Mecom Oil Co. is apparently still recognized as authoritative in relation to its facts. In Humble Oil & Refining Co. v. Luckel, 154 S.W.2d 155 (Tex.Civ.App.1941, error ref'd w.o.m.), an injunction prohibiting exploration pending the outcome of a title suit was denied. The court took the position that the legal remedy was adequate.

> If appellants enter on the land and develop it for oil, producing oil, they do so at their peril, and the profits of their developments would inure to appellees, free of contribution on their part, for the reason that, knowing the facts of appellees' title, appellants, as a matter of law, would not be good faith developers of the land. Houston Production Co. v. Mecom Oil Co., Tex.Com.App., 62 S.W.2d 75. If appellees enter upon the land pending the litigation, in an attempt to develop it for oil, and by their attempt destroy the oil value of the land, then they would be liable to appellees for the damages suffered by them, in the event they should be cast in this suit. Humble Oil & Refining Co. v. Kishi, Tex.Com.App., 276 S.W. 190.

If the land is being drained or if the trespasser must drill to satisfy leasehold obligations, does she have any acceptable alternative to taking the risk outlined in the *Luckel* case? In Whelan v. Killingsworth, 537 S.W.2d 785, 55 O.&G.R. 224 (Tex.Civ.App. 1976) a "bad faith" trespasser under *Mecom* argued for the abolition of the rule because of the "necessity to encourage full and rapid development of oil reserves as well as to protect against drainage.... " In refusing to ignore *Mecom* the court said that the district court had "ample equity powers to provide for the development of property pending the final outcome of litigation." To what equitable powers does the court refer? Would that include receivership?

Further developments in Texas have continued to muddy the waters. In Brannon v. Gulf States Energy Corp., 562 S.W.2d 219, 59 O.&G.R. 320 (Tex.1977) the Texas Supreme Court rejected the contention that "as a matter of law" a trespass was not in good faith because of the trespasser's knowledge of a prior lease and claim on the mineral interest. *Mecom* was not cited but the opinion emphasized that "no suit was filed" prior to the trespasser's entry upon the leasehold and the spudding of its first two wells. In its opinion on a motion for rehearing the court also rejected the argument that the trespasser "should be held as a matter of law not to have been in good faith in drilling the third well because it was commenced after this suit was filed." *Brannon* is a principal case in Chapter 4 *infra*.

The inconsistent holdings of *Mecom* and *Brannon* were ignored by the court in Mayfield v. deBenavides, 693 S.W.2d 500, 85 O.&G.R. 162 (Tex.App. 1985) which cited *Mecom* for the proposition that entry "with full knowledge of the pendency of an action" conclusively constitutes bad faith. It further noted that the converse, however, is not true. An entry is not automatically in good faith "because a suit has not been filed. . . . Rather, it is a circumstance to be considered along with other factors in determining the good faith of the trespasser."

3. Edwards v. Lachman, 1974 OK 58, 534 P.2d 670, 51 O.&G.R. 343, provides an extensive analysis of what constitutes good or bad faith. The case also suggests that a trespasser may start out as a good faith trespasser but my turn into a bad faith trespasser after she realizes that a well that has been drilled is bottomed on the lands of another. That creates substantial problems when it comes to measuring the damages. On remand in *Edwards*, the trial court allowed the trespasser credit for the cost of drilling from the shallower formation to the bottom of the deeper formation. The Court of Appeals reversed and gave a credit for the cost of drilling from the surface to the bottom of the deeper formation. The Supreme Court reversed the Court of Appeals and found that *Edwards I* limited the recovery of the trespasser to the benefits conferred upon the plaintiff. 1977 OK 175, 567 P.2d 73. The trial court distinguished *Champlin Refining, supra* by only finding a benefit for the drilling from the shallower to the deeper formation. The benefit was measured by the costs of drilling to the deeper formation. Is that consistent with *Champlin Refining?* For other cases dealing with how to calculate the costs that may be set-off by the good faith trespasser, see BP America Production Co. v. Marshall, 288 S.W.3d 430 (Tex.App. 2008); Carter Oil Co. v. McCasland, 207 F.3d 728, 3 O.&G.R. 659 (10th Cir. 1953).

4. Would a different standard of good faith be appropriate in a case involving hard minerals, rather than oil and gas? *Compare* Whittaker v. Otto, 248 Cal.App.2d 666, 56 Cal.Rptr. 836, 26 O.&G.R. 1 (1967) *with* Dethloff v. Zeigler Coal Co., 82 Ill.2d 393, 45 Ill.Dec. 175, 412 N.E.2d 526, 68 O.&G.R. 608 (1980), cert. denied, 451 U.S. 910 (1981).

5. Oklahoma has no per se rules regarding continued development after a challenge has been made to the title of the oil and gas lessee. *See* Dilworth v. Fortier, 405 P.2d 38, 23 O.&G.R. 424 (Okl.1964).

6. Huge Oil commences the drilling of a well while a lease is still in existence. The primary term ends and Huge ceases drilling activities. The lease does not contain any provision for extending the lease beyond the primary term unless production is achieved. Huge honestly believes that the lease has a "savings" provision allowing short cessations of drilling operations. Huge continues operations and finds oil. If Huge is later found to be a good faith trespasser can it recoup the costs expended by it in drilling the well while the lease was still in force? *See* Hunt v. HNG Oil Co., 791 S.W.2d 191 (Tex.App. 1990).

7. Are there limits on the costs the good faith trespasser may deduct from the revenue received from the sale of the hydrocarbons? Can the good faith trespasser deduct actual costs if such costs exceed the costs that would have been incurred by a reasonable and prudent operator? Bryan v. Big Two Mile Gas Co., 577 S.E.2d 258 (W.Va. 2001).

8. In Belcher v. Elliott, 312 F.2d 245, 18 O.&G.R. 14 (6th Cir.1962), ambiguous language in a conveyance was construed as reserving to Belcher the oil and gas. The grantees had, however, purported to sell the oil and gas rights to Elliott, who had commenced drilling a well. The question arose whether Elliott and associates "are entitled to a lien on the property for enhanced value by reason of drilling the well":

A Kentucky statute provides in KRS 381.460:

> If any person, believing himself to be the owner by reason of a claim in law or equity founded on a public record, peacefully occupies and improves any land, and the land, upon judicial investigation, is held to belong to another, the value of the improvements shall be paid by the successful party to the occupant, or the person under whom and for whom he entered and holds, before the court rendering judgment or decree of eviction causes the possession to be delivered to the successful party.

The general rule is set out in Kelly v. Kelly, 293 Ky. 42, 168 S.W.2d 339, 341, as follows:

> As said in 31 C.J. 319 § 27: "As a general rule in order that one may recover compensation for improvements made on another's land, even in a court of equity, it is necessary that he shall have made such improvements in good faith while in bona fide adverse possession of the land under color of title. There must be three concurrent essentials: (1) The occupant must have made the improvements in good faith; (2) he must have been in possession adversely to the title of the true owner; and (3) his possession must have been under color or claim of title."

Mr. Elliott and his associates were in possession adversely to the Belchers and had possession under color or claim of title. It could be that the drilling of the well was a considered risk of title on the part of Messrs. Elliott, Rice and Moore. But this does not necessarily eliminate good faith. Mrs. Belcher came to Kentucky from the Belcher home in Ohio and talked to her lawyer, Mr. Burke. Mr. Burke told her that Mr. Elliott was preparing to drill or had started drilling a well on the land in question. Neither the Belchers themselves, nor their attorney, gave notice to Elliott to stop the drill work on the land and did not take legal action to enjoin Elliott and associates from proceeding.

This standing by on the part of the Belchers, with knowledge of the improvements going on, together with the other elements mentioned, causes us to conclude that in equity and all fairness, under Kentucky law, the defendants-appellees are entitled to a lien upon the gas and oil rights and any moneys that may be due plaintiffs-appellants, for the amount by which the improvements have enhanced the value of the land, not exceeding the cost of the well and the connecting pipe.

Compare Swiss Oil Corp. v. Hupp, 253 Ky. 552, 69 S.W.2d 1037, 1039 (1934): "The test to be applied is . . . whether . . . the circumstances were calculated to induce or justify the reasonably prudent man, acting with a proper sense of the rights of others, to go in and to continue along the way."

9. In GREER v. STANOLIND OIL AND GAS CO., 200 F.2d 920, 2 O.&G.R. 229 (10th Cir.1952), the court, considering a case where "the proof showed no more than the expenditure of $17,780.00 for the drilling of a well with a showing of gas" by a good faith trespasser and the trial court had applied "the rule that an occupying claimant who in good faith and under color of title constructs or erects improvements upon the premises before his title fails is entitled to recover the value of the improvements to the rightful owner," held that "benefit to the rightful owner—not cost to the trespasser is the test of improvement. . . . And the burden is on the trespasser to prove that his outlay did enhance the value of the property." The court found further that "There was no proof that the drilling of the well enhanced the value of the lease." For further discussion of the right of the good faith trespasser to recover from the true owner for improvements enhancing the value of land, *see* Martin & Kramer. *Williams and Meyers Oil and Gas Law* § 226.2.

10. When there is a bad faith trespass or bad faith possession and production, is there a distinction to be made between the true owner seeking the value of the production and retaining the well itself? *Compare* Edmundson Brothers Partnership v. Montex Drilling Co., 98-1564 (La.App. 3 Cir. 5/5/99) 731 So.2d 1049, 142 O.&G.R. 266 (applying La.Civ.Code article 488, court ruled that the extraction of minerals diminishes the substance of the thing, so a possessor of minerals in bad faith does not have the right to have his expenses reimbursed) *with* Amoco Production Co. v. Texas Meridian Resources Exploration Inc., 180 F.3d 664, 143 O.&G.R. 17 (5th Cir. 1999) (Federal court applied La.Civ. Code article 497 to allow a bad faith possessor to recoup well drilling costs -- labor and material -- when the lessor elected to retain the benefits of the well thus drilled).

———

Martin & Kramer, *Williams and Meyers, Oil and Gas Law*

§ 227. Subsurface Trespass

The drilling of a well bottomed on the land of another is a trespass, although the surface location of the well is on the driller's own land. The intent of the driller of the directional well is immaterial to the commission of the trespass, which occurs when there is unauthorized entry upon the land of another. However, if the driller intentionally bottomed the well on another's land, the trespass is in bad faith, and the trespasser is liable for the value of the oil at the surface, *i.e.*, without a credit for drilling and operating costs. Even in the case of unintentional directional drilling resulting in good faith subsurface trespass, production from the well may be enjoined, at least where substantial injury results from the trespass. . . .

It seems safe to say that the plea of good faith in drilling a trespassing directional well will rarely be favorably received in court today. With the making of directional well surveys a standard oil field practice, most trespasses would seem to be intentional or at least inexcusably negligent.

––––––

NOTES

1. The important question today is not so much the liability of a subsurface trespasser as the means of proving that a trespass has occurred. A landowner may have a well-founded suspicion that a neighboring well is bottomed on his land, but how is he to prove it when the well is in the exclusive possession and control of another? Although demands for directional surveys have been stoutly resisted, the courts have held that discovery procedures authorize the making of such surveys. Rule 34, of the Federal Rules of Civil Procedure, has been interpreted as authorizing deviational and directional well surveys upon a showing by the applicant of good cause. Williams v. Continental Oil Co., 215 F.2d 4, 3 O.&G.R. 2080 (10th Cir.1954), cert. denied, 348 U.S. 928 (1955). A Texas court has reached the same conclusion under a rule permitting trial courts to " . . . entertain suits in the nature of bills of discovery, and grant relief therein in accordance with the usages of courts of equity." Hastings Oil Co. v. Texas Co., 149 Tex. 416, 234 S.W.2d 389, 393 (1950). This case also enjoined production from the well pending adjudication of the trespass question. Tex.R.Civ.P. 737. Rule 11 (Tex. Admin. Code § 3.11) now requires inclination and/or directional surveys under certain circumstances. Can a court require a party demanding a directional survey to post a bond to cover possible damage to the well? *See* Gliptis v. Fifteen Oil Co., 204 La. 896, 16 So.2d 471 (1943)

2. SWEPI, L.P. v. CAMDEN RESOURCES, INC., 139 S.W.3d 332 (Tex.App. 2004), reflects the interplay between common law causes of action and governmental regulation. SWEPI was attempting to defend an action asserting that it allowed oil and gas to be drained to a Camden well. The Railroad Commission refused to order a directional survey and made an implicit finding that the Camden well was properly drilled. The court, nonetheless, ordered the tests that SWEPI desired because the Commission has no power to resolve common law property claims. Whether or not there is a trespass can only be resolved through judicial, not administrative, action.

3. L & G OIL CO. v. RAILROAD COMM'N, 368 S.W.2d 187, 18 O.&G.R. 664 (Tex.1963), upheld the validity of Section V of Statewide Rule 54 providing that upon written request of an operator and a showing of probable cause the Commission will conduct a directional survey of a suspect well. The particular order in issue was held invalid because it authorized the complaining party instead of the Commission to make the survey.

If the owners of a producing well contract to have a directional well survey done for their own information and insert a provision in the agreement prohibiting the surveyor from releasing the results of the survey to any third person, will the surveyor be liable for telling an adjoining tract owner that the well is bottomed under his land? *See* Lachman v. Sperry–Sun Well Surveying Co., 457 F.2d 850, 43

O.&G.R. 50 (10th Cir.1972).

4. The fact that a well is bottomed outside the tract on which its drilling began is subject to what one court has called "natural concealment," leading to the conclusion that the statute of limitations did not begin to run at the time the well was completed as a producer but only after the discovery of these facts by the injured party. Alphonzo E. Bell Corp. v. Bell View Oil Syndicate, 24 Cal.App.2d 587, 76 P.2d 167 (1938). The time periods involved in *Bell* made it unnecessary to deal with Cal.Code of Civil Procedure § 349³/₄ which provides that actions based on the drilling and operation of "a well drilled for oil or gas or both from a surface location on land other than real property in which the aggrieved party has some right, title or interest or in respect to which the aggrieved party has some right, title or interest" must be brought within 180 days. As to "new wells" the cause of action "shall be deemed to have accrued ten days after the time when the well which is the subject of the cause of action was first placed on production." The statute was expressly made applicable to existing causes of action and provided that the "time for commencement of existing causes of action which would be barred by this section within the first one hundred eighty days after this section becomes effective, shall be the said first one hundred eighty days."

The statute also contains a provision on damages which seems to be broader in its application than the limitations provisions. It reads as follows:

> Measure of damages. In all cases where oil or gas has been heretofore or is hereafter extracted from any existing or subsequently drilled well in this state, by a person without right in good faith or acting under an honest mistake of law or fact, the measure of damages, if there be any right of recovery under existing law, shall be the value of the oil or gas at the time of extraction, without interest, after deducting all costs of development, operation and production, which costs shall include taxes and interest on all expenditures from the date thereof.

In STANDARD OIL CO. OF CALIFORNIA v. UNITED STATES, 107 F.2d 402 (9th Cir.1939), cert. denied, 309 U.S. 654, 673 (1940), the argument was made that the damages provision "was intended to apply to cases of so-called 'slant drilling' only, not to cases . . . involving a surface trespass." The court held that "the language employed admits of no such construction." The constitutionality of the statute was also sustained in this case.

5. PAN AMERICAN PETROLEUM CORP. v. ORR, 319 F.2d 612, 18 O.&G.R. 1061 (5th Cir.1963). The governing statute of limitations was two years but defendant had been producing from a slant well bottomed on plaintiff's land in the East Texas field for five years. It was proved that defendant had fraudulently concealed its trespass, but that plaintiff's geologist had written the company of strong suspicions before the well was completed that if it should produce it would do so because of being bottomed on plaintiff's land. An inquiry made thereafter would have revealed the trespass. The jury found on sufficient evidence that plaintiff should have made inquiry when the well was completed and plaintiff was

therefore limited to two years' recovery.

6. Defendant life insurance company made loans to pipeline companies that purchased oil illegally produced from slant wells. The loans were secured by assignments of proceeds from the sale of the oil, and defendant had extensive management powers to take over operation of the well to protect its interests. The well operator knew the well was illegal, but the pipeline company and defendant did not. Should the defendant be liable for conversion? Suppose that defendant collected sums in excess of what was due it and paid over the excess to other persons having oil payments and other interests in production. Is defendant liable for the excess it collected as well as for amounts owing to it? Is the statute of limitations tolled as to defendant when plaintiff had no reason to know of the trespass on his land? *See* Pan American Petroleum Corp. v. Long, 340 F.2d 211, 22 O.&G.R. 212 (5th Cir.1964), cert. denied, 381 U.S. 926 (1965), and Dau & Ratleff, *Pipeline and Lender Liability for Slant–Hole Production*, 43 Tex.L.Rev. 772 (1965).

SECTION 4. ADVERSE POSSESSION OF MINERALS

(See Martin & Kramer, *Williams and Meyers Oil and Gas Law* §§ 223–224.8.)

GERHARD v. STEPHENS

Supreme Court of California, 1968.
68 Cal.2d 864, 69 Cal.Rptr. 612, 442 P.2d 692, 31 O.&G.R. 28.

TOBRINER, JUSTICE. . . .

[The successors to certain undivided mineral interests brought suit to quiet title as against the surface owners who had purported to lease for oil and gas development. The trial court found that the severed mineral title had been lost by abandonment or adverse possession. In a portion of the opinion which is found at page 236, *supra,* the Supreme Court concluded that some of the mineral interests had been lost by abandonment but others had not been so lost. In the portion of the opinion which follows the court dealt with the question whether the severed mineral interests had been lost by adverse possession. Eds.]

2. Adverse Possession

The trial court concluded that the . . . defendants had acquired title to section 31 "in fee" by adverse possession. In support of that conclusion the court found that for over 20 years prior to the instigation of this litigation defendants and their predecessors had improved the Syncline Ranch, including section 31, by fencing it and conducting cattle operations on it; they had excluded trespassers from the property; they had paid all taxes assessed against section 31, although the mineral estate was not separately assessed. The court also found that for more than five years before the commencement of this action defendants had exercised dominion over the section 31 oil and gas rights by negotiating leases and receiving rentals. These acts, the court determined, were so open and notoriously adverse to the interests of plaintiffs' predecessors as to effectuate an ouster and mature a title in defendants to the entire fee in section 31, including the interests now claimed by plaintiffs.

Plaintiffs contend that no sufficient evidence supports the trial court's finding. They point out that defendants engaged in no visible activity adverse to plaintiffs' claimed interests until 1956, when oil was discovered on section 31. They argue that defendants' recording of several instruments in which they claimed to own without reservation the entire estate in section 31 and their failure to account to plaintiffs for the rentals received from the leases did not give notice to plaintiffs that their alleged cotenants were claiming adversely to them.

We shall explain that the cases hold that the owner's possession of the surface does not necessarily earn property rights, previously conveyed to another, in the oil and gas underlying the surface. We point out that defendants engaged in no *subsurface* activity sufficient to acquire a prescriptive title to the interest previously conveyed to the corporations; we further point out that defendants' *surface* activities were not adverse to plaintiffs' enjoyment of their interests.

Although the courts generally hold that adverse possession of the surface extends to the underlying mineral estate (*see* cases cited, 35 A.L.R.2d 124, 129–138; 1 Williams & Meyers, Oil & Gas Law, § 224, p. 337; 1 Kuntz, Oil and Gas, § 10.2; 1A Summers, Oil and Gas, § 138, p. 306), continued possession of the surface, following a conveyance of the oil and gas rights apart from the surface, does not establish the possessor's rights against the legal owner. (*See* cases cited, 35 A.L.R.2d 124, 154–165; 1 Williams & Meyers, op. cit. *supra,* § 224, p. 338; 1A Summers, Oil and Gas, § 138, pp. 307–312.) The reason for the rule lies in the fact that the true owner of a mineral interest would not be alerted to a hostile claim on the part of an occupant who takes no steps to penetrate the surface. (*See* Smith v. Pittston Co. (1962) 203 Va. 408, 412, 124 S.E.2d 1; 1 Williams & Meyers, op. cit. *supra,* § 224.1, pp. 340–341.)

. . .

Thus the courts have indicated that mere possession and ownership of the surface, in the absence of visible activity sufficient to impart to the true owner notice of an adverse claim, does not give rise to *adverse* title to rights in the underlying minerals. Indeed some courts have taken the extreme position that, following a conveyance of the mineral rights, subsurface activity by the surface owner on a part of a tract will not give title by constructive adverse possession to the mineral rights appurtenant to the entire tract . . . . Other courts have held that the development of part of the subsurface estate under color of title to the entire tract will mature a prescriptive title to the entire mineral estate .

Applying these rules to the instant case we conclude that no evidence supports the trial court's finding that defendants acquired a prescriptive title to the oil and gas rights. Defendants' lessees did no drilling on section 31 until shortly before the commencement of the present actions. Defendants' execution and recording of leases does not by itself constitute the actual and visible physical enjoyment of the subject rights essential to place the true owner on notice of an adverse claim.

We recognize that defendants' lessee Hess drilled a dry hole on the Syncline Ranch just outside of section 31 in 1951, but, even if this single action would suffice as a continuous and notorious user for the requisite period, defendants cannot on this basis establish title under the doctrine of constructive adverse possession. (*See* Code Civ.Proc. §§ 322, 323.) Although defendants held color of title to a complete interest in the oil and gas underlying the Syncline Ranch including section 31[26] their lessee's activity did not intrude upon the interests now claimed by plaintiffs. The California cases hold that a person in actual possession of only a part of the land to which he has color of title cannot establish ownership of the entire tract by adverse possession unless his *actual* possession infringes upon the presumptive possession of the true owners of the land. . . .[27]

Defendants attempt to sustain the trial court's finding by pointing out that because California regards the oil and gas interests here in dispute as profits *a prendre* or incorporeal hereditaments, the servient owner may extinguish such servitudes by acts other than exploring or drilling for the substances himself. "In a non-ownership state, where the grantee . . . of oil and gas rights acquires only the grantor's . . . right to drill and appropriate, a right in the nature of a *profit a prendre,* an incorporeal hereditament, . . . it would seem, although there is practically no authority, that excluding the grantee . . . would be such an interference with the essence of the right that the statute would begin to run, and the same conclusion would be reached if there were any interference with such right as would give an action to the grantee. . . . " (2 American Law of Property, § 10.7, p. 523; *see* 1 Kuntz, op. cit. *supra*, § 10.4, pp. 218–219; 1 Williams & Meyers, op. cit. *supra*, § 224.4, pp. 359–360, .)

We find no evidence, however, that could support a finding that the profit was "extinguished by the user of the servient tenement in a manner adverse to the exercise of the [incorporeal hereditament], for the period required to give title to land by adverse possession." (Glatts v. Henson (1948) 31 Cal.2d 368, 371, 188 P.2d 745, 746 (easement).) We find no exclusion of the grantee that would be such

---

[26]   A deed describing land by surface boundaries without reservation will convey the grantor's title to the oil and gas rights as well as to the surface estate. (Standard Oil Co. of California v. John P. Mills Organization (1935) 3 Cal.2d 128, 132–133, 43 P.2d 797.) A deed that purports to convey a complete interest in land therefore gives "color of title" to the mineral estate even though the deed is *ineffective* to convey that title because of a prior conveyance. Thus, in the instant case, defendants now have color of title by virtue of decrees of distribution in the Frusetta and Cornwell estates, and a 1940 quiet title decree against Carroll, an oil lessee. (Code Civ.Proc. § 322.) These instruments describe the property as the Syncline Ranch, including "all of Section 31" and make no exception in favor of the rights previously deeded to the corporations.

[27]   In view of our conclusion that defendants and their predecessors at no time utilized the rights now claimed by plaintiffs we need not consider plaintiffs' further argument that defendants could not acquire a prescriptive title against their cotenants because no evidence established the requisite ouster of plaintiffs or their predecessors.

an interference with the essence of the right as to start the running of the statute. To extinguish an easement by adverse user the use "must either interfere with a use under the easement or have such an appearance of permanency as to create a risk of the development of doubt as to the continued existence of the easement." (2 American Law of Property, § 8.102, p. 308.) Moreover, "[a]n easement cannot be acquired or extinguished by adverse use unless the party whose rights are affected thereby has knowledge of the adverse nature of such use. This knowledge may be either actual or constructive, resulting from notice either express or implied." (Clark v. Redlich (1957) 147 Cal.App.2d 500, 508, 305 P.2d 239, 244.)

The facts of the instant case bear no resemblance to those situations in which the servient owner has successfully extinguished an easement by obstructing a specifically defined right of way for the prescriptive period. If defendants had covered the surface of section 31 with permanent improvements, we would not disturb a finding that the user was adverse to the development of oil and gas. Here, however, defendants' fencing of the Syncline Ranch and their use of the surface of section 31 for cattle grazing was not inconsistent with plaintiffs' alleged right to enter upon the tract and explore for oil.[28] Indeed defendants themselves, through their lessees, were able to use the land to drill for oil and gas in spite of the surface cattle operations.[29]

3. Laches

The trial court held that plaintiffs' claims to the interests were barred by laches. The court found that subsequent to the time that plaintiffs' predecessors had acquired these claims important documents had been lost and many persons who had knowledge of the numerous transactions concerning section 31 had died. The court also found that plaintiffs have possessed a cause of action since the year 1928, when defendants executed a deed of trust to the French–American Corporation describing the entire fee title without reservation. It found that between 1928 and 1957 the value of the oil rights to section 31 greatly increased because of the efforts and expenditures of defendants and their lessees.

Whether or not laches occurred in a particular case primarily presents a question for the trial court . . ..

A defendant asserting laches on plaintiff's part must show that plaintiff has acquiesced in defendant's wrongful acts and has unduly delayed seeking equitable

---

[28]   If mineral owners try to enter to develop the minerals and surface owners refuse entry, fence the land, and threaten suit, making clear that they are making a hostile claim to the minerals and the mineral owners acquiesce in this conduct, have the surface owners acquired title by limitations? *See* General Refractories Co. v. Raack, 674 S.W.2d 97 (Mo.App.1984, application to transfer to Supreme Court denied). [Eds.]

[29]   As to defendants' practice of excluding strangers from the ranch, even if plaintiffs had notice of it, they could not, in the absence of a frustrated attempt to assert their rights, be charged with knowledge that defendants would refuse entry to one holding a legal right to use the property. (*Cf.* 1 Kuntz, op. cit. *supra,* § 10.4, p. 219.)

relief to the prejudice of defendant.[30] "[M]ere lapse of time, other than that prescribed by [statutes of limitations], does not bar relief." (Maguire v. Hibernia Sav. & L.Soc., *supra,* 23 Cal.2d 719, 736, 146 P.2d 673, 682.)

The crucial point, however, is that laches does not bar the quieting of title if the party asserting the defense fails to demonstrate that he was in adverse possession of the contested property during the period of delay. . . .

. . .

In the instant case, as we have pointed out, *supra,* prior to the 1956 drilling, defendants and their lessees did not invade plaintiffs' interests in a manner creating a potential of prescriptive title. Defendants' surface activities were entirely consistent with their ownership of the servient estate and posed no threat to plaintiffs' rights. Although defendants, by recording instruments in which they claimed ownership without reservation of all of section 31, cast a cloud on plaintiffs' title, plaintiffs' delay in seeking to quiet title cannot constitute laches. Defendants' failure to account to plaintiffs for any rentals received under the early oil leases covering section 31 cannot support a finding of laches, because in the absence of drilling on section 31 plaintiffs could not be charged with knowledge of any accrued rights.[31] In summary we find no evidence to support the trial court's finding that plaintiffs' action is barred by laches.[32]

. . .

NOTES

1. The court does not mention that the mineral owner has no cause of action against the surface occupant so long as the latter does not penetrate the surface. Since adverse possession is tied to the running of a statute of limitations for invasion of a possessory interest is that a better basis for the decision? *See* Nelson v. Teal, 293 Ala. 173, 301 So.2d 51, 49 O.&G.R. 118 (1974) and Downey v. North Alabama Mineral Development Co., 420 So.2d 68, 77 O.&G.R. 1 (Ala.1982). Can

---

[30]   The trial court found prejudice in that almost all persons who participated in, or had knowledge of, the transactions concerning section 31 had died or lost their memory. The loss of witnesses is a factor demonstrating prejudice, and the cases do not require that defendant must demonstrate that their testimony would have been favorable to him.

[31]   Plaintiffs' right to have those rentals covered in an accounting, however, may be barred by the statute of limitations.

[32]   Defendant Shell Oil Company claims that in any event certain plaintiffs' laches are demonstrated by their failure, after learning of the discovery of oil on section 31, to assert their claims for a period less than one year, during which time Shell expended considerable money in connection with the property. The trial court made no finding as to the effect of this shorter delay. [In Perpetual Royalty Corp. v. Kipfer, 253 F.Supp. 571, 25 O.&G.R. 495 (D.Kan.1965), aff'd per curiam, 361 F.2d 317, 25 O.&G.R. 504 (10th Cir.1966), cert. denied, 385 U.S. 1025 (1967), mineral interests were held to be extinguished by laches where the owners "stood silently by" from 1946 when the occupiers of the land began claiming all mineral interests, through an active development period from 1956 to 1964 when "time, energy and money" were expended in "drilling producing wells." Eds.]

one adversely possess a non-possessory or incorporeal interest? Does *Gerhard* resolve the conundrum of having actual and exclusive possession of a non-possessory interest?

2. Are Dormant Minerals Acts such as the one enacted in Indiana and discussed in Texaco v. Short *supra,* a better approach to the problem of eliminating stale claims than the traditional doctrine of adverse possession? *See* Polston, *Legislation, Existing and Proposed, Concerning Marketability of Mineral Titles,* VII Land & Water L.Rev. 73 (1972).

3. Should the same rule that applies to severed oil and gas interests be applied to severed hard rock mineral interests? In BURLINGTON NORTHERN, INC. V. HALL, 322 N.W.2d 233, 241 (N.D.1982) the court described how one adversely possessed a severed coal interest. It said:

> The acquisition of the title to severed minerals by adverse possession requires the taking of possession of the minerals by drilling, or conducting other mining activities. . . . the drilling or other mining activities must be 'sufficient to impart to the true owner notice of an adverse claim.' . . . In order to be adverse, the acts upon which the claimant relies must not only be actual 'but also visible, continuous, distinct, and hostile and of such character as to unmistakably indicate an assertion of claim of exclusive ownership by the occupant. The burden of proving adverse possession is on the person alleging it and must be established by clear and convincing evidence.

Does it appear likely under *Gerhard* or *Burlington Northern* that severed oil and gas or hard rock mineral estates can be adversely possessed in the absence of actual production for the statutory period?

4. *O* conveys one-half mineral interest in Blackacre to *X.* Thereafter *T* enters and takes adverse possession of the surface of Blackacre. Can *T* perfect limitation title against *O* 's interest in the land, both surface and mineral? Dixon v. Henderson, 267 S.W.2d 869, 3 O.&G.R. 1748 (Tex.Civ.App.1954), and Fadem v. Kimball, 612 P.2d 287, 67 O.&G.R. 45, (Okl.App.1979, approved for publication by Supreme Court), held that he could.

5. YTTREDAHL V. FEDERAL FARM MORTGAGE CORP., 104 N.W.2d 705, 13 O.&G.R. 487 (N.D.1960). *O,* owner of Blackacre in fee simple, executed a mortgage to *M* in 1934. In 1936–7 *O* conveyed perpetual, non-participating royalties to *X.* In 1942 the mortgage was foreclosed and the land conveyed to *P,* who remained in continuous possession until bringing a quiet title action in 1955. *X* was not joined in the foreclosure proceeding. Held, although he has been in possession for the statutory period of time, *P* 's possession was not adverse to the severed royalty interest of *X. P* is in the same position as *O* was, and since *O* could not claim adversely to *X,* neither can *P.* Accord: Wisness v. Paniman, 120 N.W.2d 594, 18 O.&G.R. 632 (N.D.1963).

Compare Payne v. A.M. Fruh Co., 98 N.W.2d 27, 11 O.&G.R. 225 (N.D.1959). A tax lien attached to *O* 's land in 1932 for nonpayment of taxes from

1932–5. In 1936–7 *O* conveyed perpetual, non-participating royalty to *X*. The redemption period expired and a tax deed issued in 1941 to the county, which in 1943 conveyed the land to *P*. In an action to determine title brought by *P* against *X*, held, for *P*. Although the tax deed to the county and the county's deed to *P* were invalid because *X* did not receive notice of the expiration of the redemption period, the origin of *P* 's title dates back to 1932 and possession for the statutory period under such a title is adverse to *X*. A severance of minerals is ineffective as to the county and its privies if the land is subject to a tax lien at the time of severance.

Compare also Broughton v. Humble Oil & Refining Co., 105 S.W.2d 480 (Tex.Civ.App.1937, error ref'd). *T* was in adverse possession of land when *O*, the record owner, severed the minerals by oil and gas lease. Thereafter *T* perfected a limitation title to the surface. *Held, T* also acquired limitation title to the minerals.

6. In McCoy v. Lowrie, 42 Wash.2d 24, 253 P.2d 415, 2 O.&G.R. 621 (1953), reliance was placed, in claiming title to minerals, on a "somewhat unusual statute applicable only to vacant and unoccupied lands . . . which requires no element of adverse possession but merely color of title and the payment of taxes for seven successive years . . .." The claimants argued that they had satisfied the requirement of having paid taxes, "insisting that, because they have paid all the taxes levied on the N.W. 1/4 of Section 15 . . . they must have paid the taxes on the minerals, because RCW 84.04.090 (cf. Rem.Rev.Stat. 11108), defining 'real property' for taxation purposes, states that it includes 'the land itself' and 'all substances in and under the same.' ' A portion of the opinion on this point follows:

> This contention is answered in a case decided very recently by the supreme court of Colorado, *i.e.,* Mitchell v. Espinosa, 125 Colo. 267, 243 P.2d 412 (1952). The Colorado taxing statute defining 'real property' for taxation purposes and referred to in the opinion is quite similar to our own, and real property thereunder includes lands, minerals, and quarries in and under the land. The question presented, as stated by the court, was:

>> Where a reservation of oil rights is contained in a deed to land which is thereafter assessed for taxes without change or recognition of the reserved oil interest; where the owner of said reserved oil interest took no action to cause it to be assessed; and where no assessment was made on said severed oil interest; will a tax deed issued for unpaid taxes accruing subsequent to the date of execution and recording of the deed under which the severance of oil interests was effected, convey to the grantee in the treasurer's deed the oil rights theretofore severed and reserved?

> The answer was negative, for the reason that there had never been a valid tax on the reserved oil rights. As the court said:

>> Where a separate and distinct estate consisting of "mines, minerals and quarries in and under" specific land is created by reservation thereof, a sufficient description of this property for assessment purposes requires specific reference to the severed estate. Thus there must be in the assessment a sufficient description of the estate in oil, such as "all oil and

gas beneath and underlying" the specific quarter section of land, or other description of the surface property, before it can be held that severed oil rights have been assessed for the payment of taxes.

McCoy is followed in Gilbreath v. Pacific Coast Coal & Oil Co., 75 Wash.2d 255, 450 P.2d 173, 32 O.&G.R. 493 (1969), where a concurring opinion suggests that the Washington rule as applied here should be changed by prospective legislative action. In Discussion Notes, 32 O.&G.R. 499 at 500, Professor Maxwell writes: "Certainly a strong argument can be made that the burden should be on the holder of the mineral interest to get his interest separately assessed if he is to be protected either from statutes such as the one applied here or from tax sales under more common statutes."

In McCracken v. Hummel, 43 Cal.App.2d 302, 110 P.2d 700 (1941), the "oil and gas rights" were severed by reservation but were never separately assessed. In an action to quiet a title based on a tax sale the court concluded that the assessment in the name of the surface owner "included the assessed valuation of the oil rights." In the absence of a positive assertion of the right to a separate assessment by the mineral owner the total assessment to the surface owner was valid and the tax deed based on it carried ownership of all interest in the land. *Compare* Failoni v. Chicago & North Western Ry. Co., 30 Ill.2d 258, 195 N.E.2d 619, 20 O.&G.R. 13 (1964), where severed minerals were not separately assessed yet were not acquired by a surface owner holding under color of title under a limitation statute giving ownership to vacant and unoccupied land by payment "of all taxes legally assessed thereon for seven successive years;".

7. STEED v. CROSSLAND, 252 S.W.2d 784, 1 O.&G.R. 1848 (Tex.Civ.App.1952, error ref'd). Wilson conveyed to Lindsay reserving all minerals. The deed was not recorded. Thereafter, Lindsay conveyed to Crossland by a deed purporting to grant the full fee simple title. Crossland claims title to the minerals as a bona fide purchaser under the recording act and by adverse possession. *Held,* for Wilson. Crossland is not a bona fide purchaser because he is on notice of the contents of all deeds in his chain of title whether recorded or not. Nor can he acquire title by adverse possession:

> The legal effect which a failure to record a deed has upon the rights of parties affected by such conduct depends upon and is controlled by statute; and the statute applicable here is Art. 6627, V.A.T.S. which provides that unrecorded deeds 'as between the parties and their heirs, and as to all subsequent purchasers, with notice thereof or without valuable consideration, shall be *valid and binding.*' . . . Because of this statute it is necessarily held that a deed takes effect between the parties when delivered to the grantee, not when it is recorded. . . .

> Thus, between Wilson and his grantee Lindsay, the minerals were severed from the land (and retained by Wilson) when Wilson delivered his deed to Lindsay because this deed took effect at that time between Wilson and his grantee Lindsay. This event occurred before T.S. Crossland made his purchase

and took his deed from Lindsay; and we have held that T.S. Crossland was not an innocent purchaser. Accordingly, Wilson's exception of the minerals was fully as effective as against T.S. Crossland as it was against Lindsay before him.

The consequence is that appellees' possession of the surface never extended to and included the minerals and thus appellees never acquired any title to the minerals.

Under this decision, can an operator afford to drill on land with a missing link in the chain of title?

Suppose an undivided one-half mineral interest has been conveyed by a landowner but the conveyance has never been recorded. Will a purchaser for value from one who has adversely possessed the surface of the land for the period of the statute of limitations be able to claim a full title to the minerals under a notice recording statute if the purchaser has no knowledge of the unrecorded mineral deed? See Taylor v. Scott, 285 Ark. 102, 685 S.W.2d 160, 84 O.&G.R. 22 (1985).

BARFIELD V. HOLLAND, 844 S.W.2d 759, 120 O.&G.R. 556 (Tex.App. 1992, writ denied). In attempting to resolve a dispute regarding the reservation of a severed mineral interest the court said: "Plaintiffs' duty to inquiry as to what would be conveyed by the three deeds was to examine the public records... with respect to the title to such lands, not to rely on representations made by the grantors of the deeds."

PAYNE V. WILLIAMS, 91 Ill.App.3d 336, 414 N.E.2d 836, 68 O.&G.R. 627 (1980). Payne asserted title to a mineral estate underlying a 40 acre tract of land based on a tax deed. Williams asserted that the tax deeds were void because of a lack of notice to the then-title owner. Evidence was adduced at trial that the affidavits filed in the tax sale that due diligence had been exercised to locate the owners of the mineral estate were fraudulent since the affiant had personal knowledge of who the owners were. At the time of the tax sale, there had been a severance of the mineral and surface estates, but the purchaser at the tax sale was also the owner of the surface estate. Can the present possessors, who were donees of a gift inter vivos of the mineral estate from the purchaser at the tax sale claim adverse possession under color of title? What if the present possessors were bona fide purchasers for value? The Illinois court concluded that the tax deeds were void as a matter of law and that the present possessors could not be bona fide purchasers for value since they were donees, not purchasers for value. Does the present possessor have to separately adversely possess the surface and mineral estates since there had been a merger of the estates following the tax sale? The court concluded that even if the purchaser did not have personal knowledge regarding the fraudulent concealment of notice to the title owners, the public records regarding the tax purchaser's application for tax deeds should have put him on notice that the deeds were void. Thus, there was not any merger of the surface and mineral estates which required that there be adverse possession of the mineral estate through some type of development or extraction activities.

For the impact of a lost deed on an adverse possession claim, see Witt v. Graves, 787 S.W.2d 681, 108 O.&G.R. 223 (Ark. 1990).

8. What is the effect of adverse possession of the surface after a landowner has executed an oil and gas lease? Discussion Notes, 1 O.&G.R. 447 at 450 (1952), state: "Logically, adverse possession should run as to rights retained by the lessor, both under the existing lease, as to royalties for example, and as to the reversionary interest which would divest the lessee of all title upon expiration of the lease. However, no cases have been found which expressly discuss this problem." *See* Walker, *The Nature of the Property Interests Created by an Oil and Gas Lease in Texas*, 7 Tex.L.Rev. 539 at 577–580 (1929).

In HOPE LAND MINERAL CORP. V. CHRISTIAN, 225 Mich.App. 43, 570 N.W.2d 626, 139 O.&G.R. 355 (1997), the court by implication treats the reserved rights of the lessor in a gas storage lease as real property by allowing those interests to be adversely possessed by the purchaser of the interest at a tax sale. The court does not refer to the leasehold payments as royalty, but rentals. The result may be explained by the court's treatment of Hope Land's retained interest as the possessory estate, notwithstanding the clear fact that Panhandle Eastern Pipeline was operating the gas storage facility under a valid and subsisting 1957 lease. Professor Norvell asks, "How would the defendant have notice that the gas storage company was paying to the plaintiff the storage rentals or that subsequent deeds were recorded that asserted interests contrary to his ownership?" Discussion Notes, 139 O.&G.R. 361-62.

What if the lessor's rights are not considered appurtenant to the surface or if there are non-operating interests, such as a perpetual nonparticipating royalty owned apart from any surface ownership; when would adverse possession extinguish such interest?

9. Do the principles developed in applying the concept of adverse possession to severed minerals help in determining the question whether title to severed minerals can be gained or lost by accretion and erosion? *See* Nilsen v. Tenneco Oil Co., 614 P.2d 36, 68 O.&G.R. 57 (Okl.1980), discussed in Murphree, *Oil and Gas: The Inapplicability of Accretion to Severed Mineral Estates*, 34 Okla.L.Rev. 826 (1981); *see also* J. P. Furlong Enterprises, Inc. v. Sun Exploration & Production Co., 423 N.W.2d 130, 99 O.&G.R. 393 (N.D.1988); Comment, *Accretion and Severed Mineral Estates*, 53 U.Chi.L.Rev. 232 (1986). Does the establishment of a surface boundary by acquiescence affect the ownership of severed mineral interests? *See* Sachs v. Board of Trustees of Town of Cebolleta Land Grant, 92 N.M. 605, 592 P.2d 961, 64 O.&G.R. 151 (1978), critically examined in terms of its impact as a precedent by Professor Custy in Discussion Notes, 64 O.&G.R. 158 (1980).

In KIM-GO V. J.P. FURLONG ENTERPRISES, INC., 460 N.W.2d 694, 114 O.&G.R. 71 (N.D. 1990), *on later appeal*, 484 N.W.2d 118, 114 O.&G.R. 82 (N.D. 1992), the North Dakota Supreme Court determined that parites who separately own tracts of land as a result of a river abandoning its old bed and forming a new bed acquire ownership of the old bed on an undivided basis. The second appeal applied a proportional, rather than a value-based methodology to delineate the newly-created separately-owned mineral interests. See also, Jackson v. Burlington Northern Inc., 205 Mont. 200, 667 P.2d 406, 78 O.&G.R. 479 (1983).

NATURAL GAS PIPELINE CO. OF AMERICA v. POOL

Supreme Court of Texas, 2003
124 S.W.3d 188, 168 Oil & Gas Rep. 199, 47 Tex. Sup. Ct. J. 153

OWEN, JUSTICE. In these consolidated proceedings, lessors under three oil and gas leases contend that the leases terminated because intermittently over the years there were periods of time ranging from 30 to 153 days when there was no actual production. We do not decide whether the leases terminated because even assuming they did, the lessees thereafter acquired by adverse possession fee simple determinable interests in the mineral estates that are identical to those the lessees held under the leases. Accordingly, we reverse the judgments of the court of appeals and render judgments for petitioners.

I

Two separate suits were brought in the same trial court by the same lessors against the same defendants. The first suit involved two leases; the second suit involved a third lease. The cases were not consolidated in the trial court or the court of appeals, and the court of appeals issued an opinion in each case.

In *Pool 1*, two leases were executed by J.T. Sneed and his wife in 1926 and 1936, respectively. In a separate agreement, the leases were consolidated as to a portion of the lands they covered for purposes of natural gas exploration and production. The 1926 lease at issue in *Pool 1* provided it would remain in effect for a term of ten years and "as long thereafter as oil or gas, or either of them, is produced from said land by the lessee." The 1936 lease similarly provided that it would remain in effect "so long as natural gas is produced."

A well, known as the J.T. Sneed # 1 well, was drilled on the consolidated acreage, and it produced gas until a replacement well was drilled in 1994. The replacement well has produced without interruption. But according to records from the Texas Railroad Commission, there were periods of time when there was no production from the J.T. Sneed # 1. Those periods were in August 1941, June through September 1963, July and August 1964, June 1979, March 1983, and July 1984. There is evidence that the J.T. Sneed # 1 did not produce for 122 consecutive days in the summer of 1963 and for 62 consecutive days in 1964. The other periods of non-production were shorter.

The lease at issue in *Pool 2* was executed in 1937. It provided that "[s]ubject to the other provisions herein contained, this lease shall remain in force pending the commencement and continuation of drilling operations on said land as hereinafter provided, and as long thereafter as natural gas is produced and marketed from any well on said land."

Two producing wells were drilled on the acreage covered by the lease at issue in *Pool 2*. However, there was no actual production from either of these wells in August 1959, July and August 1960, June and July 1961, June through October 1963, July and August 1964, and June 1969. The periods of no actual production ranged from 30 to 153 days. Another well was drilled on the *Pool 2* lease in 1996, and it has produced in paying quantities without interruption.

The plaintiffs in the trial court, who are the respondents in this Court, are the successors of the Sneeds' interests in all three leases, and they contend that the leases terminated due to cessation of production. They brought suit to quiet title, for trespass, conversion, and fraud, and for actual and exemplary damages.

. . .

[The trial court found in separate proceedings that both leases had terminated and that the lessees were bad faith trespassers. Eds.]

The lessees appealed both judgments. In *Pool 1*, the court of appeals held that the leases had terminated due to cessation of production, the lessees could not establish adverse possession even if they were trespassers because they had not given notice of repudiation of the lessors' title, . . . [In Pool 2, the court of appeals also found no adverse possession. Eds.]

We granted the lessees' petitions for review. Because the lessees established adverse possession as a matter of law, and resolution of that issue is dispositive, we do not reach other issues presented by the lessees' petitions.

II

In Texas it has long been recognized that an oil and gas lease is not a "lease" in the traditional sense of a lease of the surface of real property. In a typical oil or gas lease, the lessor is a grantor and grants a fee simple determinable interest to the lessee, who is actually a grantee. Consequently, the lessee/grantee acquires ownership of all the minerals in place that the lessor/grantor owned and purported to lease, subject to the possibility of reverter in the lessor/grantor. The lessee's/grantee's interest is "determinable" because it may terminate and revert entirely to the lessor/grantor upon the occurrence of events that the lease specifies will cause termination of the estate. In the cases before us today, the lessors retained only a royalty interest. When an oil and gas lease reserves only a royalty interest, the lessee acquires title to all of the oil and gas in place, and the lessor owns only a possibility of reverter and has the right to receive royalties. A royalty interest, as distinguished from a mineral interest, is a non-possessory interest.

A mineral estate, even when severed from the surface estate, may be adversely possessed under the various statutes of limitations. Once severance occurs, possession of the surface alone will not constitute adverse possession of minerals. Generally, courts across the country including Texas courts have said that in order to mature title by limitations to a mineral estate, actual possession of the minerals must occur. In the case of oil and gas, that means drilling and production of oil or gas.

In order to acquire title under a statute of limitations, that statute's requirements must be met. In these cases, we consider the three-, five-, and ten-year statutes of limitations. Suit was filed in both cases more than ten years after the last cessation of actual production. The last period of nonproduction occurred in 1984 in *Pool 1* and in 1969 in *Pool 2*. Both suits were filed in 1998.

The three-year statute of limitations says: "A person must bring suit to recover real property held by another in peaceable and adverse possession under title or color of title not later than three years after the day the cause of action accrues." The five-year statute says,

> (a) A person must bring suit not later than five years after the day the cause of action accrues to recover real property held in peaceable and adverse possession by another who:
>
> > (1) cultivates, uses, or enjoys the property;
> >
> > (2) pays applicable taxes on the property; and
> >
> > (3) claims the property under a duly registered deed.

. . .

"Adverse possession" is defined in the Civil Practice and Remedies Code as "an actual and visible appropriation of real property, commenced and continued under a claim of right that is inconsistent with and is hostile to the claim of another person." "Peaceable possession" is defined as "possession of real property that is continuous and is not interrupted by an adverse suit to recover the property."

The court of appeals concluded in these cases that the lessees' continuation of oil and gas operations and possession of the minerals after the leases terminated was not adverse because no notice of repudiation had been given to the lessors. The court reasoned in *Pool 2* that because the lessees' original possession of the mineral estate was permissive, adverse possession could not be established "unless notice of the hostile nature of the possession or repudiation of [the record title owners'] title is clearly manifested." The court of appeals employed similar reasoning in *Pool 1*.

We first consider the relationship between the parties, which guides us in determining whether the lessees' possession was adverse. The parties to the leases were not co-tenants. As discussed above, the lessors retained only a royalty interest and the possibility of reverter. The lessors had no right to possess, explore for, or produce the minerals. The exclusive right to do so was conveyed to the lessees. Accordingly, even when the lease was in effect, there was no co-tenancy. More importantly, if the leases terminated as the lessors contend, the lessees retained no interest whatsoever in the minerals. The entire mineral interest reverted to the lessors. There was no co-tenancy.

A lessee's position after a lease expires is more analogous to one holding over after the execution of a deed or after a judgment vesting title in another is entered. We have long said that "as a general rule, a party holding over after the execution of a deed or the rendition of an adverse judgment is merely a permissive tenant." In such circumstances, "possession cannot be considered adverse until the tenancy has been repudiated, and notice of such repudiation has been brought home to the titleholder." But we have said that actual notice of repudiation is not required. Rather, notice can be inferred, or there can be constructive notice.

This Court held in *Tex-Wis Co. v. Johnson* [534 S.W.2d 895 (Tex. 1976) Eds.]that a "jury [may] infer notice of a repudiation without any change in the use of the land," if there has been "long-continued use." In *Tex-Wis,* the Court cited some of its early decisions to that effect. One of those decisions was *Vasquez v. Meaders.* [156 Tex. 28, 291 S.W.2d 926 (1956). Eds.] This Court said in that case:

> It is not necessary that actual notice of adverse claim and disseisin be given to the landlord. It is sufficient if constructive notice is given, and constructive notice will be presumed where the facts show, as they do in this case, that the adverse occupancy and claim of title to the land involved in this suit has been long continued, open, notorious, exclusive and inconsistent with the existence of title in the respondent.

The Court also quoted at length from the decision in *Mauritz v. Thatcher* [140 S.W.2d 303 (Tex.Civ.App. 1940, writ ref'd.) Eds.]*:*

> It is the settled law in this state that a tenant cannot dispute the title of his landlord by setting up a title either in himself or in a third person during the existence of his tenancy until such notice of a termination thereof is given to the landlord as amounts to an actual disseizin. Limitation upon an adverse possession in a case of this kind begins to run from the time of such notice of a termination of tenancy. It is not necessary, however, that actual notice of an adverse holding and disseizin be given to a cotenant or owner. Such notice may be constructive and will be presumed to have been brought home to the co-tenant or owner when the adverse occupancy and claim of title to the property is so long-continued, open, notorious, exclusive and inconsistent with the existence of title in others, except the occupant, that the law will raise the inference of notice to the co-tenant or owner out of possession, or from which a jury might rightfully presume such notice.

. . .

In this case, the lessors asserted no claim for at least fourteen years with regard to the *Pool 1* leases, and for at least twenty-nine years with regard to the *Pool 2* lease. That is a strong circumstance tending to authorize an inference of notice of adverse possession. There is also uncontroverted evidence that the lessees' long-continued possession was "open, notorious, exclusive, and inconsistent with the existence" of title to all the minerals in the lessors.

It is important to bear in mind that the lessees were not required to give actual or constructive notice that they were no longer claiming an interest under the leases, but instead that they were claiming an interest that was inconsistent with the lessors' title to all the minerals when the leases expired. Thus, it was not the leases that the lessees must have been adverse to, but the lessors' fee title to all the minerals after the leases allegedly terminated. The lessees continued to claim rights under the leases, and it was that claim that was adverse to the lessors' fee title, unencumbered by the leases.

It is also important to recognize that the holding over of an oil and gas lessee after the lease has expired can differ from a tenant of the surface with regard to what is "open, notorious, exclusive, and inconsistent." A tenant of the surface that holds over and does nothing more than continue to occupy the premises as before, paying the same rent as before, is not in the same position as an oil and gas lessee who holds over. Surface leases do not typically contemplate that the tenant will remove permanent fixtures on or improvements to the property or consume or destroy the property itself. But an oil and gas lease contemplates that the mineral estate itself may be permanently and irrevocably depleted by removing and exhausting the minerals. An oil and gas lessee that holds over continues to physically remove and dispose of the very valuable, non-renewable minerals for its own account. Such actions are by their nature hostile to the lessor's ownership of all the minerals in place once the lease expires and the mineral estate reverts to the lessor in its entirety.

In both of the cases before us, the court of appeals relied on a decision from this Court, *Killough v. Hinds.* [161 Tex. 178, 338 S.W.2d 707 (1960). Eds.] In the *Killough* case, Hinds acquired an oil and gas lease on forty acres of land, the surface of which was owned by the Killoughs or their predecessors. Hinds and his wife were given permission by the surface owners to construct a home on the property. Hinds later sold his interest in the oil and gas lease to a third party, but he and his wife continued to live on the surface of the property. This Court held that Hinds had failed to establish title to the surface under the ten-year statute of limitations because "the erection of the barn, pig pen, chicken house and the grazing of milk cows, all of which were relied on to show adverse possession," were not "inconsistent with the permissive use and the right by which Hinds entered upon and occupied the property and built his residence, nor are they sufficient as a matter of law to afford the record owner constructive notice of a repudiation of that permissive use."

. . .

There is evidence of a "notorious act of ownership over the property, distinctly hostile to the claim of the" lessors that "indicate[s] unmistakably an assertion of claim of exclusive ownership" to show notice in the cases before us today. Assuming that the leases terminated because of cessation of actual production, an issue that we again note we are not deciding, the character of the lessors' real property interests changed dramatically and instantaneously, as did the lessees' interests, as we have explained above. The lessors had owned a mere possibility of reverter in the respective mineral interests as long as the leases covering those interests remained in effect. The lessors had no right to possess or explore for minerals. That right had been granted in toto to the lessees. The lessees had the right to explore for and remove all of the oil and gas from the premises, subject only to the obligation to pay royalties on the oil and gas that was actually produced. And if the lessees did not properly account to the lessors for that royalty, the leases would not terminate. The lessors would be relegated to bringing suit to recover the unpaid royalties. As long as the leases continued in effect, the lessees were entitled to recover and sell one hundred percent of the oil and gas, to the point of totally exhausting those valuable resources. Once the leases terminated, the lessees had no right to explore for, produce, or sell any of the oil and gas, much less one hundred percent of all that was produced. Those rights reverted to the lessors. Thereafter, it was the lessors that had the exclusive right to all the proceeds from production, subject only to an equitable accounting to the former lessees for the actual cost of production.

After the leases allegedly terminated, the lessees' continued production and sale of all the oil and gas and payment of royalty on only a relatively small percentage of the proceeds was open, notorious, and hostile to the lessors, who received payments each month of only a 1/8 royalty for more than ten years after they say the leases terminated. Moreover, the lessees drilled new wells after the time the lessors contend the leases terminated. The act of drilling wells is an act hostile to the lessors' exclusive right to explore for and remove the valuable minerals as well as the lessors' exclusive right to make the decision whether to drill and therefore impact the speculative value of the mineral estate if the well were unsuccessful.

Our conclusion is consistent with the Fifth Circuit's decision in *St. Louis Royalty Co. v. Continental Oil Co.,*[193 F.2d 778 (5th Cir.1952). Eds.] in which the facts were similar to those before us today. In *St. Louis Royalty Co.,* the lessors brought suit long after the lessees had conducted successful exploration, drilling, and production operations, contending that the lease had expired many years earlier because there was no actual production during the primary term. The lease had a 60-day drilling clause, and a well had been successfully drilled during the 60-day period after the primary term expired. The Fifth Circuit held that the well maintained the lease under the 60-day clause. But that court also held, as an alternative ground, that the lessees "by open, notorious, and adverse possession for more than five years with payment of taxes, under a claim of right brought home to plaintiff, acquired a good and perfect title, to the leasehold interest." The opinion in *St. Louis Royalty Co.* further observed that "the things [lessees] were doing were not done secretly and in a corner, but openly and in the face of all the world, with full knowledge thereof brought home to plaintiff and its predecessors in title." Our Court later described *St. Louis Royalty Co.'s* alternative holding and the facts supporting it, saying,

[T]he plaintiff and its predecessors in title, although fully informed of the lessee's claim to and operations under the lease, did not assert that the same had terminated until some ten years after the end of the primary term, during which time the lessee had developed the property by drilling seven producing wells.... The [Fifth Circuit] then went on to say that if the leases had lapsed, the plaintiff was still not entitled to recover because the evidence established as a matter of law that the defendants had perfected a limitation title to the leasehold estate.

In the cases before us today, the fact that the lessees were claiming the right and title to all of the production and the right to drill and explore for oil and gas, subject only to the royalty obligation, was open and notorious and was hostile to the lessors' claim that all title and interest reverted to the lessors and that the lessees ceased to own any interest in the minerals. As one noted treatise has concluded, "a lessee who develops and produces under a lease which previously terminated may acquire title to such lease by adverse possession."

The lessors also contend that the lessees' possession was never "adverse" because the lessees did not recognize that the leases had terminated when they continued operations and therefore lacked the requisite intent to adversely possess the mineral interests. Our decision in *Calfee v. Duke* [544 S.W.2d 640 (Tex.1976). Eds.] leads us to disagree with this contention. In that case, Calfee, an heir to J.H. Duke, occupied approximately 248 fenced acres of land. His deed from his parents did not include 24 of those acres. The other Duke heirs brought a trespass to try title suit, claiming that they and Calfee became co-tenants upon Calfee's parents' death with regard to the 24 undeeded acres. We observed that Calfee "never thought of himself as claiming adversely to anyone for the simple reason that he thought he was the rightful owner and had no competition for that ownership." We held that this satisfied the statute's requirement of adverse possession:

That being his claim of right, and it being coupled with his actual and visible possession and use, the adverse claim and possession satisfy the statutory requirements and cannot be defeated by Calfee's lack of knowledge of the deficiency of his record title or by the absence of a realization that there could be other claimants for the land.

In the cases before us today, the lessors essentially contend that the lessees should have notified them that the leases had terminated and that the fee interest in the minerals had reverted to the lessors. This is tantamount to saying that the running of limitations is suspended until the record titleholder obtains actual knowledge of what it owns. This is a novel proposition indeed. It would mean, for example, that limitations would be suspended whenever heirs did not realize that they had inherited an interest. That has never been the law in Texas. A record titleholder's ignorance of what it owns does not affect the running of limitations. The lessees' possession of the mineral estates in the cases before us today was adverse, and all the requirements of the three-, five-, and ten-year statutes of limitations were met.

Statutes of limitations require someone with a claim to assert that claim within a specified period of time, and the statutes of limitations dealing with real property are no different. The Legislature has required those claiming an interest in real property to "bring suit" within certain periods of time. Statutes of limitations are designed "to compel the assertion of claims within a reasonable period while the evidence is fresh in the minds of the parties and witnesses" and to "prevent litigation of stale or fraudulent claims." The lessors in the cases before us contend that title reverted to them a number of years ago, perhaps as long as 57 years before suit was filed under the *Pool 1* leases, and as long as 39 years before suit was filed under the *Pool 2* lease. The last occasion when title could have reverted under the *Pool 1* leases was 14 years before suit was filed, and for the *Pool 2* lease, it was 29 years. These are the types of claims that statutes of limitations were intended to foreclose.

Our decision should not be read as awarding fee simple absolute interests to the lessees in the oil and gas resources at issue. The lessees acquired the same interest that they adversely and peaceably possessed, that is, the oil and gas leasehold estates as defined by the original leases. Those interests are fee simple determinable interests in the respective properties on the same terms and conditions as the original leases. The terms that made the original mineral estates "determinable" continue to apply to the fee simple determinable interests acquired by adverse possession.

The court of appeals accordingly erred in failing to hold that the lessees acquired leasehold interests by adverse possession.

. . .

IV

As noted throughout this opinion, we have not reached the question of whether the leases terminated due to cessation of production. Specifically, it is unnecessary to decide whether the terms "is produced" or "so long as natural gas is produced" as used in the leases before us mean that the leases would terminate whenever actual production ceased, or instead, whether the leases would terminate only when production in paying quantities ceased. Nor do we decide whether the doctrine of temporary cessation of production includes or should include cessation of production for economic reasons.

The dissent contends that in deciding these cases based on statutes of limitations, we have "put[ ] the cart before the horse." The dissent proposes that these cases be remanded for a trial of fact issues that it concludes are raised under its view of the temporary cessation of production doctrine.

For reasons of judicial economy, this Court has long required that dispositive issues must be considered and resolved and that a judgment moving the case to the greatest degree of finality must be rendered. . . . The disposition of the statutes of limitations questions resolves the cases before us, and therefore, we do not reach other issues presented.

For the foregoing reasons, we hold that the court of appeals erred in failing to hold that the lessees in these two cases acquired fee simple determinable mineral estates by adverse possession. Accordingly, we reverse the judgments of the courts of appeals and render judgments for the lessees.

NOTES

1. What estate was owned by the lessors in this case after the oil and gas leases presumably terminated? Was it a fee simple absolute? a fee simple determinable? If it was a fee simple absolute does the *Pool* opinion explain why the adversely possessing former lessee only gains limitation title to a fee simple determinable estate? If the lease stops producing oil and gas who becomes the owner of the mineral estate?

2. Does the court adequately explain why the lessee has a continuing obligation to make royalty payments? The lease which conveyed the fee simple determinable estate to the lessee has presumably expired. What is the basis for the court's conclusion that the lessee must continue to pay royalty on production occurring after limitations title has been achieved?

Martin & Kramer, *Williams and Meyers Oil and Gas Law*

§ 224.4 Adverse possession of severed minerals

Surface occupancy alone, it has been seen, is not usually sufficient to establish title by adverse possession to previously severed minerals. But such exercise of dominion over minerals as may be said to constitute adverse possession thereof will, if continued for the statutory period, mature a limitation title. What are the requisites of such adverse possession?

A typical statement is that the dominion exercised over the minerals must give notice to the owners of the mineral estate that the occupier of the surface is claiming the minerals thereunder. An actual, public, notorious and uninterrupted working of the minerals for the statutory period is generally required. The mere execution, delivery, or recording of oil and gas leases or mineral deeds will not constitute adverse possession. Nor does it suffice that minerals are mined for a time with the knowledge of the owner; there must be continued adverse possession of the minerals for the statutory period.

. . .

Regarding solid minerals, there is a question whether the opening and operation of a mine is sufficient to mature limitation title to those minerals remaining untouched in their beds. Oil and gas present a similar problem. What area is adversely occupied by a single producing well? Arguably the possession extends to the area drained by the well, since the true owner has notice from a well on the surface that drainage is occurring in the sub-surface. Of course, such notice does not extend to the owner of the minerals in tracts other than the one on which the well is located.

A number of producing states have enacted statutes providing for compulsory pooling and unitization under certain circumstances. Under these statutes, production from a compulsory unit is treated as production from all tracts included in the unit for some purposes. Will such production serve the purpose of maturing title by adverse possession to a tract not drilled on but within the unit? It can be argued that the only use that can be made of a small mineral tract has been made when it is incorporated into the pooled unit, and, therefore, when a well is drilled on a unit that drains the tract, adverse possession has been taken of it. Notice to the landowner is provided by the statute, which typically permits the regulatory body to establish compulsory drilling units and forbids drilling on smaller tracts. Any landowners within the specified distance from a well has notice that his land has been included in a compulsory unit and is being used by others for mining purposes. Perhaps the argument should be accepted even apart from the notice imparted by a compulsory pooling or unitization statute. If the mineral claimant is required as a practical matter to join in a pooling agreement in order to exploit the land, has he not, by making the pooling agreement, done everything possible to mine the minerals?

NOTES

1. If a lease covering an undivided mineral interest does not survive its primary term because its unitization with other acreage is invalid under the lease clause provided, will production from wells located on other tracts in the unit established constitute "adverse possession of the unleased one-tenth mineral interest in place?" *See* Hunt Oil Co. v. Moore, 656 S.W.2d 634, 79 O.&G.R. 576 (Tex.App.1983, error ref'd n.r.e.). Should it make a difference if the pooling or unitization agreement had been executed to meet the requirements of a drilling unit established by the conservation commission? *See* Dye v. Miller & Viele, 587 P.2d 139, 63 O.&G.R. 136 (Utah 1978).

2. Would the statute of limitations begin to run at the time drilling operations were commenced on a well completed as a dry hole, where further operations were successful and the property continuously produced oil and gas thereafter?

Would the mining of coal for the statutory period perfect limitation title to oil and gas if the coal miner claimed under a grant of "all minerals"?

In Hurst v. Rice, 278 Ark. 94, 643 S.W.2d 563, 76 O.&G.R. 442 (1982), a claim was made that collection of royalties for the period of the statute of limitations under a lease by the claimant which became part of a unit producing gas constituted "adverse possession for all minerals." Held: "[s]uch a claim from gas production goes to the gas only and the drilling and production of a gas is not adverse to the mining or stripping of coal, a solid mineral. We need not decide if drilling and producing a gas or other liquid mineral is ever adverse to the drilling of any solid mineral. Nor need we decide whether the mining of one solid mineral is adverse to that mineral alone or to the mining of all other minerals. Likewise, we do not decide whether the mining or producing of mineral will be treated as adverse to all minerals when a claim is made to all minerals."

See 1 Kuntz, Law of Oil and Gas § 10.5.

3. In 1970 Sun executed a lease with a surface owner who, unknown to the parties, owns a 1/4 mineral interest. Shortly thereafter Sun assigned the lease to Oatman, but only covering the mineral rights to a depth of 5532 feet. Oatman, after conducting a title examination, determined that Sun had probably not leased all of the mineral estate and obtained a receiver's lease after having a receiver appointed to deal with the mineral interests. Oatman drilled within the depth limitations but paid Sun only 7/8 of the benefits under their agreement, asserting that the receiver's lease covered 1/8 of the mineral estate which was not covered by the 1970 Sun lease. Under Texas law a five year adverse possession statute was applicable. Tex.Civ.Prac. & Rem.Code Ann. § 16.025. Both parties first learned of the existence of a will which dealt with the 1/8th interest in question in 1977. The court determined that the statute of limitations began running in 1977 because that is when Oatman's exclusive possession was made known to Sun. The adverse possession of the 1/8th mineral interest was good as to all depths notwithstanding Oatman's lawful right to possess the mineral interests down to a depth of 5532 feet. Sun Operating Limited Partnership v. Oatman, 911 S.W.2d 749, 131 O.&G.R. 104 (Tex.App. 1995, writ denied).

NOTE: SEVERANCE BY ADVERSE POSSESSOR BEFORE LIMITATION TITLE IS PERFECTED

O is the record owner of land. *T* enters and takes adverse possession thereof. Before the statutory period has run, *T* conveys the minerals by deed or oil and gas lease to *E. T* remains in possession for the statutory period and acquires limitation title to the surface. Does a limitation title to the minerals also inure to *E?* Clements v. Texas Co., 273 S.W. 993 (Tex.Civ.App.1925, error ref'd), held that it does.

Suppose *T* conveys the land to *F,* reserving the minerals. *F* enters and continues in adverse possession, the periods of occupation of the two aggregating to the statutory period. Does a limitation title to the minerals inure to *T?* Houston Oil Co. of Texas v. Moss, 155 Tex. 157, 284 S.W.2d 131, 5 O. & G.R. 90 (1955), holds that it does. Accord, McLendon v. Comer, 200 S.W.2d 427 (Tex.Civ.App.1947, error ref'd n.r.e.).

Compare Thomas v. Southwestern Settlement & Development Co., 131 S.W.2d 31 (Tex.Civ.App.1939, error dism'd, judgment correct). Plaintiffs were record owners of $^3/_{35}$ undivided interest in surface and minerals; *X* was the record owner of the remaining $^{32}/_{35}$ interest. *X* purported to convey all of the surface to *Y* and to reserve all of the minerals. *X* and *Y* joined in an agreement putting a tenant in possession of the surface, and this tenant occupied for the statutory period. *X* and *Y* claimed title to the $^3/_{35}$ interest in surface and minerals under a statute permitting acquisition of title upon 5 years adverse possession "under a deed or deeds duly registered." *Held,* Y acquired limitation title to the surface but X did not acquire limitation title to the minerals.

> . . . [T]he surface occupancy of the tenants, being confined to that part claimed by Southwestern Settlement & Development Company [Y], matured title by limitation under the five year statute only to the land claimed by that company. Such actual possession of a part of the surface for farming and stock raising extended by construction to the limits of the claim as defined by the recorded muniments of title of the Southwestern Settlement & Development Company. It did not extend to the oil and gas estate because that estate was not claimed by the claimant of the surface. And certainly an adverse possession of the surface, claimed wholly by another, who did not claim the oil and gas, could not be extended by construction to include the oil and gas claimed by Houston Oil Company and Republic Production Company [X]. This results not from any actual severance of the oil and gas estate from the surface; for appellees could not sever the $^3/_{35}$ interest belonging to appellants. But instead it results from a segregation or severance of the claims of the respective limitation claimants, as shown by their recorded muniments of title. To mature a limitations claim there must be actual adverse possession of some part of the property claimed. Hill v. Harris, 26 Tex.Civ.App. 408, 64 S.W. 820. There was no possession of any part of the oil and gas, and so there was no possession of any part of the property comprehended within the claims of Houston Oil Company and Republic Production Company. There being no possession no limitation title could ripen in them.

Can the *Thomas* case be reconciled with *Houston Oil Co. v. Moss?*

Compare also Smith v. Nyreen, 81 N.W.2d 769, 8 O.&G.R. 595 (N.D.1957). X 's heirs went into adverse possession of land under color of title. Before the statutory period had run, one heir, *Y,* obtained a deed to the entire surface from the other heirs, who reserved $^8/_9$ of the minerals. Y remained in possession for the balance of the statutory period. *Held,* Y was a cotenant with the other heirs; his possession of the surface under claim of title to it and to 1/9 of the minerals was possession of the entire mineral estate on behalf of all the cotenants and the limitation title inures to the benefit of all the cotenants. As between the record owner and the adverse possessors, there had been no severance of the minerals. Is it significant that in *Thomas* the purported severance occurred prior to possession of the surface? *See* Ates v. Yellow Pine Land Co., 310 So.2d 772, 51 O.&G.R. 307 (Fla.App.1975), cert. denied, 321 So.2d 76 (Fla.1975). Is it important that in *Thomas* X was a cotenant with plaintiffs? *See* Thweatt v. Halmes, 265 Ark. 606, 580 S.W.2d 685, 63 O.&G.R. 157 (1979); Bell v. Lyon, 635 S.W.2d 586, 75 O.&G.R. 574 (Tex.App.1982, error ref'd n.r.e.).

For a case which follows the prevailing view that post-entry severance of the mineral estate will not prevent the adverse possessor from claiming both the surface and mineral estate, see Hurst v. Southwest Mississippi Legal Services Corp., 610 So.2d 374, 123 O.&G.R. 472 (Miss. 1992). The Mississippi Supreme

Court stated:

> Unless a severance has occurred, the party in adverse possession of the surface need not exercise active dominion over the minerals in order to perfect a valid title to both surface and minerals.... An attempted severance of mineral rights by the record title holder at any time after the adverse claimant enters possession is invalid. 610 So.2d at 379.

To the same effect, while finding no severance had occurred, is Krosmico v. Pettit, 1998 OK 90, 968 P.2d 345, 141 O.&G.R. 112.

SECTION 5. EFFECT OF DIVIDED OWNERSHIP ON OIL AND GAS OPERATIONS

A. CONCURRENT OWNERSHIP
(See Martin & Kramer, *Williams and Meyers Oil and Gas Law* §§ 502–510.)

PRAIRIE OIL & GAS CO. v. ALLEN

Circuit Court of Appeals of the United States, Eighth Circuit, 1924.
2 F.2d 566, 40 A.L.R. 1389.

PHILLIPS, DISTRICT JUDGE. This action was originally brought in the state court by Lizzie Allen against the Prairie Oil & Gas Company, hereinafter called Prairie Company, to recover damages for alleged conversion of a quantity of petroleum oil. It was properly removed to the federal court. Thereafter, Skelly Oil Company, hereinafter called Skelly Company, on motion of Prairie Company, was made a party defendant.

[In 1911, Good Land Company conveyed the tract in issue to J.C. Trout, reserving and excepting $^9/_{10}$ of the minerals therein and full and free right of ingress and egress for the purpose of exploring for and developing the minerals. In 1912, J.C. Trout, joined by his wife, conveyed this property to Lizzie Allen. Thereafter Good Land Company entered into an assignment and contract with Kay–Wagoner Oil & Gas Co., whereunder the latter covenanted to complete an oil well on the premises and to pay certain royalties to Good Land Company. Kay–Wagoner Co. assigned an undivided ¾ interest in this contract or lease to Skelly Oil Co. On June 17, 1920, Skelly commenced operations on the land which resulted in the completion of a producing well on July 12, 1920, and thereafter Skelly completed two additional wells. The reasonable and necessary expenditures of Skelly Co. in the development and operation of the property up to and including Feb. 1, 1922, amounted to $153,380.40. Prairie Company was the purchaser of the oil produced under a division order executed by Good Land Co., Kay–Wagoner Co., and Skelly Co. Eds.]

The oil produced and delivered to Prairie Company was as follows: To May 13, 1921, 25,284.71 barrels, of the market value of $57,586.82; from May 13,

1921, to April 7, 1922, 19,252.02 barrels, of the market value of $27,371.98. All oil run prior to May 13, 1921, was paid to the persons and in the proportions designated in the divisional order. One-tenth of the purchase price of all oil run after May 13, 1921, was retained by the Prairie Company.

Skelly Company, in taking possession of the land and developing the same, did not hold adversely to Lizzie Allen, but recognized that she owned the surface and one-tenth of the oil and gas.

The possession by Skelly Company, the development and operation of the property for oil and gas, the production of oil therefrom and the disposal of the same to the Prairie Company were with the full knowledge of Lizzie Allen.

. . .

On February 25, 1921, Lizzie Allen procured W.W. Croom to write a letter to Skelly Company, in which he stated that Lizzie Allen owned one-tenth of the oil and gas rights in this land, that she had never received any payment for or statement concerning her share of the proceeds, and that he desired Skelly Company to advise him its interpretation of the rights of Lizzie Allen in the premises. On February 26, 1921, Skelly Company replied thereto by letter in which it stated that Lizzie Allen was being credited with one-tenth of the gross proceeds of oil sold from the premises and charged with one-tenth of the cost of drilling the wells and equipping and operating the property, and when the receipts exceeded the costs it would account to her for one-tenth of the net proceeds monthly, and that as soon as well No. 3 was completed it would furnish a statement to her of the cost of the wells and the amount received for oil sold.

On March 16, 1921, Skelly Company mailed to Lizzie Allen a statement setting forth the expenses of development and operation of the property and the amounts received for oil sold therefrom, and thereafter continued from time to time to mail her similar statements.

On May 7, 1921, Lizzie Allen notified Skelly Company, in writing that she objected to its taking oil and gas from the land.

On May 12, 1921, Lizzie Allen made demand in writing upon the Prairie Company to surrender and deliver to her one-tenth of all the oil which had been produced from this land and received by it. The Prairie Company refused this demand.

. . .

It was expressly stipulated: "That . . . Lizzie Allen . . . is the owner of an undivided one-tenth interest in and to the oil, gas and mineral in and under said land. . . . "

. . . [The lower court] concluded that Lizzie Allen was entitled to recover from the Prairie Company $2,737.19, the value of one-tenth of the oil delivered after May 7, 1921, and from the Prairie Company and Skelly Company $5,758.68, the value of one-tenth of the oil produced and delivered prior to May 7, 1921. Judgment was entered accordingly. From this judgment Skelly Company and Prairie Company sued out a writ of error to this court.

[The court rejected the contention that Lizzie Allen was the owner of a $^{1}/_{10}$ cost-free royalty interest in the land but held instead that she was the owner of $^{1}/_{10}$ of the minerals. As to the mineral-royalty distinction and construction problems arising from the language of particular conveyances, see Chapter 6, *infra.* Eds.]

Under the provisions of the deed to Trout, Good Land Company retained the ownership of nine-tenths of the oil, gas and mineral in and under the land, and reserved the right to enter upon the land and use so much of the surface thereof as might be reasonably necessary for operating, drilling, mining and marketing the same. Having reserved the right to use the surface, Good Land Company, being a tenant in common with Lizzie Allen, would have had the right to develop the land for oil, drill oil wells thereon, and produce oil therefrom, without her consent. Tenants in common are the owners of the substance of the estate. They may make such reasonable use of the common property as is necessary to enjoy the benefit and value of such ownership. Since an estate of a cotenant in a mine or oil well can only be enjoyed by removing the products thereof, the taking of mineral from a mine and the extraction of oil from an oil well are the use and not the destruction of the estate. This being true, a tenant in common, without the consent of his cotenant, has the right to develop and operate the common property for oil and gas and for that purpose may drill wells and erect necessary plants. He must not, however, exclude his cotenant from exercising the same rights and privileges.

. . .

In the case of Burnham v. Hardy Oil Co., *supra*, the court said:

It seems to us that the peculiar circumstances of a cotenancy in land upon which oil is discovered warrant one cotenant to proceed and utilize the oil, without the necessity of the other cotenants concurring. Oil is a fugitive substance and may be drained from the land by a well on adjoining property. It must be promptly taken from the land for it to be secured to the owners. If a cotenant owning a small interest in the land had to give his consent before the others could move towards securing the oil, he could arbitrarily destroy the valuable quality of the land.

There are cases to the contrary. *See* Gulf Refining Co. v. Carroll et al., 145 La. 299, 82 So. 277; Zeigler et al. v. Brenneman et al., 237 Ill. 15, 86 N.E. 597 (probably because of an Illinois statute, *see* Murray v. Haverty, 70 Ill. 320); South Penn Oil Co. v. Haught, 71 W.Va. 720, 78 S.E. 759. We believe, however, that the rule above stated is supported by the better reason and by the weight of authority.

But counsel for Lizzie Allen say that since she did not join in the lease it was void as to her and that the Skelly Company, in going upon the land in question and developing and producing oil therefrom, was a trespasser. Counsel bases this contention upon the following language of the Supreme Court of

Oklahoma in Howard et al. v. Manning, 79 Okl. 165, 192 P. 358, 12 A.L.R. 819:

> Neither of the tenants in common is entitled to the exclusive possession of all the land to the exclusion of his cotenants, nor entitled to possession of any particular part of it. As he cannot exclude his cotenants by his own occupation of the land, he cannot, without their consent or ratification, lease either all or any particular part of the land in such a way that his lessee will have the right to the exclusive possession of all the land or any part thereof. It is well settled that a lessee of one tenant in common by a lease in which the other tenants have not joined is, as to them, a trespasser so far as he occupies any portion of the land. The lessee of one tenant in common is a trespasser as to the other tenants in common, but the lease is not void as against the tenant in common executing it.

While it is not entirely clear from the language used, we are fully persuaded the Oklahoma court, in declaring that a lease by a tenant in common was void and the lessee who entered a trespasser, referred to a lease by one tenant in common of the whole of the common property, and an occupancy by the lessee to the exclusion of the other cotenant. . . .

. . .

In the case of York et al. v. Warren Oil & Gas Co., 191 Ky. 157, 229 S.W. 114, the court after quoting extensively from the earlier case of New Domain Oil & Gas Co. v. McKinney, 188 Ky. 183, 221 S.W. 245, said:

> It will thus appear that this court is committed to the doctrine that one cotenant may lease his undivided interest in the joint real property for oil and other mineral and such lease contract is valid and binding on the cotenant making the same but not upon the other cotenants; however, the lessee becomes a cotenant in the mineral leased with the other joint owners and may, as could his grantor, enter upon the joint property and explore for, mine and market minerals, accounting to the other joint tenants for their proportion of the minerals recovered according to the rule adopted in the McKinney Case.

In the case of Compton v. People's Gas Co., 75 Kan. 572, 89 P. 1039, 10 L.R.A.,N.S., 787 the Kansas court held that where one tenant in common, owning one half of certain property, executed an oil and gas lease thereon to one party and the other tenants in common, owning the other half, executed another oil and gas lease thereon to another party, each lessee was entitled to the possession of the premises for the purpose of mining the same for oil and gas, but that neither was entitled to the exclusive possession.

In the instant case the lease to the Kay–Wagoner Company, which was later assigned to Skelly Company, was not of the whole property but of the undivided interest therein belonging to Good Land Company and the rights of Lizzie Allen were fully recognized. We therefore conclude that Skelly Company was not a trespasser as against Lizzie Allen, but that the lease to it was valid, and that upon entry it became a tenant in common with Lizzie Allen during the continuation of the lease and occupied the same relation to Lizzie Allen and the property as the Good Land Company would have occupied, had it entered and developed and operated the property for oil and gas.

The one question remaining then is: What is the proper basis of accounting between Skelly Company and Lizzie Allen? Under the general rule Skelly Company would be bound to account to Lizzie Allen for one-tenth of the net profits determined by deducting from one-tenth of the value of the gross production one-tenth of all reasonable and legitimate expenses for development and operating the property for oil and gas, but in the event of loss it could not compel her to reimburse it for any part of the loss.

In the case of New Domain Oil & Gas Co. v. McKinney, 188 Ky. 183, 221 S.W. 245, the Court of Appeals of Kentucky, after an exhaustive review of the authorities in passing upon the basis of accounting where one cotenant operated jointly owned oil property with the knowledge of his cotenant's interest, said:

> In cases of a willful trespasser the rule seems to be well settled, as will be seen from the authorities, *supra,* that the plaintiff is entitled to recover, in a suit for an accounting, the value of the mineral, without deducting any expenses incurred in mining it. But we do not understand that this stern rule should be applied in cases where a cotenant operates the mine with the knowledge of his cotenant's interest. Especially should this rule not be applied against a cotenant where the mineral involved is of a fugacious nature and liable to be exhausted by adjacent operators. In such case if one tenant is able and willing to develop the mine and extract the oil before it is entirely lost and his cotenant is not, he should be allowed to do so without incurring the penalty of accounting to his cotenant for the gross amount of oil produced; but since he may not convert, to any extent, his cotenant's interest, he must account to the latter for his proportion of the net value of the oil produced, which is its market value, less the cost of extracting and marketing. Any other rule, it seems to us, would be not only inequitable, but illogical; for, if the operating tenant could be made liable for the gross amount of the plaintiff's proportion of the oil produced, the rights of the former would be fixed according to the rules governing an operation by a willful trespasser who had no interest whatever in the mineral. We cannot believe that an exclusive operation by one joint tenant is so tainted with wrong as against his cotenant as to require the application of the same rules that should be applied to a willful trespasser. Moreover, it would be illogical to say that a cotenant could lawfully operate the joint property (as he may do, Thornton, *supra*, § 312) and at the same time be compelled to divide the gross proceeds or profits with his joint tenant, thus circumscribing his rights according to the rule applicable to a willful and malicious trespasser without pretense of title or right. There is no rule of law known to us that would render the plaintiff cotenant immune from the observance of that healthy

maxim, 'He who seeks equity must do equity,' and the equity which plaintiff in such cases is required to do is to pay his proportionate part of the actual expenses of operation and marketing, if they are not above that which is usual and customary in that locality.

. . .

While an accounting on a royalty basis has been held proper under the peculiar facts of particular cases (*see* South Penn Oil Co. v. Haught, *supra*, and discussion of rule in New Domain Oil & Gas Co. v. McKinney, *supra*), we see nothing in the facts of the instant case which take it out of the general rule.

The right of one cotenant to sue the other for the value of the use of the former's interest in the joint property which did not exist at common law prior to the enactment of St. 4 Anne, c. 16, is given by section 3804, Oklahoma Rev.Laws 1910 (section 7361, Compiled Okl.St.1921).

We therefore conclude that Lizzie Allen is entitled to an accounting from Skelly Company for the market value of the oil produced less the reasonable and necessary expense of developing, extracting and marketing the same. The royalty paid by Skelly Company to the Good Land Company is not a part of the cost of production and should not be included in the same.

The judgment of the lower court is therefore reversed, and the cause remanded, with instructions to grant the Prairie Company and Skelly Company a new trial; and it is so ordered.

NOTES

1. Since the enactment of the STATUTE OF WESTMINSTER II, c. 22 (1285), a cotenant has been subject to the law of waste. This rule is in effect in all states as a part of the common law or by local statute. Most of the cases that have considered the issue are in accord with the principal case. Is there sufficient justification for removing the concurrent owner from the common law or statutory interdiction against waste in these circumstances? Can a sound distinction be made between solid minerals and oil and gas for purposes of the law of waste? For a general discussion of the topic of concurrent ownership *see* Marla E. Mansfield, *A Tale of Two Owners: Real Property Co-ownership and Mineral Development*, 43rd Rocky Mt. Min. L. Inst. ch. 20 (1997)

The principal case was followed in P & N Investment Corp. v. Florida Ranchettes, Inc., 220 So.2d 451, 31 O.&G.R. 297 (Fla.App.1969) and Lichtenfels v. Bridgeview Coal Co., 344 Pa.Super. 257, 496 A.2d 782 (1985), *on sub. app.*, 366 Pa.Super. 304, 531 A.2d 22, 95 O.&G.R. 571 (1987). Slade v. Rudman Resources, Inc., 237 Ga. 848, 230 S.E.2d 284, 56 O.&G.R. 210 (1976), appears to rely on a statute to reach the same result although it says it is adopting "the prevailing rule because it makes sense."

Other recent cases reject claims of conversion where one cotenant has developed and failed to account to the non-developing cotenant. See In Re Hillsborough Holdings Corporation, 207 B.R. 299, 305, 136 O.&G.R. 1 (M.D.

Fla. 1997) ("Where, as here, the coal owner captures and sells gob gas [gas from a coal deposit], the gas owners are entitled to share in any profits on such sales after taking into account the cost borne by the coal owner in capturing and marketing the gas"). *See also* Mitchell Energy Corp. v. Samson Resources Co., 80 F.3d 976, 983, 134 O.&G.R. 580 (5th Cir. 1996):

> A unique legal relationship exists between cotenants. Unlike one who is not a party to the cotenancy, any cotenant has the right to extract minerals from the common property without consent or participation of the other cotenants. This right is subject only to a duty to account for the other cotenants' proportionate part of the value of the oil and gas produced, less their proportionate part of the drilling and operating expenses. Thus, the parties do agree that Samson did not convert gas by producing it and selling it. Instead, the issue is whether a tort action lies against Samson for converting the proceeds of the gas sales when it failed to pay Mitchell. We conclude that no conversion action lies.

2. Should it matter whether the complaining party was a concurrent owner of the fee or the oil and gas lessee of a concurrent owner? *Compare* Budd v. Ethyl Corp., 251 Ark. 639, 474 S.W.2d 411, 40 O.&G.R. 335 (1971), *with* Jameson v. Ethyl Corp., 271 Ark. 621, 609 S.W.2d 346, 69 O.&G.R. 19 (1980). *See also* Fife v. Thompson, 288 Ark. 620, 708 S.W.2d 611 (1986), and South Central Petroleum, Inc. v. Long Brothers Oil Co., 974 F.2d 1015, 121 O.&G.R. 421 (8th Cir. 1992). A cotenant possesses a right of accounting for revenues less expenses from production. Ark.Code Ann. § 18-60-101. Often, expert witnesses will have to be used to determine the appropriate amount of set-off for expenses.

What would be an appropriate remedy for a concurrent owner who is denied entry to the concurrently owned premises for the purpose of developing the land for oil and gas? *See* Garcia v. Sun Oil Co., 300 S.W.2d 724, 7 O.&G.R. 1256 (Tex.Civ.App.1957, error ref'd n.r.e.).

Can a cotenant of a working interest owner file a statutory lien on the cotenant's interest? John Carey Oil Co., Inc. v. W.C.P. Investments, 126 Ill.2d 139, 533 N.E.2d 851, 106 O.&G.R. 435 (1988) surveys the cases from the various states and holds that a statutory lien can be attached against the interests owned by a nonoperating co-owner.

Does the execution of a joint operating agreement affect the right of a co-tenant to drill without the consent of the other co-tenants? *See* Cox v. Fagadau Energy Corp., 68 S.W.3d 157, 162 O.&G.R. 527 (Tex.App. 2001).

3. As brought out in the principal case, several states have allowed one cotenant or coowner to veto development by another. Louisiana is one of these. Louisiana's special rules on co-ownership are found in the Louisiana Mineral Code in Articles 164 through 187. The articles provide separately for co-owners of land and for co-owners of a mineral servitude.

§164. Creation of mineral servitude by co-owner of land

A co-owner of land may create a mineral servitude out of his undivided interest in the land, and prescription commences from the date of its creation. One who acquires a mineral servitude from a co-owner of land may not exercise his right without the consent of co-owners owning at least an undivided eighty percent interest in the land, provided that he has made every effort to contact such co-owners and, if contacted, has offered to contract with them on substantially the same basis that he has contracted with another co-owner. A co-owner of the land who does not consent to the exercise of such rights has no liability for the costs of development and operations, except out of his share of production.

§175. Co-owner of mineral servitude may not operate independently

A co-owner of a mineral servitude may not conduct operations on the property subject to the servitude without the consent of co-owners owning at least an undivided eighty percent interest in the servitude, provided that he has made every effort to contact such co-owners and, if contacted, has offered to contract with them on substantially the same basis that he has contracted with another co-owner. A co-owner of the servitude who does not consent to such operations has no liability for the costs of development and operations except out of his share of production.

The severity of Article 175 is mitigated by Article 176:

§176. Co-owner of mineral servitude may act to prevent waste or destruction or extinction of servitude

A co-owner of a mineral servitude may act to prevent waste or the destruction or extinction of the servitude, but he cannot impose upon his co-owner liability for any costs of development or operation or other costs except out of production. He may lease or otherwise contract regarding the full ownership of the servitude but must act at all times in good faith and as a reasonably prudent mineral servitude owner whose interest is not subject to co-ownership.

Difficulties in applying the co-ownership articles are seen in GMB Gas Corp. v. Cox, 340 So.2d 638, 57 O.&G.R. 362 (La.App.1976) and Cox v. Sanders, 421 So.2d 869, 74 O.&G.R. 271 (La.1982).

4. If the drilling cotenant need only account for the net profits, is there any great deterrent to development in the fact that the land is held in concurrent ownership? Consider the following questions:

(a) Who bears the risk of unsuccessful development?

(b) What if the commercial discovery of oil and gas comes after the drilling of several dry holes; can the developing cotenant claim the cost of the dry holes as against his non-joining cotenant's share of production from other producing wells?

See, e.g., Connette v. Wright, 154 La.1081, 98 So. 674 (1923) (cotenant of leasehold had not agreed to drilling of wells but had signed division orders and received a share of the proceeds from a producing well; *held,* his claim to a share of the proceeds is subject to a pro rata share of the costs of development, including the costs of dry holes drilled); Moody v. Wagner, 167 Okl. 99, 23 P.2d 633 (1933). *But compare* Davis v. Sherman, 149 Kan. 104, 86 P.2d 490 (1939), wherein a cotenant in the working interest, after drilling a producing well, drilled a second well without the consent of his concurrent owner, which proved to be a dry hole. He was denied reimbursement for a proportionate share of the costs of drilling the dry hole. *Accord,* Ashland Oil & Refining Co. v. Bond, 222 Ark. 696, 263 S.W.2d 74, 3 O.&G.R. 577 (1953). *Compare* Knight v. Mitchell, 97 Ill.App.2d 178, 240 N.E.2d 16, 31 O.&G.R. 96 (1968), where recovery of costs of "fracking" a well from a non-consenting cotenant was denied when the results of the process conferred no benefit.

(c) The capital invested in an attempt to find oil and gas may not be recouped for some time even if the attempt is successful. Is the drilling and investing cotenant entitled to recover interest on his investment in an accounting with the nonconsenting cotenant? It has been held that he is not. Cox v. Davison, 397 S.W.2d 200, 24 O.&G.R. 128 (Tex.1965). *See* Haggard, *Production Attributable to a Nonconsenting Mineral Owner is Chargeable with Legal Interest on Drilling Costs*, 3 Houston L.Rev. 109 (1965).

5. There is limited authority that a cotenant (or his lessee) may develop minerals in a tract without the consent of his concurrent owners and account to such concurrent owners by payment of the usual royalty. McIntosh v. Ropp, 233 Pa. 497, 82 A. 949 (1912). Is this a better rule than that applied in the principal case? In Kentucky, a cotenant who did not join in an oil and gas lease or participate in drilling operations has been held entitled only to a share of the customary royalty on production prior to the institution of a suit to establish his interest in the premises, but in the case of production subsequent to the filing of his suit, he has been given a share of the value of the production less a share of the costs. Petroleum Exploration Corp. v. Hensley, 284 S.W.2d 828, 5 O.&G.R. 708 (Ky.1955); Gillispie v. Blanton, 214 Ky. 49, 282 S.W. 1061 (1926). Does the Kentucky position take care of any difficulties of the Pennsylvania position?

How should production be accounted for when a cotenant who has the right to execute oil and gas leases covering the entire interest in the tract secures production without a lease? *Compare* Martin & Kramer, *Williams and Meyers Oil and Gas Law*, § 329 *with* Bullard v. Broadwell, 588 S.W.2d 398, 64 O.&G.R. 336 (Tex.Civ.App.1979, error ref'd n.r.e.), which rejects the suggestion of the treatise that the producing executive owner account on the basis of the appropriate fraction of the usual royalty in the area. In Kumberg v. Kumberg, 232 Kan. 692, 659 P.2d 823, 79 O.&G.R. 534 (1983), however, the devisees for life of an interest in the "net profits" of land had their rights in production under a void lease by the landowner calculated on the basis of the usual bonus and royalty in the area. The Kansas court distinguished the *Bullard* case on the ground that the claimants in *Kumberg* "have no possessory interest in the land or the minerals; they are not cotenants; they are lifetime income beneficiaries." Is this a valid distinction? Is it relevant that the person to whom the land was devised in *Kumberg* was to share the "net profits" from the land with his siblings, the claimants, so long as any of them survived?

WAGNER & BROWN LTD. v. SHEPPARD

Supreme Court of Texas, 2008
282 S.W.3d 419, 52 Tex. Sup. Ct. J. 130

BRISTER, JUSTICE. One observer has estimated that 85 percent of the 27,000 wells drilled in the East Texas oil field in the first half of the 20th century were unnecessary -- resulting in a huge waste of money and natural resources. As one means of reducing excessive drilling, the Texas Legislature provided for voluntary pooling in 1949, and compulsory pooling in 1965.

Since then, this Court has never addressed how a pool of producing properties is affected if a lease in the pool expires. In this case, the courts below held that expiration of a lease removes those minerals from the pool and bars recovery of any costs incurred before termination. But the pooling agreement here did not depend on the continuation of underlying leases, nor was the equitable right of reimbursement for improvements necessarily extinguished by termination of the lease. Accordingly, we reverse and remand for further proceedings.

I. Background

Jane Sheppard, a CPA and retired family lawyer, owns 1/8th of the minerals underlying a 62.72-acre tract in Upshur County, Texas. C.W. Resources, Inc. leased her 1/8th interest, and along with Wagner & Brown, Ltd. leased the other 7/8ths of the minerals from other owners. Sheppard's lease had a special addendum providing that if royalties were not paid within 120 days after first gas sales, her lease would terminate the following month.

Sheppard's lease also authorized pooling with adjacent tracts. On September 1, 1996, C.W. Resources, Wagner & Brown, and mineral lessees on adjacent tracts signed a unit agreement pooling the Sheppard tract and eight others to form the W.M. Landers Gas Unit. One month later, a gas well was successfully completed and began producing, and a second well was completed in September 1997. Both wells were physically located on the Sheppard tract, but pursuant to the unit agreement proceeds and costs were split among all the tracts in proportion to acreage.

The parties agree that Sheppard's lease terminated on March 1, 1997, and since then she has been an unleased co-tenant, entitled to her share of proceeds from minerals sold less her share of the costs of production and marketing.

The dispute here concerns both the proceeds and the costs. Regarding the proceeds, the question is whether the termination of Sheppard's lease also terminated her participation in the unit (in which case she is entitled to 1/8th of 100 percent of production, as both wells are on her tract), or did not do so (in which case she is entitled to 1/8th of only 51.3 percent of production -- the proportion her tract bears to total acreage in the unit). Regarding the costs, the question is whether Sheppard should bear any costs incurred *before* her lease terminated, or any costs incurred *after* the lease terminated that relate to the unit but not her lease.

The trial court granted summary judgment for Sheppard, finding that termination of the lease also terminated her participation in the unit, that she was not liable for any costs incurred before termination, and that she was liable for costs incurred after termination only if they pertained solely to her lease; the court of appeals affirmed. For the reasons stated below, we disagree.

II. Does Termination of the Lease Also Terminate the Unit

[In this part of the opinion the court concludes that the termination of the Sheppard lease does not terminate the inclusion of the Sheppard mineral estate in the pooled unit. That part of the decision is discussed in Chap?, Section ?? *infra.* Eds]

III. Must Sheppard Pay a Share of Unit Expenses

Based on the conclusion that the Landers unit had terminated, the courts below denied Wagner & Brown's claim for leasehold, land/legal, and overhead expenses to the extent they related to the unit rather than solely to Sheppard's tract. Because we have held the unit did not terminate, Wagner & Brown properly accounted to Sheppard for both production and expenses on a unit basis.

The record reflects that these expenses were for landman fees, lease bonuses, recording fees, and title opinion expenses, part or all of which related to tracts in the unit other than Sheppard's. Additionally, "overhead" expenses (including administration, supervision, office services, and warehousing) were calculated pursuant to a standard accounting agreement relating to the unit. At trial, Sheppard produced no evidence that any of these expenses were not reasonable and necessary; to the contrary, she stipulated that many of them were.

On appeal, Sheppard suggests no basis for barring these expenses except that they did not all relate to her tract. But the evidence at trial established that two wells could not have been drilled on her tract without the pool, at least not without getting an exception from Railroad Commission that would have curtailed production. From the perspective of other interest holders in the unit, expenses to maintain Sheppard's tract were clearly necessary as all the wells were located there; the same kind of expenses to keep other tracts and leases in the unit were of similar value to Sheppard, not just as a matter of equal treatment but because future wells might be located off her tract.

Although Wagner & Brown's evidence was not contradicted, the trial court was not required to believe all of it. Accordingly, we do not hold that Wagner &

Brown conclusively established that all its charges were reasonable and necessary. But some of them certainly were, so there is no evidence to support the trial court's award of none of them. In these circumstances, we must reverse and remand for a reassessment of damages.

IV. Must Sheppard Pay Drilling Costs

Sheppard concedes that once her lease expired, she must pay her share of expenses to produce and market gas thereafter. But she argues, and the courts below agreed, that she did not have to pay any of the drilling or other costs incurred before the lease expired.

But the general rule regarding improvements is otherwise: "The principle is well established in equity that a person who in good faith makes improvements upon property owned by another is entitled to compensation therefor." Thus, when a son lost a farm because his deed from his father was oral, we held he might still seek reimbursement for improvements he built to the extent they enhanced the farm's value. [Sharp v. Stach, 535 S.W.2d 345, 351 (Tex. 1976), Eds.] As we explained in . . . [Stephenson v. Luttrell, 107 Tex. 320, 179 S.W. 260, 262 (Tex. 1915)], this rule arises in equity:

> When two persons are cotenants in the ownership of land, and one of them incurs expense in the improvement of the property which is necessary and beneficial, it is equitable that the one incurring the expense shall have contribution from his cotenant in an amount which is in proportion to the undivided interest owned by such cotenant.... The law implies a contract between him and his cotenant, authorizing him to spend for him the money which was necessarily spent. . . .

Under Texas law, only a naked trespasser cannot recover for improvements that benefit another's land

As oil and gas wells are improvements to real property, the same rule applies to them: one who drills a well in good faith is entitled to reimbursement. This rule applies even if an operator's lease is not valid, so long as the operator believed in good faith that it was. Similarly, a co-tenant who drills without another co-tenant's consent is entitled to reimbursement:

> It has long been the rule in Texas that a cotenant has the right to extract minerals from common property without first obtaining the consent of his cotenants; however, he must account to them on the basis of the value of any minerals taken, *less the necessary and reasonable costs of production and marketing.* [Brannon v. Gulf States Energy Corp., 562 S.W.2d 219, 224, 59 O.&.G.R. 320 (Tex. 1977).]

Thus, there is no question that under Texas law Wagner & Brown could have recovered a share of drilling costs from Sheppard had she signed no lease at all. The question is whether the rule should be different because there was a valid lease that was mistakenly allowed to expire. It is of course true that equity does not favor those who sleep on their rights. But it is hard to see why one who obtains a lease and then loses it by mistake is entitled to less equity than one who by mistake never had a valid lease in the first place.

It is true that C.W. Resources breached Sheppard's lease, but a breaching party is not necessarily barred from reimbursement for improvements. For example, Texas law permits recovery to builders upon substantial performance, even if they have breached their building contract. As we have explained, this rule is based on "the owner's acceptance and retention of the benefits arising as a direct result of the contractor's partial performance."

. . . we believe the equitable nature of such claims must turn on the equities in each case. Thus, for example, an operator who intentionally terminates a lease has a weaker claim to equity than one who does so by accident. One who immediately offers to reinstate an expired lease has cleaner hands than one who does not. As with other equitable actions, a jury may have to settle disputed issues about what happened, but "the expediency, necessity, or propriety of equitable relief" is for the trial court, and its ruling is reviewed for an abuse of discretion.

Based on the summary judgment record here, it appears that the trial court abused its discretion in refusing reimbursement of drilling costs. There is no question Sheppard's tract has enjoyed ten years of production from two gas wells drilled at the sole risk and expense of the defendants. Neither well was drilled by a trespasser; the defendants had title to the other 7/8ths of the minerals on her tract, and thus an unquestionable right to drill the wells where they were. Sheppard had the option to continue collecting royalties free of any drilling costs, as Wagner & Brown offered to reinstate her lease on that basis. Having chosen instead (as was her right) to be a co-tenant with full benefits to the minerals she owned, it would be inequitable to allow her to escape the burdens that come with that choice.

Further, it is well-settled that "equity abhors forfeiture." Consistent with that rule, Texas law requires that in construing mineral leases "doubts should be resolved in favor of a covenant instead of a condition" so that forfeiture is avoided. Sheppard's lease leaves no room for doubt that the operators forfeited the lease when the first royalty was delayed; but it says nothing about whether they forfeited drilling costs too on that basis. While the defendants lost forever their legal claim to Sheppard's minerals, that does not necessarily mean that they also lost their equitable claim for the improvements they had built on her tract. Again, equity might deny such a claim had the lease expired long after production had begun and the operators had recovered most of their drilling costs. But the lease here terminated at the very outset of production, so denying reimbursement would work a substantial forfeiture. Under the facts in the record here, the trial court abused its discretion by allowing Sheppard to reap all the proceeds of production without bearing any of the costs.

The court of appeals denied reimbursement on the basis that Sheppard was a royalty owner at the time drilling costs were incurred, and thus was not liable for them. But Sheppard ceased being a royalty owner shortly after production began.

To the extent working interest owners paid drilling costs out of production over the following months, Sheppard too would be liable for those costs as she became a working interest owner after her lease terminated.

Because of the absence of authoritative caselaw, the evidence presented in the parties' cross-motions for summary judgment did not focus on the equitable issues like those we have indicated above. We recognize there might be other facts not appearing in the record that would justify denying Wagner & Brown an equitable right of reimbursement for drilling or other pre-termination costs. Accordingly, in the interest of justice we reverse the court of appeals' judgment and remand the case to the trial court for further proceedings consistent with this opinion.

For the reasons stated above, we reverse the judgment of the court of appeals, and render judgment that (1) termination of Sheppard's lease did not terminate her participation in the Landers unit, and (2) Wagner & Brown may account for both production and costs on a unit basis to the extent allowed by the parties' agreements, custom in the industry, and the trial court's order that operating expenses be deducted from production on a well-by-well bases (from which order the defendants have not appealed). We reverse and remand for further proceedings in the trial court consistent with this opinion regarding (1) the reasonable and necessary costs of production to be deducted from amounts due Sheppard for costs incurred after termination of her lease, (2) whether the defendants are entitled to equitable recovery of their pre-termination costs, and if so (3) the reasonable and necessary amount of those costs.

NOTES

1. The Texas Supreme Court while allowing expenses incurred for pooled unit purposes did not resolve another important issue when it comes to co-tenant accounting. The operator had drilled two wells on the unit. Could the operator aggregate the costs incurred in both wells when doing its accounting or must it account on a per-well basis. The Court of Appeals decision, 198 S.W.3d 369, 379 (Tex.App. 2006) had required the operator to account on a per-well basis. Because the operator did not appeal that issue, the Supreme Court merely noted that it did not reach that issue in its opinion. 282 S.W.3d at 425 (n.24).

2. A case decided after *Wagner & Brown v. Sheppard,* seems to ignore the Supreme Court's conclusion that overhead and lease acquisition costs are to be used as a set-off where you have a good faith trespass. Bomar Oil & Gas, Inc. v. Loyd, 2009 Tex.App. LEXIS 5505 (July 15, 2009), *on reh'g,* 298 S.W.3d 832 (Tex.App. 2009).

3. In Earp v. Mid-Continent Petroleum Corp., 167 Okla. 86, 27 P.2d 855 (1933), the court recognized that the status of a party as a trespasser may change over a period of time. In *Earp,* a lessee of a cotenant engaged in no drilling or production operations on the leased premises during its primary term. The lessee of the other cotenants drilled a productive well. There was no joint operating agreement executed between the two separate lessees. The court found that

during the primary term of the inactive lessee's lease, the inactive lessee need not make delay rental payments to keep the lease alive. Thus, during this period of time, the drilling lessee must account to the non-drilling lessee under the normal rules relating to cotenancy accounting. The non-drilling lessee would have to make royalty payments to its lessor even though it may not be receiving any funds if the drilling and operational expenses exceed the revenue from the sale of the hydrocarbons. After the end of the primary term, however, the non-drilling lessee could not receive the benefit of production from the drilling lessee and thus its lease terminated and the lessor attained the status of an unleased cotenant as to the entirety of the fractional mineral interest owned.

ANDERSON v. DYCO PETROLEUM CORP.

Oklahoma Supreme Court, 1989.
782 P.2d 1367, 107 O.&G.R. 504.

LAVENDER, JUSTICE. This appeal arises from a grant of summary judgment in favor of Appellee, Panhandle Eastern Pipe Line Company, one of three defendants sued by Appellants, a group of working interest owners in the Yowell No. 1–26 natural gas well located in Roger Mills County.... Appellants alleged in their petition that Panhandle and El Paso were purchasing natural gas from the Yowell No. 1–26 well from Dyco (alleged to be the owner of approximately 47% of the working interest) and other working interest owners.

...

Oklahoma Law Does not Recognize a Claim for Conversion on the Facts Revealed by The Instant Record

In relation to Panhandle's motion for summary judgment, the record in this case reveals that Appellants, along with Dyco and others, own various percentages of the working interest in the Yowell No. 1–26 natural gas well. It also reveals Panhandle has been purchasing gas from the well from Dyco and other working interest owners and paying these working interest owners for the gas purchased, while Appellants have not received any of the proceeds from the sales.... The record further reveals Appellants are not parties to the agreement Dyco has with Panhandle for the sale of natural gas.... As will be explained, in these circumstances, Oklahoma law provides no tortious action for conversion in favor of working interest owners like Appellants against a purchaser of natural gas who buys gas from one or more other working interest owners in the same well.

Under Oklahoma law Appellants and the other working interest owners in the well are tenants in common. As cotenants each is entitled to market production from the well and the sale of gas to a purchaser by one or more cotenants without consent of other cotenants is lawful. Under ordinary circumstances it does not involve tortious conduct, i.e. conversion, on the part of either the purchaser or on the part of the working interest seller because each cotenant has the right to develop the property and market production under the common law.

Recently in *Teel v. Public Service Company of Oklahoma,* [767 P.2d 391 (Okla.1985)] this Court outlined the limited circumstances under which a purchaser may be liable for conversion when purchasing natural gas from a well operator. This situation initially arises when the cotenants name another cotenant to exploit the cotenancy for their mutual profit by entering into an operating agreement amongst themselves to have the operator market or sell gas produced from the well. It also requires that there be an absence of a division order. Additionally, and most importantly for purposes of disposition of the instant case, *a purchaser must have notice that a working interest owner not a party to the division order has revoked the operator's right to sell his/her share of gas under the operating agreement and the purchaser must continue to treat the situation as if it is purchasing the nonconsenting cotenant's share of gas, while failing to account to each working interest owner for his/her pro rata share of the proceeds.* [emphasis by court. Eds.] The *Teel* situation only comes into play when a purchaser purports to buy gas of an owner from an operator which is not authorized to deliver it.

Our situation is virtually the opposite of that faced in *Teel.* In *Teel* the working interest owner, by revoking the operating agreement, expressed an intent not to have the operator sell or the purchasers buy his gas. The purchasers, contrary to such intention, continued to treat the situation after revocation as if they were still purchasing gas from a working interest owner *that did not want to sell it to them.* [emphasis by court. Eds.] In contrast Appellants here take the position that anytime one cotenant sells gas to a purchaser, which as noted above the cotenant has a lawful right to do, the purchaser becomes liable to each and every other working interest owner for conversion simply by the force of the working interest owner's *desire to sell his gas and be paid in proportion to his percentage of working interest in the well, even though not a party to any purchase contract.* [emphasis by court. Eds.] We find no authority for this result under the common law of Oklahoma and we reiterate the admonition in *Teel* itself contained in the first footnote thereof, that, "[i]nsofar as *Teel* announces a rule of conversion, it is confined to the pecular scenario found [therein]." Further, as *Teel* notes, "[a] gas sales contract executed by a cotenant is limited to his/her interest" and we see nothing in the facts of this case that would indicate gas purchased by Panhandle under agreement with Dyco, on behalf of itself or itself and some of the other working interest owners, would constitute either a wrongful sale or purchase of Appellants' interest as was the case in *Teel.* Therefore, on the instant record Appellants have no claim against Panhandle for conversion of their respective percentages of working interest in the Yowell No. 1–26 natural gas well.

Such a disposition does not leave working interest owners like Appellants without a remedy. The law has been settled for some time that a producing cotenant must account to a non-producing cotenant for the market value of the production less any reasonable and necessary expenses of developing, extracting and marketing. Further, certain practices of the industry have been acknowledged by the courts to remedy situations like that apparently existent here where only certain working interest owners have sold production. These practices involve balancing in kind the production from the well by allowing cotenants like Appellants the opportunity to market gas from the well (i.e. taking a certain percentage of an over produced party's gas until any imbalance in the cotenant's takes from the well are made up), by periodic cash balancing whereby under produced cotenants receive cash from producing cotenants in proportion to their respective interests and cash balancing upon any particular gas reservoir's depletion. Instead of bringing an action for accounting or relying on one of the potential solutions set forth above, Appellants sought instead to turn what should have been largely an equitable proceeding into a tortious one not sanctioned by Oklahoma law. Accordingly, although not for the reason of federal preemption relied on by the trial court, we affirm the grant of summary judgment in favor of Panhandle because Appellants have no cause of action for conversion against it under the material facts revealed by the record.

[The court also rejected Appellant's claim under Okla.Stat.Ann., tit 52, § 23 that requires pipelines to "ratably take" gas under specified circumstances. Eds.]

NOTES

1. The problems presented in *Anderson* are neither novel nor uncommon. Frequently there will be more than one working interest owner in a well. Where natural gas is the principal product the taking of that gas can create substantial problems. As the authors have said elsewhere:

> Parties to a unit operating agreement may deal expressly with the problem of gas takes. Often an agreement will provide that the unit operator will have the right but not the duty to sell the gas for the nonoperating parties. The agreement may also provide a mechanism for balancing sales. However, when there is no agreement for balancing or no agreement for the operator to sell the gas of the others, what are the relations among the working-interest owners with respect to the gas sold and the gas remaining in the ground? One must look to statute or to common-law rules of property.

Kramer & Martin, *The Law of Pooling and Unitization*, 19–121 to 19–122. *See also,* Martin, *The Gas Balancing Agreement: What, When, Why and How*, 36 Rocky Mtn. Min. L.Inst. 13–1 (1990); Pierce, *The Law of Disproportionate Gas Sales*, 26 Tulsa L.J. 135 (1990); Kuntz, *Gas Balancing Rights and Remedies in the Absence of a Balancing Agreement*, 35 Rocky Mt. Min. L. Inst. 13-1 (1989); Smith, *Gas Marketing by Co–Owners: Disproportionate Sales, Gas Imbalances and Lessors' Claims to Royalty*, 39 Baylor L.Rev. 365 (1987).

2. For a discussion of how Louisiana dealt with the problem of working-interest co-tenants and their ability to sell gas, *see* Amoco Production Co. v. Thompson, 516 So.2d 376, 98 O. & G.R. 273 (La. App. 1987), writ denied, 520 So.2d 118 (La. 1988) discussed in Kramer & Martin, *supra* at

§19.05[2].

3. Where the working interest owners have not entered into a balancing agreement should the court attempt to cash balance or to balance in kind? In POGO PRODUCING CO. V. SHELL OFFSHORE, INC., 898 F.2d 1064, 108 O.&G.R. 441 (5th Cir.1990) the Fifth Circuit concluded that balancing in kind was the preferred method, although under certain circumstances cash balancing would be less inequitable. *Accord,* Questar Pipeline Co. v. Grynberg, 201 F.3d 1277 (10th Cir., 2000); Doheny v. Wexpro Co., 974 F.2d 130, 122 O.&G.R. 227 (10th Cir. 1992).

When the Commissioner of Conservation revises existing units and creates new units, problems of over- and under-production may occur. In HUNT OIL CO. V. BATCHELOR, 644 So.2d 191, 132 O.&G.R. 569 (La. 1994), the underproduced natural gas owners sought an order for an accounting from the overproduced owners after a revision of the units was ordered. The underproduced parties wanted to balance in cash. The Commissioner instead ordered balancing in kind. The Lousiana Supreme Court found that it was not arbitrary nor capricious for the Commissioner to prefer balancing in kind over balancing in cash and affirmed the validity of the order.

Does the Oklahoma court in *Anderson v. Dyco* tell us how an accounting is to be made? If, as the court suggests, an in kind balancing is possible, how does this "cotenancy" differ from the accounting in *Prairie Oil & Gas v. Allen supra*?

What factors should be considered in determining whether balancing should be made in cash or in kind? HARRELL V. SAMSON RESOURCES, 1998 Okla. 69, 980 P.2d 99, 142 O.&G.R. 62 found that an attempted sale of its interest by an overproduced party triggered a right of the underproduced party to demand cash balancing. If the sale had gone through and the underproduced party had remained in an underproduced status, would the underproduced party's balancing rights run with the land and bind the new owner? For discussion of this subject in Oklahoma, see Mark Christiansen, Discussion Notes, 142 O.&G.R. 80-86.

HEIMAN V. ATLANTIC RICHFIELD CO., 891 P.2d 1252, 131 O.&G.R. 59 (Okla. 1995). ARCO was the operator of a unit well. Heiman was a non-operating working interest owner. ARCO sold 100% of the gas stream. Heiman did not make separate arrangements to take in kind. Heiman attempted to sell his gas to ARCO's purchaser, but ARCO withheld its approval of the contract and the gas purchaser refused to execute the contract. Is Heiman entitled to a cash balancing? Is Heiman entitled to prejudgment interest on its share? What is the impact of the Oklahoma "Sweetheart Gas Act" (Okla.Rev.Stat. tit. 52, §541-547 (1983) superseded in part by Okla.Rev.Stat. tit. 52, §570.1 et seq. and Okla.Rev.Stat. tit. 52, §581.1 et seq.) on Heiman's right to both cash balancing and prejudgment interest?

4. Assuming that concurrent owners and their lessee or lessees decide to enter into an agreement for the joint development of possible oil and gas in their land, what are some of the problems that the document establishing their

relationship should cover? *See* the Operating Agreement, Appendix 9.

KRUG v. KRUG, 5 Kan.App.2d 426, 618 P.2d 323, 68 O.&G.R. 445 (1980), affirmed an order establishing in advance of drilling by one cotenant the terms of participation in any production by the non-drilling cotenant, allowing for deduction of reasonable costs and excluding any deduction of dry hole costs.

Although the nonconsenting cotenant does not incur personal liability for the expenses incurred by the active cotenant in exploration, an agreement for development will usually result in such liability. *See* Hill v. Field, 384 F.2d 829, 27 O.&G.R. 747 (10th Cir.1967).

EL PASO PRODUCTION CO. v. VALENCE OPERATING CO., 112 S.W.3d 616 (Tex.App. 2003, rev. denied). Under the terms of a joint operating agreement, the operator may withhold monies due from the sale of hydrocarbons owned by a non-operating working interest owner. If the operator wrongfully withholds the hydrocarbons or the funds has it converted the personal property interest of the non-operating working interest owner under *Dyco Petroleum*? Yes according to the Texas Court of Appeals because of the nature of the cotenancy relationship between the parties to the joint operating agreement.

Many of the provisions of typical operating agreements are discussed in Conine, *Property Provisions of the Operating Agreement—Interpretation, Validity, and Enforceability*, 19 Tex.Tech.L.Rev. 1263 (1988).

5. When a cotenant (or his lessee) drills for and produces minerals from a tract without the joinder or consent or ratification of the other concurrent owners, there are three possible bases of accounting for the removal of minerals in states in which such removal is considered waste. Accounting may be required on the basis of a proportionate share of the proceeds of the development without any deduction for the costs of such development or operation—that is, on the basis of the value of the oil or gas at the mouth of the well. Some courts have ruled that the non-joining concurrent owners may recover a share of the production after deducting a proportionate share of the costs—that is, on the basis of the value of the oil or gas in place. Other courts have restricted the non-joining concurrent owners to a recovery of a proportionate share of the lease proceeds when such production was by a lessee.

The latter two alternatives typically have been adopted and followed by the courts only where there has been a showing that the non-joining concurrent owners stood by in silence while the operations were carried on, the operations were conducted under the mistaken belief that the operator or his lessor owned the entire fee simple estate, or there was a risk of drainage when the operations were pursued. The non-joining cotenant may elect to waive the tort of waste and ratify the lease and share in the benefits, including bonus, rentals and royalties. *See* Martin & Kramer, *Williams and Meyers Oil and Gas Law*, § 504.2.

6. Leases executed by a concurrent owner normally purport to cover all of the minerals in the described tract. The lessee will usually insist that the lease be

executed in this fashion so that he may be entitled to the benefit of the doctrine of estoppel by deed in the event the leasing cotenant thereafter should acquire some or all of the outstanding undivided interest in the minerals. Execution of a lease in this fashion entitles a non-joining cotenant later to ratify the lease and to share in the proceeds. Among the questions which may arise from the execution of a lease by a cotenant which purports to convey all of the minerals are the following:

(a) How long may the non-joining concurrent owners postpone decision whether to ratify the lease or to disaffirm it and receive a share of the proceeds of production less a share of the costs?

(b) What constitutes a ratification of the lease?

(c) Will estoppel by deed operate if the bonus paid covers only the interest conveyed?

7. Should each cotenant share in production arising from a unit agreement entered into by another cotenant? *See* Superior Oil Co. v. Roberts, 398 S.W.2d 276, 24 O.&G.R. 77 (Tex.1966) reprinted at p. 855 *infra.*

Roberts is distinguished in Edwin M. Jones Oil Co. v. Pend Oreille Oil & Gas Co., 794 S.W.2d 442, 112 O.&G.R. 501 (Tex.App. 1990); Kelley v. Apache Products, Inc., 709 S.W.2d 772 (Tex.App.1986). *See generally,* 1 B. Kramer & Martin, *The Law of Pooling and Unitization*, §8.03 (1999).

SCHNITT v. McKELLAR

Supreme Court of Arkansas, 1968.
244 Ark. 377, 427 S.W.2d 202, 30 O.&G.R. 1.

FOGLEMAN, JUSTICE. This case involves a determination whether the trial court was correct in finding that certain instruments executed by certain McKellar heirs conveyed their interest in an undivided one-fourth working interest in oil, gas and minerals to J.H. Carmichael, Jr. and J.C. Stevens.

Appellant is the successor in interest to some of the McKellar heirs who were parties to these instruments. He filed this action for a declaratory judgment to determine the interests of Carmichael and Stevens and their respective wives under the instruments and for partition of all surface and mineral interests. . . .

[The portion of the opinion construing the instruments is omitted. Eds.]

When we apply these well known rules of construction, we must come to the conclusion that there was a conveyance of an [undivided one-fourth] interest in he oil, gas and minerals. . . .

Appellant contends alternatively that the court should have granted his prayer for partition and cites Ark.Stat.Ann. § 34–1801 et seq. (Repl.1962). This statute provides for partition on petition of any person having any interest and desiring a division of land held in common. There can be no doubt that under our construction of the contract, the interest of appellees is an interest in land. This court has held that a mineral deed placed of record constitutes a constructive severance of the minerals from the surface and makes two titles, one the surface and the other the mineral title. It has also been said that the sale of an undivided mineral interest operates as a severance of oil interest from the surface and creates two separate and distinct estates. The interest of appellees would not be an interest held in common with the owners of the surface in the sense of § 34–1801. On the other hand, the ownership of the minerals would be such an interest and the statute would be applicable to a division of the mineral interests, on the petition of the owners of the surface and an undivided interest in the minerals. This holding seems to be in keeping with the weight of authority, it being generally held that minerals, as a part of the real estate, if held in cotenancy, may be the subject of partition. *See* Annot., 173 A.L.R. 854.

Generally, partition, either by partition in kind or by sale and division of proceeds, is something to which each cotenant has an absolute and unconditional right, under both common law and statute. Freeman on Cotenancy and Partition, § 433, p. 571; Knapp on Partition, p. 27; Annot., 15 Ann.Cas. 778; 2 Williams & Meyers, Oil & Gas Law, § 506.2, p. 601. The basis for such right is well expressed in Dall v. Confidence Silver Mining Co., 3 Nev. 531, 93 Am.Dec. 419 (1867) where it was said:

> . . . As the law deems it against good morals to compel joint owners to hold a thing in common, a decree of partition may always be insisted on as an absolute right. It is not necessarily founded upon any misconduct of the cotenants or part owners. Hence, in decreeing a partition, the rights and equities of all the parties are respected, and the partition decreed so as to do the least possible injury to the several owners; and 'courts of equity,' says Mr. Story, 'may, with a view to the more convenient and perfect partition or allotment of the premises, decree a pecuniary compensation to one of the parties for owelty or equality of partition, so as to prevent any injustice or unavoidable inequality': Story's Eq.Jur., sec. 654.

The fact that interests in oil, gas and minerals are involved does not change the problem. . . .

It has long been recognized in other states that the owners of undivided interests in minerals are entitled to either partition or sale against the remaining owners, even under statutes worded as § 34–1801 was before the 1941 amendment.

A Mississippi statute provided for partition by tenants in common of any joint interest in the freehold. The Supreme Court of Mississippi held that under this statute the owner of the entire surface and an undivided one-half interest in the minerals was entitled to have a partition of the mineral interests.

The Court of Appeals of Kentucky held that the owner of the surface in fee simple and of an undivided one-half interest in oil, gas and other minerals was entitled to partition of the mineral rights under their partition statute which required the showing of a vested possessory estate in land which is jointly held in any manner.

The Texas Court of Appeals, in Henderson v. Chesley, 273 S.W. 299 (Tex.Civ.App.1925), aff'd 116 Tex. 355, 292 S.W. 156 (1927), held that the owner of an undivided one-half interest in mineral rights was entitled to partition against the owner of the fee and an undivided one-half interest in the minerals under a statute which gave any joint owner or claimant of any real estate or any interest therein the right to compel partition.

It is suggested by appellees in their brief, however, that it would be a great injustice to them to partition the mineral interest in kind and that there is no need to require this interest to be sold. It has been said that this absolute or unconditional right cannot be defeated by showing that a partition would be inconvenient, injurious or even ruinous to a party in interest.

The manifest hardship arising from division of property of an impartible nature has been generally and almost universally avoided by statutes authorizing sale of the property when its division in kind would tend to greatly depreciate its value or otherwise seriously prejudice the interests of cotenants. . . . This court has also recognized that, in any case to which Ark.Stat.Ann. § 34–1801 (Repl.1962) is applicable, the right to partition or sale is absolute, no matter how small the interest of the owner seeking partition or how great the majority who object. Overton v. Porterfield, 206 Ark. 784, 177 S.W.2d 735. . . .

Dictum in Overton v. Porterfield, *supra*, wherein partition and sale of the oil and gas leasehold were involved, expressed the thought that, in such case as was there involved, § 53–401 et seq. did not impose upon a court of equity the imperative duty to order a sale whenever a proper petition was filed. The court said that a rule adopted in Oklahoma might apply in this type of case under certain circumstances. That rule, quoted in the opinion from Wolfe v. Stanford, 179 Okl. 27, 64 P.2d 335 (1937), is that the court should be vested with discretion to grant or deny relief in order to prevent the right to partition from becoming a weapon of oppression and fraud in the hands of the financially fortunate who might use the right as a means of foreclosure of an owner of limited means. The Oklahoma Court clearly qualified this discretion as existing only in cases where there was no disagreement between the parties rendering co-ownership impractical and stated specifically that inability of a cotenant to purchase should not constitute a defense in the absence of approaching development or rapidly increasing values. The qualification was applied in Henson v. Bryant, 330 P.2d 591 (Okl, 1958).

But even in those cases wherein the doctrine that the equity court has the discretion to deny either partition or sale to prevent the remedy from becoming an instrument of fraud or oppression, it is generally held that the matter is one of defense to be pleaded and proved as such. In this case, appellees did not plead their present contention that a decree of partition or sale would constitute fraud or oppression against them, nor did they offer any evidence tending to support such a claim.

We therefore, modify the decree of the chancellor to award appellees an undivided one-fourth interest in oil, gas and minerals, to the extent of the undivided interest conveyed by the grantors in the "contracts." Appellees' interests were found by the trial court to be $^{13}/_{27}$ of ¼ of $^{7}/_{8}$. Neither party questions this finding. We affirm the decree as modified, except as to the failure to grant partition of the mineral estates. We remand the case for appropriate proceedings for partition of the mineral interests, either in kind or by sale. In its proceeding, the trial court should bear in mind the peculiar nature of the property rights involved. Upon remand, the trial court should correct its decree so as to conform to this opinion.

Affirmed as modified and remanded.

NOTES

1. In HENSON V. BRYANT, 1958 OK 231, 330 P.2d 591, 9 O.&G.R. 923, cited in the principal case, the court noted that a cause of action for partition of the mineral estate had been pleaded and proved and then commented on the defense that had been offered:

> The burden was on the plaintiffs in error to establish the defense of oppression. No other defense was urged. Henson's evidence established only a payment by him through his wholly owned corporation of what was apparently an adequate consideration for the interest ultimately conveyed to him. This is no defense. The fact that the party seeking partition created the interest, and the fact that plaintiffs in error do not desire partition, are each equally unefficacious in defense in the absence of additional considerations not urged in this action. Furthermore, since the evidence fails to disclose approaching development of oil properties, rapidly increasing value of this property, inability of the parties to purchase at a sale if partition in kind is not possible, or any other factual issue tending to establish the conclusion that oppression would result from partition of the mineral estate, the judgment must be affirmed.

The position that equitable defenses are relevant on the question of partition *vel non* is also discussed in Strait v. Fuller, 184 Kan. 120, 334 P.2d 385, 10 O.&G.R. 145 (1959).

In DE MIK V. CARGILL, 1971 OK 61, 485 P.2d 229, 39 O.&G.R. 79, it was urged that partition should be granted at the behest of a working interest owner against the owner of an overriding royalty interest "as a matter of equity" since the overriding royalty "places an undue burden upon the working interest and creates an intolerable economic condition which precludes further development." The court denied partition as between these interests: "Inequitable hardship and oppression are not elements which bestow jurisdiction upon a court of equity in a partition action. These issues are defensive issues to be pleaded and proved. . . . [T]he overriding royalty interests lacked possessory rights in the leasehold and defendants were not tenants in common in an estate in real property."

The *De Mik* analysis was followed by the Kansas Supreme Court in

Muslow v. Gerber Energy Corp., 237 Kan. 5, 697 P.2d 1269 (1985). Working interest owners and royalty interest owners are not cotenants since there is no unity of possession. The court did note that a working interest owner could defeat the partition of the entire estate by carving out royalty interests which would not be subject to a partition action.

DUNN V. PATTON, 240 Ill.Dec. 783, 718 N.E.2d 264 (Ill.App. 1999) held that the plaintiff/owner of a one-half interest in the minerals was not a tenant in common with the parties who owned the surface and other one-half of the minerals and thus the plaintiff could not compel partition.

In Witt v. Sheffer, 6 Kan.App.2d 868, 636 P.2d 195, 73 O.&G.R. 299 (1981), an order of partition was held to have properly covered leased and unleased portions of a tract which was held in common by owners of interests of fee simple duration and owners of defeasible term interests and to have appropriately extended to the related reversionary interests. The partition order appealed in the *Witt* case was not described in the opinion. If it provided for partition by sale, need it have encompassed the reversionary interests? If partition in kind of undivided interest leases is the remedy requested, ought lessors and royalty owners of the tracts subject to lease be made parties to the partition action? Texas Oil & Gas Corp. v. Ostrom, 638 S.W.2d 231, 75 O.&G.R. 390 (Tex.App.1982, error ref'd n.r.e.)?

The failure of a voluntary unitization agreement to create a tenancy in common between participating lessees was the basis of the denial of partition as a remedy for dissolution of the unit in Tenneco Oil Co. v. District Court of Twentieth Judicial Dist., 1970 OK 21, 465 P.2d 468, 36 O.&G.R. 441.

The Oklahoma Supreme Court embraced the same rationale in *Tenneco Oil*, in a situation where a compulsory pooling order had been entered. The force-pooled working interest owners failed to convince the court that they and the designated operator were common law cotenants. Schulte v. Apache Corporation, 1991 OK 61, 814 P.2d 469, 113 O.&G.R. 336.

An Oklahoma statute requires for partition of a mineral estate an evidentiary showing that a co-owner is frustrating development and that the development objectives of the plaintiff cannot be realized by a pooling order of the Oklahoma Conservation Commission. Okl. Stat. Ann., tit. 12, § 1501.

3. MOSELEY v. HEARRELL, 141 Tex. 280, 171 S.W.2d 337 (1943). The defendant in this case sought to defeat partition by alleging that plaintiff was endeavoring to acquire her interest in the land; that she would be financially unable to buy in the property at a receiver's sale; and that if her interest should be sold by the receiver it would not bring its full value, and in addition, she would be compelled to pay a large Federal income tax out of her receipts from the sale. These allegations were made for the purpose of showing that it would be inequitable to compel partition of the property. The court concluded, however, that there was no requirement for the showing of equitable grounds as a prerequisite to the exercise of the right of partition, nor may the right of partition

be defeated by the showing of inequities, since under the Texas statute, a cotenant is entitled to the remedy of partition as a matter of right. Equities were said to be relevant only to the determination of *how* the property is to be partitioned.

It may sometimes be inequitable to enforce partition of the jointly owned property; but this is one of the consequences which he assumes when he becomes a cotenant in land. If he does not provide against it by contract, he may expect his cotenant to exercise his statutory right of partition at will.

This suggestion that an agreement between co-owners prohibiting partition is valid has been recognized in other cases as well. For example, such an agreement has been implied from the provisions of an operating agreement which was to last for the duration of production under the leases which were concurrently held. Sibley v. Hill, 331 S.W.2d 227, 12 O.&G.R. 291 (Tex.Civ.App.1960). In Odstrcil v. McGlaun, 230 S.W.2d 353 (Tex.Civ.App.1950), an agreement not to partition was implied from the fact that one co-owner had been given the exclusive power to lease the entire tract of land and in Long v. Hitzelberger, 602 S.W.2d 321, 67 O.&G.R. 92 (Tex.Civ.App.1980), from the fact that the time for performance of a farmout drilling obligation had not expired. Thomas v. Witte, 214 Cal.App.2d 322, 29 Cal.Rptr. 412, 18 O.&G.R. 277 (1963), holds that an agreement not to seek partition can be implied from the fact that the property in which the cotenancy exists has been pooled or unitized.

On the other hand, there is authority that agreements not to partition must be limited to a reasonable period of time. Roberts v. Jones, 307 Mass. 504, 30 N.E.2d 392, 132 A.L.R. 663 (1940). What policy supports this limitation on freedom of contract? Are the agreements not to partition in the three cases cited in the preceding paragraph limited to a reasonable period of time?

In ROBINSON v. SPEER, 185 So.2d 730, 24 O.&G.R. 410 (Fla.App.1966), cert. denied, 192 So.2d 498 (Fla.1966), a reservation of an undivided one-half interest in oil and gas contained a provision giving the grantors "the right to choose one-half (½) in acreage" of the oil and gas in case of disagreements. The court recognized the clause as a valid provision governing method of partition which was not in violation of the Rule against Perpetuities.

The execution of a joint operating agreement between working interest owners who may be co-tenants may evince an implied intent not to seek a partition of the working interest so long as the joint operating agreement is in force. Dimock v. Kadane, 100 S.W.3d 602, 608, 615, 622, 157 O.&G.R. 91, 103, 115, 123 (Tex.App. 2003); MCEN 1996 Partnership v. Glassell, 42 S.W.3d 262, 148 O.&G.R. 84 (Tex.App. 2001, rev. denied).

4. "To the extent that equities are relevant on the issue of partition *vel non* or on the issue of the method of partition, a number of considerations have been urged upon our courts. Perhaps the most significant of these is that producing mineral lands or potentially producing lands should not be partitioned (or should not be partitioned in kind). Certainly the fact that land is, or potentially may be,

oil-producing, is persuasive to most courts that partition must be by sale rather than in kind. However, if the entire tract is producing or is known to be capable of producing, some courts have permitted a partition in kind among the cotenants. Where the land is only potentially producing, some courts have permitted partition in kind, but a number of courts have required that the partition be by sale, despite the plea of the objecting cotenant, who seeks partition in kind, that a forced sale may deprive him of his property in kind, which he desires to retain, and that by reason of his lack of funds, he is unable to protect his interest, which may be purchased by his concurrent owner or owners for a mere pittance. In one case it was suggested that in partitioning minerals in kind, it may be desirable to divide the surface acreage in a checkerboard pattern, with each owner being given more than one allotment in the several portions of the tract subject to partition." Martin & Kramer, *Williams and Meyers, Oil and Gas Law* § 506.3.

5. Is there an alternative to partition either in kind or by sale? Could a court give the defendant in a partition action the privilege of drilling under the terms of an ordinary commercial lease (by which plaintiffs would take one-eighth royalty on their undivided interest) and if defendant declined, give the same privilege to plaintiff? *See* Seeligson v. Eilers, 131 F.Supp. 639, 4 O.&G.R. 1737 (D.Kan.1955), *rev'd*, Shell Oil Co. v. Seeligson, 231 F.2d 14, 5 O.&G.R. 1307 (10th Cir.1955).

6. For an extended treatment of partition of oil and gas interests in Texas, see Dorothy J. Glancy, *Breaking Up Can Be Hard to Do: Partitioning Jointly Owned Oil and Gas and Other Mineral Interests in Texas*, 33 Tulsa L.J. 705 (1998). It observes: "Antiquated common law concerns about unities of time, title, interest and possession, which still complicate court decisions with regard to joint ownership in other jurisdictions, are virtually absent from Texas decisions regarding joint ownership law."

7. While partition is a statutory remedy or cause of action, the parties are always free to voluntarily partition a jointly-owned mineral estate on terms different than that would be required if the statute was followed. See Bixler v. Oro Management, LLC, 2004 WY 29, 86 P.3d 843, 165 O.&G.R. 858 where the Wyoming Supreme Court ordered a statutory partition and Bixler v. Oro Management, LLC, 2006 WY 140, 145 P.3d 1260 where the court accepted the parties' voluntary partition agreement.

B. SUCCESSIVE INTERESTS
(See Martin & Kramer, *Williams and Meyers Oil and Gas Law* §§ 511–515).

WELBORN v. TIDEWATER ASSOCIATED OIL CO.

United States Court of Appeals, Tenth Circuit, 1954.
217 F.2d 509, 4 O.&G.R. 385.

PHILLIPS, CHIEF JUDGE. Welborn brought this action against Tidewater

Associated Oil Company to recover damages for alleged slander of title through the procuring and recording of an oil and gas lease by Tidewater from Martha Smith nee Heck and J.B. Garrett. Tidewater interposed a motion to dismiss the first amended complaint on the ground that it failed to state a claim upon which relief could be granted. From a judgment sustaining the motion and dismissing the action, Welborn has appealed.

The material facts as alleged in the first amended complaint are these: Since on and prior to June 9, 1943, Smith has been the owner of a life estate and Garrett the owner of the remainder interest in a 98–acre tract of land situated in Coal County, Oklahoma.

On June 9, 1943, Smith, acting as the duly appointed guardian of Garrett, then a minor, and pursuant to lawful proceedings in the county court of Coal County, Oklahoma, and a public sale and confirmation thereof by such county court, did make and deliver to Welborn an oil and gas lease on such tract of land by which the interest in such land of Garrett was leased to Welborn for a primary term of ten years from June 9, 1943. Such proceedings in the county court consummated by such lease were done voluntarily by Smith and pursuant to a written petition filed by her in such county court. Such proceedings, however, purported only to effect a lease of the remainder interest of Garrett and were all done by Smith in her capacity as guardian of Garrett and not individually.

Smith is not a party to the instant action.

On November 17, 1952, George M. McDaniel procured the execution of an oil and gas lease covering such tract of land from Smith and Garrett. Such lease was recorded in the office of the county clerk of Coal County, February 11, 1953. On December 18, 1952, Tidewater procured a written assignment of such lease from McDaniel and on February 11, 1953, recorded such assignment in the office of the county clerk of Coal County.

On March 2, 1953, Welborn demanded that Tidewater release the McDaniel lease, on the ground it constituted a cloud on the oil and gas lease held by Welborn.

It is well settled that a remainderman may not make an oil and gas lease to permit immediate exploration and production without the consent of the life tenant. Likewise, a life tenant cannot drill new oil or gas wells, or lease the land to others for that purpose. A life tenant and the remainderman may lease the land by a joint lease and they may agree as to the division of the rents and royalties.[33]

---

[33]   Amarex, Inc. v. Sell, 566 P.2d 456, 58 O.&G.R. 557 (Okl.1977), construes the provisions of the Oklahoma Statutes [53 Okl.Stat. (1971), §§ 71–73] providing for appointment of a trustee to execute an oil and gas lease on the application of the life tenant where there are contingent remainders. The statutory method was held exclusive only as to the interests of contingent remaindermen not in being. Leases which had been granted by the life tenant and the contingent remaindermen in being were valid. For an analysis of the impact of the decision on Oklahoma practice see Professor Kuntz' comments in Discussion Notes, 58 O.&G.R. 565. [Eds.]

In the absence of such agreement, the life tenant is not entitled to any part of the royalties, but is entitled only to the income from such royalties.

The oil and gas lease to Welborn was wholly ineffective to permit Welborn to go upon the land during the existence of the life estate and explore for, develop, or produce oil and gas from the land.

The records in the probate court of Coal County clearly showed that Smith was the owner of the life estate and Garrett of the remainder interest. As such guardian, Smith could only give a lease on the remainder interest of Garrett. Smith as owner of the life estate did not individually, expressly consent to the developing of the land or the production of oil and gas therefrom under the Welborn lease. The lease was made pursuant to a judicial sale. A sale by a guardian under an order and confirmation of the probate court in Oklahoma is a judicial sale and the rule of caveat emptor applies to such a sale.

It follows that the most that Welborn acquired was a contingent right to go upon the land and develop it for oil and gas and produce oil therefrom after the death of the life tenant, in the event that death occurred prior to the expiration of the primary term of Welborn's lease. That primary term has now expired.

It follows that Welborn had no title which could be slandered by a lease executed jointly by the life tenant and remainderman.

While we doubt that it can be said that Smith is estopped to deny that she consented individually to the lease to Welborn, we refrain from passing on that question, because to so adjudicate would adversely affect the rights of Smith as life tenant, and she is not a party to the instant action. She would be an indispensable party to an action which would seek an adjudication that she is estopped to deny that she consented to the lease to Welborn and the immediate development of the land for oil and gas under such lease and during her life tenancy.

The judgment is affirmed.

NOTES

1. In GLASS v. SKELLY OIL CO., 469 S.W.2d 237, 39 O.&G.R. 307 (Tex.Civ.App.1971, error ref'd n.r.e.), a will provided: "I give to each of the . . . life tenants, during the period of their life tenancies, the right and power to lease said lands for oil, gas, and mineral production. . . . Each of said tenants shall collect, receive and retain as their own respective property all bonuses, delay rentals and royalties accruing from production . . . in proportion to the acreage so leased." The life tenants were held entitled to keep the bonus paid for an oil and gas lease. After their deaths, would their heirs or devisees be entitled to royalty payments?

2. In GASKILL v. UNITED STATES, 238 Kan. 238, 708 P.2d 552 (1985), the court, in response to a question certified by the Tenth Circuit Court of Appeals, considered a devise to the testator's wife of, "a life estate . . . with full rights, powers and authorities to sell convey, exchange, lease for oil and gas and

otherwise . . . dispose of any part or all of said estate during her lifetime, all without authority of or order from any Court." The will then went on devise "all of the remainder of my estate . . . unto my children . . . equally . . . in *fee simple.* . . . " The question certified was whether the life tenant had "a right under Kansas law to consume the corpus." The Supreme Court of Kansas answered the certified question, "No:"

> If the power of sale or disposal by a life tenant under the will is merely the ordinary one bestowed on a custodian of property, and no words suggesting that the remaindermen may be compelled to take less than the entire corpus are employed, the will is construed to permit the life tenant to have only the income derived from the life estate.

Would giving the life tenant the power to dispose of the mineral estate extinguish any justiciable interest by the remaindermen in challenging the life tenant's retention of the royalties? Singleton v. Donalson, 117 S.W.3d 516 (Tex.App. 2003).

3. If the life tenant and the remainderman join in the execution of a lease but do not agree upon the division of the proceeds, how should the proceeds be divided?

(a) HASKELL v. WOOD, 256 Cal.App.2d 799, 64 Cal.Rptr. 459, 27 O.&G.R. 709 (1967):

> It appears to be well established in the United States that where a mine or well producing rents or royalties was not open at the time of the death of the owner of the preceding estate of inheritance, or the opening of such mine or well has not been specifically authorized by him, in the absence of a contrary intention manifested by the instrument creating an estate for life or a different result required by a statute governing the subject, rents or royalties derived therefrom are principal, to be conserved for the remainderman, rather than income, i.e., interest derived from such rents or royalties. . . .

> Defendant . . . contends that . . . the Principal and Income Law (Civ.Code, § 730 et seq.) is governing.

> Professor Bogert assumes that the law is applicable only to trust. He also notes that the majority of courts, where not controlled by statute, hold that a life tenant is not a trustee for the remainderman, even though he owes the latter duties which are to a certain extent similar to the obligation of a trustee.

> The facts of this case negate the applicability of this law. This is not a case involving a trust or other similar fiduciary obligation. Rather, it concerns a family business transaction. The father loaned his son money to purchase land. Land was later purchased, and to satisfy his debt, the son conveyed back to the father a life estate.

> We therefore hold that . . . the life tenant is the owner of and entitled to

receive all income or interest derived from the monies on deposit or subsequently received pursuant to the oil and gas lease. Upon the termination of the life estate the corpus of the monies, which is the property of the reversioner, shall then be paid to her or to her successors or assigns.

(b) In TOLER'S ESTATE, 174 Cal.App.2d 764, 345 P.2d 152, 11 O.&G.R. 309 (1959), the court rejected a contention that the income from oil and gas leases given by an executor having the power "to lease, encumber and sell" should be "credited 95% to principal and 5% to income":

> The amounts received from oil leases, as far as the record shows, did not result in the finding of oil or gas or the withdrawal of any from the property. Its real effect was mere ground rent as distinguished from natural resource taking. Thus, there could have been no depletion of the corpus and there appears to us no reasonable cause to distinguish between the allocation of net income from the farm operations and the income from the oil lease.

Would it matter whether the amounts discussed were "bonus" or "delay rentals?" *See also* VanAlstine v. Swanson, 164 Mich.App. 396, 417 N.W.2d 516 (1987), *app. denied,* 430 Mich. 885 (1988).

(c) WHITTINGTON V. WHITTINGTON, 608 So.2d 1274, 123 O.&G.R. 244 (Miss. 1992). The Mississippi Supreme Court in a 4-3 decision interpreted various deeds to create a life estate and vested remainder in a 1/2 of 1/8 royalty. It then applied the "conventional" rule that in the absence of the application of the open mine doctrine, or specific language in the deeds to the contrary, the life tenants are only entitled to receive interest from the invested royalty payments and are not entitled to receive the corpus of such payments. The dissent argued that the clear intent of the parties showed that the life tenant was to receive the corpus, not merely the interest. Both the majority and the dissenting opinion treated the various instruments as unambiguous, although they gave diametrically opposed interpretations of that same language.

(d) HYNSON V. JEFFRIES, 697 So.2d 792, 136 O.&G.R. 595 (Miss.App. 1997). Decedent in his will established a trust and provided that all "of the income of the trust shall be paid to my said wife during her lifetime . . ." At his death, there were producing oil and gas properties. Under the will, should the royalties be paid to the widow life tenant? Or does "income" refer only to the interest earned by royalty, which should be treated as corpus? Should the common law open mine doctrine, taken up in the next case, apply? The court concluded that the provisions of the Uniform Principal and Income Act adopted in Mississippi (Miss.Code Ann. §91-17-1 et seq.) controlled the distribution of royalty payments to a life tenant and the remaindermen and superseded common law rules.

Compare SUCCESSION OF WEISS, 687 So.2d 1084, 136 O.&G.R. 61 (La.App. 1997). The Louisiana court concluded that the language of the will clearly evidenced an intent on behalf of the testator to distribute all of the royalty income to trust beneficiaries, rather than the 27-1/2% authorized by La.

Trust Code §2152.

4. *Trustee's Powers to Lease.* The power of a trustee to permit development for minerals of premises held in trust turns on several factors. If the trustee's estate is limited in duration to less than a fee—as in the frequently occurring situation of a life estate held in trust followed by a legal remainder interest—the trustee has no greater power than a legal life tenant to permit the development of the premises, absent enlargement of his powers by statute or some provision in the creating instrument. If, however, the creating instrument is construed as giving the trustee an estate in fee simple (that is, the life estate and the future interests are both equitable interests), the trustee has full power to permit the development of the premises for oil and gas.

Even though the trustee's estate may be less than a fee, in a number of cases he has been held to have power to develop or lease the premises by virtue of an express power to develop or lease granted by the creating instrument or by the implication of such a power from a granted power of sale, or from other language in the creating instrument. Absent such express or implied authority, his sole relief is by the intervention of a court of equity or by a court acting under a statutory grant of power where there is danger of drainage.

Where a trustee has developed or leased premises, the problem then arises as to the appropriate method of division of the proceeds among the successive equitable interests under the trust. The common-law rule is the same one applicable to successive legal interests; that is, normally the rentals go to the income account of the trust for the current beneficiary and the balance of the proceeds of development are attributed to the corpus of the trust, and the current beneficiary is entitled only to the income from such invested sums. Rarely have the proceeds been divided between the income and corpus on the basis of the computed values of the equitable current and remainder interests. By statute in a number of states, specific provision has been made for this situation. The typical statute provides that 27½ % of the royalties shall be attributed to the corpus account and the balance shall be attributed to the income account under the trust. The figure of 27½% was obviously borrowed from the pre–1970 Federal tax provisions concerning depletion allowances; later reduction of the depletion allowance to 22% and then to 15% has not as yet been followed by a similar reduction in attribution to corpus under all such statutes. Whatever the appropriateness of this percentage for depletion purposes, it is difficult to see that it has any specific relevance to the problem of division of the proceeds of development between the income and corpus accounts of a trust estate. If the life beneficiary has a considerable expectancy, during his early life this arbitrary rule gives him less than he would be entitled to receive under the more rational rule of calculating the relative values of the life estate and future interests; where the life beneficiary has a relatively short expectancy, this rule gives him a far greater share than he would appear to be entitled to. The rule may be justified on the basis of simplicity of administration, but if so, does not the same desideratum of simplicity of administration apply in the case of legal life estates and remainders?

See Patrick Martin & Bruce Kramer, *Williams and Meyers Oil and Gas Law* §§ 514–514.1.

MOORE v. VINES

Supreme Court of Texas, 1971.
474 S.W.2d 437, 41 O.&G.R. 82.

STEAKLEY, JUSTICE. This suit was brought by parties[34] holding remaindermen interests under the joint will of Ruby N. Vines, deceased, and Troy C. Vines, to obtain a construction of the will and a declaration of their rights as to two tracts of land located in Van Zandt County, Texas, and identified as Tract A and Tract B. Defendants to the action, and respondents here, are Troy C. Vines, the life tenant and independent executor of the decedent's estate; Pan American Petroleum Corporation, Service Pipe Line Company and four assignees through assignments originally made by Troy C. Vines.[35] The question with which we are concerned is whether the open mine doctrine is operative as to Tract A. The courts below answered in the affirmative; our view is otherwise.

The full facts of the case are available in the opinion of the court of civil appeals and only those matters material to our determinations will be stated. Troy and Ruby Vines were married in 1931. Thereafter they acquired the two tracts in question. On March 26, 1951, they executed an oil and gas lease covering both tracts. Upon a subsequent divorce in 1953, Ruby was awarded Tract A as her separate property and Troy was awarded Tract B as his separate property. After remarrying in 1958, they executed a joint will in August of 1959, which provided:

. . .

> Upon the death of either of us, it is our will that all the property owned by both of us, both real and personal and both separate and community go to the survivor to be used by such survivor during his or her life time.

> The survivor is hereby appointed independent executor of this will and is directed to act without bond.

> Upon the death of both of us the separate real estate owned by Troy Vines shall go to L.R. Vines for his life time and the separate real estate owned by Ruby Vines shall go to her heirs.

> In case L.R. Vines shall die before Troy Vines does then Troy Vines real estate at the deth (sic) of the survivor shall go to Harold David Moore, grandson of Ruby Vines. The said Harold David Moore to have said land at

---

[34]   Hoyal B. Moore and Harold David Moore, son and grandson of decedent by a previous marriage; Vida Aleen Moore Handley, daughter of decedent by the prior marriage, and her husband, joined pro forma, were involuntary plaintiffs.

[35]   Loyd Russell Vines, Harlie L. Clark, Elise M. Young and Stanton L. Young.

death of L.R. Vines. . . .

Ruby Vines died in October, 1959. The will was probated and Troy was appointed independent executor. The oil and gas lease which had been executed by Troy and Ruby in 1951 remained in effect during its ten year primary term upon timely payment of delay rentals. No production was obtained and the lease expired in March of 1961.

On April 3, 1961, Troy Vines individually executed and delivered to Pan American Corporation an oil and gas lease covering Tract B, his separate property. . . . Troy Vines was entitled to execute the lease upon his separate property and to enjoy the royalties which resulted therefrom.

On June 1, 1961, Troy Vines, individually and as independent executor, joined by the remainderman, Hoyal B. Moore, and the involuntary plaintiffs,[36] executed an oil and gas lease to Pan American Corporation covering Tract A. This tract was the separate property of Ruby Vines at her death. Production was timely obtained upon Tract A, and Pan American paid $64,500 in royalties to Troy individually and as independent executor; he has expended all but $2,150.

On October 3, 1963, Vida Aleen Handley, joined by her husband, made a conveyance by deed to Troy Vines of all her one-half interest in the royalties accruing from production on Tract A. This portion, to be paid during Troy's lifetime, remained his regardless of the applicability of the open mine doctrine to the tract. The issue with which we are concerned is the correct disposition of the remaining one-half interest in the royalty proceeds from Tract A, as between Troy, the life tenant, and the remaindermen.

Ordinarily a life tenant who dissipates the corpus of an estate is liable to the remaindermen for waste. Waste is defined as "permanent harm to real property, committed by tenants for life or for years, not justified as a reasonable exercise of ownership and enjoyment by the possessory tenant and resulting in a reduction in value of the interest of the reversioner or remainderman." American Law of Property, Vol. I, ¶ 2.16e. Mineral royalties and bonuses are part of the corpus and the life tenant is entitled only to the interest thereon. The open mine doctrine forms an exception to the general rule. Clyde v. Hamilton, 414 S.W.2d 434 (Tex.Sup.1967). Blackstone expressed the original doctrine in this manner: "To open the land to search for mines of metal, coal, § c., is waste; for that is a detriment to the inheritance: but if the pits or mines were open before, it is no waste for the tenant to continue digging them for his own use; for it is now become the mere annual profit of the land." 2 W. Blackstone, Commentaries *282. *See also* Woodward, The Open Mine Doctrine in Oil and Gas Cases, 35 Texas L.Rev. 538 (1957). We have held that the doctrine is invoked where wells were producing under a lease executed by the testator at the time the life estate came into being, and where producing wells were drilled after the vesting of the life estate but under an oil and gas lease in force and effect at the time.

---

[36]   Vida Aleen Moore Handley and her husband, joined pro forma.

Youngman v. Shular, 155 Tex. 437, 288 S.W.2d 495 (1956). It is said that the fiction will be indulged that the wells were opened by the testator or settlor where they were drilled under authority of an oil and gas lease executed by them; or, as it has been otherwise expressed, the drilling of wells under authority of an existing lease is the equivalent of an open mine. *See* 1 Kuntz, Oil and Gas, § 8.2 (1962). It is further stated in this treatise that there is uncertainty as to the result if the lease in existence at the time the life estate began thereafter terminates. The comment is made that solution of the problem turns on whether emphasis is given to the factor that wells are drilled under authority of an existing lease which is the equivalent of an open mine, it being apparent that such authority for opening mines or drilling wells no longer exists after termination of the outstanding lease; or if emphasis is given to the intention of the grantor that the land be dealt with as he had dealt with it, from which it would follow that the life tenant would be entitled to enjoy the proceeds of a subsequent lease.

The Supreme Court of Oklahoma in Lawley v. Richardson, 101 Okl. 40, 223 P. 156, 43 A.L.R. 803 (1924), reached these conclusions:

> An examination of the foregoing cases, together with the authorities holding that a life tenant has no right to lease the lands for oil and gas purposes or to open new mines on the property where the opening of the same is not authorized under a contract executed prior to the death of the owner, discloses that a life tenant takes the land in the condition in which it was when the estate vested in him, and that he is entitled to all of the rents and profits which may accrue from the lands by reason of minerals which may be produced from mines or wells existing at the time of the death of the testator, or which may be produced from mines or wells opened under authority of conveyances executed prior to the vesting of the life estate. The life tenant has no authority by his own acts to obtain a profit on income from the land which would result from an injury to the inheritance, but he is entitled to the income and profits from the land when it is produced by reason of conditions which have been fixed by the deceased prior to his death, although the production of the same may result in an injury to the inheritance.

Among the cases examined was Andrews v. Andrews, 31 Ind.App. 189, 67 N.E. 461 (1903) where the mineral lease in question was executed before the beginning of the life estate. In upholding the right of the life tenant to the rents and profits thereunder, the Court emphasized that the life tenant had not attempted to grant any new right; that the additional wells were not opened by any authority from the life tenant; that the wells were opened by virtue of a lease without reference to the wishes of the life tenant; and that the opening of the new wells under the lease was, in practical effect, the act of the testator. This same emphasis is present in Cherokee Const. Co. v. Harris, 92 Ark. 260, 122 S.W. 485 (1909) where it was said that a life tenant is not entitled by an original act of his own to obtain from the land any profit that would result in an injury to the inheritance.

The Court in Heyser v. Frankfort Oil Co., 316 F.2d 441 (10th Cir.1963), considered a comparable problem in this manner:

> Unquestionably, the occupant of an Oklahoma probate homestead, like the holder of a life estate in lands, takes the land in the condition in which it was when the estate vested. . . . The historical rule against waste by a life tenant or a homestead occupant seems to militate against the opening of new mines or the drilling of new wells, after the vesting of the estate and to confine new exploration to authority expressed or clearly implied at the time of the vesting of the estate.

The better view, in our opinion, is that the open mine doctrine is not applicable beyond the lease in existence at the time of the vesting of the life estate of Troy Vines, *i.e.,* beyond the last leasing act of Ruby Vines. The rights of Troy Vines in such respect rested on this lease and expired upon its termination. The lease executed after the vesting of the life estate was not the equivalent of an open mine at such prior time. Troy Vines as life tenant was not authorized by the will of Ruby Vines to lease the land for mineral development nor was he given enjoyment of the proceeds from any such lease. Under these circumstances we are unable to attribute an intent to Ruby Vines that the land should continue to be leased for mineral development for the benefit of Troy Vines with a resulting diminishment in the value of the interest of the remaindermen.

We are otherwise in agreement with the conclusions of the court of civil appeals. For the reasons stated, however, we reverse the judgments below and remand the cause to the trial court for further proceeding consistent with this opinion.

It is so ordered.

MCGEE, JUSTICE. I respectfully dissent.

In addition to the facts relied upon by the majority, one other fact is significant. It was stipulated by the parties that a prior oil and gas lease was executed by Troy C. Vines and wife, Ruby Vines, to Fred Herschback, dated March 12, 1943. The record reflects that the land was from 1943 until the death of Ruby Vines under an oil and gas lease. Hence it was clearly the intent of both Troy and Ruby Vines to procure production from the land almost from the time of its acquisition in 1941. Their *intention* was not merely to lease the land for oil and gas, but to secure production as soon as possible.

The common law regarding the open mine doctrine has been interpreted through a long line of cases. The key word in these and cases following is *intention*—intention determined at the death of the testator.

. . .

The open mine doctrine developed from the principle that the intention of the creator of the life estate controls. Under this doctrine, if the owner of a tract opens a mine, through lease or production before he creates the life estate, unless a contrary intention appears in the instrument creating the life estate, it should be presumed that the creator of the life estate intended to allow the life tenant to continue the exploitation of the mine and receive the royalty income therefrom.

The rights of the life tenant must be ascertained at the time of vesting of the life estate. Tract A was subject to an oil and gas lease at the time of Ruby Vines' death and the vesting of Troy Vines' life estate. Mitchell v. Mitchell, 157 Tex. 346, 303 S.W.2d 352 (1957); Comment, The Open Mine Doctrine, 8 Hous.L.Rev. 753 (1971); Guittard, Rights of a Life Tenant in Production and Use of Oil and Gas, 4 Tex.B.J. 365 (1941).

. . .

The Vines' joint will contained no expressed intention with regard to the distribution of royalties as between the life tenant and the remaindermen. In Clyde v. Hamilton, [414 S.W.2d 434 (Tex.1967)], we said:

> An exception to the rule that a life tenant is entitled to nothing but interest on the royalties and bonus is found in the 'open mine' doctrine. At common law, if mines or pits were open at the time the life estate began, it was not waste for the life tenant to continue digging for his own use; and the proceeds were regarded as a profit from the land. When the settlor of a trust or a testator had opened the mine, and he gave no directions as to the impounding or expenditure of the proceeds from the mine, the law presumed an intent that the life tenant could expend or dispose of them as this settlor or testator could. So the proceeds did not become part of the corpus to be preserved for the remaindermen. . . . " 414 S.W.2d at 439.

The law is clear in Texas that the creation of an oil and gas lease prior to the vesting of a life estate, absent language in the deed, will, or trust agreement expressing a different intention of the creator of the life estate, will give rise to the application of the open mine doctrine. This has been shown above. This is true even though production is obtained after the death of the creator of the life estate. The question here raised is the same as that posed in Mitchell v. Mitchell, 298 S.W.2d 236[Tex.Civ.App.1957, rev'd on other ground as above noted, 157 Tex. 346, 303 S.W.2d 352 (1957)]: "Must the royalty come out of the very lease made by the testatrix in order to constitute income payable to the life tenant?" 298 S.W.2d at 248. I agree with the answer given in that opinion and the comment of Professor Woodward on that opinion in 35 Texas L.Rev. 538:

> A second tract of 150 acres was leased to Sun Oil in 1935. The lease was in force in 1940, when the testatrix died, by virtue of payment of delay rentals. It continued in force until 1945, the end of the primary term, and then expired. Then in 1946 the testamentary trustees executed a new lease on this tract to Gainer. Thus was presented one of the questions posed by Judge Garwood in the *Shular* case: Does the open mine doctrine apply where a lease was in force at the time the creator of the life estate died, but thereafter expired without production, and a new lease to a different lessee is executed by the trustees: The court concluded that it does. It stated the issue in this

way: 'Whether the trustees' lease should be considered as opening a new mine in land to which the "open mine" doctrine has previously had no application, or as being only a renewal, continuing in existence a previous condition or use of the land?' It was reasoned that if production has been going on under the lease executed by the testatrix at the time of her death, there could be no question, but must the royalty issue out of the very lease executed by the testatrix? The Court examined the Kentucky case of Daniels v. Charles. There hard minerals were involved. The husband and wife executed a contract to give a lease for coal mining operations two years before the husband died. This contract was never performed and expired under its own terms about a year before the husband died. Some seven years later the wife made a lease to the same mining company that had contracted to take a lease during husband's lifetime. The court said that had the contract been executed by the husband the lease would still be in force at the time the wife executed the lease, so that the execution of the lease would have been in performance of the contract, the lease could have been attributed to the act of the husband. In such a case the wife, by virtue of her dower rights, would have been entitled to payment of 1/3 of the royalties. However, since the contract executed by the husband had expired prior to his death, there was no room for application of the open mine doctrine.

The Texas court (in Mitchell) rejected this reasoning as applied to the facts under consideration. *It began with the fundamental premise that the life tenant's rights are fixed at the moment that his life estate comes into being, and not at some later date.* (Emphasis added.)

The following considerations are also useful in a case such as this: (1) the lease to Pan American executed by Troy C. Vines on Tract A was executed within a few months of the expiration of the lease to Cities Service in which Ruby Vines had joined; and (2) the record does not indicate that Tract A could have been put to some better or different use which would prohibit oil or gas production when the lease to Pan American was executed, so there is no reason to believe that Ruby Vines' earlier manifested intention would have been different when the second lease was executed. Comment, The Open Mine Doctrine, supra at 764–765.

The judgments of the trial court and the Court of Civil Appeals should be affirmed.

DENTON, J., joins in this dissent.

On Motion for Rehearing

STEAKLEY, JUSTICE. Petitioner, Hoyal B. Moore, has filed a Motion for Rehearing urging that we render judgment in his favor against Respondent, Troy C. Vines, for the sum of $32,251.51, with interest thereon from February 27, 1970, and that this cause be remanded to the trial court solely for the purpose of permitting the determination of the amount of royalties presently held in suspense by Pan American Petroleum Corporation, and for appropriate orders for the investment of these funds during the lifetime of Vines. Petitioner says that "This money [the $32,251.51] should be paid to him now and not at the time of the death of Troy C. Vines." The argument is that the life estate of Vines in the royalties paid to and expended by him were forfeited for waste; otherwise, of course, interest on this sum will continue to accrue to Vines as life tenant. *See* Clyde v. Hamilton, 414 S.W.2d 434 (Tex.1967). There is no suggestion of the existence of statutory authority for forfeiture of a life estate under such circumstances.

It appears to be generally recognized that forfeiture of a life estate, whether of the entire estate or limited to the thing wasted, rests upon specific statutory enactments. It has been further said that forfeiture of the estate for life for waste is a drastic sanction which is granted sparingly and then strictly within statutory authority for such forfeiture. Restatement of Property, § 152 (d), comment f, at 503 (1936).

Apart from the absence of statutory authority, we are of the view that in any event the conduct of Vines under the circumstances was not so unconscionable as to support a forfeiture of his life estate to the extent sought by Petitioner.

The Motions for Rehearing of Hoyal B. Moore, Petitioner, and of Troy C. Vines, Respondent, are overruled.

NOTES

1. WHITE v. BLACKMAN, 168 S.W.2d 531 (Tex.Civ.App.1942, error ref'd w.o.m.), involved the claim of a surviving spouse to a homestead interest in premises upon which there were 41 producing oil wells at the death of the decedent. Although concluding that it was bound by authority to apply the open mine doctrine and give the surviving spouse all royalties accruing subsequent to the death of the decedent, the court evidenced its reluctance as follows:

> The application of the 'open mine theory' to the facts here will result in such a depletion and impairment of the corpus as to leave it as a skeleton or ghost for the remainderman, for the production from the wells, if continued, will appropriate all the oil from the land prior to the end of this widow's life expectancy (19.5 years). This result, a gross injustice to the married daughters (remaindermen), a testator would be helpless to limit or prevent.

For purposes of the open mine doctrine should a distinction be drawn between so-called "legal" life estates created by operation of law (*e.g.,* dower and curtesy, homestead) and so-called "conventional" life estates created by deed or will? *See* Davis v. Bond, 138 Tex. 206, 158 S.W.2d 297 (1942); Youngman v. Shular, 155 Tex. 437, 288 S.W.2d 495, 5 O.&G.R. 1069 (1956); and McGill v. Johnson, 799 S.W.2d 673 (Tex.1990).

2. Prior to the adoption of the Mineral Code [La.R.S. Title 31] effective January 1, 1975, Louisiana decisions made it clear that "[t]he usufructuary of land is not entitled to any of the economic benefits flowing from execution of mineral leases or production thereunder where the leases were executed and the

production secured subsequent to the creation of the usufruct. However these cases did not answer many questions such as: when has a well or mine been 'actually worked' within the meaning of the code; if one well has been drilled at the time the usufruct is created what is the right of the usufructuary to production from multiple sands within the well, from deeper sands subsequently penetrated, or from other wells; what are the rights of the usufructuary of a mineral right; what are the rights of a naked owner to use the surface for mineral operations; what are his obligations and liabilities for surface use? Articles 188–196 provide solutions to these and other problems." Comment to Article 188, Louisiana Mineral Code.

In brief, the Code adopts the open mine doctrine for conventional usufructs of land (in the absence of express provision in the instrument creating the usufruct) and for legal usufructs of land. One who has the usufruct of a mineral right, as distinguished from the usufruct of land, is entitled to all of the benefits of use and enjoyment that would accrue to him if he were the owner of the land. If the usufruct does not include the mineral rights or the benefit of the open mine doctrine, the naked owner of the land has all of the rights in minerals that he would have if the land were not subject to the usufruct but the naked owner's enjoyment of these rights is subject to reasonable user restrictions.

3. Certain of the numerous problems concerning development of minerals arising out of the division of the fee ownership between a life tenant and the owner of a future interest have been considered above. Where the future interests are of a tenuous character (*e.g.,* contingent remainders, executory interests) the problems are of greater difficulty. Statutes in a number of states authorize the appointment of a receiver to lease land which is subject to these tenuous varieties of future interests.

Other types of division of the fee ownership into successive interests raise problems of less complexity.

Where the fee is divided between the owner of a nonfreehold possessory interest (*e.g.,* estate for years) and the owner of a future interest, there has been some controversy over the right of the latter to develop mineral resources without the consent of the former. Clearly the owner of the possessory nonfreehold interest is not entitled to produce minerals or to share in any production from the premises, absent an express authorization, but to what extent may the owner of the future interest interfere with the enjoyment of the land by the owner of the possessory interest by drilling operations? As a matter of practice, before operations are pursued, the owner of the future interest attempts to obtain the consent of the owner of the possessory interest to the operations. The latter has such a bargaining position that he may be able to exact a substantial consideration for his consent. The consent of the tenant to the drilling operations is usually evidenced by a "Tenant's Consent Agreement," which recites that for and in consideration of a stated sum, the tenant gives "his consent and approval of all the rights, interests and privileges stipulated in the within and foregoing oil and gas mining lease and the full use and exercise thereof by the lessee, or lessee's successors and assigns, subject to the condition, however, that any and all damages or loss sustained by any of his crops and other property on said premises as a result of the exercise of said rights under said oil and gas mining lease, shall be paid him on the basis of the actual cash value thereof."

If the land has been leased or minerals severed by conveyance prior to the creation of the possessory interest of the tenant, the latter is subordinate to the rights of development of the premises, and so much of the surface as may reasonably be necessary for the development of the minerals may be employed by the operator or lessee.

"Where ownership is divided between a possessory defeasible fee interest and some future interest (*e.g.,* possibility of reverter), the weight of modern authority permits the owner of the defeasible fee interest to develop or lease the land without the joinder of the owner of the interest subsequent thereto and without the necessity of accounting to the owner of the future interest for any share of the proceeds of development of leasing. Thus if land were granted 'so long as used for Church purposes,' the owner of such premises may develop or lease the land and is entitled to the entire proceeds of such development or leasing unless and until there is a cessation of use of the premises for Church purposes. When creating instruments are inartistically drafted, difficult construction problems arise in determining (a) whether a life estate or a defeasible fee was created thereby; or (b) whether a possessory estate in fee simple defeasible was granted or reserved or a mere right of surface user, viz., an easement or, in Louisiana, a servitude." Martin & Kramer, *Williams and Meyers Oil and Gas Law* § 515.

4. It would not be unusual for the parties to grant a life tenant the right to execute leases without joinder of the future interest owners. Likewise it would not be unusual for a grantor/donor to expressly provide for the allocation of royalty and the other economic benefits between the present and future interest owners. It would even be possible for a grantor/donor to authorize the owner of a term mineral interest to lease the minerals for a period in excess of the term. See RLM Petroleum Corp. v. Emmerich, 1995 OK 50, 896 P.2d 531, 132 O.&G.R. 340; Steger v. Muenster Drilling Co., 134 S.W.3d 359 (Tex.App. 2003).

C. SURFACE-MINERAL SEVERANCE: HEREIN OF SPLIT ESTATES

See *Martin and Kramer,* Williams and Meyers Oil and Gas Law §§ 218-218.15

It is universally accepted that the owner of a unified fee simple absolute estate encompassing both the surface and mineral estate may "sever" all or a portion of the mineral estate by deed. *See e.g.,* Bodcaw Lumber Co. v. Goode, 160 Ark. 48, 254 S.W. 345 (1923); Lathrop v. Eyestone, 170 Kan. 419, 227 P.2d 136 (1951); Barker v. Campbell-Ratcliff Land Co., 1917 OK 208, 64 Okla. 249,

167 P. 468; Humphreys-Mexia Co. v. Gammon, 113 Tex. 247, 254 S.W. 296 (1923). The deed or lease severing the mineral estate may provide for the creation of express easements that give the severed mineral owner specifically defined use rights that will burden the surface estate. *See e.g.*, Columbia Gas Transmission Corp. v. Zeigler, 83 Fed.Appx. 26 (6th Cir. 2003); Smith & Marrs, Inc. v. Osborn, 2008-NMCA-43, 180 P.3d 1183; Landreth v. Melendez, 948 S.W.2d 76, 137 O.&G.R. 170 (Tex.App. 1997). Where there are such express provisions, the scope of the easement of surface use will be dependent upon the language used. In many severance deeds, however, the parties did not provide for an express easement of surface use. Courts nonetheless invariably found that in the absence of an express easement there would be created an implied easement of surface use for the benefit of the mineral estate. See Bruce M. Kramer, *The Legal Framework for Analyzing Multiple Surface Use Issues*, 44 Rocky Mtn. Min.L.Fdn.J. 273 (2007) which explores the historical development of the implied easement of surface use, including judicial avoidance of many common law implied easement requirements in order to find an implied easement of surface use. Professor Kramer posits the theory that there are two basic approaches to determining the scope of the implied easement of surface: the first is the "reasonably necessary" or uni-dimensional approach which focuses on the actions of the owner of the mineral estate; the second is the "reasonable accommodation" or multi-dimensional approach that looks at both the acts of the mineral estate owner along with the injury that may be caused to the surface estate.

The following cases reflect two different approaches to resolving the question of what is the scope of the implied easement or surface use.

SUN OIL CO. v. WHITAKER

Supreme Court of Texas, 1972.
483 S.W.2d 808, 42 O.&G.R. 256.

MCGEE, JUSTICE. . . .Earnest Whitaker is the owner of the surface estate and Sun Oil Company is the owner of a mineral leasehold estate in a 267-acre tract of land in Hockley County. Sun acquired its lease on the property on April 5, 1946, from L. D. Gann and his wife, then the owners of the fee title subject to an outstanding non-participating one-sixteenth free royalty in the west one-half of the tract. The surface estate was conveyed by Gann and his wife to Whitaker on January 2, 1948. The conveyance to Whitaker was subject to Sun's lease, and the deed expressly excepted and reserved all minerals that might be produced from the land to the Ganns, their heirs and assigns.

Sun's lease has been kept alive beyond the primary term of five years by production from eight oil wells which are producing from the San Andres formation. When production from its oil wells decreased because of diminishing pressure in the San Andres formation, Sun obtained permission from the Railroad Commission to inject fresh water into the San Andres in furtherance of a pressure maintenance program. Whitaker and his son-in-law, Doyle Henderson, are using water from the Ogallala formation for cultivating the land as an irrigated farm.

. . . Sun sought a permanent injunction enjoining the defendants from interfering with its production of not more than 100,000 gallons of fresh water per day, through an existing supply well, from the Ogallala formation underlying Whitaker's tract of land for use in producing the oil. By cross-action Whitaker sought to enjoin Sun from producing and using the fresh water to produce the oil. Whitaker also sought to recover actual damages for the water theretofore used and for crops destroyed, and, as well, exemplary damages. The case was tried to a jury and based upon the jury's verdict, judgment was rendered that Sun take nothing by its suit, that Whitaker recover the sum of $12,598.03 for actual and exemplary damages, and that Sun be permanently enjoined from producing and using the fresh water for its waterflood program. The court of civil appeals affirmed. 457 S.W.2d 96. Judgments of the courts below are reversed and judgment is rendered that the permanent injunction prayed for by Sun is granted, and all relief sought by Whitaker is denied except Whitaker is to recover the sum of $431 which has been tendered into court by Sun.

Sun's lease grants and leases the 267-acre tract to Sun "for the purposes of investigating, exploring, prospecting, drilling and mining for and producing oil, gas and all other minerals. . . . " The lease also provides: *"Lessee shall have free use of* oil, gas, coal, wood and *water from said land except water from Lessor's wells for all* operations hereunder. . . . "

The evidence shows that the water produced from Sun's well is being produced from the only available source of water on the land and that such water in being used exclusively for the benefit of the leased premises, the so-called Gann-Whitaker tract. Efforts to use available salt water, other than that produced with the oil, have failed. The waterflood operation will result in the production of additional oil, valued at $3,200,000. The evidence further shows that the Sun water supply well is equipped so that it cannot produce in excess of 100,000 gallons of water per day. Sun's water supply well is located 3,138 feet from Whitaker's water supply well on this lease.

The defendants stipulated at this trial that (1) "the waterflood process is a reasonable and proper operation for the production of oil from the San Andres Reservoir under the L.D. Gann tract"; (2) the use of "Ogallala water as the extraneous or make-up water for injection into the San Andres Reservoir under the L. D. Gann tract in conducting secondary recovery of oil by a waterflood process" is a reasonable and proper operation; and (3) "the location of the injection wells and the rates of water injection" as conducted by Sun "constitute reasonable and proper operations for the production of oil." There is, therefore, no fact issue in the case concerning the stipulated matters.

Sun relies on two legal theories upon which it bases its claim to use the Ogallala water, from its own water wells, under the Gann lease to waterflood wells on this lease. (1) As owner of the dominant estate by virtue of its oil and gas lease it has the implied right as a mineral lessee to use such part of the surface and so much thereof as may be necessary to effectuate the purposes of the lease, and

(2) it possesses an expressed contractual right to "*free use of . . . water from said land except water from Lessor's wells for all operations* hereunder. . . . "

In affirming the trial court's judgment, the Court of Civil Appeals dealt with the case as though it involved only the second of Sun's theories; the Court held that the meaning of the quoted language authorizing free use of water from the Whitaker land was doubtful and ambiguous when applied to the subject matter of the contract, and that evidence introduced on the trial supported jury findings that the parties to the lease did not contemplate or intend that large quantities of water would be used for waterflood purposes. We need not decide whether the opinion of the Court of Civil Appeals is sound inasmuch as we are satisfied that Sun has the implied right to free use of so much of the water in question as may be reasonably necessary to produce the oil from its oil wells.

The oil and gas lessee's estate is the dominant estate and the lessee has an implied grant, absent an express provision for payment, of free use of such part and so much of the premises as is reasonably necessary to effectuate the purposes of the lease, having due regard for the rights of the owner of the surface estate. The rights implied from the grant are implied by law in all conveyances of the mineral estate and, absent an express limitation thereon, are not to be altered by evidence that the parties to a particular instrument of conveyance did not intend the legal consequences of the grant.

The implied grant of reasonable use extends to and includes the right to use water from the leased premises in such amount as may be reasonably necessary to carry out the lessee's operations under the lease.

. . .

Water, unsevered expressly by conveyance or reservation, has been held to be a part of the surface estate. Fleming Foundation v. Texaco, Inc., 337 S.W.2d 846 (Tex.Civ.App.1960, writ ref'd, n.r.e.). However, that decision expressly recognized the right of the oil and gas lessee to drill water wells on said land and to use water from such wells to the extent reasonably necessary for the development and production of such minerals. The added language in the instant case that Sun was to have "free use of . . . water from said land except water from Lessor's wells for all operations" under the lease added no limitation on the implied grant except that such water should not be taken from lessor's wells.

. . .

. . . Sun has an implied right to waterflood because the waterflood operation is reasonably necessary to carry out the purposes of the lease. The reasonableness of Sun's waterflood operation stands uncontradicted in this record. Its use of Ogallala water for injection was approved by the Railroad Commission. The stipulations are conclusive under this record that the use of Ogallala water was reasonably necessary to *effectuate the purposes of the lease.*

. . .

We have concluded that there is no evidence to support the jury's finding that it is not "reasonably necessary" for Sun to use the water underlying the Whitaker farm for its waterflood project. As pointed out above, efforts to use available salt water for the waterflood project have failed, and there is no other source of usable water on the leased Whitaker tract which is available to Sun. To hold that Sun can be required to purchase water from other sources or owners of other tracts in the area, would be in derogation of the dominant estate.

. . .

Judgments of the courts below are reversed and judgment is hereby rendered granting Sun's application for permanent injunction enjoining Whitaker from interfering with its production of not more than 100,000 gallons of fresh water per day, through its supply well, from the Ogallala formation underlying the Whitaker-Gann lease tract for use in producing oil and denying all relief sought by Whitaker except that Whitaker recover the sum of $431 which had been tendered into the registry of the court.

Dissenting opinion by DANIEL, J., in which GREENHILL, STEAKLEY and DENTON, JJ., join.

DANIEL, JUSTICE (dissenting). I respectfully dissent. I adhere to that portion of the majority opinion which restates the general rule that, unless otherwise provided by contractual provisions, an oil and gas lessee has an implied right or easement to make such use of underground water as may be reasonably necessary for ordinary and customary primary drilling and producing operations. However, I completely disagree with the majority's extension of this "implied easement" doctrine so as to permit Sun also to take, consume and deplete an enormously greater quantity of water for a vastly different repressuring and secondary recovery process in the face of jury findings that the parties to the lease did not contemplate or mutually intend such use; that it would materially affect the supply which the surface owner could produce by wells for irrigation of the surface; that such use was not reasonably necessary; and that it will substantially reduce the value of the surface for agricultural purposes.

The majority fails to consider the vital distinction between the occupancy and use of the surface, including water, for ordinary drilling and production operations which do not substantially consume, diminish or destroy the surface estate and the relatively new, extraordinary and far more extensive taking of fresh water for injection into oil sands in a manner which substantially destroys and diminishes the surface estate. Water flooding is not an ordinary primary production method; it is an extraordinary and extraneous medium usually employed after ordinary primary operations have terminated.

Underground water is part of the surface estate, and unless severed by reservation or conveyance, it belongs to the owner of the surface. Whitaker, purchaser of the surface estate in 1948, had drilled wells to the Ogallala formation, a closed and isolated underground fresh water reservoir, encountered at a depth of approximately 200 feet, and this is his only source of water for domestic and irrigation purposes. Whitaker was using the water for agricultural irrigation for 5 years before Sun sought to use the same source of water for water flooding its oil sand.

. . . Although water from other sources was admittedly available by purchase, Sun proceeded to drill a 200 foot well into the Ogallala fresh water sand on Whitaker's land, from which it pumps water at the rate of not to exceed 100,000 gallons per day without the consent of or compensation to Whitaker. The fresh water is injected by pressure pumps through the two injection wells into the San Andres sand at an average depth of 4770 feet, where it mixes with the salt water and increases the reservoir pressure. According to Sun, the use of this "extraneous medium" will cause the wells on the Whitaker tract to yield an additional one million barrels of oil worth about $3,000,000. For this purpose, Sun proposes to consume a total of 4,200,000 barrels of Whitaker's Ogallala water, which will shorten the life of Whitaker's water supply by at least eight years. Whitaker owns only the surface estate and therefore receives no royalty from the oil production.

. . .

Contract Provisions

Sun correctly argues that, so long as no violation of law is involved, the parties to its lease were free to contract as they pleased; its lessor could have granted Sun the right to consume and deplete the entire Ogallala water sand and the entire surface estate of the Gann-Whitaker tract for secondary water flooding and repressuring if the lessor had so elected to contract or convey. On the other hand, any such right to destroy or substantially diminish and consume the surface estate should be clearly spelled out in the contract and not be implied from general provisions relating to substantially non-consuming and non-destructive occupancy and uses of the surface. . . .

. . .

If Sun had contemplated or intended for its lease to cover free use of water for secondary repressuring of the subsurface oil sands, it could have made this known to its lessor and so contracted in specific language, by use of a provision which the record shows to have been placed in Sun leases on other tracts as early as 1947, in the following words:

Lessee shall have free use of oil, gas, coal and water from said land except water from Lessor's wells for all operations hereunder, *including repressuring, pressure maintenance, cycling, and secondary recovery operations* . . .

In my opinion, the trial court, based upon the jury findings, and the Court of Civil Appeals have correctly decided against Sun's claim to an express grant of the right to take the fresh water under Whitaker's land for repressuring the San Andres sand by water flooding. . . .

. . .

Implied Easement

For the same reasons stated above with reference to the alleged express grant, it should be held that Sun is not entitled under the general grant of the minerals to an implied right to take, consume, and deplete the fresh water sand on the Whitaker land without compensation to Whitaker. The holding of an implied grant, irrespective of the existence or terms of an express grant, is far reaching, regressive, and without direct precedent since the days when all minerals were royal patrimonies owned by the sovereign crown or state. . . .

The rule in Texas that a severed mineral estate is dominant and the surface estate servient had its genesis in the Spanish law under which the King held separate ownership of all minerals under both private and public lands. As with all royal patrimonies, the sovereign's separate and severed mineral ownership on private lands rendered the surface estate servient and subject to any use the King might find necessary to mine for and produce the minerals on or beneath the lands of his subjects.

By the Constitution of 1866, the State released and relinquished all mines and minerals to the owners of the soil. Subsequent court decisions, while recognizing the dominant easement of private owners or lessees of severed minerals, have gradually required more regard for the servient surface estate than was previously required under the "dominio radical" (ultimate or basic ownership) of the King or the sovereign state.

This Court has led the way in working out accommodations which preserve unto the severed mineral owner or lessee a reasonable dominant easement for the production of his minerals while at the same time preserving a viable servient estate. This is particularly important in a State whose most productive resources are oil and agriculture, both of which depend heavily upon declining sources of water. Texas' 177,221 oil wells produced more than 1.2 billion barrels of oil in 1970, with a value in excess of $4 billion, and with jobs furnished for nearly 100,000 people. Approximately 41% of these wells were producing with the aid of some type of secondary recovery process. Total farm income for 1969 was in excess of $3.4 billion and furnished employment for nearly 300,000 persons. More than half of the total value of Texas crops is estimated to be produced through irrigation of 8,206,249 acres in 146 counties, which have a total of 83,115 irrigation wells. In addition, by 1967 there was an estimated 272,000 fresh water ponds, lakes and reservoirs of less than 200 acre-feet of storage capacity covering approximately 457,000 surface acres. Section 5.140 of the Texas Water Code permits these reservoirs to be constructed and the impounded water to be used for domestic and livestock purposes without obtaining permits from the Texas Water Rights Commission. The State has expressed its concern for conservation and protection of these small surface reservoirs and the ground waters of the State through Acts of the Legislature.

The majority opinion today applies directly to 4,200,000 barrels of fresh water to be taken without compensation to the owner of only one 267 acre irrigated farm, whose usefulness will be diminished and shortened for a period of at least eight years. However, the rationale of the opinion is so far reaching that it could be applied to any and all of the ground waters of the 8,206,249 acres now under irrigation which underlie acreage covered by similar oil and gas leases, if and when all or a substantial portion of such waters can be shown to be reasonably necessary for injection into a producing oil stratum for repressuring through a secondary water flooding process. The majority is establishing a precedent for untenable conflicts in which the dominant estate may not only properly occupy and use the surface and reasonable amounts of water for primary drilling and production operations but also completely consume and destroy the surface owners' fresh water supply without compensation even if there is no express grant of free use of water for operations under the lease. No such right of total consumption of a fresh water supply and its destruction of surface uses for agricultural purposes should be implied in any oil and gas lease which is silent on the subject.

In *Getty* [Getty Oil Co. v. Jones, 470 S.W.2d 618 (Tex.1971)] we recognized, once again, that a severed mineral estate carries with it a dominant easement for the purposes of the mineral lease, but our opinions on original submission and rehearing should have dissipated any idea that the right of the owner of the mineral lease to use of the surface is absolute or unfettered. We not only reemphasized the limitations on the rights of the owner of the mineral lease to uses which are reasonably necessary and which are made with due regard for the rights of the surface owner, but we also held that the selection of a use which unduly interferes with existing uses by the surface owner, if there are reasonable alternatives available to the mineral lessee, will be held unreasonable and, consequently, not reasonably necessary to effectuate the purposes of the lease. The question is divided, therefore, into two parts: (1) Does Sun's use of Ogallala water underlying Whitaker's tract unduly interfere with Whitaker's surface use? (2) Does Sun have a reasonable alternative?

The Ogallala formation is a depletable and isolated underground reservoir in which the water is not replenished except by such surface water as may percolate down into the reservoir. It is the only source of water available to Whitaker for domestic and irrigation purposes, and the water was being so used by him before Sun entered upon its waterflood project. Sun proposes to produce from its single well and to consume 4,200,000 barrels of the water, which use will shorten the life of Whitaker's water supply by at least eight years. This will cause a substantial decrease in the value of the land. Sun admitted that it "would be physically possible to purchase water to use in waterflooding the L. D. Gann lease involved in this case", and one of its expert witnesses testified that in his opinion it would be economically feasible to purchase water for water flooding the lease; that in his opinion the cost would be about $42,000.00 for the necessary 4,200,000 barrels; and after paying for the water and the oil royalty there should be a net value of $2,000,000 in additional oil recovered from the water flooding project.

In the present case, there was evidence of probative value in support of the existence of a reasonable alternative practiced by operators in the area by which Sun could obtain water at an economically feasible cost for the conduct of its secondary recovery operations without consuming Whitaker's water and diminishing the value and utility of his surface without compensation, and this supports the jury finding that "it is not reasonably necessary for Sun Oil Company to use water from the Ogallala formation underlying the Whitaker farm to waterflood the L. D. Gann lease."

The jury findings compel the conclusion that Sun's use of 4,200,000 barrels of the Ogallala water underlying the Whitaker tract (1) constitutes an undue interference with Whitaker's reasonable use of the surface, and that (2) Sun has a reasonable alternative. The judgments of the courts below correctly denied Sun's prayer for an injunction and correctly awarded injunctive relief and damages to Whitaker.

McFARLAND v. TAYLOR

Arkansas Court of Appeals, 2002
76 Ark.App. 343, 65 S.W.3d 468, 158 O.&G.R. 209

BIRD, JUDGE. Appellants, Wayne McFarland and Phillip Pittman, are the lessees of mineral rights in land owned by appellee Bennie Taylor and his neighbors in Union County. Since acquiring their leases in 1998, appellants used a road across appellee's land for access to a well that was located on land to the west of appellee's property. In 2000, appellee blocked this road. Rather than use another road, appellants filed suit for an injunction directing appellee to remove the obstruction across the disputed road. After the chancellor refused to issue an injunction, appellants filed this appeal. Because the chancellor did not abuse his discretion in denying appellants' petition for an injunction, we affirm.

After living nearby for about twenty years, appellee purchased this tract of land in 1983 for the purpose of building a house there. Appellee built a house, but it was destroyed by fire before he could occupy it. According to appellee, there were no wells on the adjoining land at that time but that, soon afterward, the oil well to the west was constructed and he gave its operators permission to use his road, which extends from Arkansas Highway 275 to the western boundary of appellee's property. Different oil-well operators used the road over the next decade. At trial, appellee testified that he gave express permission to use his road to all operators of the well but that he cautioned them that their use could continue only until he withdrew his permission. At some point, another road leading from Highway 15 to the well was built.

In 1998, appellants obtained assignments of the mineral leases and began using the Highway 275 road. Appellee testified at trial that, as before, he gave appellants conditional approval to use this road until he withdrew his permission. Appellants also made some improvements to the Highway 15 road.

In 1997 or 1998, appellee's son and daughter-in-law, Brent and Chelsea Taylor, and their small daughter, moved into a mobile home on appellee's land. Mr. and Mrs. Taylor testified that there is a significant amount of traffic at all hours of the day and night on the Highway 275 road, which is used as their driveway, and that they were concerned about the safety of their three-year-old daughter. Mrs. Taylor also testified that she was worried about her own safety, because her husband works the 3:00 p.m. to 11:00 p.m. shift. Appellants did not dispute that they use the Highway 275 road at night. These concerns prompted appellee to withdraw his permission for appellants to use the Highway 275 road. His blocking of the road prompted appellants to file this lawsuit.

Although appellants testified that the Highway 15 road could not be used as an alternate route without much improvement at great expense, appellee presented evidence to the contrary. Gordon Height, an engineer with the Arkansas Highway and Transportation Department, testified that the Highway 15 road could be adequately improved for between $ 1,000 and $ 1,500. Sam Jean, who has twenty-five years' experience in "dirt work" and who has previously worked on the Highway 15 road, testified that he could make that road usable for heavy trucks for no more than $ 1,500. In his letter opinion, the chancellor stated that he was impressed with Mr. Jean's "experience, his knowledge, and his forthrightness."

The chancellor made the following findings in his letter opinion, which was incorporated in his order denying appellants' petition for an injunction:

> If the issue before me was whether it was reasonable for the [appellants] to use the Highway 275 road once or twice a month for an eighteen wheeler and/or a workover rig, I would have no difficulty in concluding that such a limited amount of use would be reasonable. However, taking into account the testimony from the July hearing as to the considerable amount of traffic over the road at all hours of the day and night and taking into account that workover rigs will have to visit the well sites in order to make the wells operational and keep them operational, and taking into account that Highway 15 road can be made fully usable for not more than $ 1,500.00, it is my conclusion that the most reasonable ingress and egress for [appellants] to their oil and gas properties is from Arkansas State Highway 15.

. . .

Appellants argue that the chancellor erred in considering whether it would be reasonable to use an alternate route across appellee's land because it was not used as a mobile-home site before appellants began production under their lease. According to appellants, a comparison of the reasonableness of using an alternate route can be made only when there is a pre-existing use by the surface owner. Appellee responds that appellants are factually and legally incorrect. We agree with appellee.

We are not prepared to hold that, as a matter of law, a mineral owner is always entitled to choose between two or more means of access to the minerals, without regard to necessity or to the harm it may cause the surface owner, if the surface owner's use did not predate the mineral owner's use. The respective rights of mineral and surface owners are well settled. The owner of the minerals has an implied right to go upon the surface to drill wells to his underlying estate, and to occupy so much of the surface beyond the limits of his well as may be necessary to operate his estate and to remove its products. Diamond Shamrock Corp. v. Phillips, 256 Ark. 886, 511 S.W.2d 160 (1974). His use of the surface, however, must be reasonable. Id. The rights implied in favor of the mineral estate are to be exercised with due regard for the rights of the surface owner. See id. (citing Getty Oil Co. v. Jones, 470 S.W.2d 618 (Tex. 1971)).

In Martin v. Dale, 180 Ark. 321, 21 S.W.2d 428 (1929), the Arkansas Supreme Court made it clear that, in all circumstances, the mineral owner's use must be necessary and the potential harm to the surface owner must be considered:

> It is not questioned that Lenz, as agent for the trustee to whom the lease was given, had the right of access to the lands covered by the lease; but this is a right which arose out of necessity, and not as a matter of convenience. In other words, while the right of entry was implied, this right did not authorize Lenz to enter as he pleased; it was his duty to do so in the manner least injurious to his grantor, and if a means of ingress existed when the lease was taken, and which continued to be available, this entry, and no other, should have been used, although it was not the most convenient.

180 Ark. at 324, 21 S.W.2d at 429.

In any event, appellee demonstrated that the road in question had been used for residential purposes for many years. Appellants have apparently based their argument on the incorrect factual assumption that the Highway 275 road was built as part of the oil and gas operations near appellee's land. Appellee, however, testified that this road has been used for decades as access to a barn and a potato shed. He said that he has been familiar with this land since 1965 and that this road had been used many years before there were any oil wells on the adjacent land. Appellee also testified that, when he bought his land in 1983 for the purpose of building a house there, the road was not being used for access to an oil well. Appellee further stated that appellants and their predecessors had used the Highway 275 road with his permission. He said he had informed them that they could use it until he told them "to quit." Clearly, appellee's testimony would support a finding that this road had been in residential use for many years before the nearby oil production began.

Therefore, the chancellor was correct in considering whether it was necessary for appellants to use the Highway 275 road and whether it would be reasonable to require them to use the alternative Highway 15 road. He found that the "most reasonable ingress and egress for [appellants] to their oil and gas properties is from Arkansas State Highway 15." Generally, what is reasonable is a question of fact. Although we review chancery cases de novo, we will not reverse a chancellor's finding of fact unless it is clearly erroneous. In light of the testimony credited by the chancellor and discussed in his letter opinion, his finding of fact in this regard is not clearly erroneous.

Based on the evidence presented, we cannot say that the chancellor abused his discretion in denying injunctive relief to appellants.

Affirmed.

NOTES

1. Can the approach taken by the Texas Supreme Court in *Whitaker* be reconciled with the approach taken by the Arkansas Court of Appeals in *McFarland*?

2. In Getty Oil Co. v. Jones, 470 S.W.2d 618, 39 O.&G.R. 657 (Tex. 1971), the Texas Supreme Court seemingly adopted the multi-dimensional, reasonable accommodation, doctrine when determining the scope of the implied easement of surface use. The court in *Getty Oil* required the lessee to accommodate the surface owner's use of center-pivot irrigation systems that could not operate with the wellhead equipment placed on the surface. Even though it would require the oil and gas lessee to expend funds in excess of the amount needed to safely and efficiently produce the hydrocarbons, the court found that "where there is an existing use by the surface owner which will otherwise be precluded or impaired, and where under established practices in the industry, there are alternatives available to the lessee whereby the minerals can be recovered, the rules of reasoanble usage of the surface may require the adoption of an alternative by the lessee. 470 S.W.2d at 622.

3. If the jurisdiction has caselaw utilizing both the "uni-dimensional" and "multi-dimensional" tests what kind of jury instruction should be given by the trial judge. *See* Trenolone v. Cook Exploration Co., 166 S.W.3d 495 (Tex.App. 2004); Davis v. Devon Energy Production Co., 136 S.W.3d 419, 160 O.&G.R. 897 (Tex.App. 2004).

4. Occasional cases have found an excessive or negligent user of surface easements by a mineral owner or lessee, but such instances are relatively rare. *See e.g.,* Brown v. Lundell, 162 Tex. 84, 344 S.W.2d 863, 14 O.&G.R. 611 (1961), imposing liability for negligently allowing salt water, produced from a well and disposed into an open pit, to escape from the open pit by percolation and seepage into the subsurface stratum so as to pollute the underground water supply, thus permanently impairing the value of the land:

> We agree that the owner-operator of the lease has the right to use so much of the land, both surface and subsurface, as is reasonably necessary to comply with the terms of the lease contract and to carry out the purposes and intentions of the parties. . . . It does not follow, however, that the operator may use either the surface or the subsurface in a negligent manner so as to damage the landowner.

5. Even under the reasonably necessary test, the lessee may not engage in

activities that are unreasonable, excessive or negligent. In Oryx Energy Co. v. Shelton, 946 S.W.2d 637, 135 O.&G.R. 618 (Tex.App. 1996), the court suggested that even in the absence of a finding of negligence a mineral lessee may be liable to the surface owner for damages caused by excessive use as evidenced by oil spills and leaks and abandoned pipes.

6. One use that is treated as being per se excessive is where the use of the surface of Blackacre is for the benefit of the mineral estate under Whiteacre. In oil and gas leases there may be provisions relating to surface use that do allow the lessee to use the surface for the benefit off other mineral estates. The court in Caskey v. Kelly Oil Co., 737 So. 2d 1257, 143 O.& G.R. 32 (La. 1999) found that language of a lease allowing use of the land for operations "on any adjacent lands"' was not limited to adjoining premises also owned by the lessor. An "adjacent lands" clause promotes the efficient development of oil and gas fields and the state's public policy of developing mineral resources, and numerous commentators have recognized the validity of a contractual clause granting easements or surface rights in the leased property in connection with operations on other premises. In Salvex, Inc. v. Lewis, 546 So.2d 1309, 104 O.&G.R. 493 (La.1989), writ denied, 551 So.2d 1323 (La.1989), the lessor sought a servitude of pipeline passage from an "enclosed lease" over land of the lessee no longer under lease. The court allowed a statutory servitude of passage by road to the nearest public road but found a servitude of pipeline passage beyond the contemplation of the Civil Code. In Pittsburg & Midway Coal Mining Co. v. Shepherd, 888 F.2d 1533 (11th Cir.1989), the court approved as a statement of law the proposition that "in Alabama, as well as in every other jurisdiction, a simple severance without express grants, gives to the mineral owner no right to use the surface of one tract to aid in the mining of another tract, even though the miner owns the minerals under both." The court went on to find in the documents before them "specific grants to use the surface" in relation to operations on other lands which were sufficient to allow the uses that would not have been available by implication based on mere severance of the minerals from the surface. In Kysar v. Amoco Production, 73 Fed.Appx. 349, 160 O.&G.R. 585 (10th Cir. 2003), *on certified question to New Mexico Supreme Court*, 2004-NMSC-025, 135 N.M. 767, 93 P.3d 1272, 160 O.&G.R. 595, *decided*, 379 F.3d 1150, 160 O.&G.R. 623 (10th Cir. 2004), the courts provide an exhaustive review of the rights of the mineral owner to use the surface after a portion of a lease has been pooled.

7. Surface use issues are often very fact-specific. Surface owner installs an irrigation system known as a center pivot system which requires surface clearance in order to operate at peak efficiency. A mineral lessee drills a well and places it in a subterranean cellar so as not to interfere with the surface owner's irrigation system. The same lessee seeks to drill an additional two wells and proposes to construct earthen ramps that will allow the system to operate without having to dig out the cellars. The surface owner alleges the ramp system will alter the water distribution system causing substantial injury. Fasken v. Darby, 901 S.W.2d 591,

134 O.&G.R. 302 (Tex.App. 1995, no writ).

8. More than 10 states have adopted some type of surface owner protection or damages act. While Indiana was the first state to enact such a statute, it was North Dakota in 1975 that enacted the first comprehensive statute. N.D.Cent. Code §§ 38.11.1-01 to 38.11.1-10. In addition, the National Commissioners of Uniform State Laws drafted the Model Surface Use and Mineral Development Accommodation Act dealing with both hardrock mining and oil and gas operations in 1990. The Model Act's history and development is explored in Wenzel, The *Model Surface Use Act and Mineral Development Accommodation Act: Easy Easements for Mining Interests*, 42 Am.U.L.Rev. 607 (1993).

There is not a uniform approach taken in the several states that have adopted some form of surface use legislation. Colorado does little more than legislatively adopt the reasonable accommodation doctrine that had been previously adopted by the Colorado Supreme Court in Gerrity Oil & Gas Corp. v. Magness, 946 P.2d 913, 138 O.&G.R. 1 (Colo. 1997). Colo.Rev.Stat. § 34-60-127. The Texas statute is even more limited because it only applies to subdivision developers who may designate potential oil and gas well sites when they file their subdivision plat. Tex.Nat.Res.Code Ann. §§ 92.001-92.007 (West 2010). The states which have most recently adopted such statutes have typically imposed upon the mineral owner the duty to make payments for surface use rights regardless of what easements they may have. N.M.Stat.Ann. §§ 70-12-1 et seq. (2007); Wyo.Stat.Ann. §§ 30-5-401 et seq. Under the New Mexico, North Dakota and Wyoming statutes, a strict liability regime replaces the "reasonable accommodation" common law regime. *See generally*, John Wellborn, *New Rights of Surface Owners: Changes in the Dominant/Servient Relationship Between the Mineral and Surface Estates*, 40 Rocky Mtn.Min.L.Inst. 22-1 (1994); Lowe, *Eastern Oil and Gas Operations*, 4 Eastern Min.L.Inst. 20-18 (1983); Pearce, *Surface Damages and the Oil and Gas Operator in North Dakota*, 58 N.D.L.Rev. 456 (1982).

The Oklahoma Surface Damages Act was interpreted in Houck v. Hold Oil Co., 1993 OK 166, 867 P.2d 451, 128 O.&G.R. 93 to apply to wells drilled subsequent to the Act's passage even if the lease was executed prior to the Act. The Act requires willful acts in order for the surface owner to recover treble damages. The Act sets up a damages regime based on the basic concept of diminution in fair market value. *See also* Schneberger v. Apache Corp., 1994 OK 117, 890 P.2d 847, 131 O.&G.R. 587, where the Oklahoma Supreme Court answered a certified question from the U.S. District Court limiting the amount of damages to the diminution in value rather than the cost of remediation where the cost of remediation was grossly disproportionate to the diminution in the value of the surface.

9. Would a change in the relationship between severed mineral and surface owners constitute either a taking of the mineral owner's property interest constituting the owner's implied or express easement to use the surface, or an

impairment of an obligation of contract? In Murphy v. Amoco Production Co., 729 F.2d 552, 81 O.&G.R. 321 (8th Cir. 1984), the court upheld the validity of the North Dakota statute against such constitutional challenges.

10. Musser Davis Land Co. v. Union Pacific Resources, 201 F.3d 561, 145 O.&G.R. 282 (5th Cir. 2000) held that a lease granting the lessee the exclusive right to explore for oil and gas without defining or limiting in any way the term "exploration'" bestowed upon the lessee the right to conduct seismic exploration to determine the presence of subsurface trapping mechanisms favorable to oil and gas production. The court also held that in the absence of a contrary provision in the oil and gas lease or other contract between the parties, the lessee is entitled to the ownership of the seismic data it develops pursuant to its prudent and reasonable geophysical operations incidental to its exercise of the exclusive right to explore and produce oil and gas under the lease. Query: does the lessee owe royalty on money received for sale of seismic data?

D. OTHER DIVIDED INTERESTS
See Martin and Kramer, *Williams and Meyers Oil and Gas Law* §§ 516–519.

———

1. *Interests arising from the marital relationship.*

Entry into the marriage relationship by a man and woman may have the effect of giving one spouse an interest in the realty owned by the other at the time of the marriage or thereafter acquired by the other during coverture. The common law marital interests of the spouse in the realty owned by husband and wife (dower and curtesy) have been severely limited by statute in many states, but in a substantial number of states, these interests (or a statutory substitute therefor) continue to be important. In other states, realty owned by a husband and wife during marriage as their "homestead" may be subject to certain rights of the non-owning spouse.

During marriage, to the extent that the non-owning spouse continues under modern law to have an interest in the realty owned by the other spouse, the joinder of the non-owning spouse in a conveyance of a mineral interest or in a lease is generally required. In the absence of such joinder, the interest granted by the owning spouse may be defeasible in whole or in part upon the survival of the owner by the non-owning spouse, or the non-owning spouse may be entitled to have a share of the proceeds of leasing or development sequestered during the joint lives of the husband and wife pending the contingency of the survival of the owning spouse by the non-owning spouse.

Several of the states in which oil and gas is produced have the community property system whereunder property acquired during marriage other than by gift, inheritance, devise or bequest, or other than that which is acquired with separate funds of one spouse, is owned by the community of husband and wife. Since the husband has considerable powers of management, operation and disposition of community property, he may develop lands owned by the community. Some community property states require joinder of the spouses in a conveyance of community real property while in others the husband alone may convey such property.

2. *Easements, servitudes.*

When the fee simple title to land is burdened by an easement (*e.g.,* a right of way), the owner of the easement is not entitled to share in the proceeds of development of the premises, but such development may be enjoined to the extent that it will interfere with the enjoyment of the easement. In the case of oil and gas production, the fee owner or his lessee may not drill at such a place on the surface as will interfere with enjoyment of the easement, but if he can locate his well on the surface so as not to interfere with the easement, the well may be bottomed on a portion of the tract beneath the right of way. The easement owner himself, of course, has no right to remove the oil and gas.

Difficult problems arise with a degree of frequency concerning the construction of instruments granting a right of way as to whether a mere easement was granted or whether a fee simple was granted in the land included within the right of way.

Development of land burdened by equitable servitudes restricting the use of the land may be enjoined by the persons entitled to the benefit of the servitude where such development violates the restriction. This is of particular importance in residential subdivisions of cities which have been restricted for "residential uses only," where subsequent development indicates that a producing formation underlies the development. Absent agreement of all of the benefited parties, development may be enjoined at the suit of any person entitled to the benefit of the agreement. Drilling activities may, of course, be curbed also by zoning ordinances.

The ownership of oil and gas underlying roads and streets has created substantial problems. *Compare* Village of Kalkaska v. Shell Oil Co., 163 Mich.App. 534, 415 N.W.2d 267, 95 O.&G.R. 545 (1987), aff'd, 433 Mich. 348, 446 N.W.2d 91, 107 O.&G.R. 18 (1989), *with* Town of Moorcroft v. Lang, 779 P.2d 1180 (Wyo.1989). *See also,* Inversiones Del Angel, S.A. v. Callon Petroleum Company, 883 F.2d 29 (5th Cir.1989); City of Shreveport v. Petrol Industries, Inc., 550 So.2d 689 (La. App. 2d Cir.1989).

3. *Security interests.*

Premises which are subject to a mortgage or a deed of trust usually may not be developed by the owner or by his lessee (unless the lease is senior to the security interest) without the consent of the secured creditor. Absent such consent, development of the land for minerals will generally be said to constitute voluntary waste which is enjoinable by the secured creditor or for which he may recover damages, and typically such "waste" will occasion an acceleration of the mortgage debt under the terms of such security agreements.

Before land can be developed by a lessee of the mortgagor, it is usually necessary, therefore, to obtain a subordination agreement from the mortgagee. Such an agreement will normally provide specifically for the allocation of the proceeds of the lease between the mortgagor and the satisfaction of the debt for which the mortgage or deed of trust is security.

A mortgage, deed of trust, or vendor's lien is entitled to priority over any subsequent interests created by the landowner, and such interests may be cut off by foreclosure.

If the premises have been leased prior to the execution of the mortgage or other security interest by the lessor, the royalty or production interests of the mortgagor pass under the later executed mortgage or deed of trust without express mention therein. In the event of foreclosure of the mortgage, mineral interests created before the mortgage was executed (if unaffected by recordation statutes) are not cut off by the foreclosure.

Chapter 3

THE OIL AND GAS LEASE–A CLOSE LOOK
AT ITS MORE IMPORTANT CLAUSES
SECTION 1. THE HABENDUM
AND DELAY RENTAL CLAUSES

(*See* Martin & Kramer, *Williams and Meyers*
Oil and Gas Law §§ 602-602.6, 603-604.12, 631-635.)

A. INTRODUCTION

1.

Dated December 30, 1857
Deed Book P, p. 357
$1 in hand

Pennsylvania Rock Oil Company

   to

E.B. Bowditch and E.L. Drake

Demise and let [all the lands owned or held under lease by said company in the County of Vanango, State of Pennsylvania] to bore, dig, mine, search for and obtain oil, salt water, coal and all minerals existing in and upon said lands, and take, remove and sell such, etc., for their own exclusive use and benefit, for the term of 15 years, with the privilege of renewal for same term. Rental, one-eighth of all oil as collected from the springs in barrels . . . Lessees agree to prosecute operations as early in the spring of 1858 as the season will permit, and if they fail to work the property for an unreasonable length of time, or fail to pay rent for more than 60 days, the lease to be null and void.

2.

The State of Texas
Nacogdoches County

This indenture . . . witnesseth that the said party of the first part have this day agreed to grant to the party of the second part their heirs & assigns the exclusive privilege of mining operations upon a certain tract or parcel of land herein described, to-wit: . . .. [A]nd the said party of the first part for themselves, their heirs, executors & administrators do hereby bargain, sell & convey and release in fee simple to the said party of the second part, their & etc. forever, the rights and privileges upon the said premises for the uses and purposes

*aforementioned without law or hindrance in any manner in and for the consideration hereinafter named to-wit: [a cash consideration of $300] & also for the consideration of the one twelfth part of all products of sd. lands in the way of minerals or oil that may hereafter be saved, procured or found upon said lands aforenamed by said parties, their heirs or assigns free of any expense to the party of the first part & the said party of the second part agree on their part to put the works in operation for mining within five years next ensuing or this shall be void.
. . .]*

3.

In consideration of the sum of one dollars, . . . R.W. Bradford . . . of the first part, hereby grant and guaranty unto the Federal Oil Company, . . . second party, all the oil and gas in and under the following described premises, together with the right to enter thereon at all times for the purpose of drilling and operating for oil and gas, ... The first party shall have the one-eighth part of oil produced and saved from said premises, ... In case no well is commenced within one day from this date, then this grant shall become null and void, unless second party shall thereafter pay at the rate of eight dollars and seventy-five cents ($8.75) for each month such commencement is delayed, in advance. . . .

4.

It is agreed that this lease shall remain in force for a term of three years from this date, and as long thereafter as oil or gas, or either of them, is produced from said land by lessee. . . . If no well is completed on said land on or before the 17th day of January, 1917, [1 year from date of execution of the lease] this lease shall terminate as to both parties, unless the lessee on or before that date shall pay or tender to the lessor, . . . the sum of Three Hundred Eighty and no--100 dollars, which shall operate as a rental and cover the privilege of deferring the completion of a well for twelve months from said date. In like manner and upon like payments or tenders the completion of a well may be further deferred for like periods of the same number of months successively.

5.

That the grantor, . . . has granted and conveyed unto the grantee all the oil and gas in and under the following described tracts of land: . . . To have and to hold unto and for the use of the grantee, his heirs and assigns, for the term of five years from the date hereof, and as much longer as oil or gas are produced in paying quantities, ... Grantee agrees to commence a well on said premises within one year from the date hereof or thereafter pay the grantor as rental $16.50 each three months in advance from quarter to quarter to the end of this term, or until said well is commenced, or this grant is surrendered as herein stipulated, ...

––––––

Although there is great variation in the provisions of oil and gas leases, these five excerpts typify the major types of leases formerly or now in use in the United

States. These excerpts are taken from Moses, *The Evolution and Development of the Oil and Gas Lease*, 2 Sw.Legal Fdn. Oil & Gas Institute 1 at 7 (1951). *See also* on the evolution of the oil and gas lease form, Martin & Kramer, *Williams and Meyers Oil and Gas Law* §§ 601-601.5. The structure of the habendum clause in coal leases usually differs from that found in oil and gas leases and stems from the nature of the industry. *See* Norvell, *The Coal and Lignite Lease Compared to the Oil and Gas Lease*, 31 Ark.L.Rev. 420, 436 (1977). Form 1 is from the lease under which America's first well, the Drake well, was drilled in Titusville, Pennsylvania, in 1859, to the substantial depth of 69½ feet. Form 2 is from the document under which the first commercial producer was drilled in Texas in 1866. The Drake lease was for a long primary term but contained no "thereafter" clause. This form has passed completely from current usage as it was found unsatisfactory by both lessors and lessees. The former desired some assurance of a periodic return from the land in the form of rentals or royalties, and the latter desired to insure that their interest would last so long as oil or gas was produced from the land. Form 2 (a grant of the oil and gas in fee) was also unsatisfactory to lessors inasmuch as it did not provide for a periodic return pending the commencement of production. Form 3 is a "no-term" lease. Some of the reasons for the general abandonment of this form by the industry are indicated in the case and note which follow. Form 4 is an "unless" lease, which is the lease form most commonly employed today. Form 5 is an "or" lease, which was once widely utilized and is still used in California but is only occasionally employed in other states.

———

FEDERAL OIL CO. v. WESTERN OIL CO.

Circuit Court of the United States, D. Indiana, 1902.
112 F. 373.

[Defendants Bradford in 1901 executed and delivered to complainant a lease the essential elements of which are set forth in the third example of lease forms at the beginning of this chapter. On October 22, 1901, complainant mailed to the lessor a check for $8.75 as required by the lease, which the lessor refused to accept, stating that he would no longer be bound by the lease. The lessors executed a lease of the same premises to Western Oil Co. which entered upon the premises and began to drill for gas. Complainant seeks to have its title to the lease quieted and to enjoin defendants from in any manner asserting any right or title adverse or hostile to the leasehold title of the complainant. To this bill the defendants filed their demurrer. Eds.]

BAKER, DISTRICT JUDGE. . . . The only consideration yielded at the time of making the lease was $1 in hand paid by the complainant to the lessors. As will be seen later, there was no binding promissory consideration on the part of the complainant for the execution of the lease. The bill, which is verified, alleges that the leasehold interest claimed to have been acquired exceeds $2,000 in value. The cash payment, if actually made, was merely nominal, and it is quite apparent from a consideration of the terms of the whole lease that the lessors would not have executed it for any such paltry consideration. If there was no further consideration which the lessee was bound to yield to the lessors, a court of equity would be bound to refuse the enforcement of the lease. The consideration would be so trifling, compared with the value of the leasehold interest, as to shock the moral sense. . . . Where, as in this case, the only consideration is prospective royalties to arise from exploration and development, failure to promptly explore and develop the demised premises renders the agreement nudum pactum, and works a forfeiture of the lease, for it is of the essence of such a lease that the work of exploration shall be commenced and prosecuted with promptness. The smaller the tract of land demised, the more imperative is the need of prompt exploration and development, because the lessor is entitled to his royalty as promptly as it can be had, and delay endangers the drainage of the oil and gas from the demised premises through wells in its immediate vicinity. Leases of coal, stone, and other like materials are corporeal hereditaments, and constitute an essential part of the land itself, and are capable of present, absolute grant, while oil and gas are of a fugitive and volatile nature, a grant of either of which creates only an inchoate right, which will become absolute only upon its reduction to possession. A lease to mine for oil or gas is a mere incorporeal right to be exercised in the land of another. It is a profit a prendre, which may be held separate and apart from the possession of the land itself. Except to the extent of $1, the lessee has yielded no consideration for the lease; nor is it bound by any enforceable promise or covenant, for the breach of which the lessors would have a right of action to compel the payment or yielding of any further consideration whatever. The lessors' right to a portion of the oil can only arise when it has been produced and saved on the premises. The lessors have a right to a royalty for gas only when it is found in sufficient quantities to transport, and then only for the product of each well when actually transported. The complainant, it is true, agrees that the second well shall be completed within 90 days after the first well, and that a well shall be drilled each 90 days thereafter until seven wells are sunk; but the complainant does not agree that it will ever drill the first well. Doubtless the lessors expected the exploration and development of the demised premises to commence at once, but the language of the lease is that, "in case no well is commenced within one day from this date, then this grant shall become null and void," unless the complainant should thereafter pay in advance at the rate of $8.75 for each month such commencement is delayed. Thus it is seen that the complainant is under no obligation enforceable at law ever to commence the drilling of the first well, and it is under no obligation to pay for failure to commence or complete the sinking of the first well. The provision that the second well shall be completed within 90 days after the first well is sunk is not enforceable at law, because the complainant is not bound ever to commence or complete the first well. The complainant is under no obligation to pay the monthly rental of $8.75. The lessors could maintain no action to recover the same if the complainant should refuse to continue payment. Such a lease is without consideration, and must be held a nudum pactum and void. A lease so unfair, inequitable, and against good conscience no court ought to enforce.

And for another reason the court cannot enforce this lease: The lease expressly provides that the complainant shall have the right to remove all its property from the demised premises at any time, and may cancel and annul this contract, or any part thereof, at any time. It is a well-settled rule of law that a lease which is determinable at the will of one party is equally determinable at the will of the other party. . . .

And for a still further reason the court must refuse to enforce this lease: The court will not decree that one party shall specifically perform a contract which the other party, at its option, may refuse to carry out. It is of the essence of a decree that it should be mutually binding and conclusive on both parties. It would be an idle formality for the court to enter a decree against the defendants in this case, for the reason that the complainant has the right to render the decree ineffective at any moment that it pleases. . . .

For these reasons the demurrer is sustained and the bill dismissed for want of equity, at the costs of the complainant, and the temporary restraining order heretofore granted is dissolved.

NOTES

1. The instrument involved in the principal case attempted to create a so-called "no term" lease. How would you describe the lease that resulted from the court's decision?

2. Is the no-term lease any different from a lease which includes a "surrender" clause containing the following language:

> At any time or times Lessee shall have the right to surrender this lease as to the whole or any part of the acreage covered by this lease . . .

Under *Federal Oil* would this unilateral right of surrender by the Lessee necessarily imply a concurrent right in the Lessor. Would that render the lease unenforceable? The early case authority was nearly unanimous in finding leases with surrender clauses enforceable. *See e.g.,* Lindlay v. Raydure, 239 F. 928 (E.D.Ky.1917), *aff'd,* Raydure v. Lindley, 249 F. 675 (6th Cir. 1918). *See generally,* Martin & Kramer, *Williams and Meyers Oil and Gas Law* § 680.2.

3. "The 'no-term' lease was heavily attacked in the courts. . . . Even the courts that enforce these leases, however, hold that the lessee can not by the payment of rentals put off drilling indefinitely. Consequently, this lease, like the fixed term lease, fails to provide the requisite certainty as to the rights and obligations of the parties. The amount of litigation over the enforceability of these leases undoubtedly led to their disuse. It appears improbable that any modern lease is intended to be for 'no-term.' Most 'no-term' leases found to have been created in the past two decades probably resulted from mistake or misunderstanding in the completion of the blanks in store-bought lease forms." Martin & Kramer, *Williams and Meyers Oil and Gas* Law § 601.3. An example is

Sidwell Oil & Gas Co., Inc. v. Loyd, 230 Kan. 77, 630 P.2d 1107, 71 O.&G.R. 266 (1981), where a signed lease was sent to the lessee with open blanks for the date of the lease and the date of the first rental payment. The last day of the three year primary term was inserted as the date of the first rental payment. The lease was held void. "[T]here was no meeting of the minds as to [its] material provisions."

4. Can the parties enter into a no-term lease contingent on the payment of minimal rentals? Would the lessor be able to claim that the lessee had breached an implied covenant to develop the leasehold if the lessee made no effort to develop and merely continued to pay rentals? In Dallas Power & Light Co. v. Cleghorn, 623 S.W.2d 310, 72 O.&G.R. 153 (Tex.1981), the Texas Supreme Court, without discussing the validity of a no-term lease, found that there was no implied covenant to develop because the parties had expressly agreed not to require any development by the lessee. *In accord,* Warm Springs Development Co. v. McAulay, 94 Nev. 194, 576 P.2d 1120, 60 O.&G.R. 516 (1978), where the court found that annual rental payments effectively eliminated any implied covenant to develop and kept the lease alive for as long as the lease allowed such payments to be made.

5. By a gas storage agreement, supplementing an old oil and gas lease, the lessor purported to modify and extend the lease and to grant the additional right of "introducing, injecting, storing and removing gas of any kind." The term of this agreement was for ten years "and so much longer . . . as gas is being produced, stored, withdrawn, or held in storage by the lessee" in the storage area. Payments to be made to the grantor under this agreement were $200 per year per well. The ten year term of the agreement has long since passed and the grantor has brought suit seeking a declaration that their land is free of its burdens. Does the law as to "no-term" leases form a useful analogy? *See* Myers v. East Ohio Gas Co., 51 Ohio St.2d 121, 364 N.E.2d 1369, 5 Ohio Op.3d 103, 58 O.&G.R. 341 (1977). The concurring opinion in *Myers* emphasizes that the Agreement at issue is "one of a myriad such made with owners of portions of the so-called Clinton Sands Storage Area, pieced together at the expense of more than $35,000,000 of ratepayers' money and essential to a storage scheme which in the past critical year provided heat for northeastern Ohio homes and businesses when other areas were in short supply." In an earlier case involving another of these storage agreements, the Ohio Court of Appeals considered and dismissed the law relating to "no-term" leases. Rayl v. East Ohio Gas Co., 46 Ohio App.2d 167, 348 N.E.2d 390, 75 O.O.2d 306, 54 O.&G.R. 501 (1975).

6. Modern courts in dealing with no-term leases that may be kept alive through the payment of nominal rentals have couched the analysis in terms of lack of mutuality. In ICG Natural Resources, LLC v. BPI Energy, Inc., 399 Ill.App.3d 554, 339 Ill.Dec. 214, 926 N.E.2d 446 (2010) invalidated a 99-year coalbed methane lease where the lessee had no obligation to explore for and produce CBM. There were no rental payments provided for in the lease. Where

even nominal rental payments are provided for, the courts have found no lack of mutuality and upheld the lease. *See* Northup Properties, Inc. v. Chesapeake Appalachia, L.L.C., 567 F.3d 767 (6th Cir. 2009) and Snyder Bros. v. Peoples Natural Gas Co., 450 Pa.Super. 371, 676 A.2d 112, *app. denied,* 546 Pa. 683, 686 A.2d 1312 (1996). But there are contrary views to maintaining a lease indefinitely through the payment of nominal rentals. Hite v. Falcon Partners, 2011 Pa.Super. 2.

B. THE PRIMARY TERM

1. DELAY RENTALS--THE "UNLESS" LEASE

If no well be commenced on said land on or before one year from the date hereof, this lease shall terminate as to both parties, unless the lessee on or before that date shall pay or tender to the lessor or to the lessor's credit in the _____ Bank at _____, or its successors, which shall continue as the depository for rental regardless of changes in the ownership of said land, the sum of _____ Dollars, ($_____) which shall operate as a rental and cover the privilege of deferring the commencement of a well for twelve months from said date. In like manner and upon like payments or tenders the commencement of a well may be further deferred for like periods of the same number of months successively. All payments or tenders may be made by check or draft of lessee or any assignee thereof, mailed or delivered on or before the rental paying date. It is understood and agreed that the consideration first recited herein, the down payment, covers not only the privilege granted to the date when said first rental is payable as aforesaid, but also the lessee's right of extending that period as aforesaid, and any and all other rights conferred. Should the depository bank hereafter close without a successor, lessee or his assigns may deposit rental or royalties in any National bank located in the same county with first named bank, due notice of such deposit to be mailed to lessor at last known address.

PHILLIPS PETROLEUM CO. v. CURTIS

United States Court of Appeals, Tenth Circuit, 1950.
182 F.2d 122.

PHILLIPS, CHIEF JUDGE. Phillips Petroleum Company brought this action against Joe W. and Lois B. Curtis to quiet title to an oil and gas lease. From an adverse judgment, Phillips has appealed. The question presented is whether, under the facts, which are not disputed, Phillips was entitled under equitable principles to be relieved from a termination of the lease due to its failure to pay the delay rental on or before the time stipulated in the lease.

On October 4, 1946, Joe W. Curtis and Lois B. Curtis gave an oil and gas lease on the SW1/4 SW1/4, S. 12, T. 3 N., R. 3 W., Garvin County, Oklahoma, to James T. Blanton. The lease is what is known in oil and gas terminology as an "unless" lease. It contained this provision:

If no well be commenced on said land on or before the 4th day of October, 1947, the lease shall terminate as to both parties, unless the lessee on or before that date shall pay or tender to the lessor, . . . the sum of Forty and No/100 Dollars . . ..

In like manner and upon like payments or tenders the commencement of a well may be further deferred for like period of the same number of months successively.

Such lease was numbered 51365-A.

At the time they executed the above-mentioned lease, the lessees [lessors? Eds.] owned an undivided 1/3 interest in the minerals in such tract of land. Shortly after they executed such lease they sold an undivided 1/24 interest in such minerals, subject to such lease. Thereafter, under the terms of the lease, the rental was to be apportioned and the amount due the lessees [lessors? Eds.] on October 4, 1948, was $11.67.

On January 22, 1948, Phillips acquired such lease from Blanton by assignment. The delay rentals for October 4, 1947, were duly paid. No well was commenced on any portion of such lease on or prior to October 4, 1948, and no delay rental was paid or tendered to the lessees [lessors? Eds.] or for their credit in the depository bank named in the lease on or prior to October 4, 1948, because of an inadvertent error made by an employee of Phillips. Promptly upon discovering such error, about December 1, 1948, Phillips informed the lessees [lessors? Eds.] thereof and tendered to them the amount of the rental, which they refused. Phillips continued to tender the amount of the rental, with legal interest, but the lessees [lessors? Eds.] refused to accept same.

On January 2, 1946, Mabel M. Arnett executed an oil and gas lease to Phillips, covering an undivided 2/3 interest in the SE1/4 of the NE1/4, the SW1/4 of the SW1/4, the S1/2 of the NW1/4 of the SW1/4, the SW1/4 of the NE1/4 of the SW1/4 and the NE1/4 of the NE1/4 of the SW1/4, all in S. 12, T. 3 N.,R. 3 W., Garvin County, Oklahoma. Such lease was numbered 51365.

On January 2, 1948, Joe W. and Lois B. Curtis executed and delivered to Phillips a lease covering an undivided 1/3 interest in the SE1/4 of the NE1/4, the S1/2 of the NW1/4 of the SW1/4, the SW1/4 of the NE1/4 of the SW1/4 and the NE1/4 of the NE1/4 of the SW1/4, all in S. 12, T. 3 N., R. 3 W., Garvin County, Oklahoma. Such lease was numbered 51365-B.

The two leases from Joe W. and Lois B. Curtis covered a 1/3 interest in the 120 acre tract and the lease from Arnett covered the remaining 2/3 interest in the 120 acre tract.

Phillips paid $200 per acre for leases Nos. 51365-A and 51365-B and $50 per acre for lease No. 51365.

Phillips is a large producer of petroleum products and a large purchaser and holder of oil and gas leases. It handles approximately 36,000 delay rental payments each year. It has set up and maintains a comprehensive and systematic set of records and accounts which is administered by trained and competent personnel for the purpose of servicing its oil and gas leases and paying delay rentals thereon. Under its system errors are exceedingly rare.

On September 15, 1948, an employee of Phillips, in the performance of his duties, made an entry in its lease records that lease No. 51365-A was held by "Production," because of his inadvertent conclusion that a well located on the NW1/4 of the SE1/4 of the NE1/4 of Section 12, which was commenced by Phillips on July 31, 1948, and completed by it on September 9, 1948, as a commercial producer, was a commencement and completion of a well on lease No. 51365-A. The error of such employee did not constitute gross negligence.

Phillips was ready, willing, able and desirous of paying such rental, if necessary, to keep such lease in force and would have timely and properly paid such rental, but for the mistake of its employee.

It will be observed that the error was made by an employee of Phillips, subject to its direction and supervision, and not by an independent agent, such as the United States Post Office Department or a depository bank.

. . .

Under an "unless" lease the lessee has an option to continue the lease in force by paying the stipulated rental on or before the designated date, and it is subject to termination at his will, which privilege he may exercise by a failure to commence a well or pay the rental within the time stipulated in the lease, in which case the lease automatically terminates.

When a contract is optional in respect of one party thereto, it will be strictly construed in favor of the other party; and an "unless" lease is construed "most strongly" against the lessee and in favor of the lessor.

Time is of the essence of the provision in an "unless" lease for the payment of delay rental; and the payment of the delay rental on or before the specified date is a condition precedent, which must be performed in order to continue the lease in force and give the lessee the right to defer drilling operations; and such a lease is automatically terminated at the end of the original term, upon failure of the lessee either to commence a well or to pay the delay rental on or before the specified date, by reason of the agreement of the parties.

While in certain decisions the Supreme Court of Oklahoma has referred to the effect of the failure to pay delay rental on or before the specified date as a forfeiture of the lease, it has expressly held that the effect is not a forfeiture, but merely a termination of the lease in accordance with the agreement of the parties and that the equitable principles with respect to relief from forfeitures have no application; and in Williams v. Ware, 167 Okl. 626, 31 P.2d 567, at page 570, the court expressly held that 23 Okl.St.Ann. § 2, has no application to an "unless" lease which has terminated by failure to pay delay rental within the specified time.

The case of Brunson v. Carter Oil Company, D.C.Okl., 259 F. 656, was decided May 31, 1919, by the United States District Court for the Eastern District of Oklahoma. The facts in the case were these: On May 18, 1916, A.C. Flowers and Cordie Flowers executed and delivered to Carter an "unless" oil and gas lease covering certain lands in Stevens County, Oklahoma. Thereafter, the Flowers conveyed the land to Brunson. On May 5, 1917, Brunson mailed to Carter a copy of his deed from the Flowers. Upon receipt of such deed Carter wrote to Brunson advising him that an entry of the change of ownership had been noted on its rental sheet showing that Brunson had acquired the land and was entitled to the rentals thereafter to accrue. However, through inadvertence and mistake of the clerk charged with the duty of making such notation on the rental sheet, such notation was not made. Carter failed to pay the rental due to Brunson on the lease, May 19, 1917, due to such error. However, on April 15, 1918, Carter prepared a check in the sum of $90, the amount of rental due on the lease May 19, 1918, payable to the order of the First National Bank of Duncan, Oklahoma, for deposit to the credit of the original lessors, and forwarded such check to the bank on April 20, 1918. The bank deposited the check to the credit of the original lessors. The court held that since Carter had paid a substantial consideration for the lease and had clearly manifested its intention to keep the lease alive by payment of rental, a termination of the lease would amount to a forfeiture and that Carter was entitled to equitable relief.

The case of Harvey v. Benmo Oil Co., 272 F. 475, was decided April 11, 1921, by the United States District Court for the Eastern District of Oklahoma. In that case the lessee under an "unless" lease, on June 7, 1920, forwarded a letter and a check to cover the delay rental by registered mail, properly addressed, with postage duly prepaid thereon, to the American State Bank at Beggs, Oklahoma, the depository bank. The delay rental was due June 16, 1920. Through delay in the mails the letter and check did not reach the bank until June 18, 1920. The court held that a termination of the lease on account of the failure to pay the delay rental, under the circumstances, would be in the nature of a forfeiture and that the lessee was entitled to equitable relief.

In two cases, Brazell v. Soucek, 130 Okl. 204, 266 P. 442, and Oldfield v. Gypsy Oil Co., 123 Okl. 293, 253 P. 298, the Supreme Court of Oklahoma held that the lessee under an "unless" lease was entitled to equitable relief against the termination of the lease by reason of the lessee's failure to pay the delay rental within a specified time, and as applied to the particular facts in those cases, approved the principles laid down in the Brunson and Harvey cases. In the Brazell case, the lessee left a check with the depository bank, drawn on such bank, to pay the delay rental, advised the cashier of the bank when the delay rental was due and instructed him to deposit the check to the credit of the lessor. Sufficient funds were on deposit to the credit of the lessee to cover such check. Because of a misunderstanding as to the date when the rental was due, the cashier failed to deposit the check in time.

In the Oldfield case, the lessee forwarded a draft by registered mail to the

Stillwell National Bank, the depository bank, to cover the rental payment in ample time for it to reach the depository bank in the usual course of mail, before the due date of the rental. The postal clerk erroneously delivered the registered letter to the First National Bank of Stillwell, which signed the return registered receipt. Due to such error the draft did not reach the depository bank in time.

It will be observed that in the Brazell and Oldfield cases, the failure of the check and draft, respectively, to be deposited in the depository bank to the credit of the lessor within the time specified in the respective leases, was due to the mistake of an independent agency not under the supervision and control of the lessee and was not due to the mistake of an agent or employee of the lessee in the performance of his duties, acting under the supervision and direction of the lessee.

In Gloyd v. Midwest Refining Co., 10 Cir., 62 F.2d 483, we held that where the lessee, under an "unless" lease mailed a check payable to the order of the lessor in ample time for it to reach him in the usual course of mail, before the delay rental was due, and the check failed to reach the lessor before the due date because of unusual delay in the mails, the lessee was entitled to equitable relief against the termination of the lease.

The Harvey case and the Gloyd case fall in the same category as the Brazell and Oldfield cases.

In no case, so far as we have been able to discover, has the Supreme Court of Oklahoma held that the lessee was entitled to equitable relief from the termination of an "unless" lease by reason of failure to pay the delay rental within the time specified in the lease, where the failure, although unintentional, was caused by a mistake of an employee or agent of the lessee in the performance of his duties, acting under the direction, supervision and control of the lessee.

Moreover, in New England Oil & Pipe Line Co. v. Rogers, 154 Okl. 285, 7 P.2d 638, 641, the court distinguished the Harvey case and the Brazell case on the ground that in those cases the failure of the delay rental payments to reach the lessor on or before the time they were due, was caused by the mistake of an independent agency and was not due to the mistake or negligence of the lessee and expressly repudiated the Brunson case. The court said:

> Harvey v. Benmo Oil Co., *supra,* and Brazell v. Soucek, *supra,* are clearly distinguishable from the instant case, as therein the mistakes, errors, etc., by reason of which the rentals when due did not reach the parties entitled thereto, were not the mistakes or negligence of the lessee . . .. The principle applied in the Brunson Case, *supra,* to the effect that, where down payments for what is termed an 'unless lease' is substantial, and is a consideration for the execution of the lease, and also for the option to renew or extend the term from time to time by drilling or commencing a well, as the case may be, or with a stipulated sum called rental for the privilege of deferring for a definite time the drilling or commencement of a well, the failure to drill or pay works a forfeiture of a valuable right conferred under a valid mutual

contract based on a valuable consideration, has never been applied by this court. On the other hand, this court has many times before and since the Brunson Case held that in such cases, upon failure to drill or pay within the time provided by the option clause, the lease automatically terminates, and that the equitable rules against forfeiture do not apply.

Accordingly, we conclude that under the law of Oklahoma, denial of equitable relief, harsh as it may seem, was proper under the facts and circumstances of the instant case.

Affirmed.

NOTES

1. Does the unless lease create a fee simple determinable estate? What is the future interest that is retained by the lessor if the lessee owns a fee simple determinable? Do fee simple determinable estates automatically terminate upon the occurrence or non-occurrence of the limitation? What is the nature of the limitation under the delay rental provisions of an "unless" lease? If the lessee continues in possession of the premises after the limitation has occurred, what is the lessee's status?

2. The *Curtis* case specified the date upon which rentals were to be tendered if drilling had not commenced. Suppose a lease provides:

This Agreement, made and entered into this 10th day of January, 1974.

If no well be commenced on said land on or before one year from the date hereof, this lease shall terminate, unless, on or before such date, lessee shall pay [rentals] . . ..

Suppose further that the lease is executed on February 15, 1974, and delivered March 20, 1974. Is a tender by February 15, 1975 timely? By March 20, 1975? *See* Hughes v. Franklin, 201 Miss. 215, 29 So.2d 79 (1947); Greer v. Stanolind Oil & Gas Co., 200 F.2d 920, 2 O.&G.R. 229 (10th Cir.1952); Moses, *The Date in the Oil and Gas Lease*, 40 Tulane L.Rev. 97 (1965).

3. Small errors can be fatal. In Young v. Jones, 222 S.W. 691 (Tex.Civ.App.1920), the lessee tendered $73.29 when he should have paid $76.25. *Held,* the "unless" lease terminated.

4. Contemporary leases seem invariably to provide for payment of rentals to a depository bank to the credit of the lessor. The usual practice is for the lessee to mail to the depository bank a check to the order of the bank in advance of the due date of rentals. The check is accompanied by written instructions as to the persons whose accounts are to be credited with deposits. For typical letters of transmittal and accompanying documents *see* Gulf Production Co. v. Perry, 51 S.W.2d 1107 (Tex.Civ.App.1932, error ref'd). Most cases treat the depository bank as the agent of the lessor to receive payment and hence error or dereliction by the bank cannot cause termination of the lessee's interest. *See* Martin & Kramer, *Williams and Meyers Oil and Gas Law*§ 606.5.

The anniversary date of an oil and gas lease was July 1. The depository bank received a draft from the lessee by mail for rentals on June 30 but did not deposit same to the lessor's account until July 7. The lessor contended that the lease had terminated by failure to make timely payment of rentals. What result? *See* Kronmiller v. Hafner, 75 S.D. 439, 67 N.W.2d 353, 4 O.&G.R. 276 (1954).

In Miske v. Stirling, 199 Mont. 32, 647 P.2d 841, 75 O.&G.R. 517 (1982), a time draft given by the lessee to the lessor to cover a "bonus rental" was paid when the "collecting bank," holding the draft as "agent" of the lessor received funds from the lessee and "dedicated" them to payment of the draft.

5. A lease was executed by plaintiff individually and by plaintiff's brother individually and as attorney in fact for other named concurrent owners of the described tract. The lease authorized payment of delay rentals by deposit "to the credit of lessor" in a named bank. Rentals in a correct total amount were paid by deposit in the bank to the credit of plaintiff's brother "Individually and as Attorney-in-Fact for" the named persons for whom he had acted in the execution of the lease and, in addition, for plaintiff. Plaintiff sued to cancel the lease as to his interest for nonpayment of delay rentals. Is a clause declaring that the bank is the agent of the lessor for receipt of rentals a basis for saving the lease? *See* Trad v. General Crude Oil Co., 468 S.W.2d 612, 40 O.&G.R. 93 (Tex.Civ.App.1971), *error dism'd w.o.j.*, 474 S.W.2d 183 (Tex.1971); *cf.* Wagner v. Mounger, 253 Miss. 83, 175 So.2d 145, 22 O.&G.R. 601 (1965).

6. What if the mails fail? Would it matter if the rentals were mailed in sufficient time to arrive? Would modern overnight delivery services render this issue moot? Would the failure of the delivery system, whichever one is chosen, be imputed to the lessor or lessee? *See generally,* Ballard v. Miller, 87 N.M. 86, 529 P.2d 752, 49 O.&G.R. 449 (1974); Corley v. Olympic Petroleum Corp., 403 S.W.2d 537, 25 O.&G.R. 73 (Tex.Civ.App. 1966, writ ref'd n.r.e.).

What if the lease provides that the rental payments are effective upon mailing? In Gillespie v. Bobo, 271 Fed. 641 (5th Cir. 1921), the court held the lease terminated where the lessee made a minor error in the address leading to a late arrival of the delay rental check notwithstanding the leasehold provision making the rental payment effective upon mailing. *See also* Baker v. Potter, 223 La. 274, 65 So.2d 598, 2 O.&G.R. 1072 (1952). Should the result depend on whether the "unless" lease is treated as a fee simple determinable estate? Is the allocation of fault in the misdelivery appropriate for fee simple determinable estates? Should the same rules apply to options to renew or extend a lease where the lessee mails the payment in sufficient time to arrive prior to the deadline but the check is lost in the mail? *See* APC Operating Partnership v. Mackey, 841 F.2d 1031, 98 O.&G.R. 324 (10th Cir.1988).

7. A check for the delay rentals was delivered to the lessor on the due date, August 12, 1946. The check was deposited by the lessor somewhat more than a month later, and when it reached the drawee bank, payment was refused for lack of funds in the lessee's account. The lessee had a small but active account, and

with the exception of four days from the time he gave the check until the time it was presented for payment, he had on deposit sufficient funds to pay the check. A bank officer had agreed to honor the delay rental check when presented even if the lessee's account should be overdrawn thereby. By mistake of the bank, payment was refused despite this agreement. As soon as the mistake was discovered, three days after presentment of the check, the drawee bank telephoned the correspondent bank to correct the mistake, but the check had already been returned to the lessor. Has the lease terminated? *See* Hamilton v. Baker, 147 Tex. 240, 214 S.W.2d 460 (1948).

8. Can lessor stand by, knowing that the proper rental has been deposited to his credit in the wrong bank and is available to him on demand and then successfully claim that the lease is terminated? *See* Carroll v. Roger Lacy, Inc., 402 S.W.2d 307, 25 O.&G.R. 225 (Tex.Civ.App.1966, error ref'd n.r.e.).

9. The Bankruptcy Code of 1978, 11 U.S.C. § 362(2) provides that the filing of a petition "operates as a stay . . . of (3) any act to obtain possession of property of the estate or of property from the estate or to exercise control over property of the estate." Will the filing of a bankruptcy petition by the lessees preserve "unless" leases if rentals are not paid as provided in the leases? *Cf.* In re Trigg, 630 F.2d 1370, 68 O.&G.R. 489 (10th Cir.1980), decided under the Bankruptcy Act of 1898, holding that the lessee's bankruptcy did not stop the bankrupt's lease from terminating automatically on failure to pay rentals. The Bankruptcy Rule considered in the *Trigg* case provided that the filing of a petition . . . "shall operate as a stay of the commencement or continuation of any court or other proceeding against the debtor . . . ." The relationship of sec. 108(b) of the Bankruptcy Code, providing for extension of time "to cure a default," to delay rental payments is discussed in Good Hope Refineries, Inc. v. Benavides, 602 F.2d 998, 1003, 64 O.&G.R. 37, 44 (1st Cir.1979), cert. denied, 444 U.S. 992 (1979): "When a debtor or a trustee fails to exercise or renew an option by paying the agreed price, there is no contractual 'default' to be cured." *See* Baker and Schiffman, *Effect of Bankruptcy Law on Specific Oil and Gas Insolvency Problems*, 35 Sw. Legal Fdn. Oil & Gas Inst. 187 (1984).

10. Article 133 of the 1975 Louisiana Mineral Code provides for termination of an oil and gas lease when an "express resolutory condition" occurs. Such a terminating event would happen under an "unless" rental clause if delay rentals were not properly paid. HARRY BOURG CORP. V. DENBURY ONSHORE LLC, 129 Fed.Appx. 55, 163 O.&G.R. 733 (5th Cir. 2005) (not selected for publication in the Federal Reporter)(well settled in Louisiana law that when the "unless" drilling clause is used the lessee does not covenant to drill or pay. The failure of the lessee to drill or pay a stipulated sum of money ipso facto terminates the lease).

11. Did you understand how the respective leases described the lands that were subject to the lease? Would the use of governmental surveys make Phillips Petroleum culpable in not making the correct payment? Should the culpability of

the lessee be relevant to determine whether the lease should terminate upon the failure to make a timely and accurate delay rental payment? What is the lessee's employee or employees acted in good faith in setting up the payment schedule for delay rentals? If the unless clause creates a fee simple determinable estate that automatically terminates is the good or bad faith of the lessee relevant?

Leases sometime contain provisions, either independent of the delay rental clause, or intertwined with the delay rental clause that read as follows:

> *If oil or gas is not being produced from the leased premises, on the first anniversary date of this lease, this lease shall then terminate as to both parties unless on or before such anniversary date Lessee shall pay or tender (or shall make a bona fide attempt to pay or tender, as hereinafter stated) to Lessors to to the credit of Lessors in the _____ Bank, the sum of _____ (herein called rentals). . . . If Lessee shall on or before the anniversary date, make a bona fide attempt to pay or deposit rental to a Lessor entitled thereto according to Lessee's records or to a Lessor who, prior to such attempted payment or deposit has given Lessee notice, . . and if such payment or deposit shall be ineffective or erroneous in any regard, Lessee shall be unconditionally obligated to pay to such Lessor the rental property payable. . . .and this lease shall be maintained in the same manner as if such erroneous or ineffective rental or deposit had been properly made, provided that the erroneous or ineffective rental payment or deposit be corrected within thirty (30) days after receipt by Lessee of written notice from such Lessor of such error..*

Is this an unless lease if the lease contains such a clause? If so, has a fee simple determinable estate been created? Are the provisions relating to erroneous payments repugnant to the creation of a fee simple determinable estate? Is the language used in the provision language creating a limitation on the estate granted?

KINCAID v. GULF OIL CORP.

Court of Appeals of Texas, 1984.
675 S.W.2d 250, 82 O. & G.R. 356.

CANTU, JUSTICE. This is an action filed by appellants, as lessors of an oil and gas lease covering 25,866.79 acres of land located in Zavala and Uvalde Counties seeking declaratory relief against appellees, lessees and assignees of interests in said lease, based upon appellant's belief that delay rentals had not been timely paid as required by the lease. Appellees are Gulf Oil Corporation, Omega Minerals, Inc., Major Petroleum Corporation, and CRB Oil & Gas, Inc. The case was tried to the District Court of Uvalde, Texas, on an agreed statement of evidence.

The facts which give rise to this appeal may be summarized as follows. On

March 1, 1974, Gulf Oil Corporation (Gulf) entered into two separate leases. The leases both covered Kincaid lands. The first lease covered 25,814.19 acres and was designated by Gulf as internal lease number 212112. The following parties were named as lessors in the lease: E.D. Kincaid, Jr. and William Alex Kincaid, Independent Executors and Testamentary Trustees under the Last Will and Testament of Frank T. Kincaid, Jewel Armstrong Kincaid, widow of Frank T. Kincaid, Jewel Frances Garwood and husband, Roy H. Garwood, Jr., and Elizabeth Ann Maner and husband, James Ray Maner. This lease, for purposes of the opinion, will hereinafter be referred to as the F.T. Kincaid Lease. The second lease covered 25,866.79 acres and was designated by Gulf as internal lease number 212111. The following parties were lessors in the lease: E.D. Kincaid, Jr. and William Alex Kincaid, Individually and as Independent Executors under the Last Will and Testament of E.D. Kincaid, Deceased, and Mrs. Adaline Kincaid, Individually and as President of A.V.K. Ranch Company, Inc. This lease, for purposes of the opinion, will hereinafter be referred to as the E.D. Kincaid Lease. It is the E.D. Kincaid lease which is the subject matter of this appeal.

The aforementioned leases are identical in all respects with the exception of the named lessors and the tract or property descriptions. Both E.D. Kincaid, Jr. and William Alex Kincaid are named as Lessors under both leases; however, in each lease they are acting in somewhat different capacities. Although E.D. Kincaid, Jr. and William Alex Kincaid act in the stated capacities for each estate, the estates are maintained separately and independently of each other. Moreover, the leases cover contiguous tracts of land in Zavala and Uvalde Counties, Texas.

Both the E.D. Kincaid Estate lease and the F.T. Kincaid Estate lease contained an "unless" type clause as well as other provisions dealing with erroneous payment of delay rentals. . . .

[The leases contained a delay rental clause similar to the one that precedes this case. Eds.]

It is noteworthy that both leases provide (1) that First State Bank of Uvalde is the depository bank for payment of delay rentals (2) that delay rentals are due and payable on or before March 1 of each year of the lease term; and (3) that delay rentals are $1.00 per acre. Moreover, it is important to note that the two leases were prepared by an attorney hired by appellants.

From 1975 through 1978, Gulf paid annual delay rental payments under both leases by checks payable to the First State Bank of Uvalde (the Bank). Attached to these checks were credit allocation instructions.

On May 30, 1979, Gulf entered into a letter agreement with Major Petroleum Corporation (Major) covering 15,000 acres which was subsequently assigned to Omega Minerals, Inc. (Omega). CRB Oil & Gas, Inc. eventually acquired all of Omega's interest under this letter agreement. In 1979, pursuant to the terms of the leases, Gulf exercised its option to extend the primary term under the leases for five years.

In January, 1980, Gulf made the decision to pay the delay rental due on the F.T. Kincaid lease and to hold payment of the rental due on the E.D. Kincaid lease. The decision to delay payment was based on conversations between Gulf and Omega that drilling operations were being undertaken which might eliminate the delay rental obligations under the E.D. Kincaid lease.

In February, 1980, a check payable to the First State Bank of Uvalde in the amount of $25,815.19 for the delay rental obligation on the F.T. Kincaid lease was issued by Gulf and received by the Bank for the heirs of the F.T. Kincaid Estate. The lessors under the F.T. Kincaid lease were credited for the payment. E.D. Kincaid, Jr., William Alex Kincaid, E.D. Kincaid, III and Warren Neill, Vice President of Operations of the Bank, were all aware of the delay rental payment on the F.T. Kincaid lease. Thereafter, several inquiries were made by E.D. Kincaid, Jr. and William Alex Kincaid as to whether the delay rental under the E.D. Kincaid lease had been received by the Bank.

On February 29, 1980, Gulf was informed that drilling operations under the E.D. Kincaid lease had ceased. Consequently, Gulf's legal department made the decision to pay the delay rental on the E.D. Kincaid lease on or before March 1, 1980. It was decided that Omega would make the delay rental payment on behalf of Gulf because Omega's agent was closer to Uvalde. Moreover, it was uncertain whether Gulf's bank had the facilities to wire transfer any monies after banking hours or on weekends and holidays. [The court judicially noted that March 1, 1980 fell on a Saturday. Eds.]

Thereafter, Rosamond Whitehead, an Omega employee, was instructed to issue two checks; one in the amount of $25,815.19 and the other in the amount of $25,867.79. She also received instructions to rent a plane, fly to Uvalde, take the checks to the Bank and inform the Bank that she was making the payment on Gulf's behalf. In preparing the checkstubs and the delay rental receipts, Whitehead used the F.T. Kincaid Estate lease for purposes of identifying the lessors who were to receive the payment. It is significant that previously, Gulf had only forwarded to Omega the F.T. Kincaid lease and had not forwarded the E.D. Kincaid lease.

Whitehead thereafter called the Bank and informed Charles Reed, Sr., Vice President of Trusts for the Bank, that an Omega agent would be delivering a delay rental check to the Bank that day on behalf of Gulf for some Kincaid properties. Because Whitehead was concerned that the check would not be delivered until after banking hours, she arranged with Reed to deliver the check to his home after banking hours if necessary.

On the afternoon of February 29, 1980, several calls were made between Gulf and the Bank. The substance of these calls was to confirm that (1) the Bank was aware that Omega was acting on Gulf's behalf; and (2) to inform the Bank that a telegram would be sent to the Bank to confirm same. The telegram was subsequently sent.

It is undisputed that Reed was aware that a rental payment was to be delivered to the Bank. E.D. Kincaid III was also informed that "an oil company was bringing a lease check to the Bank." Moreover, it was presumed by Neill that the payment to be made was for the E.D. Kincaid lease. However, since the rental was not due until March 1, the Bank was instructed by E.D. Kincaid III and Ed Vaughan, the Bank's attorney, not to accept payment after business hours on February 29th.

Thereafter, written instructions were left for Tom Rothe, the bank officer on duty on March 1, that someone was going to be delivering a check for the Kincaids and to accept it and hold it until Monday, March 3rd.

On March 1, 1980, during regular banking hours, Omega's agent delivered a check payable to the First State Bank of Uvalde in the amount of $25,867.79 and a delay rental receipt to Rothe. Rothe accepted the check and signed the delay rental receipt. On March 3, 1980, the check and the receipt were delivered to Reed. William Alex Kincaid, on March 3, 1980, looking at the check but not reading the check or the receipt, concluded that the rentals for the E.D. Kincaid lease had been paid. He, thereafter, informed E.D. Kincaid, Jr. that the delay rental for the E.D. Kincaid Estate lease had been paid.

Thereafter, E.D. Kincaid III informed Neill that the check was payable to the F.T. Kincaid Estate lease and told Neill not to credit the E.D. Kincaid Estate. He then orally instructed Vaughan, to return the check to Omega. This instruction was subsequently reduced to writing on March 7, 1980. Moreover, a letter was written to Omega by Vaughan on March 3, 1980, concerning the check. Upon receipt of this letter, Royis Ward of Omega, telephoned Vaughan and advised him that a response letter would be forthcoming. On March 5, 1980, Ward sent a letter to the Bank instructing the Bank to deposit the check delivered on March 1, 1980, to the lessors under the E.D. Kincaid Estate lease. Carbon copies of this letter were sent to Vaughan, E.D. Kincaid III and Gulf.

In February, 1981, Gulf tendered a delay rental check in the amount of $25,867.79 to the Bank for payment of the delay rentals under the E.D. Kincaid Estate lease. It is noteworthy that neither the delay rental check tendered by Omega to the Bank on March 1, 1980, nor the February 1981, check have been cashed by the Bank or returned to Omega nor have the lessors under the E.D. Kincaid Estate lease received credit for either check consistent with their instructions.

. . .

In essence, appellants' contentions seem to challenge the following: (1) the trial court's construction of the lease; and (2) the trial court's failure to adjudge that said lease *ipso facto* terminated.

It is well settled that with the usual "unless" lease, a failure of the lessee either to begin a well or to pay the delay rentals, ipso facto terminates the lease on the date set out for the action and the estate reverts to the lessor without the necessity of re-entry, declaration of forfeiture or legal action. W.T. Waggoner

Estate v. Sigler Oil Co., 118 Tex. 509, 19 S.W.2d 27, 30 (1929). Moreover, an oil or gas lease, providing for the production of oil and payment of royalties within a given time unless delay rentals are paid conveys a determinable fee so that if the lessee fails to drill or to pay delay rentals, his lease is terminated. Humble Oil & Refining Co. v. Harrison, 146 Tex. 216, 205 S.W.2d 355, 360 (1947). In applying these rules, some cases have required a strict compliance by the lessee with the terms of the lease relating to the payment of delay rentals, holding that a small deficiency in the amount of the payment or a failure to make the payment until a short time after its due date terminates the lease. *Id.* at 360-361. The application of the rule has been relaxed in some cases, however. *Id.* at 361. In Perkins v. Magnolia Petroleum Co., 148 S.W.2d 266 (Tex.Civ.App.--Galveston 1941, writ dism'd judgmt cor.), for example, it was held that a lessee sufficiently complied with the requirements of the payment of delay rentals by making a joint deposit in the depository bank of the total amount due under separate leases to different lessors where the lessors were disputing the extent of their respective interests. *See id.* at 269. And in Buchanan v. Sinclair Oil & Gas Co., 218 F.2d 436 (5th Cir.1955), it was held that lessors who received, without objection, smaller delay rentals than that provided for in the subject lease could not successfully contend that failure of lessee to pay the stipulated rental had terminated the lease.

We find it unnecessary, however, to rely on the doctrine of equitable estoppel enunciated in the *Humble Oil & Refining Co.* and *Buchanan* cases inasmuch as Paragraph 5 of the lease in question clearly distinguishes it from a majority of the leases at issue in earlier decisions in which a failure to make proper payment of the rental was held to have brought about a termination of the rights of the lessee. *See* Woolley v. Standard Oil Co. of Texas, 230 F.2d 97 (5th Cir.1956).

In construing an oil and gas lease, like any other instrument, the primary and all important consideration is the intention of the parties as gathered from the instrument. . . . We must also bear in mind that if its language is reasonably susceptible of a construction argued for by one of the parties, that prevents a forfeiture, such construction is to be preferred to one resulting in a forfeiture. . . . . Moreover, even where different parts of the instrument appear to be uncertain, ambiguous or contradictory, yet, if possible, the court will harmonize the parts and construe the instrument in such way that all parts may stand, and will never strike down any portion unless there is an irreconcilable conflict where one part of the instrument destroys, in effect, another part. . . .

Further, the rule of construction for ambiguous instruments provides that if a contract is ambiguous or its meaning is doubtful, it will be construed most strongly against the party who drafted it. . . .

Considering the lease as a whole and giving effect to every provision of the instrument, we think that the "unless" clause in paragraph 5 operates as a special limitation on the term of the estate acquired by the lessee, notwithstanding the general limitation of five years provided by paragraph 2. . . . However, by its very terms, the provision of paragraph 5 (i.e., "or shall make a bona fide attempt to pay or tender, as hereinafter stated") operates as a limitation upon the "unless" provision for termination because of non-payment of delay rentals. It is, therefore, obvious that the parties intended to alleviate the harsh result ensuing from the rule of automatic termination upon the failure of a lessee to meet the stipulated conditions in an ordinary and unmodified "unless" type of lease. We know of no appellate court in Texas which has had occasion to construe a lease such as the one in the instant case; however, the United States Court of Appeals for the Fifth Circuit in the aforementioned case of Woolley v. Standard Oil Co. of Texas upheld a practically identical lease provision.

There remains for determination whether appellees did, in fact, make a bona fide attempt to pay the delay rental as provided for in the E.D. Kincaid lease. The uncontroverted evidence shows that on March 1, 1980, Gulf, acting through Omega, sent the full amount of the E.D. Kincaid rental to the depository bank, the agent of the lessors, and at all times thereafter there was on file with the bank, the Omega check made out for the correct rental amount. On February 29, 1980, Gulf made the decision to pay the delay rental on the E.D. Kincaid lease because a dry hole on said lease had recently been plugged and abandoned. It was agreed that Omega, acting for Gulf would deliver the check to the Bank since Omega was closer in proximity to the Bank. Numerous telephone calls were made on behalf of Gulf and Omega in order to ensure a prompt delivery of the rentals. Thereafter, on March 1, 1980, the payment was timely made. The payment was for the full amount of the E.D. Kincaid rental and the check was at all times thereafter on file with the Bank. Although the check was made out to the proper payee, i.e., the depository bank for lessors, the credit allocations on the receipt were incorrect. When Omega learned of the mistake on March 3, 1980, it promptly replied by letter to the Bank instructing the Bank to deposit the check in the E.D. Kincaid account. The record standing thus, we think it plain that appellees did, make a bona fide attempt to pay the rental as contemplated by the lease.

The judgment of the trial court is affirmed.

NOTES

1. What result if the *Curtis* principles were applied? Was Gulf any less culpable than Phillips Petroleum? Should culpability have any impact on finding whether a fee simple determinable estate has terminated?

2. Does the court fully appreciate the conflict between the "good faith performance" provisions of the unless clause and the automatic termination character of the standard unless clause? Can the parties create a "fuzzy" limitation that will automatically terminate the lease? Is the court allowing the parties to create a new estate in land? Is that allowed in a non-oil and gas context?

3. In the absence of a specific provision relating to the good faith of the lessee, should a court allow a good faith attempt or substantial performance to excuse the automatic termination of a lease? See Burlington Resources Oil & Gas Co. v. Cox, 133 Ohio App.3d 543, 729 N.E.2d 398 (1999).

4. Did the court apply a canon of construction to assist it in determining the intent of the parties? Did the court construe the lease in a light most favorable to the lessor? the lessee? Was there anything unusual about the drafting of this particular lease instrument?

5. What if the parties negotiate a delay rental and drilling clause which combines elements of the traditional "unless" and "or" form clauses? Should equitable defenses be allowed if the lessee fails to meet the drilling or rental payment provisions? *See* Phyfer v. San Gabriel Development Corp., 884 F.2d 235, 106 O.&G.R. 502 (5th Cir.1989).

6. What if the parties changed the delay rental clause to the following: "It is agreed that the royalty in each year commencing with the year of August 1, 1979, shall not be less than Two Thousand ($2,000.00) Dollars and if in any lease year the royalty does not equal said sum of $2,000.00, the lessee agrees to pay the lessor the difference between said sum of $2,000.00 and the royalty received; said payment to be made within thirty (30) days of the close of the lease year." There was a producing well on the lease, but in the 1992-93 lease year the royalties paid to the lessor were less than $2,000.00. The lessee inadvertently fails to pay the difference within 30 days of the end of the 1992-93 lease year. Is the clause to be treated as an automatically terminating limitation (condition, in the court's terminology) or merely a covenant which will allow the lessee to bring an action for damages? M & C Oil, Inc. v. Geffert, 21 Kan.App.2d 267, 897 P.2d 191, 132 O.&G.R. 263 (1995) concluded that the clause was only a covenant and that forfeiture was not warranted by the inadvertent delay in making the minimum royalty payment.

HUMBLE OIL & REFINING CO. v. HARRISON

Supreme Court of Texas, 1947.
146 Tex. 216, 205 S.W.2d 355.

HART, JUSTICE. [Suit by Humble Oil & Refining Co. and others against D.J. Harrison for removal of cloud from and to quiet title to certain oil and gas interests. Humble alleged in its petition that Harrison cast a cloud on its title by contending that the leasehold estate had terminated as to Harrison's interest by reason of the asserted failure of Humble to pay to him the correct amount of delay rentals. A 3/4 interest in the 1074.4-acre tract here involved was the property of Mrs. Lottie Otto. By a series of "unless" type leases providing for semi-annual delay rental payments, this tract was leased to Humble's assignor. In 1944, Mrs. Otto, joined by her husband, conveyed an interest in the minerals, subject to the existing leases, to Harrison. A copy of the deed was supplied by Harrison to Humble. In a portion of the opinion which is omitted the court discussed the construction of the deed from the Ottos to Harrison and concluded that its effect was to convey 1/2 of the minerals (viz, 2/3 of the Ottos' 3/4 interest in the minerals) and 1/2 of the proceeds of the existing leases, including rentals. The supervisor of Humble's land rental division construed the instrument, however, as conveying 1/2 of the Ottos' interest in the rentals, and before the next rental payment date, March 1, 1944, when $750 was payable as rental under the Otto leases, $750 was remitted to the depository bank with instructions to credit $375 to the Ottos and $375 to Harrison. Before the rental anniversary date, Harrison was notified by the depository bank of the amount of the deposit, and he made no objection as to the amount of the deposit to his credit in the bank and made no request for any additional payment. The next rental payment which fell due on May 8, 1944, was a payment of $1639.50, and again Humble deposited this sum in the depository bank with instructions to distribute 1/2 to the Ottos and 1/2 to Harrison. Again Harrison was notified by the bank of the deposit and made no objection as to the amount of the payment and no demand for an additional payment. Eds.]

On June 10, 1944, Harrison for the first time made known to Humble that he contended that the delay rentals placed in the bank to his credit by Humble were insufficient. On that date he wrote a letter to the Needville bank, sending a copy to Humble, in which he acknowledged receipt of the deposit slips dated February 23, 1944, and April 28, 1944, and concluded as follows:

These rentals have not been paid in accordance with the terms of the respective leases, and for that reason these leases have lapsed and are null and void as to my interest. This letter will serve to notify you that I refuse to accept the rentals referred to.

The next rental-paying date for the 874.4-acre tract was November 8, 1944, and prior to that date Humble deposited with the Needville bank to Harrison's credit the sum of $1093. On November 7, 1944, Humble wrote to Harrison, stating that it had made the deposit to Harrison's credit, with the explanation that it was willing to make what it contended was an overpayment in order to satisfy Harrison's contention and to keep the lease in effect. Harrison promptly replied on November 8, 1944, by a letter in which he referred to his letter of June 10, 1944, and made the demand that Humble execute releases to him, "since these leases are null and void as to my interest."

On February 28, 1945, Humble made a similar tender to Harrison of what it considered an overpayment of the rental on the 200-acre tract, by depositing $500 in the Needville bank to Harrison's credit, and notified Harrison of this deposit. Harrison replied on March 8, 1945, stating that "these leases are null and void as to my interest," and demanding that Humble forward to him executed releases covering his mineral interest. Humble, on each rental-paying date thereafter, has tendered, and Harrison has refused to accept, payments of $1093 on the 874.4-acre tract and $500 on the 200-acre tract. Humble has also paid the Ottos and the Paddocks respectively, all rentals due and payable to them.

(1) Construing the mineral deed as a whole, we conclude that the proper construction to be placed on the deed is that Harrison is granted an undivided one-half interest in all of the minerals and an undivided one-half interest in all royalties, bonuses and rentals. . . .

. . . While an ambiguity is created by the reference in the mineral deed to the leases executed by the Ottos and the rentals payable thereunder, we conclude that, considering the deed as a whole, the intention of the parties was that Harrison was granted the right to receive one-half of the entire rentals and not merely one-half of the rentals payable to the Ottos. . . . It follows that Harrison was entitled to receive one-half of the semiannual rental of $2186 payable on the 874.4 acres, or $1093, and one-half of the annual rental of $1000 payable on the 200 acres, or $500, and that in tendering $819.75 and $375 Humble did not tender to Harrison the amounts which he was entitled to receive under the mineral deed to him from the Ottos.

(2) The question remains whether Humble's lease is terminated as to Harrison's interest because Humble mistakenly advised the depository bank to credit Harrison's account with a less amount than he was entitled to receive and to credit to the Ottos' account a greater amount than they were entitled to receive, at the time of each payment. It is undisputed that Humble deposited in each instance a total amount which paid the delay rentals in full; their only mistake was in misconstruing the mineral deed and erroneously dividing the delay rentals between the Ottos and Harrison. There is no evidence that Humble did not act in good faith, nor does the evidence show that Humble acted negligently. The evidence shows affirmatively that Humble has been willing to make overpayments in order to keep the lease alive. While we have concluded that the construction of the mineral deed adopted by Humble's agent, a licensed attorney, was erroneous, there were provisions in the deed which make its meaning ambiguous and the construction adopted by Humble's agent was not without reasonable foundation. The letter sent by Harrison to Humble with the copy of the deed was not, when considered with the deed, unambiguous, because it likewise was subject to the interpretation that one-half of the royalties *payable to the Ottos* had been assigned to Harrison. Although Harrison was notified, by receiving the deposit slips, that Humble construed the mineral deed to entitle him to receive only one-half of the Ottos' share of the rentals, he made no protest and no demand for an additional payment, but remained silent until after the rental-paying dates had passed.

It is well settled in this state that the lessee under "unless" leases, such as those involved in this case, has a determinable fee, and that if he fails to drill or to pay delay rentals his lease is terminated. . . . In applying this rule, some cases have required a strict compliance by the lessee with the terms of the lease relating to the payment of delay rentals, holding that a small deficiency in the amount of the payment or a failure to make the payment until a short time after its due date terminates the lease. . . . The application of the rule has been relaxed in some cases, however. In Perkins v. Magnolia Petroleum Co., Tex.Civ.App., 148 S.W.2d 266, writ dismissed, judgment correct, it was held that a lessee sufficiently complied with the requirement of the payment of delay rentals by making a joint deposit in the depository bank of the total amount due under separate leases to different lessors, where the lessors were disputing as to the extent of their respective interests. In Miller v. Hodges, Tex.Com.App., 260 S.W. 168, it was held that the lease continued in effect and that the lessee was excused from making payments of delay rentals after the lessor brought suit to cancel the lease, during the pendency of the suit. In Mitchell v. Simms, Tex.Com.App., 63 S.W.2d 371, it was held that where the lessors received a payment of the delay rental after its due date, the lessors became estopped to assert that the lease had terminated because of delay in the payment.

In the present case, Humble's failure to pay to Harrison his full share of the rentals originally payable to the Ottos is due primarily to a misconstruction of an ambiguous mineral deed to which Humble was not a party, and, secondarily, to a failure on the part of Harrison, after he knew that Humble had misconstrued the deed, to notify Humble of the proper construction. Each of the leases executed by the Ottos contained the provision that "no change in the ownership of the land or part thereof, the minerals or interests therein, shall impose any additional burden on Grantee." It would be an imposition of an additional burden on the lessee to require that the lessee determine at its peril the proper construction of an ambiguous instrument thereafter executed by the lessors, conveying a part of their interest in the minerals and the royalties, bonuses and delay rentals. Where, as in this case, the lessee has in good faith made a mistaken construction of the lessors' partial conveyance of their interests and lessee has made a payment in accordance with such construction, of which the assignee has notice, the duty rests on the assignee to notify the lessee of its mistake so that the lessee will have an opportunity to make a proper payment of the delay rentals. Where the assignee, instead of giving the lessee such notice, remains silent, we hold that the assignee is estopped to assert that the lease has terminated as to his interest on the ground that the lessee has failed to pay to him a sufficiently large share of the delay rentals.

While we do not know of any case presenting the exact situation involved here, we think the general principles of equitable estoppel are applicable. In Burnett v. Atteberry, 105 Tex. 119, 145 S.W. 582, 587, this Court said:

> An estoppel may arise as effectually from silence, where it is a duty to speak, as from words spoken. One may be induced to act to his injury on account of the silence of one interested in a transaction, and when such course of action is permitted with the knowledge of the interested party or induced by silence or tacit acquiescence, the doctrine of estoppel may be invoked.

. . .

We hold that Harrison is estopped to assert that Humble's lease has been terminated because of its failure to make sufficient payments of delay rentals to him. Accordingly, the judgments of the district court and the Court of Civil Appeals are reversed and judgment is rendered in favor of petitioners.

[SMEDLEY, JUSTICE, concurred on the ground that Humble correctly construed the mineral deed from the Ottos to Harrison and made the appropriate delay rental payments to keep the lease alive. Eds.]

NOTES

1. Could Humble have avoided a construction of the deed and preserved its lease by drawing a check to the credit of the Ottos and Harrison jointly and mailing the check to the depository bank with instructions to credit the payees' accounts as their interests should appear?

2. Could Humble have avoided a construction of the deed and preserved its lease by filing an interpleader action and depositing the rentals into the registry of the court? *See* Citizens Nat. Bank of Emporia v. Socony Mobil Oil Co., 372 S.W.2d 718, 20 O.&G.R. 89 (Tex.Civ.App.1963, error ref'd n.r.e.); HNG Fossil Fuels Co. v. Roach, 99 N.M. 216, 656 P.2d 879, 76 O.&G.R. 88 (1982).

3. Did the court give any weight to the fact that the initial "erroneous" delay rental payment had been accepted by Harrison? Should a lessor be estopped from asserting that a lease has terminated because it is "a day late or a dollar short" when the lessor accepts the tender that is made? The cases are divided on the subject. *Compare* Buchanan v. Sincalir Oil & Gas Co., 218 F.2d 436, 4 O.&G.R. 400 (5th Cir. 1955)(estoppel would apply) *with* Miller v. Kellerman, 228 F.Supp. 446, 20 O.&G.R. 523 (W.D.La. 1964), *aff'd*, 354 F.2d 46, 24 O.&G.R. 586 (5th Cir. 1965), *cert. denied*, 384 U.S. 951 (1966)(estoppel would not apply). Could a lessor who receives an inadequate check one month prior to the deadline do nothing and then claim that the lease terminated automatically. *See* Schwartzenberger v. Hunt Trust Estate, 244 N.W.2d 711, 56 O.&G.R. 271 (N.D. 1976).

4. If a lessor attacks the continued validity of a lease must the lessee continue to make payments during the period of time that the litigation is ongoing? If the lessee wins is the lessor entitled to receive the rentals? Should the lessee be entitled to additional time to achieve production if the primary term has expired during the time the litigation was active? *See* HNG Fossil Fuels Co. v. Roach, 103 N.M. 793, 715 P.2d 66, 89 O.&G.R. 113 (1986).

5. If a lease provides for a delay rental of $1.00 per mineral acre and a mutual mistake as to the extent of the lessor's ownership results in the lessee applying the proportionate reduction clause so as to underpay the rentals, does the principle of the *Harrison* case require notice to the lessee of the mistake before the lease can be terminated? *See* Schwartzenberger v. Hunt Trust Estate, 244 N.W.2d 711, 56 O.&G.R. 271 (N.D.1976).

If the lessee could determine by examination of the record that the apparent equal co-ownership of lessors on the face of the lease was erroneous, will payments made on the basis of the appearance of the lease to the apparent co-owners and accepted by them trigger the special limitation of the rental clause? *See* Norman Jessen & Assoc., Inc. v. Amoco Prod. Co. 305 N.W.2d 648, 69 O.&G.R. 478 (N.D.1981). If an accurate examination of the record would have revealed that the lease actually covered the 80 mineral acres originally contemplated in the leasing transaction rather the 60 mineral acres which became the basis for payment and acceptance of rentals, should the lease terminate under the rental clause? Hold 80 acres? Hold 60 acres? Borth v. Gulf Oil Exploration and Production Co., 313 N.W.2d 706, 74 O.&G.R. 60 (N.D.1981), adopted the theory of Kugel v. Young, 132 Colo. 529, 291 P.2d 695, 5 O.&G.R. 951 (1955), that the proffer of a rental for less acreage than covered by a lease amounted to an offer to keep under lease only the lesser acreage which, when accepted, continued the lease only to that extent. These cases are discussed and criticized in Maxwell, *Some Comments on North Dakota Oil and Gas Law--Three Cases from the Eighties*, 58 N.D.L.Rev. 431, 446 (1982).

6. In LEDFORD v. ATKINS, 413 S.W.2d 68, 26 O.&G.R. 644 (Ky.1967), an assignment of an oil and gas lease provided for the payment of delay rentals before a given date with the further proviso that failure to make the payments before such date would cause the assignment to be void. The assignee was ill and in a comatose condition until shortly before the rental date and seriously ill through the rental date; as a result proper rentals were not paid by the time specified in the assignment. (Rentals were presumably paid in time to keep the underlying lease alive; no issue of its termination was presented.) The court affirmed a judgment that the assignment remained in effect. "To forfeit the lease under these conditions and especially so where the lessee [assignee] stands to lose a substantial investment already made in developing the lease, in our opinion, would be a gross injustice."

A dissenting judge felt the assignment should be enforced as written: "When grown men have seen fit to place such a requirement in the contract, without condition, qualification or provision against unforeseen circumstances, I do not feel that a court should modify it." Is there any merit in this position?

BRANNON v. GULF STATES ENERGY CORP.

Supreme Court of Texas, 1977.
562 S.W.2d 219, 59 O.&G.R. 320.

DANIEL, JUSTICE. The question in this case is whether an oil and gas lease terminated for nonpayment of delay rentals where there was a late payment and acceptance by the lessor of a check designated as "lease rental." More specifically, the issue is whether parol evidence was admissible to vary the written designation of the late payment from a "rental" to a bonus for a new lease.

Petitioners, M. J. Brannon, Jr., Otis Thompson, Patricia A. Elliott, and Henry W. Elliott III claim under a lease dated November 20, 1973, from Clara Odessa Martin covering 202 acres in Coleman County. The first delay rental of $202.00 due November 20, 1974, was not timely paid. However, Petitioners allege that on January 17, 1975, Respondent, Gulf States Energy Corporation, while holding an interest in said lease, made a late written tender of the rental to the lessor by letter and check. Both of these designated the payment as "lease rental," and the check was accepted by the lessor. Respondent, Gulf States, countered with the allegation that the check was in fact a bonus paid in advance for a new lease which Mrs. Martin executed and delivered to Gulf States on July 9, 1975. . . .

On November 20, 1973, Mrs. Clara Odessa Martin executed an oil and gas lease to Mary Linn Elliott covering the 202 acres mentioned above. It was a five year lease providing for termination at the end of each anniversary date unless an annual rental of $202.00 was paid or the lease was otherwise perpetuated by drilling operations or production. In the meantime, Mrs. Elliott had also acquired an adjoining 1900 acre lease known as the "Evans lease." Early in 1974, Mrs. Elliott negotiated the terms of a sale of the Martin and Evans leases to Royal Russell, sole owner and stockholder of Gulf States, whereby she would assign the leases for a cash consideration and retain a 1/16th overriding royalty. Prior to Mrs. Elliott's execution of the assignments, Royal Russell negotiated an agreement with Master Drillers, Inc., on February 18, 1974. It referred to the separate tracts as one lease and recited that Master was interested in purchasing said lease and "acting in a capacity of 'Warehousing' said lease for and on behalf" of Gulf States, and that the two companies were interested in jointly developing same. . . .

At Russell's request, on February 20, 1974, Mrs. Elliott executed assignments on the Martin and Evans leases to Master Drillers, Inc., "effective as of February 18, 1974." . . .

. . . On October 22, 1974, the Internal Revenue Service seized the Martin and Evans leases along with other property of Master Drillers to enforce a lien for taxes.

Mrs. Martin's annual lease rental was not paid on November 20, 1974. She so advised Mrs. Elliott after Christmas in December of 1974. Because of this jeopardy to her overriding royalty, Mrs. Elliott testified that she called Royal Russell and asked him to pay the rentals. Russell remembered a call from Mrs. Elliott about the Evans lease but denied any mention of the Martin lease. In any event, by letter dated January 17, 1975, addressed to Mrs. Martin, with a copy mailed to Mrs. Elliott, Royal Russell transmitted a Gulf States check payable to Clara Odessa Martin in the sum of $202.00, on which was typed "Lease Rental." The letter read in part as follows:

"Dear Mrs. Martin:

Enclosed please find check # 6240 in the amount of $202.00 which is for lease rental on the following described lease:

(Description)

(s) Thank you. Sincerely,

    GULF STATES ENERGY CORPORATION

(s) Royal Russell

    (s) President RR/mdm Enclosure

cc: Mrs. Mary Elliott"

The check was received, accepted, endorsed, and deposited in her bank account by Mrs. Martin. On January 24, 1975, Mrs. Elliott wrote Mrs. Martin, "I trust you have received your rental money, since Mr. Russell advised me that he was mailing the check."

On May 27, 1975, at a public sealed-bid sale by the Internal Revenue Service, petitioner Thompson bought all rights and interests of Master Drillers in the Martin and Evans leases. Royal Russell attended the sale and submitted a lower bid on behalf of Gulf States. Thompson conveyed one-half of his purchase to petitioner Brannon by quitclaim deed dated June 9, 1975. Subsequent delay rentals, due November 20, 1975, were tendered by them to Mrs. Martin but were declined by her.

On July 9, 1975, Mrs. Martin executed a new ten-year lease on the 202 acres to Gulf States. It made a first well location on the tract in late July of 1975 and drilled three wells in spite of the recorded conveyances to Thompson and Brannon and written notices from their attorneys protesting any operations on the property by Gulf States. This suit was filed by Petitioners Brannon and Thompson on September 17, 1975, seeking a declaratory judgment that the lease of November 20, 1973, from Clara Odessa Martin to Mary Linn Elliott was in full force and effect; that the subsequent lease from Mrs. Martin to Gulf States be declared void; and that Gulf States be ordered to vacate the premises. . . .

Gulf States' claim to a superior title rests upon parol testimony of Mrs. Martin and Royal Russell that the $202.00 "lease rental" paid by Gulf States to Mrs. Martin on January 17, 1975, was not in fact a rental due under the Martin-Elliott lease of 1973, but a bonus for a new lease from Mrs. Martin to Gulf States executed by Mrs. Martin on July 9, 1975. In support of this direct contradiction of the written terms of the rental letter and check, there was other parol evidence to the effect that shortly before Christmas of 1974, Mrs. Martin orally agreed to make a new lease to Gulf States in consideration of $202.00 and a promise to drill a well. This evidence is summarized in narrative form by the Court of Civil Appeals.

Petitioner-plaintiffs have contended throughout this litigation that the parol evidence rule rendered inadmissible the testimony of Mr. Russell and Mrs. Martin contradicting the written Gulf States rental letter and the $202.00 rental check of January 17, 1975, which was accepted and endorsed by Mrs. Martin. If they are correct, then admission of such parol evidence was erroneous . . .

Contractual Nature of the Letter and Check

That a written instrument cannot be varied by parol evidence is a rule of substantive law. . . .. Yet it is true that the rule applies only to contractual or jural writings evidencing the creation, modification, termination or securing of a particular right or obligation. . . ..

We consider the letter and the check relating to "lease rentals" as contractual in nature because late payment and acceptance of annual rentals provided for in an oil and gas lease has the effect of reviving the lease as though it had never terminated. Mitchell v. Simms, 63 S.W.2d 371 (Tex.Comm.App.1933, holding approved) . . .

In Simms, a rental due on June 1, 1929, was paid in December, over six months late. The payment was accepted by the lessor, and it was held that such acceptance estopped the lessor from "claiming that the lease had terminated before payment was made." The court said: "The practical result . . . is the same as if Simms and wife (the lessors) had by binding agreement in December, revived the lease." By the same token, the written tender of lease rentals by Gulf States to Mrs. Martin two months late and her acceptance, endorsement, and deposit of the check to her account had the same effect as a binding agreement to revive the only existing lease on which any "rentals" were due. The transaction was contractual in nature. Furthermore, there is no ambiguity in the term "lease rental.". . . .

Conclusions and Instructions on Remand

Since the written lease rental payment and acceptance were contractual and unambiguous, it was error to allow the terms and effect thereof to be varied by parol evidence. We hold that the Martin-Elliott lease of November 20, 1973, is in full force and effect and that it is superior to the Martin-Gulf Coast lease of July 9, 1975.

. . .

As to ownership of the Martin-Elliott oil and gas lease, we hold that it is vested in M. J. Brannon, Jr., and Otis Thompson, subject to (1) the 1/16th overriding royalty of Patricia A. Elliott and Henry W. Elliott III, (2) the right of Gulf States to reimbursement of one-half of the lease rental paid on January 17, 1975, (3) the rights, if any, of Gulf States under its agreement of February 18, 1974, with Master Drillers, and (4) an offset in favor of Gulf States for the amount of Brannon and Thompson's share of any reasonable and actual expenses found to have been made by Gulf States in drilling, operating and producing the wells on the Martin-Elliott lease when an accounting is made between the parties for the production of oil, gas and other minerals therefrom.

If, upon remand, Gulf States asserts no interest in the Martin-Elliott lease or it is found by the court to have no interest therein, then Gulf States will be entitled to an offset of the actual amount of its reasonable expenditures in drilling and developing the leasehold estate if Gulf States is again found to have drilled and developed the 202 acre lease in good faith belief in the superiority of its lease of July 9, 1975. Petitioners insist that as a matter of law Gulf States cannot be a trespasser in good faith because of its own revival and knowledge of the Martin-Elliott lease. We disagree. Under certain circumstances, one can be completely mistaken in his claim of superior title and yet be a trespasser in good faith. The question is usually one of fact. . . .

Although Petitioners' adverse claim had been known and asserted by letters, no suit was filed against Gulf States prior to its entry upon the leasehold and spudding its first two wells. It consulted legal counsel before incurring expenses on the lease. On retrial, substantially the same evidence as to belief in the superiority of its title would support a jury finding that Gulf States conducted its operations and made its expenditures in good faith. However, the amount of the expenditures should not be a mere estimate as on the present trial. Rather, it should be based on a detailed accounting of the actual expenditures. . . ..

Accordingly, the judgments of the lower courts are reversed and the case is remanded to the trial court for further proceedings in accordance with this opinion.

On Motion for Rehearing

Both petitioner, Brannon, and respondent, Gulf States Energy Corporation, have filed motions for rehearing. . . .

Brannon . . . argues that none of the evidence on good faith development by Gulf States is admissible in view of our holding that such evidence could not vary the written instruments evidencing payment of rentals by Gulf States on the 1973 Martin lease. While parol evidence was not admissible to vary the written contract, it is admissible and should be heard on retrial for the sole purpose of determining whether Gulf States believed in good faith in the superiority of its 1975 lease when it drilled the three wells on the Martin lease. Brannon contends that Gulf States should be held as a matter of law not to have been in good faith in drilling the third well because it was commenced after this suit was filed. This particular argument is overruled because Gulf States entered upon the Martin tract and began its drilling program prior to the filing of this case. The question of Gulf States' good faith is still one of fact to be decided upon remand. . . .

NOTES

1. What is the court's basis for finding that the lease may have been revived by the lessor's acceptance of the late tendered delay rental check? *See* Harrell v. Saline Oil & Gas Co., 153 Ark. 104, 239 S.W. 731 (1922).

2. If the standard *unless* clause creates a fee simple determinable estate are the equitable doctrines of estoppel, waiver and revival applicable? For example, if the lessor accepts a late payment of delay rentals should she be estopped from asserting that the lease has terminated? *See* Mitchell v. Simms, 63 S.W.2d 371 (Tex. Com. App. 1933). If detrimental reliance is required in order to assert estoppel as an affirmative defense, how can the lessee claims such reliance where the terms of the "unless" clause create a limitation that will automatically terminate the lease. *See* O'Hara v. Coltrin, 637 P.2d 398, 71 O.&G.R. 487 (Colo.App.1981).

Would a lessor be estopped if a depository bank placed to her credit an inadequate or a late rental? Would the designation of the bank as the agent of the lessor be determinative? Would the lessor's actual knowledge prior to the anniversary date that the amount tendered by the lessee was inadequate affect the result? *Latham v. Continental Oil Co.*, 558 F.Supp. 731, 76 O.&G.R. 514 (W.D. Okl. 1980) said that it should. Should the court be influenced by the payment of a large bonus shortly before the erroneous delay rental payment was due? *See* Discussion Notes, 76 O.&G.R. 520 (1983).

3. Is the court's treatment of Gulf States as a possible good faith trespasser consistent with *Houston Production Co. v. Mecom,* discussed at p. 91 *supra.* Although Gulf States entered the premises prior to the onset of litigation did they continue to produce after litigation had begun? Were other options available to Gulf States other than continued drilling and production? Was there any evidence that the hydrocarbons were being drained from the leasehold estate?

CHANGE IN OWNERSHIP CLAUSE

The rights of either party hereunder may be assigned, in whole or in part, ... No change in the ownership of the land, or any interest therein, shall be binding on Lessee until Lessee shall be furnished with a certified copy of all recorded instruments, all court proceedings and all other necessary evidence of any transfer, inheritance, or sale of said rights.

Absent such a clause as the above, what is the effect of the assignment of a lessor's interest on the delay rental clause? Absent such a clause is the lessee's interest freely assignable? Given the clause, what is the effect of an unrecorded conveyance by the lessor of which the lessee has no notice? What is the effect of a recorded conveyance by the lessor? Is the burden of checking the records to ascertain the persons entitled to delay rental payments placed on the lessee? The following case bears on these questions.

GULF REFINING CO. v. SHATFORD

Circuit Court of Appeals of the United States, Fifth Circuit, 1947.
159 F.2d 231.

[The assignment clause of an oil and gas lease executed on July 6, 1938, provided that either lessors or lessees may assign their interest "in whole or in part" but that any change in ownership of the land or assignment of rentals or royalties shall not be binding on the lessee "until after the Lessee has been furnished with a written transfer or assignment or a true copy thereof." On November 17, 1944, the lessors conveyed to Shatford a 1/6 mineral interest in the land covered by the lease. By subsequent assignments, Shatford reduced his holdings to a 1/8 interest. No notice of any kind was given to the lessee by Shatford until May 3, 1945, when Shatford wrote a letter advising Gulf that he was the owner of "one-eighth of the royalty" in the land covered by the lease, and suggesting that he was entitled to a portion of the rental. Gulf replied promptly with a request that Shatford furnish photostatic copies or certified copies of his deeds, observing that "we will be in a position to change our records to reflect your ownership." In the succeeding month and a half, nothing more was heard from Shatford. On June 25, 1945, eleven days before the rental anniversary date, Gulf mailed to the depository bank two checks in payment of the delay rentals with instructions that they be credited to the accounts of the original lessors and certain assignees who had provided Gulf with copies of the assignments under which they claimed. No provision was made for payment of any part of the rentals to Shatford. On June 26, 1945, the day after Gulf had mailed its checks to the depository bank, Gulf received from Shatford photostatic copies of deeds showing his acquisition of an interest in the minerals. Gulf, by letter of June 28, 1945, acknowledged receipt of such copies and advised Shatford that "we have changed our records to reflect your interest in the above lease." Having already paid delay rental for the ensuing year, Gulf made no payment to Shatford. Because of Gulf's failure to pay to Shatford an additional amount as his proportionate share of delay rental, Shatford's claim for cancellation was sustained by the lower court. Eds.]

McCORD, CIRCUIT JUDGE. . . .Gulf was not required to wait until the last minute of the last day to make the payment that would keep its lease alive for another year. Each year Gulf, acting in accordance with the *"on or before"* provision of the lease, paid the delay rental about ten days "before" the anniversary date.[37] The crucial issue is whether Gulf paid the rental to the ones who, under the lease, were entitled to receive it at the time it was paid? Gulf was required to pay lessors their proportionate shares of delay rental. However, Gulf was not bound by any transfer or assignment of interests until it had received a *"written transfer of assignment or a true copy"*. This provision in the lease is a reasonable and valid one, written in clear and unambiguous language. Shatford, who bought his way in as a lessor after execution of the lease, was bound by this provision, but he did not diligently comply with it. He waited almost six months before he made a move. Then he only wrote a letter. After he was advised that Gulf, as it had a right to do, required a copy of his deeds, he made no move for another month and a half. During all this time while Shatford was doing nothing, Gulf's time was running out and the anniversary date was approaching. In late June, about the time of year it always forwarded rental payments to lessors' agent bank, Gulf moved to safeguard its interests and paid the rentals to lessors' agent bank for deposit to Beaird, the Parks, and Powell, the persons whose "written" evidence of ownership it had. Gulf did not direct the lessors' agent bank to deposit anything to the credit of Shatford because Gulf had not received a copy of any transfer or assignment in his favor. Had Shatford acted diligently and sent

---

[37] We need not discuss whether Gulf could have paid delay rentals months or years in advance of anniversary dates and thereby insulated itself from liability to assignees who had bought interests in the minerals or gave notice of purchases subsequent to such advance payments. Certainly, the payment in the case at bar, made just a few days before the deadline, was within the letter and spirit of the *"on or before"* provision of the lease. Indeed the early payment was a reasonable and prudent, safe business procedure.

copies of his deeds prior to the time Gulf paid the rental money, he would have received a share in proportion to his holdings. Contrast Shatford's conduct with that of Powell who received his one-sixth interest on the same day Shatford received his. Powell sent in certified copies of his deed, and when the rental was paid, Powell received his proportionate share. Payment to Powell certainly discloses the forthright honesty and good intent of Gulf to pay rental on time to the persons entitled to receive it.

To now permit Shatford to cancel the lease so far as it affects his one-eighth interest would be to reward him for his negligence and punish Gulf for living up to its contract. The rental money had been sent to the lessors' agent bank prior to the time Gulf received copies of Shatford's deeds. Shatford should not now look to Gulf for his portion of that rental money; he should look to those persons who received it--the persons from whom he secured his interest.

The judgment is reversed and the cause is remanded with direction to enter judgment for Gulf Refining Company.

Reversed and remanded.

NOTES

1. ATLANTIC REFINING CO. v. SHELL OIL CO., 217 La. 576, 46 So.2d 907 (1950). A mineral lease executed on January 30, 1943, included an "unless" type drilling and rental clause and an assignment clause providing that a change in ownership was not to be binding on lessee or impair the effectiveness of any payments made, until the lessee "shall have been furnished, forty-five days before payment is due, a certified copy of recorded instrument evidencing any transfer," sale or other change in ownership. The court commented as follows on the action brought by plaintiff to obtain recognition of the validity of its lease:

On the 14th day of May, 1943, Furlow [the lessor] conveyed to defendant Shell Oil Company, Inc., a one-half interest to the minerals in and under the above-described property. There was agreement between the parties however that Shell was not to participate in the rentals, and to evidence that intention certain words and provisions of the printed form to the contrary were stricken out on the typewriter and the instrument was initialed by the vendor in the margin opposite these deletions, so that the printed form appeared thus: ' . . . Conveys unto Grantee One Half (1/2) of all of the royalties, ~~rentals~~ and other benefits, ~~including money rentals payable for drilling operations~~ accruing under any valid oil, gas or mineral lease or servitude on said property which has heretofore been filed for record . . ..' While this instrument transferring a half interest in the minerals to the Shell Company was duly recorded in the Conveyance Records of Caldwell Parish on June 7, 1943, neither the plaintiff nor its assignor Nesbitt was ever served with a certified copy thereof.

The delay rentals for the years 1944 and 1945 were deposited to Furlow's credit with the depositary named in the contract of lease, the Ouachita National Bank. It appears, however, that on December 21, 1945, some 40 days before the anniversary date, the plaintiff, instead of depositing the entire amount of the delay rentals for the ensuing year to Furlow's account, deposited in the Ouachita National Bank the $120 with instructions that $60 thereof was to be credited to Furlow's account and $60 to the account of the Shell Company. This action was taken by plaintiff pursuant to a title opinion by its attorneys, dated March 12, 1945, and based on an examination of an abstract procured by plaintiff and prepared by the Clerk of Court of Caldwell Parish. The abstract contains a correct copy of the mineral conveyance to Shell as it is recorded, without showing the deletions or making reference thereto.

. . .

Plaintiff contends, in which it is supported by the opinion of the trial judge, that it was entitled to rely on the Conveyance records; that contracts must be construed with reference to the words which they contain, and, in construing the same, words which have been erased are not to be considered, citing Corpus Juris, Vol. 13, Contracts, Sec. 505, p. 539; 17 C.J.S., Contracts, § 317; and consequently that the mineral deed should be construed without consideration of the deleted words, and, so construed, that it conveyed one-half of the delay rentals to the Shell Oil Company; that therefore the payments made by plaintiff maintained its lease in force and the subsequent lease from Furlow to Reese was null.

The fallacy of this conclusion lies, we think, in the fact that the plaintiff had no occasion to rely on the public records or to construe the contract, as no certified copy of the recorded deed evidencing the transfer of mineral interest from Furlow to Shell was delivered to plaintiff and it was not required to interpret the intention of the parties or to act in reference thereto; and as no demand for payment had been made on plaintiff by Shell; nor had pressure been brought to bear by anyone which would have caused plaintiff to turn to the public record in determining a course of action. In fact, there was no occasion for delivery of a certified copy of the recorded deed in view of the agreement between Furlow and Shell that the delay rental was to be received in its entirety by Furlow. Plaintiff paid the whole delay rental to Furlow for the years 1944 and 1945, and although it received from its attorney a 'Final Opinion on Title' on March 12, 1945 (more than ten months before the succeeding delay rental payment became due on January 30, 1946), without further inquiry, either to Furlow or to Shell, proceeded to make the rental payment on December 21, 1945 (some 40 days before the due date) by depositing the amount in the named depositary, Ouachita National Bank, half to the credit of Furlow and half to the credit of Shell.

. . .

It is to be observed that in the title opinion upon which plaintiff bases its good faith in making the 1946 payment as it did, plaintiff was advised that inasmuch as it had not been served with a certified copy of the conveyance from Furlow to the Shell Company at least 45 days before payment was due, as required under section 8 of the contract of lease, the first and second rental payments could properly have been made in their entirety to Furlow and '*subsequent rental payments may be made in their entirety, at the election of the Atlantic* [plaintiff], *to L.C. Furlow until a certified copy of this mineral conveyance is served upon the Atlantic,* or the Atlantic may divide the rental payment, even if it has not received a certified copy of this deed, and pay 1/2 to Furlow and 1/2 to the Shell Oil Company, Incorporated.' (Emphasis ours.) Thus it may be seen that the plaintiff was not acting on the record--for, after relying on the title opinion and after this controversy arose, an examination of the original document disclosed the fact of the deletions and was sufficient to warrant the conclusion on the part of plaintiff that 'it would appear that all rental payments should have gone to Mr. L.C. Furlow'--but acted at its own peril in choosing an insecure method without further inquiry, when it was protected by its own contract; and furthermore, made no effort to straighten out its difficulties with Furlow, the real party in interest.

. . .

We therefore conclude that under the very terms of the contract between the plaintiff and Furlow, no drilling operations having taken place within the first twelve months or within the extended periods for the years 1944 and 1945, and plaintiff having failed to pay the required amount for the year 1946, the lease was terminated by its own terms. . . .

For the reasons assigned the judgment of the lower court is annulled and set aside, and it is now ordered, adjudged and decreed that there be judgment in favor of the defendants dismissing plaintiff's suit at its cost.

A dissenting opinion by Hawthorne, J., urged that the rental was correctly paid, reasoning as follows:

The real problem in this case is whether the paragraph containing the deleted words is sufficient expressly to reserve to the landowner, Furlow, the right to receive all the delay rental under the lease previously granted by him to Nesbitt, assignor of Atlantic.

In my opinion plaintiff herein is a third party insofar as the mineral deed is concerned, and in construing the deed, as such third party, was under no duty to consider or give effect to the words deleted by the parties thereto. Since the instrument in its entirety is free from ambiguity, when read without the deleted words, the lessee, Atlantic Refining Company, was not required to read the deleted words in connection with the instrument as written in order to determine the intent of the parties. Furthermore, in my opinion, this court has no cause to consider the deleted words in determining the intent of the parties and the meaning of the instrument. Our Civil Code in Title IV, Of Conventional Obligations, Section 5, of the Interpretation of Agreements, Article 1945, par. 3, charges us 'That the intent is to be determined by the words of the contract, when these are clear and explicit and lead to no absurd consequences'. The words of this agreement are clear and unambiguous and lead to no absurd consequences, and by them Furlow conveyed unto Shell one-half of all the royalties *and other benefits* accruing under any valid oil, gas, and mineral lease on the property which had theretofore been filed for record. Certainly the delay rental is a benefit accruing under a valid oil and gas lease. By this clause conveyance was made of one-half of the amount of the delay rental, and I cannot read therein the specific reservation of the delay rental which is necessary to circumvent the general conveyance clause under the jurisprudence. If this instrument does not embody the intention of the parties thereto, the remedy is to have the instrument reformed to express that intention, and not to read into the instrument a provision that is wholly absent by implying a positive intent from words which were deleted from the instrument entirely.

. . .

Atlantic paid the rental correctly to continue the lease in effect, and for this reason I respectfully dissent.

McCALEB, J., dissenting from the judgment of the court, made the following comments:

I readily concede that plaintiff acted at its peril in voluntarily taking notice of the transfer from Furlow to Shell and, if the delay rentals tendered by it were not in accordance with the provisions of that conveyance, it must suffer the consequences of its mistake. But I demur to the ruling that plaintiff's payment was erroneous. On the contrary, Furlow conveyed to Shell one-half of 'all the royalties, and other benefits, accruing under any valid oil, gas or mineral lease . . ." This grant of all royalties and other benefits must be considered as including rentals as they are not otherwise mentioned in the conveyance. Obviously, then, payment of one-half of the delay rentals to Furlow and the other one-half to Shell was in conformity therewith.

The only way it can be deduced that the transfer to Shell did not include rentals is to notice that rentals, including delay rentals, have been deleted from the printed form used for the transfer and to interpret that, by this deletion, the parties intended that Furlow was to retain the right to receive all rentals. Taking cognizance of this deletion would be appropriate in an action between the parties to the instrument . . . but a third person (like plaintiff), relying on the public record, was not required to investigate deletions which formed no part of the contract and which were not written, so far as he was concerned. . . . To hold otherwise is to say that the peril to which plaintiff subjected itself extended beyond the public records and that it had the onus of determining correctly an unexpressed intention of the parties to the instrument. To this, I cannot subscribe and accordingly I respectfully dissent.

2. *A,* owning an undivided 1/2 interest, *B* owning an undivided 1/4 interest,

and *C*, owning an undivided 1/4 interest in 80 acres, joined as co-lessors in a lease of Blackacre providing for $1 per acre rental and further providing that "No sale or assignment by Lessor shall be binding on Lessee until Lessee shall be furnished by registered U.S. Mail with a certified copy of recorded instrument evidencing same." *C* thereafter acquired the interest of *A* by a duly recorded instrument, but no notice thereof was given to the lessee. *B* and *C* partitioned Blackacre, and *C* was allotted 60 acres and *B* 20 acres thereof. Subsequently *C* conveyed the 60 acres to *X* by a duly recorded instrument, and *X* furnished a copy thereof to the lessee. On the next rental anniversary date, the lessee paid the delay rental to the depository bank with instructions to credit the payment as follows: to *A*, $40; to *B*, $20; and to *X*, $20. Thereafter *X* filed suit for the cancellation of the lease, alleging improper payment of the delay rental and alleging that he had complied with the provisions of the lease when he furnished defendant with a certified copy of the deed whereby he acquired from *C* the segregated portion of the leased premises, comprising the 60-acre tract. How should the case be decided? *See* Pearce v. Southern Natural Gas Co., 220 La.1094, 58 So.2d 396, 1 O. & G.R. 690 (1952).

3. Notice of assignment to plaintiff of an undivided interest in leased premises was not given to defendant lessee in the manner specified by the assignment clause of the lease covering the premises. Subsequently lessee caused an abstract of title to be prepared for the purpose of preparing a royalty division order, which abstract included a copy of the mineral deed in favor of plaintiff. Moreover a pooling and unitization agreement was prepared by defendant and signed by plaintiff. Lessee paid delay rentals to the original lessor without giving effect to the assignment to plaintiff. Is the lease subject to cancellation insofar as it covers plaintiff's undivided interest in the minerals? *See* Garelick v. Southwest Gas Producing Co., 129 So.2d 520, 15 O.&G.R. 740 (La.App.1961).

The heirs of the lessor filed under Louisiana law, an Act of Correction, affecting a lease held by Amoco. The lease had two specific provisions, one of which authorized Amoco to make delay rental or shut-in royalty payments to the credit of a bank, notwithstanding the death or incapacity of the lessor and the second which required the lessees to give written notice to the lessor in the case of a change in ownership. The heirs claimed that because a shut-in gas royalty payment was wrongfully made to the decedent after the Act of Correction was properly filed, the lease would automatically terminate. What result? Lapeze v. Amoco Production Co., 842 F.2d 132, 97 O.&G.R. 552 (5th Cir. 1988).

4. A section of land, Blackacre, was leased by *L* to *E* by an "unless" lease providing for $1 per acre per year delay rental with rental and assignment clauses like those on p. 182 and p. 206. Thereafter *L* assigned his entire interest in the NW1/4 of Blackacre to *A;* timely and proper notice thereof was given *E.* Consider the consequences of these non-cumulative situations:

(a) On the next ensuing rental anniversary date of the lease, *E* tendered $640 to *L* as rental; no payment of rental was tendered to *A*.

(b) On the next ensuing rental anniversary date of the lease, *E* tendered $160 to *A* as rental; no payment of rental was tendered to *L*.

(c) Before the next rental anniversary date of the lease, a well was commenced on the NW1/4 of Blackacre; no rental was tendered on or before the next rental anniversary date to *L* or to *A*.

(d) Production was obtained from a well drilled on the NW1/4 of Blackacre; the primary term of the lease then expired without drilling on the balance of Blackacre.

See Martin & Kramer, *Williams and Meyers Oil and Gas Law* §§ 409.4, 606.4, 677.4.

5. Can the parties avoid the problem in *Shatford* and in Gulf Production Co. v. Continental Oil Co., 139 Tex. 164, 164 S.W.2d 488 (1942) by placing a provision delaying the effective date of the assignment until a 30 or 60 day period has passed? In the *Continental Oil* case the payments were made three years in advance.

6. Lessees have avoided the problem of mistakes in delay rental payments by negotiating so-called "paid-up" leases whereby all of the delay rentals are paid at the time of the execution of the lease. While obviously exposing the lessee to a larger cash payment at the time of lease execution, it avoids the problem of having the lease terminate during the primary term.

_____

Well Commencement Clause

The typical lease excuses payment of delay rentals if a well is "commenced" on the land before the anniversary date. The following is a typical example of such a clause. As one would expect, there are numerous factual issues that arise from having to determine whether or not the lessee has complied with the commencement of drilling operations requirement. Additionally, either by express language or by judicial implication, the lessee will be required to exercise diligence in the continuation of such operations. Similar factual issues are raised when it comes to applying various leasehold savings provisions that may require the lessee to engage in drilling or reworking operations. Leasehold savings clauses are discussed at Section 3 *infra*.

This lease shall terminate . . . unless Lessee shall make or tender the payments hereinafter provided or shall within said period commence drilling operations for a well for oil and gas on the leased land and prosecute the drilling of such well with reasonable diligence until oil or gas is found in quantitites deemed paying by Lessee or until Lessee deems that futher drilling would be unprofitable or impracticable in which event Lessee may abandon the well . . . 3 Martin & Kramer, *Williams and Meyers Oil and Gas Law* §605.2 (from a California lease form).

_____

HALL v. JFW, INC.

Kansas Court of Appeals, 1995.
20 Kan.App.2d 845, 893 P.2d 837.

JEAN F. SHEPHERD, District Judge, Assigned:

John Hall, lessor, and JFW, Inc., lessee, entered into an oil and gas lease on August 3, 1990. The lease provided in pertinent part:

If no well be commenced on said land on or before August 3rd, 1991, this lease shall terminate as to both parties. . . .

If the lessee shall commence to drill a well within the term of this lease . . . the lessee shall have the right to drill such well to completion with reasonable diligence and dispatch.

The lease was altered after it was recorded. The altered lease stated it was entered into on August 13, 1990, and stated the well must be commenced on or before August 13, 1991.

The lease also contained a delay rental clause, but JFW did not timely tender payment.

JFW performed the following activities on the Hall lease:

10-22-90: Title opinion

3-30-91: Measured and staked location for well

4-2-91: Surveyed elevation of site

6-25-91: Received KCC approval of intent to drill

June/July 1991: Talked with geologist

7-18-91: Received bid for drilling mud

7-20-91: Reached verbal agreement with Duke Drilling to drill well; tentative date set for late July or early August

7-24-91: Restaked location for well

More than three days before 8-6-91: Told Duke Drilling to get rig on lease.

8-10-91: Signed written contract with Duke Drilling

Agent of Duke Drilling dug drilling pits and leveled location

8-11-91: Drilled water supply well

8-12-91: Prepared rotary hold and run-around

8-12-91: Duke Drilling picked up surface casing

8-14-91: Duke Drilling moved drilling rig onto lease and spudded well

8-14-91 to 8-20-91: Well dug to 3,000 feet; production casing installed and cemented

8-20-91 to 9-3-91: Cement allowed to cure

9-3-91: Ready to move completion rig onto lease.

Prior to completion of the well, Hall sought a determination that the lease had terminated and requested a temporary restraining order preventing JFW from entering the lease. On September 3, 1991, a temporary restraining order was issued. The trial court later denied a temporary injunction and determined the well had been commenced by August 3, 1991. This court reversed the judgment after finding the trial court prematurely had decided the merits of the case.

Upon remand, the parties completed discovery and filed motions for summary judgment. The trial court found JFW had a "firm commitment" from Duke Drilling prior to the commencement deadline and therefore determined JFW had commenced the well prior to expiration of the lease, and entered summary judgment for JFW.

Hall argues the trial court erred in finding a well had been commenced prior to the lease termination date. He argues the preliminary steps taken in anticipation of drilling were not sufficient to constitute commencement of the well under the terms of the lease. We agree. . . .

This court is not bound by the trial court's interpretation of the written contract. Simon v. National Farmers Organization, Inc., 250 Kan. 676, 680, 829 P.2d 884 (1992). The rules of construction for oil and gas leases are well established and mirror the rules for construction of contracts generally.

> [T]he intent of the parties is the primary question; meaning should be ascertained by examining the document from all four corners and by considering all of the pertinent provisions, rather than by critical analysis of a singleor isolated provision; reasonable rather than unreasonable interpretations are favored; a practical and equitable construction must be given to ambiguous terms; and any ambiguities in a lease should be construed in favor of the lessor and against the lessee, since it is the lessee who usually provides the lease form or dictates the terms thereof. Jackson v. Farmer, 225 Kan. 732, 739, 594 P.2d 177 (1979).

Unambiguous contracts are enforced according to their plain, general, and common meaning in order to ensure the intentions of the parties are enforced. The intent of the parties is determined from the four corners of an unambiguous instrument, harmonizing the language therein if possible. Ambiguity does not appear until the application of the pertinent rules of interpretation to the face of the instrument leaves it genuinely uncertain which of two or more meanings is the proper meaning. . . .

When analyzing whether a lease has terminated because the lessee has not timely commenced a well, our Supreme Court has looked to the language of the controlling instrument, reading it as a whole, to determine the nature and extent of the lessee's obligation. See Shoup v. First Nat'l Bank, 145 Kan. 971, 976, 67 P.2d 569 (1937) (an escrow agreement required not only timely commencement but continued operations with due diligence to completion).

Our courts have considered clauses which require the lessee to commence operations for drilling in a number of instances. [The court then discussed a series of Kansas cases including A & M Oil, Inc. v. Miller, 11 Kan.App.2d 152, 154-55,

715 P.2d 1295 (1986), Herl v. Legleiter, 9 Kan.App.2d 15, 668 P.2d 200 (1983), Phillips v. Berg, 120 Kan. 446, 243 Pac. 1054 (1926), Baughman v. Ault, 115 Kan. 553, 223 Pac. 815 (1924) and Hennig v. Gas Co., 100 Kan. 255, 257, 164 Pac. 297 (1917).]

There is authority in Kansas which hints that something less that actual drilling might satisfy a requirement of the "commencement of drilling." The Supreme Court has apparently assumed something less than spudding might constitute commencement of drilling operations. However, the lease in this case at hand required that the lessee actually "commence to drill" before the expiration of the lease in order to avoid termination.

Professor David Pierce offers the following analysis based on the limited authority implying that something less that actual drilling might be sufficient under some leases:

> In Herl v. Legleiter the Kansas Court of Appeals suggests something less than actual drilling may be sufficient to satisfy a commencement clause. However, it appears where something less than actual drilling is being relied upon, the lessee should be required to demonstrate what amounts to an irrevocable commitment to conduct operations, to completion, on the leased land. The best evidence of this, absent actual drilling on the premises, is an enforceable contract with a third party to drill a well on the leased land. However, the lessee runs a risk when something less than an appropriate rig is in place on the lease. 1 Pierce, *Kansas Oil and Gas Handbook* § 9.34(1991).

Allowing an irrevocable commitment to conduct operations to completion to satisfy a lease requiring the commencement of drilling, and not merely operations for drilling, too broadly states the law of Kansas. Here, actual drilling had not been commenced by the lease anniversary date, whether August 3 or 13. The only actions taken on the leased property prior to August 13 were the staking of the well location, an elevation survey, the signing of a written contract with Duke Drilling, an agent of Duke Drilling digging drilling pits and leveling the location, and the drilling of a water supply well.

An analysis of the plain terms of the lease reveals that the trial court erred in granting summary judgment for JFW. The lease provides that the lessee must "*commence to drill a well within the term of this lease.*" (Emphasis added.) Under the plain terms of the lease, actual drilling was required prior to the termination date.

JFW alleges no fraud, mistake, or duress and was free to contract on its own terms. JFW is now bound by those terms.

JFW argues that its activities up until September 3, 1991, were sufficient to extend the lease. The problem with JFW's argument is that JFW fails to distinguish the activities performed prior to the lease termination date from those occurring after that date, which are relevant only in determining if the lessee used due diligence in completing the well and not in determining whether the well was timely commenced.

JFW extensively reviews cases from other jurisdictions which would support its position that minimal activities on the leased property are sufficient to constitute commencement. These cases, however, are not persuasive, given current Kansas law. Significantly, the Shoup court declined the lessee's invitation to adopt a more liberal interpretation of the commencement clause in conformance with other jurisdictions.

JFW also argues a finding that the well had not been timely commenced would be grossly inequitable and would not constitute a reasonable interpretation of the lease. Kansas courts have refused to apply equity in cases involving commencement. See 1 Pierce, *Kansas Oil & Gas Handbook* §9.32 (1991); Shoup, 145 Kan. at 980, 67 P.2d 569 (Allen, J., dissenting.).

The case is reversed and remanded for entry of judgment in favor of Hall.

NOTES

1. Why does the court ignore equitable considerations in interpreting the well commencement provisions of the lease? Is it because the primary term is considered a fee simple determinable estate which automatically terminates upon the failure to comply with the limitation?

2. The court uses several canons of construction to assist it in interpreting the lease? Identify them and assess whether they help or hinder the court in determining the intent of the parties? Should the courts apply the same canons of construction for oil and gas leases that they use for contracts? for deeds? See generally, Bruce Kramer, *The Sisyphean Task of Interpreting Mineral Deeds and Leases: An Encyclopedia of Canons of Construction*, 24 Tex.Tech L.Rev. 1 (1992).

3. If the clause requires diligent operations does that mean that the lessee must engage in "continuous" operations. See generally Amoco Production Co. v. Carruth, 512 So.2d 571, 98 O.&G.R. 84 (La.App. 1987), *writs refused,* 516 So.2d 366 (La. 1988) criticized by Professor Martin at Discussion Note, 98 O.&G.R. 93.

4. Which of the following activities would excuse the payment of delay rentals under a well commencement clause if such activities took place prior to the anniversary date:

(a) Obtaining a drilling permit from the regulatory agency.

(b) Signing a drilling contract with an oil well drilling contractor. 1 David Pierce, *Kansas Oil and Gas Handbook* § 9.34 (1991).

(c) Making on-site preparations including building a board road, clear trees and bulldozing, or staking out the well site. Ridge Oil Co. v. Guinn Investments, Inc. 148 S.W.3d 143 (Tex. 2004); Oelze v. Key Drilling, Inc., 135 Ill.App.3d 6, 90 Ill.Dec. 1, 481 N.E.2d 801, 87 O.&G.R. 277 (1985); Petroleum Energy, Inc. v. Mid-America Petroleum, Inc., 775 F.Supp. 1420, 116 O.&G.R. 227 (D.Kan. 1991); Breaux v. Apache Oil Corp., 240 So.2d 589. 37 O.&G.R. 221 (La.App.

1970); Kaszar v. Meridian Oil & Gas Enterprises, Inc., 27 Ohio App.3d 6, 499 N.E.2d 3, 92 O.&G.R. 453 (1986).

(d) Erecting a rig and drilling platform at the drill site.

(e) Erecting a rig and drilling platform at the drill site, digging a slush pit and spudding in the well.

(f) Suppose in (d) that additional preparations for drilling, such as digging slush pits, are deliberately slowed down so that the lessee can examine the electric log on an offset well in which it has an interest. *See* Stoltz-Wagner & Brown v. Duncan, 417 F.Supp. 552, 55 O.&G.R. 315 (W.D.Okla. 1976).

(g) Suppose in (e) that lessee had failed to secure the required drilling permit from the state conservation agency. *See* Gray v. Helmerich & Payne, Inc., 834 S.W.2d 579, 582, 125 O.&G.R. 418 (Tex.App. 1992, writ denied); Goble v. Goff, 327 Mich. 549, 42 N.W.2d 845 (1950).

(h) Suppose in (e) that lessee has no rig of its own and makes no effort to get one until suit is filed after the end of the primary term. *See* Herl v. Legleiter, 9 Kan.App.2d 15, 668 P.2d 200, 78 O.&G.R. 19 (1983).

(i) Suppose in (e) the lessee commences drilling with a shallow drilling rig and drills a few feet a day for several weeks before commencing drilling in earnest. *See* Bunnell Farms Co. v. Samuel Gary, Jr. & Assocs., 47 P.3d 804, 153 O.&G.R. 1 (Kan.App. 2002); Wilds v. Universal Resources Corp., 1983 OK 35, 662 P.2d 303, 76 O.&G.R. 291; Son-Lin Farms, Inc. v. Dyco Petroleum Corp., 589 F.Supp. 1, 82 O.&G.R. 584 (W.D.Okl. 1982).

(j) Spudding a well on adjacent surface acreage intending to slant drill under the lease, but not penetrating the acreage covered by the lease. *See* A & M Oil, Inc. v. Miller, 11 Kan.App.2d 152, 715 P.2d 1295, 88 O.&G.R. 453 (1986).

See Martin & Kramer, *Williams and Meyers Oil and Gas Law* §§ 606.1, 618.1.

5. Must a lessee under a well commencement clause complete the drilling of the well during the primary term of the lease? Lessee drills Well #1 within the time period authorized by the commencement clause. The well produces some gas but eventually is plugged. The primary term expires prior to the plugging of the well. Well #2 is commenced shortly after the plugging of Well #1. Has the lease terminated? McClain v. Ricks Exploration Co., 1994 OK CIV APP 76, 894 P.2d 422, 132 O.&G.R. 164.

————

2. DELAY RENTALS—THE "OR" LEASE

Lessees agree to commence a well on said premises within _____ years from the date hereof, or pay lessor _____ cents an acre per annum, payable quarterly in advance from the _____ day of _____, 19__, until said well is commenced or this lease surrendered.

WARNER v. HAUGHT, INC.

Supreme Court of Appeals of West Virginia, 1985.
174 W.Va.722, 329 S.E.2d 88, 85 O.&G.R. 199.

McGRAW, JUSTICE. The appellants, plaintiffs in four of the actions below, assert several points of error in support of their request for reversal of the circuit court's final order. Upon the findings and conclusions which follow, the ruling of the circuit court is reversed.

I

During the month of November, 1979, the appellants leased, by separate instruments, various tracts of land in Pendleton County to D. & H. Oil Company, for oil and gas exploration and development. Each lease provides for a primary term of ten years. Further, in each lease, the lessee agreed to pay an annual delay rental, in advance, until a well yielding a royalty to the lessors is drilled on the premises. The leases contain no provision setting forth a remedy or course of action in the event of late payment or nonpayment of the delay rental. Each of the appellants, however, alleges that the agent for the lessee represented to them at the time of the execution of each lease, that if the terms were not complied with, including the rental provision, the lease would be null and void. Each lease does contain a surrender clause which permits the lessee, or its successors or assigns, to cancel the lease at any time, upon the payment of one dollar to the lessor. In May of 1980, all of the subject leases were assigned by D. & H. Oil Company to the appellee, Haught, Inc., and shortly thereafter, recorded in the Pendleton County Clerk's office.

In 1981, the appellee failed to make the delay rental payments on any of the subject leases when due. After the expected payments became overdue by a month or longer, all but one set of appellants separately mailed to the appellee or the original lessee, D. & H. Oil Company, by regular mail, notices of cancellation advising that the lessors considered the leases to be null and void due to the failure to pay delay rentals when due. On January 21, 1982, prior to the dates of some of the above-mentioned notices, the appellee mailed delay rental checks in the appropriate amounts to all of the appellants. The checks were back-dated to the respective due dates under each lease. The appellants, in all instances, refused the delay rental checks.

In February and March of 1982, each set of appellants initiated a separate civil action in the Circuit Court of Pendleton County, seeking a declaratory judgment declaring their lease forfeited and abandoned due to the appellee's failure to make timely payment of the delay rental, and removing the lease as a cloud upon the title to their real estate. The appellee answered the civil actions setting forth that the appellants had no right to the relief sought, principally because they had failed to comply with the provisions of West Virginia Code § 36-4-9a (1985 Replacement Vol.). Ultimately, counsel for both sides submitted motions for summary judgment, with supporting affidavits and memoranda, on behalf of their respective clients. On February 9, 1983, the circuit court issued a memorandum order denying the appellants' motion and granting the appellee's motion, thereby dismissing the appellants' complaints. It is from that final order that the appellants have brought this appeal.

II

The central issue presented to this Court is whether the lease cancellation provision of West Virginia Code § 36-4-9a (1985 Replacement Vol.) applies to the oil and gas leases which are the subject of this appeal. The statute provides, in pertinent part, that:

> Except in the case where operations for the drilling of a well are being conducted thereunder, any undeveloped lease for oil and/or gas in this State hereafter executed in which the consideration therein provided to be paid for the privilege of postponing actual drilling or development or for the holding of said lease without commencing operations for the drilling of a well, commonly called delay rental, has not been paid when due according to the terms of such lease, or the terms of any other agreement between lessor and lessee, shall be null and void as to such oil and/or gas unless payment thereof shall be made within sixty days from the date upon which demand for payment in full of such delay rental has been made by the lessor upon the lessee therein, as hereinafter provided, except in such cases where a bona fide dispute shall exist between lessor and lessee as to any amount due or entitlement thereto or any part thereof under such lease.

> No person, firm, corporation, partnership or association shall maintain any action or proceeding in the courts of this State for the purpose of enforcing or perpetuating during the term thereof any lease heretofore executed covering oil and/or gas, as against the owner of such oil and/or gas, or his subsequent lessee, if such person, firm, corporation, partnership or association has failed to pay to the lessor such delay rental in full when due according to the terms thereof, for a period of sixty days after demand for such payment has been made by the lessor upon such lessee, as hereinafter provided.

> The demand for payment referred to in the two preceding paragraphs shall be made by notice in writing and shall be sufficient if served upon such person, firm, partnership, association, or corporation whether domestic or foreign, whether engaged in business or dissolved, by United States registered mail, return receipt requested, to the lessee's last known address.

> A copy of such notice, together with the return receipt attached thereto, shall

be filed with the clerk of the county commission in which such lease is recorded, or in which such oil and/or gas property is located in whole or in part, and upon payment of a fee of fifty cents for each such lease, said clerk shall permanently file such notice alphabetically under the name of the first lessor appearing in such lease and shall stamp or write upon the margin of the record in his office of such lease hereafter executed the words "canceled by notice"; and as to any such lease executed before the enactment of this statute said clerk shall file such notice as hereinbefore provided and shall stamp or write upon the margin of the record of such lease in his office the words "enforcement barred by notice."

As the record indicates, none of the appellants complied with, nor do they claim to have complied with, the "demand for payment" requirements of the statute.[38] Rather, the appellants contend that the above-quoted provision applies only to "or" type oil and gas leases. Further, the appellants assert that the leases in question are not "or" type leases, but are "unless" type leases, and therefore, beyond the purview of the statute.

A brief review of the general characteristics and primary distinctions between these two classical types of oil and gas leases is warranted at this point. To begin, the "or" and "unless" nomenclature stems from the effect and obligations created under the drilling and rental clauses of typical leases.[39] In an "or" lease, the lessee covenants to do some alternative act, usually to drill a well or to pay periodic rentals, to maintain the lease during its primary term. Simply put, the lessee must "drill or pay". Conversely, the lessee in an "unless" lease does not covenant to drill a well or pay rentals. However, if the lessee does neither within the time intervals specified therein, the lease automatically expires by its own terms. In typical form, "if" no well is drilled, the lease terminates "unless" rentals are paid. *See* 3 H. Williams, *Oil and Gas Law* §§ 605 & 606 (1984); R. Donley, *The Law of Coal, Oil and Gas in West Virginia and Virginia* §§ 80 & 83 (1951).

As further elaborated upon by one authority:

The result is that the "unless" clause is construed as a clause of special limitation whereas the "or" clause is construed as a clause of condition. The consequences of this construction of the two clauses include the following:

(1) If the lease contains an "unless" clause, no affirmative act is required on the part of the lessee if he wishes to terminate the lease before the expiration

---

[38]   In some cases, notices of cancellation were sent, but apparently not as an attempt to comply with the statute. In any event, in all instances the delay rentals were tendered within the sixty-day period provided under the statute.

[39]   We note, however, that it is not the labeling, but rather the legal effects and obligations created by the language of the lease which is of primary importance. This point is of particular relevance in cases such as this one simply because the language of many leases does not track the form contained in classical "or" and "unless" clauses.

of the primary term; failure to commence a well or to pay rentals on or before the rental paying date will cause the lease to expire. On the other hand, if the lease contains an "or" clause, some affirmative act is required, *viz.,* a surrender of the lease, if the lessee wishes to terminate the lease before the expiration of the primary term.

(2) Termination of the lease is automatic by operation of the "unless" clause; in the case of the "or" clause, some action by the lessor is required to effect a termination.

(3) Since the termination of the lease by operation of an "unless" type drilling and rental clause is automatic, equitable rules against forfeiture are not applicable to a determination of whether the lease has expired. On the other hand, since the termination of the lease by operation of an "or" type drilling and rental clause requires entry by the lessor, and, in the case of a dispute, judicial determination of the rights of the parties, equitable rules against forfeiture are applicable to a determination of whether the lease should be canceled. 3 H. Williams, *supra* at § 606 (footnote omitted).

Returning to the immediate inquiry, whether the leases involved in the instant case are, in effect, "or" leases or "unless" leases, we begin with the pertinent language. Each lease contains the following drilling and rental clause:

> The said Lessee covenants and agrees to pay rental at the rate of $1.00 per acre, per year, plus $2.00 Bonus per acre the first year . . . in advance, . . . until, but not after, a well yielding royalty to the Lessors is drilled on the leased premises, and any rental paid for time beyond the date of completion of a gas well shall be credited upon the first royalty due upon the same and all rentals shall cease after the surrender of this lease as hereinafter provided for.

The surrender clause referred to above, as contained in each lease, states that:

> Upon payment of one ($1.00) Dollar at any time, by the party of the second part [the lessee], or by its successors and assigns it or they shall have the right to surrender this lease for cancellation, after which all payments and liabilities thereafter to accrue under any by virtue of its terms shall cease and determine, and this lease becomes absolutely null and void.

The appellants cite to the wording in the drilling and rental clause, "rental . . . in advance, . . . until, but not after, a well yielding royalty to the Lessors is drilled. . . . " This language, according to the appellants, creates, in effect, an "unless" lease which automatically terminates upon failure to pay the delay rental. . . .

To begin, an "unless" type lease places no obligation upon the lessee. However, in the instant leases the terms clearly provide that the lessee covenants and agrees to pay rental. This language does not create a special limitation upon the primary term, but rather an absolute obligation to drill a well or pay rental in advance of each year of the primary term. Failure to do one of these acts is an actionable breach of contract.

Additionally, as previously quoted, the subject leases contain a surrender clause permitting the lessee to voluntarily surrender the leases, thereby terminating further obligations. In delineating the distinctions between "or" and "unless" leases, Professor Donley has discussed the relevance of the "surrender clause," where he observes that:

> With [the unless] type of clause the lessee does not need the protection of a surrender clause in order to escape liability for failure to drill. . . . The lessee may simply do nothing and let his leasehold estate terminate and there is an end to the relationship between the parties. However, in the drill or pay type of clause, the lessee does need the device of a surrender clause in order that he may avoid liability for nonperformance of either of his alternative promises where it appears that the field is a nonproductive one. R. Donley, *supra,* at § 73.

We conclude that each lease in question, by its written terms, clearly obligated the lessee to do one of three acts during the primary term: (1) drill a productive well; (2) pay delay rentals; or (3) surrender the lease. Under the effect of these alternative obligations, the leases would not automatically terminate by the failure to seasonably tender the delay rentals. However, another facet of this case, other than the written terms of the leases, is also relevant to the question of the applicability of West Virginia Code § 36-4-9a (1985 Replacement Vol.). Closely related to the issue of the meaning of the express language of the subject leases is the question regarding the effect of the alleged oral statements to the appellants by the lessee's agent to the effect that failure to pay rentals on time would render the leases null and void. If admissible, and found to have been actually made, these oral statements would, in effect, change the character of these leases to resemble "unless" leases, automatically terminating upon failure to pay the rentals on time.

. . .

The appellants' allegations of the parol representations made to them, which, for the purposes of the summary judgment motions were deemed to be true, raise factual issues relating to the possibility of mistake or material misrepresentation.

. . .

Therefore, because these leases may, upon remand, be found to include an automatic termination provision added by the parol representations, we specifically address whether the notice and demand requirements provided under section 36-4-9a are applicable to such "unless" type leases. For the following reasons, we find "unless" leases to be unaffected by the notice and demand provisions.

First of all, the statute itself, in spite of its initial reference to "any" lease, manifests an intent that it was designed to deal with the rental collection problems inherent in the "or" type lease. In the final paragraph of the statute, the legislature declared that, "The continuation in force of any such lease after demand for and failure to pay such delay rental . . . is deemed by the legislature to be opposed to public policy against the general welfare." This "continuation in force after failure to pay" problem the legislature sought to alleviate has never been a problem in "unless" type leases, which automatically extinguish upon failure to pay.

Prior to the enactment of section 36-4-9a in 1943, when a lessee was in default on rental obligations under an "or" type lease that did not contain a forfeiture clause for the benefit of the lessor, generally, the lessor's sole remedy was a suit, for past due rentals, and the unproductive lease remained in force. In some cases, where the facts were such that a court could find that equitable forfeiture or a declaration of abandonment was appropriate, the lessor could have the lease terminated. The latter two remedies, however, were the exception rather than the rule. Thus, the statute provided an expeditious means, without resort to judicial process, to require lessees in "or" type leases to pay the rentals due under the lease if they did not wish to have the lease canceled under the statute.

Additionally, there is a distinct notion of inconsistency in requiring, in a lease which obligates the lessee to do nothing, notice and demand before *automatic* termination. In this regard, we note that even when contained within "unless" leases themselves, such notice and demand requirements have been held to be void as being "inconsistent and repugnant" or "irreconcilable" with the special limitation intent of the leases. *See, e.g.,* Clovis v. Carson Oil & Gas Company, 11 F.Supp. 797 (E.D.Mich.1935); Richfield Oil Corporation v. Bloomfield, 103 Cal.App.2d 589, 229 P.2d 838 (1951); McDaniel v. Hager-Stevenson Oil Company, 75 Mont. 356, 243 P. 582 (1926); Lewis v. Grininger,198 Okla. 419, 179 P.2d 463 (1947); Waddle v. Lucky Strike Oil Company, Inc., 551 S.W.2d 323 (Tenn.1977).

Accordingly, we hold that under circumstances that do not permit forfeiture or indicate abandonment, an oil and gas lease binding the lessee to drill a well on the leased premises within a certain period, or, in lieu thereof, make periodical payments of delay rental, and containing no clause of special limitation which would effect an automatic termination of the lease for failure of the lessee to perform one of the specified obligations, is not terminable due to nonpayment of the rental without the lessor's compliance with the notice and demand provisions under West Virginia Code § 36-4-9a (1985 Replacement Vol.).

. . .

III

The next question presented in this appeal is whether the circuit court erred in ruling that equity would neither require nor permit a forfeiture of the subject oil and gas leases. Although admitting that the application of equitable principles in contract law is rare and discouraged, the appellants assert that, under the circumstances presented, this is perhaps a case where it would be proper. More precisely, the appellant's claim for seeking the equitable remedy of forfeiture centers upon the fact that the delay rental checks for the next annual period under each lease were approximately two months late.

. . .

Though the payments were late, they were tendered within the time required after demand under West Virginia Code § 36-4-9a (1985 Replacement Vol.). Further, the record indicates that the first two annual payments under these leases were apparently made on time. Without the development of clear and convincing circumstances to support a forfeiture in equity, the parties should initially pursue their remedies at law.

. . .

IV

Finally, the appellants ask us to determine whether the circuit court erred in finding that the appellee had not abandoned the leases in question.

The doctrine of abandonment recognizes that, upon a proper showing of circumstances indicating intent by the lessee to abandon, the law may effect a surrender of the leased estate. . . . We further note that the element of intent is the principal distinguishing factor between abandonment and forfeiture. Abandonment "rests upon the intention of the lessee to relinquish the premises, and is therefore a question of fact for the jury; while a forfeiture does not rest upon an intent to release the premises, but is an enforced release." Garrett v. South Penn Oil Co., 66 W.Va. 587, 596, 66 S.E. 741, 745 (1909) (quoting *Thornton on the Law of Oil and Gas* § 137).

. . .

The record indicates that the parties never came to terms upon the question of whether intent to abandon ever existed. Although they were in apparent agreement on some pertinent facts, for instance, that the rental checks were indeed late, and that the appellants did intend to repossess the leased estates, they remained in disagreement over the fundamental issue of the lessee's intent.

. . .

Accordingly, we hold that the circuit court erred in granting the appellee's motion for summary judgment on the appellants' claim of abandonment. The appellants have a right to introduce evidence at trial to support their allegations of abandonment.

For the foregoing reasons, the final ruling of the circuit court is reversed, and the case is remanded for further proceedings consistent with this opinion.

Reversed.[40]

---

[40] Under some circumstances a statutory "rebuttable legal presumption" of abandonment operates in West Virginia. *See* Berry Energy Consultants and Managers, Inc. v. Bennett, 175

NOTES

1. The "or" form is also illustrated by lease form 5 set forth at the beginning of this chapter. Numerous variations in the clause are found: the lessee covenants to drill "or" to do something else. The language "or pay rentals or surrender the lease" led to claims by lessors in early cases that the lease was in effect at the will of the lessee and hence at the will of the lessor, or that the lease was void for want of mutuality. In most cases this claim was unsuccessful. *See, e.g.,* Rich v. Doneghey, 71 Okl. 204, 177 P. 86, 3 A.L.R. 352 (1918). Other variants are: "or forfeit the lease"; "or pay rentals or forfeit the lease."

Suppose that a clause giving the lessor a right "to declare a forfeiture" upon lessee's failure to pay rentals appears in a lease which is otherwise in the "unless" form. Does this transform the instrument into an "or" lease? *See* Schumacher v. Cole, 131 Mont. 166, 309 P.2d 311, 8 O.&G.R. 565 (1957).

2. If the lessee under an "unless" lease does nothing at all during the primary term his rights terminate. If there is no forfeiture clause or statute of the kind presented in Warner v. Haught an extremely ambiguous situation may result. Usually, the lessor in this situation is restricted to a suit to recover the rental but ideas such as the "equitable remedy of forfeiture" and abandonment discussed in *Haught* may be invoked. If there is a forfeiture clause the lessor can act to terminate the lease. The court's view of equity may, however, require notice and demand for performance before the forfeiture is judicially declared. Can the lessee then tender rental and avoid termination?

The abandonment theory was successfully used in Wohnhas v. Shepherd, 54 Ohio Op. 436, 119 N.E.2d 861, 3 O.&G.R. 1352 (Ohio Com.Pl.1954). There the lessee covenanted "to drill or pay rentals or surrender." There was no forfeiture clause. Before the five years of the primary term had passed, the lessor sued to quiet his title, alleging that the lease had terminated, the lessee having failed to obtain production or to pay rentals. *Held* for plaintiffs. Defendant's predecessor had drilled a dry hole on nearby land in 1951, after which all operations on the premises in question had ceased. The lessee believed, and so testified, that he supposed the leases were automatically cancelled when he failed to pay the rentals called for in the leases at the time they became due and payable.

There is no forfeiture clause in the lease in question and therefore the mere non-payment of rental is not sufficient to warrant a court in decreeing a forfeiture. In fact, under the circumstances as presented by the evidence in this case, unless Mr. Iles and the St. Clair Oil Company abandoned this lease, warranting a cancellation thereof for that reason, plaintiffs are not entitled to any relief in this proceeding. This presents an inquiry as to the status of the law on the subject of abandonment. . . . The Court finds upon a consideration of all of the evidence in this case that the defendants, Iles and St. Clair Oil Company abandoned all their right, title and interest in and to

W.Va. 92, 331 S.E.2d 823, 85 O.&G.R. 624 (1985). [Eds.]

the lease in question in this case and that the same should be and hereby is canceled. The Court is of the opinion that the law pertaining to the question of abandonment is fairly uniform in the various states and that abandonment is recognized in Ohio as a ground for cancellation.

Is the concept of "abandonment" an appropriate tool for dealing with the problem raised in this case? *See* Girolami v. Peoples Natural Gas Co., 365 Pa. 455, 76 A.2d 375 (1950).

3. Consider the following clause which was a part of a lease with a three year term "and so much longer as oil or gas was produced in paying quantities:"

> Lessee to commence a well on said premises within twelve months from this date or pay to Lessor Twenty five and 00/100 Dollars ($25.00) each year, payable yearly thereafter until said well is commenced or this lease surrendered . . .. This lease shall become null and void for failure to pay rental for any period when same becomes due and payable, provided however that Lessee or his assigns is given 10 days written notice of his failure to pay said rentals and they are not paid within said 10 days.

Is the lease an "unless" lease or an "or" lease? *See* Mossgrove v. All States Oil & Producing Co., Inc., 24 Ohio App.2d 128, 265 N.E.2d 299, 37 O.&G.R. 448 (1970).

————

3. EFFECT OF OPERATIONS DURING THE PRIMARY TERM

If prior to discovery of oil or gas on said land Lessee should drill a dry hole or holes thereon, this lease shall not terminate if lessee commences additional drilling or reworking operations within sixty (60) days thereafter or commences or resumes the payment or tender of rentals on or before the rental paying date next ensuing after the expiration of three months from date of completion of dry hole.

————

SUPERIOR OIL CO. v. STANOLIND OIL & GAS CO.

Supreme Court of Texas, 1951.
150 Tex. 317, 240 S.W.2d 281.

GRIFFIN, JUSTICE. On March 3, 1944, J.O. Dodson, et al., owners executed to P.W. Anderson an oil and gas lease on a section of land located in Scurry and Borden Counties for a primary term of 10 years. They used a printed form described as "C-88 R-Producers' 88 Special-Texas Form." The dispute between the parties is determined by the proper construction of the primary term, rental and dry hole provisions of the lease, which are, respectively:

> It is agreed that this lease shall remain in force for a term of 10 years from this date, said term being hereinafter called "Primary Term", and as long thereafter as oil or gas, or either of them is produced from said land by the lessee.

. . .

If no well be commenced on said land on or before the 3rd day of March, 1945, this lease shall terminate as to both parties, unless the lessee on or before that date shall pay or tender to the lessor, or to the lessor's credit in the Snyder National Bank at Snyder, Texas, or its successors, . . . which shall continue as the depository, regardless of changes in the ownership of said land, the sum of Three Hundred Twenty and no/100 Dollars, which shall operate as rental and cover the privilege of deferring the commencement of a well for twelve (12) months, from said date. In like manner and upon like payments or tenders the commencement of a well may be further deferred for like periods of the same number of months successively. And it is understood and agreed that the consideration first recited herein, the down payment, covers not only the privilege granted to the date when said first rental is payable as aforesaid, but also the lessee's option of extending that period as aforesaid, and any and all other rights conferred.

Should the first well drilled on the above described land be a dry hole, then and in that event, if a second well is not commenced on said land within twelve months thereafter, this lease shall terminate as to both parties, unless the lessee on or before the expiration of said twelve months shall resume the payment of rentals in the same amount and in the same manner as hereinbefore provided. And it is agreed that upon the resumption of the payment of rentals, as before provided, that the last preceding paragraph hereof, shall continue in force just as though there had been no interruption in the rental payments.

On August 21, 1944, Anderson assigned this lease to Richfield Oil Corporation (hereafter called "Richfield"), which began drilling a well on January 10, 1945, and completed it as a dry hole on February 3, 1945. On January 28, 1946, Snyder National Bank received from Richfield a check for $320, which the letter of transmittal described as being "in payment of delay rentals for the period of February 3, 1946 to February 3, 1947 due" under the Anderson lease. By endorsement on this letter, the bank acknowledged receipt of the check, stating that it "has been credited according to your instruction", sent a copy to "J.O. Dodson et al., Snyder, Texas", and returned the original to Richfield. The same procedure was followed when, on January 29, 1947, Richfield sent a check for $320 to the bank to pay delay rentals due under the lease "for the period of February 3, 1947 to February 3, 1948," and again when, on January 19, 1948, Richfield sent a check for $320 to the bank to pay delay rentals due under the lease "for the period of February 3, 1948 to February 3, 1949."

Deciding to sell its Texas oil and gas leases, Richfield published notice inviting bids for them. The leases in Borden and Scurry Counties were described as Tract 5 and included the section covered by the Anderson lease. This notice, dated October 15, 1948, stated that all bids received would be publicly opened at

Richfield's Midland office and the highest cash offer would be accepted; it stated that the original leases, assignments, title opinions and all title data in Richfield's hands were available for inspection by prospective bidders, at its Midland office; that each bidder must satisfy himself regarding the leases without regard to information or assistance furnished by Richfield; and that lease assignments would be without warranty of any kind. The Superior Oil Company (hereafter called "Superior") and Intex Oil Company (hereafter called "Intex") bid in the Tract 5 leases together. On December 1, 1948, Richfield assigned the Anderson lease to Intex, "without warranty of any kind", and on December 30, 1948, Intex assigned an undivided interest in it to Superior.

On December 15, 1948, Richfield delivered to Intex papers relating to the lease and these included the last rental receipt above mentioned showing payment of delay rentals for the period of February 3, 1948 to February 3, 1949; and on January 14, 1949, it delivered to Superior the other two receipts showing such payment for the periods from February 3, 1946, to February 3, 1947, and from February 3, 1947, to February 3, 1948.

On February 5, 1949, Superior and Intex tendered payment of the annual rental to the Snyder bank but the tender was refused, under lessors' instructions, because payment had not been made by February 3. Before March 3, 1949, Superior and Intex began drilling a second well on the lease land.

On August 9, 1947, J. O. Dodson et al. executed to one Jordan a warranty deed conveying to him the surface estate and one-half the minerals in the section covered by the Anderson lease. This deed also contained a power of attorney authorizing Jordan, when the land was not under lease, to execute oil and gas leases on all or any part of the land covering their interest as well as his own.

On March 9, 1949, Jordan executed to Stanolind Oil & Gas Company (hereafter called "Stanolind") an oil and gas lease on the whole section.

Thereafter Stanolind and Jordan, respondents, filed this suit against Superior, Intex and others, petitioners, to recover title and possession of both surface estate and minerals in the land and to enjoin them from drilling an oil well or otherwise trespassing thereon. Under our conclusions, it is sufficient to say that, when the pleadings were all in, the issue was whether the original Anderson lease, under which petitioners claim, was still in effect when petitioners started drilling a second well on the land after February 3, but before March 3, 1949.

A trial court judgment for respondents was affirmed by the Court of Civil Appeals. 230 S.W.2d 346.

The Court of Civil Appeals held that the language of the lease was free of ambiguity; that it "plainly and expressly required commencement of a second well or payment of rental within twelve months after the date of the dry hole"; that, as the dry hole was drilled February 3, 1945, payment of rental within twelve months did not keep the lease alive for thirteen months; that, after the dry hole, commencement of a well or payment of rental every twelve months was required; that neither action could keep the lease alive for more than twelve months.

Writ of error was granted on two points, the first being that the Court of Civil Appeals "erred in holding that petitioners did not have the right until March 3, 1949, to pay the rental, or to commence a well, under the lease that provides 'and it is agreed that upon the resumption of the payment of rentals . . . the last preceding paragraph hereof' (said March 3 rental paragraph) 'shall continue in force just as though there had been no interruption in the rental payments.' " The second point complains that the Court of Civil Appeals erred in holding that their lease had terminated for failure to pay rentals.

A majority of the court is unable to agree as to the meaning of the "dry hole clause", and the construction to be given to it with regard to the payment of delay rentals. The writer of this opinion and some members agree with the construction of the Court of Civil Appeals, while other members of the court disagree. However, a majority of the court are in agreement that this provision of the lease is ambiguous, and, that being ambiguous, the construction placed thereon by the parties to the lease (Richfield and the lessors) for three years and long prior to this litigation is the construction to be given to the lease by this court, and that Intex and Superior are bound by such construction.

At best the meaning of the lease is doubtful and it is capable of two constructions. This being true, the lease contract is ambiguous. . . .

In construing the lease we must find the intention of the parties. . . .

The trial court, by its Finding of Fact No. 12 and Conclusion of Law No. 6, found that the lease was ambiguous, and the parties had given it a construction. There can be no question but that Richfield (Intex and Superior's immediate grantee) and the Dodsons, lessors, construed the "dry hole clause" to fix February 3rd as the anniversary paying date. For three years, February 3, 1946 to February 3, 1947; February 3, 1947 to February 3, 1948 and February 3, 1948 to February 3, 1949, Richfield sent to the Snyder National Bank the rental payments which the Dodsons accepted, accompanied by a letter specifically stating that such payment covered a twelve months period, beginning February 3, 1946 and ending February 3, 1947, the same being true of the two succeeding consecutive years. The parties have construed this lease, and since it is ambiguous, the courts will follow the construction given by the parties.

This court said in Lone Star Gas Co. v. X-Ray Gas Co., 139 Tex. 546, 164 S.W.2d 504; loc. cit. (5-7) 508:

> The original parties to the contract, and also the trial court, construed such contract as contended for by defendant; while the Court of Civil Appeals, with considerable difficulty, construed it as contended for by plaintiffs. If there is any doubt as to the meaning of a contract like the one before us, the courts may consider the interpretation placed upon it by the parties themselves. In this instance the acts of the parties themselves indicated the

construction they mutually placed upon the contract at the time, including the acts done in its performance, and same is entitled to great if not controlling weight. Courts will generally follow the interpretation of the parties to a doubtful contract. The practical construction placed upon a contract by the parties themselves constitutes the highest evidence of their intention that whatever was done by them in the performance of the contract was done under its terms as they understood and intended same should be done. . . .

Superior and Intex were bound by this construction placed on the lease by the parties.

Prospective bidders were advised that they should carefully read and consider the documents of record affecting the land, examine the title status and familiarize themselves with all of the conditions relative thereto. Prospective bidders were notified that each bidder must satisfy himself on all matters concerning the leases independently of any information or assistance furnished by Richfield.

Before the rentals were due on February 3, 1949, Intex and Superior had in their possession the Richfield files which contained the rental receipts reciting that the annual rental payments for the three years preceding February 3, 1949 were paid on or before February 3rd of each of said years for the privilege of deferring commencement of a well from February 3rd of one year to February 3rd of the following year, and disclosing that the lessors and Richfield had construed the lease to require rentals to be paid on or before February 3rd of each year after the dry hole. Superior and Intex were invited by Richfield to look at their records before they bought. . . .

Not having paid the delay rentals by the date they were due under the "dry hole" provisions of the lease, as construed by Richfield and the lessors, the determinable fee title held by Superior, *et al.,* automatically came to an end. . . .

We agree with the Court of Civil Appeals that after the Dodsons sold an undivided one-half interest in the minerals and all of the surface on August 9, 1947, by deed giving Jordan a power of attorney to lease all the minerals, Jordan's lease to Stanolind dated March 9, 1949, was valid and binding; and that the attempted ratification of petitioners' lease by one Boyle, who had acquired his interest in the minerals afterwards, (December, 1948) was of no force and effect.

The judgment of the Court of Civil Appeals is in all things affirmed.

BREWSTER and SMITH, JJ., dissent.

BREWSTER, JUSTICE (dissenting). I am convinced that the lease involved here is not ambiguous and that when Superior and Intex, as assignees, began a second well on the land before March 3, 1949, the Anderson lease was valid and subsisting.

The rental paragraph, quoted in the majority opinion, provided that if no well was commenced on the land on or before March 3, 1945, the lease should terminate unless lessee had sooner paid lessors $320 or deposited that amount to their credit in the Snyder bank; that by such payment lessee could defer commencement of a well until on or before March 3, 1946; and that upon like payments thereafter commencement of a well could be deferred to March 3 of each succeeding year throughout the primary term. In short, lessee could hold the lease for 10 years without ever commencing a well by paying $320 on or before March 3, of each year.

Under the dry-hole paragraph, if lessee shouldered the more expensive burden of drilling a well, he was relieved of his obligation to pay the annual rental on the succeeding March 3; but if the well proved to be a dry hole he was required either to commence a second well within 12 months thereafter or pay lessors $320, evidently to insure that his failure to bring in a well had not discouraged him to the point of abandoning the lease. When he had resumed payment of rentals, *"as before provided"*, that is, within 12 months after drilling a dry hole instead of an oil well, then "the last preceding paragraph hereof (the rental paragraph) shall continue in force *just as though there had been no interruption in the rental payments.*" I assume that this meant that the entire rental paragraph, not just part of it, should "continue in force."

So, when Richfield completed the dry hole on February 3, 1945, it was required to pay $320 rental to lessors by Feb. 3, 1946, which it did by depositing that amount to lessors' credit in the Snyder bank. When that was accomplished, the effect of the dry hole as well as the force of the dry-hole paragraph was spent and the "last preceding paragraph" (the rental paragraph) was again operative, to "continue in force just as though there had been no interruption in the rental payments." Under its clear and explicit terms Richfield could keep the lease alive for 10 years without ever attempting to drill any well by paying $320 on or before March 3, of each year, beginning March 3, 1945. Therefore, the rental paragraph can be continued "in force just as though there had been no interruption in the rental payments" only by holding that after Richfield completed the dry hole and resumed payment of rental within 12 months thereafter, it could keep the lease alive by paying the annual rental on or before March 3, 1947, and on or before March 3 thereafter during the term of the lease in lieu of drilling another well. I think the accelerated payment made on Jan. 28, 1946, met the dry-hole condition and also paid the rental due by March 3, 1946, because only thus can effect be given to the entire language of both the dry-hole and rental paragraphs.

The majority holding penalizes lessees for attempting to develop the lease, which development doubtless was the primary purpose of the lessors, by drilling a well which proved to be a dry hole and which surely cost lessees many times the $320 they otherwise would have been bound to pay by March 3, 1945. Rather than abandoning the lease after drilling the dry hole they showed their faith in it by paying the $320 within 12 months thereafter exactly as the lease required. Having done that, they had the right to pay rentals thereafter, beginning on or before March 3, 1947, *"just as though there had been no interruption"* in their payment. Otherwise, the last sentence of the dry-hole paragraph would mean nothing.

That Richfield may have treated the rental as due by Feb. 3, of each year, after "resumption of the payment of rentals" cannot defeat the rights acquired by its assignees, Superior and Intex, when they purchased the lease, because, there being no ambiguity in the instrument, "resort cannot be had to an interpretation given by the parties to the terms of the agreement in order to prove a construction contrary to the plain meaning." Highland Farms Corp. *et al.* v. Fidelity Trust Co., 125 Tex. 474, 481, 82 S.W.2d 627, 630.

I would render the judgment for petitioners.

SMITH, J., joins in this dissent.

NOTES

1. Is a "dry hole" clause for the benefit of the lessor or the lessee?

BAKER v. HUFFMAN, 176 Kan. 554, 271 P.2d 276, 3 O.&G.R. 1662 (1954). An "unless" lease executed in 1948 provided for a primary term of 10 years. One producing oil well was drilled on the lease, but after April 4, 1951, the lessee failed to produce or market any oil therefrom. No other drilling was done and no rentals were paid after the drilling of the initial well. On May 20, 1952, the lessor made written demand upon the lessee for the release of said lease. Defendant failed to comply with this demand and on April 10, 1953, the lessor filed this action to cancel the oil and gas lease and to quiet title to real estate. The defendant demurred to the evidence presented by plaintiff. *Held,* the demurrer was properly sustained:

> It was plaintiff's contention that her evidence disclosed the lease had expired by its terms by reason of defendants' failure to produce oil in paying quantities after April, 1951. Unless the lease in question had expired by its own terms, the trial court did not err. . . . This was not an action for forfeiture of an oil and gas lease, nor for damages for the breach of contract where excuses of nonperformance might be pleaded, nor was it an action for cancellation of an oil and gas lease based upon a breach of an implied covenant to develop. The action was one to cancel the lease which, as alleged, had expired by its own terms because of the lessees' failure to produce oil and gas in paying quantities continuously during the primary term. Plaintiff argued that the purpose of making an oil and gas lease is for the exploration, development and production of oil and gas, and when the lessee elects to begin development and obtains oil and gas then production must continue in paying quantities if the lease is to remain in force, irrespective of whether the primary term has expired. It is true that once the lessee of an oil and gas lease undertakes to develop the leased premises, the implied covenant to fully develop the leased premises with reasonable diligence applies even during the primary term. . . . However, plaintiff has neither alleged nor shown by her evidence that defendants have breached

that covenant. . . .

. . . We find nothing in the contract providing for a forfeiture or cancellation of the lease for the failure of the lessee to produce oil and gas in paying quantities during the primary term of the lease. The failure of lessees to produce or failure, alone, of production in paying quantities during the primary term of the lease did not result in a defeasance *ipso facto.* To hold otherwise would be to read into the lease express provisions which did not exist, and this we cannot do.

Compare TEXAS CO. v. DAVIS, 113 Tex. 321, 254 S.W. 304, 255 S.W. 601 (1923). In 1901, a tract of land in Texas was leased for oil and gas development. The lease provided that it should be null and void if operations for drilling of a well were not commenced and prosecuted within two years from the date of execution of the lease, and further provided that the lessee might prevent such forfeiture by payment of delay rentals from year to year until such well was commenced. It was agreed that completion of a well should operate as a full liquidation of all rental during the remainder of the term, which was for 25 years from the time of discovery of oil or other minerals and as long thereafter as oil, water, gas or other minerals were produced in paying quantities. The lessee drilled four wells on the premises within 2 years from the date of execution of the lease. Approximately 1000 barrels of oil were marketed from one of the wells, but the other three wells were dry. Operations on the premises ceased in 1904. Plaintiffs are the successors to this lessee. In 1913 defendant acquired a lease on this tract from the landowners and in 1918 drilled a well thereon which was richly productive of oil. In 1919 plaintiffs instituted this suit for the recovery of the working interest in this property. *Held* for defendant:

The title and rights conveyed to Underwood [the first lessee] or his assigns were to be held and used for no other purpose than mineral exploration, development, and production. When that purpose was no longer prosecuted by Underwood and his assigns, their title and estate instantly terminated. . . .

We are convinced: First, that Underwood and his assigns took only a determinable fee under the grant from Arnold and wife, which terminated long ago; and, second, that abandonment of the purpose for which Underwood and his assigns were invested with their title and rights in and to the minerals and land were necessarily fatal to the maintenance of the suit of defendants in error. The evidence is undisputed which conclusively shows that those under whom defendants in error claim entirely and permanently abandoned the exploration and development of the 76-1/2 acres and the production of minerals thereon. Their estate at once terminated without the need of a conveyance. All the oil, gas and other minerals remaining in the land reverted to Arnold and wife or their assigns, and defendants in error had no title to support their suit. . . .

Since the payment of rentals authorized merely delay in drilling, the clause concerning it was no longer operative, once a well was duly commenced and completed. Nothing could have been further from the minds of the parties, as disclosed by the entire writing, than that after the land was found to contain oil or gas, in paying quantities, it must remain wholly unproductive in the event the lessee elected to pay $10 per annum. . . .

. . . The clause was plainly not intended to defeat the dominant purpose of both contracting parties, which was the production of minerals for mutual profit.

The *Davis* doctrine was discussed in the second round of litigation in Rogers v. Ricane Enterprises, 884 S.W.2d 763, 130 O.&G.R. 414 (Tex. 1994), *rev'g*, 852 S.W.2d 751, 130 O.&G.R. 391 (Tex.App. 1992), *on remand from*, 772 S.W.2d 76, 108 O.&G.R. 331 (Tex. 1989). After finding that the assignment of a small part of a larger lease had not automatically terminated, the court was faced with the argument that the assignees had abandoned the assigned portion of the lease. The sole producing well ceased production in 1961 and no further work was done on the assigned acreage until 1979. The base lease was kept alive from production on non-assigned tracts. The lower courts on remand found that the lease had been abandoned based on *Davis*. The Supreme Court, however, disagreed. Only where there is an expressed purpose stated in the lease, or in this case an assignment, that is abandoned, will the interest terminate. As the court concluded: "The assignment in this case does not, by its express terms, specify a purpose for the assignment and does not contain any language limiting the duration of the assignment to 'as long as' oil and gas is produced. 884 S.W.2d at 767. What is left of the *Davis* doctrine? Isn't the purpose of the assignment the production of oil and gas, much as the court implied in *Davis*? Was the dissent correct when it said: ". . . the purpose of this contract was for exploration, development, and production of oil, gas, and minerals, and because the jury determined that the purpose of the assignment was abandoned, I would apply Davis to conclude that the assignment automatically terminated." 884 S.W.2d at 770 (Hightower, J. dissenting). Upon remand from the Supreme Court, the Court of Appeals further remanded for a new trial as to a group of defendants who had purchased oil from the trespassing lessee. Rogers v. Ricane Enterprises, 930 S.W.2d 157, 135 O.&G.R. 178 (Tex.App. 1996).

2. What is the effect of the deletion of the delay rental clause from a lease with an habendum clause for a "one year definite term unless extended by production" which also contains a dry hole clause providing for new drilling or the resumption of rentals "within twelve months from the expiration of the last rental period for which rental has been paid" if the first well "be a dry hole?"

CITIZENS BY-PRODUCTS COAL CO. v. ARTHALONY, 170 Ind.App. 1, 351 N.E.2d 57, 54 O.&G.R. 469 (1976), faced with such a lease and a producing well drilled within twelve months of the dry hole but beyond the one year primary term, reasoned that the purpose of a dry hole clause was to relieve the lessee drilling a dry hole from rental payments. Thus, "since all provisions for a rental period or rental were deleted from the lease, the purpose of the clause is defeated." Held: the lease had terminated.

In HOOD v. ADAMS, 614 S.W.2d 574, 69 O.&G.R. 504 (Tenn.App.1981, cert. denied), the lease had a primary term of ten years but was modified by deletion of the unless delay rental clause and the addition of a clause requiring drilling within 90 days to avoid termination. The document also had a dry hole clause providing that the lease would not terminate if drilling was again commenced before the next rental date or rentals paid at that time. A dry hole was drilled within the ninety day period and the lessor sued for cancellation. The ten year primary term did not sustain the lease. The lease as modified was terminable at the will of either party. "Where a purported oil and gas lease creates an option with no obligation either to drill or to pay rent, the lease is terminable by either party in the absence of intervening equities." Held: the lease had terminated.

The lessor in the *Hood* case was apparently paid a "one dollar cash bonus." Does this explain the result?

3. How dry does a hole have to be to come within a lease clause calling for a "dry hole?"

In SLATS HONEYMON DRILLING CO. v. UNION OIL CO., 239 F.Supp. 585, 22 O.&G.R. 186 (W.D.Okl.1965), the court had before it an agreement calling for a contribution on the abandonment of a test well as a "dry hole" and found:

> that the well involved in this litigation is deemed to be a dry hole under the agreement of the parties entitling the plaintiff to the prescribed dry hole money; that the plaintiff conducted itself as a prudent and reasonable operator in connection with attempting to make the well capable of producing gas to the market and under the circumstances was not restricted in point of time in this connection and in demanding dry hole money except to be reasonable and prudent under the circumstances. The Court further finds that the plaintiff was supported by reliable data and advice and also acted in good faith, without fraud or bad motive in deciding not to lay approximately five miles of pipe line at its own expense but rather to plug and abandon the well as a dry hole.

See 3 Martin & Kramer, *Williams and Meyers Oil and Gas Law* § 614.1.

4. When is a dry hole completed? Can a reworking operation on an old well result in a dry hole for purposes of lease extension? *See* Burns v. Louisiana Land & Exploration Co., 870 F.2d 1016, 106 O.&G.R. 547 (5th Cir.1989).

5. In HARREL v. ATLANTIC REFINING CO, 123 F.Supp. 70, 3 O.&G.R. 2060 (E.D.Okl.1954), the lease contained the following dry hole clause:

> Should the first well drilled on the above described land be a dry hole, then, and in that event, if a second well is not commenced on said land within twelve months from the expiration of the last rental period for which rental has been paid, this lease shall terminate as to both parties, unless the lessee on or before the expiration of said twelve months shall resume the payment of rentals in the same amount and in the same manner as hereinbefore

provided.

The rental date of the lease in question was June 15 and rentals were paid on each anniversary date up to and including June 15, 1952. A dry hole was completed in August 1952 and no rental was paid on June 15, 1953. The lessors brought suit claiming that the lease had terminated. The court held for the lessee, reasoning as follows:

> Plaintiffs assert that even though a test well was drilled by the defendant in August of 1952, it was incumbent on the defendant to pay delay rentals prior to June 15, 1953, in order to keep the lease in effect. Such an interpretation appears to be contrary to the express wording of the lease provision which states, "Should the first well drilled on the above described land be a dry hole, then and in that event, if a second well is not commenced on said land within the twelve months from the expiration of the last rental period for which rental has been paid, this lease shall terminate . . ." In order to ascribe any meaning to this just-quoted provision it must be construed as furnishing an alternative to the lessee insofar as drilling the first well is concerned; that is, an alternative of either drilling such test well or paying the delay rental payment. Admittedly, the dry hole drilled by the defendant in August of 1952 was the first well drilled on the leased land and the "expiration of the last rental period for which rental has been paid" fell on June 15, 1953. Thus, the defendant had twelve months from such expiration to either drill a second well or *resume* delay rental payments. Naturally, if the drilling of the first well did not take the place of a delay rental payment, it would be nonsensical to provide in the lease that if a second well was not drilled within twelve months from the expiration date of the last rental payment that the delay rental payment should be resumed. There can be no resumption unless prior thereto a delay rental payment had been excused; and, the express wording makes it abundantly clear that the twelve months grace earned by the lessee in drilling the first well was to be counted from the time the last paid rental expired, that is, June 15, 1953.

Suppose a second dry hole is completed in May, 1954. Must lessee tender rentals on or before June 15, 1954, to keep the lease alive?

Suppose a well is commenced June 1, 1954, and completed July 1, 1954. Should lessee tender rentals on June 15, 1954? On June 15, 1955? *See also,* Rogers v. Heston Oil Co., 1984 OK 75, 735 P.2d 542, 92 O.&G.R. 175.

6. COLBY v. SUN OIL CO., 288 S.W.2d 221, 6 O.&G.R. 354 (Tex.Civ.App.1956, error ref'd n.r.e.). The dry hole clause provided:

> If at any times during the primary term and prior to discovery of oil or gas on said land lessee shall drill a dry hole or holes thereon, or if at any time or times during the primary term and after discovery of oil or gas the production thereof should cease from any cause, this lease shall not terminate if on or before any rental paying date next ensuing after the expiration of three months from the date of completion of dry hole or

cessation of production, Lessee commences additional drilling or re-working operations or commences or resumes the payment or tender of rentals.

The lease, dated July 21, 1942, was for a primary term of seven years. A dry hole was completed on May 29, 1948. No rental was tendered on July 21, 1948, but production was secured by another well prior to the end of the primary term. In a suit asserting that the lease had terminated lessor argued that "the lease terminated by its own terms upon the completion of the dry hole at a time when there was no rental paying date next ensuing after the expiration of three months from the date of completion of the dry hole." Rejecting this argument, the court held that the lease did not terminate and was perpetuated into the secondary term by production. What is the defect in lessor's construction of the dry hole clause?

7. In CHEVRON OIL CO. v. BARLOW, 406 F.2d 687, 33 O.&G.R. 377 (10th Cir.1969), a well was completed as a dry hole on September 8, 1967 in the last month of a five year term ending on September 18, 1967. Since there was no production the lessor took the position that the lease terminated at the end of the specified term. Paragraph 5 of the lease provided that:

> If at any time or times after the primary term or within three months before expiration of the primary term, all operations and all production hereunder shall cease for any cause, this lease shall not terminate if lessee shall commence or resume drilling or reworking operations of the production of oil or gas within three months after such cessation.

Relying on this clause the lessee entered the land and commenced drilling on November 16, 1967, completing another dry hole on December 10, 1967. On the theory that another three month period then began to run a third well was commenced on February 21, 1968 and completed as a producer on March 15, 1968. The court held that the lease had terminated:

> [W]e agree with the trial court's conclusion that the word "operations" in paragraph 5 refers to operations in connection with production rather than operations related to drilling. We are convinced that had the parties, experienced as they were in matters pertaining to oil and gas leases, intended that paragraph 5 should apply to the drilling of dry holes, they would have chosen more expressive words and not relied upon the word "operations" to convey their intent.

Many types of clauses dealing with the effect of operations during the primary term are collected and discussed in 3 Martin & Kramer, *Williams and Meyers Oil and Gas Law* §§ 611 *et seq* .

———

C. EXTENSION OF THE LEASE BEYOND THE FIXED PRIMARY TERM

1. DRILLING OPERATIONS

a. DISCOVERY AND MARKETING

It is agreed that this lease shall remain in force for a term of three years from this date, and as long thereafter as oil or gas, or either of them, is produced from said land by lessee.

BALDWIN v. BLUE STEM OIL CO.

Supreme Court of Kansas, 1920.
106 Kan. 848, 189 P. 920.

MARSHALL, J. The plaintiffs commenced this action to cancel oil and gas leases given by them to W.S. Thompson, on certain property in Greenwood county. Judgment was rendered in favor of the plaintiffs on their motion for judgment on the pleadings, and the defendants appeal.

The questions argued are that the court committed error in sustaining the plaintiffs' motion for judgment on the pleadings, in rendering judgment canceling the leases, and in refusing the motion of the defendants to set aside the judgment and permit them to amend their answer.

The defendants became the owners of the leases by regular assignments from W.S. Thompson. Two leases were given, each identical with the other except in the description of the property leased. Each lease was dated January 17, 1916, and contained the following provisions:

It is agreed that this lease shall remain in force for a term of three years from this date, and as long thereafter as oil or gas, or either of them, is produced from said land by lessee. . . .

If no well is completed on said land on or before the 17th day of January, 1917, this lease shall terminate as to both parties, unless the lessee on or before that date shall pay or tender to the lessor, or to the lessor's credit in the First National Bank at Eureka, Kansas, or its successors, which shall continue as the depository regardless of changes in the ownership of said land, the sum of three hundred eighty and no/100 dollars, which shall operate as a rental and cover the privilege of deferring the completion of a well for twelve months, from said date. In like manner and upon like payment or tenders the completion of a well may be further deferred for like periods of the same number of months successively. And it is understood and agreed that the consideration first receipted herein, the down payment, covers not only the privileges granted to the date when said first rental is payable as aforesaid, but also the lessee's option of extending that period as aforesaid, and any and all other rights conferred.

No well had been completed under either lease on January 17, 1919, and none was commenced until about December 7, 1918. The answer alleged as excuses for not commencing to drill before the 7th day of December, 1918, that it was necessary to have an adequate supply of water to furnish power for running the machinery; that the defendants were dependent on water from rainfall that collected in ponds and creeks in the vicinity of the lands described; that until November 7, 1918, rain did not fall in a quantity sufficient to supply water for drilling; that they immediately thereafter began preparations for drilling a well, but on account of excessive rainfall the roads became muddy so that fuel for the engine and casing for the well could not be transported over the public roads; that a blizzard intervened which prevented the defendants from carrying on their work; and that their employés became sick and were compelled to cease work and thereby interfered with the drilling of the well. The answer also alleged that it was impossible for them to obtain coal and casing on account of the action of the United States government in taking charge of coal supplies, in promulgating rules and regulations governing the use of coal for drilling purposes in the vicinity of these lands from about December, 1917, to May, 1918, in taking charge of all casing, and in prohibiting the purchase, sale, and use of casing in drilling wells in territory not proven, which included these lands.

1. The principal question argued by the defendants is that on account of the excuses given by them, the leases should not be forfeited by reason of the failure to complete a well before January 17, 1919. The leases provided that they should remain in force for three years from their date, and that they should terminate in one year if no well was completed within that year; they also provided for an extension of the terms of the leases for another year, and still another year, by the payment of the rental named therein; but the leases terminated absolutely at the end of three years from their date, unless a well was drilled producing oil or gas. The leases did not contemplate that any circumstance or condition should excuse the lessee from performing the conditions of the leases on his part. By their terms ample time was given in which to drill a well. The lessee was compelled to take notice of the climatic conditions and of the topography of the country in the vicinity of the lands; he was likewise compelled to take notice of the powers of the government over coal and iron industries. The lessee with knowledge of these things contracted positively that he would do certain work within a certain time, and that after that time his rights in the premises should cease, unless gas or oil should be produced from the land by the lessee. Neither was produced.

This is not an action for a breach of contract where excuses for its nonperformance might be pleaded. It is an action to cancel leases that by their terms had expired on account of the lessee's nonperformance of their conditions. By paying the rent provided for in the leases, the lessee could neglect to drill until the leases expired, but the lessee, if he failed to produce oil or gas from the lands, could not, by paying rent, extend the leases one day beyond the time fixed for their expiration, nor could excuses for nonperformance extend the leases beyond that time. Doornbos v. Warwick, 104 Kan. 102, 177 P. 527, supported this conclusion.

. . .

3. The defendants pleaded a tender of $860 on the 16th day of January, 1919, as rental under the leases, which tender was refused by the plaintiffs. . . . The tender did not extend the leases beyond January 17, 1919.

No error appears, and the judgment is affirmed.

All the Justices concurring.

NOTES

1. Some of the misfortunes which beset the lessee in the *Baldwin* case would be anticipated in many modern lease forms in the *force majeure* clause. *See* the clause included in Appendix 1, *infra. See also* Martin & Kramer, *Williams and Meyers Oil and Gas Law* §§ 683; 683.2; Kirkham, *Force Majeure–Does it Really Work?* 30 Rocky Mt.Min.L.Inst. 6-1 (1984). The *force majeure* clause is discussed in detail in Chapter 3. Section 1, Subsection f *infra*.

2. Suppose in *Baldwin* that the "unless" clause had provided that "If no well is commenced on said land . . . this lease shall terminate unless . . .." Could the lessee successfully contend that having commenced a well, he was entitled to complete it, even though completion should occur after the end of the primary term, and that completion as a producer would extend the lease into the secondary term? *Compare* Simons v. McDaniel, 1932 OK 36, 154 Okl. 168, 7 P.2d 419, and Vincent v. Tideway Oil Programs, Inc., 1980 OK CIV APP 23, 620 P.2d 910, 68 O.&G.R. 516, *with* Walker, *The Nature of the Property Interests Created by an Oil and Gas Lease in Texas*, 8 Tex.L.Rev. 483 at 523-26 (1930).

3. If a lessee files for bankruptcy and the bankruptcy court fails to instruct the trustee in bankruptcy to take actions to preserve the lease before the anniversary date, would the lease's force majeure clause allow the lessee to retake possession upon the dismissal of the bankruptcy case? *Compare* Gilbert v. Smedley, 612 S.W.2d 270, 69 O.&G.R. 312 (Tex.Civ.App. 1981, writ ref'd n.r.e.) *with* Webb v. Hardage Corp., 471 So.2d 889, 86 O.&G.R. 324 (La. App. 1985).

4. A lease executed and delivered on December 13, 1958 contained the following provisions:

1. Lessors, in consideration of Ten Dollars and other valuable consideration ($10.00) in hand paid, and of the royalties herein provided, and of the agreements of lessee herein contained, hereby grants, leases and lets exclusively unto the lessee for the purpose of investigating, exploring, prospecting, drilling and mining for and producing oil, gas and all other minerals, laying pipe lines, building tanks, power stations, telephone lines and other structures thereon to produce, save, take care of, treat, transport, store and own said products, including salt water, and to house his employees, the following described land in Hansford County, Texas, to-wit: [description of land].

2. This lease is a commencement lease under which the lessee or his assigns

are obligated to commence the drilling of a well on the leased premises within one hundred twenty (120) days from the date hereof, said well to be drilled to a depth sufficient to test the Hugoton Field or about 3000 feet. Such well shall be completed within one hundred twenty (120) days after commencement thereof. If such well is not commenced and completed within such time, then this lease shall terminate as to both parties. If however, production is obtained, then this lease will remain in force for as long thereafter as oil, gas, casinghead gas, gasoline or any of them is produced.

The following events then occurred under these clauses:

Drilling operations were commenced on the property on March 10, 1959, and a well capable of producing gas in paying quantities was completed on or about April 10, 1959. The well was shut in upon completion and production was not obtained until a pipeline connection was made on September 3, 1960, from which date production continued without cessation to the date of trial and, presumably, is still continuing.

The final day of the 120 day completion period was July 9, 1959. Did the lease remain in effect? Do the clauses create limitations or covenants? *See* Fox v. Thoreson, 398 S.W.2d 88, 23 O.&G.R. 808 (Tex.1966).

5. STANOLIND OIL & GAS CO. v. BARNHILL, 107 S.W.2d 746, 749 (Tex.Civ.App.1937, error ref'd):

The facts are that gas was discovered and more than 7,000,000 feet per day could have been extracted from the well. It is not disputed that it would produce that amount of sour gas, and, if there had been an available market, the gas could have been disposed of. There was, however, no market for sour gas in the territory. Carbon black was being manufactured from sweet gas, which abounded in almost inexhaustible quantities in the same general territory, and sweet gas was more desirable for the manufacture of carbon black, as well as other commodities, than was sour gas. . . . Appellants went to the expense of $25,000 in drilling the well which was drilled. . . .

The lease under which appellant holds provides that it shall remain in force for a term of five years from its date, and so long thereafter as oil or gas, or either of them, is produced in paying quantities. Under the above definition of the term employed by the parties in the contract, appellants, although they discovered gas to an extent and in quantity that would have complied with their obligation in the lease if a market had been available for the kind of gas discovered, did not, under the circumstances of this case, discover nor produce gas in such quantities as, under the law, would be "paying quantities," within the five years provided by the lease. Their estate, consisting of a determinable fee, determinable upon their failure to produce oil or gas in paying quantities within such term, came to an end on February 4, 1935, as found by the trial court, after which they did not possess any interest whatever in the land. The lease did not exist. The term had closed, and the interest they procured by the lease was gone. It is not a question of forfeiture for failure to continue to develop the land, nor does it rest upon any other contingency. Appellants did not contract for a term which would depend upon the possibility of procuring a market for the product at some date subsequent to its express date of expiration. The lease did not provide that it should remain in force and effect for five years, and as long thereafter as there may be prospects of a market for the product, and it is not the duty of the courts to make contracts for parties but only to construe such contracts as they make for themselves.

6. WATSON v. ROCHMILL, 137 Tex. 565, 155 S.W.2d 783, 137 A.L.R. 1032 (1941). Actual production from a well ceased on May 21, 1932, at which date the tubing became stopped up and was not cleaned out. Oil could have been produced from the well, but due to the Depression and the low gravity of the oil there was no market which would justify the operation of the well. As a result no oil was produced from the well from May 21, 1932 until January, 1935. During this interval the lessee kept a watchman on the premises and paid the taxes on the leasehold interest claimed by him. In December, 1933, he cleaned out and acidized and swabbed the well. The well was again acidized in July, 1934. At some time during this interval lessee erected a new derrick over the well and installed a pump. There was testimony to the effect that lessee expended over $8,000 in the above work and repairs. In January, 1935, a market was found for oil such as could be produced from this well, and production was resumed. This was continued until lessors filed this suit in April, 1937, for a judgment declaring the lease terminated and for removal of a cloud from the title. The trial court instructed a verdict in favor of defendant, and this judgment was affirmed by the Court of Civil Appeals.

The Supreme Court concluded that there was such cessation of production as to terminate the lease as a matter of law. In dealing with defendant's claim of estoppel, the court assumed that the lessors knew of the work being performed by lessee on the leased premises but declared:

It appears, however, that the lease in question terminated by its own terms automatically upon cessation of production after the expiration of the primary term. The lease involved the conveyance of an interest in land, and any extension thereof would necessarily have to be in writing unless the doctrine of estoppel applies. . . . The conduct of lessors, relied on by lessee as forming a basis for estoppel, all occurred after the lease had fully terminated. At that time lessee knew, or in law must have known, that his lease had terminated. In our opinion, lessors' mere silence after the expiration of the lease was not sufficient to recreate the lease. . . . Consequently, even though lessors did silently permit the improvement and development of the property after the termination of the lease, they are not now estopped to plead the termination thereof.

The judgments of the trial court and of the Court of Civil Appeals (on the matter here discussed) were reversed, and judgment was rendered in favor of the plaintiffs.

7. What if the lessee stays on after the termination of the lease and production is obtained on which royalties are paid? Brown v. Haight, 435 Pa. 12, 255 A.2d 508, 33 O.&G.R. 333 (1969), decided that the "traditional oil and gas 'lease' is a determinable fee" and not a fee "subject to the grantor's right of entry" and that the lessee holding on under such circumstances was a tenant at will.

When the lessee leaves on final termination of his possessory rights can he take with him the casing and equipment he has installed? *Cf.* Sunray DX Oil Co. v. Texaco, Inc., 417 S.W.2d 424, 27 O.&G.R. 232 (Tex.Civ.App.1967, error ref'd n.r.e.), where the lease provided for removal of fixtures and equipment "at any time within the term of this lease." After termination by cessation of production the former lessee argued "that it should not lose title to the fixtures and equipment placed on the lease by reasons of its failure to remove same, and that it should have a reasonable time in which to effect such removal. Numerous cases can be cited where leases are silent as to the period of removal or provide for removal after termination, and the courts have allowed 'a reasonable time'. An implied covenant can be said to exist in such instances. Here, there can be no implied covenant, for there is an express provision in the lease contract that the right of removal exists 'at any time during the term of this lease'. When express covenants appear in a lease, implied covenants disappear." What is the relationship of "reasonable time" to the concept of abandonment in determining the right of the former lessee to remove fixtures and equipment when the lease has terminated? *See* Sec. 3, *supra.*

Would the departing lessee be allowed to destroy a producing well by the removal of casing? If the removal of the casing were not allowed, would it then belong to the lessor? *See* Patton v. Rogers, 417 S.W.2d 470, 27 O.&G.R. 211 (Tex.Civ.App.1967, error ref'd n.r.e.).

———

b. DISCOVERY

McVICKER v. HORN, ROBINSON & NATHAN

Supreme Court of Oklahoma, 1958.
322 P.2d 410, 8 O.&G.R. 951.

BLACKBIRD, JUSTICE. Plaintiffs in error, hereinafter referred to as plaintiffs, are the owners of a 40-acre tract of land in the Witcher area of Oklahoma County. In 1953, they executed and delivered to J. W. Dutton an oil and gas lease on said tract for a term of one year from October 31st, 1953. Duttom thereafter assigned the lease to the partnership of Horn, Robinson and Nathan . . . They completed a gas well on the leased land on or about May 1, 1954, but no gas from said well has even been marketed or sold.

On October 24, 1955, plaintiffs commenced this action against the defendants to quiet their title to the property (including pipe and other well equipment defendants had installed thereon) against all claims of the defendants. Their position that defendants no longer had any right, title or interest in and to said property was predicated on the theory, both that defendants had abandoned the leasehold after completing the well in May, 1954, and that the lease had expired by its own terms on account of defendants' failure to produce gas from the leased land by October 31, 1954, the end of the lease's one-year term.

The lease involved here was executed on what is called a "Producers No. 88" form, "With Pooling and Regulation Clauses." It is regular in all respects except that the provisions of the ordinary printed form with reference to delay rentals were stricken out, or obliterated, from it. With this deletion, the only remaining provisions of the lease at all material to the present controversy read as follows:

> It is agreed that this lease shall remain in force for a term of *One (1)* years from this date, and *as long thereafter as oil or gas, or either of them, is produced from said lands* by the lessee.

> In consideration of the premises the said lessee covenants and agrees:

> . . .

> And where gas only is found one-eighth of the value of all raw gas at the mouth of the well, while said gas is being used or sold off the premises, payment for gas so used or sold to be made monthly. The lessor to have gas free of cost from any gas well on said premises for all stoves . . .

> . . .

> To pay lessor for gas produced from any oil well and used off the premises one-eighth of the value of the raw gas at the mouth of the well, payment for the gas so used or sold to be made quarterly.

> . . .

> Should the first well drilled on the above described land be a dry hole then, and in that event, if a second well is not commenced on said land within twelve months from the expiration of the last period which rental has been paid this lease shall terminate as to both parties, unless the lessee on or before the expiration of said twelve months shall resume the payment of rentals, as above provided, . . .

It will be observed from the above that the "unless" provision, or only part of the lease expressly providing for its termination, is incomplete and meaningless without the delay rental provision, which, the ordinary Producers 88 form of lease contains. In connection with this part of the lease, it is also perhaps worthy of mention that the Corporation Commission's well-spacing order, in effect in the area, allows only one well on a tract the size of the one here involved. Plaintiffs cite authorities to support the proposition that the duty to market the oil and gas found on the leased premises is an implied covenant on the part of the lessee in oil and gas leases generally, or in the regular forms thereof;

but they have failed to point out an express covenant therefor in the lease before us. Defendants point out that the mere use of the word "produced" in the quoted "thereafter" clause does not make of that clause such an express provision. They argue, in effect, that the word "market" is neither included in, nor synonymous with, Webster's definition of the word "produced." They say they could have "produced" gas from the leased premises, within the ordinary meaning of the word, if they had not "shut in" the well, but had left it open to waste its gaseous product into the air. A similar argument was made in Home Royalty Ass'n v. Stone, 10 Cir., 199 F.2d 650, but rejected, because it was a Kansas case, and the court thought that previous opinions of the Kansas Supreme Court required marketing, as well as discovery, citing Ratcliff v. Gouinlock, 136 Kan. 149, 12 P.2d 798, and Tate v. Stanolind Oil & Gas Co., 172 Kan. 351, 240 P.2d 465. . . . We have been referred to no Oklahoma case, and know of none, in which a lease, like the present one, has been declared terminated at the end of its primary term, on account of the lessee's failure to produce, where, by that time, he had not only discovered oil or gas in paying quantities, but was bringing it to the surface and reducing it to possession in a manner in which it could be, but had not yet been, marketed. Of course, it is easier to see how oil can be produced by bringing it to the surface and placing it in storage facilities (though not marketed until the storage is filled) than in the case of gas where storage is not yet a practical, accepted, or perhaps possible procedure. In this connection, see 9 Okla.L.Rev. 200, Note 4. Regardless of this difference in the nature of the product, however, it cannot be disputed, as Judge Huxman recognized in his dissenting opinion in Bristol v. Colorado Oil & Gas Corp., 10 Cir., 225 F.2d 894, and stated in the majority opinion in Christianson v. Champlin Refining Co., 10 Cir., 169 F.2d 207, 210, that:

> . . . in the very nature of the oil business . . . a reasonable time must intervene between the completion of the drilling operations resulting in production and the ability to market and sell the product of a well.

And this, in the most ordinary, or usual, situations, applies more markedly to the "nature" of the gas business than the oil business.

We do not think some of the statements of the Kansas Supreme Court in the Tate Case, supra, concerning ipso facto termination of a lease for failure to produce, can be soundly applied to such termination for failure to market. Accordingly, our views are not in accord with their interpretation and application in the Home Oil Co. Case, supra. We refer most particularly to the following statement in the Tate Case, supra [172 Kan. 351, 240 P.2d 468]:

> The great weight of authority, however, appears to be in harmony with the view that actual production during the primary term is essential to the extension of the lease beyond that fixed term. This, at least, is true *unless the lease contains some additional provisions indicating an intent to extend the right to produce beyond the primary term.* (Emphasis ours.)

No valid fault can be found with the above statement, but we say it applies only to production, per se, and as that word is ordinarily defined (not including marketing). To say that marketing during the primary term of the lease is essential to its extension beyond said term, unless the lease contains additional provisions indicating a contrary intent, is to not only ignore the distinction between producing and marketing, which inheres in the nature of the oil and gas business, but it also ignores the difference between express and implied terms in lease contracts. Here, the lease contains no provision expressly requiring the marketing of any gas produced under it. (In fact, it seems to recognize the distinction between production and marketing by its provisions for the payment of royalty on only the gas that is "used or sold off the premises.") Therefore, if a covenant by the lessee to do so, is to be included within its terms, such covenant can only be an implied one. While statements that could be construed as indicative of a contrary view have been made in cases involving some difference in both facts and law, in at least one jurisdiction (see Stanolind Oil & Gas Co. v. Guertzgen, 9 Cir., 100 F.2d 299; Berthelote v. Loy Oil Co., 95 Mont. 434, 28 P.2d 187) we think the following statement at pages 355 and 356 of Vol. 2, Summers on Oil & Gas (Perm.Ed.) sets forth a correct summation of the decisions of most courts on the matter:

> *Whenever a duty to perform an act or series of acts is fixed* by contract or *implication* of law, and the time, manner, and extend *of performance is not fixed,* the law implies that such act or acts shall be performed diligently. Where, therefore, *oil gas leases do not state the time, manner, and extent of performance* of express and implied duties to test, develop and protect the land and *market the produce, the courts have of necessity tested the lessee's performance by the standards of reasonable men and reasonable diligence.* (Emphasis ours.)

In this connection, see also Bristol v. Colorado Oil & Gas Corp., *supra.* The lease before us contains no express covenant to market gas, and likewise contains no express provision as to when this should be done. It therefore follows that it cannot be held to have terminated ipso facto, or by its express terms, on October 31, 1954, on account of the lessee's failure to market gas by said terminal date of its primary term. It also follows that the defendants lessees had a reasonable time after completion of the well the start marketing its product.

After determining that if there was any agreement in the lease in question to market the gas from the well involved, it was an implied one, and accordingly that the period lessees had within which to comply with such requirement was not governed exclusively by the terminal date of the lease's primary term – but by what was reasonable under the circumstances – the next question is: Did the trial court err in holding, in effect, that, on the basis of the evidence, arrangements to market the gas were accomplished within a reasonable time?

Under the above facts, it is our opinion that the trial court's judgment for defendants cannot be held to be clearly against the weight of the evidence. Mr. Burgin testified that: "Since the first day we completed this well, we did continuous work in trying to get a gas connection to (it) . . ." While this is a conclusion of the witness, we believe the facts we have mentioned, and others unnecessary to detail, tend to support it. They also show that defendants persisted in their efforts despite the fact that they almost lost the property in the lien foreclosure proceedings herein described. Plaintiffs have failed to point out any manner in which defendants could have been more diligent than they have been, except to argue that they should have started selling gas to Oklahoma Natural when it submitted its contract of November, 1954. They point out that there is evidence showing defendants could have purchased a used gas compressor for only $8,000, but no evidence was introduced to show that defendants' compliance with all of Oklahoma Natural's requirements would not have cost the $15,000 or $20,000 testified to by defendants' witness, Mr. Nathan. Nor did plaintiffs introduce any evidence that a reasonably prudent operator would have entered into such a contact at that time, when they then had good reason to believe that another prospective purchaser would purchase the gas on terms that should prove more profitable and satisfactory not only to themselves but to the plaintiff royalty-owners, as well. Nor is there any evidence that, when defendants procured the setting aside of the receiver's sale, plaintiffs objected. There is testimony on behalf of defendants (contradicted by McVicker), however, that about the time Mr. Burgin paid the judgment indebtedness against the property, Mr. McVickers indicated to him that he would "go along" with defendants' further efforts to market the gas. While this may furnish little, if any, basis for defendants' plea of estoppel, it perhaps received some consideration by the trial court. In this connection, see Bristol v. Colorado Oil & Gas Corporation, supra, 225 F.2d at page 898.

Oil and gas lessees ·should not be allowed to hold their leases indefinitely, while no product therefrom is being marketed and diligent efforts are not being made to accomplish this, or where, despite their efforts, there is no reasonable probability they will be successful. We agree with the majority opinion in Bristol v. Colorado Oil & Gas Corp., *supra*, 225 F.2d at page 897 that the rule of reasonableness here applied is not "unlimited in the face of diligent effort" and that such a lease may be cancelled regardless of the intensity of the lessee's efforts, where there is no reasonable probability that same will be successful, or it appears that others, with less effort, would succeed where they have failed. If such lessee cannot market the product, there is no purpose to be served, under the terms of the lease, in his retaining it, and it is difficult to conceive of any reason it should not be cancelled. The fact that the lessee has at one time invested money in it certainly would not warrant his holding it indefinitely without development, marketing, or payment of rentals for its future speculative value. This case, like others of the same character, must be determined on its own particular facts. Here, there is no evidence to invoke such considerations.

In accord with the foregoing, the judgment of the trial court is affirmed.

———

NOTES

1. In TATE v. STANOLIND OIL & GAS CO., cited in the principal case, the lease at issue contained the following clause:

> If the lessee shall commence to drill a well within the term of this lease or any extension thereof, the lessee shall have the right to drill such well to completion with reasonable diligence and dispatch, and if oil or gas, or either of them, be found in paying quantities, this lease shall continue and be in force with the like effect as if well had been completed within the term of years herein first mentioned.

The well upon which the extension of the lease depended was commenced and completed during the primary term. Although production did not commence until some sixteen months after the end of the primary term, the court held that under the circumstances a finding of production within a reasonable time was justified. On the interpretation of the lease the court concluded:

> Considering the habendum and drilling clauses together it seems to us it would be wholly inconsistent to hold the lessee, under this lease, is permitted to obtain production with diligence and dispatch after the primary term has expired, if the well is commenced but not completed during the primary term, but that he would not be permitted to obtain production within a reasonable time after the primary term if a well were completed during the primary term, as, for example, on the last day of such term.

Consider the following situations in the light of the *Tate* case:

A lessee holding Oklahoma land under a lease with a *Tate* clause, commences operations on Well No. 1 during the primary term, is drilling when the term expires, and completes the well as a dry hole one month after the term expires. While drilling operations on Well No. 1 are in progress, but after the end of the primary term, lessee commences drilling on Well No. 2 which is completed in due course as a commercial producer. Is the lease extended into the secondary term? *See* Skelly Oil Co. v. Wickham, 202 F.2d 442, 2 O.&G.R. 559 (10th Cir. 1953).

Would it make any difference that the lease contained no *Tate* clause but provided instead for a secondary term "as long thereafter as oil or gas is produced or the premises are being developed or operated"? Suppose the second well was commenced 30 days after the first well was plugged and abandoned as a dry hole? *See* Statex Petroleum v. Petroleum, Inc., 308 F.2d 815, 17 O.&G.R. 778 (10th Cir. 1962).

A lease contains the following clause:

> If at the expiration of the primary term, oil, gas or other mineral is not being produced on said land, or on acreage pooled therewith, but Lessee is then engaged in drillilng or reworking operations thereon or shall have completed a dry hole thereon within sixty (60) days prior to the end of the primary term, the lease shall remain in force so long as operations are prosecuted

with no cessation of more than sixty (60) consecutive days, and if they result in the production of oil, gas or other mineral so long thereafter as oil, gas or other mineral is produced from said land or acreage pooled therewith.

Well No. 1 was placed into production in 1976. It continually produced until December 1979. Workover operations were begun on January 1, 1980. The primary term of the lease ended on January 15, 1980. In April 1980, the lessee abandoned efforts to regain production from Well No. 1. In May 1980 Well No. 2 was begun and achieved production shortly thereafter. Has the lease terminated for failing to produce in paying quantities at the end of the primary term? See Griffin v. Crutcher-Tufts Corp., 500 So.2d 1008, 94 O.&G.R. 7 (Ala. 1986).

2. Would a contract or fee simple subject to a condition subsequent analysis of the oil and gas lease habendum clause be more consistent with the discovery rule interpretation? In HUTCHINSON v. SCHNEEBERGER, 374 S.W.2d 483, 20 O.&G.R. 962 (Ky. 1964) there was a failure to obtain production by the end of the primary term under a typical "unless" lease which would expire on December 9, 1957 unless oil or gas was then being produced. The court stated:

Equity does not favor forfeitures and the intent must must be clear and unequivocal as well as the conditions therefor.

The beginning of preparation for drilling on September 21 was apparently accepted by both parties in lieu of an additional $ 10 rental. We therefore hold that there was a substantial compliance with the terms of the lease and that there was timely commencement.

The next question to be resolved is whether there was production within the intent of the lease on December 9, 1957, the date on which the term expired unless there was production.

The obvious purpose of the contract between these parties was the development of oil or gas production on this land.

Lessee had discovered oil at a shallow level in what had theretofore been recognized as a level not providing the best production and using what seems to have been good practice elected to drill deeper to "Weir Sand," the recognized best production sand. There is no showing that discontinuing operation for the winter was not also the accepted practice. So he discontinued operation until the next spring when he returned and prepared to and did pump oil at the first discovered level. The oil was there and all parties knew it. If lessee had not reasonably and seasonally returned to pump the discovered oil we might here reach a different conclusion; however, he did return seasonally and did pump the oil already discovered.

The court concluded that the lease was held by production.

Is there anything that would require a state which adopted the discovery rule to treat the habendum clause as a fee simple determinable or a fee simple on a condition subsequent? A strong suggestion for the application of the latter is interspersed with standard fee simple determinable language in Eastern Oil Co. v.

Coulehan, 65 W.Va. 531, 64 S.E. 836 (1909):

> [W]here there has been a substantial compliance with the contract, and gross injustice would be inflicted upon the plaintiff by denying him relief, relief should be granted. . . . There can certainly be no question as to the fact that the plaintiff substantially performed its contract. . . . Where there has been such substantial performance of a contract, equity may set aside or disregard a forfeiture occasioned by a failure to comply with the very letter of an agreement.

West Virginia continues to apply fee simple on condition subseqent principles to both oil and gas and mineral leases, although as in Eastern Oil Co., both fee simple determinable and fee simple on a condition subsequent language may appear in a court's discussion of why a lease has or has not terminated. In addition the traditional maxims that equity abhors forfeitures and that instruments will be interpreted to avoid finding limitations rather than covenants have been used to deny a lessor the right to claim that a lease has automatically terminated. *Compare* Christian Land Corp. v. C & C Co., 188 W.Va. 26, 422 S.E.2d 503, 123 O.&G.R. 358 (1992)(coal lease not automatically terminated) *with* McCullough Oil, Inc. v. Rezek, 176 W.Va. 638, 346 S.E.2d 788, 90 O.&G.R. 596 (1986) (oil and gas lease both a conveyance and a contract) *with* Jolynne Corp. v. Michels, 191 W.Va. 406, 446 S.E.2d 494, 129 O.&G.R. 146 (1994)(oil and gas lease creates a fee simple determinable).

Kentucky not only has adopted the discovery rule for the interpretation of the habendum clause, but has also taken a very lessee-oriented application to the discovery requirement. Thus even where the lease does not contain a "Tate" clause allowing the lease to be continued into the secondary term when a well has been commenced, the courts have implied such a clause and allowed the lessee to continue operations even though no reservoir has been discovered when the primary term ends. Little v. Page, 810 S.W.2d 339, 116 O.&G.R. 75 (Ky. 1991).

3. In JATH OIL CO. v. DURBIN BRANCH, 1971 OK 127, 490 P.2d 1086, 40 O.&G.R. 488, a lease "for a term of ten years . . . and as much longer thereafter as oil or gas is found in paying quantities" was held to be terminated by automatic "operation of the limitation provision of the habendum clause" on "cessation of production . . . from February 1964 to July 1966" which, "under the circumstances . . ., was cessation for an unreasonable period of time and was therefore more than a temporary cessation of production." The court pointed out that a lease does not terminate if the cessation of production is a "temporary . . . one that does not extend for an unreasonable time under the particular circumstances of each case." In Discussion Notes, 40 O.&G.R. 503 at 505 Professor Kuntz comments:

> [W]hen the result in the principal case is considered, it is readily apparent that "found" is no longer synonymous with "production" unless they both mean that oil or gas must be marketed. If this is so, it is necessary to search for other meaning in the McVicker case and those cases which followed it.

Perhaps those cases only related to the situation where oil or gas is discovered within the primary term but marketing does not occur until after the primary term. If this is the case, then they can stand for the proposition that the primary term is extended by implication and that the lessee has a reasonable time within which to establish production by marketing when he has made a commercial discovery.

In STATE v. AMOCO PRODUCTION CO., 1982 OK 14, 645 P.2d 468, 73 O.&G.R. 64, production ceased when the casing collapsed on the one producing well on the lease. A second well was commenced within a month and completed as a producer in the same formation. It was held that such temporary cessation did not result in the termination of the lease. In Discussion Notes, 73 O.&G.R. 69 at 70 Professor Kulp comments:

> In the principal case, the court clearly stated the principle ["regarding the effect of cessation of production after the primary term"] as follows:
>
> after production was established, lessee became vested with an estate that would endure as long as it conducted diligent operations to produce from the known formations.

Is the status of *McVicker* now more clearly defined in Oklahoma? Kentucky also holds that "production is broadly defined to include not only its obvious denotation, the pumping of oil, but also the discovery of oil during the initial term of the lease coupled with the exploitation and removal of the discovered oil within a reasonable time thereafter." Greene v. Coffey, 689 S.W.2d 603, 85 O.&G.R. 133, (Ky.App.1985, review denied by Supreme Court). Professor Kramer comments that Kentucky is "committed to the view espoused in Montana, Oklahoma, and West Virginia, that discovery of a reservoir is sufficient to meet the habendum clause obligation of the standard lease. . . . Kentucky has been very lenient in determining the amount of hydrocarbons that need be discovered or marketed in order to maintain the lease beyond the secondary term. In order to avoid having the lessee hold the lease indefinitely after discovery, the Oklahoma decisions have imposed an implied covenant to diligently market the oil or gas." Discussion Notes, 85 O.&G.R. 138 (1985). What lease provisions would be desirable from the point of view of a Kentucky lessor to deal with such a problem?

4. What if the parties include an habendum clause that provides that "This lease shall remain in force for a term of three (3) years and as long thereafter as oil and gas or either of them is, or can be, produced." Is the "or can be" language needed in a discovery jurisdiction? Does it effectively change the limitation in a discovery plus marketing jurisdiction to merely discovering a reservoir capable of producing hydrocarbons in paying quantities?

———

c. LESSOR INTERFERENCE

GREER v. CARTER OIL CO.

Supreme Court of Illinois, 1940.
373 Ill. 168, 25 N.E.2d 805.

[The controversy here grows out of the validity of an oil and gas lease given by Bennie Irey Shaw to the Carter Oil Company on May 28, 1936, and recorded on June 24, 1936. Previously, on November 15, 1933, Bennie Irey Shaw had executed a quitclaim deed of these pfsremises to Mark Greer. This quitclaim deed was not recorded until about two years after the execution of the oil lease to Carter Oil Company. By another conveyance, subsequent to the lease, an undivided interest in the land was conveyed by Mark Greer and Bennie Irey Shaw to J.G. Burnside. The oil and gas lease was for a period of three years with a "thereafter" clause and, by its terms, the lease expired May 28, 1939, at which time there was no production of oil or gas from the leased premises. By a suit filed in October, 1938, by Mark Greer and J.G. Burnside against the Carter Oil Company, plaintiffs sought the removal of the cloud upon their title occasioned by the lease held by defendant, Carter Oil Company. The circuit court found the oil lease valid and, the three year term of the lease having expired, the court by its decree provided that additional time be given to drill a well. The plaintiffs appealed. Eds.]

GUNN, JUSTICE. [In a portion of the opinion which is here omitted the court concluded that the Carter Oil Company was an innocent purchaser of the oil and gas lease for value under the recording act with the "down payment" of the first year's rental of $14.30 operating to give it that status. Eds.]

Appellants urge that the circuit court had no power to decree that the Carter Oil Company had a reasonable time after the termination of the litigation in which to perform the terms of the lease because it had expired during the course of the litigation.

The evidence shows that Mark Greer acquired title from the owner, Mrs. Shaw, in 1933. The deed was not recorded. The Carter Oil Company lease from Mrs. Shaw was made in May and recorded in June, 1936. In May, 1937, and May, 1938, rental of $14.30 was paid the grantor, Mrs. Shaw, by depositing it in the bank at St. Elmo. The first producing well in this territory was brought in in January, 1938. The tax records show payment by Mrs. Shaw for the years 1934, 1935, 1936 and 1937. There was nothing to indicate appellants claimed prior rights until July, 1938, when the deeds were recorded. The suit to cancel the leases was started in October, 1938. A serious question was thus raised as to the validity of the oil lease. If the contention of appellants was right it was absolutely void. If, during the remaining seven months of the lease, appellees produced a well, they would be liable to account for the oil produced as though they were trespassers. . . . If they brought in a dry hole they might be liable for damages for destroying the market value of the leasehold. It was extremely hazardous for appellee, the Carter Oil Company, to move in any direction until the litigation was settled and it, of course, could not control the duration of the litigation. This situation was brought about by the grantor of both appellants and appellees, namely, Bennie Irey Shaw. It could have been avoided had appellants placed their deeds of record and given notice to the world of their rights. Under these conditions, does the principle of estoppel apply to give appellees a reasonable additional period beyond the term of their lease in which to exercise their right to drill. As affecting oil and gas leases, this is a new proposition in Illinois but the general principle is recognized. In Mills v. Graves, 38 Ill. 455, 87 Am.Dec. 314, an estoppel is defined to be an impediment or bar to the assertion of a right of action arising by means of a man's own act or where he is forbidden to speak against his own act and extends to and binds privies in blood, privies in estate and privies in law. In First Lutheran Church v. Rooks Creek Church, 316 Ill. 196, 147 N.E. 53, it was invoked to prevent the forfeiture of an estate by breach of condition subsequent induced by acts of grantor.

In Lehmann v. Warren, Webster & Co., 209 Ill. 264, 70 N.E. 600, it was said, where time was of the essence of a contract, that a party who prevents performance within the stipulated time will not be allowed to avail of the non-performance he has occasioned to avoid the agreement. In Bondy v. Samuels, 333 Ill. 535, 165 N.E. 181, 186, we said: "Estoppel may arise from silence as well as words. It may arise where there is a duty to speak and the party on whom the duty rests has an opportunity to speak, and, knowing the circumstances keeps silent. . . . It is the duty of a person having a right, and seeing another about to commit an act infringing upon it, to assert his right. He cannot by his silence induce or encourage the commission of the act and then be heard to complain."

The great weight of authority in other States holds that where the lessor brings suit to avoid an oil and gas lease and the litigation ends after the grant has expired, that the lessors are estopped to claim the lease is invalid because the term has expired, and that an additional period of time may be fixed by a court of equity in which to commence drilling operations. . . .

In most of these cases the doctrine was applied in cases where the lessor brought about the delay, but in the cases of Standard Oil Co. v. Webb, [149 La. 245, 88 So. 808], and Simons v. McDaniel, [154 Okl. 168, 7 P.2d 419], the suits were brought by the assignees or grantees of the lessor. Under these authorities, Bennie Irey Shaw would be estopped to claim a forfeiture under the circumstances of this case. Appellants derive their title from her and it was within their power to prevent a fraud being perpetrated upon third parties by recording their deeds. The lease of appellee the Carter Oil Company was recorded more than two years before the deeds of appellants were placed on record. During all this time appellants had notice that there was an outstanding title created by their grantor and suit was not begun until insufficient time was left to have the litigation terminated prior to the expiration date of the Carter Oil Company's lease. This situation, in our opinion, comes squarely within the majority rule of other jurisdictions and the principle has long been recognized in this State. We think the extension order was justified, but that the date to commence drilling should be definitely fixed and not left to construction as to what is meant by "a reasonable time." It has been suggested that the Carter Oil Company could have proceeded to drill without regard to the suit and should not be permitted to claim the benefit of an estoppel unless it was actually prevented by court order or otherwise. In the cases from foreign jurisdictions this was not required, and when the great loss the lessee might suffer by successfully drilling a well and then losing title to the land is considered, we do not think that it should be required. . . .

The decree of the circuit court is affirmed in all respects as to both appellees except as to that part of the decree fixing a reasonable period after the termination of the litigation for the drilling of the well by the Carter Oil Company. The decree should be amended so as to extend the time of drilling for the period of six months from the date the affirming order of this court is filed with the clerk of the circuit court.

The decree is reversed and the cause remanded for such purpose, only, and in all other respects the decree is affirmed.

Decree modified and affirmed.

NOTES

1. In BAKER v. POTTER, 223 La. 274, 65 So.2d 598, 2 O.&G.R. 1072 (1952), the lessor brought suit for cancellation based on the late deposit of a telegraphed rental payment. The court refused to hold that the lease had terminated and, on rehearing, held that the lessee was entitled to an addition to the primary term of a period of time equal to that taken up by the law suit. The court said that "the lessee has been deprived of his exercise of the rights granted to him by the lease by the act of the lessor in having filed and prosecuted the lawsuit against him." Similar relief was granted in Bingham v. Stevenson, 148 Mont. 209, 420 P.2d 839, 26 O.&G.R. 175 (1966), where the unwarranted repudiation of the lease by the lessor resulted in a judgment that the lessee was "entitled to have the lease extended for a period of 7 years and 9 months from date of judgment." Is there any essential difference between this relief and that awarded in Greer v. Carter Oil Co.? Suppose that the well drilled under the Greer order was dry. Could the lessee commence a second well within the six month period set forth in the decree and retain the lease if such well were productive?

2. In EASTERN OIL CO. v. COULEHAN, 65 W.Va. 531, 64 S.E. 836 (1909), the lessee began drilling operations within the primary term of the lease and struck gas in a sand at the depth of about 1240 feet, which when gauged and tested showed a capacity of about 3,000,000 cubic feet of gas per day. After striking this gas, however, the lessee decided to go deeper to the lower or Indian sand. The well was begun in ample time to have completed it to the lower sand, but time was running short and orders were given to the drillers to work on

Sunday. The lessor frightened the drillers away by suggesting they were liable to arrest and conviction for working on Sunday. The lease had been executed on August 3, 1901, for a primary term of five years, and about 30 minutes after midnight of August 2, 1906, lessor appeared on the premises and notified the drillers that the lease had expired and that they were trespassers. As a result the drillers stopped drilling, and work was not resumed until about noon of August 3rd, a loss of about twelve hours in time. Shortly after 1 a.m. of August 4, gas in immense quantities was struck in the Indian sand. Lessors promptly instituted a suit in unlawful detainer to recover possession of the premises whereupon the plaintiff lessee obtained an injunction protecting it in the possession and occupancy of the land and enjoining the lessor from in any manner interfering with it in the use, occupancy and operation of said land for oil and gas purposes under the lease, and, also, from prosecuting his action of unlawful detainer. *Held,* injunction made permanent.

The court concluded that the lease did not expire by its terms until midnight August 3, 1906.

> Our conclusion . . . is that, where a contract of lease of the character of that involved here requires of the lessee affirmative acts to be done within a certain period stipulated from the date thereof, unless there is something in the instrument itself evincing a different intention on the part of the parties thereto, the date of the instrument will be excluded in the computation of time.

The authority appears uniformly in accord with the court's conclusion in this respect. *See* Winn v. Nilsen, 1983 OK 91, 670 P.2d 588, 78 O.&G.R. 78.

3. Huge Oil is planning on drilling a well shortly before the end of the primary term. The surface owner who was not the lessor forcibly kept Huge Oil away from the proposed well site. Huge Oil immediately sought a court injunction to prevent the interference. Unfortunately, the injunction was issued after the end of the primary term. Has the lease terminated? *See* Burger v. Wood, 1978 OK CIV APP 6, 575 P.2d 977, 59 O.&G.R. 503.

Lessor leases several tracts of land to Lessee. Lessor owns the surface estate of some but not all of the tracts. The leases require drilling to commence before a certain date. As to those tracts where the surface is owned by Lessor, Lessee commences drilling operations prior to the anniversary date. The surface owner of the remaining tracts is known to the Lessee as a difficult person to deal with who has damaged Lessee's oil field equipment over disputes on other lands. Lessee begins negotiations with the recalcitrant surface owner prior to the end of the primary term. Negotiations fail and upon attempting to move heavy equipment onto the tracts immediately prior to the end of the primary term, the surface owner threatens the crew with bodily harm. The primary term expires. What result? 21st Century Investment Co. v. Pine, 1986 OK CIV APP 27, 734 P.2d 834, 92 O.&G.R. 597 found the lease was extended for a reasonable time after an injunction was issued barring the surface owner from interfering with the lessee's operations.

4. The rule in *Greer* is sometimes called the doctrine of repudiation or the lessor interference rule or the doctrine of obstruction. *See e.g.,* Jicarilla Apache Tribe v. Andrus, 687 F.2d 1324, 75 O.&G.R. 286 (10th Cir.1982); HNG Fossil Fuels Co. v. Roach, 103 N.M. 793, 715 P.2d 66, 89 O.&G.R. 113 (1986); Mission Resources, Inc. v. Garza Energy Trust, 166 S.W.3d 601 (Tex.App. 2005); Tar Heel Energy Corp. v. Menking, 621 S.W.2d 450, 71 O.&G.R. 106 (Tex.Civ.App. 1981).

Lessor, who has not been paid royalties for several months, along with an employee of the Lessee who has not been paid for several months, physically shut-in Lessee's well. Lessee did not respond to the written notice given him by the Lessor and the employee that the well had been shut-in. In the meantime the Railroad Commission of Texas ordered the well sealed because the Lessee had failed to file the appropriate monthly reports. Several months later, the Commission rescinded its shut-in order and the Lessee seeks to restart production. The Lessor sues, claiming that the lease has expired for failing to produce in paying quantities. The Lessee asserts various affirmative defenses including the "lessor interference" doctrine. What result? *See* Atkinson Gas Co. v. Albrecht, 878 S.W.2d 236, 129 O.&G.R. 243 (Tex.App. 1994, writ denied).

Lessor sends letters to the gas purchaser questioning the legal authority of the lessee to sell the gas. In a jurisdiction requiring the marketing of natural gas, should the lease terminate when the gas purchaser refuses to execute a contract or refuses to take gas under an existing contract. O'Neal v. JLH Enterprises, Inc., 862 So.2d 1021, 159 O.&G.R. 1060 (La.App. 2002).

5. The opinion in *Coulehan* further declares that equity should relieve the plaintiff from termination of his leasehold interest in order to prevent gross injustice. "A lessor should not be heard to complain of a default caused by himself, or permitted to take advantage of his own wrong." What if the lessor gives a "top lease" to a third party while the first or base lease is still ongoing? *See* Brown, *Effect of Top Leases: Obstruction of Title and Related Considerations*, 30 Baylor L.Rev. 213 (1978).

Compare GISINGER v. HART, 115 Ohio App. 115, 184 N.E.2d 240, 17 O.&G.R. 47 (1961). Lessee commenced operations for drilling four days before the end of the primary term, but lessor interfered therewith, preventing further drilling. Held: the lease terminated. It was "highly improbable, if not impossible, to produce oil and gas in paying quantities on the premises in a period of four days. It is a case of 'too little too late'."

To what extent are other equitable defenses applicable to cases of asserted termination of oil and gas leases?

VICKERS v. PEAKER, 227 Ark. 587, 300 S.W.2d 29, 7 O.&G.R. 1177 (1957). Vickers, assignor in a farm-out arrangement, sought to cancel the assignment for Peaker's failure to drill the required well by the time specified in the agreement. Held: for defendant:

> [I]t is clear that Vickers sat by and remained silent and allowed Peaker to expend at least $40,000 in testing the Smackover lime and proving the Cotton Valley formation to be productive (all of which information was valuable to Vickers), before Vickers ever indicated in any way that he was going to make any claim of any forfeiture of his assignment under which he knew Peaker was holding and claiming.

The doctrine applied by the court to save the assignment was estoppel. Are the equitable doctrines of waiver and avoidance of forfeiture equally applicable to the case? Does it ever make a difference whether the defense is estoppel, waiver, or avoidance of forfeiture?

------

2. PRODUCTION IN PAYING QUANTITIES
CLIFTON v. KOONTZ

Supreme Court of Texas, 1959.
160 Tex. 82, 325 S.W.2d 684, 10 O.&G.R. 1109, 79 A.L.R.2d 774.

SMITH, JUSTICE. This suit brought by petitioners, Lillie M. Clifton, individually and as executrix of the estate of her husband, J.H. Clifton, deceased, et al., seeks the cancellation of an oil, gas and mineral lease on the theory that after the expiration of its ten-year primary term, the lease terminated due to cessation of production.

In the alternative, and only in the event the Court should find that production had not ceased, petitioners sought cancellation of the lease (other than for 40 acres around the existing well) on the theory that the owners of the lease (the working interest) breached an implied covenant to reasonably develop the property and to "reasonably explore the same for the production of minerals therefrom . . .." It was their contention that the owners of the working interest, in the event the alternative plea should be sustained, would forfeit all rights under the lease (except as to 40 acres around the producing well) upon failure within a reasonable time, to commence and continue the drilling of wells to a depth sufficient to test all known horizons in the general area. Petitioner also sought damages because of breaches of express and implied covenants of the lease.

The lease was executed in 1940. It covers two tracts of land encompassing 350 acres owned by the Cliftons, in Wise County, Texas. In 1949, during the primary term, a well was drilled which produced both oil and gas but very little oil. The Railroad Commission classified the well as an "associated" (with oil) gas well. Other than its acidization in 1950, no other drilling or reworking operations were carried on during the intervening years until September 12, 1956, when it was successfully reworked by "sandfracting". This date was subsequent to the filing of the present case.

The judgment of the trial court, entered after a trial before the court, without the aid of a jury, contains the court's findings of fact upon the basic questions.

The court found that the existing gas well had at all material times continuously produced gas in paying quantities. Accordingly, the oil and gas lease was held to be in full force and effect, thus denying petitioners' prayer for judgment that the lease had terminated. The judgment recites a finding to the effect that petitioners were damaged by the failure of respondents to rework the existing well and to drill, but that such damages were speculative and could not be ascertained with any degree of certainty. Judgment for damages was therefore denied.

The judgment decreed "that unless on or before the expiration of 60 days after the date this Judgment shall become final, the owners of the working interest shall commence and thereafter drill with reasonable diligence and in a good and workmanlike manner a test well looking to the production of oil and/or gas on said land, the drill site to be selected by said working-interest owners and said well to be drilled to a total depth of 5600 feet below the surface of said land unless production of oil and/or gas in paying quantities be by them found at a lesser depth, this lease shall terminate and re-invest in Plaintiff and her assigns except as to the conglomerate formation found in the stratigraphic interval from 5300 to 5600 feet, from which the present well is now producing, such section being generally known as the Atoka [Morris Field] conglomerate; and as to all of such section, strata, or formation, the lease shall continue in full force and effect, as to appearing parties herein, so long as it continues producing oil, gas or other minerals conformably with the terms of the lease instrument."

Both petitioners and respondents appealed. The Court of Civil Appeals affirmed the trial court's judgment denying termination of the oil and gas lease and petitioners' claim for damages, but held that respondents were not required to drill a second well. Therefore, the judgment of the trial court requiring such drilling was reversed and rendered. 305 S.W.2d 782.

We have concluded that the judgment of the Court of Civil Appeals must be sustained. We first consider the primary question, which is: Was there any evidence to sustain the finding of the trial court that production in paying quantities had not ceased?

. . .

Petitioners base their contention that the well had ceased to produce in paying quantities upon the showing that for the period of time from June, 1955 through September, 1956, the income from the lease was $3,250 and that the total expense of operations during the same period was $3,466.16--thus, a loss of $216.16 for the sixteen months' period selected by petitioners. During the period of time indicated, some months showed a gain and some a loss. For the months of July, August, and September 1956, the total net loss amounted to $372.37. These were the months immediately following June 1956, the date respondent, Koontz, acquired his 52 per cent interest in the lease. Beginning July 1, 1956 he began making financial arrangements, securing the services of third parties, and commenced saving all oil produced from the lease to be used in reworking the well. The holding from the market of this oil accounts materially for the losses in the operations during the months of July, August, and September 1956. The record shows that for years, in view of the low allowable on gas, the oil production had made the difference between operating at a profit and at a loss. The respondent Koontz, testified that from two to three months were required to accumulate a tank of oil and that after such accumulation a sale would be made. Respondents' evidence reflects that through 1955 and 1956 there was but little variation in gas production. For the same period of time there was a great variation in oil production, resulting in a showing of a profit in months when oil sales were made.

It is undisputed that reworking operations were commenced on September 12, 1956, and that such operations resulted in an 1800 per cent increase of production. Reworking operations having thus been commenced on September 12, 1956, the evidence that there was a small operating loss for the period of time from July, 1956 through September, 1956 is not controlling in determining whether or not there had been a cessation of production in paying quantities through July 12, 1956, a date 60 days prior to the beginning of reworking operations. The question, therefore, is: Was there production in paying quantities from the existing well through July 12, 1956? Evidence, as contended for by the petitioners, as to profit or loss subsequent to July 12, 1956, is immaterial in determining whether or not there was production in paying quantities at all times through July 12, 1956, then the clause contained in the lease, which permits reworking within 60 days following cessation of production, if it ever came into operation, was complied with when reworking operations were begun on September 12, 1956, and later successfully completed. Thus, by considering only the evidence relative to production prior to July 12, 1956, we find that the lessee operated at a profit in the sum of $111.25, for the period of time beginning in June 1955 and continuing through July 12, 1956. The record shows a loss during the months of April and May, 1956. The record further shows that for the year 1954, a profit was earned each month, and that the aggregate profit was the sum of $1,575; that in 1955 the operations were profitable during nine months of the year, with a net profit of $894 for the year; and that for the first six months' period of 1956, the lease was operated at a profit of $145. These factual situations, when considered in the light most favorable to the findings of the trial court, support its finding and judgment that there was not a cessation of production in paying quantities through July 12, 1956.

Petitioners argue that it is settled under the Texas law that, after the primary term, the ordinary oil and gas lease absolutely terminates when its income no longer exceeds the cost of its operation, and that since the operations showed a loss for the months of April and May, 1956, the lease terminated. . . . The further argument is made that such established rule applies as well to a lease with a 60-day termination clause, except a period of 60 additional days is allowed in which to begin additional drilling or reworking operations. The lease under consideration does contain such a clause which is directly related to the petitioners' argument relating to the determination of cessation of production:

. . . or if after discovery of oil, gas or other mineral the production thereof should cease from any cause, this lease shall not terminate if lessee commences additional drilling or reworking operations within sixty (60) days thereafter. . . .

We agree with petitioners that if production in paying quantities ceased, the 60-day clause applies. However, the facts in the instant case compel a different result than that contended for by petitioners.

It is only in the event of a finding of cessation of production in paying quantities that the trial court would be called upon to determine whether, if within 60 days from the date of such cessation, reworking operations were begun and resulted in profitable production thereafter. After cessation of production in paying quantities, the lessee has 60 days of "grace" in which to save his leasehold, however, if production never ceased, as is the case here, the 60-day clause is not definitive of the period over which the trier of the facts must determine whether a lease is producing in paying quantities. There can be no arbitrary period for determining the question of whether or not a lease has terminated for the additional reason that there are various causes for slowing up of production, or a temporary cessation of production, which the courts have held to be justifiable. . . . We again emphasize that there can be no limit as to time, whether it be days, weeks, or months, to be taken into consideration in determining the question of whether paying production from the lease has ceased. To apply the 60-day clause as contended by petitioners would mean that the respondents would have been required to immediately commence drilling operations, upon sustaining a slight loss for one month, without regard to whether they believed the next month's production might be profitable for the reason that if they were in error and suffered another slight loss, the lease would terminate. The petitioners cite the case of Stanolind Oil & Gas Company v. Newman Brothers Drilling Company, Tex., 305 S.W.2d 169, 172, as supporting their position. That case has no application to this question since it was concerned primarily with whether the 60-day clause and the 30-day clause contained in an oil, gas, or mineral lease are cumulative in application or separate and distinct provisions. No attempt was made in that case to settle the question of over what period of time paying quantities should be determined. However, the opinion does state that "the sixty-day provision would be brought into play by a cessation of such production."

The lease instrument involved in this suit provides by its terms that it shall continue in effect after commencement of production, "as long thereafter as oil, gas, or other mineral is produced from said land." While the lease does not expressly use the term "paying quantities", it is well settled that the terms "produced" and "produced in paying quantities" mean substantially the same thing. Garcia v. King, 139 Tex. 578, 164 S.W.2d 509, 511.

The generally accepted definition of "production in paying quantities" is stated in the *Garcia* case, *supra,* to be as follows:

If a well pays a profit, even small, over operating expenses, it produces in paying quantities, though it may never repay its costs, and the enterprise as a whole may prove unprofitable.

In the case of a marginal well, such as we have here, the standard by which paying quantities is determined is whether or not under all the relevant circumstances a reasonably prudent operator would, for the purpose of making a profit and not merely for speculation, continue to operate a well in the manner in which the well in question was operated.

In determining paying quantities, in accordance with the above standard, the trial court necessarily must take into consideration all matters which would influence a reasonable and prudent operator. Some of the factors are: The depletion of the reservoir and the price for which the lessee is able to sell his product, the relative profitableness of other wells in the area, the operating and marketing costs of the lease, his net profit, the lease provisions, a reasonable period of time under the circumstances, and whether or not the lessee is holding the lease merely for speculative purposes.

The term "paying quantities" involves not only the amount of production, but also the ability to market the product (gas) at a profit. . . . Whether there is a reasonable basis for the expectation of profitable returns from the well is the test. If the quantity be sufficient to warrant the use of the gas in the market, and the income therefrom is in excess of the actual marketing cost, and operating cost, the production satisfies the term "in paying quantities". In the *Hanks* case, [24 S.W.2d 5], the trial court found that the well completed by Hanks did not produce in paying quantities within the contemplation of the terms of the lease, and this Court upheld such finding, holding that there was no evidence showing that there were any facilities for marketing the gas or any near-by localities or industries which might have furnished a profitable market therefor. The Court went further and pointed out the complete failure of the evidence to show what the gas could have been sold for at any probable market, and that there was no evidence "tending to show that the well was situated in such proximity to any prospective market which would justify the construction of a pipe line for marketing same."

In the present case we have a finding of the trial court that there was production in paying quantities, and that the lease had not terminated. The evidence supports the finding. Evidence of marketing facilities and that the gas was sold at a profit is present in the instant case, whereas the *Hanks* case, *supra,* was wholly devoid of such evidence.

Proration rules adopted by the Texas Railroad Commission are a factor in determining the productive capacity of a particular lease. The Railroad Commission may by its order, for example, permit the taking of a greater percentage of the daily volume from nonassociated gas wells to the end that adequate quantities of gas may be delivered during the winter months when the demand therefor is the greatest, and by its order reduce the percentage to be taken during the summer months when a lesser quantity of gas is needed. Relative to individual leases or gas units, the Commission has taken the position that it may by its order allow an overproduction for a period of time to meet the market demand for that period, and, in order to balance such overproduction, it takes the position that it may order that the well be cut down to a minimum volume of production.

Petitioners contend that the trial court's finding that production in paying quantities had not ceased is erroneous because the profit and loss figures heretofore considered do not include charges for depreciation as an operating expense. We are of the opinion that the trial court correctly excluded depreciation as an operating expense in determining whether and when production in paying quantities ceases. The petitioners contend that depreciation of the original investment cost should be taken into consideration as a part of the expense in operating the well. With this contention we do not agree. We do not have before us the question of whether or not depreciation on producing equipment should be charged as an operating expense, and, therefore, do not decide the question.

As the *Garcia* case, *supra,* indicates, the term "paying quantities", when used in the extension clause of an oil lease habendum clause, means production in quantities sufficient to yield a return in excess of operating costs, and marketing cost, even though drilling and equipment costs may never be repaid and the undertaking considered as a whole may ultimately result in a loss. The underlying reason for this definition appears to be that when a lessee is making a profit over the actual cash he must expend to produce the lease, he is entitled to continue operating in order to recover the expense of drilling and equipping, although he may never make a profit on the over-all operation. Depreciation is nothing more than an accounting charge of money spent in purchasing tangible property, and if the investment itself is not to be considered, as is held by this Court, then neither is depreciation.

In Transport Oil Co. v. Exeter Oil Co. Ltd., 84 Cal.App.2d 616, 191 P.2d 129, it was held that operation of an oil and gas leasehold at an annual profit of about $4,300, without deduction of reserves for depletion and depreciation and after exclusion under terms of the lease and operating agreement of overriding royalties as an operating expense, was in "paying quantities" within the habendum clause of the lease. By not including depreciation as an operating expense, we more nearly accomplish a just result for lessors and lessees alike, for, if the rule were otherwise, many leases would be terminated and the lessees' incentive to drill decreased, regardless of the magnitude of the investment.

Petitioners present two cases to support their argument that depreciation should be treated as an expense item, but they are both distinguishable. In Persky v. First State Bank, Tex.Civ.App., 117 S.W.2d 861, 863, no writ history, the Court of Civil Appeals merely held:

> We *might* take judicial knowledge that all property of this kind has some depreciation from year to year, but we could not, in the absence of testimony to support it, take judicial knowledge of the percentage or amount of such

depreciation. (Emphasis supplied.)

It is not held that depreciation is to be considered as operation expense, and, as such, is to be deducted from income. Also, we do not have pumping machinery involved in the case at bar, as was true in the *Persky* case.

The other case relied upon is United Central Oil Corporation v. Helm, 5 Cir., 11 F.2d 760, certiorari denied 271 U.S. 686, 46 S.Ct. 638, 70 L.Ed. 1151. The Court in that case held that depreciation and overhead expenses are to be considered in determining paying quantities at the time of abandonment of a lease, in an action for damages against the lessee by the lessor. However, the lease was for a period of four years rather than "so long thereafter as oil and/or gas are produced." The holding in the *Helm* case does not conflict with our holding in the present case. The case merely demonstrates the variable meaning applied to the phrase "paying quantities".

Petitioners contend that the 8 per cent overriding royalties outstanding, as shown by the record, should be excluded from the total income attributable to the working interest in determining whether or not production is being obtained in paying quantities. We do not agree. The entire income attributable to the contractual working interest created by the original lease is to be considered. Petitioners cite no Texas case, and we have found none, which supports their argument. While apparently there is no Texas decision in point on the particular question, we believe the case of Transport Oil Co. v. Exeter Oil Co. Ltd., *supra,* is authority supporting our view. In that case the basic oil and gas lease on a tract of land executed in 1921 required certain drilling operations and the payment of 162/3 per cent royalty to the lessor based upon the gross income. After several assignments of the lease in which overriding royalties have been reserved, the plaintiff in the action became the owner of the lease subject to the overriding royalty. Thereafter, the plaintiff, Transport Oil Company, assigned this lease to the defendant, Exeter Oil Company, in which assignment the defendant obligated itself to pay out of the gross income from the lease, the basic landowner's royalty, the overriding royalty reserved in previous assignments, and the override to the plaintiff totaling almost 50 per cent of the income per well. In 1942 the defendant plugged the well and abandoned the property, due to declining production. The abandonment terminated the plaintiff's interest as an overriding royalty owner, and it brought suit to recover damages. The court held that overriding royalty could not be excluded from the total income in determining whether there was production in paying quantities. *See also,* Vance v. Hurley, 1949, 215 La. 805, 41 So.2d 724.

[The court's discussion of petitioners' arguments based on the theory of implied covenants is omitted.]

The judgment of the Court of Civil Appeals is affirmed.

HAMILTON, J., not sitting.

NOTES

1. The lease in *Clifton,* like most leases, provided for a fixed primary term and a secondary term "as long thereafter as oil, gas or other mineral is produced." The lease did not provide for extension of the term "as long as oil, gas or other mineral is produced *in paying quantities.*" On what basis is the italicized phrase added to the lease by the court?

In GARCIA v. KING, 139 Tex. 578, 164 S.W.2d 509 (1942), it is said:

In order to understand and properly interpret the language used by the parties we must consider the objects and purposes intended to be accomplished by them in entering into the contract. The object of the contract was to secure development of the property for the mutual benefit of the parties. It was contemplated that this would be done during the primary period of the contract. So far as the lessees were concerned, the object in providing for a continuation of the lease for an indefinite time after the expiration of the primary period was to allow the lessees to reap the full fruits of the investments made by them in developing the property. Obviously, if the lease could no longer be operated at a profit, there were no fruits for them to reap. The lessors should not be required to suffer a continuation of the lease after the expiration of the primary period merely for speculation purposes on the part of the lessees. Since the lease was no longer yielding a profit to the lessees at the termination of the primary period, the object sought to be accomplished by the continuation thereof had ceased, and the lease had terminated.

2. The courts have interpreted *Clifton* and *Garcia* as applying a two-pronged test. The first part of the test is a "mathematical" calculation. Bruce Kramer, *Lease Maintenance for the Twenty-First Century: Old Oil and Gas Law Doesn't Die, It Just Fades Away,* 41 Rocky Mtn.Min.L.Inst. 15-1, 15-16 (1995). Operating revenue must exceed operating costs. If a profit, no matter how small is found, the requirements of the habendum clause have been satisfied as a matter of law. Skelly Oil Co. v. Archer, 163 Tex. 336, 356 S.W.2d 774, 783, 16 O.&G.R. 650 (1961); Hydrocarbon Management, Inc. v. Tracker Exploration, Inc., 861 S.W.2d 427, 126 O.&G.R. 316 (Tex.App. 1993, no writ). The second part of the test, which is only triggered where there is no profit showing, is whether a reasonably prudent operator would continue to operate the well in the manner in which it is being operated for the purpose of making a profit and not merely for speculation. Ballanfonte v. Kimbell, 373 S.W.2d 119, 120, 19 O.&G.R. 826 (Tex.Civ.App. 1963, writ ref'd n.r.e.). For either prong of the test what is the relevant time period? Should the mathematical computation also depend on what a reasonable and prudent operator would think is the appropriate period of time? Would the time frame be different for oil production than for natural gas production? *See* Everest Exploration, Inc. v. URI, Inc., 131 S.W.3d 138, 159 O.&G.R. 73 (Tex.App. 2004). Kansas appears to rely solely on the "mathematical" test to determine production in paying quantities. Reese

Enterprises, Inc. v. Lawson, 553 P.2d 885, 56 O.&G.R. 517 (Kan. 1976).

3. A statute required the plugging of wells that had ceased to produce oil or gas in "commercial quantities." In State of Ohio v. Wallace, 52 Ohio App.2d 264, 369 N.E.2d 781, 6 O.O.3d 262, 58 O.&G.R. 549 (1976), the enforcement of the statute was attacked on the ground that its language was vague, ambiguous and uncertain. *Held*, the statute is valid. "Commercial quantities" does not mean the same thing as "paying quantities." " 'Paying quantities' connotes a profit; 'commercial quantities' does not. . . . '[C]ommercial quantities' can fairly be defined as 'of sufficiently large amount to be sold by the owner to a buyer for transport elsewhere.'" For other cases dealing with the term "commercial" quantities, *see* Pan American Petroleum Corp. v. Shell Oil Co., 455 P.2d 12 (Alaska 1969), Texaco, Inc. v. Fox, 228 Kan. 589, 618 P.2d 844, 67 O.&G.R. 360 (1980).

4. Some early cases applied a subjective standard of the good faith judgment of the operator in determining whether production in paying quantities had ceased. *See, e.g.,* Gypsy Oil Co. v. Marsh, 121 Okl. 135, 248 P. 329, 48 A.L.R. 876 (1926). This standard is sometimes expressly incorporated into modern leases, for both habendum clause purposes and implied or express drilling covenant purposes. *See, e.g.,* Danker v. Lee, 137 Cal.App.2d 797, 291 P.2d 73, 5 O.&G.R. 313 (1955) (drilling covenant).

If a jurisdiction uses a good faith standard to determine whether there is production in paying quantities who bears the burden of proof? The prevailing view seems to place the burden on the party asserting that the lease has expired. *See* Superior Oil Co. v. Devon Corp., 458 F.Supp. 1063, 1071, 61 O.&G.R. 61 (D.Neb. 1978), *rev'd on other grounds*, 604 F.2d 1063, 65 O.&G.R. 368 (8th Cir. 1979); T.W. Phillips Gas & Oil Co. v. Jedlicka, 2008 PA Super 293, 964 A.2d 13, *but cf.* Positron Energy Resources, Inc. v. Weckbacher, 2009-Ohio-1208, 2009 Ohio Appl. LEXIS 985 (2009).

5. Although it is somewhat rare, the parties are free to provide an express standard of what constitutes production in paying quantities. In Renner v. Huntington-Hawthorne Oil & Gas Co., 39 Cal.2d 93, 244 P.2d 895, 1 O.&G.R. 1063 (1952), the lease defined the term as follows:

> Oil in paying quantities shall be understood to mean a well drilled from which there may be pumpted for a period of thirty (30) consecutive days a quantity of oil which shall average fifty (50) barrles of oil . . . per day.

6. The Louisiana Mineral Code has adopted the *Clifton v. Koontz* standard with the following provisions:

> **Art. 124**. When a mineral lease is being maintained by production of oil or gas, the production must be in paying quantities. It is considered to be in paying quantities when production allocable to the total original right of the lessee to share in production under the lease is sufficient to induce a reasonably prudent operator to continue production in an effort to secure a

return on his investment or to minimize any loss.

As to all other minerals, it is sufficient if a reasonably prudent operator would continue production considering the particular circumstances in the light of the nature and customs of the industry involved. In appropriate cases, such as the mining of lignite, a court may consider the total amount of production allocable to the mining plan or project of which a particular lease is a part, rather than merely the amount of production from that lease.

Art. 125. In applying Article 124, the amount of the royalties being paid may be considered only insofar as it may show the reasonableness of the lessee's expectation in continuing production. The amount need not be a serious or adequate equivalent for continuance of the lease as compared with the amount of the bonus, rentals, or other sums paid to the lessor.

See McCollam, *Impact of Louisiana Mineral Code on Oil, Gas and Mineral Leases*, 22nd Annual Inst. on Mineral Law 37 at 81 (1975); CCH, Inc. v. Heard, 410 So.2d 1283, 72 O.&G.R. 471 (La.App.1982).

7. The parties to a lease strike from the written form lease the language "whether or not in paying quantities" that follows the term "production" in the habendum clause. Does the *Clifton* two-part test apply? Should the deletion shift the burden of proof from the lessor to the lessee? In Peacock v. Schroeder, 846 S.W.2d 905, 121 O.&G.R. 525 (Tex.App. 1993, no writ) the court found that the lease required production in paying quantities and applied the two-part test. It continued to place the burden of proof to show lack of production in paying quantities on the lessor. *See also* Evans v. Gulf Oil Corp., 840 S.W.2d 500 (Tex.App. 1992, writ denied); Bachler v. Rosenthal, 798 S.W.2d 646, 114 O.&G.R. 354 (Tex.App. 1990, writ denied).

8. Does a lease terminate where there is no production for a period of time without the application of the *Clifton* two-part test? Should the court have to determine whether there was production in paying quantities over a period of time in excess of the time of zero production? Would it matter whether the well was an oil well or a gas well?

9. The habendum clause of an oil and gas lease provides that it "shall remain in force for a term of one (1) year and as long thereafter as gas is or can be produced." Does the additional language "or can be" allow the lease to be maintained in the absence of actual production if the state otherwise requires discovery plus marketing to maintain the lease in the secondary term? Would it matter if there were other leasehold savings clauses that envisioned a discovery plus marketing requirement? *See* Greer v. Salmon, 82 N.M. 245, 479 P.2d 294, 41 O.&G.R. 177 (1970); Anadarko Petroleum v. Thompson, 90 S.W.3d 550 (Tex.2002); Grinnell v. Munson, 137 S.W.3d 706, 159 O.&G.R. 1139 (Tex.App. 2004).

10. Is the existence of paying quantities an issue of law or of fact? In Morgan v. Fox, 536 S.W.2d 644, 54 O.&G.R. 345 (Tex.Civ.App.1976, error

ref'd n.r.e.), the court held that "where it is shown that a small profit has been realized from the operation of the well, it may be found as a matter of law that the well is producing in paying quantities." Does such a finding remove from the case a question of "whether or not the lessee is holding the lease merely for speculative purposes?" *See* Bell v. Mitchell Energy Corp., 553 S.W.2d 626, 58 O.&G.R. 365 (Tex.Civ.App.1977). Can a lessor prevail on a motion for summary judgment under *Clifton v. Koontz* if they do not proffer evidence on both prongs of the production in paying quantities test? Dreher v. Cassidy Limited Partnership, 99 S.W.3d 267, 158 O.&G.R. 508 (Tex.App. 2003). Who has the burden of proof on the issue of production in paying quantities? *See* Cannon v. Sun-Key Oil Co., 117 S.W.3d 416, 158 O.&G.R. 959 (Tex.App. 2003) (lessor "had the burden to prove that a reasonably prudent operator would not have continued to operate the lease in the manner in which it was operated for the purpose of making a profit and not merely for speculation during the period of cessation.").

STEWART v. AMERADA HESS CORP.

Supreme Court of Oklahoma, 1979.
604 P.2d 854, 65 O.&G.R. 530.

OPALA, JUSTICE: The dispositive issues here are (1) whether oil and gas ceased being produced in paying quantities from an Amerada Hess Corporation (Amerada) leasehold, and if this be answered in the affirmative, then (2) whether the cessation operated to terminate Amerada's leasehold estate in the premises.

Amerada, owner of oil and gas lease acquired from predecessor in title of present owners (owners), assigned a portion of its lease to Union Texas Petroleum (Union), who sank a producing well. The present owners later leased the same premises to the Rodman Corporation (Rodman) who also succeeded in drilling a producing well.

Owners brought an action to terminate Amerada's lease because oil and gas was no longer being produced in paying quantities. Amerada's own action followed. The latter suit's object was to eject Rodman from the leasehold premises and secure money judgment for the value of the oil and gas alleged to have been removed. The two causes were consolidated for trial. The trial court upheld Amerada's superior lease rights and rendered a money judgment against Rodman. On separate appeals by owners and Rodman, the Court of Appeals, in two opinions, reversed the judgments against Rodman and the owners. It held as a matter of law that Amerada's lease had expired because Union's well was not producing in paying quantities one year before and the year in which Rodman secured the succeeding leasehold. Amerada seeks our review of these decisions. We consolidated the two appeals for disposition of review on certiorari by a single opinion.

The Court of Appeals assumed that if equipment depreciation were to be considered as a production cost item for 1972 and 1973, Union's well operations were unprofitable during those two years but not so if that item were not included as an expense. The trial court viewed depreciation of equipment as not mandatorily includible in computing lifting costs, while the Court of Appeals held that depreciation of production equipment installed upon the demised premises must be considered in determining profitability of the leasehold estate....

I

The habendum clause of Amerada's oil and gas lease in suit provides that it shall remain in force, after the primary term, for as long thereafter as oil or gas is produced. The term "produced", when used in a "thereafter" provision of the habendum clause, denotes in law production in paying quantities. The phrase means that the lessee must produce in quantities sufficient to yield a return, however small, in excess of "lifting expenses", even though well drilling and completion costs might never be repaid.

The cost of drilling a producing well, i. e. the expense incurred before oil is actually lifted from the ground is not an item to be considered in computing production in "paying quantities". The lifting of oil which marks the commencement of production stage is coincidental with the completion of a well. At that stage, critical here, only those expenses which are directly related to lifting operations can be included in determining if Amerada's lease remained in force beyond its primary term.

Whether depreciation does constitute a mandatory cost item is an issue of first impression in Oklahoma. . . .

Extant case law on the national scene has given little attention to depreciation as a mandatory lifting expense item. Some case law indicates, without actually so holding, that depreciation might be considered to be in this category. Other out-of-state authority reasons that since the original investment in a well may not be considered in computing oil lifting expenses, neither should producing equipment depreciation. The rationale for not including depreciation, given by one court, was that it would bring about termination of many leases and tend to diminish the incentive to drill.

Depreciation of equipment used in lifting operations is regarded as production expense in some states. The rationale for this rule is that while depreciation of the original investment in the drilling of a well may not be *stricto sensu* an out-of-pocket lifting expense, production-related equipment does have value that is being reduced through its continued operation. We adopt this reasoning as sound and hold that depreciation should be mandatorily included as an item of lifting expense in determining whether there is production in "paying quantities". The base and the period of depreciation should be determined by reference to currently prevailing accounting standards.

II

Under a literal or strict interpretation of the "thereafter" provision in a habendum clause, uninterrupted production following expiration of primary term would be indispensable to maintain a lease in force. This would mean, of course, that any cessation of production (in the paying-quantities sense of the term), however slight or short, would put an end to the lease. Oklahoma has rejected that literal a view. Our law is firmly settled that the result in each case must depend upon the circumstances that surround cessation. Our view is no doubt influenced in part by the strong policy of our statutory law against forfeiture of estates. The terms of 23 O.S. 1971 § 2 clearly mandate that courts avoid the effect of forfeiture by giving due consideration to compelling equitable circumstances.

The "thereafter" clause is hence not ever to be regarded as akin in effect to the common-law conditional limitation or determinable fee estate. The occurrence of the limiting event or condition does not automatically effect an end to the right. Rather, the clause is to be regarded as fixing the life of a lease instead of providing a means of terminating it in advance of the time at which it would otherwise expire. In short, the lease continues in existence so long as interruption of production in paying quantities does not extend for a period longer than reasonable or justifiable in light of all the circumstances involved. But under no circumstances will cessation of production in paying quantities *ipso facto* deprive the lessee of his extended-term estate.

A decree of lease cancellation may be rendered where the record shows that the well in suit was not producing in paying quantities and there are no compelling equitable considerations to justify continued production from the unprofitable well operations. Because the trial court found that lifting expenses need not include depreciation, neither party had the opportunity to adduce proof of circumstances surrounding the stoppage of profitable production nor those factors which might afford compelling equitable considerations either in favor of or against lease cancellation.

The opinions of the Court of Appeals which set aside the trial court's judgments are vacated. The judgments of the trial court are reversed and the causes remanded for further proceedings consistent with these views.

LAVENDER, C.J., IRWIN, V.C.J., and SIMMS, DOOLIN and HARGRAVE, JJ., concur; WILLIAMS and BARNES, JJ., dissent; HODGES, J., not participating.

BAYTIDE PETROLEUM, INC. v. CONTINENTAL RESOURCES, INC.

Supreme Court of Oklahoma, 2010
2010 OK 6, 231 P.3d 1144

WATT, JUSTICE. We granted certiorari to consider an issue of first impression: whether an oil and gas lease sought to be terminated for failure to produce in paying quantities during the secondary term remains effective until the lease has been judicially cancelled? We hold that it is not the court order which terminates the lease. Rather, it is the failure to produce in paying quantities under the habendum clause during the lease's secondary term. In the instant cause, the lease terminated by its own terms before the Alfalfa County court issued its order.

Our determination that the date of termination may precede the date of adjudication and did so here, necessitates that we also address Baytide's assertion that it should not be held to the legal obligations arising based on its status as a lessee in the Unit. Specifically, Baytide insists that it should not be bound by its agreement to accept $ 13,200.00 for its equipment located on the lease.

Baytide voted to approve the method of valuation of the equipment and the values assigned thereto, totaling $ 13,200.00 for the two Ford wells. Under these facts, Baytide is bound by its agreement to accept $ 13,200.00 for the lease equipment.

Factual and Procedural History

The cause arises from a tortured procedural and factual background involving litigation in two counties and before the Corporation Commission giving rise to three appeals.

In 2000, Baytide acquired the equipment and working interests in two wells operated in Alfalfa County and known as the Ford B # 1 and the Ford B # 2. The following year, Continental obtained top leases covering the two oil and gas wells. In July of 2001, Continental filed an action in Alfalfa County seeking to have Baytide's base lease terminated for failure to produce in paying quantities. In September of the same year, while the termination cause remained pending, the Commission established a Plan of Unitization creating the Aline Oswego Unit and designating Continental as the unit operator. The Unit includes the Ford wells. Under the plan, each lessee had the option to participate. The existing oil and gas equipment on the leases was to be evaluated, and its owner given credit for the appraised value of the same.

In November of 2001, the Alfalfa County court granted Continental's request for a temporary injunction. Baytide was ordered to comply with the plan and to deliver the two wells and all operating equipment to Continental along with production and well records. The order did not require that Baytide participate in the unit's operation or set a price for the sale of the operating equipment.

The Alfalfa County court issued it's order in favor of Continental on November 19, 2002 finding that Baytide's lease had terminated under the habendum clause for failure to produce in paying quantities. Until that date and thereafter in a subsequently dismissed appeal, Baytide contended that its leases were valid. Nevertheless, the trial court found that: Continental was not claiming that Baytide's leases automatically terminated; there was a twenty-eight (28) month cessation of production during which time Baytide made no effort to restore production; Baytide presented no reasonable basis for the cessation of production; the leases expired by their own terms by reason of cessation of production for an unreasonable period of time (28 months); and the underlying lease should be canceled for failure to produce in paying quantities.

Baytide filed the present action in Garfield County in April of 2006 asserting that it was not a lessee when the Unit was formed and that its oil and gas equipment located on the Ford wells had been wrongfully converted and Continental had been unjustly enriched. The petition was amended on May 12th to add a misuse of process allegation. Through a series of orders, the Garfield County district court: quieted title in the equipment in the Unit; awarded Baytide $ 13,200 for the equipment, including the heater treater; and determined there was no evidence of misuse of process.

Baytide filed its appeal on July 18, 2008. In May of 2009, the Court of Civil Appeals affirmed finding that Baytide remained a lessee when the Unit was formed because, prior thereto, no court order had been entered terminating the lease for failure to produce in paying quantities. We granted certiorari on October 5, 2009.

Discussion

a. A court order may be necessary to adjudicate the rights of the parties when allegations are that a lease has terminated for failure to produce in paying quantities. Nevertheless, the lease is not cancelled because of the entrance of the court order. The lease terminates for the failure to produce in paying quantities under its own terms pursuant to the habendum clause.

Baytide asserts that cessation of production in paying quantities in the secondary term of a lease for an unexplained or unreasonable period of time results in the lease's expiration. Under such circumstances, Baytide contends that it is the period of time found to have exceeded the bounds of reasonableness which fixes the date of lease cancellation rather than the date a court decree enters cancelling the lease contract. Continental argues that a lease expires for lack of production in paying quantities only upon the date a court order enters cancelling the same. We disagree with Continental's argument.

Both parties find support in our pronouncement in *Stewart v. Amerada Hess Corp.,* 1979 OK 145, 604 P.2d 854. Stewart made it clear that, in Oklahoma, the cessation of production during the secondary term of a lease is not in and of itself sufficient to automatically terminate a lease. Rather, a lease remains viable so long as the interruption of production in paying quantities does not extend for an unreasonable period which is not justifiable in light of all the circumstances. Under *Stewart,* a decree of cancellation may issue where the record supports a determination that a lease is not held by production in paying quantities and no compelling circumstances justify continued production from the unprofitable well operations. Nevertheless, *Stewart* does not hold, as Continental argues, that a lease will expire for lack of production in paying quantities only upon the issuance of a judicial decree.

Smith v. Marshall Oil Corp., 2004 OK 10, 85 P.3d 830 is instructive. In *Smith,* two issues were presented. The first was whether the evidence was sufficient to support the determination that the subject leases expired due to lack of production in paying quantities. The second, involved the issue of whether the equipment left on the premises had been in place for a sufficient period of time to vest ownership in the surface owner. We answered both issues in the affirmative.

The trial court in *Smith* directed verdict in favor of Marshall Oil on August 2, 2001 finding that the oil and gas leases expired by their own terms. On certiorari, we acknowledged that in the absence of compelling equitable considerations to justify a cessation in paying quantities over a period spanning three years, the leases would terminate under the terms of the habendum clauses. Determining that no extenuating circumstances occurred justifying the failure to produce in paying quantities, we held that the leases expired by the terms of their own habendum clauses.

The leases at issue in *Smith* contained a provision allowing the lessor six months to recover equipment from the lease premises following a cessation of production. Nevertheless, the *Smith* Court held that the period in which the well ceased to produce in paying quantities indisputably triggered the six-month period for Smith to remove his equipment. The opinion goes on to provide in pertinent part at P22:

> ". . . Both Smith's actions and inaction with regard to the equipment he left on the subject leases supports a determination that ownership vested in the surface owner, under the specific lease provision . . ."

The trial court decided the quiet title issue on August 2, 2001. The period of cessation of production occurred between 1996 and 1998. If, as Continental contends, a lease expires only upon the issuance of a court order or judgment, Smith should have had six months from the date of the decree to remove his equipment from the well site. Instead, the *Smith* opinion makes it clear that the failure to present evidence regarding inclement weather to preclude removal, or any other variable for equitable consideration, Smith's failure to remove the equipment during the six-month period immediately following cessation of production, occurring some two-and-one-half years prior to the entrance of the court order, resulted in ownership of the equipment vesting in the surface owner.

Stewart provides that a lease will not expire for lack of production in and of itself. However, it does not hold that only a judicial determination may end a lease for lack of production in paying quantities. *Smith's* pronouncement makes it clear that it is the unreasonable cessation of production that causes the lease to terminate. Therefore, in conformance with *Smith* and *Stewart,* we determine that it is the failure to produce in paying quantities during the lease's secondary term rather than the entrance of a court order which terminates a lease. In the instant cause, the lease terminated by its own terms before the Alfalfa County court issued its order.

b. Baytide agreed to the valuation method and the value assigned to its property. It is bound by its agreement to accept $ 13,200.00 for the lease equipment.

* * *

Conclusion

Our prior decisions make it clear that although a court order may be necessary to adjudicate the rights of the parties when allegations are that a lease has terminated for failure to produce in paying quantities, a lease terminates not because of the entrance of a court order but for failure to produce in paying quantities under the lease's habendum clause. Furthermore, under the facts presented, where Baytide agreed to the valuation method and the value assigned to its property, it is bound by the agreement to accept $ 13,200.00 for the lease equipment.

NOTES

1. The court in both *Stewart* and *Baytide* speak in terms of the cessation of actual production in the secondary term. Is that consistent with the holding in McVicker v. Horn, Robinson & Nathan, 322 P.2d 410, 8 O.&G.R. 951 (Okla. 1958), discussed *supra* at 244 where the court concluded that a well need only be capable of producing in paying quantities in order to go from the primary to the secondary term? Should the Oklahoma Supreme Court have focused on whether the well's were capable of producing as opposed to whether there was a cessation of production?

2. One of the hallmark differences between treating the oil and gas lease as a fee simple determinable or a fee simple on a condition subsequent is whether or not the lessee is a trespasser once there is a cessation of actual or a cessation of a capability of producing in paying quantities. Does the *Baytide* opinion suggest that the prior lessee in this case was a trespasser from the moment that there was a cessation of production in paying quantities?

3. There are a number of cases dealing with the issue of depreciation and how to properly account for capital expenditures when making a production in paying quantities determination. In Lege v. Lea Exploration Company, Inc., 631 So.2d 716, 130 O.&G.R. 329 (La.App. 3d Cir. 1994), *writ denied*, 635 So.2d 1112 (La. 1994), whether there was or was not production in paying quantities turned on the accounting treatment of expenditures by the lessee for disposal of salt water. The lessee had installed a salt water disposal system at substantial expense. The lessors contended that the expenses were greater than the value of the production. The lessee contended that the system costs were in the nature of capital expenditures that should be amortized over a period of years. If the lessee's capitalization of these costs was correct, then the lease did produce in paying quantities in the relevant period. To this the lessors responded that the costs should be immediately expensed since the disposal system substituted for trucking the salt water away for disposal, and such lifting and trucking of salt water would have been properly treated as a current expense. How would you rule in this case? What factors would you consider? *See also* O'Neal v. JLH Enterprises, 862 So.2d 1021 (La.App. 2003)

4. In applying the "mathematical" test courts must determine the income or revenue stream. Historically, as in *Clifton*, one took the gross revenue from the

sale of the hydrocarbons and you excluded the lessor's royalty. One did not exclude overriding royalty payments even though the lessee does not, in theory, receive that income. La.Min.Code art. 124; Transport Oil Co. v. Exeter Oil Co., 191 P.2d 129 (Cal.App. 1948); Hininger v. Kaiser, 738 P.2d 137, 94 O.&G.R. 167 (Okla. 1987). Additional payments to the lessor, however, need be characterized as bonus or additional royalty, the former not being an operating expense while the latter is not included in lessee revenue calculations. Vance v. Hurley, 41 So.2d 724 (La. 1949).

In looking at the type of operating expenses that are deducted from the revenue the Kansas Supreme Court applied the following analysis:

These direct costs include labor, trucking, transportation expense, replacement and repair of equipment, taxes, license and permit fees, operator's time on the lease, maintenance and repair of roads, entrances and gates, and expenses encountered in complying with state laws which require the plugging of abandoned wells and prevention of pollution. Reese Enterprises, Inc. v. Lawson, 553 P.2d 885, 898, 56 O.&G.R. 517 (Kan. 1976).

How would you classify the following expenses:

(a) Depreciation of equipment including casing, tubing or the Christmas tree. *See* Mason v. Ladd Petroleum Corp., 630 P.2d 1283, 70 O.&G.R. 586 (Okla. 1981).

(b) Depreciation of equipment used in operating or lifting activities. *See* Whitaker v. Texaco, Inc., 283 F.2d 169, 13 O.&G.R. 502 (10th Cir. 1960); Bales v. Delhi-Taylor Oil Corp., 362 S.W.2d 388, 17 O.&G.R. 811 (Tex.Civ.App. 1962).

(c) Overhead expenses of operating multiple leases. *Compare* Reese Enterprises, Inc. v. Lawson, op cit. *and* Skelly Oil Co. v. Archer, 163 Tex. 336, 356 S.W.2d 774, 16 O.&G.R. 650 (1961) *with* Menoah Petroleum, Inc. v. McKinney, 545 So.2d 1216, 105 O.&G.R. 242 (La.App. 1989) and Duerson v. Mills, 648 P.2d 1276, 73 O.&G.R. 39 (Okla.App. 1982).

(d) Marketing expenses, including the construction of a pipeline. Pray v. Premier Petroleum, Inc., 662 P.2d 255, 76 O.&G.R. 449 (Kan. 1983).

(e) Cost of constructing a saltwater disposal system. Lege v. Lea Exploration Co., Inc., 631 So.2d 716, 130 O.&G.R. 329 (La.App.), *writ denied*, 635 So.2d 1112 (La. 1994).

5. What should be the accounting period for determining whether or not production was in paying quantities? *See* Cannon v. Sun-Key Oil Co., 117 S.W.3d 416, 158 O.&G.R. 959 (Tex.App. 2003); Ballanfonte v. Kimbell, 373 S.W.2d 119, 19 O.&G.R. 826 (Tex.Civ.App.1963, error ref'd n.r.e.). Does it matter that the period was unusually bad, as in the Depression, or unusually good for oil prices as in an international crisis affecting imports? Should a one year period suffice? *See* Edmundson Brothers Partnership v. Montex Drilling Co., 731

So.2d 1049, 142 O.&G.R. 266 (La.App. 1999). Is the issue relating to the relevant accounting period applicable where there has been a total cessation of production for 30 or 60 days, preceded and followed by periods of production in paying quantities? Should it matter whether the lease contains a savings provision?

The Temporary Cessation of Production Doctrine
RIDGE OIL CO. v. GUINN INVESTMENTS, INC.

Supreme Court of Texas, 2004
148 S.W.3d 143, 161 Oil & Gas Rep. 1135, 47 Tex. Sup. Ct. J. 1080

OWEN, JUSTICE. In this oil and gas case, two lessees obtained working interests under a single lease through assignments. Guinn Investments, Inc. is the lessee as to the Guinn tract, and Ridge Oil Company, Inc. is the lessee as to the adjoining Ridge tract. Ridge shut in the only two producing wells, both located on the Ridge tract, for approximately ninety days and subsequently executed new leases with the owners of the possibility of reverter of the mineral interest in the Ridge tract. Among the numerous issues presented, we hold 1) the temporary cessation of production doctrine applies when there is more than one lessee under a single lease, 2) production permanently ceased from the Ridge tract when the new leases between Ridge and the owners of the Ridge tract became effective, 3) Guinn was not conducting operations on the lease sufficient to sustain the lease at the time production permanently ceased or thereafter, and 4) Guinn is not entitled to prevail on its claims for tortious interference, fraud, or the imposition of a constructive trust.

Because the trial court did not err in granting summary judgment for Ridge, we reverse the court of appeals' judgment and render judgment for Ridge.

I

. . .

Guinn and Ridge were both lessees under a 1937 oil and gas lease. The lease covered two adjoining 160-acre tracts, which we will call the Guinn tract and the Ridge tract for ease of reference. Through various assignments, Guinn became a partial assignee under the lease, obtaining the lessee's rights and obligations with respect to the 160-acre Guinn tract. Through other assignments, Ridge also became a partial assignee, obtaining the lessee's rights and obligations with respect to the 160-acre Ridge tract. Although the record is not entirely clear, the parties agree that the possibility of reverter of the mineral interest in each tract has become separate as well. The successors to the lessor's interest in the Guinn tract have no interest in the Ridge tract, and the successors to the lessor's interest in the Ridge tract have no interest in the Guinn tract. Ridge's brief asserts that both the working interests and the mineral estates with regard to the two tracts have been "partitioned," and Guinn does not dispute that. As the court of appeals noted, "It is undisputed that Ridge Oil's lessors were not the lessors of the Guinn tract." None of the lessors are parties to this suit, with the exception of Ridge, who obtained a percentage of the possibility of reverter of the mineral estate on the Ridge tract.

At one time, there was a producing well on the Guinn tract, but it was plugged and abandoned in 1950. There has been production on the Ridge tract since 1937, and there were two producing wells on that tract at all times material to this dispute. The parties agree that until at least December 1, 1997, those wells sustained the 1937 lease as to both the Ridge and Guinn tracts. The habendum clause of the 1937 lease says:

> It is agreed that this lease shall remain in force for a term of five (5) years from this date, and as long thereafter as oil or gas, or either of them is produced from said land by the lessee, or as long as operations are being carried on.

Guinn acquired its interest under the 1937 lease in the summer of 1997. Shortly thereafter, Ridge offered to purchase Guinn's interest, but Guinn declined the offer. Ridge then decided that it would attempt to terminate the lease. Ridge told its pumper to cut off the electricity to the two producing wells on the Ridge tract and to perform no other activities on the premises until further notice. The electricity to the wells was cut off on December 1, 1997, and the wells ceased to produce on that date. . . The letters then set forth Ridge's plan to terminate the lease and obtain new leases on both the Guinn and Ridge tracts:

> However, after consulting with my attorney regarding this matter, he has advised me of another avenue that we can take to accomplish the same desired result. He advised me to take new oil and gas leases covering only [the Ridge tract] and to simply shut the two [Ridge tract] wells in for a period of 90 days which would terminate the 1937 oil and gas lease. We could then take new oil and gas leases from the mineral owners under the [Guinn tract].

In these same letters, Ridge sent a new lease to each mineral estate owner of the Ridge tract and offered to pay a $500 bonus to each upon execution of a new lease. The letters also said that Ridge would pay $50 per month to each owner for the loss of royalty proceeds while the wells were shut in, which was the average monthly royalty paid for the last six months of production. Each of the mineral interest owners of the Ridge tract accepted this offer and executed new leases effective as of March 3, 1998. On that same date, Ridge instructed its pumper to reconnect the electricity to the wells, which he did, and the wells resumed production on that date.

. . .

In Guinn's suit against Ridge and Woodward, Guinn contended that the 1937 lease had not terminated as to its tract, either because the cessation of production of the wells on the Ridge tract was temporary or because Guinn had begun operations on the Guinn tract before the lease expired and was prevented from continuing those operations by Ridge's conduct.

. . .

The court of appeals concluded that the temporary cessation of production doctrine applied, the cessation of production was temporary, and Ridge's surrender of its lease and the taking of new leases from the mineral interest owners on the Ridge tract did not terminate the 937 lease as to Guinn. . . .

Ridge petitioned this Court for review, which we granted.

II

First, we should make clear what is not at issue in this case. None of the mineral interest owners in either tract are parties to this suit, other than Ridge, who owns a portion of the mineral interest in the Ridge tract. Our determinations do not purport to adjudicate the rights of the absent mineral interest owners. The only issue among those who are parties to this suit (Guinn, Ridge, and Woodward) is whether the 1937 lease remains in effect as to the Guinn tract.

This Court has consistently recognized that when a lease covers more than one tract and provides, as the 1937 lease in this case provides, that it shall remain in force for a stated term and "as long thereafter as oil or gas, or either of them is produced from said land," production on any part of "said land" continues the lease in effect as to all land covered by the lease. This is true even when the lessee assigns interests in parts of the leased premises to different operators and there is production by only one assignee. Neither Guinn nor Ridge takes issue with these principles of law.

Ridge, however, contended in its motion for summary judgment that, when production on the Ridge tract ceased on December 1, 1997, the cessation of production immediately terminated the 1937 lease in its entirety. Ridge contended alternatively that the 1937 lease terminated when it executed new leases with the owners of the possibility of reverter of the mineral interest on the Ridge tract. Guinn countered that the cessation of production was only temporary, and therefore, under the temporary cessation of production doctrine, the 1937 lease did not terminate.

This Court has held that temporary cessation of production in paying quantities does not terminate a lease that provides it will remain in force and effect as long as oil or gas is produced. In *Midwest Oil Corp. v. Winsauer*, [159 Tex. 560, 323 S.W.2d 944 (1959) eds.] we concluded that this doctrine likewise applies to royalty deeds, and we thus said that "a temporary cessation of production in paying or commercial quantities will not cause the royalty deed to terminate" when the term of that grant was as long as oil, gas, or other minerals were produced. We reached the same conclusion in *Stuart v. Pundt*, [338 S.W.2d 167 (Tex.Civ.App.-San Antonio 1960, writ ref'd) eds.] a decision from a court of appeals in which this Court refused the application for writ of error. We reiterated in *Amoco Production Co. v. Braslau* [561 S.W.2d 805 (Tex. 1978) eds.] that only permanent cessation of production may cause the estate to terminate. More recently, we said in *Anadarko Petroleum Corp. v. Thompson* [94 S.W.3d 550 (Tex. 2002) eds.] that "a typical Texas lease that lasts 'as long as oil or gas is produced' automatically terminates if actual production permanently ceases during the secondary term," citing our decision in *Amoco v. Braslau.*

We agree with the court of appeals in the case before us today that, absent any language in a lease to the contrary, the temporary cessation of production doctrine applies when a lease covering more than one tract or interest is held by production from a well operated by a partial assignee of the lessee's rights. As noted above, the temporary cessation of production doctrine applies to leases between original lessors and lessees and to owners of nonparticipating royalty interests. It logically should apply to partial assignees of a lessee's interest as well. The central question before us today is whether there was a temporary cessation of production under the facts of this case.

III

This and other courts have held that the temporary cessation of production doctrine applies in a wide variety of circumstances. In *Winsauer*, we held that cessation of production in paying quantities for 174 days was temporary because it was caused by two successive lawsuits and subsequently an obstruction in a gas line that required the laying of a new line. In another temporary cessation of production doctrine case, *Stuart v. Pundt*, the only producing well "began to make sand, water and bottom sediment to the point where the connection was cut off and gas could not be produced." Efforts to restore the well were undertaken until it was determined that the casing in the well had collapsed. The lessee plugged and abandoned that well and drilled a new one. The cessation of production was held to be temporary, and the new well sustained the lease. In *Cobb v. Natural Gas Pipeline Company of America*, [897 F.2d 1307 (5th Cir. 1990) eds.] Natural was the lessee and also owned and operated the pipeline system into which a well maintaining the lease fed. There were three periods of non-production at issue, nine months in 1946 to 1947, three months in 1962, and another nine months in 1974 to 1975. The well did not produce during these periods because Natural's pipeline pressure was greater than the wells' pressure. After Natural added compression on the line, the well began to produce at rates higher than any in its history. The United States Court of Appeals for the Fifth Circuit held that the three extended cessations of production were temporary and that the lease had not terminated. In *Casey v. Western Oil & Gas, Inc.*, [611 S.W.2d 676 (Tex.Civ.App.-Eastland 1980, error ref'd n.r.e.) eds.] a well was not produced for two months after the lessee's contract with its pipeline purchaser expired and negotiations for a new contract were underway. The court of appeals held that this evidence supported the trial court's finding that cessation of production was temporary and, therefore, that the lease did not terminate. Accordingly, although decisions at times have said that the temporary cessation of production doctrine applies when there is "sudden stoppage of the well or some mechanical breakdown of the equipment used in connection therewith, or the like," or that the doctrine applies when the cause of a cessation of production is "necessarily unforeseen and unavoidable," the circumstances in which this and other courts have applied the doctrine have not been so limited. The court of appeals in the present case correctly concluded that "foreseeability and avoidablility are not essential elements of the [temporary cessation of production] doctrine."

Ridge contends that, when it ceased production on December 1, 1997, the 1937 lease terminated as to the Ridge tract on that date. We do not reach that contention because Ridge's alternative position in its motion for summary judgment is dispositive. Ridge asserted that at least as of March 3, 1998, the date on which new leases became effective for the Ridge tract, cessation of production from the wells on the Ridge tract was permanent with respect to the 1937 lease. We agree.

This Court acknowledged more than a half a century ago in *Superior Oil Co. v. Dabney* that parties to an oil and gas lease may validly include a provision allowing the lessee to surrender all or part of the lease. We said, "Options to surrender in contracts of lease are now regarded as valid by a practical unanimity of decision." Even if an oil and gas lease does not contain a surrender clause, the parties may mutually agree to a release, or they can effectively terminate their lease by signing a new one. When the owners of the possibility of reverter of the mineral interest in the Ridge tract executed new leases with Ridge, they effectively terminated the 1937 lease as to that tract. Production by Ridge from the Ridge tract was thereafter performed under the new, March 3, 1998 leases, not the 1937 lease. The cessation of production from the Ridge tract under the 1937 lease thereby became permanent. Guinn had no right to enter the Ridge tract to restore or obtain production, even under the 1937 lease. Ridge and its lessors, however, could not affect Guinn and its lessors' interests under the 1937 lease, and termination of the lease as to the Ridge tract could not, in and of itself, terminate the 1937 lease as to the Guinn tract. The Guinn tract, by its terms, remained in effect "as long . . . as oil or gas, or either of them is produced from said land by the lessee, or as long as operations are being carried on." The question then becomes, was there any production "by the lessee" on March 3, 1998 when the 1937 lease terminated as to the Ridge tract.

[The court concludes that Guinn's efforts on the Guinn tract were not sufficient to maintain the lease after the production ceased on the Ridge tract. Eds.]

. . .

Guinn contends that we should hold that a lessee cannot surrender or terminate a lease to destroy the rights of another partial assignee of the lessee's interest. We decline to adopt such a blanket rule of law. Even if such a rule of law might be appropriate in the context of overriding royalty interests when the underlying lease does not contain an express release provision, a question we do not address, there is a material distinction between an overriding royalty interest and that of a lessee. An overriding royalty interest is a non-participating interest. A royalty owner has no right and thus no ability to go onto the underlying property and drill or otherwise take action to perpetuate a lease. An overriding royalty interest owner is wholly dependent on the lessee to keep a lease alive. That is not true of a lessee. A lessee in Guinn's position could continue a lease in effect by drilling a well and obtaining production, or continuing operations until production is obtained, under lease provisions like those in the 1937 lease.

. . .

The trial court did not err in granting summary judgment for Ridge. The court of appeals' judgment is reversed and judgment is rendered for Ridge.

NOTES

1. In a state which treats the oil and gas lease as a fee simple determinable estate, who owns the possessory interest where there has been a cessation of production in the secondary term? Can the parties to a lease eliminate that uncertainty through leasehold provisions that extend the lease even though production has ceased?

2. If equitable defenses are not available in those states which treat production as a limitation, Watson v. Rochmill, 137 Tex. 565, 155 S.W.2d 783 (1941), would they be available in the period of time during which only a temporary cessation of production is involved? Should a court allow estoppel or other equitable defenses if the lessee proceeds under an assumption that the cessation is temporary but a court later determines that the cessation is permanent? *See* Kramer, *The Temporary Cessation Doctrine: A Practical Response to an Ideological Dilemma*, 43 Baylor L.Rev. 519 (1991).

3. Prior to *Ridge Oil Co.*, the issue of whether the temporary cessation of production doctrine should be limited to cessations caused where the lessee has no control over the stoppage or where the stoppage was unexpected was hotly debated. *Compare* Bradley v. Avery, 746 S.W.2d 341, 102 O.&G.R. 577 (Tex.App. 1988) *with* Sorum v. Schwartz, 344 N.W.2d 73, 97 O.&G.R. 336 (N.D.1984).

4. If there is a "total" cessation of production for some unspecified period of time should the temporary cessation of production apply? A series of Texas appellate cases suggests that a "total" cessation bars the lessee from arguing that there has only been a temporary cessation of production. Cannon v. Sun-Key Oil Co., 117 S.W.3d 416 (Tex.App. 2003); Ridenour v. Harrington, 47 S.W.3d 117, 149 O.&G.R 283 (Tex.App. 2001).

5. Indiana has attempted a statutory solution to the problem of determining when a cessation of production is temporary. Leases become "null and void after a period of one (1) year has elapsed . . . since operation for oil and gas has ceased, both by the non production of oil or gas and the nondevelopment of said lease. . . ." This statute was enacted in 1923 was interpreted for the first time in

Barr v. Sun Exploration Co., 436 N.E.2d 821, 75 O.&G.R. 451 (Ind.App. 1982). The court found the lease continued in effect because within a year from the cessation of production, "activities to repair, maintain and operate the oil well toward the eventual production or attempted production of oil were in evidence."

6. What are the relevant considerations in determining whether cessation of production is temporary or permanent? What factors other than time can be considered? Consider the following cases:

a) WAGNER V. SMITH, 8 Ohio App.2d 90, 456 N.E.2d 523, 78 O.&G.R. 537 (1982). A marginally productive well ceased its production of gas in 1978. Oil continued to be produced and sold through all of 1979. The lessee knew in the middle of 1978 that a hole in the casing was causing loss of production. In 1980 the lessee attempted to fix the problem by "mudding" the well. The attempt failed because of the water pressure in the well bore. In March 1981, two months after the lessor filed this action the lessee prepared to drive a smaller well casing into the existing wellbore. *Held,* the lease terminated.

b) HABY V. STANOLIND OIL & GAS CO.,228 F.2d 298, 5 O.&G.R. 1057 (5th Cir.1955). Defendant was the owner of a producing leasehold in Reagan County, Texas. The lease did not contain a "force majeure" clause. The Railroad Commission of Texas entered a shutdown order on all of the Spraberry Sand wells because of gas wastage, and defendant shut in his well pursuant to that order on its effective date, April 1, 1953. The order was later held to be void by the Supreme Court of Texas on June 10, 1953. Even though the Supreme Court had so decided the Commission had fixed the allowable of the well at zero, and, therefore, the well was shut in. Subsequent orders of the Commission provided that a well could produce only if the casinghead gas could be put to a legal use and then described four such uses. Defendant could not put the casinghead gas to any one of the four uses. On January 1, 1954, defendant resumed the production of the well. The above described shutting in of the well occurred after the expiration of the primary term of the lease. *Held,* the lease terminated.

c) KERR V. HILLENBERG, 373 P.2d 66, 17 O.&G.R. 167 (Okl. 1962). The well ceased production due to mechanical failure. The lessee attempted without success to find replacement parts for the pumping machinery. It took nearly one year to find the part. *Held,* the lease did not terminate.

d) WRESTLER V. COLT, 7 Kan.App.2d 553, 644 P.2d 1342, 73 O.&G.R. 307 (1982). Paying production based upon an old water flood project had ceased for at least three years prior to the suit for cancellation. The lessee was trying to secure a "tertiary" recovery project based "on an experimental polymer injection project being conducted on a nearby lease." Evidence was shown that even if the experiment was successful it would be several more years before it could be applied to the lessor's well. *Held,* the lease terminated.

e) MCCLAIN V. RICKS EXPLORATION CO., 1994 OK CIV APP 76, 894 P.2d 422, 132 O.&G.R. 164. In plugging an initial well and drilling a second well, the court noted that the temporary cessation of production doctrine requires the consideration of the operator's diligence in regaining production and the reasonableness of the period of time in which production is halted. A third factor may be the voluntariness of the cessation. If due to mechanical problems beyond the control of the operator, it is less likely that the lease will be terminated.

f) COBB V. NATURAL GAS PIPELINE CO. OF AMERICA, 897 F.2d 1307 (5th Cir. 1990). There were three periods of no production dating back to 1946-47. Two of the three periods of no production were 9 months in length, the third was for 3 months. There is no direct evidence of the cause of the cessations, but the lessee proffers an expert who hypothesizes that the cessations were due to low pressure on the gathering lines. The lessors offer no evidence of their own as to the cause of the cessations. The evidence satisfies the lessee's burden of proof and the temporary cessation of production doctrine applies.

6. What place does reasonable diligence by the lessee in attempting to overcome the problems that caused the cessation of production have in the determination of whether the lease has ended? *See* Gillespie v. Wagoner, 28 Ill.2d 217, 190 N.E.2d 765, 18 O.&G.R. 863 (1963) (financial trouble, bad weather--lease terminated); Feland v. Placid Oil Co., 171 N.W.2d 829, 34 O.&G.R. 416 (N.D.1969)(need for new salt water disposal facilities--lease not terminated); Waddle v. Lucky Strike Oil Co., Inc., 551 S.W.2d 323, 57 O.&G.R. 581 (Tenn.1977)(well "red tagged" by Oil and Gas Board--lease terminated); Bradley v. Avery, 746 S.W.2d 341, 102 O.&G.R. (Tex.App. 1988)(long producing well ceases production for unknown causes but is restarted within 100 days-lease terminated).

---

3. SAVINGS CLAUSES

a. DRILLING OPERATIONS CLAUSE

SWORD v. RAINS

United States Court of Appeals, Tenth Circuit, 1978.
575 F.2d 810, 61 O.&G.R. 339.

MCWILLIAMS, CIRCUIT JUDGE. This case concerns an oil and gas lease. Charles H. Sword, the owner of 160 acres of land in Meade County, Kansas, entered into an oil and gas lease for a primary term of one year commencing July 23, 1971. Sword subsequently executed an extension of the lease whereby the primary term of the lease was extended an additional three months, from July 23, 1972, to October 23, 1972. The landowners adjacent to Sword also executed oil and gas leases. All the leases, including Sword's eventually were assigned to and became vested in Wilson Rains and Arthur W. Skaer, and others, who will hereinafter be referred to as Rains.

On September 23, 1972, Rains commenced a test well on Sword's land. On October 7, 1972, a drill stem test indicated the presence of gas. Total depth was reached on October 15, 1972, on which date casing was set. By November 8,

1972, the well was completed, was flowing gas, and was ready for pipeline connection except for final open flow testing.

Rains next began efforts to market the gas. These efforts will be considered later and need not at this point be detailed. It is sufficient here to simply note that Rains made contact with Panhandle Eastern Pipe Line Company, and others, as possible purchasers of the gas. Negotiations with Panhandle ensued, resulting in the signing of a contract for the sale of the gas on June 6, 1973. Thereafter Rains commenced operations to lay a pipeline to Panhandle's gathering system, some .7 miles distant. On August 20, 1973, gas deliveries to Panhandle commenced. In the fall of 1974 a twenty-year gas purchase contract with Panhandle was finalized.

On October 23, 1973, Sword, a citizen of California, commenced the present action in the United States District Court for the District of Kansas. Jurisdiction was based on diversity. The action sought to have the oil and gas lease executed by Sword declared terminated and to quiet title to the land in Sword. It was Sword's theory of the case that the oil and gas lease had expired under its own terms because of the failure of Rains to comply with certain lease deadlines. . . .

Trial of the case was to the court. The trial court entered detailed findings of fact and conclusions of law which held, in essence, that . . . under the "continuous operation" clause in the oil and gas lease, the lease had not expired and that there had been no breach thereof by Rains. Accordingly, the trial court entered judgment dismissing Sword's claims "on the merits."

On appeal, Sword . . . seek[s] reversal of the judgment insofar as it dismissed his quiet title claim.

As previously mentioned, the oil and gas lease under consideration was for a primary period of one year, commencing July 23, 1971. On June 21, 1972, the lease was extended for a period of three months to October 23, 1972. Specifically, the extension agreement provided that " . . . said term of said lease shall be and is hereby extended, with the same tenor and effect as if such extended term had been expressed in such lease, for a period of three months from the date of the said expiration thereof and as long thereafter as oil or gas (including casing head gas) is produced from any well on the land covered by said lease . . ."

As indicated, Rains commenced drilling on the leased premises on September 23, 1972. This occurred within the primary term of the lease, as extended. By October 23, 1972, the expiration date of the primary term of the lease, as extended, Rains had completed his test drilling and had discovered gas in paying quantities. The well itself was substantially completed by November 8, 1972, and then shut down, because of adverse weather.

The habendum clause of the Sword lease provides for a primary term of one year from its date, which was later extended for an additional three months, "and as long thereafter as oil or gas, . . . and other minerals may be produced from said leased premises or operations for the drilling or production thereof are continued as hereinafter provided."

Section 7 of the lease, the so-called "continuous drilling" provision, reads as follows:

> 7. It is expressly agreed that if the lessee shall commence operations for the drilling of a well at any time while its lease is in force this lease shall remain in force and its terms shall continue for so long as such operations are prosecuted and, if production results therefrom, then so long as such production may continue.

Under the terms of the lease, then, if Rains commenced drilling a well within the primary term of the lease, as extended, which he did on September 22, 1972, the lease shall remain in force and continue "for so long as such operations are prosecuted," and if there be production, "so long as such production may continue."

Under Kansas law, in the absence of a continuous operations clause, there must be actual production within the primary period of the lease, and without such production, the lease will expire by its own terms. Home Royalty Association v. Stone, 199 F.2d 650 (10th Cir.1952). However, a "continuous operation" clause extends the lease for so long as the lessee-operator exercises due diligence in equipping the well and getting it into production, which includes the marketing of the gas. Christianson v. Champlin Refining Co., 169 F.2d 207 (10th Cir.1948). To the same effect, *see* Tate v. Stanolind Oil & Gas Company, 172 Kan. 351, 240 P.2d 465 (1952). In *Tate,* the Kansas Supreme Court spoke as follows:

> From this provision [a continuous operations clause], standing alone, it clearly appears that even though a well is only commenced during the primary term, it may be completed thereafter with reasonable diligence and dispatch. Obviously if on completion of drilling operations oil or gas is found in paying quantities the lessee, under this clause, is not expressly required to produce or market the oil or gas immediately. And, of course, that might be wholly impossible. He would, however, be required to do so within a reasonable time. But even if the drilling clause reasonably could be interpreted as requiring both production and marketing immediately upon completion of the well it is nevertheless clear the necessity therefor was extended beyond the fixed primary term.

> · · ·

> It is impossible to lay down an accurate general rule with respect to what constitutes production or marketing within reasonable time in every case. Whether either has been so obtained must be left to the particular facts of cases as they arise.

Application of the continuous operations doctrine to the instant case means that the Sword oil and gas lease continued after October 23, 1972, for so long as

Rains continued with due diligence his production efforts, which included his efforts to market the gas. In this regard the trial court found and concluded that Rains had acted within a reasonable time and with due diligence in his effort to sell the gas to a gas company. It was on this basis that the trial court concluded that the Sword lease had not expired by its own terms and accordingly refused to quiet title in Sword.

Whether Rains acted in reasonable time and with due diligence in marketing the gas is essentially a question of fact. In our view, there is substantial evidence to support the trial court's holding that Rains acted with due diligence. Certainly such finding is not clearly erroneous. It is on this basis that we affirm.

The primary term of the lease, as extended, was October 23, 1972. Rains commenced drilling a well on the leased acreage in September, 1972, and had discovered gas in paying quantities prior to October 23, 1972. The well was for all practicable purposes completed in November, 1972, and then shut down because of adverse weather. Rains then began looking for someone who might be interested in purchasing the newly discovered gas. It would have been to the monetary advantage of both Sword and Rains if the gas could have been sold in intrastate commerce. Apparently, there were no intrastate gas purchasers within reasonable proximity of the Rains well, and accordingly Rains began looking for an interstate purchaser. Contract was made with three potential purchasers, Panhandle Eastern Pipe Line Company, Michigan-Wisconsin Pipe Line Company, and Northern Natural Gas Company.

As indicated, gas sold in intrastate commerce went for a higher price than that for gas sold in interstate commerce. However, in 1970, the Federal Power Commission, acting in response to the shortage of natural gas that was being dedicated to interstate pipelines, issued new policy statements and regulations designed to attract new dedications of gas to the interstate market. By Order No. 428-B, issued July 15, 1971, the Federal Power Commission established an exemption from certain filing and rate requirements for small producers of natural gas. This Order enabled small producers to initiate sales in interstate commerce at a price in excess of the existing area rate. Rains throughout had been proceeding on the premise that as a small producer he could eventually sell his gas through open market negotiations with competing gas purchasers, free from the existing area rate. However, on December 12, 1972, the Court of Appeals for the District of Columbia ruled that the Federal Power Commission had exceeded its authority under the Natural Gas Act in exempting small producers from area rate regulation. Texaco, Inc. v. Federal Power Commission, 154 U.S.App.D.C. 168, 474 F.2d 416 (1972). It was not until June 10, 1974, that the Supreme Court vacated the order of the Circuit Court and remanded the case with direction that there be further proceedings before the commission. Federal Power Commission v. Texaco, Inc., 417 U.S. 380, 94 S.Ct. 2315, 41 L.Ed.2d 141 (1974).

It is Rains' position that he acted reasonably and with due diligence in marketing the gas. In determining whether Rains proceeded with due diligence, considerable emphasis is laid on the rather chaotic and uncertain market conditions resulting from the *Texaco* decision by the District of Columbia Circuit Court which struck down the small producers' exemption. The trial court recognized the effect of the *Texaco* decision, and, in light of all the facts and circumstances, found that Rains had acted with due diligence in executing a gas purchase agreement in June, 1973, and by commencing delivery in August 1973.

As indicated in both Christianson v. Champlin Refining Co., 169 F.2d 207 (10th Cir.1948) and Tate v. Stanolind Oil & Gas Co., 172 Kan. 351, 240 P.2d 465 (1952), what constitutes marketing within a reasonable time and with due diligence depends on the particular facts of the case at hand. In *Christianson,* where there was about fifteen months between completion and a marketing contract, the lessee was held to have acted within a reasonable time under the circumstances. In *Tate* about four months elapsed between completion and a marketing contract, and the lessee was there held to have acted within a reasonable time. The instant case falls in between *Christianson* and *Tate,* there being some eight months between completion and a marketing agreement.

The passage of time is not in itself determinative of the question of reasonable time, and due diligence, though it of course is a factor to be considered. Bristol v. Colorado Oil & Gas Corporation, 225 F.2d 894 (10th Cir.1955).[41] In *Bristol* we observed that although reasonable time and due diligence do not have the same meaning in the application of the rule of reason, both are nonetheless essential ingredients of the rule. Other factors also entered into the picture. For example, the lessee does not have to accept the first offer. It is to his benefit, as well as to the benefit of the lessor, that he obtain as high a price as possible for the gas. All that the law requires is that the lessee act with due diligence in marketing the gas. The trial court here held that Rains had so acted, and in our view, there is ample evidence to support this determination.

Sword notes that at one time Rains was considering selling his interest. Counsel argues that such negates any *continuing* effort by Rains to market. This argument is too semantical, and cannot, by itself, dispose of the entire controversy. Rains counters this argument by asserting that notwithstanding the possibility of a sale, he nonetheless was at all times continuing his efforts to market the gas. In any event, the trial court took this fact into consideration and from the totality of the evidence, and not from one isolated fact, found due diligence on the part of Rains. We are not inclined to disturb this holding.

. . .

Judgment affirmed.

———

---

[41] In *Bristol* a lapse of 7-2/3 years was held not to preclude a finding of due diligence in marketing.

NOTES

1. Drilling operations clauses sometimes provide a time limit to avoid some of the problems in *Sword.* For example,

> . . . so long as such operations continue or additional operations are had, which additional operations shall be deemed to be had where not more than sixty (60) days elapse between abandonment of operations on one well and commencement of operations on another well. Fields v. Stanolind Oil & Gas Co., 233 F.2d 625, 5 O.&G.R. 1364 (5th Cir.1956).

2. A key factual issue in applying such clauses is to determine whether the lessee has engaged in additional operations or, more frequently, additional drilling or reworking operations. For example, should preliminary activities such as staking, applying for a well permit, brush clearing and site leveling be sufficient? *See* Johnson v. Yates Petroleum Corp., 1999 NMCA 66, 981 P.2d 288, 142 O.&G.R. 309. There is substantial disagreement as to whether physical operations are required under this type of clause. *Compare* Del Ray Oil & Gas, Inc. v. Henderson Petroleum Corp., 797 F.2d 1313, 92 O.&G.R. 96 (5th Cir.1986)(applying Louisiana law) *with* Sheffield v. Exxon Corp., 424 So.2d 1297, 76 O.&G.R. 419 (Ala.1982) and Gulf Oil Corp. v. Reid, 161 Tex. 51, 337 S.W.2d 267, 12 O.&G.R. 1159 (1960).

3. Would the following definition of operations clarify the problem of what actions are sufficient to trigger the application of the savings provision:

> [O]perations for and any of the following: drilling, testing, completing, reworking, recompleting, deepening, plugging back or repairing of a well in search for on in an endeavor to obtain production of oil, gas sulphur or other minerals, excavating a mine production of oil, gas, sulphur or other mineral, whether or not in paying quantities. Ice Brothers, Inc. v. Bannowsky, 840 S.W.2d 57, 58, 121 O.&G.R. 125 (Tex.App. 1992).

4. Should there be a difference in application between a leasehold savings clause that requires the lessee to engage in "operations" rather than requiring the lessee to engage in "drilling operations?" *See* Bargsley v. Pryor Petroleum Corp., 196 S.W.3d 823 (Tex.App. 2006).

5. Does adding the term "reworking" make the savings provision narrower or broader in application? Would applying for a water discharge permit be a "reworking" operation? In Coronado Oil Co. v. U.S. Department of the Interior, 415 F.Supp.2d 1339 (D.Wyo. 2006), the court found that such an application raises a factual issue as to whether the lessee was engaged in reworking operations.

6. The widespread use of horizontal drilling and hydraulic fracturing techniques call into question the efficacy of the traditionally drafted savings provisions found in many leases. See Bruce M. Kramer, *Keeping Leases Alive in the Era of Horizontal Drilling and Hydraulic Fracturing: Are the Old Workhorses (Shut-in, Continuous Operations, and Pooling Provisions) Up to the*

Task? 49 Washburn L.J. 283 (2010).

———

b. DRY HOLE AND CESSATION OF PRODUCTION CLAUSES
SUNAC PETROLEUM CORP. v. PARKES

Supreme Court of Texas, 1967.
416 S.W.2d 798, 26 O.&G.R. 689.

GREENHILL, JUSTICE. This case principally involves the construction of the provisions of an oil and gas lease dealing with operations, or lack of them, at and after expiration of the primary term, and also the question of whether a new lease was a "renewal or extension" of a former lease so as to perpetuate an overriding royalty interest. The suit was instituted by Frank Parkes to establish that he owned an overriding royalty interest and for a money judgment for royalties alleged to be due. The parties submitted the case to the trial court, sitting without a jury, upon an agreed statement of the facts. Judgment was for the plaintiff Parkes. The Court of Civil Appeals sitting at Amarillo reformed the judgment in matters not material here and affirmed. 399 S.W.2d 840 (1966).

The first question is whether the oil and gas lease in question terminated under its own terms. That question turns upon the 60-day and 30-day clauses in the lease dealing with operations in progress at the end of the primary term, the cessation of production, the completion of a dry hole after the end of the primary term, and related questions. This Court has had some of these problems heretofore in at least three cases: Rogers v. Osborn, 152 Tex. 540, 261 S.W.2d 311 (1953); Stanolind Oil & Gas Co. v. Newman Brothers Drill. Co., 157 Tex. 489, 305 S.W.2d 169 (1957); and Skelly Oil Co. v. Harris, 163 Tex. 92, 352 S.W.2d 950 (1962). Our decision on the first question is based upon those opinions, and they will be referred to several times herein.

As considered material here, the stipulated facts are these: on April 17, 1948, one O'Hern, as lessor, executed an oil and gas lease to the plaintiff Parkes. It provided for a royalty of 1/8th and a primary term of 10 years. The lease, which was on 160 acres in Ochiltree County, Texas, provided that it could be pooled for gas only.

On May 15, 1957, some nine years later, Parkes sold and assigned his lessee's interest to L.H. Puckett for $5,600. Parkes also reserved an overriding royalty of 1/16th of 7/8ths of the production from the lease, or from any extension or renewal thereof. The assignment provided that the assignee would be under no obligation to keep the lease in force by payment of rentals, by drilling, or by development operations, and that assignee should have "the right to surrender any or all part of such leased acreage without the consent of assignor." Thereafter Puckett assigned the oil and gas lease; and it subsequently was assigned to the petitioners, Sunac Petroleum Co. et al., who were the defendants in the trial court.

The end of the primary term, April 17, 1958, was approaching. On April 14, 1958, the lessees pooled the land in question with other lands for gas purposes only. On the following day, drilling operations began on land within the 640-acre gas unit but not upon the 160 acres in question. When the primary term ended on April 17, 1958, there was no production from, or operations upon, the particular 160-acre lease, but a well was being drilled upon the gas unit.

On June 11, 1958, the well on the 640-acre gas unit was completed as an *oil* well as distinguished from a *gas* well. Oil was produced in paying quantities from that well, at least until the commencement of a second well.

On June 24, 1958, approximately 68 days after the expiration of the primary term and 13 days after the completion of the oil well on the gas unit, the defendants Sunac et al. began the drilling of a second well on the particular 160 acres in question. The well was completed as a producing oil well on July 29, 1958; and the well continued to produce thereafter.

Approximately a year after the completion of the above-mentioned well on the 160 acres, the successors in interest of the lessor "asserted that a question existed as to whether or not the said lease had been maintained in force and effect during the period from the expiration of the primary term until drilling operations were commenced" on the 160 acres. Thereafter, on August 17, 1959, the lessees (now Sunac et al.) for a cash consideration of $27,000, "procured a new oil, gas and mineral lease" from the successors of the original lessor. As will be discussed more in detail later, the new lease had different provisions from the original lease. The well on the 160 acres was still producing when the new lease was executed. The lessees Sunac et al. stopped paying the 1/16th of 7/8ths overriding royalty to the plaintiff, Parkes, about December 1, 1959; and this suit followed.

Parkes, of course was entitled to a 1/16th of 7/8ths overriding royalty under the original lease and as long as it remained alive. If it was continued in force beyond the primary term by drilling or other operations, he is entitled to prevail under the original lease. We shall first decide, therefore, whether the original lease terminated under its own provisions.

The original lease contained these provisions:

2. Subject to the other provisions herein contained, this lease shall be for a term of ten years from this date (called 'Primary term') and as long thereafter as oil, gas or other mineral is produced from said land hereunder.

. . .

5. If prior to discovery of oil or gas on said land Lessee should drill a dry hole or holes thereon, or if after discovery of oil or gas the production thereof should cease from any cause, this lease shall not terminate if Lessee commences additional drilling or reworking operations within sixty (60) days thereafter . . ... If at the expiration of the primary term oil, gas or other mineral is not being produced on said land but Lessee is then engaged in drilling or re-working operations thereon, the lease shall remain in force so long as operations are prosecuted with no cessation of more than thirty (30) consecutive days, and if they result in the production of oil, gas or other

minerals so long thereafter as oil, gas, or other mineral is produced from said land.

Paragraph 9 gave the lessee the right to pool "the gas leasehold estate" and "the lessor's gas royalty estate" with other leases to create gas units of not more than 640 acres. It is then stated:

> The commencement of a well, or the completion of a well to production, on any portion of an operating unit shall have the same effect under the terms of this lease as if a well were commenced, or completed, on the land embraced by this lease.

Paragraph 5 above set out contains two sentences which provide for the continuation of the lease beyond the primary term. The first sentence speaks of *additional* drilling or reworking operations. The second sentence speaks of drilling or reworking operations.

This Court in Rogers v. Osborn, 152 Tex. 540, 261 S.W.2d 311 (1953), held that for the first sentence in paragraph 5 of the lease, the 60-day provision, to be brought into operation, one of two events must occur: (1) there must be a dry hole before a discovery of oil and gas, or (2) there must be a cessation of production after the discovery of oil or gas. In that case, before the end of the primary term, there had been a completed well which produced some gas and a trace of oil; and the well was intermittently opened or bled to remove nonmarketable oil and waste from it. The Court concluded that the well was not a "dry hole." Moreover, since there had been no marketable production, there could not have been a "cessation of production." The holding was that the 60-day first sentence of paragraph 5 was not operative and did not prolong the life of the lease. A second well, begun and completed as a producer after the primary term, did not continue or revive the lease. Rogers v. Osborn also had a holding with regard to the 30-day second sentence of paragraph 5. It will be discussed later herein.

The above requirements for the operation of the 60-day first sentence were repeated in Stanolind Oil & Gas Co. v. Newman Brothers Drilling Co., 157 Tex. 489, 305 S.W.2d 169 (1957). In *Newman Brothers,* there was a dry hole; and because of the dry hole, the holding was that the lease was prolonged under the 60-day provision.

Under the test of *Rogers v. Osborn,* which was reiterated in *Newman Brothers,* the 60-day first sentence of paragraph 5 did not come into play to prolong the lease in this case. We agree with the Court of Civil Appeals that the completion of an oil well is not to be considered as the drilling of a dry hole. 399 S.W.2d at 843. The completion of a producing oil well on a unit authorized for gas only, but not located on a particular lessor's premises, would not, and did not, prolong the life of the oil and gas lease on the 160 acres.

The 30-day second sentence in paragraph 5 deals with drilling and reworking operations. As was pointed out in Rogers v. Osborn, the sentence does not speak of *additional* drilling (the drilling of additional wells), but it speaks of drilling and reworking operations being conducted at the time of the expiration of the primary term. An additional holding of this Court in *Rogers v. Osborn* was that the drilling or reworking must be upon the particular well being drilled or reworked at the expiration of the primary term in order for the 30-day second sentence to become operative. The opinion states that the 30-day second sentence of paragraph 5 applies to "continuous prosecution of the very operations" being conducted when the primary term ended.

In *Newman Brothers,* there was a well being drilled at the end of the primary term, and the Court held that the 30-day second sentence prolonged the life of the lease [at least] until the completion of the well as a dry hole. Since the well was completed as a dry hole at a time when the lease was in force and effect, the Court further held that the drilling of the dry hole, prior to the discovery of oil or gas, brought into effect the 60-day first sentence of paragraph 5. A second well was completed within 45 days from the plugging of the dry hole. So the lease was held not to have terminated under the specific drilling operations of the 30-day provision which resulted in a dry hole, and the dry hole in turn brought into play the 60-day sentence of paragraph 5.[42]

In the case at bar, the particular operation under way at the expiration of the primary term did not end in a producer of gas so as to perpetuate the lease under the 30-day second sentence of paragraph 5. There were no additional drilling or reworking operations on the particular well, the oil well on the 640-acre gas unit. Under *Rogers v. Osborn* and *Newman Brothers,* the drilling of another well after the primary term would not prolong the lease unless the particular operation being conducted at the expiration of the primary term resulted in a dry hole (or, under other circumstances, upon cessation of production). So the lease was not prolonged after the end of the work or final operations on the first well under the 30-day clause, and the well did not end in a dry hole or cessation of production so as to activate the 60-day clause.

The parties also cite Skelly Oil Co. v. Harris, 163 Tex. 92, 352 S.W.2d 950 (1962). In *Skelly,* both provisions similar to those in paragraph 5 of the lease here involved had 60-day time limitations; *i.e.,* the second sentence dealing with drilling and reworking also gave the lessee a 60-day period rather than 30 days. In *Skelly,* a well was being drilled at the end of the primary term on the unit with which the land in question had been pooled. The well was completed as a producing gas well, but it was capped because there was no pipeline immediately available. Within 60 days the lessee procured a pipeline connection, and production promptly commenced. The holding was that the lease did not terminate. The 60-day clause (similar to the 30-day clause in the lease in question) gave the lessee 60 days after cessation of operations within which to

---

[42] The dissent in *Newman Brothers* was of the view that the dry hole must have occurred before the end of the primary term for the 60-day provision to have applied. Under that view even if the oil well on the gas unit here were considered to be a "dry hole," it was not completed during the primary term; and hence the lease here under consideration terminated.

begin production from that well. If it had been completed as a dry hole, the opinion in *Skelly* said the lessee would have had 60 days under the provisions of paragraph 6[paragraph 5 of the lease before us] in which to commence additional drilling or reworking operations.

Our holding here is that the first oil and gas lease terminated. The drilling and completion of the particular operation (the oil well) off the lessee's premises after the primary term did not end in the production of gas so as to prolong the lease under the 30-day provision; and there was no cessation of production or dry hole to activate the 60-day sentence.

[In a portion of the opinion reproduced hereinafter, the court concluded that the second lease from the successors of the original lessors was not a renewal or extension of the original lease. Eds.]

The judgments of the courts below are reversed and judgment is here rendered for petitioners.

HAMILTON, JUSTICE. I respectfully dissent.

. . . [I]f the proper disposition of this case requires that the question of the effect of the completion of an oil well on a consolidated gas unit after the expiration of the primary term of the lease involved be resolved, and consequently the question of whether or not the 1948 lease in fact remained in force until the effective date of the new lease be resolved, then it is respectfully submitted that, under the facts of this case, the said 1948 lease did in fact remain in force.

The 1948 lease authorized the lessee to combine the 160 acres with other lands as to the gas leasehold estate. Two days before the primary term of the original lease expired, Sunac placed the 160-acre lease in a unit with other lands and commenced drilling operations. These operations were on the land within the unit but were off the land included within the original 160-acre lease. As a consequence of the lease and the gas unit, Sunac needed to complete a well that produced gas in order to hold the lease by production. If it did so, it complied with the terms of the original lease. Instead of a gas producer, however, the well was completed as a producing oil well on June 11, 1958.

Completion of the producing oil well did not continue the lease in force, but Sunac on June 24 commenced the drilling of another well that was located on the original 160-acre tract. It did this by authorization of this provision of the original oil and gas lease:

> If, prior to discovery of oil or gas on said land, lessee should drill a dry hole or holes thereon, or if after discovery of oil or gas the production thereof should cease from any cause, this lease shall not terminate if lessee commences additional drilling or reworking operations within sixty days thereafter or (if it be within the primary term) commences or resumes the payment or tender of rentals on or before the rental paying date next ensuing after the expiration of three months from date of completion of dry hole or

cessation of production. . . .

The second drilling operation resulted in a producing oil well on July 29, 1958.

The quoted provision was construed in Stanolind Oil and Gas Company v. Newman Brothers Drilling Company, 157 Tex. 489, 305 S.W.2d 169 (1957). We held in that case that a lessee, who prior to discovery of oil or gas, and within the primary term, commences drilling which results in a dry hole, may, within sixty days commence additional drilling. In the *Newman* case the additional drilling operations resulted in production, and we held that the lease did not expire. The Court of Civil Appeals in this case held that the production of oil on the lands which required the production of gas only was the completion of a producing well rather than a "dry hole" and there was no authorization for drilling the second well to keep the lease alive. With this I disagree.

The majority has held that the production of oil was not production which would hold the gas-only unit and hence it would not hold the original lease. The majority, however, holds that the completion of the oil well prevented the well from being a "dry hole". Hence we have the majority holding that under the terms of the lease, the same well may be classified as a producer for some purposes but as a nonproducer for others. It is a nonproducer, says the majority, and will not hold the lease. But it is not a dry hole because it is a producer. It is inconsistent to say that the production of oil does not keep the lease alive but to say that the production of oil keeps the completed well from being a dry hole. Consistency as well as the meaning of the lease compels the construction that the production of oil on a gas-only unit was not production and because it was not production it was a dry hole.

The pooling provision of the original lease equates non production with a dry hole. The provision concludes with this sentence:

> The commencement of a well, or the completion of a well to *production,* on any portion of an operating unit shall have the same effect under the terms of this lease as if a well were commenced, or completed on the land embraced by this lease. (Emphasis added.)

If, as the majority holds, the completion of the oil well was not production under the lease and particularly the pooling provision it had to be a dry hole. Because it was a dry hole, the *Newman* case applies, and additional drilling was timely commenced which resulted in production of an oil well. The lease did not lapse.

[The discussion of the extension and renewal clause of the assignment is reproduced hereinafter. Eds.]

I would affirm the judgments of the Court of Civil Appeals and the trial court.

SMITH and POPE, JJ., join in this dissent.

———

NOTES

1. Consider the application of the following clause to the facts of *Rogers, Newman Brothers, Harris* and the principal case:

> If, at the expiration of the primary term, Lessee is conducting operations for drilling a new well or reworking an old well, or if, after the expiration of the primary term, production on this lease should cease, this lease nevertheless shall continue as long as said operations continue or additional operations are had, which additional operations shall be deemed to be had where not more than sixty (60) days elapse between abandonment of operations on one well and commencement of operations on another well, and if production is discovered, this lease shall continue as long thereafter as oil, gas or other mineral is produced and as long as additional operations are had.

Would the results have been different? See generally, Martin & Kramer, *Williams and Meyers Oil and Gas Law* § 617.

2. The primary term of a lease on Texas land would have expired on December 9, 1978. Prior thereto lessee commenced a well which was continuously drilled until completed as a gas well on January 18, 1979. On February 17, 1979, shut-in gas royalty was tendered by lessee and rejected by lessor. A contract for the sale of the gas was concluded on June 7, 1979, and gas was continuously produced in paying quantities after that date. Suppose the lease contained the same clause as Paragraph 5 in *Parkes* and a shut-in clause the same as clause (b) on pages 300, *supra.* Would the lease continue in force? Suppose the drilling operations clause was identical to that in Note 1, *supra.* Would the lease continue in force? *See* Gulf Oil Corp. v. Reid, 161 Tex. 51, 337 S.W.2d 267, 12 O.&G.R. 1159 (1960); Skelly Oil Co. v. Harris, 163 Tex. 92, 352 S.W.2d 950, 15 O.&G.R. 653 (1962); Duke v. Sun Oil Co., 320 F.2d 853 (5th Cir.1963), reh. granted on proper anniversary date and timeliness of tender, 323 F.2d 518, 19 O.&G.R. 238 (5th Cir.1963); Mayers v. Sanchez-O'Brien Minerals Corp., 670 S.W.2d 704, 81 O.&G.R. 38 (Tex.App.1984, error ref'd n.r.e.).

3. When is a well completed within the meaning of a clause giving a lessee the right to drill a well commenced during the primary term "to completion with reasonable diligence and force with like effect as if such well had been completed within the primary term"? In LeBar v. Haynie, 552 P.2d 1107, 55 O.&G.R. 381 (Wyo.1976), it was contended that after the "Lewis Formation" was tested at 6744 feet the well was "completed" as a dry hole and the lease terminated. The original drilling rig had been removed after setting casing to 6719 feet and another rig placed over the hole during the testing. The second rig was removed on August 2 and no further work was done until August 27 when another rig was obtained and work undertaken that resulted in commercial production at 7115 feet on October 2. *Held:* the well was not completed until Oct. 2. The evidence before the trial court was sufficient to satisfy the test of a good faith intent to drill deeper carried out with reasonable diligence. Important factors were the setting of casing to a depth "900 feet below the point necessary to test the Lewis" and evidence

that the second rig was removed because it was not properly equipped to drill to the deeper "Teapot Formation."

————

c. Effect of Express Savings Provisions on Temporary Cessation of Production Doctrine

SAMANO v. SUN OIL CO.

Supreme Court of Texas, 1981.
621 S.W.2d 580, 70 O.&G.R. 64.

POPE, JUSTICE. George Samano and others (hereafter called Samano), as lessors of an oil and gas lease, sued Sun Oil Company and Tanya Oil Company (hereafter called Sun), as lessees, for a declaratory judgment that the lease had expired; because, during the secondary term, there was neither production nor any drilling or reworking operations for a continuous period of seventy-three days. The question presented by the case is whether a sixty-day limitation period for drilling or reworking operations was applicable to the secondary term of the lease. The trial court granted lessor, Samano, a summary judgment, holding that the lease had terminated. The court of civil appeals, with a divided court, reversed that judgment, 607 S.W.2d 46, holding that the sixty-day requirement in the drilling or reworking clause applied only to operations in progress at the end of the primary term. We reverse the judgment of the court of civil appeals and affirm the judgment of the trial court.

Paragraph 2 of the Samano lease is an early habendum clause which also includes a continuous drilling or reworking clause. Paragraph 2 of the lease provides: 2. Subject to other provisions herein contained, this lease shall remain in force for a term of ten years from this date, called primary term, and as long thereafter as oil, gas or other mineral is produced from said land, or as long thereafter as Lessee shall conduct drilling or re-working operations thereon with no cessation of more than sixty consecutive days until production results, and if production results, so long as any such mineral is produced.

Samano and Sun executed the lease on March 29, 1934, so the ten-year primary term ended March 29, 1944. Production in paying quantities extended the lease beyond the primary term and until May 4, 1977, when production stopped. Sun did nothing to restore production until July 15, a continuous period of seventy-three days. Sun urges that the cessation of drilling or reworking operations for the seventy-three days was a temporary cessation.

. . .

Standard rules of English show that paragraph 2 was carefully drafted and that its meaning is clear and easily breaks into three parts. These are the three divisions of the clause:

Subject to the other provisions herein contained, this lease

(1)

shall remain in force for a term of ten years from this date, called primary term, (March 29, 1944 was end of term)

(2)

and as long THEREAFTER as oil, gas or other mineral is produced from said land, (May 4, 1977 was last day of production)

(3)

or as long THEREAFTER as Lessee shall conduct drilling or re-working operations thereon with no cessation of more than sixty consecutive days until production results, so long as any such mineral is produced. (July 3, 1977, was the end of the sixty-day period).

Sun and the majority of the court of civil appeals have ignored this division and more particularly, they have ignored the word "thereafter" which was used not once, but twice, each time with a meaning and a reference to what had gone before.

The draftsman stated three distinct things which would prolong the term of the lease. The habendum clause first provided for a ten-year primary term. "Thereafter," meaning "after that," or after the duration of the primary term, the lease would continue in force as long as there was production. We are now into the secondary term, when we reach the third provision. That provision is the continuous drilling or reworking clause, and it looks back upon the two prior habendum provisions, the one for the primary term and the other one for its extension into the secondary term by production. It then states that the duration of the lease can be extended even further. We know this because the second "thereafter" now refers to what has already been stated. It was not until both methods (1) and (2) for extending the life of the lease were stated in the habendum clause that the lease provided for yet a third extension, that is, "thereafter by drilling or reworking operations." An inseparable part of this drilling or reworking clause was that there could be no cessation of drilling or reworking operations for more than sixty days not seventy-three days.

Sun quotes a number of textbooks on English grammar in support of its contention that modifiers are intended to refer to the words closest to them in the sentence. That, of course, is the correct rule. It was not observed by the court of civil appeals. The rule concerning modifiers is:

The reader naturally assumes that the parts of a sentence which are placed next to each other are logically related to each other. . . . The rule which will guide you may be stated in two parts: (1) place all modifiers, whether words, phrases, or clauses, as close as possible to the words they modify; (2) avoid placing these elements near other words they might be taken to modify. [citing Kierzek, The MacMillan Handbook of English (3d ed. 1954).]

. . . Consistent with good English, the first "thereafter" refers to the extension of the primary term; and the second "thereafter" reasonably refers to both not just one of the prior statements about duration of the lease.

If, as urged by Sun and held by the majority of the court of civil appeals, the clause requiring drilling or reworking within sixty days applies only to operations in progress at the end of the primary term; according to the grammar books, the drilling or reworking clause would have followed next and immediately after the clause stating the ten-year primary term. Instead, consistent with the rules of grammar, the drilling or reworking clause within sixty days is next to and immediately after the part of the habendum clause concerning the secondary period by production. The court of civil appeals erroneously leaps over that clause to apply the second "thereafter" exclusively to the primary term clause.

This court has already decided, and correctly so, that the drilling or reworking clause, including its express limitations by time, applies to operations in progress at the end of the primary term. While rules of good English are not always controlling, under those rules there is stronger reason for holding that the drilling or reworking clause within sixty days applies to the instance of cessation of production during the secondary term, than there is for holding that the clause only applies to the operations in progress at the end of the primary term.

The better, consistent, and more workable rule would be to apply the same rule to both parts of the habendum clause that is, to operations at the end of the primary term and also the cessation of production during the secondary term, both of which immediately are followed by the sixty-day drilling or reworking clause.

This grammatical construction of the compound sentence is also good common sense. All of the drilling or reworking clause must be applied, if it is to be applied at all. This includes the sixty-day limit which is an integral part of that clause. It means that when there is the right to drill or rework, that operation must be done in that stated time. It means that the right to drill or rework was not intended to be within the stated time in one instance but within a reasonable time in the other. The exercise of the right and the time limit to do it are both necessary parts of the whole.

The habendum has two events which maintain the lease in force, the ten-year term and production after that term. Both of those events are also terminating factors. It is not reasonable to hold that the lessor and lessee intended, as Sun says and the court of civil appeals has held, that the two terminating events stated in the habendum clause should be treated differently with respect to the drilling or reworking clause. Sun says that the drilling or reworking clause applies to operations in progress at the end of the primary term, but it is not applicable to the other terminating event, the cessation of production during the secondary term. . . . Sun says that the parties intended a specific limit of sixty days for one-half of the habendum clause, but intended an uncertain period of time which can be determined by a fact-finder for the other half of the habendum clause.

. . . The reasonable and common-sense meaning of the clause is that the whole drilling or reworking clause, including its sixty-day limit for the operations, must be applied to the whole habendum clause.

Both Samano and Sun cite many of the same cases. This is so, because this court has not previously addressed this question. But, this court has never held that a specific time limit for drilling or reworking operations, stated as a part of the habendum clause and required to keep a lease alive, applies exclusively to the operations at the end of the primary term and has no application to operations required to keep a lease alive upon cessation of production during the secondary term. Precedents about leases that state no time for drilling or reworking operations are not helpful. In those cases, of course, the rule of reasonable temporary cessation applies. . . .

. . .

Professor Kuntz has discussed this problem and agrees with this analysis:

If the clause is of the type that is combined with the habendum clause and provides "and so long thereafter as oil or gas is produced from the land or the premises being developed or operated," *it has an apparently broad purpose of preserving the lease during operations as well as during production without regard to whether the operation began during the primary term or began after its expiration.* Accordingly, such clause should be construed to preserve the lease while the lessee continues to conduct operations, regardless of when the operation began so long as it began while the lease was in effect. (Emphasis added.)

4 Kuntz, Oil and Gas § 47.4, at 121 (1972). He would construe the lease provision to permit drilling or reworking operations so long as the lease was in effect, whether it be during the primary term or during the secondary term when the lease is preserved by production. That is the common-sense meaning of the contract that would protect the mutual interests of both the lessors and lessees. The sixty-day provision is an integral part of the drilling or reworking provision affording a known time for commencing drilling or reworking operations, while the contract is in effect during the secondary period. Neither precedent nor sound reason exists for striking down that agreement.

When production stopped on May 4, 1977, during the secondary period, Sun had an express sixty days to drill or rework the well. When it failed to do so, the lease by its express terms automatically terminated. The judgment of the trial court was a correct one.

The judgment of the court of civil appeals is reversed and that of the trial court is affirmed.

Dissenting Opinion by DENTON, J., in which MCGEE and BARROW, JJ., join.

NOTES

1. Why did Sun Oil Co. interpret the savings provision narrowly? Was the savings provision designed to protect the lessor's or lessee's interest? Would it matter if Texas recognized the temporary cessation of production doctrine? *See generally* Martin & Kramer, *Williams and Meyers Oil and Gas Law* § 616.2.

2. In MCCULLOUGH OIL, INC. v. REZEK, 176 W.Va. 638, 346 S.E.2d 788, 90

O.&G.R. 596 (1986) the lease contained a cessation of production clause requiring the lessee to resume operations within 60 days of the cessation in order to keep the lease alive. The lease also contained a notice clause requiring the lessor to provide the lessee with notice of the breach of any provision in the lease. The original lessee assigned his interest which eventually was transferred to McCullough Oil. There was no production for over 6 years and McCullough and the original lessor re-leased the premises. The lessee/assignor sued claiming that he was entitled to notice before the lease terminated. The court found that the 60 day provision was a limitation and not a promise or covenant. As a limitation the fee simple determinable leasehold estate automatically terminated at the end of the 60 day period and no notice was required. *McCullough Oil* was cited with approval in Christian Land Corp. v. C. & C. Co., 188 W.Va. 26, 422 S.E.2d 503, 123 O.&G.R. 358 (1992) in support of the conclusion that a coal lease could be abandoned even if the lease did not contain a limitation causing its automatic termination. *See also* Jolynne Corp. v. Michels, 191 W.Va. 406, 446 S.E.2d 494, 129 O.&G.R. 146 (1994).

———

d. SHUT-IN GAS ROYALTY CLAUSE

Consider the following shut-in gas royalty clauses:

(a) Where gas from a gas well is not sold or used, lessee may pay as royalty $640 per well per year and while such payment is made, it will be considered that gas is being produced within the meaning of the habendum clause.

(b) If a well capable of producing gas or gas and gas-condensate in paying quantities located on the leased premises (or on acreage pooled or consolidated with all or a portion of the leased premises into a unit for the drilling or operation of such well) is at any time shut in and no gas or gas-condensate therefrom is sold or used off the premises or for the manufacture of gasoline or other products, nevertheless such shut-in well shall be deemed to be a well on the leased premises producing gas in paying quantities and this lease will continue in force during all of the time or times while such well is so shut in, whether before or after the expiration of the primary term hereof. Lessee shall use reasonable diligence to market gas or gas and gas-condensate capable of being produced from such shut-in well but shall be under no obligation to market such products under terms, conditions or circumstances which, in lessee's judgment exercised in good faith, are unsatisfactory. Lessee shall be obligated to pay or tender to lessor within 45 days after the expiration of each period of one year in length (annual period) during which such well is so shut in, as royalty, an amount equal to the annual delay rental herein provided applicable to the interest of lessor in acreage embraced in this lease as of the end of such annual period, or, if this lease does not provide for any delay rental, then the sum of $ 50.00; provided that, if gas or gas-condensate from such well is sold or used

as aforesaid before the end of any such annual period, or if at the end of any such annual period, this lease is being maintained in force and effect otherwise than by reason of such shut-in well, lessee shall not be obligated to pay or tender, for that particular annual period, said sum of money. Such payment shall be deemed a royalty under all provisions of this lease. Such payment may be made or tendered to lessor or to lessor's credit in the depository bank above designated. Royalty ownership as of the last day of each such annual period as shown by lessee's records shall govern the determination of the party or parties entitled to receive such payment.

Assume that the clauses above appear in a lease on land located in a jurisdiction following the discovery and marketing rule. The primary term of the lease would expire on July 1, 1991. A gas well is completed on May 1, 1991. When are shut-in royalties due? Suppose the payment is not tendered on the due date. What are the consequences of missing the due date? *See* Amber Oil & Gas Co. v. Bratton, 711 S.W.2d 741, 92 O.&G.R. 359 (Tex.App. 1986). Suppose payments are tendered on June 1, 1991. What would be the consequence of a second tender on June 30, 1992?

On these questions, see Kramer, *Lease Maintenance for the Twenty-First Century: Old Oil and Gas Law Doesn't Die, It Just Fades Away*, 41 Rocky Mtn.Min.L.Inst. 15-1 (1995); Beck, *Shutting-In: For What Reasons and For How Long?*, 33 Washburn L.J. 749 (1994); Maxwell, *Termination of Oil and Gas Leases-The Failure of Drafting Solutions*, 15 Sw. Legal Fdn. Oil & Gas Inst. 193 (1963).

TUCKER v. HUGOTON ENERGY CORP.

Kansas Supreme Court, 1993
855 P.2d 929, 125 O.&G.R. 301

LOCKETT, JUSTICE: This appeal arises from eight consolidated lawsuits tried to the court in which plaintiffs/appellants claimed certain of defendants' oil and gas leases had automatically terminated because the leaseholds had failed to produce gas in commercial quantities. The trial court ruled in favor of the plaintiffs in two cases and against them in six cases. Plaintiffs in those six cases appeal, claiming the trial court: (1) failed to make sufficient findings of fact to conclude the wells were producing or capable of producing gas in paying quantities; (2) erred in finding the shut-in royalty payments perpetuated the leases; and (3) erred in considering the suit as an equitable action for lease forfeiture rather than a legal action to terminate the leases. Defendants cross-appeal, claiming the trial court erred by not entering judgment in their favor on grounds the plaintiffs were estopped from claiming the leases terminated.

Plaintiffs-lessors are (1) certain landowners of the lands covered by the defendants' oil and gas leases involved in these lawsuits and (2) Plains Resources, Inc. (Plains). Defendants-lessees are (1) Hugoton Energy Corporation (Hugoton), (2) Plains Petroleum Operating Company (PPOC), (3)

Hamilton Brothers Oil and Gas Corporation, (4) Texaco, Inc., (5) Mesa Mid-Continent Limited Partnership, and (6) Mesa Operating Limited Partnership. Defendant Hugoton is the successor operator to PPOC. The other defendant-lessees owned smaller working interests in the gas units but did not operate the units.

The wells involved in this lawsuit are in the Bradshaw Field in Hamilton County, Kansas. They were initially drilled in the 1960's by Kansas-Nebraska Natural Gas Company, Inc., now KN Energy, Inc. (KN Energy). The primary term of the various oil and gas leases expired long ago. Each of the wells has produced gas only and each produced gas nearly every month until the mid-1980's. KN Energy initially operated the wells and owned a substantial interest in the associated leasehold estates. It also owned the gathering and pipeline system which transported the gas produced from those wells.

Gas from these wells is of relatively low quality. The wells have relatively low deliverability, and they produce significant amounts of water. Because the wells produce large quantities of water, they cannot be turned on and off to meet current demands. They, therefore, require a high level of maintenance.

In 1983, KN Energy formed Plains Petroleum Company, a wholly owned subsidiary. KN Energy transferred its gas units in the Bradshaw Field to this new corporation and then entered into a Gas Purchase Contract for all the gas produced from the units.

As the price of gas increased, KN's sale of gas drastically decreased because of a decline in industrial gas sales. KN lost substantial sales, for instance, when a major commercial customer, a gas-fired energy plant, switched to coal. As a result of the decline in gas sales, KN's purchases of natural gas under the gas purchase contract also declined.

On September 13, 1985, Plains Petroleum Company became independent of KN Energy. Plains Petroleum Company then formed a wholly owned subsidiary, Plains Petroleum Operating Company, on December 1, 1986, and transferred its Bradshaw Field interests and the KN Energy gas purchase contract to PPOC.

In 1986, the wells involved in this appeal encountered mechanical problems and production from the wells ceased. PPOC elected not to repair and produce those wells because of the high cost of maintenance. The wells remained off production for more than three years, each subsequently resuming production at a different time. During those periods in which the wells were not producing, PPOC tendered "shut-in" royalty payments which were accepted by the lessors.

In 1989, PPOC decided to sell its properties in the Bradshaw Field. It prepared and distributed a Sales Brochure which contained information regarding its Bradshaw Field properties, including revenue and operating expense information for the wells for the 28-month period from January 1987 through April 1989. The brochure indicated that operating expenses exceeded revenues for each of the wells during that period.

In the summer of 1989, Plains Resources determined that the leases at issue had terminated because the wells were not producing in paying quantities or were not capable of producing in paying quantities. Consequently, Plains acquired new oil and gas leases covering the properties associated with the wells. In November 1989, Hugoton became the successor in interest to the original lessees when it acquired PPOC's leases covering those same properties.

Plaintiffs brought this action to terminate Hugoton's oil and gas leases.

Trial Court Decision

After reviewing the evidence, the trial court found that all the leases involved in the lawsuit had valid shut-in royalty payment provisions; when the six wells involved in this appeal were first shut down in 1986 it was for mechanical reasons; and PPOC subsequently chose to invoke the shut-in royalty payment provisions of the leases rather than to repair the wells because PPOC would not be able to recover its costs of repair at the price being paid by and the limited market available through KN Energy.

As relevant to this appeal, the trial court concluded that when the wells were shut down and shut-in royalty payments made, there was no market available for the sale of the natural gas that could be produced from the wells. The court also concluded that prior to and after the shut-in period, four of the wells were producing in paying quantities and were capable of producing in paying quantities when shut in and that, as to two of the wells, repairs had commenced within the time frame required by the leases but that a sufficient period of time had not passed to determine whether the revenue from the sale of gas would exceed the Lease Operating Expense (LOE).

The trial court refused to terminate defendants' leases in those six cases.

As noted previously, plaintiffs in those six cases appeal, and defendants cross-appeal. Because the record does not support plaintiffs' conclusion that the trial court considered this to be an equitable action for lease forfeiture rather than a legal action to terminate the leases, we do not address plaintiffs' third claim of error.

Shut-In Royalty Payments

Plaintiffs next contend the trial court erred in concluding the shut-in royalty payments perpetuated the leases because (1) the trial court erred in finding that no market existed for the gas, and (2) the wells were not mechanically capable of producing gas.

There were three different shut-in royalty clauses and related repair clauses involved in this case. All were similar to the following:

> Where gas from a well or wells, capable of producing gas only, is not sold or used for a period of one year, lessee shall pay or tender, as royalty, an amount equal to the delay rental as provided in paragraph (5) hereof, payable annually at the end of each year during which such gas is not sold

or used, and while said royalty is so paid or tendered this lease shall be held as a producing property under paragraph numbered two hereof.

If, upon, or after the expiration of the primary term of this lease, the well or wells on the leased premises, or on the consolidated gas leasehold estate, shall be incapable of producing, this lease shall not terminate provided lessee resumes operations for drilling a well on the leased premises or on the consolidated leasehold estate within one hundred twenty (120) days from such cessation, and this lease shall remain in force during the prosecution of such operations and, if production results therefrom, then as long as production continues.

Plaintiffs first argue that shut-in royalty clauses cannot be invoked when there is a market for the gas or the wells are not capable of producing because of mechanical problems. Although the trial court concluded that at the time the wells were shut down and the shut-in royalty payments made there was no market for the gas that could be produced from the wells, plaintiffs contend that conclusion is inconsistent with the trial court's factual findings. They point out the court's factual finding that, as to each of the six wells, PPOC invoked the shut-in royalty payment provisions because it would not be able to recover its costs of repair at the price being paid by and the limited market available through KN Energy. They argue the finding that there was a limited market is inconsistent with the conclusion that there was no market and, since there was a limited market, shut-in royalty clauses could not be invoked.

As a general rule, forfeiture of oil and gas leases for breach of an implied covenant is disfavored. Whether a lessee has performed its duties under expressed or implied covenants is a question of fact. In the absence of a controlling stipulation, neither the lessor nor the lessee is the sole arbiter of the extent, or the diligence with which, the operations and development shall proceed. The standard by which both are bound is what an experienced operator of ordinary prudence would do under the same or similar circumstances, having due regard for the interests of both. In making such a determination, consideration is given only to the circumstances existing at the time without benefit of the wisdom of hindsight.

In Pray v. Premier Petroleum, Inc., 233 Kan. 351, 352, 662 P.2d 255 (1983), the lessor executed an oil and gas lease which contained the standard provision with a primary term of two years "and as long thereafter as oil and gas, or either of them, is produced from said land by the lessee, or the premises are being developed or operated." The lease also contained a shut-in royalty clause. During the primary term of the lease, the lessee began drilling operations on the property. A well capable of producing gas in paying quantities was completed.

In *Pray*, the lessee's effort to market the gas was fruitless since it was the only gas well in the area and was three miles from a gas line. The lessee tendered payment under the shut-in royalty provision. For five years, the shut-in royalty payments were paid and accepted. Thereafter, the lessors refused to accept them. Lessors then filed a quiet title action claiming the gas well was not capable of producing in paying quantities because the cost of connecting the pipeline rendered it unprofitable. The trial court concluded that the cost of transporting the gas to the nearest market should be used in determining whether the well was capable of producing in paying quantities and determined the lease had expired by its own terms.

On appeal, the sole issue was whether the costs of connection could be used in determining the well's capability of producing in paying quantities. We determined costs of connecting the pipeline should not be included and that to do so would essentially negate the effectiveness of a shut-in royalty clause.

In reversing the trial court, the *Pray* court discussed some of the typical provisions of an oil and gas lease. It noted that, generally, under the habendum clause of an oil and gas lease, oil or gas must be produced in "paying" or commercial quantities in order to perpetuate a lease beyond its primary term. Paying quantities is synonymous with commercial quantities. Although the phrase "in paying quantities" may not appear in oil and gas leases, it implicitly is a part of the habendum clause.

With regard to the "thereafter" provision, also known as the extension clause, of the habendum clause of an oil and gas lease, the court noted that it is generally accepted the phrase "in paying quantities" means production of quantities of oil and gas sufficient to yield a profit to the lessee over operating expenses, even though the drilling costs or equipping costs are never recovered and even though the undertaking as a whole may thus result in a loss to the lessee.

The court further stated that a shut-in royalty clause in an oil and gas lease enables a lessee, under appropriate circumstances, to keep a nonproducing lease in force by the payment of the shut-in royalty and that such a clause by agreement of the parties creates constructive production. As such, the clause becomes an integral part of the habendum clause.

The shut-in royalty clause was developed to protect against automatic termination of a lease where a gas well was drilled and no market existed for that gas. The reason that the shut-in royalty clause has come into growing use goes back to the inherent physical nature of natural gas. Unlike oil, it cannot be produced and stored or transported in railroad cars or tank trucks. A lessee completing a gas well consequently often had a special and quite onerous problem in finding a market outlet for the gas production. This would, at times, result in losing a lease at the end of the primary term.

The "shut-in" royalty clause applies to circumstances where "a well capable of producing a profit is drilled but for the time being no market exists." There is a mutual interest between the lessor and lessee when no market for the gas exists. However, once a market for gas is secured, that mutual interest no longer exists because the lessor's interest is in securing royalty payments from production, while the lessee's interest is divided between receiving revenues and minimizing expenses associated with that production. Thus, the "shut-in" royalty clause serves the interests of both parties only in situations where no market exists for the gas.

The fact a lease is held by payment of shut-in gas royalties does not excuse the lessee from its duty to diligently search for a market and to otherwise act as would a reasonable and prudent lessee under the same or similar circumstances.

Although a shut-in royalty clause does not normally specify the shut-in gas well must be capable of producing in paying quantities, such a requirement is implied. As noted, a shut-in royalty clause is a savings clause allowing for constructive production. Such clauses provide that upon payment of the shut-in royalty it will be considered as if gas is being produced within the meaning of the habendum clause. In order to achieve the desired result, namely a profit, production, whether actual or constructive, must be in paying quantities.

To obtain the maximum profit from its use of gas, the lessee chose not to produce gas from the wells that required constant maintenance. Because, in this case, at the time of shut-in there was a limited market available to defendants-lessees for the gas producible from the six wells at issue, the shut-in royalty clauses could not be invoked to perpetuate the leases. Thus, the trial court erred in finding the shut-in royalty clauses were properly invoked.

When a mineral lease is terminated because of failure to produce, subsequent production from the land will not extend its term or revive rights which the parties fixed in the contract. Thus, because the leases had terminated because they were not producing in paying quantities, the trial court erred in considering the productive ability of the wells subsequent to the shut-in period when determining whether the wells were producing or capable of producing in paying quantities when shut-in. The trial judge's determination that the wells were currently producing did not save the leases because they had already terminated.

Because of our holding that the existence of a limited market precluded invocation of the shut-in royalty provisions in this case, we do not reach plaintiffs' assertion that the wells were incapable of producing in paying quantities due to their mechanical problems.

Cross-Appeal

In reaching a decision on the merits, the trial court did not consider the defendants' affirmative defense of equitable estoppel. Equitable estoppel is the effect of the voluntary conduct of a party that precludes that party, both at law and in equity, from asserting rights against another who relies on such conduct. A party seeking to invoke equitable estoppel must show that the acts, representations, admissions, or silence of another party (when it had a duty to speak) induced the first party to believe certain facts existed. There must also be a showing the first party rightfully relied and acted upon such belief and would now be prejudiced if the other party were permitted to deny the existence of such facts. There can be no equitable estoppel if any essential element thereof is lacking or is not satisfactorily proved. Estoppel will not be deemed to arise from facts which are ambiguous and subject to more than one construction.

Defendants claim they continued to pay the operating expenses on the wells because of plaintiffs' acceptance of the shut-in royalty payments. They claim their reliance on the acceptance was reasonable and that plaintiffs are now estopped from terminating the leases.

In light of our holding that the shut-in royalty clauses could not be invoked, the matter must be remanded for the trial court to determine whether plaintiffs should be equitably estopped from claiming termination of the leases at issue is required.

Affirmed in part, reversed in part, and remanded for a new trial on the limited issue of equitable estoppel.

NOTES

1. Could the parties have specifically negotiated a shut-in royalty clause that would apply to a well that would not meet the production in paying quantities test? Should the court apply the *Clifton v. Koontz* two-pronged test to determine if the well is capable of producing in paying quantities? Everest Exploration, Inc. v. URI, Inc., 131 S.W.3d 138 (Tex.App. 2004); Maralex Resources, Inc. v. Gilbreath, 134 N.M. 308, 2003-NMSC-023, 76 P.3d 626, 158 O.&G.R. 441.

2. What is the appropriate time period to determine whether the well is producing in paying quantities or capable of producing in paying quantities? Should it be when the well is shut-in? after payments have been made? What if the price of gas drops between the time the well is shut in and the time that gas is physically produced? Can the lessor claim that the lease has expired because if there had been production during that period of time it would not have constituted production in paying quantities?

3. Who should have the burden of proof to show that the well is or is not capable of producing in paying quantities in order to trigger the application of the shut-in royalty clause? In Hydrocarbon Management, Inc. v. Tracker Exploration, Inc., 861 S.W.2d 427, 126 O.&G.R. 316 (Tex.App. 1993, writ denied), the court placed that burden on the lessor. The court also defined the term capable of producing in paying quantities as "a well that will produce in paying quantities if the well is turned 'on' and it begins flowing, without additional equipment or repair." *Id.* at 433-34. Is that definition broad enough to include shut-ins caused by marketplace changes? What if the only buyer in the field refuses to purchase natural gas? What if some operation or activity other than turning the valve on is required to attain production? In Blackmon v. XTO Energy, Inc., 276 S.W..3d 600 (Tex.App.—Waco 2008), the court applied the

"turn-on" test but nonetheless found that even though an amine processing plant would be needed in order to produce the natural gas, the lessee could maintain the lease by the payment of shut-in royalties? What if a lessee needs to engage in a second or third hydraulic fracing operation in order to increase production to a level necessary to meet the production in paying quantities test? During the period of time the well is shut-in, is the well capable of producing in paying quantities?

4. In a discovery plus marketing state that also treats the lease as a fee simple determinable estate does a failure to make a timely and/or accurate payment automatically terminate the lease? In Freeman v. Magnolia Petroleum Co., 141 Tex. 274, 171 S.W.2d 339 (1943) a delay in several months in tendering the $ 50.00 shut-in royalty automatically terminated the lease. The court said:

> If [lessees] had wanted to prevent lapsation of the lease for non-production, they could easily have done so by paying the fifty dollars on or before the last day of the primary term. . . . The lease lapsed as a matter of law when they so failed, and it could not be revived by their attempt to perform the condition more than four months after the contract said it should be performed. 171 S.W.2d at 342.

5. What happens if the shut-in royalty is paid to the wrong person? *See* Amber Oil & Gas Co. v. Bratton, 711 S.W.2d 741, 92 O.&G.R. 359 (Tex.App. 1986). *See generally,* Forbis, *The Shut-In Royalty Clause: Balancing the Interests of Lessors and Lessees*, 67 Texas L.Rev. 1129 (1989).

6. How do shut-in royalty clauses and cessation of production or continuous operations clauses interact? Would the lessee have 60 days from the cessation of production to make the payment assuming the shut-in clause did not have an express starting date? *See* Mayers v. Sanchez-O'Brien Minerals Corp., 670 S.W.2d 704, 81 O.&G.R. 38 (Tex.App. 1984).

7. After discovery of gas "in significant quantities" by drilling through the end of the primary term under a drilling operations clause requiring "drilling, mining or reworking" the well was equipped for oil production, and gas was flared for a year in an unsuccessful attempt to produce oil. After suit was filed by the lessor to cancel the lease of record the well was shut in and shut-in royalty was tendered under a clause similar to clause (a) above. Is the lease still in effect? *See* Doty v. Key Oil, Inc., 83 Ill.App.3d 287, 38 Ill.Dec. 922, 404 N.E.2d 346, 67 O.&G.R. 352 (1980).

8. A gas well was drilled under a lease providing for payment "at the end of each yearly period during which . . . gas is not sold or used . . . of an amount equal to the delay rental . . . this lease shall be held as a producing lease under . . . [the habendum clause]." The well was shut in for lack of a market "since it was the only gas well in the area and was three miles from a gas line." Payments under the above clause were refused. In a suit by the lessors to quiet title the trial court relied on Reese Enterprises, Inc v. Lawson, 220 Kan. 300, 553 P.2d 885, 56 O.&G.R. 517 (1976), to the effect that "the cost of connecting a gas well to a

pipeline can be used in determining the well's capability of producing in paying quantities." In Pray v. Premier Petroleum, Inc., 233 Kan. 351, 662 P.2d 255, 76 O.&G.R. 449 (1983), the Kansas court reversed:

> [T]he trial court's holding essentially negates the effectiveness of the shut in royalty clause included in this lease. Involved here is precisely the type of situation contemplated by the clause--a well capable of producing a profit is drilled but for the time being no market exists. This is much different from cases such as *Reese* where no shut-in royalty clause is involved and a determination of production in paying quantities can be made simply by looking at the performance of the well. . . . A case such as the one at bar . . . necessarily involves some speculation. This speculation should not include the cost of taking the gas to market when the parties have foreseen in the lease the possibility of a market might not exist.

How speculative can the prospects of paying production be and yet support the use of a shut-in clause to extend the lease? *See* Young v. Dixie Oil Co., 647 S.W.2d 235, 77 O.&G.R. 583 (Tenn.App. 1982), where the use of a shut-in clause was held sufficient to sustain a lease although it "was very questionable" if "the entire field might now be profitable if all three shut-in wells were put into production." The court noted that "the lessee has proceeded diligently toward" achieving production in paying quantities. *See also* Beck, *Shutting-in: For What Reasons and For How Long?*, 33 Washburn L.J. 749 (1994); Lowe, *Shut-in Royalty Payments*, 5 Eastern Min.L.Inst. 18-1 (1984).

9. What if a state statute requires the lessor to put the lessee on notice of a default before a lease can be terminated? Does the lessor have to inform the lessee of a late or inadequate payment and provide an opportunity to have the defect cured before the lease can be terminated? *See* Acquisitions, Inc. v. Frontier Explorations, Inc., 432 So.2d 1095, 78 O.&G.R. 282 (La.App. 1983).

PROBLEM

On August 1, 1980, Baker executed an oil and gas lease to Easy Oil Co. covering 25,000 acres of land in Colorado. In general form, the lease followed the pattern of the Producers 88 lease reproduced in Appendix One of this casebook. The precise wording of the clauses applicable to this problem follows:

2. Subject to other provisions herein, this lease shall remain in force for a period of ten years from the date of its execution and as long thereafter as oil, gas or other minerals are produced.

3. The royalties to be paid by lessee are: . . . a royalty of $640 per year on each gas well from which gas only is produced while gas therefrom is not sold or used off the premises, and while said royalty is so paid, said well shall be held to be a producing well. . . .

4. If no well be commenced on said land on or before one year from the date hereof, this lease shall terminate as to both parties, unless the lessee on or before

that date shall pay or tender to the lessor or to the lessor's credit in the 1st National Bank at Golden, Colo., or its successors, which shall continue as the depository for rental regardless of changes in the ownership of said land, the sum of Twenty-Five Hundred Dollars, ($2500) which shall operate as a rental and cover the privilege of deferring the commencement of a well for twelve months from said date. In like manner and upon like payments or tenders the commencement of a well may be further deferred for like periods of the same number of months successively. All payments or tenders may be made by check or draft of lessee or any assignee thereof, mailed or delivered on or before the rental paying date. It is understood and agreed that the consideration first recited herein, the down payment, covers not only the privilege granted to the date when said first rental is payable as aforesaid, but also the lessee's right of extending that period as aforesaid, and any and all other rights conferred. Should the depository bank hereafter close without a successor, lessee or his assigns may deposit rental or royalties in any National bank located in the same county with first named bank, due notice of such deposit to be mailed to lessor at last known address.

7. If prior to discovery of oil or gas on said land Lessee should drill a dry hole or holes thereon, or if after discovery of oil or gas the production thereof should cease from any cause, this lease shall not terminate if lessee commences additional drilling or reworking operations within sixty (60) days thereafter or (if it be within the primary term) commences or resumes the payment or tender of rentals on or before the rental paying date next ensuing after the expiration of three months from date of completion of dry hole or cessation of production. If at the expiration of the primary term oil, gas or other mineral is not being produced on said land but lessee is then engaged in drilling or reworking operations thereon, this lease shall remain in force so long as operations are prosecuted with no cessation of more than thirty (30) consecutive days, and if they result in the production of oil, gas or other mineral so long thereafter as oil, gas or other mineral is produced from said land.

10. It is further understood and stipulated herein that payments for gas from a well or wells producing gas only, where the same is not sold or used, as provided for above in paragraph 3, shall be deemed production and shall hold the lease in full force and effect as to 640 acres only for each well, which acreage is to include said gas well. The exact form of the acreage affected by each said gas well is to be determined by the lessee but is to be as near as possible in the form of a square as the boundaries of the land covered by the lease and location of the well or wells may permit and with each said well located as near the center of said 640 acres as possible.

In October 1990 Baker brought suit against Easy for a declaratory judgment that the oil and gas lease of Easy had terminated. Uncontroverted affidavits show the following:

The first well drilled on the land was completed as a commercial oil producer on December 4, 1986. The production from this well declined radically until it completely ceased production on May 15, 1988, when it was plugged and abandoned. Delay rentals had been properly tendered until drilling commenced on this well. From January 1, 1988 to May 15, 1988 operating expenses from the well exceeded the income received from the sale of the working interest oil.

Five other wells were drilled on additional 640 acre sections of the land:

Well No. 2: Completed September 14, 1987, as a gas well, which was immediately shut in for lack of a market. Shut-in royalties have been properly tendered ever since.

Well No. 3: Drilling commenced October 21, 1987, and completed as a dry hole November 20, 1987.

Well No. 4: Drilling commenced July 15, 1989, and completed as a dry hole August 20, 1989.

Well No. 5: Drilling commenced January 9, 1990, and completed as a dry hole March 3, 1990.

Well No. 6: Drilling commenced June 20, 1990, and completed on July 27, 1990, upon discovery of gas and gas condensate in commercial quantities. The well was temporarily shut in pending execution of a gas sales contract and connection with a pipeline, and shut-in gas royalty was tendered by Easy to Baker on August 10, 1990, but was refused. On September 29, 1990, the well was connected to a pipeline and gas and condensate were thereafter marketed from the well.

On July 1, 1988, delay rentals were tendered and accepted for the period August 1, 1988 to July 31, 1989.

Both parties move for summary judgment. How should the trial court rule?

––––––

e. LEASEHOLD SAVINGS CLAUSES IN DISCOVERY JURISDICTIONS

PACK v. SANTA FE MINERALS

Oklahoma Supreme Court, 1994
1994 OK 23, 869 P.2d 323, 128 O.&G.R. 550

SIMMS, JUSTICE: Appellants, Santa Fe Minerals and other oil and gas companies, (lessees) appeal the judgments entered by the district court in the quiet title actions instituted by appellees, Mary Lou Pack, Ann E. Watts, Robert E. Stevens, Jo E. Stevens, John V. Balzer, Jake F. Balzer, and Lydia Balzer (mineral rights owners/lessors). These separate actions instituted by Pack, Watts and Stevens (No. 74,605), and by the Balzers (No. 74,606) were consolidated for trial, and are consolidated for purposes of this opinion. The trial court entered judgment in favor of the mineral rights owners, canceling oil and gas leases and quieting title in the lessors.

The Court of Appeals affirmed, holding the leases terminated of their own terms under the provisions of the "cessation of production" clause. Certiorari was granted to consider the first impression question of whether a lease, held by a gas well which is capable of producing in paying quantities but is shut-in for a period in excess of sixty (60) days but less than one year due to a marketing decision made by the producer, expires of its own terms under the "cessation of production" clause unless shut-in royalty payments are made. We find that under such circumstances, the lease does not expire of its own terms. The opinion of the Court of Appeals is vacated, the judgment of the district court is reversed, and this cause remanded with directions to enter judgment in favor of the lessees.

The stipulated facts disclose that in both cases the mineral owners or their predecessors in interest entered into oil and gas leases with the lessees. Each of the leases contained similar provisions including a habendum clause, a shut-in or minimum royalty clause, and a 60-day cessation of production clause.

The habendum clause provides for the primary term of the lease to be for ten (10) years and "as long thereafter as oil, gas, casinghead gas, casinghead gasoline or any of them is produced." The shut-in royalty clause provides for a fifty dollar ($50.00) royalty payment per year for each well from which gas is not sold. When the royalty payment is paid, the well is deemed a producing well for purposes of the habendum clause. The cessation of production clause provides for the lease to continue after the expiration of the primary term as long as production does not cease for more than sixty (60) days without the lessee resuming operations to drill a new well.

The primary terms of each of the leases expired, and the leases continued pursuant to the habendum clause due to the wells' capability to produce in paying or commercial quantities. Each of the wells continued to be capable of producing in paying quantities up until the time of trial, but lessees have chosen at times not to market gas from the wells for periods exceeding sixty (60) days. The lessees stipulated that they chose to overproduce the wells during the winter months when the demand for gas is higher and the price for gas increases. Because the Oklahoma Corporation Commission imposed annual allowable limitations as to how much gas may be produced from the wells, the lessees curtailed the marketing of gas from the wells during the summer months when prices were lower so as not to exceed the annual allowable limits. The intention and result of this practice was to obtain the highest price for the gas and still stay within the allowable production limits set by the Oklahoma Corporation Commission. Such a practice was common with most of the gas producers in the state.

Although some marketing of the gas continued during the warmer months, such sales were exceeded by the monthly expenses, so the wells were not profitable during that period. Additionally, the Pack well was shut-in for one month during this period in order to build up pressure in preparation for an annual well test to determine its annual allowable limit.

The mineral owners filed suit in district court asserting the leases terminated by their own terms when the wells failed to produce for a sixty (60) day period and the lessees neither commenced drilling operations nor paid shut-in royalty payments. The trial court determined that an interruption in the sale and marketing of gas from the wells in excess of sixty (60) days constituted a cessation of production within the meaning of the cessation of production clause resulting in a termination of the leases. Judgment was entered accordingly, and the lessees/producers appealed that judgment.

I.

The term "produced" as used in the lease clauses means "capable of producing in paying quantities" and does not include marketing of the product.

A.

The Habendum Clause

This Court has long held that the terms "produced" and "produced in paying quantities" have substantially the same meaning. State ex rel. Commissioners of the Land Office v. Carter Oil Co. of West Virginia, 336 P.2d 1086 (Okla.1958). Therein, we construed a typical habendum clause which extended the lease past its ten-year primary term as long thereafter as oil or gas is produced in paying quantities. We held that in order to extend the fixed term of ten years "and acquire a limited estate in the land covered thereby the lessee must have found oil or gas upon the premises in paying quantities by completing a well thereon prior to the expiration of such fixed term." 336 P.2d at 1094. The Court then rejected the lessors' argument that production in paying quantities required the lessees to not only complete a well capable of producing in paying quantities but also remove the product from the ground and market it. Thus, where a well was completed and capable of producing in paying quantities within the primary term, the lease continued, so far as the habendum clause was concerned, as long as the well remained capable of producing in paying quantities, regardless of any marketing of the product.

This rule of law has been consistently upheld. See, e.g., Gard v. Kaiser, 582 P.2d 1311 (Okla.1978) and McVicker v. Horn, Robinson and Nathan, 322 P.2d 410 (Okla.1958). Perhaps one of the best explanations for the rule was given in McVicker where we stated:

> To say that *marketing* during the primary term of the lease is essential to its extension beyond said term, *unless the lease contains additional provisions indicating a contrary intent*, is to not only ignore the distinction between producing and marketing, which inheres in the nature of the oil and gas business, but it also ignores the difference between express and implied terms in lease contracts. 322 P.2d at 413 (Emphasis in original).

More recently, in Stewart v. Amerada Hess Corp., 604 P.2d 854 (Okla.1979), we reaffirmed the rule that an oil and gas lease could not be terminated under the habendum clause merely because the subject well ceased producing in paying quantities. Rather, the finder of fact must also look into the circumstances surrounding the cessation, including the "[d]uration and cause of the cessation, as well as the diligence or lack of diligence exercised in the resumption of production." 604 P.2d at 858, fn. 18. In so holding we affirmed our rejection of a literal construction of the habendum clause stating:

> Under a literal or strict interpretation of the "thereafter" provision in a habendum clause, uninterrupted production--following expiration of primary term--would be indispensable to maintain a lease in force. This would mean, of course, that *any cessation* of production [in the paying-quantities sense of the term], however slight or short, would put an end to the lease. Oklahoma has rejected that literal a view. Our law is firmly settled that the result in each case must depend upon the circumstances that surround cessation. Our view is no doubt influenced in part by the strong policy of our statutory law against forfeiture of estates. The terms of 23 O.S.1971 § 2 clearly mandate that courts avoid the effect of forfeiture by giving due consideration to compelling equitable circumstances.
>
>   . . .
>
> In short, the lease continues in existence so long as the interruption of production in paying quantities does not extend for a period longer than reasonable or justifiable in light of the circumstances involved. But under *no* circumstances will cessation of production in paying quantities *ipso facto* deprive the lessee of his extended-term estate.
>
> A decree of lease cancellation may be rendered where the record shows that the well in suit was not producing in paying quantities and there are no compelling equitable considerations to justify continued production from the unprofitable well operations. 604 P.2d at 858 (Emphasis in original) . . .

Finally, in State ex rel. Comm'rs of the Land Office v. Amoco Production Co., 645 P.2d 468 (Okla.1982), we held:

> The provision about paying quantities adds little to the term production since it is settled that production must be in paying quantities even though a lease contains no such provision . . . *It is the ability of the lease to produce that is the important factor rather than the actual production applied, as an example of ability to produce a shut-in gas well will hold a lease as long as the operator seeks a market with due diligence.* 645 P.2d at 470 . . . (Emphasis added).

Therefore, the lease in the case at bar cannot terminate under the terms of the habendum clause because the parties stipulated that the subject wells were at all times capable of producing in paying quantities. The habendum clause of these leases is satisfied.

B.

The Cessation of Production Clause

The cessation of production clause of the subject leases provides:

If, after the expiration of the primary term of this lease, production on the leased premises shall cease from any cause, this lease shall not terminate provided lessee resumes operations for drilling a well within sixty (60) days from such cessation, and this lease shall remain in force during the prosecution of such operations and, if production results therefrom, then as long as production continues.

The lessors/mineral interest owners argue the term "production" used in this clause includes removal and marketing of the product. They would require the lessee to continually market the gas from the well, and any cessation of such marketing for a period of sixty days or more would result in termination of the lease. We cannot accede to such a constrained construction of the clause as it discounts the intended meaning of production as this Court has determined in numerous cases, a few of which are cited above. In construing the provisions of a contract we are bound to consider all the provisions thereof and use each provision to help interpret the others. Thus, our understanding of the term in the cessation of production clause is influenced by how we have interpreted it in other provisions of oil and gas leases such as the habendum clause. In Hoyt v. Continental Oil, 606 P.2d 560 (Okla.1980), we noted our commitment "to the principle that production means production in paying quantities in Oklahoma when the term appears in the habendum clause of an oil and gas lease." 606 P.2d at 563. We then held:

After the primary term, the effect of the cessation of production clause is to modify the habendum clause and to extend or preserve the lease while the lessee resumes operations designed to restore operations. If the lessee fails to resume operations within the 60-day period provided in this clause neither the cessation of production clause or the habendum clause is satisfied and the lease terminates upon the expiration of the given time period.

. . .

[W]here, as here, *the primary term has expired* and the effect of the [cessation of production] provision is to modify the habendum clause . . . *there is a cessation of production if the habendum clause requires production in paying quantities and such requirement is not met.* 606 P.2d at 563 (Emphasis added).

Therefore, the cessation of production clause serves the purpose of modifying the habendum clause to cause the lease to continue for the stated time period to give the lessee an opportunity to begin drilling a new well. The term "production" as used in the cessation of production clause must mean the same as that term means in the habendum clause. Any other conclusion would render the habendum clause useless after the primary term expires, a conclusion clearly not intended by the parties to the lease. If we were to conclude that the term "production" as used in the cessation of production clause means removal and marketing of the product as the mineral interest owners assert, then after the primary term expires, the lessee must not only keep the well in working order so that it is capable of producing in paying quantities as would be required under our well-settled understanding of the habendum clause, but gas would have to be removed and marketed in order to meet the requirements of the cessation of production clause. Any cessation of marketing for a sixty day period, under the lessors' interpretation of the clause, would constitute a violation of the clause resulting in termination of the lease. Such a result ignores the express terms of the habendum clause which provide for the lease to continue after the primary term as long as the well is capable of production in commercial quantities regardless of marketing.

Such a construction further disregards the express provisions of the cessation of production clause which are intended to come into play in the event that production from the well shall cease, i.e. the well becomes incapable of producing in paying quantities. The clause allows the lessee the opportunity to begin drilling another well on the leased premises within sixty (60) days so that production in paying quantities may be realized in another well. In other words, the lease terminates when the habendum clause is no longer satisfied, but the cessation of production clause is a saving clause which gives the lessee sixty (60) days to get the lease producing again. This is seen in French v. Tenneco Oil Co., 725 P.2d 275 (Okla.1986), where we stated:

> "[T]he provision is construed as giving the lessee a fixed period of time within which to *resume production or commence additional drilling* or reworking operations in order to avoid termination of the lease . . ." Restoration of production in paying quantities within that period obviates the need to drill . . . 725 P.2d at 277 (quoting Greer v. Salmon, 82 N.M. 245, 479 P.2d 294 (1970) (Emphasis in original).

Thus, the clause does not require the lessee to market oil or gas actually extracted from the well. If the well was capable of production in commercial quantities at all times, but for a short period *had less than commercial quantities marketed from it*, the lessors would require the lessees to begin drilling operations for another well that under the facts would be unnecessary and uneconomical.

Contrary to the argument of lessors and the decision of the trial court, Hoyt v. Continental Oil, *supra*, and French v. Tenneco Oil Co., 725 P.2d 275 (Okla.1986), do not support termination of the lease under the cessation of production clause. In both cases, this Court affirmed termination of the lease based upon the terms of the cessation of production clause. However, we did so because at the time of expiration of the primary terms of those leases, the subject wells were not capable of production in paying quantities, i.e. they were not "producing" wells under either the habendum clause or the cessation of production clause. The trial court in the case at bar mistakenly concluded that "production in paying quantities" required the wells to be making a profit during the relevant months. Neither *Hoyt* nor *French* require such.

Likewise, our holding in Hamilton v. Amwar Petroleum Co., Inc., 769 P.2d 146 (Okla.1989), terminating a gas lease pursuant to the terms of the cessation of production clause does not control here. The well in *Hamilton* was capable of production in paying quantities at the time it was completed, but the evidence indicated that it was not capable of producing in paying quantities upon the expiration of the primary term. Unlike the wells in *Hoyt, French*, and *Hamilton*, the wells in the instant case have at all times been capable of producing in paying quantities.

In his treatise, Professor Kuntz observed that in jurisdictions such as Texas "where marketing is required as a part of production for purposes of the habendum clause," the shutting in of a gas well from which gas had been marketed could be a cessation of production under the cessation of production clause. 4 Kuntz, *The Law of Oil and Gas*, § 47.3(b) at p. 106 (1990) Id. However, the result is the opposite in jurisdictions such as Oklahoma that do not require marketing.

In a jurisdiction where marketing is not required to satisfy the habendum clause and the effect of the cessation of production clause is to modify that clause, the effect of a shut-in well from which neither oil nor gas has been marketed would be the same [no cessation of production] as in a jurisdiction where marketing is required, but the reason would be different. In this instance *there would be production for purposes of the habendum clause and therefore no cessation of production. If a well is shut-in after oil or gas has been marketed, the result would be different from the result reached in a jurisdiction that requires marketing for production, for the reason that production has not ceased and the cessation of production clause is not invoked.* (Emphasis added).

The cessation of production clause only requires the well be capable of producing gas in paying quantities. A gas lease does not terminate under the cessation of production clause for failure to market gas from the subject wells for a sixty (60) day period. Therefore, the lease will continue as long as the well is capable of production in paying quantities subject, of course, to any violation of any other express provisions such as the shut-in royalty clause or implied covenants such as the covenant to market.

C.

The Shut-in Royalty Clause

As noted in Part B above, the shut-in royalty clause plays a specific role in the continuation of the lease where the well is shut-in for a period of one year. In Gard v. Kaiser, 582 P.2d 1311 (Okla.1978), this Court looked at the effect such a clause has where the subject gas well was shut-in for three years due to low gas pressure. Prior to being shut-in, the well was completed and began production before the primary terms expired. Thus, the term of the lease was extended under the habendum clause as long as the well was capable of producing in paying quantities. However, because of low pressure, gas from the well could not enter a pipeline without additional equipment which lessee failed to install. Thus, gas was not marketed during the shut-in period. The lessors brought an action to cancel the leases on the grounds that the lessees failed to pay the shut-in royalty payments. They argued that according to the language of the lease, as soon as production ceased and the well was shut in, the lease automatically terminated.

The Court noted that in Oklahoma the shut-in royalty clause is " 'not required as an additional special limitation to extend the term'" of an oil and gas lease after a commercial discovery of gas satisfies the habendum clause. 582 P.2d at 1313 (quoting 4 Kuntz, The Law of Oil and Gas, § 46.3). Such a commercial discovery (capable of producing in paying quantities) extends the lease with or without the shut-in royalty clause, and the lease continues "subject to forfeiture for failure to comply with the implied obligation to market the product." Id.

We then held that "shut-in gas provisions are not to be construed as limitations or conditions which would affect termination of the leases," 582 P.2d at 1314-15. Thus, the failure to pay shut-in royalties in and of itself does not operate to cause a termination of the lease. Rather, it is the failure to comply with the implied covenant to market which results in lease cancellation.

In the situation at bar, lessors/mineral owners did not receive any royalty payments because gas was not being sold from the wells for a short period of time. The well, for all practical purposes, was shut-in for a period less than one year. Thus, if the well stayed in this shut-in state for one full year, the mineral owners would be entitled to the minimum shut-in royalty payment of fifty dollars ($50.00) as they agreed to be bound to in the lease. If the lessees chose to market gas from the wells, the mineral owners would receive royalty payments just as they have during the winter months. Either way, the mineral owners get paid what they are entitled to under the terms of the lease.

If we were to interpret the cessation of production clause to require marketing, then the shut-in royalty clause would be rendered meaningless. The lessees would not be able to shut-in the well and pay shut-in royalties to keep the lease viable because the cessation of production clause would mandate continuous marketing of gas. Thus, such a construction of the cessation of production clause would nullify the provisions of the shut-in royalty clause. The cessation of production clause cannot be interpreted in the manner asserted by the mineral owners.

II.

The Implied Covenant to Market (Doctrine of Temporary Cessation)

To this point we have held that the express provisions of the subject oil and gas leases do not require the lessees to remove and market oil or gas from the leases in order to keep the lease from terminating. However, our inquiry does not stop there for the long-standing rule is that typical oil and gas leases contain an implied covenant to market oil and gas from the subject wells. In State ex rel. Commissioners of the Land Office v. Carter Oil Co. of West Virginia, supra, we affirmed the rule of law set down in Cotner v. Warren, 330 P.2d 217 (Okla.1958) and stated:

> In other words in the absence of a specific clause requiring marketing within the primary term fixed in the lease, the completion of a well, as provided therein, capable of producing oil or gas in paying quantities will extend such term, provided that within a reasonable time the actual length of which must of necessity depend upon the facts and circumstances of each case, a market is obtained and oil or gas is produced and sold from such well. In such event if the producing and marketing thereof in such quantities from the well so completed is continued, the lease will extend until the economic exhaustion of the product. 336 P.2d at 1095.

In State v. Carter Oil, the evidence showed that due diligence was exercised in the seeking and obtaining of a market and, under the circumstances, such market was found within a reasonable time. Consequently, the lease was not terminated where the lessees completed the well which was capable of producing in paying quantities but were unable for over one year past the expiration date of the primary term to market the product due to not having a pipeline in the area in which to transport the product.

In *Cotner*, we adopted the rule concerning the implied covenant to market holding that the controlling factual finding is whether or not the temporary cessation of marketing was for an "unreasonable length of time." 330 P.2d at 219. We concluded that five to six months of *voluntary* cessation of marketing was not unreasonable under the circumstances because the operator was attempting to work out a problem with the lessees. The facts indicated that the problem was resolved when Warren bought out the rest of the partners becoming the sole lessee and immediately attempted to enter the premises to resume operations and drill another well. The lease had not expired under the facts and circumstances presented. Compare Hunter v. Clarkson, 428 P.2d 210 (Okla.1967) in which this Court affirmed the trial court's cancellation of an oil and gas lease where production and marketing from the well had been voluntarily ceased by the lessee for a period of five months without any circumstances to justify the cessation.

It, therefore, follows that the lessees in the cases at bar may voluntarily cease removal and marketing of gas from the subject wells for a reasonable time "where there are equitable considerations which justify a temporary cessation." Townsend v. Creekmore-Rooney Co., 332 P.2d 35, 37 (Okla.1958). If the temporary cessation is reasonable, the lease will continue, and the burden of proving that the lessees failed to use reasonable diligence in the operation of the well squarely rests with the lessors. The lessors have failed to meet that burden. We find that under the facts and circumstances of the case at bar the lessees' temporary cessation was for a reasonable time and justified by equitable considerations.

III.

Conclusion

In Rist v. Westhoma Oil Co., 385 P.2d 791, 792 (Okla.1963), the syllabus by the Court reads:

Where a cause is submitted upon an agreed statement of facts, it is the duty of this court on appeal to apply the law to such facts as a court of first instance and direct judgment accordingly.

We conclude, based upon the stipulated facts presented to us, that the leases did not terminate under the terms of the habendum clause, the cessation of production clause, or the shut-in royalty clause. Furthermore, under the circumstances of this case, the lease should not be canceled under the doctrine of temporary cessation because such temporary cessation was for a reasonable time.

For the above and foregoing reasons, the opinion of the Court of Appeals is VACATED, the judgment of the district court is REVERSED, and this cause is REMANDED with directions to enter judgment for the appellant lessees.

NOTES

1. Does *Pack* resolve all of the problems relating to the application of express savings clauses or implied savings doctrines, such as the temporary cessation of production doctrine, in a discovery jurisdiction where the lease is treated as a fee simple subject to a condition subsequent?

2. In HOYT V. CONTINENTAL OIL CO., 1980 OK 1, 606 P.2d 560, 66 O.&G.R. 83, the cessation of production clause stated: "If after expiration of the primary term . . . production . . . shall cease . . . [the] lease shall not terminate provided . . . lessee resumes operations for drilling . . . within sixty days." The lessor claimed that production in paying quantities had ceased for a period in excess of 60 days due to only marginal and sporadic production for over 12 months. The court said:

After the primary term, the effect of the cessation of production clause is to modify the habendum clause and to extend or preserve the lease while the lessee resumes operations designed to restore production. If the lessee fails to resume operations within the 60-day period provided in this clause neither the cessation of production clause or the habendum clause is satisfied and the lease terminates upon the expiration of the given time period. [606 P.2d at 563]

Is the holding in *Hoyt* consistent with the rationale in *Pack*?

3. In FISHER V. GRACE PETROLEUM CORP., 1991 OK CIV APP 112, 830 P.2d 1380, 118 O.&G.R. 491, cert denied, several different leases were involved that contained shut-in gas royalty and cessation of production clauses. Production was obtained in the primary term, but ceased during the secondary

term. The Court of Appeals applied a *Clifton* test to determine whether the habendum clause obligation to produce in paying quantities was complied with. It did not apply either the temporary cessation of production doctrine or the discovery rule approach. Is that consistent with *Pack*? Dean Eugene Kuntz in analyzing *Fisher* in the Oil and Gas Reporter said:

> The holding of the court that the leases in this case terminated can be justified only if the finding of the trialcourt that the well did not "produce in paying quantities" for a twelve-month period was supported by evidence that the well was not capable of producing in paying quantities during that time. The court was in error in its understanding of what constitutes production in Oklahoma. . .
>
> . . . the court recognized the Oklahoma doctrine that capability of production in paying quantities constitutes production, but reasoned that the doctrine is not applicable where marketing begins during the primary term but ceases after the primary term has expired. In Pack. . . . the Supreme Court reaffirmed the capability doctrine and applied it where marketing had begun during the primary term but had ceased after the primary term had expired. 118 O.&G.R. at 508-09

As usual, Dean Kuntz's analysis was right on point and correctly viewed the Oklahoma discovery doctrine as fundamentally different from the approach taken in discovery and marketing states. *See also* Danne v. Texaco Exploration & Production, Inc., 1994 OK CIV APP 138, 883 P.2d 210, 132 O.&G.R. 623, for a discussion of the Oklahoma law on how leases terminate upon a cessation of production in the secondary term.

4. What if the lessee testified at trial that he shut in an oil well because the prices were too low? Since the well is capable of producing oil or gas in paying quantities should the lease terminate if it is in the secondary term and contains savings provisions such as those found in *Pack*? In SMITH V. MARSHALL OIL CORP., 2004 OK, 10, 85 P.3d 830, 158 O.&G.R. 39, the court found that the lease terminated due to the lack of production in the secondary term.

f. THE FORCE MAJEURE CLAUSE

Given the dire consequences that flow from failing to pay delay rentals or meet the production in paying quantities requirement, it is not surprising that many lease, whether individually negotiated or on a pre-printed form, contain a force majeure clause. As with the other savings clauses that we have just studied, there is no uniform treatment of force majeure clauses. They range in length from a sentence or two to a page or more. Several variants of force majeure clauses can be found at Martin & Kramer, *Williams and Meyers Oil and Gas Law* § 683.1. The following clause is reasonably short but attempts to cover many of the situations that may occur during the drilling and production phases of an oil and gas development project:

When drilling or other operations are delayed or interrupted by a lack of

> water, labor or materials, or by fire, storm, flood, war, rebellion,
> insurrection, riot, strike, differences with workmen, or failure of acrriers
> to transport or furnish facilities for transportation, or as a result of some
> order, requisition or necessity of the government, or as a result of any
> cause whatsoever beyond the control of the Lessee, the time of such delay
> or interruption shall not be counted against Lessee, anything in this lease
> to the contrary notwithstanding.

This clause was the subject of litigation that led to the following opinion.

SUN OPERATING LIMITED PARTNERSHIP v. HOLT

Texas Court of Appeals, 1998
984 S.W.2d 277, 142 O.&G.R. 392

QUINN, JUSTICE. Elizabeth Holt, Robert P. Holt, Comfort Holt Winders,
Nick D. Holt, and Coy Miles Holt (collectively referred to as the Holts) sued
Sun Operating Limited Partnership, Oryx Energy Company, its General
Partner, Faulconer Energy Joint Venture--1988, Global Natural Resources
Corporation of Nevada, and Chevron U.S.A., Inc. (collectively referred to as
the Sun Parties) and others for a judgment declaring that two oil and gas leases
had terminated and for damages arising from the acquisition of product from
the lands after termination. The causes of action alleged sounded in equity and
tort, that is, they included a demand for 1) an accounting and restitution and 2)
damages for conversion and trespass. After conducting a bifurcated trial by
jury, the court entered judgment declaring that the leases had terminated and
awarding damages to Elizabeth and Robert Holt. Both the Sun Parties and Holts
appealed. Though the Sun Parties alleged seven points of error, and the Holts
nine, we need only address the first three uttered by the Sun Parties. They are
dispositive and involve 1) whether the cessation of production was temporary
which, in turn, afforded the Sun Parties a reasonable time to resume production
and 2) whether the cessation of production was excused via the force majeure
clauses contained in the leases. We reverse and remand.

Background

The dispute involved two oil and gas leases (referred to as leases number
one and two) affecting interests in property located in Hansford County. The
Sun Parties were either the original lessees or succeeded to the interest of the
original lessees under each agreement, while the Holts succeeded to the
interests of the original lessors under lease number one and Elizabeth and
Robert Holt succeeded to the interests of the original lessors under lease two.
Each document was executed in December of 1947 and contained several
provisions pertinent to this appeal. The first dealt with the lease term:

> Subject to other provisions herein contained, this lease shall remain in
> force for a term of ten years from this date, called primary term, and as
> long thereafter as oil, gas or other mineral is produced from said land, or

as long thereafter as Lessee shall conduct drilling or re-working operations thereon with no cessation of more that sixty consecutive days until production results, and if production results, so long as any such mineral is produced.

The second pertained to possible interruptions in drilling and operations caused by various specified acts. Referred to by us as the force majeure clause, it provided that: [The clause is excerpted at the beginning of this section, Eds.]

After execution of the leases, the lessee drilled wells and extracted gas from the land in paying quantities. This continued until April 11, 1983. At that time, the land encompassed by lease number one held five producing wells. That under lease two held five similar wells. Moreover, the gas obtained from each was being purchased by Panhandle Eastern Pipeline Company (Panhandle) and transported from the leases via Panhandle's transmission line 200. No other transmission line was connected to the wells.

As previously alluded to, the production of gas was interrupted on April 11th. Panhandle had begun major repairs and renovations upon line 200 which caused production from both leases to cease for more than 60 consecutive days. Production resumed, however, by September 23, 1983, the date on which the repairs were completed. All concede that the wells were *capable* of producing during the entire time.

From the foregoing interruption arose the present controversy. The Holts invoked that portion of the habendum clause directing that the lease remain in force as long as production does not stop for more than 60 consecutive days. Because it had so stopped, they argued that the leases automatically terminated. The Sun Parties disagreed with their opponents' interpretation of the events, and the results thereof, and raised various defenses. The two pertinent here concerned 1) whether they had a reasonable (as opposed to a specific) time within which to resume production, and 2) whether the cause of the interruption fell within the scope of the force majeure clause which, in turn, prevented the cessation from causing the leases to terminate.

Trial of the suit was bifurcated into separate proceedings involving liability and damages. During the former, the court submitted only one question to the jury which read:

Do you find that the failure to produce oil, gas and other minerals from the Lease Number 1 premises and the Lease Number 2 premises during the period of May 26, 1983, to August 1, 1983, was solely caused by "force majeure" as defined by paragraph 10 of Lease Number 1 and Lease Number 2?

Answer "Yes" or "No"

Answer: _____

In answering this question you should consider paragraph 10 of the leases in its entirety . . .

. . .

In connection with Question No. 1 you are instructed as follows:

(a) Before such an occurrence can constitute "force majeure," the operators of the wells located on the Lease Number 1 and Lease Number 2 premises must have exercised due diligence and taken all reasonable steps to avoid, remove and overcome the effect of "force majeure".

(b) For "force majeure" to be the sole cause of the failure to produce oil, gas and other minerals from the Lease Number 1 and Lease Number 2 premises during the period of May 26, 1983, to August 1, 1983, the alleged "force majeure" must have been the only cause of said failure and said failure cannot have been caused in whole or in part by the negligence of the operators of the wells located thereon.

(c) "Negligence" shall mean the failure to act as a reasonably prudent operator under the same or similar circumstances.

To the only question posed, the jury answered "no." Based upon this finding, the court declared that the leases were terminated. Damages were eventually awarded to the Holts once the remaining portion of the bifurcated trial was completed.

Unsatisfied, both parties appealed from the final judgment and asserted a myriad of error. However, we find several raised errors in the Sun Parties' brief dispositive of the appeal and address them. Finally, in addressing them, we do not necessarily do so in numerical sequence but rather in their logical sequence.

Point of Error One

[The court in this section reaffirmed the holding of Samano v. Sun Oil Co., 621 S.W.2d 580, 70 O.&G.R. 64 (Tex. 1981) that where the parties have expressly stated a period in which production must be reobtained after a cessation, that period controls, rather than the reasonable period that is otherwise a part of the temporary cessation of production doctrine. Eds.]

Point of Error Three

The third point raised by the Sun Parties involves the court's charge on force majeure. They contend that the instruction accompanying it was inaccurate since it imposed upon them the "burden to prove that they could not have overcome the interruption of production by the exercise of reasonable diligence." We agree and sustain the point.

The theory of force majeure has been existent for many years. Often likened to impossibility, it historically embodied the notion that parties could be relieved of performing their contractual duties when performance was prevented by causes beyond their control, such as an act of God. But, much of its historic underpinnings have fallen by the wayside. Force majeure, is now little more than a descriptive phrase without much inherent substance. Indeed, its scope and application, for the most part, is utterly dependent upon the terms of the contract in which it appears. This court recognized as much in Hydrocarbon Management, Inc. v. Tracker Exploration, Inc., 861 S.W.2d 427 (Tex.App.--Amarillo, no writ), when we said that the "lease terms are controlling regarding *force majeure,* and common law rules merely fill in gaps left by the lease." Id. at 436 (italics in original). In other words, when the parties have themselves defined the contours of force majeure in their agreement, those contours dictate the application, effect, and scope of force majeure. . . . More importantly, we are not at liberty to rewrite the contract or interpret it in a manner which the parties never intended.

Here, the trial court instructed the jury that before an occurrence may constitute force majeure, the well operators "must have exercised due diligence and taken all reasonable steps to avoid, remove and overcome the *effects* of 'force majeure.'" (emphasis added). Yet, the lease provision at issue stated that:

> When drilling or other operations are delayed or interrupted by lack of water, labor or materials, or by fire, storm, flood, war, rebellion, insurrection, riot, strike, differences with workmen, or failure of carriers to transport or furnish facilities for transportation, or as a result of some order, requisition or necessity of the government, or as the result of any cause whatsoever beyond the control of the Lessee, the time of such delay or interruption shall not be counted against Lessee, anything in this lease to the contrary notwithstanding.

In comparing the clause with the instruction given by the court, we find nothing in the former that expressly obligates the lessee to do that described in the latter. In other words, the clause says nothing about requiring the lessee to "exercise due diligence and take all reasonable steps to avoid, remove and overcome the *effects* of 'force majeure.'" (emphasis added). It says nothing about the lessee having to act reasonably to *remediate* the result caused by the force majeure event. Rather, the parties merely agreed that when certain specified acts occurred, any resulting delay or interruption in "drilling or other operations" was not to "be counted against Lessee." Given this, we choose not to rewrite the contract by interjecting such a duty.

. . .

As to the Holt's notion that public policy demands that we imply such an obligation into the clause, we say the following. Well operators should not be allowed to simply sit back and do nothing once production has ceased. Yet, that is no reason to rewrite the parties' contract when the law already exists to prevent a lessee from doing that. Indeed, the Texas Supreme Court imposed upon lessees and well operators various duties addressing the concern raised by the Holts. Those duties include the obligations to develop the premises, protect the leasehold, and manage and administer the lease. And, within the task of management and administration lies the requirement that the lessee reasonably market the mineral produced and secure the best price reasonably possible.

More importantly, if an operator chooses to do nothing once an event of force majeure occurred and terminated production, it is quite conceivable that his actions would run afoul of one or more of the aforementioned duties. At the very least, one could argue a lessee breaches his duty to manage and administer when the mineral remains in the ground despite the availability of reasonable measures to extract and sell it. The remedy for such a breach would be a suit for damages or, in extraordinary circumstances, termination of the lease.

Simply put, the policy argument voiced by the Holts has been adequately addressed via other bodies of oil and gas law. Thus, there is no need for us to provide further remedy by implying into every force majeure clause the requirement that the lessee exercise diligence to overcome the effects of force majeure once it occurs.

In sum, the Sun Parties correctly argue that the court erred in instructing that they had to use due diligence to avoid, remove, and overcome the effects of force majeure. Such was not intended by the parties, given the language in their agreement. Nor do we choose to contort their language to achieve an end that effectively works a forfeiture. *See* Kincaid v. Gulf Oil Corp., 675 S.W.2d 250, 256 (Tex.App.--San Antonio 1984, writ ref'd n.r.e.) (stating that the court must "bear in mind that if . . . language [of a lease] is reasonably susceptible of a construction . . . that prevents a forfeiture, such construction is to be preferred to one resulting in a forfeiture").

As to whether the court's error was harmful, we note several things. First, after perusing the record it appears that a major thrust of the Holts' suit involved the Sun Parties' alleged failure to use diligence in remediating the force majeure event and resulting absence of production. Second, the inaccurate instruction was submitted to the jurors as a means of assisting them in deciding the only question presented during the liability phase of the trial. Third, the Holts repeatedly emphasized, during their closing argument, the instruction and the Sun Parties' alleged failure to exercise diligence to overcome the effects of the force majeure event and regain production. Under these circumstances we cannot but conclude that the instruction probably caused the rendition of an improper judgment and that the mistake was harmful as per Texas Rule of Appellate Procedure 44.1(a)(1). We accordingly sustain point of error three.

Point of Error Two

In their second point, the Sun Parties assert that they were entitled to a directed verdict since the evidence conclusively established that the sole cause for the cessation in production was the failure of Panhandle to transport the gas. And, since Panhandle's failure to transport fell within the scope of the force majeure clause, the cessation allegedly could not result in termination of the lease.

In determining whether a directed verdict was appropriate, we must undertake a two step analysis. The first step requires us to construe the force majeure clause. Once that is done, we must then apply the evidence of record to determine if it conclusively established that the force majeure event was the sole cause of the cessation.

a. Construction of the Force Majeure Clause

1. Does the Phrase "other operations" Include Production?

According to the force majeure clause, any delay or interruption in "drilling or other operations" is "not [to] be counted against" the lessee if that delay or interruption is caused by a specified force majeure event. Thus, to successfully invoke the clause at bar, the Sun Parties must initially show that the concept of production falls within the scope of "drilling or other operations." Here, they contend that it comes within the parameters of "other operations." We agree for several reasons.

First, in construing the words of a contract, we must accord them their plain, ordinary, and generally accepted meaning unless the document provides otherwise.. Next, lay authorities commonly describe "operation" to mean 1) a process or series of acts performed to effect a certain purpose or result or 2) a process or method of productive activity. *American Heritage Dictionary Of The English Language* 920 (1976); *see Webster's New Collegiate Dictionary* 804 (1976) (defining "operation" as a method or manner of functioning). In other words, the ordinary definition of the word "operation" connotes an overall process aimed at achieving a particular end.

When that meaning is considered in the context of an oil and gas lease we cannot but conclude that the term encompasses the production of minerals. For example, such leases are executed for the purpose of developing the field, obtaining production, and paying the royalty owners. Circle Dot Ranch, Inc. v. Sidwell Oil & Gas, Inc., 891 S.W.2d 342, 346 (Tex.App.--Amarillo 1995, writ denied). That this is true is exemplified by the duties which have been implicitly assigned to the lessee. Again, they include the tasks of developing, protecting, managing, and administering the leasehold and marketing the minerals extracted. In other words, the lease itself creates a relationship between the lessor and lessee wherein the latter agrees to assume an operation composed of, among other things, exploration, development, production, marketing, and payment. And, this may be why courts in neighboring jurisdictions have deigned to include the exploration, development, production, and marketing of oil and gas within the definition of "operation."

Finally, an event mentioned in the force majeure clause as a basis for relieving the lessee from performing is the "failure of carriers to transport or furnish facilities for transportation." There can be little dispute with the proposition that an entity owning a pipeline which transports minerals from the well site is a carrier. *See* Tex. Nat. Res.Code Ann. § 111.002(1)-(2)(Vernon 1993) (stating that a common carrier includes one who owns, operates, or manages a pipeline for the transportation of crude petroleum for hire or engages in the business of transporting crude petroleum); Op. Tex. Att'y Gen. No. H-830 (1976) (stating that a pipeline transporting gas is a common carrier if it holds itself out as available to transport gas to all who may desire its services and a private carrier if it does not so hold itself out). Similarly indisputable is the proposition that at least one aspect of such a carrier's duty is to transport minerals from the well site. Before that particular duty can be performed, however, there must be production. So, before "other operations" can be delayed (for purposes of the force majeure clause) by the failure of a carrier to transport or provide transportation facilities, the phrase must of necessity and logic encompass production or the capability of the well to produce. If this were not so, then there would be little reason for the parties to have included that particular force majeure event in the clause.

Given the interrelationship between the carrier's duty to transport and the preexisting need for production, the meaning of the word "operate" when considered in the context of the oil patch, and the construction given the word by other jurisdictions, we conclude that "other operations" encompasses production under the lease before us. So, if production is delayed or interrupted because of one of the force majeure events, "the time of such delay or interruption [in production] shall not be counted against" the lessee.

2. Does the Force Majeure Clause Extend the Habendum Clause?

Both the Sun Parties and the Holts acknowledge that a force majeure clause could extend the lease term set forth in an habendum clause. However, the Holts believe that the former is not sufficiently worded to achieve that effect. Of course, the Sun Parties disagree. We too disagree with the Holts.

Normally, the duration of a lease is determined by its habendum clause. Gulf Oil Corp. v. Southland Royalty Co., 496 S.W.2d 547, 552 (Tex. 1973). However, the lease term may be affected by other provisions in the document, depending upon the intent of the parties. In assessing that intent, we look at the entire instrument. Here, the habendum clause reads:

> *Subject to other provisions herein contained,* this lease shall remain in force for a term of ten years from this date, called primary term, and as long thereafter as oil, gas or other mineral is *produced* from said land, or as long thereafter as Lessee shall conduct drilling or re-working operations thereon with no cessation of more that sixty consecutive days until *production* results, and if *production* results, so long as any such mineral is *produced.*

(emphasis added). From its terms we garner several things. First, the clause is expressly subject to other provisions in the instrument. That is, other provisions may modify the life of the lease. Second, the clause designates production (or lack thereof) as the condition generally determining when the lease terminates once the primary term has lapsed. If the lessee is producing minerals in paying quantities, the lease remains viable. Once production stops and the lessee fails to drill or rework for sixty consecutive days, the lease automatically terminates.

Yet, the habendum clause is not the only provision of the lease which addresses production and its continuation. Another exists, and it is the force majeure clause. That part of the instrument speaks to how delays and interruptions in various activities, including production, should be treated. Moreover, those delays, according to the parties, would "not be counted against Lessee." Simply put, to terminate the lease because production lapsed for reasons stated in the force majeure clause would be to "count" the delay "against the [l]essee," and that contradicts the intent of the parties as illustrated in the lease.

So too would it be tantamount to reading the provisions as requiring a forfeiture when the parties' intent to obtain such a result is hardly clear. Thus, we eschew such an interpretation of the provision, read the habendum clause in harmony with the entire lease including the force majeure clause, and conclude that the force majeure clause has the effect of extending the habendum clause of the leases before us.

Finally, Gulf Oil Corp. v. Southland Royalty Co., the case relied upon by the Holts to support a position contrary to the foregoing, is unauthoritative for several reasons. First, the court in *Gulf* recognized that the life of a lease could be affected by more than the habendum clause, depending upon the lease involved and the intent of the parties. Second, and unlike the instrument before us, the habendum clause in *Gulf* did not say that it was subject to other provisions in the lease. More importantly, the habendum clause in *Gulf*, unlike that here, contained a time definite. That is, the parties set a specific date on which the lease would end. It was not to continue as long as there was production but was to end 50 years after its execution, regardless of the presence of production. Given the clear intent of the parties that the lease would end on a particular date, the *Gulf* court felt constrained against interpreting the force majeure clause in a manner which contradicted the unequivocal intent expressed by the parties in their agreement.

3. Does the Force Majeure Clause Require that the Cause of the Cessation be Beyond the Reasonable Control of the Lessee?

The next question concerns whether the particular event causing production to cease must be outside the reasonable control of the lessee. The Holts say it does, while the Sun Parties say it does not. We conclude that the Holts are correct given the wording of the clause at issue.

Again, the matter is dependent upon the intent of the parties as garnered from the wording of the instrument involved. And, it has been held that the parties have evinced an intent that all force majeure events be outside the lessee's reasonable control where the lease names specific *force majeure* events and then follows that enumeration with a catch-all referring to acts beyond the lessee's reasonable control. For instance, in PPG Industries, Inc. v. Shell Oil

Co., 727 F.Supp. 285 (E.D. La. 1989), *affirmed*, 919 F.2d 17 (5th Cir. 1990), the court held that the cause there relied on did not have to be outside the parties' control. This was so because reference to the matter of control preceded the other *force majeure* events itemized in the clause. Id. at 187-88. Given this, the court held that the parties intended that the events listed after reference to "circumstances . . . reasonably beyond . . . control" did not have to be outside PPG's control. Id. at 287. However, great pains were taken by the court to distinguish those cases wherein the reference to control followed or ended the litany of other specified events. In those situations, "the reasonable control language and the enumerated events [were] plainly and grammatically tied together," according to the court. Id. at 288. Thus, the former served to modify the latter. And, all the itemized acts had to fall beyond the lessee's reasonable control.

This court took a similar tack in Hydrocarbon Management, Inc. v. Tracker Exploration, Inc. There, the lease specified that:

> Should lessee be prevented from complying with any express or implied covenant of this lease, from conducting drilling or reworking operations or from producing oil or gas . . . by reason of scarcity of, or inability to obtain or use transportation, equipment or material, or by reason of any Federal or state law or any order, rule or regulation of governmental authority asserting jurisdiction or otherwise by operation of *force majeure* (which term includes any other similar or dissimilar cause, occurrence, or circumstance not within the reasonable control of lessee), then while so prevented lessee's . . . need to conduct drilling or reworking operations or to produce oil or gas shall be suspended and this lease shall remain in force so long as lessee is so prevented.

861 S.W.2d at 435. Furthermore, Hydrocarbon attempted to invoke that portion of the clause involving orders issued by "governmental authority" to justify a cessation of production by one of its wells. That is, the Railroad Commission had directed Hydrocarbon to shut in the well, and because the Commission had ordered the shut-in, Hydrocarbon was allegedly entitled to invoke the *force majeure* clause to ameliorate the situation. We rejected Hydrocarbon's argument, however, since the circumstance causing the cessation was within its control.

Admittedly, the particular *force majeure* event invoked by Hydrocarbon did not expressly mention anything about the lessee's ability to control its occurrence. Nevertheless, reference to circumstances beyond the lessee's control was the last event mentioned in the force majeure clause. So, we implicitly interpreted it as modifying the preceding language. That is, in listing circumstances beyond the lessee's control last, the parties intended that the "cause[s], occurrence[s] or circumstance[s]" enumerated before it be outside the lessee's reasonable control before they could serve as *force majeure*.

Here, the parties concluded their litany of *force majeure* events by mentioning causes beyond the lessee's control. Given this, and the teachings of *PPG* and *Hydrocarbon* we hold that the juxtaposition evinced an intent that the qualification regarding control apply to each of the foregoing *force majeure* events. So, before any event can be successfully invoked as *force majeure* by the Sun Parties, it must be outside their reasonable control.

4. Does the Shut-In Royalty Clause Pretermit the Force Majeure Clause?

Like most every oil and gas lease, that at bar contains a shut-in royalty clause. It provided that "where gas from a well producing gas only is not sold or used, Lessee may pay as royalty Fifty Dollars ($50.00) per well per year, and upon such payment it will be considered that gas is being produced within the meaning of" the habendum clause. The Holts insist that the latter superseded the *force majeure* clause once production ended. That is, because shut-in royalties could have been paid by the Sun Parties in lieu of production, they could not rely upon the *force majeure* clause to maintain the lease. We disagree.

Simply put, a shut-in royalty clause does not *ipso facto* take precedence over every other clause which may affect the term of the lease. The court in Skelly Oil Co. v. Harris, 163 Tex. 92, 352 S.W.2d 950 (1962) held as much. In that case, the court recognized that alternative means existed by which the term of the lease could be maintained. One means encompassed the payment of shut-in royalties and the other the performance of drilling and reworking operations without cessation of more than 60 days. Because of the existence of these options, the lessee was not "under a duty to make such [royalty] payment[s]" as the "exclusive method of maintaining the lease in force." Id. at 953. Either provision could be utilized. The same applies when one of the options involves invocation of a force majeure clause like that before us. *See* Frost Nat'l Bank v. Matthews, 713 S.W.2d 365, 368 (Tex.App.--Texarkana 1986, writ ref'd n.r.e.) (stating that though Frost made the shut-in royalty payments, such were apparently unnecessary to avoid termination since the lease was susceptible to extension via the *force majeure* clause).

Again, *force majeure* clauses are now, for the most part, creatures of contract. Their meaning and scope are dependent upon the meaning and scope assigned by the parties via their agreement. Here, the parties ended the *force majeure* clause with the phrase "anything in this lease to the contrary notwithstanding." By including that passage in the clause, they evinced an intent to allow the lessee to rely upon the clause and its effect regardless of any other provision contained in the document. In other words, "[w]hen drilling or other operations are delayed or interrupted [by a specified *force majeure* event] the time of such delay or interruption shall not be counted against Lessee, [the shut-in royalty clause] notwithstanding."

So, to heed the position espoused by the Holts would be to implicitly read the shut-in royalty clause as superseding the *force majeure* clause despite the parties' intent to the contrary. Instead, we opt to give meaning to all the words in the force majeure clause and reject the argument that the Sun Parties were pretermitted from invoking force majeure merely because the lease contained a shut-in royalty clause.

b. Did the Sun Parties Prove as a Matter of Law That Production Ceased Solely Because of a Force Majeure Event?

[The court, however, could not find that as a matter of law that Sun lacked reasonable control over the occurrence and effect of the cessation of production. Eds.]

Accordingly, we reverse the judgment entered below and remand the cause for further proceedings.

————

NOTES

1. Would a cessation of production from an unknown cause or causes constitute a force majeure? Would it depend on whether the cessation was caused by a problem with the well or a problem with the pipeline? Does the principal case hold that the force majeure does not have to be beyond the control of the lessee? What if the parties had express language in the clause requiring the force majeure event to be beyond the control of the lessee. *See* Maralex Resources, Inc. v. Gilbreath, 2003-NMSC-023, 134 N.M. 308, 76 P.3d 626, 158 O.&G.R. 441. If the force majeure clause contains a list of events that are beyond the control of the lessee should a court infer such a requirement before the lessee's obligations are excused. *See* Moore v. Jet Investments, Ltd., 261 S.W.3d 412, 420 (Tex.App. 2008).

2. Are financial or monetary concerns likely to trigger the force majeure clause? Would it matter if the financial issues are self-generated? *See* Welsch v. Trivestco Energy Co., 221 P.3d 609 (Kan.App. 2009). Would the payment of money ever be excused under a force majeure clause?

3. What is the impact of governmental orders relating to activities by the lessee? Presuming that the express force majeure provision covers governmental orders or actions, should the force majeure clause apply where it is the lessee's actions would led to the order interfering or stopping production or other operations. *See* Moore v. Jet Investments, Ltd., 261 S.W.3d 412 (Tex.App. 2008); Atkinson Gas Co. v. Albrecht, 878 S.W.2d 236 (Tex.App. 1994).

4. What if the force majeure clause requires the lessee to notify the lessor in writing of a force majeure event? Would such a clause be enforceable by both the lessor and the lessee or solely by the lessee? See Goldstein v. Lindner, 2002 WI APP 122, 254 Wis.2d 673, 648 N.W.2d 892, 157 O.&G.R. 148.

————

SECTION 2. THE ROYALTY CLAUSE

(*See* Martin & Kramer, *Williams and Meyers Oil and Gas Law* §§ 641-660)

A. CONCEPTUAL CLASSIFICATION OF ROYALTY INTERESTS AND CONSEQUENCES THEREOF

The royalties to be paid by Lessee are: (a) on oil, one-eighth of that produced and saved from said land, the same to be delivered at the wells or to the credit of Lessor into the pipe line to which the wells may be connected; Lessee may from time to time purchase any royalty oil in its possession, paying the market price therefor prevailing for the field where produced on the date of purchase; (b) on gas, including casinghead gas or other gaseous substance, produced from said land and sold or used off the premises or in the manufacture of gasoline or other product therefrom, the market value at the well of one-eighth of the gas so sold or used, provided that on gas sold at the wells the royalty shall be one-eighth of the amount realized from such sale; where gas from a well producing gas only is not sold or used, Lessee may pay as royalty $100.00 per well per year and if such payment is made it will be considered that gas is being produced within the meaning of Paragraph 2 hereof [the primary term and thereafter clause]; . . .. Lessee shall have free use of oil, gas, coal, wood and water from said land, except water from Lessor's wells, for any operations hereunder, and the royalty on oil, gas and coal shall be computed after deducting any so used. Lessor shall have the privilege at his risk and expense of using gas from any gas well on said land for stoves and inside lights in the principal dwelling thereon out of any surplus gas not needed for operations hereunder.

. . .

It will be noted that the royalty clause quoted above provides that the royalty for oil shall be paid "in kind" whereas the royalty for gas is to be paid in money. Typically both oil and gas royalties are paid in the same manner; that is, after production is obtained, the lessor and lessee sign a "division order", which is a contract of sale to the purchaser of the gas or oil, and payment is thereafter made directly by the purchaser to the lessor and the lessee respectively. Despite this practice the typical lease form provides, as does the clause quoted above, for different treatment of the oil and of the gas royalty.

Early lease forms treated gas with disdain, frequently providing for a periodic flat sum payment which had no relationship to the amount or value of the gas produced. McGinnis v. Cayton, 173 W.Va. 102, 312 S.E.2d 765, 80 O.&G.R. 130 (1984), involved an 1893 lease which provided for a one-eighth royalty on oil, but for "$100 per year for so long as the well produced gas." Successors to the original lessors sued to reform or void the gas royalty provision. The trial court dismissed for "failure to state a claim upon which relief could be granted," but the Supreme Court held that plaintiffs were entitled to a trial on issues of "mutual mistake of fact and that the contract itself did not allocate the risk of changed conditions." A concurring opinion found a baiss for the lessors' case in broader areas of modern contract law relieving parties to contracts from apparent obligations. What if the royalty clause provides for a percentage of a fixed sum such as $ .04/MCF. Should the royalty owner get a percentage of the market value, the proceeds received by the lessee that are substantially in excess of $ .04/MCF? *See* Taylor v. Arkansas Louisiana Gas Co., 604 F.Supp. 779, 85 O.&G.R. 1 (W.D.Ark. 1985), *aff'd*, 793 F.2d 189, 90 O.&G.R. 201 (8th Cir. 1986).

SHEFFIELD v. HOGG, 124 Tex. 290, 77 S.W.2d 1021, 80 S.W.2d 741 (1934). The issue in this case was whether a lessor's retained interest in premises which had been leased for oil and gas development was subject to a county ad valorem real property tax. The court commented as follows:

> The fifth clause of the Hogg-Hamman contract states that as consideration therefor the lessors *"shall have"* a certain royalty, being one-eighth of the oil produced and one-eighth of the gas produced on the lands, the oil to be delivered in any pipe lines the lessors may designate, connected with the wells or into the lessor's private storage upon the leased land or any adjoining land; while for the gas the lessors *"shall have"* one-eighth interest in all money realized from gas marketed from said land, and for the sulphur or other minerals $1 for each ton produced and saved from the land, under quarterly cash settlements.
>
> Endeavoring to reach the true purpose and intent of parties, we can draw no substantial difference, so far as taxation is concerned, between an agreement *excepting* from a grant or a lease a certain fractional portion of minerals, or an agreement *reserving* the same portion, or an agreement that the lessor '*shall have*' or *rather shall continue to have* the same portion, or an agreement that the lessees *shall yield or shall deliver to the lessor* exactly the same portion. In either instance, the title to the specified mineral portion is intended to remain or vest, and does actually remain or vest in the lessor. It logically can make no difference, as may have been intimated in this justice's and in other far greater jurists' reasoning, whether the oil is retained by the lessor as oil and gas, readily convertible into cash on the market, or whether the lessee is given a power to sell *all* of the oil and gas, always accounting for a fixed royalty portion to the lessor. Sound principle, supported by the highest authority, goes further and compels us to accede to the proposition that dealing with oil and gas or dealing with solids in place, like sulphur, lignite, salt, coal, or lime, the lessor owning the entire fee-simple title to the land, and his assigns, who have been careful to secure to themselves, their heirs or assigns (by exception or reservation or by contract for "having" or yielding or paying, or for delivery, or by what-not similar contractual clause), the right to a portion of the proceeds or profits derived from the lessee's or his assigns' authorized sale of the minerals, throughout the duration of a determinable fee, which may be perpetual, have and own a fee-simple interest in land, or at least have a right belonging or

appertaining to the horizontal strata of the land in which the minerals are embedded. . . . We therefore hold that all the property interests of ascertainable value, secured to the lessors or their assigns under the Hogg-Hamman lease, are subject to taxation as real estate in the county wherein the land lies, as adjudged by the district court.

CALLAHAN V. MARTIN, 3 Cal.2d 110, 43 P.2d 788, 101 A.L.R. 871 (1935). In issue were the assignability of a royalty interest and whether recordation of the assignment charged a subsequent purchaser of the affected premises with constructive notice thereof. In rejecting the view that the transfer was merely a personal contract of the landowner or a contract for personal property, the court described a royalty as follows:

> The royalty return which the lessee renders to his lessor for this estate in the land is rent, or so closely analogous to rent as to partake of the incidents thereof. . . . In discussing the nature of landowner's royalties, the Supreme Court of the United States said in United States v. Noble, 237 U.S. 74, 80:

>> The rents and royalties were profit issuing out of the land. When they accrued, they became personal property; but rents and royalties to accrue were a part of the estate remaining in the lessor. As such they would pass to his heirs, and not to his personal representatives. . . .

>> It is said that the leases contemplated the payment of sums of money, equal to the agreed percentage of the market value of the minerals, and thus that the assignment was of these moneys; but the fact that rent is to be paid in money does not make it any the less a profit, issuing out of the land.

> That the right to receive future rents and oil royalties is an incorporeal hereditament, an interest in land, is recognized . . . .. Thus, the recognition of oil royalties as rents affords a basis for sustaining an assignment of oil royalty by a landowner as a transfer of an interest in real property.

GEOSTAR CORP. V. PARKWAY PETROLEUM, INC., 495 N.W.2d 61, 122 O.&G.R. 467 (N.D. 1993). GeoStar executed a contract with Parkway whereby GeoStar would find abandoned or marginally productive oil wells which would be offered to Parkway. Parkway would pay a cash finder's fee and assign an overriding royalty interest in the leases it was able to purchase. Parkway allegedly failed to convey certain interests and GeoStar sought specific performance of the promise to convey. If the royalty interests to be conveyed were only personal property, a statute created a presumption that damages would be an adequate remedy for the breach. N.D.Cent. Code § 32-04-09. An earlier North Dakota decision, Texaro Oil Co. v. Mosser, 299 N.W.2d 191, 69 O.&G.R. 81 (N.D. 1980), had concluded that a royalty interest was personal property. The court disavows that language, insofar as unaccrued royalty is concerned and treats it as real property, eliminating the statutory presumption that damages are adequate and substituting another statutory presumption that damages are inadequate. N.D.Cent. Code § 32-04-09.

ATLANTIC OIL CO. v. COUNTY OF LOS ANGELES, 69 Cal.2d 585, 72 Cal.Rptr. 886, 446 P.2d 1006, 31 O.&G.R. 440 (1968), which considered the question whether lessees from public entities were entitled to subtract the royalty from the value of the leasehold prior to assessment for ad valorem tax purposes, and concluded that they were not. The court said:

> Although it is classified under general concepts of property laws as an incorporeal hereditament and an interest in land, we conclude that the right to receive royalties is not classified as real property for purposes of taxation. Section 104 of the Revenue and Taxation Code defines real property to include "all . . . minerals . . . in the land, . . . and all rights and privileges appertaining thereto." . . . [I]n California the right to receive royalties is not a right appertaining to minerals "in the land." Rather, it is an interest in oil and gas when they are removed from the land and reduced to possession. The purpose of the qualifying phrase "in the land" is to differentiate such royalty interests in extracted minerals from interests in minerals still in the land. Only the latter, which include the right to drill for and extract oil and gas from the land, are real property for tax purposes.

Sheffield v. Hogg, *supra,* was cited to the California court but its persuasive authority was rejected on the ground that California, unlike Texas, does not adhere to "the doctrine of title to oil and gas in place." Is this a sound distinction?

Consider the use of ownership concepts in BEZZI v. HOCKER, 370 F.2d 533, 26 O.&G.R. 328 (10th Cir.1966), where the plaintiff had the right to one-half of a 1/8 royalty interest for a fixed term of twenty years. During this twenty-year period natural gas was removed from the land subject to the plaintiff's interest and reinjected and was not finally removed and sold until the plaintiff's interest had expired. The plaintiff's suit was to recover a royalty share of the proceeds of the gas so removed and sold. In affirming a decision for the defendant owner of the operating interest in the land, the court placed primary reliance on the principles of Oklahoma law which it stated as follows:

> It has been held . . . that the owner of land has a qualified title to the oil and gas in and under his land with the exclusive right to produce it, but has no absolute title thereto. . . . It is recognized that oil and gas are mobile and fugacious, and if it escapes to other lands or comes under another's control, whatever title the original owner had, is lost. . . . We think . . . Oklahoma has clearly indicated that whatever title Bezzi [the term royalty owner] may have had to the residue gas prior to January 11, 1961[the end of the twenty year term], it was lost when that gas was reinjected into the common source of supply and commingled with the virgin gas existing there, becoming subject to the law of capture.

GREENSHIELDS v. WARREN PETROLEUM CORP., 248 F.2d 61, 8 O.&G.R. 937 (10th Cir.1957), cert. denied, 355 U.S. 907 (1957). Royalty owners who had not signed division orders claimed that their gas had been converted by the purchaser under a gas sales contract executed by the operator. Consider the significance of the following comments by the court:

> After the execution of the gas purchase contracts, Warren Petroleum Corporation and Oklahoma Natural Gas Company, the purchasers under those contracts, sent instruments entitled 'Stipulation of Interest' to all of the royalty owners in the field. The documents amounted to certification of legal title and authorization for payment on the royalty interest in accordance with the terms of the gas purchase contracts. A substantial number of royalty owners executed the stipulation but others, including the plaintiff Greenshields, refused to sign.
>
> Greenshields now maintains that inasmuch as he has not signed the "Stipulation of Interest" form, there has been no transfer of title to the gas and that, therefore, the purchasers have converted the gas upon appropriation at the mouth of the well. It is true that oil and gas in natural state is not susceptible of ownership in place apart from the land in which it is found, but is personal property when reduced to possession. Whether or not title passes upon the occurrence of production must be determined from the language of the lease . . .. In the Producers 88 lease here under consideration, it is provided that the lessor shall receive a portion of the gross *proceeds* at market rate of all gas, contrasting with the provision for his receipt of one-eighth *part of all oil* produced. It is well settled that the provision concerning the payment for gas operates to divest the lessor of his right to obtain title in himself by reduction to possession and that thereafter his claim must be based upon the contract with the one to whom he has granted that right. His claim can only be for a payment in money and not for the product itself.

ATWOOD v. HUMBLE OIL & REFINING CO., 338 F.2d 502, 21 O.&G.R. 402 (5th Cir.1964), cert. denied, 381 U.S. 926 (1965), reh. denied, 381 U.S. 956 (1965). The lessors contended that they were entitled to royalty oil in kind so that the acts of the defendant lessee in processing royalty oil without their consent constituted conversion and required an accounting for net profits. The court construed the royalty clause providing for delivery to the lessors with other portions of a complex transaction to find that "the lease, when construed with these contemporaneous agreements, provides for payment by the lessee in money. The reference to delivery at the wells or to the credit of the lessor in the pipeline is apparently intended to fix the lease as the locality for determining the price at which the lessee is to make payment." Thus, the court determined there was no conversion. What if this construction had not been possible? Would a conclusion that conversion had occurred follow?

Consider what other legal consequences, if any, should turn upon the fact that the royalty clause of a lease provides for payment of royalty "in kind" or "in money." *See* Martin & Kramer, *Williams and Meyers Oil and Gas Law* §§ 659-659.1.

B. PROBLEMS OF ALLOCATION OF EXPENSES AND CALCULATION OF AMOUNT DUE

HENRY v. BALLARD & CORDELL CORP.

Louisiana Supreme Court, 1982
418 So.2d 1334, 74 O.&G.R. 280

BLANCHE, JUSTICE. Plaintiff landowners seek to recover outstanding royalty payments allegedly due under several gas leases executed between plaintiffs as lessors and defendants as lessees. Where gas is produced and then sold off the leased premises, the leases provide for royalty payments to the lessors equal to a percentage or fraction of the market value of the gas sold. Definition of the term "market value" in the context of these gas leases is at the heart of this dispute. Defendants have paid royalties based upon the price received from an interstate purchaser pursuant to a long term sales contract executed in 1961. In essence, defendants maintain that the 1961 contract price is equal to the market value of the gas under the royalty provisions of the gas leases. Plaintiffs assert that royalties are to be calculated on the basis of the *current* market value of the gas, a value greatly in excess of the 1961 contract price.

The issue presented by this litigation was set out by the court of appeal:

> Is the amount due the lessors as royalty under these leases to be based upon the prevailing market value at the time the gas was committed to the purchaser by the lessees under a long term gas sales contract, or is the royalty to be based instead upon the current market value determined on a daily basis the moment the gas is produced and/or delivered to the purchaser? Henry v. Ballard & Cordell, 401 So.2d 600 (2rd Cir. 1981), at p. 602.

As noted by the appellate court, this issue is res nova in Louisiana, although it has been the subject of considerable litigation in other jurisdictions. The magnitude of interests affected by its resolution in Louisiana mandated our decision to grant writs in these consolidated cases.

The leases at issue affect property in the Cameron Pass Field in Calcasieu Parish. The royalty provisions of the respective leases read as follows: [Two of the leases provided for "market value" of the gas sold or used, while two other leases provided for "market value" of gas sold or used off the premises and "amount realized" for gas sold at the wells. Eds.]

The leases were found to be productive of natural gas in 1961. Pursuant to its contractual obligation to diligently market the production, Ballard & Cordell executed a sale of the gas to American Louisiana (now Michigan Wisconsin) Pipeline Company, an interstate purchaser of natural gas and the *only* available market for gas from the Cameron Pass Field in 1961.[43]

---

[43]   The implied covenants which burden the lessee's interest include, among others, the

Evidence at trial conclusively established that negotiations between Ballard & Cordell and American Louisiana were conducted in good faith and at arm's length and that the resulting sales contract was quite favorable from the standpoint of both defendants and plaintiffs-lessors. The price obtained in the sale was equal to or better than prices in comparable sales made at that time. The price escalation clause, which provided for price increases over the term of the contract, was among the best contained in any such sale. The sales contract extended for a term of 20 years, a customary term for such contracts in the natural gas industry in 1961. Long term sales contracts (extending for as long as the life of the lease) were universally insisted upon by pipeline purchasers to enable them to obtain requisite financing for the construction of capital intensive pipeline facilities.

Beginning with the first deliveries of gas to American Louisiana, defendants have made royalty payments to the plaintiffs landowners based on the proceeds actually received for the sale of gas production from the leases under the 1961 sales contract. In 1976, the sales contract price first became out of line with the *current* market value of natural gas. In 1978, as the disparity between the 1961 sale contract price and prices paid by purchasers in more recent contracts continued to increase, plaintiffs filed this suit for outstanding royalties, contending the leases provide that royalty payments must be calculated on the basis of the current market value of the natural gas.

Plaintiffs' argument relies heavily on the 1934 case of Wall v. Public Gas Service Co., 178 La. 908, 152 So. 561 (1934). In *Wall,* this Court was required to interpret a mineral lease royalty clause which provided for royalties based upon "the value of such gas calculated at the market price". The controversy in *Wall* centered on whether the royalties were to be calculated on the basis of the market price in the field, or the market price where the gas was sold (a point two miles from the field). In dicta, however, the *Wall* court adopted Webster's definition of market price as "the price actually given *in current market dealings*." (Emphasis supplied in *Wall*). . . .

[T]he Third Circuit reversed the trial court judgment. Acknowledging that the market value of a thing is the price which it might be expected to bring if offered for sale in the market, the court of appeal did not agree with the conclusion of the trial court that the term "market value", standing alone, clearly means *current* market value. Instead, the court of appeal found that the parties to these lease contracts "intended for the royalties to be governed by the market value of the natural gas which prevailed in 1961 when the gas was committed to the purchaser under the gas sales contract rather than by the

---

covenant to protect the lease from being drained of hydrocarbons, the covenant to reasonably develop the leased premises, the covenant to market production obtained from the leased premises with dispatch, the covenant to use reasonable care in operations conducted on the lease, and when appropriate, the covenant to seek favorable administrative action to aid the discharge of the lessee's duties under the lease. See R.S. 31:122.

current market value, to be determined on a daily basis as gas is produced and or delivered to the purchaser."

For the reasons hereinafter assigned, we affirm the decision of the court of appeal.

The ambiguity in the language of the royalty provisions arises from the failure of the parties to the lease to expressly state whether "market value" means *current* market value. We note that the same or similar contract language has often been interpreted by the courts of other jurisdictions, and that these cases may be clearly divided according to two distinct lines of legal reasoning.

Although the majority of jurisdictions have interpreted the ambiguity in the royalty provisions against the lessee, they have done so by ignoring the practical realities of the oil and gas industry, and the obligations of the lessee to market the gas at the best possible price at the time the leases were made. These cases hold that the lessee is obligated to pay royalties based upon the current market value of the natural gas, and that this royalty obligation is unaffected by the contracts executed by the lessee for the sale of the gas.

The majority position has been most notably set forth by the Texas Supreme Court in Texas Oil & Gas Corp. v. Vela, 429 S.W.2d 866 (Tex. 1968). . . . Noting that none of the royalty owners had ever agreed to accept royalties on the basis of the price stipulated in the gas sale agreements, the Texas Supreme Court held that, regardless of the economic realities inherent in the marketing of natural gas, the royalties to which the owners were entitled "must be determined from the provisions of the oil and gas lease, which was executed prior to and is wholly independent of the gas sales contracts."

The practical and economic necessities of the oil and gas industry at the time the leases were negotiated are given little or no consideration in the line of cases represented by *Vela*. Because such factors necessarily would have bearing upon the intent of the parties concerning payment of royalties, the result reached in those cases has been greatly criticized. Among those factors demanding consideration are the following:

(1) Where the mineral lease provides for payment to the lessor of a fractional royalty interest, the lease arrangement is in the nature of a cooperative venture: the lessor contributes the land and the lessee the capital and expertise necessary to develop the minerals for the mutual benefit of both parties. "From this arises the affirmative, although implied obligation of the lessee to market or dispose of the product in a reasonable and prudent way to secure the maximum benefit possible for both parties."

(2) The ultimate objective of the royalty provisions of a lease is to fix the division between the lessor and lessee of the economic benefits anticipated from the development of the minerals. In the case of oil, a truly fungible product, the lessor is usually given a fractional part of the production, with

authority to take or dispose of it as he desires. "Most leases recognize such an arrangement is neither practicable nor realistic for gas. Consequently, they usually contemplate that the lessee will dispose of the gas (in a prudent manner) and pay the lessor the fractional part of the value which he is to enjoy from the enterprise." [44]

(3) "When most leases currently in production were executed, and certainly until quite recently, it was neither practical nor would it have been prudent for a lessee to refuse to sell the gas under a long-term contract. Such prices were generally higher than the so-called spot or short-term price."

[The court then discussed Tara Petroleum Co. v. Hughey, 630 P.2d 1269, 71 O.&G.R. 836 (Okla. 1981) which reached a similar result. Eds.]

Like the Oklahoma court in *Tara,* we believe that ambiguity in royalty provisions such as those at issue in this litigation cannot be resolved without consideration of the necessary realities of the oil and gas industry. Strong support is found for this minority position in the articles of our civil code applicable to the interpretation of contracts. See C.C. arts. 1945, *et seq.* and C.C. arts. 1964, 1966.

Article 1950 of the Civil Code directs us to endeavor to ascertain the common intention of the parties to the contract, where there is anything doubtful in their agreement. In ascertaining this intention (where it cannot be adequately discerned from the contract or agreement as a whole) the circumstances surrounding the parties at the time of contracting are a relevant subject of inquiry. In the instant cases, the known obligation of the lessee to market discovered gas reserves, and the accepted universal practice of marketing such reserves under long-term gas sales contracts provide the background against which these leases were executed.

The custom of the industry may also be considered in determining the true intent of the parties as to ambiguous contract provisions. C.C. arts. 1964, 1966; At trial, defendants presented unrefuted evidence that customary practice in the oil and gas industry required the lessee to pay "market value" royalties on gas in dollar amounts equivalent to the price received under a long-term sales contract (less permissible transportation charges), and the lessors to accept royalty payments so calculated.

---

[44]   [Harrell, *Developments in Non Regulatory Oil & Gas Law,* The 30th Annual Institute on Oil & Gas Law & Taxation, Southwestern Legal Foundation, 311 (1979)]at p. 335.   Indeed, Professor Harrell contends that:

. . . any determination of the market value of gas which admits the lessee's arrangements to market were prudently arrived at consistent with the lessee's obligation, but which at the same time permits either the lessor or lessee to receive a part of the gross revenues from the property greater than the fractional division contemplated by the lease, should be considered inherently contrary to the basic nature of the lease and be sustained only in the clearest of cases.   At p. 336.

Applying the pertinent rules of contract interpretation to the evidence presented in these cases, we find the parties to the mineral leases at issue intended that royalties based on the "market value" of the gas be computed on the basis of the price received for the gas under the 1961 sales contract.

We emphasize that plaintiffs-lessors have never contended in these cases that the 1961 gas sales contract was not made in good faith, or was unreasonable in any respect, whether as to price, contract term or otherwise. On the contrary, when the gas was discovered in 1961, because of the geographical isolation of the Calcasieu Pass Field, there was but one economically feasible market for the gas–the interstate pipeline of American Louisiana. American Louisiana would only agree to purchase the gas under a 20-year sales contract. Uncontradicted evidence at trial established that the price obtained by defendants in the 1961 sale was equal to or better than prices obtained in similar sales in the relevant market area at that time.

As noted by the court of appeal:

> The effect of holding in plaintiffs' favor would be to force the lessees to pay to the lessors many times more for their fractional royalty interests than the lessees actually received for all the gas. 401 So.2d at 608.

We do not propose to penalize defendants' good faith compliance with their lease obligations by requiring them to pay royalties based on a current, fluctuating, day-to-day market value of gas several times higher than the price received by them in a sales contract admittedly in the best interest of both lessors and lessees. Had plaintiffs shown that the purpose of the market value royalty clause was to provide them with protection as to price, regardless of what disposition is made of the gas by lessee and regardless of what price was received, then we would arrive at a different conclusion.

But plaintiffs have given us little help, except to claim that we have departed from our long-standing position in *Wall, supra.* As noted by the court of appeal, the *Wall* court was *not* concerned with whether the *current* market price, as opposed to a *past* market price was to be used in determining royalty amounts. Apparently, all parties in *Wall* agreed that the royalties were to be based upon the current market price at the time the gas was produced. None of the parties in *Wall* introduced evidence of a contrary intent.

In the instant case, only defendants have presented evidence of the intent of the parties to the mineral lease. All of this evidence indicates that the parties intended for a *past,* rather than a *current,* market value to control computation of royalties. As did the court of appeal, we find the *Wall* case clearly distinguishable on this basis.

Therefore, considering all of the circumstances set forth hereinabove which surrounded the parties at the time of contracting, the known obligation of the lessee to market discovered gas reserves, and the accepted, universal practice of marketing such reserves under long-term gas sales contracts, "market value" in the context of these leases could only mean the market value of the gas when it was marketed under the 20 year gas sales contract.

For the foregoing reasons, the decision of the court of appeal is affirmed.

AFFIRMED.

[The concurring decision of JUSTICE CALOGERO and the dissenting opinions of JUSTICES DENNIS, WASTSON and LEMMON are omitted.]

NOTES

1. Does the court find that the term "market value" is ambiguous? Does the court admit extrinsic evidence to show what the practical necessities of the oil and gas industry are in relation to a determination of what market value is?

2. Is the decision based in whole or in part on the notion that an oil and gas lease is a cooperative venture between the lessor and lessee? Does the lessor share in the potential losses if the well does not achieve payout? Should the lessee share its potential gains and/or losses from activities that affect how the gas is actually marketed?

3. The Montana Supreme Court has said:

Where there is no stipulation to the contrary in a lease of this kind, market price is understood to mean the *current* market price being paid for gas at the well where it is produced. . . Thus, market price is determined at the time the gas is produced. . . . The price to be paid is not to be an arbitrary price fixed by the lessee but the price actually given in current market dealings. Montana Power Co. v. Kravik, 179 Mont. 87, 586 P.2d 298, 306, 62 O.&G.R. 472 (1978).

Is there a different meaning ascribed to the terms market price or market value in different states? Consider your answer after reading the following case.

PINEY WOODS COUNTRY LIFE SCHOOL v. SHELL OIL CO.

United States Court of Appeals, Fifth Circuit, 1984.
726 F.2d 225, 79 O.&G.R. 244, reh'g denied, 750 F.2d 69 (1984), cert. denied, 471 U.S. 1005 (1985).

WISDOM, CIRCUIT JUDGE: This case concerns the interpretation of royalty clauses in certain Mississippi oil and gas leases. The plaintiffs are the owners of mineral rights in the Thomasville, Piney Woods, and Southwest Piney Woods fields in Rankin County, Mississippi. They leased their rights to defendant Shell Oil Company through various conveyances beginning in the mid-1960s. The cause of this controversy, as in many similar suits across the country, was the unforeseen and unprecedented rise in natural gas prices brought on principally by the actions of the Organization of Petroleum Exporting Countries (OPEC) in the early 1970s. Unfortunately for both the plaintiff lessors and lessee Shell, Shell had already committed the gas for sale under long-term contracts at pre-OPEC prices. Unsurprisingly, the lessors brought this class action to recover royalties that they allege Shell owes and has not paid. The district court found for Shell, except on one relatively minor issue, and certified this appeal so that the questions of liability could be decided before the determination of damages. We affirm in part, reverse in part, and remand.

I. Facts

The facts of this case are recounted in detail in the district court's opinion. Piney Woods Country Life School v. Shell Oil Co., 1982, S.D.Miss., 539 F.Supp. 957[74 O. & G.R. 485]. For our purposes it is enough to say that Shell began leasing activities in Rankin County in the 1960s. Shell used seven different lease forms, with three different royalty provisions.[45] The "Commercial" royalty provision provides for royalty:

> . . . on gas, including casinghead gas or other gaseous substance[s], produced from said land and sold or used, the market value at the well of one-eighth (1/8) of the gas so sold or used, provided that on gas sold at the well the royalty shall be one-eighth (1/8) of the amount realized from such sale[s]. . . .

The "Producers 88-D9803" provision calls for royalty:

> . . . on gas, including casinghead gas or other gaseous substance, produced from said land and sold or used off the premises or in the manufacture of gasoline or other product therefrom, the market value at the well of one-eighth of the gas sold or used, provided that on gas sold at the wells royalty shall be one-eighth of the amount realized from such sale. . . .

And the "Producers 88 (9/70)" provision orders the lessee"

> . . . to pay lessor on gas and casinghead gas produced from said land (1) sold by lessee, one-eighth of the amount realized by lessee, computed at the mouth of the well or (2) when used by lessee off said land or in the manufacture of gasoline or other products, the market value at the mouth of the well, of one-eighth of such gas and casinghead gas. . . .

The Commercial and Producers 88-D9803 leases provide for royalty based on "market value" except when the gas is "sold at the well[s]"; the Producers 88 (9/70) royalty is based on "amount realized" except for gas used by the lessee. Shell computed royalties in the same manner under all these provisions, however, and contends that they all have the same legal effect.

The gas from these fields is "sour" – it contains hydrogen sulfide. Before the

---

[45]   We adopt the names that the district court used to refer to the various royalty clauses. The "Commercial" provision is found in the "Paid-up Mississippi Rev. 7/17/45" and "Commercial--Form CC-78" lease forms; the "Producers 88-D9803" provision is found in the "Producers 88-D9803 (Revised 10/1/48) with Pooling Provision" and "(Mississippi) Form 0-280 Rev. 3 (8-61) 5M-Producers 88 Rev." lease forms; the "Producers 88 (9/70)" provision is found in the "Producers 88 (9/70)--Paid up with Pooling Provision Mississippi-Alabama-Florida", "Producers 88 (9/70) with Pooling Provision-Mississippi-Alabama-Florida", and "Producers 88 Revised-Alabama-Mississippi (11/56)" lease forms. Brackets in the text indicate variations between the lease forms.

gas can be put into the mainstream of commerce it must be processed. Rather than attempt to find someone to process the gas, Shell decided to do the processing itself. At its Thomasville plant, Shell treats the sour gas from the wells and recovers "sweet gas" – dry methane – and elemental sulfur.

Shell began efforts to market the gas from these fields in 1970. Shell sought buyers on the intrastate market because it wished to avoid restrictive federal regulations on interstate sales. *See* 15 U.S.C. §§ 717-717z (1982); 42 U.S.C. § 6399 (1976). After extensive negotiations with several potential buyers, Shell contracted with MisCoa, [a partnership of two Mississippi corporations, Mississippi Chemical Corporation and Coastal Chemical Corporation, eds.] of Yazoo City, Mississippi, to sell up to 46,667 thousand cubic feet (Mcf) a day to MisCoa for 53 cents per Mcf, with an increase to 54.59 cents after 15 million Mcf were delivered, and price escalation of three percent a year thereafter. On May 23, 1972, Shell contracted to sell excess gas to Mississippi Power and Light (MP & L) for 45 cents per Mcf, with escalation of one percent a year. Both contracts appear to have been the best available at the time. Both contracts provide that title to the gas passes in the field, when the gas is still sour. But in fact the buyer does not take control of the gas until it is processed and "redelivered". In the MisCoa contract, the measurements of quality and quantity that determine how much MisCoa pays are not made until the gas is "redelivered", as sweet gas, in Yazoo City. The MP & L contract provides for "redelivery" near the Thomasville plant. Both contracts state that the sale price includes "substantial consideration" for Shell's agreement to gather and process the gas and, in MisCoa's case, to assume the risk of loss during transportation to Yazoo City. Apparently, the parties agreed that title would pass at the wells so that the parties could avoid state regulations on pipelines. But the passage of title at the wells is also relevant to the royalty clauses in Shell's leases. Because the gas is supposedly sold "at the wells", Shell has paid royalties based on the actual revenues received from its sales of sweet gas and sulfur, rather than on market value. Shell deducts from these royalties a substantial portion of the costs of processing the gas.

The lessors filed this class action on December 27, 1974, alleging that Shell computed royalty payments improperly. The case was tried without a jury in November and December 1979. On May 3, 1982, the court issued its findings of fact and conclusions of law. The court found that Shell properly deducted the costs of processing from the royalty payments and properly based royalties for gas sold on the actual revenues realized since title to the gas passed from Shell to MisCoa at the wells. The court also rejected the plaintiffs' claim that Shell breached its duty to market the gas. The court did find that Shell should have paid royalties, based on current market value, for gas *used* in off-lease operations, but rejected the plaintiffs' evidence on market value and asked the plaintiffs to provide further evidence. Without explanation, the court also rejected the plaintiffs' claims for royalties on gas used by Shell at the Thomasville plant.

Upon the plaintiffs' motion the court issued a final judgment on the claims decided, and certified the case for appeal under Federal Rule of Civil Procedure 54(b).

II. Jurisdiction

Shell argues that we do not have jurisdiction over this appeal, on the ground that the claims not decided are inseparable from those certified. This is plainly incorrect. The district court's decision effectively disposed of all the issues except the amount of extra royalty owed by Shell. Litigation remains on only one claim, the royalties due on off-lease use of gas by Shell. One fact - the market value of gas - is relevant to several claims, but this does not make the claims inseparable. Alternatively, Shell argues that the district court abused its discretion in certifying the appeal. This contention is meritless. The only thing left for the plaintiffs to do in the district court is to present lengthy evidence on the market value of the small amount of gas Shell used in off-lease operations. The plaintiffs' potential recovery would not justify the expense of this proof. We think that the district court acted wisely and in the interests of "sound judicial administration" in certifying the case at this stage. . . ..

Shell also argues that the certification reserves some issues other than the amount of Shell's liability on the off-lease use claim – for example, whether "market value" means "current market value" in the context of sold gas as well as used gas. We understand this appeal to present all the issues except two factual questions: the market value of the gas and the amount of gas for which royalties are due.

III. The Meaning of the Royalty Clauses

The basic issues underlying this case are the meaning of "market value" and "sold at the wells" in a royalty clause and the propriety of deducting processing costs from the lessors' royalties. The royalty clauses prescribe different formulas for the calculation of royalties depending on whether the gas is sold or used, and on whether the sale or use is "at the well". Consideration of these distinctions guides our resolution of all the issues in this case.

In both the Commercial and Producers 88-D9803 royalty clauses, the royalty on gas "sold at the well" is based on the amount realized from sale, while on other gas the royalty is based on "market value at the well."[46] In interpreting similar provisions, courts have struggled with perceived grammatical ambiguities. In Exxon Corp. v. Middleton, 1981, Tex., 613 S.W.2d 240, for example, the court held that "off the premises" modified both "sold" and "used" in a lease form similar to the D9803; the court therefore held that "at the wells", which is used in apparent contrast to "off the premises", means "on the [leased] premises" rather than "in the field of production". Because the gas was in fact placed in the buyer's control at the time title passed, *Middleton* did not question that the gas

---

[46] By contrast, the Producers 88 (9/70) royalty clause treats all gas sold by lessee the same: royalty is based on "amount realized by lessee, computed at the mouth of the well"; only on gas used by the lessee is royalty based on market value.

was "sold" at that point. Here, however, the lessors argue that the place where title formally passes is not necessarily the place where gas is "sold" for the purposes of the royalty provisions. Resort to grammatical parsing is less instructive here than is a consideration of the purpose of the gas royalty clause, taken as a whole.

Commentators have suggested several reasons for differentiating between gas "sold at the well" and other gas in determining royalty. One author states that "gas is 'sold at the well' when the gas purchaser bears the expense of connecting his lines to the well". In such a case, the sale price is a proper measure of value. On the other hand, if the lessee pays for transportation, he should be compensated for it; the costs of transport are, therefore, deducted from the sale price to arrive at "value at the well". Note, Henry v. Ballard & Cordell Corp.: *Louisiana Chooses a Point in Time in the Market Value Gas Royalty Controversy,* 43 La.L.Rev. 1257, 1267 (1983). Similarly, Professor Owen L. Anderson suggests that "market value at the well" is determined by deducting transportation costs from the *value* (not the sale price) at the place of sale. Anderson, *David v. Goliath: Negotiating the "Lessor's 88" and Representing Lessors and Surface Owners in Oil and Gas Lease Plays,* 27 Rocky Mtn.Min.L.Inst. 1029, 1120 (1982).

These explanations are helpful, but not complete: in addition to transportation, other actions by the lessee away from the wellhead may affect the value of the gas when sold. In this case MisCoa paid, not for sour gas at Rankin County wells, but for sweet gas delivered to Yazoo City. Transportation and processing both increased the value of the gas to MisCoa. *See* Harmon, *Gas Royalty - Vela, Middleton, and Weatherford,* 33 Inst. on Oil & Gas L. & Tax'n 65 (1982). Harmon states,

> The most logical conclusion is that the oil and gas industry understood that the term "market value" in this gas royalty clause would be something less than the gross proceeds received for the sale of the gas, where the gas was sold and the delivery point was some distance removed from the field in which the wells were located. In such circumstances the royalty on gas would be based on market value – the amount for which the gas could have been sold in the field, less compression, gathering, and treating costs. When the gas was 'sold at the wells', whether the delivery point was on or off the lease, the 'amount realized' from the sale would be the basis of any royalty payment.

For reasons discussed in this opinion we disagree with Harmon's conclusion that "market value" may not exceed actual proceeds. But we do find his discussion instructive on the purpose of the distinction between gas sold at the well and gas sold off the lease. We conclude that the purpose is to distinguish between gas sold in the form in which it emerges from the well, and gas to which value is added by transportation away from the well or by processing after the gas is produced. The royalty compensates the lessor for the value of the gas at the well: that is, the value of the gas after the lessee fulfills its obligation under the lease to produce gas at the surface, but before the lessee adds to the value of this gas by processing or transporting it. When the gas is sold at the well, the parties to the lease accept a good-faith sale price as the measure of value at the well. But when the gas is sold for a price that reflects value added to the gas after production, the sale price will not necessarily reflect the market value of the gas at the well. Accordingly, the lease bases royalty for this gas not on actual proceeds but on market value.

"At the well" therefore describes not only location but quality as well. Market value at the well means market value before processing and transportation, and gas is sold at the well if the price paid is consideration for the gas as produced but not for processing and transportation.

IV. Point of Sale

Based on the foregoing discussion, we conclude that the gas sold by Shell was not "sold at the well", within the meaning of the lease, even though the sale contracts provide that title to the gas passes on or near the leased premises. Under the MisCoa contract, for example, title passes in the field but the sale price is not determined until after the gas is processed and transported to Yazoo City. The sour gas is metered in the fields, but this measurement appears to be of little significance. MisCoa effectively pays only for the amount of sweet gas delivered at Yazoo City, and pays a price commensurate with the value of sweet gas at the time the contract was made. The contract explicitly states that the sale price includes consideration for processing and for Shell's assuming the risk of loss during transportation to Yazoo City.

The District Court based its holding that the gas was sold in the fields on the provisions of the Uniform Commercial Code. The UCC applies to sales of natural gas, and therefore governs the sale contract between Shell and MisCoa. Miss.Code Ann. § 75-2-107(1) (1981). The parties to a sale contract may arrange for passage of title in any manner they choose. *Id.* § 75-2-401(1). Title may pass although the quantity of goods has not yet been determined, *id.* § 75-2-105(4), and the quality not yet brought up to contract specifications, *id.* § 75-2-501(1). There is no need to question that, under the sale contract, title passed in the fields. But the simple passage of title does not control whether the gas was "sold at the well" within the meaning of the leases. In the leases, "at the well" refers to both location and quality: gas is "sold at the well" only if its value has not been increased before sale by transportation or processing. This is plainly not the case here.

To interpret the leases otherwise would place the lessors at the mercy of the lessee. The lessors had no say in Shell's choice of where to put the passage of title. Their interests were either irrelevant or adverse to Shell's. Shell and its buyers wanted to avoid state pipeline regulations; but their decision to do so had the effect of placing the "point of sale" on the lease, thereby avoiding Shell's obligation to pay royalties based on market value. The opportunity for manipulation is apparent. Harmon, for example, counsels producers to "attempt to obtain appropriate contract amendments which would move the sales point onto the premises of each lease from which gas delivered under the contract is produced" to avoid payment of market value royalty. 33 Inst. on Oil & Gas L. & Tax'n at 95. We note as well the strange results that may occur if the determination of whether gas is "sold at the well" turns solely on the place where title passes. For example, if gas from several leases is delivered at a single point in the fields some lessors may be entitled to market value royalty while others receive proceeds royalty; similarly, gas produced from one lease through a directional well drilled on another lease would be sold "off the lease" even if delivered at the wellhead itself. *See* Holliman, *Exxon Corporation v. Middleton: Some Answers But Additional Confusion in the Volatile Area of Market Value Gas Royalty Litigation,* 13 St. Mary's L.J. 1, 47-49 & 48 nn. 185-86. If the place of delivery is controlling, it makes "[t]he happenstance of the point of delivery . . . very significant". Hoffman, *Pooling and Unitization: Current Status and Developments,* 33 Inst. on Oil & Gas L. & Tax'n 245, 265 (1982). We are convinced that Mississippi law does not allow the lessors' rights under an oil and gas lease to turn on such "happenstance", especially when the point of delivery is in the lessee's control.

Mississippi law looks beyond the formal passage of title when the interests of persons not party to the contract are at stake. State ex rel. Patterson v. Pure-Vac Dairy Products Corp., 1964, 251 Miss. 472, 170 So.2d 274, appeal dismissed, 1965, 382 U.S. 14, 86 S.Ct. 46, 15 L.Ed.2d 9. *Pure-Vac* held that a dairy could not evade state milk regulations by changing its sale contracts to provide that title passed at the dairy's loading docks in Tennessee, since the milk was delivered by the dairy to Mississippi customers who paid only for what they received in Mississippi. The court held that the sale occurred in Mississippi and could therefore be regulated. We agree with Shell that *Pure-Vac* does not require us to hold that title to the gas did not pass in the fields. In *Pure-Vac* the court rejected a sham devised specifically to evade state regulation. Here the district court found no bad faith. We do not hold that the gas sale contract was ineffective to pass title in the fields. But *Pure-Vac* does mean that agreement between a buyer and seller on a place for title to pass need not be conclusive for the purposes of laws extrinsic to the contract. Similarly, the reference in the lease to "sold at the well" need not be controlled by the point at which title passes in the sale contract. We therefore conclude that the gas was not "sold at the well" within the meaning of the Commercial and Producers 88-D9803 leases.

V. The Meaning of Market Value

Because the gas was not "sold at the well", the royalties under the Commercial and Producers 88-D9803 leases must be computed on the basis of "market value at the well" rather than actual proceeds. The district court reached the issue of what market value means in that it held that Shell owed market value royalty under all the leases for gas *used* in off-lease operations. Shell did not appeal this holding and, perhaps because of the small amount of money involved, did not appeal the district court's further holding that "market value" means value when the gas is delivered rather than when the sale contract is made. The plaintiffs contend that this holding is not before us now. Shell, however, contends that the district court's definition of market value applies only to "used" gas, not to "sold" gas. We see nothing in the district court opinion to suggest this result, and we reject Shell's argument that "market value at the well" has different meanings in different clauses of the same sentence. Shell's argument is enough, however, to put the meaning of market value before us, at least with respect to royalties on gas sold by Shell. The district court certified this appeal to dispose of all of the issues except the amount of damages owed by Shell. We affirm the district court's holding.

The meaning of "market value" has been at the fore of lease litigation ever since the price of gas began to increase at a rate much faster than the price escalation clauses of existing gas contracts contemplated. As long as the price of gas increased at a rate substantially equivalent to the inflation rates provided in sale contracts, it made little difference whether current market value or actual proceeds formed the basis of royalty payments: the check to the lessor was the same. But when the price of gas began to rise much faster than anticipated, this distinction became of the utmost moment to lessors and lessees.

Until the 1970s the authority was clear that "market value" referred to market value at the time of production and delivery rather than when the applicable sale contract was made. *See, e.g.,* Wall v. United Gas Public Service Co., 1934, 178 La. 908, 913, 152 So. 561, 563; Foster v. Atlantic Refining Co., 5 Cir.1964, 329 F.2d 485, 489-90 (Texas law); Texas Oil & Gas Corp. v. Vela, 1968, Tex., 429 S.W.2d 866.[47] This rule has been upheld more recently in, *e.g.,* Lightcap v. Mobil Oil Co., 1977, 221 Kan. 448, 562 P.2d 1, 11, cert. denied, 1977, 434 U.S. 876;[48] Montana Power Co. v. Kravik, 1978, 179 Mont. 87, 586

---

[47] In *Foster* the lease provided for a one-eighth royalty on oil and gas, "the same to be delivered to the credit of the Lessor into the pipeline and to be sold at the market price therefor prevailing for the field where produced when run." Does the language of this clause distinguish the case in a relevant fashion from *Vela* where a royalty phrased in terms of "the market price at the wells" was also enforced in terms of contemporary gas values rather than those which were operative when the gas contract was entered into? [Eds.]

[48] In *Lightcap* "market value" royalty payments were "determined by the traditional 'free market, willing buyer-willing seller' test, without regard to any governmental regulation of the producer's sales price." Matzen v. Cities Service Oil Co., 233 Kan. 846, 667 P.2d 337, 77 O.&G.R. 462 (1983), followed *Lightcap* in holding that market value was to be established at the time the gas was physically delivered but went on to say that "[s]trict adherence to *Lightcap* would require that the existence of federal regulation should be wholly disregarded when establishing market value for the purpose of computing royalties." "[M]arket value" was properly determined by "the highest federally regulated rate for any Kansas gas sold in interstate commerce from the Hugoton field without regard to 'vintaging' during the years covered by the dispute."

The Texas court has also held that while current market value was the proper basis for royalty payments "federal regulation has placed a ceiling on market value, which in this case,

P.2d 298, 302; Exxon Corp. v. Middleton, 1981, Tex., 613 S.W.2d 240, 244-45. This line of cases, to which we refer as the *Vela* rule, was uncontradicted until very recently. A recent line of cases, however, holds that "market value" is equivalent to the price assigned in the sale contract, at least as long as that contract was made prudently and in good faith. The first case so to hold was Tara Petroleum Corp. v. Hughey, 1981, Okla., 630 P.2d 1269. Tara has been followed in Hillard v. Stephens, 1982, 276 Ark. 545, 637 S.W.2d 581, 583,[49] and Henry v. Ballard & Cordell Corp., 1982, La., 418 So.2d 1334.[50]

The *Vela* rule is based principally on the doctrine that a gas sale contract is only executory until the gas is delivered, Martin v. Amis, 1926, Tex.Com.App., 288 S.W. 431, 433, and on the premise that the distinction made by the lease between market value and amount realized is meaningless under the *Tara* rule. *Tara*, on the other hand, found the *Vela* rule unfair because, when prices are rising, the *Vela* rule requires the lessee to pay the lessor an increasing percentage of the total revenues.

Shell argues that Simpson v. United Gas Pipe Line Co., 1944, 196 Miss. 356, 17 So.2d 200, repudiated the rule that a gas sale contract is an executory contract. According to Shell, *Simpson* held that all the gas on the leased premises is sold when the sale contract is made; market value therefore means market value at the time of the contract; and since Shell's sale contracts were made in good faith they are conclusive evidence of the market value of the gas at that time. This argument rests on a misreading of *Simpson*. *Simpson* held that the plaintiff lessor was entitled only to royalties based on the contract price, because the lessor had signed a division order agreeing to take royalties based on that

---

happens to coincide with the proceeds from the contract." First National Bank in Weatherford v. Exxon Corp., 622 S.W.2d 80, 71 O.&G.R. 96 (Tex.1981), discussed in Laity, 36 Sw.L.J. 185 at 193 (1982). A similar conclusion is reached in Shell Oil Co. v. Williams, Inc., 428 So.2d 798, 76 O.&G.R. 221 (La.1983), which distinguished Henry v. Ballard & Cordell Corp., 418 So.2d 1334, 74 O.&G.R. 280 (La.1982) on the ground "that the evidence [there] indicated the parties intended that the market value provision referred to a 'past' rather than a 'current' market value." *See* Veron, *In Search of Precedent in the Oil Patch: Louisiana's Market Value Cases*, 44 La.L.Rev. 949 (1984). [Eds.]

[49] *Compare* Diamond Shamrock Corp. v. Harris, 284 Ark. 270, 681 S.W.2d 317, 84 O.&G.R. 13 (1984), where the lessee relied on the price in a long term contract entered into before the lease was executed and of which the lessor had no notice; the court declined to adopt the contract price as governing royalty under a provision "that on gas sold at the wells the royalty shall be one-eighth of the amount realized from such sale." Current market was held the basis for royalty payments in part because any uncertainty should be resolved against the party that drafted the lease. [Eds.]

[50] In Scott Paper Co. v. Taslog, Inc., 5 Cir.1981, 638 F.2d 790, the appeals court affirmed the lower court's use of sales revenues (less processing costs) as a measure of "market price", but this was because the parties had agreed to this method in a division order. *Id.* at 797; *cf.* Simpson v. United Gas Pipe Line Co., 1944, 196 Miss. 356, 17 So.2d 200, discussed in the text of this opinion.

price. The court held that the lessor benefitted from the division order because the order facilitated a long-term gas contract under which the lessor "got an immediate sale for the gas". The court was concerned only with whether the division order was supported by consideration. The reference to "immediate sale" had nothing to do with passage of title but simply meant that the long-term contract made further marketing efforts unnecessary and that royalty would begin to be paid immediately.

We fully agree with the district court's conclusions that a gas sale contract is executory and that the sale is executed only upon production and delivery. Under section 75-2-105 of the Mississippi Code[51] the gas underground is future goods; no particular gas is sold until it is identified – *i.e.*, brought to the surface. The sale contracts are not transfers of an interest in land; accordingly, under section 75-2-107(1),[52] the contracts are contracts to sell and only become effective as sales when the gas is severed from the land. The logic of these provisions is clear. Under the leases, Shell has a *defeasible* interest in the gas underground. The most it could sell MisCoa or MP & L is that same defeasible interest: in effect, the right to possession of the gas if it is produced before Shell's lease terminates. The leases may terminate if Shell breaches implied or express covenants or if there is a cessation of production after the expiration of the primary term of the lease.[53] If the leases terminate, Shell would no longer have the power to deliver title to the gas. MisCoa might be able to sue Shell for a breach of *contract to sell,* but it would have no claim to the gas itself or to specific performance of the contract, because all title to the gas would have reverted to the lessors. Shell could not "sell" the gas to MisCoa, because a "sale" consists in the passing of title, *id.* § 75-

---

[51]   "(1) 'Goods' means all things (including specially manufactured goods) which are movable at the time of identification to the contract for sale other than the money in which the price is to be paid, investment securities (Chapter 8) and things in action. 'Goods' also includes the unborn young of animals and growing crops and other identified things attached to realty as described in the section on goods to be severed from realty (Section 2-107) [§ 75-2-107].

"(2) Goods must be both existing and identified before any interest in them can pass. Goods which are not both existing and identified are future goods. A purported present sale of future goods or of any interest therein operates as a contract to sell."

Miss.Code Ann. § 75-2-105 (1981).

[52]   "(1) A contract for the sale of minerals or the like (including oil and gas) or a structure or its materials to be removed from realty is a contract for the sale of goods within this chapter if they are to be severed by the seller but until severance a purported present sale thereof which is not effective as a transfer of an interest in land is effective only as a contract to sell."

Miss.Code Ann. § 75-2-107(1) (1981).

[53]   Forfeiture by the lessee has traditionally been the most common remedy for breaches of the requirements of the lease. *See* 4 H. Williams, *Oil and Gas Law* § 681 (1981). *But see* Waldman, *The Demise of Automatic Termination,* 54 Okla.B.J. 2767 (1983) (forfeiture disfavored in recent Oklahoma cases).

2-106.

We therefore conclude that the gas was not sold until it was produced. The sale contract itself provides that title passes when the gas is delivered. Accordingly, the basis of royalty should be "market value at the well" at the time of production and delivery, as the district court held.[54]

We also find the district court's decision supported on other grounds. First, the explicit language of the leases distinguishes between gas sold at the well and gas sold off the lease, and between amount realized and market value. The *Tara* rule eliminates the differences: the lessor receives a royalty based on amount realized no matter where and in what form the gas is sold. To be sure, *Tara* states that the contract price is the market price only if the contract was made prudently and in good faith, and therefore the sale price is market value only in those circumstances. 630 P.2d at 1274. But the same is true of royalty under an "amount realized" clause. If the lessee makes a sale in bad faith, he breaches both his duty to deal in good faith with the lessor and his duty to market, and the lessor will not be bound to the contract price. See Kretni Development Co. v. Consolidated Oil Corp., 10 Cir.1934, 74 F.2d 497, 500. Under *Tara,* therefore, any distinction between "market value" and "amount realized" is illusory.

We also reject the interpretation advanced by Harmon and Lowe that market value may never exceed actual proceeds. These commentators suggest that market value simply means actual proceeds less processing expenses. *See* Harmon, 33 Inst. on Oil & Gas L. & Tax'n at 69; Lowe, Developments in Nonregulatory Oil and Gas Law, 32 Inst. on Oil & Gas L. & Tax'n 117, 146-47 (1981). Certainly a lease *may* provide for royalties based on proceeds less processing. But this result may be accomplished clearly and explicitly by stating that royalty is "one eighth of the amount realized from the sale of the gas less processing and transportation expenses", or even more simply, "one eighth of the amount realized by lessee, computed at the well".[55] "Market value" as used in these leases is not equivalent to the formulations just recited, because it applies to royalties not only on gas sold at the well but also to gas *used.* There are no "proceeds" or "amount realized" for such gas. It strains language and credulity to argue that the "market value" of gas produced in 1977 and *not* sold to MisCoa is

---

[54]   Does it follow that because the interest of a lessee in an oil and gas lease is a defeasible interest in the oil and gas underground the most the lessee can sell, as the opinion puts it, "is that same defeasible interest: in effect, the right to possession of the gas if it is produced before [the] lease terminates?" Does the fact that the Uniform Commercial Code would classify the sale of gas as a contract to sell "future goods" if the gas is "to be severed by the seller" throw light on this question? Is this general line of inquiry any more relevant to the meaning of a gas royalty clause than is the specification in a gas contract between a lessee and a gas purchaser of where "title passes?" [Eds.]

[55]   The Producers 88 (9/70) royalty provision at issue in this case bases royalties on "one-eighth of the amount realized by lessee, computed at the mouth of the well". Shell and other lessees plainly knew how to draft a proceeds lease.

what the gas would have been sold for under a long-term contract made in 1972 rather than what the gas would be worth if sold on the open market when produced. Although it is conceivable that the single use of the term "market value" might require two different computational formulas when applied to two different types of gas, the more natural reading is that the term has the same meaning for both types of gas.

Moreover, Shell and those from whom it received leases by assignment were or should have been aware that "market value" had been held to mean value at the time of production both in old cases like *Wall* and in new cases like *Foster* and *Vela. Foster* was decided in 1964. *Vela* was decided by the Texas Court of Civil Appeals in 1966 and affirmed by the Texas Supreme Court in 1968. From the record it appears that no lease at issue here predates *Foster* and only a few predate *Vela.* Though not binding on Mississippi lessees, these decisions were widely discussed in the industry and should have alerted the lessees to the potential legal effect of the royalty clause. Shell certainly had ample opportunity to change the language if in fact it intended a "proceeds" lease. *Vela,* 429 S.W.2d at 871. It has long been doctrine that mineral leases are construed against the lessee and in favor of the lessor.[56] But our decision that market value means value rather than proceeds is not simply an instance of interpretation against the lessee.[57] It is rather a holding that, although the royalty clauses might have been less than lucid to laymen, they were quite readily understandable to those in the industry. Shell knew what a "market value" lease was and what a "proceeds" lease was. *See* Lightcap v. Mobil Oil Corp., 1977, 221 Kan. 448, 457, 562 P.2d 1, 8; Hillard v. Stephens, 1982, Ark., 637 S.W.2d 581, 587 (Hickman, J., dissenting). Shell "cannot expect the court to rewrite the lease to [its] satisfaction". *Foster,* 329 F.2d at 490.[58]

Shell asserts that "[r]oyalty payments based upon good faith contract prices have always been the custom in Mississippi". Brief for Appellee at 42. The

---

[56] In one oil-producing state (Oklahoma) this doctrine has not been applied in many years, but even in Oklahoma the doctrine has never been overruled. *See* Waldman, *The Demise of Automatic Termination,* 54 Okla.B.J. 2767 (1983). Elsewhere it is still clear that lessees will be held responsible for the leases they procure. This rule is justified even if some lessors have better legal advice and more of a role in negotiations today than they did at one time, because lessees still have greater expertise in and knowledge of the industry and the law surrounding it, still provide the lease forms, and still have bargaining advantages that limit the ability of lessor a to take an active role in the making of leases. *Id.* at 2773.

[57] If the price of gas declines, a market value royalty clause would benefit a lessee who has contracted to sell gas at a favorable price. *See, e.g.,* Note, 43 La.L.Rev. at 1266 n. 49.

[58] For the same reason we reject the argument, accepted in *Henry,* 418 So.2d at 1338, that the "ultimate objective of the royalty provisions of a lease is to fix the division between the lessor and lessee of the economic benefits anticipated from the development of the minerals." If the purpose were to fix royalty at a permanent percentage of lessees' profits, the lessee could certainly devise language to say so. We hold to the contrary that the market value clause serves in part to protect the lessor from bad bargains by the lessee.

district court did not make a finding to this effect and we doubt its universal truth. In a recent Louisiana case, Shell itself stipulated that market value royalty was determined by current market value. Shell Oil Co. v. Williams, Inc., 1983, La., 428 So.2d 798, 799. Of course, Shell is not bound by that stipulation in this case, but *Williams* does indicate that current market value is not unheard of as a basis for royalty in Shell's own leases, even in a state where the same assertion of "custom" was made, and where "market value" has now been held to mean contract price, *see Henry,* 418 So.2d at 1340-41. This allegation of "custom" is of course self-serving. The payment of royalties is controlled by lessees, and lessors have no ready means of ascertaining current market value other than to take lessees' word for it. The formula for determining royalty and the statements issued to lessors may be, as in this case, complex; it is likely that many lessors have never known just what basis lessees were using. For a practice to be legally relevant custom, both parties to the contract must have actual or presumed knowledge of the practice. Those not engaged in an industry will not be presumed to know that words which have common meanings outside the industry have a different meaning inside it. Market value in these leases is most easily understood to mean current value: the lessors cannot be presumed to know that Shell and other producers made a practice of basing their royalty payments on a different criterion. We will not find "custom" binding on lessors from a practice within the control and understanding only of the lessees.

The most important rationale underlying the *Tara* rule is the concern that it is unfair to require the lessee to pay increasing royalties out of a constant stream of revenues. The reasoning is as follows: the lessee has a duty to market the gas; gas is customarily sold in long-term contracts; the lessee is thus forced by his duty to the lessor to enter into a long-term contract but then sees his profits whittled away as the market price of gas rises. The *Tara* court reasoned further that its rule was not unfair to lessors, because they still receive royalties and are protected by the lessee's obligation to make a reasonable sale contract.

We appreciate that a lessee may find itself economically disadvantaged when its royalty obligations increase while its sale revenues remain constant. But this is no more than the risk assumed by every business venturer who undertakes the role of middleman. In this case it appears that Shell chose to market its gas to MisCoa at an initial price higher than any previously negotiated in the United States, but with a low annual escalation rate. Shell might instead have accepted a lower initial price but reserved the opportunity to redetermine the price at later intervals. The district court held that Shell's decision was not a breach of the lease-created covenant to market, since the choice of the MisCoa contract was a prudent and reasonable one. But prudence does not relieve Shell of its obligations under the lease. To say that the gas contract was what a "prudent operator" would have negotiated under the circumstances begs the question. The question is whether the operator should be saved from loss of profits simply because it acted prudently and the world economy acted adversely. If Congress or the Mississippi legislature wishes, it may establish a relief fund or create tax credits to assist the unfortunate lessees. It is not the function of the courts, construing and enforcing contracts under state law, to intervene on behalf of producers experienced in the petroleum industry, and thereby deprive lessors of their legitimate contractual expectations. "Stripped of all its trimmings [the "fairness" argument] is simply: We cannot comply. This is no answer . . .. The fact that the ascertainment of future market price may be troublesome or that the royalty provisions are improvident and result in a financial loss to [the lessee] 'is not a web of the Court's weaving.'" *Foster,* 329 F.2d at 490 (footnote omitted).

We conclude that the *Tara* rule is unfair to lessors. A landowner may decide to accept a royalty based on a smaller fractional share of market value, rather than to hold out for a larger share of proceeds, because of an expectation that the market value will rise. If the courts then intervene and declare that "market value" is the same as "proceeds", this expectation is destroyed. Moreover, the lessor has no means with which to persuade the lessee to renegotiate the lease to reflect the changed legal rule. This is illustrated by the following hypothetical. A landowner is offered leases by two producers. The first offers a 1/8 market value royalty; the second offers a 1/6 proceeds royalty. The landowner decides to lease to the first operator, because he thinks the market value of gas will rise enough to compensate for the lower fractional share. This is a business risk: if the price does not rise enough, the lessor loses money. If, however, the price rises as the lessor thought, the lessor has won his bet, just as the lessee has lost his gamble that the price would not rise; and the lessor ought to profit. But if the *Tara* rule intervenes, it takes away the lessor's legitimate expectation of market value royalties without any compensation at all. The lessor would be frozen into 1/8 proceeds royalties on a long-term low-price sales contract made by his lessee.

By contrast, under the *Vela* rule, which we adopt, the lessor's royalties increase according to the provisions of the lease. At some point the lessee may find continued operation so unprofitable that it is more economical to cease production. At this point the lessor has a strong incentive to renegotiate the lease, because a cessation of production will mean the end of all royalties until another lessee can be found, and the present lessee has a comparative advantage over others in that it has the necessary facilities already in place. The lessee's buyers will have an incentive to renegotiate the sale contracts rather than seek new and more expensive sources of gas.[59]

In short, the rule proposed by Shell would deprive the lessors both of their expected market value royalties and of any opportunity to recover some of these royalties through renegotiation of the lease. By enforcing the clear terms of the market value lease, we preserve those expectations and provide opportunities and incentives for the parties to make new contracts more nearly reflecting current

[59] In these renegotiations, for example, the lessor might agree to take 1/6 of actual proceeds instead of 1/8 of market value; or the buyer might consent to pay a price nearer to market price, or to compensate the lessee for royalties. *See* Harmon, 33 Inst. on Oil & Gas L. & Tax'n at 95; Anderson, 27 Rocky Mtn.Min.L.Inst. at 1112-13.

economic conditions.

Accordingly, we affirm the district court's holding that market value means current market value at the time of production.

VI. Proof of Market Value

Max Powell, the plaintiffs' expert witness on the issue of market value, based his estimates of market value on the average of the top three prices for gas sold in seven counties that he had previously selected as a relevant market area. This technique is the same one that Powell used, and that the court approved without contest, in Butler v. Exxon Corp., 1977, Tex.Civ.App., 559 S.W.2d 410. It is similar to the method used in Exxon v. Middleton, 1981, Tex., 613 S.W.2d 240. The district court, however, found Powell's testimony unpersuasive. Shell challenged Powell's data on the ground that sales of sweet gas are not comparable with sales of sour gas. The district court rejected this argument, noting that the processing costs, which reflect the price of converting the sour gas to sweet gas, are deducted from royalty. But the court found Powell's data noncomparable because of "[t]he unprecedented volume, deliverability and reserves of production" in the Thomasville-Piney Woods fields. The court devised its own formula:

> Once the relevant market area is defined and comparable sales are identified, the Court concludes that computation of market value should be made by the division of net sales receipts derived from sales of comparable gas (after necessary adjustments for variances in Btu content and compression changes) by the total volume sold.

The court made no findings on market value and directed the parties to consult the court for directions on further proof. *Id.*

Although the district court did not certify this issue for appeal, the plaintiffs now ask us to order the district court to use Powell's method for determining market value on remand. This we decline to do. Market value is a question of fact, and it is up to the factfinder to determine the probative strength of relevant evidence. The plaintiffs contend that Powell's evidence was the only evidence on the issue, but the Shell-MisCoa contract, while not *conclusive* evidence of market value, was also plainly *relevant* to the issue. The district court found that the sales used by Powell were made under conditions different enough to weaken their persuasiveness. The court was within its discretion to seek evidence of sales it considered more nearly "comparable".

A number of courts have struggled with the question of how to prove market value. *See, e.g.,* Exxon v. Middleton, 1981, Tex., 613 S.W.2d 240; Montana Power Co. v. Kravik, 1978, 179 Mont. 87, 586 P.2d 298; Weymouth v. Colorado Interstate Gas Co., 5 Cir.1966, 367 F.2d 84. The only general rule that emerges from these cases is that the method of proof varies with the facts of each particular case. In determining market value at the well, the point is to determine the price a reasonable buyer would have paid for the gas at the well when produced. Comparable sales of gas at other wells may be used to do this. Another method is to use sales of processed gas and deduct processing costs. Yet another relevant measure is the one proposed by Shell, the actual sale price of the gas less costs. "This is the least desirable method of determining market price", *Montana Power,* 586 P.2d at 303-04, but its persuasiveness is a matter for the factfinder.

Completely comparable sales are not likely to be found.[60] Sales that have some different characteristics must be considered. "[O]bjections against uncomparable sales, irrelevancies and unreliable hearsay [go] to the weight which the [factfinder] . . . should attach to the expert's opinion, and the protection to the other side as has always been true under our system [is] cross examination." *Weymouth,* 367 F.2d at 91. Especially when the court is factfinder, the court as well as counsel may question the expert opinions. But it should not dismiss fairly comparable sales out of hand because of certain incomparable qualities. The district court stated that there was a "lack of relevant evidence addressing comparability", 539 F.Supp. at 987. The evidence produced by plaintiffs may, in

---

[60]   Courts have identified a wide range of factors that may affect the price of natural gas. *See, e.g., Hugoton Prod. Co. v. United States,* 1963, 315 F.2d 868, 161 Ct.Cl. 274, listing the following relevant factors:

(a) The volume available for sale. Generally the greater the volume or reserves, the greater the price the seller could command.

(b) The location of the leases or acreage involved, whether in a solid block or scattered, and their proximity to prospective buyers' pipelines.

(c) Quality of the gas as to freedom from hydrogen sulphide in excess of 1 grain per 100 cubic feet.

(d) Delivery point.

(e) Heating value of the gas.

(f) Deliverability of the wells. The larger the volume that could be delivered from a reserve, the greater the price the seller could command.

(g) Delivery or rock pressure. The higher the pressure, the less compression for transportation is required.

Exxon v. Middleton also contains a thorough discussion of the factors affecting market value, including the following:

Market value may be calculated by using comparable sales. Comparable sales of gas are those comparable in time, quality, quantity, and availability of marketing outlets. *Vela, supra.*

Sales comparable in time occur under contracts executed contemporaneously with the sale of the gas in question. Sales comparable in quality are those of similar physical properties such as sweet, sour, or casinghead gas. Quality also involves the legal characteristics of the gas; that is, whether it is sold in a regulated or unregulated market, or in one particular category of a regulated market. Sales comparable in quantity are those of similar volumes to the gas in question. To be comparable, the sales must be made from an area with marketing outlets similar to the gas in question. Gas from fields with outlets to interstate markets only, for instance, would not be comparable to gas from a field with outlets only to the intrastate market.

the opinion of the court, have been insufficient, but it was plainly *relevant.* The plaintiffs produced evidence suggesting that, while some factors differentiate the Thomasville?Piney Woods wells from the wells upon which Powell based his data, those factors (abundance and availability) would *increase,* not decrease, the value of the plaintiffs' gas. It was, of course, for the court as factfinder to accord this testimony such weight as it saw fit, and to seek better information as well. We therefore reject the plaintiffs' request that on remand we mandate the use of the Powell formula as the exclusive measure of market value. But if, on remand, the search for better measures of market value at the well proves unsuccessful or inordinately burdensome, we think it the duty of the trier of fact to decide the question as best it can on the basis of the evidence that is presented. The plaintiffs must meet their burden of proof and produce sufficient relevant evidence to support their contentions. Having provided such evidence they should not be faced with the impossible task of establishing market value with absolute certainty or perfection. *See* 3 H. Williams, *Oil and Gas Law* § 650.3 (1981). It is Shell's responsibility to rebut the plaintiffs' evidence and provide evidence of its own to support its contentions. The decision is for the finder of fact, but the decision is made by balancing, not by erecting an insuperable barrier to relevant evidence.

VII. Processing Costs

The gas from these fields is extremely sour. It is an expensive proposition to convert the raw gas into marketable sweet gas and elemental sulfur and to transport the gas and sulfur to the buyers. Shell has passed on to the royalty owners a proportion of these costs, as determined by complex formulas that compensate Shell for expenses and capital investment. The plaintiffs argue that this is improper, since a royalty interest is "not chargeable with any of the costs of discovery and production", Mounger v. Pittman, 1959, 235 Miss. 85, 86-87, 108 So.2d 565, 566. They cite authority for the proposition that "production" includes the lessee's marketing efforts and that therefore any costs necessary to make gas marketable are to be borne exclusively by the lessee. *See* West v. Alpar Resources, Inc., 1980, N.D., 298 N.W.2d 484; Sterling v. Marathon Oil Co., 1978, 223 Kan. 686, 576 P.2d 635; Schupbach v. Continental Oil Co., 1964, 193 Kan. 401, 394 P.2d 1; Gilmore v. Superior Oil Co., 1964, 192 Kan. 388, 388 P.2d 602; California Co. v. Udall, D.C.Cir.1961, 296 F.2d 384; 3 E. Kuntz, *Law of Oil and Gas,* § 40.5 (1967); M. Merrill, *Covenants Implied in Oil and Gas Leases* 212-18 (2d ed. 1940). They argue that leases are construed against the lessee and that, in the absence of specific language authorizing deductions, no deductions may be permitted.

Shell responds, and the district court held, that production ends when the gas is extracted from the earth. Expenses incurred after production may be charged against royalty computed "at the well". 3 H. Williams, *Oil and Gas Law* § 645 (1981). Accordingly, the costs of processing and transportation may be deducted.

We agree with Shell that the specification in the leases that royalty is computed "at the well" controls. In Part III of this opinion, we discussed the purposes of distinctions between "amount realized" and "market value at the well". We concluded that "at the well" refers not only to the place of sale but also to the condition of the gas when sold. "At the well" means that the gas has not been increased in value by processing or transportation. It has this meaning in conjunction with "value" or "amount realized" as well as with "sold". The lessors under these leases are therefore entitled to royalty based on the value or price of unprocessed, untransported gas. Freeland v. Sun Oil Co., 5 Cir.1960, 277 F.2d 154, 157. On royalties "at the well", therefore, the lessors may be charged with processing costs, by which we mean all expenses, subsequent to production, relating to the processing, transportation, and marketing of gas and sulfur.

We emphasize, however, that processing costs are chargeable only because, under these leases, the royalties are based on value or price at the well. Processing costs may be deducted only from valuations or proceeds that reflect the value added by processing. Thus, processing costs may not be deducted from royalties for gas "sold at the well", because the price of such gas is based on its value before processing.

The function of processing costs in determining royalties based on "market value at the well" is to adjust for imperfect comparisons. As we discussed in Part VI of this opinion, the best means of determining the market value at the well of the plaintiffs' gas would be to examine comparable sales of *sour gas* at other wells in the area. Apparently there were no such sales. The next-best method is to examine sales of sweet gas and sulfur, to determine the market value of the products resulting from processing at the Thomasville plant. Processing costs may then be deducted as an indirect means of determining what a buyer would have paid for the sour gas at the wellhead.

Under the Producers 88 (9/70) royalty provision, "amount realized by lessee, computed at the mouth of the well", is the basis of royalty for *all* gas sold, not merely that sold at the well (as in the other leases at issue here). Under this clause, processing expenses are deducted from the amount realized from the sales of sweet gas and sulfur to arrive at the royalty basis.

The plaintiffs argue that Shell is entitled to deduct at most costs of sulfur production, but not to make deductions against residue gas. We see no basis for this distinction. The lessors are entitled to gas royalty at the well. This means royalty based on the value or price of the sour gas before it is separated into marketable constituents. The value or sale price of the residue sweet gas reflects Shell's processing costs just as surely as does the value or price of the sulfur.

We agree with the plaintiffs that the processing costs, under both the "market value" and "amount realized" provisions, must be reasonable. The plaintiffs charge that Shell's formulas allocate more of the costs to royalty interests than is justified. According to the plaintiffs, the district court failed to consider this issue. We are unable to determine whether it did. The district court's opinion makes no specific finding whether the formulas are reasonable, but the final judgment states that the court denied the claim for "improper and *excessive* charges made against . . . royalty interests". The record would support a finding of reasonability, but since this case is to be remanded we leave the issue open for further consideration by the district court.

VIII. Plant Fuel

Shell uses gas from the plaintiffs' leases for operations on other leases ("off-lease fuel") and at the Thomasville plant ("plant fuel"). The district court upheld the plaintiffs' claim for royalty on off-lease fuel; but without explanation it denied the claim for royalty on plant fuel. Shell does not dispute that plant fuel is "gas used off the lease" within the meaning of all the royalty clauses. Its response is rather than any royalties paid for use of plant fuel can simply be charged back to the lessors as processing costs. "The futility of such an exercise," says Shell, "is readily apparent. It would clearly be a wash."

It seems to us that the validity of Shell's argument depends on the accounting method used. The plaintiffs strenuously, although confusingly, insist that it will not be a "wash". They point out that, under the plant-lease split employed by Shell, only a portion of the costs of processing are borne by royalty owners. Furthermore, the working interests bear the costs on 7/8 of the gas produced. Another possibility is that the total additional royalty payments might equal the total additional charges to royalty, but that individual royalty owners might be entitled to more or less, depending on the particular well from which Shell takes the plant fuel in a given month. Finally, processing costs are not per se chargeable to market value royalty. They must be *reasonable* costs, and the market value of sour gas may be more or less in a given time period than the value of the finished products less processing expenses.

Even if Shell is right that it owes no extra royalties for plant fuel, we do not think that this would justify its practice. The lessors are entitled to have their royalty determined according to the provisions of the lease, not according to an ad hoc method adopted without their consent. The leases plainly call for market value royalty on gas used off the lease and the royalty should be determined on that basis. We therefore hold that plant fuel is gas used off the lease and the lessors are entitled to market value royalty on that gas. Shell may, however, treat the royalty payments as processing costs to be divided, as any other processing costs, among the various working and royalty interests.

IX. Attorney Fees and Prejudgment Interest

The plaintiffs requested attorney fees and costs on the theory that Shell exhibited bad faith. They also requested prejudgment interest, under Mississippi law, on the theory that Shell acted in bad faith and was guilty of conversion. The district court denied both requests. We affirm that holding. The plaintiffs made no showing that Shell acted in bad faith either before or during this litigation. Nor is there any basis for holding Shell guilty of conversion. "Conversion requires an intent to exercise dominion or control over property inconsistent with the true owner's rights." Shell neither intended to exercise nor exercised any control inconsistent with the lessors' rights. It had every right to take the gas; it simply failed to pay royalties according to the proper measure. This is breach of contract but it is not conversion.

Our affirmance on this issue is without prejudice to the right of the attorneys representing the plaintiff class to seek fees for their efforts on behalf of the class at the conclusion of this litigation.

X. Conclusion

We hold that "at the well" refers to gas in its natural state, before the gas has been processed or transported from the well. The gas sold by Shell under the MisCoa and MP & L contracts was therefore not "sold at the well" within the meaning of the leases. Accordingly, under some of the leases the plaintiffs are entitled to royalty based on market value for that gas. We hold that in the leases "market value" means the market value of the gas at the time the gas is produced and delivered. Because the leases call for royalty based on market value or price "at the well", royalty is based on the value or price of the gas before it is processed or transported. To determine the correct basis for royalty, processing and transportation costs may be deducted from values or prices established for processed and transported gas. We also hold that the plaintiffs are entitled to royalty on plant fuel, but Shell may include these royalties as processing costs. On remand, the district court may, in its discretion, seek more accurate measures of market value than the evidence of comparable sales already produced by the plaintiffs. If those efforts prove unsuccessful or unduly expensive, the court must determine market value based on the evidence already submitted. The evidence already produced by the plaintiffs on market value is clearly relevant, and Shell has the burden of rebutting that evidence and producing more accurate evidence of market value.

The judgment of the district court is AFFIRMED in part, REVERSED in part, and REMANDED for a determination of damages in accordance with this opinion.

NOTES

1. In *Piney Woods* the plaintiffs used an expert witness to prove market value. The testimony of the expert was challenged, in part, on the ground that the method he used in reaching his conclusion, based on the "average of the top three prices for gas sold in . . . 'a relevant market area,'" utilized data relating to sweet gas when sour gas was the substance actually being valued in the litigation. The defendant lost on this aspect of the challenge because the expert had deducted the processing costs of sour gas in making the computation. Nevertheless, the appellate court refused to mandate the use of the expert's data on remand. The abundance and availability of the gas to be valued bore on its comparability to the gas from the fields used in the expert's computations: "[i]t was for the [trial] court as factfinder to accord this testimony such weight as it saw fit, and to seek better information as well."

On remand for a reconsideration of damages, the trial court entered judgment for the defendant, finding that at no relevant time was the market value of the gas greater than the price paid under the gas sales contract which the lessee used as the basis for its royalty-payments. On appeal, this judgment was affirmed as to the earlier years involved and vacated and remanded for further consideration as to the later years involved. [905 F.2d 840 (5th Cir.1990)].

How should the burden of proof be allocated in the adjudication of these complex factual matters? Judge Wisdom's conclusion in the principal case (at p. 359) suggests that the trial court can seek better measures of market value than have already been furnished but goes on to direct the court, if a better basis is not found, "to determine market value based on evidence already submitted". The Wisdom opinion then places on [the defendant] "the burden of rebutting that evidence and producing more accurate evidence of market value."

Judge Smith, however, on the appeal from the judgment for the defendant on the remand interpreted this latter proposition to mean "that it was possible for the royalty owners to present relevant unchallenged evidence, yet still-not meet their burden of proof." The Smith opinion puts emphasis on the following statement in Judge Wisdom's opinion: "[t]he evidence produced by plaintiff's may, in the opinion of the court, have been insufficient, but it was plainly *relevant.*"

Judge Wisdom's opinion also says: "The plaintiffs must meet their burden of proof and produce sufficient relevant evidence to support their contentions. Having provided such evidence they should not be faced with the impossible task of establishing market value with absolute certainty or perfection. It is [defendant's] responsibility to rebut the plaintiffs' evidence and provide evidence of its own to support its contentions. The decision is for the finder of fact, but the decision is made by balancing, not by erecting an insuperable barrier to relevant evidence." *See* 3 Williams, Oil and Gas Law § 650.3 (1991), which is cited by Judge Wisdom for this proposition.

2. Can you reconcile the results in *Piney Woods* and *Henry*? Do they reflect different judicial approaches to the interpretation of a written instrument? As reflected in the *Piney Woods* decision, three states, Arkansas, Louisiana and Oklahoma have adopted the market value equals arms'-length contract price approach. Most of the other major producing states have followed what is often referred to do as the *Vela* rule that was adopted in *Piney Woods*. For a more complete analysis of those cases see Martin & Kramer, *Williams and Meyers Oil and Gas Law* §§ 650 - 650.4.

3. In these cases the royalty owners were asserting a claim for the difference between the contract price and the higher current market value. These differences began to arise because of the divergence between regulated interstate natural gas prices and unregulated intrastate natural gas prices. Typically, long-term contracts executed before the emergence of a large intrastate market were at low prices. Thus the royalty owners wanted their royalties based on these higher intrastate

market prices. But should the *Vela* rule apply where the lessee has locked in a high price in a long-term gas purchase contract and the current market price or value is substantially less? If the lease provides for payment of royalty based on market value, may the lessee calculate royalty based on that lower current market value rather than the higher proceeds received under the favorable contract? In Yzaguirre v. KCS Resources, Inc., 53 S.W.3d 368, 157 O.&G.R. 853 (Tex. 2001), the court reaffirmed its view in *Vela* that market value is to be determined at the time the gas was delivered and not at the time the gas was committed to a long-term contract in a situation where the contract price far exceeded the current market value.

4. Royalty clauses of leases typically provide, or are read to provide, that the royalty share of production (or of the proceeds or the value thereof) shall be delivered to the lessor or non operator "free of the costs of production." A question then arises concerning the costs which are to be borne by the operator alone out of its share of production and the costs which the royalty owner may be called upon to share proportionately.

The expenses incurred in exploring for mineral substances and in bringing such substances to the surface are clearly "costs of production" and are not chargeable against the usual royalty or nonoperating interest absent some express contractual agreement to the contrary. Included among such costs which the operator usually carries alone are:

(a) Costs of geophysical surveys.

(b) Drilling costs.

(c) Tangible and intangible costs incurred in testing, completing, or reworking a well, including the cost of installing the Christmas tree.

(d) Secondary recovery costs. *See* Kingwood Oil Co. v. Bell, 244 F.2d 115, 7 O.&G.R. 779 (7th Cir.1957).

5. Although the lessee bears the full costs of production, most leases provide something like the following: "*Lessee shall have free use of oil, gas, coal, wood and water from said land, except water from Lessor's wells, for any operations hereunder, and the royalty on oil, gas and coal shall be computed after deducting any so used.*" Does such a provision allow the lessee to use oil or gas for off-lease processing of lease production? In Bice v. Petro-Hunt, L.L.C., 768 N.W.2d 496, 2009 ND 124, the North Dakota Supreme Court concluded that residue gas used off of the leased premises to fuel the central tank batteries was in furtherance of the leased operations and thus the "free use" clause allowed the lessee off-lease use of the residue gas. Interpreting the "free use" clause to only allow the lessee to use residue gas on the leased premises "could lead to an absurd result because it would require lessors who have a central tank battery on their property to bear the entire burden of the 'free use' clause, notwithstanding benefits conferred on the other lessors." A contrary conclusion was reached in Roberts Ranch Co. v. Exxon Corp., 43 F. Supp. 2d 1252, 144 O.&G.R. 133, (W.D. Okla. 1997) where

the court found that a royalty clause requiring payment for gas used off the premises prevailed over an implied easement or an express free use clause.

6. Until recently, a royalty or other nonoperating interest in production has usually been subject to a proportionate share of the costs incurred subsequent to production. As *Piney Woods* puts it: "[t]he royalty compensates the lessor for the value of the gas at the well; that is, the value of the gas after the lessee fulfills its obligation under the lease to produce gas at the surface, but before the lessee adds to the value of this gas by processing or transporting it." *Supra* p. 347. Among these "subsequent to production" costs traditionally borne by nonoperating interests as well as by operating interests have been:

(a) Gross production and severance taxes. *See* Ashland Oil Co. v. Jaeger, 650 P.2d 265, 76 O.&G.R. 397 (Wyo.1982).

(b) Transportation charges or other expenses incurred in conveying the minerals produced from the well-head to the place where a buyer of the minerals takes possession thereof. *See* Johnson v. Jernigan, 475 P.2d 396, 37 O.&G.R. 240 (Okl.1970).

(c) Expenses of treatment required to make the mineral product saleable, *e.g.,* expenses of dehydration or of extracting hydrogen sulphide. *See* Holbein v. Austral Oil Co., 609 F.2d 206, 65 O.&G.R. 56 (5th Cir.1980) (dehydration – royalty clause provided for a share of "amount realized"); Compare West v. Alpar Resources, Inc., 298 N.W.2d 484, 68 O.&G.R. 499 (N.D.1980) (hydrogen sulphide-royalty clause provided for share of "proceeds").

(d) Expenses of compressing gas to make it deliverable into a purchaser's pipeline. *See* Martin v. Glass, 571 F.Supp. 1406, 78 O.&G.R. 111 (N.D.Tex.1983). *But see* Schupbach v. Continental Oil Co., 193 Kan. 401, 394 P.2d 1, 21 O.&G.R. 304 (1964).

What if the evidence shows that the compressors were installed "to increase production from the wells?" *See* Parker v. TXO Production Corp., 716 S.W.2d 644, 94 O.&G.R. 393 (Tex.App.1986); *Compare* Merritt v. Southwestern Elect. Power Co., 499 So.2d 210, 93 O.&G.R. 491 (La. App. 2d Cir.1986), where the parties stipulated that compression was "necessary to maintain a flow pressure in the gathering system." If a particular expense contributes to both pre- and post-production phases should the charges be apportioned? The opinion in *Merritt* states (at p. 213): "if compression charges are necessary in order for the well to produce, i.e. for the gas to reach the wellhead, then such charges are not deductible from royalty payments; [i]f, on the other hand, the compression charges are necessary only to push the gas from a producing well into the pipeline, then this cost is a post production or marketing cost and is therefore deductible from royalty payments."

(e) Manufacturing costs incurred in extracting liquids from gas or casinghead gas.

Caveat: Recent cases, discussed below, from several states, including

Colorado, Oklahoma, and West Virginia, have called the traditional analysis on allocation of post-production costs into question.

7. In FREELAND V. SUN OIL CO., 277 F.2d 154, 13 O.&G.R. 764 (5th Cir.1960), cert. denied, 364 U.S. 826 (1960), the question was whether a Louisiana mineral lessor's 1/8 gas royalty was to be based on 100% of the liquids extracted by an independent gasoline processing plant or on only that portion deliverable to the mineral interest owners after the processor has taken its share. The applicable portion of the royalty clause provided:

> The royalties to be paid by Lessee are: . . . (b) on gas, including casinghead gas or other gaseous substance, produced from said land and sold or used off the premises or in the manufacture of gasoline or other products therefrom, the market value at the well of one-eighth of the gas so sold or used, provided that on gas sold at the wells the royalty shall be one-eighth of the amount realized from such sale . . ..

The liquid content of the gas was not separated on the leased premises. On the contrary, it was carried with other full stream gas to the processing plant. At the processing plant, the gas first went through an inlet separator which removed the condensate from the full stream gas. As there were still heavier hydrocarbons remaining, the gas flowing out of the separator was sent to absorption towers where additional liquids were absorbed out. The residue gas was then sent through a dehydrator and was delivered to the pipe line company or returned to the recycling plant for reinjection into the reservoir. The liquids recovered from the inlet separator and absorption tower steps were then further processed through the fractionator to obtain propanes, butanes, motor fuel, and other products.

The processing plant was not owned or operated by either the lessor or lessees. It was owned by the Acadia Corporation, a corporation owned by an enterprise having extensive experience and technical competence in gas plants. The plant was an expensive and substantial industrial installation. It cost approximately $1,800,000. It was built by Acadia to handle the gas from this field after prolonged negotiations with the lessee-oil companies. Acadia, the processor, received under the processing contract as compensation for building and operating the plant 35.7% of the end products. The balance, 64.3% was returned to the lessees. The lessee-oil companies paid the 1/8 royalty on this 64.3%. What was at issue was the royalty on the 35.7% retained by Acadia, the processor.

The court concluded:

> In determining the market value of such gas at the well where there is no established criteria of a market, the Louisiana approach, which is binding on us, is to consider the end product of the extraction process as a factor. But it is a factor in reconstructing a market value at a place where in fact there was no, or little, market and consequently an appropriate deduction must be made.

. . .

To put it another way: in the analytical process of reconstructing a market value where none otherwise exists with sufficient definiteness, all increase in the ultimate sales value attributable to the expenses incurred in transporting and processing the commodity must be deducted. The royalty owner shares only in what is left over, whether stated in terms of cash or an end product. In this sense he bears his proportionate part of that cost, but not because the obligation (or expense) of production rests on him. Rather, it is because that is the way in which Louisiana law arrives at the value of the gas at the moment it seeks to escape from the wellhead.

Whether the costs of extraction (35.7%) payable to the processor are regarded as such, or perhaps may more properly be considered as an element in the retrospective reconstruction of a market value at the wellhead, the question of the reasonableness of the amount is the same. And as such, it is a question of fact on which the Trial Court's findings . . . are not shown to be clearly erroneous.

How would you draft a lease clause or contract producing a different result? Consider a provision that the lessors will be entitled to a percentage of: "all plant products, or revenue derived therefrom, attributable to gas produced by lessee from the leased premises." In Yturria v. Kerr-McGee Oil & Gas Onshore, LLC, 291 Fed.Appx. 626 (5th Cir. 2008) the court interpreted this language as requiring the lessee to calculate royalties based on the total revenue generated by the sale of the natural gas liquids without including the third-party costs of separating out the NGLs. While recognizing the general rule in Texas is that third-party processing costs involved in separating NGLs from the gas stream may be used in the net-back calculation, the court concluded that the unique language of the agreement unambiguously entitled the royalty owners to a share of 100% of the revenue generated from the NGL processing operations.

8. In MARATHON OIL CO. v. UNITED STATES, 807 F.2d 759, 90 O.&G.R. 6 (9th Cir.1986), cert. denied, 480 U.S. 940 (1987), a royalty of 12 1/2% "on the reasonable value of production [from federal lands] 'computed in accordance with the Oil and Gas Operating Regulations'" of the federal government was netted-back from the sales price of the liquified gas in Japan less "certain actual costs." Although there was evidence in the case of the market value of the gas in question at the wellhead (gas sold on a long-term contract from the same unit), the use of the net-back formula was held not unreasonable under federal law. *See also* California Co. v. Udall, 296 F.2d 384, 16 O.&G.R. 22 (D.C. Cir.1961), which required the lessee to compute royalties under a federal lease without deducting dehydration and compression costs to prepare gas for sale at a pipeline. For a discussion of royalty determinations in federal leasing situations and their relationship to state law, see Cassamassima, *Royalty Valuation Rulemaking of the Minerals Management Service Impact on OCS Operations*, 39 Inst. on Oil & Gas L. & Tax'n (1988)

9. In AMOCO PRODUCTION CO. v. FIRST BAPTIST CHURCH, 579 S.W.2d 280,

67 O.&G.R. 568 (Tex.Civ.App.1979), error ref'd n.r.e., 611 S.W.2d 610, 67 O.&G.R. 590 (1980), the royalty clause at issue provided "that on gas sold at the wells, the royalty shall be 1/8 of the amount realized from such sale." The court found the lessee liable under this "proceeds" lease on proof that some gas from the unit in which plaintiff lessor's lands were pooled was sold at a higher price than that which had been negotiated by defendant lessee for gas allocated to plaintiff's lands and that lessee showed a lack of good faith in obtaining collateral benefits from these transactions. In denying the application for writ of error the Supreme Court cautioned that "the holding should not be interpreted as implying an absolute duty to sell gas at market value under a 'proceeds' royalty provision . . .. Although in a proper factual setting, failure to sell at market value may be relevant evidence of a breach of the covenant to market in good faith, it is merely probative and is not conclusive."

The implied covenant to market obligates the lessee who is successful in finding economic deposits of hydrocarbons to market them. Suppose that a royalty clause provides for "one eighth of the proceeds from the sale of gas" and that the lessee discovers a deposit of sour gas; if the lessee constructs a plant on the lease premises to remove the hydrogen sulphide, should the royalty percentage be applied to the value of the sweet gas at the tail gate of the plant? In *West v. Alpar Resources, Inc.*, 298 N.W.2d 484, 68 O.&G.R. 499 (N.D.1980), the court construed the language of the royalty clause "most strongly against the lessee" and concluded that royalties should be paid as "a percentage of the total proceeds received by [the lessee] without deduction for the cost of extracting hydrogen sulphide and without deduction for any other cost incurred by lessee ." If the lessee had not itself invested in a gas plant but had sold the gas at the well to an entrepreneur who constructed the plant and sold the gas at the tailgate to a pipeline company, should the royalty obligations of the lessee be different? *See* *Martin v. Glass*, 571 F.Supp. 1406, 1415, 78 O.&G.R. 111 (N.D.Tex.1983).

The relationship of lessor and lessee from which the implied covenant to market springs is a relationship in which one party has transferred control of its mineral resources to another. The royalty clause purports to state the benefits that must be paid to the lessor in exchange for this transfer. A breach of the marketing covenant resulting in inadequate returns from sale of the substances produced is illustrated by *First Baptist Church*.

10. Prior to the 1990s, the *Piney Woods* approach to the deduction of post-production expenses in the calculation of royalty where the proceeds or valuation took place downstream of the wellhead was nearly universally accepted. The following case reflects a view that is at odds with that presented in *Piney Woods*. As you read it, and the recent cases it discusses, can you discern what factors led the court to interpret the same royalty language extant in *Piney Woods* markedly different than had most of the courts at the time that the *Piney Woods* decision was rendered?

———

TAWNEY v. COLUMBIA NATURAL RESOURCES, L.L.C.,

Supreme Court of Appeals of West Virginia, 2006
633 S.E.2d 22, 167 O.&G.R. 496

MAYNARD, JUSTICE. In this case, we address two certified questions from the Circuit Court of Roane County which we reformulate into the following single question:

In light of the fact that West Virginia recognizes that a lessee to an oil and gas lease must bear all costs incurred in marketing and transporting the product to the point of sale unless the oil and gas lease provides otherwise, is lease language that provides that the lessor's 1/8 royalty is to be calculated "at the well," "at the wellhead" or similar language, or that the royalty is "an amount equal to 1/8 of the price, net of all costs beyond the wellhead," or "less all taxes, assessments, and adjustments" sufficient to indicate that the lessee may deduct post-production expenses from the lessor's 1/8 royalty, presuming that such expenses are reasonable and actually incurred.

For the reasons that follow, we do not believe that the lease language set forth in the certified question permits CNR to deduct post-production expenses from the lessors' royalty payments.

I. Facts

Plaintiffs below are the owners of oil and gas ("lessors") which have been leased to Defendant Columbia Natural Resources or a predecessor in interest ("CNR"). At least since 1993, CNR has taken deductions from Plaintiffs' 1/8 royalty for "post-production" costs. These costs include CNR's delivery of gas from the well to the Columbia Gas Transmission ("TCO") point of delivery, CNR's processing of the gas to make it satisfactory for delivery into TCO's transportation line, and losses of volume of gas due to leaks in the gathering system or other volume loss from the well to the TCO line.

The post-production deductions taken by CNR include both monetary and volume deductions. CNR took deductions from royalty owners in equal amounts regardless of the distance from the well to TCO's transportation line. Even though CNR sent royalty checks to the lessors with an accounting of the purported amount of gas produced from the well, the purported price for which the gas was sold, and the purported amount of the royalty, CNR did not disclose on the accounting statements that deductions were taken.

Lessors have brought a class action suit against CNR for damages due to the allegedly insufficient royalty payments. There are approximately 8,000 Plaintiffs with 2,258 leases of varying forms and types. According to CNR, at least 1,382 leases at issue have language indicating that the royalty payment is to be calculated "at the well," "at the wellhead," "net all costs beyond the wellhead," or "less all taxes, assessments, and adjustments." CNR moved for summary judgment on the basis that the above lease language is clear and unambiguous and allows the lessee to deduct the royalty owners' proportionate share of post-production expenses, provided such expenses are actual and reasonable.

III. Discussion

It is the position of CNR that the "at the wellhead"-type language at issue in this case is clear and unambiguous and provides that the lessee may deduct the post-production costs of gas from the lessors' 1/8 royalty payments. Specifically, CNR explains that "at the wellhead" language indicates that the gas is to be valued for the purpose of calculating the lessors' royalty at the wellhead. However, the gas is not sold at the wellhead. In fact, the gas is not sold until the lessee adds value to it by preparing it for market, processing it, and transporting it to the point of sale. Thus, CNR concludes that the only logical way to calculate royalties at the wellhead is to permit lessees to deduct the lessors' proportionate share of post-production expenses, i.e., transportation and processing costs, from the total price received by the lessee.

The lessors, in contrast, assert that the "at the wellhead"-type language at issue is either silent or ambiguous on the subject of the allocation of post-production costs between the lessor and the lessee, and thus the language should be construed against the lessee. Further, because the lease language does not expressly address the allocation of post-production costs, the lessors posit that, pursuant to the lessee's implied covenant to market the gas recognized in Syllabus Point 4 of *Wellman v. Energy Resources, Inc.,* 210 W.Va. 200, 557 S.E.2d 254 (2001), the lessee must bear all costs incurred in marketing and transporting the gas to the point of sale. Thus, the lessors conclude that CNR was not permitted to deduct post-production costs from the lessors' 1/8 royalty but rather must bear all such costs itself.

Both the lessors and CNR cite for support cases from other states which indicate to us that courts are divided on the effect of "at the wellhead"-type language on the allocation of post-production costs between the lessor and the lessee. For example, in *Creson v. Amoco Production Co.,* 129 N.M. 529, 10 P.3d 853 (N.M.App.2000), the New Mexico court held that "at the well" language was sufficient to require the allocation of post-production expenses between lessor and lessee. The issue in *Creson* concerned specific language in a "Unit Agreement" which stated that royalties shall be based on the "net proceeds ... at the well." The agreement also contained a provision titled *"Royalty Owners Free of Cost"* (emphasis in the original) providing that "[t]his Agreement is not intended to impose, and shall not be construed to impose, upon any Royalty Owner any obligation to pay Unit Expense unless such Royalty Owner is otherwise so obligated." The lessors argued that post-production expenses were "unit expenses" under the Unit Agreement; thus, the lessees were not permitted to deduct those expenses from the sales price before calculating the royalties owed to the lessors. While the court recognized that some states do not permit post-production costs to be charged to the royalty owners, *citing Garman v. Conoco, Inc.,* 886 P.2d 652 (Colo.1994), it rejected this approach. Instead, the court determined that "the phrase 'net proceeds ... at the well' is unambiguous and means that Plaintiffs are entitled to royalties based on the value of the ... gas as it emerges at the wellhead." The court thus concluded that post-production, value-enhancing costs were properly included in calculating the royalty owed to the lessors.

The Colorado Supreme Court took the opposite approach in *Rogers v. Westerman Farm Co.,* 29 P.3d 887 (Colo.2001). The issue in that case dealt with the sufficiency of the lease language "at the well" or "at the mouth of the well" to determine the proper allocation of costs between the parties of the post-production expenses of gathering, compressing, and dehydrating the gas prior to its entry into the interstate pipeline. The *Rogers* court did not hinge its decision on a finding that "at the well" language was ambiguous. Instead, the court found such language to be completely silent with respect to allocation of costs. In other words, said the court, the language "does not indicate whether the calculation of market value at the well includes or excludes costs, and does not describe how those costs should be allocated, if at all, between the parties." Because it deemed the lease language silent, the court found that the lessees' implied covenant to market the gas governs the allocation of costs. The court explained that under the implied covenant to market the gas, the lessee alone must bear the costs to make the gas marketable when the gas is not marketable at the physical location of the well. The Colorado court recognized that it may be in the minority of states on this issue, citing contrary authority from Oklahoma, Texas, and Michigan, but noted that these courts generally do not recognize the lessor's implied covenant to market the gas.

This Court finds it unnecessary to adopt wholesale the reasoning of either of the courts above in answering the question before us. Instead, we simply look to our own settled law. We begin our analysis with the recognition that traditionally in this State the landowner has received a royalty based on the sale price of the gas received by the lessee. In Robert Donley, *The Law of Coal, Oil and Gas in West Virginia and Virginia* § 104 (1951), it is stated:

> From the very beginning of the oil and gas industry it has been the practice to compensate the landowner by selling the oil by running it to a common carrier and paying to him [the landowner] one-eighth of the sale price received. This practice has, in recent years, been extended to situations where gas is found[.]

"The one-eighth received is commonly referred to as the landowner's royalty." *Wellman v. Energy Resources, Inc.,* 210 W.Va. 200, 209, 557 S.E.2d 254, 263 (2001). In *Wellman,* we expressly recognized the general duty of a lessee to market the oil or gas produced. We explained:

> In *Davis v. Hardman,* 148 W.Va. 82, 133 S.E.2d 77 (1963), this Court stated that a distinguishing characteristic of [the landowner's royalty] is that it is not chargeable with any of the costs of discovery and production. The Court believes that such a view has been widely adopted in the United

States. In spite of this, there has been an attempt on the part of oil and gas producers in recent years to charge the landowner with a *pro rata* share of various expenses connected with the operation of an oil and gas lease such as the expense of transporting oil and gas to a point of sale, and the expense of treating or altering the oil and gas so as to put it in a marketable condition. To escape the rule that the lessee must pay the costs of discovery and production, these expenses have been referred to as "post-production expenses." ...

The rationale for holding that a lessee may not charge a lessor for "post-production" expenses appears to be most often predicated on the idea that the lessee not only has a right under an oil and gas lease to produce oil or gas, but he also has a duty, either express, or under an implied covenant, to market the oil or gas produced. The rationale proceeds to hold the duty to market embraces the responsibility to get the oil or gas in marketable condition and actually transport it to market.

This Court held in Syllabus Points 4 and 5 of *Wellman,*

4. If an oil and gas lease provides for a royalty based on proceeds received by the lessee, unless the lease provides otherwise, the lessee must bear all costs incurred in exploring for, producing, marketing, and transporting the product to the point of sale.

5. If an oil and gas lease provides that the lessor shall bear some part of the costs incurred between the wellhead and the point of sale, the lessee shall be entitled to credit for those costs to the extent that they were actually incurred and they were reasonable. Before being entitled to such credit, however, the lessee must prove, by evidence of the type normally developed in legal proceedings requiring an accounting, that he, the lessee, actually incurred such costs and that they were reasonable.

Accordingly, the present dispute boils down to whether the "at the wellhead"-type language at issue is sufficient to alter our generally recognized rule that the lessee must bear all costs of marketing and transporting the product to the point of sale. We conclude that it is not. . . .

We believe that the "wellhead"-type language at issue is ambiguous. First, the language lacks definiteness. In other words, it is imprecise. While the language arguably indicates that the royalty is to be calculated at the well or the gas is to be valued at the well, the language does not indicate *how* or *by what method* the royalty is to be calculated or the gas is to be valued. For example, notably absent are any specific provisions pertaining to the marketing, transportation, or processing of the gas. In addition, in light of our traditional rule that lessors are to receive a royalty of the sale price of gas, the general language at issue simply is inadequate to indicate an intent by the parties to agree to a contrary rule--that the lessors are not to receive 1/8 of the sale price but rather 1/8 of the sale price less a proportionate share of deductions for transporting and processing the gas. Also of significance is the fact that although some of the leases below were executed several decades ago, apparently CNR did not begin deducting post-production costs from the lessors' royalty payments until about 1993. Under these circumstances, we are unable to conclude that the lease language at issue was originally intended by the parties, at the time of execution, to allocate post-production costs between the lessor and the lessee.

CNR asserts, however, that when read with accompanying language such as "gross proceeds," "market price," and "net of all costs," the wellhead-type language clearly calls for allocation of post-production expenses. We disagree. First, we note that the word "gross" implies, contrary to CNR's interpretation, that there will be no deductions taken. Hence, the phrase "gross proceeds at the wellhead" could be construed to mean the gross price for the gas received by the lessee. On the other hand, the words "gross proceeds" when coupled with the phrase "at the wellhead" could be read to create an inherent conflict due to the fact that the lessees generally do not receive proceeds for the gas at the wellhead. Such an internal conflict results in an ambiguity. Likewise, the phrase "market price at the wellhead" is unclear since it contemplates the actual sale of gas at the physical location of the wellhead, although the gas generally is not sold at the wellhead. In addition, we believe that the phrase "net of all costs beyond the wellhead" could be interpreted to mean free of all costs or clear of all costs beyond the wellhead which is directly contrary to the interpretation urged by CNR. Finally, CNR also claims that the phrase "less all taxes, assessments, and adjustments" clearly indicates that post-production expenses can be deducted from the lessors' royalties. Again, we disagree. Absent additional language that clarifies what the parties intended by the words "assessments" and "adjustments," we believe these words to be ambiguous on the issue of the allocation of post-production expenses.

CNR also cites for support this Court's statement in *Wellman* that,

> the language of the leases in the present case indicating that the "proceeds" shall be from the "sale of gas as such at the mouth of the well where gas ... is found" might be language indicating that the parties intended that the Wellmans, as lessors, would bear part of the costs of transporting the gas from the wellhead to the point of sale[.]

According to CNR, this language was included in the opinion for the purpose of giving meaning to our holding in Syllabus Point 5 of *Wellman* where we stated that the allocation of post-production expenses will be permitted where expressly provided for in a lease. We find CNR's reliance on the above language to be misplaced. This Court has held that "when new points of law are announced ... those points will be articulated through syllabus points as required by our state constitution." . . . The comments relied upon by CNR are dicta insofar as they are not necessary to our decision in *Wellman.* The fact is that we simply did not decide in *Wellman* whether "at the wellhead"-type language is or is not ambiguous. Therefore, we find no merit to CNR's reliance on our language in *Wellman.*

CNR further cites for support *Cotiga Development Company v. United Fuel Gas Company,* 147 W.Va. 484, 128 S.E.2d 626 (1962), wherein this Court distinguished the wellhead or field price of gas from the price received by the lessee when the gas is marketed. However, while we did distinguish between the wellhead price and the actual selling price of gas, we did not define wellhead price, determine how it is calculated, or decide the specific question currently before us. Therefore, we find our discussion in *Cotiga* unhelpful in deciding the present issue. Accordingly, in light of the above, we conclude that the "at the wellhead" type language at issue is ambiguous because it is susceptible to more than one construction and reasonable people can differ as to its meaning.

Having found the language at issue ambiguous, the lessors urge that the language should be construed against CNR consistent with "[t]he general rule as to oil and gas leases ... that such contracts will generally be liberally construed in favor of the lessor, and strictly as against the lessee." CNR posits, to the contrary, that the lease language at issue should not be construed against it. According to CNR, many of the lessors are business entities which are as sophisticated in commercial matters as CNR. Further, says CNR, many of the lessors consulted with attorneys experienced in oil and gas law and even amended the leases prior to signing them.

We choose to adhere to our traditional rule and construe the language against the lessee. Significantly, CNR drafted the "the wellhead"-type language in dispute. Under our law, "[u]ncertainties in an intricate and involved contract should be resolved against the party who prepared it." Simply put, if the drafter of the leases below originally intended the lessors to bear a portion of the transportation and processing costs of oil and gas, he or she could have written into the leases specific language which clearly informed the lessors exactly how their royalties were to be calculated and what deductions were to be taken from the royalty amounts for post-production expenses.

It is also CNR's position that having found the disputed lease language herein ambiguous, the rules of interpretation require that the intent of the parties now be determined by the finder of fact. This is incorrect. Under our law, " '[i]t is the province of the court, and not of the jury, to interpret a written contract.' "

Accordingly, this Court now holds that language in an oil and gas lease that is intended to allocate between the lessor and lessee the costs of marketing the product and transporting it to the point of sale must expressly provide that the lessor shall bear some part of the costs incurred between the wellhead and the point of sale, identify with particularity the specific deductions the lessee intends to take from the lessor's royalty (usually 1/8), and indicate the method of calculating the amount to be deducted from the royalty for such post-production costs. We further hold that language in an oil and gas lease that provides that the lessor's 1/8 royalty (as in this case) is to be calculated "at the well," "at the wellhead," or similar language, or that the royalty is "an amount equal to 1/8 of the price, net all costs beyond the wellhead," or "less all taxes, assessments, and adjustments" is ambiguous and, accordingly, is not effective to permit the lessee to deduct from the lessor's 1/8 royalty any portion of the costs incurred between the wellhead and the point of sale.

IV. Conclusion

For the reasons set forth above, we answer the reformulated certified question as follows: . . .

Answer: No.

NOTES

1. Why was "at the well" apparently unambiguous for the court in *Piney Woods* but ambiguous for the *Tawney* court?

2. The *Tawney* decision is the most recent in a series of cases that impose the burden of many post-production costs upon the lessee, even when "at the well" or similar language is present in the lease. As noted by the West Virginia court, Colorado took a similar approach in Rogers v. Westerman Farm Co., 29 P.3d 887, 149 O.&G.R. 373 (Colo. 2001), a decision that was preceded by Garman v. Conoco, Inc., 886 P.2d 652, 132 O.&G.R. 488 (Colo. 1994) (relied on by the West Virginia court in *Wellman v. Energy Resources, Inc.*) and since followed in *Savage v. Williams Production RMT Co.*, 140 P.3d 67 (Colo.App.2005). The Colorado decisions develop a "marketable product rule" to the effect that where a lease is silent as to allocation of costs, the implied covenant to market obligates the lessee to incur costs necessary to render the gas marketable. The *Westerman Farm* court attempted to define "marketability" for the purposes of determining the fact question of when natural gas becomes marketable and thus becomes cost-bearing. The court stated that "marketability includes both a reference to the physical condition of the gas, as well as the ability for the gas to be sold in a commercial marketplace." It then provided the following guidance indicating that gas must be both in a marketable condition and at a place to be marketed: "Gas is marketable when it is in the physical condition such that it is acceptable to be bought and sold in a commercial marketplace, and in the location of a commercial marketplace, such that it is commercially saleable in the oil and gas marketplace." The duty of the lessee is both to condition the gas for market and to move it to a commerical market. Costs incurred beyond this point may be shared by lessor and lessee unless they provide otherwise. The duty to prepare natural gas for a market and to move it to a commercial market is independent of the good or bad faith of the lessee.

3. The Oklahoma cases on treatment of post-production costs between lessee and lessor include Wood v. TXO Prod. Corp., 1992 OK 100, 854 P.2d 880, 125 O.&G.R., TXO Prod. Corp. v. State ex rel. Comm'rs of the Land Office, 1994 OK 131, 903 P.2d 259, 132 O.&G.R. 189 (holding that "the costs

for compression, dehydration and gathering are not chargeable to [lessor] because such processes are necessary to make the product marketable under the implied covenant to market"), Mittelstaedt v. Santa Fe Minerals, Inc., 1998 OK 7, 954 P.2d 1203, 140 O.&G.R. 551, and XAE Corp. v. SMR Property Management Co., 1998 OK 51, 968 P.2d 1201, 141 O.&G.R. 557 (*infra* p. 750). These decisions are based on the Oklahoma court's determination that the lessee's implied duty to market involves obtaining a marketable product.

On the Kansas approach *see* Sternberger v. Marathon Oil Co., 257 Kan. 315, 894 P.2d 788, 132 O.&G.R. 65 (Kan. 1995) (the court stated that the "lessee has the duty to produce a marketable product, and the lessee alone bears the expense in making the product marketable"; the court appears to distinguish between "gathering" and "transportation" on the basis of the point at which "marketable product" is obtained). *See also* Davis v. Key Gas Corp., 124 P.3d 96 (Kan. Ct. App. 2006), review denied (lessee had an implied obligation to protect lessor against transportation costs and other expenses that would reduce lessor's royalty).

4. Notwithstanding the approach by the Colorado, Kansas and Oklahoma courts, the *Piney Woods* approach appears to be the one most widely accepted. In Hurinenko v. Chevron U.S.A., Inc., 69 F.3d 283, 134 O.&G.R. 249 (8th Cir. 1995), the court applying North Dakota law, found that a royalty clause calling for "market value at the well" was unambiguous and allowed the lessee to deduct post-wellhead costs in determining the amount of royalty due. The court distinguished West v. Alpar Resources, Inc., 298 N.W.2d 484, 68 O.&G.R. 499 (N.D. 1980) because the royalty clause in that case, "the proceeds from the sale of gas" was found to be ambiguous. Casinghead gas has no discernible market value at the well because it is only sold after processing. Therefore the "work-back" or "net-back" method of calculating royalties was "clearly appropriate" in this case. Any doubt about Northa Dakota law was resolved In Bice v. Petro-Hunt, L.L.C., 768 N.W.2d 496, 2009 ND 124, when the North Dakota Supreme Court found that the term "market value at the well" is not ambiguous and declared: "We join the majority of states adopting the 'at the well' rule and rejecting the first marketable product doctrine." Thus the lessee could deduct post-production costs from the plant tailgate proceeds prior to calculating royalty. *See also* Koch Oil Co. v. Hanson, 536 N.W.2d 702, 137 O.&G.R. 539 (N.D. 1995); Amerada Hess Corp. v. Conrad, 410 N.W.2d 124, 96 O.&G.R. 191 (N.D. 1987).

5. Texas has confirmed its approach to royalty valuation issues which focuses on the language of the specific clause. See Judice v. Mewbourne Oil Co., 939 S.W.2d 133, 133 O.&G.R. 513 (Tex. 1996) and Heritage Resources, Inc. v. Nationsbank, 939 S.W.2d 118, 134 O.&G.R. 547 (Tex. 1996). The *Heritage Resources* court said:

Market value at the well has a commonly accepted meaning in the oil and gas industry. . . Market value is the price a willing seller obtains from a willing buyer. . . There are two methods to determine market value at the well.

The most desirable method is to use comparable sales. A comparable sale is one that is comparable in time, quality, quantity, and availability of marketing outlets.

Courts use the second method when information about comparable sales is not readily available. This method involves subtracting reasonable post-production marketing costs from the market value at the point of the sale.

Would the Texas Supreme Court's rationale also apply to a royalty clause which provided for royalty based on "net proceeds at the well" where the sale took place downstream from the wellhead?

Heritage and *Judice* were followed in Tana Oil & Gas Corp. v. Cernosek, 188 S.W.3d 354 (Tex.App.—Austin 2006), review denied, where the terms "at the well" and "mouth of the well" defined the extent of the royalty obligation. The lessee sold the natural gas stream at wellhead to third party, receiving 84% of the amount realized from the sale of the processed liquids and 84% of the amount realized from the sale of the residue gas. Held: the lessors are not entitled to have their royalty calculated based on 100% of the produced natural gas volumes. The *Cernosek* court emphasized that the language found in the royalty clause governs the relationship between the lessor and lessee.

6. The New Mexico court in Creson v. Amoco Production Co., 129 N.M. 529, 10 P.3d 853, 145 O.&G.R. 324 (N.M. Ct. App. 2000), cert. denied, 129 N.M. 519, 10 P.3d 843 (2001), held that when royalty payment was governed by a unit agreement providing that royalty was to be calculated on the "net proceeds derived from the sale of carbon dioxide gas at the well," this provision allowed the working interest owners to use the netback methodology where the actual sale occurred downstream of the wellhead. ("In the absence of an express agreement to the contrary, such post-production costs generally include transportation costs, expenses of treatment such as dehydration, expenses of compressing gas so that it can be delivered into a pipeline, and other 'costs incurred in adding value to the well-head product.'" 10 P.3d at 857).

Creson was followed in Elliott Industries Limited Partnership v. BP America Co., 407 F.3d 1091, 162 O.&G.R. 611 (10th Cir. 2005), finding that the term "at the well" is unambiguous and allows for the lessee to calculate royalties using a netback methodology. *But see* Ideal v. Burlington Resources Oil & Gas Company LP, 233 P.3d 362, 2010-NMSC-022 (remanded "to the district court to resolve whether the marketable condition rule applies as a matter of law or because the parties may have intended it to apply."), and Davis v. Devon Energy Corp., 147 N.M. 157, 218 P.3d 75, 2009 -NMSC- 048.

7. What if the royalty clause provides that royalty is to be paid on the "gross proceeds at the wellhead." Is the use of the terms gross proceeds and wellhead inconsistent? Compare Schroeder v. Terra Energy, Limited, 223

Mich.App. 176, 565 N.W.2d 887, 138 O.&G.R. 361 (1997) with Judice v. Mewbourne Oil Co., 939 S.W.2d 133, 133 O.&G.R. 513 (Tex. 1996).

8. In KILMER V. ELEXCO LAND SERVICES, INC., 990 A.2d 1147 (Pa. 2010) plaintiff royalty owners claimed that the net-back method of calculating royalties under mineral leases violated Pennsylvania's Guaranteed Minimum Royalty Act (GMRA). They based their argument in part on the assertion that the legislature in 1979 intended to reflect the "First Marketable Product Doctrine" by requiring lessees to bear all the costs necessary to market the gas, which would include post-production costs. They also relied on Pennsylvania's Conservation law, which defines royalty owner as "any owner of an interest in oil or gas lease which entitles him to share in the production of the oil or gas under such lease or the proceeds therefrom without obligating him to pay any costs under such lease." The court rejected both claims, ruling instead that "the GMRA should be read to permit the calculation of royalties at the wellhead, as provided by the net-back method in the Lease . . . ."

9. Professor David Pierce has advanced the "royalty value theorem" for analysis and explanation of royalty litigation: "The royalty value theorem recognizes the inherent conflict between lessor and lessee when the lessor is compensated through a fixed fraction of the value of oil and gas. The theorem provides, 'When compensation under a contract is based upon a set percentage of the value of something, there will be a tendency by each party to either minimize or maximize the value.' To understand how the theorem works in practice, the physical and economic facts regarding the production and marketing of oil and gas must be recognized. First, oil and gas tend to increase in value as they move downstream from the point of production. Second, much of the increase in value can be attributed to costs associated with transporting, treating, aggregating, packaging, and marketing the production. The resulting downstream value of the production consists of two components: (1) the enhanced value associated with the actual investments in the production; and (2) the enhanced value of the production in its current form and location. The theorem recognizes that lessors will seek to maximize the value of their fraction of royalty by arguing for a valuation point as far removed from the point of production as possible. Lessees, in turn, will argue for a valuation point at the moment the oil and gas is extracted from the ground." David E. Pierce, *The Renaissance of Law in the Law of Oil and Gas: The Contract Dimension*, 42 Washburn L.J. 909, 927 (2004).

For other treatment of the topic of post-production costs, *see* Rachel M. Kirk, *Variations in the Marketable Product Rule from State to State*, 60 Okla.L.Rev. 769 (2008); Barry L. Wertz, *The* Tawney *Decision*, 55th Ann. Inst. on Min. Law 239 (2008); Brian S. Wheeler, *Deducting Post-Production Costs When Calculating Royalty: What Does the Lease Provide?*, 8 Appalachian J. L. 1 (2008); Byron C. Keeling & Karolyn King Gillespie, *The First Marketable Product Doctrine: Just What Is the Product?* 37 St. Mary's

L.J. 1 (2005) (describes the development of the "marketable product" doctrine as going "From Consistency to Chaos"); M. Benjamin Singletary, *Royalty Litigation on Processed Gas: Valuation, Post-Production Activities and the Marketable Condition Rule*, 55 Inst. On Oil & Gas L. Ch. 8 (2004); Edward B. Poitevent, II, *Post-Production Deductions from Royalty*, 44 S. Tex. L. Rev. 709 (2003) (survey of recent cases showing a "hodge podge pattern" of jurisprudence and observing that "considerable room remains for the states to develop their respective theories"); Scott Lansdown, *The Marketable Condition Rul*", 44 S. Tex. L. Rev. 667 (2003) (arguing that "the marketable condition rule represents an attempt to change well-established law in many states to reach a result that is counter to the specific language of the lease, based in most cases, on a 'leap of faith' from the implied marketing covenant that is logically unsupportable"); Owen L. Anderson, *Royalty Valuation: Should Overriding Royalty Interests and Nonparticipating Royalty Interests, Whether Payable in Value or in Kind, Be Subject to the Same Valuation Standard as Lease Royalty*", 35 Land & Water L. Rev. 1 (2000) (urging use of a "contract-law approach" with a nation-wide "uniform regime for royalty remittance and valuation" rather than "state-by-state, interest-by-interest, and clause-by-clause basis").

10. How appropriate is it to make use of class action for resolution of contract disputes, as the *Tawney* court has? Discussing the *Tawney* decision, Professor Pierce has asked that if the lease is ambiguous regarding royalty calculation, isn't the proper response to try and ascertain the intent of the parties to resolve the ambiguity consistent with their free will as contracting parties? He comments:

> Instead of seeking to ascertain the intent of the parties, the court selects a single rule of construction to resolve the matter: "We choose to adhere to our traditional rule and construe the language against the lessee." The court assumes that CNR selected the particular form in each case that contained the wellhead language. If it can be proven, as a fact, that the *lessor* or the *lessor's attorney* selected the lease form containing the wellhead language, does that mean the *lessor* loses? Because the court makes this rule of construction outcome determinative, it would seem imperative to due process that the lessee have the opportunity to inquire into how each lease came into being.

David E. Pierce, *Developments in Nonregulatory Oil and Gas Law: Beyond Theories and Rules to the Motivating Jurisprudence*, 58 Inst. On Oil & Gas L. §1.04 [3][a] (2007).

The courts have split on the propriety of the class action tool in royalty disputes. For two cases rejecting class action, *see* Union Pacific Resources Group, Inc. v. Neinast, 67 S.W.3d 275, 284, 153 O.&G.R. 395 (Tex. App.-- Houston [1st Dist.] 2001, no writ) (reversed class certification order because the class failed to meet the predominance standard because there was not a single implied covenant to market that governed each lease regardless of the terms of the royalty clause or other relevant leasehold provisions); Stirman v. Exxon Corp., 280 F.3d 554, 154 O.&G.R. 107 (5th Cir. 2002) (class action for nationwide class of natural gas royalty owners should not have been certified because the class representatives did not meet the typicality requirement and the plaintiff did not sustain its burden of proof to show that common issues of law or fact predominated in light of possible variances in applicable state laws); and Bowden v. Phillips Petroleum Co., 247 S.W.3d 690 (Tex. 2008) (denied class certification to two sub-classes of royalty owners; the court found a lack of predominance of common issues because there were express provisions in a number of leases that dealt with the issue of marketing which would negate the existence of an implied covenant to market). In addition to *Tawney* allowing class proceedings, *see* Seeco, Inc. v. Hales, 330 Ark. 402, 954 S.W.2d 234 (1997) (royalty owners entitled to be certified as a class in an action seeking unpaid royalties against gas producer and producer's local distribution company on various grounds including failure to enforce take or pay provisions, breach of implied covenant to market, fraud, and unjust enrichment). *Compare* Lewis v. Texaco Exploration & Production Co., 698 So. 2d 1001, 138 O.&G.R. 68 (La. Ct. App. 1997) *with* Chevron USA, Inc. v. Vermilion Parish School Board, 377 F.3d 459, 160 O.&G.R. 13 (5th Cir. 2004), aff'g 215 F.R.D. 511 (W.D. La. 2003). And *see* John Burritt McArthur, *The Class Action Tool in Oilfield Litigation*, 45 U. Kan. L. Rev. 113 (1996).

11. Class action royalty proceedings often raise choice of law issues concerning the nature of the royalty interest, be it lessor's royalty, overriding royalty, or non-participating royalty (or some hybrid interest). Should a court apply the law of the place of making of a contract (*lex loci contractus*), the place of performance, the place of production (*lex rei sitae*), or the law of the forum (*lex fori*)? Would it make a difference if royalty is characterized as a matter of contract or of property?

Farrar v. Mobil Oil Corp., 234 P.3d 19 (Kan.App., 2010): The court applied *lex rei sitae,* asserting that "the rights to future royalty payments pursuant to such leases are essentially part of the lessor's realty interests. . . . Whether the lessor's rights to royalty are technically classified as personalty or realty, it is difficult to imagine a more intimate contact with Kansas than the construction and enforcement of instruments that license the exploration for Kansas minerals."

Ideal v. Burlington Resources Oil & Gas Company LP, 233 P.3d 362, 2010-NMSC-022: The court held that forum law/site-of-property law applied to class action involving royalty claims. Stating that in New Mexico a grant or reservation of the underlying oil and gas, or royalty rights therein is a grant or reservation of real property, the court ruled: "Because the land is located in New Mexico, New Mexico law properly applies."

Weber v. Mobil Oil Corp., 2010 OK 33; 2010 Okla. LEXIS 36: The court held that forum law/site-of-property law applied to a class action involving royalty claims, stating, "Royalty to be derived from future mineral leases is real property and contracts involving such royalty are governed by the law of the situs."

Because a class action may involve production in non-forum states, is it appropriate for the forum court in state "A" to assert jurisdiction to adjudicate royalty claims arising from production from hundreds or thousands of wells in states "B" and "C"? The class action device allows a district court in state "A" to issue rulings on the law of states "B" and "C" arising from real estate and production in those states with no effective way for a litigant to obtain authoritative adjudication from the courts of the state where the oil and gas is located. *See*, for example, Wallace B. Roderick Revocable Living Trust v. XTO Energy, Inc., 679 F.Supp.2d 1287 (D.Kan., 2010) (class action certification issues for royalty claims under implied covenant to market for three states, Colorado, Kansas and Oklahoma).

————

C. THE OBLIGATION TO PAY ROYALTY UPON THE RECEIPT OF TAKE-OR-PAY OR SETTLEMENT MONIES

KILLAM OIL CO. v. BRUNI

Texas Court of Appeals, 1991.
806 S.W.2d 264, 118 O.&G.R. 280 (writ denied).

BUTTS, JUSTICE. This is an appeal from a summary judgment entered in favor of the plaintiffs, Fred Bruni, Ernest Bruni, and Ernesto Ramirez, as Trustees under the Bruni Mineral Trust ("Trust").

In 1974, the Trust, as lessor, drafted and entered into an oil and gas lease with the defendants, as lessee. At the time the lease was executed, Killam and Hurd operated as a partnership under the name of Killam & Hurd, Ltd. That relationship ended, and the lease was subsequently assigned to Killam Oil and to Hurd Enterprises, with each entity owning a one half interest. The lessees completed nine producing gas wells on the leased premises. The gas produced from two wells was sold to a pipeline under a Gas Purchase Contract ("the Contract") executed on November 24, 1981, between the lessees, as seller, and United Texas Transmission Company ("UTTCO"), as buyer. Under the contract, UTTCO was entitled to make up payments within five contract years for gas not taken.

The Gas Purchase Contract contained a "take-or-pay" provision obligating UTTCO either to take a specified annual quantity of gas or pay Killam and Hurd for the gas not taken. In 1986, Killam sued UTTCO when it failed to either take or pay for the minimum quantity as required under the contract. Hurd settled without suit for $2.8 million and Killam settled for $4 million. Each recovered the same amount after costs, expenses, and attorney fees were deducted.

The Trust sued Hurd and Killam seeking a royalty share of the settlement proceeds received from UTTCO. The Trust alleged four alternative claims in its petition. The Trust first alleged a breach of marketing duty, breach of duty of good faith and fair dealing, conversion, and fraud. The second claim was that the Trust was entitled to a royalty share because the take-or-pay provisions constituted a constructive sale of the gas. In the next two claims, the Trust alleged unjust enrichment and equitable reformation, respectively. Both sides moved for summary judgment, the issue being whether the Trust was entitled to receive a royalty portion of the take-or-pay settlement proceeds. The trial court granted summary judgment in favor of the Trust determining that the Trust had the right to collect a royalty share, thereby concluding that, as a matter of law, the gas royalty clause was applicable to the settlement payment. The judgment was by severance. Both Killam and Hurd appealed.

Killam and Hurd argue that under the terms of the lease and Texas law the Trust was not entitled to royalty on the settlement arising from UTTCO's breach of the "take-or-pay" provision of the contract.

Hurd brings four points of error: under the Bruni lease royalty is not due on payments for gas not produced and sold; case law holds that producers owe no royalty on take-or-pay payments; the trial court erred in denying Hurd's motion for summary judgment because Hurd had no liability to pay royalties on take-or-pay payments under the express terms of the Bruni lease; and, the trial court erred in granting summary judgment that Hurd owed royalty on payments made to settle take-or-pay liability arising from a contract between the producers and the pipeline covering the Bruni lease. Killam raises three points of error asserting that the court: erroneously granted the Trust's motion for summary judgment; erred in ruling as a matter of law that the Trust was entitled to a royalty share of the settlement received by Killam; and erroneously denied Killam's motion for summary judgment because, as a matter of law, the royalty clause of the lease did not apply to a settlement received by Killam arising from UTTCO's breach of the gas purchase contract.

The party moving for summary judgment has the burden of showing that no genuine issues of material fact exist and that it is entitled to judgment as a matter of law. When both parties file motions for summary judgment and one motion is granted and the other is overruled, the trial court's judgment becomes final and appealable. On appeal, the court of appeals should determine all questions presented including the propriety of the order overruling the losing party's motion for summary judgment.

The issue whether a standard royalty clause applies to settlement of a take-or-pay provision has not been directly addressed by the Texas courts. However, we are guided by the principles of law enunciated in other cases construing oil and gas leases, particularly the royalty clauses.

In construing the provisions of an oil and gas lease, the court must determine the intention of the parties, as expressed in the lease.. Unless a conflict or ambiguity exists, the lease alone is deemed to express the parties' intent. Extrinsic evidence cannot be considered if the lease clearly discloses the parties' intention, or if the lease is susceptible of only one legal meaning.

The royalties to which a lessor is entitled must be determined from the provisions of the oil and gas lease. The pertinent provision of the royalty clause in the lease involved here provides:

> The *royalties* to be paid by lessee are: . . . (b) on gas, including casinghead gas and all gaseous substances, *produced* from said land and *sold or used off the premises* or in the manufacture of gasoline or other product therefrom, the market value at the mouth of the well of one-eighth of the gas so *sold or used* provided that on gas *sold* at the wells the royalty shall be one-eighth of the amount realized from such; . . . (emphasis added).

In Garcia v. King, 139 Tex. 578, 164 S.W.2d 509 (1942), the court held that the term "produced", as used in the lease to allow for an extension beyond the primary term, meant "produced in paying quantities". The Garcia court emphasized that in order to understand and properly interpret the language used by the parties we must consider the objects and purposes intended to be accomplished by them in entering into the lease. The object of a mineral lease is to secure development of the property for the mutual benefit of the parties. Consequently, it has become well established under Texas law that the term "production" as used in oil and gas leases means the actual physical extraction of the mineral from the soil.

Killam and Hurd cite Monsanto Co. v. Tyrrell, 537 S.W.2d 135 (Tex.Civ.App.–Houston [14th Dist.] 1976, writ ref'd n.r.e.) and Exxon Corp. v. Middleton, 613 S.W.2d 240 (Tex. 1981) and maintain that royalties are not due to the lessor on settlement proceeds resulting from the breach of take-or-pay provisions in a gas purchase contract. We agree.

The issue confronting the court of appeals in *Monsanto* was whether an advance payment for gas production constituted "recovery from production." The oil and gas lease in *Monsanto* provided for the initial royalty to increase when the lessee recovered its total drilling costs from a stated percentage of production. The lessee entered into a gas purchase contract whereby the purchaser made an advance payment to the lessee as partial payment for the gas committed to the purchaser. The lessor contended that the advance payments should be applied as an immediate credit to the recovery of well costs, thereby triggering the increased royalties. The court ruled that the advance payments were not "recovery from production" and would, therefore, not be accounted for by the lessee as recovery of expenses until the gas was actually produced from the lease.

The royalty clause in the leases in Exxon Corp. v. Middleton provided for royalty to be based upon a percentage of the gas produced from the land as related to either the amount realized for the sale of minerals at the well or market value if the minerals were sold off the premises. The gas was delivered to a plant within the field upon which the leases were located but not on the leased premises. The lessor contended that the gas was sold when the gas contracts became effective. That court disagreed, holding that the term "produced" as used in the lease meant a physical extraction from the land and the term "sold" meant delivered. The lessee's obligations under the royalty clause were unaffected by its gas contracts. The court reasoned that the parties could have provided for royalties to be calculated on the amount realized for a sale off the premises, if they so intended.

Recently in Diamond Shamrock Exploration Corp. v. Hodel, 853 F.2d 1159 (5th Cir. 1988) the court ruled that gas royalty clauses [like the present one] do not apply to take-or-pay payments because "royalties are not owed unless and until actual production, the severance of minerals from the formation, occurs."

In the present case, the lease entitled the Trust to royalty payments on gas actually *produced.* This is not a suit by the Trust to recover royalties on the gas actually taken by UTTCO under the contract or on gas sold by Killam and Hurd on the spot market. The dispute between UTTCO and Killam and Hurd arose when UTTCO neither took the gas nor paid as required under the contract. The Trust contends that the settlement payments received by Killam and Hurd from UTTCO might have included underpayment for gas sold on the spot market. However, the gas contract was independent of the lease. The Trust, as drafters of the lease, could have specifically included a provision allowing for royalties to be paid upon proceeds received by Killam and Hurd from settlements of disputes arising from a breach of take-or-pay provisions in gas contracts. The parties knew how to and did provide for royalties to be paid based upon the gas produced and sold. In so doing, the Trust unambiguously limited its right to royalty payments only from gas actually extracted from the land. Moreover, the gas not actually produced remains in the ground and the Trust will be entitled to royalties upon the subsequent extraction.

We hold, as a matter of law, that the Trust is not entitled to royalties on the settlement proceeds arising from the take-or-pay provision of the contract between Killam, Hurd, and UTTCO. Therefore, we need not address the Trust's contention as to what the proceeds might have represented. This is because under a standard lease, take-or-pay payments do not constitute any part of the price paid for produced gas, nor do they have the effect of increasing the price paid for gas that was taken. These payments are made when gas is *not produced,* and as such, bear no royalty.

The summary judgment granted by the trial court in favor of the Trust is reversed. Judgment in favor of Killam Oil and Hurd Enterprises is hereby rendered.

———

NOTES

1. Do you agree with the following analysis of *Bruni*?:

The court approached the issue as a simple case of construing a written instrument. It applied the traditional Texas principales and canons of construction, relying in large part on the express language of the royalty clause. It likewise found the royalty clause language unambiguous so that extrinsic evidence or other factors could not be admitted to ascertain the intent of the parties. The key to the court's analysis was a rigorous look at the royalty clause to determine what events triggered the royalty payment obligation.

The two key terms that needed to be defined were "produced" and "sold." If the gas was neither produced nor sold there would be no royalty payment obligation.

Bruce M. Kramer, *Liabiity to Royalty Owners for Proceeds From Take-or-Pay and Settlement Payments*, 15 East.Min.L.Inst. 14-1 (1994). *See also* Discussion Notes, 118 O.&G.R. 309-310.

2. Prior to *Bruni* two other courts had similarly denied royalty owners a share of take-or-pay or settlement monies. *See* State of Wyoming v. Pennzoil, 752 P.2d 975, 100 O.&G.R. 359 (Wyo. 1988) and Diamond Shamrock Exploration Co. v. Hodel, 853 F.2d 1159, 103 O.&G.R. 38 (5th Cir. 1988). Is there a difference between advance royalty payments not triggered by production and take or pay payments? *See* Cheyenne Mining & Uranium Co. v. Federal Resources Corp., 694 P.2d 65 (Wyo. 1985). *See generally*, Kramer, *Royalty Obligations Under the Gun – The Effect of Take-or-Pay Clauses on the Duty to Make Royalty Payments*, 39 Inst. on Oil & Gas L.&Tax'n 5-1 (1988).

3. Is the *Bruni* interpretation of when gas is produced consistent with the *Vela, Middleton* and *Piney Woods* discussed earlier whereby production is to be valued at the time that it is physically extracted from the ground?

FREY v. AMOCO PRODUCTION CO.

Louisiana Supreme Court, 1992.
603 So.2d 166, 113 O.&G.R. 478.

COLE, Justice. Frederick J. Frey and other owners of gas royalty interests ("Frey") under a mineral lease ("Lease") commenced suit against their mineral lessee, Amoco Production Company, in the United States District Court for the Eastern District of Louisiana to recover a royalty share of the proceeds received by Amoco in settlement of the take-or-pay litigation ("Settlement Agreement") arising under the "Gas Purchase and Sales Agreement" ("Morganza Contract") between Amoco and its pipeline purchaser, Columbia Gas Transmission Corporation.[61] The Lease's royalty clause provides Frey a "royalty on gas sold

---

[61] In 1981, Amoco and Columbia entered into an exclusive long-term gas purchase and

by the Lessee [of] one-fifth (1/5) of the amount realized at the well from such sales."

The take-or-pay royalty claim was tried on undisputed facts pursuant to opposing motions for partial summary judgment. After trial, the district judge granted partial summary judgment in favor of Amoco, closely following the reasoning of Diamond Shamrock Exploration Co. v. Hodel, 853 F.2d 1159 (5th Cir. 1988). The district court determined the sale of gas cannot occur absent physical production and severance of the gas, and therefore, under Louisiana law, take-or-pay payments do not constitute part of the sale price of natural gas. . . . Frey appealed to the United States Court of Appeals for the Fifth Circuit. A three judge panel of that court reversed. Frey v. Amoco Production Co., 943 F.2d 578 (5th Cir. 1991). The Court of Appeals decided, *inter alia,* take-or-pay payments are part of the "amount realized" from the sale of gas under the Lease, and thus such payments, received by the lessee in settlement of the take-or-pay dispute with its pipeline purchaser for gas not taken, are subject to the lessor's royalty. The court, relying on Louisiana law, reasoned the payments "constitute economic benefits that Amoco received from granting Columbia the right to take gas from the leased premises, a right Amoco got through the Lease." The court thus determined "it would be contrary to the nature of the Lease as a cooperative venture to allow a benefit by any name that is attributable to the gas under the leased premises to inure exclusively to the lessee." On Petition for Rehearing, however, the Court of Appeals withdrew that portion of its opinion regarding Frey's entitlement to a royalty interest on the proceeds of the take-or-pay settlement and certified to us the following question:

Question Certified

Whether under Louisiana law and the facts concerning the Lease executed by Amoco and Frey, the Lease's clause that provides Frey a 'royalty on gas sold by the Lessee of one-fifth (1/5) of the amount realized at the well from such sales' requires Amoco to pay Frey a royalty share of the take-or-pay payments that Amoco earns as a result of having executed the Lease and under the terms of a gas sales contract with a pipeline-purchaser. Frey v. Amoco Production Co., 943 F.2d at 578 *op. withdrawn, in part, on reh'g, ques. certified,* 951 F.2d 67 (5th Cir. 1992) (per curiam).

In exercising the certification privilege granted by Rule XII of the Rules of the Supreme Court of Louisiana, the Court of Appeals maintained it continued

---

sales contract. The contract required Columbia to take delivery of a specified minimum of gas based upon the deliverability of the wells drilled by Amoco over each contract year or to pay for the minimum quantities even if it did not take delivery of the gas. In 1983, Columbia's failure to abide by the "take-or-pay" provision engendered litigation between Amoco and Columbia regarding the latter's alleged $265 million take-or-pay liabilities. The litigation was compromised by a Settlement Agreement effective July 1, 1985, and the gas purchase contract continued in effect as amended.

to believe in the propriety of its ruling but felt compelled to defer to the Louisiana Supreme Court on this critical point of Louisiana law. As noted by the federal court, this issue is *res nova* in Louisiana and, as evidenced by the filing of numerous amici curiae briefs, the extent of the interests affected by its resolution considerable. The controversy centers around Frey's alleged entitlement to a royalty share of the $66.5 million in take-or-pay amounts paid by Columbia to Amoco under the Settlement Agreement. The parties characterize $45.6 million of the total as a "recoupable take-or-pay payment" and the remaining $20.9 million as a "non-recoupable take-or-pay payment." Not at issue is a $280.2 million payment made to Amoco under the Settlement Agreement for past and future price deficiencies in natural gas as Frey has already received royalty on that entire amount.

Response To Certified Question

Having granted certification, we respond by i) examining the legal precepts which govern the issues raised by the certified question, and ii) applying the legal precepts to the precise language of the royalty clause to determine whether the proceeds received by Amoco in settlement of the take-or-pay litigation with Columbia constitute part of the "amount realized" from gas sold by Amoco to Columbia. Although the Fifth Circuit disclaimed any intention that we confine our reply to the precise form or scope of the question certified, we do not deem prudent an excursion far beyond the bounds of the precise question certified.

For the reasons hereinafter assigned, we find the take-or-pay payments under the facts of this case and the royalty clause at issue are subject to the lessor's royalty in favor of Frey.

I. The Legal Precepts

A. Louisiana Mineral Law

Ownership of land does not include ownership of oil or gas. The vesting of title to fugitive minerals, such as oil or gas, occurs when the minerals are reduced to possession at the wellhead. Thus, with respect to oil and gas, possession marks both the vesting of title and mobilization.

Although the right to explore and develop one's property for the production of minerals, and to reduce minerals to possession and ownership, belongs exclusively to the landowner, the landowner may convey, reserve or lease this privilege. In this manner, rights in minerals may be considered "separable component parts of the ownership of land." A mineral lease, one of the manners in which mineral rights are segregated from ownership, *id.,* is a contract by which the lessee is granted the right to explore for and produce minerals. Mineral leases are construed as leases generally and, wherever pertinent, codal provisions applicable to ordinary leases are applied to mineral leases. Accordingly, where the Louisiana Mineral Code, neither expressly nor impliedly provides for a particular situation, resort is made to the Louisiana

Civil Code or other laws, either directly or by analogy. However, where there is conflict between the provisions of the Civil Code or other laws, the Mineral Code prevails. We begin with the Mineral Code.

Article 213(5) of the Mineral Code defines royalty, as used in conjunction with mineral leases, as:

> [A]ny interest in production, or its value, from or attributable to land subject to mineral lease, that is deliverable or payable to the lessor or others entitled to share therein. Such interests in production or its value are "royalty," whether created by the lease or by separate instrument, if they comprise a part of the negotiated agreement resulting in execution of the lease. "Royalty" also includes sums payable to the lessor that are classified by the lease as constructive production.

Although Article 213(5) provides a standard for interpretation of the mineral lease, the rather expansive definition of royalty is not dispositive of the lessor's right to a royalty share of take-or-pay payments. The principle of freedom of contract is expressly recognized by the Mineral Code, and therefore, the accessorial right of royalty may not be defined absent reference to the oil and gas lease in which it appears. Accordingly, this Court must give consideration to the fundamental principle that the lease contract is the law between the parties, defining their respective legal rights and obligations, as well as the rules for interpretation of contracts as laid down in the Civil Code, Title IV at Chpt. 13, art. 2045 *et seq.* Disinclined to rewrite a mineral lease in pursuit of equity, we are nonetheless cognizant the terms of a mineral lease are neither intended to, nor capable of, accommodating every eventuality. *See* Martin, *A Modern Look at Implied Covenants To Explore, Develop, and Market Under Mineral Leases,* 27 Inst. on Oil & Gas L. & Tax'n 177,194 (1976).

The purpose of interpretation is to determine the common intent of the parties. Words of art and technical terms must be given their technical meaning when the contract involves a technical matter, and words susceptible of different meanings are to be interpreted as having the meaning that best conforms to the object of the contract. A doubtful provision must be interpreted in light of the nature of the contract, equity, usages, the conduct of the parties before and after the formation of the contract, and other contracts of a like nature between the same parties. When the parties made no provision for a particular situation, it must be assumed that they intended to bind themselves not only to the express provisions of the contract, but also to whatever the law, equity, or usage regards as implied in a contract of that kind or necessary for the contract to achieve its purpose. To these basic concepts, we add one other. In Louisiana, a mineral lease is interpreted so as to give effect to the covenants implied in every such lease.

Under the facts before us, a search for the parties' specific intent relative to the obligation to pay royalty on the take-or-pay proceeds would prove fruitless. Surely neither Amoco nor Frey contemplated at the time the Lease was executed in 1975 the demand for gas would fall and pipelines would be financially unable to comply with their obligations under long-term gas purchase and sales contracts. It is even more unlikely the parties contemplated producers would receive take-or-pay payments in settlement of gas contract litigation. Accordingly, we look not at the parties' intent to provide expressly for take-or-pay payments, but rather at the parties general intent in entering an oil and gas lease, *viz.,* the lessor supplies the land and the lessee the capital and expertise necessary to develop the land for the *mutual benefit* of both parties. In this manner, the royalty clause is given an expansive reading, reflecting the mutuality of objectives and sharing of benefits inherent in the lessee-lessor relationship. Consequently, we endeavor to ascertain the meaning of the royalty clause in a manner consistent with the nature and purpose of an oil and gas lease, having due regard for: 1) the function of a royalty clause; and 2) the lessee's implied obligation under Mineral Code Article 122 to market diligently the gas produced.

1. The Attributes of Royalty

As stated, the royalty clause is construed not in the abstract but in reference to the economic and practical considerations underlying the royalty interest and with due regard to the relationship between a lessor and lessee. The lessor-lessee relationship ensues from a synallagmatic contract in which the obligation of each party is the *cause* of the other. Where royalty is conferred by the lease, royalty is the reason or cause for the lessor to obligate himself thereto. Stated differently, royalty is the compensation or "consideration" the lessee pays to the lessor to secure the privilege of exercising the right to explore and develop the property for production of oil and gas.

By virtue of the beneficial relationship between lessee and lessor, the former avoids having to pay up front for the privilege of exploration, and the latter, assuming a passive role, is guaranteed participation in any eventual yield accruing from the lessee's entrepreneurial efforts, unconstrained by financial and operational responsibilities. *See* Martin, 27 Inst. on Oil & Gas L. & Tax'n at 194 n. 62. Inherent in the concept of lease as a bargained-for exchange is the recognition a lessor would not relinquish a valuable right arising from the leased premises without receiving something in return. In Wemple, [Wemple v. Producers' Oil Co., 145 La. 1031, 83 So.2d 232 (1919) Eds.] lessor sued lessee to recover a one-eighth royalty on natural gasoline the lessee extracted from casinghead gas by the use of a separator. At the time the Lease was executed in 1909, the parties were unaware of this procedure. We determined the casinghead gas fell under the oil clause of the Lease, reasoning the parties would not have contemplated a lease where the lessee could extract a valuable substance and "give nothing in return."

This Court also had occasion to discuss the nature of a royalty interest in the context of ascertaining the meaning of "market value" of gas under a mineral lease. *See* Henry v. Ballard & Cordell Corp., 418 So.2d 1334, 1339

(La. 1982). There, we reasoned the ambiguity in the royalty provision could not be resolved without consideration of the practical and economic realities of the oil and gas industry at the time the leases were negotiated and the obligations of the lessee to market the gas at the best possible price at the time the leases were made. Adopting the reasoning of Professor Thomas Harrell, we announced the rule that a lease arrangement is in the nature of a cooperative venture in which the lessor contributes the land and the lessee the capital and expertise necessary to develop the minerals for the mutual benefit of both parties. Id. at 1338. . . . We also concluded the ultimate objective of the royalty provision of the lease is to fix the division between the lessor and lessee of the economic benefits anticipated from the development of the minerals. We then quoted with approval Professor Harrell's contention:

> [A]ny determination of the market value of gas which admits the lessee's arrangements to market were prudently arrived at consistent with the lessee's obligation, but which at the same time permits either the lessor or lessee to receive a part of the gross revenues from the property greater than the fractional division contemplated by the lease, should be considered inherently contrary to the basic nature of the lease and be sustained only in the clearest of cases.

In light of *Henry*, we conclude an oil and gas lease, and the royalty clause therein, is rendered meaningless where the lessee receives a higher percentage of the gross revenues generated by the leased property than contemplated by the lease. The lease represents a bargained-for exchange, with the benefits flowing directly from the leased premises to the lessee and the lessor, the latter via royalty. An economic benefit accruing from the leased land, generated solely by virtue of the lease, and which is not expressly negated, *see* La.Rev.Stat. § 31:3, is to be shared between the lessor and lessee in the fractional division contemplated by the lease.

2. Implied Covenant To Market

Despite the purely contractual relationship between the lessor and lessee, the respective parties' obligations can not be determined absent reference to the covenants implied in every oil and gas lease. *See* Pearson & Watt, *To Share or Not To Share: Royalty Obligations Arising out of Take-or-Pay or Similar Gas Contract Litigation*, 42 Inst. on Oil & Gas L. & Tax'n 14-1, at § 14.04[1] (1991). These covenants address matters not expressly covered by the lease, including protection of the lessor's interest, and assist the court in ascertaining the duties incident to the relationship of lessor and lessee. Martin, 27 Inst. on Oil & Gas L. & Tax'n 177, 195. At the heart of the implied obligations in Louisiana is the notion the parties consent to perform these obligations in order "to effectuate the basic objectives of an oil and gas lease."

In Louisiana, the implied covenants originate not in the general principle of cooperation found in the law of contracts, *see* 5 Williams and Meyers, *Oil and Gas Law* § 802.1 (1991), but rather as particularized expressions of Civil

Code Article 2710's's mandate that the lessee enjoy the thing leased as "a good administrator." La.Rev.Stat. § 31:122, comment; La.Civ.Code art. 2710. The duty to act as a "reasonably prudent operator," imposed on the mineral lessee by Article 122 of the Mineral Code, is thus an adaptation of the obligation of other lessees to act as "good administrators." La.Rev.Stat. § 31:122, comment.

Article 122 of the Mineral Code provides:

> A mineral lessee is not under a fiduciary obligation to his lessor, but he is bound to perform the contract in good faith and to develop and operate the property leased as a reasonably prudent operator for the mutual benefit of himself and his lessor. Parties may stipulate what shall constitute reasonably prudent conduct on the part of the lessee.

The legislature intended to incorporate within Article 122 the existing jurisprudence on the subject, and accordingly, the mineral lessee's obligation to act as a "good administrator" or "reasonably prudent operator" is clearly specified in four situations. Relevant for our purposes is the implied obligation to market diligently the minerals discovered and capable of production in paying quantities in the manner of a reasonable, prudent operator. Encompassed within the lessee's duty to market diligently is the obligation to obtain the best price reasonably possible. *See, e.g.,* Tyson v. Surf Oil Co., 195 La. 248, 196 So. 336 (1940). *See also* La.Rev.Stat. § 31:122, comment; Martin, 27 Inst. on Oil & Gas L. & Tax'n at 191. . .

Regarding the fulfillment of the implied covenants, including the duty to market diligently, the lessee's conduct must conform to, and be governed by, what is expected of ordinary persons of ordinary prudence under similar circumstances and conditions, having due regard for the interest of both contracting parties. Our analysis is complicated by the paucity of litigation dealing with the implied obligation to market diligently, as well as the recognition that the lessee's conduct must be evaluated with due regard for the facts known at the time the Morganza Contract was executed. Consequently, an examination of the take-or-pay provision, a clause found in nearly all gas purchase contracts, including the contract between Amoco and Columbia, is essential to resolving the issue before us.

Under a take-or-pay provision, the pipeline-purchaser commits to take or, failing to take, to pay for a minimum annual contract volume of natural gas which the producer has available for delivery. Williams & Meyers, *Manual of Oil & Gas Terms* 1233 (1991). Where gas is paid for but not taken, the contract normally permits the purchaser to "make-up" the deficiency by taking an excess amount of gas ("make-up gas") over a specific term and, in turn, to receive a refund or credit. *See id.*

At the time of execution of the Morganza Contract, long-term gas contracts containing take-or-pay provisions were standard in the industry, as are take- or-pay clauses. *See* Henry, 418 So.2d at 1336; Kramer, *Royalty Obligations Under the Gun–The Effect of Take-or-Pay Clauses on the Duty To*

Make Royalty Payments, 39 Inst. Oil & Gas L. & Tax'n 5-1, at § 5.02 (1988). In the past, long-term contracts were universally insisted upon by pipeline purchasers to enable acquisition of financing for the construction of capital intensive pipeline facilities. Additionally, take-or-pay provisions allow the pipeline flexibility in the amount of gas taken, assuaging the difficulties caused by the cyclical nature of demand and the absence of an open market for natural gas. *See* Johnson, *Natural Gas Sales Contracts,* 34 Inst. on Oil & Gas L. & Tax'n 83, 111 (1983). Indeed, because gas ordinarily can not be stored upon production, the only economic means of transporting gas is via pipeline. *See* 1 B. Kramer & P. Martin, *The Law of Pooling and Unitization* § 5.01[3] (3d ed. 1991) ("A producer without a contract with a nearby pipeline is unable to sell its gas whatever the price.").

The producer also benefitted from a guaranteed minimal annual cash flow and was protected from decline in demand, minimizing the likelihood it would be unable to recoup the substantial costs associated with operation and maintenance. *See* Medina, McKenzie, & Daniel, *Take or Litigate: Enforcing the Plain Meaning of the Take-or-Pay Clause in Natural Gas Contracts,* 40 Ark.L.Rev. 185, 191 n.16 (1986), and the authority cited therein; . . . Further, because take-or-pay clauses made it more likely the purchaser would take at a more consistent level, drainage to the reservoir was minimized while the ultimate recovery, and the resultant economic benefits to the lessor and lessee, were maximized.

At the time the Morganza Contract was executed, a lessee who failed to execute with a pipeline purchaser a long-term gas sales contract containing a take-or-pay clause would likely be deemed to have acted imprudently. By the same token, the producer who failed to renegotiate a long term gas contract in the face of the pipeline's financial inability to perform fully under a long-term gas purchase contract, given the market conditions caused by the decline in demand and the rise in producer-pipeline litigation, would also likely be deemed to have acted imprudently.

B. Louisiana Sales Law

Louisiana law determines if and when a sale of minerals occurs. *See, e.g.,* Henry, *supra* (Gas is "sold" for purposes of determining "market value" at the time the gas sales contract is executed). As stated, oil and gas in place are insusceptible of ownership apart from the soil of which they form a part. Equally well-settled is the principle ownership of gas vests at possession, i.e., at the wellhead. These precepts, however, do not in any manner strike discord with the Civil Code's express recognition that future things may be the object of a contract. *See* La.Civ.Code art. 1976; Indeed, the sale of a future thing is specifically provided for in Article 2450 of the Civil Code, which article states, "A sale is sometimes made of a thing to come: as of what shall accrue from an estate, of animals yet unborn, or such like other things, although not yet existing." La.Civ.Code art. 2450. *See* H. Daggett at § 60 (the sale of a royalty interest is the conveyance of a future thing). Such a contract is not aleatory, but certain, since the price is to be paid for a specific future object.

Because the sale of a future thing is subject to the suspensive condition the object of the contract actually materialize, the sale is an exception to the general rule that ownership, and risk, are immediately transferred to the vendee upon agreement as to the object and the price thereof. *See* La.Civ.Code arts 1767, 2456, 2457 and 2471. Rather, the contract is executory, and ownership and risk remain with the vendor, until if and when the thing which is the object of the contract actually materializes. Should the thing fail to come into existence, there is a failure of cause and the parties are relieved of their respective obligations. If, however, the condition is fulfilled by the materialization of the object of the contract, the effects are retroactive to the inception of the obligation, and therefore, the transfer of ownership deemed to have taken place at the time of the contract. *See* La.Civ.Code art. 1775 (fulfillment of a condition has effects that are retroactive to the inception of the obligation); 4 Aubry & Rau, *Droit Civil Francais* § 302 (La.St.L.Inst. trans. 1971); S. Litvinoff, *Obligations* § 62.

We also note one need not own a thing in order to perfect a sale. The sale of a thing belonging to another is not absolutely null, but only relatively so, and such nullity is in the interest of the purchaser.

II. Application of Precepts

Frey maintains the take-or-pay payments received from Columbia under the Settlement Agreement constitute part of the price, or total revenues, received by Amoco in return for Columbia's purchase of gas under the contract. Additionally, Frey contends the payments constitute economic benefits flowing from the Lease, to which Frey is entitled a royalty share under Mineral Code Article 122 and *Henry, supra*.

On the other hand, Amoco contends the clear language of the Lease requires a "sale" before the royalty obligation is triggered, and no sale of gas is possible unless and until gas is produced, i.e., reduced to possession at the wellhead. *See* La.Rev.Stat. §§ 31:7 and 31:213(5). Characterizing a take-or-pay payment as a payment for volumes of gas *not* produced or sold, Amoco reasons the payment can not be deemed consideration for prior gas production, and therefore, the gas royalty clause remains untriggered. Amoco further contends as it assumed all the risk and bore entirely the expense of development, the take-or-pay proceeds received in settlement of its dispute with Columbia should inure solely to its benefit, and royalties are due only on the gas if and when it is eventually produced. Finally, Amoco points out Frey has received a royalty on all amounts received by Amoco from Columbia in payment of the gas produced and sold from the leased premises, including gas taken by Columbia in recoupment of its recoupable take-or-pay settlement payment to Amoco.

While at first blush Amoco's argument may appear the better of the two, recourse to fundamental principles dictates a different conclusion. Applying Louisiana law to the royalty clause, we conclude, as did our brothers on the federal bench, the take-or-pay payments made to Amoco by Columbia in settlement of the take-or-pay litigation form part of the "amount realized" by Amoco from the sale of gas to Columbia and are therefore subject to the lessor's royalty clause in favor of Frey. The take-or-pay payments are a portion of the price paid to Amoco by Columbia for the gas actually delivered to Columbia under the Morganza Contract. More important, the proceeds constitute economic benefits which are derivative of Amoco's right to develop and explore the leased property, a right conferred by and dependent upon the Lease between Amoco and Frey. Our interpretation of the royalty clause is consonant with the intention of the parties in entering an oil and gas lease.

At the outset, we make three significant observations. First, Amoco does not dispute there was, at all relevant times since 1982, continuous production from the leased property. The take-or-pay litigation arose from Columbia's failure to take or pay for the full quantities of gas specified in the contract, not from Columbia's failure to take *any* gas. Secondly, our decision does not turn on whether a "sale" of gas occurred between Amoco and Columbia for purposes of the gas purchase contract, but whether a "sale" occurred between Columbia and Amoco so as to trigger the royalty clause of the Lease. Lastly, we decline to accept Amoco's cramped characterization of the take-or-pay payment, i.e., the payment was for gas *not* produced or sold. This description refused, Amoco's argument collapses in upon itself like a house of cards. . . .

Next, Amoco's characterization of the take-or-pay payment as a payment in *lieu* of production disregards the fact the right *not* to take gas is merely a corollary of the right to take gas. *See* Frey, 943 F.2d at 584 n.5 for a similar argument. Stated differently, Columbia would not have bargained for the right *not to take gas* although paying as if it had, without having the *right to take gas.* The obligation to take, or to pay if not taken, is alternative, and thus Columbia's primary obligation under the gas purchase contract could be satisfied in one of two ways.

Finally, it is a myopic eye which perceives the lessor as sharing none of the risks associated with bringing the gas to the ground, *i.e.,* pre-production. One obvious pre-production risk shared by both lessee and lessor is the risk of drainage from the reservoir. *See* La.Rev. Stat. §§ 31:8 (rule of capture), 31:122 and the comment thereto; The take-or-pay clause, assuring the production of a relatively constant amount of gas, acts to protect the reservoir from drainage. *See also* Johnson, 34 Inst. on Oil & Gas L. & Tax'n at 109 (The take-or-pay provision influences the rapidity and extent of development of the gas reservoir); Furthermore, both the lessee and lessor share the risk of an erroneous market forecast by the lessee, the lessor's royalty being dependent on the producer-pipeline contract.

A. Gas Purchase Contract As the Sale of Gas

The Frey-Amoco Lease explicitly predicates Amoco's obligation to pay royalty on the *sale* of gas. In contrast, royalty on oil and miscellaneous minerals is triggered by *production*. The discrepancy in "triggering" events is indicative of the physical and economic dissimilarity between oil and gas, the latter incapable of being stored or transported by the lessor. Moreover, the variance of language supports Frey's contention production is not a prerequisite to a sale. *See* La.Civ.Code art. 2050. Had the parties desired to condition the payment of royalties on *production* of gas, the Lease could easily have so provided. *See* La.Rev.Stat. § 31:3.

We determine the gas purchase and sales contract between Amoco and Columbia was a sale of a specified future thing, *viz.,* natural gas. *See* La.Civ.Code art. 2450. Columbia secured the right to take and pay for a minimum quantity of gas over a specified term *or* to pay as if it had accepted delivery of the required minimum volume. Prior to reducing the gas to possession at the wellhead, Amoco possessed only an exclusive right to explore and develop the property for the production of minerals and to reduce them to possession and ownership. *See* La.Rev.Stat. §§ 31:6, 31:15, 31:114; When the gas reached the surface of the ground, and the suspensive condition thus materialized, the executory sale of a future thing, taking place between Amoco and Columbia by virtue of the execution of the Morganza Contract, was perfected retroactive to the execution of the contract. *See* La.Civ.Code arts 1767, 1775, 2450, 2471. *See also Henry,* supra (market value of gas is determined at the time the gas sales contract is executed although gas has not yet been delivered); Miller v. Nordan-Lawton Oil & Gas Corp. of Texas, 403 F.2d 946 (5th Cir. 1968) (for purposes of shut-in royalty clause, market is obtained when the lessee executes a gas purchase contract); Callery Properties, Inc. v. Federal Power Comm'n, 335 F.2d 1004, 1021 (5th Cir. 1964), *rev'd on other grounds sub. nom.,* United Gas Improvement Co. v. Callery Properties, Inc., 382 U.S. 223, 230, 86 S.Ct. 360, 364, 15 L.Ed.2d 284 (1965) (take-or-pay payments constitute a sale sufficient to establish FPC jurisdiction over the transaction). Consequently, the sale of gas occurred at the time the gas was committed to the pipeline, *i.e.,* at the execution of the Morganza Contract. Despite Amoco's lack of ownership of the gas at the time the Morganza Contract was executed, La.Rev.Stat. §§ 31:6, 31:7, 31:15, 31:114, the sale was nonetheless valid. *See* La.Civ.Code art. 2452;[62]

---

[62]   Moreover, because there *was* production at all relevant times, and because we find the Morganza Contract constituted the sale of a future thing for purposes of the Lease, we need not address Amoco's contention that "production" only occurs when the minerals are actually physically severed from the ground. *See, e.g.,* Union Oil & Gas Corp. of Louisiana v. Broussard, 237 La. 660, 671, 112 So.2d 96, 99 (1958) (royalty owner's right to share in the minerals accrues if and when they are produced); Union Oil Co. of California v. Touchet, 229 La. 316, 325, 86 So.2d 50, 53 (1956) ("It is well settled that the right of the owner of a royalty interest is restricted to a share in production if and when it is obtained.").

1. Take-or-pay payment: "price" for gas delivered

Having determined the event triggering the obligation to pay royalty on gas occurred, *viz.* the sale of gas, we address our conclusion that the take-or-pay payments are part of the "amount realized" by Amoco from the sale of gas to Columbia. We interpret the "amount realized" by Amoco from the sale of gas to Columbia to encompass both: 1) the *total* price paid by Columbia for the natural gas delivered, and 2) the "economic benefits" derived from the lessee's right to develop and explore, a right conferred by the lease.

Total revenue under a gas purchase contract is a function of quantity *and* price. Because the producer is willing to negotiate a lower price in exchange for the guarantee the pipeline will either take or pay for a specific minimum quantity of natural gas, the take-or- pay provision effectively lowers the price the producer charges the pipeline per unit of gas. *See* 4 Williams & Meyers at § 724.5. Consequently, absent the take-or-pay provision, the price of gas, and thus the royalty owed thereon, would be higher. This conclusion is also supported by the affidavit of Dr. David Johnson, Professor of Economics at Louisiana State University, submitted in evidence in conjunction with Frey's Motion for Partial Summary Judgment. Application of this theory to the case before us leads to the conclusion that the price of gas taken under the contract includes not only the contract price paid per unit of gas delivered, but also the sums paid in the form of take-or-pay payments made in settlement of the Morganza Contract litigation. To simplify the equation, the actual price paid by the pipeline per unit of gas is determined by dividing the total quantity of gas delivered by the total amount paid to the producer, the latter including take-or-pay payments. Viewed in this light, take-or-pay payments effectively increase the price of gas actually delivered to the pipeline. Failure to characterize these payments as part of the total price paid for gas sold under the contract is to disregard the obvious economic considerations underlying the take-or-pay clause.

2. Economic benefits constituting part of the "amount realized"

Although we find the take-or-pay payments in this case constitute part of the price paid by Columbia for the gas taken, we choose not to rest our decision on this conclusion alone, anticipating the theoretical difficulties inherent in classification of take-or-pay settlement proceeds as part of the price paid for gas delivered where, for instance, no gas is taken by the pipeline-purchaser. Moreover, the term "amount realized" connotes the sum total, the whole, or the final effect of the economic benefits obtained by Amoco in the exercise of the rights granted by the synallagmatic contract of Lease, and, is composed, in part, of the advantages flowing to Amoco by virtue of the sale of natural gas under the Morganza Contract.

The parties enter into a mineral lease in expectation of making a profit, and toward that end, incur reciprocal obligations. In exchange for a royalty interest, Amoco receives the exclusive right to explore and develop the leased premises. *See* La.Rev.Stat. § 31:114. Indeed, all of the rights of Amoco, in and to the Frey property and the minerals thereunder, derive from the lease executed between Frey and Amoco. By virtue of rights granted by Frey to Amoco, Amoco negotiated, and eventually renegotiated, the Morganza Contract with Columbia, allowing Amoco to sell the gas produced from Frey's property.

The benefits which accrue to Amoco under the Morganza Contract are derivative of the rights transferred to Amoco by Frey. Clearly, but for the Lease there would be no Morganza Contract, no Settlement Agreement, and ultimately no take- or-pay payments made to Amoco. *Henry*, supra, is authority for this determination. Therefore, even if we failed to find the take-or-pay proceeds constitute part of the price received by Amoco for the sale of natural gas, the payments nonetheless are economic benefits which accrue to the lessor under the rationale of *Henry* . . .

As we have stated, the duty before us is not to divine the intent of the royalty clause in the abstract. Rather, the process reflects our appreciation of the cooperative nature of the lease arrangement as well as an understanding of the economic and practical considerations underlying the royalty clause. Retention by Amoco of the entire take-or-pay payment would permit Amoco to receive a part of the gross revenues from the property greater than the fractional division contemplated by the Lease. *See* Henry at 1338 n. 10 (citing Harrell, 30 Inst. on Oil & Gas L. & Tax'n at 336). Such a result can not be countenanced by this Court.

B. Right to Take-or-Pay Proceeds Derived from Lease

Further support is found in Article 122 of the Mineral Code. Amoco is bound to market diligently minerals discovered and capable of producing in paying quantities with due regard for Frey's interests. Upon segregation and conveyance of his right of ownership in the minerals to Amoco, Frey assumed a passive role, and presently possesses merely an accessorial right dependent upon the continued existence of the lease. In contrast, Amoco's superior position accords it exclusive control over the development and management of the property for the production of minerals.

Cognizant the majority of commentators urge the courts to defer to lessee marketing decisions, we do not take lightly our duty to scrutinize the lessee's implied covenant to market diligently in light of the practical, economic, and environmental concerns. *See* Martin, 27th Oil and Gas Inst. at 177; McCollam, 50 Tul.L.Rev. at 810; 5 Williams and Meyers at § 856.3. After careful consideration, we find Article 122 of the Mineral Code, which governed and sanctioned Amoco's decision to secure a market for natural gas via a long-term gas contract under the then-existing conditions, likewise governs Amoco's royalty obligations regarding take-or-pay payments made in settlement of the gas contract litigation. See *Henry*, supra. In so doing, we recognize the virtually perfect alignment of interests existing among the lessee, lessors, and society regarding certain limited aspects of the lease, including resolution of the pipeline-purchaser's financial inability to comply with the take-or-pay provisions in the long term gas contract. *See, e.g.,* Pierce, *Lessor/Lessee Relations in a Turbulent Gas Market, Oil and Gas Law and Taxation,* § 8.02[1] (1987) (regarding the decision to produce or defer). *But see* Donohoe, *Implied Covenants in Oil & Gas Leases & Conservation Practice,* 33 Inst. on Oil & Gas L. & Tax'n 97, 98 (1982) (lessor/lessee relationship is "complex one of mutual interest mingled with antagonism.")

More specifically, were a producer to force a pipeline-purchaser to comply with the long-term gas contract, despite the decline in price and market, it is not unlikely the pipeline would be faced with insurmountable financial problems and, eventually, forced into insolvency. The producer who forces compliance with the contract in the face of the pipeline's financial ruin, would ultimately frustrate its own primary objective of assuring a market for the gas. A financially insolvent pipeline will not purchase *any* gas, and a royalty interest is worthless if no gas is sold. Confronted with the potential loss of its market for gas, the prudent producer consents to settlement and is thus assured an uninterrupted market for gas.

Indeed, in brief to this Court, Amoco acknowledges the failure to renegotiate the gas purchase contract would have resulted in a "fire-sale" disposition of the Morganza gas on the spot market. Because the duty to market is a continuing one, Amoco should not be able to enjoy the benefits of the settlement while refusing to share the benefits with Frey. Assuming Amoco acted reasonably in settling the Morganza Contract litigation, and there is no evidence to suggest otherwise, the take-or-pay payments received by Amoco as a result of the settlement inure to the mutual benefit of Amoco and Frey.

The practical considerations sustain our position. For example, "if lessors did not share in take-or-pay payments, lessees would have an incentive to compromise volume gas prices under their contracts or settlements with pipelines in exchange for favorable take-or-pay terms." See also Mineral Code art. 109, comment, for an analogous situation involving the owner of the executive right's exercise of his rights vis a vis the royalty, or non-executive holder. Accordingly, we find justice best served by the decision we reach today.

C. Conflicting Jurisprudence

We are aware of the federal and state court jurisprudence holding contrary to the position we assert herein. See e.g, Gerard J.W. Bos & Co., Inc. v. Harkins & Co., 883 F.2d 379 (5th Cir. 1989) (payments under a settlement for cancellation of a take or pay gas contract are not subject to royalty); *Diamond Shamrock,* supra (no royalty due on take-or-pay payments received in ordinary course of business); Piney Woods Country Life School v. Shell Oil Co., 726 F.2d 225, 234 (5th Cir. 1984), *cert. denied,* 471 U.S. 1005, 105 S.Ct. 1868, 85 L.Ed2d 161 (1985) ("a gas sale contract is executory and . . . no particular gas is sold until it is identified–i.e., brought to the surface."); Killam Oil Co. v.

Bruni, 806 S.W.2d 264, 267 (Tex.App.-San Antonio 1991), writ denied (no royalty is due on settlement proceeds resulting from breach of take-or-pay provision in gas purchase contract); Wyoming v. Pennzoil Co., 752 P.2d 975, 980 (Wyo. 1988) (royalties are due "only upon physical extraction of the gas from the leased tract"). See also ANR Pipeline Co. v. Wagner & Brown, 44 FERC § 61,057 (1988) (take-or-pay payments do not violate maximum lawful price (MLP) provisions of the NGPA); Kramer, 39 Oil & Gas L. & Tax'n at § 5.01 n. 1 (commentators split evenly over whether royalty obligations are owed upon receipt of take-or-pay monies). We need hardly state that such law is only persuasive in our jurisdiction. Nor does this mark the first time we decline to follow a majority view. See, e.g., *Henry,* supra. Most important, the mineral law of Louisiana evolved not from the common law, but from the Civil Code, richly steeped in our civilian heritage. See La.R.S. 31:2, comment.

Finally, in responding to the question certified, we have deliberately refrained from distinguishing recoupable payment from nonrecoupable payment. We note in passing, however, Columbia has recouped in cash the entire $45.6 million recoupable take-or-pay settlement payment made to Amoco. Clearly, Frey is entitled to a royalty payment for the gas actually taken, and has, in fact, received such a payment. However, it may be in cases such as this the lessor is also entitled to compensation for the lost time value of money. We leave to the federal courts the intricacies of the accounting, unpersuaded the complexity of the task should in some manner influence our decision. In this case, the settlement agreement establishes the price per unit of gas recouped during the make-up period, and thus we pretermit discussion of the issues raised should the original contract price be higher or lower than the price of the gas at the time of recoupment. We do note, however, courts are free to provide for such contingencies in their judgments and thus to protect the producer and royalty owner from overpayment or underpayment of royalty, respectively.

Conclusion

Accordingly, our answer to the question certified to us is in the affirmative. Pursuant to Rule XII, Supreme Court of Louisiana, the judgment rendered by this court upon the question certified shall be sent by the clerk of this court under its seal to the United States Court of Appeals for the Fifth Circuit and to the parties.

NOTES

1. Should the obligation to pay royalty on take or pay or settlement proceeds depend on how the payments are characterized? A Tenth Circuit opinion made the following classifications:

First, royalty payments are not due under a "production"-type lease unless and until gas is physically extracted from the leased premises. Second, nonrecoupable proceeds received by a lessee in settlement of the take-or-pay provision of a gas supply contract are specifically for non-production and thus are not royalty bearing. Third, any portion of a settlement payment that is a buy-down of the contract price for gas that is actually produced and taken by the settling purchaser is subject to the lessor's royalty interest at the time of such production, but only in an amount reflecting a fair apportionment of the price adjustment payment over the purchases affected by such price adjustment. Harvey E. Yates Co. v. Powell, 98 F.3d 1222, 1231, 135 O.&G.R. 100 (10th Cir. 1996).

What happens if there is a "buyout" of the contract and then a second contract executed with lower prices? In re Century Offshore Management, Inc., 111 F.3d 443, 136 O.&G.R. 40 (6th Cir. 1997).

2. Does the Harrell principle of cooperation used by the court in *Frey* extend to how the lessee and the gas purchaser characterize any settlement monies that come from a breach of a take-or-pay clause? See Bruce Kramer, "Royalty Obligations Under the Gun–The Effect of Take-or-Pay Clauses on the Duty to Make Royalty Payments," 39 Inst. on Oil & Gas L.&Tax'n 5-1 (1988). Arkansas has also embraced the Harrell principle, in SEECO, Inc. v. Hales, 341 Ark. 972, 22 S.W.3d 157, 148 O.&G.R. 155 (2000). Does the principle of cooperation apply to a non-lease relationship? *See* JN Exploration & Production v. Western Gas Resources, Inc., 153 F.3d 906, 141 O.&G.R. 93 (8th Cir. 1998) (a seller of natural gas who was entitled to a percentage of the "net proceeds" from the sale by the processor/purchaser to third parties claimed a share of a lump sum take-or-pay settlement payment; the court refused to extend the "Harrell rule" to the context of a U.C.C. contract situation.).

3. Consider the right of the lessor to share other payments received by the lessee: (1) cash and an overriding royalty received for an assignment of the lease; (2) payments to a lessee by a gas purchaser for gathering and transporting gas through the lessee's pipeline. *See* Maxwell, *Oil and Gas Royalties?A Percentage of What?*, 34 Rocky Mt.Min.L.Inst. 15-1, at 15-35 (1988).

D. OVERRIDING ROYALTY CALCULATION
FOLLOWWILL v. MERIT ENERGY CO.

United States District Court for the District of Wyoming, 2005
371 F.Supp.2d 1305, 162 O.&G.R. 872

DOWNES, DISTRICT JUDGE. This matter comes before the Court on the Defendants' Joint Motion for Partial Summary Judgment on Plaintiffs' Wyoming State Law Claims. The Court . . . FINDS and ORDERS as follows:

Background

All Plaintiffs, who are citizens of Colorado, own overriding royalty interests (ORRI) carved out of federal oil and gas leases covering lands located in various counties in Wyoming Plaintiffs bring this action against Defendants for alleged violations of the Wyoming Royalty Payment Act (WRPA) and common law. Specifically, Plaintiffs assert the following claims for relief: (1)

unjust enrichment resulting from Defendants failure to pay the proper ORRI amount due them in accordance with the provisions of WRPA; (2) an accounting of all the production, sales of production, production revenues, and appropriate expenses deductible and deducted from and allocable to their ORRI payments; (3) breach of duty to pay Plaintiffs their share of the production revenues; (4) breach of the implied covenant of good faith and fair dealing; (5) injury to property rights; (6) all statutory remedies provided for violation of WRPA; (7) failure to properly report production, sales, and deduction information as required by WRPA; (8) preliminary and permanent injunction restraining Defendants from failing to pay Plaintiffs the amount due them from the production revenues; (9) refusing to give Plaintiff information about the wells and production revenues as required by WRPA; and (10) punitive damages.

All of the Plaintiffs' ORRI are derived from leases issued by the United States Department of the Interior, Bureau of Land Management (BLM) and are located on federal land in Wyoming. The BLM form mineral leases all contain the following language:

> This oil and gas lease is issued ... subject to the provisions of the Mineral Leasing Act and subject to all rules and regulations of the Secretary of Interior now or hereafter in force, when not inconsistent with any express and specific provisions herein, which are made a part hereof.

The leases also contain language that the United States, as lessor, is to be paid a "12½ % royalty on the production removed or sold from the leased lands computed in accordance with the Oil and Gas Operating Regulations (30 CFR Pt. 221)." Defendants' payments of royalties due to the United States under the federal leases are comprehensively regulated by the federal government pursuant to the Mineral Leasing Act, the Federal Oil and Gas Royalty Management Act, 30 U.S.C. §§ 1701 et seq., the Royalty Simplification and Fairness Act, 30 U.S.C. § 1735, and implementing regulations of the United States Department of the Interior, Minerals Management Service (MMS).

Plaintiffs Dorman Followwill and C. Dennis Irwin, Jr. acquired their interests in these leases as a result of an agreement dated February 9, 1973, with C & K Petroleum Inc. Pursuant to the terms of the agreement, Followwill and Irwin agreed to provide exploratory geologic services and C & K Petroleum agreed to provide certain cash compensation and, in addition, "to assign to Followwill and Irwin overriding royalties under the following terms and provisions: With respect to all leases acquired hereunder by C & K ... within a period of five (5) years from the date of completion by Followwill and Irwin of their services hereunder, ... C & K will assign to Followwill and Irwin jointly in equal shares an overriding royalty equal to one and one-half (1½ %) per cent of 8/8ths of all the oil and gas produced, saved and marketed under the terms of such leases ...." The agreement further stated, "Said overriding royalties are to be computed and paid in the same manner as the corresponding lessor's royalty."

The subsequent Assignments of Overriding Royalty to Plaintiffs Followwill and Irwin, recorded August 11, 1976 and December 13, 1976, assign "an overriding royalty equal to 1½ % of all the oil, gas, casinghead gas, and other hydrocarbon substances which may be produced, saved and sold *under the terms of that certain oil and gas lease listed above* ...." (Emphasis added.) The Assignments further state that "the obligation to pay any overriding royalties or payments out of production of oil created herein, which, when added to overriding royalties or payments out of production previously created and to the royalty payable to the United States, aggregate in excess of 17 ½ %, shall be suspended when the average production of oil per well per day averaged on the monthly basis is 15 barrels or less."

The remaining Plaintiffs ORRI were reserved or created when the subject federal leases were assigned to another party, at which time the assignor (each Plaintiff) retained an ORRI as partial consideration for the lease assignment. By this process, the assigned leases ultimately reached the Defendants. The assignments were made using a standard BLM assignment form which includes a section where the assignor can specify what ORRI he or she is reserving. That section refers the assignor to Item 4 of the General Instructions which provides:

Overriding royalties or payments out of production-Describe in an accompanying statement any overriding royalties or payments out of production created by assignment but not set out therein. If payments out of production are reserved by assignor, outline in detail the amount, method of payment, and other pertinent terms.

None of the Plaintiffs/assignors specified any other terms regarding the method of calculating their ORRI which would be different from the method of valuation for the lessor's royalty as set forth in the federal leases. Moreover, the BLM assignment forms contain the exact same language establishing a 5% cap on overriding royalties as that contained in the Assignments to Plaintiffs Followwill and Irwin.

Most of Plaintiffs' claims hinge upon the applicability of WRPA to their overriding royalty interests in the federal leases. Defendants' motion seeks summary judgment as to Plaintiffs' WRPA claims and all claims dependent upon WRPA. Defendants contend that Plaintiffs' state-law claims which rely on WRPA fail for several reasons. First, Defendants argue that WRPA has no application to this case because the parties have, by specific language in written executed agreements, made explicit reference to federal procedures and methods of royalty computation. Second, Defendants argue that the Wyoming legislature did not intend that the provisions of the WRPA apply to a situation such as that presented by the facts of this case. Finally, Defendants argue that Plaintiffs' WRPA claims are preempted by the comprehensive federal royalty scheme.

Discussion

In Wyoming, royalties and overriding royalties are to be computed in accordance with WRPA, "[u]nless otherwise expressly provided for by specific language in an executed written agreement." Wyo. Stat. § 30-5-305(a). Defendants contend that the plain language of the leases and assignments at issue in this case establish federal regulations as the basis on which to determine the value of production used to calculate the various royalties to be paid out of production, thereby negating application of WRPA.

"An assignment of an oil and gas lease is a contract." *Wolter v. Equitable Resources Energy Co.,* 979 P.2d 948, 951 (Wyo.1999). Therefore, the Court will examine the reservation language in accordance with the general principles of contract interpretation. Likewise, the assignment of an overriding royalty is also a contract subject to the general principles of contract interpretation. The "prime focus" of the Court in interpreting a contract is to determine the parties' intent. . . .

Because the assignments at issue do not specifically state that the overriding royalties are to be computed and paid in the same manner as the corresponding federal lessor's royalty, the Court finds the assignments ambiguous on that point. However, WRPA does not require that the specific language necessary to avoid WRPA applicability be in the assignment itself. Therefore, it is appropriate for the Court to consider evidence in addition to the assignments in order to determine the intention of the parties.

In the case of Plaintiffs Followwill and Irwin, they expressly agreed that any overriding royalties they received would be computed and paid in the same manner as the federal lessor's royalty. . . .

In the case of the remaining Plaintiffs, the Court concedes that the absence of an agreement like that entered into by Followwill and Irwin makes the analysis more problematic. But, it is more plausible in the eyes of the Court to apply the same reasoning in determining the parties' intent. There is simply no evidence that the parties' intent was anything other than the use of federal regulations to value the various royalties to be paid on production.

The original federal leases expressly state that the lessor's royalty on production removed or sold from the leased lands shall be computed in accordance with federal regulations. The subsequent assignments at issue reserve an overriding royalty of the lessee's interest in the federal lease. Historically, an important aspect of an "overriding royalty" (whether reserved by the lessor in a lease or created by a subsequent instrument executed by the lessee or the lessee's successor) is that it is a "royalty," that is, "in the absence of an express agreement to the contrary it is free of costs of which the lessor's royalty is free and it is subject to the costs to which the lessor's royalty is subject." Williams & Meyers, Manual of Oil & Gas Terms at 768 (1994). WRPA did not exist at the time of these assignments, so the parties could not have intended that the definitions contained in WRPA would apply to the ORRI being reserved. In light of the historical understanding of an overriding royalty, and the absence of an express agreement to the contrary (which is specifically contemplated by General Instruction number 4 on the BLM assignment form), the parties intended that any overriding royalty would be computed and paid in the same manner as the federal lessor's royalty. A further indicator of this intent is that each of the assignments at issue specifically refer to a limit on overriding royalties or payments out of production when aggregated with the royalty payable to the United States, as did the assignments to Plaintiffs Followwill and Irwin.

When considering all of the evidence, it is clear to the Court that the parties' intent at the time the assignments were made was to compute any ORRI in the same manner as the federal lessor's interest. The federal oil and gas leases, from which Plaintiffs' ORRI inextricably derive, contain specific language expressly providing for the computation of overriding royalties in accordance with federal regulations. Accordingly, WRPA is inapplicable when determining whether Defendants correctly calculated the value of Plaintiffs' ORRI. All of Plaintiffs' claims based upon a calculation of their ORRI in accordance with WRPA must be dismissed. THEREFORE, it is hereby

ORDERED that the Defendants' Joint Motion for Partial Summary Judgment on Plaintiffs' Wyoming State-Law Claims (Docket # 80) is **GRANTED**. Plaintiffs' First, Second, Fifth, Sixth, Seventh, Eighth, Ninth, and Tenth Claims for Relief are **DISMISSED.**

NOTES

1. The principal case is an illustration of overriding royalty, the calculation of which is linked to the calculation of the lessor's royalty. The overrides here are fairly typical of the manner in which overrides arise in a variety of oil and gas transactions. Why would override owners want to tie the calculation of their royalties to those of the lessor? Why would they want them un-tied in this instance? Does one need to know more about the Wyoming Royalty Payment Act to answer this?

2. Wyoming's Royalty Payment Act, Wyo. Stat. Ann. § 30-5-301 et seq., is a remedial measure to protect royalty owners by defining what are the costs of production that are borne solely by the working interest owner. Included in the list of costs of production are exploration, development, enhanced recovery, drilling and completion, pumping or lifting, recycling, gathering, compressing, pressurizing, heater treating, dehydrating, separating, storing or transporting costs. In Morris v. CMS Oil & Gas Co., 2010 WY 37, 227 P.3d 325, the court did not disturb a lower court ruling that the WRPA may supplant common law remedies for failure to make accurate and timely royalty payments. The burden of proof of showing untimely or inaccurate payments was on the plaintiff. The opinion explores the reporting or check stub requirements of the WRPA as well as the need to pay directly or into an escrow fund, the royalties that come due within 6 months of initial production and within 60 days of subsequent

production. The Act has been interpreted as not adopting the "marketable product" common law rule of Colorado and Oklahoma, so that gathering charges incurred after gas has been dehydrated, and therefore marketable, still fall within the statutory definition of gathering and are not chargeable against the royalty interest. Wold v. Hunt Oil Co., 52 F. Supp. 2d 1330, 145 O.&G.R. 422 (D. Wyo. 1999). The venue provisions of the Act have been interpreted to allow a court to have jurisdiction over a class action lawsuit involving production from counties outside of the county where the lawsuit is filed. BP America Production Co. v. Madsen, 2002 WY 135, 53 P.3d 1088, 156 O.&G.R. 193. On the Wyoming act generally, *see* Brandin Hay, *Wyoming's Royalty Payment Act*, 31 Land & Water L. Rev. 823 (1996); Rickey Turner, *Royalty Dethroned: Wyoming's Approach to Gathering Costs, Cabot Oil & Gas Corp. v. Followwill, 93 P.3D 238 (Wyo. 2004)*, 5 Wyo. L. Rev. 665 (2005)

BOLDRICK v. BTA OIL PRODUCERS

Court of Appeals of Texas,Eastland, 2007
222 S.W.3d 672

JOHN G. HILL, JUSTICE. James P. Boldrick appeals from a final judgment that denied his summary judgment motion while granting the summary judgment motion of appellee BTA Oil Producers, a partnership. The judgment, among other things, declared that certain overriding royalty interests claimed by Boldrick are not payable to him until such time that nonconsent penalty provisions of a September 1973 joint operating agreement have been fully recouped by consenting parties and, accordingly, that BTA is not required to cause the payment of those overriding royalty interests until those funds are received by BTA. Boldrick contends in five points on appeal that the trial court erred in granting BTA's summary judgment motion and denying his for the following reasons: (1) his overriding royalty interests are not subject to the nonconsent penalty provisions of the September 1973 joint operating agreement as between him and BTA; (2) his overriding royalty interests are not "subsequently created interests" as that term is used in the joint operating agreement for BTA's benefit; (3) the court has misconstrued the effects of one division order presented to it and failed to recognize the language of the division order that applies to the well in question; (4) even if the interest BTA relinquished to Chevron during payout pursuant to Paragraph 12 of the operating agreement includes the overriding royalty interests claimed by Boldrick, BTA is not excused from the specific language of its overriding royalty grant to Boldrick's predecessor in interest; and (5) any obligation or lack of obligation of BTA for drilling or development of the oil and gas leasehold estate is not a controlling issue in the case. We construe all of these points as a single issue: whether the trial court erred in granting summary judgment for BTA while denying summary judgment to Boldrick. We affirm.

The facts in this case are undisputed. On September 15, 1973, Texaco as operator and Ben J. Fortson and Exxon as nonoperators entered into a joint operating agreement for the exploration and development of their leases and interests for oil and gas with respect to all of Section 51, Block 34, of the H & TC Ry. Co. Survey, Ward County, Texas. On February 4, 1977, Texaco and Sabine Production Company entered into a sublease agreement with respect to this same property. This sublease was subject to the 1973 joint operating agreement. BTA and Sabine shared a sublease interest.

After a test well, the 7706 JV-P Stallings No. 1 Well, was drilled and was paid out as defined in the February 4, 1977 agreement, BTA executed an assignment of overriding royalty interest, pursuant to the terms of a February 11, 1977 letter agreement, to Sabine, Carroll M. Thomas, Clyde R. Harris, and R.G. Anderson. Boldrick was the successor of Harris's interest by virtue of an assignment to him from Harris. BTA's assignment of an overriding royalty interest provided in part that "[s]aid overriding royalty interests shall be free and clear of all costs of development and operation" and "[t]his Assignment shall not imply any leasehold preservation, drilling or development obligation on the part of Assignor."

Subsequently, Chevron USA, Inc., the operator under the 1973 operating agreement and an owner of an undivided interest, proposed the drilling of the Stallings Gas Unit 2H Well; but BTA elected "non-consent status" as that term is defined in Paragraph 12 of the operating agreement. The Stallings Gas Unit 2H Well was drilled and completed.

Chevron initially made payments to Boldrick on production from the Stallings Gas Unit 2H Well but, later, requested that the funds paid be returned because Chevron contended that its division order was the result of a mistake. Neither BTA nor Boldrick are currently receiving any payments on production from the Stallings Gas Unit 2H Well.

Boldrick and others sued BTA and Chevron/Texaco for money damages, alleging breach of contract, unjust enrichment, and conversion all because its share of the overriding royalty interest, which, according to the assignment was to be free and clear of all costs of development and operation, was being used for the benefit of the defendants, including BTA. As a counterclaim, Chevron sought a declaratory judgment that it has no obligation to pay the overriding royalty interest claimed by Boldrick and the other plaintiffs, while BTA sought declaratory judgment that it has no obligation to account to the plaintiffs for the overriding royalty interest and that, because it has not received any of the proceeds attributable to the share of oil and gas claimed by the plaintiffs, it has no obligation to account to the plaintiffs for the overriding royalty interest. As noted, BTA and Boldrick are the only remaining parties to the lawsuit.

Paragraph 31(b) of the joint operating agreement provides that any subsequently created interest shall be specifically made subject to all terms and provisions of the operating agreement. It defines a subsequently created interest so as to include the creation, subsequent to the joint operating agreement, of an overriding royalty created by a working interest owner out of its working interest.

Inasmuch as BTA, a working interest owner, created Boldrick's overriding royalty out of its working interest subsequent to the operating agreement, the overriding royalty is subject to all terms and provisions of the operating agreement.

Paragraph 31(b) further provides that, where such a working interest owner elects to go nonconsent under Paragraph 12 of the joint operating agreement, the subsequently created interest shall be chargeable with a pro rata portion of all costs and expenses under the operating agreement in the same manner as if it were a working interest. Consequently, inasmuch as BTA elected to go nonconsent, Boldrick's overriding royalty became chargeable with a pro rata portion of all costs and expenses under the operating agreement in the same manner as if it were a working interest. Inasmuch as the use of the proceeds that would have come to Boldrick under his overriding royalty to meet the costs and expenses under the operating agreement is mandated by the operating agreement, such a use could not constitute a breach of contract between Boldrick and BTA that was subject to the operating agreement and could not constitute unjust enrichment or conversion. Any issue as to whether BTA must reimburse Boldrick for any costs and expenses paid with the proceeds of Boldrick's overriding royalty if and when it receives proceeds representing its working interest in the well in question was not determined in the trial court, and we do not address that issue in this appeal.

Boldrick contends in point one that his overriding royalty interest is not subject to the nonconsent penalty provision of the operating agreement as between him and BTA. In point two, he contends that his overriding royalty interest is not a "subsequently created interest" as that term is used in the joint operating agreement for BTA's benefit. He groups these two points for purposes of argument in his brief. . . .

Boldrick takes the position that the documents of record, including title opinions, clearly show why his interest is not a subsequently created interest. He refers us to provisions of the 1977 sublease agreement that prohibited any assignment by Sabine without the written consent of Texaco and that, if Texaco ever reacquired Section 51 from Sabine or its assigns, that interest would be free and clear of overriding royalty interests, while noting that Texaco agreed to the creation by BTA of the overriding royalty interest to his predecessor in title and that Texaco never reacquired its interest from BTA. We fail to see how any of these facts show that Boldrick's overriding royalty interest was not subject to the nonconsent penalty provisions of the joint operating agreement as between him and BTA or that it is not a "subsequently created interest" as that term is used in the joint operating agreement. . . . We disagree with his contention that Chevron/Texaco's consent to the assignment by BTA of Boldrick's overriding royalty interest means that that interest, which was created subsequent to the 1973 joint operating agreement, was not a subsequently created interest under the terms of that agreement. . . .

Boldrick urges in point three that the division order relied upon by the trial court did not apply to the well as to which BTA was a nonconsenting party. The division order, executed by Boldrick's predecessor, provides that BTA has no obligation to disburse funds that it has not received. Boldrick suggests that the division order did not apply to the well in question nor did it relate to the same formation. However, the division order by its terms is applicable to all wells located in the described area, an area that includes the well in question. Even if this division order was not effective, it would not change our conclusion that, considering all the documents in question and any other summary judgment evidence, BTA has no present obligation to pay Boldrick for his overriding royalty interest that is currently being used as provided by Paragraph 31(b) of the joint operating agreement. Boldrick relies upon a Chevron division order that Chevron contends was a mistake. For the reason stated, we agree that Chevron's division order was a mistake. We disagree with Boldrick's assertion that BTA, in a capacity as purchaser or assignee, has any current duty to pay him his overriding royalty.

Boldrick insists in point four that, even if the interest that BTA relinquished to Chevron during payout pursuant to Paragraph 12 of the joint operating agreement includes the overriding royalty interest claimed by him, BTA is not excused from the specific language of its overriding royalty grant to him. In point five, he asserts that BTA's obligation or lack thereof for drilling or development of the oil and gas leasehold estate is not a controlling issue of the case. He argues these two points together.

Boldrick refers us to Paragraph 12 of the joint operating agreement. That paragraph provides that, when operations are commenced for the drilling of a well by consenting parties, the nonconsenting parties relinquish to the consenting parties all of their interests in the well and share of production until the proceeds equal the total of the costs and penalties that are provided in the paragraph. The paragraph notes that proceeds that apply to the total due before the interests revert back to the nonconsenting party do not include overriding royalty interests. However, as previously noted, Paragraph 31(b) provides that, where those overriding royalty interests are subsequently created interests as defined, each is chargeable with a pro rata portion of all costs and expenses as if it were a working interest. If there is any conflict between the two paragraphs, Paragraph 31(b) provides that it is applicable "[n]otwithstanding anything herein to the contrary." We find nothing in Paragraph 12 that conflicts with our holding in this case.

Boldrick refers us to language in Paragraph 13 of the joint operating agreement stating that "[e]ach party shall pay or deliver, or cause to be paid or delivered, all royalties due the owners of ... overriding royalties ... as shown by Exhibit A." We first note that Paragraph 13 deals with "RIGHT TO TAKE PRODUCTION IN KIND" and that this case does not involve the taking of production in kind. We also note that Boldrick's royalty interest is not included in either the original "Exhibit A" attached to the joint operating agreement or the amendment to "Exhibit A" that was part of the 1977 sublease transaction. Again, we note that, in the event that Paragraph 13 conflicts with Paragraph 31(b), we must give effect to Paragraph 31(b) because of its language of "[n]otwithstanding anything herein to the contrary." . . .

Boldrick asserts that the trial court erroneously concluded that BTA owned no interest and had conveyed its interest back to the operator. The only citation he gives in support of his assertion is the 1977 sublease agreement between Sabine and Texaco. In its opinion letter written to the parties, the trial court merely noted that, under the operating agreement, the interest in a producing well of a nonconsenting party is relinquished to the consenting parties, is accomplished without deeds of conveyance, and reverts upon the recoupment of penalties attributable to nonconsent. This expression by the trial court is consistent with Paragraph 31b of the joint operating agreement.

Boldrick refers us to a number of cases that he indicates show that for at least some purposes the nonconsenting owner continues to have an ownership interest pending payment of expenses and penalty. These include *United States v. Cocke,* 399 F.2d 433 (5th Cir.1968) (carrying party, not carried party, has right to certain federal income tax deductions); *Dorsett v. Valence Operating Co.,* 111 S.W.3d 224 (Tex.App.-Texarkana 2003), *rev'd on other grounds,* 164 S.W.3d 656 (Tex.2005) (nonconsenting party relinquishes right to share of production revenue until consenting party receives designated share of expenses); *R.R. Comm'n of Tex. v. Olin Corp.,* 690 S.W.2d 628 (Tex.App.-Austin), *writ ref'd n.r.e.,* 701 S.W.2d 641 (Tex.1985) (Texas Railroad Commission has authority to order a nonconsenting party, which has a carried interest pending payment of specified costs, to plug well where operator lacks the funds to do so). Whatever interest BTA might continue to have pending its payment of the costs of development of the well in question and the penalties provided due to its nonconsenting status, it does not change the fact that Boldrick's interest as the holder of an overriding royalty interest assigned to it by BTA out of BTA's interest is chargeable with a pro rata share of all costs and expenses to be received by the consenting parties and applied to the costs of production. In oral argument, Boldrick relied upon the case of *Seagull Energy E & P, Inc. v. Eland Energy, Inc.,* 207 S.W.3d 342 (Tex.2006). We find that case to be distinguishable. In *Seagull,* the court held that one selling one's oil and gas working interest remains liable to the operator under the operating agreement unless released by the operator or the terms of the agreement. The court based its ruling on the fact that the operating agreement did not deal specifically with the issue of an assignment of a working interest to a third party. In the case at bar, the joint operating agreement does have a specific provision that deals with what happens to an overriding royalty interest created by a nonconsenting party. We conclude that the trial court did not err in granting BTA's motion for summary judgment and in denying Boldrick's motion for summary judgment.

The judgment is affirmed.

NOTES

1. Typically, an overriding royalty is a fractional interest in the gross production of oil and gas under a lease, in addition to the usual royalties paid to the lessor, free of any expense for exploration, drilling, development, operating, marketing and other costs incident to the production and sale of oil and gas produced from the lease. However, in the principal case, the override was carved out of an interest that under a joint operating agreement could be subject to nonconsent penalty if the lessee failed to participate in the drilling of a well. Does the decision turn on the words of the instrument creating the overriding royalty or the words of the joint operating agreement? Was the assignment of the override a contract to pay or a conveyance of an interest in real property subject to conditions? What inferences should be drawn from the provision stating "[t]his Assignment shall not imply any leasehold preservation, drilling or development obligation on the part of Assignor"? How would you draft an instrument for an override that would require payment of royalty when the assignor goes non-consent as in *Boldrick*?

2. It has been held that parties could agree to the conveyance of overriding royalty on property they did not own. Scrivner v. Sonat Exploration Co., 242 F.3d 1288, 150 O&GR 275 (10th Cir. 2001) (agreement included an appraisal procedure to provide for an equivalent value to the override assignments if the assignor was unable to convey the overrides).

3. There is no uniformity as to the size of an overriding royalty, but commonly it will not be larger than 1/8 of 7/8, and much smaller fractions are frequent. By the express terms of the instrument creating the overriding royalty or by the terms of a later instrument executed by the owner of the override (e.g., a unitization agreement), it may be provided that the amount of an overriding royalty shall vary depending on the amount of production (a variable royalty) or that the overriding royalty shall terminate, temporarily or permanently, if production falls below a given level or if production is not profitable after payment of the overriding royalty.

E. REMEDIES FOR NONPAYMENT
(*See* Martin & Kramer, *Williams and Meyers Oil and Gas Law* §§ 656-656.7)

CANNON v. CASSIDY
Supreme Court of Oklahoma, 1975.
542 P.2d 514, 52 O.&G.R. 533.

SIMMS, JUSTICE: Can an oil and gas lease be cancelled for lessees' failure to pay accrued royalty when that failure is in violation of the express terms of the lease but such remedy is not expressly provided by the lease?

The trial court answered in the negative, granting judgment to lessees. The Court of Appeals reversed the trial court, cancelling the leases and quieting title thereto against lessees. We Grant Certiorari, Reverse the Court of Appeals, and

Reinstate the District Court's judgment in favor of lessees.

The oil and gas leases in question provided for quarterly payments to lessors of one-eighth (1/8) royalty on gas sold. The leases do not provide for forfeiture in the event of lessees' failure to pay accrued royalties.

Lessees did not pay lessors royalties for gas produced and marketed for approximately eleven months. Lessors brought this action for cancellation of the oil and gas leases based upon the non-payment and they also sought to quiet their title to the real property. Lessors pled that the failure of lessees to pay accrued royalties breached both the express and implied covenants of the lease.

Trial was to the court on lessors' motion for judgment on stipulated facts. Briefs and exhibits were presented by the parties. The parties stipulated that: " . . . from August, 1971, until July, 1972, gas was produced from the subject tracts and sold to Cities Service Oil Company but lessees failed to account and remit to plaintiffs the proceeds attributable to their royalty interest derived from such sales in the amount of $1,693.62."

The trial court gave judgment for lessees based on the finding that cancellation of an oil and gas lease for lessees' failure to pay royalty as provided in the lease will not lie without an express provision in the lease which authorizes cancellation thereof for non-payment of royalties.

Lessees submit that the trial court's ruling is correct and in support thereof they cite the case of Wagoner Oil & Gas Co. v. Marlow, 137 Okl. 116, 278 P. 294 (1929), wherein the ninth syllabus by the Court states that:

> Failure to pay royalty or for injury to the land as provided by the lease will not give the lessors sufficient grounds to declare a forfeiture, unless by the express terms of the lease they are given that right and power.

Lessees further contend that cancellation of the lease would be harsh and inappropriate as lessors had a speedy and adequate remedy at law for money damages.

In their attempt to exempt this matter from the scope of the holding in *Wagoner, supra,* lessors present the novel argument that lessees' non-payment of royalties constituted more than a breach of the express terms of the lease. They submit that the non-payment was additionally a breach of the implied covenant to market.

However, lessors are unable to present any persuasive authority to the Court in support of this assertion. Primarily they rely on the suggestion by Earl A. Brown in his two volume work, Law of Oil and Gas Leases, 2nd Ed., at § 6.02 that the implied covenant to market should be seen as a two-pronged obligation including both the sale of the products and payment to lessor of his share of the proceeds.

Cancellation, of course, is recognized as the proper equitable remedy for a breach of implied covenants where justice will be thus served. Coal, Oil and Gas Co. v. Styron, Okl., 303 P.2d 965 (1956).

From their premise that failure to pay royalties is a breach of the implied covenant to market, lessors' argument concludes with the claim that our decision in Townsend v. Creekmore-Rooney Company, Okl., 332 P.2d 35 (1958) is direct authority supporting cancellation of the leases involved herein. However, the facts in *Townsend, supra,* did not concern the payment of royalties. There we held only that the unexplained *cessation* of marketing of oil or gas from the leases for an extended period of several months was prima facie sufficient to justify cancellation as a breach of the implied covenant to market.

Neither Mr. Brown's opinion or lessors' arguments persuade us to adopt the notion that payment of royalties comes within the ambit of the implied covenant to market.

Consequently, we do not depart from the rule expressed in *Wagoner, supra,* that lessee's failure to pay royalty as provided by the lease will not give lessors sufficient grounds to declare a forfeiture unless by the express terms of that lease they are given that right and power.

We note in passing that the overwhelming majority of jurisdictions which have considered this issue are in accord.

. . .

Through stipulation of the parties lessors' damages were agreed to amount to $1,693.62. As lessees correctly contend, lessors had a plain, speedy and adequate remedy at law available to them which, if pursued, would have fully compensated them for this loss. This Court has repeatedly held that in such circumstances equity will not grant relief to a litigant who fails to pursue his remedy at law. . . .

Therefore, the judgment of the Court of Appeals as modified and corrected by the Order of Amendment is Vacated and the Judgment of the Trial Court Granted in favor of lessees is Affirmed.

HODGES, V.C.J., and DAVISON, IRWIN, LAVENDER, SIMMS and DOOLIN, JJ., concur.

WILLIAMS, C.J., and BARNES, J., dissent.

NOTES

1. The normal remedy of the lessor for nonpayment of royalties is a damage action at law for breach of the express covenant to pay royalty on production. *See* Hafeman v. Gem Oil Co., 163 Neb. 438, 80 N.W.2d 139, 7 O.&G.R. 41 (1956). If the royalty is payable in kind, a royalty owner also has a tort action for conversion available. This remedy will be unavailable where the lessor has executed a division order, or failed to provide necessary storage for royalty oil. In the latter case the lessee may be viewed as having implied authority to dispose of the royalty oil.

If neither lessee nor lessor effectively bestir themselves to get a division order signed and royalty paid, is interest due on late royalty payments? *See*

Schaffer v. Tenneco Oil Co., 278 Ark. 511, 647 S.W.2d 446, 77 O.&G.R. 233 (1983).

2. Should punitive/exemplary damages be available to the lessor for improper or inadequate payment of royalty? In Exxon Mobil Corp. v. Ala. Dept. of Conservation and Natural Resources, 986 So.2d 1093 (Ala.,2007), the court held that "punitive damages cannot be awarded … absent a finding of fraud." The jury had awarded the state punitive damages of $11.8 billion for underpayment of royalty based on fraud, reduced by the trial judge to $3.5 billion. The fraud determination was overturned by Alabama Supreme Court as not supported by substantial evidence. What facts would be necessary to establish that breach of an obligation to pay royalty constitutes a fraudulent act?

The lease in question in Exxon Mobil Corp. v. Ala. was drafted by the state and was noteworthy for a provision stating that "[i]n case of ambiguity, this lease always shall be construed in favor of LESSOR and against LESSEE."

3. Before 1975, lease cancellation was often available in Louisiana as a remedy for nonpayment of royalties even in the absence of an express forfeiture provision. *See, e.g.,* Fontenot v. Sunray Mid-Continent Oil Co., 197 So.2d 715, 27 O.&G.R. 140 (La.App.1967), writ ref'd, 250 La. 898, 199 So.2d 915, 27 O.&G.R. 150 (1967). The Louisiana cases were based on an extension of Civil Code provisions concerning ordinary leases to oil and gas leases.

The Louisiana Mineral Code (LSA-R.S. 31:137-31:143), effective January 1, 1975, has made cancellation available only under very limited circumstances. The remedy is now available only where the lessee fraudulently withholds royalties due, or does not either pay the royalties or give a good reason for failure to pay them in response to a written notice sent by the lessor. Even when cancellation is available as a remedy it is only to be granted where the remedy of damages "is inadequate to do justice." *See e.g.,* Wegman v. Central Transmission Inc., 499 So.2d 436, 96 O.&G.R. 113 (La.App. 1986), writs denied, 503 So.2d 478 (La.1987). In those situations where cancellation is available but not granted, or where the original failure to pay was willful and without reasonable grounds, the lessor is entitled to double the amount of royalties due, interest on that sum from the due date, and a reasonable attorney's fee. In all other cases, the lessor's damages are limited to interest on the royalties and a reasonable attorney's fee if the interest is not paid within thirty days of a written demand for it. A written notice of failure to pay is always a prerequisite to a judicial demand for damages or cancellation of the lease. La.Rev.Stat. 31:137. *See* Rivers v. Sun Exploration & Production Co., 559 So.2d 963, 969, 110 O.&G.R. 243 (La.App. 1990). A clerical error of the lessee which delayed royalty payments was found justifiable and not a basis for cancellation in Nunez v. Superior Oil Co., 644 F.2d 534, 69 O.&G.R. 45 (5th Cir.1981). Where shut-in payments were designated as "royalty" in the lease clause involved, the Mineral Code was applied to require notice as a prerequisite to cancellation of a lease for non-payment. *See* Acquisitions, Inc. v. Frontier Explorations, Inc., 432 So.2d 1095, 78 O.&G.R.

282 (La.App.1983).

Because the Mineral Code articles 137-141 only specify remedies for nonpayment of royalty to lessors, they failed to address interest owners, such as overriding royalty owners, who are not lessors. The Mineral Code was amended in 1982 (R.S. 31: 212.21 to 31: 212.23) to provide similar remedies and procedure for obtaining payment by a royalty owner other than a mineral lessor and by the purchaser of a mineral production payment, though without lease cancellation since that would virtually never be in the interest of such an interest owner. Because they are parallel provisions, Articles 212.21-.23 will be applied or interpreted in the same way as those of Articles 137-140. *See* Cimarex Energy Co. v. Mauboules 40 So.3d 931, 170 Oil & Gas Rep. 572, 2009-1170 (La. 4/9/10).

Payment of lease royalty is classified as "rent" under the Louisiana Mineral Code, La. R.S. 31:123, and is thus subject to a three year prescription (statute of limitations) under Louisiana Civil Code article 3494. Quality Environmental Processes, Inc. v. Energy Development Corp., 835 So. 2d 642 (La. App. 2002), writ denied, 834 So. 2d 439 (La. 2003); Terrebonne Parish School Board v. Mobil Oil Corporation, 310 F.3d 870, 883-84, 157 O.&G.R. 1167 (5th Cir. 2002).

4. Should the oil and gas lease relationship be governed by traditional landlord-tenant statutory or common law principles? In Tyson v. Surf Oil Co., 195 La. 248, 196 So.2d 336 (1940), the court held that the provisions of the Civil Code providing for a landlord's lien were applicable to an oil and gas lease. The court observed:

> This court has . . . firmly established the rule that mineral leases would be construed as leases and the codal provisions applicable to ordinary leases would be applied thereto insofar as they may be; 'Until the Legislature shall have passed laws specifically applicable to the industry of mining, which is a new one in this state, the parties engaged in those pursuits and the courts of the state will adhere to the jurisprudece on the subject and treat mineral contracts as leases.'

> Under the same authorities and, as provided in Article 2705 of the Civil Code, the mineral lessor is entitled to a lien, privilege and right of pledge upon the property placed on the leased premises by the lessee.

In California, as well, ordinary landlord-tenant law has been applied to oil and gas leases. *See* Renner v. Huntington-Hawthorne Oil and Gas Co., 39 Cal.2d 93, 244 P.2d 895, 1 O.&G.R. 1063 (1952). Is the application of landlord-tenant law to oil and gas leases conceptually improper? Does it lead to improper or undesirable results in such cases? *See* Casenote, 4 U.C.L.A.L.Rev. 485 (1957). In Cross v. Lowrey, 404 So.2d 645, 71 O.&G.R. 221 (Ala. 1981), however, the court decided that an oil and gas lessee was not governed by "landlord and tenant law, which prevents a tenant from disputing or challenging his landlord's title to the leased premises."

5. The lessor may be able to bring an equitable action requiring the lessee to account for unpaid royalties. The principal benefit of an accounting is that the burden of showing the amount due shifts from the lessor to the lessee. *See* Phillips Petroleum Co. v. Johnson, 155 F.2d 185, 194 (5th Cir.1946), cert. denied, 329 U.S. 730 (1946).

In KELLER v. MODEL COAL CO., 142 W.Va. 597, 97 S.E.2d 337 (1957) the lessor sought cancellation of a coal lease, inter alia, for failure of lessee to pay royalties due. The court stated:

> In addition to the allegations relating to the lack of development, the small amount of royalty received and the improper use of the property, the plaintiff has alleged "the need for discovery of whether the true and accurate amount of coal shipped from said premises is the same as that reported to your plaintiff has been paid the full amounts due by defendant" [*sic*]. And in the prayer of the bill, plaintiff prays for "an accounting of the tonnages of coal removed from said premises, the information for which was not furnished monthly to your plaintiff, and is peculiarly within the knowledge of the corporate defendant herein, discovery of the amount of coal transported, by truck and by railroad weight, or otherwise, the amount of custom coal removed and sold, and a court of law being inadequate to render the plaintiff the relief he seeks and to which he is entitled, that this Court of Equity takes cognizance of the complete matters between and among the parties to this suit". The matter of discovery is often necessary to an accounting, and an accounting is likewise often necessary in order to determine whether there is anything due under such a contract as is here involved. Furthermore, it can hardly be gainsaid that courts of equity in most instances afford more convenient and more thorough procedure for determining satisfactorily such matters. . . .

> . . . The allegations and prayer are, when read together and considered as a whole, sufficient to require the Court to retain jurisdiction for the purpose of enabling the plaintiff to have the discovery and accounting prayed for by him as a part of the relief sought in this suit. The defense that the plaintiff has an adequate remedy at law may not be definitely determined at the present stage of this suit, and even if it does now or eventually appear that he has or may have an adequate remedy at law, such facts will not, according to the authorities cited, deprive a court of equity of jurisdiction in such a matter.

6. Conventional remedies are usually of little use to the average lessor, as they all require that he know the accuracy of the royalty paid by the lessee. The lessor often has no way of obtaining this information, and the size of his investment may make an investigation impractical. How does the lessor check up on the lessee in such a situation?

7. As seen in the discussion of the Wyoming Royalty Payment Act, *supra* page 380, state legislatures have become increasingly involved in the issue of

royalty payments and remedies for breach of the royalty obligation. For example, in Colorado the Oil and Gas Conservation Commission has exclusive jurisdiction over royalty disputes arising under a state statute which gives lessees or first purchasers either 60 or 90 days after the hydrocarbons are sold to make royalty payments. Failure to comply with the time requirements allows the royalty owner to sue for damages based on a set interest rate, costs and attorney's fees. The royalty owner, however, must provide written notice to the payor prior to the filing of the administrative action. West's Colo.Rev.Stat.Ann. § 34-60-118.5. For other state statutes regulating the royalty payment process, *see e.g.,* Arkansas Code Ann. § 15-72-305, Okla.Stat.Ann., tit. 52, § 540, Utah Code Ann. § 40-6-9.

8. Should the parties to a lease be able to provide that failure to make timely and/or accurate royalty payments would cause the lease to automatically terminate? The following case answers that question.

HITZELBERGER v. SAMEDAN OIL CORP.

Texas Court of Appeals, 1997.
948 S.W.2d 497, 137 O.&G.R. 149 (writ denied).

DAVIS, CHIEF JUSTICE. This case concerns the construction of an oil and gas lease and a unit agreement. Robert Hitzelberger sued Samedan Oil Corp., et al. (collectively "Samedan") for a declaration that his lease terminated due to the late payment of royalties and for other relief. After a bench trial based entirely upon submitted documents, the trial court found in Samedan's favor and declared the lease to be in effect. From this decision Hitzelberger now appeals. Because the trial court erred as a matter of law in finding that Hitzelberger's lease did not terminate, we reverse and render in part and reverse and remand in part.

In 1990, NCNB Texas National Bank signed an oil and gas lease with Massad Oil Company covering tracts of land in Navarro County. Through intervening conveyances Hitzelberger became the successor to NCNB, while Samedan succeeded Massad Oil. Samedan, wanting to expand its area of development, asked Hitzelberger to sign a unit agreement to pool his tracts with other surrounding tracts to form the South Kerens Unit. Hitzelberger agreed to sign if the royalty provisions in his lease survived the unit agreement. Samedan accepted this condition and altered its unit agreement in an attempt to satisfy Hitzelberger's request. Subsequently, Samedan developed a producing well within the unit, but not located on Hitzelberger's land. The first sale of oil from this well occurred on June 12, 1992. Samedan paid an initial royalty to Hitzelberger 120 days later on October 10, 1992. After timely making two monthly royalty payments, Samedan failed to make the monthly royalty payments due in January and February of 1993 in accordance with the royalty provisions of Hitzelberger's lease.[63] Thereafter, Hitzelberger notified Samedan

---

[63]   Samedan admitted that a clerical error prevented the January and February royalty

that the lease terminated because of the late payment of royalties. Samedan refused to release the lease; whereupon, Hitzelberger brought this action. After a bench trial, the court made the following conclusions of law:

1. The lease is unambiguous.

2. The unit agreement is unambiguous.

3. The primary term of the lease was three years from May 2, 1990, the date of execution of the lease.

4. Samedan made timely payment of royalties for production.

5. If Samedan failed to make timely payments of royalties for production during the primary term of the lease, then that failure to pay royalties does not cause the lease to terminate.

6. In the alternative, Samedan's failure to make timely payments of royalties for production does not cause termination of the lease during the term of the unit agreement.

7. The lease has not terminated, and remains in full force and effect.

8. Samedan did not commit fraud.

Based on these conclusions, the trial court decided the lease had not terminated because: (1) under the lease's habendum clause, the lease cannot terminate during the primary term, even if the royalty payments were late; (2) the unit agreement amended the lease, so that the lease cannot terminate for late royalty payments during the term of the unit agreement; or (3) Samedan timely paid the initial royalty by tendering payment within 120 days of the first production from the "leased premises"--a well under Hitzelberger's tracts.

. . .

Hitzelberger's first point of error alleging the trial court erred by entering judgment for Samedan, instead of him, is supported by multiple sub-points. Some of these sub-points include: (1) the course of dealings and the parties' own interpretation showed that both parties understood the withholding of royalty payments would terminate the lease; (2) the lease treated production from anywhere in the unit as "production from the leased premises," and required termination for the late payment of a royalty from a well anywhere in the unit; (3) the unit agreement treated production from anywhere in the unit as "production from the leased premises," and the unit agreement changes made by Samedan preserved the royalty provisions of Hitzelberger's lease; and (4) the "initial royalty payment" was actually made, thus subsequent payments were due monthly without regard to the well location within the unit. Hitzelberger's second point of error attacks the trial court's conclusion that

---

checks from being sent to Hitzelberger. This clerical error resulted when a Samedan employee put a hold on Hitzelberger's payments until he returned a signed division order. Hitzelberger's lease did not require him to sign a division order to receive royalty payments. Samedan does not dispute this.

Samedan made timely payment of royalties. This point of error also attacks the court's findings that (1) the production from the unit in October 1992 did not constitute production from Hitzelberger's "leased premises," (2) the first production from Hitzelberger's "leased premises" occurred on December 28, 1992, and (3) within 120 days Samedan timely tendered a royalty check for the December production. In his ninth point of error, Hitzelberger argues the trial court erred when it concluded that late royalty payments during the primary term do not cause the lease to terminate. Hitzelberger's tenth point of error complains of the trial court's alternative finding that Samedan's failure to make timely royalty payments does not cause termination of the lease during the term of the unit agreement. Additionally, Hitzelberger argues in his eighth point that if the lease and unit agreement are not interpreted as he claims, then their meaning is at least ambiguous.

An oil and gas lease is a contract and must be interpreted as a contract. *TSB Exco v. E.N. Smith, III Energy Corp.*, 818 S.W.2d 417, 421 (Tex.App.--Texarkana 1991, no writ). Whether a contract is ambiguous is a question of law for the court to decide. This determination is made by looking at the contract as a whole in light of the circumstances present when the parties entered the contract. If a contract is worded in such a manner that it can be given a definite or certain legal meaning, then it is not ambiguous. On the other hand, a contract is ambiguous when its meaning is uncertain and doubtful or it is reasonably susceptible to more than one meaning. If there is no ambiguity, the construction of a written instrument is a question of law for the court.

As with any contract, the ultimate goal in interpreting a lease is to determine the parties' intent. When construing a lease to seek the intention of the parties, we consider all the provisions of the lease and by harmonizing, if possible, those provisions which appear to conflict by using the applicable rules of construction. However, no single provision taken alone will be given controlling effect; rather, all the provisions must be considered with reference to the whole lease. Because the parties to a lease intend every provision to have some effect, we will not strike down any portion unless there is an irreconcilable conflict. We also give terms their plain, ordinary, and generally accepted meaning unless the lease shows that the parties used them in a technical or different sense. If after applying the established rules of construction, a lease remains reasonably susceptible to more than one meaning, extraneous evidence is admissible to determine the true meaning of the lease. Otherwise, we will enforce an unambiguous lease as written; and, in the ordinary case, the writing alone will be deemed to express the intention of the parties.

This lease construction dispute regards how Hitzelberger's lease interacts with the unit agreement. The habendum clause in Paragraph 2 of Hitzelberger's lease provides:

Subject to the other provisions hereof, this lease shall be for a term of Three years from this date (called "Primary Term") and as long thereafter as oil and gas, or either of them, is produced in paying quantities from said land or lands with which said land is pooled hereunder and the royalties are paid as provided.

The crux of this dispute centers around the royalty clause contained in Paragraph 3 of Hitzelberger's lease. Paragraph 3 provides:

(g) Within 120 days following the first sale of oil or gas produced from the leased premises, settlement shall be made by Lessee or by its agent for royalties due hereunder with respect to such oil or gas sold off the premises and such royalties shall be paid monthly thereafter without the necessity of Lessor executing a division or transfer order. If said initial royalty payment is not so made under the terms hereof, this lease shall terminate as of 7 A.M. the first day of the month following expiration of said 120-day period. After said initial royalty payment, with respect to oil or gas produced during any month, if royalty is not paid hereunder on or before the last day of the second succeeding month, this lease shall terminate at midnight of such last day.

Paragraph 4 of the lease contains a typewritten sentence, which provides "[t]his is a paid up lease and all references to delay rentals shall be disregarded." Paragraph 5 of the lease grants the Lessee the right to pool or combine the lands covered by the lease or any part thereof. This pooling provision in Hitzelberger's lease provides:

For the purpose of computing the royalties and other payments out of production to which the owners of such interests shall be entitled on production of oil and gas, or either of them, from any such pooled unit, there shall be allocated to the land covered by this lease and included in such unit (or to each separate tracts within the unit if this lease covers separate tracts within the unit) a pro rata portion of the oil and gas, or either of them, produced from the pooled unit after deducting that used for operations on the pooled unit. Such allocation shall be on an acreage basis, thus, there shall be allocated to the acreage covered by this lease and included in the pooled unit (or to each separate tract within the unit if this lease covers separate tracts within the unit) that pro rata portion of the oil and gas, or either of them, produced from the pooled unit which the number of surface acres covered by this lease (or in each such separate tract) and included in the unit bears to the total number of surface acres included in the pooled unit. Royalties hereunder shall be computed on the portion of such production whether it be oil and gas, or either of them, so allocated to the land covered by this lease and included in the unit just as though such production were from such land. In the event only a part or parts, of the land covered by this lease instrument is pooled or unitized with other land, or lands, so as to form a pooled unit, or units, operations on or production from such unit or units, will maintain this lease in force only as to the land included in such unit or units.[The court noted that the unit created in this case was not created pursuant to this clause since it was over 1000 acres in size but cited the provision because of its definition of leased premises. Eds.]

We cannot look to the circumstances present when the oil and gas lease was signed because neither Hitzelberger nor Samedan is an original party to the lease. Nevertheless, we find this lease to be unambiguous as it can be given a definite or certain legal meaning. Construing these lease provisions together, we conclude that Samedan owed Hitzelberger an initial royalty payment within 120 days from the first sale of oil or gas produced from anywhere within the pooled unit in which Hitzelberger's tracts may be a member. After this initial royalty payment, monthly royalties are due on or before the last day of the second succeeding month without Hitzelberger executing a division order. It is also clear that the failure to meet the initial royalty payment or monthly royalty payments will terminate the lease. Furthermore, the paid-up provision in the lease did not require Samedan to make delay rental payments during the primary term of the lease.

Samedan does not dispute the unambiguous nature of this lease. However, Samedan argues it is clear that the lease could not terminate for late royalty payments during the primary term. Samedan claims that "the paid-up nature of the lease gave Samedan an absolute and irrevocable right to the oil and gas from Hitzelberger's tracts for a three-year period." Thus, according to Samedan it "need do nothing" to keep this lease in force during the primary term, including paying royalties. We believe the words clearly express an intention that delay rentals need not be paid during the primary term to keep the lease in force. Samedan's interpolation that royalty payments need not be made during the primary term goes beyond the clear expressed intent of the typewritten language in Paragraph 4.

In an attempt to bolster this theory, Samedan draws our attention to the language in the habendum clause granting the secondary term, which provides "as long thereafter as oil and gas, or either of them, is produced in paying quantities from said land or lands with which said land is pooled hereunder and the royalties are paid as provided." Samedan contrasts this language with that describing the primary term to show that the twin conditions of production in paying quantities and payment of production royalties applies only to the secondary term. The basis for Samedan contending that the royalty payment condition applies only to the secondary term is the absence of a comma separating the royalty payment condition from the secondary term. Thus, Samedan asserts that timely royalty payments are not necessary during the primary term. We believe Samedan's focusing on the absence of a comma places too great a weight on too frail a reed. To accept this argument we must ignore the intent expressed in the whole document and focus on punctuation in one sentence. We decline to follow this interpretation.

Samedan asserts that if we construe Paragraph 3(g) to apply during the primary term, then we violate a rule of construction because it will make the paid-up provision in Paragraph 4 meaningless. Id. However, this is not the case. Paragraph 4 states that delay rentals are paid-up. Because this provision certainly and definitely applies to delay rentals and not royalty payments, we will apply this unambiguous provision as written. No delay rentals are due within the primary term. However, royalties become due under Paragraph 3(g) when production begins and the late payment of royalties will result in termination. This construction allows us to harmonize and give effect to the habendum, delay rental and royalty payment provisions.

Samedan further urges this Court to classify Paragraph 3(g) as a covenant, instead of a condition. An important distinction exists between a condition and a covenant. Rogers v. Ricane Enterprises, Inc., 772 S.W.2d 76, 79 (Tex.1989); Parten v. Cannon, 829 S.W.2d 327, 329-330 (Tex.App.--Waco 1992, writ denied). The distinction between conditions and covenants lies in the appropriate remedy for their breach. Breach of a condition results in automatic termination of the leasehold estate upon the happening of the stipulated events. Breach of a covenant does not automatically terminate the estate, but instead subjects the breaching party to liability for monetary damages, or in extraordinary circumstances, the remedy of a conditional decree of cancellation. Doubts should be resolved in favor of a covenant instead of a condition. The language used by the parties to an oil and gas lease will not be held to impose a special limitation on the grant unless it is clear and precise and so unequivocal that it can reasonably be given no other meaning. We find the clear and precise language of Paragraph 3(g) to be a special limitation or condition. Thus, a breach of Paragraph 3(g) results in automatic termination of the lease.

Next, Samedan attempts to support its claim that the lease could not terminate for late royalty payments during the primary term by arguing that the habendum clause plays the dominant role in determining the duration of the lease. Relying upon Gulf Oil Corporation v. Southland Royalty Company [496 S.W.2d 547, 44 O.&G.R. 637 (Tex. 1973) Eds.], Samedan professes that when the habendum clause provides "a plain and certain answer to the question of when the lease terminates, we cannot change that answer with words elsewhere in the lease not certainly directed to the same question." In light of this authority, Samedan asserts that Paragraph 3(g) of Hitzelberger's lease "does not specifically direct itself toward amending the habendum." We believe Samedan's reliance on *Gulf Oil Corporation* for this proposition is misplaced. In *Gulf Oil Corporation*, the lease's habendum clause granted the lessee a term not to exceed 50 years. The lessors sought to enforce this habendum clause, whereas the lessee argued that the 50-year term should be delayed due to proration orders of the Railroad Commission. Attempting to find a lease provision to support its contention, the lessee looked to the force majeure or excuse clause. However, the Supreme Court noted that "the lease is silent about the lessee's rights if some of its wells are not allowed to produce to full capacity." Thus, the court did not find the force majeure or excuse clause was certainly directed to the termination of the lease. Samedan insinuates that any language affecting the term of a lease must directly refer to the habendum clause. This is not correct. The Supreme Court sought to determine if the language was certainly directed toward the question of termination, not certainly directed toward the habendum clause.

Samedan also relies upon Clark v. Perez to support its argument that the habendum controls. 679 S.W.2d 710, 714 (Tex.App.--San Antonio 1984, no writ). The court in that case stated that "the habendum clause will control unless properly modified by other provisions, and the fixed term therein stated should not be extended by words found elsewhere in the lease not certainly directed to the modification of the habendum clause." However, we disagree with this conclusion and decline to follow the theory that the habendum clause must be modified. The language found elsewhere in the lease must modify the duration of the lease, not the habendum clause.

Furthermore, the habendum clause is not the ultimate determiner of the duration of an oil and gas lease. Although the duration of the estate granted is traditionally determined by the habendum clause, this need not be the case. The lease may provide elsewhere for the modification of the term stated in the habendum. It is always a question of resolving the intention of the parties from the entire lease. The labels given to such clauses as "habendum" and "granting" are not controlling, and we should give effect to the substance of unambiguous provisions. We reject Samedan's claim that the habendum clause controls unless another provision specifically refers to it. Moreover, by using the phrases "[s]ubject to the other provisions hereof" and "the royalties are paid as provided," the habendum clause in Hitzelberger's lease indicates an intent to be modified by other provisions in the lease. Therefore, we find the trial court erred as a matter of law in relying upon this argument to conclude that the lease cannot terminate during the primary term. However, the trial court did not err in concluding that this lease was unambiguous.

Because we find the lease to be unambiguous, we cannot consider extrinsic or parol evidence. Also, the interpretation given to the unambiguous lease by Hitzelberger and Samedan is of no consequence. Only where a lease is first found to be ambiguous may we consider the parties' interpretation. Where the meaning of the lease is plain and unambiguous, a party's construction is immaterial. Thus, we will enforce this unambiguous lease as written without regard to extrinsic evidence, parol evidence or the parties' interpretation. We overrule that part of Hitzelberger's first point of error relating to the parties' interpretation of the lease.

. . .

[The court also found that the terms of the unit agreement were not ambiguous. Eds.]

Thus, the unit agreement amends the individual oil and gas lease's provisions relating to operations, but leaves the lease provisions relating to royalty payments in effect. Article 3.3 supports this interpretation by declaring that all oil and gas lease provisions not contradicting uniform operations shall remain in effect. Although Samedan may incur additional administrative expenses in complying with the royalty provisions of the individual leases, this interpretation does not interfere with the uniform operations of the unit.

The unit agreement also defines what constitutes production from the leased premises. Article 3.4 provides production from any part of the unit will be considered production on each tract, except for the purpose of determining royalty payments. However, Article 6.1 specifies all oil and gas produced and saved will be allocated among the tracts in accordance with the respective participation effective when the oil and gas is produced. Moreover, the amount of oil and gas allocated to each tract in the unit will be deemed for all purposes to have been produced on such tract. We do not find that Articles 3.4 and 6.1 irreconcilably conflict. Article 3.4 concerns activities that will hold the leases of the individual tracts by production and operations on one tract in the unit; whereas, Article 6.1 more broadly defines what is production from the "leased premises" and can be applied to royalty payments. Therefore, production allocated from any tract in the unit will constitute production from the "leased premises" on the other tracts in the unit and invoke the royalty provisions in the individual oil and gas leases. The unit agreement amends individual oil and gas leases to the extent necessary to conduct uniform operations, but leaves in effect the individual lease's royalty provisions.

Samedan achieved production from a tract within the unit on June 12, 1992. Even though this production came from a tract other than Hitzelberger's, it constituted production from his "leased premises" under his lease and the unit agreement. Thus, Samedan was required to pay royalties as set forth in Paragraph 3(g) of Hitzelberger's lease to maintain its lease on Hitzelberger's tracts. Samedan failed to make two monthly royalty payments in accordance with Paragraph 3(g). Although the law does not favor forfeitures, we apply the unambiguous language of Hitzelberger's lease and the unit agreement to declare that the lease covering Hitzelberger's tracts terminated at midnight on January 31, 1993. We sustain Hitzelberger's second, ninth and tenth points of error.

. . .

Because Hitzelberger's lease terminated during the primary term due to late royalty payments, the unit agreement did not amend the lease's royalty provisions, and Samedan failed to timely pay royalty payments to Hitzelberger, we find that the trial court erred as a matter of law in relying upon any of the three alternative grounds to enter judgment for Samedan. We reverse and render judgment that Hitzelberger's lease terminated at midnight on January 31,

1993. We remand all other issues to the trial court for further proceedings consistent with this opinion.

NOTES

1. Have the parties to this lease created a royalty clause similar in effect to the "unless" type of delay rental clause? What happens if the royalty check is mailed out two weeks ahead of the deadline but does not arrive due to a delay in the mails? Should the court apply the caselaw relating to delay rentals?

2. Did the court apply any canons of construction in interpreting the royalty clause? Did Samedan argue that the canon of construction that typewritten language should prevail over the printed form language applied? Was this a "paid-up" lease? Did Samedan have to make any delay rental payments or drill in order to keep the lease alive during the primary term? Did the typewritten delay rental clause have any direct impact on the royalty payment obligation?

3. Automatic termination of the lease for nonpayment of royalty may not always be to the advantage of the lessor. The lease at issue in Stream Family Ltd. Partnership v. Marathon Oil Co., 27 So.3d 354, 2009-561 (La.App. 3 Cir. 12/23/09), writ of certiorari and/or review denied, 31 So.3d 1064, 2010-0196 (La. 4/16/10) provided as follows:

> Untimely or improper payment for royalties shall constitute an express resolutory condition of this lease with respect to LESSEE's rights. Except in instances of willfully or persistently late or improper payment LESSOR shall give twenty-one (21) days written notice of LESSEE's failure to make timely or proper payment of royalties as a prerequisite to a successful judicial demand for dissolution of the lease. In the instance of willfully or persistently late or improper payment, LESSOR need not give such notice and the lease shall resolve immediately.

The lessor had consented to an assignment of the lease on the condition that the lessee not be relieved of any obligations *under the lease*. Thereafter, the lessee had no direct involvement with the lease. The assignee produced a well on the lease but paid no royalties to lessor from 1998 to 2004. Lessor was aware of the nonpayment but made no demand for payment until 2004. The lessor brought suit against the original lessee and the assignees. The claim for royalties unpaid prior to three years before suit had prescribed. The court ruled that the reference to termination for "willfully or persistently late or improper payment" included termination for nonpayment. Because the lease had thus terminated prior to 2001, the assignor/lessee could not be held liable for nonpayments after the termination of the lease as its continued liability only pertained to obligations under the lease. The provisions of Mineral Code Articles 137–141 have no application to royalty obligations when a lease specifies that the lease will terminate automatically upon specified circumstances of non-payment of royalty. This is covered by Article 133 of the Mineral Code, which provides "A mineral lease terminates . . . upon the occurrence of an express resolutory condition." See note 3, p. 413 supra.

———

F. FREE GAS CLAUSE AND RELATED PROVISIONS

Lessor shall have the privilege of using free of charge, at his own risk and expense, gas from any gas well on said land for stoves and inside lights in the dwellings on said land and appurtenant structures, and also of using such gas for pumping water for irrigation or other use upon the premises by paying therefor at the same rate which Lessee receives for such gas.

Lessee shall permit Lessor to use, at Lessor's risk and expense, for surface operations upon the leased land, water developed by Lessee when and so long as not required by it in its operations hereunder, but not so as to interfere with any of Lessee's operations hereunder.

———

The provision in the above-quoted clause entitling the lessor to purchase gas has been added to the free gas clause in some areas of the country as a result of the increased demand for an economical supply of gas for farming and related operations. In Kansas the regulatory commission has encouraged the provision of irrigation gas to farmers. And in Oklahoma there has been legislation (subsequently held to be unconstitutional) designed to require lessees to make gas available for this purpose. Texas legislation of 1954 eliminated one ground for the opposition by lessees to making gas available for irrigation or other agricultural purposes by providing that such action should not subject any person to regulation as a gas utility. Free gas clause provisions are very prevalent in the Appalachian Basin area and sometimes provide for payments in lieu of the provision of free gas equal to an agreed-to volume of natural gas.

A clause entitling the lessor to water developed by the lessee appears in some leases in arid areas.

———

Although the economic importance of the free gas clause is relatively slight as compared to the royalty provisions of an oil and gas lease, there has been a substantial crop of litigation over the construction and application of these clauses in leases, particularly in Kentucky. Among the issues which have been litigated are: (1) the measure of damages for breach of a covenant to supply free gas; (2) whether the clause is an element in calculating damages for failure of a lessee to drill a well; (3) the quantity of gas to which the lessor is entitled; (4) whether the covenant "runs with the land"; (5) what premises are entitled to the benefit of the free gas; (6) the right of the lessee to remove casing and plug a well from which free gas has been provided; (7) right of the operator to regulate flow and pressure of gas to lessor by installations near the well; (8) apportionment of the benefit of the clause; (9) whether the right to free gas is extinguished by an attempt to sever it from the dominant premises; (10) the right of the lessee to use compressors to accelerate the flow of gas from the wells if such use interferes with the taking of free gas; (11) whether gas must be furnished the lessor even though there is no production from the leased premises; (12) whether lessor's taking gas when there is no other production from the premises amounts to production under the habendum clause of the lease; (13) whether the lessor is entitled to gas produced from an oil well; and (14) whether landowners whose lease has been pooled are entitled to free gas from the pooling unit although the well is not located on their premises. The cases are collected and discussed in 3 Martin & Kramer, *Williams and Meyers Oil and Gas Law* §661 *et seq.*

G. DIVISION AND TRANSFER ORDERS
Martin & Kramer, *Williams and Meyers Oil and Gas Law*
§ 701. Introduction

Oil and gas leases, operating agreements, mineral and royalty deeds, and other instruments affecting title to oil and gas may contain fairly specific provisions concerning rights to the proceeds of oil and gas leases and of development operations. A person charged with responsibility for making distribution of such proceeds should examine the relevant instruments and make distribution in accordance with his construction of their provisions.

Not infrequently, however, such instruments affecting oil and gas rights contain ambiguities, inconsistencies, or lacunae. The validity of certain claims may be in dispute. The interests of unknown parties may be involved. Under such circumstances, how should distribution be made? The person responsible for the distribution could be expected under such circumstances to seek a means of protecting himself against liability in the event of an improper distribution. And even though the person making distribution is unaware of any such ambiguity, dispute, or other difficulty, he could be expected to seek protection in the event of some future claim that payment had been made improperly.

Increasingly, state statutes attempt to deal with the proper payment of royalties and the impact of the execution of division orders on the payment obligations, although in some cases the results have been mixed.

There is, then, an obvious need for protection of the distributor of such fund against liability for improper payment. To meet this need, instruments known as division orders and transfer orders are employed.

In brief, a division order is simply a direction and an authorization to a person who has (or will have) a fund for distribution among persons entitled thereto as to the manner of distribution. A transfer order is a direction and authorization to change the distribution provided for in a division order.

The function of these instruments may be illustrated by the following. Blackacre is leased by Lessor to Lessee. Lessee drills a well and discovers commercial production. Lessee seeks to sell the oil produced to Purchaser, who is ready and willing to buy the oil. Purchaser might examine the lease and ascertain therefrom that Lessee had authority to sell the oil produced; under such circumstances Purchaser could accept the oil and pay Lessee therefor. But what if Lessee was without authority to sell Lessor's oil? Or what if the lease had terminated prior to production? Would not Purchaser be liable to the landowner for taking the oil from Lessee and making payment to Lessee? Purchaser would, therefore, seek to secure the execution of a division order by Lessor and Lessee in which Purchaser is given specific directions as to the payment to be made to each party.

Then, at some later date, Lessor (or Lessee) transfers some portion or all of his interest in the premises to Assignee, and Assignee requests that future payments be made to him. Purchaser may accept Assignee's word that the assignment has been made, but caution dictates that he demand an instrument executed by the assignor of such interest which specifically authorizes the making of future payments to Assignee. For this purpose Purchaser demands the execution of a transfer order. This order instructs Purchaser to give credit, effective at a specified date and hour, for oil or other products received in accordance with the division of interests recited in the order.

The lessee who obtains production may be an integrated operator who takes the product of the premises, treats and refines it, transports it to market, and sells the products to the ultimate consumer. Or the lessee may sell the product of the lease to some other person or corporation. The lessee may or may not have authority under the lease or other operative instrument to dispose of the royalty oil or gas. But whether the lessee utilizes the product off the premises or sells the product to another, and whether or not the lessee has authority to dispose of that portion of the product attributable to a royalty or other interest in the premises, it is customary to secure the execution of a division order before the product of the premises is removed therefrom. The addressee of the division order, that is, the person given directions and authorization concerning the payments to be made, may be the lessee or it may be some other person or corporation.

The procedures followed in the preparation and execution of division orders have been summarized by Professor Masterson as follows:

A purchaser of production from an oil or gas well or wells must attempt to determine to whom and in what proportion royalty payments, as well as payments to owners of the leasehold estate, should be made. The usual procedure in such a situation follows. An abstract of title including all instruments affecting the land in question is secured. This abstract is then examined by a lawyer for the purchasing company, who, based upon his examination, prepares a title opinion, which lists ownership subject to compliance with title requirements, which requirements are listed in the title opinion. This type of opinion is called "a division order opinion," or a "pipe line opinion." An investigator in the employ of the purchasing company, who is usually a member of the land department of the company, attempts to procure whatever title curative matter is necessary to meet the requirements. During this time, the purchasing company buys the oil and gas, but holds up payments pending clearance of title and distribution and execution of division orders. After title is cleared, then division orders are prepared by said company listing title as it is set forth in the lawyer's opinion. To expedite matters, it is customary to prepare a separate division order for each owner, which instrument lists only his interest. In this connection, sometimes title will be cleared as to some interests but not as to all. Often in this situation division orders will be procured from those whose title is clear, and payments as to them will start.

NOTES

1. May the lessee insist on a division order being signed as a prerequisite to the payment of royalties? Blausey v. Stein, 61 Ohio St.2d 264, 400 N.E.2d 408, 65 O.&G.R. 27 (1980), held that a lessee may do so; *but see* TXO Production Corp. v. Page Farms, Inc., 287 Ark. 304, 698 S.W.2d 791, 87 O.&G.R. 1 (1985): "To begin with, the oil and gas lease did not require the lessors to sign such an order. [Lessee] submitted testimony that division orders are recognized by custom and usage as being required in the oil and gas industry, but there is no proof that [lessees] were so familiar with the oil and gas business that their knowledge and acceptance of the particular usage must be presumed. Absent such proof, they were not bound by any such custom or usage." In Hull v. Sun Refining & Marketing Co., 789 P.2d 1272, 109 O.&G.R. 49 (Okl. 1989), it was argued that the common law included the "recognized custom and usage of the oil and gas industry that royalty owners [be required to] execute division orders before receiving royalty payments." The court held that under 52 O.S.Supp.1985 § 540 "the only condition justifying suspension of royalty payments under § 540 arises when title is questioned." The opinion notes, however, that § 540 was amended in 1989 to include a passage on division orders which provides in part: "A division order is executed to enable the purchaser of the production from the leasehold to make remittance directly to the interest owners for their royalty interest, and is not intended to and does not relieve the lessee of any liabilities or obligations under the oil and gas lease." This provision is limited, in its terms, to "division orders executed on or after July 1, 1989." The court expresses no opinion on the impact of this language "on future causes presenting the issue of execution of division orders." What is your opinion? *See* 4 Martin & Kramer, *Williams and Meyers Oil and Gas Law* § 704.8.

2. SNIDER v. SNIDER, 208 Okl. 231, 255 P.2d 273, 2 O.&G.R. 711 (1953). A father and son who owned respectively one-fourth and three-fourths of the minerals in a particular tract signed the following instrument:

Whereas, the Undersigned, Elias F. Snider and Louis Fort Snider, are owners of the surface right and an undivided mineral right, interest and title in and to the following described real estate, to wit:

The Southwest Quarter of Section 10 in Township 7 North, Range 17 West of the Indian Meridian, Kiowa County, Oklahoma,

and whereas it is desired to provide for the payment of royalty from actual production of oil, gas, or any other mineral thereon.

Now, Therefore, it is agreed between the parties that without regard to the actual ownership of said land and right in said undivided mineral interest that all royalty shall be paid and divided among the parties as follows:

That there shall be paid to the said Elias F. Snider three-fifths of all such royalty payments and to the said Louis Fort Snider two-fifths of all such royalty payments. The wives of the respective parties being Nellie R. Snider the wife of Elias F. Snider and Clara Snider the wife of Louis Fort Snider have joined in the execution of this stipulation and agreement to avoid any question arising as to homestead, community or any other arising questions as to rights in said real estate.

After the death of the father the son brought suit to cancel this instrument and to quiet title in the son to three-fourths of the minerals. In considering the appeal from a judgment cancelling the instrument the court said:

It is certain that this stipulation is not a conveyance of mineral interest. Nowhere does there appear a granting clause of any kind in the stipulation. Nor is it an assignment of royalties in perpetuity. There is no word of assignment in the instrument, no duration stated, no clause making the contract binding upon the heirs, personal representatives, or assigns of the parties.

The purpose of the instrument and the intention of the parties with reference thereto is easily seen within the four corners of the instrument itself. It was a stipulation between father and son designed to authorize the payment by the oil company of accruing royalties on a different basis than that of ownership by the parties. In practical effect it is a division order and as such subject to termination by either party. Whether there was or was not consideration for the execution of the stipulation would not affect the right to terminate upon notice where no specified duration is set forth in the division order and no proof is made as to duration.

3. SMITH v. LIDDELL, 367 S.W.2d 662, 18 O.&G.R. 696 (Tex.1963). The grantee of a 1/4 undivided interest in the minerals in 152 out of 223 acres of land sued to establish his interest in 1/4 of the 1/8 royalty reserved in a lease on the entire 223 acres, basing his claim on a clause in the lease providing:

Since, however, T.M. Smith, under date of August 30th, 1933, conveyed to H.B. Finch, Trustee, one-fourth (1/4) of all the minerals owned by him under the land covered by this lease, . . . lessee is hereby especially directed by the said T.M. Smith to pay to the said H.B. Finch, Trustee, one-fourth (1/4) of all royalties apportioned above herein to the said T.M. Smith . . ..

The court said:

The . . . lease neither created nor extended the rights of Finch. . . . Since, by the terms of the trust deed, Smith did not convey to Finch an interest in the 71.446 acres, we hold that he was under no obligation to direct Humble, the lessee, to pay to Finch 1/4 of the royalty apportioned to Smith under the terms of the lease. The direction to Humble to pay to Finch certain royalty is merely a division order. . . . Since Smith was under no contractual obligation to see that Liddell was paid 1/4 of 1/8 of all production, he had the right to revoke and he did.

4. DALE v. CASE, 217 Miss. 298, 64 So.2d 344, 2 O.&G.R. 962, 37 A.L.R.2d 811 (1953). Case conveyed to Dale a fractional interest in the minerals in a tract then subject to an oil and gas lease by a deed reciting that "This deed shall not participate in present oil and gas lease, but shall participate in any and all such future leases." Production was obtained under this lease and a division order was executed by all parties in interest, including both Case and Dale, and by the terms of the division order, a share of the royalty was paid to Dale for about two years. Case then sued to recover the full royalty due under the lease and to require an accounting for royalties improperly paid to Dale prior to the date of the filing of the suit. The chancellor construed the mineral deed as not entitling Dale to any share of the royalty paid under the lease in effect at the time of the conveyance, and held (1) that plaintiff was not entitled to a decree against the oil companies for the amount of royalties paid to Dale prior to the filing of the suit, but, (2) that he was entitled to a decree against Dale for the amount which had been paid to him. *Held,* affirmed.

Finally, the appellants' attorneys contend that, when Case and his wife signed the division orders, and failed to register a protest on account of the failure of the division orders to show Case's ownership of the royalty rights under the lease in the mineral interest conveyed to Dale, they thereby assented to Dale's interpretation of the reservation clause in the mineral deed. But, as we have already stated, in our opinion the meaning of the reservation in the mineral deed which Dale accepted is clear, and Dale was not entitled to the royalty payments to be made under the existing lease. Dale acquired no greater rights as a result of Case's signing of the division orders. Dale was not entitled to profit by the mistakes made by the operating companies in the preparation of the division orders or by Case's failure to discover the mistakes before signing the orders. The fractions appearing on the reverse side of the division orders which Case signed might have puzzled the mind of a bank president, such as Higdon, who was accustomed to dealing in figures of that kind; and we do not think that Case, whose formal education had not extended beyond the seventh grade, lost the rights reserved to him in the mineral instrument because of his failure to detect the error in the royalty interest set opposite his name on the reverse side of the division orders sent to him by the oil companies. Case, after signing the division orders, was estopped from asserting a claim against the oil companies for the amounts paid to Dale prior to the filing of the bill of complaint, but Case was not estopped from asserting his rights against Dale and his grantees who were not entitled to the royalties which were being paid to them.

5. MINVIELLE V. SHELL OIL CO., 321 F.Supp. 884, 39 O.&G.R. 280 (W.D.La.1971), aff'd, 445 F.2d 862, 39 O.&G.R. 285 (5th Cir.1971). Lessors signed a division order which contained a paragraph reserving their rights to show a greater participation in production than was stipulated in the order. Although a dispute as to the lessee's performance of obligations under the lease was compromised by an agreement in which the lessors agreed that there were "no outstanding express or implied obligations under the . . . Lease" which had not been met, the lessors brought an action against the lessee "to cancel the lease and to recover royalties allegedly wrongfully paid to the State as owner of certain water bottoms." The court found that the obligation of the lessor to pay royalties was included in the compromise agreement but went on to comment:

> The cases are legion to the effect that an oil and gas lease can be cancelled for failure to pay royalty only if there has been an arbitrary refusal or an unreasonable delay. . . . In the instant case, Shell at all times paid royalties in accordance with the agreements which had been made. There was no arbitrary refusal to pay royalties. Thus plaintiffs are not entitled to a cancellation of the lease.

6. In MADDOX V. GULF OIL CORP., 222 Kan. 733, 567 P.2d 1326, 60 O.&G.R. 13 (1977), cert. denied, 434 U.S. 1065 (1978), defendant had been given conditional permission by the FPC to increase its gas prices in 1965, subject to refund in case the FPC failed to give final approval. Until 1972, when an FPC opinion authorizing the increases became final, defendant properly withheld royalty payments attributable to the increases. In the fall of 1972 defendant paid the "suspense royalties", but although the fund had been commingled and used by defendant in its business operations, interest was not paid. Plaintiffs, in a class action, sued to recover this interest. More than one-half of the royalty owners had signed division orders expressly authorizing defendant to withhold without interest royalties in dispute. The trial court held that the waiver of interest provision in the division orders was ineffective. The supreme court agreed:

> This court has said a division order is an instrument required by the purchaser of oil or gas in order that it may have a record showing to whom and in what proportions the purchase price is to be paid. Its execution is procured primarily to protect the purchaser in the matter of payment for the oil or gas, and may be considered a contract between the sellers on the one hand and the purchaser on the other. (Wagner v. Sunray Mid-Continent Oil Co., 182 Kan. 81, 92, 318 P.2d 1039 and authorities cited therein.) Generally speaking, a division order is not a contract between sellers themselves, especially in view of the fact that each of the parties having an interest in production may in fact execute separate division orders. (Wagner

v. Sunray Mid-Continent Oil Co., *supra.*)

It was the duty of Gulf under the lease contracts it had with its royalty owners to market the gas at the best prices obtainable at the place where the gas was produced. The insertion in the division orders of matters contrary to the oil and gas leases, or contrary to the law, cannot be unilaterally imposed upon the lessor by the lessee or the purchaser. Here the unilateral attempt by Gulf in the division orders to amend the oil and gas leases, and thereby deprive the royalty owners of interest to which they were otherwise entitled, was without consideration. Therefore, the provisions in the division order regarding waiver of interest are null and void as determined by the trial court.

7. SIMPSON v. UNITED GAS PIPE LINE CO., 196 Miss. 356, 17 So.2d 200 (1944). The royalty provision in the lease covering the land here in question provided that "The grantor shall be paid 1/8th of the value of such gas, calculated at the rate of the market price at the well." A gas well having been drilled, a gas purchase contract was made by the lessees with defendant's assignor which obligated the lessees as sellers to sell to the purchaser all gas from this well, and provided that the buyer should pay for the gas 4 cents per thousand cubic feet for the first five years, 4.5 cents per thousand cubic feet for the next five years, and for the succeeding five years the price was to be determined by mutual agreement or arbitration, but not to be less than 5 cents. Lessors and lessee executed a division order which provided that all interests were to be paid in accordance with and under the terms of the gas purchase contract. The rapid decline of available gas supplies in the field in which this well was located caused a rise in the market price of gas to 9 cents per thousand cubic feet. Plaintiff lessors sought to recover royalty based on the market price of the gas, as provided in the lease, rather than on the basis of the contract price in the gas purchase contract. *Held* for defendant.

[T]his division order establishes and fixes the gas price to be paid by appellee to appellant. . . . The division order was supported by sufficient consideration. . . . It is evident the purchaser had to expend considerable money to comply with that contract. It is natural that it would not have done so unless and until the gas owners had agreed to the gas prices stipulated therein. That contract fixed the prices to be paid by the purchaser for the succeeding fifteen years. It took a chance on these prices. The gas owners had a guaranteed price for that period. They got an immediate sale of the gas and the arrangement imposed upon others the expense of saving and marketing the gas, as well as the risk in the fluctuating price. It appears that the market price of gas went up about January, 1940. Suppose it had gone down. Or suppose it yet goes down. Appellee is bound to pay the contract price. . . . The method adopted here appears to be the usual method in such cases, and "the division order is a contract between the carrier, or purchaser, and the parties signing."

GAVENDA v. STRATA ENERGY, INC.

Supreme Court of Texas, 1986.
705 S.W.2d 690, 88 O.&G.R. 568.

SPEARS, JUSTICE. This oil and gas case concerns the effect of division and transfer orders. The issue is whether division and transfer orders are binding until revoked when an operator who prepares erroneous orders under pays royalty owners, retaining part of the proceeds for itself. The trial court held the orders were binding until revoked and rendered summary judgment for the operators. The court of appeals affirmed. 683 S.W.2d 859. We reverse the judgment of the court of appeals in part and remand the cause to the trial court.

In 1967, the Gavenda family conveyed land in Burleson County to the Feinsteins. The Gavendas reserved a fifteen-year one-half non-participating royalty interest:

> PROVIDED, HOWEVER, the Grantors herein except from this conveyance and reserve unto themselves, their heirs and assigns, for a period of fifteen (15) years from the date of this conveyance, an undivided one-half (1/2) non-participating royalty of all of the oil and gas in, to and under that produced from the hereinabove described tract of land. Said interest hereby reserved is a non-participating royalty and shall not participate in the bonuses paid for any oil and/or gas lease covering said land, nor shall the Grantors participate in the rentals which may be paid under and pursuant to the terms of any lease with reference to said land.

The Feinsteins later sold the land, subject to the Gavendas' oil and gas reservation, to Billy Blaha. In 1976, Blaha executed an oil and gas lease for a 1/8th royalty.

After various conveyances with overriding royalty interests taken, Strata Energy, Inc. and Northstar Resources, Inc. each acquired a working interest in the lease. They drilled one producing oil and gas well in July, 1979 and another in February, 1980. Strata and Northstar entered into a joint operating agreement naming Strata the lease operator and providing that Strata would disburse all royalties from production. The agreement also provided that Strata's actions were made on behalf of both Strata and Northstar.

Strata hired an attorney to perform a title examination, and he erroneously informed Strata that the Gavendas were collectively entitled to a 1/16th royalty, rather than the actual 1/2 royalty. Following the attorney's erroneous report, Strata prepared the division orders and disbursed the proceeds. When various Gavendas died and royalty ownership changed, Strata prepared and sent transfer orders to the new royalty owners reflecting the same collective 1/16th royalty. The Gavendas signed these division and transfer orders and received the disbursements. The Gavendas were underpaid by 7/16th royalty, 7/16th of gross production, and Strata and Northstar kept at least part of the underpayment.

On discovering this error, the Gavendas on September 29, 1982 revoked the division and transfer orders. Two days later, their royalty interest terminated under the express terms of their reservation. Later in 1982 the Gavendas filed suit to recoup more than 2.4 million dollars in underpaid royalties owed them under the deed reservation.

Both sides filed motions for summary judgment. The trial court and the court of appeals held for Strata and Northstar, maintaining the division orders were binding until revoked. The court of appeals, however, reversed summary judgment and remanded as to Victor Gavenda's estate. The court of appeals held that there were fact issues whether the division and transfer orders encompassed Victor Gavenda's estate.

Both parties have appealed to this court. The Gavendas bring three points of error, contending the rule that division orders are binding until revoked does not apply when there is unjust enrichment, and therefore they should be allowed to recover royalty deficiencies from Strata and Northstar.

Strata and Northstar contend the division and transfer orders were also binding on Victor Gavenda's estate. They, however, do not dispute: (1) the deed reserved a 1/2 royalty, or 1/2 of gross production; or (2) the amount of the royalty underpayment $2,435,457.51 plus interest. The only issue then is whether under these facts division orders are binding until revoked.

Division orders provide a procedure for distributing the proceeds from the sale of oil and gas. They authorize and direct to whom and in what proportion to distribute funds from the sale of oil and gas. 4 H. Williams, Oil and Gas Law § 701 at 572 (1984). Transfer orders are later changes in the division order's distribution.

In Texas, division and transfer orders do not convey royalty interests; they do not rewrite or supplant leases or deeds. Exxon Corp. v. Middleton, 613 S.W.2d 240, 250 (Tex.1981); Phillips Petroleum Co. v. Williams, 158 F.2d 723, 727 (5th Cir.1946). The weight of authority clearly supports this rule. Williams, *supra*, § 707 at 612.

The general rule in Texas, though, is that division and transfer orders bind underpaid royalty owners until revoked. Exxon Corp. v. Middleton, 613 S.W.2d 240, 250 (Tex.1981); Chicago Corp. v. Wall, 156 Tex. 217, 293 S.W.2d 844, 847 (1956). One principle underlining this rule is detrimental reliance. 4 H. Williams, *supra*, § 707 at 614; 3 A. Summers, Oil and Gas § 590 at 139-140 (1958).

Detrimental reliance explains why purchasers and operators are usually protected by the rule that division orders are binding until revoked. In the typical case, purchasers and operators following division orders pay out the correct total of proceeds owed, but err in the distribution, overpaying some royalty owners and underpaying others. If underpaid royalty owners' suits against purchasers and operators were not estopped, purchasers and operators would pay the amount of the overpayment twice--once to the overpaid royalty owner under the division order and again to the underpaid royalty owner through his suit. They would have double liability for the amount of the overpayment. Chicago Corp. v. Wall, 293 S.W.2d 844 (Tex.1956). Exposing purchasers and operators to double liability is unfair, because they have relied upon the division order's representations and have not personally benefited from the errors.

Generally, the underpaid royalty owners, however, have a remedy: they can recover from the overpaid royalty owners. Allen v. Creighton, 131 S.W.2d 47, 50 (Tex.Civ.App.-Beaumont 1939, writ ref'd). The basis for recovery is unjust enrichment; the overpaid royalty owner is not entitled to the royalties. *See* 4 Williams, *supra,* § 707 at 613 (1984); 3 A. Summers, *supra,* § 590 at 139-40 (1957).

In Exxon v. Middleton, 613 S.W.2d 240 (Tex.1981), we held the division orders were binding until revoked, even though there had been no detrimental reliance. To provide stability in the oil and gas industry, we held for the distributors of the proceeds because they had not profited from their error in preparing the division order?in short, because there was no unjust enrichment.

In *Middleton,* Sun and Exxon, the well operators, sold the gas off the premises to purchasers under long term contracts. They then prepared division orders and distributed to the royalty owners a 1/8th royalty based on the contract price. The leases, however, provided for gas sold off the premises for a 1/8th royalty based on market value at the well. The royalty owners sued for their royalty deficiencies – the difference between a 1/8th royalty based on market value at the well and a 1/8th royalty based on contract price.

Exxon and Sun could not have relied on the division orders' representations in making the long-term gas contracts, for they had executed these contracts before the royalty owners had signed the division orders. *Id.* at 249. More importantly, however, Exxon and Sun did not benefit from the discrepancy between the leases and the division orders. They paid out 1/8th of what they received from the purchaser – 1/8th of the contract proceeds. Moreover, at the time of these transactions, long-term gas contracts were the best arrangement Exxon and Sun could obtain for the royalty owners.

When the operator, however, prepared erroneous orders and retained the benefits, we held that the division orders were not binding. Stanolind Oil & Gas Co. v. Terrell, 183 S.W.2d 743 (Tex.Civ.App.- Galveston 1944, writ ref'd). In *Terrell*, Stanolind Oil & Gas prepared the division orders and distributed the bonus and royalty accordingly. Stanolind Oil & Gas deducted the gross production tax from the bonus, although the lease provided that there would be no deductions. Despite the lease provision, Stanolind was shifting the gross production tax from itself to Terrell. It was profiting from its own error in drawing up the division order. There was unjust enrichment.

Applying the law to this case, we hold that the division and transfer orders do not bind any of the Gavendas. Strata both erroneously prepared the division and transfer orders and distributed the royalties. Because of its error, Strata underpaid the Gavenda family by 7/16th royalty, retaining part of the 7/16th royalty for itself. It profited, unlike the operators in Exxon v. Middleton, at the royalty owner's expense. It retained for itself, unlike in Chicago Corp. v. Wall, part of the proceeds owed to the royalty owners. Therefore, Strata is liable to the Gavendas for whatever portion of their royalties it retained, although it is not liable to the Gavendas for any of their royalties it paid out to various overriding or other royalty owners.

Strata also argues that it is not responsible for the loss because the attorney who prepared the title opinions was not Strata's agent but rather an independent contractor. We cannot agree. The attorney-client relationship is an agency relationship. The attorney's acts and omissions within the scope of his or her employment are regarded as the client's acts; the attorney's negligence is attributed to the client. Texas Employers Insurance Ass'n v. Wermske, 162 Tex. 540, 349 S.W.2d 90, 93 (1961). We note that Strata and Northstar have filed a third party cross-action against the attorney.

The Gavendas also request attorney's fees under Tex.Rev.Civ.Stat.Ann. art. 2226 (Vernon Supp.1985), now codified in the Civil Practice and Remedies Code, ch. 959, § 38, 1985 Tex.Sess.Law Serv. 7123-7124 (Vernon). Attorney's fees are recoverable in suits for royalty payments owed under oil and gas leases. Atlantic Richfield Co. v. Manges, 702 F.2d 85, 87 (5th Cir.1983); Gerdes v. Mustang Exploration Co., 666 S.W.2d 640, 645 (Tex.App.-Corpus Christi 1984, no writ). Because we see no distinction between allowing attorney's fees on underpaid royalty suits based on leases and those based on deed reservations, we hold the Gavendas can recover their attorney's fees.

We reverse the judgment of the court of appeals in part and remand the cause to the trial court to determine the amount of royalties owed by Strata and Northstar to the Gavendas, prejudgment interest due thereon, and attorney's fees. Although the total amount of the royalty deficiency is undisputed, we cannot tell from the record what portion of the royalty deficiency was retained by Strata and Northstar.

NOTES

1. Is it accurate to say, as the Texas court does in distinguishing *Middleton* from the situation in the principal case, that in *Middleton* "Exxon and Sun did not benefit from the discrepancy between the leases and the division orders"?

2. The opinion notes that the defendants in this case "have filed a third party cross-action against the attorney" who construed the reservation of a "one-half (1/2) non-participating royalty" as entitling its owners to a "1/16 royalty." Was there a rational basis for such a construction? *See* Chapter 5, *infra*.

3. On remand in the principal case the trial court entered judgment for the plaintiffs "for $2,014,540.47 in royalty deficiency, $2,437,114.03 in interest and $750.47 in royalty fees." Strata Energy, Inc. v. Gavenda, 753 S.W.2d 789, 790, 109 O.&G.R. 87 (Tex.App. 1988) (the opinion on appeal from the judgment on

remand). One issue on appeal was "whether appellants are entitled to any 'offsets and credits for that portion of the Gavendas' royalties disbursed to others' (other than the $335,756.75 royalty paid to Billy Blaha, the mineral owner and lessor)." The payments made by the lessees to others than Billy Blaha were made to overriding royalty owners and to assignees of portions of the working interest. The appeals court determined that these interests had nothing to do with the Gavenda's 1/2 "non participating royalty" interest in "all the oil and gas" in the leased tract and thus defendants received no credit for payments made to these interests.

Suppose that a title opinion on the land involved in the principal case had, when ordered by a potential lessee after the conveyance by the Gavenda family, interpreted the language of the Gavenda's term royalty to give them, for its fifteen year term, 1/2 of the oil and gas produced from the land free of the expense of production. What factors would enter into the calculations of the potential oil and gas lessee when deciding whether or not to lease the land with a view to developing it?

4. Several states have enacted statutes regulating the use of division orders. Wyoming's statute provides that a "division order that alters or amends the terms of an oil and gas lease or other contractual agreement is invalid to the extent of the alteration or amendment and the terms of the oil and gas lease or other contractual agreement shall take precedence." Wyo.Stat. § 30-5-304. Another approach simply provides that: "Royalty payments may not be withheld because an interest owner has not executed a division order." Would either the Wyoming or North Dakota statutes take care of the problem in *Gavenda*?

Several other states which have enacted division order statutes which, under certain circumstances, require the royalty owner to execute a division order before interest begins to accrue on the unpaid amount. Tex.Nat.Res.Code §§ 91.401 et seq. The Texas statute initially provides: "Such a division order does not amend any lease or operating agreement . . ." but then states: "Division orders are binding for the time and to the extent that they have been acted on and made the basis of settlements and payments." *Id.* §§ 91.402(c)(2), (g). Does this change the state of the law regarding division orders in Texas?

LOUISIANA LAND AND EXPLORATION CO. V. PENNZOIL EXPLORATION AND PRODUCTION CO., 962 F. Supp. 908, 915, 138 O.&G.R. 322 (E.D. La. 1997): The court found that the division orders in question might have effects on the underlying contractual obligations to pay overriding royalty or working interest payments through the doctrine of estoppel despite a Louisiana statute limiting certain effects of division orders (La.Rev.Stat. § 31:138.1). To be required to make sizable interest payments would affect the lessee to its detriment.

5. May the parties expressly change or waive the provisions of a division order statute? For example, may the parties waive the right to receive statutory interest for late or under-payments of royalty by executing a division order that contains such a waiver provision? *See* In re Tulsa Energy, Inc., 111 F.3d 88, 135

O.&G.R. 579 (10th Cir. 1997)(interpreting Okla.Rev.Stat.Ann. tit. 52, § 570.10).

_____

SECTION 3. MISCELLANEOUS CLAUSES

Although the nature and content of an oil and gas lease may turn upon the particular printed form which a lease hound has purchased at the nearest stationer's shop, when the parties to a leasing transaction are reasonably sophisticated or are adequately represented by counsel, the leasing negotiations may be characterized by serious bargaining over a number of miscellaneous provisions. Some of these provisions are treated elsewhere in this book, *e.g.,* the dry hole, shut-in-gas well, drilling operations, Mother Hubbard, entirety, and pooling or unitization clauses and express drilling and restoration covenants. There may be an interplay between an express provision of the lease and implied covenants. Space does not permit of an extended discussion of the numerous other miscellaneous clauses which may be the subject of bargaining in the leasing transaction beyond the brief mention of the following:

1. *Covenants of title and subrogation clauses.* An oil and gas lease may be made without any covenants of title. If the lessor does not desire to make any he must be careful in the selection of the language employed so as to avoid statutory implied covenants of title arising in some states from the employment of particular words of art. Typically a lessee will insist on a covenant of warranty (this apparently being the only covenant of title customarily included in oil and gas leases) together with a subrogation clause, a common form being as follows:

> *Lessor hereby warrants and agrees to defend the title to said land and agrees that Lessee at its option may discharge any tax, mortgage or other lien upon said land, either in whole or in part, and in event Lessee does so, it shall be subrogated to such lien with the right to enforce same and apply rentals and royalties accruing hereunder toward satisfying same.*

Consider what benefits, if any, may accrue to the lessee by virtue of the inclusion of a covenant of warranty in the lease. It is relatively rare that any substantial cash consideration is paid upon the execution of a lease. What damages may the lessee recover upon breach of the covenant of warranty? Is the covenant of value in your jurisdiction in terms of the application of the doctrine of estoppel by deed? If rentals and/or royalties have been paid to the lessor prior to the discovery of his failure of title, does the inclusion of this covenant in the lease afford substantial assistance to the lessee seeking to recover such payments?

2. *Drilling and operating restrictions.* Depending on the situation at the time of the leasing transaction as to the existence of improvements on the leased premises or as to the utilization of the surface for agricultural or grazing purposes, a variety of restrictions on the lessee's operations may be imposed. The following are typical:

> *When requested by lessor, lessee shall bury lessee's pipe lines below plow depth.*

No well shall be drilled nearer than 200 feet to the house or barn now on said premises without written consent of lessor.

Lessee shall pay for damages caused by lessee's operations to growing crops on said land.

The lease may or may not contain express provisions concerning the easements granted the lessee, his right to use the lessor's water wells in the development of the premises and any special obligations concerning the maintenance of fences by the lessee, or the restoration of the premises by the lessee upon the termination of the lease to the conditions prevailing before the commencement of operations.

3. **Release of record.** To facilitate the quieting of the lessor's title after the expiration, forfeiture or termination of a lease, a clause is frequently included in the lease providing that:

The owner(s) and holder(s) of this lease at the time it expires or is terminated shall discharge this lease of record.

Even absent such a clause, in a number of the producing states, there is a statutory requirement, on demand of the lessor, that the lessee release of record a lease that has expired, terminated or been forfeited.

4. **Removal of fixtures.** A typical provision of oil and gas leases provides that:

Lessee shall have the right at any time during or after the expiration of this lease to remove all property and fixtures placed by Lessee on said land, including the right to draw and remove all casing.

The Removal of Fixtures clause is designed to permit the lessee to remove fixtures after the lease has terminated or when it can no longer be operated at a profit. Usually there is no right to draw and remove the casing while the well is capable of producing in paying quantities. The phrase "at any time" in this clause is generally construed to mean "within a reasonable time."

WILLISON V. CONSOLIDATION COAL CO., 536 Pa. 49, 637 A.2d 979, 128 O.&G.R. 134 (1994), held that the express lease provision authorizing the lessee to remove fixtures "at any time" meant that the lessee, "was not required ... to continue operating its well until gas production was exhausted or until parties other than Consol came to view the well as lacking in profitability." Consol had acquired the 1901 lease involved in this case in 1987 "for the purpose of plugging the well, to allow the extraction of valuable coal which would otherwise have had to remain in place to provide physical support for the operating well." The lessors sought to prevent removal of fixtures since some of the gas produced was used by them on site, without charge, for domestic purposes.

In TRINIDAD PETROLEUM CORP. v. PIONEER NATURAL GAS CO., 416 So.2d 290, 73 O.&G.R. 538 (La.App.1982), writ denied, 422 So.2d 154 (1982), a current lessee moved a saltwater pump left on the premises by a prior lessee to a salvage yard for repairs. This act was held to be a conversion in spite of the new lessee's belief that the "custom" of the industry gave ownership of such equipment to successor lessees. The conversion claim was made in an unsuccessful suit maintaining the continuing validity of the prior lease. That document had no "removal of fixtures" provision. What, if anything, should the new lessee have done in relation to the pump in these circumstances?

5. *Surrender clause.* The inclusion of this type of clause is of especial importance in "or" form leases, but it is also frequently found in "unless" type leases. In simple form, the clause provides:

> *Lessee shall have the right at any time without Lessor's consent to surrender all or any portion of the leased premises and be relieved of all obligation as to the acreage surrendered.*

By virtue of this clause, a lessee may be able to retain that portion of leased acreage which appears most promising while relieving himself of obligations as concerns payment of rentals, protection, exploration or development of that portion of the leased premises which appears to him to be least promising. A lessee may also shed itself of a lease it expects not to be profitable. In SHANKS V. EXXON CORP., 674 So. 2d 473, 136 O.&G.R. 579 (La. Ct. App. 1996) the court held that an oil and gas lessee who releases the lease is thereby relieved of responsibility to the lessor for well costs previously accrued. The leased tract was included in a unit drilled by another company. Believing the well would never achieve payout, the lessee executed a release of the lease. The lessor later sought to hold the surrendering lessee liable for the well costs that were incurred by the unit operator before the surrender but that were deducted from the lessor's share of production after the lease had been surrendered. The liability for well costs attributable to a well drilled on someone else's land was not necessarily created by virtue of the leases. Rather, the unitization order, including the leased tracts at issue or portions thereof in the compulsory unit, affected the lessee's obligation for well costs, as an "owner"of a tract included within the unit. A non-operating owner or lessee, who does not consent to operations within a compulsory drilling unit by a unit operator, has no liability for the costs of development and operations, except out of his share of production. The liability of a non-consenting, non-operating owner for well costs only arises as there is production. The lessee's liability for well costs accrued only as there was production from the unit well and only to the extent of its proportionate share of production. The lessee's entire proportionate share of production was applied to the payment of well costs during the existence of the leases. Thus, the lessee had paid all well costs for which it was liable by law. For later treatment of same issues, see Shanks v. Exxon Corp., 984 So.2d 53, 168 Oil & Gas Rep. 479, 2007-0852 (La.App. 1 Cir. 12/21/07).

Chapter 4

COVENANTS IMPLIED IN OIL AND GAS LEASES
SECTION 1. INTRODUCTION

We saw in the previous chapter that an oil and gas lease commonly imposes no affirmative obligation to drill in search of oil and gas. The lease is usually drawn so that the operator has a choice of drilling or paying rental. However, in certain circumstances the landowner will not be satisfied with rental in lieu of drilling. For example, suppose a producing oil well is brought in on an adjoining tract, close enough to drain the leasehold. Suppose further that the operator owns the lease on that tract as well. Or suppose oil is discovered on the leasehold itself, and one producing well is keeping the lease in force on a section of land. The landowner is understandably desirous of further development, particularly if government spacing regulations permit one well for every 40 acres. Again, suppose that a section of land has been held under lease for twenty years by ever-dwindling production from the northwest quarter. Production on other land to the southeast indicates the possible presence of other deposits under the leasehold.

In all these cases landowners have sought court action to force the operators to drill, although the leases have said nothing about a drilling obligation. Despite this silence relief has been granted, and one of the questions we face is, on what grounds?

Clearly an oil operator is in business to search for, produce and market petroleum products. The oil and gas lease is the instrument by which land is secured by the operator for these purposes. Land is the raw material and the lease is the contract of purchase. Thus, while the lease may be silent, it is not difficult for the courts to find drilling obligations there by merely looking past the printed page to the underlying purposes of the instrument.

The imposition of unexpressed drilling duties is aided also by the fact that everything cannot be written down in advance in an oil and gas lease. The drilling program depends on subsurface conditions not usually known when the lease is signed. So the courts have reasoned that the parties to a lease, though having no specific drilling program in mind, intend exploration and development to be undertaken as become necessary and proper. In the more technical phraseology of the cases, the lessee has the implied duty to operate the premises in an ordinary, prudent manner, having regard for the interests of both himself and the lessor. What the lessee *should* do is determined by what a hypothetical ordinary, prudent operator *would* do, which (for better or worse) may be left to the jury.

For analytical purposes, it is helpful to divide the general duty to operate the premises in an ordinary, prudent manner into a number of specific duties. The cases and textbooks speak of the covenant to drill offset wells, the covenant to reasonably develop, and the covenant to market the product, to mention a few. It is well to remember, however, that no enumeration of the specific implied duties of the lessee is exclusive. The law gives the lessor a remedy for any harm resulting from the lessee's breach of the ordinary, prudent operator standard, whether or not the harm falls into one of the categories of specific duties already recognized.

Martin & Kramer, *Williams and Meyers Oil and Gas Law*

§ 803. Are Covenants Implied "In Law" or "In Fact"?

The question is sometimes discussed, though seldom litigated, whether covenants are implied "in fact" or "in law." Eminent authorities have disagreed: Mr. A.W. Walker, Jr. believes that covenants are implied in fact;[64] Professor Maurice Merrill believes very firmly they are implied in law.[65]

Before we enter into this controversy, it may be well first, to define what is meant by implication in fact and implication in law and second, to inquire what difference it makes.

A covenant is implied in fact when its existence is derived from the written agreement and the circumstances surrounding its execution. A covenant is implied in law when it is added to the contract by a court to promote fairness, justice and equity. Mr. Walker thus describes a covenant implied in fact: "Covenants will be implied in fact where necessary to give effect to the actual intentions of the parties as reflected by the contract or conveyance construed in its entirety in the light of the circumstances under which it was made and of the purposes sought to be accomplished."[66] Professor Merrill says that " . . . the real basis of the doctrine of implied covenants in oil and gas leases [is] to be found in a theory of enforcing that conduct which, under the circumstances, fair dealing between lessor and lessee fairly demands that the latter pursue. . . ."[67]

[Consequences of distinction]

Does it make any difference whether covenants are implied in fact or in law? Judging from the reported cases, the answer seems to be, not so often in the past, but increasingly the distinction is significant, as the use of class action is accepted by the courts for contract litigation. The decisions disclose four

---

[64]   Walker, *The Nature of the Property Interests Created by an Oil and Gas Lease in Texas*, 11 Texas L.Rev. 399, 402–6 (1933).

[65]   Merrill, Covenants Implied in Oil and Gas Leases §§ 7, 220 (2nd Ed.1940).

[66]   Walker, *supra* n. 64 at 402.

[67]   Merrill, *supra* n. 65. Professor Merrill's interrogatory has been recast in declaratory form, but we have no doubt that his question was rhetorical.

consequences that may depend upon the distinction.

First, the statute of limitations applicable to an action for breach of implied covenant may be determined by the classification of the covenant as implied in fact or implied in law. Several cases in point hold the covenant to be implied in fact and give the lessee the benefit of the longer period of limitation applicable to written contracts. These reject the contention that since covenants are implied by courts to accomplish justice and are not part of the contract made by the parties the shorter limitation period applicable to contracts not in writing should govern implied covenants.

Second, it is said that the continued liability of the original lessee, after his assignment of the lease, depends upon whether the covenants in the lease are implied in fact or in law. n6 If implied in fact, the lessee remains liable; if implied in law, he does not, since liability is predicated on a relationship that has terminated. Here, again the cases seem to adopt the position that the covenant is implied in fact, and hence the lessee remains liable for performance of the covenant. The importance of this rule is minimized by the fact that lessees can and often do provide expressly for the termination of their liability upon assignment of the lease.

Third, venue of an action for breach of covenant may be determined by whether the covenant is implied in law or in fact. Thus, where venue of an action for breach of a *written* contract to perform an obligation lies in the county of performance, venue for implied covenant actions lies in the county where the leasehold is located, because the covenant is implied in fact, is therefore part of the writing, and expresses an obligation to be performed in a particular county.

Fourth, the increased acceptance of class actions by courts in royalty litigation has led some courts to distinguish among types of implied covenants. If a covenant is implied in fact, then the words of the contract (such as a clause of a lease) and related documents (such as a division order or letters between royalty owner and lessee) and actions of the parties may be subject to examination. Such inquiries do not lend themselves readily to class action treatment. If a covenant is implied in law, a court may be more willing to pay little regard to the specific words of a contract or may place the burden on a defendant lessee to show that the parties have contracted out of a court imposed duty. This has been implicit in some class action royalty cases involving the implied covenant to market but has been addressed directly by the New Mexico Supreme Court in *Davis v. Devon Energy Corp.*, 218 P.3d 75. In this case, royalty owners claimed breach of an implied covenant to market; gas producers, they asserted, had improperly deducted from royalty payments the costs of making coalbed methane gas marketable and sought class action certification for money damages. In *Continental Potash, Inc. v. Freeport-McMoran, Inc.*, 115 N.M. 690, 858 P.2d 66, 80 (1993), *cert. denied*, 510 U.S. 1116 (1994) , the New Mexico Supreme Court adopted the analysis of a Texas case as to "whether an implied covenant exists in the context of mining law" and concluded: "implied covenants are not favored in law, especially when a written agreement between the parties is apparently complete. ... The general rule is that an implied covenant cannot co-exist with express covenants that specifically cover the same subject matter." Indicating that implied covenants must be implied in fact, the court stated: "[a]n implied covenant must rest entirely on the presumed intention of the parties as gathered from the terms as actually expressed in the written instrument itself." Now, in *Davis*, the court distinguished between two types of implied covenants--those that are implied in fact and those that are implied in law:

> In implying a covenant, courts may either be effectuating the parties' intentions by interpreting the written terms of an agreement and analyzing the parties' conduct, or they may be stating that a duty imposed by law creates an obligation on one or more of the parties to the agreement. Depending on the nature of the implied promise, the court may or may not be required to interpret the parties' agreement and effectuate their intentions.

The district court relied on *Continental Potash* in denying class certification, finding that individualized issues predominated over common ones and that the class litigation would not be a superior method for resolving the plaintiffs' claims because managing the various claims for damages would be difficult. Reversing on this point, the New Mexico Supreme Court held that *Continental Potash* was inapplicable to these class actions; the district court's conclusion that the marketable condition rule has been incorporated in the existing duty to market was sufficient to certify the classes. Although cautioning that "we make no decision in this opinion regarding the existence or the applicability of the marketable condition rule," the court said the primary issue to be litigated on remand was whether the costs deducted by the defendants were necessary to make the CBM gas "marketable."

[Summary]

In summary, we believe there is a large element of truth on both sides of the implied "in law"/implied "in fact" controversy. If a court thinks the implication of a covenant is inappropriate under the circumstances, it will steer close to the words of the contract and find no covenant in fact arising from them. If on the other hand the court thinks the lessee should be under a duty in the case (or class of cases) before it, it may justify implying one on the basis of equity and justice.

We do not mean to suggest that the concepts of implication "in law" and "in fact" are no more than forensic tools in supporting conclusions previously reached. It is probably true that the two concepts are mild derivatives of the polar positions sometimes taken with respect to the function and province of courts in interpreting and enforcing contracts. One position cleaves to the literal words of the contract, permitting a loss to lie where it falls in the absence of express agreement shifting the loss. The other position regards the words of the contract as the mere design of the structure, with a good deal of detail work to be filled in by the courts. The latter position accords a more active role to the courts in supplying missing intention and in redressing the balance of an unequal bargaining position. Courts seem to swing between the two extremes, but in most producing states, the incidence of decision lies close to the middle.

———

One fruitful development in the law of contracts in recent years has been the discussion of the concept of "relational contracts." Consider the "implied in law" versus "implied in fact" distinction in relation to the following excerpt:

Meyers and Crafton, *The Covenant of Further Exploration – Thirty Years Later*, 32 Rocky Mt. Min. L. Inst. 1–1, 1–19 to 1–22 (1986)

[2] Relational Contracts, Opportunistic Behavior, and the Oil and Gas Lease

To understand the problem facing the oil and gas lessor, one must understand the concept of a relational contract. The relational contract can be compared with the classic commercial contract in which "the parties generally are able to reduce performance standards to rather specific obligations." By contrast:

> a contract is relational to the extent that the parties are incapable of reducing important terms of the arrangement to well-defined obligations. Such definitive obligations may be impractical because of inability to identify certain future conditions or because of inability to characterize complex adaptations adequately even when the contingencies can be identified in advance.[68]

The typical relational contract involves a situation in which an asset (or something of value) is managed by the performing party, with the income (or return on capital) of the passive party solely dependent on the performing party's actions. Clearly this is the situation faced by the lessor of an oil and gas lease.

Because the obligations in relational contracts are not specified with particularity, it is often difficult to determine whether or not behavior by the relational promissor falls within the parameters of the contract as contemplated by the parties. Thus, relational promisees are often victimized by what is now called opportunistic behavior. As defined by Professor Muris, opportunistic behavior "occurs when a performing party behaves contrary to the other party's understanding of their contract, but not necessarily contrary to the explicit terms of the agreement, leading to a transfer of wealth from the other party to the performer . . ."

[3] Implied Covenants as a Response to Opportunistic Behavior

While different types of private contractual mechanisms have been developed to monitor and prevent the type of opportunistic behavior that can exist

---

[68] Goetz & Scott, *Principles of Relational Contracts*, 67 Va.L.Rev. 1089, 1092 (1981).

in a relational contract, these devices are costly. As Professor Muris notes:

> Resources spent to implement or prevent opportunistic behavior do not help produce a commodity or service that the contracting parties mutually value. Accordingly, the elimination or reduction of these expenditures will improve the wealth of society by freeing the resources' productive use. The challenge for the law is to help produce these laws without imposing still higher costs in the process.[69]

The judicial implication of covenants into oil and gas leases is nothing more than a response to the problem of lessees acting opportunistically. Given the relational nature of the oil and gas lease, and the "generic agency" between the lessor and lessee, the courts have recognized that without the imposition of judicially created implied covenants to effectuate the controlling intention of the parties as manifested in the lease, a lessee would have the ability to act opportunistically, that is, to manipulate the contract so as to maximize its wealth at the expense of the lessor.

Whether the judicial response be a covenant of reasonable development, a covenant to protect from drainage, or a covenant of further exploration, the rationale is the same. In effect, the implied covenants and the prudent operator standard seek to eliminate lessee opportunism by requiring the lessee to act "for the common advantage of both lessor and lessee."

It is noteworthy that the standard implied into oil and gas leases by the courts is clearly the same as that implied into non-oil and gas relational contracts. The only difference is the terminology utilized, viz., the double duty of "good faith" and "best efforts."

NOTES

1. Should "best efforts" be the same as "reasonable prudent operator"? The authors' statement that relational contracts do require "best efforts" rather than good faith exercise of business judgment is certainly open to question. *See Zilg v. Prentice–Hall, Inc.*, 717 F.2d 671 (2d Cir.1983), *cert. denied*, 466 U.S. 938 (1984).

2. How does breach of an implied covenant under a standard established by a court to control "opportunistic behavior" differ from a tort? On implied covenants generally, see Patrick H. Martin, *Implied Covenants in Oil and Gas Leases—Past, Present & Future*, 33 Wash.L.J. 639 (1994).

In reading the cases that follow, consider carefully the nature of the loss the lessor is complaining about. Has he suffered permanent loss of recoverable hydrocarbons? Mere delay in their recovery? Has he even proved that hydrocarbons are in place beneath his land? Consider also the relief that properly repairs plaintiff's loss.

———

[69]   Id. at 524.

SECTION 2. IMPLIED COVENANT TO PROTECT FROM DRAINAGE (THE OFFSET WELL COVENANT).

(*See* Martin & Kramer, *Williams and Meyers Oil and Gas Law* §§821–826.3)

SUNDHEIM v. REEF OIL CORP.

Supreme Court of Montana, 1991.
247 Mont. 244, 806 P.2d 503, 114 O.&G.R. 42.

MCDONOUGH, JUSTICE. Plaintiffs, Noel Sundheim, Bertha Sundheim and Leona Johnson (hereinafter referred to as Sundheims) appeal from an order of the Fifteenth Judicial District, Roosevelt County, granting summary judgment in favor of defendants, Reef Oil Company, the estate of Frank Hiestand, and Woods Petroleum Corporation. We affirm in part and reverse in part.

. . .

The Sundheims are the owners of mineral interests in land located in Roosevelt County in northeastern Montana. In May of 1967, they entered into oil and gas leases covering these interests. Those leases were entered into with W.C. Kaufman and they had ten year primary terms.

On March 10, 1969, the leases were assigned to Woods Petroleum Corporation. In 1974, Woods Petroleum entered into a Dry Hole Contribution Agreement with Anadarko Production Company, which had a lease on land adjacent to the Sundheim leasehold. Woods agreed to provide financial assistance in drilling a well on a lease held by Anadarko. In return, Anadarko promised to share all well data and structural information obtained through drilling the well. Anadarko found oil and the well initially produced 596 barrels of oil a day.

Woods Petroleum decided to drill its own well on the Sundheim leasehold. This well, Sundheim No. 1, was completed in March of 1975 and had an initial production of 419 barrels per day.

In 1976, production on the Sundheim No. 1 well began to decline. In an effort to increase production Woods installed a pumping unit on the well. Production, however, continued to decline and in 1977 the sucker rods broke and production ceased.

Following this breakdown, Woods Petroleum considered whether further investment should be made to rework the well and reestablish production. Woods determined that no more than $20,000.00 would be spent on reworking the well. According to an internal memorandum, its engineers determined that low production was caused by reservoir conditions. No mention is made of the possibility that the well was not producing large enough amounts of oil due to mechanical problems. Some work was then done on the well; however due to Woods' belief that the decline in production was caused by reservoir conditions, the well was cemented in, on July 20, 1977.

Following this discontinuance, Reef Oil Corporation contacted Woods

Petroleum. Reef Oil was interested in acquiring the open well bore, tubing, wellhead, tanks and other equipment still intact and left at the well in return for Reef Oil assuming responsibility for operating and eventually plugging the well. Woods agreed to this arrangement and in July of 1978 the Sundheim No. 1 well was "sold" to Reef Oil. Before this "sale" took place Reef Oil entered into new oil leases with the plaintiffs. These leases provided for a term of three years and contained provisions for annual delay rentals of $1.00 per acre per year.

According to deposition testimony, Reef Oil did not have the financial ability to redrill the well. Apparently, it was Reef Oil's intent to acquire the well and then attempt to interest others to develop the prospect. Reef Oil retained E. Earl Norwood, a certified petroleum geologist to do an in-depth geologic analysis. Norwood reported the decline in production was caused by mechanical problems and was not caused by reservoir conditions. This conclusion is supported by another engineer, Robert M. Watkins. According to Mr. Watkins, the well's low production probably resulted because it was not backflushed with fresh water to clear out salt accumulations which restricted the flow of oil.

Reef Oil did not have the financial ability to rework the well. Consequently, nothing was done with it until the early part of 1980, when Reef Oil assigned the Sundheim leasehold to Frank Hiestand. In 1981, Heistand worked out a farmout arrangement with a Canadian Oil Company who drilled a new well, the Sundheim No. 2 well, which was 300 feet to the south of the Sundheim No. 1 well. This new well yielded only salt water. According to the Sundheims, as much as 145,000 barrels of oil were drained from their leasehold by adjacent oil wells during the time period between 1977 and 1981.

On January 30, 1986 the Sundheims filed a complaint against Reef Oil Corporation, Woods Petroleum Corporation, Frank Hiestand and American Penn Energy. American Penn Energy was dismissed by stipulation of the parties. In their amended complaint against the remaining three defendants, the Sundheims set forth three separate counts alleging that the defendants breached the implied covenant to protect their leasehold from drainage, the implied covenant to reasonably and prudently develop the leasehold, and the implied covenant to reasonably and prudently operate the existing Sundheim No. 1 well. Frank Hiestand subsequently died and Andy Hiestand, the personal representative of his estate was substituted as a party-defendant.

On February 17, 1987, Reef Oil moved for summary judgment. Reef's motion was granted on May 18, 1987. In its memorandum in support of its order of summary judgment, the District Court held that the Sundheims were barred from bringing an action for breach of the implied covenant to protect from drainage because they had not served written notice upon Reef Oil and demanded it drill an offset well. The court further held that Reef Oil did not breach the implied covenant of reasonable development because the Sundheims accepted delay rentals and because this covenant does not arise until after production has been obtained on the leasehold. The District Court dismissed Sundheims' third cause of action on the ground that the implied covenant to prudently operate a well is not an independent cause of action.

. . .

I. Reef Oil Corporation and Frank Hiestand.

As stated above, the Sundheims based this lawsuit on three separate allegations. In summary, their amended complaint alleged that the defendants breached the implied covenants of protection, to reasonably develop the leasehold and to prudently operate the existing well. As a result, Sundheims allege the defendants are liable for damages. The District Court granted summary judgment in favor of defendants Reef Oil and Frank Hiestand on each of these counts. We will discuss each issue separately.

A. The Implied Covenant to Protect from Drainage.

Each of the Sundheims' allegations are based upon covenants which have been implied by common law, into oil and gas leases. The purpose of the implied covenants is to fully effectuate the intentions of the parties to a lease. Obviously, the primary intention is to produce oil and gas for a profit and to obtain royalties for the lessor. *U.V. Industries Inc. v. Danielson* (1979), 184 Mont. 203, 602 P.2d 571. To insure this result is obtained the courts have implied, through the express terms of oil and gas leases, certain duties which must be performed by the lessee. These duties include the covenant to protect from drainage and the covenant to reasonably develop the leasehold, both of which are at issue in the case now before us.

The Sundheims maintain that Reef Oil Corporation and Frank Hiestand breached the covenant to protect their leasehold from drainage. In their brief, the Sundheims estimate that as much as 145,000 barrels of oil were drained from their leasehold by adjacent oil wells. They further argue that Reef Oil and Hiestand, in accordance with their duty to protect, should have drilled offset wells in order to capture the oil before it was drained. Because these defendants failed to drill such wells, they breached the covenant to protect, and are therefore liable for damages for the value of the oil drained from Sundheims' leasehold.

The District Court did not reach the merits of the Sundheims' argument because it found that they were barred from asserting their claim due to the fact that they failed to give written notice of the drainage or demand that an offset well be drilled. In reaching its conclusion, the court relied upon *U.V. Industries Inc. v. Danielson* (1979), 184 Mont. 203, 602 P.2d 571, where we stated:

> The offset drilling rule generally requires the lessor . . . to serve written notice or demand upon the lessee or its assigns to drill an offset well as a precondition to the latter's duty to drill. Supra, 602 P.2d at 584.

Based upon this language, the District Court interpreted *U.V. Industries* very narrowly to hold that the Sundheims could not bring an action for breach of the implied covenant to protect. All parties to this action agree that no written notice or demand was served.

In *U.V. Industries* the plaintiffs brought a lawsuit alleging that the defendants failed to protect their leasehold from drainage and were liable for damages. The case involved four defendants. Two of the defendants owned an interest in a well, which was located on adjoining property. Apparently, this well caused the drainage to the plaintiff's leasehold.

Faced with these facts, we held that written notice or demand was not required to be given the two defendants who held interests in the adjoining well. We based this conclusion upon the fact that these two defendants, through their ownership of the draining well, obviously knew of the necessity to protect the plaintiff's leasehold. Thus, application of the notice requirement was unnecessary. Two other defendants who held no such interest, were not deemed to have any knowledge of the drainage. Therefore, it was held that they must be given notice of their obligation to protect before any duty on their part would arise.

The District Court interpreted *U.V. Industries* to hold that written notice or demand is always a prerequisite to a lawsuit for breach of the covenant to protect, unless the defendant owns the adjoining well which is draining the plaintiff's property. The Sundheims, on the other hand, maintain that under *U.V. Industries,* written notice or demand is only required when the defendant does not have knowledge of the drainage. According to Sundheims, a notice requirement in a situation when a defendant already has knowledge would be wholly unnecessary. Because there is evidence which tends to establish that Reef Oil and Hiestand knew of the draining of their leasehold, the Sundheims further argue that the District Court erred in granting summary judgment based upon lack of notice.

We agree with the Sundheims' interpretation of *U.V. Industries.* There is no sound reason to require a lessor to give a lessee notice of drainage in an action for damages, if the lessee already has such knowledge. Under the implied covenant to protect, a lessee is charged with the duty to act as a reasonable operator and to protect the leaseholder's interest. This duty requires the lessee to manage the leasehold in such a manner as to bring profit to both himself and the lessor. Obviously, if the lessee failed to drill an offset well, while knowing the leasehold was being drained, he would fail to meet this duty. In such a circumstance, he should not be excused for the sole reason that the lessee failed to comply with a requirement that is unnecessary.

. . .

Clearly, the reasonable notice requirement is satisfied when the lessee has knowledge of drainage. . . .

[A] person is deemed to have constructive notice when he is in possession of all of the relevant facts and circumstances. . . . Therefore, we now clarify our earlier holding in *U.V. Industries* and hold that before a lessee's duty to drill an offset well arises, he must have reasonable notice of the necessity to protect the leasehold. Such notice can either be express, from the lessor, or constructive, gained from the surrounding circumstances. Further support of this holding is found in the general rule that oil and gas leases are to be construed liberally in favor of the lessor and against the lessee. . . .

This rule is only applicable to those cases in which a lessor is seeking relief in the form of damages, however. If he is seeking other forms of relief, such as forfeiture of the lease, he must give written notice in order to give the lessee a chance to cure the breach. See Summers, Oil and Gas Volume 2 § 412 (1990).

The burden of proof is upon the lessor to establish the fact that the lessee knew of the drainage. Therefore, upon remand, the Sundheims must prove that both Reef Oil Corporation and Frank Hiestand knew or should have known that the Sundheim leasehold was being drained by the adjacent wells.

[The court next discussed the implied covenant to reasonably develop the leasehold].

It is undisputed that the delay rental payments were made by Reef Oil and were accepted by the Sundheims. Therefore, by the terms of the contract, Reef Oil was not impliedly required to engage in further development of the Sundheim leasehold. The rental clauses contained in the lease were presumably bargained for and are supported by consideration. We will not look beyond these express provisions in order to impose a duty upon Reef Oil which is in contravention of their terms. In short, the rental clause contained in the Sundheim leases relieves Reef Oil of all drilling obligations save that of offset drilling. Williams & Meyers, Oil and Gas Law § 835.1 (1990). See also Hemingway, Law of Oil and Gas § 8.10 2d Ed. (1983). The duty to drill an offset well is imposed through the covenant to protect, not through the covenant to reasonably develop. Therefore, there was no breach of the development covenant and summary judgment on this issue was properly granted.

C. Prudent Operator/Reasonable Man Standards.

The Sundheims maintain that Reef Oil and Frank Hiestand, in addition to breaching implied covenants to the lease, also violated the prudent operator standard. This standard is best understood through analogy to the reasonable man standard of tort law. Simply stated, the prudent operator is a reasonable man engaged in oil and gas operations. He is a hypothetical oil operator who does what he ought to do not what he ought not to do with respect to operations on the leasehold. Williams and Meyers, Oil and Gas Law § 806.3 (1990).

Sundheims argue that Reef Oil and Frank Hiestand breached this obligation because they did not develop the leasehold. Instead, they took the lease for speculative purposes, and while they were searching for an individual to commence drilling operations substantial drainage occurred. The District Court granted summary judgment in favor of the defendants on this count. It held that the prudent operator standard was not in itself an independent cause of action and the prudent operator standard is applied in conjunction with and serves to define the other implied covenants.

We agree with the District Court. We have previously held that the

Sundheims have presented a valid cause of action for breach of the implied covenant to protect. The duty to act as a prudent operator underlies this contractual obligation. In order to prevail on this issue, the Sundheims must prove at trial that, under the circumstances of their case, Reef Oil and Frank Hiestand did not act as reasonably prudent operators to prevent drainage. Proof of this issue necessarily entails a showing that actual drainage occurred and that an offset well would have produced oil in paying quantities. Williams and Meyers, Oil and Gas Law § 822 (1990).

Proof of the second element of the above test would necessarily establish that the defendants failed to act as reasonably prudent operators and, assuming drainage can be established, that they breached the covenant to protect. This rule of law is supported by the rationale that a reasonably prudent operator would drill an offset well if he could gain a profit and if it was necessary to protect the value of the lease. On the other hand, no breach of this standard, could be established if it cannot be shown that such a well could be drilled at a profit to the lessee. Accordingly, in such circumstances there would be no breach of the covenant to protect.

It is clear from the facts of this case, that the Sundheims allegations of breach of the covenant to protect and breach of the prudent man standard are one in the same. Proof of one necessarily entails proof of the other. Therefore, the District Court correctly granted summary judgment dismissing the allegation alleging breach of the prudent operator standard.

[The remaining issues are omitted]

Conclusion

The District Court's order awarding summary judgment in favor of Reef Oil Corporation and the estate of Frank Hiestand on the issues of breach of the covenant to develop and the prudent operator standard are affirmed. Summary judgment dismissing Woods Petroleum Company is also affirmed. The District Court's orders granting summary judgment to Reef Oil and Frank Hiestand on the issue of breach of the covenant to protect and its order granting Rule 11 sanctions is reversed. This case is remanded for proceedings consistent with this opinion.

TURNAGE, C.J., and HARRISON, HUNT, WEBER, TRIEWEILER and BARZ, JJ., concur.

NOTES

1. Notice the phrase production "in paying quantities" that we first encountered in the habendum clause reappears in an implied covenant case. Does it mean the same thing as it did when applied to habendum clause cases? The following trial judge's instructions regarding drainage were upheld in Good v. TXO Production Corp., 763 S.W.2d 59, 104 O.&G.R. 306 (Tex.App. 1988, writ ref'd):

In answering Special Issues, you are instructed that a reasonably prudent operator has the duty to prevent substantial drainage of gas and condensate from Section 229 if a reasonably prudent operator would drill a well to prevent such drainage. However, such a duty arises only if a reasonably prudent operator with knowledge of the risks involved would protect from such drainage by drilling a well and also unless [sic] a reasonably prudent operator would have a reasonable expectation of producing gas and condensate in paying quantities. The term "paying quantities" as used in this charge, means production of gas and condensate in such quantities as would give the operator a reasonable profit after deducting the costs of drilling, completing, operating, taxes and marketing. You are likewise charged that the term "reasonably prudent operator" as used in this charge, means an operator of ordinary prudence, that is having neither the highest or [sic] the lowest prudence, but on the contrary an operator of average prudence and intelligence, acting with ordinary diligence under the same or similar circumstances.

How does this definition differ from the use of "in paying quantities" in the habendum clause cases? Why should this difference exist?

2. In determining whether the proposed protection well is likely to be profitable, should the court take into account *diminutions* in production, resulting from the operation of the proposed offset well, at other wells operated by the lessee on this leasehold? What of diminished production at wells operated by the lessee on other leaseholds? For consideration of these and related issues, see a series of articles by Professor, now Judge, Stephen F. Williams, *Implied Covenants' Threat to the Value of Oil and Gas Reserves*, 36 Inst. on Oil & Gas L. & Tax'n, ch. 3 (1985); *Implied Covenants in Oil & Gas Leases: Some General Principles*, 29 U.Kan.L.Rev. 153 (1981); *Implied Covenants for Development and Exploration in Oil & Gas Leases—The Determination of Profitability*, 27 U.Kan.L.Rev. 443 (1979).

3. Repressuring and enhanced recovery operations are typically conducted by means of reservoir-wide unitization of the tracts involved. In the absence of unitization, however, special problems may arise. Consider a repressuring operation, without unitization but with the following features: lessee turns one of several wells on plaintiff's lease into a gas injection well; as a result of the program some oil is drained away from lessor's tract, but an approximately equal quantity is drained toward it; aggregate recovery on plaintiff's leasehold is increased substantially beyond what it would have been in the absence of the program. Does the drainage away from lessor's tract violate the protection covenant? If not, is it because of the matching counter-drainage, or because of the basic soundness of the repressuring program taken as a whole, or for some other reason? *Compare* Carter Oil Co. v. Dees, 340 Ill.App. 449, 92 N.E.2d 519 (1950), with Ramsey v. Carter Oil Co., 74 F.Supp. 481 (E.D.Ill.1947), *aff'd*, 172 F.2d 622 (7th Cir.1949), *cert. denied*, 337 U.S. 958 (1949).

4. It has been typical to think of the implied protection covenant as being met by the drilling of an offset well, as in the principal case. Courts have recently

recognized, however, that an alternative means of meeting the obligation might sometimes be pooling or unitization of the lease tract with the draining tract. *See, e.g.,* Amoco Production Co. v. Alexander, 622 S.W.2d 563, 72 O.&G.R. 125 (Tex.1981) (*infra*); Pierce v. Goldking Properties, Inc., 396 So.2d 528 (La.App.1981), *writ denied,* 400 So.2d 904, 69 O.&G.R. 263 (1981). In such a case, as no well would be drilled, naturally the plaintiff need not prove that a well would meet the profitability test. Will pooling that is done in good faith insulate the lessee from liability for drainage? In addition to *Pierce v. Goldking Properties,* see Southeastern Pipe Line Co. v. Tichacek, 997 S.W.2d 166, 143 O.&G.R. 179 (Tex 1999) for an affirmative answer. The lessee's implied covenant to protect against drainage through pooling was explored in Eagle Lake Estates, L.L.C. v. Cabot Oil & Gas Corp., 330 F. Supp. 2d 778, 161 O.&G.R. 734 (E.D. La. 2004), *See* note 2 page 950 *infra*.

5. On the notice issue in the principal case compare Article 136 of the Louisiana Mineral Code:

Art. 136. Written notice; requirement and effect on claims for damages or dissolution of lease.

If a mineral lessor seeks relief from his lessee arising from drainage of the property leased or from any other claim that the lessee has failed to develop and operate the property leased as a prudent operator, he must give his lessee written notice of the asserted breach to perform and allow a reasonable time for performance by the lessee as a prerequisite to a judicial demand for damages or dissolution of the lease. If a lessee is found to have had actual or constructive knowledge of drainage and is held responsible for consequent damages, the damages may be computed from the time a reasonably prudent operator would have protected the leased premises from drainage. In other cases where notice is required by this Article damages may be computed only from the time the written notice was received by the lessee.

6. Should the production in paying quantities standard be applicable in a case in which the lessee is the party causing the drainage to the adjacent tract (i.e. is a common lessee)? Consider the next case.

———

FINLEY v. MARATHON OIL COMPANY

United States Court of Appeals, Seventh Circuit, 1996.
75 F.3d 1225, 137 O.&G.R. 10

POSNER, CHIEF JUDGE. The appeal in this diversity suit presents interesting questions, exotic in this circuit, of oil and gas law, as well as questions concerning the application of amendments made in 1993 to the rules governing pretrial discovery in federal district courts.

The Finleys own two adjacent parcels of land in southern Illinois. At different times they leased oil and gas rights in the two parcels to Marathon Oil Company. The leases provided that the Finleys would receive one-sixth of any oil produced and Marathon would keep the other five-sixths as compensation for the risk and expense of producing. Later the Finleys entered into a "communitization" agreement with Marathon, consolidating the two leases into one but not altering the terms of the leases in any respect relevant to this appeal.

Immediately to the north of the Finleys' two parcels is a parcel owned by the heirs of McCroskey, who have leased their oil and gas rights to a different oil company. The underground formation from which the oil in the Finleys' land is pumped extends under the McCroskey parcel. The oil is extracted from the formation by the method known as secondary recovery, which involves injecting water into the formation with the aim of forcing the oil into producing wells and thence up to the surface. Marathon drilled an injection well at location T-2 on the Finleys' property and over a period of many years poured hundreds of thousands of barrels of water down it in an effort to extract additional oil. The Finleys presented evidence, vigorously contested by Marathon, that the injection caused oil to migrate from the plaintiffs' part of the formation to the part under the McCroskey parcel and that this migration, or "drainage" as it is called in the trade, would have been prevented if only Marathon had drilled another producing well on the Finleys' property between the injection well and the boundary with the McCroskey parcel. The Finleys claim that in failing to prevent this drainage Marathon broke its contract with them and committed a breach of fiduciary duty as well. The trial judge dismissed the latter claim before trial; the jury brought in a verdict for Marathon on the breach of contract claim; and so the suit was dismissed and the Finleys appeal.

The lease required Marathon to prevent, so far as it was commercially practicable (that is, cost-justified) to do so (an important qualification), the drainage of oil from the lessor's property to that of adjacent landowners. The requirement was explicit in certain agreements made between Marathon and McCroskeys' heirs—agreements of which the Finleys, we may assume, were third- party beneficiaries. It is also an implicit corollary of the implied duty of an oil and gas lessee to operate the leasehold prudently rather than wastefully. Fransen v. Conoco, Inc., 64 F.3d 1481, 1487 (10th Cir.1995); Tidelands Royalty "B" Corp. v. Gulf Oil Corp., 804 F.2d 1344, 1348-49 (5th Cir.1986); Williams v. Humble Oil & Refining Co., 432 F.2d 165, 171-72 (5th Cir.1970); Amoco Production Co. v. Alexander, 622 S.W.2d 563, 568 (Tex.1981). Both the general duty of prudent operation and the subsumed duty to avoid drainage illustrate the office of the common law of property and contracts in interpolating into a contract or conveyance provisions that the parties would almost certainly have agreed to had the need for them been foreseen. Although oil and gas litigation is not frequent in Illinois, there is no doubt that the common law of Illinois recognizes the duty to avoid drainage. See Carter Oil Co. v. Dees, 340 Ill.App. 449, 92 N.E.2d 519, 523 (1950); Ramsey v. Carter Oil Co., 74 F.Supp. 481, 482 (E.D.Ill.1947) (applying Illinois law); Geary v. Adams Oil & Gas Co., 31 F.Supp. 830, 835 (E.D.Ill.1940) (ditto).

Whether Marathon complied with its duty to avoid drainage was a question of fact that, if we set to one side for a moment the Finleys' objection to two of the trial judge's evidentiary rulings, the jury was entitled to resolve as it did in favor of Marathon. Marathon presented expert evidence that there was no recoverable oil in the portion of the Finleys' property in which the injection well at T-2 was drilled and so there could not have been a diversion of oil to the McCroskey leasehold as a consequence of the pumping of water into that well.

Had Marathon been the lessee of the McCroskey oil and gas rights and had deliberately diverted oil from the Finleys' properties to the McCroskey property because it had a better deal with the heirs—had Marathon, in other words, been the common lessee of the two properties—it might have been guilty of conversion. Ramsey v. Carter Oil Co., *supra*, 74 F.Supp. at 483. Indeed, even without such a nefarious purpose, diversion by a common lessee might be deemed a per se violation of the duty to avoid drainage, making the lessee liable for the loss to the lessor without regard to the cost to the former of avoiding the drainage. E.g., Cook v. El Paso Natural Gas Co., 560 F.2d 978, 982-84 (10th Cir.1977); Geary v. Adams Oil & Gas Co., *supra*, 31 F.Supp. at 834-35. Or might not. The per se rule illustrated by *Cook* and less clearly by *Geary* has been severely, and as it seems to us justifiably, criticized, 5 Howard R. Williams & Charles J. Meyers, Oil and Gas Law §§ 824.1, 824.3 (1995), and has been rejected in many jurisdictions. See, e.g., Tidelands Royalty "B" Corp. v. Gulf Oil Corp., *supra*, 804 F.2d at 1353-54; Williams v. Humble Oil & Refining Co., *supra*, 432 F.2d at 172-73; Rogers v. Heston Oil Co., 735 P.2d 542, 547 (Okla.1984). *Rogers* states clearly what we take to be the correct view, that as in other cases of fiduciary duty the common lessee must treat the lessor as well as he would treat himself, but not better. Maybe, despite *Geary*, which is after all just the opinion of one district judge, and not a recent opinion at that, Illinois would adopt this view and reject *Geary*.

We need not pursue these byways. Marathon was not the lessee of the heirs' oil and gas rights, had therefore no incentive to divert the oil to them, had in fact an incentive not to do so, was found not to have done so, and at worst merely blundered (if drainage there was, as apparently there was not, and if the drainage could have been prevented at a cost less than the value of the oil drained away) rather than committed a deliberate wrong. We shall see in a moment that Marathon's incentive to avoid drainage may have been impaired. But the Finleys neither remark this nor presented at trial any evidence that Marathon was acting deliberately rather than mistakenly, or indeed that it had committed any wrong more serious than a garden-variety breach of contract; for an oil and gas contract does not make the lessee the fiduciary of the lessor in the absence of special circumstances not shown here. O'Donnell v. Snowden & McSweeney Co., 318 Ill. 374, 149 N.E. 253, 255 (1925); Kirke v. Texas Co., 186 F.2d 643, 648 (7th Cir.1951) (applying Illinois law); Norman v. Apache Corp., 19 F.3d 1017, 1024 (5th Cir.1994). So the Finleys could not have obtained punitive damages, Mijatovich v. Columbia Savings & Loan Ass'n,

168 Ill.App.3d 313, 119 Ill.Dec. 66, 69, 522 N.E.2d 728, 731 (1988); Naiditch v. Shaf Home Builders, Inc., 160 Ill.App.3d 245, 111 Ill.Dec. 486, 496, 512 N.E.2d 1027, 1037 (1987); Amoco Production Co. v. Alexander, *supra*, 622 S.W.2d at 571, which was the main reason they had for dressing their claim in fiduciary-duty breeches.

The basis for the claim that Marathon was the Finleys' fiduciary is that the "communitization" agreement was the equivalent of a unitization agreement, with Marathon corresponding to the unit operator. Many cases do hold that the unit operator, that is, the manager of the unitized field, is the fiduciary of the owners of the field, Fransen v. Conoco, Inc., *supra*, 64 F.3d at 1487; Leck v. Continental Oil Co., 971 F.2d 604, 607 (10th Cir.1992); Young v. West Edmond Hunton Lime Unit, 275 P.2d 304, 308- 09 (Okla.1954); 6 Williams & Meyers, *supra*, § 990, p. 869, and while other cases think it enough that the participants should have a duty of good faith toward each other, e.g., Amoco Production Co. v. Heimann, 904 F.2d 1405, 1411-13 (10th Cir.1990); Rutherford v. Exxon Co., U.S.A., 855 F.2d 1141, 1145-46 (5th Cir.1988), Illinois is in the former camp. Carroll v. Caldwell, 12 Ill.2d 487, 147 N.E.2d 69, 74 (1957). Unitization refers to the exploitation of an oil or gas field that lies under the surface of a number of landowners as if it were under joint ownership, in order to optimize the rate, and minimize the cost, of production. It eliminates the temptation to socially wasteful conduct that would otherwise exist, when an oil field is owned in common, because each operator has an incentive to remove as much oil as fast as he can, even though the cost of production for the field as a whole might be less and the total amount recovered greater if the field were exploited by a single owner. Unitization is especially needful when secondary recovery is the method of getting the oil out of the ground, because the injection of water is likely to flood out parts of the formation, so that some of the lessors will no longer have any recoverable oil beneath their land, yet their sacrifice will have benefited the lessors as a whole. 6 Williams & Meyers, *supra*, § 901, p. 4.

Unitization makes the owners of the rights in the unitized field joint venturers, and joint venturers owe fiduciary duties to one another. Carroll v. Caldwell, *supra*, 147 N.E.2d at 74; Newton v. Aitken, 260 Ill.App.3d 717, 198 Ill.Dec. 751, 756, 633 N.E.2d 213, 218 (1994); Meinhard v. Salmon, 249 N.Y. 458, 164 N.E. 545, 546 (1928) (Cardozo, C.J.). It is only a slight further step, and one taken by a number of jurisdictions as we have noted, to regard the manager of the joint venture, or in the case of oil-field unitization the unit operator, as the fiduciary of the joint venturers. But this is not a case of unitization. The two leases were owned by the same people and operated by the same producer, Marathon. The communitization agreement merely formalized the ownership and operating arrangements. It was no more a joint venture than when one supplier sells to one dealer.

What is true is that a conflict of interest is built into any royalty arrangement, or any other arrangement in which one party to a contract receives a percentage of gross revenues. The Finleys were entitled to one-sixth of all oil produced under the lease, and this was the equivalent of a 16.67 percent royalty on the gross revenues (minus selling costs) from the sale of all the oil. It was a matter of indifference to them, but intense interest to Marathon, how much the oil cost to produce. Suppose the oil had a value of 100, and cost 90 to produce. Production would then be profitable from the standpoint of the property as a whole and of course from the Finleys' standpoint. But Marathon would lose money because it would bear the entire cost and that cost (90) would exceed its share of the revenues (83.33). A similar conflict is created by royalty arrangements between publishers and authors and by contingent-fee contracts. It is not, however, a conflict that has prompted the courts to impose a fiduciary duty on the cost-bearing party to such contracts, Marathon in this case. Lawyers are deemed fiduciaries of their clients but this is so whether or not the lawyer is being paid on an hourly or a contingent basis. We noted earlier that Illinois has expressly declined to make the oil and gas lessee a fiduciary of the lessor.

This analysis shows, however, that the Finleys' breach of contract claim is not so groundless as it might at first appear to be. One might think it irrational for Marathon to divert oil from property under lease to it to property leased to a competitor. But if there was recoverable oil in the vicinity of T-2 yet not enough to cover the cost of extracting it after dividing it with the Finleys in accordance with the terms of the lease, Marathon would have a disincentive to drill another producing well even if the consequence of not doing so was to divert some recoverable oil to another producer. The Finleys do not argue the point, however; nor is there evidence that Marathon yielded to this temptation to short-change them.

The district judge was right, therefore, to throw out the claim of breach of fiduciary obligation, and the only other question raised by the appeal is whether he abused his discretion in refusing to permit the introduction of two pieces of rebuttal evidence at the trial of the contract claim. One was a set of charts prepared by the Finleys' expert showing a striking correlation between the injection of water into the well at T-2 and the production of oil from the wells on the land owned by McCroskey's heirs. The other was a jar containing oil taken from the injection well itself shortly before trial and purporting to demonstrate that there was recoverable oil in the part of the formation that was proximate to the well, contrary to the testimony of Marathon's experts.

. . .

The jar was to be the piece de resistance of the Finleys' case. The ground had been carefully laid in the cross-examination of Marathon's three experts. The Finleys' lawyer had asked each of them what would happen if you drew liquid out of the injection well at T-2. One said that you would get clear water, another that you would get dirty water, and the third that, because the water that had been pumped into the injection well had come from elsewhere in the oil field and therefore contained traces of oil, and oil is lighter than water, you would at first get oil and later, after all the small amount of oil had been drawn off, water. Although Marathon had stopped injecting water months earlier because it had concluded that there was no recoverable oil in the vicinity of the well, the previously injected water had not yet all drained off and one of the Finleys had, shortly before trial, gone to the injection well, opened a tap, and filled a small jar (the kind used for preserves). The top third or so of the jar, which was exhibited to us at the oral argument, consists of a dense black substance, plausibly oil, and the bottom two-thirds or so of what appears to be murky water. The Finleys wanted to show the jar to the jury and argue that it impeached Marathon's experts. The trial judge refused to permit this on the ground that in and of itself the jar proved nothing.

We think that he was entitled to reach this conclusion and we add that judges should be cautious in allowing the introduction of physical exhibits, especially as rebuttal evidence, without firm foundations. Physical exhibits ("demonstrative evidence") are a very powerful form of evidence, in some cases too powerful, as we learn in Julius Caesar from Antony's masterful demagogic use of Caesar's blood-stained toga and slashed body to arouse the Roman mob. After hearing a welter of confusing and contradictory testimony, perhaps of a technical nature, as here, or being led through a maze of inscrutable documentation, also as here, the jury is invited to resolve its doubts on the basis of a simple, tangible, visible, everyday object of reassuring familiarity. "Seeing is believing," as the misleading old saw goes. The trial judge must make sure that the jury is not misled concerning the actual meaning of the object in the context of the litigation. The tiny jar with its film of oil meant nothing in itself. Because oil will float to the top of a column of water, and because the injection well, which was more than 1,000 feet deep, contained a very long column of water, even a tiny residue of oil would, coalescing at the top of the column, produce a few ounces of oil, all the jar contained. This would be no evidence at all that the formation under the well contained enough recoverable oil to have justified Marathon's drilling another producing well. It would not contradict the third expert but indeed would corroborate his testimony; probably not the second either; and the first was merely guessing that the residue was too small even to dirty the water—a mistake, but an irrelevant one.

It is true that had the jar been allowed into evidence Marathon would have had an opportunity to rebut along the lines just suggested. That procedure would have been a proper one for the district judge to adopt but it was also proper for him to insist, as a precondition to admitting the jar, that the Finleys lay a proper foundation. They should have asked their expert to "sponsor" the jar, explaining how it showed that Marathon's experts should not be believed. Of course, it is quite possible that no reputable expert would have so testified.

We find no reversible error. The judgment for the defendant is therefore AFFIRMED.

NOTES

1. As usual for an opinion penned by Posner, J., the decision in *Finley v. Marathon* fairly sparkles with economic analysis. Posner observes that the royalty owners are indifferent to costs. Is this fact not the source of much of our implied covenant litigation? Note further that royalties are a cost from the perspective of the lessee. Should we not distinguish, however, between lessor royalty and royalty pre-existing the lease - which we would deduct in our calculation of profitability in evaluating what a prudent operator would do - and overriding royalty carved out of the lease? Can you explain why?

2. In COOK V. EL PASO NATURAL GAS CO., 560 F.2d 978, 58 O.&G.R. 206 (10th Cir.1977), discussed in *Finley*, the lessee was the sublessee of plaintiff owner of an overriding royalty reserved in the sublease agreement. The lessor on both tracts was the United States, and thus it faced no drainage. The United States prohibited the drilling on the tract at issue because of the existence of potash deposits on the land. The defendant lessee sought from the appropriate state agency to have the proration unit increased from 320 acres to 640 acres to embrace the subject tract, but this application was denied by the agency. The plaintiff nevertheless brought suit for compensatory royalty, claiming that the lessee was in breach of a duty to protect against drainage. The court of appeals affirmed a trial court judgment for plaintiff, which it summarized as follows:

> The court concluded that since the law implies a duty on an oil and gas lessee to protect the leased premises from offsetting adjoining land drainage of oil and gas, and since this is a covenant which runs with the land and the owner of an overriding royalty interest has standing to invoke the implied covenant to protect against drainage, and where a common lessee exists for two abutting oil and gas leases, the common lessee is obligated to protect its lessor from oil and gas drainage from a well located in the other lease. *This is not controlled or limited by the test as to whether a reasonable prudent operator would in the circumstances drill an offset well.*

> The court further concluded that the common lessee is under a duty to prevent drainage *regardless of whether or not the drilling of an offset well on nonproducing land would satisfy the standards of the prudent operator rule.* [Emphasis added by eds.]

Is this rule of strict liability for the common lessee justified? Is there a problem with unjust enrichment to the lessee, or does the lessee simply have to pay royalty twice on the same oil or gas? *Compare* Breaux v. Pan American Petroleum Corporation, 163 So.2d 406, 20 O.&G.R. 476 (La.App.1964), *cert. denied*, 246 La. 581, 165 So.2d 481 (1964) *with* Phillips Petroleum Company v. Millette, 221 Miss. 1, 72 So.2d 176, 74 So.2d 731, 4 O.&G.R. 38 (1954); R.R. Bush Oil Co. v. Beverly–Lincoln Land Co., 69 Cal.App.2d 246, 158 P.2d 754 (1945).

3. In ROGERS V. HESTON OIL CO., 1984 OK 75, 735 P.2d 542, 92 O.&G.R. 175, the Supreme Court of Oklahoma stated:

We further hold that where the lessee is the owner of an adjoining lease which is draining the lessor's tract, the "prudent operator" rule applicable to the lessee's implied covenant to protect against drainage is whether the lessee as a prudent operator would drill a protection well on the lessor's tract *if the draining lease were owned by a third party, and not the lessee.* (Emphasis in original.)

4. Judge Posner's comparison of the glass jar evidence in *Finley* to Caesar's toga brings home vividly the importance of evidence and expert testimony in establishing damages in a drainage or other implied covenant case. Another such case is KERR-MCGEE CORP. V. HELTON, 133 S.W.3d 245, 160 O.&G.R. 240 (Tex. 2004.) which held that the expert witness testimony in a drainage case was incompetent to prove damages. The trial court ruled in favor of the lessor and awarded damages in excess of $1 million dollars, and the appeals court affirmed. However, the Texas Supreme Court reversed. Looking to the United States Supreme Court standards in Daubert v. Merrell Dow Pharms., Inc., 509 U.S. 579, 590 (1993), the Texas court said that if the expert's testimony is not reliable, it is not evidence. The reliability requirement focuses on the principles, research, and methodology underlying an expert's conclusions. Under this requirement, expert testimony is unreliable if it is not grounded "in the methods and procedures of science and is no more than subjective belief or unsupported speculation." The court found unreliable the plaintiff's expert witness, who had testified that a proposed hypothetical well would have produced at approximately the same rate as other offset wells. Even if the witness had relied on data generally used by petroleum engineers to estimate production, there was too great an analytical gap between the data and the expert's conclusions. From the court's opinion: "When asked if the production from these wells would tell him [the plaintiff's expert] what the hypothetical well would have produced, he answered: 'No, it does not.' Further, [he] was asked, '[Y]ou simply do not have any factual basis for projecting the production of that hypothetical well, do you?' He responded, 'That is correct'."

5. Some cases have declared that the implied protection covenant is negated by an express covenant requiring lessee to protect against draining wells within a specified distance of the leasehold, where the draining well is farther away. Is it logically sound to infer from an express duty to protect against wells within 150 feet from the leasehold an intention that no duty shall exist to protect against wells 151 or more feet away? Apart from logic, do you think the typical lessor understands this to be the effect of an express offset well clause? How should a court treat an express offset well clause requiring the lessee to protect against wells within 150 feet when the spacing rule of the state, at the time the lease was executed, forbade drilling closer than 330 feet from leaselines?

Judge Wisdom of the Fifth Circuit considered these questions in WILLIAMS V. HUMBLE OIL AND REFINING CO., 432 F.2d 165, 38 O.&G.R. 212 (5th Cir.1970), reh. denied, 435 F.2d 772, 38 O.&G.R. 245 (5th Cir.1970), cert.

denied, 402 U.S. 934 (1971). The question arose in Louisiana; after noting that there were no decisions in point in that state, Judge Wisdom reviewed the jurisprudence of other states, notably California, Mississippi and Texas, observing that the Supreme Court of Texas in 1966 had held in Shell Oil Co. v. Stansbury, 410 S.W.2d 187, 25 O.&G.R. 578 (Tex.1966), that an express offset covenant does not limit "the lessee's obligation to protect against drainage when the lessee was also the operator whose drilling on adjacent property was causing the drainage." (432 F.2d at 175.) In concluding that Louisiana would adopt the same rule, he proffered three reasons:

(a) "First, lease provisions in derogation of implied obligations are to be strictly construed." (432 F.2d at 177.)

(b) "Second, when the lessee inserts an express offset provision into a lease, he is, of course, attempting to limit the implied responsibility to protect against drainage that he would otherwise have under the law of the jurisdiction. As a general proposition, when the average landowner reads a lease form containing the usual offset provision, it is reasonable for him to assume that the lessee is shouldering an added responsibility. Although there is no objection to enforcing a clause stating clearly that the lessee is not obligated to offset a well more than 150 feet from the boundary line, 'to accomplish this through indirection, by appearing to confer a benefit while taking one away, smacks of fraud or unfair dealing.' The element of unfairness is most striking in those leases in which the offset distance has been fixed at a figure equal to or even shorter than that permitted by the Conservation Department's spacing rules; those clauses, if enforced, would—because of the impossibility of there ever being a well within that distance—effectively relieve the lessee of all obligation to protect against drainage. Thus because of the indirection of the usual offset provision and the consequent possibility of misinterpreting its legal effect, such clauses should be, at the very least, strictly construed." (432 F.2d at 177–8, citations omitted.)

(c) Third, the lease in question was executed in 1933 when 150 feet may have represented a sound estimate of the distance at which drainage occurs. But in the 1960's, when the case arose, it is known that drainage reaches much farther and to hold the lessor to the earlier knowledge "might be wholly unfair." (432 F.2d at 178.)

6. What is the effect of accepting delay rentals on lessor's right to enforce the protection covenant? In addition to the discussion in *Sundheim* above, consider the following passages from ROGERS V. HESTON OIL CO., note 3 *supra*, 735 P.2d at 545–46:

We next consider whether an oil and gas lessor, by accepting payment of delay rentals, even if timely made in accordance with the terms of the lease contract, with knowledge on the part of the lessor that oil or gas is being drained from his premises at the time that payment is made, thereby waives his right to recover damages for such drainage during the time covered by the delay rental payment. Upon further consideration of our holding in Carter Oil Co. v. Samuels, [181 Okl. 218, 73 P.2d 453 (1937)], we now conclude that *Carter* should be overruled, and that payment and acceptance of delay rental under the terms of the lease does not in and of itself constitute a waiver of an implied covenant on the part of the lessee to protect the lessor from drainage.

An implied covenant to protect the leased premises from drainage by surrounding wells is a covenant implied in fact to carry out what the parties must have intended which the courts read into the lease as an integral part thereof. Such a covenant becomes a part of the lease only where its inclusion in the lease is not inconsistent with other terms of the lease. It is not a covenant implied in law, and in the absence of a statute on the subject, it does not arise out of public policy. . . .

By the terms of the covenant thus implied, the lessee has a duty to protect the land from drainage by adjoining wells so long as the drilling of a protection well or wells will, in the judgment of a reasonably prudent operator, be a profitable undertaking, . . . having due regard for the interests of both lessor and lessee.

The next question is whether the implied covenant of protection against drainage, thus included within the lease agreement, is a separate and independent covenant from the delay rental covenant, so that the payment and acceptance of delay rentals under the written terms of the lease by the lessor with knowledge of drainage by adjacent or adjoining producing leases in and of itself constitutes a waiver of the implied covenant.

The "unless" clause of the lease relates to the contractual obligation of the lessee to develop the lease. The implied covenant to protect against drainage relates to an entirely separate subject, the obligation of the lessee to protect against drainage. Each is a separate clause, relating to a different subject matter. We see no basis in reason or logic to construe the implied covenant as an imposed modification upon the development clause, particularly where it has been established that the implied covenant may only be imposed where it is not inconsistent with other provisions of the lease. In the early case of Eastern Oil Co. v. Beatty, 71 Okl. 275, 177 P. 104 (1918), we first determined that during the period of the lease when the "unless" clause was operative, the payment of delay rentals under the "unless" clause delayed the lessee's obligation to protect against drainage in the same manner and to the same extent as it delayed the lessee's obligation to develop. This holding was the basis for our determination in Carter Oil Co. v. Samuels, *supra.* We now believe, and hold, that the two clauses are separate and distinct, and that neither is a modification of the other, and that Eastern and Carter and the cases subsequently likewise so holding should be and are overruled.

Our conclusion is further fortified by what we perceive to be a fundamental injustice which may arise were we to treat the two clauses as interdependent. In the case at bar, the defendant lessee owns the lease upon plaintiff lessor's tract, and upon adjacent and adjoining leases, including the one on which drainage is alleged to exist of plaintiff's tract.

Compare ORR V. COMAR OIL CO., 46 F.2d 59 (10th Cir.1930). Counsel for the plaintiff contended that, under the lease, the acceptance of rentals excused the obligation for primary development but not the obligation to drill off-set wells:

> The defendant was obligated, under an implied covenant of the original lease, to exercise reasonable diligence to protect the Orr tract from drainage through wells on adjoining tracts. . . .

> The drilling of an off-set well on the Orr tract would have performed such implied covenant and also the covenant for primary development. Therefore, plaintiff was not entitled to both the drilling of an off-set well and to the payment of delay rentals. He was not entitled to have that which would have performed the covenant for primary development and also to that which was to be paid in lieu thereof. Plaintiff, having accepted delay rentals with full knowledge of the production from Bechtel No. 3—3W and the necessity for off-set drilling, if any there was, could not consistently insist upon the drilling of an off-set well during the period for which such delay rentals were accepted.

What is the theory that supports the *Orr* result? Waiver, estoppel, or election of remedies?

Under the view of the *Orr* case, it is the acceptance of rentals not the tender of rentals that precludes suit for breach of the offset covenant. That is, the lessee does not have an option to drill the offset well or to pay delay rental, but the lessor has the option to accept the rental or to sue for damages. Can the lessee omit payment of rentals and thereby terminate liability for breach of an implied covenant?

Should acceptance of royalty payments preclude suit for breach of implied covenants? *See* Colpitt v. Tull, 1950 OK 199, 204 Okl. 289, 228 P.2d 1000; Scilly v. Bramer, 170 Pa.Super. 276, 85 A.2d 592 (1952); *but cf.* Mooers v. Richardson Petroleum Co., 146 Tex. 174, 204 S.W.2d 606 (1947).

AMOCO PRODUCTION CO. v. ALEXANDER

Supreme Court of Texas, 1981.
622 S.W.2d 563, 72 O.&G.R. 125.

CAMPBELL, JUSTICE. This is an action by royalty owners for damages because of field-wide drainage. The trial court, after a jury verdict, rendered judgment for the Alexanders, lessors, for actual and exemplary damages against Amoco, lessee. The Court of Civil Appeals reformed the trial court's judgment and affirmed the judgment as reformed. 594 S.W.2d 467. We modify the judgment of the Court of Civil Appeals and affirm the judgment as modified.

The Hastings, West Field, in Brazoria County, is a water-drive field. Water and oil are in the same reservoir. Because water is heavier than oil, the water moves to the bottom of the reservoir driving the oil upward. As oil is removed, water moves up to fill the space.

As the oil is produced, the oil-water contact (a measure of the reservoir water level) gradually rises until the wells begin to produce water along with oil. As the wells are produced, the fluid from the wells contains increasingly higher percentages of water. When the wells produce almost all water, the wells are abandoned. The wells are then said to be "watered out" or "flooded out."

The Hastings, West Field, reservoir is not horizontal. It is highest (closer to the surface) is the southeast part. It is lowest in the northwest. Hence, the reservoir dips downward gradually from the southeast to the northwest. Leases on the higher part of the reservoir are called "updip leases" and on the lower, "downdip leases." The Alexanders' leases with Amoco are downdip. Amoco, with 80% of the field production, also has updip leases. Exxon, Amoco's chief competitor in the field, owns leases generally updip from the Alexanders and downdip from the remainder of the Amoco leases.

In water-drive fields, such as the Hastings, West Field, natural underground conditions and production of oil updip work to the disadvantage of downdip leases. As the oil is produced, the oil-water contact rises. The greater the production from updip leases, the sooner the wells on downdip leases will be "watered out" because of the waterdrive pushing the oil to the highest part of the reservoir. The downdip leases, therefore, are the first to water out. Moreover, production anywhere in the field will cause the oil-water contact to rise and move from the downdip leases to the updip leases. This is field-wide drainage.

The Alexanders' theory of this lawsuit is that Amoco slowed its production on the Alexander–Amoco downdip leases and increased production on Amoco updip leases causing the Alexander–Amoco downdip leases to "water out" much sooner. Oil not produced from the Alexander leases will eventually be recovered by Amoco as the water pushes the oil to the Amoco updip leases. Their theory of liability is that Amoco owed the Alexanders an obligation to obtain additional oil production from the Alexander leases by drilling additional wells and reworking existing wells to increase production. If Amoco had fulfilled that obligation, additional oil would have been produced from which the Alexanders would have been paid 1/6th royalty. The Amoco updip leases pay 1/8th royalty.

The Alexanders contend they pleaded two legal theories of recovery: (1) in contract, breach of Amoco by its implied obligation to take such steps as a reasonably prudent operator would have taken to protect the Alexander leases from drainage; and (2) in tort; for "intentional acts and omissions" undertaken by Amoco "for the purpose of increasing Amoco's production from its updip leases" and the deliberate waste of the Alexanders' royalty oil. The jury found:

(1) Amoco failed to operate the Alexander leases as a reasonably prudent operator.

(2) Amoco operated its leases under an intentional policy of maximizing its profits by producing less oil from the Alexander leases than would have been produced by a reasonably prudent operator, while increasing the drainage of oil from the Alexander leases by Amoco production on other leases.

The jury awarded actual and exemplary damages.

We must determine whether:

(1) Amoco had a duty to protect from field-wide drainage, or a duty not to drain the Alexander downdip leases by its operations updip.

(2) Amoco had a legal duty under the Alexander leases to apply to the Railroad Commission for permits to drill additional wells at irregular locations, to obtain the permits, and drill the wells.

(3) The trial court erred in admitting testimony that the Railroad Commission would have granted exception permits to allow Amoco to drill additional wells on the Alexander leases.

(4) The Alexanders are entitled to recover exemplary damages.

Field–Wide Drainage

Whether Amoco had a duty to protect the Alexander downdip leases from field-wide drainage, or a duty not to drain the leases by its updip operations has not been considered by the Texas courts. The Court of Civil Appeals held Amoco had a duty to protect the Alexanders from field-wide drainage.

An oil and gas lessee has an implied obligation to protect from local drainage. Local drainage is oil migration from under one lease to the well bore of a producing well on an adjacent lease. Local drainage depends upon production from wells in a specific area in a field. It will begin, increase, or decrease according to production. Local drainage may be in several directions in one field and can be prevented by drilling offset wells. Field-wide drainage in a water-drive field, however, is relatively independent of the location of particular wells. It depends on the water-drive and production from all wells in the field. Protecting from field-wide drainage, therefore, is more difficult than protecting from local drainage.

Amoco urges the Court of Civil Appeals correctly held the drainage in this case was field-wide but the court erred in holding the law imposes an obligation upon Amoco to prevent field-wide drainage, or an obligation not to drain the Alexander leases by its updip operations. Amoco recognizes the obligation to protect from local drainage, but states the Court of Civil Appeals was in error in extending that obligation to require a lessee to protect his lessor from field-wide drainage. Amoco argues this imposes a new implied obligation never previously held to exist.

. . .

Amoco contends the Court of Civil Appeals' holding expands the offset drilling obligation beyond the point of fairness and workability by including within it the obligation to offset field-wide or regional drainage. Field-wide drainage affects all leases in the field; and if the duty exists, each lessee may be required to drill offset wells. The drilling of offset wells increases field-wide drainage and sets off a chain reaction[;] the drilling of each additional well would trigger a field-wide obligation to drill more offsets and each drilling would further accelerate the field-wide drainage. Amoco argues, therefore the end result of carrying out the obligation would be self-defeating.

Amoco also says that updip leases enjoy a natural advantage over downdip leases. If the natural drainage is to be offset, the only valid way is through field-wide regulation by the Railroad Commission regulating rates of production to protect correlative rights in the field.

The implied covenant to protect against drainage is part of the broad implied covenant to protect the leasehold. The covenant to protect the leasehold extends to what a reasonably prudent operator would do under similar facts and circumstances. "As is true of the other implied duties, it is not easy to separate the duty from the standard of performance. The lessee is required generally to do what a prudent operator would do. Protection of the leased premises against drainage is but a specific application of that general duty." 5 E. Kuntz, A Treatise on the Law of Oil and Gas § 61.3 (1978). The covenant to protect from drainage is not limited to local drainage. It extends to field-wide drainage. Oil lost by field-wide drainage is just as lost as local drainage oil. The methods of safeguarding from the loss may be different and protecting from local drainage may be easier. However, it is no defense for a lessee to say there is no duty to act as a reasonably prudent operator to protect from field-wide drainage.

A lessor is entitled to recover damages from a lessee for field-wide drainage upon proof (1) of substantial drainage of the lessor's land, and (2) that a reasonably prudent operator would have acted to prevent substantial drainage from the lessor's land. In Shell Oil Co. v. Stansbury [410 S.W.2d 187 (Tex.1966)], this Court held a reasonably prudent operator would have drilled a well on the lessor's land to protect from drainage. However, because of the complexity of the oil and gas industry and changes in technology, the courts cannot list each obligation of a reasonably prudent operator which may arise. The lessee must perform any act which a reasonably prudent operator would perform to protect from substantial drainage.

The duties of a reasonably prudent operator to protect from field-wide drainage may include (1) drilling replacement wells, (2) re-working existing wells, (3) drilling additional wells, (4) seeking field-wide regulatory action, (5) seeking Rule 37 exceptions from the Railroad Commission, (6) seeking voluntary unitization, and (7) seeking other available administrative relief. There is no duty unless such an amount of oil can be recovered to equal the cost of administrative expenses, drilling or re-working and equipping a protection well, producing and marketing the oil, and yield to the lessee a reasonable expectation of profit.

Clifton v. Koontz, 160 Tex. 82, 96–97, 325 S.W.2d 684, 695–96 (1959).

The Court of Civil Appeals has not imposed a new obligation upon Amoco. The jury, in finding that Amoco failed to operate the Alexander leases as a reasonably prudent operator, has determined that Amoco failed in its duties under the implied covenants to protect the leasehold.

Amoco argues the Court of Civil Appeals did not consider that Amoco has obligations to all of its lessors in the field. Anything it does to maintain or increase production from updip leases may accelerate the water drive and expose Amoco to liability to downdip lessors. If Amoco fails to maintain or increase updip production, it is exposed to liability from the updip lessors. Amoco argues the Court has placed it between contrary obligations from which there is no escape. The fulfilling of one obligation necessarily causes the breach of the other.

The conflicts of interest of Amoco, as a common lessee, cause us concern. The Alexander leases provided for 1/6th royalty while Amoco's updip leases provided for 1/8th royalty. There is no economic incentive for Amoco to increase production on the Alexander lease because it will eventually recover the Alexanders' oil updip. Money invested in the Hastings, West Field, will have a longer productive life if invested updip. The greater the updip production the sooner Amoco's competitor Exxon will water out. Money spent updip will yield greater returns than money spent downdip because of higher daily production. With downdip operators out of production Amoco can produce its upper sands without competition and can begin production from its lower sands where it does not have significant production competition.

These conflicts would not occur if Amoco was not a common lessee (lessee common to downdip and updip lessors). If the Alexanders were the only Amoco lessor, their interests would more nearly coincide. Amoco's interest would be to capture the most oil possible from the Alexander leases before they watered out.

Amoco's responsibilities to other lessors in the same field do not control in this suit. This lawsuit is between the Alexanders and Amoco on the lease agreement between them and the implied covenants attaching to that lease agreement. The reasonably prudent operator standard is not to be reduced to the Alexanders because Amoco has other lessors in the same field. Amoco's status as a common lessee does not affect its liability to the Alexanders.

Duty to Apply for Administrative Relief

The Railroad Commission rules in the Hastings, West Field, prohibit the drilling of a well nearer than 660 feet to any other well and nearer than 330 feet to any property line or lease line. The rules allow the Railroad Commission to grant drilling permits as an exception to the spacing regulation. These exceptions are commonly referred to as Rule 37 permits. This rule provides:

> [T]he Commission in order to prevent waste or to prevent the confiscation of property will grant exceptions to permit drilling within shorter distance than above prescribed whenever the Commission shall determine that such

exceptions are necessary to prevent waste or to prevent confiscation of property.

The Alexanders contend Amoco should have drilled replacement wells in the extreme updip corner of each lease. The wells would be within 50 feet of the lease line and 200 feet apart. The wells could not be drilled unless the Railroad Commission granted Rule 37 permits. Amoco did not apply for the permits.

The Court of Civil Appeals held that when Amoco determined the leases were watering out, prudent operation demanded drilling replacement wells unless it would be economically unfeasible. If Rule 37 permits were required, Amoco should have applied for them in furtherance of its duty to prudently operate the leases. Because Amoco failed to apply, the Court of Civil Appeals held, the Alexanders were entitled to show the exceptions most likely would have been granted and they suffered damages because of Amoco's failure.

Amoco states the holding of the Court of Civil Appeals amounts to the imposition of an implied covenant obligating a lessee to seek exceptions to regulations limiting the drilling and production of wells. Amoco argues there is no Texas authority for imposing an obligation to seek administrative relief and there is no duty to seek administrative relief.

We disagree with Amoco's argument that there is no duty to seek administrative relief. Amoco owed the Alexanders the duty to do whatever a reasonably prudent operator would do if the Alexanders were its only lessor in the field.

The duty to seek favorable administrative action may be classified under the implied covenants to protect the lease, or to manage and administer the lease. Regardless of the category, the standard of care in testing Amoco's performance is that of a reasonably prudent operator under similar facts and circumstances.

We do not agree with the Court of Civil Appeals if its holding means that in every case of field-wide drainage the lessee must seek Rule 37 exceptions. There may be facts where the prudent operator would not seek administrative relief. The probability that the Railroad Commission will grant or deny the permit is a consideration to be made by the prudent operator. The jury, from evidence justifying granting or denying the permit, can determine if a reasonably prudent operator would have applied for the permit.

The jury found Amoco failed to operate the Alexander leases as a reasonably prudent operator. Does this finding, based in whole or in part on Amoco's failure to apply for Rule 37 permits, establish liability for failure to drill the replacement wells? If it does not, what remedies do the Alexanders have? They have no rights in the management or operation of the oil leases. Amoco, as operator, could quickly determine when the wells began watering out. Amoco, because of its conflicting interests, had no economic incentive to protect the Alexander leases. The downdip lessors, after their leases have watered out, have no opportunity to capture the oil updip.

It is the failure to act as a reasonably prudent operator that triggers the loss. If the Railroad Commission denies the Rule 37 permits, after a reasonably prudent application, the operator has no liability for not drilling the wells. We hold that an operator, who fails to act as a reasonably prudent operator by not seeking Rule 37 permits, is liable for loss caused by the failure to drill the wells.

Testimony that the Railroad Commission would Grant Rule 37 Permits

The Alexanders' two expert witnesses testified, in response to hypothetical questions, that the Railroad Commission would have granted approval to drill the replacement wells. Amoco contends the admission of this evidence is reversible error.

The jury question was whether Amoco operated the leases as a reasonably prudent operator. In answering this question the jury had to determine whether a reasonably prudent operator would have requested Rule 37 permits. The jury was not asked whether the Railroad Commission would have granted them.

There was other evidence on which the jury could base its decision. Amoco knew that: this was a water-drive field and the Alexander leases were downdip; these leases would water out first; the leases began watering out; a 25% increase in field production, allowed by the Railroad Commission, would speed up the watering out of the leases; there was a process by which permits could be granted; drainage was occurring as early as 1972; Amoco's action updip was accelerating the drainage; Amoco could recover the Alexander oil on its updip leases; Amoco had no economic incentive to drill replacement wells on the Alexander leases; the Alexander leases would be the first to require additional replacement wells.

There was evidence that the Railroad Commission granted twenty-two Rule 37 permits to Exxon and Amoco updip in the same fault block section of this field beginning in 1975. The jury could also consider that Amoco, in opposing a 1973 Rule 37 application on the adjacent Pearland lease, represented that the Pearland lease had 2.9 million barrels of oil originally in place. This meant that in 1973 the lease had already produced all of its recoverable oil originally in place. However, in the course of this litigation, Amoco revealed that the calculation of its own reservoir engineers showed that 3.884 million barrels of oil were originally in place under the Pearland lease.

. . . [W]e need not decide whether it was error to admit the answers to the hypothetical questions in this case because the answers are cumulative to other evidence presented to the jury. . . .

Exemplary Damages

Amoco argues that exemplary damages are not recoverable because the Alexanders failed to plead and prove a tort allowing recovery of exemplary damages. We agree.

The Alexanders alleged "a breach by Amoco of both its express and implied covenants . . . under the lease contracts . . . to protect said leases from drainage and to operate said leases as a reasonable and prudent operator" and their "royalty interest under Leases A and B has been wasted and damaged . . .."

First, Amoco argues that exemplary damages are not recoverable because a breach of the implied covenant to protect against drainage is an action sounding in contract and not in tort. The rights and duties of the lessor and lessee are determined by the lease and are contractual. The lease constitutes the contract.. . .

In Texas Pac. Coal & Oil Co. v. Stuard, 7 S.W.2d 878, 882 (Tex.Civ.App.—Eastland 1928, writ ref'd), the Court of Civil Appeals held that "the implication to develop after drilling the exploratory well is a part of the written contract and is governed by the four-year statute of limitation." In Indian Territory Illuminating Oil Co. v. Rosamond, 190 Okl. 46, 120 P.2d 349, 354 (1941), the Oklahoma Supreme Court held that "the implied covenant to protect against drainage is a part of the written lease as fully as if it had been expressly contained therein . . . " and applied the statute of limitations relating to actions on written contracts. We hold that the implied covenant to protect against drainage is a part of the lease and is contractual in nature.

Exemplary damages are not allowed for breach of contract. . . .. Even if the breach is malicious, intentional or capricious, exemplary damages may not be recovered unless a distinct tort is alleged and proved. . . .. We hold that a breach of the implied covenant to protect against drainage is an action sounding in contract and will not support recovery of exemplary damages absent proof of an independent tort.

[The court rejected a contention that Amoco's behavior was actionable as waste and therefore in a category where exemplary damages were permissible.]

. . .

Trial of Common Lessee Cases

The courts that have considered the common lessee problem have considered facts different from this case. In those cases the common lessee was causing the drainage by production on adjacent or adjoining land. The drainage was caused by production independent of water-drive and was local drainage. However, those decisions are analogous because of the common lessee. Professors Williams and Meyers have put these cases in three categories. *See* 5 H. Williams & C. Meyers, Oil and Gas Law § 824 (1980); Meyers & Williams, *Implied Covenants in Oil and Gas Leases: Drainage Caused by the Lessee*, 40 Texas L.Rev. 923 (1962). First, there are cases which state the lessee was causing the drainage but place no significance on that fact. *See, e.g.,* Billeaud Planters v. Union Oil Co. of Cal., 245 F.2d 14, 18–19 (5th Cir.1957); Gerson v. Anderson–Prichard Prod. Corp., 149 F.2d 444, 445–46 (10th Cir.1945); Chapman v. Sohio Petroleum Co., 297 S.W.2d 885, 886–87 (Tex.Civ.App.—El Paso 1956, writ ref'd n.r.e.). Second, other cases state that the lessee caused the drainage but hold this fact does not alter the ordinary rules of liability for failure to protect from drainage. Hutchins v. Humble Oil & Ref. Co., 161 S.W.2d 571, 573

(Tex.Civ.App.—Galveston 1942, writ ref'd w.o.m.); accord, Tide Water Associated Oil Co. v. Stott, 159 F.2d 174, 177 (5th Cir.1946). Third, there are cases holding the liability of the lessee is increased when the lessee is causing the drainage. *See, e.g.,* Cook v. El Paso Natural Gas Co., 560 F.2d 978, 982–84 (10th Cir.1977) (reasonable prudent operator rule inapplicable in common lessee case, proof of drainage all that is required); Bush Oil Co. v. Beverly–Lincoln Land Co., 69 Cal.App.2d 246, 158 P.2d 754, 758 (1945) (immaterial whether protection well would be profitable if drainage caused by lessee's affirmative act); Phillips Petroleum Co. v. Millette, 221 Miss. 1, 72 So.2d 176, 183 (Miss.1954) (lessee strictly liable for substantial drainage caused by own affirmative acts).

This Court in Shell Oil Co. v. Stansbury, 410 S.W.2d 187, 188 (Tex.1966), expressly overruled the *Hutchins* case, *supra,* and held that an express offset provision does not limit the lessee's obligation to protect from drainage when the lessee is the one causing the drainage. In drainage cases, Texas courts place upon the lessor the burden to prove that substantial drainage has occurred and that an offset well would produce oil or gas in paying quantities. *Clifton v. Koontz, supra.*

The judgment of the Court of Civil Appeals is modified to prohibit the recovery of exemplary damages and affirmed as modified.

NOTES

1. Would it be sensible to compromise between the *Cook* (note 2 p. 460 supra) and *Alexander* views of drainage by the lessee by shifting the burden of proof to the lessee where it is responsible for the drainage? See Seacat v. Mesa Petroleum Co., 561 F.Supp. 98, 76 O.&G.R. 459 (D.Kan.1983); Haken v. Harper Oil Co., 1979 OK CIV APP 32, 600 P.2d 1227, 65 O.&G.R. 33; cf. Dixon v. Anadarko Production Co., 1972 OK 165, 505 P.2d 1394, 44 O.&G.R. 397 (further development case; lessee was engaged in operations on adjoining lands); Elliott v. Pure Oil Co., 10 Ill.2d 146, 139 N.E.2d 295, 7 O.&G.R. 228 (1956).

2. Does the field-wide character of the drainage provide any reason for modifying the principles governing the covenant against drainage? Or is it merely a circumstance likely to be accompanied by other problems—that the proposed offset wells would violate the spacing rules and that the lessee would have to obtain administrative relief before drilling the proposed wells? Would Amoco's "chain-reaction" argument have any weight at all if one were confident that the spacing rules allowed only the economically sound number of wells and that the regulatory agency would defend them unflinchingly?

3. Can you formulate the court's standard as to how likely administrative relief must be for the lessee's "reasonable prudent operator" duty to encompass an obligation to seek such relief? Is a decision such as that in the principal case likely to increase the pressure on the regulatory agency to grant exceptions to its well-spacing rules?

4. The court states that a lessee's "duties" to protect from field-wide

drainage include, among other things, a duty to seek voluntary unitization (the only kind available in Texas). If the lessee secures unitization, are there circumstances where that would not fully discharge its duty? (Suppose the lessee by compulsory pooling establishes a unit that encompasses plaintiff's tract, but he might conceivably have obtained an exception to drill on that tract, and, had he done so, plaintiff's royalty would have been greater.) Suppose lessee is unable to secure unitization because the lessor-plaintiff has refused consent to a "fair" plan of unitization; would that effort discharge the duty or estop lessor? *Compare* Young v. Amoco Production Co., 610 F.Supp. 1479, 85 O.&G.R. 376 (E.D.Tex.1985); Rush v. King Oil Co., 220 Kan. 616, 556 P.2d 431, 442–43, 57 O.&G.R. 192, 210–11 (1976); Tide Water Associated Oil Co. v. Stott, 159 F.2d 174 (5th Cir.1946), *cert. denied*, 331 U.S. 817 (1947).

In U.V. INDUSTRIES, INC. v. DANIELSON, 184 Mont. 203, 602 P.2d 571, 64 O.&G.R. 454 (1979), the court found liability, under the implied protection covenant, in lessee's failure to apply for compulsory unitization.

Compare Trahan v. Superior Oil Co., 700 F.2d 1004, 78 O.&G.R. 297 (5th Cir.1983). Plaintiffs claimed that at a hearing of the Louisiana Conservation Commissioner, held to determine the production unit boundaries for a well drilled by lessee on another tract, lessee failed to present certain evidence that would have resulted in the unit boundaries being formed so as to include part of lessor's tract. There was no allegation of fraud, bad faith, or intentional deception. The court held that on these facts the rule against collateral attack on Commissioner orders precluded any recovery against lessee under the implied protection covenant. However, it suggested that subsequent developments (new wells, etc.) might create a duty on the lessee's part to seek relief from the Commissioner; lessee's inaction could then entail liability, for there would be no order of the Commissioner being collaterally attacked. *Id.* at 1025. *Trahan* was distinguished in Eagle Lake Estates, L.L.C. v. Cabot Oil & Gas Corporation, 330 F.Supp.2d 778, 161 O.&G.R. 734 (E.D. La. 2004). Here the plaintiffs complained of drainage that took place between the date of first production and the application of the defendant for a unitization hearing and made no claims that the defendant had not fairly represented the plaintiff's interests at the unitization hearing. The Commissioner's order in *Eagle Lake Estates* did not deal with the subject matter of plaintiffs' complaint and thus there was no collateral attack on the order.

5. Under what circumstances might the lessee discharge the obligation to pursue administrative relief simply by notifying the lessor of the problem and the possible remedy? *See* Weaver, *Implied Covenants in Oil and Gas Law Under Federal Energy Price Regulation*, 34 Vanderbilt L.Rev. 1473, 1541–48 (1981). Would such notice ever suffice if the lessor lacked standing to pursue the administrative remedy? *Compare* Amoco Production Co. v. Ware, 269 Ark. 313, 602 S.W.2d 620, 624, 68 O.&G.R. 416, 422–23 (1980) (in exonerating lessee against claim that failure to appeal from decision of conservation commission

violated its implied protection covenant, court notes ability of lessor to appeal).

6. Suppose a unit well, in which a part of the leased tract is participating, drains oil or gas from another part of the lease? Could this give rise to a violation of the implied covenant against drainage? How might this be anticipated in drafting the lease? *See* Continental Oil Co. v. Blair, 397 So.2d 538, 69 O.&G.R. 464 (Miss.1981).

7. The *Alexander* court's rejection of punitive or exemplary damages is not the invariable result, even in Texas. In W.L. Lindemann Operating Co., Inc. v. Strange, 256 S.W.3d 766 (Tex.App.—Ft. Worth, 2008 pet. denied), a lessor and co-working interest owner pled fraud against the operator in order to assert punitive damages in a case where she was essentially asserting fraudulent drainage. The court found sufficient evidence to support the jury's verdict that the operator made misrepresentations regarding why a well had suffered a diminution in production that were relied upon by the lessor. Thus, the lessor was entitled to punitive or exemplary damages in addition to actual damages caused by the drainage of oil from her lease.

In SPAETH V. UNION OIL CO., 710 F.2d 1455, 77 O.&G.R. 142 (10th Cir.1983), involving drainage by the lessee, the federal court construed Oklahoma law to permit punitive damages when there was a "reckless and wanton disregard of another's rights." It said that the evidence showed such a disregard, but, objecting to the size of the punitive damage award ($3 million), it observed that "the dispute was essentially between two oil companies" and that the "trial was a contest between experts and was completely devoid of emotional overtones." Actual damages were $17,542. It remanded to the trial court to "order a remittitur of a substantial amount in punitive damages." On remand, the trial court ordered a remittitur of $1 million, leaving a net punitive damage award of $2 million. The court of appeals affirmed. 762 F.2d 865 (10th Cir.1985).

---

5 Martin & Kramer, *Williams and Meyers Oil and Gas Law*

§ 825. Remedies for Breach of the Protection Covenant: Introduction

The courts have employed a variety of remedies to enforce the implied covenant to protect from drainage. Among them have been outright cancellation of the lease, conditional cancellation of the lease, cancellation of the lease combined with damages for past loss, mandatory injunction to drill a protection well, damages in the amount of the loss, both past and prospective. The decree of cancellation may extend to all or to only part of the leasehold, depending on the circumstances. This is not to say that every jurisdiction permits the lessor a free choice of remedies. A number of cases hold that cancellation of the lease is not ordinarily available, that the usual remedy is damages. Another case, apparently confined to the situation of breach of the offset covenant when the lease is being held by rentals during the primary term, limits relief to cancellation and bars recovery of damages.

———

What is the proper measure of damages? Royalty on the amount of oil or gas drained away? Royalty on the amount of oil or gas the protection well would have produced? Can prospective as well as past damages be awarded? Is the measure of damages so unsatisfactory that equitable relief of forfeiture should always be available?

———

SECTION 3. IMPLIED COVENANTS OF REASONABLE DEVELOPMENT AND FURTHER EXPLORATION
(*See* Martin & Kramer, *Williams & Meyers Oil and Gas Law* §§ 831–847).

———

5 Martin & Kramer, *Williams and Meyers Oil and Gas Law*
§ 832. Elements of Cause of Action for Breach of Development Covenant

　　. . . In the typical development case, there is no permanent loss of recoverable minerals; rather, the dispute is over the rate of production from the leasehold. Even if the rate of production could be increased by drilling additional wells, and although those wells might be profitable, it does not follow that a prudent operator would drill the additional wells, because his overall profit from the leasehold might thereby be reduced. Thus, a field developed on a 40–acre basis may produce the oil in place twice as fast as the same field developed on an 80–acre basis, but the doubled drilling and operating costs of the former development pattern could reduce the lessee's profit considerably below that to be made on the 80–acre pattern. The covenant of reasonable development must strike a balance between the lessor's interest in rapid production of the minerals and the lessee's interest in reduced drilling and operating costs. It is this weighing of interests that causes the law of reasonable development to be somewhat amorphous when compared to the law of offset drilling.

　　The obligation imposed on the lessee by the implied covenant of reasonable development may be generally stated as follows: upon securing production of oil or gas from the leasehold, the lessee is bound thereafter to drill such additional wells to develop the premises as a reasonably prudent operator, bearing in mind the interests of both lessor and lessee, would drill under similar circumstances. This duty arises in the primary term of the lease as well as in the secondary term, but ordinarily it arises only after production is obtained on the leasehold; it cannot arise when the lease is being held by delay rentals under an express rental clause of the lease.

———

WHITHAM FARMS, LLC v. CITY OF LONGMONT

Colorado Cou rt of Appeals, 2003
97 P.3d 135, 160 O.&G.R. 444

Opinion by CARPARELLI, JUDGE. In this action based on breach of the implied covenant to develop in an oil and gas lease, plaintiffs, Whitham Farms LLC and Life Bridge Christian Church (collectively, Whitham Farms) and defendant City of Longmont appeal the judgment entered in favor of defendants North American Resources Co., now known as EnCana Energy Resources, Inc. (NARCO); Taku Resources LLC; Geronimo Energy Partners, LLC; SOCO Wattenberg Corporation; Barry L. Snyder; and Murray J. Herring. We affirm.

The three oil and gas leases at issue relate to 310 acres of real property located in Weld County, Colorado. NARCO and the remaining defendants are the lessees of the three leases and NARCO is the operator of those lease rights.

In 1990, the City of Longmont purchased the surface and mineral estates of approximately 25 acres of the property that was encumbered in 1981 by one of the oil and gas leases. In 1995, Whitham Farms purchased the surface and mineral estates of 150 acres of the property encumbered by the same lease.

In 1982, NARCO's predecessor in interest drilled the only well that exists on the property. NARCO recompleted that well as a producing well in 1997.

In 1999, Whitham Farms demanded that NARCO release the oil and gas lease or any acreage not necessary to support the existing well. Whitham Farms cited § 38-42-104, C.R.S.2002, as support for its demand, which requires that when an oil and gas lease "becomes forfeited or expires by its own terms," the lessee must record a written surrender of the lease. Longmont made the same demand in 2001.

After NARCO refused to terminate the lease, Whitham Farms filed this lawsuit seeking declaratory relief. In its answer, Longmont joined in Whitham Farms' complaint and also requested that the oil and gas lease be terminated.

At trial, the sole issue was whether NARCO had breached the lease's implied covenant of reasonable development. The parties agreed that the court could rule based on the pretrial briefs and stipulated facts.

Whitham Farms argued that it was not economically prudent for NARCO to develop the oil and gas reserves and, therefore, that NARCO had breached the covenant of reasonable development and, as a result, Whitham Farms was entitled to equitable termination of the lease.

NARCO agreed that it was not economically prudent for it to develop the oil and gas reserves, but argued that it had not breached the covenant of reasonable development, was not obligated to develop the resources at the present time, and was entitled to continue its leasehold until development became economically prudent.

The trial court ruled that Whitham Farms and Longmont failed to establish a breach of the implied covenant of reasonable development. Accordingly, it entered judgment against Whitham Farms and Longmont, and in favor of the remaining defendants.

I. Covenant of Reasonable Development

On appeal, Whitham Farms and Longmont contend that the trial court erred when it ruled that the oil and gas lease would not terminate for breach of the implied covenant of reasonable development unless Whitham Farms and Longmont established that the drilling of an additional well or wells would be economically viable. We disagree.

A. Purpose of Oil and Gas Lease

The purpose of an oil and gas lease is to make the mineral estate profitable to both parties through the exploration, development, and production of resources located under the leased premises. *See Davis v. Cramer,* 808 P.2d 358, 360 (Colo.1991)("[t]he fundamental purpose of an oil and gas lease is to provide for the exploration, development, production, and operation of the property for the mutual benefit of the lessor and lessee"). . ..

The primary consideration in such lease transactions is the royalty derived from the development of the resources. As a result, development is a recognized expectation of both parties. Gary B. Conine, *The Future Course of Oil and Gas Jurisprudence: Speculation, Prudent Operation, and the Economics of Oil and Gas Law,* 33 Washburn L.J. 670 (1994).

B. Implied Covenants

Consistent with the expectations of both parties, Colorado courts recognize four implied covenants: to conduct exploratory drilling; to develop after discovering resources that can be profitably developed; to operate diligently and prudently; and to protect the leased premises against drainage. *Mountain States Oil Corp. v. Sandoval,* 109 Colo. 401, 125 P.2d 964 (1942); *Gillette v. Pepper Tank Co.,* 694 P.2d 369 (Colo.App.1984). In addition, a division of this court has recognized that the covenant to conduct exploratory drilling includes both exploration before discovering an initial reservoir and later exploration for additional reservoirs in unproven areas. *Gillette v. Pepper Tank Co., supra.*

The covenant of reasonable development protects the lessor's expectation that upon finding exploitable resources, the lessee will develop those resources for the mutual profit of both parties. When it is possible to develop known resources profitably and a lessee fails to act diligently to do so, the lessee is deemed to have breached the covenant, and a lessor may seek equitable termination of the lease. *N. York Land Assocs. v. Byron Oil Indus., Inc.,* 695 P.2d 1188 (Colo.App.1984).

C. Prudent Operator Standard

When, as here, there is a proven field of oil or gas, courts have held that a lessee is required to further develop the lease when there is a reasonable expectation that one or more new wells would generate enough revenue to cover the cost of development and return a reasonable profit. *See, e.g., Chenoweth v. Pan Am. Petroleum Corp.,* 314 F.2d 63 (10th Cir.1963)(applying Oklahoma law); *see also Gillette v. Pepper Tank Co., supra.* Thus, when a prudent operator would have a reasonable expectation of such economic viability and a lessee is not developing the field, it is proper to conclude that the lessee has breached the covenant of reasonable development and to grant an equitable termination to the lessor.

D. Speculation

Whitham Farms and Longmont argue that it was not necessary that they prove that the drilling of additional wells would be economically viable. Instead, they argue that the lease should be terminated because additional wells would *not* be economically viable and a prudent operator would *not* further develop the field.

In support of this argument, they refer to cases in which courts have concluded that the lessee was holding the lease for the purpose of speculation. For example, in *North York Land Associates v. Byron Oil Industries Inc., supra,* the trial court found that a prudent operator would not explore or develop the leasehold at the current time or in the foreseeable future and, thus, cancelled the lease. On appeal, a division of this court agreed. The division noted that under the prudent operator standard, the lessee was not obligated to explore, develop, or drill on the leasehold any further. Nonetheless, the division held that a corollary to the prudent operator standard controlled: a lessee may not retain its lease based on its production of oil on a small portion of the property and the mere speculative and remote hope that the economics might change and non-viable mineral holdings might become profitable at some unspecified time in the future.

The division stated that "If the exploration of the surrounding area shows very poor potential for profitable wells, then the lessee must surrender the lease rather than hold it on the mere speculative and remote hope that the economics might change in favor of profitable drilling." *N. York Land Assocs. v. Byron Oil Indus., Inc., supra,* 695 P.2d. at 1191.

Relying on *Sauder v. Mid-Continent Petroleum Corp.,* 292 U.S. 272, 54 S.Ct. 671, 78 L.Ed. 1255 (1934), the division concluded that cancellation is appropriate when the lessor wants to develop the land for purposes other than the production of oil and gas. *N. York Land Assocs. v. Byron Oil Indus., Inc., supra.*

E. Questions of Fact

Here, to determine whether there has been a breach of the implied covenant to reasonably develop, it is necessary to resolve several questions of fact. First, has a reservoir of oil or gas been found? Second, if there is a reservoir, is it possible to develop it profitably? Third, if it can be profitably developed, has the lessee developed it? Fourth, if profitable development of the reservoir is not likely at the current time, is there a potential to develop it in the foreseeable future? And, fifth, is the continuation of the lease consistent with the covenant of reasonable development, or is the lessee holding the lease rights for speculation?

F. Burden of Proof

Before we discuss the trial court's findings regarding these factual issues, we address Whitham Farms' contention that it was NARCO's burden to prove that its continued holding of the lease was not for speculation. We disagree.

The burden of proving a prima facie case for recovery on a civil claim is on the plaintiff. The burden of proof generally rests upon the party who asserts the affirmative of an issue, and "thus, the party seeking to change the status quo bears the burden of proof."

With regard to implied covenants in oil and gas leases, the "general rule is that the lessor has the burden of proof to show that the lessee did not act in good faith and as a reasonably prudent, similarly situated businessm[a]n." *Alumet v. Bear Lake Grazing Co.,* 119 Idaho 946, 952, 812 P.2d 253, 259 (1991); *see also Sanders v. Birmingham,* 214 Kan. 769, 776, 522 P.2d 959, 965 (1974)("[t]he burden of proof is upon the lessor to establish by substantial evidence that the covenant has been breached by the lessee"); *Durkee v. Hazan,* 452 P.2d 803 (Okla.1968).

We are persuaded by the rationale of these cases, which are consistent with Colorado authority, and, therefore, decline to adopt the burden-shifting approach urged by Whitham Farms and adopted by the Oklahoma Supreme Court in *Doss Oil Royalty Co. v. Texas Co.,* 192 Okla. 359, 137 P.2d 934 (1943)(once the lessor proves, by a preponderance of the evidence, that additional development would not be profitable, the burden shifts to the lessee to prove why in equity and good conscience the undeveloped portion of the lease should not be surrendered).

G. Findings of Fact

Whether an implied covenant in an oil and gas lease has been breached is a question of fact for determination by the trial court. *See Mountain States Oil Corp. v. Sandoval, supra.* On review, this court must accept a trial court's findings unless they are clearly erroneous. *Gillette v. Pepper Tank Co., supra.*

Here, it is undisputed that: (1) the lease is in the Denver-Julesburg Basin, a field that has been explored, proven, and developed; (2) it has been possible to profitably develop the oil and gas within this leasehold; (3) NARCO further developed the lease into another geological formation as recently as 1997, approximately three years before Whitham Farms filed suit; (4) the existing well is currently producing; and (5) this well has economic value.

Experts for both sides opined that, at the time of trial, a prudent operator would not drill additional wells. Neither expert expressed an opinion whether additional development of the oil and gas would be profitable within a reasonable time or in the foreseeable future.

The trial court found that NARCO "acted reasonably in developing the lands as recently as 1997 and has not ruled out the possibility of further development should such development become profitable." It also found that "the parties are subject to a pooling agreement that specifically provides that one well is sufficient to validate NARCO's lease as to all the lands within the unit." The court thus concluded that Whitham Farms and Longmont failed to establish a breach of the implied covenant of reasonable development. We conclude that the trial court's findings and conclusions were not clearly erroneous. Because Whitham Farms and Longmont did not present any evidence that the resources could not be profitably developed within a reasonable time, they did not met their burden of proving that NARCO was speculating or that the lease should be cancelled on that basis.

[The court's discussions of the effects of a pooling agreement and the accommodation doctrine are omitted].

IV. Longmont's Appeal

In its appeal, City of Longmont contends that the lease should be limited to the existing well. Relying on *Graefe & Graefe, Inc. v. Beaver Mesa Exploration Co.,* 635 P.2d 900 (Colo.App.1981), it argues that one well, drilled on the property twenty years ago, should not justify NARCO's "holding" the entire property. We disagree.

In *Graefe,* the lessee breached an express provision in a lease and was required to restore a well and surrounding acres to production. The lease was cancelled as to the remaining 2840 acres based on breach of the implied covenants of further exploration and development. In contrast, the trial court here found no breach of the express terms or the implied covenants of the lease. Moreover, the parties stipulated that the leasehold and the mineral estates of the three leases have been pooled so that the production from the existing well is deemed to be production for each of the three leases.

Accordingly, we discern no error by the trial court in failing to limit the lease to support of the existing well, and we conclude that *Graefe & Graefe, Inc. v. Beaver Mesa Exploration Co., supra,* does not dictate a different result.

The judgment is affirmed.

Judge ROTHENBERG and Judge VOGT concur.

NOTES

1. A concise statement of the implied covenant of reasonable development is found in the West Virginia case of United Methodist Church v. CNG Development Co., 222 W.Va. 185, 663 S.E.2d 639, 643 (2008). Here the court quoted and applied the following standard from an earlier case:

> . . . a lessor cannot require further development of the premises, after the lessee has acquired a vested interest in the minerals by the completion of a

paying well, except upon proof to the effect that operators for oil and gas of ordinary prudence and experience in the same neighborhood under similar conditions have been proceeding successfully with the further development of their lands or leases, and the further fact that additional well would likely inure to the mutual benefit of both lessors and lessee.

The court ruled that a trial court may consider the equitable remedy of partial rescission in fashioning relief to be awarded upon proof sufficient to establish a breach of implied covenant of development in connection with an oil and gas lease dispute. A lessee should, however, be permitted a reasonable time period to undertake efforts to further develop leased property before the trial court took evidence of a breach of implied covenant of further development or of extreme hardship to owner.

2. As the principal case notes, some courts have employed a burden shifting approach once the lessor proves, by a preponderance of the evidence, that additional development would not be profitable. The effect is to require the lessee to undertake additional exploration, as contrasted with development of proven formations. *See* for example, Davis v. Ross Production Co., 322 Ark. 532, 910 S.W.2d 209, 133 O.&G.R. 213 (Ark. 1995): "If the lessee-operator contends there is nothing to be gained by continued development, the lessee has lost nothing by cancellation of the nonproducing portion of the lease." A Colorado court has recognized an exploration covenant, as seen in *Gillette v. Pepper Tank Co., infra.* For another recent case in which a court found that a lessee satisfied the implied covenant of reasonable development with respect to the drilling of wells, *see* Meisler v. Gull Oil, Inc., 848 N.E.2d 1112 (Ind. App. 2006).

3. On the implied covenant to develop in Arkansas, as distinguished from the implied covenant to explore, *see generally* James W. McCartney & John C. LaMaster, *The Implied Covenant of Exploration in Texas and Arkansas*, 13 *U.Ark.Little Rock* 25 (1990). While Arkansas the courts do not expressly adopt an exploration covenant, the authors conclude that the "implied covenant of development . . . is defined and applied so broadly in Arkansas that it has become, in effect if not in name, very similar to the traditional definition of the implied covenant of exploration." *Id.* at 49. The reason for this is that the prudent operator standard is applied to prevent the lessee from holding acreage for speculation. *See* Skelly Oil Co. v. Scoggins, 329 S.W.2d 424, 12 O&G.R. 163 (Ark. 1959) and Byrd v. Bradham, 655 S.W.2d 366, 78 O.&G.R. 1 (Ark. 1983), both cited in the principal case. The Arkansas court has also followed the anti-speculation approach in Crystal Oil Co. v. Warmack, 855 S.W.2d 299, 127 O.&G.R. 288 (Ark. 1993) where a lease was found to have terminated for nondevelopment thus "activating" a top lease.

4. UNION GAS & OIL CO. v. FYFFE, 219 Ky. 640, 294 S.W. 176 (1927). Lessor, who was entitled to a royalty of $100 per year on each well producing gas, sued for breach of the reasonable development covenant. Discussing the

claim, the court said:

> It seems reasonable to us that, if one gas well would take out all of the gas within ten years, ten gas wells might take it out in one year, and, if one gas well only should be drilled on a tract of land, in such a case the lessor would receive his $1,000 in ten years, whereas, if ten wells should be drilled on the same tract, he would receive his $1,000 in one year. We see no other difference in the operation so far as the lessor is concerned, but there would be a great difference so far as the lessee is concerned. If one well can be drilled for $3,000, and the gas from it can be marketed through a period of ten years, a fair income may result to the operator, but, if ten wells must be drilled in one year at a cost of $30,000, and the gas is exhausted at the end of one year, it appears to us that the business of producing gas would be exceedingly hazardous.

Is it appropriate for courts to give the lessor relief against a lessee that has adopted a development strategy that maximizes its profit on the tract? Recall that the lessee bears 100% of the costs and receives, typically, 7/8 of the return. Is its decision on the scope of development likely to maximize the excess of benefits over costs? If lessees as a class could expect to be able to maximize that value, what would be the impact on the maximum size of the bonus that potential lessees would be willing to offer? May there be a social waste in requiring lessees to drill wells more densely than they would choose on a profit-maximizing basis?

The royalty provision of the lease in *Union Gas & Oil Co. v. Fyffe* reflects the era when wells were drilled to find oil; gas, in the absence of a pipeline system for its transportation and distribution, was typically flared. In the current era, is such a clause subject to attack on grounds of mutual mistake? *See* McGinnis v. Cayton, 173 W.Va. 102, 312 S.E.2d 765, 80 O.&G.R. 130 (1984).

5. The usual action for breach of the reasonable development covenant is based on the alleged failure of the lessee to drill the proper number of wells on the leasehold. Assuming that the number of wells that have in fact been drilled are sufficient to prevent drainage to other tracts of land and will eventually recover all the producible oil and gas in place, what is the loss the lessor complains of?

The answer, we suggest, is loss of interest on the capital represented by the unproduced royalty oil and gas. If that is correct, the measure of damages should be, not the amount of royalty that would have been forthcoming from more wells, but interest upon such royalty. See Mission Resources, Inc., v. Garza Energy Trust, 166 S.W.3d 301, 321 (Tx. App., Corpus Christi-Edinburg, 2005)("The amount lost by appellees is the interest income on the delayed royalties. That is the amount measured by the jury's instruction. The jury's award of damages did not force Coastal to pay royalties twice; it awarded appellees interest on the royalties appellees would have received if Coastal had not breached the implied covenant to develop the leasehold.").

The courts have not generally adopted the interest rule as the measure of damages and accordingly have had to face the additional question of how to prevent a double recovery when the oil or gas on which royalty has already been paid under a court order is actually produced later on.

COTIGA DEVELOPMENT CO. v. UNITED FUEL GAS CO., 147 W.Va. 484, 128 S.E.2d 626, 17 O.&G.R. 583 (1962), reversed a trial court decision adopting the interest rule. The court stated:

> We cannot agree, however, that the trial court was correct in its holding as a proposition of law that mere interest is the proper measure of the damages sustained by Cotiga, though it is readily conceded that the trial court had some basis for that holding in view of certain language contained in the opinion in Grass v. Big Creek Development Co., 75 W.Va. 719, 84 S.E. 750, to which reference will be made subsequently herein. It is difficult to discern from the relatively few decisions of appellate courts any clearcut rule or rules by which a trial court may be guided in such a situation. A brief review of decisions by other appellate courts in similar or analogous cases may be helpful. In doing so, we bear in mind that this is not a case involving a claim for damages for drainage. In such a situation, the gas or oil is completely lost to the lessor and the measure of damage is the "diminution of the royalties by reason of such drainage. . . ." . . . Cases of this nature are clearly distinguishable from the present situation because in the present case the gas is still in place, and has not been lost to Cotiga. Such distinction has been made by this Court previously. Grass v. Big Creek Development Co., 75 W.Va. 719, 730–731, 84 S.E. 750, 754.
>
> The royalty rule, substantially as contended for by Cotiga, has been approved in Illinois. Daughetee v. Ohio Oil Co., 263 Ill. 518, 105 N.E. 308. The Illinois royalty rule thus announced was considered and disapproved by this Court on the ground that such a rule would result in double payment of royalties to the lessor. Grass v. Big Creek Development Co., 75 W.Va. 719, 84 S.E. 750. The royalty rule has been approved and applied in Kentucky with one qualification: an oil and gas lessor must first make a demand upon the lessee to develop and market the oil or gas and liability for payment of royalties is not retroactive but accrues only upon such demand. . . . In a Texas case the court indicated that the royalty rule would be applied in an action against the lessee for failure to develop oil and gas in accordance with the terms of the lease, but the court also suggested that credit should be given to the lessee for royalties thus paid when the mineral is actually produced and marketed. Texas Pac. Coal & Oil Co. v. Barker, 117 Tex. 418, 6 S.W.2d 1031. . . .
>
> . . .
>
> Freely recognizing the difficulty of formulating a rule for the proper measure of damages in cases of this nature, in the light of the comparative paucity of precedents available for guidance, and recognizing also that a lessor suffering damages in such a situation should be compensated for such

loss but only once, it is our judgment that the fairest and most equitable rule is the one suggested by the Texas case previously referred to herein. The Court accordingly holds that Cotiga is entitled to recover from United Fuel royalties on the gas which should have been marketed from the leasehold during the period in question computed on "the rate received" by United Fuel; but that United Fuel shall have the right to offset and take credit for such sum, dollar for dollar rather than on a cubic foot basis, in the settlement and payment for royalties for gas next thereafter marketed from the leased premises. That is to say, damages thus computed in this case shall be deemed *pro tanto* the equivalent of the payment in advance of royalties on gas not actually produced and marketed from the premises, and United Fuel shall be entitled to credit for such sum, dollar for dollar, without interest thereon, when such gas shall be actually produced and marketed at the "rate received" by United Fuel when such gas is ultimately extracted and marketed.

6. The importance of the expert witness appears in GENNUSO v. MAGNOLIA PETROLEUM CO., 203 La. 559, 14 So.2d 445 (1943). Plaintiff, seeking cancellation of a lease for non-development proved that he was the owner of royalty in a 184–acre tract of land under lease to defendant; that the one well drilled on the tract twenty months before suit produced its allowable of approximately 120 bbls. daily; that defendant had recovered his investment and was beginning to show profit. Plaintiff offered no expert testimony. Defendant's defense rested largely upon the expert testimony of an employee who stated that in his opinion an ordinary, prudent operator would not drill another well. *Held,* cancellation denied; uncontradicted expert testimony shows an ordinary, prudent operator would not drill.

7. A checklist of factors may be used in determining whether an oil and gas lease has been reasonably developed. See Edmundson Brothers Partnership v. Montex Drilling Co., 98-1564 (La.App. 3 Cir. 5/5/99) 731 So.2d 1049, 1055, 142 O.&G.R. 266. The court there said the jurisprudence has articulated six factors which are especially pertinent in considering whether a lessee has breached his development obligation Those six factors are:

1) Geological data;

2) Number and location of wells drilled on or near the leased property;

3) Productive capacity of existing wells;

4) Cost of drilling compared with profit reasonably expected;

5) Time interval between completion of last well and demand for additional operation; and

6) Acreage involved in the lease under consideration.

8. Conservation laws and regulations affect implied covenants: (1) a spacing order determines the maximum density of drilling on a lease and thereby limits the effect of the implied protection and development covenants, these covenants

typically operating only to the extent that lessee has not drilled the number of wells permitted under the order; (2) a proration order determines the maximum allowable production for each well and thereby affects proof of probable profit from new wells. *See generally* Kramer & Martin, *Pooling and Unitization,* §20.04.

Suppose the leasehold consists of 640 acres and contains one well. The spacing order permits one well per 320 acres but the proration order is based on 640–acre units, so that two wells on 640 acres have the same total allowable as one well on 640 acres. Is lessee liable for breach of covenant for failure to drill a second well, assuming it could be drilled at a profit? (Consider again the issue of how profitability is to be calculated.)

In the second half of the 20th century regulatory commissions adopted regulations requiring wider spacing than in earlier years. It is not surprising, therefore, that current litigation involving the drainage or the reasonable development covenant is greatly reduced. However, litigation persists.

In RUSH v. KING OIL CO., 220 Kan. 616, 556 P.2d 431, 57 O.&G.R. 192 (1976), for example, the conservation rules provided for 20–acre spacing, and the lessee had drilled on that basis. However, the conservation rules allowed the lessee to drill additional wells, whose allowables would be taken out of allowables for existing wells on the leasehold. The court held that on these facts the spacing order was no defense to a claim for breach of the implied covenant of further development. The commission's attribution of acreage to a well "does not constitute a finding that the well will adequately and effectively drain the tract assigned to it. . . ." 220 Kan. at 626, 556 P.2d at 440–41, 57 O.&G.R. at 206–07.

9. Concerning the effects of conservation laws and regulations on implied covenants, see Ruyle v. Continental Oil Co., 44 F.3d 837, 131 O.&G.R. 315 (10th Cir. 1994) and Fransen v. Conoco, Inc., 64 F.3d 1481, 131 O.&G.R. 331 (10th Cir. 1995). The plaintiffs in both cases were the owners of mineral interests in a certain section of land in Oklahoma; some of the plaintiffs had leased their acreage to Conoco but others of the plaintiffs had not. They claimed that defendant Conoco had failed to protect the section from drainage and had failed to develop as a prudent operator. The section was the subject of an existing Corporation Commission unit and there was already a unit well for which Conoco was the operator. The appeals court held for the defendant in both cases. Only one well was permitted by the Corporation Commission on the section in issue, and no prudent operator would drill a well that was prohibited by law. The plaintiffs therefore could not establish an essential element of their claims for breach of the implied covenants of development and protection without avoiding, evading or denying the effect of a Corporation Commission order.

———

GULF PRODUCTION CO. v. KISHI

Commission of Appeals of Texas, 1937.
129 Tex. 487, 103 S.W.2d 965.

SMEDLEY, COMMISSIONER. Appellees recovered judgment against appellant for damages on account of appellant's failure, according to a jury's finding, to develop with reasonable diligence for the production of oil two tracts of land leased by appellees and their predecessors in title to appellant. The Court of Civil Appeals reversed the trial court's judgment and rendered judgment in favor of appellant but, pending action on motion for rehearing, certified to the Supreme Court the following question:

> Was appellees' petition, wherein they attempted to plead a cause of action based upon an implied covenant to drill wells in development of the leased premises in excess of the number agreed upon and stipulated for in the two leases, subject to appellant's general demurrer? In other words, since, by the express language of the two leases, provision was made for the development of the leased premises for oil by stipulating the number of wells to be drilled after the bringing in of the discovery well, and appellees plead affirmatively that all these wells had been drilled, were both leases, or either of them, subject to an implied covenant for additional development?

. . .

Appellant completed a producing well on the tract of land covered by the first lease on or about May 2, 1921, and thereafter prior to January 20, 1927, it drilled to completion fifteen wells on that tract. All but three of such wells produced oil in large quantities. It completed its first producing well on the other tract, the 20 acres, on or about October 5, 1920, and thereafter prior to January 1, 1927, it completed on that tract six wells, all of which produced oil in large quantities. The foregoing facts as to the wells drilled by appellant are set out in appellees' petition, and copies of the two leases are made parts of the petition and attached as exhibits. It is not alleged that the wells were not drilled successively within the time stipulated in the leases. The petition alleges that reasonable diligence in the development of the first lease required the drilling of an average of fifteen wells per year from January 20, 1927, to the time when the suit was filed, and that reasonable diligence in the development of the other lease required the drilling of an average of five wells per year during that period. The prayer of the petition is for the recovery of damages in an amount equal to the royalties for the oil that would have been produced in the period from four years immediately preceding the filing of the suit to July 31, 1931, if the leased land had been developed with reasonable diligence.

We agree with the conclusion expressed by the Court of Civil Appeals that "the conditions of the leases prescribing the number of wells to be drilled by appellant necessarily excluded the implied covenant plead by appellees for further development, thereby denying as a matter of law appellees' contention that appellant rested under an implied covenant to drill wells in the development of the leased premises in excess of the number called for by the terms of the two leases."

The argument first presented by appellees in support of the implication of a covenant for further development by the drilling of additional wells after the completion of the wells expressly provided for by the terms of the leases proceeds from the premise that a covenant for reasonable development of land leased for the purpose of producing oil or gas is always implied, and that it always exists unless there is in the lease an express agreement that the lessee shall not use reasonable diligence to develop the premises or an express stipulation or covenant so absolutely irreconcilable with the implied covenant as to destroy it. The true rule is that the implied covenant arises only out of necessity and in the absence of an express stipulation with respect to development of the leased premises. When the parties have not exercised their right to make express provision for the development of the land after production is obtained, a covenant for reasonable development is implied in order that the purpose for which the lease is made, the production of oil and gas with payment of royalty to the lessor, may be accomplished.

. . .

Associate Justice Pierson, in Freeport Sulphur Company v. American Sulphur Royalty Company of Texas, 117 Tex. 439, 450, 6 S.W.2d 1039, 1041, 60 A.L.R. 890, hereinafter more fully discussed, introduced his discussion of the important subject of implied covenants with the statement that: "The court cannot make contracts for parties, and can declare implied covenants to exist only when there is a satisfactory basis in the express contracts of the parties *which makes it necessary to imply* certain duties and obligations in order to effect the purposes of the parties in the contracts made." (Our italics.)

. . .

It follows that the existence of an implied covenant for development is not to be assumed and that it becomes necessary first to examine the leases under consideration to ascertain whether the parties have expressly agreed or stipulated as to the number of wells to be drilled in the development of the premises.

The first two paragraphs of the first lease relate to exploration. The lessee binds itself to begin operations for the drilling of the first well within ninety days from the date of the lease. In the event oil in paying quantities is not found in the first well, the lessee is given the right to drill additional test wells, commencing each such additional well within sixty days from the completion of the preceding well, with provision that failure to commence the second or additional wells within such time shall automatically operate as a forfeiture of the lease. Thus the parties expressly contracted with respect to the drilling of wells for exploration and there was no field for the operation of an implied agreement for the beginning of an exploratory well or wells. . . .

The agreement for development after the discovery of oil in paying quantities is found in the third paragraph. The lessee is required, in order to prevent the termination of its rights under the lease, to begin the drilling of another well within sixty days after the discovery of oil and thereafter to continue to drill additional wells (beginning each additional well within sixty days after the completion of the last prior well) until a total of twelve wells shall have been drilled on the leased premises. Nothing is said as to the drilling of other wells after the completion of the twelve.

In the lease covering the 20–acre tract express provision is also made for the drilling of exploratory wells. The fourteenth paragraph of the lease provides that, if oil in paying quantities is found on the leased premises, additional wells shall be drilled until as many as four producing wells are drilled, the additional wells to be begun with not more than ninety days' interval between the completion or abandonment of one and commencement of work on another, and that failure to drill such additional wells shall terminate the lease.

It is our opinion that the parties expressed, in the third paragraph of the first lease and in the fourteenth paragraph of the second lease, their agreements and intentions as to the number of wells required to be drilled by the lessee in the development of the premises covered by the two leases and as to the time when the wells should be drilled; in other words, that these two paragraphs contain the agreements of the parties upon the subject of development and the subject of diligence of development. This being true, there is no necessity for the implication of a covenant for development with reasonable diligence. To imply such covenant would be to make an agreement for the parties upon a subject about which they have in their written contracts expressly agreed.

. . .

So here, if it be conceded that the drilling of the fifteen wells on the one tract and the six wells on the other tract did not constitute reasonable development of all of the leased land, still there was no implied obligation for further development, because by the terms of the leases the parties expressly stipulated upon the subject of development. They deemed it advisable to set out in the leases the number of wells required to be drilled in the development of the property after the discovery of oil instead of leaving that matter to implication and to ultimate decision by a court or jury. This having been done, the obligations with respect to development imposed by the leases are to be determined by the express stipulations of the leases. . . .

The first proposition formally presented by appellees in their brief is that the express stipulations as to drilling wells relate to the estate granted, and not to the same subject as that to which the implied covenant for development relates, and therefore do not exclude the implied covenant. The proposition confuses the remedy or the result of failure to develop with the subject matter. The subject of the express stipulations as to the drilling of wells is development. The parties agreed upon the number of wells that the lessee would be required to drill and made performance more certain by stipulating that the lessee's failure to drill the wells should work the termination of its estate. The fact that the remedy or the penalty for noncompliance with the express stipulations of the leases is termination rather than liability for damages does not change the subject of the stipulations. They still relate to development. An express covenant may be, and often is, accompanied by a condition subsequent giving a right to terminate the estate of the covenantor on its breach. Tiffany's Real Property (2d Ed.) § 76, pp. 264, 265. A covenant for development, express or implied, may be by express language made a condition as well as a covenant. Brewster v. Lanyon Zinc Co. (C.C.A.) 140 F. 801, 812. It was held in Cole Petroleum Co. v. United States Gas & Oil Co., 121 Tex. 59, 41 S.W.2d 414, 86 A.L.R. 719, that the implied covenant for development came within the meaning of an express stipulation for forfeiture of an assignment of an oil and gas lease for failure of the assignee to fulfill any of the covenants of the agreement. Thus a condition subsequent may be attached even to an implied covenant for development or, differently stated, an implied covenant for development may be made by express agreement a condition as well as a covenant, but no change is thereby made in the subject to which the covenant relates, and it remains a covenant to the performance of which the covenantor is bound until the termination of the estate is caused by a breach. Similarly, the stipulations in the leases in the instant case, whether they be classified as special limitations, conditional limitations, or conditions subsequent, represent the agreement of the parties with respect to development and the subject to which they relate is development, the drilling of wells.

Because of the peculiar nature of the estate created by an oil and gas lease, the implied covenant for development is not a true covenant; that is, it does not impose a continuing, enforceable duty. The lessee may elect to permit the lease to terminate by ceasing to devote the premises to the purpose of oil and gas exploration, development, and production and thereby rid himself of the implied obligation, but the obligation continues as long as the lease is kept alive. See the valuable article by A.W. Walker, Jr., on The Nature of the Property Interests Created by an Oil and Gas Lease in Texas (11 Texas Law Review, pp. 408, 409). The express stipulations for development contained in the leases and the covenant for development implied in the absence of express stipulation have in a practical sense the same purpose, the development of the leased premises, and the obligation imposed by the express stipulations, like that implied, exists only while the lease is kept alive. For this additional reason the express stipulations and the implied obligation cannot be said to relate to different subjects.

Appellees' second proposition argues, in an effort to create a field for the operation of an implied covenant, that the express stipulations as to drilling wells do not state fully the duty of the lessee to develop but relate exclusively to development to be done within a specified period less than the entire term, and that there is an implied covenant for reasonable development during the remainder of the period for which the leases may endure. We find nothing in the leases suggesting that the express stipulations do not state fully the lessee's duty as to development. They specify the number of wells that must be drilled in the development of the property. The provisions that each well must be commenced within sixty or ninety days from the completion of the last prior well were intended to accomplish the diligent drilling of the wells required and not to divide the terms of the leases into separate periods of development.

. . .

Answering the question certified, it is our opinion that appellees' petition, in so far as it attempted to plead an implied covenant to drill wells in development of the leased premises in excess of the number of wells agreed upon and stipulated for in the two leases, was subject to appellant's demurrer. Since, by the express language of the two leases, provision was made for the development of the leased premises for oil by stipulating the number of wells to be drilled after the bringing in of the discovery well, and appellees pleaded affirmatively that all such wells had been drilled, the leases were not subject to implied covenants for reasonable development.

Opinion adopted by the Supreme Court.

NOTES

1. Does the reasonable development covenant obtain while a lease is being held in force by the payment of delay rentals? Refer again to the facts of the *Sundheim* case, *supra*. In addressing the issue, the court gave the following analysis, 806 P.2d at 509:

> As their second issue, the Sundheims argue that the District Court erred in granting summary judgment on their claim that Reef Oil and Frank Hiestand breached the implied covenant to reasonably develop the leasehold. The District Court held that this covenant was not breached because the Sundheims accepted delay rental payments. . . .
>
> In short, the Sundheims maintain that their acceptance of delay rental payments does not excuse the defendants from fulfilling their duty to reasonably develop their leasehold. They argue that because oil was discovered and pumped in paying quantities from the Sundheim No. 1 well, Reef Oil and Frank Hiestand had a continuing duty to further develop the leasehold. According to Sundheims, the payment of delay rentals has no bearing on this duty.
>
> We need not dwell upon the legal intricacies of the Sundheims' argument. The issue can be resolved through resort to the clear wording of the leases. Paragraph 2 of these leases states:
>
>> It is agreed that this lease shall remain in force for a term of three years from date and as long thereafter as oil or gas of whatsoever nature or kind or either of them is produced from said land or premises pooled therewith or drilling operations are continued as hereinafter provided. If prior to discovery of oil or gas on said land or on acreage pooled therewith, lessee should drill a dry hole or holes thereon, or if after discovery of oil or gas production thereafter should cease for any cause, this lease shall not terminate if lessee commences additional drilling or

reworking operations within sixty (60) days thereafter or (if it be within the primary term) *commences or resumes the payment or tender of rental on or before the rental paying date next ensuing after expiration of three (3) months from the date of completion of a dry hole or cessation of production.* [Emphasis added by the court. Eds.]

Paragraph 4 of the leases states:

If operations for the drilling of a well for oil or gas are not commenced or if there is no oil or gas being produced on said land or on acreage pooled therewith as hereinafter provided on or before one year from the date hereof, this lease shall terminate as to both parties unless the lessee on or before that date shall pay or tender to the lessor . . . the sum of $320.00 which shall operate as a rental and cover the privilege of deferring the commencement of operations for drilling of a well for twelve months from said date. In like manner and upon like payments or tenders the commencement of operations for drilling of a well may be further deferred for like periods of the same number of months successively.

According to these two sections of the Sundheim leases, Reef Oil had a duty to explore for or produce oil and gas from the Sundheim leasehold. If it failed to adequately carry out this duty, the terms allowed the Sundheims to terminate the lease. However, Reef Oil could, instead of engaging in drilling or production operations, make delay rental payments of $320.00 per year. Upon payment of this sum, Reef Oil was excused from the drilling and operating requirements contained in the lease.

See also State ex rel. Com'rs of Land Office v. Couch, 1956 OK 183, 298 P.2d 452, 6 O.&G.R. 346; Link v. State's Oil Corp., 229 S.W. 693 (Tex.Civ.App.1921).

For a general treatment of the subject of the relationship of express provisions to implied covenants, *see* Jacqueline L. Weaver, *When Express Clauses Bar Implied Covenants, Especially in Natural Gas Marketing Scenarios*, 37 Nat.Res.J. 491 (1997).

2. If a commercial well is completed on the leasehold, does the lessee have the duty to use due diligence to develop during the primary term or may he defer development until the primary term expires? *See* Berry v. Wondra, 173 Kan. 273, 246 P.2d 282, 1 O.&G.R. 1099 (1952); Exxon Co. v. Dalco Oil Co., 609 S.W.2d 281, 68 O.&G.R. 703 (Tex.Civ.App.1980, *error granted and judgments of both courts below set aside and cause dism'd as moot, due to settlement of all matters by the parties*).

3. Does the duty of reasonable development obtain when the lease is held in force by the payment of royalty under a shut-in gas royalty clause? *See* Amerada Petroleum Co. v. Doering, 93 F.2d 540 (5th Cir.1937)

4. In DAWES v. HALE, 421 So.2d 1208, 75 O.&G.R. 233 (La.App.1982), a

lease of 40 acres provided that in case of termination or cancellation of the lease for any cause, the lessee should have the right to retain around each producing well the number of acres allocated to each well under any relevant spacing or proration rules, or, if there were none, 40 acres around each well. Lessee drilled three wells in the northwest corner of the tract, each of which was producing two barrels of oil per day. If the lessor made out a case of violation of the implied covenant of reasonable development, should the acreage retention clause entitle the lessee to retain the entire tract? The court said no.

In GOODRICH V. EXXON COMPANY, USA, 608 So.2d 1019, 123 O.&.G.R. 438 (La. App. 1992), lessors sought partial cancellation of an oil and gas lease for the lessee's failure to develop as a prudent operator. The trial court ordered the horizontal cancellation of the lease, less and except those areas located within 40 acres, as provided for by the acreage retention clause of the lease, of the bottom hole of each producing well to the base of the deepest producing sand as well as areas held by unitized production. The appeals court upheld this determination even though the units were in some instances smaller than 40 acres. *Dawes v. Hale* was limited to its facts.

5. Should a court give effect to a lease clause that expressly leaves all development to the discretion of the lessee? Consider the following case:

OLIVER V. LOUISVILLE GAS AND ELEC. CO., 732 S.W.2d 509, 94 O.&G.R. 240 (Ky. App. 1987) (discretionary review denied by Kentucky Supreme Court). The plaintiff was owner of two tracts of land, the "Judd" tract and the "Skaggs" tract. The lessee entered into two "oil, gas and gas storage" leases with plaintiff's predecessors. The following language appeared in both the Judd lease of 1963 and the Skaggs lease of 1957:

> Habendum—To Have and to Hold unto and for the use of the Lessee for the term of 20 years from the date hereof and as much longer as oil or gas is produced in paying quantities or as the property continues to be used for the underground storage of gas.

> Rents and Royalties— . . . Lessee may, but is not obligated or required to drill the leased premises.

An annual rental fee was also provided in each lease. Seven wells were drilled on the Judd tract and none on the Skaggs tract. The plaintiff brought suit claiming that the leases should be cancelled except as to storage rights. In rejecting the claim, the court stated:

> While it is true that under some circumstances there exists an implied covenant in oil and gas leases that a reasonable attempt will be made to explore and develop the resources, there is no room for an implied covenant where the lease agreement itself makes the matter of development discretionary with the lessee. Warren v. Amerada Petroleum Corporation, 211 S.W.2d 314 (Tex.1948). . . .

Appellant requests that this Court adopt the reasoning of Cowden v.

Broderick and Calvert, Inc., 131 Tex. 434, 114 S.W.2d 1166 (1938), which held that the language 'lessor agrees that all other development shall be at the discretion of the lessee' did not leave the development to the complete option of the lessee. However, no Kentucky court has held that an implied covenant cannot be negated under any circumstance. Moreover, the Texas court noted that the implied covenant arises only in *absence* of express stipulation. The case is factually distinguishable from the case at bar, as therein the court determined that the purpose of the contract was to further production of oil and gas, and reasoned that leaving further development to the uncontrolled will of the lessee would prevent the accomplishment of the purpose for which the lease was made.

Such is not the case here. Accordingly, as the agreement clearly determines the rights of the parties, negating any implied covenant to drill or develop, the judgment of the Green Circuit Court is affirmed.

Compare COWDEN v. BRODERICK & CALVERT, 131 Tex. 434, 114 S.W.2d 1166, 117 A.L.R. 61 (1938). The lease in this case provided the following:

In the event of production on adjoining land, the lessee agrees to drill proper and necessary offsets along property lines; lessor agrees that all other development shall be at the discretion of the lessee.

The court held that this provision precluded application of the prudent operator standard to the implied covenant to develop, but went on to say:

The lessor, in agreeing that all other development should be at the discretion of the lessee, did not leave the development to the complete option of the lessee and his assigns, to develop or not to develop at their pleasure. . . . A construction of the lease that would leave further development to the option or uncontrolled will of the lessee and his assigns should be avoided, because it would tend to prevent the accomplishment of the purpose for which the lease was made, that is the production of oil and gas with payment of royalty to the lessor.

6. Should an implied covenant to continue production apply to prevent a lessee from plugging a well and removing its well equipment? See Willison v. Consolidation Coal Co., 536 Pa. 49, 637 A.2d 979, 128 O.&G.R. 1 (1994). The lease was granted in 1901 and provided for annual payments of $300 per year for each gas well, later reduced to $50.00, and for free gas for domestic use for the lessors. It also allowed the lessee "to remove all machinery and fixtures placed on said premises, and further, upon the payment of $50.00, at any time . . .[lessee] shall have the right to surrender this lease for cancellation . . ." In 1987, the oil and gas lessee sold its rights to Consol, which also held the rights to mine coal under the Willisons' land. Consol acquired the oil and gas lease for the purpose of plugging the well, to allow the extraction of valuable coal which would otherwise have had to remain in place to provide physical support for the operating well. The lessors sought to enjoin the plugging of the well and wanted to keep the lease alive to continue their supply of free natural gas. The

trial court found that the lease did not allow the abandonment of a producing well. The court said that there an implied covenant to continue to operate a well as long as it is profitable to do so. The court found then a breach and forfeiture and ordered the lessee to sell the equipment used at the well-site to the lessors for market value. The Pennsylvania Supreme Court reversed. The lessee contended on appeal that the decisions below erroneously relied upon implied provisions, perceived interests of the public, and equitable considerations to reach a result that was inconsistent with the express meaning of the lease in question. The Pennsylvania Supreme Court reversed:

> The decisions below may indeed reflect the commonly followed approach in other jurisdictions, as well as the approach outlined in various treatises on oil and gas law. It is clear, however, that construing the present lease in terms of presumed purpose and other external factors yields a result that is contrary to the terms of the lease. The lease expressly gave Consol a right to remove its equipment "at any time." That right has been extinguished by the courts below. [637 A.2d at 981].

In connection with *Willison*, consider again Judge Posner's comments in *Finley, supra*, concerning the differing incentives of lessor and lessee because the lessor bears no costs of production.

———

5 Martin & Kramer, *Williams and Meyers Oil and Gas Law*
§ 834. Remedies for Breach of the Covenant of Reasonable Development

As in protection covenant cases, the courts have recognized three separate remedies for breach of the covenant of reasonable development. In some cases, the lease has been cancelled outright, save for a small area surrounding the existing producing wells. In other cases the more moderate conditional decree of cancellation has been entered. Under this form of remedy the lease is cancelled unless a specified number of wells are drilled within a fixed period of time. Cancellation may be based on the theory that the implied obligation to develop is a condition subsequent, for breach of which the lessor may reenter. More often, however, it is granted on the equitable ground that the legal remedy is inadequate. Where this is the ground, the alternative or conditional decree has been favored. However, prolonged failure to develop may result in a decree of outright cancellation. Notice and demand may be a prerequisite to cancellation, either by judicial decision or by express lease provision.

———

In addition to equitable relief by way of cancelling the lease, the remedy of damages is also available for breach of the covenant, and the fact that proof of damages in such cases is inherently imprecise has been no obstacle to recovery. Devising a measure of damages that properly compensates the lessor for his loss has been a difficult task, and may account for the willingness of many states to award the equitable relief of cancellation freely. For discussions of the proper measure of damages for breach of drilling obligations *see* Maxwell, *Appropriate Damages for Breach of Implied Covenants in Oil and Gas Leases*, 42 Sw. Legal Fdn. Oil & Gas Inst., 7–1 (1991); Emanual, *Remedies for Breaches of Implied Covenants and Express Obligations to Drill in Oil and Gas Agreements*, 7 Eastern Min. L. Inst. ch. 16 (1986); Mansfield, *Relief from Express Drilling Obligations in an Uneconomic Market: The Federal Response and the Doctrines of Force Majeure, Impracticability, and the Prudent Operator*, 22 Tulsa L.J. 483 (1987); Maxwell, *Damages for Breach of Express and Implied Drilling Covenants*, 5 Rocky Mtn. Min. Law Institute 435 (1960); Scott, *Measure of Damages for Breach of a Covenant to Drill a Test Well for Oil and Gas*, 9 U.Kan.L.Rev. 281 (1961).

GILLETTE v. PEPPER TANK CO.

Colorado Court of Appeals, Div. II, 1984.
694 P.2d 369, 83 O.&G.R. 271.

PIERCE, JUDGE. From a judgment entered after trial to the court in an action for determination of the validity of an oil and gas lease held by defendants, both defendants and plaintiffs appeal. . . .

The lease provides for a 4–month primary term and as long thereafter as oil or gas is produced. . . . Also, a unitization agreement, executed in 1963, affects certain portions of the leased lands. The lease in question was first executed in 1951 by Donald P. and Miles T. Gillette, and was assigned, principally, to defendant, Pepper Tank Company (Pepper) in 1959. The lease covers approximately 3,360 acres of land.

Prior to the assignment, a number of successful wells were drilled in 1952–1953, and one well was drilled thereafter in 1957. After the assignment, one well was drilled in 1972; it was plugged and abandoned the same year. There have been no drilling efforts of any nature on the leased property since that time. The major development effort of Pepper was a water-flood operation commenced in 1963 and abandoned in 1971. Presently, all the wells have been abandoned with the exception of Gillette well # 10, which continues to produce at a marginal rate.

Lessors alleged the lease had been terminated during its secondary term because of defendants' failure to produce oil or gas in paying quantities. They further alleged that defendants had breached certain implied covenants of the lease and requested a decree cancelling the lease and quieting title in the lessors.

Upon finding violations of the implied covenants . . . the trial court granted conditional cancellation of the lease. The terms of the cancellation were that, if within 60 days Pepper filed with the court a plan of development for the non-producing areas, the cancellation would be ineffective; but if Pepper failed to submit a plan, in order for the cancellation to be effective, lessor, Underwood, was required to file a plan of development. Additionally, the court ruled that if

Pepper made all necessary repairs upon the pits, filed an engineer's report indicating completion of such repairs, and thereafter properly maintained such pits, the cancellation would be ineffective as to the producing area.

We affirm that conditional decree except as to part of the acreage involved.

I.

A.

Pepper contends that the court's findings as to breaches of the implied covenants to drill, develop, and operate diligently are not supported by the evidence. We disagree.

Colorado recognizes, in general, four implied covenants in oil and gas leases: to drill; to develop after discovery of oil and gas in paying quantities; to operate diligently and prudently; and to protect leased premises against drainage. Mountain States Oil Corp. v. Sandoval, 109 Colo. 401, 125 P.2d 964 (1942); see also Graefe & Graefe, Inc. v. Beaver Mesa Exploration Co., 635 P.2d 900 (Colo.App.1981). The basis for these covenants is founded on the concept that in the secondary term of an oil and gas lease:

> "[t]he work of exploration, development, and production should proceed with reasonable diligence for the common benefit of the parties, or the premises be surrendered to the lessor." Brewster v. Lanyon Zinc Co., 140 Fed. 801 (8th Cir.1905).

Reasonable diligence is, "whatever, in the circumstances, would be reasonably expected of all operators of ordinary prudence, having regard to the interests of both lessor and lessee." *Brewster, supra.*

This obligation to explore, develop, and produce, once production is acquired, includes both an implied covenant of reasonable development and an implied covenant of further exploration. H. Williams & C. Meyers, Oil & Gas Law §§ 831–847 (1983). The implied covenant of reasonable development requires a determination that additional development will be profitable. This determination rests on proof that, more probably than not, production of oil or gas will be found in paying quantities. The implied covenant of further exploration does not need such proof, but rather requires the lessor to show unreasonability by the lessee in not exploring further under the circumstances.

Among the circumstances relevant to determining whether there has been a breach of the promise of further exploration are: the period of time that has lapsed since the last well was drilled; the size of the tract and the number and location of existing wells; favorable geological inferences; the attitude of the lessee toward further testing of the land; and the feasibility of further exploratory drilling as well as the willingness of another operator to drill. H. Williams & C. Meyers, *supra.* The distinction between the two covenants is helpful in our analysis of Pepper's contention of insufficient evidence concerning what a prudent operator in the position of Pepper would have done.

The trial court found one well currently operating, a well drilled and abandoned in 1972, and an abandoned water-flood project. It also found deliberate failure to clear title, and speculative holding by Pepper, while finding some interest by third parties in drilling and developing the lease. Relying on *Mountain States Oil Corp., supra,* the court concluded that Pepper had violated the implied covenants to drill and to develop in a known area.

Whether implied covenants are breached is primarily a question of fact. *Mountain States Oil Corp.,* supra. Therefore, the findings of the trier of fact must be accepted on review unless they are clearly erroneous. . . . Here, the findings by the court support its determination of violation of the implied covenant to further explore. *Brewster, supra*; Sauder v. Mid–Continent Corp., 292 U.S. 272, 54 S.Ct. 671, 78 L.Ed. 1255 (1934).

As to the implied covenant to operate prudently, the trial court found improper maintenance and discharge of water, damage to the surface, and a poorly conducted water-flood operation. Again, a breach of this implied covenant is a question of fact, and therefore, the trial court's finding will not be disturbed on review as the record supports its decision. . . .

. . .

C.

Pepper further contends that the court erred in not segregating the unitized land from the non-unitized portion of the lease. We agree in part and disagree in part.

Unitization relieves the lessee of the obligation of the implied covenant for reasonable development for each tract separately. However, the implied covenants are not abrogated as to the lessee's obligation concerning the unit as a whole. Thus, implied covenants, including the implied covenant to operate the lease in a prudent manner as well as to develop the land reasonably after discovery, are recognized as to the unit. . . . .

Whether Pepper has breached the implied covenants of reasonable development and further drilling on the unitized portion of the lease must be considered in view of the entire unit, because, according to the terms of the unit agreement, production on any portion of the unitized area extends the lease as to the unitized portion of the lease. . . . Thus, if Pepper complies with the trial court's order concerning repairs in the SW 1/4 of the SE 1/4 of Sec. 9, the trial court may find the production from Gillette well # 10, even though marginal, sufficient to hold the entire unit and show no breach of the implied covenants as to the unitized area. . . .

The trial court did not consider the entire unit; we therefore reverse as to the remaining portions of the lease affected by the unitization agreement, namely: S 1/2 of Sec. 4, the E 1/2 of the SE 1/4 of Sec. 9, and the NW 1/4 of the SE 1/4 of Sec. 9.

II.

A.

Lessors argue that conditional cancellation is not appropriate, but rather the court should have granted an absolute cancellation. We disagree.

Breach of the implied covenants generally leaves no adequate remedy of law. . . . Thus, the district court may decree cancellation in whole or in part. . . .

. . .

B.

Lessors also argue that there is no need for requiring lessor Underwood to submit a plan of development. Again, we disagree.

The trial court properly relied upon Humble Oil & Refining Co. v. Romero, 194 F.2d 383 (5th Cir.1952) in fashioning equitable relief, under the present circumstances, including the requirement that Underwood file a plan of development with the court.

. . .

The judgment is reversed as to the following portions of the lease affected by the unitization agreement: the S 1/2 of Sec. 4, the E 1/2 of the SE 1/4 of Sec. 9, and the NW 1/4 of the SE 1/4 of Sec. 9. The trial court is to reconsider its findings regarding this portion of the lease in accordance with the views expressed in this opinion. The conditional decree as to the remainder of the lease is affirmed.

BERMAN and STERNBERG, JJ., concur.

―――――

SUN EXPLORATION AND PRODUCTION CO. v. JACKSON

Supreme Court of Texas, 1989.
783 S.W.2d 202, 107 O.&G.R. 383 (rehearing overruled, 1990)

RAY, JUSTICE. After reargument, our prior judgment and opinions dated July 13, 1988 are withdrawn, and the following opinion is substituted.

The issue in this cause is whether there exists in Texas oil and gas leases an implied covenant to explore, independent of the implied covenant of reasonable development. Sun Exploration and Production Company (the successor in interest to Sun Oil Company) and Amoco Production Company (together referred to as Sun) brought an action for declaratory judgment and an injunction against the Jacksons. Sun sought to establish the validity of an oil, gas, and mineral lease covering a 10,000 acre tract in Chambers County, known as the Jackson Brothers Ranch. Sun also sought a permanent injunction enjoining the Jacksons from denying Sun Oil entrance onto the leased property. The Jacksons counterclaimed alleging breach of implied covenants to reasonably develop and explore the entire lease, and seeking cancellation of the lease. Based on the jury's findings, the trial court rendered judgment for the Jacksons, unconditionally cancelled that portion of the lease on which Sun had not drilled extensively and conditionally cancelled the lease below the depth to which Sun had drilled (8480 feet) in a developed area. The court of appeals affirmed the unconditional cancellation and reversed and remanded as to the conditional cancellation. 715 S.W.2d 199, rehearing overruled, 729 S.W.2d 310. We reverse the judgment of the court of appeals and remand for the trial court to determine attorney's fees and whether Sun and Amoco may be entitled to any injunctive relief.

In March of 1938, Ocie R. Jackson and other interested members of the Jackson family executed an oil, gas, and mineral lease to Sun covering the 10,000 acres of the Jackson Brothers Ranch. Sun owns a majority of the working interest under the lease. Amoco owns a small minority of the working interest, which it acquired many years ago. The Jacksons own the entire surface and retain a majority of the outstanding nonparticipating royalty interest. In 1941, Sun drilled its third well on the Jackson Brothers Ranch, resulting in the discovery of a reservoir commonly known as the Oyster Bayou Field. Production from that reservoir continues to this date. The Oyster Bayou Field produces from a small part of the 10,000 acre tract. The Jacksons complain that only the Oyster Bayou Field on the Jackson lease has been developed by Sun. The Jacksons assert that Sun has neglected to explore and develop the larger remaining part of the lease.

Sun and Amoco sued to enjoin the Jacksons from interfering with their right to enter the lease and the Jacksons counterclaimed against Sun and Amoco for breach of implied covenants. The jury found that Sun had not failed to reasonably develop the Jackson lease, but that Sun had failed to reasonably explore the portions of the lease that were outside the Oyster Bayou Field. Based on the jury's verdict, the trial court rendered judgment for the Jacksons. The judge determined the remedy: unconditional cancellation of one portion of the lease and conditional cancellation of another portion of the lease.

There are three generally recognized implied covenants in Texas: "(1.) to develop the premises, (2.) to protect the leased premises, and (3.) to manage and administer the lease." Amoco Product. Co. v. Alexander, 622 S.W.2d 563, 567 (Tex.1981). The Jacksons would like us to recognize an implied covenant of further exploration.

This court has held that no implied covenant of further exploration exists independent of the implied covenant of reasonable development. Clifton v. Koontz, 160 Tex. 82, 98, 325 S.W.2d 684, 696 (1959). In *Clifton*, the court held that the covenant of reasonable development encompassed the drilling of all additional wells after production on the lease is achieved. 160 Tex. at 96–97, 325 S.W.2d at 695–96. By "additional wells" the court meant both additional wells in an already producing formation or stratum, or additional wells in "that strata different from that from which production is being obtained." 160 Tex. at 96, 325 S.W.2d at 695. The critical question was whether the lessor could prove a reasonable expectation of profit to lessor and lessee. . . . Therefore, under *Clifton* if a party could prove that a reasonably prudent operator would have drilled the well, that well fell within the implied covenant of reasonable development, without regard to whether the well was classified as exploratory or developmental.

In answer to question one, the jury found that Sun did not fail to reasonably develop the Jackson lease. This finding is dispositive of the case. The law of Texas does not impose a separate implied duty upon a lessee to further explore the leasehold premises; the law recognizes only an implied obligation to reasonably develop the leasehold. Because the jury determined that Sun has not failed to reasonably develop the Jackson lease, the court of appeals should have rendered judgment for Sun. In failing to do this, the court erred.

In analyzing the issues and instructions submitted to the jury, the court of appeals found:

(1) The terms "explore" and "develop" as they appear in the trial court's questions and instructions were used interchangeably;

(2) Question one refers only to the drilling of additional wells to test the formation situated inside the Oyster Bayou Field; and

(3) Question two refers to the drilling of additional wells to test potentially productive formations situated outside the Oyster Bayou Field.

715 S.W.2d at 203.

This analysis distorts the meaning of the questions and instructions which were submitted to the jury. The court of appeals found that question one and the instruction that accompanied it, limited the jury's attention to the matter of drilling additional development wells to test the formation situated inside the Oyster Bayou Field. In this, the court was mistaken. The language of question one is in no way limited to activity within the confines of the Oyster Bayou Field. Instead, the issue inquires whether Sun "failed to reasonably develop the Jackson lease," not merely the Oyster Bayou Field.

Likewise, the language of the instruction which accompanied question one does not limit the jury's attention to the Oyster Bayou Field. The relevant language of the instruction advises the jury that:

the term "to reasonably develop" means the development which a prudent operator would do with respect to any known producing formation of the lease.

Contrary to the finding of the court of appeals and the position advanced by the Jacksons, the quoted language of the instruction does not limit the jury's attention to the Oyster Bayou Field; it asks the jury to focus upon any known producing formation of the lease. The analysis offered by the court of appeals and the Jacksons suggests that the language of this instruction asked the jury to focus only upon formations which are *currently* producing hydrocarbons on the Jackson lease, but this is not what the language of this instruction does.

The instruction is worded broadly and allows the jury to include in its deliberations actions taken with regard to any formation on the lease that is currently producing or that has been determined to be productive of hydrocarbons but is not producing now. In addition, the wording of this instruction allowed the jury to consider Sun's actions over the years in discovering, producing and depleting several one-well reservoirs located outside the Oyster Bayou Field. Indeed, it was essential to word the instruction so as to permit such matters to be considered by the jury because the language of question one required the jury to determine whether Sun had reasonably developed the entire lease. Thus, in connection with question one, the jury was instructed to consider all activities undertaken by Sun with regard to the development of any formation capable of producing hydrocarbons, situated anywhere on the lease, and not simply those activities conducted by Sun inside the Oyster Bayou Field.

For these same reasons, the construction of question two proposed by the court of appeals as focusing the jury's attention on the reasonableness of Sun's drilling activities outside the Oyster Bayou Field is likewise erroneous. In reality, these two questions, when considered together, first asked the jury to determine whether Sun had reasonably *developed* the entire Jackson lease. In answering this question, the jury was required to consider activities affecting any formation on the Jackson lease known to be capable of producing hydrocarbons, whether such formations were currently producing or not. These issues then asked the jury to determine whether Sun had reasonably *explored* all of those portions of the Jackson lease not known to be capable of producing hydrocarbons. The construction of the jury's answers to these two questions advanced by the court of appeals has created an ambiguity in the jury's verdict when, in fact, none exists. The clear and unambiguous finding of the jury is that Sun has reasonably developed the entire Jackson lease.

Having determined there is no implied covenant to explore, independent of the implied covenant of reasonable development, and that the jury found that no breach of the covenant to reasonably develop occurred, we hold the lease remains valid. It is thus unnecessary for us to pass upon the validity of the remedies of conditional and unconditional cancellation.

We affirm that part of the court of appeals judgment that overrules Sun's motion for remand of the cause for consideration of disqualification or recusal.

The judgment of the court of appeals is reversed and the cause remanded to the trial court for a determination of attorney's fees and injunctive relief, to which Sun and Amoco may be entitled.

SPEARS, J., files a concurring opinion in which COOK, J., joins.

GONZALEZ, J., files a concurring opinion in which DOGGETT, J., joins.

NOTES

1. Which is the better approach for a court to follow, the Colorado or the Texas approach? Why? Consider the implications for the attorney drafting an oil and gas lease—is it more practical to provide express obligations to explore or to negate the implication (or imposition) of such obligations by a court? Consider

the following observation by Professors Goetz and Scott:

> [T]he courts' tendency to treat state-created rules as presumptively fair often leads to judicial disapproval of efforts to vary standard implied terms by agreement. Indeed, as the system for implying terms into contracts becomes more highly refined, the opting-out burden of those who prefer a different allocation of risks grows weightier. Furthermore, conflicting applications of the parol evidence and plain-meaning rules betray widespread judicial uncertainty over the proper method of interpreting agreements that intermingle express and implied terms. These and other related problems have exposed a central question that remains unresolved: To what extent do implied and express terms, and standardized and individualized forms of agreement, function in antagonistic rather than complementary ways? Goetz and Scott, *The Limits of Expanded Choice: An Analysis of the Interactions Between Express and Implied Contract Terms*, 73 Calif.L.Rev. 261, 262 (1985) [footnotes omitted].

2. No treatment of the implied covenant of further exploration would be complete without consideration of the late Dean Meyers' seminal and controversial article, *The Implied Covenant of Further Exploration*, 34 Tex.L.Rev. 553 (1956). The Fifth Circuit heeded Dean Meyers' call for the implication of the covenant in Sinclair Oil and Gas v. Masterson, 271 F.2d 310, 11 O.&G.R. 632 (1959), cert. denied, 362 U.S. 952 (1960), finding that the Texas Supreme Court had anticipated such a case in *Clifton v. Koontz,* 160 Tex. 82, 325 S.W.2d 684, 10 O.&G.R. 1109 (1959). In a subsequent case from the Fifth Circuit involving the same lands as the *Masterson* case, Judge Brown made the following comments that indicate the lack of deference subsequently given the *Masterson* decision by the Texas state courts:

> The magnitude of these operations and the interest at stake in these lands [the Masterson Ranch] is revealed in our prior decision, Sinclair Oil & Gas Co. v. Masterson, ... This case has been severely criticized as not well considered. Brown, The Law of Oil & Gas Leases, 1958; 1966 Cumulative Supplement, p. 243 et seq. As an Erie proposition, its vitality is severely sapped, if not extinguished, by the express repudiation of it by the Court of Civil Appeals, "the first writing Texas court," . . . in Felmont Oil Corp. v. Pan American Petroleum Corp., 334 S.W.2d 449 (Tex.Civ.App.1960, error refused n.r.e.), because, that Court stated, we had ignored Clifton v. Koontz. . . .. Perhaps the Fifth Circuit has been overruled again. Weymouth v. Colorado Interstate Gas Co., 367 F.2d 84, 102 n. 57, 25 O.&G.R. 371, 393 n. 57 (5th Cir.1966)

Other treatments of the implied covenant of further exploration, in addition to the Meyers and Crafton article excerpted above, include Martin, *A Modern Look at Implied Covenants to Explore, Develop, and Market under Mineral Leases*, 27th Sw. Legal Fdn. Oil & Gas Inst. 177 (1976); Williams, *Implied Covenants for Development and Exploration in Oil and Gas Leases—The Determination of*

Profitability, 27 Kan.L.Rev. 443 (1979); Williams, *Implied Covenants in Oil and Gas Leases: Some General Principles*, 29 Kan.L.Rev. 153 (1981); Weaver, *Implied Covenants in Oil and Gas Law under Federal Energy Regulations*, 34 Vanderbilt L.Rev. 1473 (1981); Allison, *Explorvelopment: A Theoretical Hybrid Searching for Fertile Legal Soil in an Unfertile Economy*, 39 Sw. Legal Fdn. Oil & Gas Inst. 9–1 (1988).

A California court has been the most recent to show reluctance in embracing a covenant of further exploration. LUNDIN/WEBER CO. V. BREA OIL CO., 117 Cal. App. 4th 427, 11 Cal. Rptr. 3d 768, 160 O.&G.R. 855 (2004)("where parties have chosen not to extend the obligation to explore for oil or gas beyond the discovery and development of paying quantities, a court should not insert obligations in direct conflict with the limitation expressed by the parties.).

3. The Louisiana approach to implied covenants does not distinguish sharply between further development and exploration. As summarized in the Comments to Article 122 of the Louisiana Mineral Code:

Further exploration: As noted, Williams and Meyers characterize the covenant of further exploration as being separate from the covenant of reasonable development. This is a defensible view. However, historically in Louisiana the obligation of further exploration can be viewed as an evolutionary offshoot of the obligation of reasonable development. See Carter v. Arkansas Louisiana Gas Co., 213 La. 1028, 36 So.2d 26 (1948). Although the jurisprudence does not make a clear distinction between the obligation of further exploration and the obligation of reasonable development, the distinction nevertheless exists. See Sohio Petroleum Co. v. Miller, 237 La. 1015, 112 So.2d 695 (1959); Middleton v. California Co., 237 La. 1039, 112 So.2d 704 (1959); Carter v. Arkansas Louisiana Gas Co., *supra*; Nunley v. Shell Oil Co., 76 So.2d 111 (La.App. 2d Cir.1954).

The obligation to explore and test all portions of the leased premises after discovery of minerals in paying quantities was first announced by the Louisiana Supreme Court in Carter v. Arkansas Louisiana Gas Co., *supra*. The lessor sued for a cancellation of the lease as to the entire lease premises. A main fault traversed the tract and it was admitted that full development had taken place on one side of the fault. Lessor demanded development on the other side of the fault. Based upon the evidence presented, the court felt the lessor had borne the burden of proving that a reasonable, prudent operator would further explore by drilling wells on the undeveloped side of the fault. Partial cancellation was granted as to the acreage on the undeveloped side of the fault. However, the court noted that had the plaintiff appealed from the judgment of the lower court awarding partial cancellation, he would have been entitled to cancellation of the entirety of the lease. The importance of the *Carter* case, however, stems from the fact that the court, in its application of the obligation of reasonable development, went further and imposed upon the mineral lessee the duty of further exploration after initial production in paying quantities had been obtained. The court quoted with approval language from the Oklahoma case of Fox Petroleum Co. v. Booker, 123 Okl. 276, 282, 253 P. 33, 38 (1926):

> The principle, as we understand it, is that development of every part of the lease is an implied condition. Therefore, whether the undeveloped portion be a single tract remote from the rest, or a considerable portion of a very large tract . . . or the east 100 acres of a tract of 160, it is an implied condition that the lessee will test every part.

The jurisprudence since the *Carter* decision has recognized that the obligation of further exploration is embodied in our law. Middleton v. California Co., supra; Sohio Petroleum Co. v. Miller, *supra*; Reagan v. Murphy, 235 La. 529, 105 So.2d 210 (1958); Wier v. Grubb, 228 La. 254, 82 So.2d 1 (1955); Eota Realty Co. v. Carter Oil Co., 225 La. 790, 74 So.2d 30 (1954); Nunley v. Shell Oil Co., *supra*. Federal cases applying Louisiana law are to the same effect. Cutrer v. Humble Oil & Refining Co., 202 F.Supp. 568 (E.D.La.1962); Romero v. Humble Oil & Refining Co., 93 F.Supp. 117 (E.D.La.1950).

The basic similarity between the obligation of reasonable development and the obligation to further explore is that in both instances there must be discovery in paying quantities to make the obligations operative. Some cases have been rather liberal in holding that the lessor has borne the burden of proving that a reasonable, prudent operator would further explore the lease premises. In Nunley v. Shell Oil Co., *supra*, and Romero v. Humble Oil & Refining Co., *supra*, the courts accepted testimony that lessor had a firm offer from an experienced person to take a lease with a drilling obligation on the unexplored portion of the premises. In both of those cases, it appeared that the defendants had taken the position that even if a well were drilled it would be unsuccessful. The courts seem to have viewed this as tantamount to asserting that the acreage had been condemned. In response to this it was stated that if the acreage had been condemned there was no reason why the lessee should be permitted to sit on the undeveloped acreage in the presence of an offer to drill an exploratory well. Differing circumstances are found in Middleton v. California Co., *supra*, and Saulters v. Sklar, 158 So.2d 460 (La.App.2d Cir.1963). In the *Middleton* case considerable development had taken place, several million dollars in royalties had been paid to the lessor, and lessee presented evidence that it was engaged in a program of seismic exploration of the untested acreage on a 4,600 acre lease. In view of the large size of the lease, the amount of the royalties paid, and the demonstrated plans of the lessee, the court refused to give relief to the plaintiff lessor. In Saulters v. Sklar, *supra*, defendant lessee's response to the plaintiff's demand for further exploration was decidedly evasive. The error committed by the lessee in the *Nunley* case of taking the position that the acreage was condemned was avoided by stating that it felt that the acreage might be valuable for a deep test in the future but that lessee did not deem the present was the time to make such a test. The principal distinction which can be found between *Saulters* and *Nunley* is that in one case the lessee took a categorical position that further exploration would be fruitless and in the other the lessee indicated a willingness to consider making a deep test at a future date though that date was undefined. Additionally, the court appears to have given some weight to the fact that the offer to drill on which plaintiff relied was apparently obtained in close proximity to the time of the trial and for that specific purpose.

An extensive discussion of the Louisiana implied obligation of reasonable development/further exploration is found in Noel v. Amoco Production Co., 826 F.Supp. 1000, 1004-09, 125 O.&G.R. 38 (W.D.La. 1993). See generally Thomas A. Harrell, *A Mineral Lessee's Obligation to Explore Unproductive Portions of the Leased Premises in Louisiana*, 52 La.L.Rev. 387 (1991). The plaintiff in *Noel* failed to meet the burden of persuading the court that a reasonably prudent operator would have investigated the undeveloped portions of the leased premises for the presence of potentially profitable oil and gas deposits, and that her lessee failed to conduct such operations.

4. Notwithstanding the fact that the Louisiana Mineral Code Comments indicate that the Louisiana courts were following Oklahoma precedent in adopting a "test every part" approach, the Oklahoma Supreme Court has expressed a rejection of the implied covenant of further exploration. In MITCHELL v. AMERADA HESS CORP., 1981 OK 149, 638 P.2d 441, 449, 72 O.&G.R. 104 the Oklahoma court declared: "We thus hold there is no implied covenant to further explore after paying production is obtained, as distinguished from the implied covenant to further develop." But the court has apparently not disturbed the established Oklahoma doctrine that achieves much the same effect by shifting the burden from the lessor to prove a breach of the implied covenant of reasonable development after the passage of a reasonable period of time to the lessee to show why a prudent operator would continue to hold acreage without additional drilling. See Dixon v. Anadarko Production Co., 1972 OK 165, 505 P.2d 1394, 44 O.&G.R. 397; Crocker v. Humble Oil & Refining Co., 1965 OK 126, 419 P.2d 265, 25 O.&G.R. 681; Lyons v. Robson, 1958 OK 232, 330 P.2d 593; Trawick v. Castleberry, 1953 OK 142, 275 P.2d 292, 4 O.&G.R. 63. See also Doss Oil Ry. Co. v. Texas Co., 1943 OK 154, 192 Okl. 359, 137 P.2d 934.

Compare Sonat Petroleum Co. v. Superior Oil Co., 710 P.2d 221, 88 O.&G.R. 605 (Wyo.1985) where the Wyoming court expressly rejects the Oklahoma burden-shifting approach.

5. The Kansas Deep Rights Act, Kan.Stat.Ann. §§ 55–223 to 55–229 creates a presumption in favor of the lessor in a claim on an implied development or exploration covenant, if:

(a) At the time such action is commenced there is no mineral production pursuant to such lease from a subsurface part or parts of the land covered thereby with respect to which such relief is sought and (b) initial oil, gas or other mineral production on the lease commenced at least 15 years prior to the commencement of such action. . . .

The presumption may be overcome by the lessee by a preponderance of the evidence. The Act is upheld against various constitutional attacks in Amoco Production Co. v. Douglas Energy Co., 613 F.Supp. 730, 85 O.&G.R. 446 (D.Kan.1985). See also Lewis v. Kansas Production Co., Inc., 40 Kan.App.2d 1123, 199 P.3d 180 (2009).

6. The implied covenants of reasonable development and further exploration, and the related concerns about a lessee holding on for speculation, are sometimes said to find their analogue in Aesop's fable of the "dog in the manger":

A DOG lay in a manger, and by his growling and snapping prevented the oxen from eating the hay which had been placed for them. "What a selfish Dog!" said one of them to his companions; "he cannot eat the hay himself, and yet refuses to allow those to eat who can." *See* for example, BLAKE V. TEXAS CO. 123 F.Supp. 73 (D.C.Okl. 1954): "Continued failure by the defendant to drill additional wells on the lease in question doubtless will ripen into a breach of its implied obligation. The defendant cannot forever choose to delay further development and at the same time prevent others from drilling. Should the total time of inactivity measured in years become unconscionable in and of itself, the plaintiffs will be entitled to oust the 'dog in the manger'."

———

NOTICE, DEMAND AND JUDICIAL
ASCERTAINMENT CLAUSES

(*See* Martin & Kramer, *Williams and Meyers Oil and Gas Law* §§ 682–682.5).

Many leases contain clauses similar to the following:

In the event Lessor considers that Lessee has not complied with all its obligations hereunder, both express and implied, Lessor shall notify Lessee in writing, setting out specifically in what respects Lessee has breached this contract. Lessee shall then have sixty (60) days after receipt of said notice within which to meet or commence to meet all or any part of the breaches alleged by Lessor. The service of said notice shall be precedent to the bringing of any action by Lessor on said lease for any cause, and no such action shall be brought until the lapse of sixty (60) days after service of such notice on Lessee. Neither the service of said notice nor the doing of any acts by Lessee aimed to meet all or any of the alleged breaches shall be deemed an admission or presumption that Lessee has failed to perform all its obligations hereunder.

1. Do such clauses simply state expressly what the law otherwise requires? In Lewis v. Kansas Production Co., Inc., 40 Kan.App.2d 1123, 199 P.3d 180 (2009) the court ruled: "Since there has been no compliance request, and no evidence of abandonment or futility in requesting compliance, cancellation is not appropriate. [Lessee's] actions have been sluggish, but before the extreme remedy of forfeiture could be imposed, under these facts, there should have been a demand for compliance. Absent such a demand, under these facts, [lessee] should be given a reasonable time to comply with the lease." Outright termination of the lease for breach of the implied covenant to explore and develop minerals by the trial court was reversed. The appropriate remedy for breach was conditional cancellation and grant of reasonable time to lessee to drill a well and begin production.

2. In a jurisdiction normally requiring notice and demand as a prerequisite to the cancellation remedy, should the requirement be excused if the lessor can demonstrate that such a demand would have been futile? *See* Amoco Production Co. v. Douglas Energy Co., 613 F.Supp. 730, 85 O.&G.R. 446 (D.Kan.1985); Goodrich v. Exxon Corporation, 642 F.Supp. 150, 92 O.&G.R. 536 (W.D.La.1986). How would the lessor prove the futility of a demand? Does, or should, the law require notice before the commencement of a suit for damages? What should a letter demanding development contain? *See* Taussig v. Goldking Properties Co., 495 So.2d 1008, 94 O.&G.R. 265 (La.App.1986).

3. Do notice and demand or judicial ascertainment clauses apply to the delay rental, habendum or royalty clause of the lease? Consider the comments of Judge Wisdom in Williams v. Humble Oil & Refining Co., 432 F.2d 165 at 178–180, 38 O.&G.R. 212 at 236–237 (5th Cir. 1970), on the operation of a notice and demand clause:

"Notice provisions were originally inserted in mineral leases primarily to protect the lessee against a forfeiture of the lease for breach of some express or implied obligation. Because forfeiture or cancellation is an extraordinary remedy that terminates the lessee's interest it is fair to require that notice, demand, and an opportunity to cure the breach be given the lessee. Thus whenever the lessor has sought cancellation, the Louisiana courts have required strict compliance with the notice requirements of the lease. ...

"On the other hand, an action for damages threatens no such harsh consequences as termination of the lessee's interest. There is less reason to insist upon notice and demand. Indeed, in the ordinary drainage action notice to the lessee would be superfluous. The harm has already been committed, and no action by the lessee to repair the breach could adversely affect the lessor's right to compensation for past harm. In many cases the fact of injury is already known to the lessee, who has superior knowledge of his own operations, long before it becomes known to the lessor. The lessor's inability to give notice of that which he does not know should not in all fairness bar him forever from recovery for damages already incurred. The oil industry was undoubtedly conscious of these considerations when the standard notice provision in the instant lease was drafted. Therefore we hold that failure to comply with a notice provision drafted in this language and in these circumstances was not then intended to and does not now bar an action for damages."

Article 136 of the Louisiana Mineral Code (R.S. 31:136) declares that "If a mineral lessor seeks relief from his lessee arising from drainage of the property leased or from any other claim that the lessee has failed to develop and operate the property leased as a prudent operator, he must give his lessee written notice of the asserted breach to perform and allow a reasonable time for performance by the lessee as a prerequisite to a judicial demand for damages or dissolution of the lease." The Comment to this article indicates that it was intended thereby to overrule *Williams v. Humble Oil & Refining Co.,* insofar as it relieves the lessor of the necessity to put the lessee in default. On the operation of Article 136, see generally Keith B. Hall, *Implied Covenants: Claims under Article 122,* 57th Ann. Inst. on Min. Law 172 (2010).

SECTION 4. OTHER IMPLIED COVENANTS

(*See* Martin & Kramer, *Williams and Meyers Oil and Gas Law* §§ 853–861.5).

McDOWELL v. PG & E RESOURCES CO..

Court of Appeal of Louisiana, Second Circuit, 1995.

658 So.2d 779, 136 O.&G.R. 320, writ denied, 661 So.2d 1382.

HIGHTOWER, JUDGE. This is an appeal from a judgment cancelling two oil and gas leases after the district court concluded that the lessees breached the implied covenant to diligently market production. We reverse.

Facts

By two separate contracts in 1978 and 1980, W. Howard McDowell, mineral rights owner, granted another party oil and gas leases involving approximately 133 acres in Jackson Parish. Plaintiffs ("the McDowells") are successors to the original lessor, while defendants presently own all the leasehold or working interests.

The McDowell No. 1 Well, drilled on one of the leased tracts, serves as the unit well for a 640-acre gas unit designated by the Louisiana Office of Conservation and encompassing all lands covered by both leases. One of the defendants, PG & E Resources Company ("Resources"), has been the operator of that well since August 1989.

Prior to March 1990, Resources and its predecessor operator combined the "wet gas" produced by the McDowell well with "dry gas" from the Breedlove No. 1 Well, located on lands not covered by the leases. This mixture met the quality standards necessary for transmission of the McDowell gas through a pipeline operated by United Gas Pipeline Company ("United Gas") for ultimate sale. In early 1990, however, when Breedlove stopped producing, United Gas refused to accept the unmixed wet gas from McDowell.

Thus, in March 1990, unable to transport and sell the McDowell gas, the operator faced a shut-in well predicament.[70] Thereafter, Resources endeavored to reestablish a market through steps that included:

- Pursuing repeated requests for United Gas to grant "liquid exceptions" that would allow transmission to be resumed through that company's pipeline.

- Striving to secure dry gas to mix, as before, with McDowell's wet gas, by engaging in reworking operations on Breedlove No. 1, from March 9, 1990 until April 3, 1990 (costing approximately $25,000), and, thereafter, recompletion operations on that same well until July 10, 1990 (costing $25,287).

- For that same purpose, on September 2, 1990, spudding Breedlove No. 2 (after staking the well in May, and then encountering site and timber removal problems) to eventually result in the abandonment of a dry hole at the end of October 1990 (costing $273,532).[71]

---

[70] In January (1991) of the year immediately succeeding the shut-in of the well, Resources submitted shut-in royalties to the lessors, all in accordance with the lease.

[71] In addition to the sizeable expenditures made, trial testimony reflects the considerable confidence of Resources personnel that they could reestablish production from Breedlove. Likewise, in light of a substantial earlier investment in the valuable McDowell well, Resources wanted McDowell back "on stream" as soon as possible. Testimony also indicates that, absent an effort first to reestablish Breedlove and thus save the cost of another pipeline, Resources arguably would not have acted in the best interest of the lessors or

- Concurrently with these activities, contacting two gas carriers, Crystal Oil Company ("Crystal") and Tex/Con, about purchase arrangements despite the existence of a "buyer's market" in 1990.

- By August 1990, contemplating an agreement to sell gas from the McDowell Well and another well, Ferguson, to Tex/Con.

- On September 4, 1990, agreeing to sell only the Ferguson gas to Tex/Con because, if the then-ongoing drilling operations for Breedlove No. 2 proved successful, no sale of McDowell gas to Tex/Con with the attendant pipeline construction expense would have been necessary.

- When the Breedlove No. 2 drilling proved unsuccessful, immediately undertaking to: (a) complete negotiations for the sale of McDowell gas to Tex/Con; (b) secure necessary pipeline rights-of-way; and (c) build the required pipeline. (Testimony indicates the Tex/Con price to have been more attractive than the Crystal offer.)

- By December 4, 1990, reaching agreement for a three-well sale of gas (including McDowell) to Tex/Con.

- On April 28, 1991, placing the McDowell Well back on line, after construction of the pipeline (costing $82,236).

In May 1991, after actual production resumed, the McDowells again began receiving royalties on the minerals sold from their property.

About one month prior to the resumption of production, however, the McDowells executed in favor of Jim C. Shows, Ltd., another lease on the same lands covered by the 1978 and 1980 contracts. In March and May 1991, considering the prior agreements no longer valid, the new lessee mailed letters demanding a release of defendants' contracts. When Resources refused to comply, the McDowells brought suit seeking a judicial declaration that, as a result of a 90-day cessation of production, the two older leases had expired by their own terms. Defendants countered that these instruments remained in effect by virtue of force majeure provisions and the payment of shut-in royalties. After trial, upon discerning a breach of the implied covenant to market diligently as envisioned by Article 122 of the Louisiana Mineral Code, *see* LSA-R.S. 31:122, the district court ordered the two leases cancelled. Defendants now appeal . . . .

Discussion

I. Did the Leases Expire "By Their Own Terms?"

Plaintiffs instituted suit asserting that the leases "expired by their own terms" under Paragraph 6 thereof. Careful reading of the agreement, however, does not support that proposition.

Paragraph 5 provides that, in a shut-in or force majeure situation, with the payment of shut-in royalties, the lease continues in effect during such shut-in

---

lessees of McDowell.

period "as though production were actually being obtained. . . ." The trial court found that, around March 11, 1990, when United Gas Pipeline refused to accept the "wet" gas from McDowell, a shut-in situation came into existence. Defendants, as stated, thereafter paid shut-in royalties in accordance with the contract.

In brief, plaintiffs contend that the gas purchase offers, eventually secured from Crystal and Tex/Con, contravene a finding that no market existed and preclude application of the shut-in clause. This argument, however, ignores the explicit language of the contract. According to Paragraph 5, absent the availability of a pipeline, only a market "at the well" would negate a shut-in situation.[72]

Essentially, the lease contemplates the continuous need for transporting gas from the well-head site. Thus, the wording of Paragraph 5 comports with the general concept that a shut-in provision serves to address equitably the transportation difficulties inherent in gas marketing, and, also, to balance the competing interests of the contracting parties. See Davis v. Laster, 242 La. 735, 138 So.2d 558 (1962) . . .. Basically, because natural gas ordinarily cannot be stored upon production, a pipeline provides the only economic means of transportation. . . .

Both preliminary gas purchase offers (from Crystal and Tex/Con) would have necessitated that the operator construct a pipeline, and, thus, did not end the shut-in situation. Before that could occur, under Paragraph 5, the "market" had to be brought to the well. Hence, the mere existence of potential buyers within the involved area did not suffice. Moreover, Resources' efforts to connect the nearby markets to the well, including, e.g., its attempts to reinstitute the Breedlove mixing process, are more appropriately evaluated by examining the reasonableness of its business decisions in light of Article 122's implied obligation of diligent marketing. Cf. Williams and Meyers, Oil and Gas Law, § 632.4 (1993); Forbis, *The Shut-In Royalty Clause: Balancing the Interests of Lessors and Lessees*, 67 Tex.L.Rev. 1129 (1989).

In its reasons for judgment, the trial court referred to the 90-day cessation clause enunciated in Paragraph 6 of the lease.[73] That provision applies,

---

[72] In pertinent part, Paragraph 5 of the lease states:

If Lessee obtains production of minerals on said land or on land with which the leased premises or any portion thereof has been pooled, and if, during the life of this lease either before or after the expiration of the primary term, all such production is shut in by reason of force majeure or the lack either of *a market at the well* or wells or of an available pipeline outlet in the field, this lease shall not terminate but shall continue in effect during such shut-in period as though production were actually being obtained on the premises . . ., and [at indicated times] Lessee shall pay or tender . . . to the royalty owners [the appropriate shut-in payment]. [Emphasis added by the court.]

[73] In pertinent part, Paragraph 6 of the lease states:

This lease will continue in full force and effect within or beyond the primary term as

however, only "if production previously secured should cease for any cause. . . ." Very importantly, in a Paragraph 5 shut-in situation, production does not "cease" but continues constructively. See Davis, *supra* (observing that the very purpose of a shut-in royalty clause is to maintain the lease as though the gas had not ceased flowing). Thus, the 90-day cessation of production provision never applied "by its own terms."

Neither can the Paragraph 6 provision be "transposed" or "imported" into Paragraph 5 to provide arbitrarily a 90-day limitation with respect to shut-in or force majeure situations. . . .

II. Did Resources Breach the Implied Covenant of Diligent Marketing?

The trial judge's decision did not center upon a breach of Paragraph 6, but instead essentially relied upon the diligent marketing covenant implied in all oil and gas leases. In any consideration of that general duty, the issue evolves into matters of factual circumstances, reasonable and diligent efforts, industry standards, prudent operations, etc. . . . . There are few cases dealing with this obligation in Louisiana.

First of all, we agree with defendants' well-stated position that the trial court should never have reached the implied covenant question. To cancel a lease for breach of an implied covenant, the plaintiff is required to place the defendant formally in default prior to seeking judicial intervention. . . . Such a "putting in default" never occurred in the present case. Nor is this merely a preliminary matter to be addressed by exception prior to answer. Instead, without a putting in default, no cause of action is disclosed. . . . This is especially true here, where plaintiffs' petition did not rely upon the breach of an implied covenant, thus presenting defendants with no reason, certainly no reason before answer, to raise a "putting in default" objection.

Plaintiffs' pleadings and the new lessee's demand letters merely sought cancellation and release of the leases as having expired "by their own terms." These, or similar devices, cannot substitute for a placing in default with respect to the alleged breach of an implied duty. . . . . Furthermore, the default requirement is designed to serve two purposes in the present context: (1) To provide notice that the lessor considers the lessee's actions (or inaction) as violative of the implied obligation to market, and (2) to afford the lessee a reasonable opportunity to perform that obligation. Id. Yet here, before the March demand letter, Resources had signed the gas purchase contract with

---

long as any mineral is produced from said land hereunder or from land pooled therewith. . . . [I]f *production* previously secured *should cease* from any cause after the expiration of the primary term, this lease shall remain in force so long thereafter as Lessee either (a) is engaged in drilling operations or reworking operations with no cessation between operations or between such cessation of production and additional operations of more than ninety consecutive days; or (b) is producing oil, gas, sulphur or other mineral from said land hereunder or from land pooled therewith. [Emphasis added by the court.]

Tex/Con and had measures well under way toward constructing the necessary pipeline. And, of course, before the May letter, production had actually resumed.

Even approaching the diligent marketing issue on its merits, however, the record before us still does not disclose a breach of the implied covenant. Certainly, there is no authority for simply imposing the 90-day cessation period from Paragraph 6, as did the trial judge, to fix a precise time span within which a lessee must effectuate a marketing relationship. Nor does any of the testimony concern industry standards as to what reasonably should be expected under the circumstances shown.

Under LSA-R.S. 31:122, a lessee's conduct will be evaluated by what is expected of ordinary persons of ordinary prudence under similar circumstances and conditions, having due regard for the interest of both contracting parties. . . . As explained by Kramer and Pearson, *The Implied Marketing Covenant in Oil and Gas Leases: Some Needed Changes for the 80's*, 46 La.L.Rev. 787 (1986), the most that can be required of a lessee (even if payment of shut-in royalties cannot hold a lease indefinitely) is an effort to market gas within a reasonable time, i.e., the lessee must conduct himself as a reasonable and prudent lessee. Also, where the interests of the lessor and the lessee are aligned, as here, the greatest possible leeway should be extended to the lessee in his decisions about marketing gas. Williams and Meyers, Oil and Gas Law, § 853, et seq. (1993).

Thus, what constitutes a reasonable time for obtaining a market, of necessity, will be measured by the operator's exercise of due diligence under the circumstances. See Acquisitions, Inc., supra; Bristol, supra; cf. Lelong, supra; Risinger, supra. Here, following the shut-in situation that developed in March 1990, Resources instituted continuous and simultaneous efforts to reestablish a market. Notwithstanding plaintiffs' arguments about self-dealing, had the reworking or recompletion of Breedlove No. 1 or the drilling of Breedlove No. 2 proved successful, McDowell gas sales could have immediately recommenced without the cost of a pipeline. Over the thirteen months in question, Resources never stopped its endeavors to resume such sales, and even pursued several avenues at great expense. Evaluating these actions in light of the existing circumstances, we cannot conclude that the operator did not act for the mutual benefit of both the lessor and the lessee. Neither can we discover any breach of the reasonably prudent operator standard. . . .

Cancellation obviously is a harsh remedy. . . . Consequently, the right to dissolve a mineral lease will be subject to judicial control according to the factual circumstances of each case. . . . In order to cancel a mineral lease the breach must be shown to be substantial. . . . As observed by Kramer and Pearson, *supra*, at 825:

For breaches of marketing that are essentially challenges to the reasonableness of the actual marketing decision, . . . only "under extraordinary circumstances" should the lessor be entitled to a decree of cancellation. . . . [C]ancellation . . . would be too harsh a penalty for a mistake in business judgment.

In the present matter, plaintiffs have simply not shown facts warranting a cancellation. Neither is any self-dealing or fraud involved so as to call for "punishing" a wrongdoer. To the contrary, as demonstrated by the record, third parties are actually paying for this litigation in an obvious effort to supplant Resources and, in the process, inflict a substantial loss on the lessees.

Conclusion

For the reasons assigned, the district court judgment is reversed and set aside, and plaintiffs' demands are now dismissed. All costs, here and below, are assessed to plaintiffs-appellees.

BROWN, JUDGE, dissenting.

In an oil and gas lease, a lessee undertakes to manage and develop the property for the mutual benefit of himself and the lessor. Frequently, the interests of the lessee conflicts with those of his lessor. In these circumstances, the lessee must exercise its power in fairness and good faith. . . .

The primary purpose of an oil and gas lease is clearly to obtain production and income. Obviously, the lessee has an obligation to market the product once discovered. In fact, the lessee has a duty to obtain the best price possible. Frey v. Amoco Production Co., 603 So.2d 166 (La.1992). Ordinarily, the interests of the lessee and lessor coincide. Where the interests of the two diverge, as in this case, and the lessee lacks incentive to diligently market the gas, his business judgment must be questioned.

The majority opinion is solely premised on the factual conclusion that the shut-in provisions of the leases were applicable because there was no market "at the well." Without such a shut-in situation, the leases automatically expired with the cessation of production for 90 days. Thus, I will address the application of the shut-in clause based on the loss of market at the wellhead.

Prior to March 1990, the McDowell gas was either sold to United Gas, transported through the United Gas pipeline for sale to others, or transported through the gathering system of Crystal for sale to Highland Energy Company or others. Testimony given by a Resources manager of marketing, Bobby R. McAlpin, indicated that at the time the Breedlove well failed, United Gas was merely transporting the McDowell gas to the ultimate buyer. Thus, the key event alleged to have extinguished the market was the loss of the means to transmit the gas; however, other means for transmitting the gas were quickly identified. The Breedlove well stopped producing sometime between January and March 1990. In March 1990, United Gas informed Resources that the gas from the McDowell well did not meet the company's quality specifications. United Gas refused to accept any further production from the McDowell well. Thus, production ceased from the McDowell well.

Resources chose to solve the problem by reestablishing the dry gas production from the Breedlove well. Efforts to rework the Breedlove well commenced immediately. Resources simultaneously searched for a carrier to accept the McDowell gas in its "wet" state. On April 24, 1990, Crystal Oil Company ("Crystal") tentatively agreed to accept the McDowell gas. Resources chose, however, not to respond to the Crystal offer, hoping that the negotiations then underway with Tex/Con, another gas carrier, would produce a better price.

Efforts to rework the Breedlove well from March to June 1990 failed. Thereafter, Resources immediately began the more elaborate task of recompleting the Breedlove well. About the same time, Resources began the research necessary to identify landowners over whose property a pipeline to Tex/Con would have to be built if negotiations with that carrier were successful. In August 1990, Tex/Con tentatively agreed to accept production from various Resource wells, including the McDowell well. When an agreement with Tex/Con was finalized, however, Resources had withdrawn the McDowell well. Breedlove recompletion efforts failed in September 1990 and the Breedlove No. 1 well was abandoned. Resources chose to drill a second well on the Breedlove tract. This second well resulted in a dry hole and work on the Breedlove No. 2 well was abandoned in October 1990. The date of withdrawal of the McDowell gas from the Tex/Con agreement coincided with the commencement of drilling of the Breedlove No. 2 well.

When the second Breedlove well failed, defendant began to take the steps necessary to complete an agreement with Tex/Con to transmit the production from the McDowell well. Resources negotiated with landowners concerning rights-of-way from the McDowell well site to Tex/Con's pipeline and approved the purchase of these rights in December 1990. The numerous working interest holders, however, did not grant their approval until mid-February 1991. The rights-of-way were acquired, the feeder pipeline was built and the McDowell well resumed production in April 1991.

Royalty checks to plaintiffs had stopped at the end of May 1990. Thereafter, plaintiffs tried to contact Resources to learn why the payments had stopped; however, Resources failed to respond. When Resources tendered shut-in royalty checks in January 1991, plaintiffs again sought an explanation. In February 1991, Resources responded, explaining that it considered the wells shut-in under the terms of the leases.

Evidence shows that Resources initially pursued and remained aware of alternative means to transmit the McDowell gas to an ultimate buyer. Resources chose to devote its efforts to restoring production in an unrelated oil and gas unit in another field which benefited only Resources' interest rather than to avail itself of the means to readily restore the McDowell well to a producing status.

Resources had no economic incentive to immediately hook up the McDowell well through a feeder pipeline to Tex/Con. Resources' interest was in reestablishing production on its Breedlove lease.

Under these circumstances there was a market at the well site and the shut-in provisions were not applicable. . . .

NOTES

1. Should the implied covenant to market be governed by the prudent operator standard or one that demands less of the lessee? *Compare* Kramer & Pearson, *The Implied Marketing Covenant in Oil and Gas Leases: Some Needed Changes for the 80's*, 46 La.L.Rev. 787 (1986) *and* Weaver, *Implied Covenants in Oil & Gas Law Under Federal Energy Price Regulation*, 34 Vand.L.Rev. 1473, 1515–19 (1981) *with* Scott Lansdown, *The Implied Marketing Covenant in Oil And Gas Leases: The Producer's Perspective*, 31 St. Mary's L.J. 297 (2000) (arguing that the implied marketing covenant has been misapplied with negative impacts on the oil and gas industry), and Martin, *A Modern Look at Implied Covenants to Explore, Develop, and Market under Mineral Leases*, 27th Sw. Legal Fdn. Oil & Gas Inst. 177, 190–92 (1976). Consider also this statement from ROBBINS V. CHEVRON U.S.A., INC., 246 Kan. 125, 785 P.2d 1010, 108 O.&G.R. 42:

> It is not the place of courts, or lessors, to examine in hindsight the business decisions of a gas producer. One learned treatise on the subject, 5 Williams & Meyers, Oil and Gas Law § 856.3 (1989), states:
>
> > The greatest possible leeway should be indulged the lessee in his decisions about marketing gas, assuming no conflict of interest between lessor and lessee. Ordinarily, the interests of the lessor and lessee will coincide; the lessee will have everything to gain and nothing to lose by selling the product. p. 411.
>
> The treatise cautions against second-guessing an operator's marketing decision:
>
> > There is great risk that close judicial supervision of the lessee's conduct in selling gas will inhibit his exercise of his best judgment to the detriment of both landowner and operator. Scrutiny of lessee's actions by judges (or, worse, juries) in the light of *after-acquired* knowledge will tend to encourage the operator to take the least hazardous and perhaps least profitable course of action. It is unnecessary to impose this conservatism on the operator when his interest in selling the gas is fully identified with that of his landowner. pp. 412–13. (Emphasis supplied by court. Eds.)

2. To what extent is the lessee obliged to incur additional expense in order to market the product, as by constructing a pipeline, installing pumps or a booster plant? *See* Rhoads Drilling Co. v. Allred, 123 Tex. 229, 70 S.W.2d 576 (Tex.Comm.App.1934).

3. In much of the 1980s, a large portion of the gas supply was committed to contracts in which the price to be paid by the pipeline-purchaser was too high for

the purchaser to resell. Suppose a lessee agrees to reduce the price in one contract in exchange for the pipeline's explicit agreement to buy gas produced by the lessee from other leaseholds? Suppose the lessee agrees to a price reduction in the expectation that the pipeline will be more ready to accommodate him in possible future transactions relating to other leaseholds? *See* Amoco Production Co. v. First Baptist Church, 579 S.W.2d 280, 67 O.&G.R. 568 (Tex.Civ.App.1979), *error ref'd n.r.e.*, 611 S.W.2d 610 (1980).

4. Does an express shut-in-gas royalty clause negate an implied covenant to market gas? In addition to the treatment of the issue in the principal case, *see* Davis v. Cramer, 808 P.2d 358, 113 O.&G.R. 201 (Colo.1991); Berry Energy Consultants & Managers, Inc. v. Bennett, 175 W.Va. 92, 331 S.E.2d 823, 85 O.&G.R. 624 (1985). What would the plaintiffs in the principal case need to show on remand to prevail on this point?

5. Can the lessor's signing of a division order limit the lessee's obligations under the marketing covenant? *See* Cabot Corp. v. Brown, 754 S.W.2d 104, 99 O.&G.R. 154 (Tex.1987).

6. What is the relationship between the royalty clause and the implied covenant to market? This question was partly taken up in Chapter 3 but other questions might be raised. If the royalty clause provides expressly that the lessee will pay royalty on the basis of the proceeds received, and the lessee markets the gas and pays on the basis of the proceeds, can the lessor nevertheless complain that the proceeds received should have been higher? What if the lessee makes a contract with a gas purchaser in advance, then acquires a lease that will be subject to that gas purchase contract? Is there a duty to disclose such an arrangement to the lessor? On these questions, *see* Ainsworth v. Callon Petroleum Co., 521 So.2d 1272, 99 O.&G.R. 320 (Miss.1987); Diamond Shamrock Corp. v. Harris, 681 S.W.2d 317, 84 O.&G.R. 13 (Ark. 1984). *See also*, Williamson v. Elf Aquitaine, Inc., 138 F.3d 546, 139 O.&G.R. 64 (5th Cir. 1998)(well established in Mississippi that implied covenants are inapplicable when an issue is expressly covered by the language in a lease).

Return also to Chapter 3 for the treatment of the implied duty to market in *Tawney* at page 369 *supra* (quoting from *Wellman*) and in the Colorado cases. The court in GARMAN V. CONOCO, INC., 886 P.2d 652, 659, 132 O.&G.R. 488 (Colo. 1994) stated:

> Conoco argues that the implied covenant to market exists separately from the allocation of marketing costs. We disagree. Implied lease covenants related to operations typically impose a duty on the oil and gas lessee. Accordingly, the lessee bears the costs of ensuring compliance with these promises. The purpose of an oil and gas lease could hardly be effected if the implied covenant to drill obligated the lessor to pay for his proportionate share of drilling costs. In our view the implied covenant to market obligates the lessee to incur those post-production costs necessary to place gas in a condition acceptable for market.

Consider the comments on *Garman's* employment of the implied covenant to market to transform Colorado law by one of the editors of this book:

> Most remarkably, the Colorado Supreme Court, in *Garman v. Conoco, Inc.,* did not even believe that the language creating the royalty obligation was at all relevant to define that obligation. The language of the assignment creating the overriding royalty interest was not deemed to be a critical or even necessary factor because the court was going to rely on extrinsic factors to create the obligation on behalf of the lessee to place all production into a marketable condition or form. Thus, the court's sense of justice and fairness led it to apply the implied covenant to market as the defining principle for determining the lessee's royalty obligation. The court did note that once the natural gas was placed into a marketable condition, post-marketable condition expenses incurred that enhance the value of the natural gas may be used in a netback methodology to calculate the royalty owed. . . .
>
> Cases applying the extrinsic approach to the "at the well" issue change the question from "what are production costs?" and "what are post-production costs?" to "when is a marketable product achieved?" Most of the courts that have adopted the extrinsic approach have not provided a clear definition of a marketable product. . . . As stated elsewhere, "It appears that the Colorado Supreme Court has done nothing less than fashion a new rule for the purpose of enhancing royalty values throughout Colorado." Extrinsic factors, namely the relationship between the lessee and the lessor as defined by the court, without regard to the language of the lease, now control the royalty obligation.

Bruce M. Kramer, *Interpreting the Royalty Obligation by Looking at the Express Language: What a Novel Idea?*, 35 Tex. Tech L. Rev. 223 (2004)

7. Suppose that the lease shut-in royalty clause provides "If Lessee obtains production of minerals on said land and if during the life of this lease either before or after the expiration of the primary term, all such production is shut in by reason of force majeure or the lack either of a market at the well or wells or of an available pipeline outlet in the field, this lease shall not terminate but shall continue in effect during such shut-in period as though production were actually being obtained on the premises . . ." if specified shut-in royalty payments are made. May the lessee shut in a producing well because the market is no longer a favorable one and the lessee reasonably believes that a better price can be obtained a year or two years later? That is to say, a market may be available but a better market will be available if the lessee waits to produce the gas. To put the question another way, is there a right to market in the manner of a reasonable prudent operator that would cause us to interpret "lack of a market" as "lack of a favorable market"? For some guidance in a case where there was no shut-in royalty clause *see* Gazin v. Pan American Petroleum Corp., 367 P.2d 1010, 16 O.&G.R. 1009 (Okl.1961).

8. What implied duties does a lessee have to share information with its lessors or to represent them in law suits that might affect the lessors' interests? Consider the exceptional case in which the lessor showed substantial drainage, but encountered difficulty in receiving damages because of judicial doubts about the scope or existence of implied covenant duties, HECI EXPLORATION CO. V. NEEL, 982 S.W.2d 881, 143 O.&G.R. 142 (Tex. 1998), The lessee (HECI Exploration) learned that another company (AOP) was producing excessively from a well on adjacent acreage. The lessee complained to the Railroad Commission three times and secured regulatory action to curb the wrongful acts of the adjacent producer. That producer nevertheless continued to overproduce. The lessee brought suit against the adjacent producer alleging that the overproduction damaged the reservoir and resulted in lost reserves in the lessee's well on the lessor's tract. The court in that suit granted the lessee permanent injunctive relief against the adjacent producer, and a jury awarded the lessee $1,719,956 actual damages and $2,000,000 punitive damages. The lessee never informed the lessors (Neels) of the overproduction, did not notify them of the legal action, and did not pay them a share of their recovery. The lessors did not learn of their lessee's suit against the adjacent producer for lost production until May 1993. The plaintiffs then brought a claim for royalty on the recovery had by their lessee. The court of appeals ruled that the lessors did not have a claim for royalty on the money obtained by their lessee. The court of appeals said, however, that the royalty owners could have a claim for damages based either on a theory of unjust enrichment or a breach of a duty/implied covenant to notify the lessors of a claim that they could have against the adjacent producer. The court observed:

> A lessee who recovers for the entire reduction in reservoir value due to the neighboring producer's misdeeds without having to carve out a royalty share of production profits at the royalty owner's expense. A lessee who recovers for the damage to its own interests but does nothing regarding the lessor's interest leaves the injury to that portion of the leasehold unredressed.

The supreme court reversed on both theories. The supreme court initially divided the duty/implied covenant cause of action into two separate issues--(1) is there a duty to notify the lessor of the need for suit, and (2) is there a duty to notify the lessor of the lessee's intent to sue. The court did not answer the question of whether there is an implied covenant to notify of the need to sue because it found that that cause of action was barred by limitations since the discovery rule did not apply. The court did, however, find that there was no implied covenant to notify of the lessee's intent to sue. In rejecting this covenant, the court made the following observations about implied covenants:

> This Court has not lightly implied covenants in mineral leases. ... Our decisions have repeatedly emphasized that courts "cannot make contracts for [the] parties." A covenant will not be implied unless it appears from

the express terms of the contract that "it was so clearly within the contemplation of the parties that they deemed it unnecessary to do so, or it must appear that it is necessary to infer such a covenant in order to effectuate the full purpose of the contract as a whole as gathered from the written instrument." ... A court cannot imply a covenant to achieve what it believes to be a fair contract or to remedy an unwise or improvident contract. ...

The necessity requirement was not met because the litigation commenced by HECI Exploration would not be binding on the Neels since they had an independent cause of action. The Neels would not be collaterally estopped from filing their own action since they were neither parties nor privies to the HECI Exploration action. Finally, the supreme court found that HECI Exploration was not unjustly enriched, even if the damage recovery included the injury to the entire reservoir interest, not just its own. The court rejected the unjust enrichment claim and observed:

> The fact that HECI may have recovered more than it was entitled is a matter about which AOP could have complained in the suit against it, but that does not give rise to a cause of action for unjust enrichment in favor of the Neels. HECI and the Neels had independent causes of action against AOP. Notwithstanding any excess recovery by HECI, the Neels would have been entitled to recover from AOP the full amount of damage attributable to their interest had they timely sued AOP.

The court did not deal with the impact of the implied covenant to protect against drainage on the factual circumstances in *Neel*. It is reasonably clear that HECI Exploration had a duty under the implied covenant to prevent drainage to stop the illegal production by the adjacent lessee. But the duty was not extended to sharing in the benefits received by fulfilling the implied covenant obligations.

9. If the lessee sells gas at the maximum lawful price under federal or state regulation, can the lessee later be held liable for breach of the marketing covenant for not including a provision in the gas purchase contract for price redetermination upon deregulation of the price? *See* Davis v. CIG Exploration, Inc., 789 F.2d 328, 92 O.&G.R. 631 (5th Cir.1986)

---

5 Martin & Kramer, *Williams and Meyers Oil and Gas Law*

§ 861. Implied Covenant to Conduct Operations With Reasonable Care and Due Diligence: Introduction

The implied covenant to conduct operations on and manage the leasehold estate with reasonable care and due diligence is a catchall obligation covering those acts or omissions not comprehended by the more specific implied covenants dealt with in the preceding sections. It is merely another formulation of the general duty of the lessee to act in such manner as to accomplish the purposes of the lease agreement. Four types of disputes fall within the ambit of the covenant . . .: [(1) claims that operations on the land have been carelessly conducted causing damage to the royalty interest; (2) claims that premature abandonment of the lease has damaged the royalty interest; (3) claims that lessee has failed to maximize the recovery from the land by using advanced production techniques; (4) claims that the lessee failed to seek favorable action by the regulatory commission that would have benefited the royalty interest; and (5) claims that lessee as holder of multiple leases failed to produce a fair share from each.]

———

For example, a court found the lessee in breach of its implied covenants on account of its failure to employ fire flooding. *See* Waseco Chemical & Supply Co. v. Bayou State Oil Corp., 371 So.2d 305 (La.App.1979), *writ denied*, 374 So.2d 656, 65 O.&G.R. 351 (1979). In Sinclair Oil & Gas Co. v. Bishop, 1967 OK 167, 441 P.2d 436, 30 O.&G.R. 614, the lessor claimed that the lessee should have produced oil from a shut-in well on the lease to prevent drainage by neighboring wells that were producing oil. The lessee asserted as a defense that the reservoir was primarily gas and that it would have to flare much gas to produce any oil; thus, it was not acting imprudently by waiting for a market for the gas before producing oil. The court agreed with the lessee that it was not acting imprudently in not producing but that to prevent drainage the lessee should have gone to the Corporation Commission to stop the wasteful flaring of gas on the adjacent land; there was an implied covenant to protect the leasehold that was to be fulfilled administratively. *See generally,* Donohoe, *Implied Covenants in Oil and Gas Leases and Conservation Practice*, 33 Sw. Legal Fdn. Oil & Gas Inst. 97 (1982).

———

Return for a moment to the rationale given for the implication of covenants at the beginning of this chapter. Are they implied in law or in fact? Are they the result of the relationship between the parties? Are they a device to control opportunistic behavior? Now consider whether the implied covenants approach should be extended to other areas of oil and gas law. For example, consider whether the purchaser of gas under a long-term contract owes implied obligations to the gas producer. *See* Williams Natural Gas Co. v. Amoco Production Co., 1991 WL 58387 (Del.Ch., 1991). Do the parties to a joint operating agreement owe implied obligations to one another regarding development of the contract area? *See* Texstar North America, Inc. v. Ladd Petroleum Corp., 809 S.W.2d 672, 117 O.&G.R. 376 (Tex.App.1991). Does the operator of a compulsory unit owe implied covenant duties to the non-operators? *See* Parkin v. State Corporation Commission, 234 Kan. 994, 677 P.2d 991, 80 O.&G.R. 39 (1984).

———

SECTION 5. REMEDIES FOR BREACH OF EXPRESS DRILLING AGREEMENTS

(*See* Martin & Kramer, *Williams and Meyers Oil and Gas Law* §§ 883–885.5).

This section is included in a chapter on implied covenants because (1) express covenants often provide a substitute for implied covenants and (2) doctrinally, the problems of remedies for breach of express and implied covenants are closely related.

JOYCE v. WYANT

United States Court of Appeals, Sixth Circuit, 1953.
202 F.2d 863, 2 O.&G.R. 693.

MILLER, CIRCUIT JUDGE. Appellant filed this action in the district court to recover damages resulting from the failure of the appellees, assignees of the lessee under an oil lease from the appellant covering land located in Caddo Parish, Louisiana, to drill three wells on the leased property. The District Judge sustained the motion of the appellees to dismiss the complaint for failure to state a cause of action, 105 F.Supp. 979, from which ruling this appeal was taken.

The material provisions of the lease which give rise to the present controversy are as follows:

Lessor in consideration of Lessee's obligation to drill four wells on the premises herein leased as hereinafter set forth, of the royalties hereinafter provided, and of the agreements of Lessee hereinafter contained, hereby grants, leases and lets unto Lessee for the purpose of exploring, prospecting, drilling and mining for and producing oil, gas and other minerals . . ., the following described land . . .:

The term of this lease is 60 days from date hereof and as long thereafter as Lessee complies with his obligations hereunder, and he produces oil, gas or other minerals in paying quantities from the leased premises.

The royalties to be paid by Lessee are: . . .

The actual consideration for this lease is that Lessee hereby obligates himself to drill four wells on the leased premises at such locations as he might choose; . . .

Lessee shall begin actual drilling of a well, hereinafter referred to as "initial well" on the leased premises within 60 days from date hereof. . . . Following completion of the "initial well," whether producer or a dry hole, Lessee shall in succession drill . . . three other wells on the leased premises within 60–day intervals following the completion of the immediately preceding well drilled by Lessee hereunder. . . .

Lessee shall have the right at any time during or after the expiration of this lease to remove all property and fixtures placed by Lessee on said land, including the right to draw and remove casing. . . .

In case of cancellation or termination of this lease for any cause, Lessee shall have the right to retain under the terms hereof ten acres of land around each well producing such tract to be designated by Lessee in as near a square form as practicable.

The complaint set out the execution of the lease on July 6, 1950, the completion of the initial well on or about August 22, 1950, and the failure of the appellees up to November 6, 1950, to begin the drilling of a second well and their refusal thereafter to drill the second well, on the ground that the initial well produced insufficient oil to make further operations profitable.

In support of the motion to dismiss, appellees contended that the lease expired by its terms following the expiration of 60 days after the completion of the first well, during which time there was no production of oil, gas or other minerals in paying quantities from the leased premises, with the result that there was neither the right nor the obligation upon their part to drill the remaining three wells. Appellant contended that there was a contract obligation on the part of the appellees to drill the remaining three wells, even though the first well proved unproductive. The trial judge upheld the contention of the appellees. For a more detailed statement of the lease provisions and the reasons of the District Judge for sustaining the motion, *see* Joyce v. Wyant, D.C., 105 F.Supp. 979.

It appears to be the well settled rule in oil and gas leases that a lease containing a definite primary term, with the provision that the lease will terminate at the end of such term *unless* the lessee or his assignee performs some additional act provided by the lease, such as the payment of money or additional drilling, does not place a binding duty upon the lessee to do anything. Such a lease is usually described as an "unless" lease. The "unless" clause is regarded as a limitation on the lessee's estate or the period of the grant. Unless the lessee performs the additional act, which he is not obligated to perform, the lease automatically terminates at the expiration of the primary grant. Such a lease is distinguished from the so-called "or" lease where the lessee is obligated either to drill a well or pay rental, and can be held in default upon failure to do so. . . .

The lease under consideration in this case provides for a definite primary term of 60 days, without use of either the "or" provision or the "unless" provision with respect to an extension thereafter. Instead it uses the phrase "and as long thereafter as Lessee" performs certain acts. The District Judge construed it as an "unless" lease. Appellant contends that it is not an "unless" lease, relying chiefly on the fact that the term "unless" is not used. The contention has but little merit. The District Judge recognized that it was not the typical "unless" lease, in that the term "unless" was not used, but held that it fell within that classification in that the legal effect of the lessee's failure to drill within the prescribed time was the same as under the usual form of an "unless" lease. . . .

Appellant contends, however, that the habendum or term clause, which was given strong consideration by the District Judge, relying partly upon J. J. Fagan & Co. v. Burns, 247 Mich. 674, 679, 226 N.W. 653, 67 A.L.R. 522, is in conflict with other provisions of the lease which state that the actual consideration for the lease is the lessee's obligation to drill four wells and which require the other three wells to be drilled even though the initial well proves to be a dry hole. Appellees attempt to reconcile the apparent inconsistency. If such an inconsistency exists it results in requiring a construction of the lease in the light of all of its provisions. The logical way to have provided for the effect now contended for by appellant would have been to have made the primary term of sufficient length to cover the sixty-day intervals between the successive drillings. The relatively short primary term of 60 days, during which only one well was required to be drilled, the provisions pertaining to termination of the lease without any provision for liability on the part of the lessee for failure to drill the remaining wells, the right in the lessee to remove all property placed by him on the land, including the removal of casing, at any time during the lease, and his right to retain ten acres of land around a producing well in the event of the cancellation of the lease "for any cause" lead us to the same conclusion as was reached by the District Judge. As pointed out in Logan v. Tholl Oil Co., *supra* [189 La. 645, 180 So. 475], where the lease contained an "as long as" clause, the main consideration for such a lease is the development of the land for oil, gas and the obligation on the part of the lessee is either to develop with reasonable diligence or give up the lease.

The District Judge based his ruling partly upon the case of Fogle v. Feasel, 201 La. 899, 10 So.2d 695, where the factual situation was very similar to the present case. Appellant contends that the ruling is more closely controlled by the case of Fite v. Miller, 192 La. 229, 187 So. 650, 122 A.L.R. 446, and 196 La. 876, 200 So. 285.

While the ruling in Fogle v. Feasel, *supra,* was based upon two separate grounds, one of which is not applicable to the present case, it is nevertheless also an authority on the other point, which is the point in issue here. . . .

We recognize that there were certain forfeiture provisions in the lease in that case, upon which the Court relied, which are not present in this case. That factual difference does not invalidate the rule which the Court relied upon in reaching its conclusion, namely, that the various provisions of the lease, considered in their entirety, indicated that the only penalty contemplated by the parties for the violation of any of the provisions of the lease was that in such event the lease was to become null and void. In our opinion, the provisions in the present lease, considered in their entirety, do not indicate an intention to hold the lessee liable for damages upon failure to drill.

In the case of Fite v. Miller, *supra,* the principle [*sic*] question before the Court was one of damages, not a construction of the lease. As shown by the opinions in that case, the Court referred several times to the unconditional obligation on the part of the lessee to drill the well which he refused to drill. No question seems to have been raised about the existence of such an obligation under the terms of the lease. The opinions do not set out the literal provisions of the lease, but appellant has undertaken to supply us with the information that the lease contained provisions as to its duration which were similar to the provisions of the lease in the present case, in that it was a lease for a term of five years "and as long thereafter as oil, gas or other mineral is produced. . . ." However, we gather from the opinion, although not definitely stated, that the obligation to drill came into existence during the primary term of five years, making it unnecessary for the Court to construe or to give any effect to the "as long thereafter" provision. In any event, it is not discussed. In the present case it is the legal effect of such a phrase that determines whether there was an obligation to drill on the part of the lessee after the expiration of the primary term, which is the real question involved. We are of the opinion that the District Judge was correct in considering the ruling in Fogle v. Feasel, *supra,* more applicable to the present case than the ruling in Fite v. Miller, *supra.*

The judgment is affirmed.

FISHER v. TOMLINSON OIL CO.

Supreme Court of Kansas, 1974.
215 Kan. 616, 527 P.2d 999, 51 O.&G.R. 28.

FONTRON, JUSTICE. This action is brought to recover damages for breach of a contract to drill an oil well. Essential facts are not in dispute. On August 16, 1971, the Union Gas System, Inc. and Field C. Benton, hereafter collectively referred to as Union Gas, owned oil and gas leases known as the Glasscock leases which covered two quarter sections of land in Elk County. On that date Union Gas entered into an agreement with L. B. Fisher, the plaintiff herein, to assign to Fisher the oil rights under said leases subject to a reservation of an undivided one-sixteenth of a seven-eighths overriding royalty. It was agreed that the assignment was to be escrowed and was to be delivered to Fisher when a producing oil well was obtained. The agreement further provided that Fisher should commence drilling operations for an oil test on or before August 16, 1972, and continue to a depth sufficient to test the Mississippi limestone, commonly found at from 1900 to 2000 feet, unless oil was found in paying quantities at a lesser depth or impenetrable substances were encountered at a lesser depth; that on completion of a producing oil well the escrow agent should deliver the assignment to Fisher; that should the first test be a dry hole, Union Gas would extend the agreement an additional year; that if no production was obtained on or before August 16, 1972, or during the extension period, the assignment was to be returned to Union Gas; and that Fisher should have the right to assign the agreement in whole or in part.

Pursuant to the foregoing agreement Union Gas executed an assignment of the Glasscock leases to Fisher and deposited the same with the escrow holder.

Conversations were thereafter had between Mr. Fisher and Mr. Robert J. Gill, vice president of the Tomlinson Oil Co., Inc., (hereafter called defendant or Tomlinson) with respect to an assignment to Tomlinson of the Union Gas contract and the leases covered thereby. The discussions culminated in an agreement between the two parties which was evidenced by a letter dated October 27, 1971, addressed to Mr. Fisher by Mr. Gill on Tomlinson's behalf, reciting that Fisher was to execute an assignment of the interests referred to in the Union Gas contract and deliver the same to the escrow holder. In return, Tomlinson agreed to drill the leases before August 16, 1972, and if oil was found to return to Mr. Fisher or his assigns a one-fourth working interest in the leases after all expenses of drilling, testing, etc. were paid. Fisher manifested his acceptance of the terms set out by Mr. Gill in the letter by attaching his signature under date of 12/27/71. Terms of the agreement were substantially reiterated in a letter from Fisher to Gill dated February 17, 1972, and Gill, by a notation thereon dated March 8, 1972, acknowledged that the letter correctly expressed the agreement. Drilling was not commenced by Tomlinson prior to August 16, 1972. To the contrary, Mr. Gill advised Mr. Fisher by phone on August 11 that Tomlinson did not desire to drill the well and wished to be relieved from its obligation. Fisher refused to release Tomlinson and advised the company he would hold it liable for damages.

The present action was filed October 3, 1972. . . . Judgment was entered for plaintiff in the amount of $8500, this being within the amount required to drill a well, as was stipulated by Tomlinson. The defendant has appealed, alleging that the cost of drilling a well was an improper measure of damages and, consequently, summary judgment was premature.

The points are interrelated, centering around the basic issue of what measure of damages should be applied to Tomlinson's breach of its agreement with Fisher.

Questions regarding the measure of damages for breach of a drilling contract have come before this court on a number of occasions. Our latest expression is found in Denman v. Aspen Drilling Co., 214 Kan. 402, 520 P.2d 1303, where Justice Fromme, speaking for the court, said:

> There is considerable confusion in the law concerning proof of damages for breach of drilling contracts. The confusion arises from some of the terminology used in the Kansas cases, as well as in the cases of other states. (See cases collected in Anno: 4 A.L.R.3d 284–313.) It should be understood that the rule as to the measure of damages for breach of contract is the same in drilling contracts as it is in other contracts. . . . (p. 404, 520 P.2d p. 1306.)

After reviewing prior decisions of this court dealing with damages arising out of broken drilling agreements, Justice Fromme went on to say:

> The measure of damages in all these cases was the same, *i.e.,* the damage which arose naturally from the breach itself. The evidence necessary to establish the measure of damages may be different but the measure of damages remains the same.

So in the present case we must ask ourselves if the evidence introduced to establish the cost of drilling a well is the best evidence obtainable under the circumstances of this case to show the natural and ordinary consequences of the breach and to enable the court to arrive at a reasonable estimate of the loss which resulted. . . . (pp. 405, 406, 520 P.2d p. 1306.)

Under the particular circumstances present in *Denman,* we held that the cost of drilling a well was not the best evidence obtainable to establish the natural and ordinary consequences of the breach. The facts in that case disclosed that Denman held two blocks of leases which he assigned to the Aspen Company, reserving an overriding royalty; that Aspen, which had agreed to drill a well on each of the two blocks, refused to drill the second well after the first had come in dry, and turned the leases back to Denman who thereupon negotiated a drilling agreement with Thunderbird Drilling Company, containing no reservation of royalty; that Thunderbird proceeded to drill a well on land covered by the second block of leases, and struck oil. Denman sued for the cost of drilling a well. In a case such as that, opined this court, evidence of the value of the lost royalty interest would appear to be proper to establish the ordinary and natural consequences of the breach, not the cost of drilling a well.

In the present case the circumstances are different; sufficiently different, we believe, to justify the trial court in accepting the costs of drilling as the best evidence available to measure damages, and in entering judgment on that basis. In so doing the court followed our decision in Gartner v. Missimer, 178 Kan. 566, 290 P.2d 827, where we said:

> The lease was given to appellants in consideration for their drilling a Bartlesville well within sixty days, a second Bartlesville well within ninety days, an Arbuckle well within six months and a second Arbuckle well within one year if there was production from the first Arbuckle well. There was nothing ambiguous about this contract; it required the drilling of at least three wells. Appellants were not excused from drilling the second Bartlesville well if the first Bartlesville well came in dry, nor were they excused from drilling the first Arbuckle well if the two Bartlesville wells were dry holes, but they did not have to drill a second Arbuckle well if the first Arbuckle well was a dry hole. In that event it made no difference whether the two Bartlesville wells were producers or dry holes. The appellants were bound by the contract to drill a well to the Arbuckle lime and if they breached their contract, they were answerable in damages that were the natural and ordinary consequence of their breach (15 Am.Jur., Damages, § 43) to the appellees who in turn were entitled to have a well drilled to test the Arbuckle lime in and under their land. . . . The cost of drilling a well, or of completing a well which had been started but later abandoned by the driller has been generally held to be the proper measure of damages. 58 C.J.S. Mines and Minerals § 226(b), p. 613; Fite v. Miller, 192 La. 229, 187 So. 650, 122 A.L.R. 446, anno. 458. (pp. 570, 571, 290 P.2d p. 831.)

The similarity of the circumstances revealed in this case and in *Gartner* is quite marked. In *Gartner,* as here, the acreage under lease was in close proximity to producing wells; in *Gartner,* as here, no well was subsequently drilled on the leased acreage; in *Gartner,* as here, the evidence going to damages related to the cost of drilling, there being testimony in *Gartner,* and a stipulation in this case.

This case differs from *Denman* in that here, unlike the situation in *Denman,* the Glasscock leases were not re-negotiated and no well was subsequently drilled on that acreage. Here also, quite unlike *Denman,* we have no evidence going to the value of an overriding royalty interest which could be used as a measure of damages.

We believe the similarities and the differences we have noted distinguish the instant action from *Denman,* bringing this case within the rationale of *Gartner.* The rule expressed in *Gartner* finds support in the language of two subsequent decisions, In re Estate of Stannard, 179 Kan. 394, 295 P.2d 610, and Cookson v. Western Oil Fields, Inc., 10 Cir., 465 F.2d 460. In the *Cookson* case, the federal court of appeals, although holding that the lessee had substantially complied with the terms of its drilling contract, commented on the *Gartner* and *Stannard* cases in this wise:

> In both of the cited cases, [*Gartner* and *Stannard*] the breach was the failure of the defendant to perform the essence of the contract—the adequate oil exploration—before the time of termination of the agreement. Plaintiffs in those cases were aggrieved in that they had given good consideration to receive information as to whether they had valuable oil and gas deposits, and did not receive that information in return. Therefore, they were entitled to receive that for which they had bargained, or to be put in a position equivalent to their having received it. . . . (p. 463.)

The rule espoused by this court in *Gartner* seems to be in accord with the generally prevailing view in this country. In an annotation contained in 4 A.L.R.3d, Drilling Contract—Damages, § 5, p. 290, the text recites:

> The cost of sinking the well contracted for, or of completing one partially sunk if the defendant has partially performed, has been adopted by a number of courts as the general measure of damages for breach of a contract to drill an oil or gas well.

Cases from a substantial number of jurisdictions are cited in support of this textual statement.

The strong thrust of Tomlinson's discontent with the summary judgment entered in plaintiff's favor is simply this: that Fisher sustained no damage from Tomlinson's breach of contract and that Tomlinson was precluded from making any showing to that effect. During a wordy colloquy which occurred at the pretrial hearing the defendant informed the trial court that it wished to present the testimony of certain geologists to the effect the Glasscock leases would not produce oil and a test well would turn out to be dry, making Fisher's overriding royalty interest valueless. For this reason, Tomlinson argues, Fisher was not damaged and the summary judgment entered in his favor was premature.

We think the situation is not quite as Tomlinson would depict it. For example, the lease which Union Gas assigned to Fisher would have been extended for an additional year had a well been drilled, and the acreage would thus have been available for further exploration. Equally as important, Fisher was entitled to the information for which he had bargained, that is, whether there were oil deposits under the land covered by the leases—or at least he was entitled, as was said in *Cookson,* "to be put in a position equivalent to [his] having received it." It is generally recognized throughout oil and gas producing areas that oil is where you find it, and the surest way to find out whether oil is present is to drill for it. This is exactly what the defendant refused to do despite his contractual obligation.

In the *Gartner* case, which went to the jury, we held that the trial court did not err in refusing to give an instruction directing a verdict for nominal damages. We are of the opinion that nominal damages would not have been appropriate in this case and that the trial court did not err in applying the *Gartner* rule as to damages and in entering judgment for the stipulated cost of drilling a well.

The judgment is affirmed.

NOTES

1. DAVENPORT V. DOYLE PETROLEUM CORP., 1940 OK 92, 187 Okl. 40, 100 P.2d 445. Plaintiff, owner of 160 acres of land, executed an oil and gas lease on 120 acres thereof to defendant, Doyle Petroleum Corporation. The lease was the ordinary printed Form 88 (Producers, Oklahoma) reciting one dollar and other good and valuable consideration for a five-year term and as long thereafter as oil or gas was produced in paying quantities. If no well was commenced on or before May 18, 1938, the lease was to terminate unless lessee paid lessor the sum of $120, as rental for privilege of deferring commencement of actual drilling for another 12 months. In like manner commencement of a well could be deferred for successive 12 month periods during the primary term. There was added the following typewritten clause:

> Notwithstanding the preceding printed part of this lease, lessee agrees within 60 days from date to begin the actual drilling of a test well for oil or gas on or offsetting the above land, and prosecute such drilling with due diligence to a depth of 3,000 feet unless oil or gas in paying quantities is found at a lesser depth. For failure to commence, prosecute and/or complete said test well, this lease at option of lessor, shall terminate.

Plaintiff sued for damages for breach of the covenant to drill.

> Defendant rightly contends it had the option, under the printed portion of this lease, to drill or pay rentals, and unless it did so the lease would terminate. Not so, under the typewritten paragraph, because it was therein expressly provided, ". . . this lease, *at the option of lessor,* shall terminate." This shift of the option to terminate the lease from the lessee to the lessor assumes added significance when we consider the paragraph begins *"Notwithstanding"* (meaning, without prevention or obstruction from or by,

in spite of, despite)" . . . the preceding printed part of this lease . . .

Can it be said by any reasonable construction, lessee was not thereby obligated to drill a test well? And, further, can it be said lessee had any option other than to drill the specified test well either on the 120 acres covered by the lease or on land off-setting said 120 acres? "Notwithstanding" the option given defendant in the printed "unless" portion of the lease, the agreement contained in the typewritten paragraph positively obligated defendant to commence actual drilling within 60 days either upon or offsetting plaintiff's land. This, plaintiff alleges, was the principal inducement for execution of the lease, and on demurrer must be considered true.

The question is, whether the provision authorizing plaintiff to terminate the lease, at his option, for failure to commence drilling limits his right to termination alone and precludes him from seeking damages for lessee's breach of his express covenant to drill.

. . .

We perceive no conflict between the typewritten paragraph and the printed portion of the Producers 88 Form lease. If defendant had drilled the well offsetting plaintiff's land and discovered oil or gas in paying quantities, then it would have been necessary to immediately drill on plaintiff's land. If the offset had been drilled to 3,000 feet either before or subsequent to expiration of one year and neither oil or gas was found in paying quantities, defendant had the option to timely pay $120 delay rental or let the lease terminate as therein provided. If defendant had exercised its option to drill on plaintiff's land and diligently continued, but was unable to reach 3,000 feet before the expiration of one year, no delay rental would be necessary to continue the lease. Likewise, if oil or gas had been discovered in paying quantities prior to expiration of one year. But when defendant failed to commence actual drilling as agreed in the lease contract, either upon or offsetting plaintiff's land, its positive covenant was breached, whereupon plaintiff could terminate the lease at his option. But plaintiff, instead of terminating the lease, affirmed it and commenced this action for damages for breach of the covenant to drill. This plaintiff was entitled to do, and the demurrer to his petition should have been overruled.

Is the *Davenport* case consistent with the *Joyce* case?

2. Farmout agreements are the type of contract in which the question of whether there is an express drilling obligation is most likely to be raised. For a case on whether the well specified in the agreement was optional or obligatory, *see* Petrocana, Inc. v. Margo, Inc., 577 So.2d 274, 115 O.&G.R. 84 (La.App. 3d Cir.1991). For additional discussion of this topic, *see* Lowe, *Analyzing Oil and Gas Farmout Agreements*, 41 Sw. L.J. 759, 782 (1987); Schaefer, *The Ins and Outs of Farmouts: A Practical Guide for the Landman and the Lawyer*, 32 Rocky Mt. Min. L. Inst. 18–1, 18–16 (1986); Scott, *How to Prepare an Oil and Gas*

Farmout Agreement, 33 Baylor L. Rev. 63, 67 (1981).

3. In *Denman* [discussed in *Fisher, supra*] should the court have required proof that plaintiff would have sold his royalty and the price at which he would have sold? *See* Riddle v. Lanier, 136 Tex. 130, 145 S.W.2d 1094 (1941); Whiteside v. Trentman, 141 Tex. 46, 170 S.W.2d 195 (Com.App.1943).

4. In a farm-out situation, is the proper measure of damages the value to the assignor of the information that would have been forthcoming if the well had been drilled? *See* Hoffer Oil Corp. v. Carpenter, 34 F.2d 589 (10th Cir.1929), cert. denied, 280 U.S. 608 (1930).

5. Would a liquidated damage clause solve the problem? For an affirmative answer, *see* Presnal v. TLL Energy Corp., 788 S.W.2d 123, 110 O.&G.R. 529 (Tex.App. 1990).

6. Should all the difficulties in measurement of damages be avoided in these cases by giving the promisee specific performance of the contract to drill?

SECTION 6. REMEDIES FOR BREACH OF OBLIGATION TO RESTORE CONDITION OF PREMISES

(*See* Martin & Kramer, *Williams and Meyers Oil and Gas Law* §§ 218.12, 673.5).

An oil and gas lease or an instrument severing minerals from the surface may by express provision impose an obligation on the lessee or mineral owner to restore the condition of the premises after cessation of mineral operations. In the absence of an express lease provision imposing a duty on the lessee the question may arise whether this obligation is implied in the lease or deed. This next case concerns an express lease provision.

CORBELLO v. IOWA PRODUCTION

Louisiana Supreme Court, 2003
850 So.2d 686, 157 O.&G.R. 1120

JOHNSON, JUSTICE. This matter arises from a suit filed by landowners against Shell Oil Company to recover damages in trespass after expiration of a surface lease which was granted to Shell, for unauthorized disposal of saltwater on the property, and for the poor condition of the leased premises. After a trial on the merits, the jury awarded damages to the landowners. . . .

Facts and Procedural History

[In 1929, Ferdinand and Eva Heyd granted an oil and gas mineral lease in favor of Shell Oil Company. The lease covered 320 acres of land in Calcasieu Parish. Shell operated the mineral lease until 1985, when it transferred its interest in the lease to Rosewood Resources, Inc. In 1961, Shell obtained a separate lease from plaintiffs, Mr. Heyd's four surviving children, on 120 acres within the 320 acres covered by the oil and gas lease, known as the Iowa Field. Thereafter, Shell built an oil terminal on a five acre parcel within the leased acreage which it operated until 1993.]

The 1961 surface lease expired May 10, 1991. On May 9, 1991, plaintiffs sent Shell a letter regarding the lease's termination date. The letter notified Shell that it had breached the lease agreement by disposing of saltwater on the property and by failing to maintain the property as provided in the lease. For approximately sixteen or seventeen months, plaintiffs and Shell attempted to resolve these issues. [In May 2000, the case was tried before a jury. The Third Circuit Court of Appeal affirmed the jury's awards of $33 million to restore the leased premises and $16,679,100.00 for Shell's illegal disposal of saltwater on the leased premises. Shell filed a writ of certiorari.]

Law and Analysis

Restoration of the Leased Premises

We begin our discussion with Shell's assignments of error regarding the damage award for failure to reasonably restore the leased premises. Shell argues that the court of appeal erred in affirming the $33 million breach of contract award for failure to reasonably restore plaintiffs' property for a number of reasons. First, Shell argues that the court of appeal erroneously held that damages for a breach of a contractual obligation of restoration in a lease can be disproportionately greater (in this case, 300 times greater) than the market value of the property. Our first inquiry, then, is whether plaintiffs are entitled to the amount of damages needed to restore the property, without regard to market value, or whether the damage award should be tethered to the market value of the property.

Shell argues that the legal principles that restrain immovable property damages in tort and specific performance cases to the market value of the property should also apply in cases of damages for breach of contract. Shell contends that the amount of damages for breach of the contractual obligation of restoration of the property must be rationally or reasonably related to the market value of the property. Otherwise, awarding an amount which is disproportionately greater than the market value, according to Shell, would give plaintiffs a windfall because plaintiffs would be in a better position than they would be in if Shell had performed under the contract. That is, if Shell had complied with its obligation under the contract to restore the property to its original state upon termination of the lease, plaintiffs would be in possession of property worth $108,000.00. Thus, Shell argues, plaintiffs' damage award for Shell's breach of its contractual obligation should be reasonably or rationally related to the amount of $108,000.00.

In furtherance of its position, Shell argues that the amount of damages recoverable under a breach of contract claim must be controlled by the parties' mutually agreed upon expectations. La. C.C. art.2045. Here, Shell maintains that the parties bargained for a restoration obligation limited by "reasonableness." Shell argues that to award plaintiffs damages in the amount of $33 million (300 times greater than the value of the property) is unreasonable. Shell contends that no rational, objective lessor would expect that the lessee would be required to restore the property at a cost disproportionately greater than the value of the property itself. Likewise, no rational, objective lessee would expect that such a reasonable restoration obligation would require the lessee to expend sums to restore the property that would be grossly disproportionate to what it would cost to purchase the property outright. For this reason, Shell argues, the contractual obligation of reasonable restoration is confined by an "economic balancing process which balances the cost of perfect restoration against the value of the use to which the land is being put." La. R.S. 31:22 (comments).

Plaintiffs counter that this case is governed by the principle that "the contract is the law between the parties." Plaintiffs maintain that the parties bargained for, among other things, reasonable restoration of the property to its original condition, in exchange for Shell's use of the land for production of oil and gas for profit. Plaintiffs point out that the agreement did not in any way limit Shell's cleanup obligation to the value of the land. Plaintiffs maintain that the measure of damages for Shell's breach is not the damage to the "value" of the land, but the cost to the innocent landowner of doing what Shell promised but failed to do, restore the land to its original condition. Therefore, plaintiffs argue that the damage award should not be reasonably or rational related to market value of the property. Instead, the damage award should be that which was bargained for, the amount necessary to restore the property. . . .

After careful review of the applicable law and facts of this case, we disagree with the arguments presented by Shell and find that the damage award for a breach of contractual obligation to reasonably restore property need not be tethered to the market value of the property.

We agree with plaintiffs that the agreement in this case, like other contracts, is the law as between the parties. Since a contract establishes the law between the parties, the purpose of contract interpretation is to determine the common intent of the parties. . . ..

Section 8 of the surface lease entered into by Shell and plaintiffs reads:

Lessee agrees to indemnify and hold lessor harmless from any and all loss, damage, injury and liability of every kind and nature that may be caused by its operations or result from the exercise of the rights or privileges herein granted. **Lessee further agrees that upon termination of this lease it will reasonably restore the premises as nearly as possible to their present condition.** (Emphasis added by the court)

We first note that the language of the contract itself does not limit Shell's liability for reasonable restoration to the market value of the property. If Shell wanted such a limitation, it could have bargained for such; it did not do so. . . .

Shell correctly notes that our courts have consistently restrained property damage awards in tort cases. . . .Generally, three approaches have been followed by the Louisiana courts in arriving at the amount of damages to property: (1) the cost of restoration if the thing damaged can be adequately repaired; (2) the difference in value prior to and following the damage; or (3) the cost of replacement new, less reasonable depreciation, if the value before and after the damage cannot be reasonably determined, or if the cost of the repairs exceeds the value of the thing damaged. . . . We held in *Roman Catholic Church v. La. Gas Serv. Co.,* 618 So.2d 874 (La.1993), which involved tortious damage to immovable property, that "... if the cost of restoring the property in its original condition is disproportionate to the value of the property or economically wasteful, unless there is a reason personal to the owner for restoring the original condition or there is a reason to believe that the plaintiff will, in fact, make the repairs, damages are measured only by the difference between the value of the property before and after the harm."

We find that damages to immovable property under a breach of contract claim should not be governed by the rule enunciated in *Church.* We find that the contractual terms of a contract, which convey the intentions of the parties, overrule any policy considerations behind such a rule limiting damages in tort cases. We recognize that in some cases, as in the instant case, the expense of restoration of immovable property can be extremely high. However, while we find it logical in tort cases to tether the amount of damages by balancing the amount to be paid by the negligent tortfeasor against the goal to restore the plaintiff, as closely as possible, to the position which he would have occupied had the accident never occurred, this same logic should not be extended to breach of contract cases. . . .

Thus, we decline to set forth a rule of law, suggested by Shell, that in cases of breach of a contractual obligation of restoration in a lease, the damage award to plaintiffs must be tethered to the market value of the property. To do so would give license to oil companies to perform its operations in any manner, with indifference as to the aftermath of its operations because of the assurance that it would not be responsible for the full cost of restoration. The landowner would receive a damage award tethered to the market value of the property and would be left with partially clean land and potential liability to others who are affected by this unclean land. In the end, it is the oil companies, not plaintiffs, who would get a windfall. Accordingly, we find Shell's argument that plaintiffs' damage award in this case should be rationally related to the market value of the property to be without merit. . . .

The remaining question, then, is whether the jury's award for restoration was reasonable. In determining damages, the trier of fact is accorded much discretion. The assessment of damages by jury is a determination of fact. . . . Upon review, we conclude that the court of appeal did not err in affirming the $33 million award in restoration damages.

We find that the fact that the contamination of the groundwater, for which the plaintiffs recovered $28 million in restoration damages, is a public injury as well as private injury, does not prevent plaintiffs from collecting damages for cleanup of the groundwater. . . .

Conclusion

Based on the above analysis, . . .the court of appeal's decision in all other respects is hereby affirmed.

VICTORY, J., dissenting in part. I dissent from the majority opinion affirming the $33,000,000.00 breach of contract award for Shell's failure to reasonably restore plaintiff's property. The contract between the parties requires only that Shell "will reasonably restore the premises as nearly as possible to their present condition." The jury was clearly wrong in determining that $33,000,000 was a "reasonable" amount of money to restore a piece of property worth $108,000.00. Further, $28,000,000 of that $33,000,000 was awarded to install a groundwater recovery system to protect the Chicot Aquifer from contamination. However, plaintiffs failed to offer any reliable, scientific evidence demonstrating a substantial threat to the aquifer, and its expert did not test the aquifer for contamination. In any event, since $28,000,000 of the award is to protect the aquifer, the landowner should be required to use the money to install the groundwater recovery system.

NOTES

1. *Corbello* was followed in Hazelwood Farm, Inc. v. Liberty Oil & Gas Corp., 844 So. 2d 380, 158 O.&G.R. 35 (La. App. 2003), where the court held the landowner was entitled to damages measured by the cost to restore the property even though the damage was already present when landowner purchased the property. The liability was based on a lease granted by the mineral owner, and the court found that the present landowner was a third party beneficiary of the lease provision. For critical commentary on *Corbello, see* Mary Beth Balhoff, *Corbello v. Iowa Production and the Implications of Restoration Damages in Louisiana: Drilling Holes in Deep Pockets for Thirty-three Million Dollars*, 65 La. L. Rev. 271 (2005); Thomas M. McNamara, Patrick W. Gray, and Amy E. Allums, *Current Issues in the Scope of the Obligation to Restore Property after Oil & Gas Operations*, 56 Oil & Gas Inst. Ch 4 (Matthew Bender 2005); Robert L. Theriot, *Duty to Restore the Surface (Implied, Express, and Damages)*, 52nd Ann. Inst. on Min. Law 141 (2007).

2. The extent of an *implied* duty to restore the surface in Louisiana law was the controversy taken up in TERREBONNE PARISH SCHOOL BOARD V. CASTEX ENERGY, Inc., 893 So. 2d 789, 161 O.&G.R. 747. (La. 1/19/05). Comments to Article 122 of the Louisiana Mineral Code and some case authority indicated that lessees may have a duty to restore the surface but the scope of the duty was uncertain. *Castex* concerned marshland leased by a school board prior to the adoption of the Mineral Code, which lease specifically authorized the dredging of canals for access to well sites and contained no provisions addressing an obligation to restore the surface. The plaintiff school board asserted the lessee had an implied duty to restore the land because of damage arising from the dredging

of two canals and a slip. The plaintiff sought monetary damages calculated using the *Corbello* measure, i.e. cost to restore the land to its original condition at the time of the leasing. Instead, the trial court awarded actual restoration. On appeal, the Louisiana Supreme Court ruled that "in the absence of an express lease provision, Mineral Code article 122 does not impose an implied duty to restore the surface to its original, pre-lease condition absent proof that the lessee has exercised his rights under the lease unreasonably or excessively." The court's analysis looked to the Civil Code articles on leases, which do not impose a strict obligation to return leased property in an unchanged condition but instead allow for deterioration of the leased property because of necessary "wear and tear." The specific rights granted in the lease are relevant to determining necessary "wear and tear," and a lessor may be considered to have given assent to the "wear and tear" normally involved in exercising the rights granted. The express grant of the right to dredge canals constituted consent to or approval of the changes necessarily incident to dredging.

The Louisiana Supreme Court amplified and reiterated its *Castex* ruling in Broussard v. Hilcorp Energy Co., 24 So.3d 813, 820, 2009-0449 (La. 10/20/09). Landowners sued lessees and others to recover for negligent operation of the leases that resulted in contamination of the property. Lessees responded with exceptions of prematurity, vagueness, improper cumulation of actions, and lack of subject matter jurisdiction. Upon the lessees' appeal the Supreme Court held that owners' statutory obligation under Mineral Code Article 136 to provide pre-suit written notice of claim that lessee failed to develop and operate the property as a prudent operator did not apply. The rationale was that such claims do not come within the prudent operator duties contemplated by Articles 122 and 136, the court stating that "claims for remediation/restoration are not related to any duty encompassed by Article 122."

3. Act 312 of 2006 of the Louisiana legislature, codified at La. R.S. 30:29, 29.1, and 2015.1(L), establishes procedures for judicial resolution of claims for environmental damage to property arising from activities subject to the jurisdiction of the Department of Natural Resources, Office of Conservation. A litigant alleging environmental damage must provide notice of the claims to the state of Louisiana through the Commissioner of Conservation of the Department of Natural Resources and the attorney general. Once it is determined that environmental damage exists, the legally responsible party is to submit a plan for remediation which is then reviewed by the department and the parties. Once a plan is approved and adopted by the court, the court will order implementation of the plan and the court and the Department will have oversight to ensure compliance with the plan. The Act does not preclude an owner of land from pursuing a judicial remedy or receiving a judicial award for private claims suffered as a result of environmental damage, except as otherwise provided in the act. The procedures for implementing Act 312 have been the source of controversy: M.J. Farms, Ltd. v. Exxon Mobil Corp., 07-2371 (La.7/1/08), 998 So.2d 16; Duhon v. Petro E, LLC, 34 So. 3d 1065, 2009-

1150 (La.App. 3 Cir. 4/7/10); Tensas Poppadoc, Inc. v. Chevron USA, Inc., 10 So. 3d 1259, 2008-1266 (La.App. 3 Cir. 2009); Germany v. ConocoPhillips Co., 980 So.2d 101 (La. App. 3 Cir. 3/5/08). See Michael R. Phillips, *Act 312 Updates*, 57th Ann. Inst. on Min. Law 237 (2010).

The Louisiana Mineral Code, in tacitly recognizing the implied easement (servitude) of a mineral servitude owner to use the surface of the land, imposes a duty of restoration on the servitude owner. a. Rev. Stat. § 31:22. Dupree v. Oil Gas & Other Minerals, 731 So. 2d 1067, 142 O.&G.R. 258 (La. Ct. App. 1999) held that this article of the Mineral Code imposed liability upon the servitude owners for damage to the land caused by the lessee of the servitude owners when the lessee of the servitude owners was bankrupt.

4. The case authority in most states generally denies the existence of a duty to restore absent express language. *See e.g.* Oxy USA, Inc. v. Cook, 127 S.W.3d 16, 158 Oil & Gas Rep. 526 (Tex. App.--Tyler 2003) (where there is no express duty to restore the surface, the court will not imply such a duty). By a law enacted in 1959 in South Dakota, a lessee may be required to restore the condition of the premises unless the requirement is waived by the surface lessee. See S.D. Codified Laws § 45-9-15.1. Somewhat similar requirements concerning the restoration of leased land to its previous condition after the cessation of operations are imposed by statutes in Alberta, Illinois (225 Ill. Comp. Stat. 725/6.), and Kansas (Kan. Stat. Ann. § 55-132a).

5. The usual measure of damages for permanent injury to the land or structures thereon as a result of the wrongful conduct of the defendant is the difference in value thereof before and after such wrongful conduct. Attempts to treat injuries to land as temporary and therefore value damages as the cost of restoration or remediation have been rebuffed where it cannot be proved that the remediation is "actually or economically feasible." *See, e.g.*, Meade v. Kubinski, 277 Ill. App. 3d 1014, 661 N.E.2d 1178 (1996) (reviews various factors, including permanency of the injury, the cost of restoration, and the diminution in value of the real property to ascertain whether the traditional measure of damages giving only the diminution in value should be awarded); North Ridge Corp. v. Walraven, 957 S.W.2d 116, 139 O.&G.R. 649 (Tex. App.--Eastland 1997, writ denied) (jury award of cost of remediation damages in the amount of $509,000 was reversed where that cost made remediation economically infeasible); Kraft v. Langford, 565 S.W.2d 223 (Tex. 1978).

6. Some courts are following an approach like that of *Corbello supra*. See McNeill v. Burlington Resources Oil & Gas Co.,143 N.M. 740, 182 P.3d 121, 2008 NMSC 22 (N.M. 2008), which rules that the distinction between temporary and permanent damages to land is no longer viable, so that a jury can consider both the cost to repair and diminution in value in measuring damages to land. The court expressed its confidence that "when directed to fashion an award of damages by the most reasonable measure available, a properly instructed jury will recognize sham claims and avoid awarding windfalls."

Chapter 5

TITLE AND CONVEYANCING PROBLEMS ARISING FROM TRANSFERS BY FEE OWNERS AND LESSORS

SECTION 1. INTRODUCTION

Martin & Kramer, *Williams & Meyers Oil and Gas Law*

§ 301. Definition and Analysis of Terms

The permissible interests that may be created in oil and gas are best identified and understood by beginning with fee simple absolute ownership of land. *A,* the owner in fee absolute of Blackacre, has the same rights, privileges, powers and immunities with regard to the minerals therein as he has in the surface, subject of course to regulation under the police power of the state. This totality of interest may be granted or reserved separate and apart from the surface, and such severance of minerals from surface interest creates what is called here a *mineral estate,* being the most complete ownership of oil and gas recognized in law. The owner of the mineral estate (*B*) has the same rights, privileges, powers, and immunities as *A* had before him: with respect to the minerals, *B* stands in the shoes of *A.*

What are *B* 's rights, powers, privileges and immunities in the mineral estate in Blackacre that are significant to oil and gas conveyancing? First *B* has the right to sell all or a part of his mineral estate. Second, *B* has the right to explore for and develop the minerals on Blackacre for himself. This second right is called here the *development right,* and the easements appurtenant to the mineral estate for access thereto are called *exploration* and *development easements.* The development right is not commonly exercised by owners of mineral estates because of the expense and risk involved, but operators sometimes buy mineral estates rather than oil and gas leases. Third, *B* has the right to execute oil and gas leases to operators to secure the exploration and development of the minerals. Leasing is the typical arrangement between landowner and operator for the exploitation of the mineral estate. The right to lease is called here the *executive right* or the *leasing right.*

Apart from creating a *leasehold estate* in an operator by execution of an oil and gas lease, what other interests can *B* create in the mineral estate of Blackacre? The most obvious interest is a fractional share of the mineral estate itself; that is, *B* can part with an undivided interest in the mineral estate, thus substituting another person for himself as to a share of the estate. As to his undivided interest, *B* 's transferee again has all the rights, powers, privileges and immunities in his share of Blackacre that *B* has in his share.

[Non-executive interest]

Often the parties to an oil and gas conveyance do not desire to create concurrent, undivided interests in the mineral estate. They may prefer that the executive right, or perhaps other of the ownership interests be lodged in the hands of one person. Thus, by grant or reservation, the executive right may be placed in one person, the other taking a *non-executive* interest in the oil and gas. To understand the nature of a non-executive interest in oil and gas, it is first necessary to investigate the usual interests created by an oil and gas lease. They are:

Bonus

One form of bonus, the cash bonus, has been defined as " 'a premium paid to a grantor or vendor' and strictly is the cash consideration or down payment, paid or agreed to be paid, for the execution of an oil and gas lease." The cash bonus is usually fixed on a per acre basis.

A second form of bonus, sometimes called an oil bonus or a royalty bonus, is payable out of production. An oil or royalty bonus usually takes the form of a production payment, which can be described as a right to a certain sum of money or a fixed number of units out of an agreed fraction of any oil or gas produced. While instruments creating production payments vary considerably in terminology, two characteristics predominate in all: (1) the sums to be paid or the units to be delivered are fixed in amount, and (2) they are to be derived from the production of the minerals. In other words, the payment is to be made only when the minerals are produced and no personal liability exists except to pay out of production.

Production payments reserved by a lessor in the lease could be classified as a distinct interest in the minerals or as landowner's royalty, but the cases that have considered the nature of these interests have treated them as bonus, payable to the owner thereof rather than the royalty owner where the ownership differs.

One court has suggested that a royalty in excess of the usual 1/8 landowner's royalty is bonus. For present definitional purposes, such excess or overriding landowner's royalty will be defined as "royalty" and not as "bonus." Such definition does not necessarily control the disposition of the interest as between a royalty owner and a bonus owner, where the two differ.

Rentals

"The term 'rental' as used in oil and gas leases refers to the consideration paid to the lessor for the privilege of delaying drilling operations." The delay rental clause takes two common forms. In the "or" form lease, the lessee has the duty to pay rental or, in the alternative, to drill. If he fails to do either, the lessor has, at his option, the right to receive the rentals or to forfeit the lease. Ordinarily, the lessee may avoid liability by exercising the right, given by most such leases, of surrendering the lease prior to the rental due date. The "unless" form lease functions in the same manner fundamentally. The lessee must drill or pay to keep the lease. However, no affirmative action is necessary on his part to avoid liability for rent, because failure to drill or to pay the rental automatically ends the lease. Thus the lessee can never be liable for rent, but he keeps his lease in precisely the same manner as does the lessee under the "or" form lease.

Royalty

"The term 'royalty' in the strict sense is held to mean a share of the product or the proceeds therefrom, reserved to the owner for permitting another to use the property." While the word "royalty" has sometimes been used to designate other lease interests besides the landowner's share of production, confusion will be minimized by the adoption of a narrow definition. Thus when the term is used in connection with the lessor's interests under an oil and gas lease it will mean that portion of the oil produced for which the lessee must account to the lessor under the terms of the lease, exclusive of oil payments but inclusive of excess or overriding landowner's royalty.

Reversionary Interests

The fixed term, "thereafter" clause lease ordinarily formerly provided that the lease should last for a certain period of time, such as five or ten years, and so long thereafter as oil or gas was produced. This type of lease terminated at the end of the primary term if there was no production at that time, or if there was, when production ceased. This reversionary interest is usually called a possibility of reverter. In recent years this form has been modified so that drilling or reworking operations in progress at the expiration of the primary term will extend the lease, and similarly, if production ceases after the end of the primary term, such operations will also extend the lease. The drilling clause in a lease also creates a reversionary interest, whether it be of the "or" or "unless" type, for a failure to drill or pay rental under either clause enables the lessor to recover the full mineral estate. This reversionary interest is usually called a possibility of reverter in the "unless" lease and a right of re-entry (or power of termination) in an "or" lease. In some states, the various implied covenants, (*e.g.*, to prevent drainage, to explore further, and to develop reasonably) are regarded as both covenants and conditions, for the breach of which the lessor may, in certain cases, terminate the lease. Lastly, in Texas, oil and gas leases may be subject to an implied special limitation that the premises be devoted to oil and gas purposes. Opinion varies as to the exact nature of the reversionary interests, but it is sufficient for our purposes to recognize that the effect of the termination of the lease is (assuming no conveyance in the meantime) to reinvest in the lessor the same interest in the oil and gas that he had before executing the lease, *viz.*, the executive right and the right to transfer the mineral estate or lesser mineral interests.

Benefit of Covenants

Lastly most oil and gas leases contain express covenants (*e.g.*, promise to pay royalty) and implied covenants (*e.g.*, to use due diligence to prevent drainage). The benefit of these covenants is an interest owned by a lessor.

[Varieties of transfers]

If *B,* the owner of the mineral estate in Blackacre, executes a lease, he may nonetheless thereafter convey an undivided interest in the mineral estate, subject to the lease. The effect of this conveyance is to transfer a proportionate share of lease benefits (bonus, if not already paid; rentals; royalty; benefit of covenants) plus an undivided interest in the possibility of reverter or other reversionary interests. When the lease terminates, *B* 's transferee then owns a concurrent, undivided interest in the mineral estate, with executive and development rights. In short, the transferee then owns exactly the same interest he would have owned had there never been a lease.

However, as indicated previously the parties may not wish to create a concurrent interest, with both parties having executive rights. They may create non-executive oil and gas interests as follows:

(1) *B* could convey an interest to endure only for the life of the existing lease. Thus his grantee would have no right to drill (because the lessee owns this right exclusively) and no right to execute future leases.

(2) *B* could convey a *royalty interest,* which is a right only to receive a certain part of the oil produced from Blackacre free of exploration and production costs. A royalty owner has no right to explore and develop, to execute leases, or to receive bonus and rental. While a royalty interest can be limited to endure only for the life of an existing lease, it may be made perpetual without giving the royalty grantee any share of the executive rights. In practical effect, an owner of a perpetual royalty interest owns a share of the royalty provided for under an existing lease and under any future lease. If the mineral estate owner should develop for himself, the royalty owner will also take a share of that production, free of costs. Royalty interests as defined here are also called non-participating royalties. This term, tautological under our definition, seems to have been used to distinguish "participating royalty," which was an interest that participated in bonus and delay rentals or in the executive power, or in both. Because of this ambiguity and because the term "participating royalty" has become obsolete, we have adopted another name for such interest, as indicated below.

(3) *B* could convey a *non-executive mineral interest,* which is defined here as the right to royalty and to either bonus or rental or both, under existing or future leases, the owner of which has no development right and no executive rights, *i.e.,* no right to exploit the minerals himself and no right to join in the execution of leases. It is our contention that a non-executive mineral interest is nothing more than a royalty interest that also shares in bonus and rental.

This description of royalty and non-executive interests has proceeded on the assumption that at the time of transfer an oil and gas lease existed on Blackacre and that the transfer created a right to share in benefits under the existing lease or under it and future leases. Of course, the same interests may be created prior to the execution of a lease on Blackacre, operating then as a present interest with enjoyment postponed, instead of operating in part under a present lease and in part under future leases. It is also true that a non-executive mineral interest does not necessarily share in all three monetary benefits under a lease: an instrument could exclude such owner from bonus or rental just as he could participate in bonus and rental and be excluded from royalty.

Apart from the Rule against Perpetuities, any of the interests above may be created for any period of time. Usually, however, they take one of three forms: (1) perpetual duration; (2) duration for a fixed period of years and so long thereafter as oil or gas is produced; (3) duration for a fixed period of years and no longer.

[Summary]

In summary we have identified and described the following interests in oil and gas that are commonly created by landowners in sales or trades:

1. *Mineral interest.* Created by mineral deed or by reservation. Duration usually in perpetuity, or for a fixed period with a thereafter clause, or for a fixed period only. Owner has same rights as landowner before severance. If concurrent, undivided interest created by grant or reservation of fractional mineral interest, concurrent owners share all rights, including right to develop or lease.

2. *Royalty interest.* Created by royalty deed or by reservation. Duration varies as in mineral interests; some royalty interests endure for life of existing lease only. Owner has right to receive a certain part of the oil or gas, as, if and when produced, free of costs of production; has no rights to develop or lease.

3. *Non-executive mineral interest.* Created by grant or by reservation in a deed with specific language that governs the sharing of bonus, rental and royalty and excluding one party from participation in execution of leases. Owner has rights as spelled out in the creating instrument; has no right to develop or execute leases.

ALTMAN v. BLAKE

Texas Supreme Court, 1986
712 S.W.2d 117, 91 O.&G.R. 346

KILGARLIN, Justice. Hazel Altman, Duncan B. Clark, and Ellen Dixon Clark present this court with one issue: whether a mineral interest conveyed in a deed by W.R. Blake, Jr. to his father is a one-sixteenth royalty interest or a one-sixteenth interest in the mineral fee. In a suit filed by Altman and the Clarks against the heirs of W.R. Blake, Sr., both sides moved for summary judgment. The trial court construed the deed to convey a one-sixteenth non-participating royalty interest to Blake, Sr., and rendered judgment for his heirs. The court of appeals affirmed the trial court judgment. 703 S.W.2d 420. Finding that the deed conveyed to Blake, Sr. a one-sixteenth interest in the mineral fee, we reverse the judgment of the court of appeals.

W.R. Blake, Jr. was the owner of the surface and the unsevered mineral estate of a 348-acre tract of land in Hockley County, Texas. On May 30, 1938, Blake, Jr. executed a mineral deed to his father, W.R. Blake, Sr. The granting clause of that lease provided:

W.R. Blake, Jr. ... does hereby grant, bargain, sell, convey, transfer, assign and deliver unto W.R. Blake, Sr., of Lubbock, Texas, hereinafter called Grantee (whether one or more) an undivided one-sixteenth (1/16) interest in and to all of the oil, gas and other minerals in and under and that may be produced from the following described land situated in Hockley County, State of Texas, to wit:

> All of that certain tract or parcel of land being all of Labors 15 and 16, League 2, Jones County School Land, Hockley County, Texas.

But does not participate in any rentals or leases, containing 348 acres, more or less, together with the rights of ingress and egress at all times for the purpose of mining, drilling, exploring, operating and developing said lands for oil, gas and other minerals, and storing, handling, transporting and marketing the same therefrom, with the right to remove from said land all of Grantee's property and improvements.

On January 14, 1939, Blake, Jr. executed a warranty deed conveying the 348-acre tract to D.A. Clark. The warranty deed described the property conveyed as:

> All that certain tract or parcel of land being all of Labors 15 and 16, League 2, Jones County School Land, Hockley County, Texas, consisting of 348 acres, more or less, save and except one-sixteenth (1/16) of the minerals, non-participating, which has previously been sold.

The 348-acre Hockley County tract of land is now under an oil and gas lease, and a producing well has been completed. The oil and gas lease reserves to the lessors a one-eighth royalty. At issue in this case is the apportionment of that royalty between the Blake, Sr. heirs and the Clark heirs.

The heirs of D.A. Clark: Altman, Duncan Clark, and Ellen Clark, filed suit against the heirs of W.R. Blake, Sr.: W.R. Blake, III and Jacqueline Blake Beatty, to determine the interest conveyed to Blake, Sr. under the 1938 deed. The Clark heirs alleged that Blake, Sr. received a one-sixteenth ownership interest in the severed mineral estate and was therefore entitled to one- sixteenth of the one-eighth reserved royalty. The Blake heirs contended Blake, Sr. received a one-sixteenth royalty interest under the deed and therefore was entitled to a one-sixteenth royalty, or one-half of the one-eighth reserved royalty.

Neither the Clark nor the Blake heirs assert that the 1938 Blake deed is ambiguous. The construction of the deed is thus a question of law for the court. The primary duty of the courts in interpreting a deed is to ascertain the intent of the parties. But it is the intent of the parties as expressed within the four corners of the instrument which controls. In seeking to ascertain the intention of the parties, the court must attempt to harmonize all parts of a deed, since the parties to an instrument intend every clause to have some effect and in some measure to evidence their agreement.

The 1938 Blake deed first conveys "an undivided one-sixteenth (1/16) interest in and to all the oil, gas and other minerals in and under and that may be produced from" the 348-acre tract in Hockley County. The deed then purports to limit the interest conveyed, stating that the grantee "does not participate in any rentals or leases." Finally, the deed conveys to the grantee "the rights of ingress and egress at all times for the purpose of mining, drilling, exploring and developing said lands."

Obviously had the inserted language denying the grantee the rights to make leases and to receive delay rentals not been included in the deed, no one would dispute that the instrument conveyed only an undivided one-sixteenth interest in the mineral fee. The Blake heirs argue that precisely because this limiting language was included, the deed conveyed a royalty interest. They contend that the effect of the inserted language was to strip the estate first granted of its characteristics as a mineral fee and to leave only royalty.

There are five essential attributes of a severed mineral estate: (1) the right to develop (the right of ingress and egress), (2) the right to lease (the executive right), (3) the right to receive bonus payments, (4) the right to receive delay rentals, (5) the right to receive royalty payments. By its express terms, the 1938 Blake deed reserves to the grantor attributes (2) and (4): the right to lease and the right to receive rentals.

This court has before recognized that a mineral interest shorn of the executive right and the right to receive delay rentals remains an interest in the mineral fee. Delta Drilling Co. v. Simmons, 161 Tex. 122, 338 S.W.2d 143 (1960). In Delta Drilling, this court held that the grantor of an undivided one-eighth interest in a severed mineral estate had reserved the exclusive right to lease the property along with the right to receive all delay rentals. Nonetheless, we concluded that the grantee was the owner of a mineral interest and therefore entitled only to one-eighth of the royalty reserved in an oil and gas lease and not to a one-eighth royalty.

The courts of appeals have consistently applied the logic and the rule of Delta Drilling. See Grissom v. Guertersloh, 391 S.W.2d 167 (Tex.Civ.App.--Amarillo 1965, writ ref'd n.r.e.); Etter v. Texaco, 371 S.W.2d 702 (Tex.Civ.App.--Waco 1963, writ ref'd n.r.e.). In both Etter v. Texaco and Grissom v. Guetersloh, the courts of appeals were presented with deeds in which the grantor reserved the executive right and the right to receive delay rentals. Despite these reservations, the courts concluded that the deeds conveyed mineral interests and not royalty interests to the grantees.

The Blake heirs argue that this case can be distinguished from Delta Drilling, Etter, and Grissom. They assert the 1938 Blake deed retains to the grantor not only the right to lease and the right to receive delay rentals, but also the right to receive bonus payments and the exclusive right to develop the minerals. First, the Blake heirs contend that a party can "participate" in a lease in at least two ways: (1) by executing a lease and (2) by receiving any bonus given as consideration for execution of the lease. Thus, they argue that the language in the 1938 Blake deed, "but does not participate in any ... leases," prohibits the grantee from executing any oil and gas lease and from receiving any bonus. We find this argument unpersuasive.

The Blake heirs cite us to no authority which would support their definition of participation. If "participation" includes the right to receive bonus payments, why would it not encompass all lease benefits, including the rights to receive royalties and delay rentals? The parties to the 1938 deed thought it necessary to expressly reserve to the grantor the right to retain delay rentals. This suggests they did not define "participation" so broadly. The 1938 deed does not reserve the right to receive bonuses, and it is not necessarily included within the term "participation." The Blake heirs certainly do not contend that they cannot participate in the royalty reserved under the lease.

The Blake heirs also argue that despite the express language of the 1938 deed, the grantee did not receive the right of ingress and egress because that right necessarily was reserved to the grantor as part of the power to lease. Whether this contention is correct or not, it cannot distinguish this deed from those construed by the courts in Delta Drilling, Etter, and Grissom. In all of those cases, the exclusive right to lease was retained by the grantor. If the right to lease includes the exclusive right of ingress and egress, that right was retained by the grantors in Delta Drilling, Etter, and Grissom. The Blake heirs rely on Watkins v. Slaughter, 144 Tex. 179, 189 S.W.2d 699 (1945), to support their position that this deed conveys a royalty interest rather than a mineral interest. The deed in Watkins reserved to the grantor, Slaughter, a one-sixteenth interest in and to all of the oil, gas and other minerals in and under and that might be produced from the land. The deed then granted Watkins the exclusive right to lease and to collect bonus and delay rentals. Unlike in this case, however, the deed in Watkins unequivocally stated that the grantor should "receive the royalty retained herein only from actual production." This court placed great emphasis on the parties' designation of Slaughter's retained interest as a royalty interest. There is no such direct language of royalty in the 1938 Blake deed. We believe the intent of the parties as expressed by the terms of this deed was to convey a mineral interest.

This case is much more analogous to Delta Drilling, Etter, and Grissom, and is controlled by those decisions. We are aware that some jurisdictions and authorities have held that a deed purporting to grant a mineral interest, but reserving all the attributes of mineral ownership--the right to lease, the right to receive bonus, and the right to receive delay rentals--should be interpreted as conveying only a royalty interest.

However, as we stated, we can find nothing to distinguish this deed from those interpreted by this court in Delta Drilling and by the courts of appeals in Etter and Grissom. The common law in Texas has been that a conveyance such as the one in this case is a conveyance of minerals and not of royalty. We recognize the necessity for stability and certainty in the construction of mineral conveyances.

Therefore, we hold that the 1938 Blake deed conveyed to W.R. Blake, Sr. a one- sixteenth interest in the mineral fee. We reverse the judgment of the court of appeals and render judgment that the Blake heirs, W.R. Blake, III and Jacqueline Blake Beatty, are entitled to one-sixteenth of the one-eighth royalty reserved under the oil and gas lease.

NOTES

1. What does the court mean when it says that there are 5 *essential* elements of a severed mineral estate? Are they truly essential or are they merely component parts of the whole? To use the Holmesian aphorism are they 5 sticks in a bundle of sticks we call a mineral estate? Which, if any, of the 5 is necessary to make the interest a mineral interest, rather than some other type of interest? What other sticks might the court have identified? One of our authors has used a variation of the "bundle of sticks" analogy in discussing the mineral estate. His "mineral bug," a chalk board creature, has a development head, a leasing power leg, a royalty leg, a rental leg, and a bonus leg all attached to an expense-bearing body. In the *Altman v. Blake* conveyance a full mineral interest is conveyed and then some attributes of that interest are verbally stripped away. What mineral elements are removed from the *Altman* "mineral bug?" How far would the surgical process have to go to change the basic nature of the creature? See *French v. Chevron, infra.*

2. Can the owner of a mineral interest transfer the development right to Able, the executive right to Baker, the right to receive bonus payments to Cassie, the right to receive delay rentals to Danielle and the right to receive royalty to Ellen? Who is the owner of the mineral estate? If Baker exercises the executive right to lease to Rex Oil, will Able be able to develop oil and gas at the same time as Rex Oil? If so, are Able and Rex cotenants who share with one another? If not and if each cannot produce oil and gas, does this not suggest that the court's list is seriously flawed? Perhaps the "executive right" stick is not a separate stick from the "development" stick. See note 1 after *French v. Chevron infra.* Perhaps, too, the right to receive bonus should be regarded as correlative to the executive right.

3. Other jurisdictions follow the principle that the component parts of the mineral estate can be separately conveyed. *See e.g.*, Mull Drilling Co., Inc. v. Medallion Petroleum, Inc., 809 P.2d 1124, 115 O.&G.R. 52 (Colo.Ct.App. 1991); Westbrook v. Ball, 222 Miss. 788, 77 So.2d 274, 4 O.&G.R. 363 (1955); Antelope Production Co. v. Shriners Hospital for Crippled Children, 236 Neb. 804, 464 N.W.2d 159, 114 O.&G.R. 337 (1991); Boley v. Greenough, 2001 WY 47, 22 P.3d 854, 154 O.&G.R. 133.

4. Authority exists in almost all states for the severance of land into two estates, a surface estate and a mineral estate. *See e.g.*, Bodcaw Lumber Co. v. Goode, 160 Ark. 48, 254 S.W. 345 (1923); Lathrop v. Eyestone, 170 Kan. 419, 227 P.2d 136 (1951); Barker v. Campbell-Ratcliff Land Co., 1917 OK 208, 64 Okla. 249, 167 P. 468; Humphreys-Mexia Co. v. Gammon, 113 Tex. 247, 254 S.W. 296 (1923). It is also clear that the owner of a "unified" estate conveys both the surface and mineral estate unless specific reference is made to one or the

other. Harris v. Currie, 142 Tex. 93, 176 S.W.2d 302 (1943); McCoy v. Lowrie, 42 Wash.2d 24, 253 P.2d 415, 3 O.&G.R. 621 (1953).

5. In the Appalachian Basin where coal mining activities antedated oil and gas operations, the Pennsylvania courts have taken the position that the owner of the unified estate may divide it into three separate estates, the surface, the minerals, and the right to subjacent support. Consolidation Coal Co. v. White, 2005 PA Super. 155, 875 A.2d 318; Hetrick v. Apollo Gas Co., 415 Pa.Super. 189, 608 A.2d 1074, 120 O.&G.R. 70 (1992).

6. If Able, the owner of the unified fee simple absolute estate, executes and oil and gas lease to Baker and retains a possibility of reverter in the mineral estate, has there been a severance of the surface and mineral estate. Clearly as to Baker's fee simple determinable estate in the minerals there has been a severance but in Pounds v. Jurgens, 296 S.W.3d 100 (Tex.App. 2009), the court in the context of determining whether a sale of the surface for non-payment of taxes also conveyed the possibility of reverter, concluded. that there was not a severance. Thus, the purchaser of the surface estate at the tax sale received not only the surface estate but the possibility of reverter in the minerals.

———

By the reason of the civil law antecedents to the contemporary jurisprudence of Louisiana, the law of oil and gas in that state, unique in a number of important respects, for many years was characterized by a number of ambiguities and uncertainties. The Louisiana Mineral Code, effective January 1, 1975, did much to clarify that law. Among other things the Code clearly adopted the nonownership theory both for solid minerals and for oil and gas and clearly and explicitly reaffirmed the application of the rule of capture and the doctrine of correlative rights.

More important for present purposes, the Code utilizes the generic term *mineral rights* to describe three basic interests which may be created by a landowner, namely the *mineral servitude,* the *mineral royalty,* and the *mineral lease,* as well as a variety of other interests (*e.g.,* a production payment). Mineral rights are denominated *real rights,* which are subject either to the prescription of nonuse for ten years (*prescription liberandi causa*) or, in the case of a mineral lease, to special rules of law governing the term of their existence (*viz.,* limiting the primary term of the lease to ten years). Mineral rights are characterized as incorporeal immovables having a situs in the parish or parishes (counties) in which the land burdened is located; they are alienable and heritable; and sales, contracts and judgments affecting them are subject to the laws of registry.

———

Martin & Kramer, *Williams & Meyers Oil and Gas Law* § 216

A *mineral servitude* is defined by Article 21 as "the right of enjoyment of land belonging to another for the purpose of exploring for and producing minerals and reducing them to possession and enjoyment." There is no obligation on the servitude owner to exercise it, but if he does, he may use only so much of the land as is reasonably necessary to conduct his operations, and he is obligated, insofar as practicable, to restore the surface to its original condition at the earliest reasonable time. In substance, then, a mineral servitude is the equivalent of a mineral interest in states other than Louisiana.

A *mineral royalty* is defined by Article 80 as "the right to participate in production of minerals from land owned by another or land subject to a mineral servitude owned by another. Unless expressly qualified by the parties, a royalty is a right to share in gross production free of mining or drilling and production costs." The royalty owner lacks executive rights and may not conduct operations to explore for or produce minerals. In substance, then, a mineral royalty is the equivalent of a royalty interest in states other than Louisiana.

A *mineral lease* is defined by Article 114 as "a contract by which the lessee is granted the right to explore for and produce minerals." Two or more noncontiguous tracts of land may be included in a single lease, and operations on the land burdened by the lease or land unitized therewith sufficient to maintain the lease according to its terms will continue the lease in force as to the entirety of the land burdened. This definition, it is seen, would be equally appropriate in states other than Louisiana.

[Prescription liberandi causa]

Perhaps the most important feature of the Louisiana law for present consideration is that usually the mineral servitude and the mineral royalty are subject to the operation of the civil law concept of prescription *liberandi causa* on ten years nonuse. Thus Article 27 provides that a mineral servitude is extinguished by prescription resulting from nonuse for ten years, and Article 85 provides that a mineral royalty is extinguished by prescription from nonuse for ten years.

Prescription of a mineral servitude is codified by Articles 28-61 and prescription of the mineral royalty is codified by Articles 86-100 of the Mineral Code. The details of the code provisions are generally the same for these two types of mineral rights and in our discussion we shall indicate the similarities and divergences.

Prescription commences both for a mineral servitude and a mineral royalty "from the date on which it is created."

[Interruption of prescription by use]

Articles 36 and 87 provide that prescription of nonuse running against a mineral servitude or a mineral royalty is interrupted by the production of any mineral covered by the act creating the servitude or royalty. Prescription is interrupted on the date on which actual production begins and commences anew from the date of cessation of actual production. Articles 38 and 88 then make it clear that the production requirement is satisfied if there has been production which is actually saved; it is not necessary that the production be in paying quantities. Production from a conventional or compulsory unit including all or part of the tract burdened by a mineral servitude or a mineral royalty interrupts prescription, but if the unit well is on land other than that burdened by the servitude or royalty, the interruption extends only to that portion of the tract included in the unit. (Articles 37 and 89).

Prior case law had established that the running of prescription against a mineral royalty or mineral servitude might be interrupted by constructive production when a well capable of paying production was shut-in by a lessee of land burdened by a mineral royalty or servitude. This rule is preserved by the Code. A tested shut-in well capable of producing minerals in paying quantities on the tract burdened or on a unit that includes all or some part thereof will interrupt prescription. If the shut-in well is on a portion of the unit other than the tract burdened, then prescription will be interrupted only as to the portion of the tract included in the unit. (Articles 34 and 90). If the land, or part thereof, burdened by a mineral servitude or a mineral royalty is included in a unit on which there is a well shut-in prior to the creation of the unit, located on other land within the unit, and capable of paying production, prescription is interrupted on and commences anew from the effective date of the order or act creating the unit. (Articles 35 and 91).

For interruption of prescription of a mineral servitude it is sufficient that there be "good faith *operations* for the discovery and production of minerals" (Article 29; emphasis supplied) as distinguished from the *production* required by Article 87 for the interruption of prescription of a mineral royalty.

By good faith is meant that the operations must be

(1) commenced with reasonable expectation of discovering and producing minerals in paying quantities at a particular point or depth,

(2) continued at the site chosen to that point or depth, and

(3) conducted in such a manner that they constitute a single operation although actual drilling or mining is not conducted at all times.

Interruption of prescription of the mineral servitude commences on the date actual drilling or mining operations ("spudding-in" of a well) are commenced, as distinguished from the date preparatory operations (*e.g.*, surveying, clearing a site) are commenced. (Article 30). And prescription commences anew from the last day on which actual drilling or mining operations are conducted. If prescription commences anew on cessation of operations, it may later be interrupted by the resumption of good faith operations to complete the well or mine or place it in production. (Article 32). And if prescription has commenced anew after production has ceased, it may be interrupted by good faith operations to restore production or to secure new production from the same well or mine. (Article 39). The cessation of the good faith operations after interruption of prescription will cause prescription to begin anew from the last day on which operations are conducted in good faith to secure or restore production in paying quantities with reasonable expectation of success. (Article 41).

[Interruption of prescription by acknowledgment and contractual extension of prescription]

It has been established in Louisiana that accrual of prescription may be interrupted by an acknowledgment of the servitude or royalty which is made for the purpose of interrupting the running of prescription. And it was also established that the period of prescription could be extended for a term less than ten years. The litigation on the matter was most extensive. Several cases indicated a reluctance to find intent to interrupt prescription by acknowledgment. Frequently arising was the question whether the joinder of a landowner in a lease (or a counterpart of a lease) had the effect of interrupting prescription by acknowledgment. Clearly such joinder did not necessarily have that effect; this was a question of intent. Similarly, if the landowner joined the owner of a mineral servitude in authorizing pooling or unitization of leased premises by a lessee, the agreement might be treated as an acknowledgment made with intent to interrupt prescription, which then began to run anew.

. . .

[Suspension of prescription]

Suspension of prescription (as distinguished from interruption of prescription) is dealt with in Articles 58-61 as concerns mineral servitudes and Articles 97-100 as concerns mineral royalties. Interruption of prescription means that the running of the prescriptive period stops, and if the running begins anew, the prescriptive period is ten years from the date of resumption. Suspension means that for some period of time the prescriptive period does not run, but if the period of suspension comes to an end the prescriptive period begins to run anew as if the date of resumption immediately followed the last date immediately preceding the suspension. Thus if there were five years of nonuse prior to suspension, the prescriptive period following suspension would be five more years.

Articles 59-61 and 98-100 deal with suspension of prescription by obstacle. If the owner of a mineral servitude is prevented from using it by an obstacle which he can neither prevent nor remove, the prescription of nonuse does not run as long as the obstacle remains. This preserves a general principle of Louisiana law which had been applied only in rare instances; most of the prior jurisprudence had dealt with defining things that were not obstacles. An obstacle as to any mineral constitutes an obstacle as to the running of prescription as to all minerals covered by the act creating a servitude. Article 61 makes it clear that a compulsory unitization order does not constitute an obstacle.

The major difference in the effect of obstacle on mineral servitudes and on mineral royalties is tied to the fact that "production" is required to interrupt prescription of a mineral royalty whereas "good faith operations" will suffice to interrupt prescription of a mineral servitude. Hence to suspend prescription of a mineral royalty there must be an obstacle "to actual production" as distinguished from an obstacle by which "the owner of a mineral servitude is prevented from using it." Thus it would not be sufficient to suspend prescription of a mineral royalty that there were an obstacle to drilling or reworking operations although such an obstacle would suffice to suspend prescription of a mineral servitude.

. . .

[Effect of multiple mineral servitudes or mineral royalties]

When servitudes or mineral royalties are created on separate, noncontiguous tracts by a single instrument, or when servitudes or mineral royalties are created by separate instruments on contiguous parts of a single tract, the servitudes or mineral royalties are separate for each tract (or parts of the tract covered by the individual instruments). (Articles 64, 73, 103). Drilling operations or production otherwise sufficient to interrupt prescription will have that effect only as to the specific tract or tracts upon which such operations are pursued. This gives rise to a number of problems. Where a road divides a tract on which a servitude is created, the question whether a single mineral servitude or royalty is created may turn upon the ownership of the bed of the road. One tract may be subject to a servitude or royalty affecting that tract only and also may be subject to a servitude or royalty affecting a larger parcel in which the tract is included. And where servitudes or mineral royalties are created in separate contiguous tracts forming a single continuous body of land, a question may arise whether the instrument created single or multiple interests.

[Divisibility of mineral servitude and mineral royalty]

Prior to the enactment of the Louisiana Mineral Code there had been some continuing controversy over the question of whether mineral servitudes and mineral royalty were divisible or indivisible. It appears to have been the general view that the indivisibility principle was applicable, but this rule had been "functionally circumvented in several instances." Thus it was held that a mineral servitude owner might divide the "advantage" of the servitude by assigning to another person all of his interest in a designated portion of the land affected by the servitude and create a situation whereby user of a portion would not prevent prescription of the remainder. And several cases involving unitization had indicated that mineral servitudes were divisible in the event of inclusion of less than all of the premises burdened by the servitude in a compulsory unit.

Article 62 of the Mineral Code specifies that "the rights and obligations of the owner of a mineral servitude are indivisible" except as otherwise provided in the succeeding articles. (A parallel provision relating to the mineral royalty is found in Article 101). The succeeding articles then declare that:

A single mineral servitude is created by an act that affects a continuous body of land although individual tracts or parcels within the whole tract are separately described.

An act creating mineral servitudes on noncontiguous tracts of land creates as many mineral servitudes as there are tracts unless the act provides for more.

The division of a tract burdened by a mineral servitude does not divide the servitude.

The owners of several contiguous tracts of land may create a single mineral servitude in favor of one or more of them or of a third party.

Co-owners of land constituting a continuous whole may partition it and reserve a single mineral servitude in favor of one or more of them.

A single mineral servitude is established on a continuous tract of land notwithstanding that certain horizons or levels are excluded or the right to share in production varies as to different portions of the tract or different levels or horizons.

The conveyance or reservation by a mineral servitude owner of a portion of his rights does not divide the mineral servitude but creates only a co-ownership, except that if a person other than the servitude owner acquires all of the rights granted by the act creating the servitude in a specific geographic area, the servitude is divided.

Execution of a lease or other contract for use or development of a portion of a tract burdened by a mineral servitude does not divide the servitude.

Unitization of a portion of a tract burdened by a mineral servitude does not divide the servitude.

As has been indicated, Articles 3 and 72 permit the parties to alter by contract certain of the legal rules applicable to mineral servitude, and Article 103 makes this freedom of contract applicable also to mineral royalties. The parties may fix the term of a mineral servitude or shorten the applicable period of prescription of nonuse; but they may not provide for a period of prescription greater than ten years. They may restrict the rules of use regarding interruption of prescription, provided that they may not be made less burdensome; the parties may agree that an interruption of prescription resulting from unit operations or production shall extend to the entirety of the tract burdened by the servitude regardless of the location of the well or of whether all or only part of the tract is included in the unit.

[Limitation on continuation of lease without drilling or mining operations or production]

For many years it remained uncertain whether an oil and gas lease was subject to liberative prescription. Few occasions arose for testing the question. Such production or drilling operations on the land as would suffice to extend the life of the lease into the secondary term would usually interrupt prescription, if applicable to leases. The standard use of relatively short primary terms in oil and gas leases (ten years or less) resulted in termination of the lease by its express provisions if there were no production from nor drilling operations upon the land at the expiration of the primary term, so the prescription question was rarely raised.

The question of the applicability of prescription *liberandi causa* to oil and gas leases was resolved by *Reagan v. Murphy*[74] decided in 1958. The Supreme Court

---

[74]   235 La. 529, 105 So.2d 210, 10 O.&G.R. 514 (1958), noted and criticized, 19 La.L.Rev. 207 (1958).

held that prescription *liberandi causa* was inapplicable to oil and gas leases. It declared that an oil and gas lease produces merely personal rights and obligations between the parties rather than real rights and hence the lease is not subject to prescription *liberandi causa.*

Article 115 of the Mineral Code specifically codifies the rule of *Reagan v. Murphy* by declaring that "The interest of a mineral lessee is not subject to the prescription of nonuse." The article proceeds to specify that "the lease must have a term" and it then imposes a limit on the duration of the primary term:

> Except as provided in this article, a lease shall not be continued for a period of more than ten years without drilling or mining operations or production. Except as provided in this article, if a mineral lease permits continuation for a period greater than ten years without drilling or mining operations or production, the period is reduced to ten years.

The Comment to this article argues that this provision, although it may be described as new, is consistent with the public policy underlying the system of prescription applicable to other mineral rights.

> . . . The net effect of this limitation in combination with the first sentence [of the article] is to free the mineral lease of the use rules applicable to servitudes while accomplishing the end of prohibiting all basic forms of mineral rights from remaining outstanding for periods greater than ten years without some form of development. . . . This leaves the matter of what form of drilling or mining operations or production will maintain the mineral lease within the discretion of the contracting parties. Established custom in this regard indicates that there is virtually no danger to the basic philosophy of a system of terminable mineral rights in permitting this freedom.

SECTION 2. CONVEYANCING PROBLEMS IN SEVERING MINERALS
A. FORMALITIES

Martin & Kramer, *Williams & Meyers Oil and Gas Law*
§ 220. Conveyancing Formalities and Requirements

For an instrument effectively to create a mineral, royalty or leasehold interest it must satisfy the usual requirements of the jurisdiction for an effective conveyance. Among such requirements found in one or more jurisdictions are the following:

(a) The instrument must contain appropriate words of grant.

(b) There must be a memorandum in writing sufficient to satisfy the statute of frauds, and in some states the writing must be sealed.

(c) Delivery and acceptance of the instrument is required.

(d) The usual requirements of the jurisdiction applicable to conveyances of realty concerning acknowledgment of the instrument and privy acknowledgment by the wife of the grantor or joinder by a spouse in certain instances are applicable to conveyances of the interests here considered.

(e) A special requirement of recordation or listing the interest for taxation within a given period of time may be applicable to mineral deeds.

The question of whether consideration is required for an oil and gas lease has been the subject of some dispute. There is, however, no dispute that consideration is required for a valid contract for a lease. Some courts and writers have assumed the requirement of consideration for a lease. Certainly in Louisiana a "serious" consideration is required. More frequently, however, the oil and gas lease has been viewed primarily as a conveyance rather than as a contract and in accordance with rules applicable to conveyances, other courts and writers have declared that consideration is not required.

Apart from the question whether consideration is requisite to a lease, the payment of consideration may be important in a number of contexts. If consideration has been bargained for but has not been tendered or paid, a conveyance or lease may be set aside for failure of consideration. Gross inadequacy of consideration may be such evidence of fraud in the light of the surrounding circumstances as to lead a court to rescind or cancel a conveyance. And in any event, whether or not consideration is requisite to the validity of the particular conveyance in question, payment of some consideration is desirable so that the grantee or lessee will be entitled to the status of a "purchaser" under applicable recording statutes and may be necessary for him to have the benefit of estoppel by deed. Payment of consideration may also be important to prevent the implication of a resulting use or trust or to prevent the conveyance from being voidable at the behest of creditors. Further, the validity of an unexecuted contract to convey may turn on the existence of consideration.

Occasionally the question arises as to whether bonus may be recovered by the lessee from the lessor in the event of failure of title. Certainly, if failure of title amounts to a breach of the covenant of title in the lease, the lessee should be entitled to recover the bonus paid.

Except in Louisiana, the courts do not generally appear concerned with the adequacy of the consideration and hence a consideration of one dollar is usually considered sufficient for the lease or other conveyance.

In summary it may reasonably be argued that except in Louisiana, consideration is not requisite to the validity of such conveyances as are here considered but in view of the contrary authority, however erroneous it may be, and the subsidiary advantages of the payment of consideration, it would be unwise for a grantee to fail to pay some consideration, especially for an oil and gas lease.

NOTE

Problems that can be characterized as conveyancing formalities are frequently involved in appellate oil and gas litigation. *See generally* United States Pipeline Corp. v. Kinder, 609 S.W.2d 837, 69 O.&G.R. 151 (Tex.Civ.App.1980, error ref'd n.r.e.) (an agreement to sell unproduced gas in the ground is an interest in realty governed by the statute of frauds and void in the absence of an adequate description of the acreage involved); St. Romain v. Midas Exploration, Inc., 430 So.2d 1354, 77 O.&G.R. 115 (La.App.1983) (a Louisiana oil and gas lease signed by the lessor is valid without the lessee's signature if acceptance by the lessee can be inferred from words or actions); Page v. Fees-Krey, Inc., 617 P.2d 1188, 67 O.&G.R. 520 (Colo.1980) (land recording law is applicable to the reservation of an overriding royalty in an assigned federal oil and gas lease); Christy v. Petrol Resources Corp., 102 N.M. 58, 691 P.2d 59, 82 O.&G.R. 555 (App.1984) (the term, "net profits interest" does not always describe an interest in land but the nature of interests so designated will vary with the provisions of particular transactions); Jones v. Alpine Investments, Inc., 1987 OK 113, 764 P.2d 513, 102 O.&G.R.339 (unincorporated association lacks capacity to hold title to mineral interest); Davis v. Griffin, 298 Ark. 633, 770 S.W.2d 137 (1989) ("words of conveyance" necessary for conveyance of "oil and gas interests"); Eagle Gas v. Doran & Associates, Inc., 387 S.E.2d 99 (W.Va.1989) (acceptance of delay rentals cured failure to deliver lease); Carlile v. Carlile, 1992 OK 57, 830 P.2d 1369, 119 O.&G.R. 78 (delivery requirement met under contested fact situation); Washington v. Slack, 813 P.2d 447 (Mont.1991) (misdescription in mortgage instrument not destructive of priority); Energetics, Ltd. v. Benchley, 189 Mich.App. 247, 471 N.W.2d 641 (1991) (last recordation, not lease termination, is event from which time runs under statute requiring recording of severed mineral rights within twenty years); Exxon Corp. v. Breezevale Limited, 82 S.W.3d 249, 157 O.&G.R. 785 (Tex.App. 2002, rev. denied) (Statute of Frauds applies to agreement relating to production sharing contract); Bixler v. Oro Management Co., LLC, 2004 WY 29, 86 P.3d 843 (common law doctrine of merger applies to agreement executed prior to excution of warranty deed).

B. INTERPRETATION OF THE WORD "MINERALS"

(*See* Martin & Kramer, *Williams & Meyers Oil and Gas Law* § 219)

MOSER v. UNITED STATES STEEL CORP.

Supreme Court of Texas, 1984.
676 S.W.2d 99, 82 O.&G.R. 143.

On Motion for Rehearing

CAMPBELL, JUSTICE. Our opinion of June 8, 1983, is withdrawn. The Motion for Rehearing is denied.

This is a suit to quiet title to an interest in uranium ore. We must determine whether uranium is included in a reservation or conveyance of "oil, gas and other minerals." The trial court awarded title to the defendant mineral owners, and the court of appeals affirmed the trial court judgment. 601 S.W.2d 731. We affirm the judgments of the courts below, and hold that uranium is a part of the mineral estate.

The Mosers, plaintiffs, and the Gefferts, defendants, own neighboring tracts of land in Live Oak County. Prior to 1949, the boundary between the Mosers' land and that of the Gefferts was a winding road. In 1949, the road was straightened and, as a result, no longer represented the true boundary between the two ranches. The new road separated a 6.77 acre tract of the Geffert ranch on the Mosers' side of the road and a 6.42 acre tract of the Moser ranch on the Gefferts' side of the road. To avoid crossing the highway to reach their tracts, the Mosers' predecessor in title and the Gefferts executed similar deeds conveying the surface estates of the isolated tracts to the other party. The 1949 deeds contain identical language reserving:

> [A]ll of the oil, gas, and other minerals of every kind and character, in, on, under and that may be produced from said tract of land, together with all necessary and convenient easements for the purpose of exploring for, mining, drilling, producing and transporting oil, gas or any of said minerals.

Substantial quantities of uranium were discovered on the 6.77 acre tract. The Mosers, as surface owners of the 6.77 acre tract, sued the Gefferts to establish ownership of the uranium. The Gefferts, as owners of the mineral estate under the 6.77 acre tract, counterclaimed to establish that uranium is one of the "other minerals" reserved from the conveyance of the surface.

At trial, the parties offered conflicting evidence on the depth of the uranium deposits and the effect its removal would have on the surface. Special issues were submitted based on the test set out by this Court in Reed v. Wylie, 554 S.W.2d 169 (Tex.1977) (*Reed* I): if substantial quantities of the mineral lie so near the surface that extraction, as of the date of the severance of the surface and mineral estates, would necessarily have destroyed the surface, the surface owner has title to the mineral. The jury found there would have been no substantial surface destruction at the time the deed was executed. The trial court accordingly held the uranium was a part of the mineral estate retained by the Gefferts in the 1949 deed.

After the Mosers' appeal to the court of civil appeals, but prior to final disposition by that court, we rendered our decision in Reed v. Wylie, 597 S.W.2d 743 (Tex.1980) (*Reed* II). The court of civil appeals held that *Reed* II should govern the appeal. In *Reed* II, we modified the rule of *Reed* I by holding that a substance "near the surface" is a part of the surface estate if it is shown that any reasonable method of production, at the time of conveyance or thereafter, would consume, deplete, or destroy the surface. 597 S.W.2d at 747. A deposit within 200 feet of the surface was held to be "near the surface" as a matter of law. In addition, we held if a surface owner establishes ownership of a substance at or near the surface, the surface owner owns the substance beneath the tract at whatever depth it may be found. *Id.* at 748. The *Moser* court of civil appeals found, as a matter of law, that at the time of trial the only reasonable method of mining uranium from the tract was by in-situ leaching or solution mining, a process which it found did not result in substantial destruction of the surface. 601 S.W.2d at 734.[75] Accordingly, the court of civil appeals affirmed.

We have previously attempted to create a rule to effect the intent of the parties to convey valuable minerals to the mineral estate owner, while protecting the surface estate owner from destruction of the surface estate by the mineral owner's extraction of minerals. *See* Reed v. Wylie, 597 S.W.2d 743 (Tex.1980); Reed v. Wylie, 554 S.W.2d 169 (Tex.1977); Acker v. Guinn, 464 S.W.2d 348 (Tex.1971). In so doing, we decided that determinations of title should be based on whether a reasonable use of the surface by the mineral owner would substantially harm the surface. Application of this rule has required the determination of several fact issues to establish whether the owner of the surface or the mineral estate owns a substance not specifically referred to in a grant, reservation or exception. *See* Reed v. Wylie, 597 S.W.2d 743, 750 (Spears, J., concurring). As a result, it could not be determined from the grant or reservation alone who owned title to an unnamed substance. Determining the ownership of minerals in this manner has resulted in title uncertainty. We now abandon, in the case of uranium, the *Acker* and *Reed* approach to determining ownership of "other minerals" and hold that title to uranium is held by the owner of the mineral estate as a matter of law.

In Texas, the mineral estate may be severed from the surface estate by a grant of the minerals in a deed or lease, or by reservation in a conveyance. *See* Humphreys-Mexia Co. v. Gammon, 113 Tex. 247, 254 S.W. 296 (1923). This severance is often accomplished by a grant or reservation of "oil, gas and other minerals." Consequently, Texas courts have had many occasions to construe the scope of the term "other minerals." We have determined that some unnamed substances have been impliedly conveyed or reserved in mineral conveyances by cataloging each, on a substance-by-substance basis, as part of the surface or mineral estate as a matter of law. *See, e.g.,* Sun Oil Co. v. Whitaker, 483 S.W.2d 808 (Tex.1972) (fresh water not included in mineral estate reservation of "oil, gas, and other minerals"); Heinatz v. Allen, 147 Tex. 512, 217 S.W.2d 994 (1949) (devise of "mineral rights" held not to include limestone and building stone); Atwood v. Rodman, 355 S.W.2d 206, (Tex.Civ.App.--El Paso 1962, writ ref'd n.r.e.) ("oil, gas, and other minerals" did not include limestone, caliche, and surface shale); Union Sulphur Co. v. Texas Gulf Sulphur Co., 42 S.W.2d 182 (Tex.Civ.App.--Austin 1931, writ ref'd) (solid sulphur deposits conveyed by ordinary oil and gas lease); Praeletorian Diamond Oil Ass'n v. Garvey, 15 S.W.2d 698 (Tex.Civ.App.--Beaumont 1929, writ ref'd) (gravel and sand not intended to be included in lease for "oil and other minerals"); Reed v. Wylie, 597 S.W.2d 743,(Tex.1980) (near surface lignite, iron and coal is part of the surface estate as a matter of law).

---

[75]   We note that under *Reed* II, however, the correct time frame for determining the existence of a surface destructive method of mineral extraction was not solely at the time of trial (the time used by the court of civil appeals), or solely at the time of conveyance, but the time of conveyance or *thereafter*. *See* Reed v. Wylie, 597 S.W.2d 743, 747 (Tex.1980).

In making these determinations of ownership, our courts have considered a number of construction aids. We have refused to employ the *ejusdem generis* rule of construction to limit the term "oil, gas and other minerals" to hydrocarbons. Southland Royalty Co. v. Pan American Petroleum Corp., 378 S.W.2d 50 (Tex.1964). Likewise, we have acknowledged that the scientific or technical definition of a disputed substance is not determinative of whether it is a mineral, because the term "other minerals" would then "embrace not only metallic minerals, oil, gas, stone, sand, gravel, and many other substances but even the soil itself." Heinatz v. Allen, 147 Tex. 512, 217 S.W.2d 994, 997 (1949). Such a construction would eliminate any distinction between the surface and the mineral estates. We have, however, approved of considering whether the substance is thought to be a mineral within the ordinary and natural meaning of the term. *See* Heinatz v. Allen, 217 S.W.2d at 997; Psencik v. Wessels, 205 S.W.2d 658, 660-61 (Tex.Civ.App.-- Austin 1947, writ ref'd). The knowledge of the parties of the value, or even the existence of the substance at the time the conveyance was executed has been found to be irrelevant to its inclusion or exclusion from a grant of minerals. *See* Cain v. Neumann, 316 S.W.2d 915, 922 (Tex.Civ.App.--San Antonio 1958, no writ). Accord Barden v. Northern Pacific Ry., 154 U.S. 288, 314, 14 S.Ct. 1030, 1033, 38 L.Ed. 992 (1893) ("[T]he knowledge or want of knowledge at the time [of the grant] by the grantee in such cases, of the property reserved in no respect affects the transfer to him of the title to it.") In Acker v. Guinn, 464 S.W.2d 348, (Tex.1971), we quoted with approval Professor Eugene Kuntz' theory that the proper focus when construing an implied grant of minerals is the general, rather than the specific, intent of the parties. We adopted the view that the general intent of parties executing a mineral deed or lease is presumed to be an intent to sever the mineral and surface estates, convey all valuable substances to the mineral owner regardless of whether their presence or value was known at the time of conveyance, and to preserve the uses incident to each estate. *Id.* at 352; Kuntz, The Law Relating to Oil and Gas in Wyoming, 3 Wyo.L.J. 107, 112 (1949).

Professor Kuntz suggested the apparently irreconcilable conflict between the rights of the surface owner to preserve the integrity of his surface fee and the right of a mineral owner who takes a mineral under an implied grant to extract his mineral could be compromised as follows:

> The rights of the surface owner to subjacent support and his right to the use of the top-soil in its place would have to be respected, and at the same time, the owner of the mineral fee should have a right of extraction. Since the right of extraction could only be exercised by destruction of the surface owner's enjoyment, it could only be accomplished with compensation for the damages to the surface estate. . . . Specific mention of the substance, however, together with the usual provisions for extraction, would demonstrate a specific intention to make the surface right subject to the rights of access for purposes of extraction and would not make the mineral owner accountable for necessary damage flowing therefrom.

Kuntz, The Law Relating to Oil and Gas in Wyoming, 3 Wyo.L.J. 107, 115 (1949).[76]

We now hold a severance of minerals in an oil, gas and other minerals clause includes all substances within the ordinary and natural meaning of that word, whether their presence or value is known at the time of severance. Heinatz v. Allen, 147 Tex. 512, 217 S.W.2d 994 (1949); Cain v. Neumann, 316 S.W.2d 915, (Tex.Civ.App.--San Antonio 1958, no writ). We also hold uranium is a mineral within the ordinary and natural meaning of the word and was retained in the Gefferts' conveyance of the 6.77 acre tract to the Mosers. We continue to adhere, however, to our previous decisions which held certain substances to belong to the surface estate as a matter of law. *See, e.g.,* Heinatz v. Allen, 147 Tex. 512, 217 S.W.2d 994 (1949) (building stone and limestone); Atwood v. Rodman, 355 S.W.2d 206 (Tex.Civ.App.--El Paso 1962, writ ref'd n.r.e.) (limestone, caliche, and surface shale); Fleming Foundation v. Texaco, 337 S.W.2d 846 (Tex.Civ.App.--Amarillo 1960, writ ref'd n.r.e.) (water); Psencik v. Wessels, 205 S.W.2d 658 (Tex.Civ.App.--Austin 1947, writ ref'd) (sand and gravel); Reed v. Wylie, 597 S.W.2d 743 (Tex.1980) (near surface lignite, iron and coal).

Having established that the mineral owner has title to uranium, we now must determine the issue of reasonable use of the surface estate by the uranium owner. The mineral owner, as owner of the dominant estate, has the right to make any use of the surface which is necessarily and reasonably incident to the removal of the minerals. *See* Sun Oil Co. v. Whitaker, 483 S.W.2d 808 (Tex.1972); Getty Oil Co. v. Jones, 470 S.W.2d 618, 627 (Tex.1971); Humble Oil & Refining Co. v. Williams, 420 S.W.2d 133 (Tex.1967). This is an imperative rule of mineral law; a mineral owner's estate would be worthless without the right to reach the minerals. A corollary of the mineral owner's right to use the surface to extract his minerals is the rule that the mineral owner is held liable to the surface owner only for negligently inflicted damage to the surface estate. General Crude Oil v. Aiken Co., 162 Tex. 104, 344 S.W.2d 668 (1961); Stradley v. Magnolia Petroleum Co., 155 S.W.2d 649 (Tex.Civ.App.--Amarillo 1941, writ ref'd); 1 H. Williams & C. Meyers, Oil and Gas Law §§ 217, 218.8 (1981).

Restricting the mineral owner's liability to negligently inflicted damage to, or excessive use of, the surface estate is justified where a mineral is specifically conveyed. It is reasonable to assume a grantor who expressly conveys a mineral which may or must be removed by destroying a portion of the surface estate anticipates his surface estate will be diminished when the mineral is removed. It is also probable the grantor has calculated the value of the diminution of his surface in the compensation received for the conveyance. This reasoning is not compelling when a grantor conveys a mineral which may destroy the surface in a conveyance

---

[76]　For a discussion of what is characterized as the "divergence between the approach of *Acker* and that advocated by Professor Kuntz in the article upon which the Texas court relies" see Maxwell, *Colorado Oil and Gas Conveyancing in Context*, 56 Colo.L.Rev. 495, 496 (1985).[Eds.]

of "other minerals."

We hold the limitation of the dominant mineral owner's liability to negligently inflicted damages does not control in a case such as this, in a general conveyance of "other minerals." When dealing with the rights of a mineral owner who has taken title by a grant or reservation of an unnamed substance such as this, liability of the mineral owner must include compensation to the surface owner for surface destruction.[77]

This holding does not affect the right of the mineral owner to enter the surface estate and use so much of the surface as is reasonably necessary to remove the minerals. Ball v. Dillard, 602 S.W.2d 521, 523 (Tex.1980); Harris v. Currie, 142 Tex. 93, 176 S.W.2d 302, 305 (Tex.1943). As in the case of the mineral owner who takes under a specific grant, the mineral owner under a grant of "other minerals" is restricted in his use of the surface estate by the dictates of the "due regard" or "accommodation doctrine." This rule is applied when the surface owner and mineral owner are attempting to use the surface estate for two conflicting and incompatible uses. We held, in Getty Oil Co. v. Jones, 470 S.W.2d 618 (Tex.1971), that even though the mineral owner held the dominant estate, he must exercise his rights of surface use with due regard for the rights of the surface owner.

Because of the extent of public reliance on our holdings in Acker v. Guinn, 464 S.W.2d 348 (Tex.1971), and Reed v. Wylie, 597 S.W.2d 743 (Tex.1980), and because of an inability to foresee a coming change in the law, the rules announced in this case are to be applied only prospectively from the date of our original opinion, June 8, 1983. *See* Sanchez v. Schindler, 651 S.W.2d 249, 254 (Tex.1983).

We affirm the judgments of the courts below. We hold as a matter of law the Gefferts are the owners of the uranium in the 6.77 acre tract.

RAY, JUSTICE, dissenting.

. . . I would hold that the new compensation rule applies in this case, so that the Mosers are entitled to compensation from the mineral owners for the destruction of their surface estate.

NOTES

1. In FRIEDMAN v. TEXACO, INC., 691 S.W.2d 586, 85 O.&G.R. 368 (Tex.1985), the court dealt with an appeal from a determination that the surface

---

[77] This holding does not affect the statutory duty of the mineral owner, or his lessee, to reclaim the surface after surface mining. *See* Texas Uranium Surface Mining and Reclamation Act, Nat.Res.Code Ann. §§ 131.001 to --.270. *See also* Surface Mining Control and Reclamation Act of 1977, 30 U.S.C. § 1201 *et seq.* (1977). The Texas Act requires an operator who is surface mining uranium to submit and effect a reclamation plan or lose his performance bond and face civil and criminal penalties. Tex.Nat.Res.Code. Ann. §§131.101 to --.270. The Code provides the surface owner with an opportunity to have the land classified as unsuitable for surface mining if he believes the land cannot be reclaimed to the Code's strict standards. *Id.* §§ 131.038, --.039, --.047. The Code also demands all reclamation efforts to proceed as contemporaneously as practicable with the surface mining operation. *Id.* §131.102(b)(14).

owners of a 640 acre tract owned the uranium on the tract, affirming and holding that *Moser* was "prospective in application only" applying "to those severances of the surface and mineral estates occurring after June 8, 1983," the date of the Texas Supreme Court's original opinion in *Moser* :

> In 1939, Adele K. Friedman and her mother leased the oil, gas and other minerals on the land to Magnolia Petroleum Company, predecessor of Mobil Oil Corporation. In 1959, Friedman and her husband conveyed the land to T.J. Martin, subject to a mineral reservation. The deed expressly reserved the right, title and interest in 'the oil, gas and other minerals, in and under said tract of land. . . .' The deed contained no definition of 'other minerals' or other indication as to what 'other minerals' included. In 1977, Martin entered into a mining lease with Texaco, Inc. The leased substances expressly included uranium.

> . . .

> The problem in this case is that at the time Friedman conveyed the surface estate to Martin in 1959, the *Acker* and *Reed* opinions had not been written. Thus, Friedman argues that the rules announced in *Moser* should apply at the time of her conveyance, and that therefore, she owns the uranium as the mineral estate owner. However, at the time of Martin's mineral lease to Texaco, in 1977, this court had written its opinion in *Acker* (1971), and that same year, 1977, issued its opinion in *Reed.* Thus, Texaco urges that it had a right to rely on those decisions when it entered into its lease with Martin.

> . . .

> Friedman insists that even if *Moser* cannot be applied retroactively, that she at least had a right to rely on the law in effect at the time of her conveyance to Martin. The only case prior to *Acker* specifically dealing with uranium is Cain v. Neumann, 316 S.W.2d 915 (Tex.Civ.App.--San Antonio 1958, no writ). *Cain* did not purport to announce a rule that uranium would always be included in a conveyance of 'other minerals.' Rather, the decision that uranium was included in a particular mineral lease was based on an examination of special provisions in that lease. 316 S.W.2d at 922. Further, as we have previously noted in *Acker,* there is no indication in *Cain* whether mining the uranium would destroy the surface. *See Cain,* 316 S.W.2d at 921-22, *reviewed by Acker,* 464 S.W.2d at 352. We conclude that the Texas law on whether uranium was included in a conveyance or reservation of 'minerals' was unsettled prior to our decision in *Acker,* and no definitive rule existed.

> We therefore hold that with respect to uranium, the rules of *Acker* and *Reed* will apply to determine the effect of severances of 'other minerals' from the surface estate for severances prior to June 8, 1983. These rules will apply even in cases where there has been no reliance on *Acker* and *Reed* because the minerals conveyed in a pre-*Acker* severance should not depend on the fortuitous event of reliance on subsequent law by a subsequent lessee or purchaser. For severances of 'other minerals' from the surface after June 8,

1983, title to uranium will be determined by the rules announced in *Moser*. If mineral and surface estates severed prior to June 8, 1983 have merged, and are subsequently reserved after June 8, 1983, the rules announced in *Moser* will apply thereafter.

Applying the rules to this case, the rules of *Acker* and *Reed* control the title to uranium under the 1939 lease of 'oil, gas, and other minerals' from Friedman to Magnolia. Therefore, in accordance with the jury findings, the uranium did not pass to Magnolia under this lease, but was retained by Friedman, and was later conveyed to Martin under the 1959 deed. Thus Friedman has no interest in the uranium.

. . .

JUSTICE RAY again dissented, joined by JUSTICE MCGEE:

. . .

The majority opinion represents yet another attempt to harmonize the recent line of cases which deal with the question of what constitutes a mineral for purposes of reserving or conveying 'oil, gas and other minerals.' [*Moser, Reed I, Reed II, Acker*] This line of cases is irreconcilable and with each attempt to rectify our prior mistakes, we only compound the problem. This time, the majority attempts to correct the mistake at the expense of the Friedmans, who are the legitimate owners of the uranium.

The law should be concerned about the rights of the Friedmans as well as those of Texaco. When the Friedmans conveyed the surface estate in 1959, Heinatz v. Allen, 147 Tex. 512, 217 S.W.2d 994 (1949), governed the question of what constitutes a mineral. In that case, this court held that the term 'minerals' is comprised of all substances 'within the ordinary and natural meaning of the word.' *Id.* at 997. Also, the case of Cain v. Neumann, 316 S.W.2d 915 (Tex.Civ.App.-San Antonio 1958, no writ) predated the 1959 conveyance. That case, written by Chief Justice Pope before his tenure on this court, holds that uranium is a mineral covered by an 'oil, gas, coal and other minerals' lease. *Id.* at 922. Thus, in 1959 when the Friedmans conveyed the surface estate and retained the mineral estate, the uranium was part of the mineral estate and owned by the Friedmans"

Consequently, T.J. Martin did not possess the right to convey uranium or any other mineral to Texaco in 1977. The cases that Texaco purportedly relied on when it entered into its 1977 lease, *Acker v. Guinn* and *Reed* I, do not concern uranium. In my opinion, the case law precedent is stronger for the Friedman's position than it is for Texaco. The majority has taken the Friedmans' property and awarded it to Texaco without compensating the Friedmans, whom I believe to be the true owners of the minerals.

I also oppose the majority opinion because it bases the determination of whether uranium is a mineral on the severance date of the surface and mineral estates. On the one hand, if a severance takes place after June 8, 1983, then the uranium is part of the mineral estate. On the other hand, if the severance occurred before June 8, 1983, then the status of uranium depends upon the fact findings required by *Acker* and *Reed.* Consequently, for the vast majority of surface and mineral estate severances, a great deal of uncertainty remains with respect to the title of uranium. In my opinion, it is essential for surface estate and mineral estate owners to be able to rely on a title examiner's opinion regarding the ownership of uranium. Under the majority opinion, however, a title examiner cannot safely determine title to uranium in a pre-June 8, 1983 severance, unless the surface estate and mineral estate owners resort to litigation.

Accordingly, I would hold that uranium is and always has been a mineral. As Justice Spears opined in his concurrence in *Reed* II: 'Whatever the rule, it should be such that the ownership of the substance in question can be ascertained from examining the instrument of grant or reservation alone.' 597 S.W.2d 743 at 751. While the rule that I espouse accomplishes certainty of title, the majority's rule fails to achieve this goal. Thus, I would reverse the judgments of the lower courts and render judgment that the Friedmans own the uranium on this tract of land." Compare Schwarz v. Texas, 703 S.W.2d 187 (Tex.1986), interpreting reservation of minerals in state patent.

2. Would it make sense to abandon the *Reed* and *Acker* approach only in the case of uranium as to post-*Moser* conveyances? Is the statement of the court in *Moser* abandoning "in the case of uranium . . . the *Acker* and *Reed* approach to the ownership of 'other minerals" ' and concluding "that title to uranium is held by the owner of the mineral estate as a matter of law" merely the beginning of a process that will encompass other substances in the meaning of "minerals" for conveyancing purposes? Various approaches to the problem and their advantages and disadvantages from a policy point of view are discussed in Maxwell, *The Meaning of 'Minerals'? The Relationship of Interpretation and Surface Burden*, 8 Tex.Tech.L.Rev. 255, 284 (1976); see also Note, *Interpretation of 'Other Minerals' in a Grant or Reservation of a Mineral Interest*, 71 Cornell L.Rev. 618 (1986).

Note that the *Moser* opinion also states that it is holding that "other minerals . . . includes all substances within the ordinary and natural meaning of that word, whether their presence or value is known at the time of severance." Martin v. Schneider, 622 S.W.2d 620, 80 O.&G.R. 110 (Tex.App.1981), held that a reservation of royalty in oil, gas and "other minerals" applied to uranium. The surface destruction test was not applicable since a "non-participating royalty owner is not entitled to produce the minerals." *Compare* Hobbs v. Hutson, 733 S.W.2d 269, 104 O.&G.R.316 (Tex.App.1987, writ denied), where it was argued that the "surface destruction test" was not applicable to disputes "between only royalty [and] mineral claimants" but the court found that the transaction under consideration involved a severance of minerals from the surface estate.

The Texas Supreme Court in Plainsman Trading Co. v. Crews, 898 S.W.2d 786, 129 O.&G.R. 280 (Tex. 1995) held that the *Acker/Reed* tests should be applied to a pre-1983 severance of a royalty interest, overruling Martin v. Schneider, 622

S.W.2d 620, 80 O.&G.R. 110 (Tex.App. 1981). The key issue for the court was that the royalty interest was carved out of the mineral estate. If the mineral estate did not contain the surface-mineable minerals, one could not create a royalty in minerals one did not own. 898 S.W.2d at 789-90. The dissent stated: "If the surface owners mine surface minerals, the [royalty owners] should collect their royalties from them. If the mineral estate owners produce subsurface minerals, such as oil or gas, the [royalty owners] should likewise collect royalties from them. Surface destruction is simply irrelevant?" 898 S.W.2d at 791 (Gammage, J. dissenting). Should it matter whether the royalty was created by deed or lease? *See also* Farm Credit Bank of Texas v. Colley, 849 S.W.2d 825, 122 O.&G.R. 113 (Tex.App. 1993, writ denied).

3. Should the *Moser* or *Acker/Reed* approach be used to deal with the definition of a "mineral" in a context other than that of a deed between two non-governmental entities? *See* Schwarz v. State, 703 S.W.2d 187, 87 O.&G.R. 602 (Tex. 1986) where the court considered other factors in interpreting a state patent reserving "minerals". For a contrary approach that does not treat a state patent differently from a private deed *see* Oklahoma ex rel. Commissioners of Land Office v. Butler, 753 P.2d 1334, 98 O.&G.R. 140 (Okla. 1987)(reservation by state of "oil, gas and other mineral rights" does not include coal).

The Texas Tax Code authorizes ad valorem taxation of "real property" which is defined to include "land," "a mine or quarry," and "a mineral in place" Tx.Tax Code §1.04. The taxpayer owns and leases land that it uses, in part, to extract limestone. After years of not separately assessing the value of the limestone, the County Appraisal District changes its position and substantially increases the appraised valuation of the taxpayer's property interest. Should the limestone be taxed as a "mineral?" A majority of the Texas Supreme Court concluded that applying the *Moser* ordinary and natural meaning test to limestone was appropriate and since an earlier decision had found that limestone belonged to the surface owner, Heinatz v. Allen, 147 Tex. 512, 217 S.W.2d 994 (1949), it was not a mineral subject to separate ad valorem taxation. Gifford-Hill & Co., Inc. v. Wise County Appraisal District, 827 S.W.2d 811, (Tex. 1991). Justice Gonzalez, in dissent argued that the interpretation of a taxation statute should not be bound by the interpretational rules or canons as applied to private deed transactions when to apply those rules or canons would defeat the purposes of the taxation statutory scheme. 827 S.W.2d at 819-20 (Gonzalez, J. dissenting).

In North Dakota a series of statutory enactments dealing with the definition of the term "minerals" has created substantial uncertainty as to how the term is to be interpreted. N.D.Cent. Code § 47-10-24 and its predecessors, 1955 N.D.Sess.Laws ch 235, § 1, and 1975 N.D.Sess.Laws ch. 244, § 1.

In Wyoming, a statute withdrew local governmental authority to regulate the "extraction or production of the mineral resources." Would a city be able to regulate sand and gravel or limestone quarries located within the city? Would it matter if another Wyoming statute expressly gave a state agency the power to regulate such extractive activities under its definition of a mining operation? Should the court apply the *Moser* plain meaning approach? *See* River Springs Limited Liability Co. v. Board of County Commissioners, 899 P.2d, 1329, 134 O.&G.R. 650 (Wyo. 1995), interpreting Wyo.Stat.Ann. §§ 18-5-201 and 35-11-103(e)(ii).

4. The *Moser* case cites the Texas Uranium Surface Mining and Reclamation Act which imposes a "statutory duty of the mineral owner, or his lessee, to reclaim the surface after surface mining" and states that this duty is not affected by the liability of the owner of a mineral "unnamed" in a conveyance to compensate "the surface owner for surface destruction." Should the existence of such a statute have a bearing on the remedies available in a situation such as *Moser*? Is it a legislative declaration that development is appropriate and affords the surface owner an exclusive remedy? Might the statute be said to be a legislative substitution of the specific remedy of reclamation for damages? Could the surface owner waive the statutory remedy in favor of damages? Should such a reclamation statute have an impact on the construction process involved in determining the meaning of "other minerals" in a grant or reservation? Should it matter whether the statute was passed prior to or subsequent to the date of the conveyance in question? Note that *Reed II* as stated by the *Moser* opinion considered that the impact on the meaning of the words "other minerals" of "the existence of a surface destructive method of mineral extraction was not solely at the time of trial . . ., or solely at the time of conveyance, but the time of conveyance or *thereafter.*" (footnote 4 to *Moser*).

5. What test should apply if a hardrock mineral is specifically mentioned? For example if a deed grants a ½ interest in the "granite," should it matter how the granite is extracted or is it clear that the grantee receives a mineral estate limited to granite? In WILDERNESS COVE LIMITED V. COLD SPRING GRANITE CO., 62 S.W.3d 844, 160 O.&G.R. 908 (Tex.App. 2001) the court refused to apply *Moser* or any other surface destruction test.

6. SPURLOCK V. SANTA FE PACIFIC RAILROAD CO., 143 Ariz. 469, 694 P.2d 299, 83 O.&G.R. 241 (App.1984, review denied). The plaintiff in a suit for conversion of helium derived its title to the lands from which the helium was taken through a deed which reserved to the defendant railroad "all oil, gas, coal and minerals whatsoever." The defendant counterclaimed to quiet its title to the mineral state. Held: "all minerals" is not ambiguous and extrinsic evidence to determine the intent of the parties to the deed is not relevant. "[T]he grantor retains ownership of all commercially valuable substances separate from the soil, while the grantee assumes ownership of a surface that has value in its use and enjoyment. . . . The grantor further retains ownership of mineral substances that are unknown at the time of the conveyance." Minerals retained as a matter of law under the general reservation of all minerals include "helium, nitrogen, potash, industrial clay, and petrified wood." Whether the parties to the conveyance severing the surface and mineral estates knew "of the existence or value of these minerals at the time of conveyance is irrelevant to the question of ownership." Such knowledge is,

however, "relevant to the mineral owners rights to burden the surface estate in the course of developing and producing these minerals." Thus, [w]ith respect to minerals specified in the conveyance or minerals commercially known to exist at the time of the conveyance, reasonable destruction of the surface estate is permissible. However, . . . with respect to substances which were unknown or had no commercial value at the time of the conveyance, . . . [t]he holder of the mineral estate owns such substances, but his development of these resources must not interfere with the surface owner's estate. Only in this way can the general intention of the parties to create and enjoy two co-existing, individually valuable estates be given effect." Does this reasoning create an ownership situation like that found in the nineteenth century English case, Hext v. Gill, LR 7 Ch. 699 (1872), which held that a bed of china clay was included in a reservation of "mines and minerals" but that the owner of the surface was entitled to an injunction against the owner of the minerals to prevent destruction or serious injury to the surface? Such a result does not deny all elements of ownership to the "owner" of the minerals. He cannot remove them but neither, of course, can the surface owner. If there is to be any movement the parties must strike a bargain.

7. WEST v. GODAIR, 542 So.2d 1386, 101 O.&G.R. 477 (La.1989). A reservation of "an undivided one-half (1/2) interest in and to all of the minerals of every nature or kind situated in, on and under the hereinabove described property" was held by the trial court, in an accounting action brought by the grantors, to cover "all pit run, field dirt, wash gravel, topsoil, and sand mined or removed from the property." [Material other than sand and gravel was apparently considered surface reasonably destroyed by the mining operation]. The Louisiana Court of Appeal reversed, finding that evidence "that gravel and sand were commercially exploitable within 25 miles of the property involved at the time the mineral servitude was established" was outweighed by "the fact that there were no negotiations whatsoever between the parties concerning the scope of the reservation." The court noted that "the customary reservation of minerals is meant to apply to oil and gas" and went on to hold that "applying the interpretation which least restricts ownership of the land conveyed the [grantors] failed to establish by a preponderance of the evidence that the reservation of minerals set forth were intended to include sand, gravel, topsoil, and pit run." West v. Godair, 538 So.2d 322, 323 (La.App. 1989). The Supreme Court of Louisiana, without opinion, reversed the judgment of the appeal court and reinstated the judgment of the trial court. Should the fact that the reservation covered an undivided one-half interest be of importance in a situation such as this one? *See also* CONTINENTAL GROUP, INC. v. ALLISON, 379 So. 2d 1117 (La. Ct. App. 1979) (reservation of all mineral rights construed as including solid minerals and as allowing strip-mining of lignite coal; case remanded for determination of whether projected exercise of the servitude by stripmining of lignite will be in accord with law); *reversed*, 404 So. 2d 428, 71 O.&G.R. 285 (La. 1981), cert. denied, 456 U.S. 906 (1982) (agreeing with the court of appeal's conclusion that the reservation of "all mineral rights" included the right to explore for and exploit lignite coal, a solid mineral, but concluding that the right to exploit

lignite had prescribed for ten years nonuser notwithstanding the production of oil and gas).

A somewhat different attitude toward the "intent of the parties" is evident in MILLER LAND & MINERAL CO. v. HIGHWAY COM'N, 757 P.2d 1001 (Wyo.1988), where the reservation was "all mineral and mineral rights" and the substance at issue was gravel. The court opined that "the only reliable rule which surfaces from the confusing and inconsistent approaches taken by those courts attempting to ferret out the subjective intent of the parties is that the word 'mineral' means what the court says it means." The question to be determined was "not whether the grantor intended to reserve gravel" but "whether or not gravel is a mineral." The court went on to hold that "gravel is not a mineral" unless it is " 'rare and exceptional in character or possesse[s] a peculiar property giving [it] special value.' " The quote is from Heinatz v. Allen, 147 Tex. 512, 217 S.W.2d 994, 997 (1949), and goes on to give examples of special value such as "sand that is valuable for making glass and limestone of such quality that it may profitably be manufactured into cement." One concurring opinion preferred "that we simply hold that gravel is not a mineral." Would this be preferable?

Does a reservation of "all coal and other minerals" unambiguously include oil and gas? Does it reserve all mineral substances? In McCormick v. Union Pacific Resources Co., 14 P.3d 346 (Colo. 2000), the Colorado Supreme Court concluded that such a reservation clearly included oil and gas, but did not answer the question of whether it reserved all mineral substances. *In accord*: Anschutz Land & Livestock Co. v. Union Pacific R.R. Co., 820 P.2d 338, 94 O.&G.R. 408 (10th Cir. 1987)(applying Utah law).

8. In *Moser* the deed was treated as unambiguous so that no parol or extrinsic evidence was admitted in order to ascertain the intent of the parties. That appears to be the majority view, although from state to state whether the minerals in a grant or reservation of "oil, gas and other minerals" include such metallic substances as gold or copper differs. *Compare* Panhandle Cooperative Royalty Co. v. Cunningham, 495 P.2d 108, 41 O.&G.R. 383 (Okla. 1972)(copper, silver, gold or other metallic substances not included; instrument unambiguous) *with* Kunkel v. Meridian Oil, Inc., 114 Wash.2d 896, 792 P.2d 1254, 110 O.&G.R. 330 (1990)(oil and natural gas not necessarily included in grant of "minerals" as instrument is ambiguous).

Would it better to treat the term "minerals" as ambiguous and allow parol evidence? In Smith v. Nugget Exploration, Inc., 857 P.2d 320 (Wyo. 1993), the court had to interpret the term "surface grazing rights" in a quitclaim deed. After applying several canons of construction and citing *Miller Land, supra*, the court concluded that the term was ambiguous in the context of the language used in the entire deed. How much less ambiguous is the term "minerals" than the term "surface grazing rights?"

9. Variables affecting the construction of instruments creating extractive rights include the following:

(a) Form of the instrument, *viz.,* whether it would be described as a mineral deed or as a lease;

(b) Designation of the substances mentioned in the instrument, *e.g.,* "all the coal mineral and mining rights," "oil, gas and other minerals," "oil, gas and other hydrocarbons;"

(c) The nature of the provision, if any, concerning surface easements and surface rights of grantor and grantee;

(d) Surface user, or the execution of instruments relating to surface user, prior in time to the instrument creating extractive rights;

(e) The conduct of the parties prior and subsequent to the execution of the instrument creating extractive rights;

(f) Knowledge in the community at the time of the execution of the instrument of the existence (or possible existence) of particular minerals in the community;

(g) The state of the art of mineral extraction at the time of the execution of the instrument creating extractive rights.

10. What policy considerations are applicable to the interpretation of the word "minerals" and the admission of extrinsic evidence in cases involving this interpretation question? A somewhat different problem is raised by a deed which grants "all the surface rights" and reserves "phosphate and phosphate rock with the right to . . . remove such deposits . . . .." Is the fact that the grantor is in the phosphate industry relevant? Is it relevant that at the time of the conveyance "there was no oil and gas development in Bear Lake County," the site of the land involved? *See* Stucki v. Parker, 108 Idaho 929, 703 P.2d 693, 86 O.&G.R. 457 (1985).

11. Does the word "gas" in a reservation of "oil and gas" include hydrocarbon gas only or does it extend to other gaseous substances such as carbon dioxide? Aulston v. United States, 915 F.2d 584 (10th Cir.1990) and Hudgeons v. Tenneco Oil Co, 796 P.2d 21 (Colo.App.1990), cert. denied, both say yes. Should the answer to such a question be approached differently in the context of a reservation of oil and gas required by a statute of the United States and such a reservation in a private conveyance?

MULLINNIX, LLC v. HKB ROYALTY TRUST

Wyoming Supreme Court 2006.
2006 WY 14, 126 P.3d 909, 162 O.&G.R. 584

KITE, JUSTICE. The dispositive issue in this case was whether deeds reserving "oil rights" which were executed in the 1940s in Campbell County, effectively reserved gas rights without a specific reference to "gas." In these consolidated appeals, Mullinnix, LLC and John W. Hickman, Fred J. Boyce and Lane Boyce (hereinafter referred to as Hickmans) contest the district court's order quieting title in gas rights in the appellees. The district court examined extrinsic evidence of the trade usage of the term "oil rights" at the time and place of the execution of the deeds and concluded the term, as used in real estate documents, did not include the gas rights. In the Mullinnix case, the district court also concluded that a document entitled "Declaration of Interest" executed in 1968, long after the deed was executed, by the grantees in the deed in question did not operate as a waiver or to estop them from asserting their full interest in the gas estate. We agree with the district court's conclusions and, therefore, affirm.

Issues

Case numbers 05-80 and 05-81 were consolidated for a bench trial and, also, on appeal. Appellants Mullinnix and Hickman filed a single brief. They articulate the issues on appeal as follows:

1. Was the Court's decision that reservations of "oil rights" exclude gas contrary to the evidence introduced at trial[?] Did the Court intentionally ignore relevant evidence and rely on inadmissible evidence in reaching its conclusion? (Mullinnix and Hickman cases)

2. In analyzing evidence to decide that a reservation of "oil rights" excluded gas, did the Court fail to follow the precedent and process set forth by the Court in Hickman v. Groves, [2003 WY 76,] 71 P.3d 256 [(Wyo. 2003)]? (Mullinnix and Hickman cases)

3. Did the Court err by allowing an expert opinion from attorney Edward Halsey interpreting deeds when that was the job of the Court, and Mr. Halsey refused to apply the decision of Hickman v. Groves to his analysis? (Mullinnix and Hickman cases)

4. Did the Court erroneously exclude evidence of conduct of parties to the deeds and their successors in interest demonstrating that they considered a reservation of "oil rights" to include "gas" as well? (Mullinnix and Hickman cases)

. . .

13. Did the Court err by interpreting a reservation of "oil and commercial gravel rights" to be a reservation of "oil", but not "oil rights"? (Hickman case)

14. Should the reservation of "oil rights" include coalbed methane gas? (Mullinnix and Hickman cases)

. . .

Groves phrases the appellate issues as follows:

1. Whether the District Court's determination that the reservation of "oil and commercial gravel rights" did not include gas or coalbed methane gas, is supported by the evidence.

2. Whether the District Court followed the process outlined in Hickman v. Groves, 2003 WY 76, 71 P.3d 256 (Wyo. 2003), in interpreting the warranty deed.

Facts

Case No. 05-81

On October 14, 1944, Jerry Hickman and Effie Hickman executed a warranty deed conveying real property located in Campbell County to Ed Willard, but reserving "to the grantors one-half of all oil and commercial gravel rights" in the property. Hickmans are the successors in interest to Jerry and Effie Hickman; and Bernice Groves, James Drake, and Edra June Drake (hereinafter referred to as Groves) are the successors in interest to Mr. Willard.

On July 20, 2001, Groves filed an action seeking to quiet title to all coal bed methane gas (CBM)[78] underlying the subject real property. Hickman filed a counterclaim seeking a declaration that they owned one-half of all of the gas, including CBM, underlying the property pursuant to the reservation of "oil rights" contained within the warranty deed. Hickmans contended the term "oil rights" had a particular meaning when the deeds were executed in 1944 in Campbell County, which included gas and they filed affidavits supporting that contention. The district court granted summary judgment in favor of Groves, ruling that the warranty deed was unambiguous and, as a matter of law, the reservation of "oil rights" did not include a reservation of gas rights. Hickman v. Groves, 2003 WY 76, P3-4, 71 P.3d 256, 256-57 (Wyo. 2003).

Hickmans appealed the summary judgment, and this Court reversed and remanded for a trial, finding a question of fact was raised concerning whether "oil rights" had a particular trade usage at the time the deed was executed. We ruled, although the term "oil rights" is unambiguous on its face, facts had been alleged showing that the trade usage of the term "oil rights" included gas and, therefore, suggested the true intent of the grantors was to reserve gas, as well as oil. Id., ¶10, 71 P.3d at 259. We reasoned: "In interpreting unambiguous contracts involving mineral interests, we have consistently looked to surrounding circumstances, facts showing the relations of the parties, the subject matter of the contract, and the apparent purpose of making the contract." Id., ¶6, 71 P.3d at 258. Consistent with our decisions in Newman v. RAG Wyoming Land Co., 2002 WY 132, 53 P.3d 540 (Wyo. 2002), and McGee v. Caballo Coal Co., 2003 WY 68, 69 P.3d 908 (Wyo. 2003), the court must focus on the general intent of the parties, concentrating on the purpose of the grant or reservation "in terms of the respective manner of enjoyment of surface and mineral estates and the exploitation of the mineral resources involved." Hickman, P7, 71 P.3d at 258. We, therefore, remanded the case to the district court for a trial to consider extrinsic evidence to resolve the issue of fact regarding the trade usage of the

---

[78]   In our earlier decisions, we thoroughly discussed the properties of coal bed methane and concluded it is chemically no different than other types of natural gas. See, e. g., Newman v. RAG Wyoming Land Co., 2002 WY 132, ¶¶9-10, 53 P.3d 540, 543 (Wyo. 2002) (discussing the chemistry of coal and coalbed methane).

term "oil rights" at the time and place of execution of the deeds. Id., P16, 71 P.3d at 262.

. . .

Discussion

Interpretation of Terms of the Deeds

In Hickman, we concluded a genuine issue of material fact existed regarding whether the term "oil rights" as used in the Hickman/Willard deed had a particular trade usage at the time and place of the deed's execution. Hickman, ¶10, 71 P.3d at 259. Quoting 11 Samuel Williston, A Treatise on the Law of Contracts, § 32: 7 (4th ed. 1999), we stated: "...circumstances known to the parties at the time they entered into the contract, such as what that industry considered to be the norm, or reasonable or prudent, should be considered in construing a contract, while the parties' statements of what they intended the contract to mean are not admissible." Hickman, ¶13, 71 P.3d at 260. Thus, we directed the district court to consider the circumstances surrounding execution of the deed to determine whether "oil rights" was a term of widely known custom and usage in Campbell County in the 1940s which included the "gas rights" without specifically mentioning the word "gas." Hickman, ¶¶10-11, 16, 71 P.3d at 259-60, 262.

After the trial, the district court ruled Mullinnix and Hickmans did not satisfy their burden of proving use of the term "oil rights" in deeds in Campbell County in the 1940s had a particular trade usage which included the "gas rights." As properly recognized by the district court, the party asserting a particular trade usage of a term has the burden of proving the existence of the trade usage. Mountain Fuel Supply Company v. Central Engineering & Equipment Company, 611 P.2d 863, 869 (Wyo. 1980). The Restatement (Second) of Contracts § 222 (1981) gives guidance in defining a "usage of trade" as: "a usage having such regularity of observance in a place, vocation, or trade as to justify an expectation that it will be observed with respect to a particular agreement."

The grantors in both deeds, the Hickmans and the Rothwells, were involved in the ranching business when they conveyed the property. Mullinnix and Hickmans attempted to prove that ranchers, who may not have been highly educated or sophisticated, often referred to their entire bundle of minerals rights as "oil rights" without distinguishing between oil and gas. Consequently, they maintained the use of the term "oil rights" in the deeds was meant to include the gas, as well as the oil. The appellees agreed that, in the 1940s, people may have used the term "oil rights" in casual conversation to mean a broader variety of mineral rights. Nevertheless, they claimed, in formal documents such as deeds, landowners (including ranchers) were more specific and described with particularity the interests being conveyed and/or reserved. Thus, according to the appellees, the term "oil rights" was used in the deeds to mean simply that--oil and not gas.

In deciding whether Mullinnix and Hickmans had satisfied their burden of proof, the district court considered the understanding of persons who had occasion to negotiate land transactions at the time the deeds were executed. In that regard, the district court found:

. . .

5. During the 1940s, the production of oil was the primary consideration for all concerned in northeastern Wyoming. At the time, natural gas was not considered a commercial product in northeastern Wyoming primarily due to the lack of pipelines and associated production and storage infrastructure. Gas then produced in northeastern Wyoming was solely the by-product of the production of oil and was customarily "flared," that is, simply burned off as a by-product of the oil production process.

6. While the general term "oil rights" was undoubtedly used during the 1940s in Campbell County during informal discussions to refer to the bundle of rights associated with surface and sub-surface holdings, warranty deeds recorded during the period habitually referred with more exacting specificity to those substances being reserved by the grantor in a conveyance. Common language employed to reserve interests included language such as "reserving unto the grantor one-half of the oil and gas rights"; or "reserving unto the grantor, all oil, gas, and other minerals." Where the grantor sought to reserve only an interest in the oil, language such as "the first parties reserve an undivided one-half interest in all oil that may be found in or under the surface of said land"; "reserving unto the grantors one-half of all the oil rights"; or "grantors reserving, however, an undivided one-half interest in all oil and minerals (not gas) in, and under or appertaining to said premises," was employed.

7. The better weight of credible evidence presented at trial includes that, where the 1940s grantor intended to reserve an interest in oil and gas, language in a deed reservation referring to both "oil" and "gas" was customary and was the language expected by those examining deeds for title purposes.

8. Similarly, the weight of credible testimony at trial established that the reasonably prudent party to a conveyance would not have relied on the common parlance to provide that a reservation of "oil rights" in a deed included gas or other minerals. Instead, the credible testimony presented indicated that a 1940s era reservation of "oil rights" *in a deed* probably would not have included a reservation of gas or other minerals. (emphasis in original).

The testimony and documentary evidence presented at the trial supports the district court's findings. Several landmen and attorneys who worked in the minerals industry during that time testified at the trial about their experiences in Campbell County in the 1940s and 1950s. Obviously, persons who dealt in the minerals trade during that era were elderly at the time of the trial in 2004. Still, the witnesses testified people knew the difference between oil and gas at that time. They also testified the term "oil rights" was sometimes used as a colloquialism or short-hand in casual conversation to mean the broader bundle of mineral rights.

Nevertheless, the witnesses consistently testified that, when used in legal documents such as deeds, parties did not routinely use the term "oil rights" to mean "oil and gas" or the entire bundle of mineral rights. Instead, the interests at stake in a conveyance or reservation were described with particularity. Numerous documents of conveyance offered into evidence showed the use of specific descriptions of different mineral interests. For example, the trial evidence included several deeds of that era which included specific references to "oil and gas" or specifically exempting "gas." Furthermore, the attorneys and the landmen testified that, if they encountered a deed which included the term "oil rights" during that time, they would have taken some type of action to correct or "cure" what they perceived as a problem with the title.

Mullinnix and Hickmans attempted to prove the grantors in each deed, i. e. Jerry and Effie Hickman and James and Vida Rothwell, were simple ranchers with limited education, suggesting that they would have used the colloquial or slang term "oil rights" to mean all of the minerals or, at least, the "oil and gas" in their deeds. Thus, they offered testimony about the experiences and education of the grantors. The relevance of that evidence to the ultimate inquiry, i. e. whether "oil rights" had a particular trade usage which included gas rights, was not, however, shown at trial. Simply because the grantors were ranchers of limited education and/or experience and may have used the term "oil rights" to mean oil and gas in casual conversation, does not mean that they would use the term "oil rights" in a legal document to mean "oil and gas" or the entire bundle of mineral rights. In fact, the evidence presented at the trial expressly refuted that leap in logic. An oil and gas lease executed by the Rothwells prior to the deed at issue in this case contained separate specific references to their oil and gas interests, indicating the Rothwells understood well the distinction between their interests in the different hydrocarbons. In addition, an 83-year-old Campbell County rancher, Charles Christensen, testified landowners in the 1940s may have referred to their mineral rights as oil rights in casual conversation, but they had a very good understanding of the scope and nature of the property they owned and were very specific when describing the property they bought or sold in formal documents. He also provided an analogy which is instructive to our analysis in this case. He testified that ranchers often referred to their cattle herds generically as "cows" in conversation, but when they were buying or selling the livestock, they would particularly describe them as "steer calves, heifer calves, cows," etc. in the bills of sale.

Substantial evidence supported the district court's conclusion that the term "oil rights" in the deeds did not include "gas." Mullinnix and Hickmans did not meet their burden of proving the term "oil rights" was used with such regularity in deeds in the 1940s in Campbell County to mean both oil and gas that a person intending to include oil and gas in a conveyance would have used the language "oil rights."

On appeal, appellee Pennaco specifically requests we revisit decisions in which we stated that, even when the language of a deed is unambiguous, the court should consider the "surrounding circumstances" in determining the meaning of its terms. See e. g., Caballo Coal Company v. Fidelity Exploration & Production Company, 2004 WY 6, ¶11, 84 P.3d 311, 315 (Wyo. 2004); Newman, ¶11, 53 P.3d at 544, McGee, ¶12, 69 P.3d at 912. Pennaco argues this interpretive procedure introduces too much uncertainty into real property title. It insists deeds should be interpreted differently than typical contracts because persons other than the parties to the deeds rely upon them. In order to remedy this situation, Pennaco urges us to announce that, henceforth, so long as the language is not ambiguous on its face, the court should establish the "plain meaning" of the language as a matter of law. It argues such a procedure would foster certainty in real estate law and allow persons examining the title to rely upon their understanding of the plain meaning of the recorded documents in determining where title reposes. The district court's decision letter after the bench trial in this case indicates it shares Pennaco's view.

In responding to this issue, we start with a reminder that the ultimate goal of our interpretation of any contract, including a deed, is to discern the intention of the parties to the document. In doing so, we look first to the plain meaning of the words of the deed. This has long been the law in Wyoming.

Pennaco and the district court insist allowing extrinsic evidence of the "surrounding circumstances" of a deed in order to determine the meaning of its terms is a new development in Wyoming law. However, a careful examination of our case law reveals it has long been the law that we look to the meaning of terms at the time of execution of an unambiguous deed. . . . If we were to adopt Pennaco's position, real property documents would be interpreted in accordance with the meaning of terms as the court understands them at the time and place of interpretation of the document rather than at the time and place of the execution of the document.

The case of Boley v. Greenough, 2001 WY 47, 22 P.3d 854 (Wyo. 2001), illustrates the importance of examining a conveyance of mineral interests at the time and place of execution. Thirty years after the Greenough parents had conveyed royalty interests to their children, a dispute arose because one provision of the assignments used the term "overriding royalty" when the grantors were not leaseholders or overriding royalty owners at the time of the assignments. Some of the confusion in Boley resulted from the evolution of the term "overriding royalty" in oil and gas law. In resolving the dispute, we emphasized the importance of interpreting the language of a conveyance at the time and place of its execution in order to effectuate the intentions of the parties to the conveyance. Boley, ¶¶14-22, 22 P.3d at 858-60.

The district court's decision letter seems to suggest, by considering extrinsic evidence of the "surrounding circumstances" of a deed's execution, we endorse a violation of the parol evidence rule. Those statements indicate a misunderstanding of the parol evidence rule, which is a rule of substantive law rather than a rule of evidence. It originated in the doctrine of merger, which states: "All provisions in a contract are merged into the deed when executed and delivered except those covenants which are deemed to be collateral to the sale. Thus, the deed regulates the rights and liabilities of the parties."

> The parol evidence rule has been stated in many ways but the basic notion is that a writing intended by the parties to be a final embodiment of their agreement may not be contradicted by certain kinds of evidence. A writing that is final is at least a partial integration. If the writing is final and also complete, it is a total integration and may not only not be contradicted by the type of evidence in question but may not even be supplemented by consistent (non-contradictory) additional terms. If it is final and incomplete it may be supplemented by consistent additional terms.

Longtree, Ltd. v. Resource Control International, Inc., 755 P.2d 195, 204 (Wyo. 1988), quoting, J. Calamari and J. Perillo, Law of Contracts, § 3-2 at 135-36 (3d ed. 1987). Consequently, the function of the parol evidence rule is to prevent parties from supplementing or contradicting the terms of the contract. See Restatement (Second) of Contracts § 231; E. Allan Farnsworth, Contracts, §§ 7.2 through 7.7 (3d ed. 1999). Once the terms of the agreement are identified, the parol evidence rule ceases to operate. The rule does not prohibit use of extrinsic evidence of the circumstances surrounding the execution of the deed to interpret the meaning of its terms. Id. By allowing evidence of the circumstances surrounding execution of the deed, courts are more apt to arrive at the parties' true intention at the time of the execution of the deed.

The proper role of the parol evidence rule was recognized by the Wyoming legislature when it adopted the Uniform Commercial Code, Wyo. Stat. Ann. §§ 34.1-1-101 through 34.1-10-104 (LexisNexis 2005). The Uniform Commercial Code specifically allows evidence of "usage of trade" to be considered in interpreting a contract. See § 34.1-1-205; Century Ready-Mix Company v. Lower & Company, 770 P.2d 692, 696-97 (Wyo. 1989). "Custom and usage of a particular place or trade can be proved to give to the words of a written contract a meaning different from that which would be given to the words by their more general usage" without violating the parol evidence rule. 6 Arthur Linton Corbin, Corbin on Contracts § 579 (2002). In Hickman, this Court quoted, at length, Williston's esteemed treatise on contracts, which discusses the proper application of evidence of custom and usage in determining the meaning of contract terms. In light of Pennaco's argument and the district court's decision letter, we think parts of that discussion bear repeating.

Historically, it has been recognized that familiar words may have different meanings in different places and that every contract will therefore have a relation to the custom of the country where it is made. ...

In subsequent years, numerous cases were decided where words with a clear normal meaning were shown by usage to bear a meaning which was not suggested by the ordinary language used. This is not only true of technical terms, but of language which, at least on its face, has no peculiar or technical meaning or significance.

Therefore, evidence of usage may be admissible to give meaning to apparently unambiguous terms of a contract where other parol evidence would be inadmissible. Thus, circumstances known to the parties at the time they entered into contract, such as what that industry considered to be the norm, or reasonable or prudent, should be considered in construing a contract, while the parties' statements of what they intended the contract to mean are not admissible.

It is currently the widely-accepted rule that custom and usage may be proved to show the intention of parties to a written contract or other instrument in the use of phrases of a peculiar technical meaning which, when unexplained, are susceptible of two or more plain and reasonable constructions. Parol evidence may be admitted to establish a technical meaning where certain provincialisms and technicalities of science and commerce have acquired a known, fixed and definite meaning different from their ordinary meaning by legal custom or usage. Thus, in the interpretation of technical terms used in a contract, it is proper to consider the meaning given to those terms in the course of prior dealings between the parties, as well as by business or trade custom or usage....

. . .

The correct rule with reference to the admissibility of evidence as to trade usage under the circumstances presented here is that while words in a contract are ordinarily to be construed according to their plain, ordinary, popular or legal meaning, as the case may be, if in reference to the subject matter of the contract, particular expressions have by trade usage acquired a different meaning, and both parties are engaged in that trade, the parties to the contract are considered to have used them according to their different and peculiar sense as shown by such trade usage. Parol evidence is admissible to establish the trade usage, and that is true even though the words are in their ordinary or legal meaning entirely unambiguous, since, by reason of the usage, the words are used by the parties in a different sense.

Hickman, ¶¶12-13, 71 P.3d at 260-61, quoting 12 Samuel Williston, A Treatise on the Law of Contracts, § 34: 5 (4th ed. 1999). See also Caballo Coal Company, ¶¶11-12, 84 P.3d at 316-17. Thus, we continue to recognize the importance of allowing the use of extrinsic evidence to interpret a contract, including custom and usage in a particular place or trade at the time of execution of the contract, in order to arrive at the plain meaning of the agreement, with the goal of more closely effectuating the parties' true intent.

If we were to accept Pennaco's invitation to adopt a definition of the term "oil rights" to be applied in every legal document coming before Wyoming courts, there might be greater predictability in resolving disputes over the meaning of that certain term. It would not, however, serve to effectuate the intent of the parties to documents and would undermine this Court's deed interpretation jurisprudence which has developed over more than one hundred years. Upon review of our numerous cases involving interpretation of real property interests, we certainly cannot say that such jurisprudence has caused excessive litigation or confusion as predicted by Pennaco.

In Newman, we surveyed the ways other jurisdictions addressed the problems associated with determining coalbed methane ownership. Newman, ¶¶20-27, 53 P.3d at 546-49. We discussed the advantages and disadvantages of various interpretation models and, ultimately, rejected any rigid rule of law established by the courts without regard to the parties' intent. Instead we returned to our longstanding procedure designed to give "effect to the general intent of the parties to the conveyance with regard to the exploitation of mineral resources." Newman, ¶27, 53 P.3d at 549. We were not convinced there is any better way to resolve disputes over property ownership than by trying to ascertain the intentions of the parties to the conveyance by looking to "the facts and circumstances surrounding the execution" of deeds. We continue to believe that is the best approach and, therefore, decline Pennaco's invitation to revise the law of deed interpretation in Wyoming.

. . .

Affirmed.

NOTES

1. Is the Wyoming court's approach likely to lead to consistent results when it comes to deed interpretation issues? Is the court's explanation regarding the admission of extrinsic evidence persuasive? Is the approach taken by the court similar to that taken by the Texas Supreme Court in *Moser*?

2. The court refers to Newman v. RAG Wyoming Land Co., 2002 WY 132, 53 P.3d 540, 156 O.&G.R. 314 and McGee v. Caballo Coal Co., 2003 WY 68, 69 P.3d 908, 157 O.&G.R. 677 as supporting the court's approach to the interpretational issue. Both *Newman* and *McGee* deal with the somewhat thornier issue of whether coalbed methane gas (CBM) is included or excluded from a grant or reservation of "coal" or "coal and mining rights." The issue of the ownership of CBM gas has been hotly contested in the past few decades, encompassing development plays in the Appalachian Basin and the Intermountain West regions. The results from the various jurisdictions have not been consistent. The results may depend upon whether the term "coal" or "coal or other minerals" or "gas" is the subject of the dispute. See Continental Resources of Illinois, Inc. v.

Illinois Methane, LLC, 364 Ill.App.3d 691, 301 Ill.Dec. 887, 847 N.E.2d 897 (2006); Cimarron Oil Corp. v. Howard Energy Corp., 909 N.E.2d 1115 (Ind.App. 2009); Central Natural Resources, Inc. v. Davis Operating Co., 201 P.3d 680 (Kan. 2009); Harrison-Wyatt, LLC v. Ratliff, 267 Va. 549, 593 S.E.2d 234 (Va. 2002); Energy Development Corp. v. Moss, 214 W.Va. 577, 591 S.E.2d 135 (2003).

3. The issue of whether minerals are reserved or granted where a governmental body is a party to the transfer raises additional issues. Some of those issues are discussed in Amoco Production Co. v. Southern Ute Indian Tribe, 526 U.S. 865, 142 O.&G.R. 437 (1999), a principal case found in Chapter 8, Section B. Many of these cases involve various public land laws that transferred ownership of public domain lands in the western part of the United States that either granted or reserved some type of mineral estate. *See e.g.*, Bedroc Ltd., LLC v. United States, 541 U.S. 176, 159 O.&G.R. 857 (2004)(Pittman Underground Water Act of 1919's reservation of "valuable minerals" held not to include sand and gravel); Watt v. Western Nuclear, Inc., 462 U.S. 36, 79 O.&G.R. 596 (1983)(Stock-Raising Homestead Act's reservation of "all the coal and other minerals" held to include gravel); New West Materials LLC v. Interior Board of Land Appeals, 398 F.Supp.2d 438, 162 O.&G.R. 415 (E.D.Va. 2005), *aff'd*, 216 Fed.Appx. 385, 167 O.&G.R. 550 (4th Cir. 2007)(reservation of oil, gas and all other mineral deposits in Small Tract Act patents includes sand and gravel).

C. THE MOTHER HUBBARD (OR COVER-ALL) CLAUSE

It is the intention of the lessor and lessee that this lease shall also include, and there is hereby included and leased for the purposes and consideration herein stated, all the land owned and claimed by lessor adjacent or contiguous to the land above described, whether in the same or different surveys.

Martin & Kramer, *Williams and Meyers Oil and Gas Law* § 221

The situation giving rise to the common practice of inclusion of such cover-all clauses in deeds and leases is the uncertainty which sometimes exists as to the quantum or nature of the interest of the grantor or lessor in certain premises or the adequacy of a particular description, which may be copied from an earlier instrument in the chain of title, to describe the premises owned or claimed by the grantor or lessor. The problem of the adequacy of a particular description was especially acute in Texas for two reasons. First, it is known that deficiencies exist in many early surveys in this state and a description based on such surveys may be inadequate or inaccurate. Second, for many years Texas law did not provide for compulsory pooling and unitization and, although there was a rule of the regulatory commission governing well spacing, a well permit for a tract smaller than the normal spacing unit could frequently be obtained by reason of certain exceptions to the spacing rule. If the particular description in a deed or lease failed to cover an adjoining strip owned by the grantor, the grantor or lessor would retain his interest in such strip. Under some circumstances he might be able later to obtain a permit for a well on such strip, and the lessee or grantee who intended to obtain and believed that he had obtained an interest in all of the premises of his lessor or grantor might suffer drainage to a portion of the premises which erroneously was believed to have been leased or conveyed to him.

Situations giving rise to a need for cover-all clauses may arise in all states for a variety of reasons. "Variances between survey lines and the lines of the tract actually owned by a landowner sometime result in this way: the landowner in fencing his tract, or in some other way indicating its boundaries, will put the fence or other monument at a convenient place, which place varies from the description in the instrument under which he claims ownership. For example, there may be a row of trees three hundred feet east of the actual east boundary line, which trees make convenient substitutes for fence posts. Thereafter the landowner matures title to the strip by adverse possession. Still later the landowner executes an oil and gas lease which carries forward the description in the deed, and thus usually fails to pick up the strip to which landowner has acquired title by adverse possession. It may be possible for such a description to include the strip under the doctrine of agreed boundary or the doctrine of boundary by acquiescence however, it is never safe to assume that either doctrine applies where the problem before the attorney is preparation of instruments, or advising as to the construction thereof, short of the law-suit stage."

A cover-all clause may be employed in a mineral or royalty deed or in an oil and gas lease.

J. HIRAM MOORE, LTD. v. GREER

Supreme Court of Texas, 2005
172 S.W.3d 609, 171 Oil & Gas Rep. 163, 48 Tex. Sup. Ct. J. 662

JEFFERSON, CHIEF JUSTICE. Mary Greer, her three sisters, and their widowed mother partitioned an 80-acre tract into four 20-acre tracts, designated 1 through 4. The land is all in the I. & G. N. R.R. Survey No. 6, A-232 ("the Railroad Survey"), in Wharton County. Each sister received title to the surface and minerals in one tract and one-fourth of a non-participating royalty interest in each of the other three tracts. Greer received Tract 3.

In 1988, the two sisters who owned Tracts 1 and 2 leased their minerals to Larry K. Childers. The SixS Frels # 1 Well was completed on an adjacent 106-acre tract in the Wm. Barnard Survey No. 14, A-801 ("the Barnard Survey"), and in 1991 that tract was pooled with Tracts 1 and 2 and four other tracts, at a specified horizon, to form the 350-acre SixS Frels Gas Unit. The following schematic drawing depicts Tracts 1-4 and the SixS Frels Gas Unit:

After 1991, Greer was thus entitled to receive 1/4 of the royalty for each of Tracts 1 and 2 from the SixS Frels # 1 Well. There was no production-- hence no royalty due Greer--with respect to Tracts 3 and 4.

In May 1997, Greer and her sister leased the minerals in Tracts 3 and 4, respectively, to J. Charles Holliman, Inc. The following September, Greer executed a royalty deed to Steger Energy Corp. At the time, there was still no production with respect to Tracts 3 and 4, and despite her lease to Holliman four months earlier, Greer was unaware of any drilling activity planned for the future. Greer's royalty deed to Steger consisted of nine numbered paragraphs in small print on a single page. The first paragraph conveyed all mineral royalties -

> that may be produced from the following described lands situated in the County of Wharton, State of Texas, to wit:

> All of that tract of land out of the AB 801 SEC 14/W M BARNARD # 14 SURVEY, Wharton County, Texas known as the MEDALLION OIL - SIXS FRELS UNIT. Grantor agrees to execute any supplemental instrument requested by Grantee for a more complete or accurate description of said land. Reference is made to this unit(s) for descriptive purposes only and shall not limit this conveyance to any particular depths or wellbores. In addition to the above described lands, it is the intent of this instrument to convey, and this conveyance does so include, all of grantors [sic] royalty and overriding royalty interest in all oil, gas and other minerals in the above named county or counties, whether actually or properly described herein or not, and all of said lands are covered and included herein as fully, in all respects, as if the same had been actually and properly described herein.

The first quoted sentence, a specific grant, describes land "known as the . . . SIXS FRELS UNIT " in the Barnard Survey. As already noted, the SixS Frels unit comprised tracts in both the Barnard Survey and the adjacent Railroad Survey, but Greer owned no interests in the Barnard Survey. [1] Greer's only royalty interests in the SixS Frels unit were in Tracts 1 and 2, both of which were in the Railroad Survey. But the fourth sentence, a general grant, refers to all Greer's interests in Wharton County, thus including not only her royalty interests in Tracts 1 and 2 in the SixS Frels Unit, but her interests in Tracts 3 and 4 as well.

During September and October, Steger acquired other royalty interests in Wharton County, and in December it sold twenty-five such interests, including the one acquired from Greer, to J. Hiram Moore, Ltd. for $ 360,000, which was market value. At that time, there was no production from Greer's Tract 3, nor was it pooled with any producing property.

Two years later, in December 1998, Kaiser-Francis Oil Co., successor to the working interest in Tract 3 that Greer conveyed to Holliman, pooled about 313 acres, including Tracts 1-4, at a different horizon than the SixS Frels Gas Unit, for production from the Greer # 1 Well which had been completed in Tract 3. Moore claimed all royalties with respect to the interests partitioned to Greer in Tracts 1-4, and when Greer disputed the claim, Kaiser-Francis suspended payments for those tracts.

Moore sued Greer to determine their respective rights, and Greer counterclaimed for declaratory relief as well as rescission and reformation based on mutual mistake and fraud. Moore moved for summary judgment, contending that it had acquired all of Greer's royalty interests in Wharton County by purchasing her royalty deed to Steger. Greer responded that she had intended to convey to Steger only her interests in the SixS Frels Unit in the Barnard Survey. In her supporting affidavit, she stated: "I did not intend to convey any other property. I specifically did not intend to convey any of my interest in the I & GNRR Co. Survey No. 6, Abstract 232 Wharton County, Texas." The trial court granted Moore's motion for summary judgment and severed Greer's claims for rescission and reformation. Those claims remain pending.

The court of appeals reversed the summary judgment with this explanation:

Here the question is not whether the property [claimed by Moore] was described specifically enough [in Greer's royalty deed to Steger], but whether the "catch-all" language is sufficient to effect a conveyance of a significant property interest that Greer contends she had no intention of conveying by this deed. *Jones v. Colle [727 S.W.2d 262, 30 Tex. Sup. Ct. J. 315 (Tex. 1987)]* sets forth the longstanding rule in Texas that a clause, like the one at issue here, can only convey small interests that are clearly contemplated within the more particularly described conveyance, and they are not effective to convey a significant property interest not adequately described in the deed or clearly contemplated by the language of the conveyance. Because the interest in Tract 3 was a substantial one, we hold that the rule disallowing such "cover-all" clauses to effectively convey a substantial property interest is the controlling law in this case. *72 S.W.3d 436, 441.*

We may construe the deed as a matter of law only if it is unambiguous. *See Westwind Exploration, Inc. v. Homestate Sav. Ass'n, 696 S.W.2d 378, 381, 28 Tex. Sup. Ct. J. 603 (Tex. 1985).* Citing *Holloway's Unknown Heirs v. Whatley, 133 Tex. 608, 131 S.W.2d 89, 92 (Tex. 1939)*, Moore argues that the deed is unambiguous and that the general description establishes that the parties intended the deed to convey all of Greer's royalty interests in the county. Pointing to a line of cases in which our courts have recognized the validity of geographic grants, Moore contends that the general description falls into that category of conveyances and thus enlarges the specific grant.

Greer, on the other hand, contends that she intended a specific conveyance only. She argues that the second grant does not enlarge the first. Citing *Jones v. Colle,* and *Smith v. Allison,* 157 Tex. 220, 301 S.W.2d 608 (Tex. 1957), she argues, and the court of appeals agreed, that the language following the specific grant was intended to convey only small unleased strips of land adjacent to the described property. *72 S.W.3d at 441.*

In *Smith v. Allison,* 157 Tex. 220, 301 S.W.2d 608, 611 (Tex. 1956), we held that a deed was ambiguous when its general description conveyed a significantly greater interest (surface and minerals in land included within the specific description) than the specific grant (minerals only) and when the amount paid for that conveyance appeared to relate only to the mineral interest specifically described. Accordingly, we noted that "the deed under question contained material inconsistent provisions that rendered it uncertain as to the property conveyed."

Because the deed was ambiguous, it was correctly submitted to the jury, and we affirmed the judgment on that verdict.

We face a similar problem here. The specific description in Greer's deed points to a survey in which Greer apparently owns no interest. The deed purports to convey "all of that tract of land out of the AB 801 SEC 14/W M BARNARD # 14 SURVEY, . . . known as the MEDALLION OIL - SIXS FRELS UNIT." As previously noted, Greer owns a 1/4 nonparticipating royalty interest in Tracts 1 and 2, which were pooled in the SixS Frels Unit; however, neither tract is in the W M Barnard Survey. Therefore, the specific description either does not describe any royalty interests owned by Greer, or it incorrectly describes her royalty interests in Tracts 1 and 2 that are part of the SixS Frels Unit by stating that they are in the W M Barnard Survey instead of the I. & G. N. R.R. Survey. The general description conveys "all of grantors [sic] royalty and overriding royalty interest in all oil, gas and other minerals in the above named county or counties, whether actually or properly described herein or not, and all of said lands are covered and included herein as fully, in all respects, as if the same had been actually and properly described herein." The deed in effect states that Greer conveys nothing, and that she conveys everything. We cannot construe this deed as a matter of law.

Given the deed's ambiguity, the trial court erred in granting summary judgment. A jury should therefore hear evidence and determine the parties' intent. *See Columbia Gas Transmission Corp. v. New Ulm Gas*, 940 S.W.2d 587, 589, 40 Tex. Sup. Ct. J. 42 (Tex. 1996). Accordingly, we affirm the court of appeals' judgment[79] and remand to the trial court for further proceedings consistent with this opinion.

JUSTICE HECHT, concurring.

The dissent discerns no principle in the Court's decision, but there is one, and a very venerable one at that: hard cases make bad law. The specific grant of royalty interests in Mary Greer's deed to Steger Energy Corp. described property in a survey she did not own. One might suppose that the wrong survey was referenced by mistake, but no, Greer now tells us under oath: the specific grant conveyed nothing, which is "specifically" -- her word -- what she intended. Her purpose all along, if she can be believed, was to take Steger's money and convey nothing in return. Steger's successor in interest, J. Hiram Moore, Ltd., who bought royalty interests at market value, including Greer's deed, argues that by the literal terms of

---

[79]   In doing so, we express no opinion on the court of appeals' holding that "a clause, like the one at issue here, can only convey small interests that are clearly contemplated within the more particularly described conveyance, and they are not effective to convey a significant property interest not adequately described in the deed or clearly contemplated by the language of the conveyance." 72 S.W.3d at 441 (citing *Jones v. Colle,* 727 S.W.2d 262, 30 Tex. Sup. Ct. J. 315 (Tex. 1987)).

the general grant in that deed, it acquired all of Greer's royalty interests in Wharton County. Those interests substantially exceeded the interest she would have conveyed had the deed referenced the adjoining survey. If Moore is right, the record does not reflect whether it still paid market value for all of Greer's interests, as it says it did for every other interest it bought.

Moore argues that general grants must always be read literally, or land titles will become uncertain, and chaos will descend. Greer argues that general grants can never include more than small strips adjacent specifically described property, or unsophisticated, perhaps careless, grantors will be duped out of property they never intended to convey. We have squarely rejected Greer's argument in two cases, and the argument is at least inconsistent with three others. But we stopped short of endorsing Moore's argument in *Smith v. Allison*. There the grantor specifically conveyed a mineral interest in two adjoining quarter-sections -- 320 acres -- then added, with this general language:

> any and all other land and interest in land owned or claimed by the Grantor in said survey or surveys in which the above described land is situated or in [sic] adjoining the above described land.

As it happened, the grantor owned the surface and minerals in 1,440 acres adjoining the two specifically described quarter-sections. We held that the general grant did not unambiguously convey all of the grantor's interest in the 1,440 acres and affirmed a judgment on a jury verdict finding that it was not her intent to do so.

The dissenting opinion argues that the general grant was not given effect in *Smith* because it literally included the surface estate as well as the minerals and was therefore repugnant to the rest of the deed that conveyed only mineral interests. The argument is certainly a reasonable one, but I doubt seriously that *Smith* would have been decided differently if the general grant had read, "any and all other mineral interest". The inclusion of the surface estate in the general grant was troublesome, but so, too, was the inclusion of the mineral interest in two-and-one-quarter sections that could have been described as easily as the two quarter-sections in which interests were specifically conveyed if the parties had ever had the remotest notion that the additional acreage was to be part of the transaction.

Situations in which general grants cannot be given effect have not arisen frequently. The Court in *Jones v. Colle* thought that it presented such a situation, but I agree with the dissenting opinion that *Jones* misread *Smith* and that in any event the result in *Jones* is not inconsistent with our other decisions. But while it only rarely happens that general grants cannot be given literal effect, I am not prepared in the unusual circumstances of this case to adopt a rigid rule that always construes general grants literally. The dissent poses five situations in which, I agree, a general grant should be construed according to its terms, [10] but it is just as easy to pose other circumstances in which it will seem highly unlikely that the parties fully intended what they actually said, and unjust to hold one of them to it. We should not use this case to make bad law. As long as a rule that gives effect to general grants with a few exceptions seems to manage the cases that arise, I would not change it simply because it could, possibly, prove unworkable. *Smith* did not destabilize land titles. Neither will this case.

With these few additional thoughts, I join in the Court's opinion.

JUSTICE OWEN, dissenting, in which MEDINA joined.

This grant is unambiguous. It purports to grant Greer's interest in a specific section of a specific survey, and it also purports to grant all of Greer's royalty interests in Wharton County, whether described in the deed or not. As it turns out, Greer does not own what she purported to convey in the specific grant. But she does own royalty interests in Wharton County, and she unequivocally conveyed all those royalty interests in the general granting section of this deed. I would give effect to this grant unless and until the deed is reformed or rescinded.

The Court's reasoning for failing to enforce the deed as written boils down to this and only this: "The deed in effect states that Greer conveys nothing, and that she conveys everything. We cannot construe this deed as a matter of law." [1] But the deed itself contemplated that the specific grant might not "actually and properly describe[]" the royalty interests that Greer owned in Wharton County, and therefore, the deed expressly provided that Greer was conveying all her royalty interests in Wharton County "whether actually or properly described herein or not, and all of said lands are covered and included herein as fully, in all respects, as if the same had been actually and properly described herein."

If the Court were to faithfully apply our precedent -- which is considerable -- it would give effect to what the written words unmistakably say in this royalty deed. This Court's decision in *Smith v. Allison* does not support the result reached today. The general clause in *Smith v. Allison* purported to convey fee simple title to all land adjoining the parcels specifically described, while the specific descriptions as well as the habendum clause and warranty clause limited the conveyance to minerals only. The general grant was therefore in conflict with and repugnant to the specific grant, and we have long held that when that occurs, the specific grant "will ordinarily control."

I have several questions for the Court: Would the Court hold that the deed in this case, including the general grant of all royalty interests in Wharton County, is ambiguous if:

1) a metes and bounds description of the land in which royalty interests were conveyed had been used as the specific description, but the description did not close;

2) there were three tracts of land specifically described in which royalty interests were granted, and all were effective, but there was one other royalty interest in Wharton County, not specifically described, that Greer owned;

3) there were three tracts of land specifically described in which royalty interests were granted, but one failed because Greer owned no interest in that one tract;

4) there were 100 tracts of land specifically described in which royalty

interests were granted, but one failed because Greer owned no interest in that one tract; or

5) there were 100 tracts of land specifically described in which royalty interests were granted, and all were effective, but there were three other royalty interests in Wharton County, not specifically described, that Greer owned?

What principal of law does the Court announce today that will give stability and predictability in construing deeds, wills, oil and gas leases, liens, and deeds of trust? I can discern none. We are told only that when there is at least one specific grant and it fails, an unambiguous general grant is rendered ambiguous. Accordingly, there will be trials, sometimes years after the grantors and grantees have passed on, to determine what a conveyance meant.

We have received a number of amicus briefs in this case. . . . [They] tell us that the failure to give effect to the plain meaning of the deed before us will lead to severe adverse consequences including the failure of previously certain titles and security interests and a proliferation of litigation that can only be resolved by a trial to determine the meaning of "ambiguous" instruments. I fear that these amici are correct in their assessment of the damage today's decision will inflict on the stability and predictability of titles. Geographic grants are commonly used in large acquisitions as well as small, personal transactions in which individuals cannot afford to have deed records scoured and legal descriptions prepared and reviewed by lawyers. But all lay people know what they mean when they say, "I intend to convey all the royalty interests I own in Wharton County." Because the Court does not give effect to the intent that is plainly and directly expressed in the deed before us, I dissent.

NOTES

1. SMITH V. ALLISON, 157 Tex. 220, 301 S.W.2d 608, 7 O.&G.R. 484 (1956), which is discussed in the principal case illustrates the difficulty presented when a cover-all clause, if given literal application, will result in conveying an area considerably larger than the area included within the particular description. This situation gives rise to a reasonable doubt whether such consequence was intended by the parties to the conveyance. The problem then is to find a theory which will justify the admission of extrinsic evidence of intent in the context of an action involving the construction (as opposed to the reformation) of a particular instrument.

The grantor in this instance owned 1760 contiguous acres falling within three sections (123, 124 and 145). The deed conveyed an undivided one-half interest in the minerals in the SE 1/4 and the NW 1/4 of Section 124 and included a broadly phrased cover-all clause. Claimants through the grantee claimed a one-half mineral interest in the NE 1/4 of Section 124 by operation of the cover-all clause; in other words, they claimed that the deed had the effect of conveying 1/2 of the minerals in three quarter-sections although the particular description was of two quarter-sections only. As the court observed in the course of its opinion, a literal reading of the cover-all clause would give the grantee an even greater interest: a ½ mineral interest in the two quarter sections specifically described and the entire interest (mineral and surface) in an additional 1440 acres.

The mere recitation of the facts gives rise to the conviction that it is doubtful that the cover-all clause was intended by the parties to have its literal meaning; the problem faced was how to justify inquiry into intent of the parties. That this was a difficult problem for the court is illustrated by the following:

a. The trial court and the Court of Civil Appeals viewed the instrument as ambiguous and declared that because of such ambiguity the intent of the parties should be ascertained from surrounding circumstances.

b. The initial opinion of the Supreme Court found the cover-all clause was ambiguous.

c. A concurring opinion by Justice McCall denied that the instrument was ambiguous but concluded that a cover-all clause will bring within a conveyance only small strips of land bordering the described tract or tracts which may not be included because of faulty description or which may have been acquired by adverse possession.

d. On motion for rehearing, the majority appeared to accept the position enunciated by Justice McCall's concurring opinion.

e. Two dissenters appeared to argue that there was a latent ambiguity arising from application of the description of the property in the deed to the lands owned by the grantor, and extrinsic evidence is admissible to resolve the ambiguity.

2. The problem relating to cover-all or Mother Hubbard clauses is not confined to Texas as the following two cases illustrate:

BARNETT V. GETTY OIL CO., 266 So.2d 581, 43 O.&G.R. 204 (Miss.1972). On July 23, 1958, J. D. Haynes executed an oil, gas and mineral lease to Roger A. Mateer covering his 1/16 interest in the S 1/2 of the NW 1/4 of Section 17, Township 1 North, Range 15 East. The ten year primary term of the lease expired without production on July 23, 1968. On May 27, 1968, J. D. Haynes executed a "top lease" to Getty Oil Co. for a primary term of three years to commence on July 23, 1968. The lease specifically described only two hundred acres but recited that it covered two hundred and forty acres; the lease included a cover-all clause. Included in the specific description was the S 1/4 of the NW 1/4 of Section 17. Testimony offered in the trial court established that it was the intention of the parties to include the S 1/2 (instead of the S 1/4) of the NW 1/4 of Section 17 and that through a scrivener's error 1/2 became 1/4 . This lease was recorded on October 31, 1968.

On July 10, 1968, J.D. Haynes conveyed to appellant Barnett his 1/16 interest in the minerals in the S 1/2 of the NW 1/4 of Section 17, and this instrument was recorded on July 16, 1968; the court's opinion does not give the precise language but indicates that the conveyance was "subject to any valid and subsisting oil, gas and mineral lease."

Production having been obtained by Getty Oil Co. under its lease, appellant Barnett claimed he was the owner of an unleased 1/16 mineral interest in the 80 acres of the S 1/2 of the NW 1/4 of Section 17; Getty Oil Co. claimed that Barnett was entitled only to a royalty on production. Barnett appealed from the Chancellor's dismissal of his original bill of complaint with prejudice. *Held,* appellant's mineral interest is subject to the Getty lease and cause remanded for an accounting of the royalties to which appellant Barnett is entitled.

Appellant claimed to be a bona fide purchaser for value without notice of the lease from Haynes to Getty Oil Co., and hence entitled to the protection afforded by Mississippi Code §§ 867-869, the applicable "notice" type recordation statute. The court concluded, however, that:

> [T]hese sections are inapplicable when the grantee (the appellant in this case) accepts a conveyance which is expressly made subject to any valid and subsisting lease or leases. . . . The fact that the appellant in the case at bar did not have actual notice is not essential.

> The appellant chose his own mineral deed and sent it to the grantor Haynes, and, if he had desired, the appellant could have made his mineral acquisition subject to only valid and subsisting leases of record. The failure of the appellant to draft the instrument to make his acquired interest subject to only recorded leases, as he now contends, is totally one of his own doing and he is now bound by the established law of this state pertaining to the enforcement of rights in accordance with the written agreement which he selected."

The court then turned to the appellant's contention concerning the construction of the lease to Getty Oil Co.:

> The appellant argues that the mineral lease from Haynes to Getty Oil Company described a portion of the property in dispute as the south quarter of the northwest quarter(S 1/4 of NW 1/4) of Section 17, Township 1 North, Range 15 East; that this lease was clear and unambiguous and that the chancellor was in error in allowing any proof as to the intentions of the parties to said instrument since the instrument should speak for itself. The appellant's position in this case is that even if the lease from Haynes to Getty Oil Company was a valid and subsisting lease, the registry statutes are applicable and the Court must therefore accord to the appellant the protections of said statutes as a bona fide purchaser for value without notice of said mineral interest covering the north half of the south half of the northwest quarter (N 1/2 of the S 1/2 of the NW 1/4) of said Section 17. The appellant submits that the only evidence offered was the self-serving statements of the employees of Getty Oil Company which were nothing more or less than an effort to vary the terms of an instrument which was plain and unambiguous on its face. The appellant points out that the appellees took the position that the coverall clause in the said lease was sufficient to cover the south half (S 1/2) as well as the north half of the southwest quarter (N 1/2 of SW 1/4) of said Section 17. However, the appellant urges that this Court in the case of Continental Oil Company v. Walker, 238 Miss. 21, 117 So.2d 333 (1960), stated that such clauses are not looked upon with favor by the courts and that the purpose of

said clauses is to cover only irregular or omitted strips of land. The appellant argues that this is not the situation in the case at bar. The appellant urges there is no proof in the record as to the intentions of the lessor, J. D. Haynes, and the only evidence offered by the appellees in this case to show intent was that of Getty Oil Company's personnel.

Where the intentions of the parties to an instrument appear clear and unambiguous from the instrument itself, the court should look solely to the instrument and give same effect as written. If, however, a careful reading of the instrument reveals it to be less than clear, definite, explicit, harmonious in all its provisions, and free from ambiguity throughout, the court is obligated to pursue the intent of the parties, and, to determine the intent, must resort to extrinsic aid.

The trial court had before it a lease which specifically described only two hundred acres, but recited that it covered two hundred and forty acres and contained a 'coverall' clause. Under these circumstances the conflict between the acreage recital of two hundred and forty acres and the description of only two hundred acres had rendered the instrument ambiguous on its face. The trial court accordingly correctly permitted evidence which conclusively showed that it was the intention of the parties to said lease that the lease cover the property described in the earlier lease of July 23, 1958, but through a scrivener's error the subject land was specifically described as the south one-fourth of northwest one-fourth (S 1/4 of NW 1/4) of said Section 17 instead of the south one-half of the northwest one-fourth (S 1/2 of NW 1/4) of said Section 17. The case of Continental Oil Company v. Walker, supra, is not contrary to the above mentioned rule of law as the appellant argues. In that case this Court expressly recognized the importance of the acreage recital in construing an oil, gas and mineral lease containing a 'coverall' clause. Therefore, there was no error in permitting evidence as to the intent of the parties.

WHITEHEAD V. JOHNSTON, 467 So.2d 240, 85 O.&G.R. 227 (Ala.1985). The area described in an oil and gas lease containing a Mother Hubbard clause was five acres and the adjacent area claimed under the clause was a one acre strip to which the lessor claimed ownership by adverse possession. The *Smith* case was cited to the court for the proposition that an ambiguity existed and "parol evidence should have been allowed to ascertain the intent of the parties." *Smith v. Allison* was found "distinguishable from the case at hand because in *Smith* the tract of land sought to be included pursuant to the Mother Hubbard clause was substantially larger than the area included within the *particular description.*" Summary judgment for the party claiming under the clause was affirmed. A dissenting opinion characterized the treatment of the Mother Hubbard clause in the majority opinion as " *absolutism.*"

———

SECTION 3. GRANTS AND RESERVATIONS OF FRACTIONAL INTERESTS

(See Martin & Kramer, *Williams & Meyers Oil and Gas Law* §§ 308-319)

DUDLEY v. FRIDGE

Supreme Court of Alabama, 1983.
443 So.2d 1207, 80 O.&G.R. 1.

ALMON, JUSTICE. This appeal involves a mineral royalty interest deed. The grantors brought this action to have the deed construed as advocated by the grantors or, failing this, to have the deed reformed. The trial court denied relief, holding that the deed was to be construed as advocated by the grantees and that the evidence did not support reformation of the deed.

The plaintiffs are A. Bruce Dudley, Jr., John N. Horner, J.H. Spencer, R.H. McLeod, and Larry U. Sims. In 1971 these five men purchased 100 acres of land in Mobile County with one-half the mineral rights. This 100-acre tract was subject to a lease referred to herein as the Daws lease, which reserved a 1/8 royalty to the lessor. The plaintiffs later purchased 20.5 acres in an adjoining section with full mineral rights and no existing lease.

In 1974 Larry Sims heard that Harris Anderson was knowledgeable in oil and gas matters and was handling some oil business in the area for a friend of Sims's. Sims called Anderson and went to his office to learn about oil and gas transactions.

According to Sims, a man from Mississippi offered plaintiffs a proposal whereby he would lease the 20.5 acre parcel from them for a 1/8 royalty plus $1000 per acre bonus if the plaintiffs would also convey five royalty acres on the 100-acre tract for $2,400 per acre, or $12,000. A royalty acre is defined as a 1/8 royalty on the full mineral interest in one acre of land. Sims asked Anderson for advice on the proposal and, according to Sims, Anderson offered to better the proposal by offering a 3/16 royalty and $1000 per acre bonus on the 20.5 acres, provided that the plaintiffs deeded five royalty acres to Anderson.

The plaintiffs did in fact execute a lease with a 3/16 royalty on the 20.5 acres and a royalty deed relating to the 100-acre parcel. Defendant Harris Anderson was named as the lessee in the former instrument and the grantee in the latter. Both instruments are dated December 2, 1974. Because the meaning and application of the royalty deed are at issue in this case, we shall set the deed out in full:

ROYALTY DEED

KNOW ALL MEN BY THESE PRESENTS that [the plaintiffs and their wives], (hereinafter called Grantor), for and in consideration of the price and sum of TEN AND MORE ($10.00 and More) DOLLARS and other valuable considerations, cash in hand paid by *Harris G. Anderson* (hereinafter called Grantee), has granted, bargained, sold and conveyed, and does by these presents grant, bargain, sell and convey, unto the said Grantee the mineral royalty interest only, as hereinafter set out affecting and relating to the property described below:

An undivided one-tenth (1/10) royalty interest only in and to that part of the minerals owned by Grantor in, on or under the following described property: [The one hundred acre tract], said interest being subject to the present oil, gas and mineral lease to S.B. Daws, and to be subject to any and all further leases at Grantor's option.

This sale and transfer is made and accepted subject to an oil, gas and mineral lease now affecting said lands, but the royalties hereinabove described shall be delivered and/or paid to the Grantee out of and deducted from the royalties reserved to the Grantor arising out of said lease. This sale and transfer, however, is not limited to royalties accruing under the lease presently affecting said lands, but the rights herein granted are and shall remain a charge and burden on the land herein described and binding on any future owners or lessees of said lands and, in the event of the termination of the present lease, the said royalties shall be delivered and/or paid out of the whole of any oil, gas or other minerals produced from said lands by the owner, lessee or anyone else operating thereon.

The grantor herein reserved [sic] the right to grant future leases affecting said lands so long as there shall be included therein, for the benefit of the grantee herein, the royalty rights herein conveyed; and the grantor further reserves the right to collect and to retain all bonuses and rentals paid for or in connection with any future lease or accruing under the lease now outstanding.

TO HAVE AND TO HOLD said royalty rights unto the said Grantee, forever; and the said Grantor hereby agrees to warrant and forever defend said rights unto the said Grantee against any person whomsoever lawfully claiming or to claim the same.

WITNESS the signature of Grantor, this the 2nd of December, 1974.

[Signatures and Notarizations]

Indorsement as required by Alabama Law: This instrument prepared by <u>Larry U. Sims.</u>"

Sims took most of the language in this deed from a royalty deed form given him by Anderson, although he made changes throughout the form. The paragraph containing the property description and the description of the interest conveyed was entirely inserted by Sims. The portion of the deed most acutely in dispute is the description of the interest conveyed as a "1/10 royalty interest."

Anderson assigned his interest in the royalty conveyed by the deed to the other named defendants. In January 1976 the plaintiffs executed a new lease on the 100-acre parcel to AMAX Petroleum Corporation. AMAX was an assignee of Daws's interest as lessee under the Daws lease, which was due to expire in March 1976. The AMAX lease recites on its face that it supersedes the Daws lease. In the AMAX lease, the plaintiffs reserved to themselves a 1/4 royalty.

The dispute which precipitated this case arose in 1977 when the Getty Oil Company began drilling in the area as assignee of the AMAX lease. Getty apparently proposed to pay defendants 1/10 of 1/8 royalty, whereupon Anderson wrote to AMAX and Getty in June and July 1977 claiming 1/10 of 1/4 royalty. Sims contested this claim, saying that defendants were only entitled to 1/10 of 1/8 royalty. Getty then wrote to Anderson, stating that "Mr. Sims advises in substance that your claim to an additional 5 net royalty acres under the December 2, 1974 royalty deed is not in accordance with the intent of the transaction," and suspended payment of the disputed 1/10 of 1/8.[80]

Plaintiffs filed this suit on November 2, 1977, seeking by the first cause of action of the complaint to have the court:

> declare that the Plaintiffs sold to Defendant, Harris G. Anderson, five royalty acres only; that Defendant Anderson's subsequent assignees be bound by such declaration; that said royalty deed be so construed by order of this Court; and that said assignments also be so construed.

The second cause of action of the complaint, as amended, averred that "through fraud, mutual mistake of the parties, or a mistake of one party which the other party at the time knew or suspected," (*see* Code 1975, § 35-4-153) the deed did not "truly express the intention of the parties as to the exact royalty interest conveyed." The averments continued that defendants were insisting upon an interpretation that plaintiffs conveyed to Anderson "a one-tenth interest in any and all royalties received by Plaintiffs instead of the true intent of the agreement that Plaintiffs convey to Defendant Anderson five permanent royalty acres." This count requested the court to revise and reform the deed to express the intention urged by the plaintiffs, and further prayed that the assignments be construed as conveying only the 5 royalty acres intended to be conveyed.

The defendants answered, denying that the intent was to convey only five royalty acres and stating that the deed granted them a 1/10 royalty interest "not only in the existing S.B. Daws lease, but by its express terms a one-tenth (1/10) royalty interest in any additional oil, gas and mineral lease which the Grantors might execute on the minerals which they own." They alleged that they were entitled to 1/10 of the 1/4 royalty because the AMAX lease "provides for a 1/4 royalty payment to mineral royalty owners including the Defendants." As to the second cause of action, the defendants denied that there was any fraud, mutual mistake of the parties, or mistake of one party which the defendants at the time knew or suspected. They further alleged that, under § 35-4-153, Code 1975,

> even if the allegations of the Second Cause of Action are true, which

---

[80] The disputed royalty amounts to 5 royalty acres for the following reason: Plaintiffs owned 1/2 the minerals in the 100 acres. This equaled 50 mineral acres within the terms of the oil and gas industry, one mineral acre being defined as the full mineral rights in one acre. Because a royalty acre is defined as a 1/8 royalty on one mineral acre, plaintiffs owned 50 royalty acres under the Daws lease. When plaintiffs conveyed a 1/10 royalty interest to Anderson while the Daws lease was in effect, 1/10 of these 50 royalty acres equalled 5 royalty acres. The AMAX 1/4 (2/8) royalty lease may be viewed as creating 100 royalty acres. Thus, contend the defendants, their 1/10 royalty interest amounts to 10 royalty acres under the AMAX lease.

Defendants deny, any revision of the royalty deed by the Court would prejudice the rights of the Defendants other than Harris Anderson, which Defendants acquired their interests as third parties in good faith and for value.

Sims and Anderson gave depositions, which were introduced as exhibits at trial. The trial court heard the case without a jury. Only Sims and Anderson testified. The trial court entered a judgment finding that the evidence did not support either the construction of the deed requested by the plaintiffs or reformation of the royalty deed; that there was no fraud, mutual mistake, or mistake of one party which the other at the time knew or suspected; and that the royalty deed truly expressed the intent of the parties. The court therefore ordered that the plaintiffs' request for relief be denied, that the royalty deed is a valid and binding deed on the parties, and that the defendants are entitled to 1/10 of the 1/4 royalties payable under the AMAX lease. Plaintiffs appeal from this judgment.

Plaintiffs argue first that the deed unambiguously supports their position that the rate of the royalty interest was fixed at 1/10 of 1/8. They say this is so because the deed states that the royalty interest conveyed was subject to the Daws lease, which provided a 1/8 royalty. They assert that the subsequent statements in the deed, that the interest was not limited to royalties accruing under the Daws lease, make the deeded royalty interest a permanent royalty interest rather than a term royalty interest which would terminate with the expiration of the Daws lease. Under this interpretation, the phrases "the rights herein granted," "the said royalties," and "the royalty rights herein conveyed" would mean the 1/10 of 1/8 permanent royalty upon which plaintiffs insist. They point to the phrase added by Sims, "and to be subject to any and all further leases at Grantor's option," and state that this gives them the option to extend to defendants the benefit of the increased royalty of the AMAX lease, an option which they have not exercised.

The defendants state that the last-quoted phrase reserves to the plaintiffs the right to lease or not to lease the minerals at their option, but not the right to withhold from defendants the benefits of a lease more favorable than the Daws lease. The phrases lifted from the last three paragraphs of the deed, in the defendants' view, are taken out of their context, which clearly shows that the 1/10 of royalty attaches to whatever royalty plaintiffs reserve in any lease. Finally, defendants point out that the deed mentions neither a 1/8 royalty nor 5 royalty acres.

We find no merit in plaintiffs' position that the deed unambiguously grants only five royalty acres, or a 1/10 of 1/8 royalty. This argument hinges on the plaintiffs' contention that the phrase "and to be subject to any and all further leases *at Grantor's option* " means that plaintiffs could choose whether or not to extend the benefits of a more favorable lease to defendants. This is not a reasonable interpretation of the language. With no additional consideration, why would the plaintiffs ever extend additional royalties to the defendants? A much more plausible and natural reading of the language is that the plaintiffs could choose whether or not to execute leases in the future. This retains executive control over the leasing of the minerals in the plaintiffs.

If the plaintiffs do not have the option whether to grant or withhold the benefits of future leases, the fact that the defendants' interest is to be "subject to" future leases just as it is "subject to" the Daws lease at the time of the conveyance defeats the argument that only the Daws 1/8 royalty is meant when the deed later refers to "the said royalties" and "the royalty rights herein conveyed." This reading is consistent with the provisions in the deed that the sale is not limited to the royalties accruing under the Daws lease and that the plaintiffs shall include the defendants' royalty rights in any future leases. Instruments are to be construed as a whole so as to harmonize their parts whenever possible.

The plaintiffs' second allegation of error is that, assuming the deed does not unambiguously support their interpretation, it is latently ambiguous and should be construed in their favor. They assert that the trial court erred in failing to find as a matter of law that the deed is ambiguous and in failing to construe it in the context of the surrounding circumstances to hold that the intent was to convey only five royalty acres. The plaintiffs argue that several items of evidence prove this intent.

The exhibits which plaintiffs cite are: 1) a sheet with two computations, allegedly in Anderson's handwriting, one of which multiplies 2400 (the alleged price per royalty acre) times 5 (royalty acres) to equal 12,000, and the other computing a price of 20,500 dollars for the 20.5 acres leased; 2) a draft signed by Anderson for $32,500 payable to plaintiffs and reciting that it is for an "O G & M lease covering 20.5 acres in Sec. 36, T 1S, R 1W," and "Royalty Deed 5 AC [Sec.] 35"; and 3) a letter from Sims to Anderson. This letter bears the date December 2, 1974, and reads:

> My group has today sold five acres of royalty on minerals in Section 35, Township 1 So., Range 1 West, to group of investors represented by you as their broker. This letter is to confirm my group's consent to contact you to review any leasing arrangement contemplated by my group involving the five acres made the subject of the royalty deed if the current effective lease expires. It is understood that my group has the legal right to make any and all decisions as to such leasing but we will be glad to have your advice in that regard.

Sims testified that he dictated this letter in Anderson's presence at the time Anderson brought him the draft and he delivered the signed deed and lease to Anderson. Sims testified that Anderson smiled or smirked when he looked at the deed, causing Sims to ask, "Now, Mr. Anderson, are you sure that we are conveying the five royalty acres and nothing else?" Anderson replied, according to Sims, "Yes, sir. That's our deal, five royalty acres, no more, no less, but how about giving me a letter saying that you will, your group will contact me if they ever re-lease." Plaintiffs argue that Anderson's behavior shows that he knew or suspected that the deeded interest would increase with a more favorable lease, contrary to the stated intent.

Defendants respond that the computations were made on the basis of five royalty acres simply because that was the only way to figure the price at the time.

Anderson testified:

> A royalty interest can be increased from one to two to three to whatever royalty you might acquire on a lease or you reserve on a lease. So at the time I acquired this, it is true I bought five royalty acres. I paid for five royalty acres but that is one-tenth of the fifty acres that they owned. If they were prudent and good businessmen enough to acquire a lease on there that gives one-fourth, well, that they are--we are not, we are non-executive royalty owners. We have nothing to say and he made it very clear in the deed, very plain in the deed, we have nothing to say about any future leases or any future royalties. That is entirely up to the executive mineral owners. We are at his mercy so to speak, now. But it's a kind of unwritten thing that he must protect us too. He can't freeze us out. So, when we acquired the one-tenth interest that would go to any interest that he might get in the future.

Anderson further testified that rather than saying defendants' interest increased to ten acres, one could just as accurately say that "each acre of ours became twice as valuable as it was on the day that we acquired it." He said that if one viewed the situation as plaintiffs owning 45 royalty acres under the Daws lease and defendants owning 5, under the AMAX lease plaintiffs' interest doubled to 90 royalty acres and defendants' to 10, rather than, as plaintiffs would have it, their interest increasing to 95 royalty acres and defendants' share remaining at 5 royalty acres.

We set out this evidence relating to the possibility of a latent ambiguity because proof of a latent ambiguity arises when collateral matters outside the writing show the meaning of the document, unambiguous on its face, to be uncertain. The trial judge, considering the matters set out above, determined that the evidence did not support the construction requested by the plaintiffs. Plaintiffs insist that this judgment is due to be reversed as a matter of law, citing the rule that whether an instrument is ambiguous is a question of law.

Plaintiffs' argument does not establish that the trial court is due to be reversed. While Sims's testimony and the above-referenced exhibits may be said to cast some doubt on the meaning of the deed, this doubt is not strong enough, in light of the recitals on the face of the deed that the defendants' royalty interest applies to future leases just as to the Daws lease and of Anderson's testimony regarding the "five royalty acres only" controversy, for us to hold that the trial court erred as a matter of law in not finding the deed ambiguous and not giving it the construction urged by plaintiffs. Moreover, even if we were to hold that the trial court erred in not finding a latent ambiguity, the remainder of the court's judgment makes it clear that the court, as finder of fact, found the facts not to support plaintiffs' interpretation of the deed. This finding would be supported by the presumptions of the *ore tenus* rule, and the facts, as set out above, support the trial court's conclusion.

The plaintiffs' final allegation of error is that the trial court erred in light of the evidence in finding no basis for reformation of the deed. The chief aspect of the evidence not already discussed above pertinent to this argument regards testimony by Sims that, after this controversy arose, he searched the probate records of Mobile

County and found some thirty or forty royalty deeds prepared by Anderson. Plaintiffs introduced one of these into evidence. It describes various parcels in numbers of royalty acres and specifies after the list of parcels that "It is the intention of Grantor by this royalty deed to convey to Grantee, its successors and assigns Five (5) royalty acres."

Sims testified that all the deeds prepared by Anderson similarly specified the number of royalty acres being conveyed. He and the other plaintiffs argue that Anderson, upon reading the deed prepared by Sims, was aware or suspected that Sims had made a mistake in failing to limit the conveyance to a certain number of royalty acres. Sims further testified that he described the interest conveyed as a 1/10 royalty interest because Anderson told him that would describe the five royalty acres being conveyed, 1/10 of fifty royalty acres being five acres.

Defendants respond that to have a deed reformed, a party must produce clear, convincing, and satisfactory evidence. The above discussion shows that the evidence was in such dispute that the trial court cannot be held in error for holding the plaintiffs did not sustain their burden of proof.

For the reasons stated, the judgment of the trial court is due to be, and is hereby, affirmed.

Affirmed.

TORBERT, C.J., and FAULKNER, EMBRY and ADAMS, JJ., concur.

NOTES

1. The principal case states that "A royalty acre is defined as a 1/8 royalty on the full mineral interest in one acre of land." Although this is a recognized meaning of the term, Williams and Meyers (in § 320.3) describe a royalty acre as "the full lease royalty on one acre of land. Thus, the owner of 50 royalty acres in a 100 acre tract is entitled to 1/2 of royalty, or 1/16 royalty where the lease provides for 1/8 royalty." If the *Williams and Meyers* definition is accepted, would the plaintiffs in the principal case gain anything from a finding that they had conveyed, in defendant Anderson's words, "five royalty acres, no more, no less?" If the tract of land described in the conveyance at issue contains exactly one hundred acres and plaintiffs own a one-half interest in it, does the language used to describe the interest conveyed, "1/10 royalty interest," convey the same interest as would a conveyance of "five royalty acres" under the *Williams and Meyers* definition? If the language of "royalty acres" is not to be used, what language would accomplish the result for which the plaintiffs contend? The defendants?

2. WADE V. ROBERTS, 346 P.2d 727, 11 O.&G.R. 529 (Okl.1959). *R* conveyed to *E* a tract of land containing 32 acres, reserving to *R* "an undivided 5/32 interest amounting to an undivided five (5) acre interest" in the minerals. The tract later increased in size by accretion so that the 5/32 interest amounted to 7.385 mineral acres. *R* claimed he was entitled to 5/32 of the enlarged tract (*i.e.* to 7.385 mineral acres); *E* contended that *R*'s reservation was limited to 5 mineral acres. *Held,* for *E*. Suppose the conveyance is in terms of mineral acres alone: "5 mineral acres in

[land described by metes and bounds with a river as the southern border], containing 100 acres, more or less." The river moves to the south and new land is created between the river and the described property, so that the acreage with the river as southern boundary is now 200 acres. Does the grantee still own 5 mineral acres in the 200 acre tract or 1/20 undivided interest in 200 acres, amounting to 10 mineral acres? *See* Arnold v. State, 750 P.2d 1137, 97 O.&G.R. 610 (Okl.App. 1987), cert denied. Kimball, *Accretion and Severed Mineral Estates*, 53 U.Chi.L.Rev.232 (1986), takes the position that "a fixed boundary rule for severed mineral estates is the better alternative."

3. HILD v. JOHNSON, 2006 ND 217, 723 N.W.2d 389. The owner of a section of land conveys an undivided 382.76/582.76 interest in the minerals in the section. The denominator of this fraction is apparently based on the belief that because 57.24 acres of the 640-acre section is beneath the Little Missouri River, the grantor only owned 582.76 acres. Subsequently, it is determined that the river is not navigable at the time of statehood and thus, grantor owns the full 640 acres. The grantees thereupon contended that the deed conveyed 382.76/582.76 x 640 acres, while the grantor asserts the deed only conveys 382.76 mineral acres. The court holds that the mineral deed expressly conveys an undivided interest, expressed as a fraction, in the minerals in all of Section 21. The grantees therefore acquired the stated fractional interest in all of the described land.

―――――

BODY v. McDONALD

Supreme Court of Wyoming, 1959.
79 Wyo. 371, 334 P.2d 513, 10 O.&G.R. 103.

BLUME, CHIEF JUSTICE. This is an action in the nature of a declaratory judgment and to quiet title to a three-fourths mineral interest in the lands hereinafter described. The court substantially granted the prayer of the plaintiffs and appellees herein, and from the judgment so entered the defendants and appellants herein have appealed to this court. The facts as disclosed by the pleadings and the evidence herein are substantially as follows: On May 29, 1914, George Edwards and Lena B. Edwards, husband and wife, being then the owners of the property herein described, made and executed a warranty deed to W.W. McDonald in and to [described property]. The deed contained the following reservation:

> Excepting and reserving to the said parties of the first part, their heirs and assigns forever an undivided one-fourth (1/4) interest in and to all oil, petroleum and other oil products now located upon or in the said lands, or that may hereafter be taken therefrom. Each of the said parties of the first part is to have and does hereby retain a one half interest in the reservation hereby made.

Thereafter on June 11, 1914, the aforesaid W.W. McDonald joined by his wife, Maggie McDonald, executed and delivered to Albert G. Cheney and Charles H. Body a warranty deed to the aforesaid property. In that deed the grantors granted, bargained, sold and conveyed unto the grantees the aforesaid property, and it contained the following:

> Excepting and reserving to the said parties of the first part [the McDonalds], their heirs and assigns, forever, an undivided one-fourth (1/4) interest in and to all oil, petroleum, and other oil products now located upon or in the said lands, or that may hereafter be taken therefrom. Each of the said parties of the first part is to have and does hereby retain a one half interest in the reservation hereby made.

The deed also contained the statement that the grantors "have good and lawful right to sell and convey the same. And the said parties of the first part [the McDonalds] will and their heirs, executors and administrators shall Warrant and Defend the same against all lawful claims and demands whatsoever."

At the same time the grantees in the last mentioned deed, namely, Body and Cheney, executed a mortgage in favor of William W. (W.W.) McDonald to secure the sum of $5,000. In that mortgage the mortgagors stated that they were well seized of the said premises, in and of a good and indefeasible estate, in fee simple. The mortgage further contained the following statement:

> provided that this mortgage hereby expressly excepts and reserves unto George Edwards and unto Lena B. Edwards, his wife, their heirs and assigns, forever, an undivided one-fourth interest in and to all oil, petroleum and other oil products now located upon or in the said lands, or that may hereafter be taken therefrom, in the proportion of one half interest in the said exception and reservation hereby made, to each of said parties, to wit: George Edwards and Lena B. Edwards;

The evidence of Charles H. Body herein shows that he knew of the reservation in the deed of the Edwardses to the McDonalds and that he, the said Charles H. Body, had been in possession of the premises since the time that he and his associate received a deed to the premises above mentioned. It further appears herein that one of the plaintiffs, Ruth Henry, is the successor in interest of Albert G. Cheney, and that Mabel I. McDonald Weaver and Anna M. Adams Wise are the successors in interest of W.W. McDonald and his wife, Maggie McDonald. Charles H. Body died on May 22, 1957, after the judgment was entered in the case below, and left his widow, Ruby Body, as his sole heir and devisee; and upon motion filed in this case, dated May 9, 1958, she was substituted as one of the plaintiffs and respondents herein.

The trial court entered judgment, stating in part as follows:

> The court concludes that either by estoppel or resulting trust the reservation contained in the said deed of June 11 and 12, 1914 is held by the Defendants for the use and benefit of George Edwards and Lena B. Edwards as such interest was set forth in the deed of May 29, 1914, if said interest is presently outstanding and claimed by the said George Edwards and Lena B. Edwards or their successors in interest.

The court further quieted the title of three-fourths of the mineral interest in and to the lands in the plaintiffs Charles H. Body and Ruth Henry and enjoined and debarred defendants herein from setting up any claims of right, title or interest in and to three-fourths of the mineral interest in the lands above mentioned. As stated before, the defendants herein have appealed from this judgment.

The plaintiffs herein contend that as they are the owners of three-fourths of the minerals contained in the lands above described and since one-fourth of the mineral rights is outstanding in favor of Edwards and his wife (or their privies), the defendants herein have no mineral interest in the lands whatsoever.

The defendants, on the other hand, while not questioning the one-fourth mineral interest outstanding in favor of Edwards and his wife, claim that they are the owners of one-fourth of the minerals contained in the lands aforesaid according to the reservation in the deed by McDonald and his wife to Albert G. Cheney and Charles H. Body.

1. It may be conceded for the purposes of this case that the deed from the McDonalds, dated June 11, 1914, to Body and Cheney is plain and unambiguous and cannot be varied by parol evidence. By that deed the McDonalds conveyed to the grantees in that deed the whole of the lands above described excepting only a one-fourth mineral interest. In other words, the McDonalds conveyed under a warranty of title, aside from the surface of the land, three-fourths interest in and to the minerals contained therein. Having thus vested the title to three-fourths of the minerals in and to the grantees, the McDonalds and their successors in interest are estopped from claiming that the grantees and their successors in interest have less than three-fourths of the mineral rights in the land. In 31 C.J.S. Estoppel § 10, it is said:

> Estoppel by deed is a bar which precludes a party to a deed and his privies from asserting as against the other and his privies any right or title in derogation of the deed, or from denying the truth of any material fact asserted in it. . . . The doctrine of estoppel by deed is applied in order to avoid circuity of action, and to compel the parties to fulfill their contracts.

Again in 31 C.J.S. Estoppel § 13, it is stated:

> A person who assumes to convey an estate by deed, or his successor, is estopped, as against the grantee, or those in privity with him, to assert anything in derogation of the deed; he will not be heard, for the purpose of defeating the title of the grantee, to say that at the time of the conveyance he had no title, or that none passed by the deed, nor can he deny to the deed its full operation and effect as a conveyance. A warrantor of title may not question the validity of the title warranted, nor may he assert an outstanding hostile title.

See also 19 Am.Jur. Estoppel §§ 5 to 10 inclusive.

A situation similar to that appearing in this case is discussed in a number of cases which sustain the judgment of the court rendered in the present action. Brown v. Kirk, 127 Colo. 453, 257 P.2d 1045; Merchants & Manufacturers Bank v. Dennis, Miss., 91 So.2d 254; Salmen Brick & Lumber Co., Limited v. Williams,

210 Miss. 560, 50 So.2d 130; Garraway v. Bryant, 224 Miss. 459, 80 So.2d 59, 61 A.L.R.2d 1387; Murphy v. Athans, Okl., 265 P.2d 461; Benge v. Scharbauer, 152 Tex. 447, 259 S.W.2d 166; Duhig v. Peavy-Moore Lumber Co., Inc., 135 Tex. 503, 144 S.W.2d 878. Syllabus 3 in Salmen Brick & Lumber Co., Limited v. Williams, *supra* [210 Miss. 560, 50 So.2d 130], states as follows:

> Where grantor had property conveyed to it with one-half mineral exception and subsequently said grantor conveyed property by warranty deed conveying fee-simple title to property but containing exception clause almost identical to that used in original deed to grantor, exception clause in second deed merely described interest owned by grantor, and therefore effect of second deed was to convey to grantee exactly what grantor had received under its original deed and grantor retained for itself none of mineral interest by such deed.

In the *Duhig* case [135 Tex. 503, 144 S.W.2d 878], syllabus 1 is similar to the foregoing and syllabus 2 states as follows:

> Where warranty deed, executed by grantor to whom realty had been conveyed with reservation of one-half undivided interest in mineral rights, expressly retained undivided one-half interest in mineral rights and purported to convey all other interest in the realty, covenant of general warranty in the deed operated as an 'estoppel' denying to the grantor and those claiming under him the right to set up a claim to one-half undivided interest in mineral rights.

The court mentioned the fact that the grantor in the second deed should not be permitted to hold the mineral interest reserved by his deed and to require his grantee to seek redress in a suit for breach of the warranty. The court quoted the rule relating to after-acquired property and stated at 144 S.W.2d 880:

> . . . What the rule above quoted prohibits is the *assertion* of title in contradiction or breach of the warranty. If such enforcement of the warranty is a fair and effectual remedy in case of after-acquired title, it is, we believe, equally fair and effectual and also appropriate here.

The other cases above mentioned are similar in effect and establish we think conclusively that in the case at bar the defendants in this case cannot question the fact that the plaintiffs herein own three-fourths of the mineral interest in and to the lands above described, and that since one-fourth of the mineral interest is outstanding in favor of Edwards and his wife (or their privies), defendants under the facts herein own no mineral interest in the lands in question here.

. . .

3. As heretofore stated Charles H. Body testified that when he and Cheney received the deed of June 11, 1914, from the McDonalds he knew that one-fourth mineral interest in the lands was outstanding in favor of George and Lena B. Edwards under the deed of May 29, 1914. So counsel for the defendants contend that the doctrine of estoppel cannot be applied herein because it is inapplicable when the parties had full knowledge of the facts. That seems to be the crux of the contention of counsel for the defendants herein. They think that this case is

distinguishable from the cases cited in point one of this opinion, including the *Duhig* case, by reason of that fact. It is true that matter was not mentioned in those cases, but inasmuch as purchasers of real property usually examine the title thereto, it is highly probable that the purchasers in those cases had knowledge of an outstanding mineral interest just as is true in the case at bar. While conjecture, it is not unlikely that the matter was not considered worthy of attention either by counsel in those cases or by the courts rendering the decisions. Counsel cite us to numerous authorities involving an estoppel in pais or equitable estoppel. These authorities are not in point. Learned counsel have failed to distinguish between an estoppel in pais and an estoppel by deed. In the case of McAdams v. Bailey, 169 Ind. 518, 82 N.E. 1057, 1059, 13 L.R.A.,N.S., 1003, 124 Am.St.Rep. 240, the court said:

> Lord Coke observes that although estoppels are odious, yet warranties are favored in law, being part of a man's assurance. 2 Institutes, 219. It is a mistake to liken an estoppel by deed to an estoppel in pais. . . .

We have seen that the McDonalds, by the deed of June 11, 1914, granted to Cheney and Body unconditionally and without qualification three-fourths of the mineral interest in the lands involved herein and warranted the title thereto. Counsel for defendants would whittle down that grant and warranty--since the Edwardses' interest is outstanding--from three-fourths to one-half of the mineral interest. No authority that this may be done has been cited and we have found none. A discussion in point contrary to the contention of defendants is found in the case of Ayer v. Philadelphia & Boston Face Brick Co., 159 Mass. 84, 34 N.E. 177, 178, a decision written by Justice Holmes, although that case involved an after-acquired title. In that case it was held that when a man gives a second mortgage but covenants and warrants as against the first mortgage and afterwards obtains title through a foreclosure of the first mortgage, his title thereby acquired inures to the benefit of the second mortgagee. In that case the parties knew that there was a first mortgage outstanding but that made no difference. Said the court in part:

> A subsequent title would inure to the grantor when the grant was of an unincumbered fee, *although the parties agreed by parol that there was a mortgage outstanding,* and this shows that the estoppel is determined by the scope of the conventional assertion, not by any question of fraud or of actual belief. But the scope of the conventional assertion is determined by the scope of the warranty which contains it. Usually the warranty is of what is granted, and therefore the scope of it is determined by the scope of the description; but this is not necessarily so, and when the warranty says that the grantor is to be taken as assuring you that he owns and will defend you in the unincumbered fee, it does not matter that by the same deed he avows the assertion not to be the fact. The warranty is intended to fix the extent of responsibility assumed, and by that the grantor makes himself answerable for the fact being true. In short, if a man by a deed says, 'I hereby estop myself to deny a fact,' it *does not matter that he recites as a preliminary that the fact is not true.* The

difference between a warranty and an ordinary statement in a deed is that the operation and effect of the latter depend on the whole context of the deed, whereas the warranty is put in for the express purpose of estopping the grantor to the extent of its words. The reason 'why the estoppel should operate is that such was the obvious intention of the parties.' Blake v. Tucker, 12 Vt. 39, 45. (Emphasis supplied.)

That discussion was cited with approval in McAdams v. Bailey, supra. If the force and effect of a warranty could not be affected or diminished by knowledge of an outstanding title in these cases, neither could it be in the case at bar. The contention here made is overruled.

4. Counsel for defendants contend and pleaded that the action herein is barred by the statute of limitations. In this case Charles H. Body has at all times been in possession of the lands involved herein. In 54 C.J.S. Limitations of Actions § 124, p. 36, it is stated:

A cause of action to quiet title or for the removal of a cloud on title has been said to be a continuing one, and never barred by limitations while the cloud exists. The statute of limitations does not begin to run in favor of defendant in an action to quiet title, or remove a cloud, so that he may invoke its running for the prescribed period as a defense, until some assertion of right or title to the premises has been made by him and until such assertion of right or title to the premises has been brought to the knowledge of plaintiff.

In the case at bar it does not appear when a controversy arose or when the defendants claimed any mineral interest as involved herein. For aught we know the defendants claimed no mineral interest in this case until they were served with process. Hence, it is clear under the foregoing authorities that the contention herein that the action is barred by the statute of limitations must be overruled.

From what we have said it is apparent that the judgment must be and is affirmed.

Affirmed.

NOTES

1. Does it matter that the fraction of minerals purportedly reserved is the same as the fraction outstanding in considering the application of the *Duhig* doctrine? *Compare* Gilbertson v. Charlson, 301 N.W.2d 144, 69 O.&G.R. 73 (N.D.1981), where one hundred percent of the minerals were conveyed by the language of the deed, fifty percent were reserved, and thirty-six and two-thirds percent were outstanding, and *Duhig* was not applied, *with* Sibert v. Kubas, 357 N.W.2d 495, 84 O.&G.R. 350 (N.D.1984), where one hundred percent of the minerals were conveyed, fifty percent of the minerals were outstanding and fifty percent were reserved, and *Duhig* was applied. The *Gilbertson* exception to *Duhig* has been further limited in Acoma Oil Corp. v. Wilson, 471 N.W.2d 476, 116 O.&G.R. 501 (N.D. 1991). Would it affect your analysis if the parties claiming that *Duhig* was not applicable were family members? Cotenants? *See also* Hartman v. Potter, 596

P.2d 653, 64 O.&G.R. 355 (Utah 1979), where the grantors of one hundred per cent of the minerals, owned only fifty percent and reserved seventy-five percent and the court ignored *Duhig*, reaching "the strange conclusion that the [grantor's] deed was to be construed as reserving 3/4 of what they owned . . . and as conveying . . . 1/4 of [what they owned]." Martin & Kramer *Williams and Meyers Oil and Gas Law*, § 311, n.17.1 Recent *Duhig* problems, including Hartman v. Potter are discussed in Ellis, *Rethinking the* Duhig *Doctrine*, 28 Rocky Mtn.Min.L.Inst. 947 (1983).

    Williams and Meyers, supra, suggest that "actual and constructive notice" to grantors and grantees "was of critical importance" to the *Hartman* court. In Scarmardo v. Potter, 613 S.W.2d 756, 69 O.&G.R. 523 (Tex.Civ.App.1981, error ref'd n.r.e.), *Duhig* was applied to a situation where the grantor of one hundred percent, owned fifty percent and reserved twelve and one-half percent. On the issue of knowledge by the grantee of the state of the title the court said: "[t]he estoppel by deed rule in *Duhig* emanates from the scope of the warranty clause and therefore the knowledge of the grantee is immaterial." 613 S.W.2d at p. 759. Should actual knowledge by the grantee of the state of the grantor's title have an impact different from constructive knowledge derived from the recordation of an outstanding interest in a third party? *See* Morgan v. Roberts, 434 So.2d 738, 79 O.&G.R. 215 (Ala.1983), where the *Duhig* doctrine was applied to a conveyance by the owner of one-half, reserving one-fourth and the court said: "Absent grantees' actual knowledge of grantors' prior conveyance of 1/2 of the mineral rights, the prior recordation of that earlier conveyance cannot excuse grantors' warranty obligations."

    2. In SALMEN BRICK & LUMBER CO. v. WILLIAMS, which is discussed in the principal case, Alexander, J., dissenting, said:

> I am unable to find support in a theory by which a court seeks gratuitously to save a grantor against an anticipated suit for breach of warranty. A warranty does not effect the conveyance. Title is acquired by the conveyance and guaranteed by the warranty. Nor is a deed void which subjects the grantor to a possible suit to enforce the warranty or for damages.

> Is it not possible that appellant intended to reserve unto itself the mineral interest which it owned? It had a right to do so. If such was its purpose, it is unjust to thrust upon it a construction which denies to it such right. It certainly used language adequate to this end. What else should it have done? Under what duty do we construe the words as an exception from the warranty, rather than as a reservation unto the grantor?

> It is important to keep uppermost in mind that this is not a suit for reformation, and there is no cross-bill by appellees. It could be true that the reservation was inadvertent and unintended. But here no mistake or fraud is alleged or shown. There was no testimony adduced. The record is documentary and intent is sought out by the detective processes of deduction. Yet, the ticket of admission to the arena of construction is ambiguity. To set out at once to create ambiguity is to reach a predetermined conclusion first and later to seek

for reason and justification.

In this sort of action, we ought to say to the appellant: We know what you meant, because we know what you said. However, we are saying to it: Regardless of what you said, we think it best to take your reserved half mineral interest and give it to appellees because it is better for you to surrender this interest than that you be exposed to embarrassment or litigation.

. . .

All the circumstances which the Court assembled to support an intent at variance with the universal meaning of the words used would be relevant in a suit to reform the deed and rewrite it in language importing a mere exception. But in this form of action, they should be held at bay since the words are not to be construed but merely defined. By definition, 'reservation' means a reservation and 'unto the grantor' means unto the grantor.

Here again the answer to an inquiry as to what the grantor meant is what it said.

3. LUCAS v. THOMPSON, 240 Miss. 767, 128 So.2d 874, 14 O.&G.R. 73 (1961). Plaintiff Thompson acquired Blackacre subject to the reservation of one-half mineral interest in a remote deed in his chain of title. He conveyed Blackacre by warranty deed to defendant Lucas, reserving one-half the minerals. The trial court's rulings in the title suit were described as follows:

> Over the objection of the defendant, the court permitted Thompson to testify that, on November 1, 1952, at the time of his execution of the deed to Lucas, he did not know that any mineral interest had been conveyed; that he thought he owned all of such rights; and that he did not discover this fact until the spring of 1959, when he was attempting to lease the land.

> In the decree, the court expressed the opinion that the parties were in good faith and believed that they each owned a one-half interest in said minerals, but without the knowledge that a one-half interest therein had been previously conveyed. A one-fourth of the mineral rights was awarded to each of the parties and their titles thereto were quieted and confirmed accordingly.

Should the judgment of the trial court be affirmed?

4. HARRIS v. WINDSOR, 156 Tex. 324, 294 S.W.2d 798, 6 O.&G.R. 1234 (1956). *R* owned 50% of the minerals by virtue of a deed from *X* reserving the other 50%; *R* conveyed to *E* by metes and bounds description followed by this clause: "being the same land described in Warranty Deed [*X* to *R*, giving a recording reference], reference to which is made for all purposes." *R* reserved a 3/8ths mineral interest "in . . . the above described premises." *Held, E* received under the deed a 1/8 th mineral interest. The subject matter of the conveyance was the 4/8ths (or 50%) owned by *R;* the prior deed reference was intended to qualify the estate being conveyed because such reference was made "for all purposes." Thus *E* was to receive 4/8ths less what *R* reserved, which was 3/8ths of 8/8ths .

Harris v. Windsor distinguishes Duhig v. Peavy-Moore Lumber Co., 135 Tex.

503, 144 S.W.2d 878 (1940), the leading case applying the rule of the principal case, on the ground that the recording reference in *Duhig* did not contain the phrase "reference to which is made for all purposes." Is this a sound distinction?

In Scholz v. Heath, 642 S.W.2d 554, 78 O.&G.R. 142 (Tex.App.1982), a prior deed reference "for all legal purposes" was "merely to further identify the area and not to qualify the estate conveyed." Compare Rutherford v. Randal, 593 S.W.2d 949, 65 O. & G.R. 76 (Tex.1980), where the deed reference was introduced with the words "for a better description of said land" but went on to say that the "interest hereby conveyed . . . is that portion that the grantor . . . received from his deceased parents." The interest received from the parents was a 1/24 interest and the granting clause used the fraction 1/240 . The court held that the deed was unambiguous and that it conveyed a 1/240 th interest: "the recital that the conveyance is the portion previously conveyed by [grantor's] parents is no more than a description of the lands wherein the mineral interests are located."

5. OPALINE KING HILL v. GILLIAM, 284 Ark. 383, 682 S.W.2d 737, 85 O.&G.R. 285 (1985) dealt with the question of the *Duhig* rule and quitclaim deeds. In reaching the conclusion that the *Duhig* rule was not invoked by a conveyance utilizing a quitclaim deed the court said:

> The opinion in the *Duhig Case* was composed for the Texas Supreme Court by a commissioner who expressed his personal opinion that when a deed purports to reserve one-half of the mineral interest, and the grantor owns only one-half of the mineral interest, the language of the deed is ambiguous but that 'when resort is had to established rules of construction and facts taken into consideration it becomes apparent that the intention of the parties to the deed' was that the grantee take the one-half interest in the minerals owned by the grantor. 135 Tex. at 506-507, 144 S.W.2d at 879-880. Apparently the 'construction' rule the commissioner would have applied was that the language of reservation used by the grantor referred only to the one-half mineral interest owned by the third party.

> The commissioner, however, noted that the members of the Texas Supreme Court for whom he was writing had expressed a rationale for decision which was entirely different. The Texas Supreme Court's decision, and thus the holding of the case, is that a grantor may not agree to warrant and defend title to a property interest and then, in a later clause in the same instrument, breach that warranty by reserving to himself some portion of that same property interest. The principle of estoppel is held to apply just as it would if the grantor in a warranty deed did not own property at the time he executed the deed but acquired it later. 135 Tex. at 507-508, 144 S.W.2d at 880-881.

> Thus, the holding of the *Duhig Case* is not at all applicable here, as Gilliam warranted nothing but conveyed by quit-claim. The chancellor's refusal to apply the holding of the *Duhig Case* was thus correct.

> The Phillips heirs would have us apply the so-called 'one step' approach, *i.e.*,

that the grantor's reservation is not a second step but is merely a reference to the already outstanding interest. That is the rationale expressed by the commissioner in the *Duhig Case* opinion. They cite Brown v. Kirk, 127 Colo. 453, 257 P.2d 1045 (1953); Garraway v. Bryant, 224 Miss. 459, 80 So.2d 59 (1955); and Murphy v. Athans, 265 P.2d 461 (Okla.1953), as having adopted that approach. While the estoppel principle is not specifically mentioned in any of those cases, the holding of each is entirely dependent upon the conveyance having been by warranty deed. In each it is noted that the grantor's reservation was construed as an exception to the warranty or intended to protect the grantor against a suit on his warranty.

In the case of a quit-claim deed containing a reservation such as the one here the reservation can have only the purpose of notifying the grantee of the mineral interest not owned by the grantor or the purpose of keeping the title to the reserved interest in the grantor. A quit-claim deed conveys the interest of the grantor in the property described in the deed. Smith v. Olin Industries, Inc., 224 Ark. 606, 275 S.W.2d 439 (1955). It is not suggested the grantor of such a deed has any duty or reason whatever to notify his grantee of the interest he does not own. In this case, the grantee, Phillips, was known by Gilliam, the grantor, to have knowledge of Long's one-half interest in the minerals, as Phillips was in the chain of title subsequent to the grant to Long and had been the grantor in the deed to Gilliam.

We look to the deed and the context in which it was made to ascertain the intent of the parties. In this case the deed and the context in which it was made show it was intended that Gilliam keep the one-half mineral interest. In view of the unlikeliness that Gilliam intended the reservation as notice to Phillips of Long's one-half interest in the minerals, the language of the reservation would be surplusage if it was not intended to keep the one-half mineral interest in Gilliam. We will not treat language of a deed as surplusage if we can attribute a reasonable meaning to it.

Soon after *Hill v. Gilliam*, the Arkansas court reached a *Duhig* result in a case with the classic *Duhig* situation: 50% of the minerals outstanding when the land is conveyed by a warranty deed providing "that the grantor retains an undivided one-half interest in and to all mineral rights." Peterson v. Simpson, 286 Ark. 177, 690 S.W.2d 720, 85 O.&G.R. 454 (1985). The court applied the "construction" rule of *Hill*, "a warranty deed which does not limit the interest in the minerals granted purports to grant 100% of the minerals." The policy basis for this rule is stated to be the "preservation of a viable recording system." A dissenting opinion rejects this idea: "If the fifty percent mineral reservation, outstanding at the time of the deed questioned here, was made in a recorded deed, subsequent grantees had constructive notice of it. If it was not recorded, subsequent grantees may be bona fide purchasers in good faith without notice. The *Duhig* rule has nothing to do with either situation." What answer can you make to this proposition of the dissent? Does the majority opinion in *Peterson* mean that the rationale of *Hill* should be rethought?

6. The conveyance in PURSUE ENERGY CORP. v. PERKINS, 558 So.2d 349 (Miss.1990), granted with warranty "2.5/32.5 interest" in two described tracts, one of which contained 7.5 acres and the other 25 acres, "containing in the aggregate 32.5 acres." The tracts were contiguous. A final clause stated "the specific intent of the grantor to convey 2.5 full mineral acres out of the above described land." The grantor owned only a 1.5/32.5 interest in the 25 acre tract but had sufficient interest in the 7.5 acre tract to cover the granted percentage of both tracts in the aggregate. It is argued that this deed conveyed only "an undivided 2.5/32.5 interest from each acre" in each of the tracts. Is there an answer to this contention in the language of the deed? Is this a situation for the operation of the *Duhig* principle?

7. ROSENBAUM V. MCCASKEY, 386 So.2d 387, 67 O.&G.R. 394 (Miss.1980), reaches the same result as *Gilliam* . What if the deed "grants" the interest claimed but "contains no warranty": does the *Duhig* rule apply? Blanton v. Bruce, 688 S.W.2d 908, 86 O.&G.R. 138 (Tex.App.1985, writ ref'd n.r.e.), holds that it does "because the deed purports to convey a definite interest in the property." In Discussion Notes, 86 O.&G.R. 148 (1986), Professor Flittie finds it "regrettable that the Texas Supreme Court did not grant writ of error" since the decision "is an undesirable extension of a rule of most doubtful merit in the first place, rendered even more doubtful by cutting it loose from its breach of general warranty base." Is there a good answer to this criticism of *Blanton?*

8. HOOKS V. NEILL, 21 S.W.2d 532 (Tex.Civ.App.1929, error ref'd). Grantors, who owned an undivided one-half interest in an acre of land conveyed all of their interest reserving a "one thirty-second part of all oil on and under the said land and premises herein described and conveyed." The court found that this language "clearly imports--in fact, it clearly states--that grantors reserve and except from the conveyance one thirty-second part of all oil on and under the land and premises thereby conveyed." Since the land conveyed was an undivided one-half of the tract described the court concluded that the reservation was effective as to a "one sixty-fourth part of all of the oil on and under the entire one acre of land." If grant is "one half (1/2) interest in the following described lots" with a reservation of "one-half interest in and to all oil and mineral rights of any kind or character in and under the above described land," what proportion of the minerals are conveyed under the principle of *Hooks v. Neill? See* Jolly v. O'Brien, 749 P.2d 1000, 98 O.&G.R. 30 (Colo.App.1987).

9. In AVERYT V. GRANDE, INC., 686 S.W.2d 632, 84 O.&G.R. 561 (Tex.App.1984, error ref'd n.r.e.), an undivided one-half interest in minerals was conveyed by a deed describing Blackacre and stating that the conveyance was "less" the undivided one half interest in minerals that had been conveyed by a previous deed described by recording references. The document then went on to reserve "an undivided 1/4 of the royalty covering all of the minerals in, to and under or that may be produced from the lands above described to the Grantor." What "royalty fraction" is reserved in this transaction? Is the case governed by *Hooks v.*

Neill? *See also* First National Bank in Dallas v. Kinabrew, 589 S.W.2d 137, 65 O.&G.R. 545 (Tex.Civ.App.1979) where an instrument identified several leases by recording references and then conveyed 1/8 of 8/8 "of all the Oil, Gas, and other minerals which may be produced, marketed, and saved from the herein described premises . . . ." The leases from which the overriding royalty was created covered less than a 100% interest in the land they described. Was the interest conveyed a 1/8 of 8/8 of production from the land described in the leases or 1/8 of 8/8 of the production attributable to the leases identified?

10. FERGUSON v. MORGAN, 220 Miss. 266, 70 So.2d 866, 3 O.&G.R. 411 (1954). *T*, who owned 50% of the minerals in Blackacre devised to the four children of a deceased son "all of my right, title and interest" in Blackacre "except an undivided one-fifth (1/5) mineral rights and interest in and under said land, which I devise to my son" who was alive. Did the son receive 1/5 of 100% of the minerals or 1/5 of 50% of the minerals?

11. Grantors, owners of a 3/4 interest in a tract of land, conveyed the tract to Grantee by a general warranty deed which reserved "one-half of whatever oil, gas and other mineral interest is owned by them in the described land at the time of this conveyance, conveying to the grantee onehalf of said interest." The heirs of Grantee claim that the "Duhig Rule" is applicable to this conveyance and "transferred the surface and a one-half mineral interest" to Grantee. Is this an appropriate application of *Duhig*? *See* Manson v. Magee, 534 So.2d 545, 103 O.&G.R. 515 (Miss.1988).

12. PRICE v. ATLANTIC REFINING CO., 79 N.M. 629, 447 P.2d 509, 31 O.&G.R. 226 (1968). The granting clause of the deed conveyed "an undivided one-half interest in the following described land." The description of the land followed. The reservation in the deed read: "Grantor[s] hereby retain unto themselves an undivided one-half of all royalty in and to the above described land."

> Appellants contend that the deed reserved to the grantors only an undivided one-half royalty in the one-half mineral interest *conveyed.* Conversely, appellees contend that the deed reserved an undivided one-half royalty interest in the *land described.* We are inclined to appellee's position. . . . [W]e think the intent of the grantors is made clear by the language used, 'one-half of all royalty in and to *the above described land.*' (Emphasis ours.) Consequently, we construe the deed as reserving to the grantors, their heirs and assigns one-half of the royalty resulting from minerals produced and marketed from the land.

What did the grantee receive under the deed as construed?

13. In BURKE v. BUBBERS, 342 N.W.2d 18, 79 O.&G.R. 571 (S.D.1984), an option to sell 801.43 acres contained the language: "The seller wants to retain 1/4 of the mineral rights." The option was exercised and the transaction carried out by a warranty deed which contained no mineral reservation. The deed was reformed to include a reservation of " 1/4 of the mineral rights." The grantor "owned 360.5675

mineral acres on or under the 801.43 surface acres to be sold; a prior owner had retained the remaining mineral rights." The trial court found that the reservation in the deed as reformed applied to "the 801.43 acres sold . . . or 200.3575 mineral acres." The Supreme Court stated:

> Here, the prior agreement of the parties, as written into the purchase option, was to complete a sale of 801.43 acres of real property, with Burkes reserving to themselves ' 1/4 of the mineral rights.' Thus, the trial court was correct in finding that the deed should be reformed to reflect a reservation of a mineral interest in the Burkes. However, the words of reservation must be understood as a reservation of one-fourth of the mineral rights actually owned by Burkes, rather than a reservation of one-fourth of the mineral rights in the total acreage sold. Since Burkes did not own the mineral rights to the entire 801.43-acre tract, but rather owned only 360.5675 mineral acres, they could not reserve one-fourth of the minerals in the 801.43 acres. One cannot reserve one-fourth of something he does not own. Burkes could reserve only one-fourth of the mineral rights actually owned by them (360.5675 acres □ 1/4 91.14 acres). Under the trial court's finding, Burkes would be reserving 55.56% of the mineral rights owned by them (200.3575 acres divided by 360.5675 acres 55.56%), and not the one-fourth mentioned in the purchase option. Therefore, the trial court's finding as to the intent of the parties concerning the amount of the reservation is clearly erroneous.

Is there a *Duhig* issue in the case that has been overlooked? *See* 1 Martin & Kramer, *Williams and Meyers Oil and Gas Law*, § 311, n.1.1.

14. SELMAN v. BRISTOW, 402 S.W.2d 520, 25 O.&G.R. 104 (Tex.Civ.App.1966) error ref'd n.r.e. 406 S.W.2d 896, 25 O.&G.R. 67 (Tex.1966). The premises in question were conveyed by Weeden to plaintiffs, reserving 1/8 th *of* the royalty, and plaintiffs conveyed the premises to defendant, reserving 1/4 th of the minerals. The trial court concluded that the royalty interest reserved by Weeden should be charged proportionately, 1/4 th to the interest of plaintiffs and 3/4 th to the interest of defendant. *Held* on this question, reversed and rendered.

> When Bristow and Austin reserved 1/4 th of the minerals, they then became entitled to 1/4 th of the usual 1/8 th royalty and were thus the owners of a 1/32 nd royalty upon the entire mineral estate. Thus, their reservation left them with more than enough minerals and royalty to satisfy the previous royalty reservation without effecting [sic] their conveyance to Selman of a full 3/4 ths of the minerals. Since the grantors undertook to convey a full 3/4 ths of the minerals and had the power to do so, they will be held to this undertaking. The deed was therefore effective to convey to Selman the surface and an unencumbered, unrestricted 3/4 ths of the minerals.

> Although Bristow and Austin had knowledge of the Weeden royalty reservation, nevertheless they chose to warrant the title to Selman, and thereby assumed the risk of failure of title.

> While the covenants of general warranty cannot be construed as enlarging the

title conveyed or impairing the grantors' title to 1/4 th of the minerals reserved to them, the warranty operates as an estoppel denying to the grantors the right to set up their 1/4 th undivided interest in the mineral estate against the grantee's title to the 3/4 ths undivided interest therein.

For the reasons stated, that portion of the judgment decreeing a proportionate reduction of the mineral estate conveyed to R.E. Selman by the 92.2 acre deed is reversed and judgment is hereby rendered granting R.E. Selman and wife, Nettie Selman, all right, title and interest to an undivided 3/4 ths of the minerals in and under the land described by the deed. . . .

See also Atlantic Refining Co. v. Beach, 78 N.M. 634, 436 P.2d 107, 30 O.&G.R. 27 (1968), to the same effect.

The *Selman* and *Beach* approach to the burdening of mineral interests by outstanding royalty interests was followed in Acoma Oil Corp. v. Wilson, 471 N.W.2d 476, 116 O.&G.R. 501 (N.D. 1991). Moen owned the surface and mineral interests covering 160 acres. Prior to 1944 by recorded deeds Moen transferred to three different persons a total of a 6.5% royalty interest. In 1944 Moen conveyed the 160 acres to Wilson by warranty deed. There were no express reservations and there was no reference to the outstanding royalty interests. Subsequent transfers were made by Wilson, whose heirs still retained a share of the mineral estate. A producing well was drilled on the 160 acres and a grantee of Wilson sought a declaratory judgment that the 6.5% royalty burden should be borne solely by the Wilson retained interests. Several of Wilson's grantees had title opinion clearly showing the outstanding 6.5% interest. The trial court's application of equitable estoppel principles was rejected by the North Dakota Supreme Court which applied the *Duhig* rule. As long as the grantor's (Wilson's) mineral estate was large enough to bear the burden of the 6/5% royalty interest, payment was to be satisfied solely out of that interest.

GORE OIL CO. v. ROOSTH

Court of Appeals of Texas, 2005.
158 S.W.3d 596, 162 O.&G.R. 573.

ARNOT III, JUSTICE. The issue in this appeal is whether the grantor's or the grantee's successors-in-interest should bear the burden of outstanding mineral and nonparticipating royalty interests. Appellees, Steve Roosth, Trustee; New Horizons Oil & Gas, Ltd.; and John D. Procter, Trustee, are successors-in-interest to the grantor, Peyton McKnight. Appellees brought suit against the leasehold interest owners and others after the leasehold interest owners failed to pay appellees the full amount to which they claimed to be entitled. Appellees sought a declaratory judgment as to their proportionate share of the royalties, sought damages for conversion, and requested attorneys' fees. Appellees contended that the reservation made by McKnight was in addition to the previous outstanding mineral and royalty interests. The defendant leasehold interest owners contended that the reservation made by McKnight was reduced by the outstanding mineral and royalty interests. The parties entered into a stipulation regarding all of the facts other than attorneys' fees. The trial court found in favor of appellees; entered judgment against the leasehold interest owners for past royalties in the amount of $ 271,210.21; . . . We modify and affirm.

. . .

Construction of the McKnight Deed

The McKnight deed, a general warranty deed from McKnight to Eagle Investment Company, provided in relevant part as follows:

> HAVE GRANTED, SOLD AND CONVEYED, and by these presents do GRANT, SELL AND CONVEY unto the said Grantee all that certain tract or parcel of land situated in Knox County, Texas, described as follows ("Property"), to-wit:

. . .

> Grantor unto himself, his heirs and assigns, reserves free of all liens a full one-eighth (1/8) non-participating royalty interest in the Property subject to any previously conveyed or reserved mineral interest as may appear of record in Knox County, Texas.

> This conveyance is made and accepted subject to all restrictions, reservations, covenants, conditions, rights-of-way and easements now outstanding and of record, if any, in Knox County, Texas, affecting the above described property.

The trial court concluded that the McKnight deed was ambiguous and that the intent of the parties to the deed was for McKnight to reserve a full 1/8 royalty interest in the property undiminished by the outstanding mineral and royalty interests that had previously been conveyed or reserved. . . .

In order to determine if the trial court erred in its construction of the McKnight deed, we must first determine the appropriate standard of review. The first issue we must address is whether the deed is ambiguous. That question is a question of law for the court and, therefore, will be reviewed de novo. Seldom have courts found deeds to be ambiguous.

A court's primary goal when construing a deed is to ascertain the true intention of the parties as expressed within the "four corners" of the instrument. The four corners rule requires the court to ascertain the intent of the parties solely from all of the language in the deed. The intent that governs, however, is not the intent that the parties meant but failed to express but, rather, the intent that is expressed. If a written instrument, such as a deed, is worded in such a way that a court may properly give it a certain or definite legal meaning or interpretation, it is not ambiguous. However, if a written instrument remains reasonably susceptible to more than one meaning after the established rules of interpretation have been applied, then the instrument is ambiguous and extrinsic evidence is admissible to determine the true meaning of the instrument.

The McKnight deed contains two "subject to" clauses. The leasehold owners correctly observe that the term "subject to" is a limiting, qualifying term and that, when used in a mineral deed, means "subordinate to," "subservient to," or "limited by."

The first "subject to" clause reads:

> Grantor unto himself, his heirs and assigns, *reserves* free of all liens a full one-eighth (1/8) non-participating royalty interest in the Property *subject to any previously conveyed or reserved mineral interest as may appear of record in Knox County, Texas.* (Emphasis added).

The leasehold owners argue that the "subject to" limitation is contained in the reservation clause and not the granting clause. The leasehold owners point out that the grant is a general grant of "all that certain tract or parcel of land." Consequently, the leasehold owners argue it is the reservation that is limited by the prior reservations and not the grant. The "subject to" clause modifies "reserves."

The leasehold owners would, harmonizing all of the language of the deed and relying on *Duhig v. Peavy-Moore Lumber Co., 135 Tex. 503, 144 S.W.2d 878 (Tex.1940)*, ascertain that the intent of the parties was that the 1/8 nonparticipating royalty interest reserved by McKnight *included* all of the other outstanding royalty interests of record.

After applying the *Duhig* principle to the first "subject to" clause, the leasehold owners address the second "subject to" clause which reads:

> This conveyance is made and accepted *subject to all* restrictions, *reservations*, covenants, conditions, rights-of-way and easements *now outstanding and of record*, if any, in Knox County, Texas, affecting the above described property. (Emphasis added)

The leasehold owners argue that this "subject to" clause makes the grant limited by the conveyance to all other reservations of record after first deducting the outstanding 1/16 royalty interest from McKnight's 1/8 royalty interest. The leasehold owners argue that the second clause is a "belt and suspenders" type of language often used by scriveners and should be subjugated to the first clause.

In effect, the leasehold owners argue that the deed is unambiguous, that the grantor reserved a 1/8 royalty interest, and that the outstanding 1/16 royalty interest should be deducted from that 1/8 royalty interest. The deed conveys a 7/8 royalty interest and the remaining mineral interest.

In contrast, appellees argue that one cannot harmonize the two "subject to" clauses under the leasehold owners' interpretation. Appellees point out that the second "subject to" clause makes the grant subject to "all...reservations." Consequently, the first "subject to" clause cannot be read to mean that the 1/8 royalty interest *includes* the 1/16 royalty interest because the second "subject to" clause clearly reads that the 1/8 royalty interest is *in addition to* the outstanding

1/16 royalty interest and all other reservations. Appellees find the deed ambiguous.

If not ambiguous, then appellees would have this court find, as a matter of law, by giving effect to all of the language and reading the deed as a whole, that the first "subject to" clause modifies the word "Property," the closest word to the clause, and not "reserves." Property is defined in the granting clause as "all that certain tract or parcel of land...described as follows ('Property')." Therefore, property as used in the reservation clause, which uses the same capital P, is the same property as used in the granting clause. McKnight would receive the full 1/8 royalty interest he intended. The second "subject to" clause makes it clear that the parties intended for McKnight to reserve a full 1/8 royalty interest and to ensure that the reservation was not later vulnerable to a *Duhig* claim.

We hold that the McKnight deed is reasonably susceptible to more than one meaning. The paragraphs quoted above appear to make both the reservation and the conveyance "subject to" the prior outstanding conveyances and reservations. Clearly, it was not the intent of the parties to the McKnight deed for both the grantor and the grantee to bear the full burden of the outstanding 1/16 prior interests.

Since the intent of the parties cannot be determined from the plain language of the four corners, we must determine whether the various canons of construction aid in construing the McKnight deed. However, the canons of construction are not helpful in this case. The canons of construction lead to conflicting results, including the following canons: construe against the grantor and in favor of the grantee, construe against the scrivener (in this case the grantee), and specific clauses control over general clauses. *See* Bruce M. Kramer, *The Sisyphean Task of Interpreting Mineral Deeds and Leases: An Encyclopedia of Canons of Construction*, 24 Tex. Tech L. Rev. 1 (1993). Consequently, we agree with the trial court that the deed is ambiguous.

Estoppel

If the McKnight deed had not stated that the "conveyance is made and accepted subject to all restrictions, reservations, covenants, conditions, rights-of-way and easements now outstanding and of record" but merely conveyed the entire premises less the grantor's reservation of a full 1/8 nonparticipating royalty interest, we might agree with the leasehold interest owners that *Duhig v. Peavy-Moore Lumber Co., supra*, and its progeny apply. Under *Duhig*, a grantor and his successors are estopped from claiming title in a reserved fractional mineral interest when to do so would, in effect, breach the grantor's warranty as to the title and interest purportedly conveyed to the grantee. . . . In this case, however, the McKnight deed contained an additional limiting clause stating that the conveyance was subject to all outstanding reservations, covenants, and restrictions. Consequently, we hold that *Duhig* does not apply and that the grantor and his successors in interest are not estopped from claiming title to the full 1/8 royalty.

Having determined that the deed is ambiguous and that appellees are not estopped by *Duhig*, we must determine whether the trial court erred in its construction of the deed based upon the stipulated evidence. When a written instrument is determined to be ambiguous, extrinsic evidence may be introduced to show the intent of the parties. In this case, other than the deed itself, the only evidence regarding the intent of the parties to the McKnight deed was an affidavit that had been filed in the county clerk's office. The affidavit was made by Thomas E. Morris, the attorney in fact for the grantee partnership in the McKnight deed. Morris filed the affidavit while his partnership still owned the property, which was foreclosed upon shortly thereafter. Morris swore that it was his understanding, both at the time of the deed and the affidavit, that McKnight was to receive a full 1/8 royalty unreduced by the previous reservations. The Morris affidavit was subsequently listed as a prior encumbrance in various deeds in the leasehold interest owners' chain of title. There is nothing else in the stipulations relating to the intent of the parties to the McKnight deed.

Based upon the stipulations before the trial court, we must uphold the trial court's findings with respect to the parties' intent. Consequently, appellees are entitled to their share of the full 1/8 royalty that was intended to be reserved by McKnight. The first and second issues are overruled. We need not reach the third issue, which relates to whether the leasehold interest owners ratified the McKnight reservation or were estopped by the inclusion in their chain of title of deeds in which the Morris affidavit was listed as a prior encumbrance.

Conclusion

The judgment of the trial court is modified to delete the award of prejudgment interest; and, as modified, the judgment is affirmed.

NOTES

1. A deed contains the following two clauses: "[Grantors convey] an undivided one-half (½) interest in and to all the oil, gas and other minerals in and under, and that may be produced from the following described land. . . It being intention of grantors herein to convey one-half of the minerals out of the interest owned by them in above-described tract of land." At the time of the conveyance the grantors own a ½ mineral interest. Did the grantors convey away all of their mineral estate or did they convey ½ of the ½ they owned or ¼ of the mineral estate. Black v. Shell Oil Co., 397 S.W.2d 877, 23 O.&G.R. 960 (Tex.Civ.App. 1965, writ ref'd n.r.e.) held that the language of the deed umambiguously conveyed to the grantee a ½ mineral estate. For another case dealing with the use of the phrase "out of" to describe the estate being conveyed see Minchen v. Hirsch, 295 S.W.2d 529, 6 O.&G.R. 1364 (Tex.Civ.App. 1956, error ref'd n.r.e.). *See generally*, Martin & Kramer, *Williams & Meyers Oil and Gas Law* § 319.

2. Bass was the owner of leased land subject to 6/14ths *of* the royalty which had been reserved by persons who had conveyed the property to him. Bass conveyed to defendant's predecessor "an undivided one-half interest in and to" said

land "subject to the Reservation contained in the following Deeds [listing the deeds in which 6/14 of the royalty had been reserved]." Is this deed ambiguous? To what share *of* the royalty paid on production is the grantee entitled, 1/2, 4/14, or 1/14? *See* Bass v. Harper, 441 S.W.2d 825, 32 O.&G.R. 486 (Tex.1969). The *Bass* interpretation of the term "subject to" was followed in Mafrige v. U.S., 893 F.Supp. 691, 135 O.&G.R. 352 (S.D.Tex. 1995) in interpreting a deed purporting to transfer some mineral interests to the United States. The court stated: "Texas courts have held that 'subject to' is a term of qualification, meaning 'subordinate to,' or 'limited by.' It can limit the estate granted, the estate warranted, or both, depending on the intent of the parties."

3. Sears was the owner of a tract of land. Sears conveyed the land to Johnson, reserving "an undivided 1/16 royalty (same being 1/2 of the usual 1/8) interest in all of the oil and gas . . . that may be produced from the land herein conveyed." Johnson conveyed the land to Williams reserving "the remaining one-half mineral interest in . . . the . . . described property." Is Johnson's deed to Williams ambiguous? What interests do Sears, Johnson and Williams have in the land? *See* Johnson v. Fox, 683 S.W.2d 214, 84 O.&G.R. 385 (Tex.App.1985).

4. Smith owned Blackacre subject to a 1/64 royalty interest in Jackson but thought Jackson and Bohannon each had a 1/64 royalty Smith conveyed to McKenney with a clause stating that 1/32 royalty had previously been reserved in Jackson and Bohannon. What is the state of the title? *See* Jackson v. McKenney, 602 S.W.2d 124, 68 O.&G.R. 98 (Tex.Civ.App.1980, error ref'd n.r.e.). What is the impact of the common law rule against exceptions or reservations in a third party? See Professor Horner's comments in Discussion Notes, 68 O.&G.R. 102 (1981). Should the findings of the trial court in dealing with conveyances of this kind be treated as factual matters to be upheld by the appellate court if they are "not clearly erroneous"? *See* Monson v. Dwyer, 378 N.W.2d 865, 88 O.&G.R. 324 (N.D.1985) and Professor Maxwell's comments in Discussion Notes, 88 O.&G.R. 328 (1986).

5. Tuer conveyed a 44.5 acre tract to Patrick by a deed which "excepted and reserved a 1/16 royalty interest." Patrick conveyed the same tract to Barrett by a deed which excepted a 1/16 royalty interest which was reserved unto Tuer" and also "excepted and reserved to [Patrick] an undivided 1/2 interest in and to all of the oil, gas, sulphur and other minerals in and under the above described land except a 1/32 royalty interest out of the minerals so reserved herein which said royalty interest is specifically conveyed by [Patrick] to [Barrett]." Then Tuer conveyed its reserved 1/16 royalty to Patrick. If the tract is leased with a 1/8 royalty reserved, how should the royalty be divided? Patrick v. Barrett, 734 S.W.2d 646, 94 O.&G.R. 578 (Tex.1987), says that the "keystone of this opinion is a clear understanding of the distinctions between an exception and reservation." Does this distinction help?

6. Stewman Ranch, Inc. v. Double M Ranch, Ltd., 192 S.W.3d 808 (Tex.App. 2006). Grantor conveys some 8,900 acres by warranty deed with the following reservation: "There is, however, excepted and reserved to the Grantors an undivided one-half (1/2) of the royalties to on the production . . . from the described lands

which are presently owned by Grantors. . . " At the time of the conveyance the grantors did not own 100% of the royalty in the described lands. Does the *Duhig* rule apply so that the grantee receives a full ½ of 50% of the royalties? The court declines, however, to apply *Duhig* because it claims there is no "failure of title." It nonetheless finds that the instrument unambiguously reserves to the grantor ½ of the percentage of royalty that the grantor owned at the time of the conveyance. Is that the same result as would have been reached had *Duhig* been applied?

———

McMAHON v. CHRISTMANN

Supreme Court of Texas, 1957.
157 Tex. 403, 303 S.W.2d 341, 304 S.W.2d 267, 7 O. & G.R. 610.

[The owners of an undivided one-sixth (1/6) mineral interest in a 240-acre tract executed a lease purporting to cover the entire mineral estate in the tract. The lease provided for a 1/8th royalty and included a proportionate reduction clause and a covenant of general warranty. Attached to the body of the printed lease contract was a typewritten clause or "rider" reading as follows:

The lessors herein reserve unto themselves their heirs and assigns, without reduction, as an overriding royalty, a net 1/32nd of 8/8 ths of all oil or gas produced and saved from the above described premises.

In an action to construe the lease and establish the rights of the parties, the trial court and the Court of Civil Appeals held that the proportionate reduction clause was applicable to the 1/32nd overriding royalty as well as to the 1/8th royalty reserved by the lease. Facts summarized by Editors.]

CALVERT, JUSTICE. At the time of the execution of the lease petitioners did not own the whole of the mineral fee estate in the 240 acres of land. They owned only an undivided 1/6th interest therein. Their undivided 1/6th interest is the arithmetical equivalent of a 16/96ths interest.

Petitioners do not question but that the proportionate reduction clause in the lease operates to reduce the normal royalty to which they are entitled from 1/8th of production to 1/6th of 1/8th, or 1/48th, of production. They contend, however, that the proportionate reduction clause has no application to the 1/32nd overriding royalty reserved by them. Respondents contend, on the other hand, that the proportionate reduction clause applies not only to the normal royalty reserved but to the reserved 1/32nd overriding royalty as well. If petitioners are correct in their construction of the lease it provides for a normal royalty of 1/48th of production and an overriding royalty of 1/32nd of production, or a total royalty of 5/96ths of production. If respondents' interpretation of the lease is correct it provides for a normal royalty of 1/48th of production and an overriding royalty of 1/6th of 1/32nd, or 1/192nd of production, a total royalty of 5/192nds. We will first address ourselves to this point of difference.

On the face of the lease the proportionate reduction clause and the overriding royalty clause present an obvious conflict. The first would require a proportionate reduction of the 1/32nd overriding royalty and the second would prohibit its reduction. One of the rules of construction for resolving conflicts requires that typewritten matter in a contract be given effect over printed matter. That rule is peculiarly applicable here. When it is applied the proportionate reduction clause and the overriding royalty clause are harmonized and the language of each is given meaning. The language of the proportionate reduction clause is given effect as requiring a reduction of the normal royalty reserved in the lease, but it is not given an effect which would render the words "without reduction" in the overriding royalty clause meaningless. To refuse to so limit the effect of the proportionate reduction clause would necessarily result in a holding that the ambiguity in the lease cannot be resolved by rules of construction, a result which both parties disavow. This construction of the lease gives the petitioners a greater royalty than the usual 1/8th of the mineral fee owned them, but parties may validly contract for a greater royalty than 1/8th of the lessor's mineral ownership. We agree that in so far as the quantum of the royalty reserved in the lease is concerned the lease is unambiguous and we hold that the quantum reserved is a 5/96 th interest.

The necessity for construction of the lease is not yet exhausted. The respondents insist here, as they did in their motion for an instructed verdict, that in the final analysis the question in the case is governed by the rule of estoppel laid down in Duhig v. Peavy-Moore Lumber Co.

But there is sound reason for declining to extend [the *Duhig* rule] to and apply it in the construction of oil, gas and mineral leases. We know as a matter of common knowledge and experience that deeds are usually prepared by the grantor or by a scrivener of his choice under his direction. He rarely prepares and executes a deed which purports to convey and which warrants title to an interest in property greater than he owns. If through carelessness or otherwise he executes a deed which purports to convey and which warrants title to an interest in property greater than he owns there is some moral justification for taking from him as much of the interest which he does own as is necessary to make good his warranty. We also know as a matter of common knowledge and experience that mineral leases are usually prepared, or standard forms completed, by the lessee. Even though a lessee knows a lessor owns less than the full fee title to the premises on which a lease is sought he often, if not usually, prepares and insists upon a lease which purports to convey the entire fee in order to make certain that no fractional interest is left outstanding in the lessor. He is protected against the possibility of being forced to pay royalty on a greater interest than that actually owned by the lessor by the inclusion of a standard proportionate reduction clause in the lease. That clause protects the lessee but it does not operate to reduce the estate which the lessor purports to convey. Klein v. Humble Oil & Refining Co., 126 Tex. 450, 86 S.W.2d 1077-1079. In many such cases, illustrated by the instant case, in which the lessor actually owns only an undivided interest in the minerals in the land described in the lease and in which there is a reservation of royalty, the lessee, by resort to the *Duhig* rule and even though owning through leases the entire 7/8ths working interest in the remainder of the minerals, could take, without paying therefor, the whole of the interest of the lessor in the minerals, including that reserved as royalty, and could, as well, recover damages from the lessor for breach of warranty. It is unthinkable and contrary to all modern human experience in the oil and gas industry to suppose that one owning an interest in the mineral fee would lease that interest for development of the mineral estate with no intention of receiving any of the returns from production of the minerals.

It is stipulated in the record that respondents had full knowledge that petitioners owned only a 1/6th interest in the minerals in and under the 240 acres of land. We may accept as true the statement in petitioners' brief that respondents themselves then owned 7/8ths of the remaining 5/6ths, or 70/96ths, of the minerals. Rule 419, Tex.Rules of Civil Procedure. Petitioners reserved a 5/96th interest, as we have held, and their warranty therefore purportedly extended to the remaining 91/96ths. It thus appears that if the *Duhig* rule were applied in the construction of the lease the respondents who own in their own right 70/96ths of the minerals could, as a matter of law, take by estoppel the entire 1/6th or 16/96ths owned by petitioners, including the 5/96 ths reserved by them, and would then have a cause of action against petitioners for damages for breach of warranty of title to the remaining 5/96ths interest.

If the warranty is enforced to the extent of its full purport the reservation will be destroyed. If the purported reservation is preserved the warranty will be breached pro tanto. This creates a latent ambiguity requiring that we repair once again to the intention of the parties for its resolution.

What did the parties intend? . . . It is evident that the parties intended the covenant of warranty to extend only to the 11/96th interest in the minerals title to which passed to respondents under the lease, and we so hold on this record as a matter of law. So holding preserves the reserved royalty and preserves the warranty for its intended purpose. There has been no breach of the warranty as we have interpreted it and the warranty cannot, therefore, be used by respondents as a vehicle for obtaining or for cutting down the royalty reserved to petitioners in the lease.

In their answer in the trial court respondents, as defendants, pleaded alternatively that the words "without reduction" were included in the overriding royalty clause as a result of a mutual mistake of the parties, and by cross-action they sought a reformation of the lease to eliminate the words. Inasmuch as the trial court instructed a verdict for respondents at the close of petitioners' evidence the cross-action was never reached for trial and no evidence was offered thereon. They are entitled to a trial of their suit for reformation. The cause must therefore be remanded.

The judgments of the Court of Civil Appeals and the trial court are reversed and the cause is remanded to the trial court for trial of respondents' cross-action.

SMITH, JUSTICE (concurring). I am of the opinion that this case should be reversed by this Court and judgment here rendered for the petitioners. I do not agree with the holding by the majority that the *Duhig* case has any bearing or application

to the present case. The *Duhig* case is not controlling, therefore, it is unnecessary to a decision here to hold, as the majority does, that it declines to extend and apply the *Duhig* doctrine in the construction of oil, gas, and mineral leases. I cannot find any sound support for the theory advanced by the majority that the *Duhig* rule applies in the construction of a deed, but not in the construction of oil, gas, and mineral leases.
. . .

It is stipulated that petitioners only owned an undivided 1/6 fee simple mineral interest in the 240 acres covered by the lease. The other 5/6 leasehold interest is owned by respondents. It was stipulated that respondents knew, at the time they undertook to acquire the lease and at the time they paid for it, that petitioners only owned an undivided 1/6 interest. The lessors (petitioners) only attempted to reserve 5/96 combined ordinary and overriding royalty out of their admitted 16/96 mineral interest in the leased premises, leaving an 11/96 interest for lessees. This is exactly what lessees bargained and paid for. This suit does not involve the ordinary 1/8 royalty. It involves only the determination of the amount of overriding royalty, and the ownership thereof.

[W]e have a plain, clear and unambiguous contractual provision whereby the respondents did covenant and agree that petitioner would receive 1/32nd of 8/8ths of all oil produced from the 240 acres, without reduction, as an overriding royalty. This contractual provision is binding upon respondents. It is a covenant running with the land. There was no breach of warranty. The *Duhig* doctrine does not in any manner nullify the contractual covenant in the lease providing for this overriding royalty.

Under my view of the record in this case the contract was sufficient without the use of the words "without reduction". Therefore, I do not believe this case should be remanded to enable respondents to try the issue of mutual mistake as to the inclusion of the words "without reduction"

GARWOOD, JUSTICE (dissenting). To the extent that it opposes the distinction now drawn between mineral leases and deeds for purposes of the *Duhig* rule, I agree with Justice Smith's concurring opinion. However, and contrary to his opinion, I think the *Duhig* rule, if not declared inapplicable to leases, would require a judgment for the lessees.

The opinion of the Court, despite its polite disclaimer of "disparagement" of the *Duhig* opinion, is, to my mind, nothing less than an open condemnation of it even as applied to deeds.

. . .

With all respect, I cannot escape concluding from the opinion that in the mind of the Court the real reason for now limiting the *Duhig* rule is less the purported distinction between deeds and leases than the conviction that the rule is simply bad law, improvidently declared. If the rule is bad law, the instant opinion seems to be the first statement of much consequence that so declares, although we have not infrequently had the *Duhig* case before us with ample opportunity to have said then what we say now against it.

I dissent from the view of the Court concerning the *Duhig* case. There is no good reason to complicate the law further in that fashion. What the final proper result of the instant suit should be need not be discussed now by me, since whatever result we reach upon the theory we now follow will necessarily be wrong, in my opinion.

My other concern about the Court's opinion is this: although my state of mind may be due to my own limitations, I must confess to some confusion as to the latter portion, in which, although having previously made clear that the *Duhig* rule will have nothing to do with mineral leases, we seem to say that, nevertheless, the lessor here has purported to convey and warrant the lessees' title to a 10/96 mineral interest, which is outstanding in third parties, and should therefore lose to the lessees his otherwise reserved 5/96 royalty interest, which he evidently did not mean to lose, and that this conflicting situation is a latent ambiguity entitling the Court to consider parol evidence of the intent of the parties.

To me what the Court appears to be saying is that, "When we apply the lease to the actual facts of ownership, we find that the necessary effect, considering the warranty (and the *Duhig* case), is to convey to the lessee without any reservation of royalty or with a smaller reservation of royalty than the lease provides; but this in turn conflicts with the express provision reserving a particular royalty to the lessor, so we have an ambiguity which develops on application of an unambiguous instrument to the actual facts." If this reasoning is not an application of the *Duhig* rule in order to produce an ambiguity, I simply fail to follow it. If it is such an application of the rule, then we seem to apply the rule while saying it does not exist as to leases.

GRESHAM v. TURNER

Texas Court of Civil Appeals, El Paso, 1963.
382 S.W.2d 791, 21 O.&G.R. 171.

PRESLAR, JUSTICE. This is an action for the construction of an oil and gas lease executed by appellants, as lessors, to appellee Fred Turner, Jr., as lessee, and dated February 27, 1947. The remaining appellees are assignees and co-owners of the lessee's interest. The sole question for determination is the amount of royalty reserved in such lease by appellants. Appellants were owners of an undivided 1/80 th mineral interest in Section 14, Block 4 1/2 , G.C. & S.F. Ry. Co. Survey, containing 922 acres, Upton County, Texas, when they executed the lease in question on a printed form in common use, which provided, among other things, as follows:

1. Lessor in consideration of Ten Dollars ($10.00) in hand paid, of the royalties herein provided and of the agreements of Lessee herein contained, hereby grants, leases, and lets exclusively unto Lessee, the following described land in Upton County, Texas, to-wit:

> Survey No. 14, Block 4 1/2 , G.C. & S.F. Ry. Co. Survey and containing
> 922 acres more or less

3. The Royalties to be paid Lessor are: (a) on oil, one eighth of that produced
and saved from said land.

The printed lease form also contained the following proportionate reduction clause:

> without impairment of Lessee's right under the warranty in event of failure of
> title, it is agreed that if Lessor owns an interest in said land less than the entire
> fee simple estate, then the royalties and rentals to be paid Lessor shall be
> reduced proportionately.

Prior to trial it was stipulated between the parties that the proportionate
reduction clause was deleted prior to execution of the lease. It was also stipulated
that the lessor owned only a 1/80th mineral interest in the 922-acre Section 14, that
the bonus consideration for the lease was $10.00 per acre on 11.52 acres, and that
payment for delay rentals was agreed by the parties to be $1.00 per acre on 11.52
acres at the time of the execution and delivery of the lease. It was also stipulated
that production of oil under the lease began during its primary term and has been
continuous, but that lessors have not executed any division orders or accepted any
of the tendered royalty payments, and that the lessees have not brought any action
to reform the lease, nor sought to vary or attack the validity of same, but have
always contended that appellants were entitled to royalty of their 1/80th part of the
1/8th royalty only.

Trial was before the court without a jury, with the plaintiffs (lessors)
contending the lease was not ambiguous and that they were entitled to a royalty of
1/8th of the total production, 1/8th of 8/8th, from Section 14. Defendant lessees
contended the lease was not ambiguous, but that it provided for lessors to receive
only a royalty of 1/8th of their 1/80th interest, and alternatively, that the lease
contained a latent ambiguity, was a product of mutual mistake, or it was a case of
constructive fraud. The four-year statute of limitations was also relied on by
defendant lessees. The trial court entered judgment that lessors were entitled only to
1/80th of the 1/8th royalty. No findings of fact or conclusions of law are in the
record, so that the basis for the judgment is not known.

We are of the opinion that the judgment is correct and that the instrument is
unambiguous. We arrive at that conclusion by finding the intention of the parties
thereto from within the four corners of the instrument, the language used by the
parties, the subject matter of the instrument, aided by the situation of the parties and
the surrounding circumstances. That, we perceive to be the correct rule of
construction of an unambiguous instrument.

We are aided in the determination of this case by the fact that the Supreme
Court of this State has previously decided a case (Gibson v. Turner, [156 Tex. 289,
294 S.W.2d 781 (1956)]) involving the same land, the issue as to the amount of
royalty to be paid under an exact lease form (except for an oil payment), but
involving different parties plaintiff and a different fraction of mineral interest

owned.

We turn now to the precise question of what royalty the parties intended. Looking to the language of the lease we note first that the royalty is "reserved". That can mean only that it is something retained out of that which was granted. What was granted? A lease of a 1/80th mineral interest. Then the 1/8th royalty reserved must come out of the 1/80th. The effect of the ordinary oil and gas lease, of which this is one, is to vest the lessee with a determinable fee to the minerals in place. Royalty is the specified portion of those minerals intended to remain and vest in the lessor. To follow the construction urged by appellants herein would be to hold that one could become vested with a 10/80th interest, 1/8th of 8/8ths, by conveying a 1/80th and making a reservation of 1/8th. The use of the word "from" can have no other meaning than that given "reserved". "The royalties to be paid Lessor are: (a) on oil, one-eighth of that produced and saved from the land." Saved from what land? The only reasonable construction is that with which the parties are dealing--the 1/80th mineral interest which was the subject of their contract. Lessor owned a 1/80th interest which he was interested in leasing, and lessee was interested in leasing said 1/80th; they made a deal and drew the lease in question to cover the subject of their deal--the 1/80th. Out of *that* 1/80 th, lessor reserved a 1/8th royalty. Another factor reflecting the intentions of the parties is that the construction which we here give to their lease would comport with ordinary business dealings, while that urged by appellants would not. We do not think it reasonable that one would make a business deal agreeing to give up 10/80th of the oil above ground for 1/80th of the same oil below the ground. The situation of the parties was such that lessee could have drilled without the lease covering lessors' fractional interest, and would have been obligated to pay lessors only 1/80th of the production. That, after deducting 1/80th of the drilling and equipping cost from lessors' share. In such a situation we do not believe the parties intended a trade whereby one would receive 10/80th for his 1/80th, plus a cash consideration and relief from the cost of drilling and equipping.

Appellants urge that this case is controlled by the case of Gibson v. Turner, *supra.* With that we do not agree. The Supreme Court, in construing the lease in that case, was dealing with a different fractional ownership from that in our case. There, the subject matter of the lease was a 9/40th mineral interest, and by the court's construction the lessors had a royalty of 5/40th. The court held, as do we, that the 5/40th was "reserved" out of the 9/40th, and it recognized that while the usual royalty was 1/8th the parties could agree on a different royalty. The court then makes the point, in support of its opinion, that it was possible to reserve the 5/40th out of 9/40th and that such could be done *without any incongruity.* Appellees urge that by such language the Supreme Court clearly intended to limit its holding to instruments in which the claimed royalty could possibly be reserved out of the interest leased. We do not speculate as to whether the Supreme Court intended such a rule, but simply point out that it makes the *Gibson* case an entirely different case from the one before us. Such language prevents the *Gibson* case from being controlling in our case, for in the instant case it would not only be an incongruity, but an impossibility, to reserve 10/80th out of 1/80th.

The judgment of the trial court is affirmed.

NOTES

1. GIBSON V. TURNER, which is distinguished by the court in the principal case, involved a lease which on its face covered 100% of the minerals in described premises although all parties realized that lessors owned only 9/40 ths of the minerals therein. The lease provided for payment of a 1/8 th royalty and included a proportionate reduction clause but prior to signature by the lessors the latter clause was deleted by running typewritten "x" through it. The court concluded that lessors were entitled to a full 1/8th[5/40th] royalty on production from the described premises. The lessees argued for a proportionate reduction of the consideration paid--the purchase price--as a result of a failure of consideration for the payment. The response of the court was as follows:

> we hold that respondents owning all the balance of the leasehold at the time the Gresham lease was executed cannot recover because they are in no danger of being evicted from the estate granted.

2. "A lease executed by a concurrent owner normally purports to cover all of the minerals in the described tract. The lessee will usually seek to have the lease executed in this fashion so that he may be entitled to the benefit of the doctrine of estoppel by deed in the event the leasing cotenant thereafter should acquire some or all of the outstanding undivided interest in the minerals." 3 Martin & Kramer, *Williams and Meyers Oil and Gas Law* § 505.

If a lease executed by the owner of 1/2 of minerals but purporting to cover all minerals provides for the payment of a 1/8 royalty, what should be the effect of:

(a) absence of a proportionate reduction clause from the lease form as initially prepared?

(b) deletion by the parties before execution of the lease of the proportionate reduction clause contained in the lease form as initially prepared?

(c) execution of a lease form which contains a proportionate reduction clause?

3. Not all opinions of intermediate appellate courts find their way to the printed reports in law libraries. Many are not certified for publication by the court, and in that event the unpublished opinion may be known only to the persons involved in the decided case. Apparently the opinion in Gresham v. Turner was to have had this fate until an inquiry by the Editors of this course book resulted in certification for and tardy publication of the opinion. Was any useful purpose served by this intermeddling by the Editors?

4. KLEIN V. HUMBLE OIL & REFINING CO., 126 Tex. 450, 86 S.W.2d 1077 (1935). *X* conveyed land to *Y* reserving a 1/8 royalty. *Y* conveyed the land to *R* with the identical reservation. *R* leased the land to *E*, reserving the customary 1/8 royalty. The lease contained a typical proportionate reduction clause. *Held, E* owns a 7/8 working interest; *X* owns a 1/8 royalty; *Y* and *R* own nothing. The court

rejected *R* 's contention that under the proportionate reduction clause he owned a royalty of 1/8 of 7/8 .

5. TEXAS CO. v. PARKS, 247 S.W.2d 179, 1 O.&G.R. 555, 2007 (Tex.Civ.App.1952, error ref'd n.r.e.). *R* leased to *E* land described as an undivided one-half interest in the E/2 of Section 208; rentals were to be paid in the sum of $160 yearly; the proportionate reduction clause provided for a reduction of rentals "if lessor owns an interest in said land less than the entire fee simple estate." Lessee tendered $80 rentals, construing the proportionate reduction clause as reducing rentals if lessor owned less than 100% of the minerals. *Held,* the lease terminated for non-payment of proper rentals, *viz.* $160. The proportionate reduction clause permits a reduction of rentals if lessor owns less than 100% of the mineral interest in "said land"; "said land" is an undivided one-half interest; therefore, since lessor owned 100% of an undivided one-half interest, rentals could not be reduced.

6. E.H. LESTER LEASING CO. v. GRIFFITH, 779 S.W.2d 226, 107 O.&G.R. 250 (Ky.App. 1989). Lessor owned a mineral interest subject to a previously conveyed 1/8th royalty interest. The lessor executed a lease reserving a 1/8th royalty interest. The lease contained a proportionate reduction clause. Is the lessee entitled to any royalty payments? a proportionately reduced royalty payment? The court applied the proportionate reduction clause, requiring not only the payment of the 1/8th royalty to the third party but a 7/64th royalty to the lessor. How did the court calculate the 7/64th fraction? For a criticism of this case see Discussion Notes, 107 O.&G.R. 254 (1991).

———

SECTION 4. CHARACTERISTICS OF MINERAL AND ROYALTY INTERESTS

(See Martin & Kramer, *Williams and Meyers Oil and Gas Law* §§ 303-307.4)

Three important questions are to be borne in mind when considering the mineral-royalty distinction:

(1) What consequences follow from the classification?

(2) What form of words will cause the interest to be classified as one or the other?

(3) Is it realistic to abandon the labels, "mineral" and "royalty" and to rely instead on the characteristics spelled out in a particular conveyance?

———

THORNHILL v. SYSTEM FUELS, INC.

Supreme Court of Mississippi, 1988.
523 So.2d 983, 99 O.&G.R. 326.

EN BANC. HAWKINS, PRESIDING JUSTICE, for the Court: The petition for rehearing is denied. The original opinion is modified in that Harris v. Griffith, 210 So.2d 629 (Miss.1968), insofar as it conflicts with our holding in this case, is overruled.

C.L. Thornhill and others have appealed from a decree of the chancery court of Jefferson Davis County finding that a mineral conveyance to Thornhill in 1945 conveyed only a non-participating royalty interest rather than an undivided mineral interest of the minerals "in place." The only issue before us is a construction of this conveyance and determining which type interest was conveyed.

Persuaded Thornhill acquired an undivided one-half mineral interest to all minerals, subject only to the right reserved by the grantors to the bonuses and delay rentals from oil and gas leases, we reverse and render judgment for the appellants.

Facts

Hardy McLeod and Joseph McLeod owned the Northwest Quarter of the Southwest Quarter (NW1/4 SW1/4) of Section 30, Township 6 North, Range 17 West, in Jefferson Davis County. On September 9, 1944, the couple executed an oil, gas and mineral lease covering the land in favor of Frank Ryba. Thereafter, on May 14, 1945, the McLeods executed a mineral deed to C.L. Thornhill conveying an undivided one-half (1/2) interest in the minerals. The deed to Thornhill appears on a standard "Form R-101 Mineral Right and Royalty Transfer" instrument. On the face of this deed was typed:

It is the intention of the grantors to convey, and they do hereby convey, twenty (20) full mineral acres of land of said tract.

Non-participating as to present or future lease rentals or bonuses.

A copy of the conveyance as it appears in the public record is attached as an appendix.

On July 28, 1948, four years after the mineral conveyance to him, Thornhill filed an application for ad valorem tax exemption on his mineral interest acquired from the McLeods. Paragraph (4) of the printed form states: "Fractional interest for which exemption is applied and nature of such interest: ..." Following this there is typed "1/2 Royalty."

After Thornhill received his interest, both Thornhill and the McLeods executed numerous instruments conveying fractional interests in the minerals and oil and gas leases. None of these conveyances are important to the issue before us.

On December 22, 1979, appellee System Fuels, Inc., spudded the A.M. Speights 30-13 Well on a 160-acre unit encompassing this forty acres. The Speights Well began producing oil on March 20, 1981. On November 11, 1980, System Fuels spudded the gas unit 30-12 on the tract, which began producing on February 6, 1981.

C.L. Thornhill and those claiming through him, the appellants here, filed suit in the chancery court alleging the McLeods, by the instrument in question, conveyed an undivided one-half mineral interest in all minerals in place, subject only to the reserved right of the grantors (as to such undivided onehalf interest conveyed) to receive all bonuses and delay rentals from the oil and gas lease in effect when the instrument to Thornhill was executed, as well as to all future oil and gas leases. The appellees, System Fuels, Inc., and others, answered, denying such ownership, and claiming that all Thornhill acquired by such conveyance was a non-participating royalty interest in the oil and gas produced, which carried with it no right to execute oil and gas leases on the land.

The chancellor found the conveyance to be a non-participating royalty interest only, and entered a decree in favor of appellees. Hence this appeal, with the sole question before us: Was this a mineral interest conveyance, or a nonparticipating royalty interest?

Law

In Mississippi we have two basic types of ownership of interests in minerals.

First, by far the most common way the holder of an interest in minerals obtains his interest is either from a reservation of or a conveyance of a fractional interest in the minerals. Such a reservation or conveyance simply creates a tenancy in common between the parties as to the minerals. Thus, if A reserves an undivided one-half interest in the minerals in his deed conveying Blackacre to B, A and B are tenants in common as to the minerals. Also, if X by a mineral deed conveys an undivided one-half interest of Blackacre's minerals to Y, X and Y are also tenants in common as to the minerals. Both parties have an equal right to go upon the land and drill for or mine minerals. Both would have to sign an oil and gas lease in order for the lessee to get a good title to all the mineral interests in Blackacre and be authorized to drill. Each would be entitled to share equally in the bonuses paid for the oil and gas lease, as well as the delay rental payments made under such lease. Finally, in event of discovery of oil or gas as between the two, they would share equally in the royalty payments made under the oil and gas lease.

It is also well settled that in such a mineral deed or reservation of the grantor may convey or reserve certain attributes of this mineral ownership. Thus, in a mineral deed the grantor may convey an undivided mineral interest, but reserve unto himself all bonuses or delay rentals, or both, as to any oil and gas lease. *Mounger v. Pittman*, 235 Miss. 85, 108 So.2d 565 (1959), *infra.*

Another type of mineral interest ownership is a royalty interest. This interest only becomes meaningful if there is a commercial production of oil and gas, at which time the royalty owner receives the agreed-upon fraction of the production. The owner of a royalty interest has no control or right of possession, and no obligation to the remaining mineral interest owners. And, unless or until oil or gas is produced commercially, no obligation of any kind is due him. A royalty interest does not share in the bonuses or delay rentals with the other mineral interest owners. He is not required to sign an oil and gas lease in order for the lessee to have the right to go upon the land and drill for oil and gas. His interest cannot be charged with any costs of drilling a well as it could be if he held a mineral interest as a tenant in common and his cotenant drilled a producing well. *Lackey v. Corley*, 295 So.2d 762 (Miss.1974).

Generally, it is quite clear just what type of ownership the parties have. A conveyance or reservation of a royalty interest will, as a rule, specifically state that the royalty owner has no right to sign oil and gas leases, and that he cannot participate in bonuses or delay rentals, and cannot be charged with drilling or exploration costs.

In this case we have a mineral deed which reserved to the grantors the rights to bonuses and delay rentals under any oil and gas leases. Did this reservation so change the character of the instrument from a mineral deed to transform it into a royalty conveyance? We hold it did not.

In the evolution of oil and gas law in this state, our courts have endeavored to accommodate to the practical needs of the parties involved with the end in view of promoting the development of our natural oil and gas resources as efficiently as possible commensurate with fairness to all parties. The words used to denote mineral ownership have occasionally had fuzzy edges. We could, of course, hold that certain words have a definite, fixed meaning at all times and under all circumstances. Circumstances occasionally show, however, that the parties did not mean what the words standing alone might appear to mean. We therefore have had to look to surrounding circumstances in addition to relying upon the words themselves. In such instances we deemed it better to let the words remain just what they are: a strong, but not necessarily conclusive indication of what the parties meant.

Here again we are confronted with the same problem. We are asked to determine whether an instrument is a mineral deed, conveyance of a "mineral estate" of the "minerals in place," or a conveyance of a "non-participating royalty interest"? As stated in Hemingway, Law of Oil and Gas:

> In probably no other area of oil and gas law, than in cases involving the mineral-royalty distinction, can examples be found of courts, on behalf of befuddled litigants, benevolently and improperly granting reformation in the guise of a judgment for title.

Section 2.7 (2nd Ed.1983).

The chancellor's problem in this case was a consequence of our decision in Harris v. Griffith, infra, wherein we benevolently took care of "befuddled litigants" to let their words accommodate what we thought the parties intended....

In *Mounger v. Pittman, supra,* we recognized that particular words in a mineral transfer should not control, but the entire instrument should be examined. We held the reservation to be of a mineral interest in place as opposed to a royalty interest. The reservation read as follows:

> We do hereby reserve for ourselves, our heirs and assigns, one-eighth of all the oil and gas which may be produced from said lands to be delivered in tanks and pipelines in the customary manner, and this shall be a covenant running with the land and all sales and other conveyances of said lands shall be subject to this reservation and agreement.

235 Miss. at 86, 108 So.2d at 566.

We then stated:

> The distinguishing characteristics of a non-participating royalty interest are: (1) Such production is not chargeable with any of the costs of discovery and production; (2) the owner has no right to do any act or thing to discover and produce the oil and gas; (3) the owner has no right to grant leases; and (4) the owner has to receive bonuses or delay rentals. Conversely, the distinguishing characteristics of an interest in minerals in place are: (1) Such interest is not free of costs of discovery and production; (2) the owner has the right to do any and all acts necessary to discover and produce oil and gas; (3) the owner has the right to grant leases; and (4) the owner has the right to receive bonuses and delay rentals.

235 Miss. at 86-87, 108 So.2d at 566.

The mineral deed in this case is on a form customarily used to convey a mineral estate with all appertaining rights as opposed to a royalty interest only. The only change made in the terms and conditions of the printed form is the typed insertion that the grantee would not receive bonuses or delay rentals from any oil and gas lease. Under conventional rules of construction, this appears a simple conveyance of an undivided one-half interest in all the minerals, with the grantor retaining the right to bonuses and delay rentals from oil and gas leases. Under ... *Mounger v. Pittman, supra,* it is well settled these rights could be separated without changing the character of the instrument from a mineral estate to a royalty interest only.

Also, under ordinary rules of construction, all that was not unequivocally and specifically reserved was conveyed by the granting clause....

Harris v. Griffith, supra, is the only case from this Court which, arguably, could support a contrary conclusion. That case, however, is distinguishable on its facts from this case. In 1937 the Griffiths executed an oil and gas lease. In August, 1944, they executed a mineral conveyance on a "Form R-101," as used in this case, conveying an undivided one-quarter interest in the minerals to Thomas O. Payne. Typed in this deed (in addition to the land description) were the following sentences:

> It is the intention of the grantors herein to convey 64 3/4 full mineral acres.
> This instrument is to be non-participating both as to bonuses and lease rentals.

210 So.2d at 631. Also, the following changes were made in the deed. In the heading the word "RIGHT" was marked through and "DEED" typed in capital letters above it. In the final conveying paragraph, all but the following first sentence thereof was marked out:

> This conveyance is made subject to any valid and subsisting oil, gas or other mineral lease or leases on said land, including also any mineral lease, if any, heretofore made or being contemporaneously made from grantor to grantee.

210 So.2d at 631.

In *Griffith* we also noted there were elements of estoppel against the parties who claimed to own a mineral interest as opposed to a royalty interest. They had paid none of the production costs for the producing well, and had not asserted any more than a royalty interest when the Griffiths had to employ an attorney, (and pay him forty percent of their mineral estate) to secure a release of an oil and gas lease from the oil company which had shut down production. When another operator took over drilling operations, over $34,000 in debts attributable to this tract had been paid by the complainants. The final operator produced a commercial well. Thereafter, in 1960 (and notarized in 1966) the defendants executed an oil and gas lease to a third party. The complainants sought to cancel the oil and gas lease as a cloud upon their title. The chancellor held the deed to Payne was only a nonparticipating royalty interest and that all executive rights, bonus and delay rentals remained with the grantor.

In *Harris* this Court first stated that parol evidence concerning the effect of the deed was not competent, because of the need for a consistent, coherent body of law on the construction of oil and gas conveyances. We added that ordinarily such instruments are not regarded as ambiguous, and that courts should construe them so as to best comport to the parties' intention appearing on the instrument itself. We quoted with approval *Richardson v. Moore*, 198 Miss. 741, 22 So.2d 494 (1945):

> In trying to solve this question, we should keep in mind certain well established principles of construction of contracts. Those applicable here are (1) the deed must be read in the light of the circumstances surrounding the parties when it was executed; (2) that the construction should be upon the entire instrument, and each word and clause therein should be reconciled and given a meaning, if that can be reasonably done; (3) that the main document and that to which it refers must be construed together; (4) that if the wording of the deed is ambiguous, the practical construction placed thereon by the parties will have much weight in determining the meaning . . ..

210 So.2d at 633.

We then stated:

> Accordingly, the Griffith-Payne deed should be analyzed in the light of the objective circumstances surrounding the parties when it was executed. Moreover, the construction must be upon the entire instrument.

210 So.2d at 633.

This Court recognized that the printed form was used to convey a fractional interest in minerals in place. We then noted that the parties showed an intent to substantially change the form to the effect the interest conveyed. We first noted that the heading had been changed, and that this change indicated an intent to convey a royalty only. We also considered the typed intention clause: "This instrument is to be non-participating both as to bonuses and lease rentals." We held this as indicative of an intent to convey something other than an interest of minerals in place and that the retention of bonuses and delay rentals was closely and materially related to the executive right. Considering the deed further, we noted the parties struck out the last printed paragraph:

> ... but for the same consideration herein above mentioned, grantor has sold, transferred, assigned and conveyed and by these presence does sell, transfer, assign and convey unto grantee, his heirs, successors and assigns, the same undivided interest (as the undivided interest herein above conveyed in the oil, gas and other minerals in said land) in all the rights, rentals, royalties and other benefits accruing or to accrue under said lease or leases from the above described land; to have and to hold unto grantee, his heirs, successor and assigns.

210 So.2d at 634.

We held that the striking of the above-quoted provision tended to show that the parties did not intend to sell the rights, rentals, royalties and other benefits under the existing lease. We then stated:

> The deletions in the title and the last printed paragraph may be considered in order to arrive at the true meaning and the intention of the parties. They are relevant factors under the circumstances of this case and the terminology of the deed in assisting the Court to reach a reasonable interpretation of the instrument.
>
> . . .
>
> In the instance case, the reservation of bonuses and rentals, *together with the other above-discussed terms of this deed,* constituted an implied retention of the executive rights. [Emphasis added by court. Eds.]

210 So.2d at 635.

We were also influenced by the fact that, as in this case, Payne, the grantee, was an experienced oil and gas investor, whereas Griffith, the grantor, lacked experience and education. Finally, this Court took into account the timing and long period of delay by the lessors under the 1960 oil and gas lease before they asserted any claim to the executive right.

There are two additional amendments to the printed form in Harris v. Griffith, not present in this case. Also, as we noted above, the defendants in Harris v. Griffith did not deem themselves to be tenants in common with the grantors when the grantors had to employ legal counsel to secure a release of an oil and gas lease, and the defendants paid none of the legal fee.... Further, the defendants had assumed none of the expenses or costs when the previous drilling had far more expense than income. As tenants in common, their interest would have been chargeable with a proportion of such costs....

The circumstances and facts of that case caused this Court to make an exception of what otherwise would unquestionably have been termed a conveyance of mineral interest in place, and not simply a conveyance of a right to royalties. Those circumstances and facts are not present in this case.

In this case the chancellor found no element of estoppel and noted the parties had not struck any of the printed portion of the mineral deed.

In our original opinion, because we found *Harris v. Griffith* clearly distinguishable (despite the very able dissenting opinion), we saw no necessity to consider overruling it.

The extensive briefs filed by the parties in connection with the petition for rehearing, as well as the views expressed in the dissenting opinion, however, have convinced us that *Harris v. Griffith* needs to be overruled, insofar as this Court construed that the instrument therein by its terms conveyed only a non-participating royalty. We are convinced that in this respect the Court erred, and that the instrument by its terms remained, despite the changes noted by the Court, a conveyance of an undivided one-half interest in the oil and gas minerals, subject only to a reservation in the grantors of the bonuses and delay rentals.

Harris v. Griffith was cogently criticized in an article published in Volume XLI, Mississippi Law Journal, Spring 1970, No. 2, p. 189, "An Analysis of the Rights and Duties of the Holder of the Executive Right," Joel Blass and Jean Rand Richey, pp. 201-205. The authors were of the view this Court erred in finding that the reservation of bonuses and delay rental "justifies the implication of an intent to retain the executive right in the grantor." They further contended this was at variance with our consistent holdings that the various incidents of ownership in a mineral estate could be separated, *Westbrook v. Ball, supra,* and the conventional rule of construction that in a conveyance of a mineral estate all is conveyed that is not specifically reserved or excepted, p. 204. The confusion the authors predicted would result from this decision have indeed come to pass. We have therefore concluded that *Harris v. Griffith,* while containing sound pronouncements of law, nevertheless erred in concluding that there was an implication of a reservation of executive rights simply by a reservation of bonuses and delay rentals. Our view is precisely the reverse: that the grant of an entire interest in minerals conveys all incidents of ownership thereto that are not specifically excepted or reserved, and that the reservation by the grantor of all rights to bonuses and delay rentals from oil and gas leases does not carry with it by implication executive rights, or any other incident of ownership.

We therefore overrule *Harris v. Griffith* insofar as it holds that an instrument such as there executed does not on its face convey an undivided one-half interest in the minerals with all rights incident thereto, with the sole exception of the reservation in the grantors of the rights to receive bonuses and delay rentals from oil and gas leases.

We must hold, therefore, that the chancellor erred in holding this conveyance to be a non-participating royalty transfer rather than of an undivided one-half interest in the minerals.

The conveyance to Thornhill was of an undivided one-half interest of the oil and gas minerals, reserving only to the grantors the right to bonuses and delay rentals from the existing and future oil and gas leases. Executive rights were not reserved. The decree of the chancery court will be reversed and judgment rendered here for the appellants.

REVERSED AND RENDERED.

HAWKINS, P.J., and PRATHER, SULLIVAN and ANDERSON, JJ., concur.

DAN M. LEE, P.J., ROY NOBLE LEE, C.J., and GRIFFIN and ZUCCARO, JJ., dissent.

ROBERTSON, J., concurs in denial of petition for rehearing with separate written opinion.

DAN M. LEE, PRESIDING JUSTICE, dissenting:

...

[W]e are called upon to determine what effect "non-participating as to present or future lease rentals or bonuses" has upon the conveyance. Did the McLeods convey a mineral interest or a royalty interest? Did Thornhill buy a mineral interest or a royalty interest? By holding that Thornhill bought mineral rights in 1945, the majority, and the concurring opinion, trample upon several well established areas of the law, including deed construction, appellate review, and most fundamentally, basic oil and gas law. In so trampling, the majority has found it necessary to overrule our decision in *Harris v. Griffith,* 210 So.2d 629 (Miss.1968), insofar as that opinion holds that retention of bonuses and delay rentals by the grantor necessarily implies retention of executive rights. Contrary to the opinion of the majority that Harris is some sort of aberration from our well established oil and gas law, Harris did what this Court has failed to do here--it applied our well established principles of deed construction, appellate review and basic concepts of oil and gas law. With all due respect to the majority, again I vigorously dissent, both to the holding in the present case and to the overruling of Harris.

. . .

In the "objective accessible world," of which the concurring opinion is much enamoured, but which both the concurring and the majority opinions ultimately ignore, the McLeods owned the surface and minerals of the 40 acres. In 1944, the McLeods leased their 40 acres in minerals to Frank Ryba. In 1945, the McLeods conveyed 20 full mineral acres to Thornhill; however, in the typed-in portion of the mineral deed, the McLeods retained the rights to all present and future bonuses and lease rentals. Appurtenant to ownership of minerals is the right to bonuses and delay rentals, along with the right to execute leases. The effect, then, of retaining the rights to bonuses and lease rentals turned what started out to be a mineral conveyance into a royalty deed. The reason this is so is because two of the rights appurtenant to ownership of minerals were retained by the McLeods, bonuses and delay rentals. But what of the right to execute leases? The right to receive bonuses and delay rentals are necessarily appurtenant to the right to execute leases. In order to understand why this is so, it is necessary to understand what a bonus is. A bonus is the cash consideration paid by the lessee for the execution of an oil and gas lease by the mineral owner. 7 *Williams & Myers, Oil and Gas Law, Manual of Oil and Gas Terms* 80. It is true that the bonus can take some other form than cash--it can be a royalty bonus or overriding royalty reserved to the mineral owner in addition to the usual mineral owner's royalty. But a bonus by any other name is still a bonus, and its purpose is to compensate the mineral owner for executing a lease.

Where the majority opinion fails in this case is when it plays a theoretical game of separating the right to execute a lease from the consideration a mineral owner is paid to execute the lease. Here, the majority says that because McLeod, in an instrument which Thornhill drew up, did not expressly retain the right to execute leases on 20 acres, the right went to Thornhill, but McLeod is still entitled to receive the bonuses on any lease Thornhill executes on the other one-half of the mineral estate. Why in the world would Thornhill want the right to execute leases while allowing the McLeods to retain the bonuses from a lease he, Thornhill, executes, when the whole purpose of the bonus is to compensate the lessor for executing the lease? In the "objective accessible world," an experienced oil and gas person, as Thornhill was, would not do that unless he, too, were playing games with the inexperienced McLeods. Furthermore, under today's majority opinion, if Thornhill executes a lease on his 20 mineral acres and there is delay in drilling, the delay rentals on Thornhill's lease would also go to the McLeods. Why would Thornhill do that?

The point is that the "executive right" the majority creates today is no more than a legal theory. An executive right unhinged from the bonus is meaningless, because by definition whoever retains the right to execute leases has the right to the resulting bonus. The bonus is nothing more, or nothing less, than consideration paid for executing a lease. Harris v. Griffith, 210 So.2d 629 (Miss.1968), was infinitely correct, in the real world, when it held that the retention of bonuses and lease rentals justifies an implication to retain the right to execute leases. Harris at 634. It is the only implication that makes any practical sense. Otherwise, both the grantor and the grantee are left with meaningless rights.

. . .

Conclusion

The majority sees this decision as a mere interpretation of words in a deed; however, the ramifications of this decision will have troublesome effects upon Mississippi's oil and gas law, not to mention the incredible injustice we do to the grantors in this case who have relied on this conveyance for 34 years. The majority has fashioned a legal theory that has little practical meaning, except to open the door for title busters and oil sharks to feast upon the unsuspecting people unversed in the legal niceties of theories and terminology. I cannot join in the creation of such a theory. I, therefore, would hold that the deed from the McLeods to Thornhill conveyed a one-half nonparticipating royalty interest and affirm the chancellor's findings.

ROY NOBLE LEE, C.J., and GRIFFIN and ZUCCARO, JJ., join in this dissent.

DEED

MINERAL ~~RIGHT~~ AND ROYALTY TRANSFER

(To Undivided Interest)

STATE OF MISSISSIPPI
COUNTY of Jefferson Davis   KNOW ALL MEN BY THESE PRESENTS:

that   Carl P. Griffith and wife, Alma R. Griffith

of Jefferson Davis County, State of Mississippi, hereinafter called grantor (whether one or more and referred to in the singular number and masculine gender), for and in consideration of the sum of Ten and No/100 Dollars $10.00, and other good and valuable considerations, paid by Thomas O. Payne

hereinafter called grantee the receipt of which is hereby acknowledged, has granted, sold and conveyed, and by these presents does grant, sell and convey unto said grantee an undivided One-quarter (1/4) interest in and to all of the oil, gas and other minerals of every kind and character in, on or under that certain tract or parcel of land situated in the County of Jefferson Davis, State of Mississippi, and described as follows:

W½ of NW¼ and N½ of SE¼ of NW¼ and 22 acres NE¼ of NW¼ South of Road, and E½ of NW¼ of SE¼ and NE¼ of SW¼ of SE¼ and E½ SW¼ of NE¼ and E½ SE¼ all in Section 2 Township 8 North, Range 19 West, and 2 acres in SE corner of SW¼ of SW¼ North of road, and 5 acres in SE corner of SW¼ of SW¼ South of road in Section 35, Township 9 North, Range 19 West. Jefferson Davis Co.

It is the intention of the Grantors herein to convey 64 3/4 full mineral acres.

This instrument is to be non-participating both as to bonuses and lease rentals.

Also, in addition to the above described land, any and all other land owned or claimed by Grantor in said section or sections in which the above described land is situated or in adjoining sections and adjoining the above described land.

TO HAVE AND TO HOLD the said undivided interest in all of the said oil, gas and other minerals in, on or under said land together with all and singular the rights and appurtenances thereto in any wise belonging, with the right of ingress and egress, and possession at all times for the purpose of mining, drilling and therefor for said minerals and the maintenance of facilities and means necessary or convenient for producing, treating and transporting such minerals and for housing and boarding employees, unto said grantee, his heirs, successors and assigns, forever, and grantor herein for himself and his heirs, successors and administrators hereby agrees to warrant and forever defend all and singular the said interest in said minerals unto the said grantee, his heirs, successors and assigns against every person whomsoever lawfully claiming or to claim the same or any part thereof.

Grantee shall have the right at any time that is not required to redeem for Grantor by payment, any prior liens or other liens on the above described lands, in the event of default of payment by Grantor, and be subrogated to the rights of the holder thereof.

This conveyance is made subject to any valid and subsisting oil, gas or other mineral lease or leases on said land, including also any mineral lease. If any, hereafter made or being such mineral interest...

WITNESS the signature of the grantor this 11th day of August 19 44

Witnesses.

Carl P. Griffith
Alma R. Griffith

ROBERTSON, JUSTICE, concurring in denial of petition for rehearing:

. . .

Today's majority and dissenting opinions are filled with much talk of mineral estate versus royalty interest. 'Tis but an object lesson that lawyers practice under a tyranny of labels. All too often we yield to the tyranny when, as here, we should stand and fight.

Not that we could live without the law's labels, for all words in the end are but labels. They are inherently incapable of exact correspondence with their object. Holmes said, more metaphorically, a word is but the skin of a living thought. My targets are careless word usage and notions of mutual exclusivity.

The realities represented by the labels "minerals" and "royalty" are complex. Problems arise when one speaks in terms of the underlying realities while another insists upon labeled communications. Such communication is impossible. When one talks in terms of underlying realities and the other insists upon the simplistic security of the labels, losses and often lawsuits loom large.

Our law recognizes many separate incidents of mineral ownership. Some of these are

A. The power to sell all or a part of the mineral estate.

B. The authority to alienate short of sale, that is, to lease (executive right);

C. The authority to go upon the land to drill and develop the minerals (right of ingress and egress).

D. The burden of participating in the cost of exploration, discovery and production;

E. The right to receive bonuses;

F. The right to receive delay rentals;

G. The right to receive (a part of) the proceeds of production;

H. Reversionary interests, to take effect at the end of the primary term of a lease or at the cessation of production; and

I. Benefit of covenants, express and implied.

These incidents of ownership may be separated and conveyed or retained as the parties see fit.

Listing the incidents of mineral ownership unmasks another myth. Mineral interests and royalty rights are never mutually exclusive categories. Royalty rights are one incident of a whole mineral estate. Conveyance of minerals without more conveys the right to receive royalties as well. Royalty is thus seen a subspecies of the mineral estate. It may be severed and conveyed, "participating" or "non-participating".

The point is as simple as it is fundamental: the owner of a (partial or whole) mineral estate has legal power to convey to another any one or more of the various incidents of his estate--and to retain the others, as he sees fit. Oil and gas economics and custom may suggest some combinations more sensible than others. The law is indifferent to his choice.

Today's is one of those cases in which the parties broke away from the law's facilities. They laid aside the labels of convenience. The instrument at issue reserved to the grantors, Hardy McLeod and Josephine McLeod, the right to receive all bonuses and delay rentals from an eight month old oil and gas lease then in effect and, as well, from all future oil and gas leases. Once they did this, the parties moved beyond the shorthand descriptives "mineral interest" and "royalty interest." It became positively dangerous to continue to speak in those terms, as this lawsuit certainly demonstrates. Those who ask after that point whether a mineral interest or a royalty interest was conveyed by McLeod to Thornhill are simply asking the wrong question. Here lies the fundamental error of Appellees, System Fuels, Inc., et al., and of my colleagues in dissent.

Our question is, which of the incidents of mineral ownership were conveyed to Thornhill on May 14, 1945, and which were retained in the McLeods? More precisely, the judicial mind has become charged to locate the authority to lease the oil and gas interest conveyed to Thornhill. Was that authority retained by the McLeods or conveyed to Thornhill?

The majority opinion, as much as I agree with it, makes the same mistake as everyone else, first on page 986 and at several points thereafter. The question is not, whether the McLeods conveyed "a mineral estate" or a "royalty interest only," but which of the various incidents of mineral ownership were conveyed and which were retained.

Two oft cited cases may be used to explain how we should construe mineral conveyances--and how we should not.

Mounger v. Pittman, 235 Miss. 85, 108 So.2d 565 (1959) is a case of consequence. Grantors reserved a one-eighth interest in the land pertaining to oil and gas. The question was whether grantor's interest was chargeable with one-eighth of the cost of production. The language at issue was

> We do hereby reserve ... one-eighth of all the oil and gas which may be produced from said lands....

235 Miss. at 86, 108 So.2d at 566. Everything else was conveyed to grantees. Though the matter could have been made more certain, construction is not difficult. The conveyance by its language gave grantees everything not reserved. Specifically, grantors reserved no right to drill or explore for oil and gas. They did reserve one-eighth of the oil and gas "which may be produced ...," that production, if any, to be generated by the parties with the right and authority to drill and explore. The words "which may be produced" could sensibly mean only "which may be produced by others than grantors." Grantors reserved the right to physical delivery of their one-eighth to themselves or as they might direct. From this it is a short step to reading the reservation as providing that grantors get their one-eighth without strings attached. The Mounger Court nevertheless offers careful definitions of "nonparticipating royalty interest" and "minerals in place". 235 Miss. at 86-87,

108 So.2d at 566. These definitions were unnecessary at the time. They have proven mischievous as they reinforce the tyranny of labels and contribute to the methodology that leads today's dissenters astray.

Lackey v. Corley, 295 So.2d 762 (Miss.1974) is wonderfully wrong and right. Grantees were not chargeable with cost of production because, as a matter of common English usage, the wording of the grant and reservation said they weren't, not because their interest was a "non-participating royalty interest." The Lackey Court reasoned wrongly that grantees' interest had "three of the four characteristics of a non-participating royalty interest," that the fourth--freedom from costs of production--would be implied to yield a "non-participating royalty interest," and that because grantees held a nonparticipating royalty interest they were not chargeable pro rata with cost of production. Lackey, 295 So.2d at 764-65. This is nonsense. Why go by way of China to cross the street? The shortest distance between two points is a straight line, and this is so in law and logic as in geometry. The *Lackey* grantees were not chargeable with cost of production because the words in the instrument, as a matter of elementary English usage, said they weren't, period.

. . .

I return to the idea of grammatical construction. The McLeods conveyed via typed wording "twenty (20) full mineral acres". The pre-printed language reflects a conveyance of an undivided one-half (1/2) interest in and to all of the oil, gas and other minerals. These two provisions may only be read as conveying to Thornhill all incidents of the minerals not reserved. The only reservations are bonuses and rentals. I read the instrument as would an English teacher having before her a manual of oil and gas terms. I am satisfied she would read the May 14, 1945, conveyance as vesting in C.L. Thornhill the legal power to lease his one-half interest in the minerals. Nor can I imagine a prudent title lawyer failing to advise a prospective lessee/client that before drilling, he had jolly well better get a lease from Thornhill as well as the McLeods.

I see the reading the Court gives the McLeods' conveyance the best it may be given consistent with the ideal of principled integrity in our law and, at once, the practical needs of the individual and his lawyer searching the land records. I join in today's decision denying the petition for rehearing of System Fuels, Inc., et al.

NOTES

1. The majority opinion in *Thornhill* is based in part on the proposition that "under ordinary rules of construction, all that was not unequivocally and specifically reserved was conveyed by the granting clause." Professor Mason of the University of Mississippi points out in his Discussion Notes on the principal case, 99 O.&G.R. 390 (1989), that the "contrary rule,that the elements of the mineral estate which are not specifically conveyed are reserved to the grantor" is relied on in the *Mounger* and *Lackey* decisions, which are both discussed in Thornhill. Professor Mason notes, however, that the issue was different in *Mounger* and *Lackey* and thus the "two rules [may logically] coexist." The aspect of ownership at

issue in *Mounger* and *Lackey* was "cost-bearing," that is, did the grantor retain "a greater share of production in the form of a cost-free royalty instead of a cost-bearing interest."

The interest before the court in *Mounger* was a reservation of "one-eighth of all the oil and gas which may be produced from said lands to be delivered in tanks and pipelines in the customary manner. There is authority in many states that this is the classic language for creating a royalty interest, an interest that does not bear the expense of production but is entitled to the named fraction of post production value, 'usually measured at the well head. *See* Kramer, *Conveying Mineral Interests -Mastering the Problem Areas*, 26 Tulsa L. J.175 (1990). *Mounger,* however, found it necessary to write at some length on "whether the reservation reserved to the grantors a non-participating royalty interest or an interest in the minerals (oil and gas) in place." The court went on to use the checklist on the distinguishing characteristics of a non-participating royalty interest set out in the opinion in the principal case and, among other characteristics lacking, found no statement "that grantors' share of production is to be free of cost of discovery and production." Result: "the interest reserved by the grantors was an estate in the oil and gas in place." Professor Mason explains the decision on the ground that if cost bearing is to be passed by a conveyance that appears to convey total ownership it must be spelled out. In other words, the incidents of ownership that are "benefits/rights" pass automatically if not reserved, the burden of costs must be spelled out. Is there a better explanation?

Lackey is described in the concurring opinion in the principal case as "wonderfully *wrong* and *right.*" The interest involved was "an undivided one-half interest in and to said oil, gas and minerals that may hereafter be produced from the said lands." This interest was excepted from a reservation of "all oil, gas and minerals." The issue was whether the exception in the reservation gave the grantee "an oil, gas and mineral estate in place or a non-participating royalty interest." The court ran the *Mounger* checklist for royalty and found no right "to do any act or thing to discover and produce oil and gas"; no "right to grant leases"; no "right to receive bonuses and delay rentals." Lacking was "specific indication [in the deed] as to whether or not the grantee will be charged with any of the costs of discovery and production." The *Lackey* court found that the interest it was analyzing had three of the four characteristics of a royalty interest. Where are they found in the language before the court? They are there because the royalty characteristics are basically negative, such as the absence of the leasing power, and, since they are not granted, they are by negative inference present. To appreciate the opinion it is useful to read its conclusion:

> Therefore, the grantee under the deed in question received a non-participating royalty interest despite the fact that the deed doe not specifically (but by implication from the entire instrument does) relieve the grantee from the cost of discovery and production. Such an exclusion of cost can be logically implied from a reading of the entire instrument because it had three

characteristics preponderating toward, and one characteristic negating, a construction that the interest of appellants is a non-participating royalty.

It is this round about approach in *Lackey* that the concurring opinion in *Thornhill* finds "wonderfully *wrong.*" The part of the opinion that is characterized as "wonderfully *right* " is harder to find. The *Lackey* opinion itself says that "Nowhere in the deed is specific indication made as to whether or not the grantee will be charged with any of the costs of discovery and production." Where do we then find, in the words of the concurring opinion, "the words in the instrument" which, "as a matter of elementary English usage, said they weren't, *period?* " They are there, but not, at first glance, "as a matter of elementary English usage." The aspect of royalty that the *Lackey* court says is specifically missing but impliedly present is really the only ingredient that is specifically present. Where is it? In the only language the court has to work with: "an undivided one-half interest in an undivided one-sixteenth interest in and to said oil, gas and minerals that may hereafter be produced from the said lands." This is is the classic language for the creation of a royalty interest nearly everywhere. Can you explain why this is so in elementary English?

The majority opinion in *Thornhill* characterizes the *Harris* case, which is overruled and put down as a situation "wherein we benevolently took care of the 'befuddled litigants' to let their words accommodate what we thought the parties intended." The issue in *Harris* is the location of the leasing power. The court held that the reservation of bonuses and rentals "constituted an implied retention of the executive rights [the leasing power]." This is the holding that was overruled. What was conveyed was characterized as "a non-participating royalty." The fraction involved was one-fourth. If we take the plain meaning of a 1/4 non-participating royalty interest, what will the grantee get? One-fourth of the gross production free of the expense of production. Note that the result of the removal of the leasing power from a conveyance that "has the effect of conveying a fractional interest in the minerals in place" has, on the face of the opinion in *Harris,* the result of changing a 1/4 mineral interest, entitled to 1/4 of the royalty reserved in a lease, to an interest-entitled to 1/4 of the gross production. Benevolence? To be fair, the *Harris* court points out that their decision "does not adjudicate the amount of royalty due." The Mississippi court continues its vigorous discussion of conveyancing in Knox v. Shell Western E & P, Inc., 531 So.2d 1181, 102 O.&G.R. 520 (Miss.1988).

The concurring opinion in *Thornhill* (at pp. 634-35) lists a number of incidents of mineral ownership. How many of these can you remove and still have something left. What will it be? *See* Maxwell, *The Mineral-Royalty Distinction--A Question of How Much*, 10 Gonzaga L.Rev. 731, 755 (1975).

2. In PRAIRIE PRODUCING CO. V. SCHLACHTER, 786 S.W.2d 409 , 112 O.&G.R. 522 (Tex.App.1990, writ ref'd), a conveyance of an undivided one half interest in oil and gas and other minerals used the classic mineral interest language, "in and under and that might be produced" and expressly set out rights of "ingress

and egress for mining, drilling and exploring for oil, gas and other minerals but reserved "delay rentals and bonus monies." Nothing was explicitly said in the document of the power to lease. In a suit by the successors to the grantor it was asserted that the grantee acquired a royalty interest with no leasing power. The appeals court reversed a holding in favor of a royalty construction and held that a reservation of bonus and delay rentals was "not inconsistent with the conveyance of a mineral interest" and the grantee had such an interest with the power to lease as one of the mineral attributes that passed by the conveyance since it was not explicitly reserved. Suppose the deed had included no development easements and had explicitly reserved the power to lease while continuing to convey a one-half interest in oil and gas and other minerals "in and under and that might be produced" from the described land. What interest would be conveyed? If it is a royalty, what fraction of the production is its owner entitled to?

3. J.M. HUBER CORP. V. SQUARE ENTERPRISES, INC., 645 S.W.2d 410, 77 O.&G.R. 357 (Tenn.App.1982), distinguishes language reserving a royalty of "one-fourth part of such oil, gas and other minerals" and a reservation of "one-half (1/2) interest in all royalties received from minerals" pointing out that while the first formulation did reserve "one-fourth" of all minerals as royalty the second reserved "but one half of the royalty."

4. In a case of first impression for Nebraska a conveyance reserving "one-half interest in oil, gas and other minerals in, on and under said real estate" with "all of the bonuses and delay rentals" explicitly given to the grantee, the court held: "that the executory right as to the grantor's undivided one-half interest remained with that interest" in spite of the "grant of bonuses and delay rentals." Antelope Production Co. v. Shriners Hosp., 236 Neb. 804, 114 O.&G.R. 337 (1991). Does a rational and workable configuration of interests result from this interpretation? Is this a valid question to ask about a judicial reading of a conveyancing instrument?

5. In a subsequent case involving the same type of form deed, the Mississippi Supreme Court concluded that parol evidence was admissible to determine the intent of the parties where mutual mistake or fraud was adequately shown on the issue of whether a mineral or royalty interest was conveyed. Bedford v. Kravis, 622 So.2d 291, 126 O.&G.R. 509 (Miss. 1993). *See also* Knox v. Shell Western E & P, Inc., 531 So.2d 1181, 102 O.&G.R. 520 (Miss. 1988).

FRENCH v. CHEVRON U.S.A., INC.

Texas Supreme Court, 1995.
896 S.W.2d 795, 134 O.&G.R. 111

Justice ENOCH delivered the opinion of the Court, in which all Justices join.

The controversy in this case is over the size of an interest conveyed by a mineral deed. Grantee's successor-in-interest claims to own a royalty interest equal to the value of 1/656.17 of all oil and gas produced from the entire tract of land involved. Grantor's successors-in-interest contend that the deed conveyed only a 1/656.17 portion of the royalty to be paid by the lessor. We agree with the grantor's successors-in-interest and consequently affirm the judgment of the court of appeals.

In 1943, George Calvert (Grantor), the owner of a 1/32 mineral interest in a 32,808.5 acre tract, deeded a fifty acre, 1/656.17 interest to Capton M. Paul (Grantee). The pertinent parts of the document, titled "Mineral Deed," read as follows:

[Paragraph I.]

That I, George Calvert, ... do grant, bargain, sell, convey, set over, assign and deliver unto Capton M. Paul, an undivided Fifty (50) acre interest, being an undivided 1/656.17th interest in and to all of the oil, gas and other minerals, in, under and that may be produced from the following described lands....

[Paragraph II.]

It is understood and agreed that this conveyance is a royalty interest only, and that neither the Grantee, nor his heirs or assigns shall ever have any interest in the delay or other rentals or any revenues or monies received or derived from the leasing of said lands present or future or any part thereof, or the renewal or extension of any lease or leases now on said lands or any part thereof. Neither the Grantee herein nor his heirs or assigns shall ever have any control over the leasing of said lands or any part thereof or the renewal or extending of any lease thereon or for the making of any lease contract to develop or prospect the same for oil, gas or other minerals, which is hereby specifically reserved in the Grantor.(Emphasis added).

Petitioner Fuller Trust, the successor-in-interest to the grantee, brought suit against Grantor Calvert's successors to construe the deed as conveying a royalty interest. Fuller Trust maintains the deed conveyed a pure fixed royalty interest of 1/656.17 of all production. Respondent Chevron claims the deed conveyed a mineral interest with a reservation of all rights stated in paragraph II. If the deed conveyed a mineral interest reduced by reservations, the grantee would receive only a 1/656.17 fraction of any royalty payable under a lease.

Both parties sought summary judgment, asserting that the deed is unambiguous and that its construction is a question of law. If the language is unambiguous, the court's primary duty is to ascertain the intent of the parties from the language of the deed by using the "four corners" rule. Luckel v. White, 819 S.W.3d 459, 461 (Tex. 1991). The trial court denied Fuller Trust's summary judgment motion, finding only that the case constituted a proper declaratory judgment action, and granted Chevron's motion, holding that the deed conveyed a mineral interest with a reservation of certain rights.

On appeal, Fuller Trust argued that the phrase "royalty interest only" indicated that the parties intended to transfer only a royalty interest in production from the land. As support, Fuller Trust cited Watkins v. Slaughter, 144 Tex. 179,

189 S.W.3d 699 (1945), in which a deed containing language similar to the Calvert-Paul deed was held to convey a royalty interest. The court of appeals concluded that under Watkins and other Texas decisions, a deed transferring a royalty interest must expressly provide that royalty be from "actual production." On the basis that the Paul deed did not include that language, the court of appeals distinguished Watkins and affirmed the judgment of the trial court.

The conflict in this case comes in reconciling language in paragraph one, which appears to convey a mineral estate, with the language in paragraph two, which explicitly states that only a royalty interest is being conveyed. A mineral estate consists of five interests: 1) the right to develop, 2) the right to lease, 3) the right to receive bonus payments, 4) the right to receive delay rentals, and 5) the right to receive royalty payments. Altman v. Blake, 712 S.W.2d 117, 118 (Tex. 1986). A conveyance of a mineral estate need not dispose of all interests; individual interests can be held back, or reserved, in the grantor. However, "[w]hen an undivided mineral interest is conveyed, reserved, or excepted, it is presumed that all attributes remain with the mineral interest unless a contrary intent is expressed." Day & Co. v. Texland Petroleum, 786 S.W.2d 667, 669 n.1. (Tex. 1990).

While the first paragraph appears to grant a mineral estate, the second paragraph of the lease specifically states that the interest conveyed is a royalty interest only. The clause in the second paragraph, beginning "and that," sets forth the consequences of the royalty interest only declaration, by going further to specifically reserve in the grantor the four components of a mineral estate other than the royalty: the rights to lease, to receive bonus payments, to receive delay rentals and to develop or prospect.[81]

We must ascertain what was meant by the language used in the conveyance, so we begin by noting the relevant canon of construction. "[I]n construing a written instrument the lawful intent of the parties must be looked to and must govern." Smith v. Brown, 66 Tex. 543, 1 S.W. 573 (1886). Because "once a dispute arises over meaning, it can hardly be expected that the parties will agree on what meaning was intended," courts use canons of construction to help ascertain the parties' intent. Southland Royalty Co. v. Pan Am. Petroleum Corp., 378 S.W.2d 50, 59 (Tex. 1964) (Calvert, C.J., concurring). The "four corners" canon of construction means that the court must look at the entire instrument to

---

[81] The court of appeals held that the deed was silent as to the conveyance of the right to develop and therefore, that right was impliedly transferred to the grantee. This conclusion is incorrect for two reasons. First, the right to develop is a correlative right and passes with the executive rights. Day & Co. v. Texland Petroleum, 786 S.W.2d 667, 669 n.1 (Tex. 1990). Second, we read the reservations clause in this conveyance as reserving the right to develop in the grantor. It states that the grantee has no control over "the making of any lease contract to develop or prospect." Consequently, we also conclude that the right to develop was reserved in the grantor.

ascertain the intent of the parties.[82]

We interpret the transfer to have conveyed a 1/656.17 mineral interest with reservation of all developmental rights, leasing rights, bonuses, and delay rentals. The conveyance grants, in essence, only a royalty interest, as stated in the second paragraph. First, the granting clause must be read in light of the rest of the document. Paragraph one states that the grantor is conveying a fifty acre interest. A "fifty acre interest" is 1/656.17 of the 32,808.5 tract, and the deed then recites that it is conveying "an undivided 1/656.17th interest in and to all of the oil, gas, and other minerals in, under, and that may be produced from the described lands." Standing alone, this would convey a 1/656.17 interest in the minerals. Paragraph two indicates that the interest in the minerals conveyed in paragraph one is a royalty interest. The remainder of paragraph two is best interpreted as explaining the consequences of the "royalty only" description. It reserves in the grantor the right to receive delay or other rentals, or any revenues from the leasing or from any renewal or extension of any lease. This reservation would be redundant and would serve no purpose whatsoever if the interests in minerals being conveyed was a 1/656.17 royalty interest, that is, 1/656.17 of all production. A grant of a royalty interest, without any further grant, does not convey an interest to the grantee in delay or other rentals, or in bonus payments, nor would it convey executive rights. The meaning of this grant is to convey an interest in the nature of a royalty--a mineral interest stripped of appurtenant rights other than the right to receive royalties. From the four corners of the document, we conclude that the parties intended to convey a 1/656.17 mineral interest with the reservations described, thus conveying only the royalty portion of the mineral interest. This harmonizes and gives effect to all portions of the deed. In other words, when a deed conveys a royalty interest by the mechanism of granting a fractional mineral estate followed by reservations, what is conveyed is a *fraction of* royalty, not a fixed fraction of total production royalty. *See, e.g.,* Brown v. Havard, 593 S.W.2d 939, 942, 946 (Tex. 1980)(distinguishing between conveyance of fraction of production as royalty and *fraction of* royalty).

The court of appeals correctly affirmed the trial court's summary judgment. We affirm the judgment of the court of appeals and render judgment by construing the Paul deed as conveying to Paul a 1/656.17 interest in the minerals and not 1/656.17 of production *as* royalty.

NOTES

1. In footnote 81 p. 642 *supra*, the court finds that the development right is "correlative" to the executive power and is reserved by the grantor. Is that consistent with *Altman v. Blake*, discussed at the beginning of this chapter that seemingly authorizes each of the parts or constituent elements of the mineral

---

[82]   For an exhaustive account of the various canons of construction used to interpret mineral deeds and leases, see Bruce M. Kramer, *The Sisyphean Task of Interpreting Mineral Deeds and Leases: An Encyclopedia of Canons of Construction*, 24 Tex.Tech L.Rev. 1 (1993).

estate to be separately conveyed or reserved? Would it be consistent with a rule that says that all constituent parts that are not specifically reserved are granted? Day & Co. v. Texland Petroleum, 786 S.W.2d 667, 105 O.&G.R. 590 (Tex. 1990). If the executive power and the development right were retained by the grantor, in what sense would the grantee have received a possessory estate? Who could sue for trespass?

2. Does Paragraph I of the *French* deed, taken alone, create a mineral interest or a royalty interest? The language used in Paragraph I includes the phrase "in and under and that may be produced from" the described land. Professor Martin has described such language as a "norm of expression," noting that "courts have established that the use of certain words will be presumed to have certain meaning and to have intended certain consequences." 37 Inst.on Oil & Gas L. & Tax'n 1-8 (1986). How would you "label" the *French* instrument if you had only Paragraph I to work with? Recall Justice Robertson's comments on the "tyranny of labels" in his concurring opinion in *Thornhill, supra.* Should this labeling yield, so far as basic mineral-royalty classification is concerned, to the stripping away of some of the elements of a mineral interest in Paragraph II? Professor Maxwell has analyzed this problem more perspicaciously than most. *See* Richard C. Maxwell, *Mineral or Royalty--The French Percentage*, 49 S.M.U. L.Rev. 543 (1996) and Richard C. Maxwell, *Oil and Gas Conveyancing--Is There Truth in Labeling*, 33 Washburn L.J. 569 (1994).

3. In view of Paragraph II of the *French* deed, is there any question as to the location of the power to develop and execute leases under that document? How does the language differ from that contained in the *Thornhill* deeds? Is the result absurd? Why would someone create a 1/656.17 mineral interest with the capacity to develop and lease? Who would lease such an interest?

4. Does the *French* court add the word "of" to an instrument which lacks it in order to reach its answer to the question presented? Compare the approach taken in Gavenda v. Strata Energy, *supra.* If the word is added, can this be explained as a result of the search for the parties' intent within the "four corners" of the instrument with the word "of" added to round out the intent discovered by this process.

5. Professor Emeritus Edwin Horner argued in an amicus brief supporting a motion for rehearing in *French* "that the court is in error in inserting the word "OF" in stating that the grantee received 1/656.17 OF the royalty, unless it is the intent and purpose to overrule Watkins v. Slaughter." (p. 8 of brief). The *Watkins* deed contained the following language:

> Together with a 15/16 interest in and to all the oil, gas and other minerals in and under and that may be produced from said land and the grantor retains title to a 1/16 interest in and to all of the oil, gas and other minerals in and under and that may be produced from said land; but it is distinctly agreed and understood that the grantor, his heirs and assigns shall not receive any part of the money rental paid on any future lease; and the grantee, his heirs or assigns,

shall have authority to lease said land and receive the cash bonus and rental; and the grantor, his heirs or assigns, shall receive the royalty retained herein only from actual production of oil, gas or other minerals on said land.

The court of appeals in *French*, in distinguishing *Watkins*, made much of the *Watkins* language providing that the grantor "shall receive the royalty retained herein only from actual production of oil, gas, or other minerals on said land." Should these words overcome the preceding language of "normative" mineral significance? Has *Watkins* been overruled sub silentio, as Professor Horner suggested? The Texas Supreme Court apparently does not think that *Watkins* has been overruled in any manner relying upon it, rather than *French*, in Temple-Inland Forest Products Corp. v. Henderson Family Partnership, Ltd. 958 S.W.2d 183, 137 O.&G.R. 575 (Tex. 1997). In *Temple-Inland* the deed also granted a 15/16 interest and reserved a 1/16 interest without using the words "from actual production." Yet the Texas Supreme Court found that the reserved interest was a 1/16 royalty interest and not a 1/16 mineral interest. If it was not a mineral interest who was the owner of the unconveyed 1/16 mineral interest?

6. A deed reserves: "an undivided 1/16 interest in and to all minerals of every kind and description, including oil and gas, in, upon and under" the land, but subsequently provides that "the right to . . . make any and all oil and gas leases upon said land is hereby granted exclusively to grantee. . . , and they shall be entitled to any and all cash bonus or bonuses. . . together with all cash rentals under such leases; but an undivided 1/16 of any and all oil and gas and other minerals developed from said land shall be owned by grantors. . . " Is the last clause relating to minerals developed the same as the *Watkins* language of actual production? See Bank One Texas, National Ass'n v. Alexander, 910 S.W.2d 530, 132 O.&G.R. 402 (Tex.App. 1995, writ denied).

7. What kind of interest is described by the following language: "a one-fourth of one-eighth royalty interest and the right to participate equally in any bonuses and delay rentals under any oil, gas and mineral lease of the lands within [described tracts]?" Does it matter whether it is characterized as mineral or royalty? *See* Brady v. Security Home Inv. Co., 640 S.W.2d 731, 75 O.&G.R. 586 (Tex.App.1982). Can a reservation of a "1/32 royalty interest" be validly coupled with a right to "participate in one-half of the bonus paid for any oil, gas or other lease and one-half of the money rentals" and a provision that the owner of these rights "shall join in the execution of any future oil, gas, or mineral lease?" *See* Elick v. Champlin Petroleum Co., 697 S.W.2d 1, 88 O.&G.R. 396 (Tex.App.1985, error ref'd n.r.e.).

EXAMPLES AND PROBLEMS

1. MCSWEYN V. MUSSELSHELL COUNTY, 632 P.2d 1095, 70 O.&G.R. 542 (Mont.1981). A contract to sell county land had a provision for a reservation to the county of " 'an undivided two and one-half percent of all oil, and other minerals lying in, under and beneath the premises." The deed as delivered reserved "an

undivided two and one-half percent royalty of all oil, gas, and other minerals lying in, and that may be produced from the premises hereinbefore described, delivered free of cost." The trial court found that the change in language from contract to deed "resulted in an unconstitutional gift or donation by the County to [the grantee]," reasoning that "all things being equal, a mineral interest in real property is more valuable than a royalty interest." Is there any truth in this reasoning?

CASTLE V. HARKINS & CO., 464 So.2d 513, 85 O.&G.R. 285 (Miss.1985), which seems to find that a conveyance in which the grantor, owner of " 1/2 interest in all minerals," reserved 63/64 of minerals ["It being clearly understood that vendee is to have 1/64 of all mineral royalties that may become due or be produced on any lease that may be made by vendor hereafter"] gave the grantee a 1/64 mineral interest which the court further described as " 1/4 of [the grantor's] royalty interest (1/4 X 1/16 = 1/64); a non-participating royalty interest calculated as 1/64 th of the whole of all oil, gas and other minerals." Does this result give the grantee the best of both worlds, mineral and royalty?

Which interest, mineral or royalty, would be susceptible of being adversely possessed? Would it matter if the jurisdiction was an ownership or non-ownership state? In Montana, an ownership state, the court presumed that a royalty interest was subject to being adversely possessed although it noted that the activities needed to adversely possess a royalty interest would be different from those needed to adversely possess a mineral interest? What type of actions would be needed to successfully adversely possess a royalty interest? Stanford v. Rosebud County, 251 Mont. 128, 822 P.2d 1074, 117 O.&G.R. 477 (1991).

Does a transfer of 50 "mineral acres" transfer a mineral interest, a royalty interest or both. *Williams & Meyers* define a "mineral acre" as "the full mineral interest in acre of land." 8 P. Martin & B. Kramer, *Williams & Meyers Oil and Gas Law* 632-33. *See* Discussion Note, 137 O.&G.R. 594-98. In Acoma Oil Corp. v. Wilson, 471 N.W.2d 476, 116 O.&G.R. 501 (N.D. 1991), the court found that a conveyance of "mineral acres" that was also described as a fractional share of an undivided parcel was not burdened by outstanding royalty interests that were not specifically reserved in the instrument.

2. ATLANTIC REFINING CO. v. BEACH, 78 N.M. 634, 436 P.2d 107, 30 O.&G.R. 27 (1968). A conveyance labeled "Mineral Deed" granted "an undivided 1/16 interest" in "all the oil, gas and other minerals, in and under, or that may be produced from" forty acres of land "together with the right of ingress and egress at all times for the purpose of developing the same." There then followed an "intention provision" stating that this "conveyance is intended only as a Mineral Deed and to convey only an undivided one half of the Royalty on the above described Forty Acres of land to wit: An undivided one sixteenth of said Minerals thereunder or which may be produced therefrom, but not so as to affect in any way the title in fee simple to said lands nor to convey any interest whatever in or to any rentals or future rentals of Oil, or Gas minerals in under or that may be produced from said lands, but same is reserved unto grantor, specifically, and the intention of

this mineral deed is to convey, and it is so understood, [an] undivided one sixteenth (1/16) of the Minerals thereunder." The instrument went on to warrant the title "in so far as it covers and includes an undivided one sixteenth (1/16) interest in and to said above described Forty acres of Minerals in and to said lands hereinabove described." The grantors "argue that the instrument is a mineral deed conveying an undivided 1/16 interest in the minerals in place, and that [the grantees] are entitled only to 1/16 of the 1/8 royalty." The grantees "contend that a proper construction of the whole instrument makes it apparent that is was intended to grant an undivided 1/2 of the royalty." In determining which meaning to give to this instrument what importance and impact should be given to: (1) the title, Mineral Deed; (2) the exploration and development easements; (3) the "intention" clause; (4) the warranty? Would conveyance by "label" have been better: 1/16 mineral interest or, alternatively, 1/16 royalty interest?

3. HNG FOSSIL FUELS CO. v. ROACH, 99 N.M. 216, 656 P.2d 879, 76 O.&G.R. 88 (1982). Interests were conveyed by two special warranty deeds to the Thomsons in 1963, each of which was described as follows: "25% of the minerals and mineral rights owned by the grantor relating to, within, upon, or underlying the real estate described on the exhibit attached hereto, . . . including, without limitation, oil, gas and all other minerals of any type or character whatsoever, non-participating." The Roaches, successors to the interests of the grantor in this instrument, leased the lands involved to HNG with a provision for the payment of $1 per acre in delay rentals annually. Delay rentals were paid to the Roaches for the first two lease years but prior to the payment date for the 1980-81 rentals the Thomsons each "claimed 25% of the rentals and bonus" due on or before September 1, 1980. To protect itself from possible multiple liability or lease termination, HNG sought and received a court order permitting it to unconditionally tender into the registry of the court the sum of $82,172.35 in full and complete satisfaction of its obligation to pay delay rentals for the year 1980-81. HNG interpled the Thomsons and the Roaches as defendants, "seeking a determination of each party's share of the delay rentals." The trial court "held that the Thomsons are entitled to share only in production of and royalties from the minerals and mineral rights on the lands in question, but do not have the right to negotiate or to execute leases, nor do they have the right to participate in lease bonuses or delay rentals." Held: the interests created in the Thomsons amounted to a "50% non participating mineral interest."

> The Thomsons have an interest in the minerals and mineral rights relating to, within, upon, or underlying the real estate. The parties agree that this interest is a mineral interest rather than a royalty interest. A royalty interest is an interest only in production of minerals. 1 H. Williams & Meyers Oil and Gas Law § 301 (1981) A mineral interest is a grant or reservation of real property. Duvall v. Stone, 54 N.M. 27, 213 P.2d 212 (1949).

> None of the parties dispute that the Thomsons' 50% non-participating interest entitles them to only 50% of the 1/8 royalty provided for in the leases,

sometimes called a 50% mineral interest or 50% *of* royalty interest. *See* Lanehart v. Rabb, 63 N.M. 359, 320 P.2d 374 (1957), *overruled on other grounds,* 93 N.M. 135, 597 P.2d 745 (1979). The issue disputed by the parties is the meaning of a 'non-participating mineral interest.'

A mineral interest includes the following incidents: the right to receive bonuses, delay rentals, and royalties; the right to execute oil, gas, and mineral leases; *Duvall v. Stone, supra;* Shepard v. John Hancock Mutual Life Insurance Co., 189 Kan. 125, 368 P.2d 19 (1962); and the right of ingress and egress to explore for and produce oil and gas; Cormier v. Ferguson, 92 So.2d 507 (La.App.1957); Jolly v. Wilson, 478 P.2d 886 (Okl.1970); 1 H. Williams & Meyers Oil and Gas Law, *supra.* A mineral interest may be created and, by appropriate language in the deed, be stripped of one or more of its normal incidents. *Shepard, supra;* Westbrook v. Ball, 222 Miss. 788, 77 So.2d 274 (1955); *see Jolly v. Wilson, supra.*

The question therefore is what incidents were removed from the Thomsons' mineral interest by the restriction that it be 'non-participating.' The term 'non-participating royalty' has a well-understood meaning in oil and gas law, entitling its owner to a share of gross production but not to bonuses, delay rentals, the executive right, or the right of ingress and egress to explore for and produce oil and gas. Federal Land Bank of Houston v. United States, 144 Ct.Cl. 173, 168 F.Supp. 788 (1958); Schlittler v. Smith, 128 Tex. 628, 101 S.W.2d 543 (1937); Arnold v. Ashbel Smith Land Company, 307 S.W.2d 818 (Tex.Civ.App.1957) (writ ref'd n.r.e.). Terminology that appears to create a 'non-participating mineral interest' usually has been construed to create a royalty interest. 1 H. Williams & C. Meyers, *supra,* at § 307.4. However, it has been held that the parties to a deed may create a mineral interest that does not share in bonuses or delay rentals and does not have the right to execute leases or the right to explore for and produce oil and gas. Swearingen v. Oldham, 195 Okl. 532, 159 P.2d 247 (1945); *cf.* Picard v. Richards, 366 P.2d 119 (Wyo.1961) (applying 'non-participating' broadly to both royalty and mineral interests). In *Swearingen,* the court relied on parol evidence to determine that the intent of the parties had been to create a mineral interest which reserved 1/16 of the 1/8 royalty to the grantor but which conveyed to the grantees the right to execute leases and to collect rentals and bonuses. In the instant case, the trial court relied on parol evidence to determine what the parties understood 'non-participating' to mean. Parol evidence may be relied upon to explain ambiguities in a written document. Maine v. Garvin, 76 N.M. 546, 417 P.2d 40 (1966). The court did not err in holding that the term 'non-participating' as used in the deed *in the instant case* means that the owner of such interest is not entitled to participate in executing leases and does not participate or share in bonuses or delay rentals.

4. ANDRUS v. KAHAO, 414 So.2d 1199, 72 O.&G.R. 455 (La.1981). An "act of sale" provided: "Sellers herein reserve an undivided one-half interest in and to all

the outstanding minerals, the other undivided one-half interest being conveyed herein to Purchasers. The Purchasers shall have the exclusive right . . . to enter into any oil, gas or mineral leases as they may deem fit . . . provided, however, that no lease be executed for less than a one-sixth royalty." The owner of the purchasers' interest leased the land and the sellers filed suit, claiming a share of the bonuses and rentals. *Held,* the executive right includes entitlement to bonuses and rentals.

> On original hearing, persuaded by plaintiffs' contention that the parties intended, as to bonuses and rentals under prospective leases, the conveyance of only a proportionate interest in keeping with the mineral conveyance, we reversed the lower court judgments and ordered the case remanded to the trial court to give plaintiffs an opportunity to prove their claims. In doing so, we overruled two prior cases of this Court, Mt. Forest Fur Farm of America, Inc. v. Cockrell, 179 La. 795, 155 So. 228 (1934) and Ledoux v. Voorhies, 222 La. 200, 62 So.2d 273 (1952).

> Upon reconsideration we now determine that our original opinion was wrong. There is ample reason to follow the jurisprudence rather than overrule it, not the least of which is to maintain consistency in the mineral law jurisprudence.

> The Louisiana Civil Code substantially predates the development of the oil and gas industry in Louisiana. Consequently, the framers of the Code did not contemplate the many and varied legal questions and problems which would arise in the course of development of that industry. With scant Codal or other statutory guidance this Court was called upon to resolve legal questions, and decide cases in which such questions arose. One such question was presented in a fact situation much like those prevailing in the case before us in the 1934 *Mt. Forest* case, and in its successor *Ledoux.* And it was there determined that a clause in an act of sale, reserving to the vendor (or granting) the exclusive right to execute mineral leases also reserved to him (or granted) the exclusive right to the bonuses and delay rentals.

> The Legislature has not to this day seen fit to pronounce different rules. On the contrary the rulings of *Mt. Forest* and *Ledoux* have since been codified in the new Mineral Code.

> The Louisiana Mineral Code was adopted by Act 50 of 1974, effective January 1, 1975. As stated by the Louisiana State Law Institute in the introduction to the Mineral Code, 'the Mineral Code is designed in large measure to supplant by way of codification the extensive jurisprudence that developed in this area of the law.' Although perhaps not directly controlling because adopted after the date of the pertinent sale in this case, it is nevertheless to be considered.

> Article 105 of the Mineral Code provides:

>> The executive right is the exclusive right to grant mineral leases of specified land or mineral rights. Unless restricted by contract it includes the right to retain bonuses and rentals. The owner of the executive right

may lease the land or mineral rights over which he has power to the same extent and on such terms and conditions as if he were the owner of a mineral servitude.

The Comments to Article 105, after outlining the facts and holdings in *Mt. Forest v. Cockrell, supra,* and *Ledoux v. Voorhies, supra,* recite:

Based on these cases and the general view with which they are harmonious, Article 105 defines the executive right as the exclusive right to lease, including the right to retain bonuses or other property given for the execution of leases and rentals or other property given for their maintenance, extension, or renewal.

Thus it is clear that Article 105 was intended to be a codification of the existing jurisprudence dealing with the subject, and as can be seen by our review of that jurisprudence, it was just that.

The Mineral Code, in Article 214, also expressly addresses the applicability of the Code to existing mineral rights. Article 214 provides:

The provisions of this Code shall apply to all mineral rights, including those existing on the effective date hereof; but no provision may be applied to divest already vested rights or to impair the obligation of contracts.

Not only have plaintiffs herein not satisfied the Court that they are being deprived of vested rights or that the obligations of their contract are being impaired, but to the contrary, under the jurisprudence in effect at the time this contract was entered into, it is clear that the jurisprudence as it stood at that time supported a result in defendants' favor. The conveyance of 'the exclusive right . . . to enter into any oil, gas or mineral leases as they may deem fit' effected also the conveyance of the right to any bonuses or rentals derived from entering into a lease, as those payments are nothing more than consideration for the exercise of the right to lease. *Mt. Forest v. Cockrell, supra.*

Although plaintiffs argue that it was never their intention to convey the right to all the bonuses and rentals to defendants, there is nothing in the record before us that supports that contention. The case is therefore in the identical posture as was *Ledoux.* There the court found persuasive and favorable to the possessor of the executive rights the absence from the agreement of any reference to bonuses and rentals notwithstanding the inclusion of a reference to royalty apportionment. In the case under consideration there was additionally in the same clause an express limitation on the mineral lease which might be engaged (minimum one-sixth royalty) without reference to any limitation upon the enjoyment of the entire bonuses and rentals in the holder of the executive rights.

As was recited in *Ledoux,* 'The parties were undoubtedly contemplating all sources of income to be derived from the transaction between them and it strikes us that if they had had the bonuses and rentals in mind as well as the royalties these would have been specified also.'

Therefore, considering the existing jurisprudence at the time the contract was entered into, as well as the Mineral Code and the express public policy interests in fostering one comprehensive body of law in the mineral law area, we find that we erred in our original opinion in overruling the prior jurisprudence and reversing the lower courts."

Is the interest retained by the "seller" in *Andrus* mineral or royalty? Does it matter? *See* Anderson v. Mayberry, 661 P.2d 535, 76 O.&G.R. 97 (Okl.App.1983), characterizing such an interest as a "non-participating mineral interest."

5. SIMSON v. LANGHOLF, 133 Colo. 208, 293 P.2d 302, 6 O.&G.R. 1011 (1956). *R* "assigned and set over . . . forty-nine percent (49%) of all oil and/or gas that may be produced, saved and marketed" from certain land to *E. Held, E* had a mineral interest.

> The 'assignment and set over,' as expressed by the parties, to the plaintiff of 49 per cent of the oil and gas that may be produced, saved or marketed from the described lands was, we think, a conveyance of these minerals in perpetuity, in situ, in fee simple. The rule applicable here is of considerable antiquity as indicated by the following quotation found in 90 A.L.R. 770, at p. 772:

>> According to the feudal law, the whole beneficial interest in land consisted in the right to take the rents and profits thereof, and accordingly the general rule has always been, in the language of Coke, that "if a man seised of land in fee by his deed granteth to another the profits of those lands, to have and to hold to him and his heirs, and maketh livery secondum formam chartae, the whole land itself doth pass. For what is the land but the profits thereof?" 1 Co.Litt. 45.' "

Does this mean that a perpetual non-participating royalty cannot be created in Colorado? *See* Corlett v. Cox, 138 Colo. 325, 333 P.2d 619, 10 O.&G.R. 912 (1958), which held that a reservation of "6 1/4 % of all gas, oil and minerals that may be produced or in other words 1/2 of the usual 1/8 royalty" constituted a "a 1/16 interest in the mineral fee estate." The court accentuated the fact that there was no oil and gas lease on the land involved at the time of the reservation? What relevance does this fact have to the conveyance under consideration? The Colorado view that refuses to recognize perpetual nonparticipating royalty interests as separate estates was used to deny the owner of such interests the right to redeem the surface estate which had been sold by the governmental authorities for non-payment of taxes. Notch Mountain Corp. v. Elliott, 898 P.2d 550, 132 O.&G.R. 1 (Colo. 1995).

The problem of perpetual non-participating royalty has been dealt with by statute in Colorado as to instruments executed on or after July 1, 1991. A royalty interest of perpetual or limited duration can be created and vests in its owner the right to receive the royalty share provided for in the creating instrument.

Colo.Rev.Stat. § 38-30-107.5.

In the construction of a fractional interest as mineral or as royalty is it significant that the fraction is greater than 1/8 ?

If a fractional interest granted or reserved exceeds 1/8 and the interest is found to be royalty rather than mineral, does the size of the fraction suggest that the interest is a fraction *of* royalty rather than a fractional royalty? *Compare* Picard v. Richards, 366 P.2d 119, 16 O.&G.R. 463 (Wyo.1961) (reservation of "a non-participating 20% royalty interest" held to reserve 1/5 of 1/8 royalty) with Arnold v. Ashbel Smith Land Co., 307 S.W.2d 818, 8 O.&G.R. 646 (Tex.Civ.App.1957, error ref'd n.r.e.) (reservation of "a one-fourth royalty" held to reserve 25% of production free of expenses). Brown v. Havard, 593 S.W.2d 939, 65 O.&G.R. 249 (Tex.1980), considered a reservation of "an undivided one-half nonparticipating royalty (Being equal to, not less than an undivided 1/16 th) of all the oil, gas and other minerals, in, to and under or that may be produced from said land" and found the language ambiguous, thus supporting a jury finding on extrinsic evidence that it reserved a 1/16 royalty. What rational meanings can this language convey? If a court does not resort to ambiguity or Lord Coke, taking language literally may result in a very large royalty. See Gavenda v. Strata Energy, Inc., *supra* p. 433 (one-half). What then? See Smith, *Methods of Facilitating the Development of Oil and Gas Lands Burdened with Outstanding Mineral Interests*, 43 Tex.L.Rev. 129, 136 (1964).

6. LITTLE V. MOUNTAIN VIEW DAIRIES, 35 Cal.2d 232, 217 P.2d 416 (1950). *R* granted to *E* "Eight and one-third per cent (8 1/3 %) of all oil, gas and other hydrocarbon substances, and minerals in, under and/or which may be hereafter produced and save[d] from" certain land. *Held, E* received a mineral interest. At the time this grant was made one-sixth royalty was common in California, and when *R* leased the land, after the grant, a one-sixth royalty was reserved. Eight and one-third percent (8 1/3 %) royalty is one-half of one-sixth royalty. Does this suggest that the court erred?

7. In addition to the usual consequences that follow the characterization of an interest as a mineral interest or a royalty interest, Kansas is almost alone in attaching two other consequences to the distinction:

(1) Under a statute a mineral deed is void unless it is recorded within 90 days of execution or listed for taxation. The statute does not apply to royalty deeds.

(2) A mineral interest may be created in perpetuity; a royalty interest in perpetuity has been held to violate the Rule against Perpetuities.

A prediction may be hazarded that unorthodox construction will continue to be forthcoming from Kansas, due to the attempt by the court to avoid the cancellation of deeds under the statute requiring recording or listing for taxation and the invalidation of instruments under the Rule against Perpetuities. Thus instruments that normally would be construed as mineral deeds may be called royalty deeds, so as to avoid the statute; and instruments normally creating royalty interests will be

construed as creating mineral interests, so as to avoid the perpetuities rule.
Martin & Kramer, *Williams and Meyers Oil and Gas Law* § 306.

Thus, in COSGROVE V. YOUNG, 230 Kan. 705, 642 P.2d 75, 74 O.&G.R. 431 (1982), a conveyance of "one-half (1/2) of the royalty in Oil and Gas produced" which specifically included "royalties to become due and owing by reason of any future oil and gas leases" was found void as a violation of the Rule against Perpetuities. A contrary position was taken in Hanson v. Ware, 224 Ark. 430, 274 S.W.2d 359, 4 O.&G.R. 325, 46 A.L.R.2d 1262 (1955). Meyers and Ray, *Perpetual Royalty and Other Non-Executive Interests in Minerals*, 29 Rocky Mtn.Min.L.Inst. 651, 663 (1983), argue that "perpetual royalty interests do not suspend the power of alienation: there are always persons in being capable of joining together to give a fee simple in the minerals." The Florida court put the matter this way: "We hold that the reservation of royalties did not violate the rule against perpetuities because it created a presently vested interest in the land. The fact that production is uncertain and may never occur does not defeat the interest." Conway Land, Inc. v. Terry, 542 So.2d 362, 365 (Fla.1989). Is there a satisfactory answer to these scholarly and judicial propositions? A dissent in *Terry* stated: "I disagree that the Terrys retained any interest in the land in question. What was retained was 'one-half of any and all royalties that may be paid or obtained from the lands aforesaid on account of any oil, minerals, or gas which may be taken from said real property.' Their interest was not a part of the value of land but was limited to sharing in monies derived from severed minerals. Upon severance the minerals become personal property. The Terrys possessed only a contingent interest in personal property."

8. Consider the following:

Oklahoma has had among the prominent producing states perhaps the greatest difficulty with the mineral-royalty distinction. Part of the trouble comes from its view that the word 'royalty' is ambiguous and may mean either royalty in the strict sense or minerals in the broader, more inclusive sense. . . .

Several influences seem to have produced this labyrinth. First the Oklahoma court has thought that many landowners used the words 'royalties' and 'minerals' interchangeably so that a transfer of royalties was really intended to transfer an interest in the full mineral title . . . .. Only where the royalty was a specified figure (*e.g.*, 1/2 of 1/8 royalty) was it clear that a nonparticipating royalty--a pure royalty--was intended. At the same time another reason existed for construing a fraction of royalties (1/3 of royalties) as a mineral interest. No court has yet worked out in full a theory for protecting the owner of such an interest from unfair dealing by the executive. Suppose '1/3 of royalties' is construed as a true royalty interest, and suppose further that the executive leases for a very large cash bonus and a 1/16 royalty. What protection does the owner of '1/3 of royalties' get? To prevent such unfair dealing, the Oklahoma court may have been persuaded to construe '1/3 of royalties' as a mineral interest, so the owner thereof could protect himself through exercise of his leasing right. Under this theory a 1/16 royalty should be construed as a true royalty, and '1/3 of royalties' under existing and future leases should be construed as a mineral interest, so that the owner has protection as to future leases. But under the view that 'royalty' is ambiguous and may have been used by the parties to mean 'minerals' both 1/16 royalty and 1/3 of royalties under existing and future leases could create a mineral interest. Since competing and inconsistent influences seem to be at work in Oklahoma, the conflict in the case authority is not surprising."

Martin & Kramer, *Williams and Meyers Oil and Gas Law* § 307.1-.2.

In HAYS v. PHOENIX MUTUAL LIFE INS. CO., 391 P.2d 214, 20 O.&G.R. 195 (Okl.1964), a reservation of "a royalty of one-sixteenth (1/16) of all oil, gas, and minerals produced from said real estate, said royalty to be delivered to Grantor, its successors and assigns free and clear of all costs of production, together with the right of ingress and egress over said property" was held to reserve royalty rather than minerals. Persuasive to the mineral construction was the reservation of the right of ingress and egress; persuasive to the royalty construction was the fact that there was an oil and gas lease on the premises at the time of the conveyance. In determining whether an interest is minerals or royalties, why should it matter that the land was, or was not, under lease at the time of the conveyance? *See* Zemp v. Jacobs, *infra*, n. 7, p. 725.

———

One of the constituent elements of the mineral estate is the so-called executive right, or perhaps more accurately the executive power. In the mineral-royalty context the ownership of the executive power may be critical in determining whether the granted or reserved interest is a mineral or royalty interest. The cases that follow deal with the more practical issue of the relationship between the owner of the executive power and the owner of the mineral. One of the authors of this text has characterized the court's approaches to this issue as falling within any of four categories: 1. No standard, 2. Good faith and fair dealing, 3. Prudent operator or utmost good faith or utmost fair dealing and 4. Fiduciary. Patrick H. Martin, *Unbundling the Executive Right: A Guide to Interpretation of the Power to Lease and Develop Oil and Gas Interests*, 37 Nat.Res.J. 312, 375-76 (1997). As you read these cases do they easily fit within Professor Martin's categories? Are the courts, even within a single jurisdiction, consistent about the standard to be applied?

———

SECTION 5 – THE EXECUTIVE RIGHT OR POWER TO LEASE
GARDNER v. BOAGNI
WHITEHALL OIL CO. v. ECKART

Supreme Court of Louisiana, 1968.
252 La. 30, 209 So.2d 11, 29 O.&G.R. 229.

HAMITER, JUSTICE. These suits arose out of a dispute over the ownership of a 29% interest in an overriding royalty granted in connection with the execution of an oil and gas lease.

[The first suit was brought by Mrs. Susan Boagni Gardner for the Susan-Alice group against the Edward-Vincent group seeking a declaratory judgment that plaintiffs were the owners of the disputed interest. The second suit was a concursus proceeding brought by the owners and developers of the lease seeking to have determined the conflicting claims to the 29% interest as between the Susan-Alice group and the Edward-Vincent group. The two suits were consolidated and the district court, after trial of the merits, dismissed the declaratory judgment action and in the concursus action it recognized the Edward-Vincent group as owners of the disputed overriding royalty interest. On appeal by the Susan-Alice group the judgments were reversed, the Court of Appeal recognizing the Susan-Alice group as the owners of the disputed overriding royalty interest.

[Upon the death of Edward M. Boagni, Sr., his four children as his heirs had become the owners in indivision of certain properties owned by him. On November 23, 1942, these four children entered into a partition agreement whereunder each of the four acquired the ownership of certain designated parts of the property. Facts summarized by the Editors.]

Further, with regard to mineral interests, the partition agreement provided:

In order that all of the parties hereto may participate to the extent hereafter set forth in the oil, gas and other mineral royalties which might accrue out of the production of any of such minerals from said lands, each of the parties hereto, as part of this agreement of partition, reserves in and to the lots hereby allotted to the others the oil, gas and mineral royalties equal to eleven per centum (11%) of one-eighth (1/8th) of all the oil, gas and other minerals produced from said lands either by the owner of the lands or of the mineral rights therein or by any lessee or others operating under any contract whatsoever. And in consideration of the transfer of certain rights to the parties hereto by Richard O. Eckart under another partition of even date herewith, all of the appearers herein transfer and assign unto the said Richard O. Eckart, his transferees and assigns, oil, gas and mineral royalty rights in and to the lots respectively allotted to them equal to seven per centum (7%) of one-eighth (1/8th) of all the oil, gas and other minerals produced from said lands; it being the intention hereof that the lot allotted to each of the said parties shall be subject to oil, gas and mineral royalty rights vested in the other parties and in the said Richard O. Eckart to the extent of a total of forty per centum (40%) thereof, leaving vested in the owner of each of said lots sixty per centum (60%) of all of such royalties to accrue under any lease or other contract affecting said land. The royalty rights so reserved by the said parties in and to the lots allotted to the others, including the rights of Richard O. Eckart, shall be and remain an obligation attached to said lands binding on any owner or owners thereof or of the mineral rights therein or lessees operating thereon, such royalties to be delivered or paid free of expense out of any oil, gas or other minerals produced from said property by any such owner or lessee, subject, however, to such proportionate deductions as may apply to the lessor's royalty under any lease affecting the land, but the owner of the fee title to each of the said lots, as herein allotted, shall have the right to grant any lease or leases affecting his or her respective lands without the concurrence of the other royalty owners therein and any and all bonuses, rentals and other considerations (except royalties) paid for or in connection with any such lease or other contract shall be payable only to the owner of the lands so leased and the other parties as royalty owners shall not participate therein. It is further provided, however, that in the event any owner should grant a lease or leases affecting his or her land as herein allotted providing for the payment of royalties on oil, gas or other minerals in excess of one-eighth (1/8th) of the whole produced from said land then the other owners of the royalty rights therein reserved or transferred to them shall participate in such excess royalties in the same percentages herein set forth; and the total royalties in which said parties shall participate shall in no event be less than one-eighth (1/8th) of the whole of the oil, gas or other minerals produced from the land.

As a result of the partition agreement Edward M. Boagni, Jr. acquired, among other lands, a certain parcel designated as Lot G in St. Landry Parish containing approximately 520 acres. By a subsequent sale from Edward Boagni, Jr., a brother (Vincent Boagni) acquired the fee and part of the mineral interest in such lot.

In December, 1958, the Edward, Jr. and Vincent Boagni interests (the Edward-Vincent group) executed a mineral lease on Lot G in favor of one Craft Thompson. The lease provided for the retention of a one-eighth royalty. Simultaneously, Thompson executed an "overriding royalty assignment" wherein he transferred to the lessors (the Edward-Vincent group) a stated percentage of his seven-eighths working interest. Eventually, production was obtained under the lease.

The Edward-Vincent group contend that the overriding royalty was assigned to them in lieu of a cash bonus; and that, inasmuch as they were entitled under the agreement of partition to retain all bonuses, they were the owners of the entire proceeds therefrom. In this connection, they urge that a distinction should be made between overriding royalty, paid instead of a cash bonus, and the ordinary royalty as rental in connection with the lease.

On the other hand the Susan-Alice group (comprised of the other heirs of Edward M. Boagni, Sr. and of Eckart) urge that the override transferred by Thompson is merely an excess royalty payment within the contemplation of the "excess royalty" provision of the partition agreement. They urge that the "(except royalties)" stipulation, when read in connection with the "excess royalty" provision, means that however royalties are taken they should be shared, in the percentage stated, by all of the royalty owners. And they assert that, since their collective portions total 29%, they are the owners of that interest in the override and are entitled to the entire amount of the proceeds deposited in the registry of the court.

In reaching his conclusion the trial judge sustained the factual and legal contentions of the Edward-Vincent group, he stating:

Prior to 1958 oil had been produced on Tract 'B' of the 1942 partition belonging to Mrs. Alice B. Heard in which the other heirs have participated as royalty owners on the basis of one-eighth. This tract was in what is known as the Opelousas Oil Field.

During the years 1958 and 1959 oil companies and lease brokers began obtaining leases in the area generally known as the Tidewater Block which was located some distance southeast of the Opelousas Oil Field.

At approximately this time, Mr. Edward Boagni, Jr., representing his wife and son, the owners of the minerals on Lot 'G' along with the heirs of Vincent Boagni, began negotiations looking toward the leasing of Lot 'G'. He testified that from his investigation of leases in this area that the prevailing royalty was one-eighth of production. He further testified that he could have received from Mr. Craft Thompson a lease on Lot 'G' providing for $100 per acre bonus and one-eighth royalty. Negotiations continued and finally Mr. Boagni obtained a proposition under the terms of which the lessee obligated himself to drill a well within ninety days and the mineral owners accepted as a bonus an overriding royalty in lieu of the cash bonus. A lease was granted, and in a separate instrument the overriding royalty was provided for.

It is my opinion that there is a difference between a lease royalty, stipulated in the lease, and an overriding royalty out of the lessee's seven-eighths interest taken as a bonus for the execution of the lease as applicable to the facts and the agreements in this case.

. . .

The [partition] agreement specifically provides that the mineral owners may retain any and all bonuses, rentals and other consideration received for the execution of the lease, except royalties. The first of these words are 'paid for or in connection with any such lease or other contract,' and those words relate to the preceding words 'all bonuses, rentals or other consideration'. These are the things which may be kept by the mineral owners, and such 'other considerations' or delayed bonus may be taken in 'other contracts'. This is what took place in the instant case.

The next pertinent language of the royalty reservation which supports my decision is 'in the event any one grants a *lease* . . . providing for payment of royalty', etc. This language in my opinion stipulates that the excess royalty in which the royalty owners may share, is such as is *provided in the lease* granted by the mineral owners. This shows that the parties contemplated that the retention of royalty interest was such as may be provided in the lease.

The overriding royalty interest involved in this suit was not provided in the lease, but was acquired specifically as a bonus or other consideration in a separate instrument.

The Court of Appeal approved the finding of the trial judge, it stating:

In exercising the executive power to grant the lease and in negotiating for the overriding royalties, Edward M. Boagni, Jr., acted honorably and openly and in accord with what he felt were the contractual rights of the Edward-Vincent parties he represented as agent. He could have accepted for them a cash bonus of $52,000 for the lease, and all parties concede that under the contract the Edward-Vincent group was entitled to keep entirely for themselves any such cash bonus. Instead, he negotiated for the overriding royalty interest at issue, which the lessee granted to the Edward-Vincent group in lieu of the $52,000 cash bonus. This overriding royalty interest was granted as a bonus or consideration for the granting of the lease by the Edward-Vincent group, which had the executive (exclusive leasing) power with regard to the tract leased.

But such court went on to adopt the contention of the Susan-Alice group that the act of partition excepted the payment of royalties from the bonus provision. It held that under the "(except royalties)" clause the executive (the party with the exclusive leasing power) could retain any bonus so long as it was not in the form of royalty, and that any royalty bonus received would have to be shared by all of the royalty owners according to their respective interests.

We disagree with the construction placed on the contract by the Court of Appeal and are of the opinion that the interpretation of the district court was correct. As we view the partition agreement, the "(except royalties)" clause merely had reference to the previously referred to one-eighth royalty reserved in a lease, not to an overriding royalty accepted in lieu of a cash bonus. And inasmuch as the override in the instant case was given and received (after negotiation) instead of a cash bonus, the owners of Lot G are entitled to the entirety thereof; all in accordance with the specific terms of the partition agreement.

It is suggested by the Susan-Alice group that the executive power held by the owners of Lot G placed them in a fiduciary capacity with regard to all of the royalty owners and that it was therefore incumbent on the executive to obtain the best "deal" for them, which in this case would require that he accept all the consideration in the form of direct royalties in and under the lease rather than take some of the benefits (the bonus) as an override.

In a similar situation the same argument was made by the royalty owner in *Uzee v. Bollinger*, La.App., 178 So.2d 508. In the opinion in that case the Court of Appeal analyzed prior decisions of this court, as well as those from the other Courts of Appeal touching on the subject, and it concluded that "Louisiana has rejected the doctrine of implied agency and fiduciary relationship as between the owner of royalty interest and the owner of the right to lease the land for mineral purposes." We agree with that conclusion.

In the instant case, under the partition agreement, all that the executive was required to do in negotiating a lease was to obtain not less than a one-eighth royalty for the benefit of the royalty owners. That was done. And having secured that advantage for the royalty owners, the executive was at liberty to negotiate, for his own benefit and his co-owners, for any additional bonuses, rentals, *etc.*, whether he accepted them in the form of cash, overriding royalty or oil payments.

For the reasons assigned the judgment of the Court of Appeal in each of the above named consolidated suits is reversed and set aside; and it is now ordered that the judgment of the district court in each of said cases be and is reinstated and made the judgment of this court. All costs in both suits are taxed to the mass on deposit in the registry of the court.

MCCALEB, J., dissents, being of the opinion that the judgment of the Court of Appeal is correct.

SANDERS, J., dissents, being of the opinion the judgment of the Court of Appeal is correct. *See* Whitehall Oil Company v. Eckart, La.App., 197 So.2d 664.

——

Louisiana Mineral Code

L.S.A. Title 31, effective January 1, 1975

§ 109. Obligation of owner of executive interest

The owner of an executive interest is not obligated to grant a mineral lease, but in doing so, he must act in good faith and in the same manner as a reasonably prudent landowner or mineral servitude owner whose interest is not burdened by a nonexecutive interest.

Comment

It is the intention of Article 109 to sustain the decisions in Vincent v. Bullock, 192 La. 1, 187 So. 35 (1939) and Humble Oil & Refining Co. v. Guillory, 212 La. 646, 33 So.2d 182 (1946), insofar as they negate any duty on the part of a landowner or a mineral servitude owner whose interest is burdened by a mineral royalty to lease for the benefit of a mineral royalty owner. No such duty is imposed on the holder of any executive interest under Article 109. The root of these decisions is in the system of prescription. The executive should not be forced to act to the detriment of his expectancy that prescription will accrue against an outstanding mineral right. However, Article 109 is intended to reverse the decisions in Whitehall Oil v. Eckart, 252 La. 30, 209 So.2d 11 (1968) and Uzee v. Bollinger, 178 So.2d 508 (La.App. 1st Cir.1965) insofar as they hold that the executive owes no duty whatsoever to the nonexecutive. It is the overwhelming majority position in the United States that the owner of an executive mineral interest owes to the owner of a nonexecutive interest some form of duty in exercising the executive right. *See* 2 Williams & Meyers, Oil & Gas Law § 339.2. The most difficult problem has been to define the duty, not to determine whether it exists. In the cited reference, the authors detail the various appellations that have been used to describe the duty.

As Article 109 states, it is intended to negate the existence of a true fiduciary relationship such as that entailed in agency, or mandate. However, it is also intended that there be a somewhat higher duty than that of ordinary care and good faith. The following passage from 2 Williams & Meyers § 339.2 at 209-210 is explanatory of the intent of Article 109 in this regard:

Where the interests of the executive and non-executive coincide, the . . .

improper exercise . . . [of the executive right] due to carelessness, inattention, indifference, or bad faith will result in liability. Where the interest of the executive and non-executive diverge, the executive will not be bound to a standard of selfless conduct, such as that imposed on trustees or guardians. He may exercise the executive right with the same self-interest in mind as if there were no outstanding royalty or non-executive mineral interest. But the executive cannot exercise . . . the executive right for the purpose of extinguishing the non-executive interest, or for the purpose of benefitting himself at the expense of the non-executive. If the conduct of the executive satisfies the ordinary, prudent landowner standard, the fact that the non-executive owner has been harmed is not actionable under this view. But if an ordinary, prudent landowner, not burdened by an outstanding non-executive interest, would have acted differently, then the executive's conduct is actionable if it causes harm. We believe this standard fairly effectuates the intent of the parties; it does not require more than can be expected of ordinary landowners and it does not permit less, especially where the 'less' is due to the executive's effort to profit at the expense of the royalty or non-executive mineral owner.

For example, if in the process of executing a mineral lease a landowner whose interest is subject to a mineral royalty negotiates for a specific understanding as to what costs are to be borne by the lessee, whatever benefits he secures for himself he should be bound to secure also for his royalty owner. Absent highly unusual circumstances, the ordinary, prudent landowner would not be likely to negotiate so that three quarters of his royalty interest would be entitled to the economic advantage of the lessee bearing certain costs and the other quarter would be burdened by such costs.

A second example might be found in situations in which the executive has a choice between a high bonus and a small royalty or a low bonus and a large royalty. In these instances, the executive should not be bound to bargain selflessly as a fiduciary but should be free to consider his own economic position in determining which way to structure the lease transaction. However, if the executive is offered a one-fourth royalty, he should not, by the expedient of denominating one-half of that proffered royalty interest as an overriding royalty, be permitted to deprive a mineral royalty owner of the right to share in that additional royalty. This, of course, assumes that the royalty deed in question is of a kind that would permit the royalty owner to share in a stated proportion of royalties, such as "one-half of all royalties accruing from the production of oil, gas, or other minerals" from the tract in question. If the deed limited the royalty owner's right to one sixty-fourth of the whole of all minerals produced, then the executive-lessor would have no motivation to engage in the subterfuge of designating a portion of the royalties as an overriding royalty as the sharing arrangement would be fixed absolutely by the royalty deed which he or his ancestor in title has given.

There are other problem areas. For example, if landowner A grants a mineral servitude covering one-half of the mineral rights to B together with executive rights over A's retained share of mineral rights, can B in negotiating a lease obtain a one-eighth royalty and a one-sixteenth overriding royalty? It is the intent of these recommendations to assure that such sleight-of-hand tricks cannot succeed. Under ordinary circumstances a lessor who can obtain a three-sixteenths royalty would not be motivated to secure any part of it as an override but would merely negotiate the full fraction as a lessor's royalty. The standard set forth in Article 109 seeks to foreclose such opportunities for unfair dealing.

§ 110. Lease in violation of obligation valid; right of nonexecutive to damages

A mineral lease granted in violation of the standard of conduct required by Article 109 is not invalid for that reason, but the owner of a nonexecutive interest may recover any damages sustained by him by a personal action against the owner of the executive right. The action prescribes one year from the date on which the lease is filed for registry.

Comment

Article 110 seeks to impose proper sanctions for violation of the duty of fair dealing imposed by Article 109 without simultaneously imposing upon mineral lessees the burden of uncertainty in taking leases from owners of executive interests as to whether the executive has violated his duty to the owner of a nonexecutive interest. To these ends, the executive is subject to personal action for violation of his duty, but the lease granted by him in violation of the duty is not invalid. As a further step toward removing uncertainty of lease titles in this regard, Article 110 provides that the action for violation of the duty established in Article 109 must be brought within one year from the date on which the offending lease is filed for registry.

NOTE

Article 109 is applied in Sparks v. Anderson, 465 So.2d 830, 84 O.&G.R. 480 (La.App.1985), upholding the sale to a coal company of lands which were subject to a reserved "sixty (60%) per cent of such royalty that may be received in any other future lease affecting coal or lignite."

LUECKE v. WALLACE

Texas Court of Appeals, 1997
951 S.W.2d 267, 138 O.&G.R. 193, no writ

JUSTICE KIDD. Appellants Jimmie Luecke and Tex-Lee Drilling & Development Company, Inc. challenge a trial-court judgment declaring that Carolyn Kay Wallace, appellee, has an interest in the oil and gas royalties and bonuses to a certain 303.055-acre tract of land (the "303-acre tract") owned by Luecke. We will affirm the trial court's judgment.

The Controversy

In 1984, Carolyn Wallace (formerly Durrenberger) and John Durrenberger were divorced. As part of their property settlement agreement, Wallace executed a deed (the "1984 deed") conveying her interest in a 303-acre tract of land to Durrenberger. In this deed, however, Wallace reserved (1) an "undivided one-half (1/2) non-participating interest in any and all oil, gas and mineral royalties reserved by" Durrenberger, his heirs, and assigns; and (2) "an undivided one-half interest in any bonus money exceeding $50.00 per acre received by" Durrenberger, his heirs, and assigns. This reservation forms the basis of this dispute.

In 1988, Durrenberger conveyed his interest in the 303 acres to appellant Jimmie Luecke. The deed from Durrenberger to Luecke (the "1988 deed") was expressly made subject to Wallace's reservation.

Early in 1992, Luecke negotiated an agreement to lease the 303-acre tract to Union Pacific Resources Company (Union Pacific). Originally, Luecke proposed to lease the 303 acres directly to Union Pacific in exchange for a 1/5th royalty and a $150 per acre bonus. In performing a title search, however, Union Pacific discovered Wallace's interest and, in a letter dated April 28, 1992, informed Luecke that, with respect to Wallace's interest, he must provide Union Pacific with "evidence of payment of 1/2 of excess bonus money above $50.00 per acre and a Ratification and Rental Division Order setting forth the interest division."

Two days after receiving this letter, Luecke and Union Pacific closed the lease agreement; however, at Luecke's request, the structure of the transaction was changed. First, Luecke leased the 303-acre tract to appellant Tex-Lee Drilling & Development Company (Tex-Lee) for a 1/8th royalty and "less than $50.00 per acre" bonus. Luecke was the president and sole owner of Tex-Lee. Tex-Lee then sold its lease to Union Pacific for the $150.00 per acre bonus and an overriding 1/5th royalty. According to a representative of Union Pacific, at the closing of the lease agreement, Luecke orally represented that he had contacted Wallace and she was "on board" with the lease of the 303 acres.

Wallace was not, however, "on board" with Luecke's lease. Indeed, according to Wallace, at the time of the closing, she had not been contacted about the lease. Only after the closing did Luecke send representatives to contact Wallace. Initially, Luecke attempted to purchase Wallace's interest in the 303-acre tract; then, when Wallace refused to sell, he unsuccessfully tried to get Wallace to sign a stipulation of her interest in the 303 acres.

On the day after the closing, Union Pacific paid Tex-Lee a $150 per acre bonus for the 303-acre tract. Five days later, Tex-Lee paid Luecke this same amount. Luecke did not, however, pay or attempt to pay Wallace anything.

Wallace brought this lawsuit for declaratory judgment of the parties' rights and duties and for Luecke's breach of duty to Wallace. Wallace alleged that, in the deed by which she transferred the 303 acres to Durrenberger, she reserved (1) a 1/2 nonparticipating royalty interest and (2) 1/2 of all bonuses in excess of $50. She contends Luecke knew of this reservation and intentionally structured the lease to avoid paying her the royalties and bonus payments she was due. Further, Wallace alleged that Luecke, as owner of the executive rights to the 303-acre tract, owed her a duty of "utmost good faith and fair dealing."[83] She argues that Luecke breached this duty when he leased the tract to Tex-Lee for a bonus and royalty that were below market value.

[The trial court granted Wallace's partial motions for summary judgment and held a trial on the merits awarding Wallace, actual damages, prejudgment interest, exemplary damages and attorney's fees. Ed's]

Discussion

Summary Judgment

[In this part of the opinion the court affirmed the trial court's summary judgment motions finding that the reservation made in the Wallace-Durrenberger instrument applied to Durrenberger's heirs and assigns including Luecke and Tex-Lee and that the reservation was a royalty interest coupled with an interest in the bonus payments. Ed's.]

Breach of Duty of Utmost Good Faith and Fair Dealing.

Luecke and Tex-Lee also argue that the trial court erred by granting partial summary judgment that Luecke, as holder of the executive right to the 303-acre tract, breached a "duty of utmost faith and fair dealing to Wallace." As noted above, the trial court's summary judgment declared that Luecke owed Wallace "a duty to obtain for her every benefit [he] obtained for himself *or* [Tex-Lee]."

An executive owes nonparticipating royalty interest owners a duty of utmost good faith. *Manges v. Guerra*, 673 S.W.2d 180, 183 (Tex.1984). This duty is a fiduciary duty arising from the relationship of the parties and not from contract. *Id.* The "duty requires the holder of the executive right ... to acquire for the non-executive every benefit that he exacts for himself." *Id.*

Luecke and Tex-Lee argue that, to establish a breach of this duty on summary judgment, Wallace would have had to establish that *she* could have made a better deal than the one-eighth royalty and the less than fifty dollar per acre bonus that Luecke received from Tex-Lee. We disagree. Wallace only needed to establish that Luecke obtained benefits for himself that he did not obtain for Wallace. The undisputed summary judgment evidence establishes the following: (1) Union Pacific offered to lease the 303-acre tract *from Luecke* for a one-fifth royalty and $150 per acre bonus; (2) Luecke restructured the transaction so that Tex-Lee would receive the $150 per acre bonus and one-fifth royalty and Luecke individually would only receive from Tex-Lee a one-eighth royalty and less than $50 per acre bonus; and (3) Luecke received from Tex-Lee all of the

---

[83]   Wallace contends that, because Luecke leased the 303-acre tract to his solely owned company, he was engaging in self-dealing and owed her a fiduciary duty.

bonus money Tex-Lee received from Union Pacific. Clearly, the lease from Luecke to his captive corporation, Tex-Lee, was a sham transaction entered into for the sole purpose of depriving Wallace of her full interest. This evidence establishes as a matter of law that Luecke, as owner of the executive right, did not obtain for Wallace every benefit he obtained for himself.

The trial court did not err in granting Wallace partial summary judgments because (1) it properly construed the unambiguous reservation in the 1984 deed and (2) it correctly concluded that Luecke breached the fiduciary duty he owed Wallace. Accordingly, we overrule Luecke and Tex-Lee's first two points of error.

Exemplary Damages

In three points of error, Luecke and Tex-Lee challenge the trial court's $25,000 exemplary damage award. Specifically, they contend both that there is no evidence to support an award of exemplary damages and that there is factually insufficient evidence to support the amount of damages awarded.

Legal Sufficiency: Award of Exemplary Damages Appropriate.

In their fifth and sixth points of error, Luecke and Tex-Lee contend there is no evidence to support the trial court's award of exemplary damages. The holder of a nonparticipating royalty interest may recover exemplary damages from the holder of the executive right for the breach of the duty of utmost good faith. *See Manges v. Guerra*, 673 S.W.2d 180, 184 (Tex.1984) (the breach of the duty of utmost good faith is a breach of fiduciary duty; a beneficiary harmed by a breaching fiduciary may recover exemplary damages). An award of exemplary damages for the executive's breach of that duty will be upheld so long as the breach was intentional, malicious, fraudulent, or grossly negligent.

As determined above, Wallace established as a matter of law that Luecke breached the duty of utmost good faith that he owed to her. After a trial to determine whether Wallace was entitled to recover exemplary damages, the trial court found that Luecke's breach of this duty "was done knowingly and willfully, with the intent to deprive Wallace of benefits Luecke knew, or had reason to know, Wallace was entitled to share." This finding that Luecke intentionally breached the duty of utmost good faith will support an award of exemplary damages. We must therefore determine whether this finding is supported by legally sufficient evidence.

Luecke argues that the evidence established he had a good faith belief that Wallace had no interest in the 303-acre tract and, therefore, there was no evidence to support the trial court's finding. After reviewing the record, however, we find ample evidence to support a finding that Luecke knew of Wallace's interest and intentionally sought to deprive her of that interest. As previously mentioned, there is evidence that Luecke originally planned to lease the 303-acre tract directly to Union Pacific for a one-fifth royalty and $150 per acre bonus. Union Pacific then notified Luecke of Wallace's reservation. Within two days of receiving this notice, Luecke restructured the lease of the 303-acre tract. This restructuring deprived Wallace of any bonus payments and minimized her share of royalty payments; at the same time, the restructuring increased Luecke's share of bonus and royalty payments. Testimony in the record indicates that Luecke falsely told Union Pacific's representatives that Wallace was "on board" with the lease of the 303-acre tract. Furthermore, there is evidence that, subsequent to his lease of the 303-acre tract, Luecke attempted to buy Wallace's interest. When Wallace refused to sell, Luecke attempted to get Wallace to sign a stipulation of interest. More than a scintilla of evidence supports the trial court's award of exemplary damages. The trial court could, therefore, reasonably conclude that Luecke knew of Wallace's interest and intentionally sought to deprive her of it. The trial court's award of exemplary damages is supported by legally sufficient evidence.

. . . We must now analyze all of the evidence in light of the *Kraus* factors to determine whether the trial court's award of $25,000 in exemplary damages is reasonable. The evidence establishes that, for his own personal gain, Luecke intentionally breached his fiduciary duty to Wallace. It further establishes that he knew that his actions would deprive Wallace of her bonus interest and substantially reduce the amount of royalties she would receive from production on the 303-acre tract.

Wallace, as a nonparticipating royalty interest owner, had no right to participate in the execution of any lease of the 303-acre tract. She was, therefore, dependent upon Luecke, the holder of the executive right, to procure a fair lease of that tract. It is because of this relationship that Luecke owed Wallace a fiduciary duty "to acquire for Wallace every benefit that he exact[ed] for himself." Instead of fulfilling his fiduciary obligation to Wallace, Luecke leased the property to Tex-Lee, his wholly owned corporation, for a small bonus and royalty. One day later, Tex-Lee leased the same property for a significantly larger bonus and royalty. Luecke, as the sole owner of Tex-Lee, received the benefit of this second transaction; but Wallace did not. We believe that this is the type of intentional self-dealing that punitive damages are meant to punish and deter. In light of the *Kraus* factors and the trial court's award of $15,153 in actual damages, we hold that the trial court's $25,000 exemplary damage award is supported by the evidence; indeed, the evidence would have supported even higher exemplary damages.

Having determined that the evidence is legally and factually sufficient to support the trial court's $25,000 exemplary damage award, we overrule appellants' fifth, sixth, and seventh points of error.

Conclusion

Having found no error, we affirm the trial court's judgment.

———

NOTES

1. Does the court apply the utmost fair dealing or fiduciary standard on the executive? Would breach of the utmost fair dealing standard support the jury's award of exemplary damages?

2. Should a court apply the fiduciary standard where there are special circumstances or where the owner of the executive power engaged in self-dealing, such as by leasing the premises to himself as was done in this case? A series of Texas cases tends to support the imposition of a fiduciary duty in those circumstances. *See e.g.,* Manges v. Guerra, 673 S.W.2d 180, 80 O.&G.R. 561 (Tex. 1984); Sauceda v. Kerlin, 164 S.W.3d 892 (Tex.App. 2005); Dearing, Inc. v. Spiller, 824 S.W.2d 728, 125 O.&G.R. 162 (Tex.App. 1992); Mims v. Beall, 810 S.W.2d 876, 118 O.&G.R. 329 (Tex.App. 1991); Mafrige v. U.S., 893 F.Supp. 691, 135 O.&G.R. 352 (S.D.Tex. 1995).

3. Professor Martin has written: "The discussion of these cases. . . should clearly indicate that the courts recognize a fairly high standard of conduct to be observed by the executive when negotiating the initial terms of the lease that will control benefits flowing to the executive and nonexecutive owners. The prudent operator standard appears adequate to control the conduct of the executive. A fiduciary standard, which apparently is prompted by the desire to punish overreaching by the executive, goes too far. Punishment is probably best left to the criminal law system. . . " Martin, *Unbundling the Executive Right: A Guide to Interpretation of the Power to Lease and Develop Oil and Gas Interests*, 37 Nat. Res. J. 311, 396 (1997). See also Ernest Smith, *Implications of a Fiduciary Standard of Conduct for the Holder of the Executive Right*, 64 Texas L.Rev. 371 (1985). In light of the facts in the principal case do you agree with Professor Martin's conclusion?

4. In SCHLITTLER v. SMITH, 128 Tex. 628, 101 S.W.2d 543 (1937), the court construed an instrument as reserving an undivided one-half interest in royalty for a term of 10 years and so long thereafter as there was production. The grantee therefore owned all bonus and rental and had the exclusive leasing power. Regarding the standard of conduct governing the exercise of the executive power, the court said:

> We think that self-interest on the part of the grantee may be trusted to protect the grantor as to the amount of royalty reserved. Of course, there should be the utmost fair dealing on the part of the grantee in this regard.

Was the standard of "utmost fair dealing" applied in the principal case?

5. In MANGES v. GUERRA, 673 S.W.2d 180, 80 O.&G.R. 561 (Tex.1984), actual and exemplary damages were allowed against the holder of one-half of the minerals and the executive rights who leased to himself for a nominal bonus in a situation where the non-executive owners were entitled to participate as to their one-half interest "in all bonuses, rentals, royalties, overriding royalties and payments out of production." Although the *Manges* case cites *Schlitter* and speaks

of "utmost good faith" it also uses the term "fiduciary," noting that the "duty arises from the relationship of the parties and not from the contract," specifically placing its approval of the exemplary damages award on the status of the executive as a "breaching fiduciary" rather than a mere contract breacher. Donahue v. Bills, 172 W.Va. 354, 305 S.E.2d 311, 79 O.&G.R. 368 (1983), also describes the nature of the duty of the executive toward the nonexecutive owner as "strict fiduciary standards." Does such language represent a tightening of the protection for nonexecutive interests? Would a standard of dealing for the executive phrased in terms of a "fiduciary" obligation reach a different result than a standard of "utmost fair dealing" in a situation where the executive entered into a joint venture agreement "worded so as to preclude any royalty payment" rather than a lease? *See* Comanche Land and Cattle Co., Inc. v. Adams, 688 S.W.2d 914, 86 O.&G.R. 150 (Tex.App.1985). Because there may be circumstances where the interests of the executive and non-executive may diverge, some have argued that the fiduciary standard should be applied, otherwise the executive will always choose an option harming the non-executive. Morse & Ross, *New Remedies for Executive Duty Breaches: The Court Should Throw J.R. Ewing Out of the Oil Patch*, 40 Ala.L.Rev. 187 (1988).

6. In Shelton v. Exxon Corp., 921 F.2d. 595, 112 O.&G.R. 180 (5th Cir.1991), the instrument which created the executive right under consideration included language giving executive the "exclusive right to enforce the obligations of ... leases and to contract and negotiate with the lessee ... with respect to each such obligation." The executive settled a dispute with a lessee over royalties by accepting an increase in "royalty fractions" from 1/6 to 9/48 thus avoiding "adverse tax consequences" to itself. One holder of a nonexecutive interest preferred a cash settlement. The court held that the "executive fulfilled its duty because it obtained for all mineral and royalty interest holders the same consideration for release of the claim, a prospective increase in royalties." This is not a case where the executive "manipulated settlement terms so that benefits usually shared by all mineral owners inure solely to the benefit of the executive." Further, Texas law does not require that the interests of the executive rights holder and those of the mineral and royalty-interest holders be one. No trust relationship existed. Are there other situations where the application of a "fiduciary" standard might make a difference? The Natural Resources Code of Texas was amended in 1985 to state that the surface owner leasing Relinquishment Act lands is under a fiduciary duty to the state. Such an owner may not lease to corporations or partnerships in which he is a principal stockholder or partner. Vernon's Texas Code Ann., § 52.187.

7. Can the executive breach his duty to the non-executive by not executing a lease? There are a series of Texas cases that hold that there can be no breach of the duty owed by the executive to the non-executive where the executive does not lease. See In re Bass, 113 S.W.3d 735, 164 O.&G.R. 834 (Tex. 2003); Aurora Petroleum, Inc. v. Newton, 287 S.W.3d 373 (Tex.App. 2009); Veterans Land Board v Lesley, 281 S.W.3d 602 (Tex.App. 2009, pet. granted). The analysis in these cases derives from language in Manges v. Guerra, 673 S.W.2d 180, 80 O.&G.R.

561 (1964) that stated that the breach of the executive duty occurs only when the executive receives benefits for herself that are not shared with the non-executive owners. Some earlier Texas cases did find a breach of the duty by failing to lease or by failing to timely lease the mineral estate. See Federal Land Bank of Houston v. United States, 168 F.Supp. 788, 10 O.&G.R. 311 (Ct.Cl. 1958); Hawkins v. Twin Montana, Inc., 810 S.W.2d 441, 117 O.&G.R. 144 (Tex.App. 1991). The Louisiana Mineral Code specifically provides that the owner of an executive right is not obligated to grant a mineral leases. La.Rev.Stat. 31:109.

8. When executive rights were first discussed by the courts, the courts grappled with classifying them. Executive rights were treated as powers of appointment, Dallapi v. Campbell, 45 Cal.App.2d 541, 114 P.2d 646 (1941), powers coupled with interests, Allison v. Smith, 278 S.W.2d 940, 4 O.&G.R. 1136 (Tex.Civ.App. 1955, writ ref'd n.r.e.) or as a statutory power in trust. See Martin & Kramer, *Williams & Meyers Oil and Gas Law* §§ 324, 338. The choice had substantial ramifications relating to the duration of the executive right, whether it could be revoked, whether it could be exercised more than once and whether the Rule Against Perpetuities would apply. *Id.* With the 1990 decision of the Texas Supreme Court in Day & Co. v. Texland Petroleum, Inc., 786 S.W.3d 667, 105 O.&G.R. 590 (Tex. 1990), the executive right was treated as a "creature of property rights. . ." whose transfer and creation is "best governed by principles of rel property and oil and gas law." 786 S.W.2d at 669.

CARLSON v. FLOCCHINI INVESTMENTS

Wyoming Supreme Court, 2005.
2005 WY 19, 106 P.3d 847, 160 O.&G.R. 930

KITE, JUSTICE. In this coalbed methane royalty dispute, the royalty owners filed an action against the mineral owners claiming they were entitled pursuant to a mineral lease executed by the parties' successors in interest to a share of an overriding royalty interest in minerals produced from certain ranch lands. Prior to trial, the district court granted summary judgment for mineral owners on royalty owners' breach of contract claim. The royalty owners' claims for breach of fiduciary duty and breach of the covenant of good faith and fair dealing proceeded to trial. After trial, the district court entered judgment for the mineral owners. Royalty owners appealed. We affirm.

Issues

Royalty owners present the following issues:

1. The district court erred in dismissing breach of contract claims against Flocchini Investments and its predecessors in interest when the undisputed evidence demonstrated that Flocchini Investments was the successor in interest to the initial signatories and where Flocchini Investments failed to share all royalties obtained on production in contravention of its contractual duties.

2. The district court's decision on Appellants' claim for breach of fiduciary duty erred by applying the wrong standard.

. . .

Facts

In 1957, Robert and Velma Wright owned the surface estate of a large ranch in Campbell County, Wyoming. Mr. Wright's sister, Alice Spielman, owned some of the minerals underlying the ranch. In May of 1957, Ms. Spielman and the Wrights entered into a "Cross Conveyance and Stipulation of Interests" concerning the minerals underlying the ranch. Pursuant to the stipulation, Mr. Wright acquired all of Ms. Spielman's interest in the minerals, including the executive right to lease the minerals, and Ms. Spielman retained a nonparticipating royalty interest in the minerals.

Between 1957 and 1979, Mr. Wright conveyed his interest in the ranch and minerals to the mineral owners and Ms. Spielman conveyed her royalty interest to the royalty owners. In 1981, the royalty owners filed suit against the mineral owners, claiming they diverted royalty payments to themselves rather than sharing those payments with the royalty owners as required by the cross conveyance and stipulation. The suit was settled when the parties reached an agreement, effective November 15, 1982, providing that the term "landowner's royalty" as used in the cross conveyance included all royalties acquired by the mineral owners for oil, gas and minerals produced from the subject ranch land, including overriding royalties, and in essence that, in exchange for dismissal of the suit, the mineral owners would divide royalty payments among the parties as intended by the cross conveyance. The settlement agreement also provided:

> The mineral owners shall negotiate all future oil and gas leases and other mineral leases in good faith and as ordinary prudent mineral owners. . . . Except as specifically hereinafter provided, should the Mineral Owners acquire an overriding royalty in the Subject Lands, the same shall be considered to be part of the "landowner's royalty."

In 1994, interest began to grow in the production of methane gas from the ranch. At that time, Durham Ranches, Inc., owned a portion of the surface estate of the ranch, the mineral owners owned a portion of the mineral interest previously owned by Mr. Wright, and the royalty owners owned the non-participating royalty interest previously owned by Ms. Spielman. The mineral owners were interested in investigating the possibilities of coalbed methane development while Durham Ranches was concerned with the effects such development would have on the ranch. Mr. Flocchini was charged with negotiating mineral leases on behalf of the mineral owners and surface use agreements on behalf of Durham Ranches. Petrox Resources, Inc. approached Mr. Flocchini concerning the production of methane gas from beneath the ranch. Mr. Flocchini and Petrox negotiated an agreement whereby the mineral owners agreed to lease the minerals to Petrox in exchange for a 15% royalty interest and Petrox agreed to compensate Durham Ranches for surface use and damages by making a lump sum payment of $ 50,000, a producing well payment of $ 1,000 per well and a 3% overriding royalty interest in all minerals under the ranch. On the basis of this agreement, the mineral owners and Petrox executed a mineral lease on August 3, 1994, providing for payment of the 15% royalty interest. By separate letter agreement, Petrox and Durham Ranches set forth the terms for surface damage compensation, including the 3% overriding royalty interest Petrox agreed to pay to Durham Ranches.

. . . Royalty owners claimed Durham Ranches was an alter ego for the mineral owners and by virtue of the letter agreement providing for payment of the 3% overriding royalty interest to Durham Ranches, the mineral owners acquired royalties that the 1982 settlement agreement required to be shared with the royalty owners. Specifically, royalty owners claimed they received only their proportionate share of the 15% landowner's royalty interest when they also should have received their proportionate share of the 3% overriding royalty interest paid to the mineral owners alter ego, Durham Ranches. Royalty owners alleged claims for breach of contract, breach of the covenant of good faith and fair dealing, breach of fiduciary duty, conversion, tortious interference, constructive fraud and fraud, alter ego liability, civil conspiracy, and attorneys fees and costs.

Discussion

1. Breach of Contract

The royalty owners claim the district court erred in granting summary judgment for the mineral owners on the breach of contract claim. Specifically, they claim Mr. Flocchini negotiated the mineral lease with Petrox so as to divert a portion of the royalties to Durham Ranches rather than sharing all royalties proportionately with the royalty owners as required by the 1982 settlement agreement. They assert the district court dismissed the claim on the sole basis that Durham Ranches was not a party to the agreement without ever addressing the real breach of contract issue, i.e., the mineral owners' liability under the 1982 settlement agreement. They agree that Durham Ranches was not a party to the 1982 agreement but contend that does not resolve the issue of whether the mineral owners, who were bound by the agreement as successors in interest to Mr. Wright's mineral interests, were in breach. They contend the agreement clearly and unambiguously obligated the mineral owners to share with the royalty owners *all royalties* paid on the oil, gas and minerals produced from the ranch lands and that by entering into a lease with Petrox that paid them more and the royalty owners less, the mineral owners violated the agreement. Specifically, the royalty owners assert Petrox initially offered to pay a royalty of 16 or 16-2/3%, meaning they would have received a higher proportionate payment. However, Mr. Flocchini countered with an offer resulting in a lower royalty payment to them and a separate royalty payment to Durham Ranches. As a result of these negotiations, royalty owners allege they obtained only a proportionate share of the 15% landowners' royalty and Mr. Flocchini's company received a 3%

overriding royalty interest. They claim the 3% overriding royalty interest was not disclosed to or shared with the royalty owners and that violated the 1982 settlement agreement. [The trial court granted the defendant mineral owners' motion for summary judgment. Eds.]

. . .

. . .our inquiry focuses on whether the evidence, viewed in the light most favorable to the royalty owners, demonstrated a genuine issue of material fact as to whether the mineral owners who were parties to the 1982 settlement agreement breached the agreement, or whether they were entitled to judgment as a matter of law on the breach of contract claim. We hold the district court properly granted summary judgment as to the royalty owners' claim because the terms of the 1982 agreement clearly and unambiguously provided the royalties to be shared were those "acquired" by the mineral owners, and the evidence was undisputed that the only party to acquire the 3% royalty interest was Durham Ranches, which was not a mineral owner.

The 1982 settlement agreement provided in pertinent part as follows:

1. Mineral Owners hereby expressly accept, approve, confirm and ratify that the term "landowner's royalty", as used in the Cross Conveyance, shall include all royalties paid upon the oil, gas and minerals produced, saved and marketed from the Subject Lands including all overriding royalties, production payments and other cost free interests based upon or measured by the production of hydrocarbons or other minerals which have been or are hereafter reserved to or acquired by Mineral Owners, except as hereinafter specifically provided.

The agreement later reiterates:

7. . . . should the Mineral Owners acquire any overriding royalty in the Subject Lands, the same shall be considered to be part of the "landowner's royalty".

The evidence presented was undisputed that the only entities that "acquired" a royalty interest in mineral production from the ranch lands were the mineral owners and Durham Ranches. The mineral owners received a 15% landowners' royalty interest. Durham Ranches received a 3% overriding royalty interest along with other payments for surface damage done to the property from coalbed methane production. The royalty owners received their proportionate share of the 15% landowners' royalty interest acquired by the mineral owners under the terms of the mineral lease. What they did not receive a share of was the 3% overriding royalty interest described in the letter agreement. However, Durham Ranches, not the mineral owners, acquired that royalty. Thus, it was not a royalty acquired by the mineral owners and they were not required under the clear terms of the 1982 settlement agreement to pay a proportionate share of it to the royalty owners. Because these facts were undisputed, we hold summary judgment was proper on the breach of contract claim.

2. The Applicable Standard For Measuring the Duty Owed by Mr. Flocchini

The royalty owners contend the district court applied the wrong standard for measuring the duty owed to them by Mr. Flocchini as a successor to the executive right under the original agreement between Mr. Wright and Ms. Spielman. They claim the court erred in relying on *True Oil Co. v. Sinclair Oil Corporation, 771 P.2d 781, 793 (Wyo. 1989)* to conclude that the fiduciary duty was defined by the 1982 agreement and was limited to acting in good faith and as a prudent mineral operator. They argue the district court's reliance on *True Oil* was misplaced because in that case True's relationship with Sinclair as agent and trustee arose from the contract itself, warranting the conclusion that the contract alone defined the obligations between the parties. In contrast, the royalty owners argue, the relationship between themselves and Mr. Flocchini arose not from the 1982 settlement agreement, but from Mr. Flocchini's position as executor of the royalty owners' interest. Thus, the royalty owners assert, the district court erred in concluding that the 1982 settlement agreement defined the relationship between the parties. The mineral owners respond that the parties' agreement controls the relationship and, therefore, Mr. Flocchini's duty as holder of the executive rights was to act in good faith and as an ordinary prudent mineral owner.

In its summary judgment decision letter, the district court stated:

The 1982 Agreement and Assignment expressly provided for and defined the duty of the Defendant as to executory authority and good faith. When parties express a standard of care, the existence and extent of their duties is controlled by that contractual expression. *See, True Oil Co. v. Sinclair Oil Corp., 771 P.2d 781 (Wyo. 1989)*. In this case [royalty owners] have attempted to separate good faith and fair dealing out of an argued fiduciary duty. As these are intertwined, and because of questions of fact relating to the question of good faith and fair dealing, Counts 2 [breach of covenant of good faith and fair dealing] and 3 [breach of fiduciary duty] present issues proper for hearing. The existence of duty is a question of law for the court to decide and, finding that a duty is present, Counts 2 and 3 should proceed.

We hold that the district court correctly concluded the 1982 settlement agreement defined the standard of care Mr. Flocchini owed to the royalty owners.

We have recognized in a number of cases that contracting parties may incorporate express terms varying the standards that would otherwise govern their relationship. Where express terms are contractually agreed upon, they control the relationship of the parties. Id. Consistent with this general principle, we concluded in True Oil Co., 771 P.2d at 793, that the rights and duties of the parties were controlled by their agreement. We cited Tenneco Oil Co. v. Bogert, 630 F. Supp. 961, 967 (W.D. Okla. 1986) for the principle, applicable in joint ventures to develop oil and gas properties, that the existence and extent of fiduciary duties is controlled by the terms of the agreement between the parties. True Oil Co., 771 P.2d at 793. Although Tenneco Oil Co. involved a joint operating agreement, the principle we cited from the case is equally applicable to the agreement at issue here in which the parties expressly agreed that mineral owners would "negotiate all future oil and gas leases and other mineral leases in good faith and as ordinary prudent mineral owners". That was the standard the parties agreed to and that was the standard that governed Mr. Flocchini's conduct.[84] Consistent with our longstanding principles of contract interpretation, we will not rewrite the parties' express and unambiguous agreement.

3. Evidence that Mr. Flocchini Violated the Good Faith, Prudent Mineral Owner Standard

Royalty owners argue next that even if the 1982 settlement agreement established the duty owed by Mr. Flocchini as that of a good faith, prudent mineral owner, the evidence presented at trial showed that he violated the duty. Essentially, royalty owners contend the undisputed evidence showed that Mr. Flocchini was guilty of self-dealing in negotiating the lease terms with Petrox. Mineral owners respond that the facts as found by the district court clearly demonstrate that Mr. Flocchini acted in good faith and as an ordinary prudent mineral owner in negotiating the lease with Petrox.

The district court made the following findings of fact relevant to the issue of Mr. Flocchini's conduct:

> 20. After Petrox had completed its negotiations with [Mr.] Flocchini, it proceeded to negotiate leases with other mineral owners in the area of the proposed development. Because of their collective 50% mineral interest, the Wright family had greater negotiating strength than any other mineral owner. The Wright family was therefore able to negotiate a 16% lease with Petrox. All other mineral owners that executed leases in favor of Petrox agreed to leases with landowners royalties equal to, or less than, the landowners royalty negotiated by [Mr.] Flocchini.
>
> 21. Two of the mineral owners that negotiated leases in favor of Petrox with respect to minerals underlying the Durham Ranch . . . are Plaintiffs in this action, and several others were professionals who purchased and sold minerals as a regular course of their business. Each of these leased their mineral interests to Petrox with landowners royalties that were equal to, or less than, the landowners royalty negotiated by [Mr.] Flocchini.
>
> 22. What a lessee would be willing to offer, and what a lessor would be willing to accept, in terms of a reserved landowners royalty in a mineral lease can vary depending upon such factors as the time of the lease, the location of the minerals, the type of mineral that the lessee intends to

---

[84] It is noteworthy that 2 Williams & Meyers, Oil and Gas Law, § 339.3, p. 222.2(1) (Revised 2002) describes the prudent landowner test as "the best compromise between the minimum standard of good faith and the rigorous standard required of a trustee." *See also* pp. 212-13 for a detailed discussion of the prudent landowner test. While citing Williams & Meyers in their brief, royalty owners fail to mention that the standard endorsed by the authors is the very one contained in the 1982 settlement agreement.

produce, the existence or absence of lease depth restrictions, the speculative nature of the proposed development, and other economic conditions at the time.

23. The amount of the landowners royalty reserved in oil leases, conventional gas leases, leases in other locations, and leases executed at other times not illustrative of the amount that an ordinary prudent mineral owner would have accepted in negotiating a lease of the type, in the location, and at the time of the Flocchini/Petrox lease.

24. The leases [Mr.] Flocchini negotiated and executed in this case were reasonable and fair to all those mineral owners and nonparticipating royalty owners that were entitled to share in the proceeds thereof. Further, the leases contain terms which an ordinary prudent mineral owner would have negotiated.

25. The effect of the methane production activity on the Durham Ranch has been profound: Fences have been broken, gates have been left open, livestock has been injured and lost, the land has been disturbed, there has been flooding during the winter, gates have frozen shut, and increased roads, traffic, dust, and debris have seriously altered the aesthetics of the ranch.

26. Given the impact of the methane production activities on Durham Ranch, and the contingent and speculative nature inherent in accepting an overriding royalty as compensation for surface use, access, and damages, the compensation [Mr.] Flocchini negotiated on behalf of Durham Ranches, Inc., in 1994 was fair and reasonable.

The district court also reached the following conclusions of law relevant to this issue:

33. By virtue of the 1982 Settlement Agreement, the Plaintiffs and the Defendant established the standard of care that would govern all negotiations between [Mr.] Flocchini and prospective mineral lessees. [Mr.] Flocchini was, pursuant to that agreement, obligated to negotiate "in good faith and as [an] ordinary prudent mineral owner."

34. In addition to the duties [Mr.] Flocchini owed to the Plaintiffs by virtue of the 1982 Settlement Agreement, he also owed corresponding duties to Durham Ranches, Inc. These mutual and corresponding duties prohibited [Mr.] Flocchini from using the bargaining advantages possessed by the mineral owners to the benefit of the surface owner, and likewise prohibited him from using the bargaining advantages possessed by the surface owner to the benefit of the mineral owners.

35. To the extent that [Mr.] Flocchini had greater bargaining power during his negotiations with Petrox than did other minority mineral owners, that greater bargaining power was derived from his ability to convey access to the surface estate. [Mr.] Flocchini owed the plaintiffs no duty to exercise that greater bargaining power for their benefit. Moreover, had [Mr.]

Flocchini used his greater bargaining power to obtain a higher landowners royalty for the mineral owners than the mineral owners could have obtained for themselves, [Mr.] Flocchini would have breached his duty to Durham Ranches, Inc.

36. Throughout his negotiation, execution, and implementation of the leases at issue in the present action, [Mr.] Flocchini: (a) acted in good faith; (b) acted as an ordinary prudent mineral owner; and (c) fulfilled all duties he owed to the Plaintiffs, including those duties both expressed and implied in the 1982 Settlement Agreement.

37. [Mr.] Flocchini did not divert royalties from the mineral owners to the surface estate owner. Rather, he simultaneously negotiated separate agreements for both the mineral owners and the surface estate owner which were unbiased and fair to each.

After examining all of the properly admissible evidence in the record, we are unable to conclude the district court's findings of fact were clearly erroneous. Giving due regard to the trial court's opportunity to assess the credibility of the witnesses, assuming Mr. Flocchini's evidence is true and giving to him every fair and reasonable inference and keeping in mind that we do not substitute ourselves for the trial court as finder of fact, we are compelled to affirm.

In reaching this conclusion, we are cognizant of the fact that Petrox's initial offer included a higher royalty payment -- 16 or 16-2/3% as opposed to the 15% ultimately paid. Had Mr. Flocchini accepted this initial offer, royalty owners' proportionate share would have been greater. However, after hearing the testimony and weighing all of the evidence the district court was persuaded that Mr. Flocchini fulfilled his duty to negotiate in good faith and as an ordinary prudent mineral owner. That we might have reached a different result based upon our review of a cold record is not sufficient grounds to warrant reversal.

NOTES

1. Are the results in *Carlson* and in *Luecke* consistent? Did they apply the same test to determine whether the executive breached its duty to the non-executive? Is there a difference between a fiduciary standard and an utmost good faith or good faith standard? Did the executive get for the non-executives all of the benefits received by the executive? How can the court find no breach of a fiduciary duty where the evidence clearly showed that the executive negotiated a lower royalty percentage than that which was originally offered by the lessee?

2. HUDGINS v. LINCOLN NAT. LIFE INS. CO., 144 F.Supp. 192, 6 O.&G.R. 1063 (E.D.Tex.1956). In his conveyance of Blackacre, R reserved "an undivided one-half (1/2) interest in the oil, gas and other minerals in, under and on said property but shall not be entitled to receive any part of any bonuses paid by leases or any part of the rentals that may be paid for the privilege of deferring the commencement of a well or drilling operations." By mesne transfers E acquired Blackacre, subject to R's reserved interest. On May 10, 1950, E leased the land to O

upon the following terms:

(1) 5 year primary term

(2) 1/8 royalty

(3) $1.00 per acre delay rental

(4) $15.00 per acre cash bonus

(5) $15.00 per acre oil payment bonus of 1/16 of 7/8 of production, if, as and when production was secured.

The court did not discuss the validity of this lease with respect to the royalty owner; either it was not questioned or its validity was assumed.

The lease from *E* was one of several acquired by *O* in the area for the purpose of making a wildcat play. In the last year of the primary term of the 1950 lease, *O* drilled a wildcat well in the area that showed promise of commercial production. A second wildcat well was planned, and *O* sought the extension of the *E* lease. *E* executed a new lease, superseding the earlier instrument, on these terms:

(1) 5 year primary term (from September 22, 1954)

(2) 1/8 royalty

(3) $1.00 per acre delay rentals

(4) $107.50 per acre cash bonus.

When requested to do so, *R* refused to ratify the lease and offered to negotiate only on the basis of a 1/4 royalty or some unspecified net profit arrangement. At all times, *R* refused to bargain on any basis that would allow a bonus but insisted on bargaining only on the basis of a larger royalty.

In an action brought by *E* to determine the rights of the parties, *R* contended that under the reservation *R* reserved a mineral interest of one-half of the minerals in place and that although *E* was entitled to all bonuses and delay rentals paid for and under oil and gas leases covering the entire mineral estate, *E* did not have the right to subject *R's* interest in the mineral estate to an oil and gas lease. On the other hand, *E* contended that the interest reserved by *R* was a royalty and that *E* had the right to subject the mineral estate to an oil and gas lease.

The court concluded that *E* had the power and authority to subject the entire mineral estate in Blackacre to oil and gas leases and that the interest reserved by *R* was to become a royalty interest under any lease thereafter executed by the grantee in his deed or his assigns.

In addition to the self-interest on the part of [*E*] that should require them to protect [*R*] as to the amount of royalty reserved in oil and gas leases, [*R*] has the right to require of [*E*] the utmost fair dealings with reference to providing for the payment of appropriate royalty in any oil or gas lease that [*E*] execute covering the lands in question. Under the facts in this case I find that [*E*] in executing the September, 1954, leases, above referred to, dealt most fairly with [*R*] with reference to the royalty provided for in said leases.

3. The role of the executive who also owns the surface estate is clearly impacted by its dual ownership. In Veterans Land Board v. Lesley, 281 S.W.3d 602 (Tex.App. 2009, pet. granted), the executive also owned the surface estate and in developing the surface placed restrictive covenants in the deeds which would prohibit the use of the surface for oil and gas exploration or production activities. That would seemingly have the practical effect of prohibiting oil and gas operations on the surface estate. The executive would benefit by increasing the value of the residential lots it was selling, which benefit would not accrue to the non-executive mineral owner. Nonetheless, the Texas Court of Appeals found no breach of the duty owed to the non-executive. At the time of printing, the Texas Supreme Court has heard oral argument in *Lesley* but has not yet published its opinion.

SECTION 6. TERM INTERESTS

(See Martin & Kramer, *Williams and Meyers Oil and Gas Law* § 331-337)

A term interest is an interest in oil and gas created by a landowner for a less-than-perpetual duration. There are fixed term interests (*e.g.,* for 20 years)[85] and defeasible term interests (*e.g.,* for 20 years and so long thereafter as oil and gas are produced). This type of interest is encountered with increasing frequency as landowners become more and more reluctant to part with potentially valuable rights in perpetuity. While the habendum clause of the defeasible term deed raises some of the same questions as the habendum clause of the oil and gas lease, it was thought better to treat the subject in this chapter, along with the other interests that landowners create for investment and speculation purposes, than to deal with it in the chapter on the oil and gas lease, which is executed for the purpose of exploration and development of the land.

Note on the Validity of Defeasible Term Interests

The next section of this Chapter deals in more depth with problems created by the common-law Rule Against Perpetuities. However, as an introduction to that material, it was thought appropriate to discuss Rule Against Perpetuities issues as they affect defeasible term interests. As the following passage suggests the Rule has created problems for parties creating certain types of defeasible term interests that may be invalid.

The *grant* of minerals or royalty for a fixed period of years (*e.g.*, 20 years) and so long thereafter as oil or gas is produced is valid under the common-law Rule against Perpetuities. The grantor has a possibility of reverter, which is not

---

[85] What property interests are created in one-half of the minerals in Blackacre when its owner, O, conveys Blackacre to A, reserving "for a period of ten years an undivided one-half (1/2) interest in and to all of the oil, gas and other minerals in, on and under" Blackacre? If A then conveys Blackacre to B "except that certain one-half undivided interest and retain unto themselves only their reversionary interest in said mineral interest" has A successfully retained any part of the minerals in Blackacre? *See* Deason v. Cox, 527 So.2d 624, 102 O.&G.R. 508 (Miss.1988).

subject to the Rule.

The *reservation* of such an interest, however, may be subject to the Rule, on the theory that a grant of land subject to a reservation of minerals for 20 years and so long thereafter as oil and gas is produced is, in effect, the transfer of the minerals to the grantee when production ceases, which shift of beneficial interest may occur beyond the permissible period of the Rule. The grantor has attempted to create a springing executory interest in the minerals not certain to vest within lives in being plus 21 years. Hence, the executory interest is void and grantor has a perpetual mineral interest. Victory Oil Co. v. Hancock Oil Co., 125 Cal.App.2d 222, 270 P.2d 604, 3 O.&G.R. 1233 (1954).

Professor Williams believed that the Rule should not be applied to reserved term interests, since they promote alienability by simplifying the mineral title rather than by fettering alienability. A number of courts have followed his views by what one of the present co-authors calls a "mangling" of property law concepts. Kramer, *Property and Oil and Gas Don't Mix: The Mangling of Common Law Property Concepts*, 33 Washburn L.J. 540, 550-59 (1994). They have done so through a variety of means. In WILLIAMS V. WATT, 668 P.2d 620, 80 O.&G.R. 162 (Wyo.1983), an interest which was to go to a grantee on expiration of a term interest for "20 years and so long thereafter as minerals continue to be produced," was characterized as a "vested remainder" rather than an "executory interest" and thus not subject to the Rule against Perpetuities. A concurring judge rejected the rationale of the majority but found the interest valid because finding it valid would comport with the "modern view as to the underlying purpose of the rule against perpetuities because alienability would be facilitated not diminished."

A number of courts have dealt with reserved defeasible term interests without mentioning the perpetuities rule and on the assumption that they were valid. For example, in MORRIS V. MAYDEN, 35 Ill.App.3d 338, 341 N.E.2d 428, 54 O.&G.R. 155 (1976), a reservation of a 1/2 mineral interest for a period governed by the continued existence of an oil and gas lease was found to be valid and no mention made of any Rule problem. A further number of courts have used the common law distinction between a reservation and an exception to uphold a reserved defeasible term interest. By treating the grantor's retained interest as resulting from a reservation, the court found that the entire interest was conveyed to the grantee followed by a second grant back to the grantor. Under that scenario, the grantee retains a possibility of reverter that for historically anachronistic reasons is not subject to the Rule. *See e.g.,* Earle v. International Paper, 429 So.2d 989, 78 O.&G.R. 217 (Ala. 1983); Bagby v. Bredthauer, 627 S.W.2d 190, 72 O.&G.R. 574 (Tex.App. 1981). However, the careful conveyancer will use two instruments to avoid the question: (1) a grant of the land, followed by (2) a grant back of the term mineral or royalty interest.

In Illinois a 1947 statute sought to include possibilities of reverter within the purview of the Rule. That might have wreaked havoc with conveyances of defeasible term interests and the basic oil and gas lease. It was necessary for the court to perform surgery to remove the fishhook left in by less careful draftsmen. The Illinois Legislature in 1947 made possibilities of reverter inalienable. Ill.Laws of 1947, p. 659; Ill--S.H.A. ch. 30, § 37b. This statute was relied on in Murbarger v. Franklin, 18 Ill.2d 344, 163 N.E.2d 818, 12 O.&G.R. 550 (1960), to prevent transfer of the reversionary interest left in the grantor of a defeasible term mineral interest. *Held,* the statute does not apply to reversionary interests created by oil and gas leases or term mineral deeds.

> [R]egardless of what appellation may be applied to the interest remaining in the lessor or term grantor of oil and gas, this court has consistently held that such reversionary interests are alienable and, to such extent, has in effect refused to apply the common-law rule to the contrary. . . . Any change in position at the present time, we believe would produce absurd results and be harmful to the titles to the oil properties in this State. An owner of land subject to an oil-and-gas lease or term deed would be unable to devise or convey, even to members of his own family, and purchasers would be discouraged from acquiring the title to mineral interests under lands subject to a lease.
>
> . . .
>
> It is true that the legislature, in 1947, enacted a prohibition against the alienability of a possibility of reverter, right of entry, or re-entry for breach of a condition subsequent. (Laws of 1947, p. 659.) But when this enactment was adopted, it did not, by force of the uses and exigencies of the country and the decisions of this court, extend to reversionary interests created by leases or term mineral deeds for oil-and-gas interests.

BEATTY v. BAXTER

Supreme Court of Oklahoma, 1953.
208 Okl. 686, 258 P.2d 626, 2 O.&G.R. 1284.

DAVISON, JUSTICE. This is a suit of equitable cognizance, wherein the owners of the reversionary interest, J.B. Beatty and Zella E. Beatty, as plaintiffs, seek, as against F.H. Baxter and some ten others, as defendants, to have the court adjudge that the determinable estates in the minerals underlying an eighty-acre of tract of land in Kay County, Oklahoma, owned by the defendants, had terminated and expired. The parties will be referred to as they appeared in the trial court, being the same as their appearance here.

At the time of his death in 1925, James S. Hubbard was the owner of a quarter section of land (160 acres). The parties here, both plaintiff and defendant, are his children or their grantees. The said Hubbard had, theretofore and in 1921, executed an oil and gas lease on the premises for a primary term of five years or as long as oil and gas was produced, the same being owned, at the time this suit was filed, by Continental Oil Company. . . . At his death, he left a will, by the terms of which he devised to one of his sons, Charles S. Hubbard, the south half or eighty acres of the quarter section. To another son, Fred B. Hubbard, was devised the north half or eighty acres, being the lands involved herein.[86] Threatened contests of the validity of the will developed.

In settlement thereof and to avoid any contest, the said Fred B. Hubbard in February, 1925, conveyed to each of his brothers and sisters an undivided mineral estate in the 80 acres which had been devised to him. The plaintiffs, herein, are the owners of the estate he retained. The defendants are the owners of the estates conveyed. Plaintiffs contend that the conveyed estates have terminated. Defendants contend that they have not. The trial court made written findings of fact and conclusions of law and, founded thereon, rendered judgment for the defendants. Plaintiffs had perfected this appeal.

Except for two, the conveyances by Fred B. Hubbard, which constitute the foundation of this lawsuit, were identical in that the habendum clause in each provided as follows:

> To Have and to Hold, All the afore granted estate, property and easements, together with all and singular the rights, privileges and hereditaments thereunder belonging or appertaining, unto the said heirs, successors and assigns, for a period of Twenty (20) years and as long thereafter as oil or gas is produced from said premises.

The other two conveyances had the phrase "or development had thereon," added.

From 1924 on, numerous wells were drilled on both the north eighty acres and on the south eighty acres. Most of them produced oil for a number of years and then were plugged and abandoned. By December, 1945, only one well remained on the north eighty which was being pumped and was producing oil. At that time, December, 1945, the lessee stopped pumping oil from that well, known as No. 3. In 1947, another well, known as No. 7, was drilled on the north eighty acres. It was completed and started producing oil in September of that year. It was some 700 to 800 feet deeper than was No. 3. A few months later and in the early part of 1948, well No. 3 was deepened and again began producing oil. The trial court found "that production was never abandoned as to the north half of said 160 acres, but that said lessee only temporarily ceased production in order to rehabilitate the well, but that said rehabilitation of the well was delayed because of the war conditions then existing. That during said temporary cessation of production the plaintiff did not assert any right or claim to said property or lease contrary to the defendants' interests. That his present claim is being made after production had begun after such cessation."

The quoted finding of fact sustained the contention of the defendants, and

---

[86]   The non-apportionment of royalties rule applies in Oklahoma [see page 737, *infra*] and therefore the owners of interests in the north 80?acres devised to Fred B. Hubbard did not share in the royalty payable on production from the south 80?acres of the 180?acre lease executed by James S. Hubbard in 1925. [Eds.]

rejected plaintiffs' argument that the cessation of production, from the 80 acres here involved, terminated the estates of the defendants, under the above quoted provisions of the habendum clauses in the conveyances from Fred B. Hubbard. During the approximate twenty-one months that oil was not being taken from the north eighty acres, it was being produced from the south eighty acres so that no question of the termination of the lease is here involved or suggested. The district superintendent of the lessee testified that the casing was never removed from Well No. 3; that it was never abandoned; that, if it had been, the casing would have been removed. At the time of the trial, after being deepened, it was producing 15 to 18 barrels of oil per day; that, during that time, the war was in progress and oil field equipment was very scarce; that under those circumstances, the fact that the casing was left in the well, indicated that there was no intention of abandoning it.

Whether or not a contract, which is effective "as long . . . as oil or gas is produced," has expired depends upon the surrounding facts in each case. Plaintiffs cite and rely upon a number of decisions of this court dealing with the interpretation of the quoted phrase but all of them are those wherein the phrase was a provision in an oil and gas lease. No case has been cited wherein the phrase was a provision in a mineral or royalty deed. Nor do we find one upon independent investigation. The cited cases constitute little authority which is applicable to the case at bar. An oil and gas lease is governed by different rules of construction from those applicable to other contracts, being construed most strongly against the lessee and in favor of the lessor. The reason therefor springs from the danger of loss of oil and gas by drainage.

The defendants herein, being grantees of royalty interests, stand in a materially different position from that of lessees. Not only is there no duty upon them to effect production, but their right to do so is doubtful. That duty rests upon the lessee and upon the success of the lessee's action in obtaining and maintaining production depends the extent of the term of the defendants' estates. . . .

The trial court found that production from the well on the land involved was only temporarily halted for rehabilitation of the well thereon and that such rehabilitation was delayed because of war conditions then existing. The testimony, in addition to establishing the facts hereinabove outlined, further showed that at that time machinery and materials were difficult to obtain and many oil companies were temporarily ceasing small production and trying to find bigger production to satisfy the war effort. The testimony further showed that casing was very scarce and, if the stoppage of production had not been only temporary, the casing would have been pulled from the well and used elsewhere. In addition, the well had been rehabilitated and was producing oil at the time of the trial. Under such circumstances the finding and judgment of the trial court was not against the clear weight of the evidence. Being so, this court will not reverse it on appeal.

. . .

The judgment is affirmed.

JOHNSON, V.C.J., and WELCH, CORN, and BLACKBIRD, JJ., concur.

HALLEY, C.J., and ARNOLD, O'NEAL and WILLIAMS, JJ., dissent.

NOTES

1. *Cf.* WILSON v. HOLM, 164 Kan. 229, 188 P.2d 899 (1948).

[W]e see no sound reason why the general principles of law, heretofore stated, governing the construction of oil and gas leases containing habendum clauses providing the estate conveyed shall continue after the expiration of its primary term so long as oil or gas is produced in paying quantities, should not be applicable to the construction of a mineral deed containing identical or similar provisions.

. . .

Obviously, since production under a lease depends in the first instance upon action or inaction on the part of the lessee and since the ultimate test as to whether an estate created by a deed has terminated depends entirely upon its own provisions, it must follow that the parties to a mineral deed, providing the estate conveyed to the grantees shall continue so long as oil is produced in paying quantities, do not contemplate that failure of a lessee to produce oil in paying quantities works a defeasance ipso facto. To hold otherwise would mean that failure of the lessee to so produce because of neglect, poor judgment, fraud, connivance with the owners of other mineral interests, voluntary abandonment of the lease, or any unjustifiable reason over which the grantees had no control, would have that result. That, however, does not mean that owners of mineral interests can sit idly by and do nothing when the lessee ceases to operate or production stops for any other reason. Neither does it mean, as appellants contend, that any cessation which is resumed at some future date cannot be deemed permanent but must be construed as temporary for that construction would result in a nullification of the defeasance clause itself. We believe proper construction of such an instrument requires the conclusion that if for any reason there is a cessation of production of oil in paying quantities on the land covered by its terms the owners of the minerals in place are required to move promptly and by their efforts actually establish that such cessation, regardless of its cause, is temporary, not permanent. In the event of their failure to do so, it is our view production as contemplated by the parties is to be regarded as having ceased, their conveyance terminates and any estate therefore held by them under and by virtue of its terms reverts to the grantors.

How would a term mineral owner go about implementing the Kansas court's suggestion? A term royalty owner?

2. In PALMER v. BRANDENBURG, 8 Kan.App.2d 154, 651 P.2d 961, 74

O.&G.R. 260 (1982), a partition decree which provided for a reservation of "all of the oil and gas which may be produced saved and marketed for so long as oil or gas or either of them be produced from the described lands [The West Half (W/2) of Section (20)] under present leases thereon [separate leases on each quarter section]" was construed as creating "a defeasible term [mineral] interest perpetuated by production on the Northwest Quarter." The basis for finding a mineral interest was "the oft stated principle that a royalty interest is a right to *share* in production. If sharing in production is a necessary element or feature of a royalty interest, as we believe it is, the reserved interest here cannot be a royalty interest." 651 P.2d at 966. The court further stated that "[c]essation of production on and termination of the lease covering the Southwest Quarter did not terminate the reserved interest as to that quarter-section." The court went on to state that they did not have before them "the question of where, as between the parties, the rights to bonus payments and delay rentals lie in the event of re-leasing of the Southwest Quarter [or] who has the power to execute an oil and gas lease, or who has drilling rights and the proper accounting for production [from the Southwest Quarter]." What is the answer to these questions?

3. Should the clause "as long as oil or gas is produced therefrom" in a term deed be construed to mean production in paying quantities? A term interest can be drafted without a "thereafter" clause. Thus, in Browning v. Grigsby, 657 S.W.2d 821, 78 O.&G.R. 548 (Tex.App.1983), the reservation of a fractional interest in oil, gas and other minerals provided "that this mineral exception and reservation shall expire on October 25, 1958, unless at that time the above named minerals or either of them is being produced from said premises in paying quantities." What is the impact of a later permanent cessation of production on such an interest?

Should the term "production in paying quantities" mean the same thing for term deeds as it does for oil and gas leases?

4. What factors should determine whether cessation of production is temporary or permanent? Are they the same that are applicable to oil and gas leases?

AMOCO PRODUCTION COMPANY v. BRASLAU

Supreme Court of Texas, 1978.
561 S.W.2d 805, 590 O.&G.R. 520.

GREENHILL, CHIEF JUSTICE. In this oil and gas case, the problem is whether term royalties expired because there was a cessation of production after the expiration of the primary terms of the term royalty deeds. Our holding is that while production did cease, there is evidence to support the trial court's finding that there was but a temporary cessation; and consequently the term royalties did not expire. As will be indicated below, the holding under the particular facts is one of first impression; and it extends our previous holdings on temporary cessation of production.

Amoco Production Company and others [Amoco] are the owners of term royalties in Frank Braslau Gas Unit in Live Oak County. They brought a declaratory judgment to determine whether the cessation of production after the primary term had caused their term royalties to expire. If they did expire, these interests reverted to the Braslaus and the Kugerls.

Trial was to the court without a jury. The judgment included findings that reworking operations on the well in question were begun and continued with diligence and good faith; that production from the gas unit was restored within a reasonable time, and that the cessation of production was temporary. The holding was that the term royalty interest did not expire.

The Court of Civil Appeals reversed. It rendered judgment that because of the cessation of production, the term royalties expired. 549 S.W.2d 260. We granted the writ of error of the Amoco group.

There is little dispute in the facts, and most of them are stipulated. By instruments dated September 2, 1946, and February 4, 1958, Frank and Morris Braslau conveyed the term royalties now owned by the Amoco group. Also on February 4, 1958, John and Anna Kugerl executed a term royalty deed to other property within the same Frank Braslau Gas Unit. Each of the grants was for a period of 15 years and as long as oil or gas was produced from the lands described. If at the end of the 15 year term there was no such production from "the lands described," the term royalties terminated.

The unit operator, Atlantic Richfield Company [Arco], drilled and completed what we will call Well Number One. During the drilling of the well, the well logs contained "positive showings" of four producing sands. We shall call the four sands A, B, C, and D, and their depths are here approximated.

Sand A[the Wilcox] was at 7,700 feet. Sand B[the Slick] was at 7,800 feet. Sand C[the Mackhank] was at 8,500 feet; and Sand D[the Massive] was at 8,800 feet.

As Well Number One was being drilled, it reached the 8,550 foot sand, Sand C. When this depth was reached, the testimony is that the well "started to kick and started to blow out." That meant to the testifying consulting geologist that these occurrences were "a very good show" of a producing sand. He further testified that the drillers "tried to balance the drilling, to hold back the pressure so they could go ahead and penetrate the sand." This meant to the witness that these occurrences gave "an indication of production."

The drillers were able to control the well, and they drilled through Sand C to the deeper Sand D at 8,800 feet.

The witness further testified, and his testimony is undisputed, that there were "probably four sands [capable of production] in the well at that time."

The well was ultimately completed in Zones B and D; *i.e.*, in the 7,800 foot [Slick] sand and in the deepest sand, the 8,800 foot [Massive] sand. The well was not then completed in the shallowest sand, Sand A, or in the 8,500 foot Sand C [the Mackhank].

The testimony is that the operators intended ultimately to produce from all four sands sequentially through the same well bore of Well Number One.

Production from Sands B and D commenced during the 15 year primary term of the term royalties in question and continued until August of 1971. At that time, the production from Zone B was depleted. On October 9, 1972, Arco, the operator, notified the owner of the working interest that production from Zone D, the deepest sand, was becoming marginal because of the high cost of handling the water produced with the oil. So at that time, Arco decided to complete the same well in the other two zones, A and C.

Production ceased from the deepest Zone D on November 13, 1972; and work was begun the next day to recomplete the well in the other two zones.

However, Well Number One was "lost" due to mechanical difficulties in trying to recomplete the well. The casing collapsed inside the well bore.

Without delay, Arco obtained permission from the Railroad Commission to "move over" 700 feet on the same "said land" in order to produce from Zone C, the Mackhank sand at approximately 8,500 feet. This Well Number Two, begun on January 12, 1973, was completed in Zone C on February 17, 1973. Commercial production was begun from Zone C on February 24, 1973, approximately 20 days after the expiration of the last of the 15 year primary terms provided for in the term royalty agreements.

Subsequently, a third well was completed on "said land" by Arco to produce from Zone A, the shallowest of the sands. Both these wells produced continuously, and were producing at the time of this suit. There is also evidence that Zone A was productive in the immediate area because of production from a well on an adjacent tract. The problem, then, is whether there was a termination of the term royalties because there was no production for 103 days or approximately three and a half months, and because production, when resumed, was from a different sand.

The owners of the term royalties contend, as the trial court held, that there was but a temporary cessation of production from known sands or zones. The owners of the reversionary interests contend that there had not been *any* production from Zones A and C; and that it is impermissible to call it a temporary cessation of production if it is necessary to drill a second well to produce from a separate zone.

Our view is that the term royalty agreement speaks in terms of production from "said land," or "the lands described," and not from any particular zone or sand. But the term royalty is in the nature of a determinable fee; and if "production" ceases, the interest terminates by its own terms. The courts have ingrafted upon that concept the holding that temporary cessation of production will not trigger the extinction of the interest.

We turn now to the cases. If there is a cessation of production after the primary term, which is not a temporary one, the estate terminates. This was the holding of Watson v. Rockmill, 137 Tex. 565, 155 S.W.2d 783 (1941), in which there was a cessation of two years and seven months. The cause was not any mechanical failure but a severe depression in the price of oil.

In Holchak v. Clark, 284 S.W.2d 399 (Tex.Civ.App.1955, writ refused), oil was "discovered" (a favorable drill stem test) during the primary term, but there was *no* production in paying quantities during the primary term. After the primary term had expired, a second well was drilled into another sand or zone, and it was productive. The holding was that the royalty interest terminated. The emphasis was upon the difference between oil "discovered" during the primary term, and oil "produced." There being no production during, or upon the expiration of the primary term, the royalty interest terminated. The opinion commented upon the absence of a "continuous drilling" provision in the instrument; *i.e.*, that if oil were discovered, it would be sufficient if the operator thereafter proceeded with diligence until production was obtained even after the expiration of the primary term. The court would not write such a provision into the contract for the parties; and the Court of Civil Appeals in this case considered the *Holchak* case and that language to be determinative of this case.

Also helpful to Braslau, *et al.,* are Gulf Oil Co. v. Reid, 161 Tex. 51, 337 S.W.2d 267 (1960), and Archer County v. Webb, 161 Tex. 210, 338 S.W.2d 435 (1960). In Gulf v. Reid, the lease had a 5 year primary term. Gulf began a well toward the end of the term and completed it as a gas well after the end of the term. The well was capped because there was no pipeline connection or ready market for the gas. Shut-in royalties were tendered four months after the end of the primary term, and a pipeline connection was secured some five months after the well was completed. By a divided court, it was held that the lease was terminated. The interest conveyed was a determinable fee; and since there was no *production* at the end of the term, and no prompt tender of the shut-in royalty, the lease terminated.

In Archer County v. Webb, cited above, the oil and gas lease was for a primary term of 10 years, and contained a provision that if gas were discovered, the lessee could pay $50 per well per year in lieu of production. A separate instrument, a term royalty, provided for a share of the production for 15 years and as long as oil or gas was *produced.* The term royalty did not contain the shut-in gas royalty provision. Gas production was obtained; the well or wells were capped; and the shut-in royalties were paid or tendered. The holding was that term royalty, a determinable fee interest, terminated because there was no actual production after the primary term. Shut-in royalty payments were not "production" after the primary term as far as the term royalty was concerned. The Archer County case and Gulf v. Reid are illustrative of this court's view of holding the parties to their agreements rather strictly.

The main cases for the opposing view are Midwest Oil Corp. v. Winsauer,

159 Tex. 560, 323 S.W.2d 944 (1959); and Stuart v. Pundt, 338 S.W.2d 167 (Tex.Civ.App.1960, writ refused). Both of these cases distinguish the *Holchak* opinion discussed above.

In Midwest v. Winsauer, the mineral interest was a term royalty, as we have here. The term royalty was for a primary term of 15 years, and as long thereafter as oil or gas were produced. There was production beyond the temporary term followed by cessation of production due to mechanical problems and litigation. The trial court had rendered a judgment that the term royalty had expired. This court held that from the facts adduced, the cessation, as a matter of law, was temporary; and that the term royalty did not terminate. The opinion says,

> Although the royalty deed does not expressly provide that the term royalty will not terminate because of temporary interruptions, we hold that such a provision is necessarily implied. 323 S.W.2d at 946.

This court then distinguished *Holchak* and another case. Referring to those cases, our opinion states,

> Neither of these cases deal with the question of cessation of production, either temporary or permanent. The question in the *Holchak* case, *supra,* was 'whether there was paying production from the land [at the end of the primary term]'. . . . The court simply held [in *Holchak*] there was no production whatever from the premises. 323 S.W.2d at 948.

The same could be said of the opinion in Gulf v. Reid and in Archer County v. Webb. Accordingly those cases are distinguished.

Finally we come to Stuart v. Pundt, 338 S.W.2d 167 (1960), an opinion of the San Antonio Court of Civil Appeals in which a writ was refused. It is regarded as the case nearest in point. While there are some distinguishing facts, we base the decision in this case upon that opinion and upon the *Winsauer* case discussed above.

In Stuart v. Pundt, the mineral interest involved was also a term royalty. There was substantial gas production during the primary term. Then the well became clogged, and gas could not be produced. As here, during the reworking operations of the well, the casing in the well collapsed. About a month after the casing collapsed, the operator obtained a permit to drill a new well on the same land. The new well was completed without delay in the same sand or zone from which the first well had produced,--and that is a distinguishing feature between that case and this case.

The holding in Stuart v. Pundt was that the mechanical difficulties and the prompt drilling of the new well constituted a temporary cessation of production; *i.e.,* that there was no termination of the mineral interest (the term royalty) even though production was from a new or different well. That court again distinguished *Holchak* and Archer County v. Webb.

We are persuaded that the holding in Stuart v. Pundt should be followed even though the new well here was completed in a sand or zone different from that from which the first well was actually produced. The production must have been from "said land," or "the lands described," as it was.

We do not have before us the extension or preservation of a term royalty during a period of further or deeper exploration. The Sand or Zone C, the Mackhank, was definitely encountered during the drilling of Well Number One to the extent that the driller feared a blowout of the well. While there was no production from that zone or sand in Well Number One, the reasoning of Stuart v. Pundt convinces us that the term royalty should not be terminated due to a temporary cessation of production when the operator was attempting to move up in the well and to obtain production in Well Number One in Zone C. The well was lost, and the operator promptly moved over and obtained production from "said land" in Zone C. As stated, the fact findings of the trial court are that there was a temporary cessation and due diligence; and there is ample evidence to support these findings.

Accordingly, the judgment of the Court of Civil Appeals is reversed, and the judgment of the trial court is affirmed.

NOTES

1. The principal case is relied on in DE BENAVIDES v. WARREN, 674 S.W.2d 353, 82 O.&G.R. 115 (Tex.App.1984, error ref'd n.r.e.), where production ceased from a well which had for thirty years produced from land in which defendant owned a 1/16 term royalty dependent on continuing production. The royalty was held to be preserved by production obtained within three months from a new well on the royalty acreage in a productive sand which had been discovered on adjacent acreage not subject to the defendant's term royalty by drilling which occurred prior to the cessation of production in the thirty year old well. A dry hold to that sand on the royalty tract preceded the drilling of the successful well. Does the holding of the De Benevides case on this point follow *Braslau* or venture into the area reserved by the Supreme Court in that case: "the extension or preservation of a term royalty during a period of further or deeper exploration"? See Professor Flittie's careful analysis of *De Benevides* on this point and on another basis of the decision: ratification by the defendant royalty owner of a pooled lease covering the royalty acreage given by the plaintiff exercising the executive right. Discussion Notes, 82 O.&G.R. 134 (1985).

2. If a habendum clause of a term deed provides "and so long thereafter as oil, gas and/or other minerals are produced therefrom, or the premises are being developed or operated", does the tender of shut-in gas royalties by the lessee pursuant to a lease clause extend the term of the mineral or royalty deed? *See* Dewell v. Federal Land Bank of Wichita, 191 Kan. 258, 380 P.2d 379, 18 O.&G.R. 624 (1963), where the lease with the shut-in royalty clause was executed under the authority of an executive right contained in the term deed itself but the lack of a provision for shut-in royalty payments in the habendum clause of the deed itself was fatal to its extension. The expansion of the *Dewell* holding to other fact situations is limited by Classen v. Federal Land Bank of

Wichita, 228 Kan. 426, 617 P.2d 1255, 68 O.&G.R. 426 (1980) discussed *infra* at Chapter 7, Section 6. Is a more favorable case for the term interest presented when its owner attempts to ratify the executive owner's lease which includes a shut in provision?

FRANSEN v. ECKHARDT

Supreme Court of Oklahoma, 1985.
711 P.2d 926, 87 O.&G.R. 326.

KAUGER, JUSTICE. This action is pending on appeal before the United States Court of Appeals for the 10th Circuit. The following question was certified by that court pursuant to the Uniform Certification of Questions of Law Act, 20 O.S. 1981 §§ 1601-1613:

> Does the completion and testing of a gas well to the point that it is capable of producing gas in paying quantities, contracting for the sale of the gas through a gas purchase contract, and commencing construction to connect the well to a pipeline satisfy the extension provision of a term mineral interest reserved in a deed providing that the primary term of the reservation be extended "in the event that . . . gas and other minerals . . . are being produced in paying quantities . . .at the time of the termination of the period above mentioned?

This Court has answered the question as follows:

> Under the terms of the warranty deed under consideration, a term mineral interest requires actual marketing for the interest to be extended beyond the primary term. The requirement for production in paying quantities is not satisfied until the gas is reduced to possession and the parties receive financial benefits from the production.

On January 22, 1952, J.E. and Esther Eckhardt, the grantors, conveyed lands in Custer County, Oklahoma by warranty deed to George and Yvonne Fransen. The grantors reserved a term interest by the following language:

> There is, however, EXCEPTED and RESERVED unto the Grantor, his heirs, executors, administrators, and assigns, an undivided one-fourth (1/4) interest in the oil, gas and other minerals in and under and that may be produced from the above described land for a period of thirty (30) years from the date of this deed, and in the event that oil, gas, and other minerals, or any of them, are being produced from said land in paying quantities at the time of the termination of the period above mentioned, then this reservation shall continue and be in full force and effect as long as such production continues;

On April 18, 1979, the Eckhardts leased the mineral interest to O.N.G. Exploration, Inc. Sometime thereafter, O.N.G. entered into an operating agreement with the Harper Oil Company by which Harper agreed to operate the working interest of the property owners. Harper began drilling operations in

January of 1981, and completed the well in September. Tests were conducted on September 17, and the well was determined to be capable of producing gas and gas condensate in paying quantities, and both flowed from the well for approximately six hours. The gas was flared, the condensate was recovered at the rate of one and a half barrels per hour, and nine barrels of condensate were stored. The well was then shut-in to await connection to a gas pipeline. Harper negotiated a gas purchase contract with Delhi Gas Pipeline Corporation on December 9, 1981, and Delhi began construction of a pipeline to the well. The pipeline was completed on April 16, 1982, and gas was produced into the pipeline on May 5, 1982. Gas and gas condensate have been sold in paying quantities since that time.

The Fransens filed an action in the district court of Custer County to quiet title and to cancel the 30-year mineral interest. The Eckhardts removed the case to the United States District Court based on diversity jurisdiction. Each party filed a motion for summary judgment. The United States District Court for the Western District denied the Fransens' motion and granted the Eckhardts' motion. The Court determined that the discovery and completion of a well which had produced, and was capable of producing, hydrocarbons in paying quantities was sufficient to extend the term mineral interest. The Fransens appealed to the 10th Circuit.

The Fransens contend that more is required than discovery and completion of a well to extend the lease mineral interest; and that the language in the deed creating and reserving the term mineral interest requires production, marketing, and economic benefit to the parties. We agree.

The critical date is January 22, 1982. On that date, if gas were not being produced in paying quantities, the deed reservation ended, and all rights in the land reverted to the Fransens. The parties agree that prior to January 22, 1982, the well was completed and found capable of producing gas in paying quantities, and that gas or gas condensate was not actually sold in paying quantities until three to four months after the deed's termination date.

The 10th Circuit certified the question of whether the term is extended because there are no controlling Oklahoma precedents. The two decisions construing "production" under Oklahoma law, Panhandle Eastern Pipeline Co. v. Isaacson, 255 F.2d 669 (10th Cir.1958) and McEvoy v. First National Bank and Trust Company of Enid, 624 P.2d 559 (Okla.App.1980) have reached different results.

In *Isaacson*, the deed contained a reservation clause similar to the one involved in this case. The primary term reserved expired on August 13, 1956. In the interval from December, 1953 to February, 1954, a well was drilled on part of the land. The well was tested in March, 1954, and gas flowed at the rate of 2,300,000 cubic feet per day. The well was equipped with casing, tubing, a separater, well head fittings, and measuring tanks. A second test was conducted in May, 1956, and a gas purchase contract was executed with Colorado Interstate

Pipeline Company. At the time of the lawsuit, the well had not been connected to a pipeline and no royalties had been paid. The 10th Circuit Court of Appeals determined that the reservation clause in the deed contained a "thereafter" clause resembling similar clauses found in oil and gas leases. The Court, while recognizing that there are dissimilarities between a deed and an oil and gas lease, and acknowledging that distinct rules of construction apply to each of the two instruments,[87] defined "production", regardless of whether it appeared in a deed or in a lease, as not requiring marketing.

The opposite conclusion was reached in McEvoy v. First National Bank and Trust Company, 624 P.2d 559 (Okla.App.1981). This case concerned a non-participating term mineral conveyance executed on October 26, 1956, with the primary term expiring on October 10, 1976. The deed conveyed an undivided 1/4 interest in the royalty in any minerals under the described property.[88] Drilling operations began on March 3, 1976, and were finished on July 10. At this time, the well was capable of producing oil and gas in paying quantities, but neither was being marketed by October 10, 1976. Gas was first sold from the well in January of 1977--oil in February of 1977, and since that time, oil and gas were marketed in paying quantities. The issue before the *McEvoy* court was whether the interest was perpetuated past its primary term when oil or gas is discovered in paying or commercial quantities in the primary term, but a market for the production is not developed until after the expiration of the primary term.

The Court of Appeals did not accept *Isaacson's* assumption that the definition of "production" is the same for both deeds and leases. The court followed Kuntz, *The Law of Oil and Gas,* § 15.8 (1962), and refused to apply to non-participating term mineral interest conveyances the rule applicable to completion leases. The court held that a well *capable* of producing in paying quantities is not the same as a well producing in *actual* paying quantities. It found that the term mineral interest terminated on October 10, 1976, when oil, gas, or other minerals were not produced in paying quantities.

---

[87] In Beatty v. Baxter, 208 Okla. 686, 258 P.2d 626, 628 (1953) the Court said: An oil and gas lease is governed by different rules of construction from those applicable to other contracts. In Jath Oil Co. v. Durbin Branch, 490 P.2d 1086, 1091 (Okla.1971) the Court said that: *"oil and gas lease cases could not be applied to term mineral interests because, in reference to provisions other than the granting clause, essentially different obligations were involved."* (Emphasis added).

[88] The deed provided:

"This grant shall run, and the rights, titles, and privileges hereby granted shall extend to grantee herein, and to the heirs, administrators, executors and assigns of the surviving grantee for a period of twenty (20) years from October 10, 1956, and as long thereafter as oil, gas, or other minerals, or either of them, is produced or mined from the lands described herein, in paying or commercial quantities. If, at the expiration of said 20 years from October 10, 1956, gas, or other minerals is not being produced or mined from said land or any portion thereof in paying or commercial quantities, this contract shall be null and void, and the grantee's rights hereunder shall terminate."

The construction of the reservation turns on the question, what is production in paying quantities? Obviously, there is a split in authority concerning whether production without marketing will extend the term interest.

The majority of the cases in Texas require marketing to extend the primary term. These cases are not applicable because they involve royalty interests, and they are distinguishable from reservations in warranty deeds, which are a mineral interests. The inherent difference between a royalty interest and a mineral interest in land, is that the owner of the royalty interest receives nothing unless gainful production is acquired, while a mineral interest owner is entitled to lease his/her share and receive renewal rentals and any other benefits available under the lease.

The Kansas courts, however, have dealt explicitly with reservations of oil and gas interests in warranty deeds. In Dewell v. Federal Land Bank of Wichita, 191 Kan. 258, 380 P.2d 379 (1963), the Court held that the reservation in a warranty deed which reserved an undivided 1/2 interest in the minerals for twenty years, and so long thereafter as oil and gas or other minerals are produced, created a determinable fee in the minerals. The Court found that the deed reservation, and an oil and gas lease executed at different times and by different parties, could not be construed together. It further held that a shut-in royalty clause in a lease does not extend the interest of the holders of the mineral rights under a reservation because shut-in royalty payments are not equivalent to production; and that the reservation was not extended beyond the primary term by the payment of shut-in royalties. In this case, no shut-in royalty payments were paid to the Fransens, nor had any benefits been received by them.

In Home Royalty Association v. Stone, 199 F.2d 650 (10th Cir.1952), the plaintiffs owned all the surface rights and one-half of the minerals. The defendant owned one-half of the minerals. The habendum clause in the mineral conveyance was very similar to the one in the present action. It provided "to have and to hold until the said grantee for the period of twenty-one (21) years from the date, and as much longer thereafter as oil, gas, or other minerals are produced". Gas was discovered and tested during the primary term. Actual marketing did not begin until after the primary term had ended. The court, applying the law of Kansas, held that a conveyance which contains only a habendum clause to set the termination point of the primary term for the production of oil or gas will extinguish the rights of the grantee unless oil or gas are discovered and marketed prior to the expiration of such term.

This Court has determined that the grantee under a terminable interest stands in a dissimilar position from that of the lessee under a lease; and we previously have recognized that oil and gas lease cases constitute slight authority when addressed to a case involving a terminable interest. This view is supported by acknowledging that an oil and gas lease is not the same as a grant or reservation of terminable interests in the minerals. Under a lease, development by the lessee is intended, and it is understood that the lessee will have the power to try to extend the lease by production. In a terminable interest, the parties do not contemplate activity by the grantee to produce the minerals for the mutual benefit of the grantee and the reversioner. The owner of the terminable mineral interest has the right of ingress and egress but this right is for his/her own benefit. The continuation of the fixed term of the terminable interest may be set by whim, by the anticipated period of the current boom, or by references to purposes other than that of stimulating discovery of oil and gas. The lessor considers operations by the lessee for the mutual benefit of the lessor and lessee. The grantor of a terminable interest does not think of any benefit but his/her own.

The purpose of a term interest is to prevent title complication by unifying the title from time to time. The advantage of title unification, in the absence of production or in the presence of exhaustion of oil and gas reserves, is important for several reasons. There is no way to determine with certainty whether the resources are exhausted. Other investors who believe that other mineral deposits may be present may want to purchase the mineral interest. Because prospective purchasers of surface interests frequently want the mineral rights included, outstanding mineral interests place practical hindrances, on and depress the price of, subsequent sales. The mineral owner owes certain duties to the perpetual royalty owner. Ultimate title unification provides certainty and extinguishes these problem obligations.

The rules of construction require that the intention of the parties, especially the intention of the grantors be discerned by examination of the entire instrument. The warranty deed states "this reservation shall continue and be in full force and effect as long as such production continues". The parties have selected an occurrence which is disassociated from their mutual interest except prescribing when one interest concludes and when the other vests in present enjoyment. While the continuation of a terminal interest may be formulated in terms of production from the premises, the parties probably thought in terms of enjoyment of benefits. A perusal of the four corners of the warranty deed, reflects that the grantors' intent was to protect their interest in an oil and gas lease which had been executed on September 5, 1957. The apparent general intention was that if speculation resulted in the production of oil and gas, the interest would continue, and if the lease were unproved, the Eckhardts would be divested of ownership so that future development would not be prejudiced.

The rules regarding production in paying quantities applicable to oil and gas leases do not apply to the reservation contained in the warranty deed. We adopt the Kuntz rationale expressed in *McEvoy,* and find that in this case, production means actual enjoyment of tangible economic benefits which result from marketing.

QUESTION ANSWERED.

DOOLIN, V.C.J., and HODGES, LAVENDER, HARGRAVE, WILSON and SUMMERS, JJ., concur.

SIMMS, C.J., and OPALA, J., dissent.

NOTES

1. LUDWIG V. WILLIAM K. WARREN FOUNDATION, 809 P.2d 660, 112 O.&G.R. 491 (Okl. 1990). Grantors of a [one-half] mineral interest "for a period of 20 years from December 22, 1961 and as long thereafter as oil and/or gas or other minerals [were] produced from the land described" brought suit to quiet title to the interest. The question for decision was stated to be: "whether a defeasible term mineral interest (as distinguished from a lease), expires when the only producing well on the property ceases production after the twenty year period set forth in the mineral deed and a new well is drilled and production thereafter re-established." The court held that it did. The conveyance by the grantor in this case created a "conditional limitation." The grantor "retains an interest and therefore termination of the conditional limitation is not a forfeiture." The lease cases illustrate the "concept that enforcing a defeasibility provision in a conveyance is a forfeiture which should be avoided. The lessee has spent heavily obtaining production from the lease he purchased. [This] large expenditure in acquiring production is one reason not to cancel his right to produce, for he should be allowed every chance to profit from what is, at the best of times, a risk and oftentimes simply a gamble."

Dissenting justices could not "agree that rules applicable to leases which are designed to enhance and encourage the profitable production of minerals do not apply with the same force of logic to defeasible mineral deeds. If the rules are fair and beneficial in the former case, why would they not be fair and beneficial in the latter?" Is the majority reasoning an answer to this argument?

2. Would you advise a client who is purchasing a term interest to include a provision such as the following in the term deed?

If at the expiration of the primary term hereof, there is a valid, recorded lease covering this interest, this interest shall in no event terminate while said lease remains in effect. This provision is cumulative of all other provisions herein.

This clause was suggested by Professor Masterson in Discussion Notes, 1 O.&G.R. 1717.

As attorney for the seller, would you resist inclusion of such a provision?

Can you devise a clause that gives even greater protection to the purchaser of a defeasible term interest?

———

SECTION 7. SPECIAL PROBLEMS RELATING TO THE RULE AGAINST PERPETUITIES

It is clear that the framework of the law of oil and gas is property law. One of the fundamental doctrines applicable to the conveyancing of interests in real property is the common law Rule Against Perpetuities. As memorialized by

Professor John Chipman Gray, the Rule is often stated in the following form: "No interest is good unless it must vest, if at all, not later than twenty-one years after some life in being at the creation of the interest." John Gray. The Rule Against Perpetuities 201 (4th ed. 1952). The Rule is, in property law terms, a rule of recent origin, having developed in the English courts beginning at the end of the 17th century. The Rule has been modified in a number of states, but its vitality continues as will be shown by the following cases involving the application of the Rule to oil and gas conveyancing issues. As you read these cases consider the following excerpt written by Professor Howard Williams, the former co-author of this casebook:

> When, however, the Rule is transposed from its original setting and is placed among property interests almost unknown until the turn of this century, it must be justified by new conditions before being permitted to annihilate the considered transactions of landowners and businessmen. In the words of Professor Leach . . . "Excessively long family settlements were the threat which produced the Rule; and the period of perpetuities was tailored to fit the needs of family gift transactions. To derive from a rule that motivated a general concept applicable to commercial transactions is a step of doubtful wisdom. Lives in being have no significance in commercial transactions, nor has the period of twenty-one years." Martin & Kramer, *Williams & Meyers Oil and Gas Law* § 325 at 64-64.1.

But as the following case suggests, the oil and gas conveyancing attorney ignores the Rule at one's peril.

––––––

TEMPLE HOYNE BUELL FOUNDATION v. HOLLAND & HART

Colorado Court of Appeals, 1992.
851 P.2d 192, 124 O.&G.R. 1 (cert. denied)

SMITH, JUDGE. In this legal malpractice action, defendants, Holland & Hart and Bruce Buell, appeal the final judgment entered on a jury verdict awarding plaintiffs, Buell Development Corporation, $3,364,011 in damages and $2,125,195 in pre- judgment interest. Plaintiffs cross-appeal the portion of the trial court's judgment denying various costs. We reverse the judgment and remand for a new trial.

The judgment at issue here arose from defendants' representation of plaintiffs in connection with the sale of stock in Kings County Development Corporation (KCDC), a California corporation, to John Rocovich. As part of the transaction, defendants drafted an option contract which provided that part of the consideration for the stock sale was an option in favor of plaintiffs to acquire from Rocovich a percentage of KCDC's minerals underlying its California farmland should KCDC ever distribute these interests to its shareholders.

After this contract was signed, KCDC instituted two lawsuits against various defendants, including plaintiffs and Rocovich. The lawsuit asserted, among other things, that plaintiffs and Rocovich had breached their fiduciary duties by entering into the foregoing stock transaction.

In 1982, KCDC settled with Rocovich, and, as part of that settlement, Rocovich conveyed all of his KCDC stock to the corporation. KCDC, meanwhile, decided to distribute its mineral interest to the corporation's shareholders within the year 1982.

While the KCDC lawsuits were pending, plaintiffs attempted to exercise their option under the contract executed with Rocovich. KCDC refused to honor the agreement between plaintiffs and Rocovich and to transfer the mineral interests, asserting that the option violated the Rule against Perpetuities and was therefore unenforceable.

In 1984, plaintiffs, who had discharged defendants as counsel in 1981, during the KCDC lawsuits, settled their dispute with the corporation. Under the terms of this settlement, plaintiffs received one-half of the mineral interests that they would have been entitled to under the option.

In 1989, plaintiffs initiated this lawsuit against defendants, asserting claims of negligence and breach of contract relating to the option contract with Rocovich. Soon after the litigation commenced, defendants sought a determination from the trial court that the option did not violate the Rule against Perpetuities.

At the evidentiary hearing on this issue, defendants attempted to establish that the option was exclusively contractual in nature and, thus, outside the Rule. The trial court disagreed, concluding that the option was subject to, and in violation of, the Rule. Moreover, the court ruled that there were no public policy reasons to allow the enforcement of the option in spite of the violation. Finally, the trial court ruled that no testimony on this particular issue would be admissible at trial.

Numerous expert witnesses for both parties testified at trial concerning the effects of the Rule against Perpetuities violation as it related to the rights of the parties under the option contract. At the conclusion of the evidence, the jury was instructed that plaintiffs had to prove two conjectural "cases within a case": (1) That plaintiffs would have lost a lawsuit brought to enforce the option as drafted by defendants; and (2) that plaintiffs would have won a lawsuit to enforce the option if defendants had not drafted it negligently.

The jury returned a verdict that plaintiffs' losses were caused solely by defendants' negligence.

I.

Defendants contend that the jury verdict in favor of plaintiffs on their claim for legal malpractice is erroneous as a matter of law. The crux of defendants' argument is the trial court's pre-trial ruling that the option contract violated the Rule against Perpetuities. Defendants argue that the trial court's ruling was erroneous, tainted the entire trial, and justifies entry of judgment in their favor. We agree that the trial court's ruling was in error and that such error did, indeed, pervade the trial proceedings. We disagree, however, that these conclusions dispose of plaintiffs' claim of legal malpractice.

As relevant here, the option provided:

In further consideration of the mutual promises of the parties as herein provided, we agree as follows with respect to the mineral interests owned by [KCDC]:

....

2. I [Rocovich] will use my best efforts to have the mineral interests ... distributed or made available to shareholders of [KCDC] as soon as possible ...

3. From and after the time of distribution, plaintiffs will have six months to purchase, at its option, 31.73576% of the mineral interests now owned by [KCDC] ..., and I [Rocovich] agree to sell to plaintiffs said interest for the sum of $305,500; if plaintiffs should not exercise its option, I [Rocovich] will, nevertheless, have the option, at any time during said six-month period, to offer plaintiffs the said 31.73576% of [KCDC] mineral interests which plaintiffs shall be required to purchase for the sum of $305,500.00.

....

This agreement will bind and enure to the benefit of the heirs, successors and assigns of the respective parties. (emphasis added)

A.

The Rule Issue

The Rule against Perpetuities is a rule of property law, the fundamental purpose of which is to keep property "unfettered," that is, free from inconvenient limitations. Restatement of Law of Property at 2129 (Introductory Note); see also Leach, Perpetuities in a Nutshell, 51 Harv.L.Rev 638 (1938). As such, the Rule operates to invalidate future interests in property, real or personal, legal or equitable, which vest too remotely, specifically, "later than twenty-one years after some life in being at the creation of the interest."

An option to purchase, such as that at issue here, may or may not be subject to the Rule. See 5A R. Powell & P. Rohan, Powell on Real Property § 767B (1992). If the option may continue for a period longer than 21 years and it creates an interest in a specific parcel of land or in any "specific identifiable thing," the option is required to comply with the Rule. See Restatement of Law of Property § 393 comment c, and § 401 comment b (1944). In such instances, the option, because it is specifically enforceable, "fetters" the property which is the subject matter of the option by imposing upon it an unfulfilled condition precedent for too long a time.

If the option is determined not to create an enforceable interest in a specific parcel of land or in any "specific identifiable thing," then no specific property is "fettered" or chargeable with performance under the option. See Restatement of Law of Property § 401. Nor is any particular property subject to a potentially objectionable unfulfilled condition precedent. Such option, therefore, while creating enforceable contract rights between the parties, "involves no fettering of any property and hence, [presents] no occasion for applying the Rule." Restatement of Law of Property § 401.

In analyzing the option here, the trial court rejected defendants' argument that the option was of this latter character, that is, "exclusively contractual." The trial court concluded, instead, that the option did implicate a specific property interest thereby subjecting it to the Rule. The crux of the trial court's conclusion was its determination that:

> The [option] does concern specific land. The [option] concerns the mineral interest which was then owned by [KCDC] ... A clearly identifiable mineral interest is involved in the [option] which can be stated with particularity and identified at the time of the [option is execution.] The [option] binds the parties to obligations regarding a particular piece of property to the exclusion of any other property....

Having found the Rule applied, the trial court turned to the question of whether the option was in violation of the Rule's terms. The trial court answered the question in the affirmative, noting that when the mineral interests would be distributed was uncertain and, hence, that "there are numerous possibilities which would result in plaintiffs' interest not vesting within the period of 21 years after Rocovich [the measuring life] dies."

Defendants contend that the uncertainty created by the unknown date of the mineral distribution, if any, was an irrelevant circumstance because the Rule had no applicability to the option. The essence of defendants' argument is, thus, that the trial court erred in its initial conclusion that the option concerned specific property.

A cursory review of the terms of the option reveals that the parties clearly identified a specific property interest as the subject matter of the option. However, based on the following principles of property and corporate law, we conclude that the terms were insufficient to create an interest in the property described.

It is fundamental that a grantor can convey no more rights in property than he himself owns. Accordingly, for the option here to have created an interest in a specific property, to wit, the mineral property identified in the option, Rocovich, the "grantor," would have to have had, at the time the option was executed, an interest, legal or equitable, in this property.

The record discloses, however, that, at the pertinent time, Rocovich's status was only that of a shareholder in KCDC. As a shareholder, Rocovich was entitled to an interest in the minerals only if and when they were distributed. However, until such time, he held no interest in specific property of the corporation. Thus, when the parties executed their contract, Rocovich's rights and interests, while, indeed, valuable and enforceable, were in KCDC's earnings and profits not in its divisible assets, such as the mineral interests.

At most, Rocovich held only a hope or chance of acquiring specific corporate property. Neither a hope or a chance, however, are sufficient to create an interest in property.

Hence, until the mineral interests were distributed, Rocovich had no rights in the mineral interests which he could convey or "fetter." The option contract could not, thus, subject this particular property to an unfulfilled condition precedent. Indeed, KCDC was equally free, both before and after the parties executed the option, to dispose of the mineral interests at issue in any particular manner it chose and to any particular person or entity.

The option contract, in sum, involved no fettering of a specific parcel of property or of any specific identifiable thing and, hence, presented no opportunity to invoke the Rule.

This is not to say that the option was unenforceable. Indeed, between plaintiffs and Rocovich, the contract created rights and obligations enforceable by one against the other, their heirs, successors, and assigns. Of critical significance, however, the contract, in the words of defendants' expert, did "nothing whatsoever to the subject matter of their contract-- property owned by a stranger to [that] contract."

The contract here is not subject to such a contingency. No interest in the minerals could arise in Rocovich or in plaintiffs unless and until the mineral interests were distributed by KCDC to its shareholders. Once the interest was created by distribution, moreover, the option to purchase it was to be exercised promptly, specifically, six months "from and after the time of distribution" or it would lapse. Hence, unlike the situation in Prime, there was no "mathematical possibility" that once created, the interest would vest beyond the period of perpetuities.

Next, plaintiffs argue that Rocovich's sale of his stock to KCDC as part of their settlement agreement "created" an interest in the minerals which triggered Rocovich's obligations under the contract and the applicability of the Rule. This argument is without merit.

Rocovich could assign his obligations to KCDC only if such assignment would not impair plaintiffs' rights. See Restatement (Second) of Contracts § 317(2) (1981). Such is clearly not the case here where the assignment, in triggering the Rule, would operate to destroy plaintiffs' rights. In conclusion, we hold that the trial court erred in ruling that the option contract violated the Rule against Perpetuities.

B.

Effect of the Erroneous Ruling

Defendants argue that the trial court's erroneous pre-trial ruling tainted the entire trial and thus constituted prejudicial error. Under the particular circumstances here, we agree.

The record reveals that, prior to trial, the trial court ruled that all testimony by defendants' experts or other witnesses concerning whether the contract violated the Rule against Perpetuities was inadmissible. Moreover, at trial the jury was instructed that:

> [Plaintiffs] must prove that if had it sued to enforce the [option] it would have lost that court action solely because of defendant's negligence....

Later, in instructing the jury as to the plaintiffs' burden in this regard, the trial court further instructed the jury:

> The court has ruled as a matter of law that the [option] violated the Rule against Perpetuities and there is no public policy reason for not applying the rule.

This latter instruction states, unequivocally, that, in drafting the option contract, defendants committed an error. When read in conjunction with the former instruction, these directives indicate that the error is significant in proving the plaintiffs' "case within a case." Moreover, the transcript discloses that extensive reference was made to the option's "unenforceability" by both the trial court and plaintiffs' counsel. Indeed, as pointed out by defendants in their brief, mention of this matter was made "10 times by the court, 56 times by the [plaintiffs'] counsel and 7 times by witnesses."

Although oftentimes these references were couched in the broader claim that defendants had not been "reasonably careful" in drafting the option, the trial court's erroneous ruling, nonetheless, as evidenced above, pervaded the trial proceedings and, most critically, influenced the instructions given to the jury. Accordingly, we conclude that defendants are entitled to a new trial on plaintiffs' claim of legal malpractice.

Defendants argue, however, that if the option is valid and enforceable, there can be no malpractice claim as a matter of law, thus entitling them to entry of judgment in their favor. We disagree.

An attorney owes a duty to his client to employ that degree of knowledge, skill, and judgment ordinarily possessed by members of the legal profession in carrying out the services for his client. One of these obligations is anticipating reasonably foreseeable risks.

Thus, although we hold here that the option did not violate the Rule against Perpetuities, the question remains whether defendants, as reasonably prudent attorneys, should have foreseen that the option, as drafted, was likely to result in litigation and whether other attorneys, in similar circumstances, would have taken steps to prevent such a result. Plaintiffs argued at trial, and presented expert testimony in support of their assertion, that the principal negligence of defendants was their not protecting plaintiffs from loss by failing to research and analyze the Rule's applicability in the option, to recognize the likelihood that a good faith dispute could occur over the enforceability of the option because of the Rule, and to take the simple step of either adding a time limitation or "savings clause" or recommending the deletion of the provision that made the option binding on heirs, successors, and assigns.

Bruce Buell testified that he had given no specific consideration to the Rule in drafting the Letter Agreement. Nor did he perform any legal research, consider the choice of law, consult with experts, or even consult with other members of his own firm on the question of whether the Rule could apply to the option. He concluded that he had no duty to do any of these things because he "was an experienced business transactions lawyer" who "knew the rule against perpetuities" and "knew when it applied," and could "spot issues like that." As a result, plaintiffs argue, Buell did not advise his clients of the real likelihood that a good faith dispute could arise over the enforceability of the option under the Rule.

On the issue of defendant's negligence, one expert attorney- witness testified that a reasonable attorney would have no reason to include a savings clause in the option and concluded that defendants met the standard of care in drafting it. However, conflicting testimony was offered by two other attorney-experts who testified that defendants had failed to meet the standard of care and should have considered the possibility that the Rule might apply to the option and should have protected it against a Rule challenge.

Perhaps the strongest testimony on this issue was that of defendant's own attorney-expert. On cross-examination, he unequivocally testified that, in his opinion, an attorney would be guilty of malpractice if he: (a) did not research the issue of the Rule in the context of this transaction, (b) failed to consider the potential for a dispute over the applicability of the Rule to the option, and (c) failed to utilize a savings clause to protect against that potential dispute.

Thus, although there was a conflict in the expert testimony as to attorney negligence in drafting the option, no witness disagreed with the premise that the option would have been protected from any Rule dispute if defendants had considered the Rule, had recognized the clear potential for dispute, and had either included a savings clause or excluded the language making the option binding on heirs, successors, and assigns.

In short, resolution of the Rule of Perpetuities issue does not conclusively resolve the issue of whether defendants met the applicable standard of care in preparing the option contract. Hence, we reject defendant's argument that judgment be entered in their favor.

The judgment is reversed, and the cause is remanded for a new trial on plaintiffs' claims for legal malpractice consistent with the views expressed in this opinion.

NOTES

1. What was the basis for the court's remand of the malpractice issue? If the interest did not violate the Rule what did the attorney do that justified a second trial?

2. In general, options relating to property interests may create Rule problems if they "fetter" the property interest for a period in excess of the Rule? *See* generally, 6 *American Law of Property* § 24.55; Richard Powell & Patrick Rohan, *Powell on Real Property* 767B (1996).

3. Was this the type of commercial transaction that Professor Williams would exempt from the application of the Rule? Did it create a remotely vesting future interest in the mineral estate? Did it affect the alienability of either the corporate stock or the mineral estate?

4. Many operating agreements include "preferential rights of purchase" provisions giving the holder the right to match third-party offers to purchase interests subject to the operating agreement? Would those violate the Rule? See cases cited in Martin & Kramer, *Williams & Meyers Oil and Gas Law* § 322 nn. 16-17 (2006). *See also* Reasoner, *Preferential Purchase Rights in Oil and Gas Instruments*, 46 Tex. L.Rev. 57 (1967).

5. WHITING OIL & GAS CORP. V. ATLANTIC RICHFIELD CO., 2010 Colo.App. LEXIS 1223. In *Whiting,* the party subject to an option to sell certain mineral interests argues that the option violates the Rule Against Perpetuities. Colorado, in 1991, joined an increasing number of states that have adopted the Uniform Statutory Rule Against Perpetuities that was developed by the Uniform Commissioners on State Laws. The Colorado statute (Colo.Rev.Stat. §§ 15-11-1101 to 1107 is generally prospective in nature but gives a trial court the power to reform a deed that pre-dates the effective date of the Act so that it will not violate the common law Rule. In *Whiting,* the court reformed the option to last only for 21 years past the death of the president of the corporation that held the option and since the option was exercised within that period it was upheld. The Colorado statute abolishes the common law Rule for transactions that occur after its effective date. At the time of the publication of this edition, the Court of Appeals decision had not yet been finalized nor had the time for an appeal to the Colorado Supreme Court passed.

LATHROP v. EYESTONE

Kansas Supreme Court, 1951.
170 Kan. 419, 227 P.2d 136.

WEDELL, JUSTICE. This was an action against various defendants to quiet title to certain lands.

The action was tried by the court on an agreed statement of facts. The only defendants now involved are Standish Hall, trustee, and The First National Bank in Wichita, a corporation, trustee. The title of the plaintiff was quieted as to all defendants except the two last mentioned. Judgment was rendered in favor of those two defendants pursuant to their respective claims based on certain written instruments executed and delivered by former owners of the fee title, and plaintiff, the present owner of the fee title, appeals.

It is agreed the plaintiff, Charles F. Lathrop, is the present owner of the fee title to the lands involved subject to such interest as the two mentioned defendants may have in the lands, if any.

The primary question presented is the nature, character and effect of the written instruments pleaded and relied on by these two appellees. For convenience appellees will be referred to as Hall and as the bank. Although neither Hall nor the bank was a grantee in the original instruments involved it is admitted they are now trustees of whatever right, title or interest may have been granted thereby. We shall first set forth the pertinent parts of the two conveyances on which Hall relies. One of them reads:

This agreement made and entered into this 22nd day of April, 1919, by and between Frank E. Eyestone and Mamie E. Eyestone, his wife, assignors, parties of the first part and The Guarantee Title and Trust Company, assignee, party of the second part.

Witnesseth, that whereas, on the 17th day of Sept. 1918, a certain oil and gas lease was made and entered into by and between the said Frank E. Eyestone and Mamie E. Eyestone his wife, lessors, and J. A. Crawford, lessee, wherein said lessors leased to said lessee for himself his heirs and assigns for oil and gas mining purposes for the terms designated therein and under the conditions thereof, the following described land located in Butler County, Kansas, to-wit: [the legal description] and which said lease is recorded in Book Misc. 41 page 53 in the office of the Register of Deeds of Butler County, Kansas, and provides that the lessee his heirs, and assigns, shall under the conditions of the premises depart of all oil produced and saved from said leased **139 land and shall pay to the lessor, his heirs or assigns, one-eighth (1/8) of all gas produced and saved from said leased land or in lieu thereof said one-eighth (1/8) of the gas, $200.00 for each gas well drilled thereon where gas only is found while the same is being used off the premises.

[Here appears paragraph of assignment to The National Refining Company.]

Now, Therefore, Be It Remembered, That, we Frank E. Eyestone and Mamie E. Eyestone, his wife, the undersigned, in and for the consideration of One Dollar ($1.00) and other valuable consideration to us in hand paid by The Guarantee Title & Trust Company, the receipt of which is hereby acknowledged have sold and by these presents do sell, assign and deliver to the said Guarantee Title and Trust Company, their successors and assigns, an undivided one-fourth (1/4) interest in and to all of the oil and gas right, title and claim, we now own or which we may hereafter be entitled under and by virtue of the above and described aforesaid lease or any oil and gas lease existing or which may hereafter exist upon the above described land or any part thereof, including all of the oil and gas, rent and royalties now accrued or accrue hereafter, also a perpetual and irrevocable right, privilege and license to enter upon said land or any part thereof and prospect for and drill wells for oil and gas therein or thereon, and also the right to use and the possession of so much of said premises as may be necessary to enable the assignee herein, their successors or assigns, to carry out the purposes and provisions of this grant, provided that this clause does not in any way interfere with the lease on the above described premises.

Provided, That the grantors herein their heirs or assigns upon payment to the grantee herein; its successors or assigns of an amount equal to three-fourths (3/4) of the expense and cost of producing and disposing of such oil and gas, shall thereupon forthwith be entitled to and shall receive from said grantee, their successors or assigns, an equal amount to three-fourths (3/4) of the net profits arising from the sale and disposition of said oil and gas as aforesaid, but in any event the grantor is not to be held for any expense in the last clause, unless it is taken from the oil and gas produced on said premises.

The other instrument relied on by Hall, dated May 5, 1919, reads in part:

Assignment of Royalty.

Know All Men by These Presents: That Ethel Johnson, a single woman, and George Johnson, a widower, of the first part, in consideration of One Dollar, and other valuable considerations, the receipt of which is hereby acknowledged, have granted, bargained, sold and conveyed and do by these presents grant, bargain, sell, and convey unto W. H. Stanley of Wichita, Kansas, party of the second part, an undivided one-fourth (1/4) in a certain oil and gas mining lease, executed by said Ethel Johnson, a single woman, and George Johnson, a widower, to E. E. Johnson, [the legal description] containing eighty acres, more or less according to government survey, which lease is dated October 30th, 1918; also

An undivided one-fourth (1/4) interest in any and all bonuses received for oil and gas lease or leases hereafter executed by first parties, or their assigns, upon said real estate, or any part thereof, and one-fourth (1/4) of the oil and gas royalties reserved to the lessors in any such lease or leases hereafter executed; to being expressly understood and agreed that when said present oil and gas lease terminates parties of the first part, or their assigns, shall have the full and exclusive right to execute another oil and gas lease or leases as herein provided; which one-fourth (1/4) of the royalties shall not be less than one-sixteenth (1/16) of the production of the lease.

The pertinent part of the conveyance on which the bank relies reads:

Declaration of Trust.

This Agreement made and entered into this 12th day of June, 1919, between Frank E. Eyestone and Mamie Eyestone, his wife, parties of the first part and The Prudential Trust Co. a corporation, organized under the laws of the State of Kansas, party of the second part, hereinafter called the Trustee.

Witnesseth, That the parties of the first part in consideration of One Dollar to them paid by the second party, the receipt whereof is hereby acknowledged, do by these presents, grant, bargain, sell and convey to the party of the second part an undivided one-half in a one-eighth oil and gas royalty reserved to them on the North Half of the Southwest Quarter of Section 8, and the Southwest Quarter of the Northwest Quarter of Section 8, in Township 23, Range 4 East of the 6th Principal Meridian, in Butler County, Kansas, in a certain oil and gas mining lease, executed by the parties of the first part to The National Refining Co., and now of record in the office of the Register of Deeds of Butler County, Kansas; also an undivided one-half interest in any and all bonuses received from any oil or gas lease or leases hereafter executed by the parties of the first part or their assigns upon said real estate or any part thereof and one-half of the oil and gas, reserved to the lessors in any such lease or leases hereafter executed; it being expressly understood and agreed that when the said oil and gas lease terminates, the parties of the first part or their assigns shall have full and exclusive right to execute another oil and gas lease or leases on said premises, subject only to the rights of the second party to whom half of the bonus received and an oil and gas royalty of one- sixteenth of the oil and gas produced and saved from said premises.. .."

As we have frequently stated the term 'royalty' is often rather loosely and inaccurately used by men in the petroleum industry, those dealing in oil and gas holdings and at times by attorneys. Some persons refer to oil and gas in place as royalty. Others refer to royalty as the landowner's share in production. We have, therefore, repeatedly held the true nature and character of the instrument is not to be determined by the name or label attached thereto but by its intent as reflected by the terms, the contents thereof.

A mineral deed is one which makes a severance, from the fee, of a present title to minerals in place. It is actually a realty conveyance. On the other hand 'royalty' in its ordinary meaning is that part of oil and gas payable to the lessor by the lessee out of oil and gas actually produced and saved. It is the compensation to the lessor provided in the lease for the lessee's privilege of drilling and producing oil or gas. It does not include a perpetual interest in and to the oil and gas in place. It is not uncommon to find 'royalty' shortly defined as 'a share' in production 'paid'. It is personal property. The lessee's interest in the oil produced, commonly seven-eighths or whatever the lease provides, is called the working interest.

The cardinal principle or test to be applied in the interpretation of such instruments, as in others, is the intention of the parties. Although all parts of the instrument are to be considered the granting clause is, of course, paramount in determining what interest was intended to be granted.

We now reach the appellant landowner's next contention. It relates to the provisions in the instruments which purport to grant an interest in royalties, rents and bonuses on other oil and gas leases which might be executed in the future. Appellant contends those covenants are personal covenants of the prior owners of the land, that they do not run with the land and hence do not bind him, a subsequent owner of the fee title.

As previously stated it is stipulated the lands described in all three conveyances are now free and clear of all oil and gas leases and there is no production on any of them. Appellant concedes such covenants, even as to future leases, would be binding on the owners of the land who made them but contends they are not binding on him, a subsequent fee title owner. It has been held an assignment by the fee title owner of an interest in royalties, rents and bonuses that accrue under an existing lease, and a lease which 'might be executed' in the future is valid between the parties thereto, both as to the existing lease and a subsequent lease where made by the same landowners. Miller v. Sooy, 120 Kan. 81, 242 P. 140. In the same case it was held that construing the instrument as being binding only with respect to the leases executed by the then landowners, and not as running with the land, the instrument was not subject to attack on the ground it violated the rule against perpetuities.

The first instrument in the instant case does not in its terms bind the 'grantors, their successors and assigns.' Although the granting clause, in both the second and third instruments, does not provide that 'We, the undersigned, for ourselves, our heirs and assigns' grant, etc., there is language in each of those two instruments which makes it appear it may have been intended to convey such interests under subsequent leases executed by the grantors, 'or their assigns.'

We need not determine whether these instruments, or any of them, were intended to be binding on subsequent fee title owners. If such was the intention when would the grant of such future interests vest? Appellant or future fee owners might never execute another lease. There is nothing in any of the instruments which imposes a duty on them to do so. Under the last two instruments, at least, the fee title owner would not be precluded from doing his own developing. Moreover there is no limitation of time within which a future lease would be required to be executed, if one were actually executed. It is, therefore, wholly problematical when, if ever, such an interest under future leases would vest. Such a grant violates the rule against perpetuities, a rule against too remote vesting. In 41 Am.Jur., Perpetuities and Restraints on Alienation, s 24, it is said: 'One of the essential elements of the rule against perpetuities is that at the time the future interest is created, it must appear that the condition precedent to vesting must necessarily happen, if it happens at all, within the period prescribed by the rule. . . . A possibility, or even a probability, that the interest or estate may vest within that time is not enough, for, it is said, the question of probabilities does not enter into the equation. If by any conceivable combination of circumstances it is possible that the event upon which the estate or interest is limited may not occur within the period of the rule, or if there is left any room for uncertainty or doubt on the point, the limitation is void.'

The foregoing statement constitutes the well recognized rule which is in harmony with our own decisions. The trial court properly held appellant's title should not be quieted insofar as an interest in and to oil and gas in place, granted by the first instrument, is concerned, but it erred in refusing to quiet appellant's title as against rights claimed under the second and third instruments. The judgment is, therefore, affirmed in part and reversed in part and remanded to the district court with directions to enter judgment in harmony with the views herein expressed.

NOTES

1. Oscar is the fee simple absolute owner of Blackacre. He executes a lease to Raider Oil reserving a 1/8th royalty. Shortly thereafter Oscar transfer to Alexis all of his interests except for the following reservation:

> It is particularly agreed, and this conveyance is made subject thereto, and said reservation and the terms and stipulations hereof relative thereto, that Alexis, its successors and assigns shall receive and be entitled to one-half the royalties payable thereunder, and all of the reversionary rights in the minerals, except there is hereby reserved to Oscar, one-half (1/2) of all royalties accruing and/or payable under the existingn leases, and, in the event of the termination, forfeiture or expiration of said leases, as and when same may, respectively, so terminate, forfeit, or expire, a perpetual non-participating free royalty interest in and too all the minerals, in, upon or under the lands conveyed hereby as follows: 1/16th on oil and gas.

Does the reserved royalty interest that burdens future production violate the Rule? Are those royalties in future production vested interests or do they vest at some unknown time in the future when the initial leases terminate? Could the interest be created in Kansas? The language above comes from a deed that was interpreted by a Texas court in Hamman v. Bright & Co., 924 S.W.2d 168, 174, 138 O.&G.R. 141 (Tex.App. 1996), *judg. vacated in aid of settlement*, 938 S.W.2d 718, 138 O.&G.R. 161 (Tex. 1997). The court of appeals decision concluded that the royalty interest was vested and not subject to the Rule. In accord: Hanson v. Ware, 224 Ark. 430, 274 S.W.2d 359, 4 O.&G.R. 325

(1955); Dauphin Island Property Owners Ass'n v. Callon Institutional Royalty Investors I, 519 So.2d 948, 97 O.&G.R. 455 (1988).

2. Kansas has adopted the Uniform Statutory Rule Against Perpetuities, Kan.Stat.Ann. §§ 59-3401, that changes the time period in which an interest must vest or fail to vest. Nonetheless, at least one Kansas appellate court decision has followed the *Lathrop* analysis rather than applying the statutory period. Fritschen v. Wanek, 22 Kan.App.2d 927, 924 P.2d 1288, 135 O.&G.R. 247 (1996).

3. Can the *Lathrop* analysis also be applied to a severed executive power? Since the power can be exercised for an indefinite period does it violate the Rule? Is it a vested interest? An early California decision suggests that the executive power of unlimited duration would violate the rule. Dallapi v. Campbell, 114 P.2d 646 (1941). But a later decision dealing with a royalty interest held in dicta that the executive power was vested at the time it was created and therefore not subject to the Rule. Keville v. Hollister Co., 105 Cal.Rptr. 238, 43 O.&G.R. 311 (1972). Should royalty interests and the executive power be treated the same?

PEVETO v. STARKEY

Supreme Court of Texas, 1982.
645 S.W.2d 770, 75 O.&G.R. 166.

SONDOCK, JUSTICE. This is a suit for a declaratory judgment. Starkey seeks to have a term royalty interest conveyed to Peveto declared void. He also seeks to have a term royalty interest in the same property conveyed to him declared valid. The trial court rendered judgment for Starkey, and the court of appeals affirmed the trial court judgment. We reverse the judgments of the courts below.

A.G. Jones and his wife conveyed an undivided three-fourths term royalty interest in several tracts of land to Peveto. The primary term of the royalty deed was "for a period of fifteen years" from April 23, 1960, and "as long thereafter as oil, gas or other minerals, or either of them is produced ... in paying commercial quantities." Jones then executed an oil and gas lease to Edge and Moehlman. The lease contained a shut-in royalty clause, but the term royalty deed did not. Thirteen years later, Peveto executed an instrument entitled "Ratification of Oil, Gas and Mineral Lease," which purported to ratify the lease executed by Jones to Edge and Moehlman. This was apparently a unilateral attempt by Peveto to acquire the benefits of the shut-in royalty clause in the Edge and Moehlman lease. Subsequently, Edge and Moehlman completed a test gas well, but due to the high sulphur content of the gas produced, the well was shut-in. Edge and Moehlman made shut-in royalty payments to Jones and Peveto for the duration of the term royalty deed.

In November of 1973, Jones conveyed to Starkey, by top deed, a three-fourths term royalty interest. This royalty deed was for a primary term of ten years and "as long thereafter as oil, gas or other minerals, or either of them, is produced ... in paying commercial quantities." The deed contained the following typed provision:

This grant shall become effective only on the expiration of the above described Royalty Deed to R.L. Peveyto [sic] dated April 23, 1960.

Four months before Peveto's deed was to expire, Jones and Peveto executed another instrument. This instrument purported to extend the primary term of Peveto's deed from fifteen years to twenty-five years.

Peveto first contends his term royalty deed was extended into its secondary term by the payment of shut-in royalties by Edge and Moehlman. We disagree. The payment of shut-in royalties does not constitute "production in paying commercial quantities." "It is now well established that the completion of a gas well capable of producing in paying quantities but shut-in due to lack of pipe line facilities or for other reasons is not considered production, or production in paying quantities, under the provisions of a term royalty deed which contains no shut-in gas well clause."

The shut-in clause contained in the lease did not modify the terms of the royalty deed. Because the royalty deed did not contain a shut-in royalty clause, it terminated at the expiration of the primary term. "It is the mineral deed, not the lease, which [must contain] the provision securing to the term mineral owners the benefits of the shut-in gas well provision."

Peveto further contends his "ratification" of the Edge-Moehlman lease enabled him to rely on the shut-in royalty clause in that lease and extended his deed past the primary term. We do not agree. The payment of shut-in royalties to those claiming under a royalty deed which contains no shut-in payment clause will not extend the interest beyond the primary term. This is so even if the royalty owners have executed an instrument ratifying the lease on the property.

Peveto next argues the royalty deed to Starkey violates the Rule against Perpetuities. Article I, section 26 of the Texas Constitution expressly provides: "Perpetuities ... are contrary to the genius of a free government and shall never be allowed ...." Tex.Const. Art. I, s 26. The Rule states that no interest is valid unless it must vest, if at all, within twenty-one years after the death of some life or lives in being at the time of the conveyance. The Rule requires that a challenged conveyance be viewed as of the date the instrument is executed, and it is void if by any possible contingency the grant or devise could violate the Rule.

The deed from Jones conveying the term royalty interest to Starkey was a standard form nonparticipating royalty deed. The printed portion of the granting clause conveyed a presently vested three-fourths royalty interest. However, following the property description, the parties inserted: "this grant shall become effective only upon the expiration of [Peveto's] ... Deed ...." This additional clause causes the Jones-Starkey deed to violate the Rule. The interest Jones conveyed to Peveto by the first term royalty deed was a determinable fee. This Court defined a determinable fee to be "an interest which may continue forever, but the estate is liable to be determined, without the aid of a conveyance, by some act or event circumscribing its continuance or extent."

All parties agree the deed from Jones to Starkey is unambiguous. Thus, the intent of the parties must be determined from the four corners of the instrument. The rights of the parties are governed by the language used and the choice of designating words is of controlling importance. The words used here postpone the vesting of Starkey's interest until some uncertain future date. A grant "effective only upon" the termination of a determinable fee cannot vest until the prior interest has terminated. A determinable fee could continue forever, and may not terminate within the time period prescribed by the Rule. The words "effective only upon" created a springing executory interest in Starkey which may not vest within the period of the Rule; therefore, the deed is void.

Because the restrictive language used in the Jones-Starkey deed prevented the grant of the interest from Jones to Starkey from vesting in interest until after Peveto's interest terminated, and since this might not occur within the period prescribed by the Rule, we hold that the instrument violates the Rule against Perpetuities.

Accordingly, we reverse the judgment of the courts below. Judgment is here rendered that the deed from Jones to Starkey is void.

NOTES

1. A common practice in the oil and gas industry is for parties to take "top leases." A top lease has been defined as "A lease granted by a landowner during the existence of a recorded mineral lease which is to be become effective if and when the existing lease expires or is terminated." Martin & Kramer, *Williams & Meyers, Manual of Oil and Gas Terms,* 1130-32. If a court applied the analysis in *Peveto* would a top lease violate the Rule? Is the lessor as the owner of a possibility of reverter conveying a portion of that interest or is the lessor conveying a springing executory interest that will not vest or fail to vest within the period allowed by the Rule? In Hamman v. Bright & Co., 924 S.W.2d 168, 133 O.&G.R. 141 (Tex.App. 1996), *judg. vacated in aid of settlement,* 938 S.W.2d 718, 133 O.&G.R. 161 (Tex. 1997), the court of appeals concluded that the language of the top lease was directly analogous to the language of the top royalty deed in *Peveto.*

2. Can top leases avoid the application of the Rule by drafting provisions showing that the parties intended to transfer a portion of the possibility of reverter rather than a springing executory interest? For several drafting suggestions see Nelson Roach, *The Rule Against Perpetuities: The Validity of Oil and Gas Top Leases and Top Deeds in Texas After Peveto v. Starkey"* 35

Baylor L.Rev. 399 (1983); Max Ernest, *Top Leasing--Legality vs. Morality*, 26 Rocky Mtn. Min.L.Inst. 957 (1980).

3. Although, as noted above, top leasing is a common and longstanding practice, there are relatively few cases dealing with the Rule problem. Most top lease cases do not discuss the Rule. In Stoltz, Wagner & Brown v. Duncan, 417 F.Supp. 552, 55 O.&G.R. 315 (W.D.Okla. 1972), the courts applied the Rule, but also applied the *cy pres* doctrine in order to validate the top lease that had an unusual provision in that it authorized the top lessee to take possession not only at the termination of the base lease but also within one year from a fixed date. The court expressly noted that the portion of the top lease allowing the top lessee to take possession at the termination of the base lease would violate the Rule.

4. Many jurisdictions have modified the remorseless application of the Rule to invalidate future interests by adopting the "wait and see" rule. Under the "wait and see" rule the validity of the future interest is not determined at the time of its creation. Instead the court will see if the interest actually vests or fails to vest within the Rule period to determine its validity. Thus where a base lease terminated at the end of its primary term and within the Rule period, a top lease was upheld, even though at the time of the creation of the top lease the base lease could have extended for a period in excess of that allowed by the Rule. Nantt v. Puckett Energy Co., 382 N.W.2d 655, 89 O.&G.R. 122 (N.D. 1986).

5. As you will see in Chapter 6, Section 2, it was reasonably commonplace for deeds in Texas to have a clause describing the interest conveyed in the event that a lease in existence at the time of the conveyance was terminated or forfeited. In Bowers v. Taylor, 263 S.W.3d 260 (Tex.App. 2007), a deed contained the following language: ". . . if said lease should be forfeited than the one fortieth (1/40) royalty interest. . . shall be cancelled and in lieu thereof. . . the [grantee] is to become vested with one-third (1/3) interest in the fee title in to the oil, gas and minerals. . ." The grantor argues that the $1/3^{rd}$ mineral interest is a springing executory interest that is invalidated by the Rule Against Perpetuities. What result applying the *Peveto* rationale?

———

Chapter 6
TRANSFERS SUBSEQUENT TO A LEASE
SECTION 1. INTRODUCTION

In this chapter we are concerned with the operating and related consequences of transfers by the lessor or by the lessee. Attention is directed first to transfers by the lessor (Sections 2 and 3), and then to various consequences of transfers by the lessee, viz., the relationship of the transferor and transferee (Section 4), the relationship of the lessor and transferee (Section 5), and the relationship of the lessor and the lessee-transferor (Section 6).

Preliminary to this discussion of consequences and regulation of transfers, we pause to note the occasions for and the nature of the transfers which may be made by lessor and lessee.

A. TRANSFERS BY THE LESSOR

Earlier chapters of this book have indicated in general the nature of the transfers which may be made by a lessor and some of the occasions for such transfers. By voluntary or involuntary conveyance or by testate or intestate succession, the lessor's interest (or some portion of his interest) in leased premises may pass to others. Consider, for example, the owner of certain premises (Blackacre) or of minerals therein who has leased his interest to Lessee. Subsequent to such lease, the Lessor may convey:

(a) his entire interest in the leased premises (Blackacre) to one person; *e.g.,* Lessor may by deed convey Blackacre to B and his heirs;

(b) his entire interest in Blackacre to several persons as concurrent owners; *e.g.,* on Lessor's death intestate, his interest may pass to his children, B and C, as tenants in common;

(c) an undivided interest in Blackacre and the lease; *e.g.,* Lessor may convey to B an undivided one-half interest in Blackacre;

(d) his entire interest in a portion; *e.g.,* Lessor may convey the northwest one-quarter of Blackacre to B and his heirs;

(e) an interest in production under the existing lease; e.g., Lessor may convey one-half of his one-eighth royalty under the existing lease covering Blackacre, or an interest in production under the existing or any future lease, *e.g.,* Lessor may convey to B a $1/16^{th}$ royalty payable on all production from Blackacre, whether under the terms of the existing lease or under the terms of any future lease which may be executed covering Blackacre;

(f) his entire interest (or some part of his interest) in all of Blackacre or in some portion of Blackacre, excepting and reserving some share of production or some share of lease proceeds; *e.g.,* Lessor may convey the northwest quarter of

Blackacre, excepting and reserving one-half of the one-eighth royalty payable under the terms of the lease covering Blackacre insofar as it covers and includes the acreage conveyed.

This enumeration, although not exhaustive of all conveyances which may be made by a lessor subsequent to a lease, is sufficiently representative to show the variety of such conveyances. In Section 2 of this chapter, attention will be directed to some of the consequences of such transfers by the Lessor.

B. TRANSFERS BY THE LESSEE

Transfers which may be made by a lessee are even more varied than those which may be made by a lessor. It will suffice here to enumerate some of the circumstances that cause a transfer to be made and to indicate the types of transfers which may be made.

(1) The original lessee may have acquired the lease without any intention of conducting exploration and development operations on the land but with the hope of conveying the premises to another person willing to engage in such operations. A lease hound, for example, learning of proposed exploration operations in wildcat territory, may rush to the area with his pad of lease forms, purchased at a stationer's shop, and seek to obtain leases from owners of minerals in the same general area. His hope is to obtain leases which will increase in value by reason of success in the exploration operations proposed for the area. If such other operations are successful, the lease hound will then seek to find a buyer of his leases. Ordinarily the lease hound under these circumstances will seek to reserve from the conveyance some nonoperating interest, *e.g.,* an overriding royalty of 1/16th of the 7/8ths working interest.

(2) One lessee may have acquired a lease covering substantial acreage and may seek to have drilling operations conducted thereon by some other operator. Thus if lessee has a leasehold covering a section of land he may be willing to transfer the leasehold insofar as it covers and includes the northwest quarter-section to an operator who, (a) as a condition precedent to receiving the transfer, must drill a well thereon, or (b) in consideration of the transfer, will covenant to drill a well thereon. By such a transfer, the lessee may be able to satisfy the drilling requirements of his lease as to the entire section of land and avoid further payment of rentals on his retained acreage. Moreover by such operations he will acquire information useful to him in reaching a decision whether (or where) to drill upon his retained acreage. Or, a lessee who has offset obligations to satisfy but lacks an available rig, may agree to transfer his leasehold insofar as it covers the drill site of the required offset well in consideration of the drilling of the offset well. In an agreement of this kind, termed a *"farm-out agreement"*, the transferor may retain some interest in production from the assigned acreage, *e.g.,* an overriding royalty or a carried interest. Farmout agreements may contain "area of mutual interest" provisions that entitle the parties to the agreement to share in any other acquired leasehold or working interests within a defined area.

(3) The lessee who has obtained production in a shallow horizon may divide

the leasehold by a conveyance of the leasehold insofar as it covers specified strata or horizons or insofar as it covers strata or horizons above or below a specified depth. Thus if a lessee with substantial overhead costs has small production from shallow horizons and is fearful that such production might not be viewed as paying production for habendum clause purposes, he may assign the leasehold insofar as it covers the shallow horizon to another operator not burdened by overhead for whom the production would clearly be "paying" production. By this means he may preserve the lease while making plans for exploration in deeper strata.

(4) A lessee may find it necessary to raise money to finance proposed drilling operations, and hence may sell certain interests to investors in the enterprise. Thus he may convey nonoperating interests (*e.g.,* oil payments, overriding royalty interests, or net profits interests) to contributors of money or services to the enterprise.

(5) Financial backing for proposed drilling may be obtained from persons owning leasehold interests in neighboring premises. Thus if the several owners of leases on Blackacre, Whiteacre, and Greenacre are faced with the problem of drilling or losing their leases at the expiration of the primary term, they may reach an agreement for the drilling of an exploratory well by the lessee of Blackacre, with some contribution to the cost of the enterprise being made by the lessees of Whiteacre and Greenacre. Under these circumstances, the owners of Whiteacre and Greenacre may sign a so-called "*Letter Agreement*" which obligates them under specified circumstances to make a payment to the lessee of Blackacre. These letter agreements may be used by the lessee of Blackacre to obtain bank or other credit required for the financing of the proposed drilling operations. Examples of such agreements are included in the Appendix. A "*bottom-hole letter*" requires the payment of an agreed consideration upon the drilling of a well to a specified depth whereas a "*dry hole letter*" provides for the payment to be made if the well is a dry hole. These letter agreements may be further described as *donation letters* or as *purchase letters,* the difference being whether the contributor is entitled to receive some interest in the premises upon which the well is drilled (a purchase letter) or is not to receive such an interest (a donation letter). Yet another form of letter agreement, the "*acreage contribution letter*", provides for the contribution of acreage to the common enterprise.

(6) Owners of leasehold interests have sought to raise money for exploration and development operations by a variety of means. In some instances partnerships or mining partnerships have been entered into by contributors of lands, services, and money. In other cases, limited partnership interests or a variety of nonoperating or operating interests have been sold in financial centers to investors attracted by the tax advantages investments in the extractive industries were said to hold for persons in upper income tax brackets. These arrangements have been varied in character.

A number of other methods of financing may be employed for substantial projects, such as offshore exploration and development. These include *advance payment financing* (also called a *forward sale*) which is akin to a production payment in that for an advance of money to be used in exploration and development the developer agrees to deliver a portion of the oil and gas produced until such time as the cumulative value of the oil and gas so delivered equals some multiple (*e.g.,* 150%) of the sum advanced. This transaction may also afford the advancing party a preferential right to purchase oil or gas produced from the project. Another method may be described as *take-or-pay financing* which involves an agreement by a purchaser to take a minimum quantity of oil or gas over a specified term at a fixed price (or at a fluctuating price which cannot be reduced below a specified level) or to make minimum periodic payments to the producer even though oil or gas is not being delivered to the purchaser. The developer may pledge the benefits of a take-or-pay contract to a lender as security for a development loan. *Throughout financing* may be employed for pipelines; producing companies commit themselves to ship quantities of petroleum products sufficient to enable the pipeline company to pay and discharge expenses and obligations which are or shall become due or payable. On the basis of such agreements the pipeline company is enabled to borrow money (frequently as much as 90 percent of the total cost) for the construction of the pipeline. Financing may be on a *nonrecourse basis* if the only recourse of a lender in the event of default is a security interest in the financed project. This form of financing has the advantage to the borrower of non-appearance on the borrower's balance sheet; it is, of course, significantly more expensive to the borrower as lenders may be expected to demand a higher rate of return than would be required for recourse borrowing which entitles the lender to pursue all assets of the borrower on the basis of the borrower's promissory note or other promise to repay the debt. Advance payment financing (or sales of production payments) are forms of nonrecourse financing.

(7) After development of a leasehold, the operator may desire to dispose of the property in order to realize his gain and to raise money for other ventures. Such a transfer may take any of the customary forms common to conveyances of land, but one form of the transaction in particular, known as the *A-B-C transaction,* was used frequently until its tax advantages were eliminated by the Tax Reform Act of 1969. Where the transaction took this form, *A,* the operator, conveyed the working interest to *B* for a cash consideration. There was excepted and reserved from this conveyance, however, a production payment which usually was larger in amount than the cash consideration paid by *B* the purchaser. Subsequently *A* sold the reserved production payment to *C* for cash. Customarily the production payment was fixed in an amount and percentage of production that would be satisfied in the space of several years. By this device B, the purchaser of the operating interest, acquired the developed leasehold for a relatively small down payment, but until the production payment was satisfied, the bulk of the production was devoted to its satisfaction.

In succeeding sections of this chapter we consider problems arising from the variety of transfers which may be made by a lessee. One final observation concerning the informality of many of these transfers should be made here.

Frequently the writing evidencing the agreement between parties to the transfer is wanting in important details of the agreement, raising problems of enforceability under the Statute of Frauds. Thus one judge commented on letter agreements in the following terms:

> These two actions . . .arise out of one of the so-called 'letter agreements' which seem to be so dear to the hearts of the petroleum industry, despite the fact that their vagueness, inexactness, omissions and lack of finality constitute a fertile breeding ground for disputes and litigation. [Hudson's Bay Oil & Gas Co. Ltd. v. Dynamic Petroleums Ltd., 26 W.W.R. 504 (Supreme Court of Alberta 1958), appeal dismissed and cross appeal allowed, 28 W.W.R. 480 (Supreme Court of Alberta, App.Div., 1959).]

But in another case a lower court decision that an agreement was unenforceable on the ground of incompleteness was reversed, one judge commenting that:

> The fact that a conveyancer would have insisted on the parties settling many matters which would be likely to arise in the relationship that is contemplated by these documents is not grounds for holding that no enforceable agreement has been made. [Calvan Consolidated Oil & Gas Company, Ltd. v. Manning, 25 W.W.R. 641, 657, 16 D.L.R.2d 27 at 43 (Supreme Court of Alberta, App.Div. 1958), appeal dismissed, [1959] S.C.R. 253, 17 D.L.R.2d 1 (Supreme Court of Canada 1959).]

SECTION 2. TRANSFERS BY THE LESSOR: HEREIN OF THE "SUBJECT-TO" CLAUSE AND THE "TWO-GRANTS" THEORY

(See Martin & Kramer, *Williams and Meyers Oil and Gas Law* §§ 340-340.5.)

HOFFMAN v. MAGNOLIA PETROLEUM CO.

Commission of Appeals of Texas, Section B, 1925.
273 S.W. 828.

STAYTON, J. Peter L. Hoffman, as plaintiff, brought this suit for the recovery of the proceeds of royalties that accrued under an oil lease on a half section of land, basing his right of action upon a deed from lessors to himself. The district court sustained general demurrers to his petition, and the parties who demurred aver as the basis of the decision, and as correct law, the point that the deed only conveyed the royalty earned by wells drilled upon a tract of 90 acres out of the larger leased tract, whereas the petition did not allege that any wells had been brought in upon this particular 90 acres. The Court of Civil Appeals was of the same opinion, 260 S.W. 950. . . .The specific inquiry is whether the deed gave plaintiff a right to participate in any royalties under the lease, unless production were had from a well upon the 90 acres.

As is shown by the allegations of the petition, the lease was in ordinary form to one R.O. Harvey as lessee; covered and conveyed the oil and gas under 320 acres out of a section of land in Comanche county; provided for a one-eighth royalty on oil, money royalty on gas, and rentals of $330 every six months, as consideration for deferring the commencement of a well, up to five years; made the rights under it perpetual during production; and negatived all obligation upon lessee to drill upon any particular part of the premises.

The deed to plaintiff is averred to have passed from lessors, Duke and wife, before the completion of any well on the half section, in consideration of $10,000 paid by him. It was delivered nine months after the date of the lease, and granted:

The following, to wit: One-half (1/2) interest in and to all of the oil, gas and other minerals in and under and that may be produced from the following described lands situated in Comanche county, Texas, to wit: A certain 90 acres" (giving metes and bounds and describing the tract as out of the section already mentioned) "together with the rights of ingress and egress at all times for the purpose of mining, drilling and exploring said lands for oil, gas, and other minerals and removing the same therefrom.

The instrument continued:

Said above-described lands being now under an oil and gas lease originally executed in favor of R.O. Harvey and now held by _____. It is understood and agreed that this sale is made subject to said lease but covers and includes one-half of all the oil royalty and gas rental or royalty due to be paid under the terms of said lease. It is agreed and understood that one-half of the money rental which may be paid to extend the term within which a well may be begun under the terms of said lease is to be paid to the said Peter L. Hoffman and in the event that the said above-described lease for any reason becomes cancelled or forfeited then and in that event the lease interest and all further rentals on said land for gas and mineral privilege shall be owned jointly by Jas. N. Duke and wife, and _____ Hoffman, each owning one-half interest in all oil, gas and other minerals in and upon said land, together with one-half interest in all future rents. . . .To have and to hold the above-described property, together with all and singular the rights and appurtenances thereto in anywise belonging unto the said Peter L. Hoffman heirs and assigns, forever.

The plaintiff contends that the conveyance to him of "one-half of all of the oil royalty . . .due to be paid under the terms of said lease" is not confined to wells upon the smaller tract. The defendants reply that, since the smaller tract is the subject-matter of the deed, the quoted language should be construed as having reference to it alone, and that any other interpretation would be unreasonable.

Before deciding the question, the court would call to mind that, if the deed is capable of the meaning contended for by plaintiff, there would be nothing unreasonable in holding under the circumstances alleged that, for so substantial a cash consideration, the lessors, having previously conveyed to lessee all of the oil, gas, and minerals under the whole half section, and retained the surface for their own consistent uses, consented to part, not only with their possibility of a reverter in the oil, *etc.*, of the 90 acres, but also with a full one-half of their right to royalties under the lease as an entirety, especially in view of the fact that the reverter was uncertain and the control over the placing of wells impossible. They may not have done that; but it would not have been wrong or inane upon its face if they had. And then again it could have been reasonable from their standpoint, and not unreasonable from that of grantee, if they had restricted their assignment of royalties to those from wells on the 90.

There are a number of rules of construction that will be noticed.

This deed, if its intention be ambiguous, is to be construed against grantors rather than against grantee; and yet, if what it purports to convey is distinctly pointed out in a manner and under circumstances showing that royalty as to wells located outside of the 90 acres was not in contemplation, although the contrary could be construed as within other language when considered separately, the specific and restricted intent would control. But the instrument must, if possible, be considered and made to speak consistently, as a whole, without the rejection of any words, and so as to declare the evident intention of the parties, and the latter is the principal rule to apply.

The deed, in the first part, conveys an undivided one-half interest in the possibility of a reverter of the oil in place under the 90 acres. For, "subject" to the lease, the one-half interest in the oil under that particular tract is conveyed to Hoffman. In the last part of the instrument it is provided that, if the lease shall be forfeited or canceled, the grantor and grantee shall share equally in all mineral interests and rights in "said land"; that is, the 90 acres. Gas and other minerals take the same course, but are not here involved.

Those passages have regard to minerals, the title to which has passed to the lessee, but which may revert to lessor and his assigns if the lease should terminate. But they are not the sole subject-matter of the conveyance. There follow other words regarding not the reverter but rents and royalties prior to the time the lease terminates, and while it is still in force, both before any well at all is drilled and after the drilling of a productive well.

The deed, after conveying the interest in oil, gas, and minerals, and referring, as the description and the allegations show, to the particular "lease" now under consideration, reads:

> It is understood and agreed that this sale is made subject to said lease, but covers and includes one-half of all the oil royalty and gas rental or royalty due to be paid under the terms of said lease. It is agreed and understood that one-half of the money rentals which may be paid to extend the term within which a well may be begun under the terms of said lease is to be paid to said Peter L. Hoffman. . . .

This is plainly a statement that the deed conveys a one-half interest in the royalty to accrue under the terms of the lease as an entirety; that is, the lease upon the whole half section.

Instead of restricting these royalties to wells on the 90 acres, the conveyance covers one-half of "all" the oil royalty under the terms of the lease. That this passage refers to an interest in the whole instead of a part of the royalty, irrespective of where the wells shall be located, is corroborated by the provisions for one-half of the delay rentals payable under the lease. These, by the provisions of the latter instrument, are to accrue before any well at all is begun, and the conveyance of them can only refer to the lease as a whole, and hence not to a particular 90 acres of it; thus showing that in this connection a segregated 90?acre tract is not intended to be the measure of the rights granted.

The Court of Civil Appeals correctly decided that, under this lease, a subdivision of the land would not affect the rights and obligations of the lessee?that the purchaser occupies no better position than his vendor does. As to the lessee the lease remained an entirety, as did his duty to pay rents and royalties. The deed said the 90 acres was subject to the lease. It was. But a conveyance of a part of the land did not subdivide the lease and turn it into two leases, one upon the 90 and the other upon the remaining 320. There was therefore no lease covering solely 90 acres, and, when the deed refers to the "lease," it can only refer to the lease on the whole 320 acres.

The words of the two instruments show this meaning, and it is one that may reasonably be viewed as the real intention of the parties.

. . .

It is suggested that there was a mistake with regard to the provisions of the deeds. If so, it may be remedied upon another trial.

. . .

We recommend that the judgment of the Court of Civil Appeals be reversed, and that the cause be remanded for new trial.

CURETON, C.J. The judgments recommended in the report of the Commission of Appeals is adopted, and will be entered as the judgment of the Supreme Court.

We approve the holding of the Commission of Appeals on the questions discussed in its opinion.

NOTES

1. Martin & Kramer, *Williams & Meyers Oil and Gas Law* § 340: "The purpose of inserting the 'subject to' clause in a deed is probably three-fold: (1) it protects the grantor against breach of warranty arising from the outstanding oil and gas lease; (2) it spells out the precise interests the grantee is to receive and thus is supposed to make the deed clearer; (3) it makes certain that rentals and royalty under the existing lease pass to the grantee, thus avoiding decisions which held that such interests do not pass without express transfer in the deed.

"It is doubtful, however, that the clause is really needed to accomplish these purposes. Protection from breach of warranty can be achieved by a mere exception in the warranty clause. Few if any states now accept the position that rentals and royalty will not pass without express mention. And while clarity of intention is gained by use of the clause in expert hands, confusion of intention is often the result of its use by the inept. Introduction of the clause into the deed creates the opportunity for conflict between it and the granting clause. The usual conflict is one of three kinds: (1) over the amount of land conveyed by the deed; (2) over the quantum of interest conveyed by the deed; or (3) over the duration of the interest conveyed by the deed."

2. In PADDOCK V. VASQUEZ, 122 Cal.App.2d 396, 265 P.2d 121, 3 O.&G.R. 582 (1953) the question presented was what was the quantum of interest conveyed, (granting clause: "3 percent of 100 percent" of oil or gas "within or underlying"; subject-to clause refers to a lease providing for a 1/8 royalty and states that deed "includes, but is not limited to 6/25 of all bonuses, rents, royalties and other benefits which may accrue . . .under said . . .Lease . . .or any other lease . . ."). What result if the "two grants" concept of *Hoffman* is followed?

3. The subject-to clause now normally contains the words "in so far as it [the lease] covers the above described land." This clause could be called the Hoffman Clause, since it makes the subject-to clause mean now what it always would have meant but for the decision in the *Hoffman* case.

4. Should a "subject-to" clause be limited in effect to defining the scope of the deed warranty? *See* Stracka v. Peterson, 377 N.W.2d 580, 88 O.&G.R. 149 (N.D. 1985).

————

GARRETT v. DILS CO.

Supreme Court of Texas, 1957.
157 Tex. 92, 299 S.W.2d 904, 7 O.&G.R. 322.

HICKMAN, CHIEF JUSTICE. This is an action in trespass to try title to a tract of land in Navarro County in which the sole controversy centers around the construction of a deed under which respondent holds an interest in the minerals in and under the land. The suit is by Mrs. Mattie Garrett and Mrs. Bee Lively, widow and surviving daughter, respectively, and sole heirs of C. S. Garrett, who died intestate before this suit was filed. The relevant provisions of the deed are as follows:

That C. S. Garrett and wife, Mattie Garrett, both of Navarro County, Texas, for and in consideration of the sum of Fifteen Thousand Dollars cash in hand paid by J. Mentor Caldwell, hereinafter called Grantee, receipt of which is hereby acknowledged, have granted, sold, conveyed, assigned and delivered and by these presents do grant, sell, convey, assign and deliver unto the said Grantee an undivided one sixty-fourth interest in and to all of the oil, gas and other minerals in and under, and that may be produced from the following described land: (Description follows)

Together with the right of ingress and egress at all times for the purpose of mining, drilling, and exploring said land for oil, gas and other minerals, and removing the same therefrom. 'Said land being now under an oil and gas lease executed in favor of I. B. Humphreys or his assigns, it is understood and agreed that this sale is made subject to the terms of said lease, but covers and includes one-eighth of all of the oil royalty, and gas rental or royalty due and to be paid under the terms of said lease.

It is understood and agreed that one-eighth of the money rentals which may be paid to extend the term within which a well may be begun under the terms of said lease is to be paid to the said Grantee and in event that the above described lease for any reason becomes cancelled or forfeited, then and in that event an undivided one-eighth of the lease interest and all future rentals on said land for oil, gas and other mineral privileges shall be owned by said Grantee, he owning one-eighth of one-eighth of all oil, gas, and other minerals in and under said lands, together with one-eighth interest in all future rents.

Respondent owns the interest which was conveyed to Caldwell by that deed.

As disclosed by that instrument, at the date of its execution, December 7, 1921, the land was under an oil and gas lease to I. B. Humphreys. That lease provided for a one-eighth royalty. No production was obtained thereunder, and it expired by its own terms. Subsequently another lease was executed, which likewise provided for a one-eighth royalty. That lease is still in existence and oil is being produced thereunder.

The trial court held that respondent is entitled to one sixty-fourth of the royalty payable under the lease now in existence, while the Court of Civil Appeals held that respondent is entitled to one-eighth of the royalty payable thereunder.

...

In construing the deed we shall be guided by the well-established rule which we recently reaffirmed in Harris v. Windsor, Tex., 294 S.W.2d 798, 799, 800, in this language:

We have long since relaxed the strictness of the ancient rules for the construction of deeds, and have established the rule for the construction of deeds as for the construction of all contracts,-that the intention of the parties, when it can be ascertained from a consideration of all parts of the instrument, will be given effect when possible. That intention, when ascertained, prevails over arbitrary rules.

Another applicable rule is that should there be any doubt as to the proper construction of the deed, that doubt should be resolved against the grantors, whose language it is, and be held to convey the greatest estate permissible under its language.

The question of immediate concern to the parties is the royalty to which the respondent is entitled under the existing lease. We shall approach the solution of that question by considering first the royalty to which it would have been entitled had there been production under the lease in existence when the deed was executed, and then determining from the language of the deed whether it was the intention of the parties that the royalty was to be the same under subsequent leases.

Should the granting clause be considered alone there would be no doubt as to the interest conveyed. It states in certain terms that the interest conveyed was 'an undivided one sixty-fourth interest in and to all of the oil . . . .' Had other language in the deed not disclosed what the parties understood 'one sixty-fourth' to mean, it would be our duty to give those words their usual meaning and construe the deed as a mineral deed to an undivided one sixty-fourth of the minerals in place. But there follows the granting clause language which clearly defines what the parties understood 'one sixty-fourth' of the minerals to mean. After reciting that the land was under an oil and gas lease, the deed provided that 'it is understood and agreed that this sale . . .covers and includes one-eighth of all of the oil royalty, and gas rental or royalty due and to be paid under the terms of said lease.' Construing all of these provisions together it is made certain that what the parties intended to convey, had there been production under the then existing lease, was a royalty of one sixty-fourth or one-eighth of the one-eighth royalty retained in the lease. The rights conveyed by the deed under the then existing lease were one-eighth of the money rentals which might be paid to extend the term within which a well might be begun and one-eighth of the one-eighth royalty.

Turning now to the rights acquired by Caldwell under that deed in the event the then existing lease should terminate, it is provided that 'then and in that event an undivided one-eighth of the lease interest and all future rentals on said land for oil, gas and other mineral privileges shall be owned by said Grantee, he owning one-eighth of one-eighth of all oil, gas, and other minerals in and under said lands, together with one-eighth interest in all future rentals.' We can discover in that language no intent to grant a less interest under a subsequent lease than that granted under the then existing lease. As pointed out above, there was granted one sixty-fourth of the minerals which the parties construed to mean one-eighth of the one-eighth royalty under the then existing lease. The provision for ownership of the minerals under future leases is that the grantee shall own 'one-eighth of one-eighth' of the minerals. Had that fraction been expressed as one sixty-fourth, it should be given the same meaning as in the granting clause which the parties understood and agreed to be a one sixty-fourth royalty or one-eighth of the one-eighth royalty. Instead of employing the fraction one sixty-fourth in defining the ownership under a subsequent lease, the provision is for one-eighth of one-eighth. Clearly, that does not denote a less interest than a one sixty-fourth, but on the contrary it emphasizes the fact that the intention was to convey one-eighth of the royalty under future leases the same as under the original lease. The court takes judicial knowledge of the fact that the usual royalty provided in mineral leases is one-eighth. The parties doubtless assumed that the royalty under future leases would be one-eighth, as it was under the lease in existence when the deed was executed.

Considering further the rights acquired by the vendee in that deed in the event the lease in existence when the deed was executed should terminate, it is provided that the vendee should acquire 'one-eighth of the lease interest and all future rentals.' That can have no other meaning than that the grantee shall have the right to lease an undivided one-eighth interest in the minerals and to receive one-eighth of all future rentals, or, at any rate, to receive one-eighth of the bonus paid for future leases and one-eighth of the rentals. The right to one-eighth of the bonus and rentals and one-eighth of the royalty left no right in the vendors in the one-eighth interest. Having all the rights incident to ownership of one-eighth of the minerals, the conclusion follows that the deed conveyed to Caldwell an undivided one-eighth of the minerals.

Construing this deed as a whole and giving effect to each and every provision thereof, we are led to the conclusion that the royalty conveyed under future leases was the same as that conveyed under the then existing lease,-that is to say, one-eighth thereof. We further conclude that having the right to receive one-eighth of the royalty, together with a one-eighth lease interest and future rentals thereon, the respondent in reality is the owner of one-eighth of the minerals in the land.

It follows that the judgment of the Court of Civil Appeals should be affirmed, and it is so ordered.

Dissenting opinion by NORVELL, J., in which GRIFFIN, CALVERT and SMITH, JJ., join.

NORVELL, JUSTICE (dissenting). In determining the intention of the parties to the conveyance in question, it is not asserted by either side that resort may be had to matters outside the written instrument itself. The meaning of the deed, i. e., the intention of the parties, must therefore be ascertained from the language contained therein. 12 Am.Jur., 748. It is clear enough that Garrett and wife intended to convey to Caldwell, respondent's predecessor in title, an undivided 1/64th interest in and to all of the oil, gas and other minerals in and under the tract of land involved. The formal granting clause expressly so provides....

NOTES

1. What canons of construction did the majority opinion use in interpreting the instruments? Did they harmonize the inconsistent provisions or did they give one clause greater weight? Would the dissenting opinion have given greater weight to the granting clause?

2. ALFORD v. KRUM, 671 S.W.2d 870, 81 O.&G.R. 189 (Tex. 1984), construed a grant of one-half of one-eighth of "all of the oil and gas . . .in and under and that may be produced from the . . .described lands" as conveying a one-sixteenth mineral interest. The deed made further reference in the subject-to clause which provided for payment of "1/16 of all the oil royalty under that lease" to the grantee. The instrument then went on to provide in the future lease clause that the grantee was to receive a "one half interest in all oil, gas and other minerals." The court ignored the future lease clause "because [where] an irreconcilable conflict exists between the granting clause and the future lease clause," the granting clause should control. A dissenting opinion found "no ambiguity in the deed that grants a one-sixteenth mineral estate so long as there is an outstanding lease and a one-half mineral estate upon the lease's termination. The fractions are different for good reason." Neither the majority nor dissenting opinions cite *Hoffman.* Were

Hoffman and *Garrett* overruled *sub silentio? See* Burney, *Interpreting Mineral and Royalty Deeds: The Legacy of the One-Eighth Royalty and Other Stories*, 33 St. Mary's L.J. 1 (2001); Herd, *Deed Construction and the Repugnant to the Grant' Doctrine*, 21 Tex. Tech L.Rev. 635 (1990); Smith, *The 'Subject-To' Clause*, 30 Rocky Mtn.Min.L.Inst. 15-1 (1984).

3. HAWKINS V. TEXAS OIL & GAS CORP., 724 S.W.2d 878, 97 O.&G.R. 399 (Tex.App. 1987). A three grant deed described in the granting clause a transfer of a "1/4 of the 1/8 royalty interest." A handwritten provision described the grantee as *not* receiving any of the lease rights, bonus or delay rentals under future leases. The subject-to clause conveyed 1/4 of the royalty under the existing lease. The future lease clause provided that after the expiration of the existing lease, each of the grantees were to individually own a 1/4 mineral interest. Where the second lease provided for a 1/6 royalty is the grantee entitled to a 1/24 or a 1/32 share? Did the grantee receive upon the expiration of the first lease, a mineral or a royalty interest? Which Texas Supreme Court decision governs, *Garrett* or *Alford?*

4. A royalty deed from Mayes to Luckel contains the following clauses:

[granting clause] I, Mary Etta Mayes ... [convey to] L.C. Luckel, Jr. an undivided one thirty-second (1/32nd) royalty interest in and to the following described property, ...

[habendum and warranty clauses] TO HAVE AND TO HOLD the the above described 1/32nd royalty interest ... unto the said L.C. Luckel, Jr.... to warrant and forever defend ... the said 1/32nd royalty interest ...

[subject-to clause] It is understood that said premises are now under lease originally executed to one Coe and that the grantee herein shall receive no part of the rentals as provided for under said lease, but shall receive one-fourth of any and all royalties paid under the terms of said lease.

[future lease clause] It is expressly understood and agreed that the grantor herein reserved [sic] the right upon expiration of the present term of the lease on said premises to make other and additional leases ... [and] shall be entitled to one-fourth of any and all royalties reserved under said leases.

[final clause] It is understood and agreed that Mary Etta Mayes is the owner of one-half of the royalties to be paid under the terms of the present existing lease, the other one-half having been transferred by her to her children and by the execution of this instrument, Mary Etta Mayes conveyed one-half of the one-sixteenth (1/16th) royalty now reserved by her.

The original Coe lease expired. The lands covered by the Mayes-Luckel deed are now under lease where the royalty reserved is 1/6. Do the grantees receive a 1/32 or a 1/24 royalty payment? What result under *Alford? Garrett?* In LUCKEL V. WHITE, 819 S.W.2d 459, (Tex.1991) the Texas Supreme Court in a 4-1-4 opinion overruled *Alford* and applied the canons of construction used in *Garrett* to give the grantee a 1/24 royalty for production from leases which provided for a 1/6th leasehold royalty. *See also* Jupiter Oil Co. v. Snow, 819 S.W.2d 466, 115 O.&G.R.

148 (Tex.1991).

5. CONCORD OIL CO. V. PENNZOIL EXPLORATION & PRODUCTION CO., 966 S.W.2d 451, 142 O.&G.R. 130 (Tex. 1998). Crosby was the owner of a 1/12th mineral interest subject to an outstanding lease. He conveyed to Southland "an undivided one-ninety-sixth (1/96) interest in and to all of the oil, gas and other minerals in and under, and that may be produced . . . . While the estate hereby conveyed does not depend upon the validity thereof, neither shall it be affected by the termination thereof, this conveyance is made subject to the terms of any valid subsisting oil, gas and/or mineral lease or leases on above described land. . . , but covers and includes one-twelfth (1/12) of all rentals and royalty of every kind and character. . ." There was no express future lease clause contained in the instrument. Crosby later conveyed to Robinson a 7/96th mineral interest. Robinson then leased to Pennzoil. Concord was a successor in interest to Southland who argued that the Crosby-Southland deed conveyed all of the 1/12th mineral interest owned by Crosby. What result under *Garrett? Alford v. Krum? Luckel v. White?* Should it matter that both Crosby and Southland were experienced in oil and gas conveyancing matters? What if the facts showed that Crosby had received the 1/12th mineral interest the day before the Southland conveyance changing only the fraction in the granting clause in preparing the Southland deed?

6. HEYEN V. HARTNETT, 235 Kan. 117, 679 P.2d 1152, 81 O.&G.R. 31 (1984). A quitclaim deed granted "an undivided 1/16 interest in and to all oil and gas and other minerals" and went on to say that if the land was "covered by a valid oil and gas or other mineral lease" the grantees would have "an undivided 1/2 interest in the Royalties, Rentals, and Proceeds therefrom, of whatsoever nature." The court found the deed ambiguous and construed it to convey an undivided 1/2 interest in the minerals. The opinion makes special mention of the "widespread confusion as to the fractional interest in the minerals required to produce a certain share of the royalties under an oil and gas lease. . . .We believe that the use of the fraction '1/16' in the initial clause of this mineral deed was simply an error commonly made in the early days of oil and gas conveyancing." The deed in *Heyen* was delivered in 1925. Would a consideration of "all of the surrounding facts and circumstances" suggest a different result in construing a deed delivered in 1986? Does *Heyen* represent a better approach to the construction of mineral deeds than *Hoffman, Luckel* or *Concord Oil? In accord,* Powell v. Prosser, 12 Kan.App.2d 626, 753 P.2d 310, 98 O.&G.R. 246 (1988); Shepard, Executrix v. John Hancock Mutual Life Ins. Co., 189 Kan. 125, 368 P.2d 19, 16 O.&G.R. 1147 (1962).

7. ZEMP V. JACOBS, 713 P.2d 25, 87 O.&G.R. 386 (Okl.App.1985, cert. denied). Land was conveyed by warranty deed reserving an interest defined as follows:

> Said first party requires said second party to keep said land under Oil or gas Lease to some responsible and reliable Oil or Gas Company at all times if demand therefor, satisfactory to said first party, at expiration of any Oil or gas lease on said land.

. . .

> The Grantors herein specifically reserve and retain an undivided one sixteenth (1/16) of all Oil and Gas produced from said land and also reserve to themselves an undivided One Half of all Oil and Gas rentals and bonuses under any Oil and Gas lease affecting said real estate or in any future Oil and Gas Lease on the above described premises . . ..

The grantors claimed a 1/16 royalty interest and the grantees brought suit to establish that the interest reserved was a 1/16 mineral interest. Held: the interest reserved was a mineral interest:

> The [grantors], having previously entered into a lease for oil and gas production, knew of the royalty available to them and chose not to reserve specifically unto themselves one half of the royalty interest. They chose instead to retain a shared executive interest via the right to participate in the leasing process. They also chose to participate equally in all bonuses and rentals and to keep for themselves one unit of every sixteen units of produced hydrocarbons. Further, while they did not specifically retain the right of access in the reservation, they also did not waive it.

> The [grantors'] language in the reservation indicates they did not intend to give up any rights. The reservation maximized the [grantors'] control to insure receipt of an undivided 1/16 of *all* oil and gas produced together with income from the lease bonuses and rentals, not only from the existing lease but also for all future leases. The [grantors], being familiar with the leasing process would also have reasonably known that there would be a cost of production to produce any oil and gas and that they would be required to share in this cost pro rata with the retained interest.

. . .

> Thus, while the actual hydrocarbon reservation for 'oil and gas produced from said land' would tend to indicate that a 'royalty' interest was created, this is negated by the [grantors] retention of rights in the critical factors indicative of a mineral interest: the executive right, or the right to grant leases, and the right to receive rentals and bonuses. . . .Consequently, we hold that the [grantors] created . . .a mineral interest and that the mineral interest has the right to receive 1/16 of the royalty under any lease existing on the realty.

Is *Zemp* a useful guide in negotiating the Oklahoma mineral-royalty "labyrinth" described earlier in Chapter 5?

8. BENGE v. SCHARBAUER, 152 Tex. 447, 259 S.W.2d 166, 2 O.&G.R. 1350 (1953). *R*, owner of 3/4 of the minerals in Blackacre, conveyed Blackacre to *E* by general warranty deed reserving a 3/8th mineral interest "but the grantee and his assigns shall have the sole power to execute all future . . .mineral leases without the joinder of the grantors herein, but said leases shall provide for the payment of three-eighths (3/8) of all bonuses, rentals and royalties to the grantor." *Held, R* owns 1/8 mineral interest and receives 3/8 of lease benefits, *E* owns 5/8 mineral

interest and receives 3/8 of these benefits.

Under the decision in Duhig v. Peavy-Moore Lumber Co., 135 Tex. 503, 144 S.W.2d 878, the effect of the deed, by reason of the outstanding 1/4th interest in the minerals and the general warranty, was that the grantee acquired by the deed the surface and a 5/8ths mineral interest, and the grantors reserved only a 1/8th mineral interest. On account of the outstanding 1/4th mineral interest in third parties the warranty was breached at the very time the deed was executed and delivered, and a 1/4th mineral interest required to remedy the breach was taken from what the grantors undertook to reserve to themselves leaving them only a 1/8th mineral interest.

The difficult question in the case arises because of the provision above quoted that the mineral leases to be executed by the grantee shall provide for payment of 3/8ths of all bonuses, rentals and royalties to the grantors. But for that provision the grantors, as owners of but 1/8th interest in the minerals as the effect of the deed, would be entitled to only a like interest, 1/8th, of bonuses, rentals and royalties under leases executed by the grantee. One-eighth of the bonuses, rentals and royalties normally would go to the grantors as owners of a 1/8th interest in the minerals and 5/8 ths of bonuses, rentals and royalties would normally go to the grantee as the owner of a 5/8ths interest in the minerals. But are not the owners of such interests in the minerals free to agree, if they desire to do so, that their fractional interests in bonuses, rentals and royalties under leases to be executed shall be in different amounts from what they normally would be?

The fractional part of the bonuses, rentals and royalties that one is to receive under a mineral lease usually or normally is the same as his fractional mineral interest, but we cannot say that it must always be the same. The parties owning the mineral interests may make it different if they intend to do so, and plainly and in a formal way express that intention. Here that intention is expressed by clear language in the deed that leases executed by the grantee under the power given shall provide for the payment of 3/8ths of all bonuses, rentals and royalties to the grantors. The provision is not an agreement that the parties to the deed shall participate in the bonuses, rentals and royalties in proportion to their ownership of mineral interests. It is rather a contractual provision that the grantors shall receive a specified part of the bonuses, rentals and royalties; namely, 3/8ths.

The warranty and its breach on account of the 1/4th mineral interest outstanding in third parties reduced the 3/8ths mineral interest which the grantors undertook to reserve from 3/8ths to 1/8th, and in so doing it left to the grantee the surface and the 5/8ths mineral interest that the deed purported to convey to him. There is seeming inconsistency between the parts of the deed that give the grantors a 1/8th and the grantee a 5/8ths mineral interest and the provision that the grantors' share in bonuses, rentals and royalties shall be 3/8ths, but there is no fatal repugnancy.

The warranty extends to what the deed purports to grant; namely, the surface and the 5/8ths interest in the minerals, and the application of the rule in the Duhig case assures the grantee of title to what the deed purports to grant to him. But the warranty does not extend to the provision in the deed as to the interest in bonuses, rentals and royalties. The deed does not purport to convey the right given by that provision.

GARWOOD, J., dissented:

Now, it seems to me that since the grantors have in their deed represented themselves to own 8/8, they have done so for all purposes, and not just for the purpose of one paragraph or clause of the instrument. The instrument is a unit and should be taken as if the grantors had said, 'We represent ourselves to own 8/8 of the minerals in this tract of land, and your commitments as well as our rights in this trade are upon the faith of the representation.' Whether the words about the contents of the leases are words of royalty or not, they are part and parcel of the entire arrangement, and this fact appears from the deed itself. It appears quite artificial to reason that the representation exists and has positive effect with regard to part of the deed but does not exist or have effect as to another part. Surely, [if the outstanding interest had been 5/8 instead of merely 2/8], it would be far fetched to believe that businessmen might agree for the grantee to have the exclusive power to lease and a substantial interest in the minerals and yet receive nothing by the deed of practical value. And yet we could not hold one way when the outstanding interest is 5/8 and another way when it is only 2/8. It is plain to me that in the instant case the same equity ('estoppel') which arises from the grantor's representation of 8/8 ownership and converts the stipulated 3/8 mineral reservation into 1/8, also justifies construing the instrument so as to reduce the so-called royalty reservation from 3/8 of the royalty to 1/8 thereof.

Suppose in *Benge* that R had reserved one-half of the minerals in Blackacre, granting the exclusive right to lease to *E,* but reserving 1/2 of all bonus, rentals and royalty under such leases to *R;* suppose further that R owned only one-half mineral interest (plus the surface) in Blackacre. Would this mineral interest owned by *E* under the *Benge* decision be something like the Cheshire Cat in *Alice in Wonderland?* " . . .and this time it vanished quite slowly, beginning with the end of the tail, and ending with the grin, which remained some time after the rest of it had gone."

9. It has been urged that the "subject-to" clause affects the duration of a grant or reservation.

(a) R conveys to E an undivided one-half interest in the minerals in Blackacre for a fixed term of 20 years, providing that "Said land now being under an oil and gas lease, it is understood and agreed that this sale is made subject to the terms of said lease." The lease is producing upon the expiration of the 20-year period and E claims his mineral interest is extended by the "subject-to" clause, which incorporated into the mineral deed the "thereafter" clause of the oil and gas lease.

Should this contention be sustained? *See* Kokernot v. Caldwell, 231 S.W.2d 528 (Tex.Civ.App.1950, writ ref'd).

(b) *R* leased land to *O* and thereafter conveyed the land to *E* by warranty deed providing, "all oil rights reserved to grantors and subject to existing leases." *O* surrendered his lease, and *E* claimed that *R* 's interest terminated, because its duration was measured by that of the lease. What result? The case authority is in conflict.

10. The "subject-to" clause has performed at least one other function, that of reviving a lease that has expired. Thus in HUMBLE OIL & REFINING CO. v. CLARK, 126 Tex. 262, 87 S.W.2d 471 (1935), when a lease terminated by virtue of a tender of delay rentals to the wrong party, the lease was revived by "ratification" when the lessor executed a mineral deed subject to the lease.

If the expired lease was not "expressly recognized in clear language" in the subsequent document, it will not be revived. McVEY v. HILL, 691 S.W.2d 67, 89 O.&G.R. 582 (Tex.App. 1985, writ ref'd).

Many printed forms now add to the "subject-to" clause the following language: "Said land is subject to an oil and gas lease in favor of _____ and it is understood and agreed that this sale is subject to said lease, *in so far as such lease may be valid.*"

GARZA v. PROLITHIC ENERGY CO., L.P.

Court of Appeals of Texas, 2006
195 S.W.3d 137, 165 Oil & Gas Rep. 332

SIMMONS, JUSTICE. Vicente Saenz and Inocencia de Saenz executed two separate deeds in favor of J. B. Claypool and Homer P. Lee. The deed in favor of J.B. Claypool is entitled "Royalty Contract," while the deed in favor of Homer P. Lee is entitled "Mineral Deed." This appeal challenges a summary judgment construing the mineral and royalty interest language contained in those deeds. The operators of the wells responsible for paying royalties to the appropriate parties interpled the funds and sought a judicial interpretation of the deeds. The trial court construed the deeds in favor of the parties claiming ownership as the grantees of the deeds (the "Claypool/Lee Claimants). The parties claiming ownership as the grantors of the deeds (the "Saenz Claimants") appeal the summary judgment, contending: (1) the summary judgment fails to give effect to all of the terms of the deeds: (2) the *Duhig* rule does not apply to the deeds; and (3) the trial court erred in admitting the Claypool/Lee Claimants' expert title opinions. We affirm the trial court's judgments.

Background

In 1938, Vicente Saenz and Inocencia de Saenz executed a Royalty Contract in favor of J. B. Claypool. The Contract conveyed "an undivided one-half (1/2) interest in and to all of the oil, gas and other minerals in and under the [Property] .

. . Together with the rights of ingress and egress at all times for the purpose of taking said minerals." The Contract further provided:

It is distinctly understood and herein stipulated that said land is under an Oil and Gas Lease made by Grantor providing for a royalty of 1/8th of the oil and certain royalties or rentals for gas and other minerals and that Grantee herein shall receive One-half (1/2) of the royalties and rentals provided for in said lease insofar only as said lease covers the land hereinabove described; but he shall have no part of the annual rentals paid to keep said lease in force until drilling has begun.

It is further agreed that Grantee shall have no interest in any bonus money received by the Grantor in any future lease or leases given on said land, and that it shall not be necessary for the grantee to join in any such lease or leases so made; That Grantee shall receive under such lease or leases one-sixteenth (1/16th) part of all oil, gas and other minerals taken and saved under such lease or leases, and he shall receive the same out of the royalty provided for in such lease or leases, but Grantee shall have no part in the annual rentals paid to keep such lease or leases in force until drilling is begun.

TO HAVE AND TO HOLD the same unto the said Grantee, his heirs and assigns, forever; and we hereby bind ourselves, our heirs, executors and administrators to WARRANT and FOREVER DEFEND all and singular the said minerals unto the said Grantee, his heirs and assigns, against all persons whomsoever lawfully claiming or to claim the same or any part thereof.

Vicente Saenz and Inocencia de Saenz also executed a Mineral Deed in favor of Homer P. Lee. The Mineral Deed conveyed "an undivided fifteen-thirty-seconds (15/32) interest in and to all of the oil, gas and other minerals in and under the [Property]. . . . together with the rights of ingress and egress at all times for the purpose of taking said minerals." The Mineral Deed contained provisions similar to the Contract, stating:

It is distinctly understood and herein stipulated that said land is under an Oil and Gas Lease made by Grantor providing for a royalty of 1/8th of the oil and certain royalties or rentals for gas and other minerals, and that Grantee herein shall receive 15/32nds of the royalties and rentals provided for in said lease; insofar as it covers the above described land; but he shall have no part of the annual rentals paid to keep said lease in force until drilling is begun.

It is further agreed that Grantee shall have no interest in any bonus money received by the Grantor in any future lease or leases given on said land, and that it shall not be necessary for the Grantee to join in any such lease or leases so made. Nevertheless, neither the Grantor, nor the heirs, administrators, executors and assigns of the Grantor shall make or enter into any lease or contract for the development of said land, or any part of same, for oil, gas or other minerals, unless each and every such lease, contract, leases, or contracts, shall provide for at least royalty of the usual one-eighth to be delivered free of cost in the pipe line, and a royalty on natural gas of one-eighth of the value of same when sold or used off the premises, or one-eighth of the net proceeds of such gas; and one-eighth of the net amount of gasoline manufactured from natural or casinghead gas. That Grantee shall receive under such lease or leases 15/32 of 1/8 part of all oil, gas and other minerals taken and saved under any such lease or leases, and he shall receive the same out of the royalty provided for in such lease or leases, but Grantee shall have no part in the annual rentals paid to keep such lease or leases in force until drilling is begun.

TO HAVE AND TO HOLD the same unto the said Grantee, his heirs and assigns forever; Grantors hereby bind themselves, their heirs, executors and administrators to Warrant and Forever Defend all and singular the said minerals unto the said Grantee, his heirs and assigns, against all persons whomsoever lawfully claiming or to claim the same or any part thereof.

The lease in effect at the time the Contract and Mineral Deed were executed, which provided for the payment of a 1/8 royalty, terminated, and the new lease provided for a 1/5th royalty. The parties filed competing motions for summary judgment. The Claypool/Lee Claimants asserted that they are entitled to 1/2 of the 1/5 royalty and 15/32nds of the 1/5 royalty, respectively. The Saenz Claimants asserted that the language limiting the royalty under future leases must be given effect, thereby limiting the royalty to be received by the Claypool/Lee Claimants to a fixed 1/16th under the Contract and 15/32nds of 1/8th under the Mineral Deed of the 1/5 royalty received under the new lease. The trial court granted summary judgment in favor of the Claypool/Lee Claimants, and the Saenz Claimants appealed.

. . .

Construction of Deed

When interpreting a deed just as in interpreting a contract, the intent of the parties is to be determined from the express language found within the four corners of the document. Construction of an unambiguous deed is a question of law to be resolved by the court. All parts of the deed are to be harmonized, construing the instrument to give effect to all of its provisions.

A. Saenz Claimants' Argument

The Saenz Claimants argue that while the granting clause of the deeds granted the Claypool/Lee Claimants an undivided one-half and 15/32nds mineral interest, respectively, the future lease clause reduced the amount of royalty payable to the Claypool/Lee Claimants under future leases. The Saenz Claimants contend that if the Claypool/Lee Claimants had decided to develop their undivided interests in the minerals, they would have been entitled to one-half and

15/32nds of the proceeds from their production, respectively. If they chose not to produce the minerals, their right to receive income from production would be reduced by the future lease clause.

B. The Claypool/Lee Claimants' Argument

The Claypool/Lee Claimants argue that the granting clauses of the deed conveyed an undivided mineral interest and that the future lease clause is nothing more than a recognition of what the royalty would be under a future lease providing for the usual 1/8 royalty. The Claypool/Lee Claimants argue that the mineral interest cannot be "transmogrified into a fixed royalty conveyance" upon the termination of the existing lease. The Claypool/Lee Claimants note that the deeds expressly warrant title to the minerals, not to a fixed royalty.

C. Analysis

The issue presented in this appeal requires us to consider two lines of cases in an area of the law in which the most recent pronouncement by the Texas Supreme Court is in the form of a plurality opinion, and only one of the justices who participated in that decision remains on the court. See Concord Oil Co. v. Pennzoil Exploration & Production Co., 966 S.W.2d 451, 41 Tex. Sup. Ct. J. 476 (Tex. 1998) The two lines of cases involve: (1) determining whether an interest conveyed in a mineral deed is a mineral ownership interest or a royalty interest; and (2) reconciling deeds containing different fractional interests in the granting clause and the future lease clause.

1. Ownership v. Royalty Interest

The first line of cases deals with determining the nature of the interest conveyed. The five essential attributes of a severed mineral estate are: (1) the right to develop (the right of ingress and egress); (2) the right to lease (the executive right); (3) the right to receive bonus payments; (4) the right to receive delay rentals; and (5) the right to receive royalty payments. Altman v. Blake, 712 S.W.2d 117, 118, 29 Tex. Sup. Ct. J. 457 (Tex. 1986). The Texas Supreme Court, however, has recognized that "a mineral interest shorn of the executive right and the right to receive delay rentals remains an interest in the mineral fee." Altman, 712 S.W.2d at 118-19.

In French v. Chevron U.S.A., Inc., 896 S.W.2d 795, 38 Tex. Sup. Ct. J. 445 (Tex. 1995), the court considered the distinction between a mineral fee interest and a royalty interest. In that case, the deed conveyed "an undivided Fifty (50) acre interest, being an undivided 1/656.17th interest in and to all of the oil, gas and other minerals, in, under and that may be produced from the following described lands...." French, 896 S.W.2d at 796. In a subsequent paragraph, the deed stated, "It is understood and agreed that this conveyance is a royalty interest only," and then listed the rights the grantee did not have, including any interest in the delay or other rentals, any control over leasing the lands, and any ability to make a lease contract to develop the land. Id. The court interpreted the transfer to have conveyed a mineral interest with reservation of all developmental rights, leasing rights, bonuses, and delay rentals. *Id.* The court noted that the reservation of the various attributes of a severed mineral estates would have been redundant if only a royalty interest was conveyed. . . .

Similarly, the deeds at issue in this case convey an undivided interest in and to all of the oil, gas and other minerals in and under the property and reserve in the grantors at least the second, third, and fourth *Altman* rights. Although the right to develop, referred to as the right of ingress and egress, appears to be expressly conveyed in the deeds, the court in *French* stated "the right to develop is a correlative right and passes with the executive rights." Accordingly, it is unclear whether the deeds reserved the first *Altman* right. Nevertheless, the phrase "in and under" refers to a mineral interest. *See generally* Laura H. Burney, *Interpreting Mineral and Royalty Deeds: The Legacy of the One-Eighth Royalty and Other Stories, 33 ST. MARY'S L.J. 1, 30-31 (2001).* Furthermore, the reservation of the second, third, and fourth *Altman* rights would have been redundant if the deeds intended to convey a royalty interest. Therefore, applying the first line of cases, we hold that the deeds conveyed a mineral interest. The question becomes whether the conflicting fractions in the deeds resulted in a conveyance of a fractional mineral interest that was greater than the royalty interest that the grantee would receive under future leases. That question leads us to the second line of cases that must be considered.

2. Conflicting Fractions

The second line of cases deals with reconciling conflicting fractions within a deed. The Saenz Claimants rely heavily on the Texas Supreme Court's decision in Luckel v. White, 819 S.W.2d 459, 461-63, 35 Tex. Sup. Ct. J. 40 (Tex. 1991). In *Luckel,* the deed at issue conveyed a 1/32nd royalty interest in the granting clause but the future lease clause stated that the grantee was entitled to 1/4th of any and all royalties reserved under future leases. The court concluded that the future lease clause was effective to convey a one-fourth interest in all royalties as to future leases, harmonizing the reference to 1/32nd by noting that one-fourth of the usual one-eighth royalty (which was the royalty under the existing lease at the time of the conveyance) is 1/32nd.

The overriding distinctions between the instant case and *Luckel* is that in *Luckel* the conflicting fractions both involved royalty interests, and the fractional royalty interest in the future lease clause was greater than the fractional royalty interest in the granting clause. In this case, the conflicting fractions involve a mineral interest and a royalty interest, and the fractional royalty interest in the future lease clause is less than the mineral interest originally granted. This may, however, be a distinction without a difference.

In *Luckel,* the Texas Supreme Court overruled Alford v. Krum, 671 S.W.2d 870, 27 Tex. Sup. Ct. J. 434 (Tex. 1984), overruled, Luckel v. White, 819 S.W.2d at 464. In *Alford,* the granting clause of the deed conveyed "one half of the one-eighth interest in and to all of the oil, gas and other minerals in and under and that may be produced from" certain land. The deed then provided that the land was subject to an existing lease and that the sale "covers and includes 1/16 of all the oil royalty and gas rental or royalty due and to be paid under the terms of said lease." The future lease clause then provided that "in the event that the said above described lease for any reason becomes cancelled or forfeited, then and in that event, the lease interests and all future rentals on said land, for oil, gas and mineral privileges shall be owned jointly by [the parties] each owning a one-half interest in all oil, gas and other minerals in and upon said land, together with one-half interest in all future rents." The Texas Supreme Court held that an irreconcilable conflict existed between the granting clause and the future lease clause; therefore, under the applicable rules of construction, the granting clause should control.

In interpreting the lease in *Luckel*, the court noted that the only significant difference between that case and *Alford* is that *Alford* dealt with the conveyance of a fractional mineral interest and the *Luckel* case dealt with a fractional royalty interest. The court stated, "That difference is not material. A royalty interest is an interest in land that is part of the total mineral estate. ... A royalty interest derives from the grantor's mineral estate and is a nonpossessory interest in minerals that may be separately alienated. The same instrument may convey an undivided portion of the mineral estate and a separate royalty interest, and the royalty interest may be larger or smaller than the interest conveyed in the minerals in place." This language in *Luckel* appears to support the Saenz Claimants' contention that the deeds in question could convey an undivided portion of the mineral estate and a separate royalty that would become smaller under future leases than the interest in the minerals in place.

In *Concord*, however, this court relied on the "two-grant" or "multiple-grant" theory to hold that a deed conveyed a 1/96 mineral interest and a separate 1/12 interest in rentals and royalties. 878 S.W.2d 191 (Tex. App.--San Antonio 1994), rev'd, 966 S.W.2d 451, 41 Tex. Sup. Ct. J. 476 (Tex. 1998). The Texas Supreme Court reversed, noting that it was not evident from the four corners of the conveyance that two differing interests were to be conveyed. The court reasoned that the deed had two references to the term "estate" indicating that only a single estate was being conveyed. In addition, the deed stated that the estate being conveyed did not depend upon the validity of and would not be affected by the termination of leases. The court noted that if two estates were conveyed, the termination of the leases would have an effect on the conveyed estate. Finally, the court reasoned, "No language in the conveyance indicates that the 1/12 interest in rents and royalties was meant to be *in addition to* or *separate from* the estate granted in the opening clause." This reasoning appears to somewhat conflict with the following language in *Luckel* which states, "The future lease clause in the Mayes-Luckel deed recites that the grantee 'shall be entitled to one-fourth of any and all royalties reserved under said leases.' This language is as effective to grant an interest as the formal 'do hereby grant, bargain, sell and convey' language of what we have designated as the 'granting' clause." Thus, *Luckel* does not appear to require language indicating two estates were conveyed or that the second estate was "in addition to" or "separate from" the first estate.

The other case frequently cited that directly deals with a conflicting fraction in a deed conveying a mineral interest, as opposed to a royalty interest, is Garrett v. Dils Co., 157 Tex. 92, 299 S.W.2d 904 (Tex. 1957). In *Garrett*, the Texas Supreme Court considered a deed that conveyed "an undivided one sixty-fourth interest in and to all of the oil, gas and other minerals in and under, and that may be produced from" certain land. The deed provided that it was subject to an existing oil and gas lease and that "it is understood and agreed that this sale is made subject to the terms of said deed, but covers and includes one-eighth of all the oil royalty, and gas rental or royalty due and to be paid under the terms of said lease." The deed further provided, "in the event that the above described lease for any reason becomes cancelled or forfeited, then and in that event an undivided one-eighth of the lease interest and all future rentals on said land for oil, gas and other mineral privileges shall be owned by said Grantee, he owning one-eighth of one-eighth of all oil, gas, and other minerals in and under said lands, together with one-eighth interest in all future rents." The court held that the deed would have conveyed a one sixty-fourth mineral interest "had other language in the deed not disclosed what the parties understood 'one sixty-fourth' to mean." The court noted that the grantee would be entitled to receive one-eighth of the one-eighth royalty under the existing lease and reasoned, "We can discover in that language no intent to grant a less interest under a subsequent lease than that granted under the then existing lease.

Although the granting clause in *Garrett* unlike *Luckel* conveys a mineral interest and not a royalty interest, *Garrett* is distinguishable from the instant case. In *Garrett*, the future lease clause conveyed all rights incident to ownership, including the right to lease and the right to receive bonus payments and annual rentals. Accordingly, the only question before that court was which fraction of the mineral interest was conveyed. Therefore, the problematic conflict between the granting of a mineral interest and a future lease provision appearing to convey a smaller royalty interest was not addressed.

The conflicting fraction problem often arises due to a 1923 case (which was later overruled) that led to the development of the "three-grant" or multiclause lease form containing: (1) the granting clause; (2) a "subject to" clause; and (3) a "future lease" clause. Because the typical royalty provided for in oil and gas leases at the time was 1/8, the conflicting fractions in the different clauses would generally be some multiple of 1/8.

Harmonizing all parts of the Contract and Mineral Deed, we conclude that the trial court properly construed the deeds by adopting the position of the Claypool/Lee Claimants. The Saenz Claimants concede that the granting clause granted a mineral interest of 1/2 and 15/32nds, respectively. The Saenz Claimants, however, contend that when the current lease was executed, the Claypool/Lee Claimants' entitlement to a portion of the royalty somehow reverted back to the Saenz Claimants. If we were to accept this contention, a reversion in interest could occur each time a subsequent future lease was executed.This position does not appear consistent with the four corners of the Contract or Mineral Deed contemplating a single conveyance with fixed rights. In order to harmonize the ownership of a fixed mineral interest with a reduced royalty interest, the Saenz Claimants argue that the Claypool/Lee Claimants could avoid the reduced royalty interest by "developing [their] undivided one-half interest in the minerals . . . ." After costs are paid, the Saenz Claimants contend that the Claypool/Lee Claimants "would [be] entitled to receive one-half of the proceeds from production." Since the Saenz Claimants exclusively control the right to lease and those leases would grant the lessee the right to develop the oil and gas in and under the Property, the Saenz Claimants fails to explain how the Claypool/Lee Claimants could protect their mineral interest's corresponding right to income from production by developing the Property leased by the Saenz Claimants to a third party.

Acknowledging the reason for the development of the three-grant or multiclause lease form and the typical royalty provided in leases at the time the Contract and Mineral Deed were executed, we can harmonize the lease provisions in such a way that the Claypool/Lee Claimants' respective 1/2 and 15/32nds mineral interests entitle them to consistently receive a 1/2 and 15/32nds interest in whatever amount of royalty is paid under the future lease clause. Our holding is consistent with the *Concord* decision because neither the Contract nor the Mineral Deed contain any language that make it evident that two differing estates were to be conveyed. This construction is bolstered by the language in the Mineral Deed which restricts the ability of the grantor to enter into a future lease that provides for less than a one-eighth royalty. This language contemplates the possibility of a future lease containing a larger royalty. The Mineral Deed then states that the grantee shall receive "under such leases" referring to the future leases containing the "usual one-eighth" royalty, 15/32nds of 1/8th. It therefore follows that if the grantor enters into a future lease with a royalty that is larger than the "usual one-eighth," the grantee would be entitled to 15/32nds of the larger royalty interest. We conclude that the trial court properly construed the Contract and the Mineral Deed.

Duhig Doctrine

In Duhig v. Peavy-Moore Lumber Co., 135 Tex. 503, 144 S.W.2d 878 (1940), W.J. Duhig, as grantee, was conveyed title to a tract of land in a deed that reserved to the grantor an undivided one-half interest in the minerals. Duhig subsequently conveyed the land to another party by a deed that stipulated that the grantor retained an undivided one-half interest in all of the mineral rights or minerals in and on the land. The question presented was whether Duhig reserved for himself the remaining one-half interest or merely excepted the one-half interest reserved to the original grantor. The court assumed that because the word "retain" means to keep what one already owns, the deed only conveyed title to the surface estate. The court noted that by adopting that construction, the warranty in the deed was breached at the very time of the execution and delivery of the deed because the deed warranted title to the surface estate and to an undivided one-half interest in the minerals. Likening the situation to the rule against after- acquired title, the court held that Duhig and those claiming under him were estopped from asserting title to the remaining one-half interest in minerals.

In *Duhig*, the problem arose because the deed purported to convey an undivided one-half interest in the minerals to the grantee; however, if the grantor "retained" a one-half interest, then the grantee would not receive any interest. Accordingly the grantor could not convey and retain a one-half interest because that exceeded the amount of minerals he owned.

In this case, the grantors owned the entire mineral estate, and the conveyances did not exceed the amount of the mineral estate that was owned. Accordingly, the warranty with regard to the mineral estate conveyed was not breached, and the *Duhig* doctrine is not applicable.

Conclusion

The trial court's judgments are affirmed.

NOTES

1. Why did the court ignore the title to the two conveyancing instruments? Should the name given the form should not be given controlling effect. In Etter v. Texaco., Inc., 371 S.W.2d 702, 705, 20 O.&G.R. 97 (Tex.Civ.App. 1963, writ ref'd n.r.e.), the court said no.

2. Is the court's attempted reconciliation of the Texas two-grant cases, including *Garrett v. Dils, Luckel v. White* and *Concord Oil Co. v. Pennzoil Exploration & Production Co.* persuasive? Are the canons of construction cited by the court useful in determining the result in this case?

3. What was the rationale of the court in rejecting the application of the *Duhig* rule to the two deeds? Does *Duhig* still have much vitality except in cases of over-conveyancing?

SECTION 3. TRANSFERS BY THE LESSOR: THE APPORTIONMENT OF ROYALTIES

CENTRAL PIPE LINE CO. v. HUTSON

Supreme Court of Illinois, 1948.
401 Ill. 447, 82 N.E.2d 624.

CRAMPTON, JUSTICE. This is an appeal from a decree of the circuit court of Wayne County in favor of appellees, in an action of equitable interpleader instituted by the Central Pipe Line Company to determine the ownership of funds in its hands. The funds were derived from the purchase of royalty oil, and a determination was also sought of the ownership of future funds to be derived from the same source.

We ascertain the basic facts to be as follows: In July, 1936, Emma Tyler, a widow, owned 114 acres of land. On July 13 of that year she leased all of the land for oil and gas; 74 acres thereof being in sections 27 and 28, and 40 acres being in section 4. Before any oil or gas wells were ever drilled upon any of the acreage, Emma Tyler in January, 1938 (as an act of voluntary division between her children), by deeds conveyed all of the land in fee. In each deed she reserved a life estate which she enjoyed until her death intestate in November, 1941. In December, 1945, wells were drilled on the 74-acre tract; no wells were ever drilled on the 40-acre tract. From those wells oil was produced, sold and marketed to an extent whereby the Central company had impounded $4726.95.

By various assignments the oil-and-gas lease as to the 74-acre tract passed to one Mitchell. By mesne conveyances the fee in the 74 acres became vested in Elsie Mae Cornstubble. A one-half of the fee in the 40 acres remained in Geneva Hutson, who is a daughter of Emma Tyler. The other one-half of that fee belonged to Lucille Coil, another daughter of Emma; this one-half became the property of Cecil Tyler subject to a reservation by Lucille of an undivided one-fourth of the minerals for 20 years or so long as oil or gas is produced. The oil-and-gas lease covering the 40-acre tract was retained by the assignee of the original lessee.

The oil-and-gas lease which Emma Tyler gave did not contain a royalty proration clause providing, in the event the leased 114 acres shall thereafter be owned in severalty or in separate tracts, that the entire 114 acres shall be developed and operated as one lease, and that all royalties accruing thereunder shall be treated as an entirety, to be divided among and paid to the separate owners in the proportion the acreage of each separate owner bears to the entire leased acreage.[89]

Errors relied upon by the appellants, and the counter-propositions advanced by appellees to negative them lead to but one issue. It is--Where a lease for oil and gas does not contain a proration clause, and the owner of the fee, subsequent to the execution of the lease, disposes of all by conveying portions thereof to others, and oil or gas is produced from some portion of the leased property after it was fractioned, does the royalty therefrom belong only to the owner of the particular portion upon which the well is located, or does the royalty belong to all the owners of all the portions upon a prorata basis? Courts of last resort of other States have been confronted with this issue, but this is the first time it has been presented to this

---

[89]The royalty proration clause referred to by the court is more commonly called an entirety (or entireties) clause. The following clause in the lease construed in Krone v. Lacy, 168 Neb. 792, 97 N.W.2d 528, 11 O.&G.R. 492 (1959), is typical:

If the leased premises are now or hereafter owned in severalty or in separate tracts, the premises, nevertheless, may be developed and operated as an entirety, and the royalties shall be paid to each separate owner in the proportion that the acreage owned by him bears to the entire lease area. There shall be no obligation on the part of the lessee to offset wells on separate tracts into which the land covered by this lease may hereafter be divided by sale, devise, or otherwise, or to furnish separate measuring or receiving tanks for the oil produced from such separate tracts. [Eds.]

court.

. . .

Where this issue has been presented to the courts of last resort in other States their decisions have not been universally harmonious. The lack of harmony is due to a fundamental difference of opinion regarding the innate character of potential oil or gas royalties. Wettengel v. Gormley, 1894, 160 Pa. 559, 28 A. 934, 40 Am.St.Rep. 733, is the earliest reported case dealing with this issue. Gormley owned three farms and gave one oil-gas mining lease. After his death each of his three children obtained a farm by devise. Subsequent to Gormley's death the lessee produced oil from one of the farms, but production was not obtained from the other two. The royalty was apportioned among the three devisees on the ground they were entitled to share therein because the oil may have been drawn from the three farms. Besides giving consideration to the fugacious nature of oil and gas, the court reasoned from the character and legal effect of the lease. An oil lease, it held, partakes of the character of a lease for general tillage, *i.e.,* agricultural rather than of one for the mining or quarrying of solid minerals. The court lacking any precedent to follow applied to the issue the doctrine of the apportionment of surface rents. The court had occasion in Wettengel v. Gormley, 184 Pa. 354, 39 A. 57, to examine its holding in the prior case. It stood by that holding, stating additionally that, inasmuch as the lease covered the three farms as a single, undivided tract, the rent (royalties) may be said to issue from each and every part. The royalties belonged to all the owners and not to the owner of any part. The court held the royalties to be personal estate which was not disposed of by Gormley in his will, hence were intestate property, to be divided among the three children in the proportion his or her share of the land bears to the whole acreage subject to the lease. The royalties were payable upon the total production from the whole acreage whenever and wherever production took place.

. . .

[The court's discussion of cases from a number of jurisdictions is omitted. It concluded that:

1. In West Virginia and Kentucky, the apportionment theory was applied in one case but the contrary doctrine was applied in others;

2. The non-apportionment theory has been explicitly followed in Arkansas, Indiana, Kansas, Ohio, Oklahoma, and Texas;

3. Two California cases appear to express approval of the apportionment theory, but both are distinguishable from the facts of the instant case;

4. The authorities cited from Louisiana were not pertinent to the inquiry in this case. [Eds.]

. . .Our careful survey of the decisions of the courts of other jurisdictions which have been confronted with the issue here, and which we have set forth in this opinion, clearly demonstrates to us that the great prevailing weight of authority is against holding unaccrued oil or gas royalties are rent, and as a consequence are to be apportioned. The fugacious oil or gas in place under a certain tract of land at a stated time is a part of the land. The fractional portion thereof, which may become royalty oil or gas in the event the underlying oil or gas is found and captured by production to the surface, ceases to be an integral part of the whole of the oil or gas under the particular land tract only when physical separation from the land occurs at the surface. Then, and only then, does the fractional portion, denominated "royalty," become personalty. In the sense of timing, it has "accrued," and may be properly labeled "rent." Before that status is reached the oil or gas is real property, and, in the absence of agreement to the contrary, the lessor owner of the tract of land conveys the oil or gas thereunder when he conveys his land. His grantee succeeds to the title to all, *i.e.,* land, oil and gas. . . .

The decision on the issue presented by this case would be the same regardless of whether the subdivision of the whole acreage came about by grant, devise or by operation of the statute controlling descent of intestate property. We agree with Professor Summers on the immateriality of how the subdivision was made; the basic controlling principle would apply in either instance.

The lease in the instant case did not provide for proration of the royalties. To accede to the view of the appellants would amount to a change in the terms of the lease from non-proration to proration without the consent of all the parties thereto. Where the writing actually expresses an agreement, equity cannot make a new one for the parties or add a provision in lieu of another without a prior accord between them for such substitution. The facts that Emma Tyler received delay rentals, and that the same were paid to her children after her death, in the proportion of their acreage, do not reinforce the contentions of appellants. Geneva Hutson was paid such rental by reason of her ownership of acreage in section 4, and she was not paid any of the rental for any interest in the 74 acres.

. . .

The decree of the circuit court of Wayne County is affirmed.

Decree affirmed.

NOTES

1. In ROBINSON v. MILAM, 125 W.Va. 218, 24 S.E.2d 236 (1942), *P* was the owner of one-fourth of the minerals in a 70-acre parcel which was a part of a 214-acre leasehold. *D* owned the balance of the minerals in the 214 acres and had an exclusive leasing power. One producing well was drilled on the 214 acres, but it was located off the 70 acres in which *P* had an interest. In lieu of further development, the lessee agreed to pay *D* a certain sum. *P* sued for a share of the royalties accruing from production from the one well on the lease and for a share of the sum paid in lieu of further development. *Held* for *P* as to the latter but for *D* as to the former. The court declared that the non-apportionment theory was followed in West Virginia, so *P* was not entitled to a share of the royalty from the producing oil well. As concerns the payment in lieu of further development the court commented as follows:

However, where a lessor and lessee agree, in lieu of development of a leasehold upon which a test well has been drilled, that the lessee will pay a certain sum, however measured, the owner of a portion of the minerals in a subdivision of the leasehold is entitled to a share of such payments, based upon his proportionate mineral ownership. . . .[T]he agreement herein . . .provides that it shall be nullified by the drilling of another well on the leasehold, in which event it necessarily follows that payment of the royalties therefrom will depend upon the location thereof.

We do not intend by what has been said herein to relax the rule as to apportionment of royalties . . ., but we believe that money to be paid by a lessee in lieu of further development is to be distinguished from royalties, the money so paid being a consideration for forbearance to enforce a covenant for reasonable development; while royalty is paid for a license to explore for minerals and when found to sever the same from the land. . . .The defendants having contracted with reference to property and an undivided interest of a portion of such property being owned by the plaintiffs, and the defendant deriving profits by reason thereof from such agreement, we believe and so hold that the plaintiffs are entitled to a proportionate share of the money arising from the agreement in lieu of further development.

2. In JAPHET v. MCRAE, 276 S.W. 669 (Tex.Com.App.1925), the court commented as follows concerning a hypothetical lease covering 1,000,000 acres of land:

One corner of that land is 75 miles from one of its other corners. The land is in several counties. Suppose the owner of the land should sell 75 acres in one corner to a person interested in oil, and who thinks he knows oil land when he sees it. He counts confidently on the one-eighth royalty attaching to that 75-acre tract. He induces the original lessee to develop his land, and a gusher is forthcoming. Under the rule [of apportionment] urged . . .the man discovering this 75-acre tract as the most valuable of the entire million acres for oil purposes would be entitled to only seventy-five one-millionths of the one-eighth royalty of the oil coming out of the gusher on the 75-acre tract. Can we say such a rule is just? A party owning another 75 acres, 75 miles distant, would get as much from the oil as the man whose 75 acres has the producing gusher. And yet this man without the gusher might not have a single drop of oil under his land, which is 75 miles distant from the gusher. Such a rule would, in our opinion, be entirely impracticable.

In GRELLING v. ALLEN, 218 S.W.2d 896 (Tex.Civ.App.1949, writ ref'd n.r.e.), the court commented as follows on the problem of apportionment or non-apportionment of royalties:

We might assume a situation in which Allen, after leasing his 88 acres of land . . .divided the entire tract into eighty-eight parcels of one acre each, and thereafter sold the eighty-eight parcels to various persons, subject to the lease covering the eighty-eight acre tract. It would hardly be logical, in the event that a producing well was drilled on one of such one acre tracts, that the owner of such tract would be entitled to all of the royalty produced from that well. We think that under such circumstances all of the owners of the one acre tracts would participate in the royalty produced from a well drilled on any portion of the subdivided large tract.

Texas came close to adopting the apportionment rule. Judge Stayton submitted a draft opinion in *Japhet* adopting the apportionment rule but it was rejected in favor of the adoption of the non-apportionment rule. Stayton, *Apportionment and the Ghost of a Rejected View*, 32 Tex. L.Rev. 682 (1954). Is the non-apportionment rule more fair? *See e.g.*, Huie, *Apportionment of Oil and Gas Royalties*, 78 Harv.L.Rev. 1113 (1965).

3. REPUBLIC NATURAL GAS CO. v. BAKER, 197 F.2d 647, 1 O.&G.R. 1142 (10th Cir.1952). Lessor executed an oil and gas lease to 640 acres of land in Kansas and thereafter conveyed to Parker an undivided one-half interest in the minerals under one of the quarter sections covered by the lease. Soon thereafter a well was drilled on the quarter section in which Parker held his mineral interest. The question presented was whether the royalty should be divided on the basis of interests in the section leased or on the basis of interests in the quarter-section upon which the well was located. For purpose of arriving at a proration formula, the Kansas Corporation Commission declared that the acreage factor of each well was to be "equal in number of acres held by production from the said well divided by 640," and the term "held by production" was construed to mean the acreage which "participates in the production upon a royalty basis and does not receive delay rentals." After completing the one well, Lessee had neither an express nor implied obligation to drill additional wells on the lease. The trial court held that the owners of the mineral interest under the quarter-section on which the well was drilled were entitled to all the royalties from such well. *Held,* "We agree with the trial court that the basic order does not have the effect of unitizing the royalty interests, and the judgment is affirmed." Huxman, Circuit Judge, dissenting, characterized the non-apportionment rule as "the harsh law of the tooth and the claw."

―――

RUTHVEN & CO. v. PAN AMERICAN PETROLEUM CORP.

Supreme Court of Kansas, 1971.
206 Kan. 639, 482 P.2d 28, 39 O.&G.R. 242.

HARMAN, COMMISSIONER. This is an action for an accounting from an oil purchasing company for a share of oil produced from leased land.

The appeal involves construction of an entirety clause in the lease under which the oil was produced. A cross-appeal concerns the validity of a separate conveyance upon which plaintiffs' claim is based.

We summarize certain background facts as stipulated by the parties prior to trial. On July 3, 1924, C.A. Mermis and his wife Paulina executed and delivered to George W. Holland as grantee an instrument entitled "Sale of Oil and Gas Royalty" in which for a consideration of $7,000 they conveyed an undivided one-fourth interest in the oil and gas minerals produced from the *west* half of a described quarter section of land in Russell county, the instrument reciting that the property was subject to a particular oil and gas lease. This instrument was not filed of record with the Russell county register of deeds until March 10, 1925--more than ninety days after its execution.

On June 13, 1936, Paulina Mermis, a widow, owner of the west half mentioned above, executed and delivered to the same George W. Holland an instrument entitled "Ratification of Mineral Deed". This instrument, which was primarily intended to ratify and confirm the July 3, 1924, document, because of possible ambiguity in the latter, was duly recorded June 15, 1936. Plaintiffs are the successors in interest to George W. Holland, who died December 22, 1946.

On June 18, 1956, the same Paulina Mermis executed and delivered to Leo J. Dreiling as lessee an oil and gas lease covering the entire quarter section for a primary term of two years. This lease was duly recorded the same day. Neither plaintiffs nor their predecessor in title ever joined in the execution of this or any other lease covering the quarter section.

Production of oil under the Dreiling lease was obtained on wells located on the *east* half of the quarter section. Stanolind Oil Purchasing Company, predecessor of defendant Pan American Petroleum Corporation, purchased the first oil produced therefrom on August 16, 1956, and production from wells on the *east* half has continued to date. No production has ever been obtained from the *west* half of the quarter section. On November 12, 1956, Stanolind obtained and issued a division order covering oil payments attributable to the production on the east half. Under this order Mrs. Mermis was to be paid all the landowner's one-eighth royalty for the production, and she was so paid by Pan American and its predecessors in title during her lifetime and until September 1, 1965. Thereafter payment was made to the executors of her estate until the land was sold to Curtis Warren and Lloyd J. Witt, then payment was made to them until May 19, 1967, when Witt died and his share was paid to the administrator of his estate until the commencement of this suit. All royalty payments since September, 1967, have been impounded by Pan American pending determination of ownership in this litigation. Plaintiffs have never received any royalty based on production under the Dreiling lease.

No claims were ever filed against the estates of Paulina Mermis or Lloyd J. Witt and the time for filing such claims has expired.

. . .

August 28, 1967, plaintiffs commenced this action by filing their petition naming Pan American as defendant. They alleged the execution of the three instruments we have mentioned; that at the time the Dreiling lease was executed the premises were owned in severalty and by reason of an entirety clause in that lease plaintiffs were entitled to 1/64 th of the proceeds of all oil produced and they asked for an accounting. Plaintiffs' claim for an apportionment of royalty is against Pan American only.

Pan American answered, asserting among other matters, the defense of laches and estoppel by conscious inaction and delay and, by way of third party practice, it brought into the action as defendants Curtis Warren, the Witt heirs and the executors of the Paulina Mermis estate so that it might be reimbursed by them if held liable to plaintiffs.

[Discussion of the invalidity of the 1924 instrument is omitted. Eds.]

Plaintiffs contend, and we think correctly so, that the 1936 instrument itself constituted a valid conveyance. . . .

. . .The trial court correctly found the existence of a valid mineral deed in and to an undivided one-fourth of the minerals in the west half of the quarter section in George W. Holland effective June 13, 1936, and the cross-appeal is not sustained.

We turn now to the principal appeal. Plaintiffs' claim to a proportionate share of royalty is based on a provision, commonly known as an "entirety clause", in the Dreiling lease. This clause provided:

> If the leased premises are now or hereafter owned in severalty or in separate tracts, the premises, nevertheless, may be developed and operated as an entirety, and the royalties shall be paid to each separate owner in the proportion that the acreage owned by him bears to the entire leased area. There shall be no obligation on the part of the lessee to offset wells on separate tracts into which the land covered by this lease may hereafter be divided by sale, devise, or otherwise, or to furnish separate measuring or receiving tanks for the oil produced from such separate tracts.

Their argument is this: Mrs. Mermis had a right to execute a lease with an entirety clause; in 1956 she did so by leasing the entire quarter section to Dreiling; when production was obtained plaintiffs were entitled to a proportionate share of the royalties by reason of the entirety clause, even though the production was only upon the east half of the quarter section.

In 3 Kuntz, Oil and Gas, § 45.4, under the heading "Correlation of royalty clause with entirety clause" we find this discussion as to the background and purpose of the entirety clause:

> Ordinarily, one important deliberate and calculated effect of the entirety clause is to overcome the nonapportionment rule which is applied in most jurisdictions. Under such rule, when the land under lease has been subdivided after the delivery of the oil and gas lease and the lessor's interest is owned in severalty, the royalty on production from a well is paid to the lessor or lessors only to the extent of their interests in the tract where the well is located, despite the fact that such well satisfies the drilling clause and the habendum clause of the lease as to all of the land described in the lease and consequently serves to hold the lease by production. Such application of the non-apportionment rule produces dissatisfaction on the part of lessors who own parts of the leased premises where there is no well. A dissatisfied lessor invariably sees an urgent necessity for further development of the lease by the drilling of a well or wells on the tract owned by such lessor. Further, even if

there are many wells and such wells are evenly distributed so as not to be objectionable to the lessors, the lessee may find the situation objectionable in that he may be required to duplicate equipment and to install additional measuring tanks or meters, where production from all of such wells is not uniform in quality and volume. If the lease provides that the lessors must bear the expense of additional equipment which is required as the result of a subdivision of the land after a lease is granted, such added expense will provide the lessors with an additional reason to be dissatisfied. In order to overcome the objections just mentioned and to maintain harmony among all of the interested parties, as well as to make certain that the obligations of the lessee to develop the property will not be increased by any conveyance by the lessor, an entirety clause is frequently inserted in an oil and gas lease. (pp. 434-435.)

See also 3A Summers, Oil and Gas (Perm. ed.) § 609.

Kansas has been among the jurisdictions adhering to the so-called non-apportionment rule (Carlock v. Krug, 151 Kan. 407, 99 P.2d 858). Our question is whether the entirety clause is applicable under the facts here.

Plaintiffs' argument is based on the assumption the term "leased premises" as used refers to and covers the entire quarter section. They contend that to hold otherwise would leave the clause with no office to perform. In support of their position plaintiffs rely almost entirely on Hoffman v. Sohio Petroleum Co., 179 Kan. 84, 292 P.2d 1107. There a half section of land was leased for oil and gas, the lease containing an entirety clause. The land was thereafter sold in separate tracts at a partition sale, subject to the existing lease. It was held the purchasers of the tract upon which no oil was produced were entitled to participate in the royalties from oil produced under the lease on the other tracts in the proportion indicated in the entirety clause.

Under its facts the Hoffman case is distinguishable from the situation here in that the oil and gas lease there was executed by the owner of the entire leased acreage prior to division of the mineral ownership and the entire acreage was subject to the lease containing the entirety clause. There can be no question as to the application of such provision to lands covered by the lease and which are separately owned at the time of the execution of the lease or subsequently. An equitable result is achieved when all the leased land is treated as a unit and royalties are paid on a proportionate basis. Here, however, the mineral interest was divided prior to execution of the lease containing the entirety clause.

We think the answer lies simply in resort to the language contained in that clause. Its premise for operation is: "If the *leased premises* are now or hereafter owned in severalty or in separate tracts". (Our emphasis.) Mrs. Mermis alone executed this lease. She was at the time owner of the east half of the quarter section (80 mineral acres) and an undivided three-fourths of the west half of the quarter section (60 mineral acres). This constituted the property embraced by the lease. As such it comprised the "leased premises" mentioned in the entirety clause. Plaintiffs or their predecessor in title did not execute or join in the lease. Mrs. Mermis could not and did not lease their mineral interest. Her execution of the lease, despite its recitation that it embraced the entire quarter section, was effective only as to the interest she owned. The "leased premises" were not the entire land (the quarter section) but only the interest therein subject to the lease. Hence the "leased premises" were not owned in severalty or in separate tracts at the time of the execution of the lease, or thereafter, and the entirety clause never became operative. The trial court took this view and we think correctly so.

. . .

We hold then that the term "leased premises" as used in an entirety clause in an oil and gas lease means the lessor's interest which is the subject of the lease.

. . .

We affirm the orders and judgment of the trial court both as to the appeal and the cross-appeal.

Approved by the Court.

SCHROEDER, JUSTICE (dissenting).

The question presented by this lawsuit is whether the plaintiffs (appellants), as the owners of an undivided interest in the minerals under a portion of the leased acreage, are entitled to share in the royalty oil produced from the leased acreage.

. . .

The "leased premises" in the lease here under consideration clearly refer to the *entire quarter section of land.* To hold otherwise would fail to give credence to the granting clause which covers:

> . . .*all that certain tract of land,* together with any reversionary rights therein, situated in the County of Russell, State of Kansas, described as follows, to-wit: [Here the legal description of the entire quarter section is given.] (Emphasis added.)

It would also fail to give credence to the "lesser interest" clause which reads:

> If said lessor owns *a less interest in the above described land than the entire and undivided fee simple estate therein,* then the royalties and rentals herein provided shall be paid the lessor only in the proportion which his interest bears to the whole and undivided fee. . . . (Emphasis added.)

The lease also contained an "entirety" clause reading in part as follows:

> . . .If the leased premises are *now* or hereafter *owned in severalty or in separate tracts,* the premises, nevertheless, may be developed and operated as an entirety, *and the royalties shall be paid to each separate owner in the proportion that the acreage owned by him bears to the entire leased area.* . . . (Emphasis added.)

. . .

Paulina Mermis had previously executed the "Ratification of Mineral Deed" and hereby knew that her grantee, George W. Holland, or his successors, owned an interest in the west eighty. In negotiating the terms of the lease she was free to limit it to the east eighty if she chose to do so. On the contrary, she chose to include the entire quarter section. The third party defendants (all defendants other than Pan American) as her successors in interest are, of course, bound by such burdens as she chose to impose upon the lands. Clearly, she had the right to execute the lease with an "entirety" clause. Any holding to the contrary would deprive her of rights possessed by her at the time of negotiating the lease.

The court's opinion overlooks the recitals in the "lesser interest" and "entirety" clauses.

Under the "lesser interest" clause Paulina Mermis agreed that if she owned a lesser interest than the "entire and undivided fee simple estate," then her royalties are to be that portion which her interest bears to the "*whole and undivided fee.*"

She applies this language to "*the above described land," and not to just her interest therein.*

The "entirety" clause recites:

> . . .If the leased premises are *now* or hereafter *owned in severalty or in separate tracts,* . . .. (Emphasis added.)

The word "now," which the court overlooks, means the time when the lease was executed. When the lease was executed Paulina Mermis and her successors in interest owned 140 of the 160 mineral acres in the lease--80 acres under the east eighty and 60 acres under the west eighty. Clearly then, under the "lesser interest" clause of the lease she was entitled to 140/160ths or 7/8ths of the 1/8th reserved royalty.

The leased premises on the date of the lease were "owned in severalty" as to the west eighty, and "in separate tracts" as between the east and west eighties. The "entirety" clause directs that "the royalties shall be paid to each separate owner in the proportion that the acreage owned by him bears to the entire leased area." Thus, under the lease the remaining 1/8th of the 1/8th reserved royalty belongs to the plaintiffs.

. . .

Under the plain and unambiguous language of the lease here in question, the plaintiffs have established a right to their proportionate share of the royalty reserved.

. . .

It is respectfully submitted the plaintiffs are entitled to their proportionate share of the 1/8th reserved royalty under the lease here in question, and the trial court should be reversed.

NOTES

1. Krone v. Lacy, 168 Neb. 792, 97 N.W.2d 528, 11 O.&G.R. 492 (1959):

The question is, may the owners of all royalty interests under a lease containing an entirety clause agree among themselves for a different division of the royalties accruing thereunder than that fixed by the entireties clause and, if they do, is such an agreement enforceable among themselves?

. . .

The entireties clause is a valid burden imposed upon the interests in the property retained by the lessors. . . .The lessors, by executing the lease, placed a restriction upon their power to alienate any part of their estate in the land covered by the lease except in accordance with the provisions of the lease itself. That they had the right to do so and that such restriction is not against public policy has already been determined by this court. Such provision, being a covenant running with the land, imposed the same burden on any interest acquired by anyone from the lessors to the lease as upon the lessors themselves.

. . .

We have come to the conclusion that by the terms of an entirety clause, such as is here contained in the gas and oil lease, the parties thereto agree that in the event the mineral interests under the leased premises become separately owned that the royalties shall nevertheless be treated as an entirety, the separate owners to participate pro rata, and that by reason of the provisions of the entireties clause, the separate owners cannot convey any interest in their royalty rights therein contrary thereto.

We recognize there are holdings in other states contrary to the foregoing. . . . [I]f the provisions of an entireties clause are not against public policy, and they are not, and the parties voluntarily enter into a lease so providing then the provisions thereof should be and, in our opinion, are in effect at all times as to all parties thereto as long as the lease remains in force and effect.

Since the entirety clause is binding on the lessors, lessees, and all holding under them, it necessarily follows that all parties affected must consent to any change in the apportionment required by the clause in order to make such change effective.

In BRUBAKER V. BRANINE, 237 Kan. 488, 701 P.2d 929, 86 O.&G.R. 351 (1985), the court repeats the language of *Krone* characterizing the entireties clause as a "restriction" on the "power to alienate" in a situation where there was no "express modification" of the lease 'entirety clause' in the conveyance of a portion of the land subject to the lease.

FOERTSCH V. SCHAUS, 477 N.E.2d 566, 85 O.&G.R. 246 (Ind.App.1985), reaches the same result as *Krone,* characterizing the entirety clause as a "covenant running with the land." The opinion in *Foertsch* points out that its ruling " . . .is not to be construed as a prohibition against parties assigning accrued royalties, as such are choses in action, and not interests in real estate." Does this mean that parties cannot assign the proceeds of potential royalties which may accrue in the future from a lease with an entirety clause?

2. Parties to mineral or royalty conveyances subsequent to a lease had no actual knowledge of the presence of an entirety clause in the lease (although they obviously had constructive notice thereof), but all parties understood that royalties were not to be apportioned. Case authority is divided on the availability of reformation of the deeds to carry out the contract in accordance with the mutual intent of the parties. *See* Martin & Kramer, *Williams and Meyers Oil and Gas Law* § 521.3.

3. Executive, the owner of the minerals in Blackacre and Whiteacre, two adjoining eighty-acre parcels, conveyed one-half the minerals in Whiteacre to *A*, retaining the exclusive leasing power, and conveyed one-quarter of the minerals in Whiteacre to *B* absolutely. Thereafter Executive executed an oil and gas lease affecting the south one-half of Blackacre and Whiteacre which covered the 50-acre interest owned by Executive and the 20-acre interest of *A*. A producing well was drilled on the leased portion of Blackacre and another upon the leased portion of Whiteacre, but the former was much more productive than the latter well. The lease contained the following entirety clause:

> If the leased premises shall hereafter be owned in severalty or in separate tracts, the premises, nevertheless, shall be developed and operated as one lease and all royalties accruing hereunder shall be treated as an entirety and shall be divided among and paid to such separate owners in the proportion that the acreage owned by each such separate owner bears to the entire leased acreage. There shall be no obligation on the part of the lessee to offset wells on separate tracts into which the land covered by this lease may be hereafter divided by sale, devise, or otherwise, or to furnish separate measuring or receiving tanks.

On what basis should the royalty on production from the two wells be divided? *See* Stroud v. D-X Sunray Oil Co., 376 P.2d 1015, 17 O.&G.R. 787 (Okl. 1962).

4. Plaintiff, the owner of a 3/4 mineral interest in Blackacre and a 1/4 mineral interest in Whiteacre, executed an oil and gas lease covering both tracts which included an entirety clause providing that:

> If the leased premises are now or shall hereafter be owned in severalty or in separate tracts, the premises, nevertheless, shall be developed and operated as one lease, and all royalties accruing hereunder shall be treated as an entirety and shall be divided among and paid to such separate owners in the proportion that the acreage owned by each such separate owner bears to the entire leased acreage.

Upon development, Whiteacre proved to be more productive than Blackacre. To what royalty is plaintiff entitled? *See* Thomas Gilcrease Foundation v. Stanolind Oil & Gas Co., 153 Tex. 197, 266 S.W.2d 850, 3 O.&G.R. 673 (1954).

5. An oil and gas lease contains the following clause:

> If this lease now or hereafter covers separate tracts, no pooling or unitization of royalty interests as between any such separate tracts is intended or shall be

implied or result merely from the inclusion of such separate tracts within this lease ... the words 'separate tract' mean any tract with royalty ownership differing, now or hereafter, either as to parties or amounts, from that as to any other part of the lease premises.

The oil and gas lease covers two separate tracts, with each tract burdened by a non-participating royalty. The lease also contains a pooling clause. A producing well is drilled on the south half of the tract. Both non-participating royalty owners ratify the lease. Should the lessee pay apportioned or non-apportioned royalties under the lease? *See* Verble v. Coffman, 680 S.W.2d 69, 70, 83 O.&G.R. 405 (Tex.App. 1984). *In accord,* London v. Merriman, 756 S.W.2d 736, 111 O.&G.R. 591 (Tex.App. 1988). *See also,* Kramer & Martin, *The Law of Pooling and Unitization* § 7.04.

SECTION 4. TRANSFERS BY LESSEE: RELATIONSHIP OF TRANSFEROR AND TRANSFEREE

(See Martin & Kramer, *Williams and Meyers Oil and Gas Law* §§ 420-420.2.)

XAE CORP. v. SMR PROPERTY MANAGEMENT CO.

Supreme Court of Oklahoma, 1998.
1998 OK 51, 968 P.2d 1201, 141 O.&G.R. 557

HARGRAVE, Justice. We granted certiorari to review the question whether the implied covenant to market under the oil and gas lease extends to an overriding royalty interest owner. The overriding royalty interest in this case was an in-kind interest granted by separate conveyance rather than reserved in the assignment of oil and gas leases.

Specifically, the question is whether the defendants/appellants (hereinafter "SMR") improperly deducted from the overriding royalties paid to plaintiffs their proportionate share of the costs incurred in gathering, processing and compressing gas produced from the subject wells. The trial judge determined as a matter of law that the gas was not in marketable form at the wellhead, and that SMR had a duty to make the gas marketable. The trial judge granted summary judgment for the overriding royalty interest owners. We hold that there is no implied covenant to market applicable in this case because no obligation was undertaken in the instrument creating the overriding royalty interest and because the interest was an in- kind interest deliverable at the wellhead.

Facts

The plaintiffs are the successors in interest to an overriding royalty interest conveyed by SMR's predecessor in title. The conveyance was a grant of overriding royalty interest executed in 1967, which provided:

Whereas, it is the present intent of J.C. Barnes Oil Company to assign, transfer, set over and deliver unto Clayton E. Lee and R.L. Beasley, in equal shares, hereinafter called Assignees, the overriding royalty interest in and to the leases described on Exhibits ... hereto attached in accordance with the terms and conditions thereof ... "An overriding royalty interest of an undivided 1/8 of 7/8 of all gas, gas condensate or other gaseous hydrocarbons which may be produced under the terms of the oil and gas leases described in Exhibits "A" and "B" attached hereto and made a part hereof, the same to be delivered to the Assignees herein, free and clear of all costs and expenses whatsoever, save and except gross production taxes or other governmental taxes properly chargeable thereto.

The parties agree that this clause created an in-kind overriding royalty interest, meaning that the overriding royalty granted was a fraction of the gas produced rather than of the gas sold. The assignment stated that it applied to all extensions or renewals of the oil and gas leases, and contained a proportionate reduction clause. The overriding royalty interests were made subject to previously existing overriding royalty interests and a production payment. The assignment contained no express provision placing a duty on the lessee to market the product.

Plaintiffs did not take their share of gas in-kind, but instead authorized SMR Property Management Company to market their share. SMR, acting as agent for itself and the other defendants, paid plaintiffs for gas produced and sold from the leased premises. SMR deducted from the amounts paid to the plaintiffs a charge for gathering and delivering the gas to an amine treatment facility where large quantities of hydrogen sulfide and carbon monoxide were removed from it. The gas was sold at the outlet from the amine treatment facility. Plaintiffs' brief states that the amine treatment facilities are located "on or near" the subject leases.

Plaintiffs sued to recover all deductions made by SMR for the gathering, delivery and treatment charges, alleging that they were entitled to be paid their overriding royalty interests on gas produced from the subject wells without any deductions for gathering, delivery and treatment of the gas. Plaintiffs sought summary judgment that as a matter of law they were entitled to receive their overriding royalty interest free and clear of any such costs or charges, arguing that the "production" process does not end until a marketable product has been obtained, and that because the gas was not marketable in its natural state, the lessee was required to bear the costs of making it marketable. They relied upon three lessor royalty cases: TXO v. Commissioners of the Land Office, 903 P.2d 259 (Okla.1994); Wood v. TXO, 854 P.2d 880 (Okla.1992) and Clark v. Slick Oil Co., 88 Okla. 55, 211 P. 496 (1922). Plaintiffs urge us to adopt the rationale of Garman v. Conoco, 886 P.2d 652 (Colo.1994), which applied the implied covenant to market under the oil and gas lease to the overriding royalty interests.

The trial court and the Court of Civil Appeals found non- marketability of the gas to be the primary issue in the case. The plaintiffs maintain that the agreement of the parties is silent as to the point at which the in-kind overriding royalty is to be delivered free of all costs and expenses and as to the required condition of the gas at the point of delivery.

. . . SMR argues that the Wood v. TXO and TXO v. Commissioners cases have no bearing in this case because those cases involved lessor royalties and the lessee's marketing covenant under the oil and gas lease. . . .

Implied Covenant

We decided the royalty owner cases based on the implied covenant of marketability under the oil and gas lease. The implied covenants in the oil and gas lease ordinarily can not be enforced by an overriding royalty interest owner. Williams and Meyers, Oil and Gas Law, § 420, p. 356-7 (1981) states the general rule:

> The owner of an overriding royalty is not entitled to the benefit of the covenants of the base lease, express or implied, in the absence of an express provision in the instrument creating the overriding royalty. The benefits of such express and implied covenants of the lease touch and concern the lessor's estate and burdens of such covenants touch and concern the lessee's estate. The assignment, either in whole or in part, of the burdened estate will not permit enforcement of the covenants which burden the assigned estate by a person other than the lessor or claimants though him of a portion or all of the benefitted estate.

This Court has said that, unless expressly assumed, implied covenants of an oil and gas leases do not extend to lease assignments with reservation of overriding royalty interest. Kile v. Amerada Petroleum Corp., 118 Okla. 176, 247 P. 681 (1925). We said that unless the contract of assignment imposed rights and obligations on the party the same as that of lessor and lessee, such relationship would not be implied. We declined to apply an implied covenant to protect against drainage in favor of an overriding royalty interest owner. We said that to hold the fundamental provisions in leasing contracts to be implied in the contract of assignment would be to stretch the rule beyond legitimate bounds where there was no express agreement in the contract that could form the basis of relief for the breach of an implied obligation. We said that such owner was limited to the terms of the assignment itself and that where there was no express obligation to drill or produce, no covenant would be implied.

In *Kile* we explained that the covenants implied in an oil and gas lease will not be implied in the transfer of an oil and gas lease with retained overriding royalty interest unless other factors are present that require that covenants be implied in order to achieve the fundamental purpose of the transaction. We pointed out that the grant of an overriding royalty interest by the owner of the lease bears no relationship to a leasing transaction and the covenants usually implied in an oil and gas lease should not be implied.

Likewise, the Supreme Court of Arkansas, in McNeill v. Peaker, 253 Ark. 747, 488 S.W.2d 706 (1973) held that covenants to reasonably develop or protect the premises from drainage by offsetting wells were not implied as a matter of law in assignments of oil and gas leases in which the assignor's reservation of an overriding royalty interest constituted the principal consideration for the assignment.

Professor Kuntz points out that the grant of an overriding royalty by the owner of the lease, as distinguished from an exception or reservation of overriding royalty, bears no resemblance to a leasing or subleasing transaction and that the relation is that of grantor and grantee, with no basis for concluding that the parties contemplated that the grantor would owe the grantee any duties regarding development. Accordingly he says, covenants usually implied in oil and gas leases should not be implied in this instance. E. Kuntz, The Law of Oil and Gas § 55.3(3).

Williams and Meyers, supra, § 420.1 recognize that there may be instances in which implied covenants should be enforceable by overriding royalty owners and Professor Kuntz states that it is possible for other features of the overriding royalty transaction to provide a basis for implying obligations. For example, where the transferor expressly covenants to develop the lease, then there is sufficient basis to invoke the implied covenants incident to the express covenant. Kuntz goes on to say that: "In the absence of an express covenant to develop and in the absence of special circumstances which reveal that development is the object of the sale, the covenants implied in oil and gas leases should not be implied in the grant of an overriding royalty interest."

The implied covenant to protect against drainage appears to be the implied covenant most often implied in favor of the overriding royalty. It has been surmised that courts that uphold a covenant to protect against drainage in favor of the overriding royalty do so because drainage is tantamount to conversion.

Louisiana and Texas have applied the implied covenant to protect against drainage to protect the assignor of oil and gas leases who reserved an overriding royalty interest. In Bolton v. Coats, 533 S.W.2d 914 (TX 1975), the Texas Supreme Court held that unless the assignment provides to the contrary, the assignee of an oil and gas lease impliedly covenants to protect the premises against drainage when the assignor reserves an overriding royalty. It is not stated whether the assignment contained any express covenants. The court stated that Bolton was entitled to the benefit of the implied covenant under his assignments if his allegations of drainage were found to be true and if the protection from drainage would have been afforded by a reasonably prudent operator under the same or similar circumstances,. . .

Some courts have begun to apply the implied covenant to market in favor of the overriding royalty owner. Garman v. Conoco, Inc., 886 P.2d 652 (Colo.1994) applied the implied covenant to market the product to the overriding royalty owners before it, with little discussion of the differences between an overriding royalty and the lessor's royalty. In a footnote the Garman court noted that some question exists whether the implied covenants under an oil and gas lease extend to overriding royalty owners, citing § 420 of William and Meyers, above. The court stated, however, that the rationale for application of the covenants to protect the lessor similarly extends to the interest of an overriding royalty owners, citing

§ 420.1 of William and Meyers, *supra*, and Bolton v. Coats, *supra*. . . .

We must disagree with our sister state's rationale as applied to the overriding royalty interest in this case. Oklahoma has viewed implied covenants differently. Implied covenants to develop an oil and gas lease grew out of the operations of oil and gas leases where production is for mutual benefit of both lessor and lessee. We made clear in Kile v. Amerada Petroleum Corp., supra, that the implied covenants of the oil and gas lease could not be enforced by the overriding royalty owner.

We discussed the nature of the overriding royalty interest extensively in Thornburgh v. Cole, 201 Okla. 609, 207 P.2d 1096, 1098 (1949), a coal mining case. We used the definition of "overriding royalty" from a California oil and gas case defining "overriding royalty" as a fractional interests in the production of oil and gas as are created from the lessee's estate, whether by reservation or by grant. We said that the definition applied equally to a coal mining lease, and illustrated the fact that such an interest "overrides" or is in addition to the royalty reserved to the land owner or lessor, and generally arises through contracts between the lessee and a third person, whereby the owner and holder of the 7/8 ths working interest provided to the lessee, contracts and agrees that some specified portion of his interest in production shall be paid to a third party, for such consideration as is agreed upon. The lessor, or party receiving the usual royalty provided for in the lease contract, having generally no interest in the working interest, is not a party to such contracts. We said that overriding royalty has been defined as: "... a certain percentage of the working interest which as between the lessee and the assignee is not charged with the cost of development or production. (citations omitted)."

Professor Kuntz posits that the word "overriding" as a further descriptive term to modify "royalty," is intended to mean that the interest is to override or be free of the burdens normally incident to the working interest out of which it is carved. Kuntz, § 63.2, p. 218. An overriding royalty is an interest in the oil and gas lease out of which it is carved, and cannot be a property interest of greater dignity than the lease itself. The overriding royalty interest is created out of the working interest in a lease. The assignee of an oil and gas lease may agree to pay the assignor a certain fraction of the assignee's share of the production. The nature of an overriding royalty interest is such that it attaches only when oil and gas are reduced to possession. Before this, the owner of an overriding royalty has no assertable right in the leasehold and the vesting of such owner's rights are dependent upon the happening of a future event or condition.

The rationale of Kile v. Amerada Petroleum Corp., supra still applies. Oklahoma has recognized that the overriding royalty interest is different from the lessor's royalty interest. An overriding royalty "overrides," or is in addition to the royalty reserved to landowner or lessor, and generally arises through contracts between the lessee and a third person. Kile v. Amerada Prod. Corp. has not been overruled and we decline to do so now. We find that the implied covenant to market cannot be enforced by the overriding royalty interest owners in this case.

In-Kind Overriding Royalty Interest

[The court in this section noted that the point of delivery for an in-kind royalty interest is at the wellhead. Thus there is no implied duty or covenant to deliver the gas in anything but its natural condition, unlike the duty of the lessee to put gas into a marketable condition where the royalty obligation is determined by a payment. Eds.]

Conclusion

Our statements in the royalty owner cases that production has not ended until a marketable product had been obtained, and the references therein to "non-operating interests," were made within the context of the lessor/lessee relationship and the lessee's implied duty, under the oil and gas lease, to market the product. That rationale does not apply in the case at bar. The Assignment whereby the royalty owners received their interests did not create any obligation to market. We find that the trial court erred determining that un-marketability of the gas at the wellhead entitled plaintiffs to summary judgment. The duty placed upon the lessee to deliver gas in marketable form arises from the lessee's implied duty, arising out of the oil and gas lease, to market the product. No such duty exists toward the overriding royalty interest owner unless such obligation is created by the assignment. Here, the obligation is merely to deliver the gas in-kind when production is obtained.

Certiorari Previously Granted; Court Of Civil Appeals' Opinion Vacated; Trial Court's Judgment Reversed.

NOTES

1. COOK V. EL PASO NATURAL GAS CO., 560 F.2d 978, 58 O.&G.R. 206 (10th Cir. 1977). A lessee (Cook) under a federal oil and gas lease assigned the lease to El Paso, retaining an overriding royalty interest. El Paso drilled a well on some adjacent lands, also leased by the federal government, in part because under federal regulations a well could not be drilled on the Cook lease since there were valuable potash deposits located there. Cook sued El Paso under the implied covenant to prevent drainage. The United States as royalty owner under both leases was indifferent to the outcome. What result? The court relied on Williams & Meyers, Section 420, but nonetheless found that the covenant ran with the land. Is that consistent with the analysis in *XAE*? Why would Cook, the original covenantor, be able to stand in the shoes of the United States, the original covenantee?

2. TIDELANDS ROYALTY B. CORP. v. GULF OIL CORP., 804 F.2d 1344, 95 O.&G.R. 604 (5th Cir.1986), *reversing*, 611 F.Supp. 795, 86 O.&G.R. 162 (N.D.Tex.1985). In connection with a sale of geophysical data to defendant Gulf plaintiff Tidelands received the right to assignments of overriding royalties on "federal offshore oil and gas leases on certain designated portions of the 2,700,000 acres covered by the geophysical data." Leases were acquired by Gulf

and overriding royalties were assigned to Tidelands. Some of these overriding royalties are in "Block 332" and defendant Gulf's "operations on adjoining property (Block 333) are, admittedly, draining gas from some of the producing sands under Block 332."

Plaintiff filed this action claiming that Gulf had breached an implied covenant to protect against drainage that arose when Gulf assigned the overriding royalty required by the sale of the geophysical data. The trial court analogized the assignment of the overriding royalty to an oil and gas lease and concluded that the implied covenant existed. On appeal, the Fifth Circuit reversed, holding that the relationship between lessor and lessee is not analogous to the relationship between an assignor of overriding royalty such as Gulf and assignee. That would be particularly true here where the plaintiff was not the owner of the development right but merely the owner of valuable geophysical information. Instead, the Fifth Circuit found that the relationship between Gulf and Tidelands was more akin to that of an executive and non-executive owner. Under Louisiana Mineral Code article 109 the executive owes the non-executive a duty to act in good faith. This duty would be met "[i]f the conduct of the grantor is supported by reasons other than merely the advantage of avoiding the royalty burden ..." The case was remanded for determination whether defendant had breached this standard.

For another unusual case where the owner of an overriding royalty sought to impose a fiduciary duty upon the working interest owner *see* Norsul Oil & Mining Co. v. Texaco, Inc., 703 F.Supp. 1520, 1549, 103 O&G.R. 446 (S.D.Fla. 1988).

3. Is the assignee of a lease subject to any duty to his assignor who has retained some interest in production (*e.g.,* an oil payment or an overriding royalty) to pay rentals during the primary term or to reassign the lease to the assignor prior to effecting a surrender of the lease to the lessor or prior to permitting the lease to expire during the primary term by nonpayment of rental? *See* Collins v. Atlantic Oil Producing Co., 74 F.2d 122 (5th Cir.1934).

4. PHILLIPS PETROLEUM CO. v. TAYLOR, 116 F.2d 994 (5th Cir.1941), cert. denied, 313 U.S. 565, 61 S.Ct. 941, 85 L.Ed. 1524 (1941).

Appellees Taylor and others executed an oil and gas lease to appellee Gray. Gray assigned the lease to appellant, retaining, as his sole consideration, an overriding royalty, payable out of the first oil and gas produced. . . .The assignee exercised his rights under the assignment, and drilled some producing wells.

The lessors and the lessee-assignor joined in this action to recover damages from the assignee for breach of an alleged implied covenant to protect the premises from drainage.

. . .

It is well settled that, under the lease in the case, the assignee impliedly covenanted to protect the lessor from drainage. To determine whether or not the assignor also has the protection of an implied covenant, the same rules govern as are applied to covenants in contracts between the lessor and lessee; and the obligations between the assignor and assignee are what the contract of assignment makes them, in the light of the provisions of the lease assigned."

. . .

Under the facts of this case, the only interest to be enforced by Gray was the overriding royalty. The oil having been drained, the only recourse by which his injury could be compensated lay in a suit for damages. His interest being a covenant running with the land, and enforceable as such, it certainly follows that his protection and the administration of justice require that the right to sue be awarded to him.

The law requires an assignee, who stands in the shoes of his lessee-assignor, to do that which an ordinarily prudent operator would do under the circumstances, having due regard for the interests of all parties. . . . For the protection of the lessor, the law imposes upon the assignee the duty, when he drills, to exercise reasonable prudence solely in order to prevent loss to the lessor of that which is rightfully his, be the loss by drainage or refusal to capture or surrender the lease. This duty, where not expressed, is imposed by implied covenant. Applying the same principle to the assignor and assignee in this case, it seems entirely reasonable to impose a similar duty.

. . .

The conclusion that the assignor has an implied covenant seems to be entirely in harmony with the related Texas decisions. The courts of that state have held that the retention of an overriding royalty, payable from the first oil produced, did not bind the assignee to drill where no duty to drill was owed the lessor, giving rise to an unmistakable inference that, had the duty been owed to the lessor, the right to the duty and its enforcement would be present in the assignor. . .

The conclusion reached by us can prejudice no one, since the implied covenant in favor of the lessor already has placed the onus of drilling or compensating for the failure so to do, and this obligation is not increased by our holding. Under any other holding, the assignor, his consideration consisting of royalty alone, might be seriously prejudiced by deprivation of his only enforcement privilege.

Accord, Bolton v. Coats, 533 S.W.2d 914, 53 O.&G.R. 379 (Tex. 1975).

Did the Oklahoma Supreme Court in *XAE* respond to the arguments made in Phillips Petroleum v. Taylor?

5. ROGERS V. WESTERMAN FARMS, INC., 29 P.3d 887, 149 O.&G.R. 373 (Colo. 2001) specifically rejected the *XAE* conclusion that the owner of an overriding royalty interest is not entitled to enforce an implied covenant. In *Rogers* the implied covenant to market was held to apply to the owners of

overriding royalty interests seeking to apply the "first marketable product" doctrine to royalty valuation issues.

6. CONTINENTAL POTASH, INC. V. FREEPORT-MCMORAN, INC., 115 N.M. 690, 858 P.2d 656, 125 O.&G.R. 565 (1993), on the other hand, applies traditional real covenant theory to reject the claim that the lessee can enforce the benefit of the implied covenant to market that runs in favor of the lessor. Only successors in interest to the covenantee/lessor can receive the benefit of the implied covenant to market. *In accord*: Elliott Industries Limited Partnership v. BP America Production Co., 407 F.3d 1091, 162 O.&G.R. 611 (10th Cir. 2005). See also Martin & Kramer, *Williams & Meyers Oil and Gas Law* § 420.3.

7. In LA LAGUNA RANCH CO. V. DODGE, 18 Cal.2d 132, 114 P.2d 351 (1941), the question presented to the court was stated as follows: "Does the interest of the holder of a fractional share in the production of oil, which is created out of the estate of the operating lessee, survive after the lessee's voluntary surrender of the leasehold by a quitclaim deed?" The majority answered the question in the negative. A dissenting judge stated the issue and his answer as follows: "Briefly stated, the issue is whether a lessee may surrender his lease to the lessor, and thereby cut off the rights of partial assignees of the lessees. To propound the issue is to answer it in the negative, unless all principles of equity and justice are to be ignored."

In evaluating these views consider:

(a) the appropriateness of distinguishing between the methods by which a lease may be terminated, *e.g.,* by failure to pay rentals, by surrender, by breach of covenant or condition, by failure to prosecute drilling operations, by failure to make payment of shut-in royalty, *etc.* See Edward v. Prince, 221 Mont. 272, 719 P.2d 422 (1986).

(b) the availability of a remedy by the owner of the nonoperating interest against the operating owner who permits (or causes) termination of the lease.

8. Is the operator of a lease liable to the owner of a nonoperating interest for negligent injury to a well or producing formation and resultant loss of a lease? *See* Whitson Co. v. Bluff Creek Oil Co., 156 Tex. 139, 293 S.W.2d 488, 6 O.&G.R. 155 (1956).

9. Should a sublessor owe a duty to act as a reasonable and prudent operator to its sublessee? Consider the following case. An assignee of a state lease executed a sublease of the working interest to a depth of 12,000 feet to one company. The assignee then assigned to another company 50% of the rights to the formations below 12,000 feet. The two owners of the deep rights drilled a well to 18,000 feet. An electric log that was run apparently showed hydrocarbons in paying quantities in the portion of the formation held by the sublessee. The sublessee asked to use the well bore of its sublessor to complete a well in the shallower formation. The well bore, however, was plugged by the two owners of the deep horizon rights. The sublessee then drilled at great expense a second well

that produced in paying quantities. The sublessee then filed suit against its sublessor and the other owner of the deep horizon rights contending that they should had an obligation to turn over the well bore to it. Should the owners of the deep rights, including the sublessor, have turned over the well? The court in Neomar Resources, Inc. v. Amerada Hess Corp., 648 So.2d 1066, 132 O.&G.R. 613 (La.App. 1994), writ denied, found that there was no obligation on the part of the deep rights owners arising from the sublessor/sublessee relationship to act as a reasonable and prudent operator toward the sublessee.

10. State law may provide for creation of a royalty interest through a state agency entering a pooling order. E.g. Okla. Stat. Ann. tit. 52, § 87.1(e); N.M. Stat. Ann. § 70-2-17(C). In such instance, a unit operator may owe a duty under state law and regulatory order to pay royalty to an interest with which the operator has no contractual relationship. The extent of that duty may be appropriately determined in the first instance by that state's regulatory agency that entered the order leading to the creation of the royalty. New Dominion, L.L.C. v. Parks Family Company, L.L.C., 216 P.3d 292, 2008 OK CIV APP 112. The court declared: "Implied covenants arise from the written lease of the parties, they are not simply created by the Supreme Court as a matter of universal justice."

───────────

SUNAC PETROLEUM CORP. v. PARKES

Supreme Court of Texas, 1967.
416 S.W.2d 798, 26 O.&G.R. 689.

GREENHILL, JUSTICE. [On April 17, 1948, O'Hern as lessor executed an oil and gas lease covering a 160-acre tract to the plaintiff Parkes for a primary term of 10 years. Some nine years later Parkes sold his interest to Puckett for a cash consideration, reserving an overriding royalty of 1/16th of 7/8ths of the production from the lease or from any extension or renewal thereof. Puckett thereafter reassigned the lease, and it subsequently was assigned to the petitioners, Sunac Petroleum Co. *et al.,* who were the defendants in the trial court. Shortly before the expiration of the primary term of the lease the lessees pooled the land in question with other lands for gas purposes only and drilling operations were commenced on the unit but not on this tract. When the primary term ended on April 17, 1958, there was no production from, or operations upon, the particular 160-acre lease, but a well was being drilled upon the gas unit. This well was thereafter completed on the 640-acre unit as an *oil* well as distinguished from a *gas* well. Approximately 68 days after the expiration of the primary term and 13 days after the completion of the oil well on the gas unit, the defendants began the drilling of a second well on the particular 160-acres in question which was completed as a producing oil well on July 29, 1958, and the well continued to produce thereafter. Approximately a year later, the successors in interest of the lessor asserted that there was a question of whether the lease was in effect, and thereafter Sunac procured a new lease from the successors of the original lessor.

Thereupon Sunac stopped paying the overriding royalty to the plaintiff, and this suit followed. The court first turned to the question of whether the original lease terminated under its own provisions, and in a portion of the opinion which is reproduced at page 289, *supra,* the court concluded that the original lease terminated at the expiration of its primary term. Eds.]

Renewal or Extension of Lease

We now turn to the question of whether the second lease from the successors of the lessor O'Hern was a renewal or extension of the original lease. The Court of Civil Appeals held that it was, relying upon its determination that the original 1948 lease was still in force when the second lease was obtained in 1959. In the light of our conclusion that the original lease had expired under its own terms, and for the reasons discussed below, we cannot agree with that conclusion.

Most of the cases holding that a subsequent lease was an extension or renewal of an earlier lease have intermixed the related questions of estoppel and constructive trust. We shall treat these questions separately.

It seems clear that the new lease was not an *extension* of the old lease. An extension, as used in this context, generally means the prolongation or continuation of the term of the existing lease. It might also encompass the enlarging of the territory or strata to be covered by the lease. The parties here did none of these things. As enlarged upon below, the lease had long since expired, and there was a new lease on the same land upon substantially different terms.

Moreover, we do not regard the new lease (and the parties refer to the second lease as a *new* lease in their stipulated facts) as a *renewal* of the old lease. The lessors and lessees entered into the new lease over a year after the old lease had expired. Ordinarily when an oil and gas lease has expired, the lessor is free to deal with the property as he sees fit. He may develop and operate it himself, or he may lease it to whomsoever he pleases and on such terms as may be agreed upon.

. . .

Since the new lease was executed under different circumstances, for a new consideration, upon different terms, and over a year after the expiration of the old lease, we hold that the new lease was not a renewal of the old lease.

Confidential Relations and Constructive Trust

There are cases from other jurisdictions which hold that a second lease is, or will be treated as, an extension or renewal of an original lease because the lessee is regarded as a trustee and is adjudged to hold the overriding royalty in the subsequent lease subject to a constructive trust in favor of the owner of the overriding royalty. As a basis for their holdings, these cases have relied on either (1) specific language in the assignment or (2) the close relationship between the parties shown by the particular facts involved.

In Probst v. Hughes, 143 Okl. 11, 286 P. 875, 69 A.L.R. 929 (1930), the lessee, before the expiration of the original lease, took a new lease which cut off the holder of the overriding royalty. The court wrote at length on relationships of trust and confidence and concluded that the new lease was a renewal or extension of the original lease. Similarly, in Howell v. Cooperative Refinery Association, 176 Kan. 572, 271 P.2d 271 (1954), a constructive trust was imposed in favor of a joint adventurer who had held an overriding royalty under the original lease. In a case tried upon a demurrer, the facts were assumed to be that the plaintiff furnished geological information to defendant for which he received his overriding royalty in the original leases. Thereafter the plaintiff furnished additional information leading to the procurement of the new lease. It was agreed between plaintiff and defendant that the new lease would be taken by a third party, and plaintiff expected that defendant would convey to plaintiff his overriding royalty interest. The new lease, as agreed, was taken by the third party after the expiration of the old lease, and it was assigned to defendant; but the defendant refused to honor plaintiff's overriding royalty. The holding was that plaintiff alleged a joint interest and ownership of the mineral estate; that plaintiff and defendant were in a position of trust and confidence; and hence the new lease would be treated as a renewal or extension of the old lease.

Another situation in which some courts have protected the holder of the overriding royalty is called a "washout" transaction, generally involving some bad faith on the part of the lessee. In this type of situation, the operator takes a new lease before the expiration of the old lease and then simply permits the old lease to expire. Oldland v. Gray, 179 F.2d 408 (10th Cir.1950); 2 Williams and Meyers, Oil and Gas Law § 420.2 (1964).

Parkes assigned the lease to Puckett, and Sunac received its interest only after numerous intermediate assignments. There is no evidence that Sunac occupied a position of confidence or trust with Parkes other than the clause relating to extensions or renewals. Sunac made a substantial effort to develop the land in question by pooling it for gas as authorized and by drilling a well. It failed, however, to keep the original lease in force because the well produced oil instead of gas. Sunac took a new lease to protect its interest only after the owners of the property raised a question as to the validity of the original 1948 lease.

The language most often pointed to by the courts as creating a fiduciary relation in this type of situation provides that the overriding royalty will apply to any extensions or renewals of the lease assigned. The assignment from Parkes to Sunac uses such phraseology. In addition, however, the assignment contains this significant provision:

> There shall be no obligation, express or implied, on the part of Assignee, its successors or assigns, to keep said lease in force by payment of rentals or drilling or development operations, and Assignee shall have the right to surrender all or any part of such leased acreage without the consent of Assignor.

Sunac was under no duty to develop the land or continue the lease in force; to the contrary, the assignment expressly gave it the right to surrender the lease at any time without Parkes' consent. We construe the "renewal or extension" provisions together with the provisions set out as relieving the lessee from the duty to perpetuate the lease, and thus the overriding royalty. This latter provision materially distinguishes the cases relied upon by Parkes.

Normally, when an oil and gas lease terminates, the overriding royalty created in an assignment of the lease is likewise extinguished. It is also generally held that the assignment of an oil and gas lease reserving an overriding royalty in the assignor does not usually create any confidential or fiduciary relationship between the assignor and his assignee.

This Court has previously held that a constructive trust arises where there is a fiduciary or confidential relationship between two or more parties and the party holding certain property would profit by a wrong or be unjustly enriched if he were allowed to keep the property. Under the facts of this case, we see no basis for a holding that there existed a confidential or fiduciary relationship. There is no constructive trust.

. . .

Moreover, the fact that Sunac might have recognized that Parkes' overriding royalty interest was outstanding as late as November, 1959, did not create a confidential relationship between Sunac and Parkes. The provisions of the two leases and the assignment are controlling here.

. . .

Estoppel

The Court of Civil Appeals concluded that by their conduct the parties treated the 1948 lease as remaining in effect until it was superseded by the new lease in 1959. On this basis the court held that Sunac was estopped to deny Parkes' claim.

It is true that Sunac continued to pay Parkes his overriding royalty through November of 1959. We do not believe, however, that this constituted a material misrepresentation on Sunac's part, or that Parkes, in reliance on these payments, acted to his prejudice or changed his position. Sunac, therefore, is not estopped to deny that Parkes' overriding royalty is applicable to the new lease.

The judgments of the courts below are reversed, and judgment is here rendered for petitioners.

HAMILTON, JUSTICE. I respectfully dissent.

I think the question as to whether or not the August 17, 1959, lease was a renewal of the 1948 lease should be determined as of the time and under the circumstances existing at the time the lessors and the petitioners, Sunac Petroleum Corporation *et al.,* entered into the new lease.

It is uncontradicted that all parties, the lessors and petitioners, as well as the respondent, treated said 1948 lease as being in force and effect at all times up until it was effectively canceled by virtue of the new lease executed by the lessors and petitioners. Although lessors had raised a question as to the validity of the lease, the lessors and petitioners chose not to have that question determined, but instead entered into a new lease to take the place of the questionable lease. At that time the question of law as to whether the old lease had expired under its own terms was one that had not been decided by this Court or any other court in this jurisdiction or any other jurisdiction. Furthermore, even if it be determined that the old lease expired under its own terms, there may have been a question of fact as to whether the lessors would be estopped to claim that the old lease had expired by having permitted the expenditure by petitioners of vast sums of money in drilling a well on the land in question and in operating said lease for more than a year without any complaint from lessors.

Instead of having these questions determined by a court of law, the lessors and petitioners settled their differences by entering into this new lease. Petitioners, without ever giving up possession of the premises, continued its operation just as it had been doing. When you add to this fact situation the fact that petitioners admitted that the new lease was a renewal of the old lease, by paying to respondent the override provided for in his assignment, for several months under the new lease, there is ample evidence to sustain the trial court's finding that the new lease was in fact a renewal of the old lease. For this reason I would affirm the judgments of the trial court and the Court of Civil Appeals.

. . .

Even though the Court has held that the 1948 lease had expired of its own terms prior to the execution of the new lease, nevertheless the new lease was a renewal of the 1948 lease in view of the position of trust and confidence and fiduciary relationship established between the parties by virtue of the language in the assignment. All that is necessary in order to clearly establish that a fiduciary relationship existed between petitioners and respondent is to note, in the light of the many authorities on such point, that respondent's overriding royalty interest expressly was reserved under the lease of April 17, 1948, or any extensions or renewals thereof, and the relative positions of petitioners, as the working interest owners, and respondent, as an overriding royalty interest owner.

There are many authorities directly holding that the creation of an overriding royalty interest or production payment by exception and reservation in an assignment of a designated lease, or any extensions or renewals of such lease, creates, in and of itself, a fiduciary relationship between the assignor (overriding royalty interest owner or production payment interest owner) and the assignee (working interest owner) and their respective successors in interest, irrespective of whether any confidential or personal relationship in fact exists between such parties. . . .

Since the opinion in the *Probst* case [Probst v. Hughes, 143 Okl. 11, 286 Pac. 875 (1930)] discloses that no confidential or personal relationship existed between the parties and that there were no unusual terms or provisions in the instrument of assignment, but that said instrument of assignment provided to the effect that the overriding royalty interest there created was applicable to any leases in extension or renewal of the assigned lease, the fiduciary relationship which the court found to exist could have arisen only from the relative positions of the parties and the "extension or renewal" provisions of the instrument of assignment.

. . .

Moreover, there are a number of authorities holding that the respective positions of an assignor-overriding royalty interest owner and an assignee-working interest owner creates a fiduciary relationship between said parties, and their respective successors in interest, merely from the relative positions of said parties, irrespective of whether or not there is any confidential or personal relationship between said parties and irrespective of whether the assignment creating the overriding royalty interest expressly provided to the effect that such interest will be applicable to renewal or extension leases.

An excellent example of the imposition by the law of such fiduciary relationship between parties to oil and gas transactions, merely by reason of the relative positions of such parties and irrespective of the existence of any confidential or personal relationship between such parties, is the long judicially recognized duty imposed upon the owner of the leasing power to exercise such power in such manner as to constitute the utmost good faith toward a nonparticipating royalty owner whose interest is subject to such leasing power.

Schlitter v. Smith, 128 Tex. 628, 101 S.W.2d 543 (1937), involved a deed to land wherein the grantor reserved "an undivided one-half interest in and to the royalty rights on all of oil and gas and other minerals in, on and under or that may be produced from the land herein conveyed." The court held that the grantor was entitled to receive one-half of such royalty as may be reserved in any oil, gas or mineral lease which may be executed by the grantee. It said: "We think that self-interest on the part of the grantee may be trusted to protect the grantor as to the amount of royalty reserved. Of course, there should be the utmost fair dealing on the part of the grantee in this regard."

Further, as reflected by the above authorities, the law applicable to oil and gas transactions goes beyond the traditional fiduciary concept and imposes a fiduciary relationship between the parties to an assignment of an oil and gas lease, in which assignment the assignor excepts and reserves an overriding royalty interest in the production under the assigned lease or any extensions or renewals thereof, merely by reason of the "extension or renewal" provisions of such assignment, irrespective of, and in the complete absence of, any confidential or personal relationship between such parties.

There are no cases to be found in this jurisdiction or in any other jurisdiction, holding a lease taken by the assignee, or by the assignee's successors in interest, after the expiration of the assigned lease and covering the same land as the assigned lease, not to be in extension or renewal of the assigned lease, and therefore not burdened with the excepted and reserved interest, where the instrument of assignment creating the excepted and reserved interest expressly provided to the effect that the excepted and reserved interest was to be applicable to leases in extension or renewal of the assigned lease. . . .

. . .

By reason of the peculiar circumstances existing at the time the new lease was taken, it is submitted that they in themselves create a relationship of trust and confidence between the parties. When the question of invalidity of the lease was raised by the lessors to Sunac, Parkes' rights were involved just as were Sunac's. If Sunac had stood its ground and suffered a lawsuit, Parkes would have been an indispensable party defendant. Parkes had just as much right to settle with the lessors as did Sunac, but Parkes had no opportunity to settle. Sunac went behind his back and sold him out by taking a new lease and thereby releasing the old one. This sort of conduct should not be allowed to cut Parkes out.

It is respectfully submitted that for the reasons and under the authorities indicated above there existed between petitioners and respondent a fiduciary relationship not only by reason of the "extension and renewal" provision, but by reason of the peculiar circumstances existing at the time the new lease was taken. Such new lease constituted a renewal of the old one irrespective of whether or not the latter lease in fact terminated prior to the effective date of the former lease.

I would affirm the judgments of the Court of Civil Appeals and the trial court.

SMITH and POPE, JJ., join in this dissent.

GILL v. GIPSON

Court of Appeals of Mississippi, 2007.
982 So.2d 415, 169 Oil & Gas Rep. 127

GRIFFIS, J., for the Court.

Facts

¶ 2. Wessie Mae Lowe was the sole owner of approximately eighty-four acres of land in Pearl River County, Mississippi. On June 16, 1970, she executed an oil, gas and mineral lease (the "lease") in favor of D.L. Royals. The lease contained a clause that if there was no oil or gas production for a period of sixty consecutive days the lease would automatically terminate. A well was located on a portion of the property and it became known as the "A. L. Lowe # 1."

¶ 3. D.L. Royals operated the well from 1970 until his death in 1989. After his death, his daughter, Brenda Gipson, operated the well. After his death, D.L. Royals' interest in the lease passed to his heirs, the appellees in this action (the "Royals").

¶ 4. Preston O. Gill is a landman. Gill attempted to negotiate the purchase of the well for his client, William C. Culp (owner of L & M Oil, Inc. and C.C.M. Enterprises, Inc.) and his prospective business joint venturer, John M. Dubose, Sr. (owner of Blue Diamond, Inc.). At that time, Gill was the exclusive agent for L & M Oil, Inc. and C.C.M. Enterprises, Inc.

¶ 5. In 1991, the Royals were approached by Dubose about acquiring the lease and were informed that Gill would contact them about the proposal. Gill and Dubose then presented the Royals an offer to purchase the lease. The offer was presented by Gill, on behalf of L & M Oil, Inc., and Dubose, on behalf of Blue Diamond, Inc. Gill had no ownership of either company. The offer presented was rejected by the Royals.

¶ 6. After speaking with Dubose and his son, John Dubose, Jr. ("Tap"), Gill was asked to present an offer on behalf of PRP, Inc. PRP, Inc. was a new corporation solely owned by Tap. After receiving permission from L & M Oil, Inc., Gill presented PRP, Inc.'s offer to the Royals.

¶ 7. On July 4, 1991, the Royals assigned their interest in the Lease to PRP, Inc. The assignment of the lease contained the following provision, which is a the center of this dispute.

> 9. *RESERVATION OF OVERRIDING ROYALTY* As a further consideration for the execution of this assignment, Assignor hereby reserves unto itself, its successors and assigns, and Assignee hereby grants and conveys to Assignor, as an overriding royalty, one-eight of eight-eighths (1/8 of 8/8) part of all oil, gas, and other minerals produced and marketed or used from the assigned acreage, and being free of all costs of operation, by this assignment under and by virtue of the lease hereby assigned or any renewals or extensions thereof.
>
> Also, if the lease hereby partially assigned should expire for any reason and Assignee or any of its employees, agents or any corporation in which Assignee or any of said persons has any interest shall secure any future lease of the assigned acreage or any production of oil, gas and other minerals in, on and under the land described in the lease hereby partially assigned, the Assignor shall receive an overriding royalty of one-eighth of eight eighths (1/8 of 8/8) of all oil, gas and other minerals produced and marketed or used from said property, being free from all costs of operation.

¶ 8. At trial, Gill described his involvement in this transaction as a "courtesy" for the Royals and PRP, Inc. Less than two months later, on August 23, 1991, PRP, Inc. assigned Gill a 2.5% carried working interest and a 1.875% net revenue interest in the lease.

¶ 9. Three years later, on August 23, 1994, PRP, Inc. assigned its interest in the lease to Trinity Oil & Gas Development, Inc. Then, on August 26, 1997, Trinity assigned its interest to Gill.

¶ 10. Gill continued production of A.L. Lowe # 1 until October of 1998. Then, for the alleged reason of increased costs, Gill ceased production of oil and gas from A.L. Lowe # 1. The well was shut in, but not plugged.

¶ 11. Gill then called Donald Wayne Lowe and asked him what did he want to do. At that time, Lowe was the current owner of the land and all mineral interests since, as Gill claimed, the lease was terminated. Lowe told Gill that he wanted a new lease. On March 31, 1999, Lowe executed an oil, gas and mineral lease to Preston O. Gill Operating Company. The lease was prepared by Gill. This lease was executed approximately three months after Gill ceased production on A.L. Lowe # 1.

¶ 12. On February 8, 2000, Gill, individually and not on behalf of Preston O. Gill Operating Company, assigned a ten percent working interest and ten percent net revenue interest in the lease to Lowe. In this assignment, Gill claimed that he was the owner of the entire working interest and net revenue interest on the A.L. Lowe # 1 well.

¶ 13. After February of 1999, the Royals received no more royalties under the 1991 lease. They were never informed that the well was to be shut down in 1999. Gill claimed that the underlying lease terminated automatically after sixty days passed without production. Gill claims that with the expiration of the original lease, the Royals' rights to royalties likewise terminated.

¶ 14. On July 28, 2001, the Royals commenced this action and filed their complaint to confirm title to overriding royalty interest in oil, gas, and minerals, to recover monies erroneously paid and for attorney's fees. The chancellor rendered a memorandum opinion that found that Preston Gill was an agent of PRP, Inc. and had breached a duty of good faith and fair dealing with the Royals. The chancellor then entered a final judgment that awarded the Royals $42,734.62 for oil royalties and $29,469.72 for gas royalties. The chancellor also awarded the Royals prejudgment interest in the amounts of $11,236.27 and $7,474,71 for the oil and gas royalties due. In addition, the chancellor awarded the Royals $10,000 in punitive damages and $10,000 in attorney's fees. It is from this judgment that Gill appeals.

. . .

Analysis

I. Whether the chancellor was manifestly wrong in his finding and interpretation of the assignment that imposed an overriding royalty in favor of the Royals.

¶ 16. Gill alleges that paragraph 9 of the Royals' assignment to PRP, Inc. should not apply to the newly formed lease between Preston O. Gill Operating Company and Donald Wayne Lowe. He states for several reasons that he is not bound by the terms of the assignment; these include: the assignment is not binding on any employees, agents or representatives of PRP, Inc.; the assignment is a personal covenant and does not run with the land; the measuring date whether Gill was an agent of PRP, Inc., is the date of the new oil and gas lease; and Gill was never an agent or employee of PRP, Inc. Gill also recognizes that there are no Mississippi cases directly on point to govern this appeal and states that this may be a matter of first impression before the courts of this State.

¶ 17. Gill claims that paragraph 9 does not bind an agent or employee of PRP, Inc. However, Gill cites us to no legal authority for this position. Nevertheless, it has no merit.

¶ 18. The chancellor's opinion addressed this issue as follows:

Defendants argue that it is immaterial to find Gill an agent of PRP, Inc. as the language of paragraph 9 of the Assignment would be against public policy for a corporation to restrict its employee or agent without their permission. However, here Gill not only had actual knowledge of the provision, he actually stepped into the shoes of PRP, Inc. by making an affirmative decision to acquire full ownership of the lease with knowledge of the provision, *thus going beyond being characterized as an agent for another in relation to the well production.*

(emphasis added). The chancellor determined that Gill's capacity as an agent or employee of PRP, Inc. had no bearing on this issue because Gill became bound by this contractual provision by virtue of the assignment of the Lease from Trinity to Gill. By the assignment to him individually, Paragraph 9 became a contractual obligation of Gill, individually, when he accepted the assignment of the lease from Trinity.

¶ 19. The chancellor's memorandum opinion does indeed reference Gill's status as an agent for PRP, Inc. Nevertheless, it is apparent that the chancellor's finding did not turn on this fact alone. The chancellor's opinion recognized that Gill "actually stepped into the shoes of PRP, Inc. by making an affirmative decision to acquire full ownership of the lease with knowledge of the provision."

¶ 20. The original lease included an automatic termination provision. The lease was to automatically terminate if the well did not produce for a period of more than sixty days. Aware of this automatic termination provision, the Royals assigned the lease to PRP, Inc. and included the following condition, reduced to its essence:

9. *RESERVATION OF OVERRIDING ROYALTY* ....

[I]f ... Assignee or any of its ... agents or any corporation in which Assignee or any of said persons has any interest shall secure any future lease of the assigned acreage or any production of oil, gas and other minerals in, on and under the land described in the lease hereby partially assigned, the Assignor shall receive an overriding royalty of one-eighth of eight eighths (1/8 of 8/8) of all oil, gas and other minerals produced and marketed or used from said property, being free from all costs of operation.

¶ 21. This provision created an implied covenant upon PRP, Inc., then Trinity, and then Gill, to continue production of the well to prevent the lease from terminating. Gill eventually obtained exclusive control over the production and development of the well, since he owned the working interest. Gill was bound by the standard of conduct known as the "prudent operator." *Southwest Gas Producing Co. v. Seale,* 191 So.2d 115, 119 (Miss.1966). The standard required that Gill's conduct be "[w]hatever, in the circumstances, would be reasonably expected of operators of ordinary prudence, having regard to the interests of both lessor and lessee, is what is required." *Id.* at 119-20 (quoting *Brewster v. Lanyon*

Zinc Co., 140 F. 801, 814 (8th Cir.1905)).

¶ 22. The chancellor found that the net effect of Gill's conduct was to terminate the interest of the Royals. Given the discretion that the chancellor's findings of facts are accorded, the evidence supports his finding. After Gill had received ownership in the working interest of the well, he maintained production for fourteen months. Then, in October of 1998, he ceased production for what he called cost overruns. While he ceased production, Gill never plugged the well. Gill's conduct allowed the underlying lease to terminate due to non-production. Gill then renegotiated a new lease with the land owner. The new lease was renegotiated and signed only four months after the previous lease and the Royals' interests had terminated. Having received a new lease, Gill then put the well back into production. The conduct and the timing of his actions support the chancellor's finding.

¶ 23. Gill also claims that paragraph 9 cannot be enforced against him because the provision is a personal covenant and does not run with the land. Gill argues that "[a]ll covenants having to do with realty and the use thereof are either real or personal.... A real covenant binds the heirs and assigns of the original covenantor, while a personal covenant does not.... Put another way, a covenant may 'run with the land,' or may simply be a matter between the grantor and the purchaser." *Vulcan Materials Co. v. Miller,* 691 So.2d 908, 914 (Miss.1997). Thus, Gill asserts that since he did not expressly agree to assume the royalty payment obligation, he is not bound to the terms of the assignment. The Royals counter that *Vulcan Materials Co.* also states that "since [the covenant in question] was a personal covenant, Vulcan contends that the provision is not enforceable to other parties unless ratified." *Id.* at 911. The court then ruled that "Vulcan is obligated to pay Miller the royalty pursuant to the royalty agreement not because it is a real covenant running with the land, but because Vulcan assumed the obligation in the deed to Granite." *Id.* at 915.

¶ 24. On August 26, 1997, Gill obtained his interest in the well from Trinity, and he continued production until October of 1998. During this time, Gill paid all royalties due the Royals. Clearly, as in *Vulcan Materials Co.,* the assignee ratified and assumed the obligation of paragraph 9. This is what the chancellor found, and we believe it to be correct.

¶ 25. Gill's brief seems to argue that there was a different entity that negotiated the new lease with Lowe. In fact, the March 31, 1999 lease was executed by Lowe and conveyed the mineral interests to the Preston O. Gill Operating Company. Less than a year later, on February 8, 2000, Gill, individually and not on behalf of Preston O. Gill Operating Company, executed a document that assigned a ten percent working interest and ten percent net revenue interest in the lease to Lowe. In this assignment, Gill claimed that he was the owner of the entire working interest and net revenue interest of the A.L. Lowe # 1 well. Preston O. Gill Operating Company may or may not be a legal entity. However, the record before us does not support a finding that Preston O. Gill

Operating Company was a legal entity that was separate and distinct from Gill.

¶ 26. Accordingly, we find that this issue is without merit. We affirm the judgment of the chancellor that awarded damages for past royalties and a continuing overriding royalty in the production of the well.

. . .

III. Whether the chancellor erred when he granted punitive damages and attorney's fees.

¶ 32. Gill claims that the Royals failed to prove they were entitled to punitive damages by clear and convincing evidence. Mississippi Code Annotated Section 11-1-65(1)(a) (Supp.2006) provides that punitive damages are awarded only if the Royals prove by clear and convincing evidence that Gill acted with actual malice, fraud, or gross negligence/reckless disregard for the rights of others.

¶ 33. Here, the chancellor had sufficient evidence to support his finding that Gill acted in "reckless disregard" of the rights of the Royals. Gill was involved in the mineral lease ever since PRP, Inc.'s first offer to purchase the assignment from the Royals. He even received a direct interest in the proceeds of the well from PRP, Inc. after the assignment had been finalized. Later, he acquired full ownership of the lease. All this occurred while he was keenly aware of the automatic termination provision.

¶ 34. With knowledge that the automatic termination provision would take effect upon non-production of the well for sixty days, Gill stopped production. After only a few months had passed, contrary to the interest of the lease and the Royals' interest, Gill contacted the landowner to negotiate and obtain a new lease for the well. Thereafter, upon the execution of a new lease, Gill resumed production from the well. The Royals claimed that they were never informed that the well was to be shut down by Gill.

¶ 35. The chancellor found that the "net effect of that course of conduct was to terminate the interest of the [Royals] with full knowledge by Gill of what that result would mean to the [Royals], notwithstanding the specific language placed in Exhibit 2 to protect their interests." In accordance with the discretion that we give to a chancellor's fact findings, we find that there was substantial credible evidence to support his award of punitive damages. Therefore, Gill's allegation that the grant of punitive damages was in error is without merit.

¶ 36. Gill also argues that attorney's fees are appropriate only in cases were punitive damages should be awarded, and because punitive damages are not appropriate in this case neither are attorney's fees. Since we have determined that the record supports an award of punitive damages, we also conclude that an award of attorney's fees was also appropriate.

¶ 40. The judgment of the chancery court of pearl river county is affirmed. All costs of this appeal are assessed to the appellant.

NOTES

1. Consider the desirability of including clauses such as the following in instruments granting or reserving nonoperating interests in production:

a. *"This reservation shall likewise apply as to all modifications, renewals of such lease or extensions that the assignee, his successors or assigns may secure."* Probst v. Hughes, 143 Okl. 11, 286 P. 875 (1930).

b. *"Should Assignees elect to abandon said lease while the same is producing oil or gas, Assignees shall, thirty (30) days prior to such proposed abandonment, give Assignors written notice thereof and Assignors shall have the right within said period to demand that Assignees assign said lease to Assignors upon Assignors paying to Assignees in cash the reasonable salvage value of the casing then in any wells which may have been drilled upon the lease, and upon receipt of such notice and payment, Assignees shall assign to Assignors said lease and the casing located in any wells situated on the lease."* Atlantic Refining Co. v. Moxley, 211 F.2d 916, 3 O.&G.R. 1298 (5th Cir.1954).

c. *"The execution of this conveyance shall not impose upon grantor any obligation with respect to developing or operating the property covered hereby, nor shall it be implied herefrom that grantor is in anywise obligated to see that the grantee receives oil aggregating the net value of $ . . ., or any lesser sum; but grantor does covenant and agree that grantee shall have the same but no greater right to require grantor to develop and operate said property in accordance with the terms of the original leases upon said property as the original lessors therein, their heirs or assigns might have or exercise."* Fleming v. Com'r, 24 T.C. 818, 4 O.&G.R. 1609 (1955).

d. *"Grantees covenant to keep or cause to be kept in full force and effect the subject oil and gas leases, and to perform or cause to be performed each and all of the covenants, terms and conditions imposed upon the original lessees or assigns, and expressly contained in such leases or expressly contained in any assignment under or through which such leases or an undivided interest therein is now held, in case any such covenant or condition in any such assignment in any wise affects the validity or continuance in force of said leases or of the interest herein conveyed to grantees, and to perform or cause to be performed all implied covenants or obligations imposed or connected with the subject leases, upon the lessee, or assigns, and to continuously operate or cause to be operated in a good and workmanlike manner in accordance with best field practices, each and all of the wells which have heretofore been drilled, or which may hereafter be drilled on the tract of land hereinabove described."* Gamble v. Cornell Oil Co., 260 F.2d 860, 10 O.&G.R. 179 (10th Cir. 1958).

Should extrinsic evidence be admitted by a court to reform an instrument to establish that pursuant to industry custom, an assignment of an overriding royalty interest would have included an extension or renewal clause? *See* CLK Company, L.L.C. v. CXY Energy, Inc., 972 So.2d 1280, 2007-834 (La.App. 3 Cir. 12/19/07) (trial court did not err in granting plaintiff a judgment which conveyed to it an overriding royalty).

2. In INDEPENDENT GAS & OIL PRODUCERS, INC. v. UNION OIL CO., 669 F.2d 624, 72 O.&G.R. 523 (10th Cir.1982), a lease assignment reserving an overriding royalty "extended the [assignor's] royalty interest to 'any modifications, extensions or renewals of the lease on subject lands.'" The test well drilled pursuant to the terms of the assignment "yielded oil and gas continuously until it was shut in because of mechanical failure in late October 1976." A new lease on the land in question was obtained by a third party on December 31, 1976 and was assigned to the original assignee in February 1977 after that party threatened to sue the third party on the ground that the third party had obtained the new lease as the original assignee's agent. The court stated Oklahoma law to be "that a lease assignment expressly subjecting lease extensions and renewals to an overriding royalty interest converts a new lease procured by the assignee into a renewal of the old one to which the reserved royalty attaches. . . .The fiduciary obligations impliedly created by the terms of such a lease assignment form the basis for the rule. Where an assignment provides that subsequent lease extensions and renewals are subject to an overriding royalty, the assignee stands as a quasi-trustee vis-a-vis the assignor and must exercise the utmost good faith in protecting the latter's interest in the leasehold. Consequently, any attempt by the fiduciary-assignee to procure rights antagonistic to those of his assignor will be defeated." The court affirmed the trial court's determination that the new lease was a renewal of the lease originally assigned to which the original overriding royalty attached.

What would be the result of the application of these principles to *Sunac's* facts in Oklahoma? Would the Oklahoma law as stated require an assignee, as a "fiduciary," to take appropriate steps to procure a new lease to protect the holder of the overriding royalty, as well as its own interests, when the producing well on the assigned lease must be "shut in because of mechanical failure" and permanent cessation of production is likely as in *Independent Gas & Oil Producers?*

In Olson v. Continental Resources, Inc., 2005 OK CIV APP 13, 109 P.3d 351, the Oklahoma Court of Appeals held that an overriding royalty interest does not survive the termination of the base lease in the absence of language in the assignment to the contrary, a fiduciary duty between the assignor and assignee or fraud. To the same effect: Ritter v. Bill Barrett Corp., 351 Mont. 278, 210 P.3d 688, 2009 MT 210.

Typical contractual provisions designed to protect nonoperating interests in production are discussed in 2 Martin & Kramer, *Williams and Meyers Oil and Gas Law* §§ 428-430.

3. Paula was in the business of assembling mineral leases for sale to developers. She would go to individual mineral owners, lease their minerals, assemble a large enough area to make it interesting to a large energy company and then sell the leases as a single package. Paula sold a package of leases to Rally Energy. The assignment called for a cash purchase price plus advance and

minimum royalty payments. These royalty payments were set at $ 4.00/acre until production was achieved. Thereafter royalty was set at a percentage of the value of minerals produced, if it was more than the acreage royalty figure. Paula assigned 10,000 acres of mineral leases. Rally after doing some exploratory work released 5000 mineral acres back to the original owners. Does Rally continue to owe minimum royalties based on 5000 acres or on the original 10,000 acres? Did Rally's obligations to Paula as to the released 5000 acres terminate when the underlying or base leases were terminated? *See* Piamco, Inc v. Shell Oil Co., 799 F.2d 262, 92 O.&G.R. 22 (7th Cir.1986).

4. With respect to the contract provisions regarding reassignment by the sublessee to protect the sublessor, consider the following provision:

> Assignor herein reserves the right to a re-assignment of the leases assigned herein if Assignee elects not to pay any delay rentals coming due. Such re-assignment shall be made within sixty (60) days of the rental payment or expiration date.

What effect should be given to this provision if during the last year of the primary term the assignee agrees with a company that has taken a top lease from the landowners not to drill the lease obtained from the assignor in exchange for that top lessee assigning the top lease to the assignee? *See* Avatar Exploration, Inc. v. Chevron, U.S.A., Inc., 933 F.2d 314, 116 O.&G.R. 459 (5th Cir. 1991).

5. Should there be an implied leasehold duty requiring the lessee to notify overriding royalty interest owners prior to exercising a right to release the underlying lease? *See* Degenhart v. Gold King Petroleum Corp., 851 P.2d 304, 122 O.&G.R. 54 (Colo.App. 1993), holding that there is no such duty and that the reservation of an overriding royalty interest does not, by itself, create a confidential or fiduciary relationship.

6. What are the assignee's duties towards the royalty owners? What if the assignee failed to pay the overriding royalty interest owners? Should the statute of limitations be tolled if the overriding royalty owner has no actual knowledge of production from the well? Would your decision be affected by the existence of a statutory production payment or division order statute requiring all oil and gas producers to make royalty payments within a specified period of time? *See* Goodall v. Trigg Drilling Co., 1997 OK 74, 944 P.2d 292, 138 O.&G.R. 118. For a somewhat different approach see Harrison v. Bass Enterprises Production Co., 888 S.W.2d 532, 130 O.&G.R. 138 (Tex.App. 1994).

One of the principal methods by which a lessee transfers his interest is through the use of a farmout agreement. While at one time farmout agreements were typically informal and incomplete, today farmout agreements are more typically based on preprinted, standardized or company-drafted forms that attempt to deal with the many complex issues relating to a farmout. A farmout agreement has been defined as:

. . . a contract to assign oil and gas lease rights in certain acreage upon completion of drilling obligations and the performance of any other covenants and conditions therein contained. It is an executory contract. It is largely used in cases where the owner of a lease is unable or unwilling to drill on a lease which is nearing expiration, but is willing to assign an interest therein to another who will assume the drilling obligations and save the lease from expiring. Often the owner of the lease retains an overriding royalty or a carried interest as his consideration. 2 Martin & Kramer, *Williams & Meyers Oil and Gas Law*, § 502.

There are many objectives that either the farmor or farmee may seek to achieve in executing a farmout agreement. *See* Lowe, *Analyzing Oil and Gas Farmout Agreements*, 41 Sw.L.J. 763 (1987); Lowe, *Recent Significant Cases Affecting Farmout Agreements*, 50 Sw.Legal Fdn. Oil & Gas Inst. ch. 4 (1999). It is also clear that the tax consequences of conveying or assigning oil and gas leasehold interests has encouraged the use of the farmout agreement. Gregg, *Oil and Gas Farmouts--Implications of Revenue Ruling 77-176*, 29 Sw.Legal Fdn. Oil & Gas Inst. 601 (1978).

STRATA PRODUCTION CO. v. MERCURY EXPLORATION CO.

Supreme Court of New Mexico, 1996.
1996 NMSC 016, 916 P.2d 822, 133 O.&G.R. 85.

FROST, CHIEF JUSTICE. Defendant-Appellant Mercury Exploration Co. (Mercury) appeals from a judgment rendered in favor of Plaintiff-Appellee Strata Production Co. (Strata) for breach of contract and negligent misrepresentation. Mercury argues on appeal that: it modified its unilateral contract offer with Strata before Strata's acceptance by performance; the trial court incorrectly interpreted the contract terms; the trial court failed to reduce Strata's damage award by the proportionate interests owned by subsequent investors; and the trial court used the incorrect measure of damages for calculating Strata's lost profits for breach of the contract. We affirm.

I. Facts

Both Strata and Mercury are engaged in the business of petroleum exploration and production. In 1991 Strata began putting together a drilling prospect in Lea County, New Mexico, which it called the Red Tank Prospect. As part of the Red Tank Prospect, Strata obtained farmout agreements from Exxon Co., Mobil Producing Texas and New Mexico Inc., and Mercury for drilling rights on three tracts of roughly adjacent land. These tracts are named the Cercion tract, the Paisano tract, and the Lechuza tract, respectively. The Mercury farmout agreement for the Lechuza tract is the contract at issue here.

A farmout agreement is an assignment of a lease and drilling rights by a lease-owner not interested in drilling to another operator interested in drilling. 8

Howard R. Williams & Charles J. Meyers, Oil & Gas Law 389 (1995). The primary characteristic of a farmout agreement is that the assignee is obligated to drill one or more wells on the assigned acreage as a prerequisite to the completion of the assignment.

Strata and Mercury entered into their farmout agreement effective August 28, 1991. Glenn Darden, a vice president of Mercury, drafted the farmout agreement. In the farmout agreement, Mercury represented that it owned or controlled all of the lease covering the Lechuza tract. The farmout agreement provided in relevant part that upon initiation of drilling a test well to a specified depth within 120 days of entering the agreement, Mercury would assign to Strata 100% of the working interest in the lease. Mercury also stated in the farmout agreement that it would assign to Strata a net revenue interest of 76.5% of the total revenue interest in the leased land. [A net revenue interest is the interest the assignee of a lease actually has in the profits of the production operation free of production costs after all overriding royalties have been paid out to the prior leaseholders. Eds.] Under the terms of the lease Mercury subdivided the Lechuza tract into four 40-acre parcels arranged in a checker- board pattern, and Strata could earn Mercury's lease rights for each parcel by successively drilling test wells on the respective parcels. Finally, a clause in the agreement noted that the agreement was on an option basis with no penalty for failure by Strata to drill the test wells other than termination of the agreement. Strata paid no consideration to Mercury for this farmout agreement.

On October 29, 1991, as part of its Red Tank Prospect development, Strata began drilling a well on the Cercion tract pursuant to its farmout agreement with Exxon. This Cercion well was a "wildcat," meaning that it was an exploratory well in an unproven territory and in fact was the first well to produce within the general area of the Red Tank Prospect tract. See 8 Williams & Meyers, supra, at 1218 (defining wildcat well). As a wildcat, the Cercion well was an extremely risky undertaking, costing approximately $600,000 to drill and complete. On November 5, Strata requested an extension of the 120-day drilling deadline, which would otherwise expire on December 11.

On November 10, 1991, Strata's title attorney, Sealy Cavin, Jr., noticed some discrepancies in a 1982 lease agreement concerning the Lechuza tract which indicated that additional, previously unknown parties might also have a working interest in the Lechuza tract. On November 26 Mercury granted Strata a 30-day extension on the 120-day drilling deadline. Cavin produced a formal drilling title opinion on December 9, 1991, which demonstrated that Mercury did not own 100% of the working interest in the Lechuza tract, nor was it able to transfer a 76.5% net revenue interest in the land which it had promised in the farmout agreement.

Strata completed the Cercion well and put it into production as a commercial well in December 1991. The location and success of the first Cercion well indicated that the Lechuza tract would also generate commercially productive wells. Strata then requested an additional 30-day extension beyond the January 11, 1992, deadline for commencing drilling under the Mercury farmout agreement, which Mercury refused. Strata therefore commenced drilling on the first 40-acre parcel of the Lechuza tract on January 10, 1992, and completed the well in early February 1992. Strata subsequently drilled wells on two of the three remaining 40-acre parcels, thereby earning the lease assignments for those parcels under the Mercury farmout agreement. The first and second Lechuza wells were commercially productive, the third was not.

Strata sued Mercury for breach of contract and negligent misrepresentation for failing to deliver 17.1875% of the working interest and 2.033258% of the net revenue interest for the Lechuza tract wells. After a bench trial, the court found in favor of Strata and awarded damages of $616,555.22. Mercury appeals from the trial court's findings in favor of Strata and challenges the court's calculation of damages. Because we affirm the trial court's award of damages for breach of contract, we need not address Mercury's claims regarding the tort of negligent misrepresentation.

II. Contract Formation

A. Mercury's Offer

Mercury first argues that its farmout agreement with Strata was a unilateral contract, which it was free to revoke or modify before Strata's acceptance. Mercury contends that Strata's discovery of Mercury's inability to transfer all of the relevant interests worked an effective modification of its offer in the farmout agreement.

The farmout agreement provided for Strata's acceptance by performance, namely drilling a test well to a specified depth on the Lechuza tract. The agreement did not provide for a penalty if Strata failed to drill a test well, except that the agreement would lapse. Strata was not obligated to take any action under the agreement. Accordingly, the farmout agreement was a traditional unilateral contract, in which the offeror makes a promise in exchange, not for a reciprocal promise by the offeree, but for some performance. In a unilateral contract, the offeree accepts the offer by undertaking the requested performance. Generally, the offeror is free to revoke or revise the offer before acceptance.

In this case the farmout agreement expressly provided that it was on an option basis, holding open the underlying unilateral contract offer for 120 days. Ordinarily, an option contract serves to make an offer irrevocable for the stated period of time. However, to be effective, the option contract generally must be supported by some consideration given in exchange for holding the underlying offer open. Strata acknowledges that it did not pay Mercury for this option contract. Accordingly, Strata must demonstrate a substitute for consideration which would serve to make the option contract binding.

B. Promissory Estoppel

Strata argues that it substantially changed its position in reliance on

Mercury's offer and that this reliance served as the consideration substitute. In essence, Strata argues that the doctrine of promissory estoppel applies in this case to make the offer irrevocable for the period stated in the option 120 days plus the 30-day extension). We agree.

The theory of promissory estoppel provides:

> A promise which the promisor should reasonably expect to induce action or forbearance on the part of the promisee or a third person and which does induce such action or forbearance is binding if injustice can be avoided only by enforcement of the promise. The remedy granted for breach may be limited as justice requires.

. . .

Accordingly, recasting the definition set out in Eavenson [Eavenson v. Lewis Means, Inc., 105 N.M. 161, 730 P.2d 464 (1986).Ed.]to reflect these considerations, the essential elements of promissory estoppel are: (1) An actual promise must have been made which in fact induced the promisee's action or forbearance; (2) The promisee's reliance on the promise must have been reasonable; (3) The promisee's action or forbearance must have amounted to a substantial change in position; (4) The promisee's action or forbearance must have been actually foreseen or reasonably foreseeable to the promisor when making the promise; and (5) enforcement of the promise is required to prevent injustice. The theory of promissory estoppel is equally applicable to option contracts otherwise lacking in consideration.

Turning to the present case, it is undisputed that Strata began drilling the first Cercion well on October 29, 1991, prior to learning of Mercury's inability to deliver 100% of the working interest and 76.5% of the net revenue interest for the Lechuza tract. The trial court found that Strata drilled the Cercion well as part of its development of the Red Tank Prospect, which included the Cercion, Lechuza, and Paisano tracts, and that Strata began drilling the well in reliance on the Mercury farmout agreement. Strata also presented evidence that the Cercion tract was part of the same geologic formation as and immediately downslope from the first Lechuza tract, indicating that a commercial strike on the Cercion tract would signify the likely probability of a commercial strike on the first Lechuza tract. Finally, the trial court found that the first Cercion well was an extremely risky wildcat well drilled in a previously untested location.

Accordingly, we conclude there was substantial evidence indicating that Strata reasonably relied on the option for accepting the Mercury farmout agreement without modification. By drilling the high-risk Cercion test well, Strata substantially and foreseeably altered its position. This reliance served to make the option in the farmout agreement irrevocable. Thus at any time prior to January 11, 1992, Strata was free to accept the original, unilateral contract offer simply by undertaking the performance required by the farmout agreement.

. . .

Accordingly, we conclude that the trial court did not err in finding that

Strata reasonably relied on Mercury's representations and substantially altered its position as a result of that reliance before learning of Mercury's title problems. As a result, the option contract became binding, and Strata was free to accept the original farmout offer by performing within the allotted time. Strata subsequently drilled test wells on three of the four 40-acre plots on the Lechuza tract in conformance with the requirements of the farmout agreement, thereby earning assignments of 100% of the working interest and 76.5% of the net revenue interest for these three plots.

III. Contract Terms

Mercury next challenges the trial court's interpretation of the terms of the farmout agreement. Mercury contends that, even if the farmout agreement created a binding contract, Mercury did not breach the contract with respect to assigning the working interest in the Lechuza tract because it never promised to deliver 100% of the working interest. Mercury points out that under the farmout agreement it only agreed to "assign to Strata 100% of Mercury's interest." (Emphasis added). Mercury acknowledges that the agreement provided: "Mercury represents that it owns or controls all of that certain lease shown on Exhibit 'A' hereof, which shall constitute the 'Contract Premises.' " (Emphasis added). Mercury contends that the term "controls" only applies to the authority to manage or oversee the property and was not equivalent to the working interest. It argues that, under the plain terms of the agreement, it never represented that it had 100% of the working interest in the land and only promised to assign all of the interest it actually owned. Mercury therefore charges that the trial court erred in interpreting the language in the agreement as a promise to assign 100% of the working interest, contrary to the plain meaning of the terms used.

In essence Mercury contends that the terms of the contract are facially unambiguous and therefore the trial court must enforce the clear language of the contract terms without referring to extrinsic evidence to determine their intended meaning. However, Mercury's view of New Mexico law is incorrect. In C.R. Anthony, 112 N.M. at 508-09, 817 P.2d at 242-43, and again in Mark V, Inc. v. Mellekas, 114 N.M. 778, 781-82, 845 P.2d 1232, 1235-36 (1993), this Court rejected the four-corners approach to contract interpretation and instead allowed courts to consider extrinsic evidence concerning the circumstances surrounding the execution of the agreement to determine if contract terms are in fact ambiguous. The question whether a contract term is ambiguous is a matter of law for the trial court to determine. The ultimate goal of this inquiry is to ascertain the intentions of the contracting parties with respect to the challenged terms at the time they executed the contract.

Even if we assume, as Mercury vigorously contends, that Mercury's representations unambiguously grant Strata less than 100% of the working interest in the Lechuza tract, the trial court was still free to consider evidence submitted by Strata that demonstrated the contract language was in fact ambiguous and which elucidated the intentions of the contracting parties at the time they executed the farmout agreement. Glenn Darden, Mercury's exploration manager, acknowledged drafting the farmout agreement on behalf of Mercury. He admitted in deposition and at trial that when he drafted the phrase, "Mercury represents that it owns or controls all of that certain lease ...," he intended to represent that Mercury owned or controlled 100% of the working interest in the lease, which it would transfer to Strata upon satisfaction of the agreement.

According to Darden's testimony, Mercury was negotiating with Strata on behalf of itself and several partners that also owned a percentage of the working interest in the Lechuza tract. In a letter dated September 10, 1991, which discussed negotiations between Mercury and Strata over the farmout agreement, Darden wrote to Strata that Mercury had obtained written approval of the farmout agreement from its partners that owned 94% of the working interest, and that approval from the partner which owned the remaining 6% of the working interest would be forthcoming. Darden acknowledged that Mercury believed that these partners along with itself controlled 100% of the working interest in the Lechuza tract and that Mercury had represented as much to Strata. Darden explained that he had mistakenly overlooked some additional parties that had retained a percentage of the working interest under the 1982 lease agreement executed by Superior Oil Co., a prior owner of Mercury's interest in the Lechuza tract. He admitted that he only realized his mistake after executing the farmout agreement, when Strata noticed the omission in its November 1991 title search and brought it to his attention.

Consequently, based on this evidence of the facts and circumstances surrounding the execution of the Mercury farmout agreement, we hold there is substantial evidence to support the trial court's findings that Mercury represented to Strata that it owned or controlled 100% of the working interest in the Lechuza tract and that it promised to assign this 100% working interest to Strata.

IV. Award of Damages

. . .

B. The Trial Court Properly Calculated Strata's Damages

Mercury's final argument is that the trial court applied the wrong measure of damages for calculating Strata's recovery. The trial court calculated Strata's damages by measuring the present value of the current and projected future production from the three wells Strata drilled on the Lechuza tract and then computing the amount of the undelivered 17.1875% working interest and 2.033258% net revenue interest based on that valuation. Using the projected oil production and a reasonable price for oil, the trial court calculated that the value of the undelivered interests totaled $616,555.22. Mercury contends that the trial court should have calculated the damages based upon the time at which Strata knew Mercury could not transfer the promised interests, which was prior to Strata's drilling the Lechuza tract wells. Mercury suggests that the court should have based the measure of damages on the market value of the land itself prior to drilling, which Mercury estimated at approximately $150 per acre, resulting in a total loss to Strata of only about $3,094. We disagree.

For cases in which profits are the inducement for entering into a contract, lost profits are the proper measure of damages for a breach of contract if they can be proven with reasonable certainty. As the federal district court in Petroleum Energy, Inc. v. Mid-America Petroleum, Inc., 775 F.Supp. 1420, 1426 (D.Kan.1991) (citation omitted), explained: The rule as to the measure of damages for breach of contract is the same in drilling contracts as it is in other contracts. A party injured by a breach of contract is entitled to recover all his damages, including gains prevented as well as losses sustained, provided they are certain and follow from the breach. A party can recover lost profits if the lost profits can be established with reasonable certainty.

Accordingly, we hold that the trial court properly relied on the value of the present and projected future oil production of the three Lechuza tract wells as the basis for measuring the amount of Strata's damages for the undelivered working interest and net revenue interest. Mercury does not challenge on appeal the actual oil production and pricing figures upon which the trial court relied nor the certainty of the projections of future production for the Lechuza tract wells. We therefore affirm the trial court's damage award of $616,555.22 in favor of Strata.

V. Conclusion

For the foregoing reasons we affirm the judgment of the trial court.

NOTES

1. Did the court ever hold that the farmout agreement was ambiguous? Does New Mexico allow parol evidence to be admitted even without the predicate finding that the instrument is ambiguous?

2. Many farmout agreements provide that the farmor has reserved a "back-in right" where upon the occurrence of a certain event, usually well payout, the farmor is entitled to convert its overriding royalty interest into a share of the working interest. If the parties do not specify how the farmee is to determine well payout, litigation is certain to follow. *See e.g.*, Mengden v. Peninsula Production Co., 544 S.W.2d 643, 55 O.&G.R. 477 (Tex. 1976); North Finn v. Cook, 825 F.Supp. 278, 125 O.&G.R. 613 (D.Wyo. 1993).

3. A farmout agreement provides that the farmee is entitled to recover "the cost of drilling, testing, completing, and equipping" the well from the proceeds of the well's production. During drilling operations, the well was cratered by an explosion, blowout and fire. A new well bore was drilled that achieved production. Can the farmee recover the costs of drilling the first well, including the costs of putting out the fire? Continental Oil Co. v. American Quasar Petroleum Co., 599 F.2d 363, 64 O.&G.R. 364 (10th Cir. 1979) found that the costs were recoverable by the farmee even though the farmee had insurance to cover the damaged well.

SECTION 5. TRANSFERS BY LESSEE: RELATIONSHIP OF LESSOR AND TRANSFEREE

(Martin & Kramer, *Williams and Meyers Oil and Gas Law* §§ 403.3, 404-410, 412-414)

BERRY v. TIDE WATER ASSOCIATED OIL CO.

United States Court of Appeals, Fifth Circuit, 1951.
188 F.2d 820.

HUTCHESON, CHIEF JUDGE. Brought to cancel, and remove the cloud of, an "unless" oil, gas and mineral lease as to a portion of the land leased which had been assigned to the defendants, the suit sought a decree adjudging that, because of their failure to drill on or develop their assigned portion, the lease as to it had terminated and was of no further effect.

Plaintiffs' primary claim was that, under Mississippi law, upon the assignment of the segregated portion of the land, it became in effect a separate lease, with the result that instead, as before, of one producing well within the primary term of the lease (here five years) sufficing to extend the lease in accordance with its terms as to all the lands described in it, there must be a well drilled on each segregated portion. They urged, therefore, that the fact that the original lessee had drilled a well on the part of the leased land retained by him could not satisfy the obligation of defendants, owners of the assigned portion, to drill on their own tract so as to extend the lease as to their portion beyond the primary term; and that since no well was drilled during the primary term on their assigned portion, the lease as to it was at an end.

[The court's discussion of four other claims by the plaintiff is omitted. Eds.]

The trial court agreed with defendants throughout. He found that the bringing in of the well by Richardson and the payment of shut-in gas royalty was a compliance with the drilling and producing obligation of the lease and that it had been extended beyond the primary term not only as to Richardson's retained part of the land but as to the assigned portions as well.

. . .

Appealing from the judgment for defendants, plaintiffs are here, conceding that in so ruling the district judge followed the rule of law prevailing in Texas and quite generally elsewhere, but insisting that the law of Mississippi is to the contrary. Urging upon us that in White v. Hunt, 193 Miss. 742, 10 So.2d 539, the Supreme Court of Mississippi has taken its stand with Louisiana in holding that, where there is an assignment of a portion of the leased premises, the assignee to hold his assigned portion is obligated to drill his own well, they insist that the judgment must be reversed and here rendered.

. . .

Appellees are here urging upon the primary point: that in decision after the decision (sic) the Mississippi courts have made it plain that, as to the law of oil and gas, Mississippi, as a newcomer, has aligned itself with Texas; that the law in

Texas, as well as generally elsewhere, is in accordance with the holding of the trial court; that the well drilled by Richardson was sufficient to extend the lease beyond the primary term; and that, therefore, their portion of the lease did not lapse because of their failure to drill on it.

. . .

In seeking to maintain their position, appellants do not at all dispute the contentions of appellees and the conclusion of the district judge: that Mississippi is as completely committed as Texas is to the ownership in place theory as regards minerals; that Mississippi, in formulating its decisions on oil, gas, and mineral questions, has evidenced great respect for Texas decisions dealing with the same or similar matters; and that the Louisiana rule announced in Roberson v. Pioneer Gas Co., 173 La. 313, 137 So. 46, 82 A.L.R. 1264, and followed *ex necessitate* in Harrell v. United Carbon Co., 5 Cir., 52 F.2d 790, a Louisiana case, is directly opposed to the rule prevailing in all others of the states whose courts have had occasion to pass upon the precise question, the rule, as it is stated in the generally recognized text books in the field of oil and gas law.

Supporting their claim, indeed basing it entirely, on their construction of White v. Hunt, 193 Miss. 742, 10 So.2d 539; they insisted below, they repeat that insistence here: that in and by that decision the Supreme Court of Mississippi, deliberately rejecting the rule prevailing in Texas and generally elsewhere, has elected to range itself with Louisiana; and that, having thus definitely established a rule of property for Mississippi, "we are bound [of course] to follow it".

We are in complete agreement with appellants' conclusion that as to rules of property in Mississippi, we must follow where the Mississippi courts lead. We are in as complete disagreement with the premise on which the further conclusion is rested, that Mississippi has laid down the rule of property for which appellants contend. That premise, stripped to its essentials is that White v. Hunt, which admittedly does not deal with the question for decision here, the obligation of assignees of assigned portions to drill thereon within the primary term, has by implication so clearly foreshadowed for Mississippi the rule of property appellants contend for here, that this court is bound to announce and approve it.

. . .

In view of the actual state of the law in Texas, and generally elsewhere, and of the general conformity of Texas and Mississippi decisions, we are in no doubt whatever that, declining to hold that the case of White v. Hunt, lays down for Mississippi the rule of property appellants contend for, we should, on the contrary, declare: that no such rule can be distilled from it; and that there is nothing in White v. Hunt which is at all out of keeping with the decision of the court below on the question presented for decision here, nothing which brings in question anything said there.

All that was for decision in White v. Hunt, all that was decided there, was that where a lease had been assigned in part, the obligation to pay the rents on the assigned portion passed under the terms of the lease to the assignee and upon his failure to pay at the time fixed in the lease it terminated as to that portion.

. . .

In White v. Hunt, neither White nor Bradshaw, his co-owner, paid anything, nor did anybody pay anything for them. What was attempted to be, but was not permitted to be, done there was, by a belated effort set on foot several months after the lease had expired for nonpayment, to avoid its effect, by having the overpayment by mistake of one Condon, as to his separate portion of the leased premises, operate as a kind of *nunc pro tunc* payment for White and Bradshaw who owned entirely separate portions of the lease. The Mississippi court quite correctly held that, under the express provisions of the lease under which White and Bradshaw held, this could not be done.

The judgment was right. It is Affirmed.

———

COSDEN OIL CO. v. SCARBOROUGH
United States Circuit Court of Appeals, Fifth Circuit, 1932.
55 F.2d 634.

HUTCHESON, CIRCUIT JUDGE. This is an appeal from a decree finding appellant, defendant below, in default in the performance of implied covenants to prosecute with reasonable diligence the development for oil and gas of a tract of 400 acres of land, assigned to it by the lessee of a tract of 10,254 acres, and directing, upon pain of cancellation, that it proceed on terms fixed in the decree to drill at least one well thereon.

. . .

The case really comes down in the end to whether plaintiff is correct in its position that abstractly and without reference to proof of lack of diligence, defendant could not have held the part of the lease assigned to it without doing some development on it, or whether defendant is right upon either of its propositions, that its portion of the lease must be looked at not as divided, but as part of the whole, or that if it is considered as divided, it cannot be made to drill upon it under the facts in this case, establishing as they do that no prudent operator, having the interests of both lessor and lessee in mind would drill at this time.

. . .

Upon the question of whether the lease is divisible or indivisible, it must be premised that it is conceded by plaintiff and defendant that the lease is indivisible as to the requirement for the fixing of the primary term by obtaining production and that there is no contention made here that as to the requirement to drill or pay rentals the finding of production in the first well did not stop the payment of rentals as to all of the assignees, and fix in each of them a determinable fee in that portion of the land assigned to him subject to become forfeit only upon the cessation of the use or upon abandonment. The contention arises upon the implied covenant to develop, the prime consideration for the leasing.

. . .

. . .Of oil and gas leases generally it may be said that ordinarily they are regarded as indivisible as to the express covenants which fix the vesting of the determinable fee, such as the drilling of a certain number of wells, when that is required, or the obtaining of production in the absence of specific requirement, and that assignees under an original lease hold their titles to their several tracts without the necessity of further rental payments, or of further compliance with these express drilling conditions when they have been complied with on any part of the lease. On the other hand, as to the implied covenant, which running with the land is imposed on each taker of any part of the lease as a consideration for his holding it, we think it is quite generally held that the contract is severable, imposing upon the holder of each segregated part the obligation to develop that part without reference to the others. Whether this is so or not as to leases in general, we think it beyond question that the implied covenant to develop under this particular lease runs with the land, and obligates each assignee as a condition to holding his part of the assigned lease, to reasonably develop it.

. . .

The clear purpose of this lease, that those holding it or any part of it should pay rentals or diligently develop for oil, the language, "If the estate of either party hereto, is assigned the covenants hereof shall extend to their heirs . . .assigns, etc.," the provision that separate payment of rentals as to portions assigned shall be sufficient to keep alive the particular portion on which rentals have been paid, show, we think, in the plainest kind of way that the parties contemplated a lease divisible as to the consideration for it, both before and after the discovery of minerals; before discovery, as to the payment of cash rentals; after discovery, as to the payment of that which in lieu of cash rentals, the development of oil from the land for the purpose of paying over to the lessor his part of the production. . . .In short, while the lease is entire as to the vesting not only in the original lessee but in all of his assigns, of a determinable fee in each as to the part of the land he owns, that determinable fee as to each owner stands or falls, is abandoned or ceases, according to his own acts, subjecting him to the obligation for damages not at all for what is being done or not being done upon the tract in general, but only for what he does. Any other construction would lead to interminable confusion.

. . .

There is ample authority for the view we take, that the lease is indivisible as to the fixing of the term; divisible as to the implied covenant to develop. It accords with reason and common sense. In taking it, however, we take it not partially, but fully. That is, we find that an assignee, after production of oil has terminated the obligation to pay rentals and has fixed in the original lessee and his assigns a determinable fee, stands as to the tract that he owns in the same position with reference to due diligence as his assignor stands to the tract retained, obligated to the same extent and no more, to do further development.

[The court concluded that the evidence showed that, as to the tract involved, there was no failure to comply with the standard of an "ordinarily prudent person." Eds.]

We do not hold that appellant may indefinitely postpone operations upon the property, depriving appellee of the opportunity to develop, or cause it to be developed. We merely hold that under the evidence before us, there being no threat of drainage, no proof by way of opinion evidence or of offers to drill the property, that development of it could be other than at a loss, no proof that an ordinarily prudent person would, under the conditions, develop it, proof merely of appellant's delay in commencing drilling operations, from October, 1928, until this suit was brought is not sufficient basis for the decree. Appellant proved that it still believes that the property under better conditions has production possibilities, and that it intends, when changed conditions shall make it appear in the interests of both lessor and lessee reasonably prudent to do so, to develop the land. The case standing thus, appellant may neither be deprived of his lease for not having drilled on it, nor required to engage in imprudent and wasteful operations thereon.

The judgment of the court below is reversed, and the cause is remanded for further proceedings not inconsistent with this opinion.

NOTES

1. The *Berry* holding, *supra*, that the habendum clause is normally indivisible so that production or drilling operations anywhere on the leased premises keeps the entire lease alive in the secondary term is the majority rule. *See e.g.,* Hurley Enterprises, Inc. v. Sun Gas Co., 543 F.Supp. 359, 75 O.&G.R. 1 (W.D.Ark.1982), *aff'd without opinion*, 696 P.2d 1001 (8th Cir. 1982); Rook v. James E. Russell Petroleum, Inc., 235 Kan. 6, 679 P.2d 158, 80 O.&G.R. 471 (1984); Cain v. Neumann, 316 S.W.2d 915, 9 O.&G.R. 1173 (Tex.Civ.App. 1958).

2. Shaw leases a 438.5 acre tract to Jefferson. Jefferson assigns 120 acres to Cowan. Production is secured on the 120 acre tract. No activity has occurred on the 318.5 acre tract for a period in excess of 20 years. Shaw's successor in interest sues to terminate the lease as to the 318.5 acre tract claiming that the original lessee and his assignees have breached the implied covenant to develop. Does the partial assignment create separate leases for purposes of the development covenant? Does much of practical moment turn on the question of divisibility so far as the development covenant is concerned. *See* Kothe v. Jefferson, 97 Ill.2d 544, 74 Ill.Dec. 43, 455 N.E.2d 73, 79 O.&G.R. 513 (1983) and 2 Martin & Kramer *Williams & Meyers Oil and Gas Law* § 409.4.

3. OAG V. DESERT GAS EXPLORATION CO., 659 N.Y.S.2d 654, 136 O.&G.R. 333 (App.Div. 1997). A partial assignment of an oil and gas lease was given to the defendants. The assigned acreage included two producing gas wells. Production in paying quantities was also extant from eight other wells. The assigned two wells were shut-in by the assignee who does not make any shut-in gas royalty payments Does the lease continue into the secondary term for the assigned acreage? The court concluded: "Where, as here, there is no provision to

the contrary, `if the assignor obtains production on the part retained, such production will not only satisfy the habendum clause of the lease as to [the] part retained but also as to the part or parts assigned."

4. In ROBERSON V. PIONEER GAS CO., 173 La. 313, 137 So. 46, 82 A.L.R. 1264 (1931), referred to by the court in the *Berry* case, heavy reliance was placed on the following clause in the lease involved:

> If the estate of either party hereto is assigned--and the privilege of assigning in whole or in part is hereby expressly allowed--the covenants hereof shall extend to their heirs, executors, administrators, successors or assigns; but no change in ownership of the land or assignment of rentals or royalties shall be binding until after the lessee has been furnished with a written transfer or assignment, or a true copy thereof; and it is hereby agreed in the event this lease shall be assigned as to part or as to parts of the above described lands, and the assignee or assignees of such part or parts shall fail or make default in the payment of the proportionate part of the rents due from him or them, such default shall not operate to defeat or affect this lease in so far as it covers a part or parts of said lands upon which said lessee or assignee thereof shall make due payment of said rental.

Forty acres of the lessee's interest in a one hundred and twenty-five acre tract had been assigned and a well drilled on the assigned acreage. As against a contention "that the drilling of the gas well . . .on the forty acres . . .kept the lease in force on the remaining 85 acres" which had been retained, the court stated:

> Our conclusion, according to our interpretation of the clause in the contract of lease allowing an assignment of it in whole or in part, and declaring the effect of an assignment as to only a part of the leased premises, is that . . .the drilling of a well by the assignees on that part of the land on which the lease was assigned to them did not have the effect of keeping the lease in force on that part of the land . . .retained.

Professor Merrill in *The Partial Assignee - Done in Oil*, 20 Tex.L.Rev. 298 at 317 (1942), comments that "It does seem . . .that this is reading far too much into so restricted a provision."

5. Article 130 of the Louisiana Mineral Code (effective January 1, 1975) provides that "A partial assignment or partial sublease does not divide a mineral lease." The official comment to the Section explains the legislative intent:

Comment

There are several cases dealing with partial assignments of leases containing a clause permitting assignment in whole or in part and providing that in the case of a partial assignment failure of an assignee to make payment of his proportionate part of the rentals will not result in termination as to the remainder of the lease. Tyson v. Surf Oil Co., 195 La. 248, 196 So. 336 (1940); Roberson v. Pioneer Gas Co., 173 La. 313, 137 So. 46 (1931); Swope v. Holmes, 169 La. 17, 124 So. 131 (1929). In all of these, the court has held that such a clause makes a lease divisible so that when there is a partial assignment, there are two leases with different sets of rights and obligations between lessor and lessee. Not only will this be true as to the rental obligation, it is true also of the effect of drilling or production on maintenance of the lease. The unarticulated premise of these cases is that in the absence of such provisions the lease would be indivisible in the sense that a partial assignment would not have the effect of creating two leases where but one existed before. It is therefore correct to say that Article 130 reflects established law insofar as assignments are concerned. As to the effect of a partial sublease, it is, again, consonant with the theory concerning the nature of the sublease to conclude that a partial sublease has no divisive effect. This is implicit in decisions such as Roberson v. Pioneer Gas Co., *supra,* in which the court had first to determine whether the transaction in question was an assignment or a sublease. Having concluded that it was an assignment, and taking cognizance of the clause regarding partial assignments, it was held that division resulted. Clearly, if the transaction had been characterized as a sublease, this result would not have followed. This conclusion is reflected in common practice in the petroleum industry. Where a portion of a lease is to be transferred to another operator in return for an obligation well, for example, the lessee will often reserve a 1/1000th overriding royalty. The economic significance of such an interest is negligible. Its purpose is solely to preserve the integrity of the original lease to avoid the effect of an assignment when the lease contains the type of clause treated in the *Tyson, Roberson,* and *Swope* cases. It should be observed that this provision, based on the concept that the lease is indivisible unless otherwise provided by contract, is harmonious with the treatment of mineral servitudes and mineral royalties, which are indivisible in the absence of special agreement to the contrary. It is, therefore, the intent of Article 130 to preserve what is understood to be established law.

6. "The non-divisibility rule is predicated upon the basis that a prudent operator might often be willing to continue development but prudent operation might require, at least temporarily, that it be conducted on one portion of the lease alone. A lessee, quite naturally, wants to drill where there is the best chance to secure production. That is his responsibility. The only obligation the lessee originally assumes with reference to development is to develop the leased premises *as a whole,* in a prudent manner. To hold that by assignment of a portion of a lease new obligations are created for the lessee and his assignee, as to each portion, is to change the contract; and if the lease is subdivided a number of times the development obligations are increased with each assignment. Obviously, such a rule would work inequity and injustice. There is no reason why a lease should be treated as 'divisible' in construing some of its covenants, and 'indivisible' in interpreting others." Brown, *Assignments of Interests in Oil and Gas Leases,* 5 Sw. Legal Fdn. Oil & Gas Inst. 25 at 43-4 (1954).

7. In what contexts will theories as to divisibility or indivisibility of a lease

be of material importance in the resolution of controversies? *See* Merrill, *The Partial Assignee - Done in Oil*," 20 Tex.L.Rev. 298 at 314 (1942); Martin & Kramer, *Williams and Meyers Oil and Gas Law* §§ 404-410.

8. Consider the desirability of including in the oil and gas lease one of the following clauses:

(a) Upon the assignment of this lease as to a segregated portion of leased premises, the rental shall be apportioned among the several leasehold owners ratably on an acreage basis, and the payment of any such rental shall be fully effective as to the portion of leased premises for which paid regardless of nonpayment by any other owner.

(b) If this lease shall be assigned or subleased by Lessee as to a particular part or as to particular parts of said land, such division or severance of the lease by Lessee shall constitute and create separate and distinct holdings under the lease of and according to the several portions of said land as thus divided, and the holder or owner of each such portion of said land shall be required to comply with and perform Lessee's obligations under this lease for, and only to the extent of, his portion of said land, provided that nothing herein shall be construed to enlarge or multiply the drilling or rental obligations, and provided further that the commencement of drilling operations on the first or any subsequent well and the prosecution thereof as herein provided, on any portion of said land, either by Lessee or any assignee or sublessee hereunder, shall be for the benefit of and shall protect the lease as a whole.

HARTMAN RANCH CO. v. ASSOCIATED OIL CO.
Supreme Court of California, 1937.
10 Cal.2d 232, 73 P.2d 1163.

PER CURIAM. Plaintiff, Hartman Ranch Company, is the owner of land subject to an oil and gas lease, which provides for a one-eighth royalty to the lessor on all oil and other substances produced. Said lease was executed on October 18, 1913, to Joseph B. Dabney as lessee. Lloyd Miley, and Buley thereafter became co-owners of the lease with Dabney. Defendant, Associated Oil Company, is in possession of said land, either as an assignee or sublessee through the aforesaid parties, and is producing oil therefrom. Said defendant is a sublessee of the land adjoining the property of plaintiff on the south. The contention of plaintiff is that the defendant by active and intensive drilling operations on this southern tract, referred to as the Lloyd lease, is draining oil from the Hartman property. Plaintiff contends that the failure of the defendant to drill additional wells on the Hartman property constitutes a breach of an implied covenant in the Hartman lease to protect the lands from drainage. Plaintiff sued to recover lessor's royalty lost through the alleged drainage. It also prayed for forfeiture of the lease for the alleged breach of said implied covenant.

The jury returned a verdict for plaintiff in the sum of $593,700 damages for loss of royalty for the 4-year period prior to March 7, 1933, when this action was filed. Upon the equitable issue of forfeiture the trial court made findings and entered a conditional decree for forfeiture, which will be described hereinafter.

The main contentions of defendant upon this appeal are: (1) That the parent lease upon which this action is brought makes express provision for the number of wells to be drilled, with which provision defendant has fully complied, and this express provision negatives the existence of an implied covenant to drill additional wells to protect from drainage; (2) that defendant is a sublessee and as such is not subject to an action by the original lessor for breach of covenants of the parent lease; (3) that the evidence is insufficient as a matter of law to establish either the fact of drainage or the amount thereof.

. . .

In the instant case the drainage of the Hartman property in possession of defendant, Associated Oil Company, is alleged to arise from the operations of said defendant on the adjoining Lloyd property to the south. We need not decide herein whether the express covenant requiring 10 wells would exclude an implied covenant generally to drill further wells to protect from drainage, due to drilling by other operators than defendant. It certainly should not be held to have been within the contemplation of the parties that one who is in possession of the Hartman leasehold, and who, as we shall hereinafter discuss, has assumed the obligations of the Hartman lease, should by its own affirmative operations on adjoining land drain oil from beneath the Hartman property. The express covenant cannot be construed as an authorization for so doing. . . .

We conclude on this branch of the case that in the circumstances shown there was an implied covenant in the Hartman lease requiring protection from drainage through operations on adjoining land by the party in possession of the Hartman leasehold.

This brings us to consideration of the question whether defendant, Associated Oil Company, is liable in damages for breach of said covenant. Defendant contends that as a sublessee it is not liable to the original lessor in damages for breach of covenants in the parent lease. This doctrine as to sublessees is well settled. It is an application of the rule, embodied in section 1465, Civil Code, that covenants running with the land bind only those who acquire the whole estate of the covenantor in some part of the property. *See, also,* section 822, Civ.Code. A sublessee is liable only to his own lessor, that is, the sublessor, since he does not acquire the whole estate, but only a portion of the unexpired term.

Respondent disputes appellant's contention that the instrument under which it holds is a sublease, rather than an assignment. Respondent further contends that, if said instrument is a sublease, appellant is liable to it by virtue of appellant's express promise in the sublease assuming the parent lease.

The generally stated distinction between an assignment and a sublease is that an assignment transfers the entire unexpired term. In the instant case the original lease was for a term of 20 years from October 18, 1913, and as much longer as the production of oil and other hydrocarbon substances should be profitable. The term of the instrument under which the defendant holds is similarly described. However, the instrument under which defendant is in possession provided for a one-fifth royalty, while the original lease provided for a one-eighth royalty. The one-fifth royalty was to be paid by defendant to Dabney (original lessee) and his associates, and the one-eighth was to be deducted therefrom by them and turned over to the parent lessor. The instrument under which defendant is in possession also gave Dabney and his associates a right of re-entry for breach of any stipulation therein.

By virtue of our decision in [several cases], we are committed to the so-called Massachusetts rule that where the transferor reserves the right of re-entry for breach of conditions he has a "contingent reversionary interest" which prevents his transfer from operating as an assignment of the whole of the unexpired term. Instead, a sublease arises. The last four citations involved oil leases.

But in the instant case the sublease to defendant contained an express promise whereby defendant assumed the parent Hartman lease. The promise of assumption is broad and unequivocal. It provides: "Said lessee [Associated Oil Company] hereby expressly assumes and agrees to perform all the obligations and covenants provided for in said parent lease and said modification thereof to be performed by the lessee in said parent lease or said modifications, provided further that nothing herein shall deprive the lessee of the right to surrender this lease and relieve itself of its obligations hereunder." Defendant contends that under certain decisions in this state the parent lessor may sue the sublessee on a contract of assumption only when the parent lessor is a party to it. In the case herein the above promise of assumption was contained in the sublease, to which the plaintiff Hartman Ranch Company, as parent lessor, was not a party.

. . .

We find no binding precedents in our decisions, and in this situation we are of the view that the better rule is that it is not necessary that the lessor be a party to the contract of assumption in order to sue the assignee or sublessee thereon.

. . .

Appellant suggests that if it is liable in an action by the lessor it may be held twice for the same obligation, since it is also liable to its promisees with whom the contract of assumption was made. This danger is not real. The rules pertaining to third party beneficiary contracts avoid such a result and will protect appellant. The recovery of damages by plaintiff herein in the amount of one-eighth royalty on the oil and gas lost through drainage and depletion will bar an action for said one-eighth by the sublessors, Dabney and his three associates.

Although our decision herein rests on the express promise of assumption by the sublessee, we are not to be understood as laying down a rule that in the absence of such a promise the lessor may not recover from a sublessee in possession the lessor's royalty percentage on oil produced from the leased premises, or on oil drained from the leased premises through operations of said sublessee on adjoining land. Under an ordinary lease the lessor cannot sue a sublessee for money rental. But where rent or royalty under an oil lease is a percentage of the oil produced, or the proceeds therefrom, it would seem, in line with certain decisions involving other royalty problems, that the lessor has a definite property right in specific property, in the royalty percentage of oil produced from the leased property or the proceeds therefrom, and that the lessee by executing a sublease rather than an assignment cannot defeat the lessor's direct right against the party by whom the oil has been produced from the leased premises. It would be but a step to hold that the lessor, upon breach of a covenant to protect against drainage, may sue the sublessee to recover his lessor's percentage upon oil not removed through wells on the leased premises, but drained from said premises by the sublessee through his wells on adjoining land.

. . .

This brings us to the third of the appellant's three main contentions--that the evidence is insufficient as a matter of law to establish either the fact of drainage or the amount thereof.

. . .

As to the fact of drainage, the record is in a state of conflict. We conclude that the defendant has not as a matter of law negatived the existence of drainage. The jury by its verdict for plaintiff on conflicting evidence concluded that there had been drainage.

[The evidence was also found sufficient to support the verdict as to the amount of damage. Eds.]

. . .

Plaintiff not only prayed for money damages for the 4-year period preceding action filed, but also for forfeiture of the lease by reason of defendant's failure to protect against drainage. . . .

[The court concluded that a conditional decree of forfeiture could not be sustained in view of failure to join as parties to the action the owners and holders of the parent lease. Eds.]

The judgment for damages is affirmed. The portion of the judgment relating to forfeiture is reversed.

NOTES

1. In RICHARDSON v. CALLAHAN, 213 Cal. 683, 3 P.2d 927 (1931), the lease contained a covenant "to keep the premises free from liens arising from drilling and producing operations under said instrument." The action was by successors of the original lessor to compel an assignee of the lease to comply with this covenant. The assignee contended that he did not "in accepting a naked assignment of the lease . . . become bound by the terms of said written lease," and that "the covenant to keep the servient tenement free from liens is a mere personal covenant not running with the land and hence not binding upon the assignee

unless specially covenanted for by him." The court concluded that the covenant in question ran with the land and was binding on defendant assignee.

2. Adjoining lessees settled a title dispute when one quitclaimed the disputed tract to the other, who agreed to drill not more than one well thereon. The agreement was designated a covenant running with the land, binding on heirs and assigns. The grantee's assignee claimed not to be bound by the promise to limit drilling to one well. In LOONEY V. SUN OIL CO., 170 S.W.2d 297 (Tex.Civ.App.1943, writ ref'd w.o.m.), the court declared that the agreement was enforceable in equity against an assignee having actual and constructive notice of the covenant, whether or not the covenant ran with the land at law. Consider the application to oil and gas litigation of the difference in the requirements for the running of covenants with the land at law and in equity.

3. What are the factors relevant to the classification of a transfer by the lessee as an assignment or as a sublease? What significance, if any, should be attached to the following:

(a) The transfer is of an undivided interest in the leasehold.

(b) The transfer is of the working interest in a subdivided portion of the leasehold.

(c) The transferor retains a power of termination or right of re-entry for breach of covenant.

(d) The transferee makes new covenants with the transferor.

(e) The transferor reserves an interest in production.

(f) The transferor reserves certain supervisory rights in the methods of operation to be pursued.

(g) The transfer is for a primary term shorter than the unexpired primary term of the original lease.

4. Should classification of a transfer as an assignment or as a sublease be viewed as relevant in the following situations:

(a) Lessor sues to recover delay rentals from the transferee.

(b) Lessor sues to recover royalties from the transferee.

(c) Lessor sues to recover damages from the transferee for breach of an express or implied offset obligation of the lease.

(d) Lessor sues for cancellation of the lease for breach of a covenant against assignments or partial assignments by lessee.

(e) Lessor sues to cancel lease as to the assigned acreage on the ground that there was no production therefrom at the expiration of the primary term.

(f) Lessor sues lessee for cancellation of the lease, not joining the lessee's transferee in the action.

(g) Lessee assigns his leasehold by an instrument containing an extension or renewal clause and providing for an overriding royalty. Thereafter assignee transfers the leasehold to defendant who, after permitting the lease to expire by nonpayment of rentals, obtains a new lease from Lessor and drills a producing well.

5. SHIELDS v. MOFFITT, 683 P.2d 530, 81 O.&G.R. 151 (Okl. 1984). An oil and gas lease contained a clause which provided: "This lease may be assigned only with the written consent of the lessors." A portion of the lease was assigned without the written consent of the lessors. The lessors brought suit against the original lessee and the assignee seeking "forfeiture and cancellation of the lease by reason of" the assignment. The court found that the clause created a perpetual disabling restraint on alienation which was void as an impermissible restraint on alienation. The opinion states: "[f]rom the clear language of the lease, it is apparent to us that the intention of the parties was to create no more than a contingent remainder in . . .lessors, and since no forfeiture was prescribed in consequence of the assignment by lessee without the consent of lessors, no forfeiture would follow." Would the inclusion of a provision for termination of the lease on assignment without written consent have changed the result? *See* the extended critical comments of Professor Hemingway in Discussion Notes, 81 O.&G.R. 159 (1984); *see also* Nowack, *Restrictions Against Alienation in Agreements Relating to Oil and Gas Interests*, 23 Alberta L.Rev. 62 (1985).

6. A provision sometimes put into an oil and gas lease for lessors is a requirement that there be no assignment of the lease without the express written consent of the lessors. When one lessee/corporation merges with another corporation, is that an assignment of the lease such that written consent of the lessors is needed? The court in Santa Fe Energy Resources, Inc. v. Manners, 430 Pa.Super. 621, 635 A.2d 648, 128 O.&G.R. 593 (1993) says no. Thus the lessors had no right to block access to the leasehold from the corporation that had succeeded to the rights of the lessee through the merger of the lessee with the plaintiff in this case.

7. While the assignment/sublease distinction invariably affects matters in the landlord/tenant area, with the exception of Louisiana and California, the courts have not relied on the landlord/tenant doctrine to analyze post-lease conveyances. Articles 127-132 of the Louisiana Mineral Code were designed to deal with the more vexing problems arising from the assignment/sublease distinction. *See generally*, Chevron, U.S.A., Inc. v. Traillour Oil Co., 987 F.2d 1138, 124 O.&G.R. 491 (5th Cir. 1993)(applying La.Min. Code article 128); Willis v. International Oil & Gas Corp., 541 So.2d 332 (La.App. 1989). In Doré Energy Corp. v. Carter-Langham, Inc., 997 So.2d 826, 2008-645 (La.App. 3 Cir. 11/5/08) the court ruled that Article 128 could be applied retroactively as the assignee/sublessee had no vested right in the lack of privity defense (an earlier case from the same court had held that the same 1927 agreement in question was a sublease; it now ruled that the jury could determine that the agreement was an assignment).

8. Where an owner of a fractional share of the working interest assigns that

interest does the assignor remain liable to the operator for joint interest billings that are incurred after the assignment? In Seagull Energy E & P, Inc. v. Eland Energy, Inc., 207 S.W.3d 324 (Tex. 2006), the Texas Supreme Court analyzed the language of a standard form joint operating agreement, applied basic rules of contract and concluded that the assignor remained liable for its share of the joint interest billings. Instead of treating the assignment as the assignment of a property interest which might insulate the assignor from liability, the court emphasized the contractual nature of the relationship and the absence of anything in either the joint operating agreement or the assignment that would absolve the assignor from liability for meeting its obligations under the joint operating agreement. See also Chieftain Intern. (U.S.), Inc. v. Southeast Offshore, Inc., 553 F.3d 817 (5th Cir., 2008) (under Louisiana Civil Code an assignee and assignor remain solidarily liable with regard to the assignor's obligations to a third party unless the third party releases the assignor).

SECTION 6. TRANSFERS BY LESSEE: RELATIONSHIP OF LESSOR (OR SUCCESSORS IN INTEREST) WITH LESSEE-TRANSFEROR

(See Martin & Kramer, *Williams & Meyers Oil and Gas Law* §§ 403.1-.2)

KIMBLE v. WETZEL NATURAL GAS CO.

Supreme Court of Appeals of West Virginia, 1950.
134 W.Va. 761, 61 S.E.2d 728.

John H. Kimble, Minnie Kimble, James Murphy, and Eva J. Murphy leased a tract of land in Wetzel County, West Virginia to the defendant on July 9, 1924, for the purpose of drilling and operating for gas. Among others, the lease contained the following provision:

> Whereas, the said first parties have this day leased to said parties of the second part two adjoining tracts of land, it is agreed that said parties are to have gas free to the amount of 150,000 cubic feet of gas per annum for heat and light in one dwelling house on or off said land after the completion of one well on either tract by making their own connection at well or nearest pipe line of second party, at their own risk.

The defendant, by writing dated February 18, 1925, assigned the lease to the Browns Run Gas Company, a corporation, which drilled a producing well on one of the tracts of land. Subsequently the leasehold was assigned to J.S. Church, Trustee. Plaintiffs herein are successors in interest to John Kimble and Minnie Kimble. [Facts summarized by Editors.]

LOVINS, PRESIDENT. Charles Kimble and Blanche Kimble brought this suit in the Circuit Court of Wetzel County against Wetzel Natural Gas Company, a corporation, seeking an injunction to require defendant to furnish plaintiffs natural gas for heating and lighting purposes free of charge, in the amount of 150,000 cubic feet a year. No proof was adduced. The trial court perpetuated a temporary injunction theretofore granted, from which decree defendant appealed.

. . .

Authorities in this and other jurisdictions seem to be in accord that a covenant to furnish gas free of charge contained in an oil and gas lease runs with the land covered by the lease.

The defendant contends that the covenant runs with the surface and not with the mineral estate where there has been a severance of the two estates, ... We think that the covenant for free gas related to the mineral estate leased by John H. Kimble, Minnie Kimble, James Murphy and Eva J. Murphy, and likewise concerned the estate in the minerals. True, there had been a severance of the surface and mineral estates, as shown by the deed to A.J. Gump and wife, but such severance was provided for in the lease by the stipulation that free gas may be furnished on or off the premises. The covenant to furnish gas for heating and lighting purposes to one dwelling is a covenant running with the mineral estate. We do not mean to say that in all instances a covenant to furnish free gas would run with the mineral estate, but in the instant case it is clear that such covenant did run with the mineral estate.

. . .

Defendant's contention that its liability is secondary and may not be enforced until there has been demand upon and refusal by defendant's assignee to furnish free gas is not tenable. The right created by the covenant being in the nature of a partial consideration for the lease, the original lessee remains liable for the breach thereof. "Although a lessee assigns the lease with the lessor's assent, he nevertheless remains liable on his express covenant to pay rent, notwithstanding rent is accepted from the assignee, unless the lessor expressly agrees to release him and substitute the new tenant in his stead." Kanawha-Gauley Coal & Coke Co. v. Sharp, 73 W.Va. 427, 80 S.E. 781, 782.

The plaintiffs may enforce the specific performance of the covenant to furnish free gas unless the lease containing it is terminated by surrender, or the covenant is vitiated by some other circumstance.

. . .

The lease here considered does not contain any express provision or language from which it may be necessarily implied that the production of gas from the leased premises is prerequisite to the validity of the covenant to furnish the lessors free gas. In the absence of such provision or language in the lease, we do not think that the validity of the covenant to furnish free gas is dependent upon the production of gas from the land covered by the lease agreement.

We conclude that the covenant for free gas in the lease here considered must be complied with so long as the lease is in force, even though the lease has been assigned, and the defendant is no longer able to effect a surrender of such lease.

. . .

. . .The answer admitted the execution of the lease containing the covenant here considered; and the production of gas on the premises. But it denied that the plaintiffs had the right to free gas because the covenant was personal to the original lessors, and denied that the well was a producing well. We think that the denials contained in the defendant's answer present no defense to the plaintiffs' bill of complaint. Therefore, the demurrer to the answer was properly sustained. The demurrer having been sustained, there was no answer to the bill, and, therefore, the rule with reference to the dissolution of a temporary injunction was not applicable.

The decree perpetuates the temporary injunction without proof. The record is not clear whether defendant elected to stand on its answer, and does not show that defendant either requested or was accorded the right to amend. No contention is made here with reference to the amendment of defendant's answer. With that in mind, we think it was not error to perpetuate the injunction without the taking of proof; but the injunction, as perpetuated, should have provided that upon a surrender of the lease the injunction would be dissolved. Accordingly, the decree of the Circuit Court of Wetzel County is modified so that upon the surrender of the lease the injunction pronounced by the trial court will be dissolved, and, as modified, the decree is affirmed.

Modified, and, as modified, affirmed.

NOTES

1. As has been noted in Chapter 4, Professor Merrill has urged that the implied covenants of oil and gas leases are implied in law whereas A.W. Walker, Jr., has contended that they are implied in fact. What significance, if any, has this disagreement to the resolution of controversies relating to the lessee's continued obligation on implied covenants of the lease after he has assigned the lease? *See* Walker, *The Nature of the Property Interests Created by an Oil and Gas Lease in Texas*, 11 Tex. L.Rev. 399 at 402 (1933).

2. An oil and gas lease contains an express covenant that the lessee will bury pipelines below plow depth. The lessee assigns the lease to a third party. The assignee breaches the covenant. Can the lessor sue the lessee/assignor for damages for breach of the covenant? Yes, according to Allain v. Shell Western E & P, 762 So.2d 709, 146 O.&G.R. 114 (La.App. 2000).

3. Consider the advisability of including in the lease one of the following provisions:

(a) It is expressly understood and agreed that this lease shall not be assigned or transferred, nor shall the premises and rights covered thereby be leased, assigned or sub-let as to any portion thereof, without the written consent of the Lessor first being had and obtained.

(b) An assignment of this lease, in whole or in part, shall, to the extent of such assignment, relieve and discharge lessee of any obligations hereunder.

(c) But it is expressly stipulated and understood that the lessee herein guarantees on the part of its assigns full and faithful compliance of all the covenants and obligations imposed on the original lessee herein.

Chapter 7

POOLING AND UNITIZATION

SECTION 1. INTRODUCTION: WELL SPACING AND ALLOWABLES

Kramer & Martin, *The Law of Pooling and Unitization*

§ 5.01 [1] Introduction

The setting of permissible production rates and amounts is one of the most difficult areas of conservation regulation. The process involves engineering aspects as well as legal concepts; it is complex and varies from state to state and perhaps within different districts in the same state. The manner in which gas marketing differs from oil marketing causes the setting of allowables for gas to be much more problematic than the setting of allowables for oil.

States regulate the rate and volume of production for two reasons. The first is the prevention of waste. This has both physical and economic aspects. Physical waste refers primarily to a reduction of the total recoverable hydrocarbons from an oil and/or gas reservoir; this may also be called subsurface waste. Economic waste refers primarily to the production of oil or gas in excess of reasonable market demand. It has been thought of, not entirely correctly, as the production of oil or gas for less than its true economic worth, and "market demand prorationing" has thus been seen as a means of propping up the price of oil. However, production in excess of reasonable market demand can also result in physical waste in the sense that produced oil with no market may be left on the ground in pits because of lack of storage or transportation facilities. A conservation agency's power to prevent waste may also be regarded broadly as encompassing powers and duties to protect the environment as well as oil and gas.

The second purpose in regulating the rate and volume of production of oil and gas wells is the protection of correlative rights. Some courts occasionally will state that protection of correlative rights is secondary to the prevention of waste, but this is more often than not an indication that the concept of correlative rights simply does not extend as far as the losing litigant has contended. The more correct and certainly preferable view is that protection of correlative rights, properly understood, is equal in weight, dignity, and importance to the prevention of waste in agency regulation. This is because the waste prevention measures restrict the right to produce and share in production from one's property under the rule of capture; unless the state affords some compensation or protection to the rights restricted, the state will be taking property without due process of law. If the state does not protect correlative rights, then it must allow the drilling and production practices that will result in waste. Prevention of waste and protection of correlative rights are more properly understood as complementary, not competing, functions of the state conservation agency. Generally, it would be incorrect to see the two purposes as conflicting. The protection of correlative rights will arise in connection with well allowables and also with well locations. In some of the cases that follow, both aspects are involved.

The relationship between prevention of physical waste and the protection of correlative rights was expressed in a 1964 study by the Interstate Oil Compact Commission as follows:

> [A] failure to protect or adjust the correlative rights of common owners of a common source of supply of oil or gas may be one of the greatest, if not the greatest, factor contributing to the physical waste of oil and gas. . . . Where, by statute or regulatory order, correlative rights are protected, so that each owner in an orderly manner is assured of the opportunity of recovering or receiving his fair share of the oil and gas from the common pool, the race is over and all have a common interest in the most efficient and economic recovery of the maximum volume of oil and gas from the common pool, which is true conservation. [Interstate Oil Compact Commission, *A Study of Conservation of Oil and Gas in the United States*, 187 (1964).]

In setting allowable production rates or volumes, a state agency may look to four basic levels of production and/or takes (purchases) for calculation and limitation. The level with which the regulatory body will be concerned will depend on whether the allowable is being set with regard to physical waste, reasonable market demand, or protection of correlative rights; the allowable may be set by looking to a combination of these.

    1. The individual well.
    2. The reservoir (or sometimes the field).
    3. A pipeline's system within the state.
    4. The entire state.

Other factors and concerns will also come into play. Special allowables have been set for wells that are marginal producers, such as stripper wells, so that they will not be abandoned prematurely. Higher allowables have been given in some instances to encourage the drilling of new wells or the use of new production techniques, such as enhanced recovery. The basis for setting allowables may change once all or a substantial portion of a reservoir has been unitized, and wells within the unit may be treated differently from the wells not subject to the plan of unitization.

Allowables may be set on a monthly, bi-monthly, quarterly, or semi-annual basis. Under the state statutes or agency regulations, there will generally be a specified period for makeup. That is, where one producer in a reservoir produces less than that producer's allowable within a given time period, say six months, it will be given the opportunity to make up this underproduction. Likewise, where another producer is overproduced, there will be a period of time in which it will be prevented from additional production until others have had their opportunity to compensate for the drainage that will have been the result of the overproduction by the overproduced well.

[2] Waste; Maximum Efficient Rate of Recovery (MER)

Oil and gas reservoirs are of different types. A reservoir primarily productive of oil may be a dissolved-gas reservoir, a gas-cap reservoir, a water-drive reservoir, or a gravity-drive reservoir. These designations refer to the primary source of energy within the reservoir that can serve to produce the oil. Gas that is produced in association with oil is called "associated gas" or "casinghead gas." Gas may also be found in reservoirs that do not contain appreciable amounts of liquid hydrocarbons, and this is referred to as "non-associated gas." Gas reservoirs may be "wet" or "dry." They are "wet" if they contain hydrocarbons that will become liquid upon production, and "dry" if their liquids content is low. The techniques for maximum recovery and the restrictions necessary to prevent waste will vary depending on the type of reservoir that is being produced. Thus, regulations based on achieving the maximum efficient rate of production will vary.

. . .

[3] Market Demand

As described in part by Williams and Meyers, under "market demand prorationing," the "commission determines what amount shall be produced in a state during a given period of time and then allocates this total amount among the producing fields in the state (field allowables) and then allocates the field allowable to the various leaseholds and wells within the field (lease and well allowables)." While this has been partially true as an element in setting well allowables in the past, it was only a part of the allowables system in the market demand states. In current practice, the market demand producing states that have been of greatest historical importance (Louisiana, Oklahoma, and Texas) no longer prorate oil production on a statewide basis. Oil well allowables are primarily set on the basis of special field rules, depth bracket yardsticks or schedules, and oil-gas ratios. These other factors have been present in the allowables systems of the producing states even when market demand has also been a factor.

The setting of allowables for natural gas poses more difficult problems than for oil. While the production of natural gas in most non-associated gas reservoirs need not be so carefully controlled as that of oil (often it is better to produce such gas reservoirs quite rapidly), the correlative rights problems are much greater. This results from the fact that the marketing of gas and oil are very different.

Oil is marketed pretty much as a fungible commodity. That is, except for variations in grade and quality of the oil, one barrel of oil is pretty much like any other and it is sold like other fungible goods: very short-term arrangements with daily fluctuations in price, with virtually anyone able to buy or sell the commodity. A buyer can shop around and buy as much or as little as the buyer likes from a particular supplier; a seller can change purchasers with a phone call and have a different tank truck company pick up the oil. It is a substance that can be transported economically in a variety of ways: pipeline, truck, rail, or vessel.

Gas, on the other hand, until recently, has been sold primarily under long-term contracts with widely varying prices (even when the gas is exactly the same) and often with a requirement that there be some minimum level of take from the producer. For example, a gas purchase contract may have a requirement that the pipeline "take" 90 percent of each well's delivery capacity or pay the producer as though the pipeline has taken that level of production. And gas can be moved economically only by pipeline. A producer without a contract with a nearby pipeline is unable to sell its gas whatever the price.

Market demand prorationing could work with oil but not with gas. If the state limits production of oil on a statewide basis, it has little or no impact on contracts, and each producer will be able to sell all the oil it produces at about the same price that every other producer receives. Moreover, because of the free movement and purchase of oil, there is little problem with imbalance of production among wells in the same reservoir.

If, on the other hand, the state limits production of natural gas, it has a dramatic impact on long-term contracts between producer and pipeline (perhaps providing a force majeure defense to the take-or-pay obligation). Producers without contracts will be unable to provide their share of the statewide market demand unless pipelines can be forced to take gas from such producers; it will lead to imbalance in takes within a common reservoir unless the state can also force the pipeline to take gas evenly throughout the common source of supply.

Because there has existed no open market for natural gas such as the market for oil, even states with market demand prorationing statutes have not tried to implement a statewide setting of allowable in the manner of oil prorationing.

. . .

§ 5.02 Well Spacing

Well spacing is concerned with the location of wells and the density of drilling into a reservoir. Spacing rules are of two types, and both may be present in one state. Rules or orders of the state conservation agency may limit the proximity of wells to property lines and to other wells; these are sometimes referred to as lineal spacing rules. For example, the Louisiana statewide spacing rule provides that wells drilled in search of oil to depths below 3,000 feet subsea shall not be located closer than 330 feet from any property line nor closer than 900 feet from any other well completed in, drilling to, or for which a permit shall have been granted to drill to, the same pool. Lineal limitations may be found in association with density spacing rules that specify the area within which a single well will be allowed. Wyoming, for example, provides a density rule of 40 acres for an oil well and 40 or 160 acres for a gas well; it has lineal rules that an oil well cannot be closer than 460 feet to the exterior boundaries of a 40–acre subdivision with the same requirement for a 40–acre gas well, and a requirement that a gas well on a 160–acre density location be at the center of the 160 acres with a 200–foot tolerance.

Spacing regulations have the effects of protecting correlative rights in areas of diverse ownership and of limiting the number of wells that may be drilled into a reservoir in a given area. This avoids the drilling of unnecessary wells. Well-spacing is done both by statewide order and by individual field or reservoir rules.

Exceptions to both kinds of spacing rules may be granted on a well-by-well basis. Other things being equal, if A is an owner of land who is adjacent to B, who has a well on his or her property, and both are over a common source of production, it will generally be to A's advantage to place its well as close to B's property as possible.

. . .

That is to say, if A is allowed to place its well close to B's property line, A will have the opportunity to produce more of the oil or gas from the common source of supply than B is able to produce. Of course, too, if there is another well on a tract to the east of A, then A would be at a relative disadvantage to that owner's well and would then want to drill another well to offset that owner's well. Now there are good reasons for granting exception well locations from time to time, but establishing relative advantage for one party in relation to another is not a sound reason for such an exception well permit. The granting of exception well permits is a frequently litigated subject. Having a relative advantage over other producers in the same reservoir can be worth millions of dollars of production for a party, and the struggle for strategic well positioning can be as hard fought as a 19th Century battle for establishment of a favorable location for a gun battery.

STACK v. HARRIS

Supreme Court of Mississippi, 1970.
242 So.2d 857, 38 O.&G.R. 1.

JONES, JUSTICE. This case, involving the drilling of an oil well as an exception from the general spacing rules, presents a question as to whether Section 6132–21(c), Mississippi Code 1942 Annotated (Supp.1968) renders the drilling of a well as an exception, absolutely and totally immune from the provisions of general rule No. 14 adopted by the State Oil and Gas Board insofar as allowables are concerned. The judge of the Circuit Court of Hinds County, Mississippi, held that it did. We disagree and reverse.

Section 6132–21(c), Mississippi Code of 1942 Annotated (Supp.1968) reads as follows:

Each well permitted to be drilled upon any drilling unit shall be drilled in accordance with the rules and regulations promulgated by the board and in accordance with a spacing pattern fixed by the board for the pool in which the well is located with such exceptions as may be reasonably necessary where it is shown, after notice and upon hearing, that the unit is partly outside the pool or, for some other reason, a well otherwise located on the unit would be nonproductive, or topographical conditions are such as to make the drilling at such location unduly burdensome. Whenever an exception is granted, the board shall take such action as will offset any advantage which the person securing the exception may have over other producers by reason of the drilling of the well as an exception, but no well drilled and completed as an exception to prescribed footage limitations for the reason that a portion of the drilling unit upon which such well is located is partly outside the pool or productive horizon shall be allocated a reduced daily production allowable whenever it shall be demonstrated to the satisfaction of the board that the productive acreage underlying such drilling unit is equal to, or more than, the reasonable minimum amount of productive acreage which would underlie such drilling unit under the minimum conditions which would permit the drilling of a well thereon so located as to comply with all applicable footage limitations; . . .

It is not denied that the well was drilled as an exception and complied with the above statute except that part which says, "Each well permitted to be drilled upon any drilling unit shall be drilled in accordance with he rules and regulations promulgated by the board . . .."

Rule 14 of the General Rules affecting the entire state is applicable to this well, there being no special rules for this particular field. Rule 14 reads, insofar as affects this issue, as follows:

(b) No well shall cross drilling unit lines unless permit is obtained from the Board after notice and hearing.

(c) Intentional deviations of short distances necessary to straighten the hole, sidetrack, junk, or correct other mechanical difficulties may be accomplished without the issuance of a permit, but the operator shall immediately notify the Board by letter or telegram of the fact thereof.

(d) Except as set forth in paragraphs (c) and (e) hereof, no well may be directionally deviated from its normal course unless authorization so to do is first obtained from the Board after notice and hearing.

(e) In the event an operator in good faith commences and proceeds with the drilling of a straight well and thereafter, for reasons acceptable to the operator, desires to directionally deviate the well, he may do so at his own risk, first notifying the Board by letter or telegram of the fact thereof. On completion of such well as a producer, the operator must immediately apply for a permit from the Board on notice and hearing for approval of such intentional deviation. Pending such approval or disapproval, the Board may assign a temporary allowable only to such well.

(f) In cases of directionally deviated drilling the Board shall have the right to assess appropriate allowable penalties to prevent undue drainage from offset properties and to adjust possible inequities caused by the directional drilling.

We need not detail the manner in which J. W. Harris, appellee, acquired the right to drill the exception well. He secured a permit from the Oil and Gas Board, employed drillers and proceeded to drill. The Smackover Sands at this point were approximately 12,600 feet to 12,800 feet below the surface. When the well had been drilled to about 10,000 feet, it was found to have drifted (or deviated) a considerable distance to the northeast. This was toward a dry hole which had been drilled in the approximate center of the forty from which the exception was carved and in which the exception well is located.

At this point, the operator, with the assistance of a drilling firm which specialized in directional drilling, changed the direction of the well, turning it approximately 180 degrees. The well then ran in a southwesterly direction practically to the point of intersection of the three drilling units upon which appellants had wells.

The permit authorized the drilling of said exception well at a point 300 feet north of the south line and 300 feet east of the west line of the forty upon which it was drilled, to-wit: the northwest quarter of northwest quarter, section 26, township 10 north, range 6 west, Wayne County. Stack and Chisholm operating as the Brandon Company had 3 producing wells, one located in the northeast quarter of northeast quarter of section 27, township 10 north, range 6 west; one in the southeast quarter of northeast quarter of said section 27 and one in the southwest quarter of northwest quarter of section 26. The operator had complied with the rules and regulations by giving notice of the fact that he was intentionally deviating the well. At one time when the bottom of the hole was located, it was in the northeast quarter of northeast quarter of section 27, being the unit where the Brandon well 27–1 was situated. Upon ascertaining this fact, the pipe was pulled and the hole plugged. The specialized directional drilling firm was rehired and redrilled the well to the east bottoming it at a point approximately 31 feet from the east line of the northeast quarter of northeast quarter and near the north lines of the two other quarter-quarter sections hereinbefore mentioned. It was practically at the junction of the four drilling units and close to the line of each of the 3 Brandon well units.

All of the wells were pumpers. Harris, the operator of the exception well, made application to the State Oil and Gas Board to approve the well as drilled with the intentional deviations and with the bottom of the hole situated as described above.

Rule 35 of the General Rules places the allowables on a depth basis. At a depth of 12,000 feet the allowables were 400 barrels daily.

The Oil and Gas Board approved the well as completed, but, provided that, because of the intentional deviations and the location of the bottom of the well, the exception well should have an allowable of only 150 barrels per day.

From this order, Harris appealed to the Circuit Court of Hinds County and secured from the circuit judge a supersedeas making bond as required by the circuit judge. On the hearing before the circuit court, the judge held that Code Section 6132–21(c) was controlling as to the allowable and that the Oil and Gas Board exceeded its authority in applying Rule 14 and reducing the allowable from 400 barrels to 150 barrels because of the intentional deviation and bottoming of the well where it was.

The circuit judge held that Section 6132–21(c), Mississippi Code 1942 Annotated (Supp.1968) controlled and under no circumstances could the Oil and Gas Board reduce the allowable.

Testimony from experts on both sides established clearly that the said exception well with the bottom located where it was, would drain from all 3 of the appellants' wells. The board was fully justified in believing this. The only defense to such drainage asserted by the appellee was that prior to the drilling of said exception well, the other wells had drained from his land. Of course, the rule of capture applied, and the oil taken by the other 3 wells was legally produced and could not be used to offset drainage from the said exception well. Plaintiff's well 27–1 was 842 feet from the bottom of the exception well; No. 27–8 was 943 feet from the hole of the exception; and No. 26–5 was 1207 feet from the bottom of said hole. It was testified by some of the experts that if the pressure equalized between the bottom of the exception well and the other wells, the drainage would extend half the distance to the other wells.

Section 6132–01, Mississippi Code 1942 Annotated (1952) declares the policy of the state as to oil and gas in these words:

It is hereby declared to be in the public interests to foster, encourage and promote the development, production and utilization of the natural resources of oil and gas in the state of Mississippi; and to protect the public and private interests against the evils of waste in the production and utilization of oil and gas, by prohibiting waste as herein defined; to safeguard, protect and enforce the co-equal and correlative rights of owners in a common source or pool of oil and gas to the end that each such owner in a common pool or source of supply of oil and gas may obtain his just and equitable share of production therefrom; . . .

In Section 6132–08, Mississippi Code 1942 Annotated (1952) under "Definitions" waste shall mean and include the following:

(3) Abuse of the correlative rights and opportunities of each owner of oil or gas in a pool due to non-uniform, disproportionate, or unratable withdrawals causing undue drainage between tracts of land or resulting in one or more owners in such pool producing more than his just and equitable share of the

production from such pool.

Section 6132–09, Mississippi Code 1942 Annotated (1952) declares that waste as defined in the act is unlawful.

Section 6132–31, Mississippi Code 1942 Annotated (1952) provides:

Owners or operators of oil or gas wells shall, before connecting with any oil or gas pipe line, secure from the board a certificate showing compliance with the conservation laws of the state and conservation rules, regulations and orders of the board. No operator of a pipe line shall connect with any well until the owner or operator of such well shall furnish a certificate from the board that such conservation laws and such rules, regulations and orders have been complied with; provided, this section shall not prevent a temporary connection of not more than seven (7) days' duration with any well in order to take care of production and prevent waste until opportunity shall have been given the owner or operator of such well to secure such certificate. The board shall have the power to cancel any certificate of compliance issued under the provisions of this section when it appears, after due notice and hearing, that the owner or operator of a well covered by the provisions of same has violated or is violating, in connection with the operation of said well or the production of oil or gas therefrom, any of the oil or gas conservation laws of this state or any of the rules, regulations or orders of the board promulgated thereunder.

The board is an arm of the state empowered by the legislature to prescribe rules and regulations for achieving in practice this policy of the state as announced in Section 1 of the Act and to enforce, maintain and carry out the said policy of the state. The powers granted to the board are in Section 613210, Mississippi Code 1942 Annotated (Supp.1968), and are as follows:

(a) The board shall have jurisdiction and authority over all persons and property necessary to administer and enforce effectively the provisions of this act and all other acts relating to the conservation of oil and gas.

. . .

(c) The board shall have the authority, and it shall be its duty, to make, after notice and hearing as hereinafter provided, such reasonable rules, regulations and orders as may be necessary from time to time in the proper administration and enforcement of this act, and to amend the same after due notice and hearing, including, but not limited to rules, regulations and orders for the following purposes:

. . .

(2) To require the making of reports showing the location of oil and gas wells; and to require the filing within thirty (30) days from the time of the completion of any wells drilled for oil or gas, of logs and drilling records.

. . .

(12) To allocate and apportion the production of oil or gas, or both, from any pool or field for the prevention of waste as herein defined, and to allocate such production among or between tracts of land under separate ownership in such pool on a fair and equitable basis to the end that each such tract will be permitted to produce not more than its just and equitable share from the pool; . . .

(13) To prevent, so far as is practicable, reasonably avoidable drainage from each developed unit which is not equalized by counter drainage.

There is no claim that the board did not have the right to make Rule 14. The only claim being, as heretofore stated, that the said section of the Code providing for full allowables renders the exception well absolutely immune from any action in the reduction of allowables regardless of what happens. We cannot condone or approve such a contention. Section 6132–21(c) contemplates, as shown by the section itself, that the well shall be drilled in accordance with the rules and regulations, and applies only in cases where the driller or operator does drill his well in accordance with such rules and regulations. If it is drilled in accordance with such rules and regulations, and there is no question about it, then the allowable cannot be reduced because it is an exception, but where it is not so drilled, the said section does not apply.

It would be a bizarre situation if the State Oil and Gas board could reduce the allowable of an oil well on a regular unit, but could not for any reason protect the equitable rights of other co-owners in the common pool by such a reduction, when the driller has violated the rules and regulations, or has intentionally deviated his well and bottomed it so that it collides with and defeats the policy of the state as announced by said Act.

The said Rule 14, sections (c), (d), and (e), shows that it is contemplated that the hole shall be, as nearly as possible or reasonable, a straight hole and shall not meander around as this hole did. Section (e) provides that if an operator in good faith commences and proceeds with the drilling of a straight well and desires to directionally deviate the well, he may do so at his own risk, first notifying the board. On completion of such well as a producer, the operator shall immediately apply for a permit from the board for approval of such intentional deviation. The board would have the right to approve or disapprove, and, if they had such rights, they certainly had the right to approve subject to a penalty such as is here involved, so as to offset any advantage that the exception well would gain over the wells adjacent.

They are charged with the duty of protecting both the public and the individual owners in oil and gas matters, and, if they should have no right to approve subject to a penalty, their only course would be to decline to approve and refuse to issue a compliance certificate which would enable the operator to dispose of his production on the market. If they failed to approve and failed to issue a certificate of compliance, the operator of the exception well would be totally deprived of any production, because he could not dispose of it. The board, in attempting to be fair to him and to the adjacent owners, sought to do equity by granting approval but reducing the allowables so that there would be compensation for the improper location of the bottom of the well.

As to "intentional" deviation, it is stated in 58 C.J.S. Mines and Minerals § 241, pages 680–681 1948) that:

> Various statutes provide for the enforcement of regulations concerning the operation of mines and wells and afford remedies for the violation thereof. Mere failure to do a required act is sufficient to render one answerable to a statute punishing "any violation" thereof. The conscious failure to observe and comply with the provisions of a statute, even though no evil intent induces the failure, is sufficient to constitute a willful violation thereof. Under a statute making miners who shall "intentionally" do certain things guilty of a misdemeanor, the word "intentionally," as so used, requires only that the act constituting the alleged violation of the statute shall be "knowingly" done.

There is no claim here that the operator did not know of this directional deviation.

In accordance with our views as hereinbefore stated, we are reversing the circuit court as to the main issue and also as to the order releasing the surety on supersedeas bond from liability. We are re-instating the order of the State Oil and Gas Board and remanding the case to the State Oil and Gas Board for such other appropriate action as may be deemed necessary, if any.

Reversed, order of Oil and Gas Board reinstated, and remanded to Oil and Gas Board.

GILLESPIE, P.J., and PATTERSON, SMITH, and ROBERTSON, JJ., concur.

NOTES

1. Should an allowable penalty be imposed for deliberately drilling at a location not permitted by the state agency? *See* Robert–Gay Energy Enterprises, Inc. v. State Corporation Commission of Kansas, 235 Kan. 951, 685 P.2d 299, 82 O.&G.R. 272 (1984). In a case such as *Stack v. Harris,* is the "penalty" actually penal in nature, or is it a limitation for the protection of correlative rights of adjacent owners? In Texas, *compare* Railroad Commission v. Sample, 405 S.W.2d 338, 25 O.&G.R. 85 (Tex.1966) *with* Harrington v. Railroad Commission, 375 S.W.2d 892, 19 O.&G.R. 830 (Tex.1964).

In SANTA FE EXPLORATION CO. V. OIL CONSERVATION COMMISSION, 114 N.M. 103, 835 P.2d 819, 120 O.&G.R. 535 (1992), the Commission held a hearing on an unorthodox location for a well and thereafter limited production from the reservoir and from each well, with a penalty imposed on the well at the unorthodox location. Appellants from the order claimed that the Commission exceeded its powers by effectively unitizing the field though not authorized to do so under the Statutory Unitization Act, NMSA 1978, §§ 70-2-11, subd. A, 70-7-1 to -21 as secondary or tertiary recovery was not involved.

The court held that the Commission was acting within its broad authority to protect correlative rights and prevent waste. The unorthodox well was located so that it could effectively drain the entire Pool. The Commission's production penalty on the well protected correlative rights.

2. Should the law guarantee a recovery of production or simply that there be a fair opportunity to produce a share of the oil or gas in a reservoir? The North Dakota Supreme Court in Hanson v. Industrial Commission, 466 N.W.2d 587, 116 O.&G.R. 294 (N.D.1991) ruled that the state agency need only give the producer the opportunity to produce in order to protect its correlative rights. The producer of a well had not shown that the Commission's denial of his request to dispose of salt water through an abandoned well had deprived him of the opportunity to produce his equitable share of the oil in a pool.

3. The Mississippi statute applied in the principal case referenced authority of the Oil and Gas Board to prevent waste. How broad a definition can be given of "waste"? Should it apply only to subsurface and surface loss of recoverable oil and gas, or should it encompass protection of all natural resources affected by oil and gas operations? That is, should the state agency be able to use its authority to prevent waste to deny a drilling permit because of an adverse impact on the environment, or would such an agency decision be beyond its authority? *See* Michigan Oil Co. v. Natural Resources Commission, 406 Mich. 1, 276 N.W.2d 141, 62 O.&G.R. 313 (1979), *aff'g*, 71 Mich.App. 667, 249 N.W.2d 135, 56 O.&G.R. 234 (1976); Gulf Oil Corp. v. Wyoming Oil and Gas Conservation Commission, 693 P.2d 227, 84 O.&G.R. 579 (Wyo.1985); Gulf Oil Corp. v. Morton, 493 F.2d 141, 47 O.&G.R. 455 (9th Cir.1973).

4. A symposium at the University of Colorado in 1985 reviewed in depth state programs for setting allowable production, and the proceedings of the seminar have been published in the University of Colorado Law Review. *Symposium: Workshop on Natural Gas Prorationing and Ratable Take Regulation*, 57 Colo. L. Rev. 149 (1985) with articles on the programs of Oklahoma, Texas, Wyoming, Kansas, Louisiana, and New Mexico. In 1992, Texas and Oklahoma began efforts to set natural gas allowables on a state-wide market demand basis.

UNION PACIFIC RESOURCES CO. v. TEXACO, INC.

Wyoming Supreme Court, 1994
882 P.2d 212, 133 O.&G.R. 549

TAYLOR, JUSTICE.   The primary question in these consolidated appeals concerns the interpretation of a contract. The contract is an operating agreement formed by four oil and gas companies who own exclusive rights to exploit minerals on certain lands in Wyoming. The contract allocates expenses and profits for natural gas exploration and production. The allocation of profit varies depending upon the geologic formation from which production is achieved. After the parties executed the contract, a natural gas well capable of significant production was discovered in one formation. Subsequently, the Wyoming administrative agency responsible for oil and gas conservation ordered that the size of the drilling unit for that formation be enlarged from its former boundaries to protect the correlative rights of an adjoining mineral rights owner and the operating rights owners. This dispute is focused on whether the order from the administrative agency superseded the parties' contract. The district court determined that the contract was modified by the administrative agency order.

We affirm.

II. Facts

On September 18, 1981, the Wyoming Oil and Gas Conservation Commission (Commission) issued two orders to regulate exploration for natural gas and associated hydrocarbons (gas) on certain Wyoming lands located in Lincoln, Sweetwater and Uinta Counties. The orders established drilling units of specified sizes where one well could be drilled to extract pools of gas that were believed to exist in distinct sedimentary formations beneath the surface. Each order also identified the area assigned for the granting of a well permit.

Included in the Commission's orders were sedimentary formations located in the Bruff Field of the Moxa Arch area of Sweetwater and Lincoln Counties. In the Frontier formation, the Commission established an irregularly sized 760-acre drilling unit which included all 640 acres of Section 15 and an adjoining 120- acre portion of Section 22 of Township 19 North, Range 112 West, 6th P.M. In the Dakota formation, the Commission established a 640-acre drilling unit which encompassed all of Section 15 of Township 19 North, Range 112 West, 6th P.M. The Dakota formation is a geologic zone which is found at a deeper depth than the Frontier formation.

Union Pacific, Amoco, Wexpro and Texaco (collectively the parties) own oil and gas working interests in the Bruff Field. Effective October 7, 1981, the parties, or their predecessors in interest, entered into an Operating Agreement to test and develop a portion of the Bruff Field. The Operating Agreement defined the "subject lands" as including the parties' interests in all of Section 15 of Township 19 North, Range 112 West, 6th P.M. (hereinafter Section 15) and the parties' interests in the portion of Section 22 of Township 19 North, Range 112 West, 6th P.M. (hereinafter Section 22) which corresponded with the Commission's Frontier formation drilling unit order. The Operating Agreement acknowledged that the Commission had established a 760-acre drilling unit for the Frontier formation and a 640-acre drilling unit for the Dakota formation. Specifically incorporated into the Operating Agreement was an agreement of Operating Provisions which was attached as Exhibit "D". The Operating Agreement also incorporated another agreement signed by the parties which is styled as a Communitization Agreement. Effective October 7, 1981, the Communitization Agreement provided exclusively for development of the Frontier formation.

The Operating Agreement designated Amoco as the operator of the wells on the "subject lands." During 1981, Amoco was required to commence drilling of a test well in Section 15 "to adequately test the Frontier and Dakota formations" unless conditions prohibited. This test well, designated the Champlin 149 Amoco "L" Well # 1 (hereinafter the L-1 Well) was located in the southeastern quarter of Section 15. The L-1 Well produced commercial quantities of gas from the Frontier formation, but was unproductive from the Dakota formation. On April 30, 1982, the L-1 Well was placed in commercial production.

In the Operating Agreement, the parties agreed to allocate the ownership of production in "working interest percentages," which were stated as:

	Frontier	Dakota
[Wexpro]	15.62501%	11.04410%
Amoco	46.71052%	53.95750%
[Union Pacific]	21.87500%	29.00420%
Texaco	15.78947%	5.99420%

Production from the L-1 Well was allocated according to these "working interest percentages" for the Frontier formation.

On June 29, 1989, Amoco proposed drilling the Champlin 149 Amoco "L" Well # 2 (hereinafter the L-2 Well). The L-2 Well would be located within the southwestern quarter of Section 15 and would be drilled to test both the Frontier and Dakota formations. Union Pacific, Wexpro and Texaco all signed "Well Authorization" agreements to signify their approval of the proposal. These agreements restated that production from the L-2 Well would be allocated according to the "working interest percentages" found in the Operating Agreement.

The L-2 Well was completed on January 23, 1990. The L-2 Well produced substantial quantities of gas from the Dakota formation. The amount of production from the L-2 Well exceeded the expectations of experts, particularly since the nearby L-1 Well had failed to produce commercial quantities of gas from the Dakota formation.

The United States Department of the Interior, Bureau of Land Management (BLM) owns the mineral rights to the portion of the "subject lands" within Section 22. On September 7, 1990, the BLM, as lessor, demanded that Texaco and Wexpro, as lessees or operating rights owners, act to protect the leased lands from drainage by the L-2 Well. In response, Texaco filed two alternative applications for orders from the Commission.

The first application sought an order to permit Texaco to drill an offset well to prevent the drainage of the gas from Section 22. The offset well would have been located in Section 22 only 800 feet away from the L-2 Well in Section 15. After conducting a hearing on November 14, 1990, the Commission denied the application for an offset well finding that a prudent operator would not drill at this location because it would interfere with the L-2 Well.

On December 13, 1990, the Commission granted Texaco's second application which sought an order to enlarge the drilling unit for the Dakota formation. The Commission ordered that the drilling unit for the Dakota formation be enlarged from 640 acres to 760 acres, which included all of Section 15 and the 120-acre portion of Section 22 which Texaco and Wexpro leased from the BLM. This enlarged drilling unit precisely corresponds with the previously existing drilling unit established for the Frontier formation on these same lands. See attached Appendix "A." The Commission determined that this enlarged drilling unit would protect the correlative rights of the BLM, Texaco and Wexpro. Texaco and Wexpro have an obligation, under terms of the Operating Agreement, to pay royalties to the BLM from their allocation of the profits from the L-2 Well.

During the November 14, 1990 hearing to determine if the Commission would grant the application for an enlarged drilling unit for the Dakota formation, Amoco presented expert testimony, accompanied by an exhibit which was admitted into evidence, disclosing the effect on the parties of an enlarged drilling unit. Amoco calculated that if the drilling unit was enlarged, Amoco's "working interest percentages" for production from the L-2 Well would decrease by 8.5197% and Union Pacific's "working interest percentages" would decrease by 4.5796%. Meanwhile, Texaco's "working interest percentages" for production from the L-2 Well would increase by 6.9483% and Wexpro's "working interest percentages" would increase by 6.1510%. However, after the Commission issued its order enlarging the drilling unit for the Dakota formation, Amoco changed its position.

On February 18, 1991, Amoco informed Texaco that it was adopting the position that the Commission's enlarged drilling unit order did not alter the terms for the allocation of production from the Dakota formation contained in the Operating Agreement. "It is Amoco's position that the Operating Agreement is binding on all signatory parties, and the entry of the Commission's Order respacing the Dakota formation in no way changes the party's contractual working interest." Union Pacific joined Amoco in claiming that the "working interest percentages" were not altered by the Commission's enlarged drilling unit order for the Dakota formation. Texaco challenged that it was apparent from reading the Operating Agreement that an enlarged drilling unit would alter the "working interest percentages" for the Dakota formation. Texaco pointed out that the Operating Agreement specifically defined "Dakota Owners" as "the working interest owners, owning the working interest in and to the Spacing Unit established for the Dakota formation." Wexpro supported Texaco's position. The parties did not seek judicial review of the Commission's order enlarging the drilling unit.

On April 17, 1991, Texaco filed an application with the Commission for a compulsory pooling order to combine the parties' interests and allocate production according to the surface acreage each party contributed to the enlarged drilling unit for the Dakota formation. On May 14, 1991, the Commission held a hearing on the compulsory pooling application. Union Pacific and Amoco protested the application arguing that the Commission lacked authority to interpret the Operating Agreement. Union Pacific and Amoco both urged the Commission to defer any action until litigation, which had been commenced in Colorado four days before the Commission's hearing, was completed. However, on June 13, 1991, the Commission issued a compulsory pooling order.

The Commission's compulsory pooling order brings together the interests of the parties in the enlarged 760-acre drilling unit on the "subject lands" in the Dakota formation. The Commission found that under the compulsory pooling order, the parties would have the following "working interest percentages" in the L-2 Well:

Amoco	45.4379%
Union Pacific	24.4246%
Wexpro	17.1950%
Texaco	12.9425%

The Commission concluded that, as a matter of law, the parties had not previously voluntarily pooled their respective interests in the Dakota formation. However, the Commission made this finding "contingent" on the outcome of litigation between the parties on the meaning of the Operating Agreement. Therefore, the parties, under the Commission's order, may apply for rescission or modification of the Commission's order after litigation over the Operating Agreement is completed. On September 10, 1991, Union Pacific and Amoco filed an action in the District Court for the First Judicial District of Wyoming challenging the validity of the Commission's June 13, 1991 compulsory pooling order. Further prosecution of this action has been suspended until a decision on this appeal is finalized.

At the time the Commission issued its compulsory pooling order, the parties were also involved in litigation over the Operating Agreement in the District Court for the City and County of Denver, Colorado. In January, 1992, however, the Colorado court exercised its discretionary power and dismissed the case, without prejudice. The Colorado court declined jurisdiction on the basis of forum non conveniens.

On January 31, 1992, Union Pacific and Amoco sought a declaratory judgment from the District Court for the First Judicial District of Wyoming holding that the Operating Agreement fixed the parties' "working interest percentages" in the Dakota formation "irrespective" of the drilling unit orders of the Commission. Texaco and Wexpro both answered and filed counterclaims seeking declaratory judgments in their favor. On June 19, 1992, Texaco filed a motion for summary judgment, which Wexpro joined. Union Pacific and Amoco responded with individual cross-motions for summary judgment.

After extensive briefing and oral argument, the district court granted a partial summary judgment in favor of Texaco and Wexpro. The district court concluded that the terms of the Operating Agreement disclosed that the parties shared production "in proportions which were equal to the proportions of net mineral acres contributed by each to the then applicable spacing units governing production from each formation." The district court determined that when the Commission enlarged the drilling unit for the Dakota formation, the interests of the parties in the proportion of net mineral acres was also altered.

After the partial summary judgment was granted, the parties stipulated to the effects of the amended "working interest percentages" and established escrow accounts to hold the disputed funds until litigation was completed. On September 22, 1993, the district court entered a judgment which required Amoco to deposit $1,620,450.09 along with future proportional shares of monthly production from the L-2 Well in the escrow accounts. The judgment also required Union Pacific to deposit $677,454.40 along with future proportional shares of monthly production from the L-2 Well in escrow accounts. Union Pacific and Amoco both filed notices of appeal.

III. Discussion

. . .

Various constructions of the Operating Agreement are offered by each party. In the Operating Agreement, Union Pacific asserts the parties unambiguously established their "working interest percentages" for any well drilled on the "subject lands" in the Dakota formation. Union Pacific declares that the express terms of the Operating Agreement do not permit subsequent modification. Accordingly, Union Pacific explains that the Commission's orders cannot supersede the Operating Agreement.

Amoco also maintains that the district court erred in ruling that the Commission's orders superseded the "working interest percentages" in the Operating Agreement. Amoco offers a policy argument suggesting limitations in the doctrine of correlative rights to protect contractual rights. Amoco argues that there is no provision in the Operating Agreement which would adjust the "working interest percentages" in the event of a change in the size of the drilling unit.

Wexpro challenges that the Operating Agreement incorporates principles of Wyoming law. Therefore, the Commission's orders, according to Wexpro, can supersede provisions in the Operating Agreement. Wexpro points out that the Operating Agreement acknowledges the drilling unit orders in place at the time the Operating Agreement was executed. Wexpro maintains that the Commission exercised its statutory obligation to prevent waste and protect correlative rights in enlarging the drilling unit for the Dakota formation.

Texaco argues that before this dispute, all the parties shared the understanding that "working interest percentages" in the Operating Agreement were determined by the amount of surface acreage each party contributed to the drilling unit. Texaco insists that the Operating Agreement does not fix "working interest percentages" for all time and all circumstances. Texaco argues that ambiguity in this language permits the court to resort to extrinsic evidence to determine the intent of the parties.

In determining if any of these varied constructions of the Operating Agreement are correct, we are guided by our established rules of contract interpretation. Contract interpretation is the process of ascertaining the meaning of the words used by the parties to express their intent. . . .

Our review of the Operating Agreement directs a conclusion that the language is unambiguous as a matter of law. . . . Accordingly, we need not consider extrinsic evidence to determine the intent of the parties. . . . However, the language of the Operating Agreement is not always simple or direct. The Operating Agreement was negotiated and drafted by professionals in a technical industry, using distinctive terminology for which customary meanings are often assumed.

The parties styled their agreement as an "Operating Agreement." The term, operating agreement, has come to have a plain meaning which reasonable persons in the oil and gas industry understand as denoting:

> An agreement between or among interested parties for the testing and development of a tract of land. Typically one of the parties is designated as the operator and the agreement contains detailed provisions concerning the drilling of a test well, the drilling of any additional wells which may be required, the sharing of expenses, and accounting methods. The authority of the operator, and restrictions thereon, are spelled out in detail in the typical agreement.

8 Howard R. Williams & Charles J. Meyers, Oil and Gas Law, Manual of Oil and Gas Terms, 837 (1991). The provisions of the Operating Agreement conform to this standard and disclose the intent of the parties to form this type of contract.

After identifying Amoco as the operator of the "subject lands," Article 2 of the Operating Agreement contains the language which the parties dispute:

2. Ownership of Production; Material and Equipment:

A. Production. Subject to Article 6 hereof [dealing with royalties], all gas and associated liquid hydrocarbons produced and saved from the Frontier formation within the Subject Lands shall be allocated in the proportions of the acreage lying outside the participating area and the acreage inside the participating area with that portion of the production allocated to the participating area to be shared by the working interest owners in the same manner as the remainder of the production in the participating area is shared. Production allocated to the acreage outside the participating area will be owned by the working interest owners and the royalty owners of that acreage. Production outside the participating area for the Frontier formation will be allocated as provided in the Communitization

Agreement attached hereto as Exhibit "A." *Allocation of production shall be owned by the parties hereto in the following "working interest percentages":*

	Frontier	Dakota
[Wexpro]	15.62501%	11.04410%
Amoco	46.71052%	53.95750%
[Union Pacific]	21.87500%	29.00420%
Texaco	15.78947%	5.99420%

The unleased Champlin acreage committed hereto (i.e., the NE/4 of said Section 15) shall be treated as though leased subject to a landowner's royalty of 15% owned by [Union Pacific]. [Emphasis added by the court.]

Article 2 initially distinguishes between the acreage located within the "participating area" and the acreage located outside the "participating area." The "participating area" is defined elsewhere in the Operating Agreement as acreage which is subject to a voluntary pooling agreement dated July 14, 1972. The acreage included in the "participating area" encompasses all of the northwestern quarter of Section 15 and a portion of the southwestern quarter of Section 15. The first three sentences of Article 2 describe the allocation of production from the acreage inside the "participating area" and the acreage outside the "participating area." After noting this distinction, Article 2 details the "working interest percentages" for the Frontier and Dakota formations.

While the parties focus on the language detailing the specific "working interest percentages," our review of Article 2, as a whole, discloses the inherent limitations of the Operating Agreement. The convoluted language of the first three sentences of Article 2 is focused exclusively on the allocation of production from the drilling unit established for the Frontier formation. The only mention of allocation of production from the drilling unit established for the Dakota formation occurs when the "working interest percentages" are detailed. Unfortunately, the parties do not explain anywhere in the Operating Agreement the precise method used to calculate the "working interest percentages" other than by the reference to surface acreage either inside or outside the "participating area."

The references in Article 2 to the Frontier formation or the Dakota formation may be considered generic; however, other provisions of the Operating Agreement illuminate the intent of the parties to relate these terms to the drilling units established by the Commission. In Article 4, the parties provided for the allocation of drilling costs and expenses in detail. The Article 4 definitions refer to the Commission's orders establishing drilling units:

A. Definition[s]:

"Frontier Owners"--the working interest owners owning the working interests in and to the Spacing Unit established for the Frontier formation.

"Dakota Owners"--the working interest owners, owning the working interest in and to the Spacing Unit established for the Dakota formation.

In the Operating Agreement, the parties refer, interchangeably, to a "drilling and spacing unit" or to a "spacing unit." The terms "spacing unit" or "drilling unit" are used in the oil and gas industry to describe the area which an administrative agency has determined one well can efficiently drain. See 8 Williams & Meyers, supra, at 357-58, 1359-62. Wyoming law refers to drilling units. Wyo.Stat. § 30-5-109(a) (1983).

The Operating Agreement contains an express reference to the Commission's orders establishing drilling units. After identifying the ownership interests of each of the parties in the "subject lands," the Operating Agreement includes a recital (hereinafter the drilling unit recital), which states:

> WHEREAS, by Order No. 11, entered September 18, 1981, in Docket No. 191-81, the Oil and Gas Conservation Commission of the State of Wyoming established the Subject Lands as a 760-acre drilling and spacing unit for the production of natural gas and associated liquid hydrocarbons from the Frontier formation; and, by Order No. 1 [sic], entered September 9, 1981 [sic], in Docket No. 190-81, the Oil and Gas Conservation Commission of the State of Wyoming established the Subject Lands on the 640-acre drilling and spacing unit for the production of natural gas and associated liquid hydrocarbons from the Dakota formation[.]

A "recital" is a formal statement in a document of some matter of fact "to explain the reasons for the transaction." . . . In the law of estoppel, a particular and definite recital provides conclusive evidence of the material facts stated. . . .

The language of the drilling unit recital is not the model of clarity. However, it is sufficient to disclose that the Operating Agreement was formed with the parties' express acknowledgement of the Commission's authority to regulate the size of drilling units in the Frontier and Dakota formations. The drilling unit recital conclusively establishes that the parties formed their agreement to correspond with the 760-acre drilling unit for the Frontier formation and the 640-acre drilling unit for the Dakota formation. These are the drilling units established by the Commission's September 18, 1981 orders. However, according to the interpretations advanced by Union Pacific and Amoco, the Operating Agreement does not acknowledge any subsequent Commission action. We disagree.

In Wyoming, the parties to a contract are presumed to enter into their agreement in light of existing principles of law. . . . These existing principles of law enter into and become a part of a contract as though referenced and incorporated into the terms of the agreement. . . . We could apply this presumption to hold that the principles of law by which Wyoming regulates the oil and gas industry became a part of the Operating Agreement. However, the parties expressly referenced and incorporated these principles of law in the Operating Agreement.

The Operating Agreement incorporates a statement of Operating

Provisions. The Operating Provisions are contained in Exhibit "D" and are referenced in the Operating Agreement so as to become a part of the contract as a whole. Kilbourne-Park Corp., 404 P.2d at 245. The Operating Provisions state terms for accounting procedures, operations of producing wells, and abandonment of wells. Operating Provision Number 16 states:

> This agreement shall be subject to all valid and applicable State and Federal laws, rules, regulations and orders, and the operations conducted hereunder shall be performed in accordance with said laws, rules, regulations and orders. *In the event this agreement or any provisions hereof are, or the operations contemplated hereby are found to be inconsistent, with or contrary to any such law, rule, regulation or order, the latter shall be deemed to control and this agreement shall be regarded as modified accordingly and, as so modified, shall continue in full force and effect.* [Emphasis added by the court.]

We hold that the plain language of the Operating Agreement incorporates principles of Wyoming law and provides for modifications of the terms of agreement when the terms of the contract are contrary to an order of the Commission. Therefore, we must determine what modifications to the Operating Agreement resulted from the Commission's December 13, 1990 order enlarging the drilling unit for the Dakota formation or from the Commission's June 13, 1991 compulsory pooling order. This determination requires an understanding of the authority which the Commission exercises.

In 1951, the legislature enacted the Oil and Gas Conservation Act, Wyo.Stat. §§ 30-5-101 to 30-5-104 and §§ 30-5-108 to 30-5-119 (1983 & Cum.Supp.1994) (hereinafter the Act) to regulate the oil and gas industry in the state. See Mark W. Gifford, *The Law Of Oil And Gas In Wyoming: An Overview*, XVII Land & Water L.Rev. 401, 415 (1982). The Act establishes the Commission, Wyo.Stat. § 30-5-103, and declares that the Commission "has jurisdiction and authority over all persons and property, public and private, necessary to effectuate the purposes and intent . . ." of the Act. Wyo.Stat. § 30-5-104(a) (emphasis added).

The Commission exercises the police power of the State of Wyoming when it issues its orders. . . . Contract rights and property rights are subject to a reasonable exercise of police power. . . . When the police power is exercised to restrict contract rights, we must determine whether the restrictions are reasonable and within the scope of the police power. . . . A valid conservation agency order directing the establishment or modification of a drilling unit does not unconstitutionally impair contract rights. . . .

The Act does not contain an express statement of purpose. However, this court has recognized that we may ascertain the intent and general purpose of an act by giving effect to every word, clause and sentence and construing all components as a whole. . . .

Our reading of the Act discloses that the purpose is to provide a comprehensive regulatory program which prevents the waste of Wyoming's oil and gas resources and protects the correlative rights of property owners. Wyo.Stat. § 30-5-102; Wyo.Stat. § 30-5-109. The Act, therefore, represents a legislative modification to the rule of capture. . . . Under the rule of capture, a land owner acquired title to all the oil and gas which the land owner could produce, even when it was proven that some of the oil and gas migrated from adjoining lands. . . . The Act permits the Commission to establish drilling units to protect the public interest by preventing waste and protecting correlative rights. . . .

The Act defines "waste" broadly as occurring under various circumstances:

(i) The term "waste" means and includes:

(A) Physical waste, as that term is generally understood in the oil and gas industry;

(B) The inefficient, excessive or improper use, or the unnecessary dissipation of, reservoir energy;

(C) The inefficient storing of oil or gas;

(D) The locating, drilling, equipping, operating, or producing of any oil or gas well in a manner that causes, or tends to cause, reduction in the quantity of oil or gas ultimately recoverable from a pool under prudent and proper operations, or that causes or tends to cause unnecessary or excessive surface loss or destruction of oil or gas;

(E) The production of oil or gas in excess of

(I) transportation or storage facilities;

(II) the amount reasonably required to be produced in the proper drilling, completing, or testing of the well from which it is produced, or oil or gas otherwise usefully utilized: except gas produced from an oil well pending the time when with reasonable diligence the gas can be sold or otherwise usefully utilized on terms and conditions that are just and reasonable;

(F) Underground or aboveground waste in the production or storage of oil, gas, or condensate, however caused, and whether or not defined in other subdivisions hereof; and

(G) The flaring of gas from gas wells except that necessary for the drilling, completing or testing of the well;

(H) The drilling of any well not in conformance to a well density and spacing program fixed by the commission or other agency, state or federal, as to any field or pool during a national emergency when casing or other materials necessary to the drilling and operation of wells are rationed or in short supply. Wyo.Stat. § 30-5-101(a)(i).

"Physical waste" is commonly understood in the oil and gas industry as referring to operational losses in oil and gas production resulting from either: surface loss or destruction of oil and gas; or, underground loss or destruction of oil and gas. 8 Williams & Meyers, *supra*, at 907. Surface loss of oil is due principally to evaporation and surface loss of gas is due principally to burning at field flares or blowing into the atmosphere. Underground loss is due to failure to recover the maximum quantity which theoretically could be produced, as by dissipation of reservoir pressure. Id.

The Act also defines "correlative rights:"

(ix) "Correlative rights" shall mean the opportunity afforded the owner of each property in a pool to produce, so far as it is reasonably practicable to do so without waste, his just and equitable share of the oil or gas, or both, in the pool. Wyo.Stat. §30-5-101(a)(ix).

Wyoming has recognized that correlative rights and the right to produce oil and gas from a pool are limited by a duty not to injure the pool and a duty not to cause waste. . . . The term "pool," as a noun, means "an underground reservoir containing a common accumulation of oil or gas, or both." Wyo.Stat. § 30-5-101(a)(iii). Under Wyoming law, each zone of a general structure, which is completely separated from any other zone in the structure, is also a "pool." Id.

The legislature authorized the Commission to "make rules, regulations, and orders . . ." and take other appropriate action to effectuate the purposes and intent of the Act. Wyo.Stat. § 30-5-104(c). In Mitchell v. Simpson, 493 P.2d 399, 401-02 (Wyo.1972), this court considered a jurisdictional challenge brought by the owner of a royalty interest who argued that he was not subject to the Commission's orders. We held that the Commission has the authority to establish a drilling unit to "prevent or assist in the prevention of waste or to protect correlative rights[.]" Id. at 402. After the drilling unit is established, the Commission also has the authority to order the pooling of all interests, including royalty interests. Id.

Larsen v. Oil and Gas Conservation Com'n, 569 P.2d 87, 89-90 (Wyo.1977) recognizes that the Commission exercises its authority in distinct stages. First, the drilling unit is established. This occurs only if the Commission makes a sufficient finding that a drilling unit is necessary to prevent waste or protect correlative rights. Id. Second, the compulsory pooling order is issued, if necessary. These stages are directed by statute. Id.

The legislature provided the Commission with the broad authority to establish drilling units:

(a) When required, to protect correlative rights or, to prevent or to assist in preventing any of the various types of waste of oil or gas prohibited by this act, or by any statute of this state, the commission, upon its own motion or on a proper application of an interested party, but after notice and hearing as herein provided shall have the power to establish drilling units of specified and approximately uniform size covering any pool.

Wyo.Stat. § 30-5-109(a). The limitation on this authority is that the acreage encompassing the drilling unit and the shape of the drilling unit "shall not be smaller than the maximum area that can be efficiently drained by one (1) well." Wyo.Stat. §30-5-109(b).

The legislature also gave the Commission continuing authority to modify its orders after a drilling unit is established:

> (d) The commission, upon application, notice, and hearing, may decrease the size of the drilling units or permit additional wells to be drilled within the established units in order to prevent or assist in preventing any of the various types of waste prohibited by this act or in order to protect correlative rights, and the commission may enlarge the area covered by the order fixing drilling units, if the commission determines that the common source of supply underlies an area not covered by the order. Wyo.Stat. § 30-5-109(d).

The plain language of this provision expressly authorizes the Commission to "decrease the size of the drilling units . . . ." Id. The legislature also authorized the Commission to "enlarge the area covered by the order fixing drilling units . . ." when data is developed that establishes the extent of the common source of supply. Id. The legislature, therefore, provided for the Commission's continuing authority to protect the public interest.

While Wyo.Stat. § 30-5-109(d) does not expressly authorize an order enlarging the size of a particular drilling unit, the Act grants the Commission implied authority to modify its orders in such a manner. . . . The legislature authorized the Commission to decrease the size of a drilling unit to prevent waste and to protect correlative rights. Wyo.Stat. § 30-5-109(d). If, based upon the evidence, the Commission determines it is necessary to increase the size of a particular drilling unit to prevent waste or to protect correlative rights, the Commission has continuing authority to modify its previous orders. . . .

The establishment or modification of a drilling unit requires the Commission to determine the amount of acreage in the unit and the shape of the unit. Larsen, 569 P.2d at 90. After the drilling unit is established, exploitation of the minerals, other than in accord with statutory requirements, is prohibited. Wyo.Stat. § 30-5-109(e). The right to exploit minerals in a drilling unit is conditioned, however, on the status of the ownership of the lands or the mineral interests. Wyo.Stat. § 30-5-109(c) and (f).

If the lands or mineral interests in a particular drilling unit are held by a single owner, that owner is entitled to an opportunity to drill for and produce, as a prudent operator, a just and equitable share of the oil or gas in a pool, subject to the conservation requirements of the Act. Wyo.Stat. § 30-5-104(d)(iv). If tracts of land or mineral interests within a drilling unit are separately owned, the Act requires another step before exploitation of the minerals may commence. The Act identifies two alternative courses of action, voluntary or compulsory pooling, permitting individual owners to exploit their rights to the minerals underlying their lands:

(f) When two (2) or more separately owned tracts are embraced within a drilling unit, or when there are separately owned interests in all or a part of the drilling unit, then persons owning such interests may pool their interests for the development and operation of the drilling unit. In the absence of voluntary pooling, the commission, upon the application of any interested person, may enter an order pooling all interests in the drilling unit for the development and operation thereof. Each such pooling order shall be made after notice and hearing and shall be upon terms and conditions that are just and reasonable. Operations incident to the drilling of a well upon any portion of a unit covered by a pooling order shall be deemed for all purposes to be the conduct of such operations upon each separately owned tract in the unit by the several owners thereof. That portion of the production allocated or applicable to each tract included in a unit covered by a pooling order shall, when produced, be deemed for all purposes to have been produced from such tract by a well drilled thereon. Wyo.Stat. § 30-5-109(f).

The term "pooling," the present participle of the verb "pool," is used to denominate the bringing together of small tracts of land for the granting of a well permit within an established drilling unit. 8 Williams & Meyers, supra, at 921-22.

Our review of the Act discloses that in Wyoming, drilling units are established for the limited purpose of controlling the density of drilling to prevent waste and to protect correlative rights. Wyo.Stat. § 30-5-109(a). . . . The Commission's December 13, 1990 order enlarging the drilling unit for the Dakota formation follows the requirements of the Act. The Commission acted under its implied authority to modify a previous drilling unit order. Wyo.Stat. § 30-5-109(d). The Commission concluded, as a matter of law, that establishing a 760-acre drilling unit for the Dakota formation would prevent waste from the drilling of an offset well in Section 22. The Commission also ruled that the correlative rights of the BLM and its lessees, Texaco and Wexpro, to produce a just and equitable share of gas from Section 22 required protection from the drainage occurring as a result of the L-2 Well in Section 15. The Commission found that the L-2 Well was capable of efficiently draining the pool underlying the enlarged 760-acre drilling unit. These findings of fact and conclusions of law are not subject to collateral attack in this appeal.

Except for the due process rights accorded "interested parties," the establishment of a drilling unit occurs without regard to ownership interests. Wyo.Stat. § 30-5-109(a). There is no reference to the allocation of ownership interests in any provision of the Act dealing with the establishment or modification of drilling units. Therefore, establishing or modifying a drilling unit does not have the effect of apportioning production from the unit. . . . We hold the Commission's order of December 13, 1990 enlarging the drilling unit for the Dakota formation did not automatically apportion production from the unit.

While the enlarged drilling unit for the Dakota formation did not result in an apportionment of production, we hold the Commission's order of December 13, 1990 superseded some of the terms of the Operating Agreement in a reasonable exercise of the police power.

Restatement (Second) of Contracts § 264 (1981) recognizes that performance of a contractual duty may be impracticable as a result of government action: If the performance of a duty is made impracticable by having to comply with a domestic or foreign governmental regulation or order, that regulation or order is an event the non-occurrence of which was a basic assumption on which the contract was made. Other jurisdictions acknowledge that when the agency charged with the responsibility for oil and gas conservation issues an order which conflicts with contractual provisions, the agency order supersedes or supplements the contractual provisions. . . .

Alston [v. Southern Production Co., 207 La. 370, 21 So.2d 383 at 384 (1945)] illustrates the operation of these principles of law. In *Alston*, the conservation agency increased the size of drilling units which had been previously established. Id. The increase in the size of the drilling unit was mandated by a federal emergency order issued during World War II which restricted the use of material for drilling after 1941. Id. at 386. The working interest owners had previously entered into voluntary pooling agreements covering the former drilling unit. However, the Supreme Court of Louisiana determined that the terms of these agreements had been superseded by the order enlarging the drilling unit. Id.

Union Pacific, Amoco, Wexpro and Texaco included in the Operating Agreement a drilling unit recital which conclusively established a basic assumption on which their agreement was made. The parties assumed that the 640 acres in Section 15 would always comprise the drilling unit for the Dakota formation. The terms of the Operating Agreement anticipated production only from the Dakota formation drilling unit established by the Commission's order of September 18, 1981. There is no other reasonable construction of the Operating Agreement.

At the time the parties entered into the Operating Agreement, the entire drilling unit for the Dakota formation in Section 22 was not included within the "subject lands." The parties had no exclusive rights to drill and produce from the Dakota formation underlying Section 22. The Communitization Agreement incorporated in the Operating Agreement and permitting development of the subject lands in Section 22 is specifically limited to development of the Frontier formation. Therefore, development of the Dakota formation in any part of Section 22 was not anticipated by the terms of the Operating Agreement.

The Operating Agreement does not provide for the effect of any future changes in the size of the drilling unit for the Dakota formation. The Operating Agreement also does not disclose the means used by the parties to arrive at the "working interest percentages" contained in Article 2. Without such information, reformation of the Operating Agreement is impossible. The language of Operating Provision Number 16, however, discloses the parties intent to permit modifications of the Operating Agreement when any provision is found to be inconsistent with or contrary to a government order. We hold that as a result of the Commission's order of December 13, 1990 enlarging the drilling unit for the Dakota formation, the inconsistent provisions of the Operating Agreement were superseded. . . . Specifically, the "working interest percentages" for the Dakota formation contained in Article 2 were rendered without further effect as of December 13, 1990.

Under Wyoming law, the Commission exercises its authority to apportion production, allocate costs, and make provisions for the drilling and operation of a well only when a compulsory pooling order is issued. Wyo.Stat. § 30-5-109(f) and (g). The Commission orders compulsory pooling of ownership interests when a drilling unit contains separately owned tracts of land or separately owned mineral interests. Wyo.Stat. §30-5-109(f). If the owners have previously entered into a voluntary pooling agreement covering the appropriate drilling unit, the Commission need not exercise its authority. Id.

Union Pacific erroneously attempts to construe the Operating Agreement as a voluntary pooling agreement for the enlarged Dakota formation drilling unit. This argument is without merit. The Operating Agreement incorporates the October 7, 1981 Communitization Agreement. The Communitization Agreement voluntarily pooled the parties' interests in Section 15 and Section 22, but only for the Frontier formation. The express language of the Communitization Agreement provides: "[T]his agreement shall include only the Frontier formation underlying said lands and the natural gas and associated liquid hydrocarbons, hereinafter referred to as 'communitized substances,' producible from such formation."

There is no plain language in the Operating Agreement disclosing an intent to voluntarily pool the parties' interests in a 760-acre Dakota formation drilling unit. This is easily explained. At the time the parties formed the Operating Agreement, the enlarged drilling unit for the Dakota formation had not been established. We hold that when the Commission issued its June 13, 1991 compulsory pooling order, there was no effective voluntary pooling agreement covering all the parties' interests in the enlarged Dakota formation drilling unit.

In the absence of a voluntary pooling agreement, the Commission is authorized to issue a compulsory pooling order. Wyo.Stat. § 30-5-109(f). Under Wyoming law, a compulsory pooling order apportions production among each separately owned tract of land in the drilling unit. Id. We offer no opinion on the validity of the Commission's compulsory pooling order of June 13, 1991 since that is the subject of separate litigation in the district court.

IV. Conclusion

"When parties make a contract and reduce it to writing, they must abide by its plainly stated terms." Colorado Interstate Gas Co. v. Natural Gas Pipeline Co. of America, 842 P.2d 1067, 1070 (Wyo.1992). However, even plainly stated terms may be the subject of significant disputes. The parties to the Operating Agreement are sophisticated corporate entities with considerable experience in forming these types of agreements. The parties also understand that their business involves a highly regulated industry. Despite these skills, the parties failed to anticipate the likelihood that a basic fact on which their agreement was premised might change. The parties never provided for the allocation of production from a well on a 760-acre drilling unit in the Dakota formation.

We affirm.

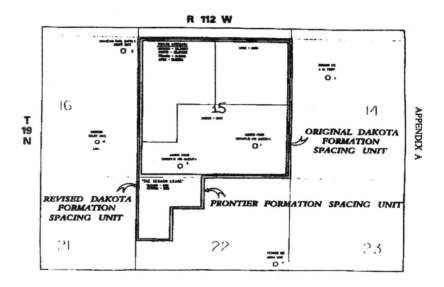

NOTES

1. The principal case introduces the distinction between spacing or drilling units and pooling. The question of the effect of a drilling or spacing unit was taken up by the Utah Supreme Court in BENNION V. GRAHAM RESOURCES, INC., 849 P.2d 569, 122 O.&.G.R. 597 (Ut. 1993). Bennion was an owner of a mineral interest that was included in a spacing unit established by the Utah Board of Oil, Gas and Mining. Graham Resources was the operator of two wells in the unit, and neither of the wells was located on property owned by

Bennion. Bennion requested that the Board order an accounting for all proceeds from the well. The Board denied the request on the basis that no pooling order had been entered for the unit and no voluntary pooling agreement existed. He sought judicial review claiming that under the statute he was a person "legally entitled" to a share of production. The statute provided, at the time, that "oil and gas proceeds derived from the sale of production from any well producing oil, gas, or related hydrocarbons in the state shall be paid to all persons legally entitled to these payments. . . ." The court rejected the claim, saying that the plain language of the statute required that a pooling order be in place before an operator has to make an accounting and payment to a nonconsenting mineral interest owner. The mechanism for assigning costs depended on the existence of a pooling order. Without a voluntary pooling order, the share of costs could only be established through a pooling order. An unleased owner is entitled under the statute to a royalty equal to the average royalty for leased royalty owners, but Bennion did not claim any entitlement under this provision. The court said that Bennion could seek a force pooling order but that the Board had no requirement to enter such an order without an application from an interested person. Query: What effective date can or must be given once a pooling order is entered?

2. In some jurisdictions, the establishment of a drilling unit results in a sharing of production. In some jurisdictions, spacing and pooling are done at the same time. *See* Kramer & Martin, *The Law of Pooling and Unitization* §5.03.

3. What is the function of a spacing unit if it does not apportion production and costs? Why have two stages in the process? Does *Bennion* suggest reasons?

4. The principal case also illustrates the complexity of the relationship between contracts, particularly joint operating agreements, and administrative agency orders. The contract may have been in existence when the order was entered, or the contract may have been agreed to after the issuance of the order. In the former, the question is raised as to the extent to which an order can modify a contract, and in the latter the question is the extent to which parties subject to an order can modify (or implement) the rights and duties vis-a-vis one another under that order. While these are distinct questions, the guiding principle in answering both is similar. That is, there is a preference for freedom of contract. A spacing, pooling, or other conservation order should extend only so far as necessary to prevent waste and protect correlative rights. When parties contract for an effect other than the effect provided for by the agency's order, they are free to do so, so long as it does not lead to waste or denial of the correlative rights of others in the common source of supply. *See* B. Kramer & P. Martin, *The Law of Pooling and Unitization*, §13.08

ACADIENERGY, INC. V. MCCORD EXPLORATION CO., 596 So.2d 1334, 119 O.&.G.R. 486 (La.App. 1992). In this case, a number of parties entered into a joint operating agreement for the drilling of a well or wells within a contract area. The agreement expressly provided for the possibility of the Commissioner of Conservation establishing a unit and well costs were to be adjusted to correspond to the Commissioner's unit. Once the well was drilled, the unit was established by the Commissioner but there were then several revisions to the original Commissioner's unit based on new geology. A number of disputes arose among the parties concerning the wells costs and the participation in a subsequent well. The trial court found that the costs and participation of the parties were to be adjusted with each revision of the unit by the Commissioner. The appeals court reversed, finding that the joint operating agreement provided for adjustment only for the first unit established by the Commissioner and thereafter well costs were not to be adjusted. The court characterized the clause as ambiguous and based its interpretation in part on the principle of interpreting against the drafter, which was AcadiEnergy. The court also stated that it would be clearly inequitable to have well cost adjustments based on subsequent revisions of the unit, after all the drilling costs had already been paid out of production, when there was no provision to adjust also the production payments received by the parties based on subsequent revisions of the unit.

SECTION 2. TEXAS AND THE PROBLEM OF THE SMALL (UNPOOLED) TRACT

Kramer & Martin, *The Law of Pooling and Unitization*, § 5.01[4]

[g] Texas

The single most important development in the law of Texas relating to allowables was the Texas Supreme Court's rejection of the allowable formulas used by the Railroad Commission for small tract, Rule 37 exception wells. As discussed in greater depth in the next section . . ., the Railroad Commission, which had no compulsory pooling power until 1965, had granted exception location wells such that a tract only a fraction of an acre in size might produce from the same reservoir as an adjacent well on 320 acres. The reason for granting a permit to drill a well at a location that was an exception to the spacing and density rules was the need to prevent confiscation and protect correlative rights. The problem that arose was the inequity that resulted when the small tract was given an allowable making it possible for it to produce a disproportionate amount of oil or gas. The typical formula for allowables was 1/3 well—2/3 acreage for gas wells, and 1/2 well—1/2 acreage for oil. The result of these formulas for allowables was to allow confiscation from the adjacent large tract; no protection was afforded the larger tract owner's correlative rights. The result was also to discourage voluntary pooling agreements.

Most thoughtful observers recognized the absurdity of the Texas allowable formulas. Professor Edwin Horner observed:

[T]o permit each tract a well is permitting the very thing that it was designed to prevent—confiscation of another's property. It would appear that it should be just as wrong for A to confiscate B's property as it is for B to confiscate A's property. . . . The justification for the original rule was to permit each mineral owner to recover his fair share of the oil and gas in place. The application of the rule as it is now being applied . . ., with the increasing amount of land to be dedicated to the spacing requirements, is permitting an owner to recover far in excess of his fair share of the oil and gas originally in place, assuming that the proper definition of 'his fair share' is that which originally underlay his tract. [Discussion Notes, 11 O.&G.R. 280–81.]

In two landmark decisions in 1961 and 1962, the Texas Supreme Court told the Railroad Commission it was time to take a different approach to allowables for small tracts.

In Atlantic Refining Co. v. Railroad Commission, 346 S.W.2d 801, 14 O.&G.R. 362 (Tex.1961) (*Normanna* case), the owners of a 320–acre tract showed that a 0.3–acre tract had $7,000 worth of gas beneath it but that under the proration formula they would be able to produce gas worth $2,500,000. The trial court upheld the commission's allowable formula. The Texas Supreme Court reversed and remanded for further proceedings. The small tract owner contended "that the Railroad Commission can take into account what a 0.3–acre tract could have produced under the rule of capture before conservation laws restricted drilling and production in fixing a proration formula, and even though such formula results in an enormous drainage from the larger tracts to the small tract, the order is valid." The Supreme Court rejected this claim. There was no justification for the wide discrepancy in allowed production. The order "does not come close to compelling ratable production; neither does it afford each producer in the field an opportunity to produce his fair share of the gas from the reservoir." While declaring invalid the order, the court declined to set a rule or standard.

After the decision in the *Normanna* case, the Railroad Commission entered the Normanna Field Order. The effect of this order was to require small tract owners to make an effort to pool on an acreage basis with adjacent owners prior to qualifying for an allowable on a special basis.

The Texas Supreme Court's opinion in the *Normanna* case was reiterated in Halbouty v. Railroad Commission, 357 S.W.2d 364, 16 O.&G.R. 788 (Tex.1962) (*Port Acres* case), *cert. denied sub nom.* Dillon v. Halbouty, 371 U.S. 888, 17 O.&G.R. 173 (1962). Here the Railroad Commission had entered the 1/3 well—2/3 acreage allocation formula for setting allowables for the Port Acres Field. The court compared two tracts in the field, one with 176 acres and one with 0.25 acre. The former would have an allowable of 4.932 MCF per acre-foot, while the latter would have an allowable of 534.551 MCF per acre-foot, or 137 times as much on an acre base and 107 times as much on an acre-foot basis. There were 20 Rule 37 wells with 0.65 percent of the total productive acreage but 14.6 percent of the allowable. The court observed:

Under the order adopted in the Port Acres Field more than 90% of the hydrocarbons recovered by the Rule 37 wells would be drained from other leases in the field. As said in *Normanna* the proration formula adopted here of 1/3—2/3 does not come close to compelling ratable production nor afford to each producer in the field an opportunity to produce his fair share of gas from the reservoir.

Continuing not to compel an allowable standard, the court said that commission personnel could devise a nearer approximation to fair share. The dissent said the court was establishing compulsory pooling by judicial decree.

Following these two decisions, the Texas courts had to address questions of their application to other existing allowable orders and of what circumstances would call for special allowable treatment by the commission. The Texas Supreme Court made it clear that the responsibility for allowables remained with the Railroad Commission, to be determined with its informed judgment, and that long-standing orders did not have to be displaced.

V–F PETROLEUM, INC. v. A.K. GUTHRIE OPERATING CO.

Texas Court of Appeals, 1990.
792 S.W.2d 508, 118 O.&G.R. 341

GAMMAGE, JUSTICE. Appellants, V–F Petroleum, Inc., and the Railroad Commission of Texas, appeal from a judgment of the district court of Travis County that reversed an order of the Commission and remanded the proceeding to the agency. By its order, the Commission denied the application of appellee A.K. Guthrie Operating Company ("Guthrie") for an amendment to the field rules of the Sara–Mag (Canyon Reef) Field. We will reverse the judgment of the district court.

In 1954, the Commission adopted for the Sara–Mag (Canyon Reef) Field an allocation formula based fifty percent upon surface acreage and fifty percent upon the number of wells producing ("50–50 allocation formula"). An allocation formula is the means the Commission uses to prorate oil production among wells in a field. A 50–50 allocation formula was common in field rules promulgated before 1961 and favored production from small tracts. . . . .

In March 1982, V–F Petroleum's predecessor in interest filed an application requesting a permit to drill a well on a substandard-sized tract within the field. Texas Railroad Comm'n, 16 Tex.Admin. Code § 3.37 (West March 1, 1990) ("Rule 37"). Guthrie protested this request for a Rule 37 exception and filed an application to amend the field allocation formula to a one-hundred percent acreage formula. Guthrie owns land on three sides of the 2.5–acre tract in which V–F Petroleum holds an interest and operates eight of the ten wells in the field. After a hearing, the Commission granted V–F Petroleum a Rule 37 exception allowing it to drill on its tract.

Shortly thereafter, the Commission held a hearing on Guthrie's application to amend the allocation formula in which V–F Petroleum intervened. The Commission denied the application on the basis that the requested allocation formula would effectively disallow a well on the V–F Petroleum tract and that V–F Petroleum had been denied an opportunity to pool. Guthrie filed a petition for judicial review of the order in the district court of Travis County. Tex.Rev.Civ.Stat.Ann. art. 6252–13a, § 19 (Supp.1990). On August 2, 1983, the district court reversed the order and remanded the proceeding to the agency. In June 1984, the Commission reopened the hearing and, after receiving additional evidence, again denied Guthrie's application to amend the allocation formula.

Guthrie then filed a second petition for judicial review in district court. The district court reversed the agency order on the basis that the 50–50 allocation formula was "illegal" because it used a per-well factor in conflict with the holdings in Railroad Comm'n v. Shell Oil Co., 380 S.W.2d 556 (Tex.1964); Halbouty v. Railroad Comm'n, 163 Tex. 417, 357 S.W.2d 364 (1962); Atlantic Refining Co. v. Railroad Comm'n, 162 Tex. 274, 346 S.W.2d 801 (1961). On appeal, this Court determined that these cases did not preclude the Commission's use of per-well factor, reversed the district court judgment, and remanded the cause for consideration of the parties' substantial evidence points. Railroad Comm'n v. A.K. Guthrie Operating Co., 742 S.W.2d 86 (Tex.App.1987, no writ).

On remand, the district court entered judgment reversing the Commission's order because it was not supported by substantial evidence and again remanded the proceeding to the agency. The court concluded further that "the present allocation formula in the Sara Mag . . . Field is illegal and has been illegal since this Court's final judgment of August 2, 1983 . . . for lack of substantial evidence." This judgment underlies the appeal now before this Court.

We first note that, when a case is remanded to the district court with instructions, that court's authority is limited to trying only those issues specified in the appellate court mandate. . . . .. In the instant cause, this Court's opinion and mandate limited the district court to a determination whether the Commission's order was reasonably supported by substantial evidence. A.K. Guthrie Operating Co., 742 S.W.2d at 88. Furthermore, our determination that the Commission order was not affected by an error of law as to the use of a per-well factor is the law of the case. . . .

In their first points of error, appellants contend that the district court erred in concluding that the Commission's order was not supported by substantial evidence. Texas Rev.Civ.Stat.Ann. art. 6252–13a, § 19(e)(5) (Supp.1990) authorizes a reviewing court to test an agency's findings, inferences, conclusions, and decisions to determine whether they are reasonably supported by substantial evidence in view of the reliable and probative evidence in the record as a whole. Although substantial evidence is more than a mere scintilla, the evidence in the record may preponderate against the decision of the agency and nevertheless amount to substantial evidence. The true test is not whether the agency reached the correct conclusion, but whether some reasonable basis exists in the record for the agency's action. . . . The agency's action, therefore, will be sustained if the evidence is such that reasonable minds could have reached the conclusion that the agency must have reached in order to justify its action. . . .

In its motion for rehearing in the Commission and in its first amended petition in district court, Guthrie asserted that the Commission's findings of fact ten, eleven, and thirteen through fifteen; and conclusions of law five, eight, and nine were not supported by substantial evidence. Accordingly, appellants here argue that these findings and conclusions are supported by substantial evidence. . . .

Finding of fact eleven states:

No opportunity is available to the owners of the 2.5 acres to pool the tract with any of the surrounding units.

a. Guthrie twice has indicated an affirmative disinclination to pool the 2.5 acres.

b. The discovery date for the field precedes the date of the Mineral Interest Pooling Act.

Conclusion of law five states:

Given the inability of the [V–F Petroleum] tract owners to pool their acreage either voluntarily or pursuant to the provisions of the Mineral Interest Pooling Act, a well on the 2.5 acre tract is necessary if they are to have a fair chance to recover their fair share of the Sara–Mag (Canyon Reef) Field.

The Mineral Interest Pooling Act ("MIPA"), Tex.Nat.Res. Code Ann §§ 102.001 et seq. (1978), permits the Commission to order mineral owners whose land shares a common reservoir to pool their interests to prevent unnecessary drilling, to protect correlative rights, or to prevent waste. . . . .. The Act's provisions do not apply to any reservoir discovered and produced before March 8, 1961. MIPA § 102.003. The reservoir here was discovered in 1954 and six wells developed during 1954 and 1955. Clearly, the provisions of the MIPA do not apply to this reservoir.

In regard to pooling generally, Mr. E.E. Runyan testified that he had twice contacted Mr. Guthrie to discuss the possibility of pooling and that each time Guthrie stated that "he could not possibly take [V–F Petroleum] into the well" and that there was no basis for negotiations on pooling. As stated previously, Guthrie owns tracts on three sides of V–F Petroleum. Runyan testified that the tract on the fourth side covers the same reservoir; that he does not know who owns the tract; that he could, but has not attempted to, learn who owns the tract; that the only drilling on that tract resulted in a dry hole; and that he is ready to pool and to bear his share of the costs.

Appellee characterizes the record as having "no evidence of any serious effort or offer to pool because V–F made none." The record does indicate that Runyan did contact Guthrie and reflects Runyan's impression that it was not profitable to pursue pooling with Guthrie. The Commission, of course, was the judge of the weight to be accorded the evidence presented. . . .

The remaining findings of fact and conclusions of law in question state:

10. If the requested change in the allocation formula is approved, no well will be drilled on the 2.5 acres.

a. No reasonably prudent operator would drill a well expecting a 304 year payout;

b. Runyan testified that he would not develop the 2.5 acre tract if the allocation formula is changed to 100% acreage.

13. A well on the 2.5 acre . . . lease could compete with upstructure wells to the west for reserves swept from undeveloped acreage as well as for the reserves underlying the 2.5 acres. Without a well, the reserves underlying the 2.5 acres will move off the lease and be recovered by one or more wells in which the owners of the 2.5 acres have no interest.

14. A 100% acreage allocation formula would restrict Runyan to producing only the reserves underlying the 2.5 acres, while Guthrie could produce from the reservoir as a whole and ultimately recover hydrocarbons from the 2.5 acre tract and from unleased acreage without competition and without regard to lease line or field-wide drainage.

15. The per-well factor in the allocation formula for the Sara–Mag (Canyon Reef) Field has not occasioned the loss of any hydrocarbons to Guthrie's leases, nor prevented Guthrie from recovering his fair share of the reservoir.

8. The per-well factor in the allocation formula of the Sara–Mag (Canyon Reef) Field has not led to any drainage of [A.K. Guthrie] acreage, nor inhibited its ability to produce its fair share of the reservoir.

9. The retention of the per-well factor in the allocation formula of the Sara–Mag (Canyon Reef) Field will permit Guthrie and Runyan both to compete for their fair share of the reservoir.

The evidence adduced at the first hearing in 1982 and at the reopened hearing in 1984 shows that the 2.5 acre tract does not have a producing well; that the oil is swept from the east across the 2.5 acre tract to the Guthrie tracts on the west; that, if no well is drilled on the 2.5 acres, the underlying oil will be produced from the well on the tract to the west; and that the current allowable allows a well to produce more oil than underlies the tract. Mr. Glenn George, a consulting petroleum engineer, testified that the amount of oil underlying the tract was approximately 8750 barrels; that the cost of drilling for and producing the oil ranged from $350,000 to $450,000; that, even at the lower cost, the value of the oil underneath the tract would not return the drilling cost; and that, during the next five years under the 50–50 allocation formula, a well would produce approximately 95,000 barrels from adjacent tracts. Mr.Runyan testified that his estimated cost was $419,300; that, under a 100% acreage formula, the well could produce eight barrels of oil per day and the payout would be 304 years; and that, in his opinion, no rational person would drill a well with a 304–year payout. He testified further that, with the 50–50 allocation formula, the well could produce sixty-four barrels per day with a one-year payout; and that he estimated that production would actually be fifty barrels per day.

Based on our review of the agency record, we conclude that findings of fact 10, 11, 13, 14 and 15; and conclusions of law 5, 8, and 9 are reasonably supported by substantial evidence. We sustain the first points of error of the Railroad Commission and V–F Petroleum.

The Commission and V–F Petroleum next argue that the district court erred in its finding that the allocation formula is illegal and has been illegal since its judgment of August 2, 1983. The judgment underlying this appeal stated, "[I]t is further ORDERED that the present allocation formula in said field is illegal and has been illegal since this Court's final judgment of August 2, 1983 for lack of substantial evidence. . . ." The parties suggest that the district court has usurped the authority of the agency in attempting to prescribe what order the agency should enter and has violated the mandate of this Court. The Commission asserts additionally that the date used has no support in the record.

We need not address these arguments. The district court couched its holding in terms of substantial evidence, the issue which this Court remanded to the district court. Having determined that the agency order is supported by substantial evidence, we need not address the Commission's point of error two or V–F Petroleum's point of error three. We note the proper inquiry is not the legality of the current formula, but whether the Commission order denying a request to amend the allocation formula is supported by substantial evidence.

. . .

We reverse the judgment of the district court and render judgment that the order of the Railroad Commission denying A.K. Guthrie's request to amend the field rules of the Sara–Mag (Canyon Reef) Field be affirmed.

NOTES

1. What factors must be considered in determining what is a "fair share"? Is a 100% surface acreage formula a fair one? Would subsurface geologic and engineering conditions make it unfair? Would the cost of the well make it unfair? Would 50 percent surface acreage and 50 percent acre-feet of productive sand be a fair formula? *See* Pickens v. Railroad Commission, 387 S.W.2d 35, 21 O.&G.R. 644 (Tex.1965) to see how the Railroad Commission responded to the *Normanna* and *Port Acres* decisions. *Compare* Railroad Commission v. Shell Oil Co., 380 S.W.2d 556, 20 O.&G.R. 888 (Tex.1964), where the court held an allowable based on 50 percent per well and 50 percent for oil acreage was not sustained by substantial evidence.

2. Is fairness to be measured on a tract basis or on a reservoir by reservoir basis? *See* BENZ–STODDARD V. ALUMINUM CO. OF AMERICA, 368 S.W.2d 94, 18 O.&G.R. 508 (Tex.1963), *rev'g*, 357 S.W.2d 809, 16 O.&G.R. 1255 (Tex.Civ.App. 1962). Here Benz–Stoddard had a 0.115–acre tract of land with ten separate reservoirs beneath. She got a permit to drill one well and complete it

in all ten. The Aluminum Company claimed this would give her more than a fair share under the 1/3—2/3 allowable formula. The evidence showed completion had been accomplished in three reservoirs and that she had recovered seven times the gas in place in all ten reservoirs. The court here held that she had a property right in all ten; on a showing that confiscation would result in any reservoir, she was entitled to a Rule 37 permit for each such reservoir. The Railroad Commission was correct in treating the ten reservoirs separately. The court acknowledged that this created a situation in which grave injustice could be done the respondent. But it said the Railroad Commission could regulate the flow of gas so that no unreasonable hardship need result: "We hold, therefore, that since the Commission is authorized to treat each completion in a separate reservoir as a separate well, it may grant multiple completions in each reservoir in which it finds there will be confiscation."

3. In RAILROAD COMMISSION v. ALUMINUM CO. OF AMERICA, 380 S.W.2d 599, 20 O.&G.R. 880 (Tex.1964) the Texas Supreme Court ruled that the *Normanna* decision was to be prospective. Too many people had relied on the proration formulas to have them overturned for existing orders.

4. RAILROAD COMMISSION v. WOODS EXPLORATION AND PRODUCING CO., Inc., 405 S.W.2d 313, 24 O.&G.R. 831 (Tex.1966), *cert. denied*, 385 U.S. 991, 25 O.&G.R. 556 (1966), held that allowables could not be set to less than market demand as a measure to protect correlative rights. Here the allowables formula (1/3 well—2/3 surface acreage) resulted in small tracts draining larger tracts, far more than 1/3—2/3 even, because when large tract wells could not meet inflated allowables, they were assigned to the other wells in the field, the small tract wells. The Railroad Commission estimated market demand as the summation of producer forecasts. But the commission had shifted to an alternative market demand calculation by limiting market demand to the result of dividing the legal deliverability of the best well in the pool by the participation factor of the pool's largest unit. Small tract owners got an injunction against this, and the Texas Supreme Court affirmed. While the commission made a powerful argument for considering correlative rights in establishing a reasonable reservoir allowable, the legislature had not given them this power, only a power to prevent waste. The commission limits were based on factors that had nothing to do with market demand. *See also* Texaco, Inc. v. Railroad Commission, 583 S.W.2d 307, 63 O.&G.R. 346 (Tex.1979).

TEXAS ADMINISTRATIVE CODE
Title 16. Economic Regulation
Part I. Railroad Commission of Texas
Chapter 3. Oil and Gas Division
Conservation Rules and Regulations
§ 3.37. Statewide Spacing Rule

(a) Distance requirements.

(1) No well for oil, gas, or geothermal resource shall hereafter be drilled nearer than 1,200 feet to any well completed in or drilling to the same horizon on the same tract or farm, and no well shall be drilled nearer than 467 feet to any property line, lease line, or subdivision line; provided the commission, in order to prevent waste or to prevent the confiscation of property, may grant exceptions to permit drilling within shorter distances than prescribed in this paragraph when the commission shall determine that such exceptions are necessary either to prevent waste or to prevent the confiscation of property.

. . .

(b) The distances mentioned in subsection (a) of this section are minimum distances to provide standard development on a pattern of one well to each 40 acres in areas where proration units have not been established.

NOTES

1. Prior to 1962 Rule 37 provided that no well shall be drilled closer than 933 feet to another well in the same producing horizon nor closer than 330 feet to a property line. Since a 20–acre tract may ordinarily be expected to measure 660 feet by 1320 feet, the 330 feet provision required that the well be precisely located at the midpoint between parallel boundary lines, and the 933 feet provision required that the well be located precisely at the midpoint of one of the two 10–acre squares making up the rectangular 20–acre spacing unit if it was to be 933 feet from the nearest diagonal offset.

2. The 40–acre spacing rule adopted in 1962 provides that no well shall be drilled nearer than 1200 feet to any well completed in or drilling to the same horizon on the same tract or farm, and no well shall be drilled nearer than 467 feet to any property line, lease line, or subdivision line. The operative effect of this order is to permit the owner of a 40–acre quarter-quarter section to drill at any location within a square measuring 386 feet by 386 feet in the precise center of the square quarter-quarter section measuring 1320 feet by 1320 feet. The owner of an 80–acre or larger tract is modestly circumscribed in his freedom of choice of a well location within the inner square of each quarter-quarter section by the requirement of 1200 feet between his wells.

Kramer & Martin, *The Law of Pooling and Unitization*, § 5.02[2][a].

Exception Wells—Texas Requirements

The Texas statutes do not specifically address exception wells. Statewide Rule 37 promulgated by the Railroad Commission provides that the commission "may grant exceptions . . . when the Commission shall determine that such exceptions are necessary either to prevent waste or to prevent the confiscation of property." [Tex. Admin. Code tit. 16, § 3.37(a)(1).] In order to obtain an exception well permit from the Railroad Commission, the following requirements have to be met:[90]

(1) Notice must be given to all adjacent lessee and unleased mineral owners (10 days);

(2) A hearing must be held at which all interested parties are allowed to be heard; and

(3) A finding from the commission must be made that the exception location is necessary to prevent waste or confiscation.

Numerous cases have been decided since Rule 37 was first adopted by the Railroad Commission in 1919. We will not attempt to go into detail on all aspects of Rule 37 exceptions but will indicate enough of the background to suggest the conditions that led to the passage of the Mineral Interest Pooling Act of 1965.

[i] Confiscation

To express it briefly, the Railroad Commission was given power that allowed it to space wells and to prorate production. However, it did not have the power to pool owners of interests so that each owner of a separate tract or an interest in a separate tract had a right to a share of production. Therefore, it followed that to prevent confiscation of the right to produce from occurring, the Railroad Commission would have to allow exceptions from spacing regulations. The owner of a tract, regardless of how small, would have a right to drill a well. The rule was commonly stated as: "Each tract of land, no matter how small, is entitled to a well as a matter of right." In Gulf Land Co. v. Atlantic Refining Co., 134 Tex. 59, 131 S.W.2d 73, 80 (1939) the Texas Supreme Court had stated: "Under one of the exceptions in Rule 37, well permits may be granted to prevent confiscation. It is the law that every owner or lessee of land is entitled to a fair chance to recover the oil and gas in or under his land, or their equivalents in kind." This principle had no application to a tract that became too small for a regular location after 1919 and after Rule 37 became effective in the area, because no confiscation would result if the owner of the land had made the tract too small for issuance of a permit. That is, a tract (or the owner of a tract) would have no right to a Rule 37 exception permit if the small tract was the result of a voluntary subdivision.

---

[90]   *See generally,* Douglass and Whitworth, *Practice Before the Oil and Gas Division of the Railroad Commission of Texas,* 13 St. Mary's L.J. 719, 722 (1982).

The second part of the equation was the amount of oil or gas the small tract owner would be allowed to produce. If the small tract were restricted to producing the amount of oil or gas that was beneath it, the right to have a well might be meaningless as the well might never pay out. It seemed reasonable that the well should have the right to produce enough oil and/or gas to recover costs and make a profit. This concept was labelled a "living allowable." Lacking the ability to engage in public utility type of cost/revenue calculations for every well, the commission adopted allowable formulas that gave some weight to acreage in the tract and some weight to each well. The disparities between small tract allowables and the amount of oil or gas in place in such a tract, and the undeniable drainage that occurred from large tracts to small tracts under the formulas, were so great that oil and gas were being confiscated from the large tracts without their owners having an opportunity to enjoy a proportionate share of production from their tracts. There was general agreement that the Texas system of small-tract exceptions and related allowables had become absurd. When the Texas Supreme Court undercut the allowable formulas being used by the commission for small tracts, it likewise undercut the foundation of the Rule 37 exceptions based on small tract concerns.

The only rational way to deal with the problem was to allow forced pooling. The Texas legislature finally enacted the Mineral Interest Pooling Act. This does not mean that the Rule 37 exceptions are no longer of any consequence.

––––––––

The next case involves an exception well that was not a small tract exception well application.

––––––––

TEXACO, INC. v. RAILROAD COMMISSION OF TEXAS

Texas Court of Civil Appeals, 1986.
716 S.W.2d 138, 93 O.&G.R. 185, writ ref'd n.r.e.

SHANNON, CHIEF JUSTICE. Texaco, Inc., seeks to set aside the judgment of the district court of Travis County which sustained an order of the Railroad Commission. TXO Production Corp. applied for and obtained a Rule 37 permit to drill a gas well in the Greasewood Field in Reeves County. Appellees are TXO and the Commission. This Court will affirm the district court's judgment.

. . .

Texaco attacks the judgment affirming the Commission's order by one point of error claiming that the Commission erred in granting the drilling permit because the permit was not required to protect correlative rights.

TXO's application sought a permit to drill a well at an exception location in Section Ten of the Greasewood Field. Texaco is the lessee of Section Nine and has a producing well in that section. Both Sections Nine and Ten are owned by the same lessor, Cornell Knight, and Section Ten like Section Nine was originally leased by Knight to Texaco and was originally developed by Texaco. Texaco's well on Section Ten recovered considerable amounts of natural gas but ultimately watered-out as the gas in the reservoir was depleted and the water encroached. At that point Texaco chose to release the acreage in Section Ten, and Knight then leased the section to TXO.

Although Section Ten is not a substandard sized tract (it is 721 acres), TXO was required to apply to the Commission for an exception to the spacing rules of Rule 37. To complete a profitable well (one which would cover drilling costs), TXO would need to drill closer to the Section Nine boundary line than the ordinary spacing rules would allow, since the remaining recoverable hydrocarbons under Section Ten were located in the far southern part of the tract, near the boundary with Section Nine. The Commission, over Texaco's protest, granted TXO the exception permit to drill in the irregular location in order to prevent confiscation and to allow TXO a chance to recover its fair share of the reserves beneath its tract.

Texaco urges that the Commission erred in granting the permit to TXO because the permit was not necessary to prevent confiscation. Texaco's contention is grounded on the argument that the lessor of Section Ten, Knight, was recovering his fair share from the well on Section Nine, and would therefore have no right to an exception permit for Section Ten. Texaco reasons that a grantor cannot create rights in a grantee which the grantor, himself, did not have and therefore TXO can have no greater right to an exception permit than Knight had.

TXO replies that oil and gas lessees are recognized as mineral owners. Such lessees have a right to protection against confiscation and a right to recover their fair share of the minerals beneath their tract, which rights are separate and distinct from the rights of their lessor. The Commission accepted this argument and granted the exception permit based solely on the rights of TXO as a lessee.

In support of the agency order, TXO refers to those opinions concerning the proposition that a mineral lessee is the owner of valuable property rights and is entitled to protection from confiscation by being allowed a fair chance to recover the hydrocarbons beneath his tract. Imperial American Resources Fund, Inc. v. Railroad Commission, 557 S.W.2d 280 (Tex.1977); Railroad Commission v. DeBardeleben, 305 S.W.2d 141 (Tex.1957); Railroad Commission v. Gulf Production Co., 134 Tex. 122, 132 S.W.2d 254 (1939); Gulf Land Co. v. Atlantic Refining Co., 134 Tex. 59, 131 S.W.2d 73 (1939). These opinions all contain language stating "[t]he basic right of every *landowner or lessee* to a fair and reasonable chance to recover the oil and gas under his property . . ." Imperial American, 557 S.W.2d at 286, (emphasis added). Similarly, the Supreme Court in *Gulf Land Co.*, supra, defined "confiscation" as "depriving the *owner or lessee* of a fair chance to recover the oil and gas in or under his land, or their equivalents in kind." 131 S.W.2d at 80 (emphasis added).

Texaco would characterize the statements from these Supreme Court opinions as "unfortunate use of loose language defining the confiscation theory."

Texaco suggests that the Supreme Court has rejected such analysis in Railroad Commission v. Williams, 163 Tex. 370, 356 S.W.2d 131 (1961). Although Texaco concedes that oil and gas lessees possess property and development rights, it insists that those rights cannot be any greater than the rights of their lessor.

In Railroad Commission v. Williams, *supra*, the Supreme Court sustained a Commission order denying a permit to drill on a 1.65 acre tract as an exception to Rule 37 because the landowner applying for the permit was already receiving his fair share of the oil beneath his tract by participating in a well on an adjoining tract. In making this determination, the Court rejected language in earlier opinions which stated that each tract is entitled to a first well as a matter of law and concluded instead that the right to a well is a right of the owner of the land rather than a vested right in the land itself.

Texaco is incorrect in its statement that the Court in *Williams* rejected the language in the opinions, quoted above and relied upon by TXO, stating the right of an *owner or lessee* to a fair chance to recover the oil and gas under his property. Although the Court in *Williams* does mention that it is the *landowner* who has the right to protection against confiscation, it also states in other parts of the opinion that "the *owners and lessees* of some small tracts are entitled to a well on their tracts as a matter of law because there is no other way to give them a fair chance to recover the oil and gas under their land." 356 S.W.2d at 136. The Court even relies upon Railroad Commission v. Gulf Production Co., supra, for this proposition.

It is true, as urged by Texaco, that the Court in *Williams* states that a grantee can have no greater rights to a permit than his grantor had. 356 S.W.2d at 137. The Court, however, limited the application of this rule when it stated that "[t]his rule is a necessary corollary to the voluntary subdivision rule." Id. An investigation of other opinions standing for this proposition reveals that those opinions involved the voluntary subdivision rule and substandard sized tracts. . . .

Unlike *Williams* and the other opinions which state that a grantee cannot have any greater rights than those which existed in his grantor, the present appeal does not involve either the voluntary subdivision rule or a substandard sized tract. Instead, it is undisputed that the tract of land involved in this appeal is of a regular size and shape and would be entitled to a well at a regular location as a matter of law. Because the rule that a grantee can have no greater rights than his grantor had is a corollary of the voluntary subdivision rule, it does not apply in this appeal. The inapplicability of the rule to this appeal is illustrated by Texaco's concession that TXO would have had a right to a permit to drill at an irregular location, had Knight conveyed his mineral estate to TXO outright rather than leasing it to TXO and retaining a royalty interest.

Imperial American Resources, supra, decided by the Supreme Court fifteen years after *Williams,* provides the strongest support for the proposition that a mineral lessee is an owner who has a right to be protected from confiscation. The

Court in *Imperial American Resources* recognized "[t]he basic right of every landowner or lessee to a fair and reasonable chance to recover the oil and gas under his property . . ." The Court in *Imperial American Resources* cites Williams for this proposition. Moreover, the facts in *Imperial American Resources* show that the permit to drill at an irregular location, which the Court sustained, is clearly based on the right of the lessee, BTA, to be protected against confiscation. The Court makes no mention of the rights of the lessor, Riggs, to an exception permit. Furthermore, the map on page 282 of the opinion indicates that Riggs was the landowner of the adjoining tract, Section eight, and was participating in a producing well on that tract. It seems likely that Riggs would not have had a right to an exception well on Tract seven since he was already recovering his fair share from the well on the adjoining tract (just as Knight is in this appeal). It seems plain that the Court in *Imperial American Resources* was relying on the *lessee's* right to protection against confiscation. Accordingly, the language concerning the right of every *landowner or lessee* to a fair chance to recover the oil and gas beneath his property is not, as Texaco claims, merely "loose language."

This Court has concluded that a mineral lessee does indeed have a property interest which is entitled to protection against confiscation. When substandard tracts and voluntary subdivisions are involved, the lessee's right may be limited by the rule that a grantor cannot create by conveyance greater rights than the grantor himself possessed. This rule, however, which is a corollary of the voluntary subdivision rule, has no application in this appeal which involves a standard sized tract. Accordingly, the Commission correctly granted an exception permit to TXO to protect it against confiscation, and the district court correctly rendered judgment sustaining the Commission's order.

The judgment is affirmed.

NOTES

1. *Voluntary Subdivision* – The principal case mentions the voluntary subdivision rule. This rule prevents the granting of Rule 37 exceptions where the landowner has segregated property for the purpose of obtaining additional drilling rights he would not have had in the absence of such segregation. In the absence of the rule, an owner could sell the mineral rights under his land to different persons, creating in each of those individuals a valuable right to a well for the prevention of confiscation. A tract may be deemed a "voluntary subdivision," although its creation was neither volitional nor intentional. Rule 37 sets forth in some detail when the voluntary subdivision rule comes into play.

2. *Century Doctrine* – In RAILROAD COMMISSION v. MAGNOLIA PETROLEUM CO., 130 Tex. 484, 109 S.W.2d 967 (1937), the court announced that where a tract of land has been voluntarily subdivided contrary to Rule 37 so that no subdivision is entitled to a drilling permit as of right, the subdivided tract may be reconstructed and granted a drilling permit as an exception to prevent confiscation. There is a recurring problem as to which subdivision gets the permit. *See* Douglass and Whitworth, *Practice Before the Oil and Gas Division of*

the Railroad Commission of Texas, 13 St. Mary's L.J. 719, 730–32 (1982).

One must still overcome the obstacle of the voluntary subdivision rule to obtain a Rule 37 exception. This is illustrated in Railroad Commission v. Williams, 356 S.W.2d 131, 16 O.&G.R. 177 (Tex.1961) where Murel Williams was denied the right to drill a well on a 3.3–acre tract as reconstituted (under the Century Doctrine after partition to two 1.65–acre tracts) because he had acquired his undivided interest in the tract from his parents who, through a well located on their 37.5–acre tract adjoining the 3.3–acre tract, were draining the gas from beneath the 3.3–acre tract. Since the parents would not have been permitted to drill a well on the 3.3–acre tract to prevent confiscation, their grantee was in no better position. The court held that the right to a well on a tract of land, however, is not a vested right in the land itself, but is a right of the owners of the land. The right of a small tract to a well is lost by merger with a larger tract. A small tract owner is not entitled to a well as a matter of right but only on a showing that waste or confiscation will result without an exception well. Williams was draining his own 3.3–acre tract by a well on the adjacent 37.5–acre tract. So he was afforded a fair and equal opportunity to recover the oil from the 3.3–acre tract; there was no confiscation. If the parents/grantors of Murel, as cotenants of the 3.3–acre tract with Smith Price, had applied for a well permit prior to the conveyance to Murel and after the H.P. Williams well had begun to drain the 3.3–acre tract, there would have been substantial evidence to support the denial of the application by the commission. The court stated:

> Murel, their grantee of the undivided one-half interest, could have no better rights to a well permit. To hold otherwise would mean that a grantor of a mineral interest could create valuable oil development rights in his grantee which he himself did not have. This rule is a necessary corollary to the voluntary subdivision rule. That rule prohibits the creation of a right to a well permit on each small tract subdivided from a large one when the owner of the larger tract did not have such right.

Compare SOHIO PETROLEUM CO. V. SCHUMACHER, 460 S.W.2d 445, 37 O.&G.R. 277 (Tex.Civ.App. 1970, no writ). The Railroad Commission granted Schumacher a permit to drill a well on a 2.668–acre voluntary subdivision of a reconstructed 17.8–acre tract that had been part of an area acquired by the county for road purposes but later abandoned by the county. The appeals court affirmed the Commission saying the evidence supported the implied finding of the Commission that the permitted well was necessary to prevent confiscation of rights of the owners of the 2.668–acre tract. The commission and court took into consideration the structural position and water drive of the tract and adjoining tract.

3. *Waste* – An exception well location may be granted to prevent physical waste. A case examining the power of the Railroad Commission to grant an exception well permit to prevent physical waste was RAILROAD COMMISSION V. SHELL OIL CO., 139 Tex. 66, 161 S.W.2d 1022 (1942) (*Trem Carr* case). Here

Trem Carr made application to the Railroad Commission for a permit to drill a second well on a 0.92–acre tract. It was to be on a 0.67–acre subdivision of the tract, and the 0.92–acre tract was in turn a voluntary subdivision from a 20–acre tract. The Railroad Commission granted the permit over the opposition of Shell Oil Company and other adjoining property owners. The sole contention here was that the drilling and operation of the well was necessary in order to prevent physical waste. The court ruled that upon a showing that in a particular field, or in a particular section of a field, on account of the peculiar formation of the underground structure or other unusual circumstances, a closer spacing of the wells is essential to recover the oil, the Commission would have authority to grant the exception, provided that it includes all those and only those coming within the exceptional situation, and providing further that it did not unduly discriminate in any other manner against producers in other areas or fields. Since there was no evidence showing any exceptional circumstances in this instance, the court held that the permit had been improperly granted by the Railroad Commission.

The requirements for unusual circumstances were met in EXXON CORP. V. RAILROAD COMMISSION, 571 S.W.2d 497, 62 O.&G.R. 105 (Tex.1978). Here the Railroad Commission permitted a recompletion in an existing well at a shallower formation even though this was only 265 feet from another well, rather than the 1,200 feet required by Rule 37. There was an existing well bore, and there was a finding that it was not economically feasible to drill at a regular location. There was a finding that the oil that would be produced could not be produced by any other existing well.

———

SECTION 3. CREATION OF POOLED UNITS

A. EXERCISE OF POOLING POWER BY LESSEE

Lessee, at its option, is hereby given the right and power to pool or combine the acreage covered by this lease or any portion thereof with other land, lease or leases in the immediate vicinity thereof, when in Lessee's judgment it is necessary or advisable to do so in order properly to develop and operate said premises in compliance with the spacing rules of the Railroad Commission of Texas or other lawful authority, or when to do so would, in the judgment of Lessee, promote the conservation of the oil and gas in and under and that may be produced from said premises. Lessee shall execute in writing an instrument identifying and describing the pooled acreage. The entire acreage so pooled into a tract or unit shall be treated, for all purposes except the payments of royalties on production from the pooled unit, as if it were included in this lease. If production is found on the pooled acreage, it shall be treated as if production is had from this lease, whether the well or wells be located on the premises covered by this lease or not. In lieu of the royalties elsewhere herein specified, Lessor shall receive on production from a unit so pooled only such portion of the royalty stipulated herein as the amount of his acreage placed in the unit or his royalty interest therein on an acreage basis bears to the total acreage so pooled in the particular unit involved.

AMOCO PRODUCTION CO. v. UNDERWOOD

Texas Court of Civil Appeals, 1977.
558 S.W.2d 509, 58 O.&G.R. 578, writ ref'd n.r.e.

McCLOUD, CHIEF JUSTICE. This case involves the cancellation by lessors of a "gas unit" designated by lessee under the pooling provisions of eight oil, gas and mineral leases. The jury found that the designation of the gas unit by the lessee was not made in "good faith".

Victory Petroleum Corporation entered into a "Farmout Contract" with Amoco Production Company whereby Victory agreed to drill a test well on Section 3, BS & F Survey, Wheeler County, Texas, and Amoco agreed to assign to Victory certain leases covering land located near Section 3. Amoco reserved an overriding royalty interest in the leases assigned. After the test well was completed as a gas well, Victory filed a gas unit declaration forming a 688.02 acre gas operating production unit designated as the Victory Petroleum Company et al., Circle Dot Ranch Gas Unit No. 1. As designated, the unit perpetuated beyond the primary term eight different oil, gas and mineral leases containing 2,252.03 acres of land.

. . .

The jury answered two issues. In special issue 1, the jury found "it was the judgment of Victory Petroleum Company, Westland Oil Development Corporation, Amoco Production Company, and L. C. Kung that it was necessary or advisable to designate the gas unit in the manner set out and described in the Unit Declaration dated May 20, 1975 in order to properly develop and operate the eight (8) oil, gas and mineral leases in question." In special issue 2, the jury found that the designation of the Circle Dot Ranch Gas Unit by Victory Petroleum Company, Westland Oil Development Corporation, L. C. Kung and Amoco Production Company "was not made in good faith".

Judgment was entered that the Unit Designation of the Victory Petroleum Company et al. Circle Dot Ranch Gas Unit No. 1 be canceled and held for naught; that the cloud upon plaintiffs' title to the oil, gas and other minerals in and under the N/2 of Sec. 1, BS & F Survey, the SE/4 and NW/4 of Sec. 2, BS & F Survey, and Sec. 81, Block M–1, H & GN RR Co. Survey, all in Wheeler County, Texas, by reason of the execution and recordation of the gas unit, be in all things removed; that the gas purchase contract by and between Victory Petroleum Company and Natural Gas Pipeline Company of America be canceled and held for naught insofar and only insofar as the contract covers the property described above; that the oil, gas and mineral leases dated May 29, 1970, executed by the Underwoods and Walsers as lessors, covering the above described property have terminated for failure of production; and, that Circle Dot Ranch, Inc., Euline S. Walser and the Estate of Donald D. Harrington, Deceased, are entitled to all royalties payable on gas and condensate heretofore or hereafter produced and sold from the existing well situated on Section 3, BS & F Survey, Wheeler County, Texas.

Amoco Production Company, Victory Petroleum Company, Westland Oil Development Corporation, L. C. Kung and Natural Gas Pipeline Company of America have appealed. We affirm.

Appellants contend the court erred in overruling their motion for instructed verdict and motion to disregard the jury's answer to special issue 2 because there is no evidence that appellants were not acting in good faith in designating the Circle Dot Ranch Gas Unit. Appellees argue the appellants "gerrymandered" the eight leases as set out in the unit designation to advance their own pecuniary interest without regard to the rights of appellees and the other mineral owners in and under the affected lands.

Appellees or their predecessors in title executed four oil, gas and mineral leases in favor of Amoco, dated May 29, 1970, with each containing a five-year primary term. The leases were assigned to Victory and each contained a pooling clause which, in part, provides as follows:

> Lessee, at its option, is hereby given the right and power to pool or combine the land covered by this lease, or any portion thereof, as to oil and gas, or either of them, with any other land, lease or leases when in Lessee's judgment it is necessary or advisable to do so in order to properly develop and operate said premises, such pooling to be into a well unit or units not exceeding forty (40) acres, plus an acreage tolerance of ten percent (10%) of forty (40) acres, for oil, and not exceeding six hundred and forty (640) acres, plus an acreage tolerance of ten percent (10%) of six hundred and forty (640) acres, for gas, except that larger units may be created to conform to any spacing or well unit pattern that may be prescribed by governmental authorities . . .

The court in Elliott v. Davis, 553 S.W.2d 223 (Tex.Civ.App. Amarillo 1977, writ filed) recently stated that the lessee must exercise "good faith" toward the lessor and royalty owners in making a pooling designation. . . . As a general rule, the question of good faith is a fact question to be determined by the fact finder . . .

. . .

Westland Oil Development Corporation and Victory Petroleum Company are "associated companies" occupying joint offices. On May 20, 1975, A. B. Rothwell, Vice–President of Westland Oil Development Corporation and Victory Petroleum Company sent the following letter to Amoco:

> We are still conducting completion operations on the Circle Dot Ranch No. 1 well in Wheeler County, Texas, although we have no doubt that we will be successful in making a Morrow Sand gas completion.

> As you are aware, the majority of your leases included under your Farmout Contract (EA 56,278) will expire at the end of their primary term on May

29, 1975. It is imperative that we file a Gas Unit Declaration of record prior to this date.

We are forwarding herewith three (3) copies of Victory Petroleum Company et al's Circle Dot Ranch Gas Unit No. 1 Unit Declaration. *This unit has been designed to hold the majority of your leases which will expire on May 29, 1975,* and also takes into consideration our recent review of your seismic records. (Emphasis added by court. Eds.)

The Victory Petroleum Circle Dot well was commenced on Section 3 in December 1974. It was completed in the Upper Morrow Formation on or about May 16, 1975, and the official potential test was conducted on May 21, 1975. The gas unit designation was signed on May 20, 1975, and filed on May 27, 1975, only two days before May 29, 1975, the termination date of the primary term of appellees' oil, gas and mineral leases.

At the time of trial, appellants had no plans to drill an additional well on any part of the 2,252 acres affected by their gas unit. Appellants excluded from the unit 90.21 acres in Section 3 where the well was located even though their records indicated the excluded acreage was probably productive, and included 45 acres in Section 81, which perpetuated beyond the primary term 602.90 acres, even though Section 81 was "lower structurally". One of appellants' witnesses testified that with the information at hand when the unit was designated, it would be "extremely stupid" to drill in Section 81, where the "seismic and subsurface" data indicated it was "lower". Their plan would be to move to a higher structural position. The Circle Dot well was drilled on a lease containing 643.23 acres. It was not necessary to bring in additional acreage in order to have a full spacing unit and receive full allowable under the Texas Railroad Commission's "spacing rules".

After May 29, 1975, the termination date of the primary term of the leases in question, appellees could have leased their land, other than Section 3 where the well was located, to other persons for a cash bonus of $75 per acre if it had not been "tied up" by the gas unit. Riney and Laxson, employees of Westland and Victory, attempted to justify the boundaries of the unit, however, the jury could have concluded from their testimony that they were not familiar with the Upper Morrow Formation and not qualified to evaluate the facts from a geological or engineering standpoint.

The jury could have properly concluded from the evidence and inferences listed above that the configuration of the unit was not established in good faith, but designated as stated in Rothwell's letter to Amoco, "to hold the majority of the leases" which would otherwise expire on May 29, 1975.

We have considered all of appellants' points of error and they are overruled. Judgment of the trial court is affirmed.

NOTES

1. *Good faith and geology.* A federal district court in North Dakota

examined the meaning of good faith and applied a good faith test to an exercise of the pooling clause that pooled private lands with a federal unit. In SOTRANA–TEXAS CORP. V. MOGEN, 551 F.Supp. 433, 77 O.&G.R. 320 (D.N.D.1982), the lessee sought to establish a federal unit that included the leased land, and the lessors refused to ratify the unit. The lessor top-leased the land to another, apparently for a higher royalty. The lessee sought a declaration that its exercise of the pooling clause was valid and the lease was still in existence under the unit. In the facts of this case, the geological evidence supported the formation of the unit; thus, the lessee had not violated his duty of good faith exercise of the clause. The court observed that "extension of the lease will almost always be a major factor in the lessee's decision to form the unit. So it can not be said with any certainty that the lessee unitized for the primary purpose of increasing oil recovery. Thus, except for extraordinary circumstances which are not present here, the courts have good reason to limit consideration to the more objective and defined determination of whether the geology supports the unit configuration."

2. Should the pooling clause of an oil and gas lease covering Blackacre be construed to permit the lessee to pool Blackacre with Whiteacre under the following circumstances:

(i) Prior to pooling, a producing well has been drilled on Whiteacre and no drilling has occurred on Blackacre; pooling occurs shortly before the expiration of the primary term of the Blackacre lease.

(ii) Prior to pooling, a producing well has been drilled on Blackacre and no drilling has occurred on Whiteacre; the lessee who seeks to pool the two tracts does (or does not) have a greater interest in production from Whiteacre than he does in production from Blackacre.

(iii) The pooling of Blackacre and Whiteacre occurs after the expiration of the primary term of the Blackacre lease but while such lease is being held under the provisions of one of its savings clauses, *e.g.*, the dry hole or continuous drilling operations clause.

Consider in this connection the following cases:

(a) IMES V. GLOBE OIL & REFINING CO., 1938 OK 601, 184 Okl. 79, 84 P.2d 1106. A community lease executed by owners of 21 lots contained a clause authorizing the inclusion of certain other lots within the terms of the lease "at any time." After producing wells were drilled by the lessee on the premises covered by the community lease, the lessee sought to permit the owners of six other lots to join in the unit and share in the royalty from the two wells. Apparently the lessee owned the royalty interest in the six lots which he sought to bring within the unit and the court was satisfied that the six lots had been condemned as valueless for oil and gas purposes. In denying the inclusion of these additional six lots in the unit, the court observed that the lessee "was virtually the agent of the lessors, and for this reason he was bound to use good faith."

(b) SOUTHWEST GAS PRODUCING CO. V. SEALE, 191 So.2d 115, 25

O.&G.R. 316 (Miss.1966). The bill of complaint alleged that the defendant operators fraudulently, in bad faith, and in violation of the duty owed by a lessee to its lessor, had gerrymandered and so constituted its drilling units as to defraud appellants in the recovery of oil belonging to them. Acting under the provision of pooling clauses in its leases, lessee included 6.35 acres of Chambers' land, 13.6 acres of Seale's land, and 20.29 acres of Johnson's land in a 40–acre pooled unit and a producing well was drilled on the Seale land. Previously lessee had drilled a dry hole on the Johnson land and apparently the Johnson lease expired; on lessee's request and on his promise to include his land in the Seale No. 1 unit, Johnson executed a new lease with pooling provisions after the drilling of the dry hole. *Held,* the chancellor was justified in holding that the lessee, Hayes, by including 20.29 acres of Johnson land (a portion of which had previously proven dry) with Seale land in Seale Unit No. 1, violated his duty to Seale of fair dealing and good faith in the formation of that unit.

(c) GILLHAM v. JENKINS, 1952 OK 150, 206 Okl. 440, 244 P.2d 291, 1 O.&G.R. 842. After drilling a successful gas well on the 80 acres under lease the lessee attempted to pool the successful gas well with an adjoining 80–acre tract to comply with wartime regulations regarding scarce materials needed to connect the well to the purchaser's gas line. The *Imes* case was cited against the validity of the pooling and the court discussed it as follows:

> In the Imes case the principal factor for consideration was not the time of development but it was the good faith of the lessees in combining additional lots to the unit after production was had. The fact of a prior discovery of oil was incidental to the proof of bad faith. . . . It appeared that the addition of the six lots was not made in good faith. The lessee sought to enrich himself at the expense of the lessor. The case is clearly distinguishable from the present case, where a real necessity and purpose existed at the time of the pooling, which was done in good faith to obtain the marketing of the gas from the well on plaintiff's land, and where the unitization was had, under compulsion, if the gas was to be marketed and benefit obtained by either party.

> . . .

> . . . We hold that the trial court was correct in its decision that, under all the circumstances of this case, it was the duty of lessee to pool or combine the involved acreage with other acreage in order to comply with the necessary Federal rules and regulations and thus secure a market for the production.

Does such a "duty" to pool depend on the existence in the lease of a pooling clause?

In Southwest Gas Producing Co. v. Seale, *supra,* the court expressed the view that the lessee was not under a duty to include and pool the Seale property in units on which wells had been drilled prior to the drilling of the first well on the Seale property. "There is no requirement of this nature in the pooling clause of the lease, or the conservation laws, or the implied covenants of the lessee."

The thrust of the plaintiffs' complaint in KINNEAR V. SCURLOCK OIL CO., 334 S.W.2d 521, 12 O.&G.R. 1185 (Tex.Civ.App. 1960, writ ref'd n.r.e.), was that the lessee failed to pool under the pooling clause and had a duty to do so. The court quoted the pooling clause, which followed a pattern typical in the industry, and said it was apparent that the paragraph did not require pooling but left it to the option of the lessee. The court could not find that pooling had become a fixed obligation of the lease.

The plaintiff in GRACE PETROLEUM CORP. V. WILLIAMSON, 906 S.W.2d 66, 135 O.&G.R. 335 (Tex.Ct.App.1995) was more successful in getting recognized a duty to pool. Here a lessee got a lease extension on a 133 acre lease based in part on a promise to place all of the lease in a unit or units. However, some 83 acres of the lease was never placed in a producing unit, and the area was apparently drained by production from adjacent units. A Texas jury found for the lessors, awarding $25,000 in actual damages for the drainage and $500,000 in exemplary damages. On appeal, the appellate court reversed as to the exemplary damages. The claim was one for breach of contract, and there was no evidence that the lessors sustained distinct damages attributable to the lessee's fraudulent misrepresentation. The apparent premise of the case is that there was a duty to pool when such a promise had been made.

(d) HUMBLE OIL & REFINING CO. v. KUNKEL, 366 S.W.2d 236, 18 O.&G.R. 344 (Tex.Civ.App.1963, error ref'd n.r.e.). The lease of Blackacre, a 145–acre tract, included a pooling clause, a continuous drilling operations clause, and a provision preserving the lease for 60 days after the completion of a dry hole. Shortly before the expiration of the primary term of the Blackacre lease on August 7, 1961, Lessee commenced drilling operations thereon which resulted in the completion of a dry hole on September 6, 1961. On November 1, 1961, Lessee executed and filed for record an oil unit designation instrument which described 160 acres of land, including Whiteacre, on which a producing well had previously been drilled, and 8.94 acres of Blackacre. The court held that the Blackacre lease was not preserved by unit production.

WILCOX v. SHELL OIL CO., 226 La. 417, 76 So.2d 416, 3 O.&G.R. 1903 (1954). Under the strict reading given the pooling clause in this case the lessee could not pool the leased tract with land on which production had already been obtained.

MALLETT v. UNION OIL & GAS CORP., 232 La. 157, 94 So.2d 16, 7 O.&G.R. 434 (1957). The lease of Blackacre, a 160–acre tract, authorized pooling "when in Lessee's judgment it is necessary or advisable to do so in order properly to develop and operate said leased premises in compliance with the orders, rules and regulations of State and Federal governmental authority, or when to do so would, in the judgment of Lessee, promote the conservation of oil and gas from said premises." Three days prior to the expiration of the primary term of the lease, the Lessee filed for record a declaration of unitization covering some 159.9 acres, including Whiteacre, on which a producing gas well had previously been drilled, and 53.3 acres of Blackacre. The court concluded that the Blackacre lease expired at the end of its primary term, declaring that:

> The language used in this lease unmistakably shows that the pooling agreement was authorized only for the purpose of development and the right to unitize with producing property was not granted. The whole tenor of the lease and the pooling agreement contemplates pooling before production.

Compare Wilcox and *Mallett* with the following two cases. Is the difference truly one of different lease language or is it a matter of judicial attitude?

GORENFLO v. TEXACO, INC., 735 F.2d 835, 81 O.&G.R. 284 (5th Cir.1984). The lessors claimed the lease was not maintained by unit operations because the lease provided only for a declaration of a unit for "production" and not for "exploration." The Fifth Circuit affirmed a district court judgment for defendants. The pooling clause in question allowed pooling before or after production, and thus operations on the pooled area for exploration before the end of the primary term were sufficient to maintain the lease.

KASZAR v. MERIDIAN OIL & GAS ENTERPRISES, INC., 27 Ohio App.3d 6, 499 N.E.2d 3, 92 O.&G.R. 170 (1985). The court ruled that a lessee who had commenced a well on the leasehold within the primary term of the lease had the power under the pooling clause at issue to pool the property after the primary term. There was nothing in the clause to indicate any limit as to when the pooling could be exercised.

(e) JONES v. KILLINGSWORTH, 403 S.W.2d 325, 24 O.&G.R. 508 (Tex.1965). By a lease executed on August 16, 1951 of tracts containing, in the aggregate, 20.55 acres, the lessee was authorized to pool or combine the leased acreage. The pooling clause provided:

> Units pooled for oil hereunder shall not substantially exceed 40 acres each in area, and units pooled for gas hereunder shall not substantially exceed in area 640 acres each plus a tolerance of 10% thereof, provided that should governmental authority having jurisdiction prescribe or permit the creation of units larger than those specified, units thereafter created may conform substantially in size with those prescribed by governmental regulations.

On July 12, 1961 the leasehold was pooled by the lessee with other premises to form a unit containing 170.86 acres; the parties to this action dealt with this unit as though it contained only 160 acres. On or about August 16, 1961, a producing well was completed on a portion of the unit other than the 20.55 acres covered by this lease. Petitioners sought a determination that the lease expired on August 16, 1961 at the expiration of its primary term. A take-nothing judgment rendered against the petitioners was affirmed by the Court of Civil Appeals. The Texas Supreme Court, three Justices dissenting, reversed and rendered. The court concluded that although the Railroad Commission *permitted* 160 proration units, it only *prescribed* 80 acre units. The pooling thus was invalid since it was for acreage larger than that prescribed by the Railroad Commission.

(f) CIRCLE DOT RANCH, INC. V. SIDWELL OIL AND GAS, INC.. 891 S.W.2d 342, 132 O.&G.R. 417 (Tex.App, 1995, *writ denied*). In a case quite factually similar to Amoco Production Co. v. Underwood, the court remanded for determining whether the lessee had acted as a prudent operator. The plaintiff lessors brought an action to cancel a lease for exercise of the pooling clause that was alleged not to be in good faith. The lessee Sidwell had drilled a well in July and August 1988 under a permit from the Railroad Commission, the application for which had designated a 643 acre drilling unit on lessors' land. In April 1989 Sidwell exercised the pooling clause to designate an irregularly shaped 668 acre unit made up out of 123 acres of the plaintiffs' land together with 545 acres covered by six other leases. This resulted in an 81% reduction in plaintiffs' royalty. Despite the defendant's admission that it was not necessary to create the pooled unit to obtain a full allowable for the well and that productive acreage from the plaintiffs' land had been excluded from the pooled unit, the trial court had granted summary judgment in favor of the defendant. The court here reversed, finding that there were fact issues concerning the application of the prudent operator standard to the conduct of the lessee. The decision is curious for it fails to mention the Underwood decision, which is a leading Texas case on good faith pooling and the facts of which are quite similar to those of this case. The court here said the issue was whether the lessee exercised the pooling option as a reasonably prudent operator would do under the same or similar circumstances by, among other things, using good faith, taking into account its and its lessors' interests. The case was remanded for determining whether the lessee had acted as a prudent operator in accordance with this standard.

(g) MISSION RESOURCES, INC., V. GARZA ENERGY TRUST, 166 S.W.3d 301, 160 O.&G.R. 144 (Tx. App., Corpus Christi-Edinburg, 2005). The Texas appeals court, following the standards applied in *Circle Dot Ranch*, affirmed an award of damages for bad faith pooling. The jury's finding of bad faith was evidently based on testimony that the lessee formed a pooled unit in a manner that financially penalized appellees, even though there were other ways to pool without creating such a penalty. The evidence, said the court, tended to prove that the lessee "did not adequately consider the financial interests of appellees in exercising its pooling power." *Mission Resources* was reversed on other grounds in Coastal Oil & Gas Corp. v. Garza Energy Trust, 268 S.W.3d 1 (Tex. 2008). The Texas Supreme Court agreed that the evidence supported the jury's finding of bad faith pooling but reversed on issues of whether actionable trespass had occurred by fracing.

3. Consider also the possibility of a lease provision requiring the lessee to drill an additional well or wells if the conservation agency authorizes a smaller unit size. The lease in PRESNAL v. TLL ENERGY CORP., 788 S.W.2d 123, 110 O.&G.R. 529 (Tex.App. 1990), contained a lease clause that authorized the lessee to pool the 115 acres of the lease to a 160 acre unit. A further clause provided that if the Railroad Commission changed its rules during the lease to allow an 80 acre

unit, the lessee had to drill a second well within 45 days or pay liquidated damages of $75,000. After the Railroad Commission changed to 80 acre units the lessee failed to drill a second well. The court reversed a trial court determination that the $75,000 was a prohibited penalty as a matter of law.

4. A lease authorized pooling into units containing no more than 660 acres, and also contained a force majeure clause. The state conservation agency, at the instance of a third party, created a *compulsory pooling* unit of 726.92 acres. Was the pooling in conformity with the lease? *See* Gordon v. Crown Central Petroleum Co., 284 Ark. 94, 679 S.W.2d 192, 83 O.&G.R. 476 (1984). Should the result be any different if the lessee itself had applied a compulsory unit larger than would be permitted to be created as a declared unit? *See* Nisbet v. Midwest Oil Corp., 1968 OK 115, 451 P.2d 687, 32 O.&G.R. 457. The lease had a clause limiting the exercise of the pooling power to 40 acres for oil and 320 acres for gas. A provision printed on the lease form stating that "should governmental authority having jurisdiction prescribe or permit the formation of larger units for maximum allowable production, any unit created hereunder may conform substantially in size with those so prescribed or permitted" was crossed out through agreement of the lessor and lessee. The lessee applied for and was granted drilling units of 640 acres. The lessor brought suit against the lessee seeking rescission of the lease; the lessor claimed the lessee had falsely represented that it would not drill or participate in units larger that 40 acres for oil and 320 acres for gas and that the lessee's applying for and obtaining the larger drilling units was in disregard of their agreement and a fraud. The court held that the striking of the language only related to the pooling power of the lessee and not to the actions of the Corporation Commission. The orders of the Corporation Commission were part of the lease, and the parties were bound thereby. *See also* Vogel v. Tenneco Oil Co., 465 F.2d 563, 43 O.&G.R. 362 (D.C. Cir. 1972).

5. The lease of Blackacre included a pooling clause authorizing the formation of a unit or units not exceeding 40 acres each in the event of an oil well, or into a unit or units not exceeding 640 acres each in the event of a gas well. Shortly before the expiration of the primary term Lessee included Blackacre in a gas unit and drilled a producing well upon lands included in the pooled unit but not upon Blackacre. The well produced both gas and gas condensate or distillate (a liquid), at a ratio of 33,000 cubic feet of gas to each barrel (5.61 cubic feet) of liquid during each 24–hour period. At the expiration of the primary term of the Blackacre lease Lessor sought a decree that the lease had terminated, alleging that the unit well was an oil rather than a gas well. What are the relevant considerations in the resolution of this controversy? *See* Diggs v. Cities Service Oil Co., 241 F.2d 425, 7 O.&G.R. 827 (10th Cir.1957).

6. What factors should be considered in the construction of a lease pooling clause as concerns such questions as the following:

(a) Whether a pooling clause authorizing pooling of the leasehold "or any portion thereof" permits pooling as to a particular gas bearing formation. *See*

Gillham v. Jenkins, 1952 OK 150, 206 Okl. 440, 244 P.2d 291, 1 O.&G.R. 842.

(b) Whether the pooling power given the lessee is exhausted by one exercise. *See* Texaco, Inc. v. Lettermann, 343 S.W.2d 726, 14 O.&G.R. 427 (Tex.Civ.App.1961, error ref'd n.r.e.).

(c) Whether the pooling power authorizes use of the leased premises in connection with unit operations of a type not anticipated at the time of the execution of the lease, *e.g.,* water flooding. *See* Miller v. Crown Central Petroleum Corp., 309 S.W.2d 876, 8 O.&G.R. 1279 (Tex.Civ.App.1958).

(d) Whether a pooling clause authorizing the lessee to create units and to enlarge or change the shape of existing units permits the lessee to change the boundaries of an existing unit so as to exclude certain premises therefrom or to include additional acreage thereby possibly diluting the existing unit acreage. *See* Grimes v. La Gloria Corp., 251 S.W.2d 755, 1 O.&G.R. 1784 (Tex.Civ.App.1952) and Expando Production Co. v. Marshall, 407 S.W.2d 254, 25 O.&G.R. 954 (Tex.Civ.App. 1966, writ ref'd n.r.e.).

(e) Whether a pooling clause authorizing the lessee to pool a royalty interest continued to authorize pooling after the royalty owner converted the royalty to a working interest? *See* Edwin M. Jones Oil Co. v. Pend Oreille Oil & Gas Co., 794 S.W.2d 442, 112 O.&G.R. 501 (Tex.App. 1990, err. denied.)

7. Should it be relevant in the exercise of the pooling clause whether all other interests in the pooled area are able to be pooled, i.e. some are pooled and some are not? What loss is sustained by a pooled interest owner when some interests in the pooled area are not pooled? Consider the following cases:

(a) UNION OIL CO. OF CALIFORNIA v. TOUCHET, 229 La. 316, 86 So.2d 50, 5 O.&G.R. 1177 (1956). The Louisiana Supreme Court declared invalid an exercise of the pooling power by a lessee when the area subject to the pooling included a lease that did not at that time authorize pooling. The court said it was "unthinkable" that the lease pooling clause gave the lessee the power to combine the lessor's land with any other land or lease in the immediate vicinity in the absence of consent by that particular landowner. Thus, the court said "the only meaning that this provision of the lease could have was that the lessor granted to the lessee authority to combine his lease with any other land or lease in the vicinity which the oil company also had authority to unitize." It should be noted of this case, however, that the court was not declaring invalid the lease or the unit. The landowner of the adjacent land had subsequently granted the pooling authority to the lessee. What was at issue was the continued validity of a royalty interest created by the lessor himself. The court's ruling meant that the effort at pooling did not interrupt prescription, and thus the royalty ceased to exist. *See also* Michigan Oil Co. v. Black, 455 So.2d 824, 84 O.&G.R. 1 (Ala.1984).

(b) HEATH v. FELLOWS, 526 F.Supp. 723, 71 O.&G.R. 542 (W.D. Okl. 1981). The lessee pooled the lessor's land with other leases, some of which did not contain pooling clauses. A successful well was begun and completed on the unit but not on plaintiff lessor's land. He sought cancellation of the lease on the ground that the pooling clause had been exercised in bad faith because of the pooling without all royalty owners joining. The court held that the unit agreement was binding on those who signed, including those where the lessee had made use of a pooling clause, stating: "For the Court to hold otherwise would in effect invalidate the pooling clause of a co-tenant which was specifically bargained for in a lease unless the other co-tenants consented to such a clause." There was no bad faith in the pooling here nor in the exercise of the power on the last day of the primary term.

(c) CELSIUS ENERGY CO. v. MID AMERICA PETROLEUM, INC., 894 F.2d 1238, 108 O.&G.R. 113 (10th Cir.1990) followed *Heath v. Fellows.* The lessees of two leases on an eighty acre tract in Oklahoma filed a declaration of pooling under the pooling clause of the leases, pooling a part interest of the tract with leases on an adjoining tract on which production had already been effected. The declaration of pooling only pertained to 37.5% in the leases as 62.5% of the interest had been assigned to a drilling partnership that was not a party to the declaration of pooling. The successors of the lessors sought to cancel the leases, claiming that the declaration of pooling was unauthorized and invalid. Subsequent to the declaration of pooling the Corporation Commission established a 160 acre spacing for a common source of supply from which the production was taking place, so there was a geological basis for the pooling, and the parties stipulated that the pooling was in good faith. The pooling clause in the appellees' oil and gas leases stated: "Lessee is hereby granted the right at any time and from time to time to unitize the leased premises or any portion or portions thereof, . . . with any other lands . . . for the production primarily of oil or primarily of gas with or without distillate." The court concluded that this granted the lessees power to pool the land, and there was nothing in the clause that restricted the lessees' power to unitize with less than 100% of the leasehold interest.

8. Who is able to exercise the pooling power contained in the lease?

(a) O'HARA v. COLTRIN, 637 P.2d 398, 71 O.&G.R. 487 (Colo.App.1981, cert. denied). The lessor claimed, among other things, that the lease had expired because it had not been properly pooled inasmuch as the pooling had been done by a sublessee and not by the lessee. The Colorado court held that the sublease/assignment distinction had no role in the exercise of the pooling clause. The lease could be assigned in whole or in part, and the sublessee assumed the power to pool when he assumed the working interest in the lease. *Compare* Doran & Associates, Inc. v. Envirogas, Inc., 112 A.D.2d 766, 492 N.Y.S.2d 504, 86 O.&G.R. 348 (1985), appeal dism'd, 66 N.Y.2d 758, 497 N.Y.S.2d 1028, 488 N.E.2d 131 (1985).

(b) PAMPELL INTERESTS, INC. v. WOLLE, 797 S.W.2d 392, 112 O.&G.R. 145 (Tex.App. 1990). Plaintiff-lessors granted a lease with a primary term of two years to Pampell that contained a pooling clause. On the day that the lease was to expire, Zeal Energy filed a unit declaration for the pooling of some of the lease acreage to form a 160 acre unit. Zeal had already commenced drilling a well on a portion of the declared unit but the operations were not on the lease in question. Plaintiff-lessors and their new lessee sought declaratory judgment from the court that the Pampell lease had expired. Pampell, which was owned by the same person who owned Zeal, asserted that Zeal filed the unit declaration as agent for Pampell. The court held that pooling clauses must be strictly complied with. The unit designation was void because it was filed by a stranger to the lease title. Even if Zeal were the agent of Pampell, the unit declaration did not comply with the lease terms because the pooling clause required the lessee to execute and record the unit designation.

9. State conservation agencies do not have occasion to pass upon the actions of a lessee in exercising the pooling power of the lease in single well drilling units. However, they often must approve the establishment of fieldwide units that are based partly on consent of the parties and partly on the compulsion of the state. In addition, the Department of the Interior must in some circumstances approve the establishment of units. In such situations, what should be the effect on the question of a lessee's good faith in exercising the pooling (or unitization) power under the lease of the state or federal agency's determination that the sharing of production under the pooling or unitization is fair and equitable?

(a) AMOCO PRODUCTION CO. v. HEIMANN, 904 F.2d 1405, 109 O.&G.R. 276 (10th Cir.1990). The Heimanns owned 48,120 acres of ranch land in New Mexico. Between 1971 and 1974, they executed three carbon dioxide and mineral leases with Amoco. Each of these three leases contained a unitization clause which granted Amoco the right to unitize the Heimanns' mineral interests with other lands in the area, subject to approval "by any governmental authority." The leases granted the Heimanns a one-eighth royalty of the net proceeds received from all oil, gas or carbon dioxide produced on their lands. In the late 1970's, Amoco embarked upon a plan to pipe carbon dioxide from northern New Mexico to its west Texas oil fields in order to enhance recovery there. Amoco therefore sought to unitize the mineral rights to approximately 1,174,225 acres of land including the Heimanns's land. The proposed agreement for the "Bravo Dome" unit allocated royalties on the basis of "surface acreage;" production was allocated according to the total surface areas contained in each tract. Amoco sought approval of the Bravo Dome unit from the New Mexico Oil Conservation Commission. The Commission found that "approval of the proposed unit agreement should promote the prevention of waste and the protection of correlative rights within the unit area" and consequently approved the unit agreement. The Heimanns and other opponents of the unit appeared before the OCC at a rehearing and presented evidence that the per-acre participation formula did not protect their correlative rights. The Commission rejected their contentions, and the Commission decision was upheld on appeal. In 1984, Amoco filed suit against the Heimanns in federal district court seeking a declaratory judgment that Amoco had properly unitized the interests covered under the leases. The district court entered a $4 million judgment for the defendant lessors based on a jury verdict that the lessee had breached a duty of good faith in exercising the unitization clause. The Tenth Circuit reversed, stating the following:

> A good faith duty is imposed where unbridled discretion is vested in an oil or gas lessee by a unitization clause. . . . If a lessee had complete discretion in unitizing an oil or gas field, the lessee might, in bad faith, combine lessor's land with less productive land, calculate a production formula which underrepresents the lessor's mineral interest, or unitize solely to avoid the termination of a lease. But where a neutral and detached agency approves a proposed unitization after undertaking an extensive and independent study of geological, physical and economic data, the agency normally will constrain such abuses by a lessee. . . .
>
> A good faith duty also may serve to assure the fair allocation of oil and gas produced by the unit. . . . Where the lessee maintains complete discretion in formulating a unitization plan, the lessee might abuse that discretion and select a participation formula which underrepresents the contribution to the unit from the lessor's land. However, where an agency such as the OCC passes upon the fairness of a proposed participation formula, concerns of lessee unfairness are ameliorated. For unless a proposed unitization plan provides for a fair participation formula, it will not win OCC approval. . . .
>
> Evaluating the statutory framework behind the OCC, we are convinced that it ameliorates the danger of lessee unfairness which gave rise to the good faith duty. Where approval of a unitization plan is finally determined by the OCC, the dangers resulting from the lessee's complete discretion which concerned this court in *Boone* are absent. . . . And where the OCC approves the participation formula after a careful and independent inquiry into the relevant geophysical and economic criteria, a fair allocation of proceeds is determined without resort to the lessee's good faith duty. Therefore, because the components of a lessee's good faith duty are necessarily encompassed within the OCC's approval criteria, it is a waste of judicial resources to conduct a second good faith inquiry here.

(b) Consider also the effect of the Bureau of Land Management approval of federal exploratory units established in FROHOLM v. COX, 934 F.2d 959, 117 O.&G.R. 74 (8th Cir.1991). Defendant lessee made use of a unitization clause in leases from plaintiff-lessors to establish two federal exploratory units that were approved by the Bureau of Land Management. The lessors brought suit against the lessee seeking a determination, *inter alia,* that the units were not formed in good faith or with geological justification. The court stated that the exercise of the unitization clauses had to be done in good faith, and ruled that establishing geological justification for unitization is tantamount to establishing good faith. The court made the following observation on the issue of the good faith of the lessee in exercising the pooling power:

> Production of oil and gas is clearly in the best interest of the lessee and

lessor. Unit agreements, such as the Antelope and Bakken agreements, are to provide for the prudent and proper development of oil and gas. A federal unit may only be established for the purpose of conserving the natural resources of any oil and gas pooled where the Secretary of the Interior determines that it is necessary or advisable in the public interest. 30 U.S.C.A. § 226(j).

10. TITTIZER V. UNION GAS CORP., 171 S.W.3d 857 (Tex. 2005); UNION GAS CORP. V. GISLER 129 S.W.3d 145 (Tex.App.-Corpus Christi, 2003). A series of related cases arose in Texas from a declaration of pooling that was inconsistent with the pooling clauses of the leases subject to the pooling. Without authorization to execute a pooling designation with a retroactive effect, the lessee's attempt to do so was ineffective. Thus the well site owners were owed full royalty on the production that occurred prior to the recordation of the pooling designation. The non-well-site owners were not entitled to royalty on the production that occurred prior to the recordation, finding that the same effective date should be applied to all leases.

11. The pooling clause in a Louisiana lease stated that the "Lessee shall execute in writing and file for record in the records of the Parish . . . an instrument identifying" the pooled acreage. The lessee executed a "declaration of unitization" of a lease tract a few days before the end of the primary term to pool it with other land on which there was a producing well. However, the lessee failed to record the same until 6 days after the end of the primary term. The court ruled that the recordation was a prerequisite to effectively pool the lease tract. The lease expired. Mobil Oil Exploration and Producing Southeast, Inc. v. Latham Exploration Co., Inc., 31 So.3d 1149, 44,996 (La.App. 2 Cir. 2/3/10).

B. VOLUNTARY POOLING AGREEMENT
THE SUPERIOR OIL COMPANY v. ROBERTS

Supreme Court of Texas, 1966.
398 S.W.2d 276, 24 O.&G.R. 77.

NORVELL, JUSTICE. This cause presents a question that has not been squarely passed upon by this Court. We find, however, that the Supreme Court of West Virginia has decided the controlling issue in the case contrary to the contentions which respondents here urge. See, Boggess v. Milam, 127 W.Va. 654, 34 S.W.2d 267 (1945) Being of the opinion that *Boggess* was correctly decided, we reverse the judgments of the courts below and render judgment that respondents (plaintiffs in the trial court) take nothing against the petitioner, The Superior Oil Company.

In stating the case, we will use the trial court designation of the parties. Plaintiffs are Newman Roberts, Rosa Lee Hartman and Olen Stafford, the only heirs of Bob Roberts, deceased. In 1947 plaintiffs owned an undivided one-half interest in a tract of one and a half acres of land being Lots 1, 2, 3, 4, 6 and 12 in Block W of the Hawley Addition to the town of Altair, Colorado County, Texas. The other one-half interest in said town lots was owned by James Craven and the defendant, Estella Todd, the only heirs of Matilda Joshua, deceased.

On September 29, 1947, James Craven (and his wife, Lovie Craven) executed an oil, gas and mineral lease to The Superior Oil Company purporting to cover all of the six town lots above described.

On October 6, 1947, Estella Todd Executed a similar lease to Superior describing the same town lots as the Craven lease.

Both leases contained the following provision: "Lessor does hereby pool and unitize the lands herein leased and all rights of the lessor therein under the terms of that certain 'Unitization and Unit Operating Agreement in the Altair Field, Colorado County, Texas' made and entered into as of the 10th day of April, 1947. Lessor ratifies, confirms and adopts said agreement and by this reference said lease and all rights of Lessor thereunder are made subject to said agreement. The terms and provisions of said agreement shall supersede the terms and provisions of this lease."

The "Unitization and Unit Operating Agreement" for the Altair Field is a detailed contract of some 37 pages. The area unitized thereby consisted of several hundred acres. Production was obtained from the unit but no wells were drilled on the six Altair town lots in which the plaintiffs own an interest, nor within a distance of 1200 feet therefrom.

It appears that although Superior offered to lease their interest in the six town lots, the plaintiffs refused such offer and never executed a lease agreement and have received no payments from the minerals produced from the Altair Field unitized area. It appears that in accounting for royalty payments. Superior considered the Craven and Todd leases as representing one and a half acres, instead of one-half of one and a half acres,-the actual undivided interest owned by James Craven and Estella Todd at the time of the execution of the leases. It appears that the royalty payment allocated to a one and a half acre interest in the unit was paid to Estella Todd and James Craven (or his heirs) up to the time this suit was filed. Superior retained the working interest share.

The two controlling factors in the litigation are (1) that no wells were drilled upon the lands in which plaintiffs hold an undivided interest and (2) that there existed no contractual relationship whatsoever between plaintiffs and Superior, the operator of the unit.

Plaintiffs do not seek to ratify or adopt the leases executed by Estella Todd and James Craven. They do not claim to be tenants in common with the royalty owners who as lessors executed leases covering lands within the unitized area. There is nothing to show that Superior has prevented plaintiffs from developing the mineral resources of the town lots in any way.

It is plaintiffs' theory that by accepting leases from Estella Todd and James

Craven, Superior became a tenant in common with them in the mineral estate underlying the town lots. Plaintiffs equate the pooling of the lots and obtaining production from the unitized area with actual production from such town lots. Of course, had Superior produced from the town lots in which plaintiffs have an undivided one-half interest, it would have to account to plaintiffs for their share of the minerals produced less the necessary and reasonable cost of producing and marketing the same. . . .

It is argued that as the Todd and Craven leases purported to cover the entire one and a half acres embraced in the six town lots and Superior used the one and a half acre factor in accounting to its royalty owners, allocating to itself 7/8 of a one and a half acre interest in and to the minerals produced from the unitized area, Superior should pay over to plaintiffs 1/2 of this 7/8 (working interest) represented by a one and a half acre share in the minerals produced from the unit, less reasonable expenses. We cannot agree with this argument. Obviously any move taken by Superior acting in connection with Estella Todd and James Craven, could not operate to place plaintiffs' property under lease or make it a part of a unitized area without plaintiffs' consent or acquiescence. No minerals were produced from their property. They had no contractual relationship with Superior or the owners of interests in the unitized lands which would give them rights in and to minerals produced from the unit but not produced from their lands. The method employed by Superior in accounting for the minerals produced from the unitized area cannot affect the situation. If, in fact, Superior has retained more than its proper share in the division of mineral proceeds between itself and the royalty owners holding lands within the unitized area, this is a matter between Superior and such royalty owners and does not affect plaintiffs in any way.

In Boggess v. Milam, 127 W.Va. 654, 34 S.E.2d 267 (1945), mentioned above, it appears that W. W. Boggess owned an undivided one-tenth interest in the minerals underlying a tract of 116 acres which had been unitized with a tract of 53 acres for mineral production purposes. The lease on the 116 acre tract and the unitization agreement were drawn so as to include W. W. Boggess but he refused to execute the instruments. A well drilled upon the 53 acre tract was a producer and Boggess contended that the unitization of the 116 acre tract in which he had an interest and the 53 acre tract which had produced effected a merger of title in and to such tracts and that he was entitled to his proportionate part of the production obtained from the unit. (11.6/169). The West Virginia Court held:

> In our opinion, the so-called unitization agreement does not effect a merger of title. Under its express terms the leases of the fifty-three acres and of the one hundred and sixteen acres not in conflict with its terms remained in effect. It consolidates only the contractual interests under the leases to the United Fuel Gas Company. Boggess having refused to become a party to the lease of the one hundred sixteen acre tract, acquired no contractual right subject to merger, and by remaining aloof, both from the lease and the

unitization agreement, neither instrument affected his interest, either favorably or unfavorably.

The Supreme Court of Mississippi followed the doctrine of the Boggess case in California Company v. Britt, 247 Miss. 718, 154 So.2d 144 (1963).

In 3 Summers Oil and Gas 492, § 612, it is stated on authority of Boggess v. Milam that:

> A unitization agreement does not effect a merger of title of the tracts involved so as to give a cotenant of a separate tract, who refuses to sign the lease of his tract or the agreement, an interest as a cotenant in the other tracts of land.

Similarly, in Myers, The Law of Pooling and Unitization, Voluntary–Compulsory, § 14.04, it is said:

> If there is no production from the tract in question the unsigned royalty interest in the absence of equitable considerations receives nothing though there is production elsewhere in the unit.

. . .

Plaintiffs have made no attempt to ratify the Todd and Craven leases or the Altair Field Unitization Agreement. It is these instruments which give rise to a claim to the mineral production from tracts not owned by plaintiffs and their cotenants, Estella Todd and the heirs of James Craven. In claiming an interest in such mineral production, plaintiffs in effect seek to claim the beneficial provisions of such contracts and repudiate the unfavorable portions thereof which would limit their claim to the same rights as those held by their cotenants who joined in the Superior lease, namely, royalty rights only. In our opinion, this position is untenable.

For the reasons stated, the judgments of the courts below are reversed and judgment here rendered that respondents (plaintiffs in the trial court) take nothing.

NOTES

1. Observe that the pooling in the principal case came about through a lease clause that incorporated by reference a unit agreement. How does this differ from the lessee's exercise of a pooling clause? There is a dearth of case law on voluntary pooling agreements; does this suggest that the most common method of pooling for the purpose of drilling a single well is through the exercise of the pooling clause? Where problems have arisen, the litigation has been, as was the issue in the principal case, over joinder of other parties. This topic is pursued below, in section D.

2. The court states that the plaintiffs in the principal case in effect sought to repudiate the unfavorable portions of the unit agreement. What are the unfavorable aspects of a pooling or unitization agreement? What does one gain by not agreeing to pooling?

3. *Superior Oil Co. v. Roberts* deals with the issue of a pooling or unitization of royalty owner interests. As we have seen in *Union Pacific Resources, supra*, it is common in the oil and gas industry for working interest owners to pool their interests using a document commonly referred to as a Joint Operating Agreement. There are several model form Joint Operating Agreements that are in widespread use including those developed by the Rocky Mountain Mineral Law Foundation and the American Association of Petroleum Landmen. The Joint Operating Agreement attempts to allocate costs and production among the working interest owners. Often the owner of the largest working interest is designated the operator for purposes of exercising control over the day-to-day operations involving drilling and production. The relationship between the operator and the non-operator is often set forth by express language in the Joint Operating Agreement. Nonetheless, there has been a substantial amount of litigation concerning the nature of that relationship. For recent scholarship on some of the problems the courts have had in defining the relationship between parties to a Joint Operating Agreement see Jonathan A. Hunter & Cheryl M. Kornick, *Operator Liability in the 21st Century: Is Being in Charge Still Worth It?*,' 51 Rocky Mtn. Min. L. Inst. Ch. 15 (2005); Wilson Woods, *The Effect of Exculpatory Clauses in Joint Operating Agreements: What Protections Do Operators Really Have in the Oil Patch?*, 38 Tex. Tech L. Rev. 211 (2005); John R. Cooney, *Recent Developments Concerning Joint Operating Agreements--Preferential Rights and Exculpatory Clauses*, 55 Inst. on Oil & Gas Law § 1.02 (2004); John Burritt McArthur, *Judging Made Too Easy: The Judicial Exaggeration of Exculpatory and Liability--Limiting Clauses in the Oilfield's Operator Fiduciary Cases*, 56 SMU L. Rev. 925 (2003; Gary Conine, *The Prudent Operator Standard: Applications Beyond the Oil and Gas Lease*, 41 Nat.Res. J. 23 (2001); Ernest Smith, *Joint Operating Agreement Jurisprudence*, 33 Washburn L.J. 834 (1994); and Hendrix and Golding, *The Standard of Care in the Operation of Oil and Gas Properties: Does the Operator Owe a Fiduciary Duty to the Nonoperator?* 44 Sw. Legal Fdn. Oil & Gas Inst. ch. 10 (1993).

C. COMMUNITY LEASE
FONTENOT v. HUMBLE OIL & REFINING CO.

Court of Appeal of Louisiana, 1968.
210 So.2d 340, 30 O.&G.R. 551.

SAVOY, JUDGE. The instant case was consolidated for purposes of trial with that of Landreneau et al. v. Humble Oil & Refining Co. et al., bearing docket number 2329 on the docket of this Court, 210 So.2d 345, and a separate decree is being handed down this date.

This is a suit for cancellation of an oil, gas and mineral lease instituted by lessor against lessee. After a trial on the merits, the district court rendered judgment in favor of defendants and against plaintiff, rejecting her demand and dismissing her suit. Plaintiff has appealed.

The oil, gas and mineral lease sought to be cancelled was executed on January 3, 1962, by Agnes Landreneau Fontenot, Semanthe Vidrine Landreneau, Calvin Landreneau, Gibbons Landreneau, Annabelle Landreneau and Charles Landreneau, as lessor, in favor of Warren L. Brown, as lessee, and subsequently assigned to defendants herein. This lease covers three separate non-contiguous tracts of land. Mrs. Fontenot, the appellant herein, owned no interest in the tract listed as No. 1 in the lease description, but owned a one-half interest in the tracts designated Nos. 2 and 3. Mrs. Fontenot prayed for cancellation of the lease as to all three tracts, and alternatively, for cancellation as to her interest in tracts 2 and 3 only. Appellants, under docket number 2329 being referred to as the heirs of Adraste Landreneau, are the owners of all of tract 1 and a one-half interest in tracts 2 and 3, and they prayed for cancellation of the lease as to all tracts. The basic facts have been agreed upon by stipulation. Thus, with minor exceptions, the questions before the Court are of law rather than fact.

On December 9, 1962, lessees commenced drilling a well which was completed as a shut-in gas well on February 9, 1963. In accordance with the terms of the lease, within ninety days from the date on which the well was shut in, lessees deposited the correct amount of shut-in royalty payment, to the credit of all lessors, in the bank designated by the lease, sufficient to maintain the lease until its next anniversary date. Two well tests were made. One in February and the other in May of 1963. Since the lease did not indicate the interest owned by lessors in the several tracts, a letter was obtained by lessees from Mrs. Fontenot and her husband stating that they owned no interest in tract 1 on which the well is situated, and did not intend, by joining in the execution of the lease, as lessor, to communitize or pool their share of the royalties payable with the other parties executing said lease as lessor. Based on this letter, lessees paid the production royalties from the two tests to the owners of tract 1 individually. A conservation unit was created effective August 1, 1963, and production began on September 4, 1963, and has continued since without interruption.

Mrs. Fontenot's primary contention is that the lease is not a joint or community lease, and production from tract 1 in which she owned no interest, did not maintain the lease as to her interest in tracts 2 and 3, therefore, the lease terminated by the failure to pay the annual rental due on January 3, 1963, and in support of this position, plaintiff cites several inapposite cases as authority for her position.

Defendants argue that although the letter from plaintiff and her husband may show that a joint community lease was not contemplated as between lessors, the lease was a joint or community lease as between lessor and lessee, under the terms of the lease and the law of this State.

The primary issue is whether the lease is a joint or community lease as between the lessee and lessor.

The lease is on a standard lease form commonly used in Southwest

Louisiana. It declares that the lessors are to be "herein called lessor (whether one or more)". It states that "lessor" for a stated consideration leases to lessee the land described as follows, to-wit: "132.8 acres in three tracts", after which the three tracts are described in detail. It further provides that on or before one year from date, the lease shall terminate unless lessee "commences operations for the drilling of A well on the land", or pay rental "for all" or part "of the land" which lessee elects to continue to hold. The lease does not specify the interest owned by the various parties.

The law is well settled that where several lessors, owning different interests in separate tracts, join in a single lease which describes all of their property and gives a total acreage figure for all of the different tracts, the lease is considered a joint lease as between lessee and lessor. . . ..

Plaintiff has apparently confused the joint lease as between lessors, and the joint lease as between lessor and lessee.

The . . . cases make it clear that regardless of the severabilities of the lease as between the lessors, it is joint "quoad the lessee" and those holding under him.

The letter signed by Mrs. Fontenot, referred to above, related only to the severability of the lease as between lessors. It did not concern or affect the joint or community status of the lease as between lessor and lessee. . . .

. . .

We are of the opinion that the subject lease was a joint lease as between lessor and lessee, and, therefore, payment of the rental on January 3, 1963, was not required since plaintiff's interest in tracts 2 and 3 was held by production from tract 1. And, since the lease was not joint or communitized between lessors, no royalty was due plaintiff from tract 1 since she disclaimed any interest therein and owned no interest therein.

The fact that the various tracts described in a lease are contiguous or non-contiguous creates no distinction in the jurisprudence in concluding that the lease is joint quoad lessee.

Plaintiff argues that the trial court erred in failing to consider the parol evidence showing that the lease was not a joint lease. In support of this argument plaintiff cites only authorities dealing with the admissibility of parol evidence relating to the severability of the lease only as between lessors. This Court has concluded that the language of the lease in this case clearly establishes that it is a joint lease as to the lessee. Plaintiff has failed to produce any authority permitting parol evidence to establish the severability of a lease as between lessor and lessee, where the language of the contract clearly indicates that the lease is a joint or community lease as between lessor and lessee.

. . .

This Court concludes that the record is clear that defendants have complied with all the terms of the lease of January 3, 1962, have made all payments called for correctly and timely, have operated the property diligently and in good faith, and that no grounds exist for cancellation of the lease.

For the reasons assigned, the judgment of the trial court is affirmed at appellant's cost.

Affirmed.

NOTES

1. The preceding case illustrates the principal issue that arises in connection with community leases. Consider the following from Kramer & Martin, *The Law of Pooling and Unitization*, § 7.03[1][a].:

> A community lease is generally defined as: "a single lease covering two or more tracts executed by the separate owners as if they were joint owners."
>
> The major legal issue that the courts have struggled with in interpreting community leases relates to the apportionment or non-apportionment of royalties that follow from production on one tract covered by the lease. In other words, the courts have had to determine whether the execution of a community lease is in effect a permanent decision to pool or unitize the lands covered by the lease for the life of the lease. This issue is sometimes discussed as whether the mere execution of the community lease creates a pooling of interests as a matter of law, or is merely an indication of an intent by the lessors to pool their interests, which can be rebutted through the admission of extrinsic evidence.
>
> In general, there have been three approaches to resolving this principal issue. The first approach, which is found in the Texas jurisprudence, suggests that merely entering into a community lease will constitute, as a matter of law, a pooling of the respective interests, unless there is contained within the lease itself express language clearly showing a contrary intent. Under this approach, extrinsic evidence could not be used to show a contrary intent on behalf of the lessors when a community lease had been executed.
>
> A second approach treats the community lease merely as evidence of an intent to pool, which gives rise to a presumption of pooling. Under this approach extrinsic evidence is admissible to determine the true intent of the lessors, and the court is free to look at matters outside the express language of the community lease. California, Louisiana, and Oklahoma with some minor differences have followed this second approach. The third approach in effect treats the community lease as a nullity by suggesting that its execution has no bearing on whether or not the interests should be pooled.
>
> Under all three approaches the ultimate test is determining the lessors' intent. The differences merely relate to the weight given the community lease document and the admissibility of extrinsic evidence to determine the true intent of the lessors. If the lessors of the segregated tracts are held to have intended to pool their interests, the royalties will be apportioned, even if the state otherwise follows the non-apportionment rule.

In determining the intent of the lessors should it be of consequence whether the property description of the lease coverage identifies the individual tracts separately? *See* Leonard v. Barnes, 75 N.M. 331, 404 P.2d 292, 23 O.&G.R. 944 (1965).

2. Among the other issues raised by the community lease are the following:

(a) Does the holder of an executive mineral interest have the power (in the absence of an express authorization) to pool nonexecutive interests by entering into a community lease? *Compare* Minchen v. Fields, 330 S.W.2d 683, 12 O.&G.R. 111 (Tex.Civ.App.1959), *aff'd in part* and *rev'd in part*, 162 Tex. 73, 345 S.W.2d 282, 14 O.&G.R. 266 (1961), *with* Le Blanc v. Haynesville Mercantile Co., 230 La. 299, 88 So.2d 377, 6 O.&G.R. 443 (1956). *Cf.* Mathews v. Sun Oil Co., 425 S.W.2d 330, 28 O.&G.R. 457 (Tex.1968).

(b) If the lessee surrenders the leasehold as it relates to a single portion of the original tract, does the former lessor of that portion remain entitled to a share of royalties from the tract that continues subject to lease? *Compare* Clark v. Elsinore Oil Co., 138 Cal.App. 6, 31 P.2d 476 (1934), *with* Duffy v. Callaway, 309 S.W.2d 853, 8 O.&G.R. 1274 (Tex.Civ.App.1958, error ref'd).

3. Texas law on community leases is seen in PARKER V. PARKER, 144 S.W.2d 303 (Tex. Civ. App. 1940, writ ref'd), and FRENCH V. GEORGE, 159 S.W.2d 566 (Tex. Civ. App. 1942, writ ref'd). In *Parker*, the owners of two separate tracts joined in the signing of a single lease covering both tracts. They described the tracts as a single tract, referred to all of the lessors as "lessor," and further provided that all royalty payments were to be made to the "lessor." Production was secured on one of the tracts and the owners on the non-producing tract sought a declaratory judgment that they were entitled to a proportional share of the royalties. The defendant owners relied on the non-apportionment rule and claimed that the lease did not cause a pooling of the two separate interests. The court's opinion concluded that the royalties should be paid to all of the lessors on an apportioned basis regardless of where the production was secured. In *French* the issue was the same, and the language of the lease was similar to that in *Parker*. The result was the same: The community lease pooled the interests as a matter of law in the proportion that the acreage covered by the respective interests bore to the acreage covered by the community lease. The holding was supported by additional leasehold language that waived the necessity of drilling offset wells to prevent drainage between the previously segregated tracts and that provided that one well producing gas would constitute full development on each 160 acres of land covered by the community lease. *Parker* and *French* were relied on in SABRE OIL & GAS CORP. V. GIBSON, 72 S.W.3d 812, 157 O.&G.R. 134 (Tex. App. 2002). The court recognized that under these decisions the separate tracts under the community lease were to be treated as effectively pooled and a single lease. Although the lessors had subsequently signed a stipulation to counteract the effects of a community lease on royalties, the lessee was not a party to the

agreement and thus was not subject to its terms. Accordingly, for purposes of lease maintenance, the lessee did not need to treat the separate tracts as separate leases.

D. OTHER PROBLEMS OF JOINDER OF OWNERS OF NON–OPERATING INTERESTS

(*See* Martin & Kramer, *Williams and Meyers Oil and Gas Law* §§ 339.3, 925–925.6).

MONTGOMERY v. RITTERSBACHER

Supreme Court of Texas, 1968.
424 S.W.2d 210, 27 O.&G.R. 774.

SMITH, JUSTICE. Petitioner, W.R. Montgomery, brought this suit to establish his right to accumulated and prospective royalty under an oil, gas and mineral lease. The material facts are undisputed. The trial court in a non-jury trial rendered judgment that Montgomery take nothing. The Court of Civil Appeals affirmed. 410 S.W.2d 925. We reverse the judgments of the courts below and remand the cause to the trial court with instructions.

In 1945 Montgomery conveyed approximately eighty (80) acres of land, designated in the record as "First Tract," to Respondents' predecessors in title but reserved for himself a non-participating royalty interest.[91] The title to the property, including the royalty interest not reserved by Montgomery, became vested in Respondents, who also owned land hereinafter designated as "Second Tract" which is contiguous to "First Tract." Montgomery owned no interest in this "Second Tract." Respondents were the owners of the executive rights under "First Tract" and "Second Tract" as well as some of the royalty rights under "Second Tract" and the royalty rights not owned by Petitioner under "First Tract."

In 1953 Respondents filed for record an oil, gas, and mineral lease dated September 4, 1951, covering both "First Tract" [80 acres] and "Second Tract." [124.19 acres]. Since Respondents held the executive rights of "First Tract," it was not necessary for Montgomery to join, nor did he, in the lease executed by Respondents. The lease contained a pooling clause whereby lessee was permitted to combine the leasehold estate with any other mineral estate in order to create

---

[91] The interest Montgomery reserved is as follows: "Out of the grant hereby made there is, however, excepted and reserved unto the Grantor herein, his heirs and assigns, forever, the equal one-half part of the usual royalty of one-eighth of all the oil, gas and other minerals on, in and under the premises hereby conveyed, such one-half part to be paid and delivered under and by the terms of the present or any future oil, gas and mineral lease or leases outstanding on said premises. But Grantor, his heirs and assigns, shall not participate in, and no reservation is here made of the lease or annual rentals and/or bonus money received by Grantee, his heirs and assigns for any future lease or leases given on said premises, and it is understood and agreed that it shall not be necessary for said Grantor, his heirs and assigns to join in any such lease or leases so made."

appropriate operating units. The lease further provided that production on any tract of land within any unitized area formed under the pooling provision should constitute full compliance with the development, drilling, and producing obligations expressed in the lease. Additionally, the lease contained the following entirety clause:

> If the leased premises are now or shall hereafter be owned in severalty or in separate tracts, the premises, nevertheless, shall be developed and operated as one lease, and all royalties accruing hereunder shall be treated as an entirety and shall be divided among and paid to such separate owners in the proportion that the acreage owned by each such separate owner bears to the entire leased acreage.

The lessee, Sun Oil Company, formed several units out of the original leased acreage by combining some of the land under the lease with land Sun held under other leases. A portion of "Second Tract" was unitized with a tract known as the Crutchfield tract, land not owned by Respondents and not covered by the lease. [Eighty acres of "Second Tract" were placed in the Crutchfield unit—making a total of 320 acres in that unit. The remainder of "Second Tract," 44.19 acres, and "First Tract" were not placed in the Crutchfield unit.] A producing well was completed on the Crutchfield tract in October, 1956, from which commercial production was begun in May, 1958. "First Tract" was placed in a unit on which a dry hole was drilled in July, 1961.

Montgomery brought this suit in May, 1964, against Sun Oil Company, the lessee, and Respondents, the holders of the executive rights, claiming by virtue of the entirety clause a share of the royalties which were accruing under the lease. Since "Second Tract" is in the Crutchfield pool, it receives a proportionate share of the royalties that are produced from the Crutchfield well. This royalty, according to the terms of the entirety clause in the lease, is "divided among and paid to such separate owners in the proportion that the acreage owned by each such separate owner bears to the entire leased acreage." Consequently, Montgomery seeks the proportion of the royalties accruing under the lease that his non-participating interest bears to the leased acreage. He contends that the entirety clause in the lease by its express terms applies to and includes his non-participating royalty interest; and since he has ratified the lease, Respondents cannot now deny the effect of the contract entered into between Respondents and Sun.

Respondents contend and the Court of Civil Appeals has held that Respondents did not have the power to bind Montgomery's non-participating royalty interest with an entirety clause; Montgomery did not properly ratify the lease; and, Montgomery did not qualify to recover as a third party beneficiary of the lease. We agree that the Respondents, the holders of the executive rights, did not have the power to bind Montgomery's non-participating royalty interest by virtue of the entirety clause alone, but do not agree with their contention that the Petitioner has not "properly" ratified the lease. We are of the opinion that the enlargement or diminishment of the rights of a prior non-participating royalty owner can be accomplished by the holder of the executive rights executing an oil, gas, and mineral lease which includes either a pooling clause or an entirety clause, provided the non-participating owner ratifies such action.

Entirety Clause

This Court has held that pooling effects a cross-conveyance among the owners of minerals under the various tracts of royalty or minerals in a pool so that they all own undivided interests under the unitized tract in the proportion their contribution bears to the unitized tract. Veal v. Thomason, 138 Tex. 341, 159 S.W.2d 472 (1942). The mere reservation of a non-participating royalty interest under a tract does not show that the royalty owner intended to give to the holder of the executive rights the power to diminish the royalty owner's interest under that tract. Consequently, pooling on the part of the holder of the executive rights cannot be binding upon the non-participating royalty owner in the absence of his consent. Minchen v. Fields, 162 Tex. 73, 345 S.W.2d 282 (1961); Brown v. Smith, 141 Tex. 425, 174 S.W.2d 43 (1943); and Nugent v. Freeman, 306 S.W.2d 167 (Tex.Civ.App.—Eastland 1957, writ ref'd n.r.e.), cited with approval in Minchen v. Fields, *supra.* We can see no distinction between the pooling clause, insofar as it has the effect of changing the aggregate ownership of the non-participating royalty owner, and the entirety clause, which, in effect, would allow the holder of the executive rights to either diminish or enlarge the ownership of that of the royalty owner. In either case, the consent of the owner must be obtained.

Respondents argue that although the lease in question covered Montgomery's non-participating interest, the entirety clause in the lease did not. They contend that since they did not have the power to diminish Montgomery's interest in "First Tract" and thereby to obtain a proportional share of any royalties accruing to that tract, they did not intend for Montgomery to use the entirety clause to share proportionally in royalties accruing to their tracts. Furthermore, Respondents claim that one of the reasons that they inserted a proportional reduction clause in the lease was to make the entirety clause operative only on those interests which they had authority to cover by the entirety clause.

We are unable to agree with Respondents' contention. The lease executed by Respondents and the original lessee explicitly described the entire tract in which Montgomery had a non-participating interest as being covered by the lease. The unambiguous entirety clause clearly indicates that it was to apply to all the interests covered by the lease. The clause points out that even if the premises are owned in severalty *at the time of the execution of the lease,* as the premises were in this case, "the [leased] premises, nevertheless shall be developed and operated *as one lease,* and *all royalties* accruing hereunder shall be treated as an entirety and shall be divided among and *paid to such separate owners in the proportion that the acreage owned by each bears to the entire leased acreage.*" (Emphasis added.) This Court has held that an "are now" entirety clause as was contained in the present lease applies to minerals held in severalty at the time of the execution of the lease. . . .

Respondents, in exercising the executive rights, had a duty to protect the non-participating royalty owner. . . . Had Respondents not intended to include Montgomery's non-participating interest within the provisions of the entirety clause, they could have easily taken affirmative steps to exclude the interest from the operations of the clause. The insertion of a proportionate reduction clause in the lease does not indicate an intention to exclude "First Tract" from the express terms of the entirety clause. The lessee, in this case, operated as if "First Tract" was covered by all of the terms contained in the lease. "First Tract" was placed in a unit and a well was drilled on that unit, even though the lessor could not pool the non-participating royalty interest. Therefore, we view it as conclusively established that in executing the lease in question, Respondents purported to bind Montgomery's interest by the entirety clause—whether or not they had the authority to do so.

Ratification

This suit was filed on May 12, 1964. Prior to the filing of suit, Petitioner claimed and made demand upon Sun Oil Company for his pro-rata share of royalties payable under the terms of the lease. The parties stipulated that, "W.R. Montgomery by his attorney informed a representative of Sun Oil Company during the last week of January, 1959, that the said W.R. Montgomery was offering to sign a ratification of the Charles E. Rittersbacher, *et al.,* lease to P.V. Hitt, dated September 4, 1951[the lease in controversy] . . .." The parties agreed that the affidavit of Montgomery would be acceptable in lieu of requiring him to appear in person as a witness. In the affidavit, Montgomery stated: "I am and have been willing to ratify the said lease and have offered and agreed with Bettis & Shepherd, Sun Oil Company's predecessor in title, as well as Sun Oil Company, to execute any division order or ratification necessary or proper with regard to any royalty interest, if such be needed." Respondents contend that this evidence does not establish actual ratification, but only amounts to an offer to make the ratification—the contention being that an offer, absent acceptance, does not meet the requirement that there must be actual ratification. Respondents further take the position that Montgomery's offer to ratify was conditional and was never fully binding upon Montgomery. This evidence demonstrates Montgomery's intention to ratify the lease, and by filing suit to enforce the lease as written, Montgomery, as a matter of law, has exercised his option to ratify the lease. We think that the manner in which he has exercised his option is analogous to the manner by which a principal can ratify the unauthorized actions of an agent—bringing a suit to enforce the unauthorized act. In such a situation it has been held that the bringing of the suit constitutes an implied ratification of the unauthorized act. . . .

This Court has never been called upon to decide the question of whether a holder of non-participating royalty has an option to make an entirety clause operative on his interest. We think that the non-participating royalty owner, so far as the existence of an option is concerned, occupies a comparable position to that of a cotenant under a lease made by his cotenant or a non-participating royalty owner under a pooling agreement made by the holder of the executive rights. As to the cotenant, it has been held that he has the right to ratify or repudiate a lease made by his cotenant which covers his interest. . . . Likewise, in the pooling area, if a non-participating royalty owner ratifies a pooling agreement, either by joining in the execution of the agreement or by accepting royalties from the pool, his interest is bound by the pooling agreement. . . . Therefore, we hold that the non-participating royalty owner has the option to ratify or repudiate a lease containing provisions which as to his interest the holder of the executive rights had no authority to insert in the lease.

Montgomery, in bringing this suit, seeks two things under the lease— royalties that have already accrued and royalties that are to accrue in the future. We have held that Montgomery has ratified the lease in question by filing suit; consequently, he is only entitled to receive royalties accruing from and after May 12, 1964, the date this suit was filed. In this connection, we point out that Montgomery, having thus ratified the lease, is as much bound thereby as if he had joined in the original execution thereof. As long as the lease is in force, he is not free to claim his full 1/2 non-participating interest under "First Tract."

We come now to the judgment to be entered. The cause must be reversed and remanded to the trial court with instructions. . . . Under our holding, Montgomery will not be entitled to royalty accruing prior to May 12, 1964, the date of filing of suit. The amount of money due Montgomery accruing under the lease subsequent to May 12, 1964, is to be determined in accordance with . . . this opinion.

Reversed and remanded to the trial court with instructions. All costs are adjudged against the Respondents.

CALVERT, CHIEF JUSTICE (dissenting). I join in the dissenting opinion filed by JUSTICE WALKER. I append the following comment.

The holding of the court is that the filing of suit by Montgomery on May 12, 1964, constituted a ratification of the lease. The parties stipulated that a well was completed as a producer on the Crutchfield Unit on October 9, 1956, and that shut-in royalty was paid from that date until May, 1958, when actual production was begun. The record reflects that Montgomery knew as early as July, 1957, that the well had been completed. The record thus reflects that Montgomery waited nearly seven years before he ratified the lease.

By agreeing that ratification has been effected by the judgment herein, I do not wish to be understood as agreeing that a non-participating royalty owner, with full knowledge of his rights, cannot lose his right to ratify through laches. *See* Nugent v. Freeman, 306 S.W.2d 167 (Tex.Civ.App.—Eastland 1957, n.r.e.). There was no plea of laches in this case.

GRIFFIN, JUSTICE (dissenting). I respectfully dissent. The entirety clause in a lease was never intended to convey and does not convey any interest owned by any land owner or mineral owner in any tract of land in any lease. It merely provides for each person to receive such part of the common production as was the ownership of the one who receives in his original tract of land or minerals.

WALKER, JUSTICE (dissenting). In my opinion petitioner has heretofore done nothing that would irrevocably bind him to the terms of the lease, and he will not be bound thereby until his tender of ratification is made effective by the judgment rendered in this case. I would hold that he is entitled to his proportionate share of royalties accruing from and after the date of judgment.

NOTES

1. Who is able to ratify a pooling agreement? Consider the ruling in FLETCHER v. RICKS EXPLORATION, 905 F.2d 890, 112 O.&G.R. 318 (5th Cir.1990). Ricks had completed a producing well on a tract it controlled. Fletcher was the lessee of an undivided 1/2 mineral interest in a nearby 29.675 acre tract. Ricks and several other lessees filed a unit designation for the well which pooled the royalty and working interests of those who either signed the agreement or ratified it. The pooled area included the 29.675 acre tract which Fletcher had leased. The lessees of the other 1/2 mineral interest were parties to the pooling agreement but Fletcher was never informed about the agreement. Fletcher eventually learned about the agreement and attempted to ratify by letter. The Railroad Commission had in the meantime determined that the 29.675 acre tract was nonproductive as to gas. Proceeds from production were distributed to all parties except for Fletcher, including those who owned working interests in the 1/2 mineral estate underlying the nonproductive tract. Fletcher argued that the filing of the unit designation or pooling agreement which included his 29.675 acre was a firm offer to join the agreement. The court did not treat the filing of the unit designation as an offer to all owners within its described area to join. Fletcher argued that under *Montgomery v. Rittersbacher* it was in the same position as a non-executive owner to a lease with a pooling clause. But the Fifth Circuit concluded that Fletcher's status as a lessee was not analogous to the status of an unleased owner and therefore refused to extend the ratification principle of *Montgomery* to a lessee. The court stated:

> The district court held that Texas law would not allow ratification of a pooling provision by the owner of an unproductive working interest. In doing so, it declined to extend the reasoning of the "cotenant" cases to cases involving "co-lessees" and refused to create a remedy heretofore unrecognized in the laws of Texas (or, perhaps, the laws of any other state, as a want of authority in the briefs indicates). The district court was correct. There is no reason to believe that the Texas Supreme Court would decide the issue any other way. The parties give us none, and we have found none.
>
> Finally, Fletcher does have (or, at any rate, had) other remedies available,

ranging from self help—he could drill his own well—to proceeding under the Mineral Interest Pooling Act.

Simply stated, Fletcher wants something for what may have once appeared to be something but is now known to be nothing. We cannot oblige him.

2. Ruiz v. Martin, 559 S.W.2d 839, 60 O.&G.R. 128 (Tex.Civ.App.1977, error ref'd n.r.e.). A, holder of entire fee in one tract and of fee subject to B's royalty interest in an adjacent tract, made a lease of both tracts. The lease contained a pooling clause but not an entirety clause. The lessee completed a gas well on the unencumbered tract. B promptly filed a written ratification of the lease. The execution of the lease was treated by the court as an offer to unitize the tract, accepted by the royalty owner's ratification. B was therefore entitled to a proportionate share of royalty from the gas well.

Would the same result follow if the lease provided that "if the lease now or hereafter covers separate tracts [defined to mean any tract with a differing royalty ownership, such as B's], no pooling or unitization of royalty interest as between such separate tracts is intended or shall be implied or result merely from the inclusion of such separate tracts within this lease." *See* Verble v. Coffman, 680 S.W.2d 69, 83 O.&G.R. 405 (Tex.App.1984).

Compare Brown v. Getty Reserve Oil, Inc., 626 S.W.2d 810, 72 O.&G.R. 588 (Tex.App.1981). The owner of executive mineral interest in Sections 27 and 29 (two 320–acre or "Spanish sections") leases both in a single lease. A well is brought in on Section 29. Under the Texas rule that without express permission an executive is not authorized to unitize (or authorize unitization by a lessee), *held,* owners of royalty in Section 29 are entitled to their share without any reduction for the lease's coverage of Section 27. Royalty owners in Section 27 presumably could secure a proportionate royalty in the Section 29 by timely ratification of the lease.

If one portion of a tract in which **A** holds a royalty interest is placed into one unit and another portion of the tract is placed in another, may **A** ratify one unit but not the other? Answering affirmatively is MCZ, Inc. v. Triolo, 708 S.W.2d 49, 91 O.&G.R. 389 (Tex.App.1986).

3. In relation to the power of the executive to pool or unitize premises subject to a non-executive interest, consider the following:

(a) Brown v. Smith, 141 Tex. 425, 174 S.W.2d 43 (1943). The court held that specific performance of a contract to purchase an oil and gas lease must be denied in an action by the contracting vendor inasmuch as the lease provided for pooling and the lessors had no power to pool the land, which was subject to an outstanding 1/32 royalty interest in a 20–acre portion of the leased premises. The court commented as follows:

> From the conclusion that Mrs. Lee's royalty interest in the 20–acre tract, and her rights incident thereto, would not be affected by the pooling agreement in the lease executed by Floyd and Ector Smith it follows, in our opinion,

that petitioners were justified in disapproving the title when respondents failed to satisfy the requirement that Mrs. Lee join in the execution and delivery of the lease. Petitioners' contract was for the execution and delivery to them of an oil and gas lease under which the royalties from all of the land, although the lease included two tracts separately owned, would be pooled and paid in proportion to the acreage owned. The lease which this suit seeks to force petitioners to accept is materially a different lease. Under it, Mrs. Lee would be entitled to a one-thirty-second royalty, one-fourth of the one-eighth royalty, in all of the oil, gas or other minerals produced from the 20–acre tract and to none of the royalty from the remaining 42.75 acres. The pooling agreement contained in the lease would affect only the royalty interests reserved in the lease by Floyd Smith and Ector Smith.

. . . When a pooling agreement which binds all of the royalty owners has been made, the lessee has greater freedom in the selection of locations for the drilling of wells on the land covered by the lease, for each of the land owners or royalty owners is entitled to his share of the royalties from all wells drilled anywhere on the land, instead of being entitled to receive only royalties arising out of the production from a certain part of the leased land. Controversies and probable liability on account of drainage are avoided. For example, would Mrs. Lee, not having joined in the lease, acquiesce in the drilling and operation of wells on the 42.75 acre tract that would drain oil from the 20–acre tract in which she owns the royalty interest?

Petitioners would suffer other disadvantages from the acceptance of a lease in which Mrs. Lee did not join. Among them may be mentioned the necessity for the construction and maintenance of separate measuring tanks on each of the two tracts for the calculation of Mrs. Lee's one-fourth of the royalties of oil and gas produced from the 20 acres and the keeping of separate books and records. In many respects the burdens and obligations of petitioners under the lease tendered by respondents would be the same as they would be under two separate leases, one affecting the 20–acre tract and the other affecting the 42.75–acre tract. In short, the lease tendered was, in a substantial particular, important to the lessee, not the lease that petitioners contracted to acquire.

(b) NUGENT v. FREEMAN, 306 S.W.2d 167, 8 O.&G.R. 40 (Tex.Civ.App.1957, error ref'd n.r.e.). *R* owned royalty in 30 acres and *E,* the executive of such land, executed a lease covering this 30 acres plus 120 adjoining acres owned by *E.* It was held that pooling did not occur because *E* had no power to pool the 30 acres, and *R* did not share in production on the 120 acres but off the 30 acres. What remedy, if any, would *R* have against *E* or the Lessee?

(c) MATHEWS v. SUN OIL CO., 425 S.W.2d 330, 28 O.&G.R. 457 (Tex.1968). The owner of executive rights in Sections 4 and 13 executed a lease covering both tracts although the non-executive interests in the two tracts were owned by different persons. The court held that production from Section 13 preserved the lease into the secondary term as to both Section 4 and Section 13, declaring:

> No pooling problem is here involved; nor is it contended that the lease was invalid when executed; nor that such lease should be cancelled in whole or in part because the operator has failed to reasonably develop the mineral potentials of the land under the lease; . . .; nor is it asserted that Mathews [the executive] has breached a duty which he as the owner of the executive rights owed to the holders of the non-participating royalty interests; . . .. It is simply argued that Brown v. Smith by analogy supports the proposition that the lease should be considered as two leases, one covering Section 13 and one covering Section 4, and that no production having been obtained from Section 4, the lease as to that tract has terminated.

> Because a lease will be considered as two leases for certain purposes, it does not follow that a single lease will be considered as two leases for all purposes whenever two or more tracts of land and diverse royalty interests are involved. It is a rule of general application that in the absence of anything in the lease to indicate a contrary intent, production on one tract will operate to perpetuate the lease as to all tracts described therein and covered thereby. . . ..

. . .

> The distinction between the inclusion of two tracts in one lease by the holder of a leasing power and the pooling of royalty interests is that the execution of a lease is an authorized act, while pooling and the cross-conveying of royalty interests is an unauthorized act.

(d) LeBlanc v. Haynesville Mercantile Co., 230 La. 299, 88 So.2d 377, 6 O.&G.R. 443 (1956). The court held that the owner of a mineral servitude, subject to an outstanding royalty, has the power to pool the land, declaring that:

> It is also well settled that the right to search and explore, which belongs to the owner of the servitude, is not given to the royalty owner; that the latter must await such time as the land has been developed, and his right is restricted to a sharing in production if and when it is obtained by the landowner or a lessee . . .. It follows that the defendant was not a necessary party to the lease . . ., the landowners having full power to enter into any lease contract they saw fit affecting the property—and that would include the power to grant a lessee the authority to pool and combine the leased acreage or any portion thereof with any lands or leases and mineral interests in the immediate vicinity—subject only to the right of the royalty owner to receive its 1/64th of the oil, gas or other minerals allocated to the acreage included in the unit.

4. Apart from pooling provisions, what other restrictions affect the executive's leasing power? May he include a shut-in royalty clause in the lease? A continuous drilling operations clause? A clause authorizing cessation of drilling

or production under specified conditions, *e.g.,* where the price of the products is less than a specified amount? A dissenting opinion by Hamilton, J., joined by two other justices, in Archer County v. Webb, 161 Tex. 210, 338 S.W.2d 435, 13 O.&G.R. 280 (1960), observed:

> With the single exception of the pooling cases, reversioners, under the majority opinion, can execute leases which define any set of circumstances as production and the effect of such lease provisions will be to extend the oil and gas lease but not to extend term interests.

5. For a voluntary pooling or unitization agreement affecting Blackacre, Whiteacre and Greenacre to be fully effective, is it necessary to secure the joinder of the following persons:

(a) The owner of an unleased undivided 1/32nd mineral interest in Blackacre?

(b) The owner of a 1/32nd royalty interest in Whiteacre?

(c) The owner of an oil payment of $100,000 payable out of 1/8 of 7/8ths of the production from Greenacre, which oil payment was reserved by the original lessee in a conveyance of the leasehold to an assignee?

(d) The owner of a mortgage executed by the owner of Blackacre prior to its lease?

(e) A judgment creditor of the lessor of Whiteacre, whose judgment was recorded in the county in which Whiteacre is located after the execution but before the recordation of the lease of Whiteacre?

(f) The surface owner of Greenacre who has no interest in the minerals?

E. COMPULSORY POOLING AND UNITIZATION

State of Louisiana
Office of Conservation
<u>Baton Rouge, Louisiana</u>
March 17, 1978
Order No. 1027

Order concerning the adoption of rules and regulations and the creation of four (4) drilling and production units for the 16,400' Tuscaloosa Sand, Reservoir A, in the PORT HUDSON FIELD, East Baton Rouge Parish, Louisiana.

. . .

Pursuant to power delegated under the laws of the State of Louisiana, and particularly Title 30 of Louisiana Revised Statutes of 1950, and after a public hearing held under Docket No. 78–166 in Baton Rouge, Louisiana, on March 14, 1978, upon the application of Amoco Production Company, following legal publication of notice and notice in accordance with rules prescribed by the

Commissioner of Conservation, the following order is issued and promulgated by the Commissioner of Conservation as being reasonably necessary to conserve the oil and gas resources of the State, to prevent waste as defined by law, to avoid the drilling of unnecessary wells and, otherwise, to carry out the provisions of the laws of this State.

DEFINITION

The 16,400' Tuscaloosa Sand in the Port Hudson Field, East Baton Rouge Parish, Louisiana, is identified as that gas and condensate bearing sand encountered in the interval between 16,418 feet and 16,793 feet in the Amoco Production Company—Georgia Pacific Corporation No. 1 Well located in Section 64, Township 4 South, Range 2 West.

FINDINGS

The Commissioner of Conservation finds as follows:

1. That the adoption of rules and regulations and the creation of four (4) drilling and production units for the 16,400' Tuscaloosa Sand, Reservoir A, in the Port Hudson Field, East Baton Rouge Parish, Louisiana, are necessary to insure orderly development, to prevent waste and to avoid the drilling of unnecessary wells.

2. That the available geological and engineering data indicate that the units outlined on the plat labeled "Amoco Production Company Exhibit No. 4 for Docket No. 78–166," a copy of which is attached hereto, are reasonable and should be adopted; that each of the said units can be efficiently and economically drained by one well located on each unit; and that creation of said units should reasonably assure to each separate tract its just and equitable share of the recoverable hydrocarbons in the reservoir.

3. That all separately owned tracts, mineral leases and other property interests within each unit created herein should be force pooled and integrated with each separate tract sharing in unit production on a surface acre basis of participation.

4. That Amoco Production Company should be designated as operator of each of the units created herein.

5. That the wells shown on the unit plat, reference Finding No. 2 above, should be designated as the unit well for the unit on which each said well is located.

6. That any well hereafter drilled to the 16,400' Tuscaloosa Sand, Reservoir A, within or without the pattern of units herein adopted, shall not be located closer than 660 feet from any unit line nor closer than 2000 feet to any other well completed in, drilling to, or for which a permit shall have been granted to drill to, the same pool.

ORDER

NOW, THEREFORE, IT IS ORDERED THAT:

1. The units shown on the plat labeled "Amoco Production Company Exhibit No. 4 for Docket No. 78–166," a copy of which is attached hereto and made a part hereof, be and are hereby approved and adopted as drilling and production units for the exploration for and production of gas and condensate from the 16,400' Tuscaloosa Sand, Reservoir A, in the Port Hudson Field, East Baton Rouge Parish, Louisiana.

The units have not been surveyed, and when a survey plat of said units showing the exterior limits thereof, the total acreage therein, and the acreage in each separately owned tract, has been submitted to and accepted by the Commissioner of Conservation or any member of his staff, insofar as it shows the exterior limits of the units, said plat shall be substituted for the above exhibit and made a part of this order by reference. In the event of conflicting claims of ownership of acreage in the units, such acreage may be so identified on the survey plat. Such identification of acreage subject to conflicting claims shall not be construed as an acknowledgement of the validity of any such claims, and shall not affect any other acreage in separately owned tracts in the units.

The survey plat shall be prepared in accordance with the requirements for unit plats and survey plats adopted by the Commissioner of Conservation. It is recognized that the exterior boundary lines of the units, as surveyed, may differ from those lines as shown on the attached plat because of the requirement that by survey the geologically significant wells be correctly located with respect to each other and to the unit boundary lines that they control.

2. The separately owned tracts, mineral leases, and other property interests within the units established herein are hereby pooled, consolidated and integrated in accordance with Section 10, Title 30 of Louisiana Revised Statutes of 1950, with each tract sharing in unit production in the proportion that the surface area of such tract bears to the entire surface area of the unit in which it is situated. Also, all operations on and production from each unit shall be considered operations on and production from each of the separate tracts within said unit and under the terms of each of the mineral leases affecting said tracts.

3. Amoco Production Company is designated as operator of each of the units created herein.

4. The unit well for each unit is designated and located in accordance with Findings 5 and 6 hereof.

5. Any future wells drilled to the 16,400' Tuscaloosa Sand, Reservoir A, within or without the pattern of units herein adopted, shall be located in accordance with Finding No. 6 hereinabove.

6. Except as they may be in conflict herewith, the provisions of all applicable Statewide Orders shall govern the exploration for and production of gas and condensate from the 16,400' Tuscaloosa Sand, Reservoir A, in the Port Hudson Field, East Baton Rouge Parish, Louisiana.

7. When additional geological and engineering information becomes

available which would indicate a required change or revision in the unit boundaries adopted herein, or which would indicate a required change or revision of other provisions of this order, the party or parties in possession of this additional information shall petition the Commissioner of Conservation for a public hearing for the purpose of considering appropriate changes.

This Order shall be effective on and after March 14, 1978.

OFFICE OF CONSERVATION OF THE STATE OF LOUISIANA

R.T. SUTTON COMMISSIONER OF CONSERVATION

FB

Amoco Prod. Co. Exhibit No. 4 for Docket No. 78–166—Attached

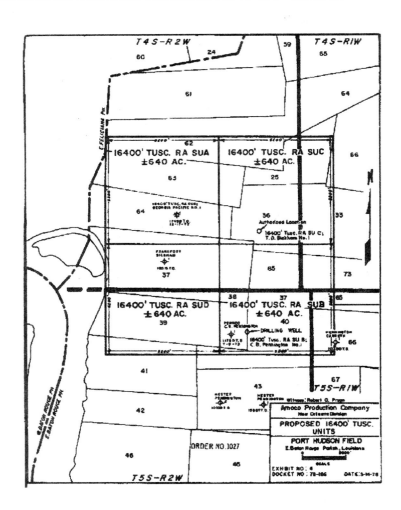

NOTES

1. The term *pooling* generally means the integration of interests in separately owned tracts for an area where a single well has been, or will be, drilled such that drilling and production costs are shared among the working-interest owners, and production is shared by all owners of rights in the mineral estate. *Compulsory pooling* generally refers to the joining together of small or irregularly sized tracts for the purpose of having sufficient acreage to receive a well permit under the relevant state or local spacing laws and regulations. There may be important differences among the states in their definitions of what constitutes a drilling unit, a spacing unit, a proration unit, or a pooled unit. When the conservation agency creates a spacing or drilling unit, there is often no automatic pooling of the interests contained within the spacing or drilling unit. Instead, if the parties cannot voluntarily agree to pool all of the interests, they must apply to the conservation agency to seek a second order to pool the interests and receive the benefits of such an order. In some states, spacing units are established and pooling accomplished in the same order. Observe how this was done in separate paragraphs in both the Findings and Order sections of Order No. 1027 *supra.*

A prerequisite of most pooling provisions is that there be two or more separately owned tracts or interests located within a spacing or drilling unit. Upon application of an authorized person, the state conservation agency will give notice to interested persons, hold a hearing on the applicant's proposal, and then enter a compulsory pooling order.

Unitization, as opposed to pooling, is the consolidation of mineral or working interests covering all or a significant part of a common source of supply. Within the unitized area there may have been many spacing, drilling, or pooled units prior to the reservoir-wide consolidation of interests. Unitization is designed to maximize production by efficiently draining the entire reservoir utilizing the best engineering techniques that are economically feasible. While unitization in many cases is in the best interests of all of the parties, voluntary unitization of all the owners has been difficult to attain. As a result, the major producing states, except for Texas, have adopted a compulsory unitization statute that allows the state conservation agency to unitize minority interests who have chosen not to join voluntarily in a unit agreement. Unlike compulsory pooling statutes, which often encompass only one statutory provision, compulsory unitization statutes cover more sections and are lengthier and more complete. The compulsory unitization process is initiated by the submission to the state conservation agency of a detailed plan. In some states, compulsory unitization can be used for specific purposes only, such as recycling operations or secondary recovery operations; but the majority of states authorize compulsory unitization in order to prevent waste, increase the ultimate amount of hydrocarbons recovered, avoid the drilling of unnecessary wells, and/or protect correlative rights. The compulsory unitization statutes of most states require a specified percentage of agreement to a unit agreement before the remainder of interests may be forced into the unit.

After development on a pattern of 640 acre units as shown above under Order No. 27, the Port Hudson Field was consolidated as a Fieldwide Unit in Order No. 1027-A-9. Th prior drilling units were replaced by a single fieldwide unit of 4779.75 acres. Although the scale of the unit makes the detail too small to read easily, the unit plat is as follows:

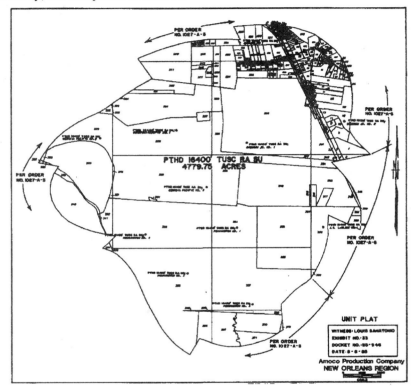

ORDER NO.   1027-A-9

2. The history of compulsory pooling begins with certain municipal zoning ordinances designed to limit drilling within the boundaries of the municipality. A number of such ordinances were enacted in the 1920's and 30's, the first of these being an ordinance enacted in Winfield, Kansas, in 1927. This was followed in the same year by the enactment of a similar ordinance in Oxford, Kansas; the latter ordinance was sustained as valid in Marrs v. City of Oxford, 24 F.2d 541 (D.Kan.1928), aff'd, 32 F.2d 134, 67 A.L.R. 1336 (8th Cir.1929), cert. denied, 280 U.S. 573 (1929). Another such ordinance was enacted in Oklahoma City, Oklahoma, in 1929, and this was followed by similar ordinances in other cities of Oklahoma and other states, *e.g.,* Texas.

These zoning ordinances were, in effect, compulsory pooling ordinances in that they limited drilling within the municipal boundaries to one well for each designated spacing unit and provided for non-drilling owners to share in the production from the well drilled for each drilling unit. On the authority of Village of Euclid, Ohio v. Ambler Realty Co., 272 U.S. 365 (1926), they were generally sustained as valid regulations under the police power, although an occasional ordinance was held invalid.

The first state compulsory pooling statutes were enacted in New Mexico and Oklahoma in 1935. In short order, most of the producing states followed the example set by those two states and enacted compulsory pooling statutes. Provision for compulsory pooling is contained in the statutes of Alabama, Alaska, Arizona, Arkansas, California, Colorado, Florida, Georgia, Idaho, Illinois, Indiana, Iowa, Kentucky, Louisiana, Maine, Michigan, Mississippi, Missouri, Montana, Nebraska, Nevada, New Mexico, New York, North Carolina, North Dakota, Ohio, Oklahoma, Oregon, Pennsylvania, South Dakota, Tennessee, Texas, Utah, Washington, West Virginia, and Wyoming. Similar statutes have been enacted in the provinces of Alberta, British Columbia, Manitoba, and Saskatchewan. Many of these statutes have been enacted in the past four decades; they evidence the growing acceptance by the industry and the legislature of the importance of this measure to prevent waste caused by excessive drilling. Still without compulsory process for pooling, however, is the important producing state of Kansas as well as several other states with lesser production. In California the pooling statutes (Calif.Pub.Res.Code §§ 3608, 3609 (West 1984)) are of very limited scope.

3. For practical purposes, the history of compulsory unitization may be said to begin in the early 1920's with Henry L. Doherty. This distinguished and far-sighted spokesman for conservation sparked much early discussion of unitization, and before many years had passed the American Petroleum Institute endorsed unit operation of oil pools. Within a short space of time, proposals for the enactment of legislation requiring unitization under certain specified circumstances came before various legislative bodies.

By Act No. 157 of 1940, § 4, provision for compulsory unitization for purposes of requiring re-cycling of gas was enacted in Louisiana, and this may be the first compulsory unitization statute. The first comprehensive statutory provision for compulsory unitization became law in Oklahoma five years later, in 1945. In 1951 a new unitization law was passed which in many respects was the same as the earlier law though containing a number of important changes. The impetus for most of the changes came from royalty owners who felt that the earlier law was too heavily weighted in favor of operators.

The Arkansas statute was enacted in 1951. Earlier, in 1948, the regulatory commission in Arkansas had ordered compulsory unitization of the McKamie–Patton field by an order which apparently met with the approval of all of the operators and 96 per cent of the royalty owners, but the order was held invalid in Dobson v. Arkansas Oil and Gas Commission, 218 Ark. 160, 235 S.W.2d 33 (1950), on the ground that the Commission was without authority to compel unitization. The legislative response was almost immediate, and a grant of authority to the Commission to compel unitization under certain specified conditions was clearly made by statute in 1951.

The Oklahoma and the Arkansas statutes, along with a model statute proposed by the Interstate Oil Compact Commission, have served as models for the several compulsory unitization statutes enacted since the midpoint of this century. Thus the Alaska and Nevada statutes are substantially identical with the Oklahoma statute, and the other unitization statutes are based in large part upon the Oklahoma, Arkansas, and model statutes.

4. When compulsory pooling is ordered by a state conservation agency, who should bear the costs of drilling the well? Should parties who are forced into a unit be required to pay the costs of a well that results in a dry hole? Should an operator be required to bear all the risks of a dry hole but then share the production of a successful well as though everyone in the unit had borne their proportionate share of costs all along? Should an operator receive some compensation for the risks it has assumed? Consider these questions in connection with the next case.

———

DAVIS OIL COMPANY v. STEAMBOAT PETROLEUM CORPORATION

Supreme Court of Louisiana, 1991.
583 So.2d 1139, 113 O.&G.R. 98.

DENNIS, JUSTICE. We are called upon to decide whether the operator of a forced drilling unit who initiated the unitization process may hold an adjacent lessee personally liable for dry hole well costs because the lessee requested the Commissioner to modify the proposed unit to include part of the lessee's leasehold. In the present case, in response to the adjacent lessee's counter-proposal, the Commissioner included parts of its leaseholds in two proposed units. Consequently, part of the land under the adjacent lessee's leases comprised 4.6859% of the acreage of one unit and 2.58931% of the other. After drilling a dry hole in each of the two units, the operator sought to hold the lessee personally liable for a proportionate share of the well costs, but the trial court ruled that such costs could not be recovered except out of production. The court of appeal reversed, holding that the lessee was personally liable because it took an active role in the unitization proceedings by submitting and arguing in support of its counterplan for unitization. Davis Oil Co. v. Steamboat Petroleum Co., 570 So.2d 495 (La.App. 5th Cir.1990). We reverse. A non-operating owner of a mineral interest, who does not consent to operations within a compulsory drilling unit by an operating owner, has no liability for the costs of development and operations except out of his share of production. Under the circumstances of the present case, in which the non-operating lessee merely introduced a counter-proposed unit plan at the Commissioner's hearing, prior to the drilling of the dry holes, only as a precaution against the uncompensated drainage of part of the land underlying its leases, the lessee did not consent to the unit operations.

Facts

Plaintiff, Davis Oil Company, and defendant, Steamboat Petroleum Corporation, held separate mineral leases on adjacent property. On March 5, 1984 Davis Oil notified the Commissioner of Conservation and all interested parties, including defendant, of its intent to request a hearing before the Commissioner to consider Davis Oil's proposed unitization plan which would create several drilling and production units in the East Manchester Field located in Calcasieu Parish. The geographic units proposed by Davis Oil did not include two adjacent tracts (the Mott tract and the Richard tract) leased by Steamboat. Steamboat filed a counterplan in which it opposed Davis Oil's plan and urged the Commissioner to adopt Steamboat's unitization plan which included parts of the tracts leased by Steamboat. A public hearing was held on May 29, 1984 at which Davis Oil and Steamboat each appeared and urged the Commissioner to adopt its unit plan. Although the Commissioner's unitization order did not totally accept either Davis Oil's plan or Steamboat's plan in toto, it did include in the units small parts of the lands upon which Steamboat held leases. According to the official unit survey plats, the Richard tract (leased by Steamboat) contributed 14.9949 acres or 4.6859% to one unit and the Mott tract (leased by Steamboat) contributed 8.2858 acres or 2.58931% to the other unit. The Commissioner's order also appointed Davis Oil as the operator of the units and authorized Davis Oil to drill unit wells.

Davis Oil drilled a well in each unit, both of which were dry holes. Steamboat was invited to participate in the drilling of each well on the basis of its interest in the units. Steamboat refused to participate. Davis Oil submitted invoices to Steamboat totaling $90,567.31 for Steamboat's proportionate share of the costs of one well and $186,409.13 for its proportionate share of the costs of the other. Steamboat refused to pay, and Davis Oil instituted this action seeking cash payments from Steamboat. Steamboat claims that it is not required to pay the costs in cash and that Davis Oil's sole remedy is reimbursement from the proceeds of production from the wells. The trial court rendered judgment in favor of defendant, Steamboat, and dismissed Davis Oil's claims with prejudice. The court of appeal reversed holding that Steamboat was liable for its share of drilling costs in cash. . . .

The Need to Allocate Well Costs Between the
Operating and Non–Operating Parties in the Unit

In order to prevent waste in the recovery of oil and gas from a producing formation located beneath separately owned or leased tracts of land and to protect the rights of each separately owned or leased tract of land overlying a common source of supply, the legislature has granted the Commissioner of Conservation the statutory authority to establish compulsory drilling units. . . . The general concept behind the establishment of drilling units is to prevent adjoining landowners or leaseholders from having to drill protective offset wells on their premises by permitting them to share production proportionately to the area of their acreage drained by the unit well. Superior Oil Co. v. Humble Oil & Refining

Co., 165 So.2d 905 (La.App. 4th Cir.), writ refused, 246 La. 842, 167 So.2d 668 (1964). As a result of these compulsory drilling units, there has arisen a need for a rule allocating the costs between the unit operator and all nonoperating parties who share in the unit's production.

The difficulty faced by judges and legislators in developing a rule that will adequately balance the rights of the unit operator and the non-operating parties whose interests have been pooled in a forced drilling unit is indicated by the substantial differences between the different state approaches to the problem. Kramer, *Compulsory Pooling and Unitization: State Options in Dealing with Uncooperative Owners*, 7 J. Energy L. & Pol'y 255 (1986); Note, *Oil and Gas– Forced Drilling Unit–Compulsory Payment In Cash of Pro–Rata Well Costs*, 39 Tul.L.Rev. 381 (1965). A rule that requires landowners or leaseholders to share in all costs and all production on the same proportional basis may work harshly on a party with little available cash. 1 B. Kramer & P. Martin, *The Law of Pooling and Unitization*, § 12.01 (3rd ed. 1989). On the other hand, a rule that requires the operating party to absorb all losses but forces the operating party to share all profits will work harshly on that risk-taker whose efforts have led to production. Id. In fact, this rule will encourage people who might otherwise contribute to the costs in advance to hold out until the results of the drilling are known, since they can get the benefits of the well without the risks. Id.

Steamboat's Liability

The statutory law in effect on the date of the compulsory unitization order provided that, when the Commissioner of Conservation requires owners of separate tracts to pool their interests and develop their lands as a drilling unit, the cost of development and operation of the pooled unit "chargeable by the operator to the other interested owners" shall be limited to the actual reasonable expenditures required for that purpose, including a charge for supervision. La.R.S. 30:10(A)(1)(c) (1950).[92] Because the law did not specify when and how such costs were or were not "chargeable" to non-operating owners, and because the parties failed to plead and prove a custom governing the question, no rule for the particular situation in the present case may be derived directly from either source of law. Accordingly, this court is bound to proceed according to equity, i.e., resorting to justice, reason, and prevailing usages. La.Civ.Code art. 4 (1987). In making wise use of these resources, however, it is appropriate and advisable for us to rely for guidance upon the substantial body of jurisprudence and doctrine. . . .

It has been noted, and is perhaps self-evident, that the operation of a mineral production unit is similar to the situation in which a tract of land or a mineral

---

[92] We note that La.R.S. 30:10 has been amended by La.Act 345 of 1984. However, because the amendment did not become effective until after the effective date of the Commissioner's unitization order issued in the present case, we do not address the provisions of Act 345.

lease is owned in indivision by several co-owners. . . . Consequently, some of the legal precepts applicable in such cases, particularly those developed in the instance in which one co-owner drills the well on the common property, with or without the consent of his co-owners, may help us to decide equitably. . . . These principles evolved from general civil law doctrine and were adapted through litigation, usage and doctrine to modern mineral law. . . .

Felicitously, several of these principles have been refined and codified as part of the Louisiana Mineral Code. After recognizing that mineral rights are susceptible of ownership in indivision and providing for when co-ownership exists, La.R.S. 31:168 et seq., the Mineral Code provides for the rights and consequences arising from co-ownership of mineral rights. La.R.S. 31:174–178. A use or possession of a mineral right inures to the benefit of all co-owners of the right. La.R.S. 31:174 (and Official Comment thereto). . . . A co-owner of a mineral servitude may not conduct operations on the property subject to the servitude without the consent of co-owners, except in specially prescribed circumstances, and a co-owner of the servitude who does not consent to such operations has no liability for the costs of development and operations except out of his share of production. La.R.S. 31:175. A co-owner of a mineral servitude may act to prevent waste or the destruction or extinction of the servitude, but he cannot impose upon his co-owner liability for any costs of development or operation or other costs except out of production. La.R.S. 31:176. Similarly, a co-owner of the lessee's interest in a mineral lease may act to prevent damage or loss of the lease and to protect the interest of all, but cannot impose upon his co-owner liability for any costs or expenses except out of production. La.R.S. 31:177.

The underlying equity considerations lead us to conclude that the Mineral Code precepts should be adopted and applied by analogy in deciding the issues raised by the present case. Accordingly, we decide that a non-operating owner or lessee, who does not consent to operations within a compulsory drilling unit by a unit operator, has no liability for the costs of development and operations except out of his share of production. This precept is consistent with an implication in this Court's opinion in Hunter Co., Inc. v. McHugh, 202 La. 97, 11 So.2d 495 (1942) and usage based thereon limiting the operating party seeking to recover a non-operating party's proportionate share of costs to the exclusive remedy of withholding it from production proceeds. . . .

Applying this principle to the present case, we conclude that Steamboat is not liable for the drilling costs incurred by Davis Oil, except out of production. Because Steamboat did not expressly or implicitly consent to Davis Oil's operations and because there was no production, the wells having been dry holes, Steamboat is not liable for any such costs. Davis Oil initiated the formation of the compulsory drilling unit without soliciting Steamboat's advice or consent. Davis Oil invited Steamboat to participate in drilling the wells, but Steamboat rejected the offer. Under these circumstances, the mere fact that Steamboat filed a modified plan with the Commissioner in an effort to prevent drainage from its lands without compensation by Davis Oil's operations did not signify or imply that Steamboat approved or consented to the operations.

The present case is distinguishable from a situation such as that presented in Superior Oil Co. v. Humble Oil & Refining Co., 165 So.2d 905 (La.App. 4th Cir.1964). In that case the court of appeal held that the owner of oil interests who had drilled a producing well before the application of an adjoining owner for the establishment of a forced drilling unit was entitled to recover in cash the proportionate share of well costs allocated to that owner and was not limited to reimbursement from the proceeds of production. We agree with the result reached by the appeals court. Under the circumstances of that case, Humble Oil plainly ratified and consented to Superior Oil's development and operations. Superior had originally drilled a well on its own lease for its own account. After completion of the producing well, Humble, an adjoining lessee, provoked a unitization hearing before the Commissioner, which resulted in an order creating a forced drilling unit composed of the leases of Superior, Humble and Shell Oil Companies, with Superior's well designated as the unit well. Superior had allegedly incurred drilling costs of approximately $300,000.00. As a result of the unitization provoked by Humble, Superior was deprived of 55% of the total production it would have received otherwise. . . .

Under the circumstances of the Superior Oil Co. case, Humble's actions in taking the initiative to form an operating unit including an already producing well resulting from Superior's successful operations reasonably can be interpreted only as implying Humble's consent to those operations. On the other hand, in the present case, Steamboat's mere participation in the regulatory unitization process in the form of a defensive request for a modification in the drilling units being formed at the initiative of Davis Oil in order to prevent uncompensated drainage of fractions of the land under Steamboat's leases does not constitute tacit consent to the operations.

In the absence of consent, to hold non-operating parties to a compulsory drilling unit personally liable for a proportionate share of unit costs in an unsuccessful venture would be tantamount to placing the less opulent party wholly at the mercy of the better situated party. . . .

Decree

For the reasons assigned, the judgment of the court of appeal is reversed and the judgment of the trial court dismissing Davis Oil's claims with prejudice is reinstated.

NOTES

1. The principal case addresses the treatment to be given to force-pooled non-operators who have not put up a share of the operating costs in advance. As this case holds, in the absence of specific legislation or regulations from the state agency, the treatment given such interests is like that given to common-law

co-tenants. *See Prairie Oil and Gas v. Allen, supra* p. 120. That is, the compulsory unit is treated as the legal equivalent of a single lease with each pooled working interest owner having a right to an undivided share of production, subject to the operator's retention of the non-operator's proportionate share of costs until payout. The effect of such an approach is to give the non-operator a "free ride," an opportunity to let the operator drill the well at its own risk and then have the non-operator responsible for no more costs than had it put up a share of costs prior to drilling. Among the states affording the non-operator a free ride are Alaska, Arizona, Indiana and Missouri. For a comprehensive survey of the differing approaches from state to state, *see* Patrick H. Martin, *Unleased and Unjoined Owners – Forced Pooling and Cotenancy Issues*, 56 Rocky Mtn. Min. L. Inst. Ch. 18 (2010).

2. As the court's footnote in *Davis v. Steamboat* indicates, the Louisiana legislature amended the pooling statute in 1984, ending Louisiana's free-ride. The 1984 amendment provided for a risk charge of 100 percent of the tract's allocated share of the cost of drilling, testing, and completing the unit well. It specified the manner of notice by the drilling party to non-operators and the method of election to participate. The penalty of the act does not apply to unleased interests, but the unleased mineral owner will have a share of costs deducted from the production. Louisiana treats the unleased owner as the owner of an 8/8 cost-bearing interest rather than providing the unleased owner with a 1/8 royalty as some states do. The 1984 act further specifies that royalty owners are to be paid their royalty regardless of the risk charge on the lessee. The penalty was increased to 200 percent in 2008 by Act 115 of 2008, codified at La. R.S. 30:10(A)(2)(b)(i). The states that follow a risk-penalty approach include Alabama, Colorado, Louisiana, Michigan, Mississippi, Montana, Nebraska, Nevada, New Mexico, New York, North Dakota, Ohio, Texas, Utah, and Wyoming, as well as the states noted below that allow working interest surrender and/or risk penalty.

3. WESTERN LAND SERVICES, INC. V. DEPARTMENT OF ENVIRONMENTAL CONSERVATION 26 A.D.3d 15, 804 N.Y.S.2d 465, 165 Oil & Gas Rep. 324 (2005). New York is a risk penalty state by virtue of N.Y. Envtl. Conserv. Law § 23-0901(3), which was revised in 2005. The New York appeals court expressed the rationale for the risk penalty provision of New York law: "It would be unfair for a nonconsenting owner or nondriller lessee to be relieved of the costs and risks associated with drilling a producing well, but at the same time reap the benefits of another's efforts in extracting oil or gas from beneath his or her land. During the period of time that the well driller is recovering these costs, the nonconsenting owner and nondriller lessees are paid a royalty of one eighth of the production attributable to his or her property." The court ruled that the Department of Environmental Conservation (DEC) could not eliminate the risk penalty and agreed with the agency that the statute left it to DEC's discretion to decide how much a nonconsenting owner is entitled to be paid after the well driller has recouped its costs and the statutory penalty and that it could not compel the well driller to transport and market the gas of a nonconsenting owner.

See also CAFLISCH V. CROTTY, 2 Misc.3d 786, 774 N.Y.S.2d 653, 161 Oil & Gas Rep. 779 (Sup. Ct. 2003). An order of the Commissioner of Environmental Conservation made pursuant the compulsory integration statute was upheld. The petitioners challenging the order asserted they should be entitled to a $7/8^{th}$ working interest in the production allocated to their tract (after the operator recouped a risk-penalty amount) rather than a $1/8^{th}$ royalty. Because the petitioner held a non-drilling leasehold interest in a tract subject to the compulsory integration, the Commissioner concluded that the petitioner was not an "owner" within the meaning of the statute.

4. Some states allow for a still more sophisticated range of options for treatment of non-operators in a unit. The Oklahoma Corporation Commission in particular has developed three distinct options that may be employed in different combinations, and some states have borrowed features from the Oklahoma precedents. In Oklahoma the non-operator may:

    (1) Participate in the cost of drilling and completing a well;

    (2) Accept a specified bonus payment and/or royalty; or

    (3) Be treated as a carried interest subject to a risk penalty.

An order providing all three is spoken of as a "three-way order." The non-operator who fails to make an election to share in costs will be deemed to have elected to accept compensation for the working interest, thereby foregoing its working-interest share of production should the well be successful. Arkansas, Idaho, Illinois, South Dakota, and West Virginia have features similar to the Oklahoma approach. The surrender of working interest may be permanent or for a limited time, depending on the statute and the exercise of discretion by the agency.

5. If a regulatory agency offers the holder of a leasehold burdened by overriding royalty a non-participating option, how should it value the leasehold and the overrides?

In O'NEILL V. AMERICAN QUASAR PETROLEUM CO., 1980 OK 2, 617 P.2d 181, 68 O.&G.R. 282, the Oklahoma Supreme Court was presented with the issue of whether the Corporation Commission had jurisdiction to modify the rights of an overriding royalty owner as part of the authority to pool. The commission had established a 640–acre drilling and spacing unit. To be pooled was a lease on 77 acres held by O'Neill, which was burdened by a 9 percent overriding royalty, a portion of which was convertible to working interest on payout. O'Neill and certain of the overriding royalty owners on this tract were given the option by the Corporation Commission to participate in the development of the well or to accept an overriding royalty of 1/16 of 7/8 on oil and 1/8 of 7/8 on natural gas with a requirement that the overriding royalty owners (except one) receive their share out of the lessee's (O'Neill's) share without regard to whether O'Neill chose to participate or take an interest. They elected the latter, reserving the right to appeal. The court agreed with the overriding royalty owners that the

Corporation Commission did not have the statutory authority to adjudicate the rights and equities of an overriding royalty interest. The authority to require pooling only extended to owners, those with the right to produce. An overriding royalty owner had no right to produce and thus is not an "owner" within the meaning of the statute. The overrides, held the court, do not come from the original lessee's interest when he chooses not to participate but are attributable to the unit operator. Although this might enable an owner of a lease to pass unbearable override burdens to third parties, the court said such a concern should be addressed to the legislature.

In NORTH AMERICAN ROYALTIES, INC. v. CORPORATION COMMISSION, 1984 OK CIV APP 14, 683 P.2d 539, 80 O.&G.R. 527, North American owned a working interest burdened by non-operating interests in excess 3/16 ths of production. The Commission established three options for all working interest owners: (1) participation; (2) relinquishment of the working interest in exchange for cash bonus of $500 per acre plus a royalty of $1/16^{th}$ of all oil or gas, this option not to be available, however, to a working interest owner whose tract was burdened by more than 3/16th non-operating interests; and (3) relinquishment of the working interest in exchange for royalty equaling $1/16^{th}$ of $7/8^{th}$ of all oil and 1/8th of 7/8th of all gas, "said fractional interest to be reduced, however, to absorb any existing non-operating interests in excess of the normal 1/8th lessor's royalty." North American challenged the order on several grounds, including a claim that it was inconsistent with *O'Neill.* The court interpreted the order as requiring the unit operator to pay the overrides, but to reduce the amounts payable to North American dollar for dollar. This reduced North American's interest to zero. Does this solve the problem of the lessee that burdens its interest in order to thwart pooling or unitization?

6. Numerous issues can arise from electing an option. If a nonoperator elects to accept a bonus for its working interest for a proposed unit well, can the nonoperator later decide to participate in well costs if a second well is drilled on the same unit? The Oklahoma Supreme Court in AMOCO PRODUCTION CO. v. CORPORATION COMMISSION, 1986 OK CIV APP 16, 751 P.2d 203, 97 O.&G.R. 593 (adoption of court of appeals opinion) rejected the Corporation Commission's policy and interpretation of the pooling statute that a compulsory pooling order pools the interests of the parties only as to the well drilled and not as to the entire unit. The court held that the statute requires that the unit be developed "as a unit." The initial pooling order establishes the rights of the parties to all wells completed in the pooled unit. Thus, where the operator of the unit drilled a second well on the same unit, a party that elected to take a bonus and a share of production cost free as to the first well could not demand to participate as a cost-bearing interest in the subsequent wells.

Is an election binding on a successor interest? In RANOLA OIL CO. v. CORPORATION COMMISSION, 1988 OK 28, 752 P.2d 1116, 98 O.&G.R. 336, the Oklahoma Supreme Court held that the acceptance of a bonus in lieu of participation in an oil and gas well operated as an assignment of the interest and prevented the bonus taker from participating as a matter of right in subsequent increased-density wells. The acceptance of the bonus effected a transfer of the working interest of the party accepting the bonus. The election of the bonus operated as a sale of the leasehold rights, and this was binding on the interest owner's successors. The court analyzed the issue in terms of risk avoidance by the party accepting the bonus and risk acceptance by the operator; to permit Ranola to participate retroactively in the three increased-density wells would divest the operator's vested interest in violation of substantive due process.

What discretion should an agency have in fashioning alternatives? *See* Application of Kohlman, 263 N.W.2d 674, 60 O.&G.R. 402 (S.D.1978); Texas Oil & Gas Corp. v. Rein, 1974 OK 9, 534 P.2d 1280, 51 O.&G.R. 69; C.F. Braun & Co. v. Corporation Commission, 1980 OK 42, 609 P.2d 1268, 65 O.&G.R. 391. Must the agency offer a non-operator the option of participating in a well only to a shallow formation, or may it instead pool multiple formations, with a requirement that the non-operator share in all costs or be subject to a risk penalty as to all costs to be recompensed out of production from any formation in the pooled area? *See* Viking Petroleum, Inc. v. Oil Conservation Commission, 100 N.M. 451, 672 P.2d 280, 79 O.&G.R. 57 (1983). Does the failure of an agency to provide an election invalidate the order? *See* Newkirk v. Bigard, 109 Ill.2d 28, 92 Ill.Dec. 510, 485 N.E.2d 321, 87 O.&G.R. 266 (1985), *cert. denied*, 475 U.S. 1140 (1986), *reh'g denied*, 477 U.S. 909 (1986).

7. A well may cost millions of dollars to drill. Payout of investment costs on the well may take a number of years. Should the operator be allowed to recoup interest on the money it has had tied up in well costs? *See* Imperial Oil of North Dakota, Inc. v. Industrial Commission, 406 N.W.2d 700, (N.D.1987); Pursue Energy Corporation v. State Oil and Gas Board of Mississippi, 524 So.2d 569, 101 O.&G.R. 63 (Miss.1988).

8. Most pooling orders provide for participation of all working interest owners on a surface acreage basis. In Humble Oil & Refining Co. v. Welborn, 216 Miss. 180, 62 So.2d 211, 2 O.&G.R. 50 (1953), the court sustained this basis for apportionment of production as against the claim that in the particular situation some premises had a higher productive capacity than others. As noted in Corley v. Mississippi State Oil and Gas Board, 234 Miss. 199, 105 So.2d 633, 9 O.&G.R. 481 (1958):

> The formula of participation based on surface acres has the merit of simplicity and certainty, but is entirely fair only in the rare cases where formations are uniform in quality and thickness throughout the unit with each tract having beneath it the same amount of reserves per acre.

Most voluntary unitization agreements and compulsory unitization orders provide for participation of all interest owners on a formula basis that incorporates multiple factors such as number of wells, porosity, acre feet of productive sand underlying each tract, and surface acreage. It often takes years to work out the formula. A mere listing of some of the numerous and complex variables in the determination of the value of a tract proposed to be included in a unit agreement may suggest the difficulties which can be encountered in negotiating an agreement:

(1) the drive mechanism available in the field;

(2) well productivity;

(3) well density;

(4) effect of prorationing;

(5) acre feet of productive formation;

(6) oil initially in place beneath a tract;

(7) extent and accuracy of information that has been obtained as a result of securing electrical logs, coring, testing;

(8) the extent of penetration into the producing formation.

———

A conservation agency in a state with a compulsory unitization statute will order unitization if the negotiated formula arrived at by the specified percentage of interest owners meets a statutory test such as that the formula be "fair and equitable." What should be found to be fair and equitable? Consider the next case that struggles with such a question and reviews other cases that have similarly struggled.

———

TROUT v. WYOMING OIL AND GAS CONSERVATION COM'N

Supreme Court of Wyoming, 1986.
721 P.2d 1047, 92 O.&G.R. 420.

BROWN, JUSTICE. By an order dated August 16, 1985, and a nunc pro tunc order dated September 1, 1985, the Wyoming Oil and Gas Conservation Commission (hereinafter Commission) approved a plan of unitized secondary recovery operations (unitization) of the Teapot Formation underlying the Mikes Draw unit area. Appellant appeals the unitization plan approved by the Commission. . . .

. . .

The Teapot Formation is located in Converse County north of Douglas. After meetings and discussions among the working interest owners concerning unitization, Intervenor, Mitchell Energy Corporation, proposed a Teapot Unit approximately 8.5 miles long by 2.8 miles wide covering approximately 7,385 acres. The appellant, Kye Trout, Jr., is the owner of working interests in leases in three wells located in the proposed unit. Mitchell Energy Corporation owns thirty-five percent of the interest in the unit area.

Counsel for appellant stated that they were not opposed to the unitization and were only opposed to the allocation formula. Appellant's witness, Guy Ausmus, a petroleum engineer, testified that they strongly supported the unit because they did not believe in waste, and if they were to stay out of the unit, that action would constitute waste. This representation by appellant's counsel was apparently considered by the Commission to be a stipulation that unitization would prevent waste. In any event, waste was not an issue before the Commission.

Before the Commission hearing on August 13, 1985, the working interest owners (operators) manifested an interest in unitization, and a technical committee was formed in January 1982 to study its feasibility. After its organization, additional meetings were held throughout the year, resulting in the technical committee report of July 1983. The operators had discussed possible formulas, including a formula proposed by appellant. The operators with interests in wells down the middle of the field would not accept the original oil-in-place and pore volume parameters favored by appellant, because they did not believe this oil to be recoverable.

In December of 1983, there was a meeting to vote on the formula for allocating unit production to the various tracts. Five different votes were taken. At this meeting a formula proposed by appellant, relying on oil-in-place as a significant factor was rejected. With respect to the Trout formula, Rob Pawlik, a petroleum-reservoir engineer, testified at the Commission hearing that this formula had no chance of receiving the requisite approval of the unit members. Appellant's proposed formula would yield more oil to him, and such oil would have to be taken from the other interest owners. Mr. Pawlik also stated that the formulas were thoroughly discussed before the meeting in December 1983, at which time the votes were taken, and he stated further that by the time of that meeting, the acceptable formula had been narrowed down by the parties and that this was the reason for only five votes. At the close of the meeting a large majority of the operators had substantially agreed on a formula.

It was determined at the Commission hearing that 82.39 percent of the operators and 93.06 percent of the royalty interest owners indicated voluntary joinder of the unit proposed by Mitchell Energy Corporation. After the hearing, the unit formula earlier favored by the operators was confirmed by the Commission, allocating unitization production based on three parameters or factors of varying weight:

Parameter	Weight
Last six months production	47.5%
Remaining proved developed producing reserves	47.5%
Original oil-in-place	5.0%

Although appellant lists three issues, his appeal is essentially a sufficiency-of-the-evidence argument. Appellant summarized his argument by stating that:

> There was no substantial evidence to support the Commission's findings. The statements of the commissioners at the close of the hearing contradict the findings placed in the order.
>
> The Commission erred in that there was no evidence whatsoever as to the effect of the formula on correlative rights or that the formula allocated oil and gas in an equitable manner. The Commission did not take into account the effect on correlative rights or the equitable distribution of oil in making its decision.
>
> The Commission's decision was arbitrary and capricious in that it was based upon a threat from the intervenor, Mitchell Energy Corporation, that it would not produce oil unless its formula were approved.
>
> The Commission erred in approving a 7,000 acre unit when the evidence was that the operator would only operate a small pilot project.

The rules for reviewing a decision of an administrative agency are well known. . . .

We examine the entire record to determine if there is substantial evidence to support an agency's findings. If the agency's decision is supported by substantial evidence, we cannot properly substitute our judgment for that of the agency, and must uphold the findings on appeal. . . . Substantial evidence is relevant evidence which a reasonable mind might accept in support of the conclusions of the agency. It is more than a scintilla of evidence. . . .

In his statement of the issue, appellant claims the Commission's decision is contrary to law and not supported by substantial evidence. He does not allege the Commission acted in excess of its jurisdiction, or that it failed to follow the required procedures.

The Commission made findings of fact and conclusions of law in support of its order. Appellant's principal contention is that the evidence was insufficient to support finding Nos. 15, 16, and 17, which state:

> 15. Kye Trout (Trout) appeared to protest the application considered herein. Trout's concern is that his leases, which were acquired in 1983, did not receive equity in the participation formula which is based 47.5 percent on the last six months' rate, 47.5 percent on remaining proved developed producing reserves, and five percent (5%) on the volume of original oil-in-place. The Commission finds that the proposed formula does allocate oil and gas in a just and equitable manner to each separately owned tract so far as can be practically determined; that the unitized operation will prevent waste, will protect correlative rights and will substantially increase the amount of produced oil and gas; and that the value of the additional reserves is much greater than the investments required to produce the additional reserves. The operator, Mitchell Energy, had discussed formulas similar to that favored by Trout, but found that an overwhelming majority of the

working interest owners would never support such a formula.

16. Many technical committee meetings and working interest owners meetings have been held since 1982 in arriving at the unitization formula and the parameters therein. The technical exhibits and data of the applicant are found to be true and accurate.

17. Mitchell Energy's application for a waterflood operation of the Mikes Draw Unit's Teapot Formation should be approved to provide for the increased recovery of hydrocarbons and for the protection of correlative rights."

A unit agreement has the effect of forming a single leasehold ownership in the unitized tract. Such agreements are economical because they allow the reservoirs of oil and gas to be produced more exhaustively as a natural unit without the constraints of artificial property lines. . . .

Unitization means that everyone who has a legal interest in the oil and gas join together in a common plan to gain effective production from a field. . . .

The Commission's authority to compel unitization in Wyoming is authorized by § 30–5–110, W.S.1977 (June 1983 Replacement). Under subsection (e)(ii), the Commission must find: "Such unit operation is feasible, will prevent waste, will protect correlative rights, and can reasonably be expected to increase substantially the ultimate recovery of oil or gas."

Correlative rights are defined by § 30–5–101(a)(ix), W.S.1977 (June 1983 Replacement), as

. . . the opportunity afforded the owner of each property in a pool to produce, so far as it is reasonably practicable to do so without waste, his just and equitable share of the oil or gas, or both, in the pool. . . .

Wyoming is in accord with other states which recognize that correlative rights and the right to produce oil from a pool are limited both by a duty not to injure the pool and a duty not to cause waste. Gilmore v. Oil and Gas Conservation Commission, Wyo., 642 P.2d 773 (1982).

Appellant disagrees with finding number 16 and states, "In fact there had not been extensive negotiations." While "extensive" and "many" are imprecise terms, the record reveals that there were negotiations involving interested parties before the unit meeting in December 1983. In our factual recitation we traced the history of negotiations among the operators beginning in January 1982 and continuing until August 1985. We cannot say that the Commission was wrong in its characterization of the negotiations.

The principal thrust of this appeal is appellant's contention that his correlative rights were not protected by the Commission's unitization order. Mr. Norris, a consulting geological engineer, was employed by the technical committee to do the geological background work prior to and part of the unitization committee work. Both Mr. Norris and Mr. Pawlik testified that the formula selected was fair and equitable to all parties for our purposes. This is equivalent to saying that the formula protects correlative rights. The purpose of the allocation formula is to assign to the various tracts a fair share of all unit production, wherever the wells are located. What is fair and equitable is to some extent in the eye of the beholder; however, both Norris and Pawlik were cross-examined regarding the basis of their conclusion that the formula was fair and equitable. Mr. Pawlik explained that original oil-in-place was not heavily weighted because such oil was not recoverable. He also thought it significant that an overwhelming majority of interest parties favored the formula agreed upon.

The Commission also heard a witness for appellant, Mr. Asmus, discuss appellant's proposed formula which gave more weight to oil-in-place. The Commission simply decided that the oil-in-place parameter "should be given only slight" weight.

Counsel for appellant also argued that the Commission itself did not believe that the approved formula set forth in the unit agreement protected correlative rights. The commissioners in effect said that where the statutory eighty percent approval is received, then the Commission has authority to approve the unit. Substantially, the same problem was present in the *Gilmore* case. There, we said:

> The operators held various meetings at which they voted on formulae to be used in allocating production under unitization. The 81 working interest owners considered a total of 71 formulae. Naturally, each of the owners wanted parameters favorable to them and wanted more weight to be given these parameters. After voting on almost 60 formulae, the owners were frustrated in their attempt to find one that would receive the statutory approval. As a result, they examined their voting records and used a computer to arrive at an equitable compromise formula that could receive the required approval. The resulting formula number 67 at issue here, received 75.89 percent approval. It appeared that no greater percentage would approve any formula proposed. Id., at 775.

Arguably, the Commission was not entirely happy with the formula favored by the majority. However, as in *Gilmore,* it was realistic. Substantially the same situation exists in this case. It was clear from the evidence that no other formula proposed would get the requisite eighty percent approval. The statutory requisite approval could not be obtained on any formula using oil-inplace as a substantial parameter. This was obvious to all concerned and had been discussed before the vote; the vote merely confirmed what all parties had known from their previous discussions.

We quoted § 30–5–110(f), W.S.1977 (June 1983 Replacement), in deciding to affirm the Commission's order approving the unit in the *Gilmore* case. About the only difference we can see in this case and *Gilmore* is that in the latter about sixty formulae were voted on, while five were voted on here. This difference under the circumstances is not significant.

> . . . [W]e do not think it would be fair or equitable to permit a very small minority of small fractional interest owners in this common source of supply

of hydrocarbons to prevent the Board and the overwhelming majority of the operating and royalty owners from pursuing a unitization program designed to obtain a maximum recovery from the field. . . . Corley v. Mississippi State Oil and Gas Board, 234 Miss. 199, 105 So.2d 633 (1958).

. . . In most instances it is impossible to use a formula which will apply equally to all persons producing from a common source. In striking a balance between conservation of natural resources and protection of correlative rights, the latter is secondary and must yield to a reasonable exercise of the former. . . . Denver Producing & Refining Company v. State, 199 Okla. 171, 184 P.2d 961, 964 (1947).

In *Gilmore,* we quoted with approval from the above cases. We then said:

There is no indication that a more equitable formula could be devised . . .. The operator's technical committee and the Commission, both having experience and expertise at unitization, settled the formula 67 as the fairest and most feasible . . .. It would appear, therefore, that there is little chance that anyone involved could devise a better formula, nor is it likely that a different formula could receive the necessary approval.

. . .

A reversal here would not give appellant a larger share of the allocation, but would send the case back to the Commission with a mandate to come up with another formula. We do not believe that even appellant has any real expectation of a different formula that would receive the required approval. If a new formula were found and agreed upon, it could spawn new lawsuits. The specter of veto by the United States Geological Survey would not disappear. It is a fact of life which we cannot ignore.

. . .

Appellant seems to expect perfection. Justice was accomplished here as much as could be under the circumstances. . . . 642 P.2d at 780–781.

What was said in the *Gilmore* case is applicable to this case. It would be virtually impossible to devise a formula acceptable to all concerned.

Appellant argues that the Commission's order was not in accordance with law because it approved the formula under threat of nonproduction. Appellant contends that the evidence shows the formula was forced upon the parties by Mitchell. Appellant also argues that the Commission's order was in error because it approved a seven thousand acre unit, and Mitchell was only planning a pilot project.

The testimony relied upon by appellant does not indicate that the Commission was threatened or blackmailed. Rather it shows that the other operators had worked over a long period of time incoming to a common agreement as to which formula should be adopted. Mitchell informed the Commission that they did not believe they could get the requisite approval for using a different formula. Appellant characterizes this statement to be a threat.

However, it appears merely a realistic appraisal of the negotiations and the work that had already been done and the fact that a majority of the operators would not agree to a different formula. Appellant does not substantiate his allegations of blackmail with competent evidence showing either a threat or Commission's acquiescence to any improper demands, nor does he cite persuasive authority. . . .

There was no reason for the Commission to approve a unitization agreement merely because the parties indicated that a different agreement would not be acceptable. The Commission had the authority to order shut-in production pending successful negotiations if it believed the allocation formula did not protect correlative rights or otherwise meet statutory criteria. . . . The Commission exercised its authority and shut-in production in the Hartzog Draw area, which was the subject of the Gilmore case.

Appellant's complaint about the Commission approving a 7,000 acre unit also lacks cogent argument and authority. Furthermore, appellant did not make this complaint to the Commission. Issues which were not raised before the administrative agency will not be considered for the first time on appeal. . . .

. . .

Conclusion

Before the Commission hearing, negotiation among interested parties demonstrated that all parties agreed that unitization was appropriate. The only problem was agreeing on a formula.

At the Commission hearing, a formula favored by appellant was considered, along with the formula proposed by Mitchell. The formula favored by appellant received only a small percentage vote, while the Mitchell formula received the necessary vote to legally form the unit. At the hearing, correlative rights was the principal matter considered. The Commission determined that the formula prepared by Mitchell was fair and equitable and that correlative rights were protected.

We must accord considerable deference to the Commission and its expertise. We cannot say that it was clearly wrong in its determination.

Affirmed.

NOTES

1. Compulsory unitization statutes typically require approval to the unitization plan by a percentage of "owners" and "royalty owners" In determining whether the requisite approval of "owners" and "royalty owners" has been achieved, how should one count overriding royalty carved out of a lessee's interest? Royalty that exists independent of a lease? Carried interests such as royalty that is convertible to working interest? Net profits interests? Other types of interest?

2. Should a conservation agency have the power to reject a proposed participation formula and require the submission of a new formula? *See* State Oil

& Gas Board v. Anderson, 510 So.2d 250, 95 O.&G.R. 467 (Ala.Civ.App.1987), *cert. denied sub nom.* Anderson v. State Oil and Gas Board, 484 U.S. 955 (1987).

3. What issues arise in connection with the termination of a compulsory unit (whether pooled or unitized)? The Kansas case of PARKIN v. STATE CORPORATION COMMISSION, 234 Kan. 994, 677 P.2d 991, 80 O.&G.R. 39 (1984), held that the Corporation Commission could not delegate the power to terminate a unitization to a majority of the owners in the unit. A 5,800–acre unit was established under the Kansas compulsory unitization statute. Under the order, operations were to continue so long as there was production in paying quantities and unit operations were being conducted. The waterflood operations were discontinued, and by 1981 there were only six producing wells in the entire unit. Royalty and mineral-interest owners applied to the Kansas Commission for dissolution of the unit, but the commission denied the application on the grounds that under the operating plan agreement, signed by 80 percent of the interest owners, the unit remained in effect until 65 percent of the working-interest owners determined that unitized substances could no longer be produced in paying quantities or that unit operations were no longer feasible. The Kansas Supreme Court held that this was an improper delegation, by the commission to the operator, of the power to dissolve the unit. *See also* Corbello v. Sutton, 446 So.2d 301, 82 O.&G.R. 79 (La.1984) (claim that unit termination and creation of new units from a previously unitized sand that utilized the current productive limits of the reservoir as the new unit boundaries violated due process barred because not timely brought); Eason Oil Co. v. Howard Engineering, Inc., 1990 OK 101, 801 P.2d 710 110 O.&G.R. 501 (rights in the well had vested on the effective entry of the 640 acre unit order and they continued with the same share of production regardless of subsequent changes to the unit configuration).

SECTION 4. POOLING OR UNITIZATION AS CROSS-CONVEYANCE OR AS CONTRACT

(*See* Martin & Kramer, *Williams and Meyers Oil and Gas Law* §§ 929–930.11).

STUMPF v. FIDELITY GAS CO.

United States Court of Appeals, Ninth Circuit, 1961.
294 F.2d 886, 16 O.&G.R. 139.

POPE, CIRCUIT JUDGE. Appellant B.J. Stumpf, plaintiff below, brought this action against Fidelity Gas Co., hereafter called Fidelity, and the other appellees, its assignees, as defendants, to procure a decree that an oil and gas lease originally given by plaintiff to defendant Fidelity had been terminated and forfeited, and to recover certain statutory and other damages because of defendants' refusal to release their claimed interest in the premises described in the lease [in Montana]. The action, originally begun in the state court, was removed to the court below on the ground of diversity of citizenship of the parties.

The lease in question was executed September 7, 1934, and provided for a so-called "primary term" of three years which could be extended by the lessee beyond the primary term by drilling a commercial oil or gas well below 2000 feet on some point on the Cedar Creek Anticline. The primary term expired September 7, 1937.[93] Under date of September 3, 1957 [1937? Eds.], Fidelity wrote to the plaintiff stating: "As you no doubt know, we have during the past year been drilling a deep test well in the Baker field, which well has recently been completed as a commercial oil well." It is the theory of the plaintiff's complaint that this statement was false. It alleges: "That no commercial oil or gas well below 2,000 feet was completed by any of defendants during the primary term of said lease as contemplated by paragraph four (4) of said lease, and said lease, therefore, by its terms, terminated and became, and still is, null and void."

The lease contained in addition to sundry provisions common in oil leases a paragraph 11[94] which authorized the lessee to pool the production under the lease with other producers and owners. Lessee was granted the right to include all of lessor's royalty interest in a unit operating agreement to that effect; and the assignee of the lessee Montana Dakota Utilities Co., another defendant, made such a unit agreement covering the lands of plaintiff and other royalty owners with itself as unit operator. This agreement became effective October 1, 1938, which was after the termination of the primary term of the lease. The agreement provided that the respective title holders party to the plan would share in the production from the unit, (called Unit No. 7), in the proportion to the number of acres held by the title holder within the area as compared with the total number of acres party to the plan. The only other reference to the unit operation contained in the lease was that referred to in footnote 93.

The substance of defendants' answer was a denial of plaintiff's assertion that no commercial oil or gas well was completed as required during the primary term of the lease; but they further allege that the lease was perpetuated by reason

---

[93] As hereafter noted, the lease, in paragraph 11, authorized lessee to include lessor's royalty interest in a "unit operating agreement". The lease also provided: "If said deep test well or a similar well shall be completed as a commercial oil or gas well below 2,000 feet, upon any lands operated jointly with the lands herein described as a so-called unit area as provided in paragraph eleven hereof, such well shall be considered as validating this lease with the same effect as though such well had been drilled on lands herein described, and production from any such well, or production allocated to the lands herein described from any well in said unit area, shall be considered as being produced on the lands herein described within the meaning of Paragraph 2 hereof."

[94] "11. If Lessee at any time shall agree with other Lessees or landowners to develop and operate this lease with other leases or tracts covered by such a co-operative or unit operating agreement as a single property, for oil and/or gas production, or in accordance with drilling and operating methods common to the several leases or tracts embraced in such agreement . . . Lessee is hereby granted the right to include all Lessor's royalty interest in such agreement."

of the commitment to the unit plan of development provided for by the unit agreement above referred to, and that the unit was now producing gas in commercial quantities.

In his reply, plaintiff alleged that the defendant Montana–Dakota Utilities Co., as assignee of the lease, knowing that it had not found oil or gas in commercial quantities below 2000 feet within the time required by the lease, "cheated, tricked, and defrauded the plaintiff by telling him that a search for commercial oil as aforesaid had been successful"; that he had no means of ascertaining for himself whether the well was a commercial one and was forced to rely and did rely upon the defendant's "aforesaid false and fraudulent declarations"; that as soon as he learned there had been no commercial oil discovered he repudiated the lease and demanded its release.

It will be observed that here there were two contracts made which, though related, were nevertheless separate and distinct. The first contractual relationship was that between the plaintiff, as lessor, on the one hand, and the defendant lessee and its assignees on the other. This was the oil and gas lease which contained the customary provisions reserving to plaintiff, as royalty to be paid by the lessee, one-eighth of the proceeds derived from the sale of all oil or gas produced.

The second was the unit operating agreement which paragraph 11 of the lease, above referred to (footnote 94, 897 *supra*), authorized the lessee to execute. In short, by that clause the lessee was granted an agent's authority to execute the unit operating agreement and extend it to cover plaintiff's lands; but the latter agreement was not a part of the lease.

It should be noted that the plaintiff's complaint, first stated in three counts, the third of which was abandoned, seeks two types of relief from the defendants. The first type is that referred to in the first and second paragraphs of plaintiff's prayer where he prays that the oil and gas lease be declared to be "forfeited, null and void", and that the defendants be required to release of record their claimed interest in plaintiff's premises. The second type is represented by the third and fourth paragraphs of the prayer in which plaintiff seeks statutory damages of $100 and special damages of $33,600. As the complaint shows, he asserts these damages resulted from the defendants' failure to release said oil and gas lease.

After the action was at issue, the defendants filed a motion to dismiss it on the ground that plaintiff had failed to join certain parties listed in the motion each of whom was a party indispensably necessary to a full and final adjudication of the controversy. Listed were some 166 persons which an affidavit in support of the motion asserted constituted a listing "of the current owners (other than Montana–Dakota Utilities Co.) of operating rights or working interests committed to the Co–Operative or Unit Plan of Development, Unit No. 7," and the "current owners (other than plaintiff) of royalty and over-riding royalty payable on gas production derived from Unit No. 7". "Unit Plan of Development, Unit No. 7", refers to the pooling or unit agreement to which we have previously referred.

The motion to dismiss was granted and the case dismissed without prejudice, and it is from that order that this appeal has been taken. In a memorandum accompanying the order for dismissal, the court stated:

> The Court's position is simply this, that under the unit agreement which was authorized by the lessor and by virtue of which the lease was amended, the royalty interest of each lessor in the unit is measured at a rate dependent upon production on all other tracts in the unit and that therefore each such royalty owner has a direct interest in this suit, the object of which is in part to free the premises involved from the unit agreement. Any judgment in this case would, it seems to the Court, necessarily affect all unit royalty owners and that a determination of this case, absent such owners, would be inconsistent with equity and good conscience.

There are cases in which it has been held that where there has been unitization or pooling for development of oil and gas leases all parties to the unit agreement are indispensable parties in a suit to try title to any parcel of land subject to a lease which has been incorporated in the unit arrangement. Leading cases to this effect are Veal v. Thomason, 138 Tex. 341, 159 S.W.2d 472, and Belt v. Texas Co., Tex.Civ.App., 175 S.W.2d 622. On the other hand, in Nadeau v. Texas Company, 104 Mont. 558, 69 P.2d 586, 593, 111 A.L.R. 874, where plaintiff and defendant were rival claimants to oil leases on the same land, the plaintiff brought suit to quiet his title to what he alleged was a prior lease; it appeared that defendant, who also claimed under lease from the same original landowner, had signed a unit and pooling agreement with other owners of lands and holders of leases. It sought to abate the suit because of failure to join as defendants the other persons who had signed the unit and pooling agreement on the ground that they were indispensable to the maintenance of the suit. The court held that it was not necessary that they be joined.

Neither the Texas cases cited nor the Montana case present facts which are just the same as those now before us. In the Nadeau case the plaintiff had never dealt with the other signers of the unit or pooling agreement and had not signed it or authorized any one to do so for him. In that respect it differs from our case.[95]

The two Texas cases cited involved actions in trespass to try title to certain lands. Involved in each case were lands upon which Veal and others had executed oil and gas leases to the Texas Company. In each case Veal's title to certain of the

---

[95] "The controversy here was whether the lease of the defendant company was valid as against the plaintiff. No attempt was made to quiet the title of the plaintiff as against the whole world, or any other person than defendant. Plaintiff might have brought such an action, but he did not. It does not appear that plaintiff or his immediate predecessors in interest had any transactions with any of the other parties prior to the inception of his title by the execution and delivery of the Reynolds lease. Hence, whatever rights may have been created by these negotiations and transactions between Ewald and the defendant, and between the latter and the various signers of this agreement, could in no wise affect the rights of the plaintiff and, therefore, they are not necessary parties to the settlement of the controversy litigated between plaintiff and defendant." 69 P.2d 593.

lands at the time of the lease was attacked by the plaintiff. Concurrently with the lease by Veal other owners of land executed other leases to the same lessee. All of them provided that all tracts so leased should be operated as one area, aggregating 6000 acres, each lessor to share proportionately in the production from the unitized block. It was provided in each lease that all these leases should be treated as one lease. In each case the court noted that the effect was to "merge all of the leases into one contract". In the *Veal* case, the Texas Supreme Court said:

> We have demonstrated that the effect of the lease contract here involved is to vest all the lessors of land in this unitized block with joint ownership of the royalty earned from all the land in such block. . . . Stated in another way, a contract affecting land in this State, which grants or reserves mineral royalty in such land, constitutes the owner of such royalty the owner of an estate in such land. 159 S.W.2d 476.

It will be noted that in the cases just referred to, the leasing to the Texas Company and the unitization of all the lands involved was accomplished by what in legal effect was a single instrument, a single contract, the terms of which made all lessors in the unitized block joint owners of interests in all the lands.[96] The Texas cases to which we have referred have been cited many times and frequently followed by other decisions in this field.

The primary respect in which the case before us differs from those is that here we have what are essentially two separate contracts. First of all Stumpf had his contract, an oil and gas lease, with Fidelity; in it he authorized the lessee or its assignee to execute another contract, namely, the unitization agreement, and that agreement was executed, as we have seen. The question then arises whether it is possible for Stumpf to procure a cancellation or declaration of nullity of the contract made with Fidelity without attacking or otherwise affecting the continued existence of the unitization agreement or questioning the validity thereof, or in any manner affecting the sundry and various persons interested therein.

It may be noted here that in his complaint Stumpf makes no attack upon the initial validity or continuing operation of the unitization agreement. He seeks an adjudication that the lease which he executed on September 7, 1934, terminated and became null and void because of the failure of the defendants to complete a commercial oil and gas well below 2000 feet within the primary term of that lease. After the defendants in their answer had set up and alleged the execution of

---

[96]   In the *Veal* case, holding that all royalty owners under the other lease contracts in the unitized block were necessary (*i.e.* indispensable) parties to the suit, the court said: "There is no escape from this conclusion, because Thomason seeks in this action to obtain a judgment freeing this land from the Texas Company lease, and if this is done by judgment which names such Company only, the royalty owners under the other leases in this unitized block will have had such royalty interest in this land, for all practical purposes, cut off and destroyed without having had their day in court." 159 S.W.2d at page 477.

the unitization agreement, copy of which was attached to their answer, plaintiff then in the reply which he was directed to file, alleged that the inclusion of the plaintiff's land in the unit agreement was void because the lease under which defendants claimed had itself terminated. (The primary term of the lease expired September 7, 1937; the unitization agreement became effective October 1, 1938; it was dated October 28, 1937.)

It is unnecessary to determine whether this portion of the reply operated to enlarge the claims under the complaint, for as noted hereafter, we are of the opinion that it makes no difference in the result here.

This brings us to the question whether the omitted parties here are indispensable within the meaning of that term and as defined in the adjudicated cases. It is not sufficient for a defendant to demonstrate that these persons would be "proper" or "necessary" parties within the meaning of Rule 19, Fed.R.Civ.P., 28 U.S.C.A. Those rules were not intended to and did not effect any alteration in the standards by which the existence of an indispensable party may be determined. Without going into any extended discussion of what constitutes an indispensable party we think it suffices to refer to the statement of the test to be applied found in State of Washington v. United States, 9 Cir., 87 F.2d 421, 427, a summary which has frequently been alluded to with approval. This court there said:

There are many adjudicated cases in which expressions are made with respect to the tests used to determine whether an absent party is a necessary party or an indispensable party. From these authorities it appears that the absent party must be interested in the controversy. After first determining that such party is interested in the controversy, the court must make a determination of the following questions applied to the particular case: (1) Is the interest of the absent party distinct and severable? (2) In the absence of such party, can the court render justice between the parties before it? (3) Will the decree made, in the absence of such party, have no injurious effect on the interest of such absent party? (4) Will the final determination, in the absence of such party, be consistent with equity and good conscience?

If after the court determines that an absent party is interested in the controversy, it finds that all of the four questions outlined above are answered in the affirmative with respect to the absent party's interest, then such absent party is a necessary party. However, if any one of the four questions is answered in the negative, then the absent party is indispensable.

So far as plaintiff's complaint seeks recovery of damages, it is clear that persons other than defendants who happen to be interested in the unit agreement would be neither necessary nor indispensable parties. It is also clear that if the plaintiff should succeed in demonstrating that the terms and conditions of his lease were not complied with by the defendants and that in consequence it became the duty of the lessee to execute a release as required by Revised Codes of Montana, 1947, Sec. 73–114, 73–115, 73–116, then the failure to execute such a release ought to entitle him to his damages, and the recovery of them should be a matter of no interest to any other party. As noted in Abell v. Bishop, 86 Mont. 478, 284 P. 525, noncompliance with the statute mentioned gives rise to three distinct causes of action, one for cancellation of the lease, one for the recovery of a statutory penalty, and one for damages. No reason is perceived why plaintiff here could not, regardless of nonjoinder with the other persons referred to, have recovery for such damages as he might be able to prove.

But apart from that consideration, we think that the decided cases would demonstrate that the persons mentioned are not indispensable parties even as to that portion of plaintiff's action in which he seeks a declaration that the lease has terminated and is now null and void.

It seems clear that if Stumpf were permitted to proceed in this case and succeeded in demonstrating that the lease in question was not extended beyond its primary term, and that hence it had come to an end, then the result of a judgment to that effect as against the defendants here would be that henceforth plaintiff would be entitled to eight-eighths or 100 percent of the oil or gas produced, or deemed produced, from his land. If the decree ran against the lease itself only and in no manner purported to deal with the continued existence of the unitization agreement, then the operator under that unit agreement would in no manner be prevented from continuing to operate and extract oil and gas in the manner therein provided.[97] The net result of the whole thing would be that the distributive shares of the proceeds of the sale of gas applicable to plaintiff's interest would be 100 percent to Stumpf instead of one-eighth to Stumpf and seven-eighths to the defendant Montana–Dakota Utilities Co.

If that is a correct analysis of the situation, then it would appear that each of the four questions posed in State of Washington v. United States, *supra,* could be answered in the affirmative. The interest of the absent parties to the unit agreement would be distinct and severable from the question of who gets the seven-eighths share in the gas produced. Such parties would not be necessary to enable the court to render justice between the parties to the original lease or their assigns. In making an award which would give 100 per cent of the proceeds from plaintiff's land to him would have no injurious effect upon the interest of any absent party; and such a determination would be consistent with equity and good conscience.

Indeed a good deal may be said to the effect that considerations of equity

---

[97]   It might be argued that if Montana–Dakota loses its seven-eighths interest in the production attributable to plaintiff's land, it will be unlikely to devote full energy to production on the unit, and thus other parties in the unit will suffer. But Montana–Dakota made this contract between itself as "vendor" and itself as operator, and obviously drew the agreement. It must be construed most strongly against it. But under any construction, Montana–Dakota as operator is firmly bound to operate the unit efficiently, to the end of securing "economical development" and the "greatest possible ultimate recovery." These obligations it cannot shed.

and good conscience argue in favor of permitting plaintiff's action to proceed. If the mere fact of the execution of the unitization agreement makes all parties interested in it indispensable, then it is obvious from the facts shown in this record that no matter how right plaintiff may be as to the facts alleged in his complaint, he is denied access to the courts, both federal and state. Some of those parties listed in the motion which the court sustained are shown to be residents of Montana and joining them would oust the court of jurisdiction. As a practical matter, some of them are shown to be deceased, and probably the determination of who are all the parties interested in the unit agreement would be beyond the means of an ordinary litigant. Nor does it appear that he could gain access to the state courts for many of these parties are shown to be non-residents of the state and it may be presumed unavailable there.[98]

An examination of the adjudicated cases whose facts most closely resemble those present here shows that the tentative conclusion suggested above, namely, that the absent parties here are not indispensable, is supported by the authorities. In Hudson v. Newell, 5 Cir., 172 F.2d 848, modified at 5 Cir., 174 F.2d 546, the plaintiffs claimed certain lands as heirs of one Hudson who during his lifetime had left a life estate therein to his wife. The wife, notwithstanding her limited interest, had made conveyances in fee of parts of the land to persons who in turn conveyed to others and they had made leases to certain oil companies reserving one-eighth interest in the royalties. Plaintiffs brought suit claiming certain of these lands and asking for a decree that they were the true owners of the land and of the mineral rights therein and of the oil and gas removed from it by the defendant oil companies who were named defendants with their respective lessors. The leases permitted unitization of the land leased together with other tracts for oil production, and persons interested in those other tracts included in the unitization arrangement were not made parties or joined in the suit. The trial court dismissed the suit because of the absence of indispensable parties. On its first opinion the court of appeals affirmed that decision and stated the facts with respect to these persons whose interest arose only through unitization of their lands with those in suit. This statement together with the reason given for the initial affirmances found in paragraph 6 of the court's opinion which was as

---

[98]   To hold that all these absent parties are indispensable would accomplish what Mr. McGee threatened to do to the bank which held a mortgage on his land which it was threatening to foreclose. As related by Professor Masterson, (*Indispensable Parties in Oil and Gas Litigation*, Sixth Annual Institute on Oil and Gas Law, Mathew Bender & Co., 1955, p. 160): "In 1925 an Iowa landowner named McGee asked a bank to extend his note which was secured by a lien on land. The vice-president with whom he was dealing refused. Thereupon McGee showed the vice-president a deed signed by McGee and which he said he was going to record right then if he didn't get his extension. This deed named the following grantees: 'Each and every member of the American Legion of Iowa, each and every member of the Independent Order of Odd Fellows of Iowa, each and every member of the Knights of Pythias of Iowa, and each and every attorney at law in Iowa.'" *See* State v. McGee, 200 Iowa 329, 204 N.W. 408.

follows:

> The question of 'unitization' and its connection with indispensable parties remains. A number of the claimants not joined have no interest unless through unitization of their lands with those in suit. In Mississippi, as elsewhere, the number of permissible oil wells is limited, and to secure production from small tracts not entitled to a well each, several tracts are united by formal recorded agreement to be served together by a single well, the production from which is to be prorated according to acreage among the several tract owners. Such units, before the filing of these suits, had been duly formed, in part from lands included in the suits. The question is whether the owners of other unitized lands acquired such an interest or title in those in suit as that they must be joined in order to grant even partial relief. The district judge held that in Mississippi the oil in place is, as in Texas, conveyed by oil leases, royalty sales, and other formal recorded contracts, and that a unitization contract vests joint ownership jointly in all the unitized reserves in place and in the royalty oil produced from the well, citing Veal v. Thomason, 138 Tex. 341, 159 S.W.2d 472.[99] While no Mississippi authorities in point are cited, we believe this to be correct, so that all these owners must be parties or the unitized lands must be excluded from any decree. No adoption by plaintiffs appears of the unitizations of their lands. 172 F.2d at page 852.

Following that decision the court granted a rehearing of the case en banc and withdrew paragraph 6 quoted above. The court held:

> As to Paragraph 6, appellants have filed a formal disclaimer of any intention to interfere with the unitizations formed before the filing of these suits to which they were not parties; and they offer so to amend their pleadings in this court or in the district court as to adopt and ratify such unitizations and to seek only to substitute themselves for their adversaries in title in these suits in such unitizations, without affecting in any manner the rights of the other parties thereto. If this is done such other parties will not be affected by such relief and are not indispensable parties to the suits. Paragraph 6 of the opinion is therefore withdrawn, and the disclaimer and agreement to ratify and adopt the unitizations above referred to will, as is therein proposed, be made a part of the mandate to the district court which shall cause the appellants to so amend their pleadings and prayers with reference to the unitized lands as to seek a decree, if they establish their title, which shall substitute themselves for their adversaries in title without affecting the rights of the other parties to the unitization agreements who are not before the court. This done, the suits may proceed without the presence of these absent

---

[99]   As we shall point out hereafter the reasoning in this portion of this quotation will be inapplicable in our case because of the express provisions of the unitization agreement here involved to the effect that this particular contract should have no such operation and should not be construed as affecting or passing title to any land.

parties. Hudson v. Newell, 5 Cir., 174 F.2d 546.

The reasons stated in the case from which we have just quoted would compel a decision here that the absent parties in this case whose interests arise solely out of the unitization agreement are not indispensable parties. As the court there observed, if there be no attack upon the unitization agreement, as such, and if the effort in the suit is merely to substitute the plaintiffs for their adversaries in title, then the other parties will not be affected by such relief, their rights will not be affected in any manner by the decree sought and the other parties would not be indispensable to the suit.

Of course the plaintiff here has not offered to adopt and ratify the unitization. In his complaint he has not attacked it. To that extent it seems clear that the validity and extent of the unitization agreement is simply not at issue in this case regardless of whether in some other action, at some other time, plaintiff might be able to make some attack upon it.

. . .

In the case of Whelan v. Placid Oil Co., 5 Cir., 198 F.2d 39, 42, the Court of Appeals referred only to the first opinion in Hudson v. Newell, *supra,* apparently unmindful of the modified decision in that case, and relying on Veal v. Thomason, *supra,* held that all mineral owners of the lands who had joined in a unitization agreement were indispensable parties who should have been made defendants in the suit. . . .

The court said:

This is true for the reason that the unitization agreement vests all the lessors of land in the unitized block with joint ownership in all the unitized reserves in place in the unit—the ownership being in the proportion which the acreage each member contributed to the pool bears to the total acreage in the unitized block.

Since the unitization agreement was not a part or parcel of the original lease itself, as it was in Veal v. Thomason, this decision is not easy to understand. Its different facts should distinguish it from the Veal case. The case has been criticized, rightly, we think, as demonstrating a misunderstanding of the Veal case and as generally unsound. It cannot be reconciled with that court's decision in Hudson v. Newell, *supra,* or with the other decisions cited above which are in accord therewith. We cannot accept the reasoning thereof.

However, even if we assumed the soundness of the decision upon the facts of that case and under the law applicable to lands in the State of Texas, it appears that for a special reason the conclusions of that case are wholly inapplicable here because of the express terms of the unitization agreement itself. Section 17 of that agreement recites as follows:

That nothing herein contained, implied or contemplated in relation hereto shall create or be deemed to create a partnership between the parties hereto, but they shall each hold their respective right, titles and interests in their respective tracts according to existing leases, permits, operating agreements and gas purchase or sales contracts and any extensions, renewals, substitutions, or modifications thereof, unchanged except as to the right of the Operator to operate Unit No. 7.

Even more significantly, it is provided in section 19 as follows:

That nothing herein contained shall be construed as affecting or passing title to any lands, leases, or permits, but the Operator shall acquire operating rights only.

It will be noted that the reason given in the quoted portion of the court's opinion, supra, is that the effect of the unitization agreement was to vest all lessors of land in the entire block with joint ownership in all acreage which each member contracted to pool. This is the theory of reciprocal conveyancing. Obviously under the present unitization agreement no such result could occur, and the stated basis for the decision in the Whelan case is wholly wanting. The quoted language of sections 17 and 19 of the instant lease furnishes a further ground for distinguishing the Veal case, supra, where, as we have noted, the court placed much emphasis upon its reciprocal conveyancing theory.

It is plain that the question of the validity, operation or extent of the unitization agreement is not in issue here. The only thing at issue is who shall receive the seven-eighths interest in the oil and gas which defendants claim under the lease. Whether plaintiff is or is not a party to the unitization agreement is simply not for adjudication in this action. If it should ultimately develop in a later suit brought by Stumpf that he is not a party to it then he will be in the position of the plaintiff in Nadeau v. Texas Co., supra, where the plaintiff had not signed the unitization agreement. If it should turn out that he is bound by the unitization agreement, then he will be in the position of the plaintiff in Hudson v. Newell, supra, and Hutchins v. Birdsong, [258 S.W.2d 218] supra, where the plaintiffs conceded their responsibility.

It appears to be the position of the defendants here that the plaintiff did attempt to attack the unit agreement. It is not clear just what is the basis for this contention. Possibly it is based upon the fact that in his reply plaintiff alleges that the inclusion of his land in the unit agreement was void because the lease itself had theretofore become void.

Disregarding any problem of pleading relating to what was formerly referred to as a departure in a reply and assuming that plaintiff had included in his complaint a count seeking invalidation as to him of the unitization agreement, we would have a situation which would still fall short of presenting a case of absence of indispensable parties. Of course it could be said that to obtain complete relief, that is to say, relief from both the lease and the unit agreement, all of the persons who are members of the unit agreement ought to be made parties; but it does not follow that such persons are indispensable parties to the granting of any relief whatever to this plaintiff. The recognized rule is stated in Rule 19(b) of the Rules of Civil Procedure as follows:

When persons who are not indispensable, but who ought to be parties if complete relief is to be accorded between those already parties, have not been made parties and are subject to the jurisdiction of the court as to both service of process and venue and can be made parties without depriving the court of jurisdiction of the parties before it, the court shall order them summoned to appear in the action. The court in its discretion may proceed in the action without making such persons parties if its jurisdiction over them as to either service of process or venue can be acquired only by their consent or voluntary appearance or if, though they are subject to its jurisdiction, their joinder would deprive the court of jurisdiction of the parties before it; but the judgment rendered therein does not affect the rights or liabilities of absent persons.

. . .

It is thus apparent that while in order to obtain complete relief in the supposititious case of a suit by plaintiff to rid himself both of the lease and of the unitization agreement, the other persons interested in the latter agreement would be deemed necessary parties, yet this does not prevent the court from proceeding to grant a portion of the relief sought and entering a judgment which, as the rule states, would not affect the rights or liabilities of the absent persons. In short, nothing would then prevent the court from determining the issue respecting the lease as between plaintiffs and defendants, making it clear that its decree in no manner affects or concerns the rights of any absent parties under the unit agreement. . . .

A multitude of cases demonstrates the general proposition that there is no rule that a court must do everything at once. . . .

It is therefore apparent that even if plaintiff had sought in this action not only a determination of the invalidity of the lease but also a determination of the invalidity of the unit agreement, the appropriate thing for the court to do (if the plaintiff was shown to be entitled thereto) would be as suggested in the Waterman case [Waterman v. Canal–Louisiana Bank & Trust Co., 215 U.S. 33, 30 S.Ct. 10, 54 L.Ed. 80 (1909)],—to grant the relief sought respecting the lease contract and shape its relief as to this in such manner as to preserve the rights of persons not before the court.

It will be apparent from what we have said that our primary difference with the trial court relates to a portion of its opinion in which it is said: "Plaintiff by his action here would withdraw his premises from the unit which would effect at least a partial rescission or cancellation of the unit agreement." In our opinion such a result is not involved in his action. As we have indicated, even if such an effort had been added to the suit, which primarily relates to the lease itself, the court ought, as suggested in Rule 19, *supra*, to proceed in the action, rendering such judgment as may be appropriate, and expressly providing that the same does not affect the rights and liabilities of the absent parties.

The judgment is reversed and the cause remanded with directions to entertain the action and to proceed therewith in such manner as may not be inconsistent with this opinion.

Upon Petition for Rehearing

PER CURIAM. The petitioners-appellees seek a rehearing on the ground that if judgment were entered in this case decreeing termination of the lease at the end of the primary term thereof, this would inevitably withdraw plaintiff's land from the unit and therefore would affect the absent unit interest owners. In consequence of this premise petitioners assert that the result envisioned by us in our opinion would be altogether impossible. The reference here is to the sentence of the opinion:

> If the decree ran against the lease itself only and in no manner purported to deal with the continued existence of the unitization agreement, then the operator under that unit agreement would in no manner be prevented from continuing to operate and extract oil and gas in the manner therein provided.

It is true that our opinion did make reference to the primary term of the lease; but it was not within the contemplation of the court that if the trial court should grant the appellant relief respecting the lease that such a judgment would decree that the lease terminated at the end of the primary term. The complaint itself shows that forfeiture or cancellation of the lease could not be effected until after December 27, 1955, the date of the notice of default, copy of which is attached to the complaint. Paragraph VII of the complaint alleges that such notice was given pursuant to paragraph 9 of the lease. Paragraph 9 of the lease, which was attached to the complaint, reads in part as follows:

> The breach by Lessee of any obligation arising hereunder shall not work a forfeiture of this lease nor be grounds for cancellation thereof in whole or in part save as herein expressly provided. . . . Should Lessee default in any of the express terms and conditions of this lease, no forfeiture or cancellation thereof shall be declared, nor any steps or proceedings to effect a forfeiture or cancellation taken unless and until Lessor shall have first given Lessee written notice, specifying the time, nature and extent of such default, and Lessee has wholly failed for a period of ninety (90) days thereafter to remedy such default.

Perhaps our opinion was at fault in not expressly quoting this provision or noting the date of the notice; but it seems to us to be clear that on the date when the unit operating agreement became effective, October 1, 1938, the lease was still in effect and even after the giving of the notice the lessee could have remedied any default and kept the lease in continuing effect. This means that the authority to enter into the unit operating agreement recited in paragraph 11 of the lease was an authority still in existence on October 1, 1938. If as a part of a working agreement between A and B, A authorizes B to bind him to another contract with C, and B does so, the contract between A and C is not itself terminated by a rescission, cancellation or forfeiture of the original agreement between A and B providing that agreement was still in effect and the authority not terminated at the time that the contract with C was executed.

The petition for rehearing is denied.

NOTES

1. Consider whether the decision in the principal case turns upon:

(a) Rejection of the cross-conveyance theory of *Veal v. Thomason.*

(b) Acceptance of the theory of *Hudson v. Newell.*

(c) The effect of Section 17 of the Unitization Agreement.

(d) The effect of the notice and demand clause of the lease.

(e) Acceptance of the decision of the Montana Supreme Court in *Nadeau v. Texas Co.* as controlling authority.

2. One commentator has observed that the difficulties presented by the Texas rule of indispensable parties are so great in situations involving pooling or unitization that undoubtedly some prospective plaintiffs have been led to "just forget the matter," declaring that:

This is not as facetious as it sounds. I am sure there have been many plaintiffs in Texas who just had to give up the question of litigating their substantive rights when faced with the procedural obstacles these pooling cases present. The title to the minerals under many a tract of land in Texas has been cured by putting the same in a large unit. Some of the best title insurance in the world, under the present law, is to place a tract of land in a large unit. Dedman, *Indispensable Parties in Pooling Cases*, 9 Southwestern L.J. 27 at 79 (1955).

3. Consider the applicability of the rationale of *Hudson v. Newell* to a situation where the plaintiff admits that his tract is unitized but the defendant claims otherwise. *See* Discussion Notes, 3 O.&G.R. 1761 (1954).

4. What is the impact of the rationale of *Veal v. Thomason* on a partition suit brought by a non-leasing concurrent owner against a leasing concurrent owner whose interest in the affected premises, Blackacre, has been pooled or unitized? *See* Douglas v. Butcher, 272 S.W.2d 553, 4 O.&G.R. 284 (Tex.Civ.App.1954, error ref'd n.r.e.).

The California Code of Civil Procedure § 872.540, relating to partition of oil and gas interests, provides:

Where property is subject to a lease, community lease, unit agreement, or other pooling arrangement with respect to oil or gas or both, the plaintiff need not join as defendants persons whose only interest in the property is that of a lessee, royalty-owner, lessor-owner of other real property in the community, unit, or pooled area, or working interest owner, or persons claiming under them, and the judgment shall not affect the interests of such persons not joined as defendants.

5. A number of cases and writers have raised the question whether a pooling

or unitization agreement has the effect of a cross-conveyance or a mere contract, *viz.,* does a pooling or unitization agreement between the owners of Blackacre and Whiteacre have the effect of conveying to the owner of Blackacre some interest in Whiteacre and of conveying to the owner of Whiteacre some interest in Blackacre, or does the agreement have the effect merely of giving each owner a contract right to share in the production from premises other than those he has contributed to the agreement. As indicated in the principal case, the so-called "cross conveyance" theory arising from the case of Veal v. Thomason has been viewed as significant in determining who are necessary parties to litigation. Consider whether the adoption of the cross-conveyance or the contract theory should be viewed as significant for the following purposes:

(a) In determining whether an executive has power to bind a non-executive owner by a pooling or unitization agreement. *See* Minchen v. Fields, 162 Tex. 73, 345 S.W.2d 282, 14 O.&G.R. 266 (1961).

(b) In determining whether a voluntary pooling or unitization agreement must comply with the requirements of the statute of frauds. *See* Discussion Notes, 1 O.&G.R. 114 (1952).

(c) In determining whether the Rule against Perpetuities applies to a lease pooling clause. *See* Phillips Petroleum Co. v. Peterson, 218 F.2d 926, 4 O.&G.R. 746 (10th Cir.1954), cert. denied, 349 U.S. 947 (1955); Kenoyer v. Magnolia Petroleum Co., 173 Kan. 183, 245 P.2d 176, 1 O.&G.R. 1126 (1952).

(d) In determining whether a venue statute applicable to "suits for the recovery of lands or damages thereto, or to remove encumbrances upon the title to land, or to quiet the title to land" is applicable to an action to determine the effectiveness of a unitization agreement. *See* Renwar Oil Corp. v. Lancaster, 154 Tex. 311, 276 S.W.2d 774, 4 O.&G.R. 697 (1955).

(e) In determining whether a conveyance of a tract by its owner which fails to mention the grantor's rights under the unit agreement in which such tract has been included will convey the grantor's interest under the unit agreement in production from tracts other than the one specifically conveyed. *See* Hoffman, *Voluntary Pooling and Unitization* 169–172 (1954).

6. As brought out in the principal case, the focus on determining whether pooling results in cross-conveyance of real property rights or is simply a contract often is joinder of parties in litigation. Rather than focusing on the classification of the rights is it not more pertinent to inquire into the effect of the litigation on non-joined parties? Is this not what the court's opinion actually does in the principal case? Consider the Louisiana case of City of Shreveport v. Petrol Industries, Inc., 550 So.2d 689, 106 O.&G.R. 60 (La. App. 2d Cir.1989). Here the City of Shreveport brought an action against a unit operator seeking to increase its share of production from a 40–acre compulsory unit within Shreveport. The claim of the plaintiff was that the true extent of the plaintiff's interest had been concealed from it and fraudulently represented as 1.5 acres rather than 13 acres, apparently based on ownership of streets and alleys. The defendant operator filed

an exception of non-joinder of parties who claimed to own the mineral interests underlying the streets and alleys. The trial court denied the exception. The appeals court reversed. The interest in the unit production of the non-joined lot owners and others currently being credited with title to minerals underlying the streets and alleys clearly would be affected by a court decision. Therefore, those parties should have been joined in the litigation.

7. Should there be a distinction between pooling that is accomplished by agreement and pooling that is accomplished by compulsion of the state in inquiring whether pooling effects a cross-conveyance or a contractual relationship? It is often said that conservation agencies have no power to affect title to property. And one could hardly say that a compulsory pooling order brings about a contractual relationship. Rather, it is a sort of forced marriage, often between unwilling participants.

8. The relationship among parties to a unit may lead to a claim of a breach of a duty arising from that relationship. It may be asserted that a unit agreement or a joint operating agreement gives rise to a joint venture in which the operator owes a fiduciary duty to the non-operators. Or it may be asserted that a unit agreement or unit order from the state creates a cotenancy and this imposes itself a high duty of one cotenant to the other cotenants. In Schulte v. Apache Corporation, 1991 OK 61, 814 P.2d 469, 113 O.&G.R. 336 plaintiffs asserted that they should be given a share of interest in certain farmouts within a compulsory unit that the operator of the unit had acquired. The court rejected as least one portion of the theory of the plaintiffs, stating:

> The plaintiffs argue the pooling order created the relationship of cotenancy in the well and that Apache breached its duties under that cotenancy relationship. In Wakefield v. State, 306 P.2d 305 (Okl.1957), the argument was made that an order force pooling mineral estates created a tenancy in common and that the parties possessed the common law rights of tenants in common. This Court rejected the argument and held that the forced pooling order did not create a tenancy in common relationship. Then in Tenneco Oil Co. v. District Court of the Twentieth Judicial District, 465 P.2d 468 (Okl.1970), we held that a voluntary unitization agreement, approved by the Commission, did not create a cotenancy. Likewise, the Commission's pooling order in the present case does not create a cotenancy.

SECTION 5. EFFECTS OF POOLING AND UNITIZATION ON EXPRESS AND IMPLIED LEASE TERMS

Kramer & Martin, *The Law of Pooling and Unitization*, §20.01-.02

The most frequently litigated issues regarding pooling and unitization relate to the effects of the inclusion of a tract of land, or a portion of the tract, in a unit upon the express and implied terms of an oil and gas lease. A closely analogous but less frequently litigated set of issues relate to the effect of such pooling or unitization on land subject to a term mineral or royalty interest.

. . .

To what extent does the fact of inclusion in a unit (either voluntary or compulsory) modify or divide the lease or term interest? In examining the cases and the statutes dealing with the effects of pooling and unitization this treatise considers four basic fact patterns that arise.

. . .

In each of the four hypothetical cases, Blackacre is the tract on which there is a lease or term interest that is the subject of the controversy. In each, Blackacre is a 160–acre tract, Gray Unit is a 640–acre unit formed by agreement or by force pooling, and Whiteacre is any other tract, a portion or all of which is in Gray Unit, but Whiteacre is not at issue in the problem. The four hypothetical cases further assume allocation of production on a surface acreage basis when the land at issue is effectively pooled.

[1] Case 1

In Case 1 all of Blackacre is in Gray Unit, and the unit well is on Blackacre. This is depicted in the following diagram:

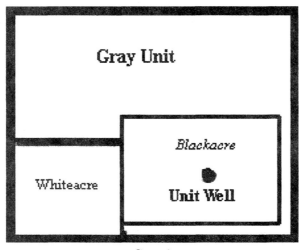

Case 1

Case 1 occasions very little controversy when the interests in Blackacre have been pooled. The unit well will maintain the lease or term interest on Blackacre, and the production will be shared among all owners of interests in Gray Unit. The only questions that might arise are whether the interests on Blackacre have been made subject to pooling. If an interest has not been effectively pooled, then that interest is not reduced by the inclusion in the unit. For example, if A is a royalty owner for of the production from Blackacre, and

A's interest has not been pooled, then A would receive 1/8th of all production from the well on Blackacre, not 1/8 x 160/640 of total production.

[2] Case 2

In Case 2, all of Blackacre is included in Gray Unit, but the unit well is on Whiteacre rather than on Blackacre. This is depicted in the following diagram:

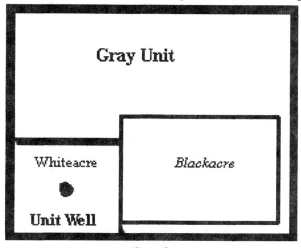

Case 2

Case 2 occasions a bit more controversy than Case 1. In most circumstances the unit well on Whiteacre will maintain a lease or a term interest on Blackacre, and the production will be shared among all owners of interest in Gray Unit. Two particular questions do arise out of Case 2. First, if an interest in Blackacre has not been pooled, is there any right of that owner to a share of production from the Gray Unit well on Whiteacre? For example, if **A** is a royalty owner for a 1/4th mineral interest on Blackacre, and **A**'s interest has not been pooled (e.g., because **A** initially refused to agree to pooling or because the owner of the executive right for **A**'s interest had no power to pool **A**'s interest), does A have a right to a share of the production from the well? The answer is clearly no. **A**'s claim to a share of production must arise by contract or by regulation; under the rule of capture there is no right to a share of production merely because the oil or gas may have migrated from beneath A's property.

The second area of controversy in Case 2 arises from interpretation of the specific language of a lease or deed for a term interest. The lease or deed may include a reference to "production from such lands as described herein" or words of similar import. Does production from a unit well on Whiteacre satisfy the requirement? The better view, as exemplified in the majority approach, holds that the unit production does maintain the interest.

[3] Case 3

In Case 3, only a portion of Blackacre is in Gray Unit and the unit well is on Blackacre. This is depicted in the following diagram:

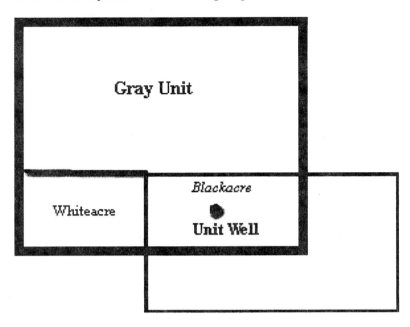

Case 3

In this circumstance, when all interests in Blackacre are pooled, the question may arise whether the unit well will serve to maintain the lease or term interest on Blackacre as to the portion not in the unit. The rule in most states is that the lease or term interest will be maintained as to both portions. There are, however, minority positions; some jurisdictions would give differing approaches whether a lease or a term interest is involved. Some jurisdictions have specifically provided by statute that a unit in a Case 3 or Case 4 situation will have the effect of dividing the lease or term interest into portions that must be separately maintained. Parties can specifically provide for a different effect from the majority rule by contract in the lease or in the grant of the term interest.

[4] Case 4

In Case 4, only a portion of Blackacre is in Gray Unit, and the unit well is on Whiteacre. This is depicted in the following diagram:

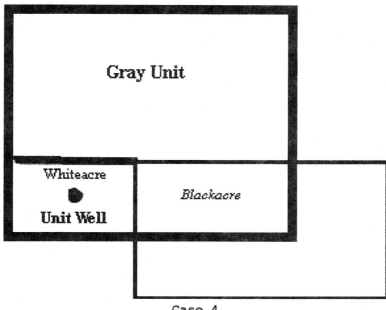

Case 4

The same question arises in this as in the Case 3 circumstances. The majority approach does not distinguish between Case 3 and Case 4, and the statements regarding the minority jurisdictions in reference to Case 3 apply with respect to Case 4. That is to say, in most jurisdictions a lease or term interest will be continued by the unit operations or production on Whiteacre. . . .

The four cases dealing with well placement and partial inclusion in a unit are only the beginning of the variations that can occur and that may raise issues about lease or term interest maintenance. Other variables that may affect the decision in a case include the following:

1. The specific wording of the habendum clause or the pooling clause of the lease or the wording of the term interest deed. For example, the lease may spell out the consequences of pooling, or the term interest deed may specify that the interest will continue for ten years and so long thereafter as there is production from "said described lands."

2. Other specific terms of the lease, such as the drilling operations clause or shut-in royalty clause.

3. Whether it is a lease or a term mineral interest or term royalty interest that is at issue.

4. Whether the pooling or unitization was the result of a community lease, inclusion of the acreage in a state-established drilling/spacing unit, a voluntary pooling or unitization, the exercise of the pooling clause by a lessee, or a compulsory pooling or unitization.

5. The specific terms of the agreement for pooling or unitization or the specific provisions of the order for pooling or unitization.

6. The specific wording of the statute under which pooling or unitization is effected.

7. Whether the pooling occurred before or after the well was drilled.

8. Whether the pooling occurred during or after the primary term of the interest in question, and whether all of the interests were pooled by the end of the primary term.

9. Whether the lessee or mineral interest owner of the interest in question has agreed to pay a share of costs for the well or has otherwise assumed liability for the drilling of the well.

Any of these variables, or combination of variables, may come into play within the four cases.

———

A. CONTINUATION OF THE LEASE UNDER THE HABENDUM CLAUSE

TRI M PETROLEUM CO. v. GETTY OIL CO.
RIDGWAY v. SHELL OIL CO.

United States Court of Appeals, Fifth Circuit, 1986.
792 F.2d 558, 91 O.&G.R. 547.

POLITZ, CIRCUIT JUDGE. In these two Mississippi diversity jurisdiction cases, consolidated for purposes of oral argument and disposition, plaintiffs-lessors appeal adverse summary judgments in suits to cancel oil and gas leases for nonproduction. Both Mississippi federal trial judges reached the same conclusion, ruling that the leases continued in force because of forced-pooling orders of the State Oil and Gas Board of Mississippi (the Board) and subsequent drilling within the pooled unit. Concluding that the district courts correctly anticipated and applied Mississippi law, we affirm both judgments.

Background

Tri M v. Getty, No. 85–4261

The material facts are not in dispute. In October 1971 Skelly Oil Company (later Getty Oil Company), acquired two oil, gas, and mineral (OGM) leases from Ora Barnes, covering land in Jefferson Davis County, Mississippi. Both leases had ten-year primary terms and would continue for "as long thereafter as oil, gas, or other mineral is produced from said land or acreage pooled therewith." Both leases contained standard clauses on pooling and drilling operations, the latter providing for continuation of the lease, despite lack of production, if drilling or reworking operations were underway at expiration of the primary term. The leases were due to expire in October 1981 unless continued by diligent drilling operations resulting in the production of oil, gas, or other minerals.

Several months before expiration of the primary term, Tri M Petroleum Company and the individual plaintiffs (hereafter "Tri M") acquired "top" leases on the lands covered by the Getty leases. In June 1981, System Fuels, Inc., seeking to develop the area but having been unsuccessful in securing approval of the various mineral interest owners, petitioned the Board for a unitization order and authority to drill a well in a 640–acre unit which included the acreage covered by the Getty leases.

On August 20, 1981, the Board force-pooled the 640–acre unit and designated System Fuels as the operator of the proposed well. Getty declined to farm out its acreage to System Fuels but initially agreed to participate in the drilling costs. Getty later refused to participate when System Fuels insisted on an agreement which provided that if the well was completed as an oil well, the size of the unit would be reduced, excluding Getty's acreage and precluding its entitlement to any production, despite its having advanced a portion of the drilling costs. When System Fuels refused to delete that clause, Getty declined to sign the agreement. System Fuels commenced drilling operations on August 25, 1981 and completed the well as a producing gas well on January 5, 1982. Getty's interest was carried to payout.

Tri M filed suit seeking cancellation of the Getty leases for nonproduction. Getty countered that the Board's forced-pooling order superseded the lease provisions and that drilling by the designated operator continued the leases in effect. The district court agreed and granted Getty's motion for summary judgment. Plaintiffs appealed.

Ridgway v. Shell Oil, No. 85–4363.

On May 2, 1974, the Ridgway plaintiffs executed an OGM lease with Shell Oil Company covering property in Rankin County, Mississippi. The lease had a five-year primary term with standard operation and production continuation clauses.

A dispute between Shell and Pursue Energy Corporation triggered this litigation. Both companies wished to drill in the same area. A voluntary accord could not be reached and both filed petitions with the Board requesting approval of a drilling unit and forced integration of all interests within the unit.

Exercising its authority under Miss.Code Ann. § 53–3–7 (1972) (prior to its amendment by Laws 1984, ch. 511, § 1; see Laws 1984, ch. 511, § 2), the Board denied Shell's application and granted Pursue's stating in its order:

> . . . All separately owned interests . . . are hereby pooled and integrated, and the persons owning the drilling rights therein and the right to share in production therefrom are hereby required to integrate their interests and said interests are hereby integrated as a gas drilling unit for the drilling,

production, development and operation of said lands as to the oil, gas and mineral rights therein.

The Ridgway property was located within the unit but was not the drilling site.

Pursue finished the drilling operation and completed a well capable of production but shut it in. Shell timely tendered shut-in royalty payments until the well was put into production. Plaintiffs instituted this action to cancel the lease, whose primary term had expired, alleging that no drilling operation by Shell had taken place. Additionally, plaintiffs contended that Shell failed to file a declaration of pooling at the end of the primary term. . . . The district court found that the lease was extended by the forced-pooling order and Pursue's drilling thereunder. It found no merit in the other allegations and granted Shell's motion for summary judgment. Plaintiffs appealed.

Analysis

We consolidated these cases on appeal because both raise the same issue: Whether the issuance of a forced-pooling or unitization order by the Board, pursuant to Miss.Code Ann. § 53–3–7 (1972), and drilling thereunder within the unit continue OGM leases on all lands located inside the unit. The controlling facts are common in both suits. Acreage under lease to Getty and Shell was included in a state-ordered pooling unit. Drilling occurred within the unit but not on the lands leased to Getty and Shell. Both refused to farm out their interests. Neither advanced drilling costs but simply "rode down the well."

Both sets of plaintiffs contend that cancellation of their leases is in order because the leases require the lessees to commence operation within the primary term, neither lessee did, and the drilling done by the Board-designated operators could not inure to the lessees' advantage. They contend that the cited Mississippi statute was not intended to affect lease obligations undertaken by the lessees.

The lessees counter that the statute-based orders of the Board preempt the provisions of the leases, and that they were foreclosed from drilling after the units were established and an operator was designated by the Board. They contend that the forced-unitization order relieved them of their personal obligation to take affirmative action by drilling and production operations in order to continue the leases beyond the primary term.

In both cases, the trial judges, schooled and skilled in the law of Mississippi, to which we duly defer, relied on the current compulsory pooling statute, its forerunners, and cases decided thereunder. The Mississippi Supreme Court had repeatedly held: "Where a unit has been properly established, it is settled that production from any of the land in the unit extends the leases upon all lands in the unit insofar as such leases cover the unitized tracts." Superior Oil Co. v. Magee, 227 Miss. 868, 87 So.2d 280, 281 (1956); Superior Oil Co. v. Berry, 216 Miss. 664, 63 So.2d 115, suggestion of error overruled, 216 Miss. 664, 64 So.2d 357 (1953). This rationale was overlaid to the forced pooling, drilling, and factual situation presented in the cases at bar.

The controlling statute, Miss.Code Ann. § 53–3–7 (1972) (applicable to these cases under Laws 1984, ch. 511, § 2), provided in pertinent part:

> (a) When two or more separately owned tracts of land are embraced within an established drilling unit, the person owning the drilling rights therein and the rights to share in the production therefrom may validly agree to integrate their interests and to develop their lands as a drilling unit. Where, however, such persons have not agreed to integrate their interests, the board may, for the prevention of waste or to avoid the drilling of unnecessary wells require such persons to integrate their interests and to develop their lands as a drilling unit. All orders requiring such pooling shall be made after notice and hearing, and shall be upon terms and conditions that are just and reasonable, and will afford to the owner of each tract the opportunity to recover or receive his just and equitable share of the oil and gas in the pool without unnecessary expense.
>
> The portion of the production allocated to the owner of each tract included in a drilling unit formed by a pooling order shall, when produced, be considered as if it had been produced from such tract by a well drilled thereon . . .

Our attention is invited to no case, and our independent research has found none, in which the Mississippi Supreme Court has applied § 53–3–7 to a drilling scenario as presented in the cases at bar. Nonetheless, we find adequate guidance from analogous Mississippi authorities.

At the outset, we note the not-unusual requirement that lease obligations in Mississippi are "construed subject to the applicable statutes for the conservation of oil and gas and the orders, rules, and regulations of the [State Oil and Gas] Board promulgated pursuant to the statutes." Frost v. Gulf Oil Corp., 238 Miss. 775, 119 So.2d 759, 762 (1960). The Mississippi Supreme Court has upheld the Board's orders and regulations in a variety of settings. . . . .

We find in the cases predating the applicable statute ample indication that the Mississippi Supreme Court would approve the continuance of the Getty and Shell leases in the instant forced-pool setting. See, e.g., Humble Oil & Refining Co. v. Hutchins, 217 Miss. 636, 64 So.2d 733, suggestion of error overruled, 217 Miss. 636, 65 So.2d 824 (1953); Superior Oil Co. v. Berry; Superior Oil Co. v. Foote, 216 Miss. 728, 63 So.2d 137, suggestion of error overruled, 216 Miss. 728, 64 So.2d 355 (1953). The *Berry* court in particular indicated that the involuntary pooling order altered the lessee's rights and concomitant obligations and would extend the leases on lands placed within the ordered unit.

Our reading of he foregoing Mississippi precedents inexorably leads us to the conclusion that the Mississippi Supreme Court would find that the Getty and Shell leases were extended as a consequence of the Board's unitization order and the subsequent drilling by the designated operator. In so deciding, we anticipate that Mississippi would reach the same conclusion as other oil-and-gas-sensitive jurisdictions. In Louisiana and Alabama it is well-established that drilling on any property within the involuntarily-pooled unit constitutes drilling on all property therein, just as production from a well located in the unit is taken as production from all property within the unit. . . ..

Plaintiffs-appellants' reliance on the decision in Wagner v. Mounger, 253 Miss. 83, 175 So.2d 145 (1965), is misplaced. *Wagner* did not involve a forced unitization order designating the only party who could drill a well. That decision is obviously not controlling, nor does it furnish meaningful guidance to our present quest.

Two further issues are presented in the Ridgway v. Shell appeal. First, was Shell obliged to file a pooling declaration, and did its failure to do so constitute a breach sufficient for cancellation of the lease? The district court concluded that the filing was unnecessary. We agree. The lease provides for voluntary pooling, in which event Shell would be required to file an appropriate notice of the pooling agreement. This was not a voluntary pooling agreement and Pursue, the designated operator in the forced-pooling arrangement, filed a copy of the Board's order in the office of the Chancery Clerk of Rankin County. Since it gave notice, this was a sufficient filing as contemplated by the lease, and Shell was not required to repeat that formality.

. . . .

For the foregoing reasons, the judgment of the district court, in each of the consolidated cases, is

AFFIRMED.

NOTES

1. The principal case involved compulsory units and a Case 4 situation. Should it matter if the pooling or unitization had come about through voluntary agreement rather than the compulsion of the state? On Mississippi law *compare* Wells v. Continental Oil Company, 244 Miss. 509, 142 So.2d 215, 17 O.&G.R. 527 (1962) *with* Texas Gulf Producing Co. v. Griffith, 218 Miss. 109, 65 So.2d 447, 65 So.2d 834, 2 O.&G.R. 1103, 2 O.&G.R. 1278 (1953). Consider the following statements from the Kansas court of appeals in Somers v. Harris Trust and Savings Bank, 1 Kan.App.2d 397, 566 P.2d 775, 60 O.&G.R. 19 (1977):

> The problem appears to be this: A leases 160 acres to B. Later the parties place eighty acres of the leased land into a unit. A producing well is brought in within the unit but not on A's included acres. Does such production extend the lease beyond its primary term as to A's eighty acres not within the unit?

> We have no Kansas authority squarely in point. The majority rule elsewhere is that where a portion of an oil and gas lease is committed to a unit, production anywhere in the unit extends the term of the entire lease . . . . The rule is based on conservation and public policy. Its rationale starts with the proposition that an oil and gas reservoir does not abide by man-made boundaries (K.S.A. 55–1302 defines a "pool" as " . . . an underground

accumulation of oil and gas in a single and separate natural reservoir characterized by a single pressure system so that production from one part of the pool affects the reservoir pressure throughout its extent. . . ."); because of the nature of the oil and gas pool and because of the growing need for conservation drilling units were established; the units were tracts of sufficient size and shape that one well properly located could sufficiently drain the oil and gas therefrom; unitization makes possible the exploitation which would be possible if the land or subsurface rights were in single ownership; reservoir energy and contents and physical equipment are not wasted; since unit operations provide an efficient and less expensive method of oil and gas recovery allowing lease termination as to outside acreage would discourage operators from unitizing appropriate acreage . . ..

. . .

The heart of the unitization agreement here is paragraph 3 of Article II. It is in very broad terms: "The production of unitized substances from the unit area through any well or wells shall be considered for all purposes as production of oil and gas from the land covered and affected by each oil and gas lease affecting any of the land within the unit area and production from any part of said unit area shall perpetuate all oil and gas leases whether or not the lands covered by any particular such oil and gas lease are productive or non-productive." Production within the unit is to be considered for all purposes as production from the land covered and affected by the lease. Additionally plaintiff's grantors were parties to the unitization agreement. . . . The majority rule already mentioned seems reasonable and just under the circumstances and one which should be adopted particularly where, as here, all affected parties have agreed to the unitization. The lease proviso for enlargement of the unit in event of agreement by all parties, including the lessors, is not inconsistent with its application. Plaintiff argues it would be inequitable to permit defendant to hold its lease on the north eighty after having done nothing on it for such a long period of time. It must be borne in mind this is not a suit by the lessors for violation of implied covenants in the lease where a multitude of factors may be involved. It is well established that the lessee's obligation of reasonable development and protection against drainage remains as to the outside acreage. These implied covenants continue to exist for the lessor's protection as to lease acreage outside the unit. . . .

We hold production within the unit, although not on the eighty acres in the unit, extended the 1949 lease beyond its primary term as to that acreage and that lease has not expired.

2. The *Somers* statement regarding the implied covenants of reasonable development and protection against drainage is reflected in other cases:

(a) BUCHANAN v. SINCLAIR OIL & GAS CO., 218 F.2d 436, 4 O.&G.R. 400 (5th Cir.1955). It was held in a Case 4 situation involving voluntary pooling that unit operations and production preserved the lease as to acreage included in the unit and as to acreage excluded therefrom, making further rental payments unnecessary to keep the lease in force during the primary term. The court commented as follows:

This is not to say, as seems to have been thought in Texas Gulf Producing v. Griffith, 218 Miss. 109, 65 So.2d 447, to be the case, that this court and the Supreme Court of Texas, in so holding, have in effect held that "the production from the unit well" (would) "keep the lease in force indefinitely as to the leased land without the unit" (and) "the lessor would be deprived of the right to drill on the leased land without the unit and deprived of any royalties, rentals, or benefits therefrom".

The exact contrary of this has been held in Texas Company v. Davis, 113 Tex. 331, and the companion case reported in that volume, and in Waggoner Estate v. Sigler Oil Co., 118 Tex. 500, and the host of cases following in their train, where it has been pointed out (1) that the estate acquired by the so-called lessee and his assigns is a determinable fee which will be lost on cessation of the use of the land for purposes of oil and gas exploration, development and production; (2) the estate of the lessee or his assigns will not survive abandonment or disuse; (3) the lessee, when the lease was taken was, and at all times afterwards remained, subject to the implied obligation to continue the development and production of oil or gas with reasonable diligence; and (4) and (5), while breach of the implied covenant will not authorize forfeiture and the usual remedy for its breach is an action for damages, in accordance with the equitable procedure established in state and federal courts, a court will, under extraordinary circumstances, entertain an action to cancel the lease in whole or in part.

In the light of these accepted, indeed imperative principles, it should be, we think it is, clear that, in urging upon us in this case the claimed inequity which would result if appellee is permitted to hold the entire lease without development by the device of including a part of it in a unit and citing the Mississippi case in support, the appellants have misconceived in nature and effect the judgment appealed from.

What and all the district judge determined, and adjudged, all that we hold and adjudge in affirming his judgment, is that, under the undisputed facts, drilling on, and production from, the unit well has kept, and is keeping, the lease in force against the claim that because there is no production from the lands described in lease No. 377, the lease has by its own terms expired or terminated. Nothing in the district judge's holding or judgment, nothing in our affirmance of it, justifies, authorizes, or will permit appellee to hold the ununitized portion of the lease against well founded legal claims, of abandonment or for damages for breach of the implied covenants, or, where there is no other adequate relief, against well-founded claims, for relief in equity.

The judgment is affirmed without prejudice therefrom to appellants' right to resort to any relief other than that specifically sought and denied in this case, to which they may be advised they are entitled.

(b) GREGG v. HARPER–TURNER OIL CO., 199 F.2d 1, 1 O.&G.R. 1685 (10th Cir.1952). A lease covered 160 acres on which two wells had been drilled, the first in 1943, a producer, and the second in 1947, a dry hole. In 1948, forty acres around the producing well were included in a 1,040–acre unit. Repeated demands for further drilling were made in 1948 and 1949. Although there was no proof that additional wells would be profitable, the court ordered conditional cancellation of the lease as to the 120 acres excluded from the unit, commenting as follows:

> Harper and Turner caused the unitization agreement excluding this 120 acre tract therefrom to be effected. This resulted in a division of the acreage of the Gregg tract, covered by the original lease. As the owner of the lease, they now stand in a different relationship to the excluded acreage than they do to the 40 acres included in the producing unit. Equity will consider the rights of both the lessor and the lessee under these circumstances. Their responsibilities as lessees become correspondingly greater toward the excluded acreage than it was before severing by unitization.

3. In 1977 the Oklahoma compulsory pooling statute was amended to provide as follows:

> In case of a spacing unit of one hundred sixty (160) acres or more, no oil and/or gas leasehold interest outside the spacing unit involved may be held by production from the spacing unit more than ninety (90) days beyond expiration of the primary term of the lease. 52 Okl.St.Ann. § 87.1(b).

No comparable provision is contained in the Oklahoma compulsory unitization statute. Could this statute be applied to leases existing at the time of the effective date of the statute? *See* Wickham v. Gulf Oil Corp., 1981 OK 8, 623 P.2d 613, 71 O.&G.R. 399. Would it apply to units arising from a despacing order? *See* Union Texas Petroleum v. Corporation Commission, 1981 OK 86, 651 P.2d 652, 75 O.&G.R. 105.

Mississippi has enacted a statute to divide leases for Case 3 and Case 4 situations. Miss. Code 1972, § 53–3–111. In Palmer Exploration, Inc. v. Dennis, 730 F.Supp. 734, 109 O.&G.R. 496 (S.D.Miss.1989), a federal court held that the statute's provision for lease division was not applicable to a voluntary fieldwide unit merely approved by the Mississippi Oil and Gas Board.

4. If a pooling clause in a lease requires that any unit agreement be filed with the county clerk, will unit production hold the lease when the lessee fails to comply with that requirement? *Compare* Sauder v. Frye, 613 S.W.2d 63, 69 O.&G.R. 147 (Tex.Civ.App.1981), *with* Petroleum Reserve Corp. v. Dierksen, 1981 OK 3, 623 P.2d 602, 69 O.&G.R. 284.

Where working interest owners join in a voluntary unitization, even though not all of their leases contain pooling or unitization clauses, will off-leasehold unit production hold the leases that contain unitization authority? *See* Heath v. Fellows, 526 F.Supp. 723, 71 O.&G.R. 542 (W.D.Okl.1981); Celsius Energy Co. v. Mid America Petroleum, Inc., 894 F.2d 1238, 108 O.&G.R. 113 (10th Cir.1990); *but cf.* Guaranty Nat. Bank & Trust of Corpus Christi v. May, 395 S.W.2d 80, 24 O.&G.R. 144 (Tex.Civ.App.1965, error ref'd n.r.e.); Michigan Oil Co. v. Black, 455 So.2d 824, 84 O.&G.R. 1 (Ala.1984).

5. One must distinguish between a *spacing* order (which merely limits where wells may be located) and a *pooling* order (which provides for sharing of production and expenses among owners in the designated tract). Where A and B own the mineral interests in the north and south halves of Blackacre, respectively, and X and Y hold leases from them, typically X's well in the north half will not serve to hold Y's lease, even though under a spacing order X's well is the only one that may lawfully be drilled on Blackacre. *See* Schank v. North American Royalties, Inc., 201 N.W.2d 419, 43 O.&G.R. 217 (N.D.1972). By contrast, an order providing pooling for Blackacre would typically result in drilling or production attributable to both halves.

6. Leases often require that continuation of the lease beyond the primary term requires that drilling operations in progress must be on the leasehold. Will such a clause be superseded by the pooling clause when the operations are on adjacent lands?

(a) MANZANO OIL CORP. V. CHESAPEAKE OPERATING, INC., 178 F.Supp.2d 1217, 151 O.&G.R. 42 (D.N.M. 2001). The primary term of a lease was to end on August 3, 1998. The lessee began drilling on an adjacent tract a directional well to be bottomed on lease land; it spudded July 27, 1998, and penetrated the subsurface of the lease land August 12, 1998. The plaintiff asserted that the lease required that operations had to be "on this lease". Held: the "adjacent property and the Lease property were 'pooled' or 'combined' in such a way as to make the provisions of paragraph five of the Lease applicable 'for all purposes' including the provisions of paragraph six of the lease which provided for an extension of time where drilling operation had begun."

(b) PIONEER NATURAL RESOURCES USA, INC., V. W. L. RANCH, INC., 127 S.W.3d 900 (Tex. App.--Corpus Christi 2004). The lease had a pooling clause with an areal limit to a unit of 320 acres. An addendum was made to the lease to permit a unit up to 380 acres. The lessee initiated a horizontal well on the 378 acre unit that was created, and the well bore did not penetrate lessor's land until after the primary term. The appeals court found that the addendum had not precluded pooling by reference to a well on "said land" and the lease had thus been maintained beyond the primary term by drilling on pooled acreage; accordingly, there was no trespass.

———

B. PUGH CLAUSE

RIST v. WESTHOMA OIL CO.

Supreme Court of Oklahoma, 1963.
1963 OK 126, 385 P.2d 791, 19 O.&G.R. 692.

Action by lessees to quiet title to oil and gas leasehold interests claimed by them in that part of lessors' mineral estate which is located below sea level. From a judgment quieting lessees' title, lessors appeal. Affirmed.

BERRY, JUSTICE. These consolidated appeals are from separate judgments entered by the District Court of Texas County in actions brought by lessees to determine whether the oil and gas leases involved had terminated as to all horizons below sea level at the expiration of their respective primary terms because of failure to obtain production at this level. They present a common question of law bearing on similar and undisputed facts.

The parties, having stipulated to the facts, each moved for judgment (on the pleadings) and the lower court gave judgment for the plaintiffs (lessees) and against the defendants (lessors). The lessors have appealed.

Plaintiffs in error here will be identified hereinafter as lessors or by name, and defendants in error will be identified as lessees or by name.

The two leases involved here cover land located in Texas County in the Guymon–Hugoton gas field. They are identical in form, differing only as to one day in date, individual lessors and land descriptions. Each was for a ten-year primary term.

By mesne assignments, lessees, Westhoma Oil Co. et al., became the leasehold owners of all horizons below sea level in each tract and other lessees became the owners of the leasehold estate above sea level. No question is raised as to the validity of these assignments.

There has been no oil or gas production nor operation of any kind as to the horizons below sea level. Lessees, Westhoma et al., contended in the lower court that their leasehold interest in the below-sea level leaseholds were extended beyond the primary term by unit production obtained by separate lessees from horizons above sea level. The trial court entered judgment for these below-sea level lessees and quieted their title to the below-sea level leasehold.

. . .

As we view the record, the respective claims of the parties involve, in the final analysis, the task of construing the terms of their written contract.

The lease contract provided, in so far as expiration and extension, as follows:

> 2. This lease shall remain in force and effect for a term of ten (10) years from the date of its delivery (hereinafter sometimes called the 'primary term') and as long thereafter as oil, gas, or other minerals are or can be produced from any well on said premises and as long as hereinafter

otherwise provided in the event of consolidation.

. . .

9. Lessee is expressly granted the right and privilege (which lessee may exercise at any time either before or after production has been obtained upon this premises or any premises consolidated herewith) to consolidate the leasehold estate, created by the execution and delivery of this lease, or any part or parts thereof, with the mineral leasehold estate or estates or parts thereof in any other lands upon which this lessee shall at the time own valid and subsisting lease or leases, located and situated near the property described herein, provided that any resulting consolidated estate shall not cover and include more than 2560 acres, . . . and it is agreed:

(a) This lease shall thereafter continue in full force and effect for all purposes as to the premises covered hereby and included in any such consolidation of estates, so long as gas is or can be produced from any well located on any part of the land included in such consolidation (whether on lands covered hereby or not) except as herein otherwise provided, or so long as oil is or can be produced from any well drilled on a portion of the land covered hereby.

(b) During the primary term hereof and until production has been obtained, from this or the consolidated leasehold, the lessee shall be privileged to pay the annual delay rentals stipulated herein on any part of this leasehold included in a consolidation of estates or not so included, and thereby continue this lease in full force and effect as to the part or portions thereof upon which rentals are so paid, but this lease, insofar as it covers any tract or tracts not included in a consolidation of estates held in force by production as herein provided, shall terminate at the expiration of the primary term hereof, unless oil, gas, or other minerals is or can be produced from a well or wells thereon.[100]

. . .

13. Notwithstanding anything in this lease contained to the contrary, it is expressly agreed that if lessee shall commence drilling operations upon these premises or upon premises consolidated herewith at any time while this lease is in force, this lease shall remain in force and its term shall continue so long as such operations are prosecuted and, if production

---

[100]   The term "Pugh" clause is frequently applied to language in a lease pooling clause providing for severance of a lease in the event less than all of the leasehold is included in a unit. In this instance the "Pugh" clause is embodied in the language of Section 9(a) of the lease ["except as herein otherwise provided"] and the final fifty-three words of Section 9(b). The clause is said to be named for Lawrence G. Pugh of Crowley, La., the first to use this clause in Louisiana in response to Hunter Co. v. Shell Oil Co., 211 La. 893, 31 So.2d 10 (1947). As is shown by the 1943 leases construed in the principal case, similar provisions were in use in other states prior to 1947. The "severance" provisions of such clauses vary in important respects. *See* Williams and Meyers § 669.14.

results therefrom, then as long as production continues."

The identical lease provisions involved in these appeals were considered in Rogers v. Westhoma Oil Co., 10th Cir., 291 F.2d 726, rehearing denied, 291 F.2d 732. In a divided opinion, the Federal Court reversed the trial court (U.S. District Court, Kansas) and held in effect that the lease contemplated both horizontal and vertical separations; that such was the intent of the parties; that such a separation of "parts" had been accomplished by lessees and that failure to explore the severed part below sea level within the primary term, effected or resulted in a termination of the lease as to the below-sea level leasehold and that unit production from the horizons above sea level did not operate in law to extend the lease as to below-sea level horizons.

The lessors urge this Court to adopt the reasoning and holding in the Rogers case, *supra,* and the lessees contend against it.

. . .

It is urged by lessors, that the subject leases "are not standard form leases commonly used in this area" and that "they have been judicially described as unusual lease forms" and that they do not contain the usual "thereafter" clause; and that because of this the "Pugh" clause contained in Secs. 9(a) and 9(b) show a different intent of the parties than that usually expressed in customary leases. They admit that while such language gives rise to no problem in case of a subsequent vertical separation of the leased land into parts, they also urge that it contemplates and recognizes a partial consolidation of horizontal structures.

Our research on the "Pugh" clause has thrown little light on its origin, except it seems to have first appeared in 1941. When we consider the circumstances existing and surrounding the execution of the lease we must acknowledge the state of affairs in this country at the time this form of lease was promulgated and entered into. Both leases were executed in January, 1943. This country was then mobilized and at war. The exploration for oil and gas was under Federal supervision so as to conserve the Nation's warehouse of steel and other strategic materials. With this background in mind, we will consider the purpose in view and the meaning of the words used to determine the intention of the parties.

. . .

Paragraph 9 grants to lessees "the right to consolidate the leasehold created by the execution and delivery of this lease or any part or parts thereof . . ." and in the event of any such consolidation it is agreed: (a) "This lease shall thereafter continue . . . as to the premises covered . . . in any such consolidation . . . so long as gas is or can be produced from any well located on any part of the land included in any such consolidation . . . . (b) . . . but this lease, insofar as it covers any tract or tracts not included in a consolidation . . . shall terminate at the expiration of the primary term hereof . . . ."

There is nowhere contained any language that purports to recognize or show intention that these terms are to apply or even recognize other than the customary application of vertical severance. Certainly the parties could have made reference to partial consolidation of separate horizontal structures by appropriate terms. But they say nothing as to depths, levels or strata.

The words "Tract or tracts", "Premises", "lands", and "leasehold estates" do not import to our minds other than their common meaning. . . . To us the contract terms are clear (Par. 9(b)) which require the payment of delay rentals by lessees as to part or portions. Can it be successfully maintained that this language makes provision for and anticipates different conditions of compliance as between lessees' assignors and lessors? We think not. The conduct of the parties indicates, and the briefs of lessors admit, that delay rentals paid by the owners of the above-sea level leasehold served to extend the lease year to year during the primary term as to all horizons. We are, therefore, unable to conclude that delay rentals on "parts" connotes an horizontal severance and that production from the upper stratum would not serve to extend the lease as to all strata. . . . Thus it seems clear to us that the parties entered into a lease agreement for a primary term of ten years with the term to be extended on production from the area described or from unit production of the area with no thought in mind of a severance as to horizontal divisions. This accords with Federal purpose to limit unrestrained development; with the plain terms of the lease and with common practice in the trade at the time.

To reach a different conclusion we would, in our view, be obliged to rewrite the contract and to change the obligations, rights and duties of the parties so that the lessors' rights were enlarged and the lessees' duties increased at the instant of the assignment of an interest in that part of the leasehold below sea level. . . .

. . .

The judgment of the trial court is free from errors. It is accordingly affirmed.

NOTES

1. The court's opinion written by Circuit Judge Breitenstein in ROGERS v. WESTHOMA OIL COMPANY, 291 F.2d 726, 14 O.&G.R. 921 (10th Cir.1961), included the following observations concerning the operation of the Pugh clause:

> The trial court held that the Pugh clauses were written in "surface sounding terms" and do not "specifically or clearly designate underground horizons" and concluded that the provision of the Pugh clauses terminating the leases at the end of the primary terms as to ununitized nonproducing portions apply only to "partial unitization of less than all of the surface acreage covered by the leases." As all the surface acreage was unitized, and as there was production from each unit, the decision was that the leases were continued beyond the primary terms as to all horizons.

> Admittedly, the Pugh clauses apply to vertical divisions of the leased premises. The query is whether they apply to horizontal divisions. In determining the issue we are aided by no decisions which are directly in point. . . .

We must arrive at the intent of the parties. The Pugh clauses are for the protection of the lessors to prevent lease continuation as to ununitized portions which are nonproducing. We find nothing in the leases which confines the application of the Pugh clauses to surface acres and vertical divisions. It is common knowledge that leases are divided both vertically and horizontally and that unitization is ordinarily on the basis of a common source of supply. While the inclusion of all surface areas in consolidations protects lessors from the hardships resulting, in the absence of a Pugh clause, from partial vertical consolidation, recognition of this fact does not solve the problem. A lease can provide for protection against continuation both of unconsolidated vertical divisions and of unconsolidated horizontal divisions. Considering these leases as a whole, we believe that a reasonable interpretation requires the conclusion that it was the intent of the parties to prohibit lease continuation as to unproductive portions without a consolidation whether such portions were the result of horizontal or vertical divisions. As the below-sea-level horizons were not included within any consolidations and as there was no production therefrom, the leases terminated as to such horizons at the end of the primary period.

In his opinion for the court, denying the petition for rehearing, 291 F.2d at 732, Circuit Judge Breitenstein made the following observations:

Counsel for Westhoma insist that the opinion of the court places an unreasonable interpretation on the leases because no rational purpose can be accomplished by applying the Pugh clauses to ununitized horizontal divisions. We disagree. An important purpose of the Pugh clauses when considered in connection with vertical divisions is to prevent lease continuation beyond the primary term as to unproductive areas not included within a unit and to obviate the necessity of an action for breach of the implied covenant to develop further in order to terminate the lease as to such areas. Exactly the same reason applies so far as horizontal divisions are concerned.

A Texas intermediate appellate court has adopted the *Rist* view in preference to that of *Rogers*. *See* Friedrich v. Amoco Production Co., 698 S.W.2d 748, 90 O.&G.R. 121 (Tex.App.1985).

2. Corporate lessees had struggled hard to obtain a federal rather than a state court decision in Kansas on the application of the Pugh clause to unitization of fewer than all formations underlying a tract of land. Rogers v. Westhoma Oil Co. involved twenty-seven actions, twenty-four of which had been removed from the Kansas state courts. Williams, *The Role of Federal Courts in Diversity Cases Involving Mineral Resources*, 13 Kansas L.Rev. 375 at 383–84 (1965):

[W]hen the diligent efforts of corporate lessees to get a federal rather than a state court adjudication of the operation of the Pugh clause led to a Pyrrhic victory, the Tenth Circuit adopting a construction of the clause unfavorable to the lessees, the same lessees embarked upon equally strenuous efforts to get a state court ruling on the matter in Oklahoma in advance of a federal court ruling which would undoubtedly follow the Tenth Circuit decision in *Rogers*. The friends of the court were numerous; in this case they represented lessees rather than lessors. Under these circumstances, of course, the lessors were equally anxious to get a federal court ruling in Oklahoma. On motion of the lessees, the United States District Court for the Western District of Oklahoma stayed further proceedings in that court pending the determination of the question by the Oklahoma courts. The plaintiffs-lessors in those federal cases sought and obtained from the Tenth Circuit a writ of mandamus directing the district court to proceed with the trial of the case and the court of appeals held that the trial court abused its discretion in granting the stay order.

It was clear from the opinion that Judge Breitenstein was reluctant to give the Oklahoma state courts a crack at his opinion in *Rogers*. . . .

Despite the action of the Tenth Circuit in issuing a writ of mandamus, the corporate lessees prevailed in the race, which combined diligence in the state court and delaying tactics in the federal courts. The court of appeals' opinion granting the writ of mandamus was dated August 2, 1962. The opinion denying the respondents' motion for modification was dated September 11, 1962. Certiorari was denied on December 17, 1962, and rehearing was denied on January 21, 1963. The opinion of the Oklahoma Supreme Court in Rist v. Westhoma Oil Co. was dated May 21, 1963. Victory of the corporate lessees in this race was profitable indeed, since the Oklahoma Supreme Court in *Rist* rejected the view of the Tenth Circuit as to the meaning of the Pugh clause and adopted the construction sought by the lessees.

3. Kramer & Martin, *The Law of Pooling and Unitization*, § 9.01:

The interpretation of Pugh clauses is often difficult because they are frequently added to a lease with little thought or analysis of their effect on each of the other provisions of the lease. As observed by one Louisiana attorney, the reason it is so difficult to incorporate the idea behind the Pugh clause in a simple paragraph is that

> the Pugh Clause affects, modifies, or amends virtually every one of the principal operations and lease maintenance provisions of the lease contract. Specifically, it modifies the provisions of the lease relating to term, the amount of and time at which rental payments may be made, the pooling clause, the effect of operations prior to discovery, the provisions pertaining to the manner of lease maintenance after production, the royalty provisions, and the implied obligation to develop. [Gremillion, *The 'Pugh Clause' and Some of Its Ramifications*, 11 La. B.J. 123, 125–126 (1963).]

The difficulties of interpreting the language of a Pugh clause are not made any easier by addition of the theoretical question of whether the Pugh clause "divides" a lease that is otherwise by its nature "indivisible."

There can be numerous variations as to how and when a partial division of a lease occurs under a lease clause limiting the effects of partial pooling or unitization. The lease clause may take effect only when the unit well is off the lease tract, or it may take effect regardless of the unit well's location. The clause may provide for a vertical division or for a horizontal division (division by depth or individual formation that has been pooled). The clause may have the effect of dividing the lease only when the pooling is the result of the lessee's exercise of the pooling power of the pooling clause of the lease, or it may divide the lease regardless of the manner in which pooling was achieved, whether by the lessee under the pooling clause or by force-pooling or unitization by the state conservation agency. It must be emphasized that a Pugh clause has only the effects that are specified by the clause itself, no less and no more. It is a limitation on the rights otherwise provided for by the lease, so it should not be read as derogating from those rights unless the parties have clearly so provided. But when the lessor and lessee have negotiated for a particular Pugh clause, it should be given the effects clearly provided for, even though other Pugh clauses commonly do not provide for a similar effect.

4. As to what events should trigger a Pugh Clause, consider the following cases.

BANNER v. GEO CONSULTANTS INTERN., INC., 593 So.2d 934, 119 O.&.G.R. 69 (La. App. 1992): the plaintiff landowners sought cancellation of leases granted to GEO in 1983 that covered 1220 acres. The leases had a five year primary term. GEO assigned the leases to Devon, who drilled a producing gas well. In 1990 Devon reassigned to GEO except for a 160 acre square around the producing well. The court found that this designation of 160 acres was a unit for purposes of the Pugh clause which provided: "Anything to the contrary herein notwithstanding, it is provided that if any portion of the lands held hereunder should be unitized in any manner with other lands, then unit drilling or reworking operations on or unit production from any unit shall only maintain this lease as to the land included in such unit." Did the court properly determine that this was a unit?

The rationale of *Banner v. GEO Consultants* was rejected in WILL-Drill RESOURCES, INC. v. HUGGS INC., 32,179 (La.App. 2 Cir. 8/18/99), 738 So.2d 1196, *writ denied*, 99-2957 (La. 12/17/99), 751 So.2d 885, 143 O.&G.R. 238. The Commissioner of Conservation established a drilling unit that was entirely within the land covered by the lease. The Pugh clause provided in part that "if . . . a drilling and/or production unit be created and established, pooling and combining a portion of the lands covered by this lease with other lands, lease or leases in the vicinity thereof, then" the clause was triggered and the lease would be severed as to acreage outside the unit. The lessors contended that the Pugh clause was triggered by the unit. The lessee and overriding royalty owners whose interests were carved out of the lease contended that the Pugh clause was not triggered by the unit because the unit did not actually pool and combine a portion of the lands covered by the lease with other lands as specified by the clause. The appeals court ruled that the clause had not been triggered. Because the unit formed consisted entirely of leased land, the Pugh Clause was not activated. The Pugh Clause operates to prevent a landowner-lessor's royalty interest from being diluted by inclusion of only a portion of his leased land in a unit with land held by other interests.

The Pugh clause in FEDERAL LAND BANK OF SPOKANE V. TEXACO, INC., 820 P.2d 1269, 118 O.&G.R. 84 (Mont. 1991) provided:

> Should any of the lands hereunder be included in a unit or units, during the primary term, either by written consent of the lessor or as a result of action by any duly authorized authority having jurisdiction thereof, then operations on or production from a well situated on lands embraced in such unit, shall serve to maintain this lease in force as to that portion of the leased premises embraced in such a unit, but shall not so maintain the lease or the remainder of the leased premises not embraced in such a unit .
> . .

The field was spaced on a 320 acre spacing pattern by the Oil and Gas Conservation Board of the Department of Natural Resources and Conservation for the State of Montana for the field where the lease tract lay. The Pugh clause modified the habendum clause and was not inconsistent with it, said the court. The inclusion of the land in a spacing unit triggered the operation of the Pugh clause held the Montana court on a certified question.

SCHWATKEN V. EXPLORER RESOURCES, INC., 34 Kan.App.2d 873, 125 P.3d 1078 (2006). A clause (a form of Pugh clause) drafted by lessors provided that "at the end of the primary term, this lease shall expire as to all lands located outside of a producing unit." At the end of the primary term, the lessee was engaged in operations but no unit had been formed. The court rejected the lessors' claim that the lease terminated under this paragraph. Other than the pooling clause. nowhere else in the lease was a portion of the leasehold referred to as a "unit." The paragraph in question conflicted with habendum clause; it was construed against lessor and habendum clause controlled.

SANDTANA, INC. V. WALLIN RANCH CO., 2003 MT 329, 318 Mont. 369, 80 P.3d 1224. The court described as a "Pugh Clause" a lease addendum that allowed retention beyond the primary term only of the "leased premise contained within the governmental Section in which a producing well is located." Pooling seems not to have been involved or necessary under this clause; it was given the effect of dividing the lease into portions that had to be separately maintained. The producing well in one section was sufficient only to hold the lease in that section.

5. Under what circumstances, if any, is a Pugh clause applicable to a compulsory unit? In BIBLER BROTHERS TIMBER CORP. V. TOJAC MINERALS, INC.,

281 Ark. 431, 664 S.W.2d 472, 81 O.&G.R. 24 (1984), the lease had been held by delay rentals until the last year of the primary term. A portion was then unitized, followed by production and the payment of royalties. Lessor sought cancellation as to remainder, on the grounds of non-payment of delay rentals. Lessor relied on a clause authorizing pooling and unitization and providing that in the event that part of the leased acreage should be "included in a unit created hereunder, then the remaining portion of the lands embraced by this lease shall be subject to delay rental payments as provided in Paragraph 4[providing for payment of delay rentals]." Does non-payment of delay rentals terminate the unincluded portion of the lease? Does the provision in question qualify as a "Pugh clause"?

See also Lowman v. Chevron U.S.A., Inc., 599 F.Supp. 14, 83 O.&G.R. 39 (M.D.La.1983), *aff'd*, 748 F.2d 320 (5th Cir.1984); Mathews v. Goodrich Oil Co., 471 So.2d 938, 86 O.&G.R. 60 (La.App.1985); Pearson v. Larry, 505 So.2d 913, 94 O.&G.R. 522 (La. App. 2d Cir.1987). Are there are sound reasons why leases would treat exercise of the pooling clause and pooling by the state agency differently and why the courts should recognize a distinction between the two? *See* Kramer & Martin, *The Law of Pooling and Unitization*, § 9.05.

6. Where a Pugh clause authorizes payment of delay rentals by the lessee for the non-unitized portion of the leasehold, does payment of rentals negate any lessee's duty under the implied development covenant? Does it make any difference if the provision for delay rentals covers a period after expiration of the primary term? *See* Morrison v. Conoco, Inc., 575 F.Supp. 876, 80 O.&G.R. 90 (M.D.La.1983).

7. A lease contained a 90–day drilling operations clause and also provided:

This lease shall continue in force and effect beyond the primary term and any continuous drilling operations in progress at the expiration thereof, as elsewhere provided herein, only as to that acreage for which production royalties are payable, even though there be one or more wells located upon the property described in this lease.

Shortly before expiration of the primary term lessee commenced a well; subsequently, but also before expiration of the primary term, the area on which the well was located was included in a compulsory unit that was partially composed of a portion of the leasehold (*i.e.*, a Case 4 situation). Lessee completed the well as a producer shortly after expiration of the primary term. Within the 90–day period it commenced operations on another well (on another unit partially composed of a portion of the leasehold, *i.e.*, also Case 4), which it also brought in as a producer. Was any land outside the first unit held beyond the expiration of the primary term? ROSEBERRY V. LOUISIANA LAND & EXPLORATION CO., 470 So.2d 178, 85 O.&G.R. 568 (La.App.1985), held that it was not. The court argued that drilling operations in the first unit would only hold acreage in the first unit; " . . . to find that drilling operations conducted within the [first] unit would hold all of the leased acreage whereas production from the [first] unit would not, results in an absurd consequence." Does it? Under the court's view, must lessee have a drilling rig ready to start work on non-unitized portions whenever a partial unitization appears imminent? A dissent argued that the continuous drilling operations clause held the entire acreage; only after its effects were exhausted would the Pugh clause come into play.

Compare EGELAND V. CONTINENTAL RESOURCES, INC., 2000 ND 169, 616 N.W.2d 861, 145 O.&G.R. 469. A pool was established by the North Dakota Industrial Commission for the exclusive purpose of drilling horizontal wells. Two leases were at issue. Both had the following Pugh clause:

> Notwithstanding anything contained herein to the contrary, no part or portion of the lands covered by this lease shall be committed to any unit plan of development or operation without the prior express written consent of the Lessor, her successors or assigns[.]

> A producing well, or well capable of production, will perpetuate this lease beyond its Primary Term ONLY as to those lands as are located within, or committed to, a producing or spacing unit established by Government authority having jurisdiction.

The lessee applied for and obtained compulsory pooling orders for each of five spacing units. The lessors brought a claim that the leases terminated, asserting that the lessee's application for compulsory pooling breached the Pugh clause, and that the lessee's drilling operations were ineffective to maintain the leases after the primary term because of the Pugh clause. The court rejected these contentions. Applying to the state agency for pooling was not a voluntary pooling that required consent of the lessors. The North Dakota court followed the majority rule that "governmental pooling and unitization orders do not divide a lease, and production anywhere on the pooled acreage holds all leases that may be wholly or partly in the unit." Although the Pugh clause modified the leases by providing that production on the leases extended the leases only as to the specific lands sharing in production, the Pugh clause did not negate the additional language in the leases allowing the leases to be extended by continuous drilling operations. The court concluded the Pugh clauses did not terminate the leases as to those lands not committed to a producing spacing unit, and the leases could still be extended by continuous drilling operations.

8. KYSAR V. AMOCO PRODUCTION CO., 135 N.M. 767, 93 P.3d 1272, 160 O.&G.R. 595 (2004). Responding to a certified question from the Tenth Circuit, the New Mexico Supreme Court ruled that a Pugh Clause incorporated in a lease was intended to limit the effects of a communitization undertaken by the lessee. The clause provided: "[T]he commencement, drilling, completion of or production from a well, or any portion of a unit created hereunder, shall not have the effect of continuing this lease in force insofar as it covers the land not included within such unit, and no unit shall be created which covers and includes land in more than one Section." As a consequence, the lessee did not have an implied right to use lease acreage not within the unit to gain access to

the unit well, which was off the lease tract. See also Kysar v. Amoco Production Co., 379 F.3d 1150, 160 O.&G.R. 623 (Tenth Cir. 2004).

9. Act No. 786 of 1991 of the Louisiana legislature amends and reenacts R.S. 30:129(B) to provide that each lease entered into by the Louisiana Mineral Board after August 1, 1991, shall contain a Pugh clause. Such a clause shall provide that the commencement of operations for the drilling of a well, the conducting of reworking operations, or production of minerals, on any portion of a unit which embraces all or any part of the property covered by such lease shall maintain the lease in effect under the terms of the lease only as to the part of the leased property embraced by the unit. The clause may provide that the acreage outside the unit(s) may be maintained by any means covered by the lease, but if by rental payments, then such payment may be reduced proportionately to the amount of acreage included in the unit as it bears to the total acreage in the lease, provided that the rental per acre on the outside acreage shall not be less than one-half of the cash payment paid for the lease per acre nor shall the lease on the non-unitized acreage be extended more than two years beyond the primary term. As a drafting exercise, draft a clause that meets this statutory requirement. Would you need to consider the impact of such a clause on other provisions of the lease? In what circumstances would you think it appropriate to include such a clause in a lease for a private lessor?

C. CALCULATION OF ROYALTIES AND PRODUCTION PAYMENTS

Martin & Kramer, *Williams and Meyers Oil and Gas Law*

§ 951. The Effect of Pooling and Unitization Upon Oil and Gas Leases: Royalty Clause

The effect of the royalty clause of the lease is modified by pooling or unitization in each of Cases I through IV. By the terms of the typical lease, the lessor is entitled to a royalty on production from Blackacre but is not entitled to a royalty on production from other tracts. By virtue of the pooling or unitization, the lessor becomes entitled to a royalty on a pro rata share of the production attributable to Blackacre under the terms of the participation formula of the agreement, whether the production is from Blackacre itself or from other tracts included within the agreement. Prior to the agreement, the lessor was entitled to a royalty measured by the full production from Blackacre; after the agreement he is entitled to a royalty measured only by a pro rata share of the production from Blackacre. Prior to the agreement, the lessor was not entitled to a royalty on production other than from Blackacre; after the agreement he is entitled to a royalty measured by a pro rata share of production from other tracts included in the unit with Blackacre.

If Blackacre is a twenty-acre tract and all of Blackacre is included within a forty-acre unit upon which one well is drilled (Cases I and II), the lessor of

Blackacre who was entitled under the terms of the lease to a 1/8 royalty on production from Blackacre will be entitled to a 1/8 royalty on an apportioned share of the production under the unit. Similarly, if Blackacre is a 200–acre tract, twenty acres of which are included in a forty-acre unit on which a producing well is drilled (Cases III and IV), the lessor of Blackacre, who under the terms of the lease was entitled to a 1/8 royalty on production from Blackacre, will thereafter be entitled to a 1/8 royalty on an apportioned share of the production under the unit. Assuming the participation formula in both instances is on an acreage basis, the lessor will be entitled under the agreement to a 1/8 royalty on 1/2 of the oil produced by the well, whether the well is located on Blackacre or on a portion of the unit other than Blackacre. In effect, then, the lessor under the agreement will have a 1/16 royalty on production from the unit.

PUCKETT v. FIRST CITY NATIONAL BANK OF MIDLAND

Texas Court of Appeals, 1985.
702 S.W.2d 232, 90 O.&G.R. 144, writ ref'd n.r.e.

MCCLOUD, CHIEF JUSTICE. The principal issue in this oil and gas case is which method of calculating royalties should be used in a "split stream" sale of gas produced from a voluntarily pooled unit. A split stream sale results when, as in the instant case, there is more than one owner of the working interest, and the individual working interest owners sell the gas allocated to them to different gas purchasers at different prices. Two methods of calculating royalties appear to have evolved: (1) the "weighted average" method whereby the royalty owners are paid on the basis of the sale of all gas produced and sold from the unit by all working interest owners; and (2) the "tract allocation" method whereby royalty owners are paid on the basis of the amount received by their individual lessee from the sale by their lessee of the proportion of gas from the unit allocated to the tract in which the royalty owner owns an interest.

This suit arose from disputes over gas produced from two gas units known as the Fort Stockton Gas Unit No. 3 (FSGU–3) and the Fort Stockton Gas Unit No. 6 (FSGU–6). Midland National Bank, now known as First City National Bank of Midland, acquired an oil and gas lease on certain lands owned by the Pucketts. In 1968, pursuant to the lease, the Bank pooled 130.57 acres of the leased land with other acreage held by the Bank, Gulf Oil Corporation and other leasehold owners to form FSGU–3. The Bank pooled an additional 62.84 acres with other acreage held by the Bank, Gulf and other leasehold owners in 1969. This unit is known as FSGU–6. The Bank has sold its share of the gas produced from the units to Northern Natural Gas Company and LoVaca Gathering Company. Gulf has sold its share of the gas produced to LoVaca Gathering Company.

In 1978, Gulf sued LoVaca and the Bank alleging that LoVaca erroneously paid the Bank for gas that should have been credited to Gulf. The Bank filed a third party action against the Pucketts, the royalty owners, seeking reimbursement for royalties mistakenly paid the Pucketts. The Pucketts counterclaimed against Gulf and the Bank alleging that their 3/16 royalty interest was underpaid. The Bank sought indemnity and, alternatively, contribution from Gulf. The Bank also filed a third party action against Northern seeking indemnity, and Gulf cross-claimed against the Bank seeking indemnity. Before trial, LoVaca settled all claims and was no longer a party to the lawsuit, and Gulf had been fully reimbursed by LoVaca and the Bank on the original claim.

In a nonjury trial, the trial court held that the Bank, the Pucketts' lessee, properly applied the "tract allocation" method in calculating the royalties paid to the Pucketts by the Bank. The trial court held that the Pucketts were entitled to a 3/16 royalty on the proceeds from the gas "allocated" to the Bank on an acreage basis. The trial court found that the Bank was entitled to recover from the Pucketts royalties mistakenly overpaid in the amount of $409,110.63. The trial court, however, found that the Bank had underpaid the Pucketts' royalties in the amount of $67,823.94 on a 5.5904 percent interest in FSGU–3 transferred by the Bank to Gulf. The Bank was awarded contribution from Gulf for $45,493.24 of this underpayment. The Pucketts were credited with the $67,823.94 underpayment and $329,592.62 in royalties due the Pucketts but withheld by the Bank pending the conclusion of the controversy. The trial court's judgment ordered the Pucketts to pay the Bank the remaining $11,694.07 in excess royalties. The Bank recovered nothing against Northern for indemnity. The Pucketts' claims for additional royalties based on the weighted average method and for attorney's fees were denied. The Pucketts and Gulf appeal.

Royalty Calculation

Under their lease, the Pucketts receive a 3/16 royalty from the production of three wells: FSGU Well 3–1 and FSGU Well 3–2 on FSGU–3 and FSGU Well 6–1 on FSGU–6. FSGU Well 3–1 and FSGU Well 6–1 were completed in 1969. Originally, the Bank authorized Gulf to sell its share of production. Gulf sold the gas to LoVaca at intrastate prices. Starting on September 1, 1971, the production from FSGU Well 3–1 and FSGU Well 6–1 was sold in a split stream arrangement: the bank sold its share of production to Northern at interstate prices, while Gulf continued to sell its share of production to LoVaca. When FSGU Well 3–1 was plugged back to the Fusselman formation and recompleted in 1975, the Bank arranged to sell its share of production from the well to LoVaca. The Bank has continued to sell its share of production from FSGU Well 6–1 to Northern. FSGU Well 3–2 was completed in 1974. The Bank has continually sold its share of production from this well to Northern. From August 1973 to the date of trial, the intrastate prices paid by LoVaca exceeded the interstate prices paid by Northern. This led to the Pucketts' claim that their royalties had been underpaid.

The Pucketts first argue that the oil and gas lease and the division orders provide that their royalties should be based upon proceeds received from the sale of "all" gas from the two units involved by all working interest owners. To the extent that Gulf sold gas from the unit, the Pucketts seek to hold Gulf jointly and severally liable.

The Pucketts urge that the tract allocation method, used by the working interest owners and approved by the trial court, is inconsistent with the terms of the lease and the division orders. We disagree.

The oil and gas lease between the Pucketts and the Bank provides in part:

5. Lessee is hereby granted the right to pool or unitize this lease, the land covered by it or any part thereof with any other land, lease, leases, mineral estates or parts thereof for the production of oil, gas or any other minerals. . . . The entire acreage pooled into a unit shall be treated for all purposes, except the payment of royalties on production from the pooled unit, as if it were included in this lease. *In lieu of the royalties herein provided, lessor shall receive on production from a unit so pooled only such portion of the royalty stipulated herein as the amount of his acreage placed in the unit or his royalty interest therein on an acreage basis bears to the total acreage so pooled in the particular unit involved.* (Emphasis added by the court. Eds.)

The royalty clause of the lease provides for payment by the lessee to the Pucketts of "the market value at the mouth of the well of 3/16 of the gas so sold or used." The two division orders each state, in part:

Until further written notice, you are hereby authorized to account to each of the undersigned for his interest in said gas in accordance with the division of interest which is correctly set out herein, subject to the terms and conditions hereinafter set forth:

1. Settlement hereunder shall be made on the basis of the proceeds derived from sales of such production and upon the volume computations made by the purchaser(s) thereof.

The operating agreements, signed by the working interest owners, provide in part:

9. *Marketing production*

Each party hereto shall have the right to take in kind or separately dispose of *its proportionate share* of the production from the Unit Area, exclusive of production used in development and producing operations and in preparing and treating oil for marketing purposes and production unavoidably lost. Upon any sale each party shall execute the division order or sales contract applicable to its own interest and shall receive the proceeds of the sale directly from the purchaser thereof. In event any party shall fail to make the arrangements necessary to take in kind or separately dispose of its proportionate share of the production, Operator shall have the right for the time being, subject to revocation at will by the party owning the same, to purchase such production or to sell the same to others at not less than the market price. . . . (Emphasis added by the court. Eds.)

The unit designations for FSGU–3 and FSGU–6 expressly provide for the allocation of production to each tract forming a part of the units stating:

The parties further agree that production of gas . . . from any portion of the hereinafter described unit shall be *allocated* between the tracts comprising such unit on the acreage basis hereinafter set forth, and that each of their interests in such production shall be paid on the basis of such allocation. (Emphasis added by the court. Eds.)

In their division orders with the Bank, the Pucketts ratified, confirmed and adopted the establishment of the units. The division orders state in part:

7. The production which is the subject of this division order is obtained from a unit, and the establishment of such unit is hereby ratified, confirmed and adopted.

The Pucketts specifically ratified and adopted the establishment of the units. See Exxon Corporation v. Middleton, 613 S.W.2d 240 (Tex.1981).

The oil and gas lease provides that the Pucketts are to be paid royalties by the Bank (their lessee) for the sale of gas marketed by the Bank. We find nothing in the pooling clause that expands the Pucketts' rights or imposes on the Bank any greater obligation regarding the marketing of the gas. The pooling clause specifies that the "lessor shall receive on production from a unit so pooled" its proportionate royalty. It does not say "on proceeds from the sale of all production" as urged by the Pucketts. The division orders signed by the Pucketts are addressed to the Bank. The division orders provide that "settlement . . . shall be made on the basis of the proceeds derived from sales of such production." This obviously means the sale of production by the Bank. There is nothing in the lease or division orders requiring Gulf or the Bank to pay the Pucketts' royalty based on the price that Gulf obtains for the benefit of its lessors for its allocated gas.

We disagree with the Pucketts' argument that the oil and gas lease and the division orders require that royalties be calculated on the basis of the proceeds received from the sale of "all gas" produced and sold from each unit. The Bank is the party obligated to market the gas once it is produced. . . .. Under the Pucketts' theory, the Pucketts would be bound by the efforts of Gulf and the other working interest owners in marketing their proportion of the gas produced, even though Gulf and the other working interest owners have no contractual obligations with the Pucketts. Likewise, under the Pucketts' theory, the lessors of Gulf would be bound by the efforts of the Bank in marketing the gas allocated to the Bank. We hold that the lease and division orders provide, as found by the trial court, that the Pucketts should receive a 3/16 royalty on the gas allocated on an acreage basis to their lessee.

The Pucketts urge us to follow Shell Oil Company v. Corporation Commission, 389 P.2d 951 (Okla.1963) (commonly referred to as the *Blanchard* case), a compulsory pooling case. There, the Oklahoma Supreme Court did adopt a "weighted average" method of calculating royalties. However, it is clear that the holding of the court was based upon a state statute, 52 Okla.Stat. sec. 87.1(d)(4) (1961), which provided:

In the event a producing well, or wells, are completed upon a unit where there are, or may thereafter be, two (2) or more separately owned tracts, any royalty owner or group of royalty owners holding the royalty interest under a separately owned tract included in such spacing unit *shall share in the one-eighth (1/8) of all production from the well or wells drilled within the unit.* . . . (Emphasis added by the court. Eds.)

In holding that the statute specifically provided that royalty owners should receive a royalty interest on "all" production, the court stated:

Under paragraph 4, above, it is specifically provided that any royalty owner shall share in the 1/8th of all production from the well in the proportion that his acreage bears to the entire acreage of the unit. . . . While it is true that under their lease agreements with their lessees, Waldo Blanchard, O.S. Black and W.R. Blanchard have authorized their lessees, respectively, to dispose of the gas produced from their *separate tracts,* the statute supersedes their lease contracts and directs that each lessor will share in 1/8th of *all production from the unitized area.* (Emphasis in original opinion).

The *Blanchard* case is clearly distinguishable. The instant case is a voluntary, not compulsory, pooling case. More importantly, Texas has no statute similar to the Oklahoma statute.

Another landmark case in the area of "split stream" sales of gas is Arkansas Louisiana Gas Co. v. Southwest Natural Production Co., 221 La. 608, 60 So.2d 9 (1952). There, the Supreme Court of Louisiana, when confronted with the question of the proper allocation of royalties, adopted the "tract allocation" method. That case, however, also involved the interpretation of statutes and a pooling order entered by the Commission of Conservation. We do note that the Texas involuntary pooling statute, Texas Mineral Interest Pooling Act, Tex.Nat.Res.Code Ann. sec. 102.001 et seq. (Vernon 1978), is similar to the Louisiana statute and arguably adopts the tract allocation method in an involuntary pooled unit. Section 102.051 states:

(a) For the purpose of determining the portions of production owned by the persons owning interests in the pooled unit, *the production shall be allocated to the respective tracts within the unit* in the proportion that the number of surface acres included within each tract bears to the number of surface acres included in the entire unit. (Emphasis added by the court. Eds.)
. . ..

It is noteworthy to point out that in both the *Blanchard* case and Arkansas Louisiana Gas Co. v. Southwest National Production Co., supra, the issue was whether or not the applicable statute changed the method of calculating royalties provided for in the respective oil and gas leases. While the lease terms were not discussed in either opinion, it is clear that each court felt, in the absence of some superseding statutory provision, that the royalties should be paid on the "portion allocated to the tract in which he has an interest." Arkansas Louisiana Gas Co. v. Southwest National Production Co., supra, at p. 10.

Next, the Pucketts argue that the pooling of their acreage resulted in a cross conveyance between all royalty owners under Veal v. Thomason, 138 Tex. 341, 159 S.W.2d 472 (1942), and, therefore, each royalty owner should be paid royalties based on the weighted average method. We disagree. The court in *Veal* was concerned with necessary parties. The court did not address the question of how to calculate royalties in a split stream sale of production from a pooled unit. Moreover, the Pucketts' lease specifically states that the entire acreage pooled in a unit shall be treated for all purposes, "except the payment of royalties on production from the pooled unit," as if it were included in the lease. The language of the lease shows an express intent not to effect a cross conveyance as to payments of royalty.

The Pucketts also rely upon TXO Production Corporation v. Prickette, 653 S.W.2d 642 (Tex.App.—Waco 1983, no writ), and assert that the court adopted the "weighted average" approach. *TXO* was a venue case. The issue was whether the lessor had established a cause of action. The court held that the lease, its pooling provision and the division order established a right to royalties, and when no royalties were paid, the royalty owner had a cause of action against the operator who was bound by the division order. The issue was not the calculation of royalties when the gas was being sold by two working interest owners at different prices to different gas purchasers. The division order provided that, "effective 1st sales," the operator would "give credit as set forth herein for all proceeds derived from the sale of gas." Also, there is no discussion in *TXO* of gas balancing which permits the working interest owners to equalize the production.

. . .

That portion of the judgment denying the Pucketts' attorney's fees is severed and reversed, and that cause is remanded. In all other respects, the judgment of the trial court is affirmed.

NOTES

1. Among the questions raised by a "split connection" is whether each lessee owes royalty only to his own lessor, for sales attributable to him pursuant to the agreement of the lessees, or whether each lessee owes each lessor royalty for every unit of gas sold. Under the "tract allocation" method, each lessee is liable only to his own lessor. For some of the consequences, *see* Symposium, *Workshop on Natural Gas Prorationing and Ratable Take Regulation*, 57 Colo.L.Rev. 149, 363–70 (1986). So far as the standard 1/8 is concerned, the *Blanchard* case, discussed in *Puckett,* adopts the second view, i.e. each lessee owes each lessor royalty for every unit of gas sold. Is that compelled by the Oklahoma statute quoted in the *Puckett* opinion? What are the *Blanchard* rule's effects so far as imposing on lessees risks of other lessees' insolvency? Its effects on administrative costs in royalty payment? *See* Mosburg, *Practical Effects of the 'Blanchard' Case*, 35 Okla. Bar Ass'n J. 2331 (1964). Subsequent legislation has changed the effect and impact of the *Blanchard* rule. One amendment to the basic conservation act has made the first purchaser(s) liable to royalty owner(s) in any

tract for the payment of proceeds from the sale of production, rather than the seller of the production as seemed suggested under the prior wording quoted above from the *Blanchard* case. Okla. Stat. Ann. tit. 52, § 87.1(e). Each royalty owner is to share in all production from the well(s) drilled within the unit, or in the gas well rental provided for in the lease on the tract, to the extent of the royalty interest owner's interest in the unit. This is not limited to a one-eighth.

2. Under the *Blanchard* doctrine, the liability for royalty commences, for all lessees, at the moment of the first sale. Under the "tract allocation" method, there are at least two possible approaches: (1) that each lessee becomes liable to his lessor from the moment of the first sale, and thereafter continuously with respect to every unit of gas sold, or (2) that each lessee becomes liable for royalty only when sales allocated to him are made. All three alternatives were presented before the Corporation Commission in the *Blanchard* case. See Mosburg, *supra*, at 2333. For a decision adopting the first possible version of the "tract allocation" method, *see* TXO Production Corp. v. Prickette, 653 S.W.2d 642 (Tex.App.1983), which is discussed and distinguished in the *Puckett* opinion.

3. Where there is a split-stream sale, the purchaser(s) from the lessees may take natural gas at different rates from the lessees so that the lessees get "out-of-balance" on their gas production; one lessee will be over-produced and the other will be under-produced. Generally the lessees will have the opportunity to get back in balance. Consider the following in connection with royalty obligations in a unit where the lessees are out of balance. Martin, *The Gas Balancing Agreement: What, When, Why and How*, 36th Rocky Mt. Min. L. Inst. 13–1, 13–42 to 13–45 (1990) [footnotes omitted]:

i) Payment of Royalty by Overproducer

Should an overproducer pay on 1) volume of gas actually sold by that overproducer or 2) the volume of gas produced by the overproducer that the producer is entitled to take in the absence of any balancing volumes? An overproducer may be reluctant to pay royalty on the overproduction volume even though selling it, just as the overproducer may be reluctant to treat such volume as income for tax purposes. This is not simply a matter of greed; it is due in part to the very real possibility of the overproducer having to balance with the underproducer in cash later and then having no practical prospect of recouping an overpayment of royalty to the overproducer's own royalty owner. In a real sense the overproduction can represent a liability of the overproducer whether there is a gas balancing agreement or not.

ii) Payment of Royalty by Underproducer

Under a "proceeds" type royalty provision of a lease, the underproduced lessee may assert that no royalty is owed as there are yet no proceeds on which to pay royalty. If one takes the position that the underproduced working interest owner has no production, then one is faced with the question of lease maintenance. Can the lessee rely on the shut-in royalty clause? This of course depends on the wording of the clause. There are variations. Some state simply that if gas is not produced, then the lessee may pay shut-in royalty. Others state if the well is shut-in for lack of a market the lessee may maintain the lease by paying shut-in royalty. Typically, the shut-in payment is much lower than royalty on production would be, whatever the price being received. Leases also provide that the lease may be maintained by other means. Can the underproducer take the position that the operator's actions in producing are operations that maintain the lease even though the lease's shut-in well clause does not apply? Some lessees have been unwilling to pay royalty on take-or-pay payments and settlements. Will we have litigation over the underproduced lessee who after receiving a payment for cash balancing from an overproducer asserts that no royalty is owed on such payments as it is not for production?

Assuming that an underproduced party is to pay royalty even though gas is not being sold by the underproduced party, what should be the value or price at which royalty is paid? Under a "value" royalty provision, the underproduced lessee may pay the lessor a royalty based on the overproducer's price received. If the price goes up when the underproducer actually sells its share of the gas, must the lessee pay its royalty owner the difference between the actual sale value and the value used in the earlier royalty payment? If the price goes down when the underproducer actually sells its share of the gas, may the lessee recoup from royalty owner the difference between the actual sale value and the value used in the earlier royalty payment? When a lessee wishes to have the possibility of recoupment in this latter situation, it may be advisable to include a notice to the royalty owner that royalty payment is being made subject to such a recoupment.

4. A lease of Blackacre provides for payment of a 1/8th royalty but the lease of Whiteacre provides for payment of a 1/6th royalty. Blackacre and Whiteacre are pooled and production is obtained from a unit well. To what royalty payment are the lessors of Blackacre and Whiteacre entitled? Does it matter whether the unit well is on Blackacre or on Whiteacre? Does it matter whether the pooled unit is formed in accordance with the provisions of a lease pooling clause or is a compulsory unit?

Suppose the printed lease on Whiteacre provided for 1/8th royalty and a typewritten rider provided for an additional 1/16th? *See* Veverka v. Davies & Co., 10 Kan.App.2d 578, 705 P.2d 558, 86 O.&G.R. 471 (1985).

5. The Blackacre lease provided for reduction of royalty when a well on the premises was "not capable of producing 50 barrels of oil per day when operated by a competent operator utilizing efficient and practicable methods and equipment for producing oil and gas." A voluntary unit was formed under which the production allocated to Blackacre was less than 50 barrels of oil per day per well. What royalty is due Lessor? *See* Beene v. Midstates Oil Corp., 169 F.2d 901 (8th Cir.1948); McLachlan v. Stroube, 324 S.W.2d 279, 11 O.&G.R. 98

(Tex.Civ.App. 1959, writ ref'd n.r.e.) (sublease that provided for a sliding scale overriding royalty based on production); Marathon Oil Co. v. Kleppe, 407 F.Supp. 1301, 54 O.&G.R. 535 (D.Wyo.1975), *aff'd*, 556 F.2d 982, 580 O.&G.R. 600 (10th Cir. 1977)(injection wells to be counted in calculating average production for purposes of sliding-scale royalty).

6. A lease clause provides that changes in ownership will be ineffective until the lessee is furnished with evidence of the change. Lessor A transfers his interest to B, and no notice is given. The operator pays A. Does the Oklahoma statute entitle B to royalty with respect to the same gas? *See* Olansen v. Texaco Inc., 1978 OK 139, 587 P.2d 976, 62 O.&G.R. 193.

7. Leases often provide that proceeds are to be used for computing royalty for sales at the well, market value for sales away from the well. Under Exxon Corp. v. Middleton, 613 S.W.2d 240, 67 O.&G.R. 431 (Tex.1981), a sale is "at the well" if transfer occurs on the lease premises. Suppose transfer takes place on unit but off lease? Is it constructively at the well for purposes of this clause? *See id.*

8. Under a joint operating agreement, one or more parties may go "non-consent" to drilling additional wells in the unit or operating agreement area, subject to a "non-consent penalty". The operating agreement may provide a clause such as this: "During the period of time Consenting Parties are entitled to receive Non-Consenting Party's share of production, or the proceeds therefrom, Consenting Parties shall be responsible for the payment of . . . all royalty . . . and other burdens applicable to Non-Consenting Party's share of production . . . ." The rights of a lessor as a third party beneficiary against the consenting parties under the joint operating agreement are discussed in In re Moose Oil & Gas Company, 347 B.R. 868 (Bankr. S.D. Tex. 2006).

D. EFFECTIVE DATE
TEXACO INC. v. INDUSTRIAL COM'N

Supreme Court of North Dakota, 1989.
448 N.W.2d 621, 109 O.&G.R. 25.

ERICKSTAD, CHIEF JUSTICE. Texaco Inc. appealed from a district court decision affirming a compulsory pooling order issued by the North Dakota Industrial Commission for the development and operation of an oil and gas well in the NE 1/4 of Section 25, Township 153 North, Range 96 West in McKenzie County. We affirm.

The NE 1/4 of Section 25 is located in the Keene–Silurian Pool. Texaco holds mineral leases covering 29/32 of the mineral interests in the NE 1/4 of Section 25, and Harley Thompson owns an undivided 3/32 of the mineral interests, which are unleased. On December 10, 1987, Texaco completed a well in the NE 1/4 NE 1/4 of Section 25. The Commission had previously set the proper spacing unit for the Keene–Silurian Pool at one well per 160 acres, and the

160 acre spacing unit for Texaco's well is the NE 1/4 of Section 25.

In August 1987 Texaco and Thompson failed to agree to terms for Texaco to lease Thompson's mineral interests in the NE 1/4 of Section 25 or terms for a joint operating agreement for drilling and operating the well. On July 15, 1988, Thompson applied to the Commission for a compulsory pooling order for the mineral interests in the spacing unit covered by the NE 1/4 of Section 25. The Commission entered an order pooling all interests in the spacing unit effective from the date of first operations. The Commission's order also required Thompson to "reimburse the operator for [his] proportionate share of the reasonable actual cost of drilling and operating [the] well, plus a reasonable charge for supervision." Texaco appealed to the district court, which affirmed the Commission's order. Texaco has now appealed to this court.

Texaco contends that the Commission erred as a matter of law in ordering compulsory pooling retroactive to the date of first operations. Texaco argues that it is not "just or equitable" for Thompson, "who has contributed nothing to the drilling of the well and has incurred no risk whatsoever throughout the entire extended process of drilling, completing and operating the well," to "now receive a retroactive share of the working interest in all of the oil and gas produced from Texaco's own well located on Texaco's solely owned leasehold." Texaco asserts that in ordering compulsory pooling retroactive to the date of first operations, "the Commission has taken a portion of Texaco's working interest in its own well, drilled on a tract in which Texaco owns 100% of the working interest, and given it to Thompson, resulting in an unlawful confiscation of Texaco's property and a taking of its property without due process of law."

Section 38–08–08, N.D.C.C., authorizes compulsory pooling:

38–08–08. Integration of fractional tracts.

1. . . . In the absence of voluntary pooling, the commission upon the application of any interested person shall enter an order pooling all interests in the spacing unit for the development and operations thereof. Each such pooling order must be made after notice and hearing, and must be upon terms and conditions that are just and reasonable, and that afford to the owner of each tract or interest in the spacing unit the opportunity to recover or receive, without unnecessary expense, his just and equitable share."

In assessing whether Section 38–08–08, N.D.C.C., authorizes the Commission to order compulsory pooling retroactive from the date of first operations, we are guided by our well-established rule of statutory construction that statutory provisions must be construed as a whole to determine the intent of the Legislature. See, e.g., County of Stutsman v. State Historical Society, 371 N.W.2d 321 (N.D.1985).

Section 38–08–08, N.D.C.C., is part of our Oil and Gas Conservation Act [ch. 38–08, N.D.C.C.], which was enacted in 1953. The Act recognizes the public's interest "to foster, to encourage, and to promote the development, production, and utilization of . . . oil and gas . . . in such a manner as will prevent waste; . . . provide for . . . a greater ultimate recovery of oil and gas . . . and [protect] . . . correlative rights of all owners." Section 38–08–01, N.D.C.C.

In furtherance of that public interest the Act modifies the "rule of capture" by authorizing the Commission to set spacing units for a common source of supply "[w]hen necessary to prevent waste, to avoid the drilling of unnecessary wells, or to protect correlative rights." Section 38–08–07(1), N.D.C.C. A spacing order shall specify the location of the permitted well on the spacing unit in accordance with a reasonably uniform spacing unit. Section 38–08–07(3), N.D.C.C. Subject to limited exceptions, after a spacing unit is established only one well may be drilled upon the unit. Section 38–08–07(4), N.D.C.C. No drilling activities may be commenced until an application for a permit to drill is approved and a permit is issued. Section 43–02–03–16, N.D.A.C.

If a wildcat well is drilled on land not covered by a spacing order, the Commission must docket a spacing hearing within thirty days and thereafter issue a temporary spacing order. Section 43–02–03–18(3), N.D.A.C. A temporary spacing order shall remain in force until a proper spacing order is issued, and during the period between the discovery of minerals and the issuance of a temporary spacing order, no drilling permits for an offset well may be issued unless approved by the enforcement officer. Id. Thus, as a practical matter, in a spacing unit in which no pooling agreement has been entered, the provisions of the Oil and Gas Conservation Act preclude an owner of a working interest adjacent to a producing well from recovering his minerals by drilling a well on his land.

The Nebraska Supreme Court recognized the effect of that denial when it affirmed an order allowing compulsory pooling retroactive to the date of commencement of production under a statute similar to our compulsory pooling statute:

> Under the common law rule of capture, appellees would have been entitled to all oil produced and appellants only remedy would have been to drill its own well. With the adoption of the Oil and Gas Conservation Act, a landowner could no longer so protect his interest. It became necessary to get a drilling permit and the act contemplates that there shall be only one well if that one can adequately pump out the oil in the pool. Here the appellants are entirely dependent for protection on the pooling order allocating to them a share in the production and the costs of production of appellees' well. The several sections of the act consistently stress the protection of correlative rights. They are clearly designed to protect adjoining landowners under whose lands a pool may extend. To do so in a fair, reasonable, and adequate manner, and to permit an adjoining owner to obtain, recover, and receive his just and equitable share, the pooling order may be made retroactive to the time production started and, insofar as costs are concerned, to the start of drilling operations. Unless the order may be made effective retroactively, it

may on occasion verge on the confiscatory. Application of Farmers Irrigation District, 187 Neb. 825, 194 N.W.2d 788, 791–792 (1972).

The Nebraska decision does not mandate retroactive pooling in every case but requires that "[a]ll pertinent factors . . . must be considered . . . [and the parties' rights must be resolved] upon an equitable basis." Id., supra, 194 N.W.2d at 792. Our compulsory pooling statute, like Nebraska's statute, does not mandate retroactive pooling but requires the Commission to balance the competing interests of the mineral owners and operators to achieve a "just and reasonable" result and "afford to the owner of each . . . interest in the spacing unit the opportunity . . . to receive . . . his just and equitable share." Section 38–08–08(1), N.D.C.C.

We agree with the Nebraska Supreme Court's rationale that unless the Commission can issue pooling orders retroactive to the date of first operations, an adjoining landowner may not receive his just and equitable share in a pool, thereby confiscating his property without due process. *See also* Ward v. Corporation Commission, 501 P.2d 503 (Okla.1972).[101] We therefore conclude that, giving effect to both the spacing and pooling provisions of ch. 38–08, the Commission may, within the guidelines of Section 38–08–08, N.D.C.C., issue compulsory pooling orders retroactive to the date of first operations.

Accordingly, we review the Commission's decision under the guidelines of that statute, which require a compulsory pooling order to be "just and reasonable" and to "afford to the owner of each . . . interest in the spacing unit the opportunity to . . . receive . . . his just and equitable share." Our review of the Commission's decision is governed by Section 38–08–14(4), N.D.C.C., which requires affirmance "if the commission has regularly pursued its authority and its findings and conclusions are sustained by the law and by substantial and credible evidence." . . .

In this case, in August 1987, Texaco negotiated with Thompson to lease Thompson's mineral interests in the NE 1/4 of Section 25 or to execute a joint operating agreement for the well. Those negotiations broke down because Thompson wanted a commitment from Texaco that Texaco would never transfer

---

[101]   In *Ward*, supra, 501 P.2d at 507, the Oklahoma Supreme Court said:

At the moment production commences, resulting pressure differentials in the common source of supply portend, in greater or less degree, drainage from all parts of the unit toward the producing unit well. This drainage is occurring from areas where oil and gas lessees are prohibited from doing anything to protect their leased premises from drainage. With the purpose of § 87.1[Oklahoma's spacing unit statute] to prevent the drilling of unnecessary wells before it, the Commission will not, except in extreme cases, make an exception to the rule that permits one producing well only on each spacing (drilling) unit. To impose this denial without granting the right to participate in production of the unit well, as of the time the non-drilling owners were prohibited from drilling, is the taking by the State of their property without due process in violation of the Fourteenth Amendment to the Constitution of the United States.

its interest to a specified company. An order establishing proper spacing for this land was in effect when this well was completed on December 10, 1987. At that time neither Texaco nor Thompson, as an "interested party," had applied for a compulsory pooling order under Section 38-08-08(1), N.D.C.C. Thompson applied for a compulsory pooling order in July 1988, and the order was entered in October 1988. Without a retroactive order, Texaco would receive the benefit of Thompson's minerals from the beginning of operations and Thompson would, as a practical matter, be precluded from recovering his minerals during that time by drilling an offset well. Under these circumstances, we believe the Commission's decision that retroactive pooling was just and reasonable in this case is supported by substantial and credible evidence. We therefore affirm the Commission's decision to allow compulsory pooling retroactive to the date of first operations.

Texaco also argues that Thompson must pay his proportionate share of the expenses of the well before he can share in the production.

The Commission order requires Thompson to reimburse Texaco for his proportionate share of the reasonable actual cost of drilling and operating the well, but does not describe how, or when, Thompson is to pay his share of the expenses. This issue was not raised or decided by the Commission. Issues which were not raised before the Commission may not be raised for the first time on appeal to this court. . . . . This issue is therefore premature, and we express no opinion as to the proper method of payment of the expenses for the well.

The district court decision is affirmed.

VANDE WALLE, LEVINE and MESCHKE, JJ., and VERNON R. PEDERSON, SURROGATE JUSTICE, concur.

VERNON R. PEDERSON, SURROGATE JUSTICE, sitting in place of GIERKE, J., disqualified.

NOTES

1. In EXXON CORP. v. THOMPSON, 564 So.2d 387, 111 O.&G.R. 471 (La.App. 1990), writ denied, 568 So.2d 1054 (La.1990) the Louisiana court upheld the conditioning of a grant of allowables upon the producer's agreeing to escrow the proceeds of production and then pay out the proceeds in accordance with the unit eventually adopted. The effect of such a conditioning of allowables is to allow the effective date of the unit to relate back to the filing of the pre-application notice that initiates the hearing for the creation of compulsory units in Louisiana. The court recognized that after a pre-application notice (when the adjoining landowner is thus deprived of his right to explore), the Commissioner has the authority to condition an allowable on compliance with a unit yet to be created. The policy of conditioning allowables was a reasonable exercise of the authority of the Commissioner to protect correlative rights. On the background of the litigation *see* Williams, *Can a Louisiana Unit Order be Effective Retroactively?* 49 La. L. Rev. 1119 (1989). *See also* the discussion of the problem of the effective date of unit orders in Kramer & Martin, *Pooling and*

Unitization, § 13.03[3](2001).

In RAILROAD COMMISSION v. PEND OREILLE OIL & GAS CO., Inc., 817 S.W.2d 36, 113 O.&G.R. 557 (Tex.1991), the Texas Supreme Court validated a practice of the Railroad Commission similar to the Louisiana method of avoiding a retrospective pooling order. The Commission had received an application from a well operator for a statewide rule 10 exception to permit down-hole commingling and an application for force pooling under the Mineral Interest Pooling Act by an adjacent owner. An interim order was issued by the Commission of May 7, 1984 to regulate the private interests in the well as if the proposed unit were formed and to prevent the party seeking the force pooling order from economic injury. The interim order gave an allowable for the well as though there had been pooling, calculated on a surface acreage basis and required a portion of the money to be escrowed. For a time the operator complied with the escrow requirement but ceased to do so when the Commission issued a final order (September 9, 1985) changing the method for calculating allowables from surface acreage to productive acre feet. Then on August 20, 1987 the Commission issued a final order pooling two sands as if they were a common reservoir, with an effective date of May 7, 1984. The appeals court ruled it was unconstitutional for the order to be effective as applied to the period September 9, 1985 through August 24, 1987 as there had been no special allowable set up to account for the new allocation basis. Pend Oreille Oil & Gas Co., Inc. v. Railroad Commission, 788 S.W.2d 878 (Tex.App. 1990). In reversing this aspect of the lower court decision, the Texas Supreme Court observed:

> [T]he problem with making a pooling order prospective only is that a pooling applicant's lands are being drained during the inevitable lag from an application for a force pooling order to the final order. To rectify this situation and avoid the proscriptions of *Buttes* [Buttes Resources Company v. Railroad Commission, 732 S.W.2d 675, 104 O.&G.R. 66 (Tex.App. 1987, error ref'd n.r.e.] and *American Operating* [American Operating Co. v. Railroad Commission, 744 S.W.2d 149, 101 O.&G.R. 305 (Tex.App. 1987), writ denied], the commission has adopted the practice of issuing an interim order if it appears that the pooling application has merit. These interim orders require the operator of the existing unit to escrow the proceeds that would belong to the successful pooling applicant if the parties were force pooled.

The appeals court erred by treating allowables as if they were vested property rights. Operators do not have a vested property right in assigned monthly allowables. Because allowables are not vested property rights, the final pooling order was not necessarily unconstitutional. Although the operator had a vested property right in its fair share of the oil and gas in place under its lease, this right was subject to state regulation. But the court did not reach the question whether the effectiveness of the interim order was abrogated by the entry of the final order in the productive acreage hearings and as a consequence arbitrarily took from the operator its fair share of the production already produced so that the final MIPA order was arguably unconstitutional; instead, the court ruled that the operator was estopped to complain about the retroactive effective date of the final MIPA order. The interim order expressly stated that a final MIPA order would supersede it. The parties were on notice that the commission was contemplating the entry of a pooling order and that the date of the final order would be made effective to the date of the interim order. The operator had the duty to obtain any modification of the interim order necessary to protect any vested rights that might have been interfered with by the entry of the final order in the productive acreage hearing.

2. The effective date of a unit order and the lessee's implied covenant to protect against drainage was explored in EAGLE LAKE ESTATES, L.L.C. V. CABOT OIL & GAS CORP., 330 F. Supp. 2d 778, 161 O.&G.R. 734 (E.D. La. 2004) When plaintiffs sued lessee for drainage occurring between well completion and defendant's application for unitization, the defendant asserted that plaintiffs could have sought from the Commissioner of Conservation an earlier effective date for the unit. The court rejected this defense because "making the effective date of a unitization order retroactive is antagonistic to the principles of Louisiana's mineral law... . This is because under Louisiana's theory of mineral ownership, making an order effective retroactively may be viewed as taking the property of one individual and giving it to another, since it requires an individual or entity who has reduced minerals to his possession during the period prior to unitization (and to which he is entitled under the law) to share those minerals (or the proceeds therefrom) with an individual or entity not otherwise entitled to them." The plaintiffs alleged that the defendant breached its obligation to protect against drainage by failing to initiate the unitization process sooner; they asserted that a prudent operator would have applied earlier, not that the defendant should have sought an earlier effective date to the unit order. The court commented: "Although any party involved in the unit can apply for unitization, it is the duty of the lessee to do so. In most situations the lessee has superior and exclusive knowledge that make it, as a prudent administrator, the appropriate party to undertake unitization, and in this case, plaintiffs have specifically alleged that defendants failed to provide them with production information, which they argue resulted in their simply not being in a position to initiate unitization."

3. COWLING V. BOARD OF OIL, GAS AND MINING, 830 P.2d 220, 118 O.&G.R. 582 (Utah 1991). The Utah Supreme Court ruled that a pooling order could not be made effective prior to entry of the spacing order. Celsius completed a well April 19, 1983; the well was then put into production in November 1983. It executed a declaration of pooling on April 19, 1983, to pool three leases covering 110 acres, all of which were leased by Adra Baird. Adjoining these tracts Celsius had a lease from the Bureau of Land Management (BLM). The Utah Board of Oil, Gas, and Mining dismissed a Celsius application for a spacing order in 1983 because Celsius had not acquired sufficient data to show the actual area drained. In January 1985,

Celsius again applied to the Board for a well spacing and drilling unit order for the pool drained by the well. On March 28, 1985, the board issued an order for a 300-acre unit, revised June 24, 1985, to 200 acres. The unit consisted of the 110 acres of the Baird leases and 90 acres of BLM land. The order was made retroactive to April 1, 1983. The Board ruled the Bairds' $230,000 in royalties that had been paid from first production to the time of the order being entered had to be shared with the BLM. On appeal, the order was overturned. The court ruled that the statewide spacing order was not sufficient to overcome the rule of capture. Until a field spacing order is entered, a correlative right is a right to an undifferentiated and unquantifiable interest in an oil or gas pool beneath one's land. The court's holding that the order could not be retroactive to the initial production when there was no spacing unit was based in part on its interpretation of the Utah pooling statute. Because the statute authorizes pooling orders to be entered only with respect to established drilling units, and because a pooling order that pools working interests must take into account the costs of drilling, the statutory scheme impliedly contemplates that pooling orders shall be retroactive to the date of first production but only if a spacing order was then in effect.

Cowling was followed by the Utah court in HEGARTY V. BOARD OF OIL, GAS, & MINING, 2002 UT 82, 57 P.3d 1042, 157 Oil & Gas Rep. 346. The court upheld the Board's determination that certain landowners were not entitled to retroactive pooling to protect their correlative rights. The landowners knew or should have known that two unit wells were planned to be on or near their property and the operator had made good faith offers to lease from the owners, which offers the owners declined. The court commented: "[A] good law, like a good parent, does nothing for a person that he or she can do independently, and it is not good law to cure one inequity by creating another." The landowners had, since 1995, substantially the same opportunity to seek state spacing and pooling that they exercised in 1999; they made a passive choice to allow their land to be drained until they took action. The court would not upset the reasonable expectations of the other owners within the relevant area.

4. *Equitable Pooling:* "The term 'equitable pooling', probably originating in a treatise [by Hoffman] on pooling and unitization, has come to be used by some courts and writers to describe the consequences of a series of Mississippi cases, which held that spacing regulations based on general conservation statutes, lacking compulsory pooling provisions, had the legal effect of pooling the land included in a drilling unit. The term 'equitable' may have been applied to this kind of pooling because it is neither voluntary (by agreement) nor compulsory under a specific provision of a statute. 'Judicial pooling' might be more appropriate since the Mississippi court effected pooling probably not contemplated by the parties, nor by the legislature that passed the conservation statute." Martin & Kramer, *Williams and Meyers Oil and Gas Law* § 906. Is the effect given spacing orders through "retroactive pooling" in the principal case and in states such as Oklahoma (as in the *Ward* decision described in the principal

case) a variation on "judicial pooling"?

5. The Texas cases, especially Ryan Consolidated Petroleum Corp. v. Pickens, 266 S.W.2d 526, 3 O. & G.R. 1148 (Tex.Civ.App.1954), *aff'd*, 155 Tex. 221, 285 S.W.2d 201, 5 O. & G.R. 99 (1955), *cert. denied*, 351 U.S. 933 (1956), make clear that owners of mineral rights who are barred from drilling by virtue of the spacing rules and the issuance of a permit to another are not entitled to share in the permitted well's production (in the absence of voluntary or, since 1965, compulsory, pooling). In jurisdictions such as Texas, what saves the constitutionality of the state's precluding such owners from drilling? The availability of Rule 37 exceptions? The availability of compulsory pooling (since 1965)? Does the *Pend Oreille* decision cited above call such cases into question?

SECTION 6. EFFECTS OF POOLING AND UNITIZATION UPON TERM INTERESTS IN OIL AND GAS

A. DEFEASIBLE TERM INTERESTS

(Refer again to the four diagrams and the discussion at pp. 896-99.)

PANHANDLE EASTERN PIPE LINE CO. v. ISAACSON

United States Court of Appeals, Tenth Circuit, 1958.
255 F.2d 669, 9 O.&G.R. 363.

BREITENSTEIN, CIRCUIT JUDGE. The trial court quieted the title of appellees Isaacson and Johnson, as the owner and oil and gas lessee respectively, to an undivided one-fourth interest in the minerals underlying certain land in Beaver County, Oklahoma. The unsuccessful defendants in that action have perfected separate appeals which present identical issues.

The land involved is the East Half of Section 21 and the West Half of the Southwest Quarter of Section 22, Township 6 North, Range 22 ECM. On August 13, 1941, O.F. Neal conveyed the land to Elmer Hall by deed which contained the following provision:

> Grantors hereby except for themselves, their heirs and assigns, an undivided one-fourth interest in and to all of the oil, gas and other minerals and mineral rights in, upon and under and that may be produced from said land, for a period of fifteen years from the date hereof and as long thereafter as oil, gas and/or other minerals are produced from said land or the premises are being developed or operated, provided such production is first had within the fifteen year period, together with the right of ingress and egress at all times for the purpose of mining, drilling and exploring and operating said land for oil, gas and other minerals and removing the same therefrom, with the right at any time to remove any and all equipment in connection therewith.

> Title to the oil, gas and other minerals and mineral rights hereby excepted is to remain vested in the said grantors, their heirs and assigns, for and during the period aforesaid.

Neal and Hall are both dead. Whatever rights Neal had under the quoted provision have passed to Isaacson, trustee, who, on April 11, 1946, made an oil and gas lease on the reserved one-fourth interest to Johnson.

The executrix of the Hall estate and the Hall heirs, appellants in No. 5744, on August 11, 1954, entered into two oil and gas leases with Panhandle Eastern Pipe Line Company, [Panhandle] the appellant in No. 5742. One lease covered the Northeast Quarter of Section 21 and the other lease covered the West Half of the Southwest Quarter of Section 22. Each lease contained appropriate provisions whereby upon the expiration of the term mineral interest the entire interest would be covered.

On September 28, 1955, The Texas Company, appellant in No. 5743, took an oil and gas lease from the Hall heirs on an undivided three-fourths interest in the Southeast Quarter of Section 21. A similar lease covering in an appropriate manner the other one-fourth interest was given by the Hall heirs to The Texas Company on February 28, 1956.

In the period December, 1953, to February, 1954, a well was drilled to the Morrow Sand on the Northeast Quarter of Section 22 by United Producing Company. [United] On a test made in March, 1954, this well, known as the Kiser well, flowed gas at the rate of 2,300,000 cubic feet per day. The well was equipped with casing, tubing, separator, well-head fittings, and measuring tanks. Since the initial test, no gas has flowed from the well except that released on a second test made in May, 1956. While the well has never been connected to any pipe line, the gas which might be produced from the well is dedicated to Colorado Interstate Pipe Line Company by a gas purchase contract. United has paid shut-in royalties to the owners of the minerals covered by the lease on the land where the Kiser well is located.

Acting under the power given to it by Title 52 O.S.1951 § 87.1, the Oklahoma Corporation Commission, [Commission] by appropriate order on May 25, 1956, established 640–acre drilling and spacing units for the production of natural gas from the lower Morrow Sand underlying Sections 21 and 22 and other lands. The Commission's order contained the following provision:

> That all royalty interests within any spacing unit shall be communitized and each royalty owner within any unit shall participate in the royalty from the well drilled thereon in the relation that the acreage owned by him bears to the total acreage in the unit.

No royalties for production from the Kiser well have been paid to anyone. While the Oklahoma statute [Title 52, O.S.1951 § 87.1(d).] permits both voluntary and compulsory pooling of working interests within an established drilling and spacing unit, no such pooling has been agreed to or ordered.

The fifteen-year primary term of the one-fourth mineral interest reserved by the Neal–Hall deed expired on August 13, 1956, unless the "thereafter" clause of the reservation extended the term. The pertinent language states that the reserved one-fourth interest is for a period of fifteen years and "as long thereafter as oil, gas and/or other minerals are produced from said land or the premises are being developed or operated, provided such production is first had within the fifteen year period."

Under the Commission's drilling and spacing order, Section 21 and Section 22 each constitutes a separate unit. The Neal–Hall deed covered land in each section. The shut-in Kiser well is located in Section 22 but on land owned by strangers to this controversy.

Three basic questions are presented, *viz.*:

1. Is the reserved mineral interest extended beyond the primary term by a well located off the deeded land but within a unit which has been established by a valid drilling and spacing order?

2. If the above question is answered in the affirmative, does the shut-in Kiser well satisfy the requirements of the "thereafter" clause?

3. If both of the above questions are answered in the affirmative, is the mineral estate extended beyond the primary term so far as the land located in Section 21 is concerned?

In the consideration of each of these questions recognition must be given to the rule that the intent of the parties controls the interpretation of the deed. This intent is to be ascertained, if possible, from the entire instrument.

The first question relates to the effect of the drilling and spacing order. The Oklahoma statute permitting the Commission to order the unitization of private property for the prevention of waste of underlying valuable minerals is a lawful exercise of the police power of the state. The Commission found that there was a "common source of supply of natural gas" underlying the lands covered by its order and that "in the interest of securing the greatest ultimate recovery of natural gas from the pool, the prevention of waste and the protection of correlative rights," the drilling and spacing order should be entered. Each unit consisted of one governmental section. On Section 22 the permitted well was specifically designated as the then existing Kiser well.

Gas produced by the Kiser well comes from the common pool. To the extent that gas is captured from the portion of the common pool underlying the East Half of the Southwest Quarter of Section 22, that land is being developed for its mineral potential even though the gas is taken from a well located on another part of Section 22.

While this court has said in Simpson v. Stanolind Oil & Gas Co., 10 Cir., 210 F.2d 640, 642, that the purpose of the Oklahoma statute here under consideration has no relation to the termination or continuation of oil and gas leases, that decision has no application here. The spacing order did not extend the primary term. It merely set the stage for the extension of the term by an activity which met the requirements of the "thereafter" clause and which occurred somewhere within the spacing unit. To hold otherwise would be to eliminate the possibility of an extension of the primary term by a well on Section 22 to the Morrow Sand.[102]

Panhandle asserts that under the "thereafter" clause the production has to be from the deeded land to extend the term. The answer is that gas produced from the common pool is production from the entire unit, including the deeded land.

Texas Company, taking a somewhat different approach, argues that the extension of the primary term could be accomplished only by (1) actual payment of royalties pursuant to the spacing order or (2) voluntary or compulsory pooling of the seven-eighths working interest. As neither happened within the primary term, Texas Company concludes that such term was not extended. The basis for the contention is that payment of royalties could only occur after production and that, under the applicable statute, production from the land of one owner within a spacing unit is production from the land of another owner within that unit when a pooling order is entered.[103] The argument is answered by the Oklahoma decision in Wood Oil Co. v. Corporation Commission, 205 Okl. 537, 239 P.2d 1023, 1027, wherein it is held that the right to participate in production from a well completed prior to the entry of a spacing order arises as a matter of law upon the entry of such order. In the case under consideration the order was entered on May 25, 1956, within the primary term. The basic rights of the parties were fixed as of that date. The failure to obtain either a voluntary pooling agreement or compulsory pooling order before the expiration of the primary date does not affect those rights.

We conclude that a reasonable interpretation of the "thereafter" clause and the legal effect of the Commission order combine to result in an extension of the term if the Kiser well satisfies the requirements of that clause.

The trial court held that the Kiser well is a "commercial" well and that United had "exercised due diligence in attempting to obtain a market for gas produced" from that well. It is conceded that the well is shut-in and that no gas from it has ever been marketed.

Appellants say that the term "production" in the mineral reservation does not mean "discovery" of gas and that the oil and gas lease rule regarding reasonable time and due diligence in marketing has no application. They concede that if the instrument was an oil and gas lease rather than a deed, the primary term would have been extended.[104] Oklahoma has recognized the difference between

---

[102]   Section 87.1(d), *supra,* prohibits the drilling of a non-permitted well into a common source of supply after the entry of a spacing order covering such common source.

[103]   Section 87.1(d), *supra,* provides in part: "The portion of the production allocated to the owner of each tract or interests included in a well spacing unit formed by a pooling order shall, when produced, be considered as if produced by such owner from the separately owned tract or interest by a well drilled thereon."

[104]   *See* Bristol v. Colorado Oil and Gas Corporation, 10 Cir., 225 F.2d 894.

the two types of instruments.[105]

The reservation in the Neal–Hall deed contains a "thereafter" clause which closely resembles similar clauses found in many oil and gas leases. Admitting the differences between a deed and an oil and gas lease and admitting that different rules of construction apply to the two classes of instruments, no good reason appears for not giving the same definition to the word "production" when it appears in one or the other.

. . .

The Kiser well discovered gas in paying quantities. All facilities necessary to deliver the gas to a committed market have been installed except for the pipe line connection. The gas has been reduced to possession and control. Temporarily it is being stored underground. To say that marketing is required to satisfy the "thereafter" clause is to read into the mineral reservation of the deed something that is not there. Under the applicable Oklahoma law the production provision has been satisfied.

The remaining question involves the land in Section 21 which constitutes a separate unit under the drilling and spacing order. Appellants contend that the term is not extended as to land outside the unit on which the Kiser well was drilled. Reliance is placed on Louisiana cases holding that the mineral term of one portion of an estate is not extended by production in a pooling unit wherein another portion of the estate is located.[106] Oklahoma decisions involving leases are to the contrary.[107] The reasoning in the lease cases is applicable to a deed. In Kunc v. Harper–Turner Oil Company, the Oklahoma Supreme Court said:

> We think the pooling by the Corporation Commission does not have the effect of creating two separate leasehold estates and that the unitization of a portion of a lease with other lands does not affect the terms of the lease, whether production is from the portion of the unit from the lease under consideration or from another portion of the unit. [297 P.2d 376]

The terms of the Neal–Hall deed control. It does not divide the mineral interest into segments. If the requirements of the "thereafter" clause are met as to one part of the reserved mineral interest they are met as to the entire interest.

The mineral reservation in the Neal–Hall deed provided for an extension of

---

[105]   In Beatty v. Baxter, 208 Okl. 686, 258 P.2d 626, 628, it was said: "An oil and gas lease is governed by different rules of construction from those applicable to other contracts, being construed most strongly against the lessee and in favor of the lessor. The reason therefor springs from the danger of loss of oil and gas by drainage."

[106]   Elson v. Mathewes, 224 La. 417, 69 So.2d 734; Childs v. Washington, 229 La. 869, 87 So.2d 111; Jumonville Pipe and Machine Co. v. Federal Land Bank of New Orleans, 230 La. 41, 87 So.2d 721.

[107]   Kunc v. Harper–Turner Oil Company, Okla., 297 P.2d 371, 376;   Trawick v. Castleberry, Okl., 275 P.2d 292, 293–294; Godfrey v. McArthur, 186 Okl. 144, 96 P.2d 322, 324; cf. Gregg v. Harper–Turner Oil Co., 10 Cir., 199 F.2d 1.

the fixed term "as long thereafter as oil, gas and/or other minerals are produced from said land or the premises are being developed or operated." Commercial quantities of gas have been tapped by a well located within a validly established drilling unit which includes part of the deeded land. Development is complete except for the pipe line connection. Until that is available the gas is stored underground. There is nothing to indicate any intent of the parties to terminate the mineral reservation at the end of the fixed period under such circumstances. On the record before us it is clear that the decision of the trial court is correct.

On each appeal the judgment is affirmed.

NOTES

1. In WHITAKER v. TEXACO, INC., 283 F.2d 169, 13 O.&G.R. 502 (10th Cir.1960), a Case 4 situation involving compulsory pooling under the Oklahoma statute, the court held that the lease was preserved in its entirety into the secondary term by unit operations and production. The opinion in this case by Circuit Judge Breitenstein relied primarily on his earlier opinion in Panhandle Eastern Pipe Line Co. v. Isaacson, *supra.* Circuit Judge Murrah, concurring in *Whitaker,* declared that:

> I agree that this case is ruled by Panhandle and, being bound by the ruling of that case as this court's expression of Oklahoma law, I must concur in the affirmance of the judgment. If I were free, however, to forecast Oklahoma law on the precise point, I should not hesitate to hold that the drilling of the well in question operated to extend the lease beyond its primary term only as to that portion which was included in the established spacing unit on which the well was drilled.

Consider whether in a jurisdiction which has held in accord with *Panhandle Eastern* in a case involving defeasible term interests, the holding of *Whitaker* necessarily follows. Consider also whether in a jurisdiction which has held in accord with *Whitaker* in a case involving an oil and gas lease, the holding of *Panhandle Eastern* necessarily follows.

In FOX v. FELTZ, 1984 OK CIV APP 60, 697 P.2d 543, 84 O.&G.R. 157, *cert. denied* (1985), the court held in a Case 4 situation that off-tract production held the defeasible term interest.

2. Consider the significance of the following cases dealing with the effect of pooling and unitization upon term interests in oil and gas:

(a) SOUTHLAND ROYALTY CO. v. HUMBLE OIL & REFINING CO., 151 Tex. 324, 249 S.W.2d 914, 1 O.&G.R. 1431 (1952). A community lease was executed by all persons having mineral interests in a contiguous 250–acre leasehold which included Blackacre and Whiteacre. Thus joining in the lease were the owners of both the term mineral interest and the possibility of reverter in Blackacre. Before the expiration of the primary term of the mineral interest, the lessee began production from wells drilled on Whiteacre. Reasoning from its conclusions as to the effect of a unitization agreement between lessor and lessee, the court concluded that in Case 2, the term of the mineral interest was extended by production from the unit but off Blackacre. The unitization being voluntary in nature, arising from the execution of a community lease, the court based its decision on the implied intent of the parties to modify the provisions of the term mineral deed.

(b) SPRADLEY v. FINLEY, 157 Tex. 260, 302 S.W.2d 409, 7 O.&G.R. 650 (1957). *A,* the owner of a 29.43 mineral acre interest in Blackacre (comprising some 117.7 acres) which had been granted for a term of fifteen years "and so long thereafter . . .", leased that interest to lessee–1 in 1945. In the same year, *B,* who owned the balance of the minerals in Blackacre and the reversion after *A* 's interest, executed a lease to lessee–2 covering both Blackacre and Whiteacre (two additional tracts comprising some 129 acres). After describing the premises, the latter lease recited as follows: "LESS 14.43 acres sold out of the above tract to . . . and less 15 acres sold out of said tract to . . ., leaving herein a total of 216.7 acres of land conveyed in this lease." This language was followed by a typical "Mother Hubbard" clause. Each of the above mentioned leases contained a pooling clause authorizing the lessee, at the lessee's option to pool the premises and further providing that "If production is found on the pooled acreage it shall be treated as if production is had from this lease, whether the well or wells be located on the premises covered by this lease or not."

Lessee–1 and lessee–2 thereafter assigned the leases to Skelly Oil Co., and the latter then executed a unitization agreement covering some 699.79 acres which included all of Blackacre and Whiteacre. A producing well was drilled on Whiteacre before the expiration of the primary term of *A* 's mineral interest. The court held that production on Whiteacre served to extend the life of the term mineral interest in Blackacre into the secondary term, declaring that:

> The case falls within the rule announced by the Southland Royalty case. By the execution of the leases, the Spradleys [*B*] and the owners of the interests covered by the deeds [*A*] agreed that production of minerals from a tract included in a pool unit would be regarded during the life of the lease as production from all of the other tracts included in such unit. The condition upon which the grant in the mineral deed was to extend beyond the 15–year term was thus modified by agreement of the parties, and was fulfilled by production of minerals from the land with which the 20.4 and 97.3 acre tracts [Blackacre] were unitized.

(c) WILLIAMSON v. FEDERAL LAND BANK OF HOUSTON, 326 S.W.2d 560, 11 O.&G.R. 299 (Tex.Civ.App.1959, error ref'd n.r.e.). Defendant conveyed four tracts of land, retaining a 1/16th royalty interest which was terminable, however, if there was no production from said land within a period of twenty years. During this twenty-year period each of the four tracts was separately leased by the reversioner-executive, each lease including a broadly phrased pooling clause. Unit production (Case 4) was obtained from a unit including three of the tracts but not from the unit including the fourth tract. Relying on Spradley v. Finley, the court held that the term interest was preserved after the expiration of the twenty-year period as to the acreage excluded from the producing unit.

3. Kansas has reversed itself as to the effect of off-tract unit production where the unit is formed voluntarily by a lessee (or lessees), under separate leases from the term and reversionary interests. In SMITH v. HOME ROYALTY ASS'N, INC., 209 Kan. 609, 498 P.2d 98, 42 O.&G.R. 589 (1972), it considered a Case 2 situation where separate leases of Blackacre, each apparently containing a pooling clause, were executed by the owner of a defeasible term one-half mineral interest and by the reversioner (holder of the remaining one-half mineral interest). The lessee pooled Blackacre with Whiteacre and obtained production from a well on Whiteacre. The court held that unit production from Whiteacre did not hold the defeasible term interest into its secondary term, saying that the "lease executed by [term interest owner] had no relation or connection with the oil and gas lease executed by the [reversioner]."

In CLASSEN v. FEDERAL LAND BANK OF WICHITA, 228 Kan. 426, 617 P.2d 1255, 68 O.&G.R. 426 (1980), however, it expressly disapproved its holding in *Smith*, "to the extent in conflict with [the *Classen* opinion]." The Land Bank by a single deed conveyed three areas, known as Tracts 1, 2 and 3, to the Classens, retaining in each a defeasible one-fourth mineral interest. (The Classens later conveyed their interest in Tract 2 to the Friesens, considered in note 4 below.) By separate leases dated November 28, 1950, both the Land Bank and the Classens leased their respective interests in Tracts 1 and 3. Evidently pursuant to authorization in the leases, the lessee formed a unit consisting of Tract 1 and certain other land, and obtained production on the other land.

The court held that Tract 1 was held by the unit production, overruling *Smith* on the ground that to do so would encourage oil and gas production, thereby helping resolve, the court said, the nation's "frightening and progressive energy crisis due principally to a shortage of petroleum reserves."

However, the court held that the unit production did not hold Tract 3 (which was outside the unit), relying on *Smith* and Friesen v. Federal Land Bank of Wichita, 227 Kan. 522, 608 P.2d 915, 66 O.&G.R. 8 (1980). Justice Herd dissented from this aspect of the opinion, arguing that production attributable to Tract 1 pursuant to the majority view should also perpetuate the reservation on the remaining acreage described in the deed, under Baker v. Hugoton Production Co., 182 Kan. 210, 320 P.2d 772, 8 O.&G.R. 714 (1958). The *Baker* decision held that when a deed creates a defeasible term mineral interest in two or more tracts of land, development or operation on any one tract will extend the interest beyond the primary term as to all other tracts unless a contrary intent is expressed in the deed.

Should *Classen* be applied retroactively to revive a defeasible term mineral interest that had expired under the law as set forth in *Smith*? *See* Kneller v. Federal Land Bank of Wichita, 247 Kan. 399, 799 P.2d 485, 110 O.&G.R. 473 (1990).

4. In the above cases the holders of both the term and the reversionary interests were bound by the unitization, either because of state compulsion or by their consent. Absence of consent complicates matters. For example, *Friesen v. Federal Land Bank of Wichita, supra,* considered whether the Land Bank's term interest in Tract 2 was extended by the production of the unit into which Tract 1 had been joined. It concluded that it did not, since the Friesens [the Classens' successors as owners of the reversionary interest in Tract 2] were not party to the leases nor to the unitization agreements covering Tract 1. Justice Herd dissented, as he had from the parallel aspect of the *Classen* decision.

A federal court applied *Classen* and *Friesen* in EDMONSTON v. HOME STAKE OIL & GAS CORP., 629 F.Supp. 620, 88 O.&G.R. 458 (D.Kan.1986), *aff'd*, 854 F.2d 1323 (10th Cir. 1988) after a certified question was answered by the Kansas Supreme Court in Edmonston v. Home Stake Oil & Gas Corp., 243 Kan. 376, 762 P.2d 176, 101 O.&G.R. 254 (1988). The case involved a term mineral interest created in 1956 covering two tracts. From 1962 to 1973 there was a producing well on Tract *A*. In 1968, a portion of Tract *A* was included in a unit formed under the Kansas Compulsory Unitization Act with the unit production not on the tract; the unit production continued to 1984. A well was commenced on *B* in 1983 but not completed until 1985. The unit production continued the interest as to Tract *A* but not *B*; the land covered by the term interest was divided by the unit. The interest ceased as to *B* when the non-unit production ceased in 1973. Had the well on *B* commenced production in 1984 before the cessation of the unit production, the interest would have continued as to *A*. The district court described *Classen* as adopting a rule of divisibility, under which a term mineral interest owner's voluntary pooling of a portion of the land in which he possesses the interest severs that land and interest from that not pooled; only the former were perpetuated beyond the primary term by off-tract production on the pooled acreage. The court concluded that the fact that compulsory unitization was involved called for no different treatment than *Classen.* Property rights were modified by the statute only to the extent necessary to conform to the provisions and requirements of the statute. The statute and order effected no transfer of ownership, and the court noted that the Kansas legislature did not change the statute following *Classen.*

But compare SHORT v. CLINE, 234 Kan. 670, 676 P.2d 76, 79 O.&G.R. 552 (1984). Part of the tract was pooled pursuant to a pooling agreement to which the owner of the reversionary interest was not a party, and unit production from off-tract wells followed. For some time there was also on-tract unit production; this ceased, however, in 1966. In addition, there was production on non-unitized portions of the tract, which production ended some time prior to November 10, 1977. Plaintiff acquired the unit leaseholds in 1973 and became unit operator. On November 10, 1977 he acquired the reversionary interest in the tract. Plaintiff argued that as there was no on-tract production, the term interest ceased. Relying heavily on the view that plaintiff acquired the leaseholds subject to the pooling agreement, and was obliged by it to operate the leases for the benefit of all lessors, the court held it would be inequitable to allow plaintiff to bring the term interests to an end on the basis of cessation of on-tract production.

For a broader decision finding that production from a voluntary unit in a Case 2 situation held a defeasible term interest, despite the reversionary owner's non-consent to the unit, *see* Shelton v. Andres, 106 Ill.2d 153, 87 Ill.Dec. 954, 478 N.E.2d 311, 86 O.&G.R. 266 (1985), *aff'g*, 122 Ill.App.3d 1089, 78 Ill.Dec. 430, 462 N.E.2d 549, 82 O.&G.R. 46 (1984).

5. If *R* conveys to *E* a defeasible term interest, should *R* be viewed as subject to a duty to join in a pooling or unitization agreement the effect of which may be to preserve the defeasible term interest into its secondary term? In this connection should it be significant whether *E* does or does not possess executive rights? Whether the jurisdiction follows the Texas or the Louisiana rule as to the executive's power to authorize pooling?

MINCHEN v. FIELDS, 162 Tex. 73, 345 S.W.2d 282, 14 O.&G.R. 266 (1961). Petitioner was the owner of a defeasible term non-executive 67.98–mineral acre interest in a 284.74–acre parcel which had been included by the executive in a lease covering 802.6 acres. The lease provided for an oil payment bonus. Production was obtained from a well on a portion of the leasehold other than the 284.74–acre parcel in which petitioner owned a defeasible term mineral interest but no well was drilled on the 284.74–acre parcel during the primary term of petitioner's interest. Upon the expiration of the primary term of the defeasible term interest this action was brought by the respondents (the executive and owner of the reversionary interest following petitioner's defeasible term mineral interest) to determine petitioner's interest. The trial court and the Court of Civil Appeals held that petitioner's defeasible term interest expired at the end of its primary term for failure of production on the 284.74–acre parcel and that petitioner was not entitled to any share of the royalty or the oil payment bonus paid by the lessee. *Held,* affirmed except insofar as petitioner was denied a recovery of any portion of the oil payment bonus and as to that the cause was remanded for the purpose of rendition of a judgment for petitioner for a pro rata share of the oil payment bonus. The court concluded that all owners of mineral interest in the 802.6 acres were entitled to share in the oil payment bonus in proportion to the mineral acres owned by them. On the power of the executive to pool non-executive interests the court had the following comments:

> By his second point petitioner contends that the Court of Civil Appeals erred in not holding that the execution of the oil and gas lease by Fields and wife constituted a unitization or pooling of all the mineral interest under the 802.6 acres, and all owners of mineral interests share in the production from any part of the 802.6 acres in proportion to their ownership. We agree with the Court of Civil Appeals that the act of Fields in executing one lease covering the 802.6 acres did not unitize or pool all of the mineral interests in said land. Brown v. Smith, 1943, 141 Tex. 425, 174 S.W.2d 43; Nugent v. Freeman, Tex.Civ.App.1957, 306 S.W.2d 167, wr. ref., n.r.e. As said by the

Court of Civil Appeals, "the reason of the rule is that where mere executive rights are conferred or reserved, there is no intention evidenced to vest authority to convey a royalty interest reserved or the royalty interest attributable to the minerals leased and to hold that such holder can unitize or pool the interest would allow him to convey such royalty interest because a unitization of the royalty and minerals under different tracts effects a cross-conveyance to the owners of minerals under the various tracts of royalty or minerals so that they all own undivided interests under the unitized tract in the proportion their contribution bears to the unitized tract. Veal v. Thomason, 138 Tex. 341, 159 S.W.2d 472." [330 S.W.2d 687] We overrule petitioner's second point.

6. Consider how a unitization agreement may be ratified by a person who has not signed the agreement itself. *See* the Court of Civil Appeals opinion in Minchen v. Fields, 330 S.W.2d 683, 12 O.&G.R. 111 (Tex.Civ.App.1959), which found there had been no effective ratification.

B. MINERAL SERVITUDES AND MINERAL ROYALTIES IN LOUISIANA

Kramer & Martin, *The Law of Pooling and Unitization*, § 20.03[6]
[footnotes omitted or renumbered and edited]

[b] Liberative Prescription of Mineral Servitude/Royalty

[i] Introduction

Louisiana is a non-ownership-in-place jurisdiction. That is, under the Louisiana Mineral Code the owner of land does not own the fugacious minerals under the surface but only the right to produce the minerals. The landowner cannot convey the minerals apart from the land, but he or she can convey the right to produce the minerals as a servitude burdening the land. Such a servitude is a real right, but it is subject to rules of liberative prescription. The principal such prescriptive rule is that the servitude must be exercised or prescription otherwise interrupted within ten years from the servitude's creation or it will expire from non-use. A mineral royalty can be created burdening the land or burdening the mineral servitude; it, too, is subject to a rule of prescription if not interrupted through exercise (or otherwise) within ten years.

The Mineral Code requires that to interrupt the ten-year liberative prescription one must commence efforts to obtain production in paying quantities in good faith from the servitude tract or land unitized with the servitude tract.[108] .

---

[108]   La. Rev. Stat. Ann. § 31:29. Use of a mineral servitude must be "by the owner of the servitude, his representative or employee, or some other person acting on his behalf." La. Rev. Stat. Ann. § 30:42. When the servitude owner has signed a pooling or unitization agreement, the operator will be operating for the servitude owner. If it is a compulsory unit upon which one is relying for interruption or prescription, the Mineral Code effectively

. .

The Mineral Code specifies the effects of inclusion of all or part of a tract burdened by a mineral servitude or mineral royalty in a compulsory or conventional (voluntary) unit. The effect is limited to the subject of liberative prescription on the mineral right. A mineral lease is not subject to the regime of liberative prescription, and thus these Mineral Code effects on a servitude or royalty have no application to a mineral lease.[109] However, one should note that if a mineral servitude terminates through the operation of a rule of prescription, a mineral lease created from it will also terminate.

The Mineral Code became effective January 1, 1975. Some mineral rights may still exist that can raise questions whether the application of the Mineral Code's provisions would impair vested rights if the new provisions change the law as it was in 1974.[110] By and large, however, the Mineral Code clarifies but does not change the existing case law on the effect of inclusion of all or a portion of a mineral servitude or mineral royalty in a unit.

[ii] Effect When Unit Well is On Servitude or Royalty Tract

When only a portion of the tract burdened by a servitude or royalty is in a unit and the unit well is on the tract burdened by the servitude or royalty, the unit well will serve to interrupt prescription for the entire servitude or royalty (unless there is an agreement providing to the contrary) if the operations or production are otherwise sufficient to interrupt the prescription of non-use.[111] The Mineral

---

provides that the unit operator will be acting for the servitude owner, obviating any need for the servitude owner to adopt the operations of the operator. La. Rev. Stat. Ann. § 30:47. Because a mineral royalty is a passive right, it does not matter who obtains the production that serves to interrupt liberative prescription.

[109] There is a different treatment of partial inclusion of a lease in a unit when the unit well is not on the leasehold [Hunter Co. v. Shell Oil Co., 211 La. 893, 31 So.2d 10 (1947); LeBlanc v. Danciger Oil & Refining Co., 218 La. 461, 49 So.2d 855 (1950) which follow the majority rule], and the analogous circumstances for a mineral servitude or royalty.

[110] *See generally* Day, *Applicability of the New Mineral Code to Existing Mineral Rights*, 22 L.S.U. Min. L. Inst. 205 (1975). Several cases since the Mineral Code have addressed the question of the applicability of the unitization provisions of the Mineral Code to pre-existing rights. See Allied Chemical Corporation v. Despot, 414 So.2d 1346, 73 O.&G.R. 319 (La. App. 1982); White v. Evans, 457 So.2d 159, 82 O.&G.R. 503 (La. App. 1984); Sandefer & Andress, Inc. v. Pruitt, 471 So.2d 933, 86 O.&G.R. 504 (La. App. 1985); and Adobe Oil and Gas Corporation v. MacDonnell, 480 So.2d 961, 87 O.&G.R. 533 (La. App. 1985).

[111] La. Rev. Stat. Ann. §§ 33, 34, 37. Note that a dry hole will serve to interrupt prescription as to a mineral servitude. La. Rev. Stat. Ann. § 31:30: "An interruption takes place on the date actual drilling or mining operations are commenced on the land burdened by the servitude or, as provided in Article 33, on a conventional or compulsory unit including all or a portion thereof. . . ." This is true even if the unit well is off the servitude tract (though there is the divisive effect described below if only a portion of the servitude tract is in the unit). The Mineral Code preserves the effect of the holding in Barnwell, Inc. v. Carter, 220 So.2d 741, 33 O.&G.R. 247 (La. App. 1969), writ ref., 254 La. 140, 222 So.2d 885, 33

Code follows the holding and rationale of the Louisiana Supreme Court in *Trunkline Gas Co. v. Steen.* [249 La. 520, 187 So.2d 720, 25 O.&G.R. 195 (1966).]

In *Trunkline Gas* a dry hole was drilled on servitude acreage included in a compulsory conservation unit; the well was held to interrupt prescription as to the entire servitude, even though only a portion of the servitude acreage was in the unit. In this case, unlike the wells in *Childs v. Washington,* [229 La. 869, 87 So.2d 111, 6 O.&G.R. 44 (1956).] and *Jumonville Pipe and Machinery Co. v. Federal LandBank,* [230 La. 41, 87 So.2d 721, 6 O.&G.R. 61 (1956).] the operations on the unit were located on the portion of the servitude tract included in the unit. The commissioner's orders in no way conflicted with the private contractual rights and therefore had no effect. . . . Thus, the servitude was not divided into two servitudes, each requiring a separate user, by the partial inclusion of the servitude in a unit here.

It is possible to enter into a community or joint lease on multiple servitudes. The effect of such pooling will be that the tracts are all pooled so that drilling (or other act that will suffice to interrupt) upon any of the servitudes will interrupt prescription as to all. [Hall v. LeMay, 191 So.2d 720, 25 O.&G.R. 509 (La. App. 1966).]

The operations or other actions that satisfy the Mineral Code articles will be an interruption of prescription rather than a suspension. [Lavergne v. Savoie, 221 So.2d 71, 33 O.&G.R. 292 (La. App. 1969).] That is to say, prescription begins anew with an interruption, so that the owner of the rights will normally have ten more years in which to make a use of the servitude or royalty. The Mineral Code does not appear to require that a landowner consent to the pooling in order for the operations or production under a voluntary pooling or unitization to have the effect of interrupting prescription. [*Compare* Alexander v. Holt, 116 So.2d 532, 13 O.&G.R. 170 (La. App. 1959).] However, the unitization must be effected prior to the running of prescription. [Baker v. Chevron Oil Co., 260 La.1143, 258 So.2d 531, 41 O.&G.R. 485 (1972).]

[iii] Effect When Unit Well is Off Servitude or Royalty Tract

The general divisive effect of the Mineral Code articles dealing with pooling and unitization is that when only a portion of a servitude or royalty is included in a unit and the unit operations are not on the tract burdened by the servitude or royalty then the unit operations or production will serve to interrupt prescription only for the portion included in the unit. [i.e. Case 4]. Article 33 dealing with drilling operations states the operation of the rule as follows:

> Operations conducted on land other than that burdened by a mineral
> servitude and constituting part of a conventional or compulsory unit that

---

O.&G.R. 255 (1969). Note, however, that a mineral royalty requires actual production or a shut-in well capable of producing in paying quantities to interrupt prescription, La. Rev. Stat. Ann. § 31:85, 90.

includes only a part of the lands burdened by the servitude will, if otherwise sufficient to interrupt prescription according to Articles 29—32, interrupt prescription only as to that portion of the tract burdened by the servitude included in the unit provided such operations are for the discovery and production of minerals from the unitized sand or sands.

There are similar provisions providing the same divisive effect with respect to shut-in wells [Art. 34] and producing wells. [Art. 37] There are analogous provisions with the same divisive effect for mineral royalties. [Art. 89 (production), Art. 90 (shut-in well)] The divisive or segregative effects of the Mineral Code apply only vertically, not horizontally. That is to say, even though an order or agreement for pooling or unitization may apply to a specified formation, prescription will be interrupted for all depths for that part of the servitude or royalty tract included in the surface boundaries of the unit. [See White v. Frank B. Treat & Son, Inc., 230 La.1017, 89 So.2d 883, 6 O.&G.R. 640 (1956).]

The Mineral Code follows the prior cases of *Childs v. Washington,* [229 La. 869, 87 So.2d 111, 6 O.&G.R. 44 (1956).] and *Jumonville Pipe and Machinery Co. v. Federal Land Bank.* [230 La. 41, 87 So.2d 721, 6 O.&G.R. 61 (1956).] The rationale given by the Louisiana Supreme Court in those cases was that an order of the Commissioner of Conservation that included only a portion of a tract burdened by a mineral servitude operated as though the parties had entered into an agreement dividing the servitude. The *Childs* court stated:

It logically follows . . . that if the landowner and the mineral owner can by agreement extend the servitude as to a portion only of the acreage without by such act interrupting the tolling of prescription as to the servitude on the remainder of the tract . . . and if the mineral owner can divide the advantages of a servitude by assigning to a third person all of his interest in a designated portion of the land affected by the servitude and create a situation whereby user of a portion would not hold the remainder . . . then clearly when the Commissioner of Conservation, acting for the State in the exercise of its police power, by his orders and after due hearing, included within a unit or units portions of a tract as to which owners had failed to agree to pool their interest for development purposes, his act produced the same result as if the parties had formed a drilling unit by convention and for development purposes had reduced the servitude to that extent. [87 So. 2d at 114.]

The rationale of *Childs* and *Jumonville* is faulty in that the commissioner and the conservation statutes have no "intent" to divide a servitude in the sense that parties agree to divide a servitude. They are not to have any effect on private rights except to the extent necessary to prevent waste and/or protect correlative rights. Since the Commissioner of Conservation is not even aware of title in establishing units, except peripherally, the portion of a tract burdened by a servitude or royalty included in a unit has nothing to do with an intent to divide the servitude or royalty. If this rationale were sound, it would seem unsound to have a different effect when the unit well was located on the tract burdened by the servitude. A better basis for the ruling was given by Justice McCaleb in a concurring opinion in *Childs.* He stated:

> While it can hardly be gainsaid that the extraction of oil from the portion of the encumbered land contained in the drilling units constituted a user of the servitude in a real sense, it was only because of the unitization orders that this user was effective in law forasmuch as no drilling explorations were ever conducted on any part of the land subject to the servitude. This being so, it seems only proper to conclude that interruption of prescription extends only to that portion of the land covered by the servitude from which the oil is actually withdrawn or where the user has taken place—that is, the part included within the drilling unit.

Whatever the rationale in *Childs* and *Jumonville,* the rule of the two cases is now embodied in statute and will not change through judicial reconsideration of the soundness of the rationale for the rule.

The pre-Mineral Code jurisprudence regarding the effect of partial inclusion of a mineral royalty interest in a unit was unclear, though there is no doubt about the approach decided upon in the Mineral Code. [Art. 89] This lack of clarity before the Mineral Code was brought out in a case involving a question of the application of the Mineral Code to interests and rights arising before the Mineral Code. In *Adobe Oil and Gas Corp. v. MacDonell,* [480 So. 2d 961, 87 O.&G.R. 533 (La. App. 1985).] the court ruled that Mineral Code Article 89, providing that ten-year liberative prescription is not interrupted by production from a unit well when the well is not located on the royalty tract, is applicable to royalty rights created prior to the 1975 effective date of the Mineral Code. In *Adobe,* the plaintiff oil company instituted a concursus proceeding to determine ownership of rights to production attributable to certain acreage included in a unit created in 1980. Claimants in one group were successors in interest to royalty deeds dated 1954 and 1955, and in another group the claimants were owners of the land. The lands covered by the royalty deeds had been partially included in a unit in 1956, the unit well for which was not on the lands burdened by the royalties. The landowner group asserted that the royalty interests had prescribed on the acreage outside the 1956 unit under ten-year liberative prescription. The royalty deed group asserted that prescription had not accrued. The appellate court ruled that Article 89 of the Louisiana Mineral Code provides that production from a unit interrupts prescription only as to that portion of the tract included in the unit if the unit well is on land other than that burdened by the royalty. This provision is applicable to rights arising prior to the effective date of the Mineral Code in 1975 unless it would impair vested rights. The state of the pre-Mineral Code jurisprudence was conflicting, the court said, and thus the royalty owners had no vested rights that would prohibit the retroactive application of the Mineral Code article.

It should be observed that simply drilling through a unitized sand will not be sufficient to satisfy the requirements of Article 33 of the Mineral Code. This article preserves the rule of the case of *Matlock Oil Corp. v. Gerard.* [263 So. 2d 413, 43 O.&G.R. 184 (La. App. 1972), writs denied, 265 So.2d 241, 43 O.&G.R. 196 (La. 1972).] In this case, several servitude tracts were involved that had been included in a unit that encompassed the Lower Hosston Sand. A well on a unit tract but outside the servitude tracts was drilled through the Lower Hosston Sand within ten years from the date of last operations for the servitude tracts. Drilling through this unitized sand would not interrupt prescription because it was not a good-faith effort to obtain production from the unit sand. No bona fide attempt was made to test this formation.

Drilling of a well through a unitized sand at a location not on the mineral servitude within the ten-year prescriptive period was not sufficient to interrupt liberative prescription when no effort was made to evaluate the unitized sand or produce from it until after the prescriptive period had run, held the court of appeal in *Malone v. Celt Oil, Inc.* [485 So. 2d 145, 90 O.&G.R. 66 (La. App. 1986).] The actions to obtain production from the unitized sand subsequent to the running of prescription were not part of a single operation commenced within the ten-year period in a good-faith effort to obtain production from the servitude tract or from land unitized with it.

The result in *Malone* may be compared with another case from the same court, *Bass Enterprises Production Co. v. Kiene.* [437 So. 2d 940, 78 O.&G.R. 456 (La. App. 1983).] In 1968, the Graveses sold land (110 acres) to Kiene, reserving a mineral servitude for one half of the minerals. In 1974, the Graveses leased their one half to Durham, and Kiene leased his interest to Bradco. No drilling took place on the servitude tract within the ten-year liberative prescription period provided by Louisiana law. However, in 1974 a well was spudded on a nearby tract. It was drilled under a permit to go to the Smackover Formation. After the well went to 10,746 feet, another company took it over and drilled to 11,520 feet. The Smackover sand was not productive. The company tested shallower sands, including the McFearin Sand. About 40 acres of the servitude tract were included in a McFearin Sand unit with the land on which this well was testing. The well was perforated, acidized, and fractured in the McFearin Sand, but it was unsuccessful. It was thereafter plugged and abandoned. In 1979 another well was drilled, which successfully produced from the Cotton Valley Formation, and the same 40 acres in the McFearin unit were included in the Cotton Valley unit. The operator of the well filed a concursus proceeding to determine to whom to pay the share of production attributable to the 40 acres. Kiene, as landowner, claimed that the Graves' servitude had prescribed because of lack of good faith drilling operations or production in the years since it was created. The servitude owners contended that the efforts to obtain production from the McFearin Sand were sufficient operations in good faith to interrupt prescription as to the portion of the servitude tract included in the McFearin unit. The court ruled that the operations to test the McFearin Sand were a good faith effort to secure production in paying quantities from land with which a portion of the servitude had been unitized. Under Article 33 of the Louisiana Mineral Code, this is sufficient to interrupt prescription of that portion of the servitude included in the unit. . . .

Article 61 is another provision of the Mineral Code that is important to note in connection with pooling and unitization. It provides: "Issuance of a compulsory unitization order establishing a unit that includes all or part of a tract burdened by a mineral servitude does not constitute an obstacle to its use." This article draws upon the case of *Mire v. Hawkins,* [249 La. 278, 186 So.2d 591, 25 O.&G.R. 160 (1966).] which held that designation of non-drilling areas within drilling units created by Conservation Commissioner did not constitute an obstacle to use of the mineral servitude on lands within the non-drilling areas. Thus, liberative prescription running against the mineral servitude was not suspended. The order is not an obstacle "for it does not prevent the use of their servitude, it merely controls the method of user." This, the court said, is through operation of law, and servitude owners take their interest subject to regulation.

[iv] Agreement for Contrary Effect Permitted

The Mineral Code permits parties to avoid the divisive effects of Articles 33, 34, 37, 89, and 90 through agreement of the landowner and the owner of the servitude or royalty. Article 75 of the Mineral Code gives this authority:

> The rules of use regarding interruption of prescription on a mineral servitude may be restricted by agreement but may not be made less burdensome, except that parties may agree expressly and in writing . . . that an interruption of prescription resulting from unit operations or production shall extend to the entirety of the tract burdened by the servitude tract regardless of the location of the well or of whether all or only part of the tract is included in the unit.

Application of the authority recognized in Article 75 was seen in the case of *White v. Evans.* [457 So. 2d 159, 82 O.&G.R. 503 (La. App. 1984).] On October 2, 1934, A sold a servitude for one half of the minerals on Blackacre (a tract of 120 contiguous acres) to *B*. A leased Blackacre and 299 other acres to Placid in 1942, and four months later *B* executed a co-lessor's agreement covering its one-half mineral servitude. In 1943 and 1944 Placid executed voluntary pooling agreements for two 640–acre units, one of which included 40 acres of Blackacre. *A* and *B* both signed the pooling agreements. On the unit that included the 40–acre portion of Blackacre, a well was spudded on June 7, 1944, (i.e. within 10 years of the servitude's creation), but the well was not on the servitude property. The well was successful and continued to produce at the time of the dispute. Landowners-plaintiffs, the successors to *A*, brought suit to declare that the production from the well did not interrupt prescription as to the 80 acres of the land burdened by the servitude that were not in the unit. The appellate court concluded that the pooling agreements should be interpreted as intending to interrupt prescription against the contiguous 120–acre servitude whether in or out of the unit. . . .

Several aspects of the rules of prescription, including a claim of an agreement satisfying Article 75, were taken up in *Sandefer & Andress, Inc. v. Pruitt.* [471 So. 2d 933, 86 O.&G.R. 504 (La. App. 1985).] The court held that operations or production on a unit would not interrupt prescription on part of a servitude tract not included within the unit, even if the rights were in existence prior to the enactment of the Louisiana Mineral Code. In this concursus proceeding to determine the owners of the rights to minerals in a 60–acre portion of a 260–acre tract, the landowners (the Pruitts) asserted that a mineral servitude that had been owned by certain mineral claimants had expired from liberative prescription accruing by ten years non-use. The court concluded the wells in question had not produced for a period of more than ten years and that none of the 60–acre portion of the tract in question was in a unit for which there was production. The court held it was the law both before and after the adoption of the Louisiana Mineral Code in 1974 that unit operations and production would interrupt prescription only as to the portion of the servitude tract included in the unit if the unit well were not on the tract burdened by the servitude. While a voluntary pooling agreement may expressly provide that operations or production on a voluntary unit will interrupt prescription as to any affected servitude tract in its entirety, the pooling agreements in question here did not provide for this effect. Nor could agreements subsequent to the extinguishment of the servitude in question have the effect of reviving the servitude.

SECTION 7. UNIT OPERATIONS AND NON–JOINING AFFECTED INTERESTS

A. OBLIGATIONS OWED BY LESSEE TO NON–JOINING LESSORS

(See Kramer & Martin, *The Law of Pooling and Unitization* § 23.01).

TIDE WATER ASSOCIATED OIL CO. v. STOTT

Circuit Court of Appeals of the United States, Fifth Circuit, 1946.
159 F.2d 174, *cert. denied*, 331 U.S. 817 (1947).

LEE, CIRCUIT JUDGE. This is a suit by oil and gas lessors for damages to their leases alleged to have resulted from lessees' recycling operations on neighboring lands.

The appellees, Elizabeth Jansing, Virginia Young Stott, and Mae Young Lubben, plaintiffs below, executed separate oil and gas leases in 1935 and 1937, covering their respective tracts of 25, 109.35, and 72.04 acres in the John Adams Survey, Anderson County, Texas. The appellants, Tide Water Associated Oil Company and Seaboard Oil Company of Delaware, defendants below, now own these leases except for an assignment to Haynes B. Ownby Drilling Company of 48.49 acres of the original 72.04 acres of the Lubben lease and of 69.36 acres of the original 109.35 acres of the Stott lease. Appellants, in the assignments to Ownby, reserved an overriding royalty and extensive controls over operations upon, and marketing of the products from, the assigned leases.

Including 88 acres of the appellees' land on which appellants still hold leases directly, appellants, during the period involved, held oil and gas leases covering 2215 acres underlaid with wet gas in a common reservoir of approximately 7355 acres. Except for the leases assigned to Ownby and those affecting two very small tracts, appellants owned leases entirely surrounding the lands of appellees on the north, east, and south.

Early in 1939 two other operators in the common pool commenced recycling operations, and in December of the same year appellants adopted the practice. In order profitably to conduct recycling operations it is necessary to unitize or pool tracts of substantial size, and beginning in mid–1939 appellants approached appellees with various proposals for unitization of their tracts with other lands and to have appellees participate in the recycling operations on the same basis as the royalty owners under other tracts subject to appellants' leases. Although both sides carried on negotiations in good faith, appellees declined to unitize and participate on this basis.

Appellants drilled a well on each of the three tracts and have produced from such tracts all of the oil that could be produced and all of the gas for which there was a market. The royalties due on these products and on the distillate or condensate separated from the gas at the well have been paid to the appellees.

Since recycling is not practicable on any of these tracts alone, appellants operated the wells on the three tracts, extracting condensate in separators at the well and selling the remaining semi-dry gas to the Lone Star Gas Company, the only market for gas in the field. Under regulations for prevention of waste, the gas could not be flared, and consequently production of condensate from the wells on the tracts was limited by the market available for the residue gas.

Under the recycling operations conducted on all other leases held by appellants (and substantially all other leases in the field) the "wet" gas is produced from withdrawal wells, processed through a gasoline plant which removes a higher proportion of liquid hydrocarbon than is possible by the use of simple separating devices at or near the well, and the remaining "dry" gas is returned under pressure through injection wells to the common reservoir. Since residue gas is not wasted but is returned to the reservoir for future use, a much higher production of gas is permitted for extraction of liquid hydrocarbons under this method of operation. The higher pressure of the dry gas at the injection wells forces the dry gas towards the points of lower pressure at the withdrawal wells, and gradually the dry gas spreads and the wet gas is withdrawn until the field becomes a dry-gas field, no longer useful for production of liquid petroleum products. In this process the dry-gas areas gradually extend from the injection wells.

Plaintiffs-appellees sued to recover damages from lessees to the extent of the royalty fraction of condensate which might have been removed from the wet gas under their lands which has now been replaced by dry gas as a result of the recycling operations conducted upon other leases in the field. The court below found that under appellees' three tracts the wet gas had been replaced by dry to the extent of 3.0 acres under the Jansing tract, 21.9 under the Lubben, and 9.0 under the Stott, with additional dry acreage under those portions of the Lubben and Stott tracts affected by the Ownby assignments. Upon these findings, after trial without jury, the court concluded that appellees had been damaged to the extent that the wet gas under their lands had been replaced by dry gas and gave judgment for the plaintiffs upon this basis.

. . .

It is conceded that a reasonable and prudent operator would not have drilled an additional well upon any of the appellees' three tracts; that the lessees were producing from these tracts all of the mineral products which could be produced in the absence of recycling; and that recycling was not practicable in the absence of unitization, which the lessors had refused. The appellants, therefore, have fulfilled their implied covenant to protect the premises from drainage.

The appellees contended, however, that there is an additional implied covenant of an oil and gas lease obligating the lessee not to injure his lessor's lease. They contend, in effect, that, even though the lessees are not liable under the implied covenant to protect from drainage, they are liable because the drainage has been effected through operations by the lessees themselves on other premises. [Most of the discussion of this contention is omitted. Eds.]

. . . The duty of an oil and gas lessee to operate the leased premises in a reasonable and prudent manner cannot prevent his operating the adjoining properties to the mutual advantage of the lessee and lessor there concerned.

. . . [T]he implied covenant of the lessee to do nothing to impair the value of the lease to the lessor is a duty "to use reasonable care in good faith to protect appellees from damage caused" by drilling on the lessee's adjoining land. . . . In the present case appellants, as lessees, repeatedly offered to the lessors plans for unitization of their three tracts in order that the appellees might participate in the recycling operations. Appellees contend, however, that no plan submitted for unitization and recycling was fair: The principal bases of contention are (1) that the contracts submitted computed royalties upon a different basis from those used by other operators in the field, and that the difference was prejudicial to their interests; (2) that the royalties offered were only 1/2 of 1/8 of the amount (less taxes)received by the recycling plant for the liquid hydrocarbons recovered in the plant; (3) that the appellants refused to make such royalties retroactive to commencement of the first recycling operations; and (4) that appellants refused to except condensate trapped in separators at the well from the effect of the unitization agreement.

The trial court found that the plan offered to the appellees was to have them participate in recycling operations "in the same manner and on the same basis as other landowners and royalty owners in the field." The record supports the trial court in this finding of fact, and there is no evidence to show that the prevailing plan was unfair to the royalty owners in general or to the appellees in particular.

When the lessee used gas for the manufacture of gasoline or other such products, the royalty provided for the lessor in the Lubben and Stott leases was 1/8 of the market value of the gas, and that in the Jansing lease was 1/8 of the net proceeds from the sale of the products after deducting the cost of manufacture. The trial court found that the prevailing royalty paid to royalty owners for wet gas in the Long Lake Field during the time involved in this suit was 1/2 of 1/8 of the amount (less taxes) received by the recycling plant for the products recovered in the plants from the wet gas. There is no evidence to show that this royalty was unfair; on the contrary, as expressed in Armstrong *et al.* v. Skelly Oil Co. *et al.,* 5 Cir., 55 F.2d 1066, 1068:

> . . . Appellees were under no obligation to erect a plant to treat this gas. When they did so they were entitled to deal with the lessor the same as a stranger would have done. Had the gas been sold to an extraction plant, the lessee, under the universal custom of the trade, would have received returns identically the same as those made by appellees.

> The method used to ascertain the value of the gas taken is fair. It would not be just to settle with the seller of the gas at the market price for the full amount delivered. That would leave nothing to cover shrinkage in extraction and the cost of manufacture, including overhead charges, and a fair return on the investment.

The refusal of the appellants to make royalties upon production from the recycling plant retroactive to the date of commencement of the first recycling operations was a refusal on their part to pay dual royalties to the appellees for the period prior to the actual effective date of the appellees' unitization and participation. We see nothing unreasonable or unfair in the appellants' refusal to concede upon this point, particularly since the appellants had approached the appellees to secure their participation before the recycling plant commenced its operations.

Nor were appellants unfair in insisting that the appellees' condensate (secured by separators near the wells) should not be excepted from the effect of the unitization. The rights of the various royalty owners within the given unit must necessarily be upon the same basis, and the appellants were fully justified in insisting that the appellees participate (if at all) upon the basis common to their other royalty owners within the unit upon their proportionate part. Moreover, as the plaintiffs' very cause of action shows, the effect of recycling operations is to make certain areas "dry" while others remain "wet;" the purpose of unitization is to equalize the rights of those royalty owners nearest the injection wells and those nearest the withdrawal wells, and this reason extends to condensate produced from separators fully as much as to the products produced in the plant. To except the condensate, in fact, would give rise to unfairness as between the royalty owners within the unit.

As the plan offered by the appellants to the appellees, for unitization of their tracts and participation in the recycling operations, was reasonable and fair in all respects, the appellants amply fulfilled any duty of fair dealing which may have been imposed upon them by the lessor-lessee relationship. The leases did not authorize unitization, and appellees were entitled to refuse, as they did, to unitize and to participate in the recycling, but the appellants were not thereby precluded from operating, and were well within their rights in proceeding to operate, their other leases to the best interest of the lessees and such lessors.

That appellants by their recycling operations acted for the mutual protection of themselves and of their lessors, including the appellees, is clearly shown by the record. Prior to the commencement of the recycling operations by appellants, two companies, one owning acreage within the producing area, were operating recycling plants. The evidence is uncontradicted that in about six or seven years, or about 1946, the withdrawal of wet gas and the injection of residue dry gas by the two companies would result in the replacement of wet gas with dry gas under a portion of appellants' acreage, including a portion of appellees' tracts of land, and that in about ten years, or about 1949, the recycling operations of the two companies would withdraw all of the wet gas from the entire reservoir and replace it with injected residue dry gas. Appellants and appellees, therefore, were alike threatened with loss of a common property right with respect to which no recovery in damages could be had. In such circumstances mutual cooperation to protect mutual interests was necessary and as binding upon the appellees as upon the appellants.

In short, the appellees may not refuse to cooperate with their lessees for their mutual protection in the adoption of the practicable customary method or plan universal in the Long Lake Field offered them by appellants and at the same time assert and demand damages. The contention of the appellees in such situation is both unique and untenable. Any damage which they suffer is damnum absque injuria and in nowise are such damages chargeable to appellants. . . .

. . . a tenant in common, particularly one holding a minor interest in the oil and gas, is not to be allowed, by withholding his consent to the development of the boundary in which his interest lies, to prevent the development of an adjoining tract under a unitization agreement to which he has been given an equal opportunity to become a party and in which his cotenants have all joined. Boggess v. Milam, 127 W.Va. 654, 34 S.E.2d 267, 269, 270.

Appellees may correct the situation for the future by participating in recycling operations "on the same basis as other royalty owners in the field."

. . .

The judgment appealed from is reversed, and the cause is remanded for further proceedings not inconsistent with the views herein expressed.

Reversed and remanded.

NOTES

1. Professor Maurice Merrill observed that "I have no quarrel with the view that a lessor who has declined to enter into a unitization program upon reasonable

terms cannot complain that the result of his lessee's participation in the unit with respect to other tracts adversely affects recovery from his own tract, or that he cannot allege a breach of the covenant for protection based on failure to take steps to prevent the very thing which must take place if unitized operation is to be successful. But I do think it is equally clear that the lessor who declines to accept unreasonable terms is in the same position as one who is excluded arbitrarily from inclusion in a unit and that such a one may maintain successfully that his lessee's conduct is a violation of implied covenant obligations." Merrill, *Implied Covenants, Conservation and Unitization*, 2 Okl.L.Rev. 469 at 479 (1949). The reasonableness of the terms offered the lessors in *Stott* is discussed in Merrill, *Unitization Problems: The Position of the Lessor*, 1 Okl.L.Rev. 119 at 128 (1948).

2. On the issue of fairness of the plan, what would be the effect of approval of the plan for unitized operations by a state regulatory commission? *See* the cases discussed at note 9 p. 853 *supra* involving the exercise of the pooling or unitization clause by the lessee.

3. Note the court's observation that "a much higher production of gas is permitted for extraction of liquid hydrocarbons" than for the gas from the wells on the plaintiffs' tracts. The fact that the state regulations allow such disparity for production is the reason for the drainage, is it not? Thus, the state may encourage, through the setting of production allowables, the unitization of oil and gas fields even when the conservation agency lacks the authority to force unitization. The suit in the principal case was against the lessee, of course, and not the Railroad Commission.

4. CORLEY v. MISSISSIPPI STATE OIL AND GAS BOARD, 234 Miss. 199, 105 So.2d 633, 9 O.&G.R. 481 (1958). *Stott* was followed in this Mississippi case in which the allowable orders of the Mississippi Oil and Gas Board were challenged directly. In *Corley,* the Mississippi court held it was proper for the Board to allocate production in accordance with surface acreage regardless of whether there were producing wells thereon. The case involved the Bailey Oil Pool of the Soso Field. Some 99.5 percent of working interests and 95.5 percent of royalty interests entered voluntary agreements for a reservoir-wide unit, and the Board approved these. Then Gulf, the operator, applied for special rules for the reservoir. There had been 64 wells on a pattern of 40–acre drilling units, with a total of 3442.02 acres assigned to them. The maximum efficient rate of recovery was fixed at 9,600 barrels per day, allocated to the producing wells on the statutory basis of surface acreage. The Board determined that the field was located under an additional 1170.33 acres on which there were no producing wells. The board then created two types of spacing or drilling units: 1) 24 units in which small fractional interests had not become parties to the agreements—these remained as 40–acre units (total of 921 acres and 24 wells); 2) the residual acreage was made part of the fieldwide unit, and the additional acreage (1170.33 acres) was attributed or allocated to it (total of 3691.34 acres and 40 wells). The

effect was to reduce the daily allowable for the field and distribute it on the basis of surface acreage. The plaintiffs complained that the order cut their allowables from 187 b/d to 104 b/d. Here the board simply increased the size of the field and reduced the MER for each surface acre. The Board, the court said, was not acting beyond its powers: "[W]e do not think it would be fair or equitable to permit a very small minority of small fractional owners in this common source of supply of hydrocarbons to prevent the Board and the overwhelming majority of the operating and royalty owners from pursuing a unitization program designed to obtain a maximum recovery from the field."

B. OBLIGATIONS OWED TO OWNERS ADJACENT TO UNIT OPERATIONS

BOYCE v. DUNDEE HEALDTON SAND UNIT

Court of Appeals of Oklahoma, Division No. 1, 1975, cert.denied.
560 P.2d 234, 56 O.&G.R. 565.

(Released for Publication by Order of Court of Appeals.)

ROMANG, PRESIDING JUDGE. These companion cases are suits for damages to three oil wells alleged to have been caused by the defendants when they forced water into a nearby well, in an effort to effect a secondary recovery of oil by means of a technique called waterflooding. The cases were consolidated for trial before the Honorable Kenneth Shilling in Carter County, and a jury returned a verdict in favor of the plaintiffs in both cases. The defendants have appealed from a judgment based on those verdicts.

. . .

In 1958 engineering studies were commenced to determine the feasibility of flooding wells in the Healdton field with water under pressure to stimulate further production of oil.

As a result of these studies, Sinclair Oil and Gas Company submitted a plan to the Oklahoma Corporation Commission under which Section 4, Township 4 South, Range 3 West in Carter County would be treated as a unit for the purposes of this project. On February 26, 1963 the Corporation Commission found that it was necessary to create the unit to prevent waste and increase the ultimate recovery of oil and gas from the common source of supply. This unit was named the Dundee Healdton Sand Unit, and was one of the defendants in these cases. The Corporation Commission named Sinclair Oil and Gas Company as operator of the unit, but that company has since merged with the defendant Atlantic Richfield Company which thereupon became the operator.

All of the plaintiffs' wells are located in the next section north of the section covered by the unitization order, and within 200 feet of the north boundary of the unit.

Injection of water into the various Healdton Sands in accordance with the order of the Corporation Commission was begun in the unit in June, 1964. The plan of unitization approved by the Corporation provided for the use of four injection wells near the north boundary of the unit. Injection Well A–4 was nearest the plaintiffs' leases. Shortly after water injection using Well A–4 was started in 1969, the plaintiffs' wells started producing water rather than oil. Plaintiffs then brought these actions seeking damages for the loss of oil production from the wells and the cost of plugging the wells, which was alleged to have been extraordinarily increased by the presence of water therein.

The jury awarded the plaintiffs interested in the Spears lease $16,125.00 and the plaintiffs interested in the Wells lease $3,000.00.

The defendants seek reversal of the judgment rendered below on the grounds that the trial court erred in its instructions to the jury, that these suits constitute a collateral attack upon the order of the Corporation Commission in contravention of 52 O.S.1971, § 111, that the plaintiffs cannot maintain this suit until they have exhausted the administrative remedies available to them, that the court erred in denying to the defendants the defenses of assumption of risk and estoppel, that the court erroneously instructed the jury on the defense of consent, and that the form of verdict submitted to the jury was improper.

The trial court instructed the jury that the Dundee Healdton Sand Unit was lawfully created and authorized to inject water into the producing formations lying under the sections covered, but that even though such operations were lawful in every respect and properly carried on without any negligence, if such operations resulted in an unreasonable interference with the peaceful occupation and enjoyment of their property by the owners of adjacent property, such adjacent owners might recover for such damages as they may have sustained by said operation.

It is the validity of this premise, stated explicitly in the instructions and applied to the fact in this case, which is disputed by the defendants in their first proposition.

The defendants argue that the owner of real property owns only such minerals as he is able to capture and bring to the surface; that the legislature has the power to regulate the manner and the extent to which this capture is accomplished; that the legislature has given the Corporation Commission the power to do this; the Corporation Commission by its order having determined that waterflooding operations within the unitized area would not materially adversely affect the plaintiffs' properties, the plaintiffs cannot now recover for damage to their wells.

. . .

The plaintiffs do not dispute the validity of the order of the Corporation Commission, nor do they contend that the defendants were negligent in any way. They simply assert that the lawful operation of the defendants' waterflood project and the resultant migration of water into the formation from which their wells were producing became a private nuisance when it destroyed the productivity of their wells, and that the order of the Corporation Commission does not insulate the defendants from liability under such circumstances.

In Fairfax Oil Co. v. Bolinger, 186 Okl. 20, 97 P.2d 574 (1939), an oil well was drilled in an area in Oklahoma City properly zoned for such an operation. Since it was properly authorized by city ordinance, the defendant in that case argued that it was not a nuisance per se, and it was not liable to the plaintiff for damages alleged to have been caused by vibrations emanating from the drilling operation in the absence of either negligence or some unusual, unreasonable or improper use of the property. The Supreme Court held, however, that the common law nuisance doctrine has been modified by Section 23 of Article 2 of the Constitution of the State of Oklahoma, and that

> . . . a legalized use of property becomes a nuisance per accidens if that use substantially damages the property of another.

In Gulf Oil Corp. v. Hughes, Okl., 371 P.2d 81 (1962), a case in which the waterflood operations of the defendant oil company were alleged to have damaged the plaintiffs' water wells, the Supreme Court of the State of Oklahoma followed its holding in Fairfax Oil Co. v. Bolinger, *supra,* and upheld an instruction in the trial court very similar to the instruction sub judice, noting that a similar instruction had been approved in British–American Oil Producing Co. v. McClain, 191 Okl. 40, 126 P.2d 530 (1942).

In 1971, the United States Court of Appeals for the 10th Circuit was faced with a dispute remarkably similar to the present case in Greyhound Leasing & Financial Corp. v. Joiner City Unit, 10 Cir., 444 F.2d 439 (1971). That action concerned waterflooding for secondary recovery in the Joinder City Field in Carter County, Oklahoma. The plaintiff's wells were excluded from the unit at the request of their predecessors in interest, who were present and participating actively in hearings before the Corporation Commission. The court in a well-reasoned opinion, upheld the plaintiffs' recovery in that case.[112] The court said, at page 441,

> The several Oklahoma Supreme Court decisions which have considered encroachments into the water or oil and gas strata underlying a plaintiff's property have applied a modified private nuisance doctrine. This is essentially the common law doctrine as altered by a provision in the Oklahoma Constitution which the Oklahoma courts have said removes the common law elements of carelessness or unreasonableness.

---

[112]   Predecessors of the named plaintiffs in *Greyhound Leasing* had participated in the engineering work preliminary to the unitization. At a later time, before approval of the unit, on their insistance the proposed unit boundaries were redrawn by the proponents to exclude the leases owned by plaintiffs' predecessors. Thus the proposal submitted to, and approved by, the Corporation Commission did not include plaintiffs' leases. The defendants argued that they agreed to the exclusion of plaintiffs' leases to avoid further delay in unitization. You are encouraged to speculate as to the reasons plaintiffs' predecessors sought to be excluded and the reasons defendants consented to the exclusion. [Eds.]

The Constitutional provision to which the court alluded was Article 2, Section 23, which provides, in part,

> No private property shall be taken or damaged for private use, with or without compensation, unless by consent of the owner . . ..

The court considered and rejected the defendants' arguments that Oklahoma's nuisance doctrine should be modified by reason of the statutory authority and actions of the Corporation Commission, and concluded that the finding of the Corporation Commission that Unit operations on the portion of the common source of supply within the unit "will have no material adverse effect upon the remainder of such common source of supply" did not authorize extension of these operations outside the unit. The court further said, at page 443,

> We also find no Oklahoma authority which would modify the rule there prevailing as to private nuisances . . . by reason of the exercise of regulatory jurisdiction by the Commission.
>
> . . .
>
> The Oklahoma statutes relative to oil and gas production, and the regulatory body and its rules and regulations present a comprehensive regulatory plan (citations omitted) and machinery which has operated successfully for many years. However, we find nothing in the statutes or decisions which place within this machinery the power or duty to adjudicate an action such as this where money damages are sought. . . .

The court said further at page 445,

> [T]he defense of assumption of risk was not available to the defendant. The plaintiff did not place itself in a place of danger as the doctrine contemplates. *See,* 73 A.L.R.2d 1378. The fact that the possibility of the intrusion of salt water was apparent to all concerned is not enough. The plaintiff was not in a position to avoid the possibility, the wells were in existence before the advent of the unitization, and nothing could be done defensively by the plaintiff alone as a practical matter. The text authorities cited by the defendant are not to the contrary as general statements of the rule. The fact that it wished to be advised of the rates and place of injection under the waterflood does not demonstrate that the plaintiff either assumed the risk or consented to the intrusion.

In the case before this court, the trial court instructed the jury that the consent mentioned in Section 23 Article 2 may be given orally, or in writing. The defendants objected to this instruction and requested an instruction to the effect that such consent may be inferred from the conduct of the plaintiffs and their operator, Tomlinson. The conduct apparently relied upon as showing such consent was the ratification of the unitization order by certain of the plaintiffs who owned an interest in property included in the unit, and the participation of Tomlinson in the studies which preceded the order of the Corporation Commission.

Bearing in mind that it was only when the defendants' activity commenced to damage the plaintiffs' wells that it became a nuisance, we find nothing in the conduct of the plaintiffs which could be interpreted as showing their consent to such damage. Ratification of the unitization order clearly is not such consent, Greyhound Leasing & Financial Corp. v. Joiner City Unit, *supra*, but even if it were, we think the instruction of the court would have been broad enough to include it.

Since ratification of the Corporation Commission's order did not constitute consent to the damages sustained, the trial court was correct in refusing to differentiate, in the forms of verdict submitted to the jury, between those plaintiffs who had ratified and those who had not.

Affirmed.

REYNOLDS and BOX, JJ., concur.

NOTES

1. The principal case, like its Tenth Circuit predecessor, Greyhound Leasing & Financial Corp. v. Joiner City Unit, 444 F.2d 439, 40 O.&G.R. 60 (10th Cir.1971), is founded on a "modified private nuisance doctrine," rather than trespass. What is the consequence of the theory? How appropriate would a remedy of injunction be? Consider the approach of the Arkansas Supreme Court in the next note.

2. JAMESON v. ETHYL CORP., 271 Ark. 621, 609 S.W.2d 346, 69 O.&G.R. 19 (1980). Ethyl held leases entitling it to remove brine (salt water) from about 15,000 acres in the Kerlin Brine Field (or about 90% of the field). Jameson owned a 95–acre tract in the field, entirely surrounded by Ethyl's leases. The parties had negotiated unsuccessfully for a number of years over a proposed lease or other extraction terms. Ethyl commenced operations, which included drilling injection wells designed to cause the brine to move toward the center (and Ethyl's extraction wells) and to prevent an influx of evidently inferior brine which would have rendered the project commercially impracticable. As a result, the bromine content (the valuable portion of brine) beneath the Jameson tract "has been substantially reduced, if not totally exhausted from a commercial perspective." 609 S.W.2d at 349, 69 O.&G.R. at 22. [Query: Would the brine beneath the Jameson tract ever have been usable from a commercial perspective, standing alone? The court notes that drilling by Jameson would have been impractical "unless she was financially prepared and willing to get into the bromine extraction business," because of the expenses of transporting brine over long distances. *Id.*] Ethyl filed for a declaratory judgment to establish the legality of its extraction process, and Jameson counterclaimed for damages and an injunction. Rejecting the Rule of Capture, the court said:

The underlying reason for adoption of the rule of capture by Arkansas and other states was the acknowledged impracticality of tracing ownership of a transient substance which migrated from lands of one owner to lands of someone else. . . .

While Arkansas' unitization laws are not, as previously noted, involved in this case, we do believe that the underlying rationale for the adoption of such laws, i.e., to avoid waste and provide for maximizing recovery of mineral resources, may be interpreted as expressing a public policy of this State which is pertinent to the rule of law of this case. Inherent in such laws is the realization that transient minerals such as oil, gas and brine will be wasted if a single landowner is able to thwart secondary recovery processes, while conversely acknowledging a need to protect each landowner's rights to some equitable portion of pools of such minerals. A determination that a trespass or nuisance occurs through secondary recovery processes within a recovery area would tend to promote waste of such natural resources and extend unwarranted bargaining power to minority landowners. On the other hand, a determination that the rule of capture should be expanded to cover the present situation could unnecessarily extend the license of mineral extraction companies to appropriate minerals which might be induced to be moved from other properties through such processes and, in any event, further extend the bargaining power of such entities to reduce royalty payments to landowners who are financially unable to "go and do likewise" as suggested by Ethyl.

The laws of trespass and nuisance and the rule of capture each evolved out of circumstances designed to balance the relative rights and responsibilities of the parties and the interests of society in general. As noted in the *Young* case, *supra* [Young v. Ethyl Corp., 521 F.2d 771, 53 O.&G.R. 111 (8th Cir.1975)], a great deal of technology and geological understanding has developed since the 1912 *Osborn* decision [Osborn v. Arkansas Territorial Oil & Gas Co., 103 Ark. 175, 146 S.W. 122 (1912)]. As envisioned in the *Young* case, which we consider to be persuasive, we are unwilling to extend the rule of capture further. By adopting an interpretation that the rule of capture should not be extended insofar as operations relate to lands lying within the peripheral area affected, we, however, are holding that reasonable and necessary secondary recovery processes of pools of transient materials should be permitted, when such operations are carried out in good faith for the purpose of maximizing recovery from a common pool. The permitting of this good faith recovery process is conditioned, however, by imposing an obligation on the extracting party to compensate the owner of the depleted lands for the minerals extracted in excess of natural depletion, if any, at the time of taking and for any special damages which may have been caused to the depleted property. By this holding we believe that the interests of the owners and the public are properly protected and served.

This matter was submitted to the Chancellor only on the issue of liability and on a remand the Chancellor must determine damages. In this regard we point out the language set forth in the case of Whitaker & Company v.

Sewer Improvement District No. 1, 229 Ark. 697, 318 S.W.2d 831 (1958), where we said:

> A court of equity should be as alert to afford redress as the ingenuity of man is to cause situations to develop which call for redress.

In other words, sometimes a court of equity must devise a formula of damages to fit a particular situation.

What rule of damages would best reconcile the interests in (1) giving the extracting parties optimal incentives to invest in extraction and (2) assuring "fair" treatment for passive owners?

3. Consider the discussion of damages in *Greyhound Leasing & Financial Corp. v. Joiner City Unit, supra.* Having concluded that the replacement of oil beneath plaintiff's land by salt water as a result of defendant's water injection program constituted an actionable nuisance under Oklahoma law, the court turned to the problem of the appropriate measure of damages:

> The measure of damages in the trial court's instruction is also challenged. The court told the jury that the usual measure for damages to real estate was a before and after comparison. The court then instructed the jury in effect that this could be computed on the difference before and after of the recoverable reserves. This is sufficiently in conformance with the before and after rule, and there was testimony of expert witnesses upon which the jury could so arrive at a dollar amount. . . .

> As to the amount of damages the defendant urges that they are excessive and that proper deductions for costs of production and related matters were not made. More particularly the defendant asserts that there should have been deducted production costs. Such an adjustment does not appear separately, but there was testimony given which included such costs and testimony as to the loss which was not challenged on this point. There is sufficient evidence in the record of the net amounts, and a method to reach them, upon which the jury could base its verdict. Further as to the point of excessiveness, the verdict was well within the spread of the experts' testimony.

4. BAUMGARTNER v. GULF OIL CO., 184 Neb. 384, 168 N.W.2d 510, 34 O.&G.R. 235 (1969), cert. denied, 397 U.S. 913 (1970). In this action at law for willful trespass and conversion, the plaintiff sought to recover the value of oil claimed to have been displaced and swept from land under lease to him by defendant into its waterflood unit recovery wells. The court commented as follows:

> We have reached the conclusion that where the primary recoverable oil has been exhausted, all interested parties in the field must be offered an opportunity to join in any unitization project to recover secondary oil on a fair and equitable basis, and if any interested party refuses to join he should not be permitted to capitalize on that refusal. To hold otherwise would discourage unitization and encourage rather than avoid waste. Consequently,

we hold where a secondary recovery project has been authorized by the commission the operator is not liable for willful trespass to owners who refused to join the project when the injected recovery substance moves across lease lines.

. . . Under the facts herein the most that plaintiff should have a right to recover is what he can prove by a preponderance of the evidence he could have obtained through his own efforts if he had drilled, developed, and operated his property outside the unitization project; that is, as if no unitization had occurred. There is evidence that any operations by plaintiff would have resulted in an economic loss. If the testimony of his manager of operations is accepted, then the profit he could have realized from his own operations for both primary and secondary recovery would have been $12,224. This would be the limit of plaintiff's just and equitable share for oil displaced from Section 16 by Kenmac in the absence of other evidence.

. . .

For the reasons enumerated, the judgment herein is reversed the cause remanded to the district court for retrial under the rule of damages enunciated above. . . .

A dissenting judge concurred in the above quoted statement of principle in the majority opinion but declared:

[B]ut if this principle is to be applied, then plaintiff has no basis for recovery whatsoever and neither does plaintiff's lessor. Quite conclusively, plaintiff would have lost money had he made an attempt to recover his share of the primary oil and he could not have shown a profit without the efforts of Kenmac in producing secondary oil, a project in which he refused to join. If, therefore, he is permitted to recover, he is then being permitted to capitalize on his refusal to cooperate and to join in Kenmac.

5. Trespass was also alleged in RAILROAD COMMISSION v. MANZIEL, 361 S.W.2d 560, 17 O.&G.R. 444, 93 A.L.R.2d 432 (Tex.1962). This was an action to set aside a Railroad Commission order permitting water injection into an irregularly spaced well, as part of a secondary recovery program. It was alleged that the injection was a trespass that would cause premature destruction of plaintiffs' producing wells. Judgment was rendered dissolving the permanent injunctions granted by the trial court against the Railroad Commission and the injecting operator.

In considering the legal consequences of the injection of secondary recovery forces into the subsurface structures, one authority, Williams and Meyers, *supra* [Oil and Gas Law, § 204.5, note 1], has stated:

What may be called a "negative rule of capture" appears to be developing. Just as under the rule of capture a landowner may capture such oil or gas as will migrate from adjoining premises to a well bottomed on his own land, so also may he inject into a formation substances which may migrate through the structure to the land of others, even if it thus results in the displacement under such land of more valuable with less valuable substances (*e.g.,* the displacement of wet gas by dry gas).

Secondary recovery operations are carried on to increase the ultimate recovery of oil and gas, and it is established that pressure maintenance projects will result in more recovery than was obtained by primary methods. It cannot be disputed that such operations should be encouraged, for as the pressure behind the primary production dissipates, the greater is the public necessity for applying secondary recovery forces. It is obvious that secondary recovery programs could not and would not be conducted if any adjoining operator could stop the project on the ground of subsurface trespass. As is pointed out by amicus curiae, if the Manziels' theory of subsurface trespass be accepted, the injection of salt water in the East Texas field has caused subsurface trespasses of the greatest magnitude.

The orthodox rules and principles applied by the courts as regards surface invasions of land may not be appropriately applied to subsurface invasions as arise out of the secondary recovery of natural resources. If the intrusions of salt water are to be regarded as trespassory in character, then under common notions of surface invasions, the justifying public policy considerations behind secondary recovery operations could not be reached in considering the validity and reasonableness of such operations. . . . Certainly, it is relevant to consider and weigh the interests of society and the oil and gas industry as a whole against the interests of the individual operator who is damaged; and if the authorized activities in an adjoining secondary recovery unit are found to be based on some substantial, justifying occasion, then this court should sustain their validity.

We conclude that if, in the valid exercise of its authority to prevent waste, protect correlative rights, or in the exercise of other powers within its jurisdiction, the Commission authorizes secondary recovery projects, a trespass does not occur when the injected, secondary recovery forces move across lease lines, and the operations are not subject to an injunction on that basis. The technical rules of trespass have no place in the considerations of the validity of the orders of the Commission.

Does the "negative rule of capture" theory followed in *Manziel* preclude the recovery of damages by a party who proves injury?

In Coastal Oil & Gas Corp. v. Garza Energy Trust, 268 S.W.3d 1 (Tex. 2008) (p. 75 *supra*), the court found that where fluids injected in a hydraulic fracturing operation migrated across property lines there was no trespass. While discussing *Manziel* and the negative rule of capture, the *Coastal Oil* case said little about the impact of a governmental permit on whether a trespass occurred. In FPL Farming Ltd. v. Environmental Processing Systems, 305 S.W.3d 739 (Tex.App.—Beaumont 2009), the court relied heavily on *Manziel* and *Coastal*

Oil to support its determination that a party in receipt of a governmental permit did not commit a trespass even though the evidence before the governmental agency was that the injected waste water stream would cross property lines. *See also* FPL Farming, Ltd. v. Texas Natural Resource Conservation Commission, 2003 Tex.App. LEXIS 1074 (Tex.App.—Austin, Feb. 6, 2003, pet. denied). But in Berkley v. Railroad Commission of Texas, 282 S.W.3d 240 (Tex.App.—Austin 2009). the court opined on the impact of a Commission order authorizing the injection of brine on a common law trespass action: "Specifically, securing a permit does not immunize the recipient from the consequences of its actions if those actions affect the rights of third parties. Nor does it authorize the recipient to act with impunity *viz* third parties. Rather obtaining a permit simply means that the government's concerns and interests, at the time, have been addressed." 282 S.W.3d at 242.

6. NUNEZ V. WAINOCO OIL & GAS COMPANY, 488 So.2d 955, 91 O.&G.R. 237 (La.1986). The Commissioner of Conservation established a drilling and production unit which included land owned by plaintiff Nunez. Defendant Wainoco had a lease on property adjacent to plaintiff's property, which was also in the unit. A directional survey revealed the well had penetrated plaintiff's property at a subsurface location. Plaintiff brought suit against defendant Wainoco claiming a trespass and seeking removal of the well and damages. The trial court ruled this was a collateral attack on an order of the Commissioner of Conservation and dismissed it. Plaintiff then filed suit against the Commissioner, Wainoco and other defendants. Plaintiff moved for summary judgment on the issue of his right to an injunction. The trial court granted summary judgment in favor of the Commissioner of Conservation, dismissing him as defendant, and granting partial summary judgment in favor of the other defendants, affirming the Commissioner's refusal to order removal of the well. The Court of Appeal affirmed the dismissal of the Commissioner from suit but ruled that the Commissioner's order could not authorize drilling on unleased property without consent of the landowner and remanded for a determination whether a trespass took place and, if so, whether it was in good faith or bad faith. Nunez v. Wainoco Oil & Gas Co., 477 So.2d 1149 (La. App. 1st Cir.1985). The Louisiana Supreme Court reversed. The enactment of conservation regulation statutes, held the court, supercedes the general concept of ownership of the subsurface of land. When a unit has been created by order of the Commissioner of Conservation, a legally actionable trespass has not occurred:

> [W]e conclude that the established principles of private ownership, already found inadequate in Louisiana to deal with the problems of subsurface fugacious minerals . . . need not necessarily be applied to other property concepts, like trespass, within a unit created by the Department of Conservation. Unitization, which creates rights and interests in a pool of hydrocarbons beyond the traditional property lines, effectively amends La.Civ.Code art. 490 and other private property laws in the interest of conserving the natural resources of the state and, in effect, of protecting

private property interests, or "correlative rights," of nondrilling landowners. By prohibiting an individual landowner in the unit from drilling wells on their own tracts, by forcing them to share production, and by limiting the amount of hydrocarbons that can be produced, the exercise of the Commissioner's power to unitize necessarily results in infringement on the usual rights of ownership. Unitization has also resulted in changes in the legal relationships between landowners and lessees within the unit. . . .

Therefore, we hold that the more recent legislative enactments of Title 30 and Title 31 supercede in part La.Civ.Code Ann. art. 490's general concept of ownership of the subsurface by the surface owner of land. Thus, when the Commissioner of Conservation has declared that landowners share a common interest in a reservoir of natural resources beneath their adjacent tracts, such common interest does not permit one participant to rely on a concept of individual ownership to thwart the common right to the resource as well as the important state interest in developing its resources fully and efficiently.

One should note footnote 29 of the opinion in which the court acknowledges that damages may be required to be paid even if there is no actionable trespass if there is damage or measurable inconvenience. In the *Nunez* case, there was no occasion to apply this since the intrusion was two miles beneath the surface with no observable consequences to the landowner and the landowner was receiving his proportionate share of production.

7. RAYMOND v. UNION TEXAS PETROLEUM CORP., 697 F.Supp. 270, 101 O.&G.R. 267 (E.D.La.1988). The *Nunez* decision was followed by a Federal district court in this case. Plaintiff landowners brought a claim against defendants for trespass, alleging that defendants injected salt water into a disposal well on neighboring property which migrated to the property (subsurface) of plaintiffs. The land of plaintiffs and the land on which the injections took place were within a force-pooled unit established by the Commissioner of Conservation. However, the injections, which were permitted by the Commissioner of Conservation, were into a different formation than the one which was force-pooled, and some of the injected water was from other land not pooled with plaintiffs' land. Plaintiffs contended the defendants were unlawfully using plaintiffs' subsurface property for disposal of salt water to the extent that the salt water was produced from lands not owned by the plaintiffs and not pooled with plaintiffs's land. Plaintiffs did not challenge the order of the commissioner permitting the salt water injection and did not seek to enjoin the injection; they sought rentals for the use of their subsurface. The court held for the defendants. There was no legally actionable trespass. Applying *Nunez,* the court ruled that the invasion of salt water under plaintiffs' land was part of a disposal operation authorized by the Commissioner. As such it was not unlawful and did not constitute a legally actionable trepass. While damages would be available upon a proper showing under the *Nunez* standard, there was no evidence here that the injection of the salt water caused

harm. Does *Raymond* go beyond *Nunez?*

8. CALIFORNIA CO. v. BRITT, 247 Miss. 718, 154 So.2d 144, 19 O.&G.R. 36 (1963) was an action for damages by the owners of an unsigned, fractional mineral interest, against the operator of a voluntary unitization of an oil field, using a pressure maintenance program which had been approved by the state regulatory agency. Plaintiffs had been offered, but had refused, an opportunity to join the unit. *Held,* there is no liability for the alleged trespass upon plaintiffs' mineral interest resulting in the alleged wrongful displacement of oil from their property.

> Since there was no invasion of appellees' mineral interest by drilling a well on it, and all of California's activities have been in accord with the requirements of the conservation act and orders of the board, this is one of those cases, somewhat unusual today, where the law of capture applies . . .

> . . . Since California has not trespassed upon appellees' land, had not drilled a well on it, and has violated no right of appellees, it is not liable to them in tort. . . .

9. *Compare* the Mississippi approach with the decision in TIDEWATER OIL CO. v. JACKSON, 320 F.2d 157, 18 O.&G.R. 982 (10th Cir.1963), cert. denied, 375 U.S. 942 (1963). This case concerned a secondary recovery waterflood program instituted by Tidewater in 1946. About 1956 Tidewater notified the Jackson brothers, owners of a lease on the Barrier land, of its intention to water flood its properties adjacent to the Barrier lease and proposed a cooperative water flood program, but the Jacksons refused to join the venture. When water injection by Tidewater flooded some of their wells, the Jacksons complained to the Kansas Corporation Commission which held there was no waste nor injury to correlative rights and concluded that there was no justiciable issue for action by the Commission; the Commission's memorandum and order were affirmed by the Supreme Court of Kansas. Jackson v. State Corporation Commission, 186 Kan. 6, 348 P.2d 613, 12 O.&G.R. 185 (1960). This suit was then commenced in federal court to recover damages for the flooding of the Jackson wells. The court concluded that the issue of tort liability for the acts done by Tidewater, in the exercise of its rights granted by the Commission to water flood its properties, survived the determination of the Commission. It affirmed the trial court's judgment for compensatory damages based upon estimated loss of profits; there had been a battle of experts in the trial court which credited the testimony of an expert witness for the Jacksons "as against the countervailing and equally competent testimony of Tidewater." With one judge dissenting, the trial court's judgment for exemplary damages was reversed inasmuch as the acts complained of were committed under color, and in accordance with Kansas law, and they were lawfully undertaken and lawfully done in the interest of conservation. The dissenter urged that Tidewater evinced a reckless disregard of the rights of the Jacksons and properly were held liable for punitive damages.

Did the Federal court, applying state law, effectively overrule the Kansas

Supreme Court by treating as tortious actions that were clearly authorized by state law? Consider the appropriate treatment to be given to the Kansas Corporation Commission's determination that there was no injury to correlative rights. *See also* Snyder Ranches, Inc. v. Oil Conservation Commission, 110 N.M. 637, 798 P.2d 587 (1990).

10. Should a distinction be made between surface and subsurface use of land for unit operations? Should a unit operator be able to make use of surface or subsurface water for unit operations? The cases are gathered and discussed in Kramer & Martin, *The Law of Pooling and Unitization*, § 20.06[1].

———

Chapter 8

PUBLIC LANDS

SECTION 1. FEDERAL LANDS

In 2000, the Federal Government owned more than 660 million acres of land-almost thirty percent of the United States. Over 587 million acres is labelled "public domain" land, viz., land which has never left federal ownership. The remaining 60 million acres are labelled "acquired lands," viz., lands acquired through purchase, condemnation, gift or exchange. The public domain constitutes almost 68 percent of Alaska, 82 percent of Nevada, 63 percent of Idaho and 64 percent of Utah. In several states the percent of federal lands is less than 1 percent. In addition, the United States has retained the right to all or specified minerals in over 66 million acres.

The United States leases onshore public lands under the provisions of several different statutes including the Mineral Leasing Act of 1920, 41 Stat. 437 et seq., 30 U.S.C. § 181 et seq., as amended by the Federal Onshore Oil and Gas Leasing Reform Act of 1987, Pub.L. No. 100-203, 101 Stat. 1330-256, (which govern the leasing of the public domain), the Acquired Lands Act of 1947, 61 Stat. 913 et seq. (1947), 30 U.S.C. §§ 351-359 (leasing of acquired lands), and the Tribal Leasing Act, 25 U.S.C. § 396 (leasing of certain Indian lands). The leasing of offshore public lands is governed by the Outer Continental Shelf Lands Act of 1953, 30 U.S.C. §§ 1331-1356 (leasing of submerged lands of the United States).

The several acts relating to the leasing of public lands differ in several respects. Not only do they cover different parts of the public domain, but they also differ on the leasing and operational aspects of oil and gas development. This section will highlight five different areas of federal public lands oil and gas law: 1). Onshore oil and gas leasing and development, 2). Unique issues affecting federal oil and gas development, 3). Federalism issues affecting those who operate on federal lands, 4). Offshore (OCS) leasing and development, and 5). Indian oil and gas leasing.

The major secondary authorities on the development of oil and gas on federal lands are Rocky Mountain Mineral Law Foundation, *The Law of Federal Oil and Gas Leases* (2010); 1 Kramer & Martin, *The Law of Pooling and Unitization* ch. 16; L. Hoffman, *Oil and Gas Leasing on Federal Lands* (1957); L. Hines, *Unitization of Federal Lands* (1953). For a good overview of federal natural resources law see G. Coggins & R. Glicksman, *Public Natural Resources Law* (2004 with supplements).

ONSHORE OIL AND GAS LEASING
AND DEVELOPMENT

Prior to 1920 federal oil and gas deposits were covered by the general placer mining laws of the United States. This was principally the Mining Law of 1872. Mining Resources Act of May 19, 1872, ch. 152, 17 Stat. 91 (codified and amended at 30 U.S.C. §§22-24, 26-30, 33-34, 37, 39-42. Oil and gas were definitively placed within the ambit of the 1872 Mining Law through the enactment of the Oil Placer Act of February 11, 1897, 29 Stat. 527. A private developer need only make a proper placer location and upon submission of proof of a valuable discovery the developer was entitled to a federal patent. The United States retained no interest in the oil and gas and received no royalties from its production. Owing in part to the dependence of the U.S. Navy on oil for fueling its ships, the United States began to withdraw lands from the placer mining process in the early 1900's. President Taft withdrew over three million acres of federal lands in Wyoming and California from the application of the mining law. This executive withdrawal was later approved by Congress which enacted the Pickett Act of 1910. 43 U.S.C. §§141-42 (repealed in 1976). In the meantime, the Supreme Court approved the executive withdrawals in United States v. Midwest Oil Co., 236 U.S. 459 (1915).

In 1920 Congress decided to develop a coherent system for leasing and developing federal oil and gas reserves. The passage of the Mineral Leasing Act of 1920, 41 Stat. 437 et seq., 30 U.S.C. §§ 181 et seq. governed federal oil and gas leasing until the enactment of the Federal Onshore Oil and Gas Leasing Reform Act of 1987. Pub. L. No. 100-203, 101 Stat. 1330-256 (1987). Today, the Bureau of Land Management within the Department of the Interior has primary, but not exclusive, jurisdiction over onshore oil and gas leases. The Minerals Management Service within the Department is primarily responsible for collecting royalty and other revenue from federal oil and gas leases.

The Mineral Leasing Act ostensibly covers "lands owned by the United States." But as a practical matter the Act has been limited to public domain lands which are those lands which the United States has gained by cession or conquest and which have not otherwise been disposed of under any of the public lands laws. 1 Law of Federal Oil and Gas Leases ch. 3 (2006). The Act also excludes from its coverage lands within incorporated cities, national parks and monuments and acquired lands.

Competitive and noncompetitive leases. Under the original 1920 Act a bifurcated system was created for the leasing of oil and gas. Lands located within a known geologic structure (KGS) of a producing oil or gas field could be leased only through a competitive bidding procedure with a maximum unit size of 640 acres. 43 C.F.R. §§ 3100.3-1, 3120.2-3. The Secretary of the Department of the Interior had discretionary authority to lease lands within known geological structures (KGSs), so that a prospective mineral developer could not force the Secretary to lease lands that fell within a designated KGS. Udall v. Tallman, 380

U.S. 1, 22 O.&G.R. 715 (1965). A KGS could be designated at any time prior to the issuance of a noncompetitive lease, even if an applicant had filed for a noncompetitive lease. Udall v. King, 308 F.2d 650, 17 O.&G.R. 277 (D.C. Cir. 1962). Likewise no competitive lease could be issued prior to the administrative determination that a KGS existed. Barash v. Seaton, 256 F.2d 714, 9 O.&G.R. 460 (D.C. Cir. 1958). One of the weaknesses of the KGS system was the lack of clear criteria by which lands were placed within a KGS. The Act did not define a KGS and the regulations which defined the term were broadly worded. Interior Department regulations defined a KGS as "technically the trap in which an accumulation of oil or gas has been discovered by drilling and determined to be productive, the limits of which include all acreage that is presumptively productive." 43 C.F.R. § 3100.0-5(1). In Arkla Exploration Co. v. Texas Oil & Gas Corp., 734 F.2d 347, 81 O.&G.R. 486 (8th Cir.1984), *cert. denied*, 469 U.S. 1158 (1985), the Eighth Circuit found that the KGS process as it was being applied was so arbitrary that it reversed the BLM's decision to allow a noncompetitive lease because the area was known to be underlain with hydrocarbons notwithstanding BLM's refusal to designate the area as a KGS.

The Secretary at his own initiative, or at the request of any interested person, would authorize the Bureau of Land Management to publish a notice of the proposed sale. The lease, if issued, would be for a five year primary term to the highest responsible, qualified bidder on the basis of the largest bonus offered with a royalty rate of not less than 12-1/2%. The Bureau could choose to reject all qualified bids, but it had to provide the highest bidder with an explanation of why the bid was rejected. Southern Union Exploration Co., 79 IBLA 90, 92 (1984).

If the lands to be leased were not within a KGS a noncompetitive lease procedure was followed. Noncompetitive leases took two forms: "over the counter" leases, which were issued to the first qualified applicant, and "simultaneous filing" leases, which were issued to the winner of a government lottery. The lottery system was used for lands not within a KGS, lands otherwise available for leasing under the Mineral Leasing Act of 1920 and lands which had been previously leased (but whose prior leases had terminated) and lands which had been specifically designated by the BLM Director. Specially designated lands typically were lands determined to generate substantial interest by prospective bidders because of some unique characteristic. If lands were taken out of the "over the counter" system and placed in the lottery system, but did not draw any bids they would be returned to the "over the counter" system. In the mid-1980's over the counter leases constituted approximately 62% of all outstanding federal oil and gas leases. A complete review of the pre-1987 system is given in 1 Law of Federal Oil and Gas Leases, ch's 3-7 (2006).

The pre-1987 leasing system was heavily criticized. Congress in response to this criticism enacted the Federal Onshore Oil and Gas Leasing Reform Act of 1987. The following legislative history suggests some of the weaknesses of the system which Congress sought to correct.

H.R. Rep. 100-378, Amending the Mineral Lands Leasing Act of 1920 to Reform the Onshore Oil and Gas Leasing Program, 7-10 (Oct. 15, 1987)

PURPOSE

The purpose of H.R. 2851 is to provide for the orderly exploration and development of the Nation's onshore oil and gas resources in a manner which ensures a fair return to the public.

BACKGROUND AND NEED FOR LEGISLATION

The Department of the Interior's onshore oil and gas leasing program consists of a competitive and noncompetitive system. The noncompetitive component was set up to emphasize oil and gas resource development by offering lands of known potential less expensively and for longer terms. Lands within a "known geological structure of a producing oil and gas field" (KGS) are offered through the competitive leasing program. Under this system, leases of not more than 640 acres may be awarded to the highest qualified bidder for five-year primary terms conditioned upon a production royalty rate of at least 12 1/2 percent and annual rental of $2 per acre per year.

Lands outside of a KGS are leased noncompetitively with tracts not previously leased issued "over-the-counter" to the first applicant. Tracts covered by leases that have expired or have been relinquished and which are not within a KGS are offered through the simultaneous leasing system (referred to as the "lottery") with a lessee randomly selected from many applicants. These noncompetitive leases, up to 10,240 acres in size, have 10-year primary terms with a fixed 12 1/2 percent production royalty rate and annual rentals of $1 per acre per year during the first through fifth years and $3 per acre per year thereafter except for OTC leases where the rental is fixed at $1 per acre per year throughout the lease term.

As of September 30, 1986, 102,885 federal onshore oil and gas leases covering 92,730,783 acres were outstanding. The vast majority of these leases, 93 percent, were issued on a noncompetitive basis. During the course of fiscal year 1986, 1,263 competitive and 7,746 noncompetitive federal onshore oil and gas leases were issued. Of the total amount of outstanding leases, only 18 percent were in production at the end of fiscal year 1986 and represented about 11 percent of total U.S. oil and gas production. These producing leases covered 12,700,808 acres of public land. Federal onshore oil and gas leasing accounted for $865 million in royalties, rents and bonuses during fiscal year 1986 with $424 million of this amount distributed to 28 states. The seven states of Alaska, California, Colorado, Montana, New Mexico, Utah and Wyoming received 94 percent of this amount. Wyoming and New Mexico combined received 61 percent of the total state share distribution.

Various problems in the federal onshore oil and gas leasing system have been the subject of numerous investigations, studies and Congressional hearings.

Fraud and abuse has long been associated with the lottery used to issue the vast majority of leases on a noncompetitive basis. On occasion, the system has been subject to manipulation and due to continuing deficiencies in making geological determinations relating to oil and gas structures, lands which should have been issued by competitive leasing to the highest bidder were instead issued noncompetitively for a minimal filing fee. Another major problem involves the so-called "40 Acre Merchants" who obtain leases which contain no known oil or gas resources, divide them into parcels of less than 40 acres, and peddle them using false promises of high return to unsuspecting citizens. There has also been a growing conflict between oil and gas development and other multiple use values in national forests and public lands primarily in the West. These conflicts can generally be attributed to the lack of proper land planning and the failure to consider potential developmental consequences prior to lease issuance.

Due to these situations, there currently exists uncertainty over whether the noncompetitive leasing system can withstand fraud and abuse, the Bureau of Land Management's ability to determine which tracts of land should be made available for leasing under the competitive or noncompetitive systems and the propriety of reserving environmental review of leasing decisions until after lease issuance. This uncertainty threatens the stability of the program and calls into question its ability to contribute to the energy needs of the Nation as well as provide for a fair rate of return to the public from the development of federally owned resources.

OVERVIEW OF LEGISLATION

The "Federal Onshore Oil and Gas Leasing Reform Act of 1987" provides for a "two-tier" leasing system. All lands available for leasing would first be offered competitively and issued to the highest qualified bidder with the minimum acceptable bid set at $2 per acre. Lease sales would primarily be conducted by oral bidding although sealed bids could be submitted. A non-refundable bidding fee of at least $75 must be paid.

The bill provides for lease sales to be held for each state at least on a quarterly basis and the public could make confidential expressions of interest about specific lands. Leases could not be issued in units larger than 2,560 acres except in Alaska where the maximum lease size would be 5,760 acres. Competitive leases would have five-year primary terms.

Lands for which no bids are received, or no bids at or above $2 per acre, would then become available for leasing during a one-year period to the first person making application for the lease and upon the payment of a non-refundable application fee of at least $75. Noncompetitive leases would have 10-year primary terms. If at the expiration of the one-year period no lease application is pending, or if a lease terminates, expires, is cancelled or is or is relinquished, the land covered by the lease would again be available under the competitive leasing program.

All leases would be conditioned upon the payment of a royalty rate of not less than 12 1/2 percent of production value and rentals of not less than $1 per acre per year for the first through fifth years of the lease and not less than $3 per acre per year for each year thereafter. A minimum royalty of not less than $3 per acre in lieu of rental would be payable on or after the discovery of oil or gas in paying quantities.

The bill would require that at least 60 days before offering lands for lease and 30 days before substantially modifying the terms of any lease the Secretary of the Interior would provide public notification through a Federal Register notice. The Secretary would also periodically notify the public of pending drilling permit applications. These notices would include information on lease terms and maps or descriptions of the affected lands.

Prior to approving a permit to drill, the legislation requires that a plan of operations must be approved by the Secretary. The plan would cover the proposed surface-disturbing activities within the lease area. A bond in an amount sufficient to ensure reclamation must also be posted. The legislation directs the Secretary to prohibit the issuance of new oil leases to any person who has failed to comply with reclamation requirements in the past.

Under the legislation, the Secretary of the Interior could not issue any oil and gas leases on public domain national forest lands without the consent of the Secretary of Agriculture.

Provision is made to authorize the Secretary to disapprove any lease assignment of less than 640 acres outside of Alaska or less than 2,560 acres within Alaska. Current law lease cancellation provisions are clarified.

The land use planning provisions of the bill provide that oil and gas leases may be issued only if leasing of the affected land has been evaluated and approved in a land use plan meeting the requirements of the section. Affected areas are public lands and public domain national forest lands on which the public has expressed substantial interest in oil and gas leasing or which the Secretary finds there is a high potential for oil and gas recovery. Where a land use plan has been completed, or there has been substantial progress toward its completion, it could be amended in compliance with the requirements of the section. The Secretary of the Interior and the Secretary of Agriculture would be required to publish in the Federal Register a list of affected plans and their completion dates which in no case shall be later tan January 1, 1991. After the specified date, no oil and gas lease may be issued until the plan is effective.

Land use plans affected by this legislation must include consideration of the potential oil and gas resources including a narrative description indicating known oil and gas reserves and lands already under lease; an analysis of the most likely social, economic and environmental consequences of exploration and development; and an identification of specific protective stipulations and the specific areas for which they apply. The Secretary is authorized to use no surface occupancy stipulations only where recovery of oil and gas from an area is feasible without surface occupancy.

The bill stipulates that oil and as leases could not be issued on lands in wilderness study areas. Oil and gas exploration could be allowed in these areas by means not requiring the construction or improvement of roads if this activity is found to be compatible with the preservation of the wilderness environment.

Simultaneous applications filed prior to enactment may be processed as could competitive lease bids. Over-the-counter applications filed prior to September 15, 1987, could be processed. The Secretary could hold one or more lease sales in accordance with the legislation prior to regulation promulgation. A number of measures are authorized to combat fraud and abuse in oil and gas leasing.

COMMITTEE VIEWS

The Committee believes the problems which have plagued the federal onshore oil and gas leasing system over the years can be addressed only through a comprehensive reform of the current program.

The legislation seeks to address these problems by requiring that all lands available for leasing be first subjected to competitive leasing. This will allow market forces, rather than geology and administrative ineptitude, to be the determining factor in the leasing program while ensuring that the public receives a fair return on the disposition of federal resources. However, subject to the competitive test, H.R. 2851 would maintain a noncompetitive "second tier" leasing system in an effort to encourage exploration and wildcatting. Under the leasing regimes proposed by the bill, the lottery would be abolished.

With respect to the "40 Acre Merchants," the legislation would provide the Secretary of the Interior with the authority to disapprove lease assignments of less than 40 acres. In addition, H.R. 2851 would establish strict penalties for fraudulent sales schemes.

In an effort to resolve many of the growing conflicts between oil and gas leasing and other land uses and values, the bill would also provide for more consistency in, and the quality and timeliness of, the manner by which the Bureau of Land Management and Forest Service management plans consider oil and gas leasing issue. The Committee intends for the legislation to promote a complete evaluation of potential exploration and developmental consequences of oil and gas leasing during the land planning stage. The current practice is to reserve these considerations until lease issuance. The basis for section 5 of H.R. 2851 is the Bureau of Land Management's most recent supplemental program guidance for oil and gas leasing. Establishing a statutory underpinning for oil and gas leasing evaluation during the land use planning stage will ensure consistent application and provide the Forest Service with a specific directive relating to oil and gas leasing.

The bill would also give the Forest Service statutory consent authority for leasing on national forest lands, authorize surface managing agencies to take the steps needed to assure the adequate reclamation of drilling operations and prohibit further leasing of lands under consideration for wilderness designation. These matters are as significant and important in ensuring the integrity of the federal onshore oil and gas leasing program as is the reform of the leasing mechanism.

. . .

A Brief Overview of the Leasing Process After 1987

The Reform Act covers both the public domain and acquired lands if the lands are "known or believed to contain oil and gas deposits." 30 U.S.C. § 226(a). The lands which are available for leasing, with the exception of tar sand areas, must be offered for leasing by competitive bidding at an oral auction. 30 U.S.C. § 226(b)(1)(A). Only lands which remain unsold after the competitive bidding process can be leased under a revised noncompetitive bidding procedure. Id. at § 226(c)(1). If the lands placed into the noncompetitive bidding process are not leased within two years, they are returned to the competitive bidding process. When enacted, the Reform Act did not change the existing length of the primary terms for competitive leases, which was 5 years, and for non-competitive leases, which was 10 years. This led to further statutory changes with the enactment of the Energy Policy Act of 1992, Pub. L. No. 102-486, which allow the Secretary to provide for competitive leases with a 10-year primary term. 30 U.S.C. § 226(e); 43 C.F.R. § 3120.1-1 (2010).

In order to lease public lands the putative lessee must meet certain eligibility standards. 43 C.F.R. Subparts 3101-3102 (2010). These standards include citizenship criteria and promises to comply with the Act's anti-fraud, reclamation and diligent development requirements. The bidder must verify her eligibility and there are criminal penalties for falsifying information on the offer sheets. The Reform Act limits individual leased tracts to 2560 acres except in Alaska where the limit is 5760 acres. No person or legal entity can control more than 246,080 acres of federal leases in a single state.

The Reform Act requires that lease sales must be held quarterly in each state where lands are available for leasing. 30 U.S.C. § 226(b)(1)(A). The traditional discretion afforded the Secretary regarding the amount and timing of making lands available for leasing was left undisturbed. Interior regulations authorize the Department to itemize the sale parcels in a Notice of Competitive Lease Sale (NCLS). The NCLS must be published at least 45 days prior to the proposed sale and must include a map of the area to be lease and the terms of the leases. All parcels must be offered through oral bidding and no written bids are accepted. The winning bid will be the highest oral bid which equals or exceeds the national minimum acceptable bid. 43 C.F.R. § 3120.5-2 (2000). See 1 Law of Federal Oil and Gas Leases ch. 7 (2010).

The national minimum acceptable bid was set by statute at $2.00 per acre with the Secretary having the power to establish a higher minimum after December 21, 1989. 30 U.S.C. § 226(b)(1)(B); 43 C.F.R. § 3120.1-2 (2000). The successful bidder must pay the minimum bonus bid, the total amount of first year rentals and an administrative fee of $ 75.00 on the day of the auction.43 C.F.R. § 3120.5-2(2000). If the high bidder fails to meet the requirements the parcel is not given to the second highest bidder, but is to be reoffered under a new competitive bidding process. Id. at § 3120.5-3. The Reform Act also changed the prior practice of having sealed bids for competitive leases by requiring that most leases be subject to oral bidding. 30 U.S.C. § 226(b)(1)(A).

The Reform Act does not prohibit BLM from receiving information or formal nominations of tracts to be offered from interested parties. 43 C.F.R. § 31201-1(e). There is no standardized form by which these expressions of interest may be filed. BLM has the authority to include any lands it determines are available for leasing in any particular auction.

The fundamental rules regarding how BLM engages in noncompetitve oil and gas leasing was not changed by the Reform Act. However, only lands that go unsold after being subject to a competitive bidding process can be sold through the noncompetitive system. 30 U.S.C. § 226(c)(1). Within two years after the lands went unsold they are available to the first qualified applicant. BLM has retained its simultaneous lottery system by declaring that all applications filed on the first business day after a competitive oral auction are to be treated as if they were filed simultaneously. 43 C.F.R. § 3110.2. BLM regulations were amended to coordinate the competitive and noncompetitive bidding procedures and to set additional minimum requirements on those who want to make noncompetitive applications. 43 C.F.R. Subpart 3110.

Onshore Oil and Gas Lease Terms. Competitive and noncompetitive oil and gas leases must provide for an annual rental payment of $ 1.50 per acre for the first five years of the lease, increasing to $ 2.00 per acre for any subsequent year. 30 U.S.C. § 226(d). Competitive leases are granted for a five year period but may be extended by production or drilling operations that are ongoing at the end of the five year term. This extension is for a period of no more than two years and requires the lessee to engage in diligent drilling operations. 43 C.F.R. § 3107.1. If production ceases in the extended term the lease will not terminate if the lessee engages in reworking or drilling operations within 60 days of being notified that the lease is not capable of producing in paying quantities and continues those operations with reasonable diligence. Id. at § 3107.2-2. Federal oil and gas leases must also provide for a minimum royalty of 12-1/2 percent on the amount produced. 30 U.S.C. § 226(b)(1)(A). Although there are different lease forms used for Indian and Outer Continental Shelf Lands Act leases, BLM does provide for a model lease form that cannot be varied from in the absence of approval from the authorizing officer. Form 3100-11 (Oct. 2008) available at http://blm.gov/blmforms/forms/index.htm.

Oil and Gas Operations. The Reform Act also affected operations by federal oil and gas lessees. While the BLM retains control over the mineral estates, other federal land management agencies such as the Forest Service now exercise control over surface operations on their lands. 30 U.S.C. § 226(g). Operations are initiated by the filing of an application for a permit to drill (APD) with the authorized officer. 43 C.F.R. § 3162.3-1. Regulations set forth a panoply of operational requirements, including reporting requirements. 43 C.F.R. Subpart 3160. The APD must contain a drilling plan and a discussion of all surface-disturbing activities. The permit cannot be issued until thirty days have expired from the publication of notice to the public that the APD has been filed. The operator must seek governmental permission for subsequent operations that were not included within the scope of the APD. Performance standards are imposed relating to safety and environmental hazards including well plugging, protection of freshwater bearing aquifers and blowout prevention. The operator must also file monthly reports of operations on BLM issued forms. The Act also requires the posting of a bond sufficient to cover the costs of reclaiming the site and the affected environment. 30 U.S.C. § 226(g).

STATE ex rel. RICHARDSON v. BUREAU OF LAND MANAGEMENT

United States Court of Appeals, Tenth Circuit, 2009
565 F.3d 683, 170 O.&G.R. 477

Before LUCERO, ANDERSON and O'BRIEN, CIRCUIT JUDGES

LUCERO, CIRCUIT JUDGE: This litigation concerns the environmental fate of New Mexico's Otero Mesa, the largest publicly-owned expanse of undisturbed Chihuahuan Desert grassland in the United States. From 1998 to 2004, the Bureau of Land Management ("BLM" or "the Agency") conducted a large-scale land management planning process for federal fluid minerals development in Sierra and Otero Counties, where the Mesa is located. Ultimately, the Agency opened the majority of the Mesa to development, subject to a stipulation that only 5% of the surface of the Mesa could be in use at any one time. Invoking the National Environmental Policy Act ("NEPA"), the Federal Land Management Policy Act ("FLPMA"), and the National Historic Preservation Act ("NHPA"), the State of New Mexico and a coalition of environmental organizations led by the New Mexico Wilderness Association ("NMWA") challenged in federal district court the procedures by which BLM reached this determination. NMWA also challenged BLM's decision not to consult with the Fish and Wildlife Service ("FWS") under the Endangered Species Act ("ESA") regarding possible impacts of the planned development on the Northern Aplomado Falcon.

The district court rejected these challenges, save for the plaintiffs' argument that BLM erred in beginning the leasing process on the Mesa before conducting additional analysis of site-specific environmental impacts flowing from the issuance of development leases. Discerning serious flaws in BLM's procedures, we affirm the district court's conclusion that NEPA requires BLM to conduct site-specific analysis before the leasing stage but reverse its determination that BLM's plan-level analysis complied with NEPA. Moreover, we affirm its conclusion that BLM complied with public comment provisions in FLPMA, and we vacate as moot the portion of the district court's order addressing NMWA's ESA claims.

I

Within Sierra and Otero counties in southern New Mexico lie the northern reaches of the richly biodiverse Chihuahuan Desert. Among the several habitats comprising this desert ecosystem is the Chihuahuan Desert grassland, much of which has depleted to scrubland over the past century and a half. A New Mexico State University biology professor identifies this grassland as the most endangered ecosystem type in the United States. The Otero Mesa, which BLM seeks to open to oil and gas development upon conclusion of the planning process that is the subject of this litigation, is home to the endangered Northern Aplomado Falcon, along with a host of other threatened, endangered, and rare species. Only a few, unpaved roads traverse the Mesa. Lying beneath it is the Salt Basin Aquifer, which contains an estimated 15 million acre-feet of untapped potable water. Recognizing the importance of this valuable resource, the state of New Mexico and many citizens and environmental groups have sought to prevent development.

A

BLM manages some 1.8 million acres of surface land and 5 million acres of subsurface oil, gas, and geothermal resources in Sierra and Otero Counties. This includes the 427,275-acre Otero Mesa. Until recently, these resources were managed under the terms of a 1986 resource management plan (the "RMP"), which contained no overall guidance on the management of fluid minerals development, leaving management decisions to be made on a case-by-case basis. Because the area saw relatively little oil and gas exploration, BLM relied on the plan without incident for a decade and issued few development leases during this time.

This state of affairs was upended in 1997, when a Harvey E. Yates Company ("HEYCO") exploratory well struck natural gas on the Otero Mesa. The strike occurred on a parcel designated the Bennett Ranch Unit ("BRU"). Oil and gas companies quickly responded by nominating over 250,000 acres in the area for federal leases. BLM determined that under the terms of then-existing internal policy, the increased development interest required the Agency to issue a management plan specifically governing fluid mineral resources. Accordingly, BLM asked existing leaseholders to voluntarily suspend their leases and began the process of amending the RMP to address possible oil, gas, and geothermal development. The stated goals of the amendment process were to determine which public lands in Sierra and Otero Counties should be available for leasing and development and to direct how leased lands would be managed.

Amending a resource management plan is a "major federal action" whose potential environmental impacts must be assessed under NEPA. Consequently, in October 2000, BLM issued a "Draft Resource Management Plan Amendment and Environmental Impact Statement for Federal Fluid Minerals Leasing and Development in Sierra and Otero Counties" (the "Draft EIS"). As NEPA requires, the Draft EIS analyzed several possible alternative management schemes for oil and gas development in the area. Of the five alternatives identified, three were fully analyzed in the Draft EIS. The other two were eliminated without further analysis.

Both eliminated alternatives would have increased the level of environmental protection for the entire plan area beyond the level provided under existing management or any of the fully analyzed alternatives. One would have done so through a blanket ban on minerals development leasing; the other, through a "no surface occupancy" ("NSO") stipulation allowing minerals development only by slant drilling from non-BLM lands. These alternatives were "considered initially but eliminated prior to further analysis" based on the conclusion that adopting a plan which so limited development would be arbitrary and capricious under FLPMA's multiple-use mandate. BLM also discounted one of the three alternatives analyzed in the Draft EIS: the "No-Action Alternative," or the option of taking no new planning action. After fully analyzing its likely impacts, BLM determined that the No-Action Alternative was not in compliance with its own policies.

Thus, BLM was left with two possible management schemes, "Alternative A" and "Alternative B." Of the two, Alternative A placed fewer restrictions on development, and BLM selected it as the preferred alternative. Alternative A opened 96.9% of the plan area but placed limitations on possible development, subjecting 58.9% of the area to a combination of NSO stipulations, controlled surface use stipulations, and timing stipulations. Of particular relevance to this litigation, Alternative A subjected 116,206 acres of the Otera Mesa and 16,256 acres of the adjoining Nutt Desert Grasslands to an NSO provision allowing surface disturbance only within 492 feet of existing roads. BLM crafted this NSO restriction "[t]o protect portions of the remaining desert grassland community by minimizing habitat fragmentation."

Also relevant to this litigation, the Draft EIS analyzed the potential impact on groundwater in the plan area only in general terms, without identifying or discussing specific aquifers such as the Salt Basin Aquifer. The Draft EIS concluded that in the construction phase of development:

The possibility for degradation of fresh water aquifers could result if leaks or spills occur from pits used for the storage of drilling fluids, or if cathodic protection wells associated with pipelines are installed in a manner that allows for the commingling of shallow surface aquifers. However, since impacts would occur only if the governing regulations fail to protect the resource, the impact is not quantifiable.

As for the production phase, the Draft EIS was equally cursory. It stated that "[p]roduction of an oil and gas well typically would not have a direct impact on groundwater resources" because regulations require that "[a]ll oil and gas wells must have a casing and cement program . . . to prevent the migration of oil, gas, or water . . . that may result in degradation of groundwater." . Finally, the Draft

EIS concluded that disposal wells, which are "used for the disposal of waste [by injection] into a subsurface stratum," would not lead to significant impacts because applicable casing and cement construction requirements and aquifer criteria would be followed and would prevent contamination.

B

Among the species for which the Chihuahuan Desert grasslands provide habitat is the Northern Aplomado Falcon ("Aplomado Falcon" or "Falcon"), listed as an endangered species since 1986. Although Falcons have only "sporadically" been seen in the United States in recent decades, the presence of breeding Falcons just across the border in Mexico led biologists to believe that the Falcon might be poised to repopulate portions of the plan area. Repopulation by the Falcon would depend on the preservation of suitable grassland habitat.

In June 2003, during the ongoing resource management plan amendment process, BLM concluded that revisions to the management plan were "likely to adversely affect" the Falcon. Accordingly, it requested in writing that FWS begin formal consultation, pursuant to § 7 of the ESA, regarding whether BLM's proposed action might jeopardize the Falcon's continued existence. Three months later, the Agency reversed course, retracted its determination that the RMP revisions were "likely to adversely affect" the Falcon, and informed FWS of its conclusion that formal consultation was therefore unnecessary. FWS concurred in this revised determination, thus ending the formal consultation process and the agencies' study of likely effects on the Falcon.

C

Three years after issuing the Draft EIS, in December 2003, BLM issued a Proposed Resource Management Plan Amendment ("RMPA") and Final EIS. Rather than selecting from among the alternatives analyzed in the Draft EIS, however, the abstract of the Final EIS explained that BLM had selected "a modified version (as a result of public input) of preferred Alternative A described and analyzed in the Draft RMPA/EIS."

This "modified version" of Alternative A ("Alternative A-modified") differed in a crucial respect from Alternative A: Rather than limiting surface disturbances to areas within 492 feet of existing roadways, Alternative A-modified would instead limit disturbances to *any* 5% of the surface area of a leased parcel at a given time, regardless of location. In addition to the 5% disturbance cap, Alternative A-modified required "unitization," a management scheme under which different operators cooperate in exploration and well development with the goal of minimizing surface impacts. "Unitization" was a new creation, never previously used by BLM in managing surface resources. Although the sections of the Final EIS describing the management plan itself were modified to reflect these new requirements, the sections describing the plan's impacts on vegetation and wildlife were not substantially modified, because the EIS concluded that the changes "do not significantly alter . . . the analysis of the environmental consequences."

Alternative A-modified did offer greater protection of the Otero and Nutt grasslands in one respect: It prohibited development on 35,790 acres of "core habitat" for five years pending further study and development of an adaptive management strategy. Thus, BLM presented the new alternative as responsive to the concerns of both industry and the environmental community. The Agency reiterated in response to public questions that it was unnecessary to analyze the impacts of A-modified because the overall "impact assessment," judged based on the "anticipated level of surface disturbance," "remained essentially the same" as under Alternative A. Based on this conclusion that the same or less surface acreage would be disturbed under Alternative A-modified, BLM reasoned, there was no substantial change from an environmental standpoint. Regarding groundwater concerns, the Final EIS added a discussion of the effects of leasing on specific basins, including the Salt Basin Aquifer, but again concluded that "the impacts on groundwater resources are expected to be minimal," adding that "[t]ypically, natural gas wells make little water and the water produced can be disposed through the use of evaporation ponds."

D

[Governor Richardson of New Mexcio objected to the BLM changes on a number of grounds, but for the most part, BLM rejected his proffered amendments to the RMP. Eds.]

E

In April 2005, the State of New Mexico filed suit against BLM, raising claims under NEPA, FLPMA, the NHPA, and the Administrative Procedure Act ("APA"), seeking declaratory and injunctive relief (the "New Mexico suit"). On May 20, BLM scheduled for July 20 a competitive oil and gas lease auction covering a 1600-acre parcel within the Bennett Ranch Unit (the "BRU Parcel"), adjacent to the parcel on which HEYCO found natural gas triggering the cascade of lease nominations that led to the RMPA process. Six days later, a coalition of environmental groups filed a second suit (the "NMWA suit"). As amended, this suit raised claims under NEPA, the ESA and FLPMA.

BLM went ahead with the July 20 auction, and HEYCO, the sole bidder, purchased the lease. During the course of litigation, however, BLM agreed not to execute the lease until resolution of the case. HEYCO has continued to prepare for the possibility of drilling, obtaining permits to build a pipeline to service wells on this lease and others it holds nearby.

* * *

II

[The court finds that New Mexico has standing to challenge the BLM decision because of the alleged environmental injury to lands located within New Mexico. The court also finds that challenges to the BLM decision not to consult with the Fish and Wildlife Service under the ESA has been rendered moot but the decision to re-introduce the Aplomado Falcon into the ecosystem. Eds.]

III

Turning to the merits of those issues over which we have jurisdiction, we first consider the plaintiffs' NEPA claims. The centerpiece of environmental regulation in the United States, NEPA requires federal agencies to pause before committing resources to a project and consider the likely environmental impacts of the preferred course of action as well as reasonable alternatives. By focusing both agency and public attention on the environmental effects of proposed actions, NEPA facilitates informed decisionmaking by agencies and allows the political process to check those decisions. The requirements of the statute have been augmented by longstanding regulations issued by the Council on Environmental Quality ("CEQ"), to which we owe substantial deference.

Before embarking upon any "major federal action," an agency must conduct an environmental assessment ("EA") to determine whether the action is likely to "significantly affect[] the quality of the human environment." If not, the agency may issue a "finding of no significant impact" ("FONSI") stating as much. . But if so, the agency must prepare a thoroughgoing EIS, as BLM did here, assessing the predicted impacts of the proposed action on all aspects of the environment, including indirect and cumulative impacts. In addition, an EIS must "rigorously explore and objectively evaluate" all reasonable alternatives to a proposed action, in order to compare the environmental impacts of all available courses of action. For those alternatives eliminated from detailed study, the EIS must briefly discuss the reasons for their elimination. At all stages throughout the process, the public must be informed and its comments considered.

NEPA is silent, however, regarding the *substantive* action an agency may take--the Act simply imposes *procedural* requirements intended to improve environmental impact information available to agencies and the public. Even if scrupulously followed, the statute "merely prohibits uninformed--rather than unwise--agency action." *Robertson v. Methow Valley Citizens Council*, 490 U.S. 332, 351, 109 S. Ct. 1835, 104 L. Ed. 2d 351 (1989).

As with other challenges arising under the APA, we review an agency's NEPA compliance to see whether it is "arbitrary, capricious, an abuse of discretion, or otherwise not in accordance with law." An agency's decision is arbitrary and capricious if the agency (1) "entirely failed to consider an important aspect of the problem," (2) "offered an explanation for its decision that runs counter to the evidence before the agency, or is so implausible that it could not be ascribed to a difference in view or the product of agency expertise," (3) "failed to base its decision on consideration of the relevant factors," or (4) made "a clear error of judgment." Deficiencies in an EIS that are mere "flyspecks" and do not defeat NEPA's goals of informed decisionmaking and informed public comment will not lead to reversal.

When called upon to review factual determinations made by an agency as part of its NEPA process, short of a "clear error of judgment" we ask only whether the agency took a "hard look" at information relevant to the decision. In considering whether the agency took a "hard look," we consider only the agency's reasoning at the time of decisionmaking, excluding post-hoc rationalization concocted by counsel in briefs or argument. "A presumption of validity attaches to the agency action and the burden of proof rests with the appellants who challenge such action." We review the district court de novo, applying the APA standard of review to the agency's actions without deferring to the district court's application of that standard.

A

According to the State and NMWA, NEPA requires BLM to complete a supplemental EIS specifically analyzing the likely environmental effects of Alternative A-modified before adopting that alternative as the new management plan for the area, and its failure to do so was arbitrary and capricious. An agency must prepare a supplemental assessment if "[t]he agency makes *substantial changes* in the proposed action that are *relevant to environmental concerns*." When "the relevant environmental impacts have already been considered" earlier in the NEPA process, no supplement is required. In a guide to NEPA published in the Federal Register, the CEQ states that a supplement is unnecessary when the new alternative is "qualitatively within the spectrum of alternatives that were discussed in the draft" and is only a "minor variation" from those alternatives.

Rather than offer additional environmental analysis of Alternative A-modified, BLM concluded in the SEIS that no further analysis was necessary because the same or less surface area would ultimately be developed under Alternative A or A-modified. For this reason, BLM determined that the change from Alternative A to Alternative A-modified was within the scope and analysis of the Draft EIS and did not substantially alter the environmental consequences as required to trigger the *§ 1502.9* supplementation requirement. BLM and IPANM continue to argue that Alternative A-modified was within the scope of the previous analysis, although for different reasons than a similarity in the final number of acres likely to be developed.

In its ruling, the district court found that the question of whether Alternative A-modified would lead to greater habitat fragmentation than Alternative A was a factual dispute. [27] It then found that there was sufficient evidence in the record to support BLM's prediction; thus, the failure to conduct additional analysis in the SEIS was not arbitrary and capricious. The court also found that actual habitat fragmentation under Alternative A-modified was dependent on factors that could not be analyzed at the planning stage.

1

As described above, Alternative A and Alternative A-modified differ primarily in the restrictions they place on surface disturbances on the Otero Mesa. Alternative A proposed a qualitative restriction on development: Disturbances would only be allowed near existing roads. Thus, they would remain contiguous rather than scattering across the landscape. By contrast, A-modified imposes a quantitative restriction: Disturbances may occupy only five percent of the Mesa at any one time.

By arguing that a difference in the degree of habitat fragmentation did not require a fresh impacts analysis, BLM neglects the fundamental nature of the environmental problem at issue. As is well documented in the record before us, the location of development greatly influences the likelihood and extent of habitat preservation. Disturbances on the same total surface acreage may produce wildly different impacts on plants and wildlife depending on the amount of contiguous habitat between them. BLM's analysis of Alternative A assumed the protections of large contiguous pieces of habitat from development. Alternative A-modified muddied this picture, doing away with any requirement of continuity of undisturbed lands. Although A-modified also requires developers to work together to minimize impacts--potentially increasing the continuity of surface developments--BLM provided so little explanation of this "unitization" restriction that it is impossible to tell whether it would create the same clustering of impacts as would the proximity restriction in Alternative A.

Moreover, this is not a case where components of fully-analyzed alternatives were recombined or modified to create a "new" alternative whose impacts could easily be predicted from the existing analysis. Nothing in the Draft EIS so much as hinted at a percentage-based surface occupancy restriction for the Otero Mesa, and there is no direct or reliable way to compare the fragmentation effects of that restriction to the effects of the restrictions analyzed in the EIS.

More generally, we cannot accept that because the *category* of impacts anticipated from oil and gas development were well-known after circulation of the Final EIS, any change in the location or extent of impacts was immaterial. . . . The situation at hand is no different. NEPA does not permit an agency to remain oblivious to differing environmental impacts, or hide these from the public, simply because it understands the general type of impact likely to occur. Such a state of affairs would be anathema to NEPA's "twin aims" of informed agency decisionmaking and public access to information.

BLM's unanalyzed, conclusory assertion that its modified plan would have the same type of effects as previously analyzed alternatives does not allow us to endorse Alternative A-modified as "qualitatively within the spectrum of alternatives" discussed in the Draft EIS. Because location, not merely total surface disturbance, affects habitat fragmentation, Alternative A-modified was qualitatively different and well outside the spectrum of anything BLM considered in the Draft EIS, and BLM was required to issue a supplement analyzing the impacts of that alternative under *40 C.F.R. § 1502.9(c)(1)(i).*

2

BLM and IPANM also argue that even if the changes in fragmentation impacts between Alternative A and A-modified require further environmental analysis, such analysis was impracticable until the leasing stage because the overall level of development could not be sufficiently predicted at the RMPA stage. All environmental analyses required by NEPA must be conducted at "the earliest possible time." Because the record reveals that BLM conducted an internal analysis of the fragmentation impacts of Alternative A-modified in 2004, we are convinced that such analysis was possible. Accordingly, we hold that NEPA requires BLM to release a supplemental EIS thoroughly analyzing its newly minted alternative at the planning stage.

3

Finally, BLM asks that we hold any error in its analysis to be harmless. The Agency contends that because members of the public had access to the SEIS and record of decision and were allowed to comment on each of these, the purposes of NEPA were fulfilled without further analysis. . . .

B

Aside from the need to analyze the specific land use plan BLM eventually selected, NMWA also charges that BLM analyzed an unduly narrow range of alternatives during the EIS process. The Agency disagrees, arguing that Alternatives A and B and the No-Action Alternative were representative of the full range of reasonable planning alternatives for the area.

The "heart" of an EIS is its exploration of possible alternatives to the action an agency wishes to pursue. . Every EIS must "[r]igorously explore and objectively evaluate all reasonable alternatives." Without substantive, comparative environmental impact information regarding other possible courses of action, the ability of an EIS to inform agency deliberation and facilitate public involvement would be greatly degraded. While NEPA "does not require agencies to analyze the environmental consequences of alternatives it has in good faith rejected as too remote, speculative, or impractical or ineffective," it does require the development of "information sufficient to permit a reasoned choice of alternatives as far as environmental aspects are concerned." . It follows that an agency need not consider an alternative unless it is significantly distinguishable from the alternatives already considered.

We apply the "rule of reason" to determine whether an EIS analyzed sufficient alternatives to allow BLM to take a hard look at the available options. The reasonableness of the alternatives considered is measured against two guideposts. First, when considering agency actions taken pursuant to a statute, an alternative is reasonable only if it falls within the agency's statutory mandate. . Second, reasonableness is judged with reference to an agency's objectives for a particular project.

* * *

Applying the rule of reason, we agree with NMWA that analysis of an alternative closing the Mesa to development is compelled by *40 C.F.R. § 1502.14*. Excluding such an alternative prevented BLM from taking a hard look at all reasonable options before it. While agencies are excused from analyzing alternatives that are not "significantly distinguishable" from those already analyzed, the alternative of closing only the Mesa--which represents a small portion of the overall plan area--differs significantly from full closure. As discussed above, the lands at issue are extraordinary in their fragility and importance as habitat. Although the record indicates that most development interest in the plan area focuses on the Mesa, so too does the interest in conservation, as expressed by the public during the comment process. Yet Alternative B, the alternative that would conserve the largest portion of the Mesa, was a far cry from closure. Given the powerful countervailing environmental values, we cannot say that it would be "impractical" or "ineffective" under multiple-use principles to close the Mesa to development. Accordingly, the option of closing the Mesa is a reasonable management possibility. BLM was required to include such an alternative in its NEPA analysis, and the failure to do so was arbitrary and capricious.

B

[The court finds BLM's decision not to consider wilderness designation under the National Wilderness Act was reasonable given the constraints that such designation places on land use. Eds.]

C

The State contends that BLM's analysis of the environmental impacts of the various alternative management plans failed to sufficiently consider a crucial impact: possible contamination of the Salt Basin Aquifer (the "Aquifer"). BLM concluded in the Draft and Final EISs that any impacts of development on the Aquifer would be "minimal," and it defends that conclusion on appeal. The State argues that this determination is arbitrary and capricious because it is unsupported by evidence in the record.

New Mexico is correct that the EISs devote little analysis to the Aquifer-- undisputably an important water resource. But insignificant impacts may permissibly be excluded from full analysis in an EIS. Thus, unless BLM's decision that impacts would be "minimal" was itself arbitrary and capricious, no further analysis was required regardless of the Aquifer's value as a freshwater resource.

In order for a factual determination to survive review under the arbitrary and capricious standard, an agency must "examine[] the relevant data and articulate[] a rational connection between the facts found and the decision made." We consider only evidence included in the administrative record to determine whether an agency decision had sufficient evidentiary support.

Our first inquiry is whether BLM "examined the relevant data" regarding the likelihood of injection into, and resulting contamination of, the Aquifer. Strikingly, BLM points to *no* record evidence explaining (1) how much wastewater a natural gas well "typically" produces, (2) whether it is reasonable to believe that wells in the plan area will be "typical," or (3) how much wastewater can practicably be disposed of through evaporation. Upon our careful review, the evidence in the record instead tends to support New Mexico's view that nontrivial impacts are possible.

[O]n this record we are wholly unable to say with any confidence that BLM "examined the relevant data" regarding the Salt Basin Aquifer before determining that impacts on the Aquifer would be "minimal." The record is silent regarding the source of BLM's determination that injection (and thus, contamination) is unlikely, and it does provide some support for a contrary conclusion. Though we do not sit in judgment of the *correctness* of such evidence, where it points uniformly in the opposite direction from the agency's determination, we cannot defer to that determination.

BLM also argues that state and federal injection well and water-quality regulations are designed to prevent the feared contamination. But the existence of these regulations does not preclude the possibility of contamination, even if the protections are intended to prevent such an outcome. Contravening the inference that existing protections are always 100% effective, the record contains evidence that, despite this regulatory scheme, groundwater contamination from gas wells has happened frequently throughout New Mexico in the past. Thus, the mere presence of these regulations cannot make up for BLM's failure to demonstrate that it "examined relevant data" supporting a finding that impacts on the Aquifer will be minimal.

We accordingly hold that BLM acted arbitrarily by concluding without apparent evidentiary support that impacts on the Aquifer would be minimal. Of course, BLM is not precluded from making the same determination once again if it provides an evidentiary basis for doing so.

D

. . . Despite granting the Agency the full measure of respect and deference warranted by the arbitrary and capricious standard of review, we must reverse.

IV

We now reach the sole issue appealed by defendant-intervenor IPANM: Whether NEPA requires BLM to produce an EIS analyzing the specific environmental effects of the BRU lease before issuing that lease.

As discussed above, after issuing the Final EIS and adopting Alternative A-modified as the new management plan for the area, BLM opened bidding for a lease on the BRU Parcel. The BRU Parcel is adjacent to the HEYCO exploratory well that struck gas and led to the outpouring of lease nominations that triggered the RMPA process. Not surprisingly, HEYCO purchased the lease. In the district court, the State successfully argued that BLM was required to produce a site-specific EIS addressing the environmental impacts of an oil and gas lease on the BRU Parcel before issuing it. IPANM contends on appeal that NEPA requires no more than (1) an EIS at the RMPA stage and (2) a later EIS when HEYCO submits an APD. In other words, the parties dispute how the environmental analysis of drilling in the plan area should be "tiered" as planning progresses from the large scale to the small.

Oil and gas leasing follows a three-step process. "At the earliest and broadest level of decision-making, the [BLM] develops land use plans--often referred to as resource management plans . . . ." Next, BLM issues a lease for the use of particular land. The lessee may then apply for a permit to drill, and BLM will decide whether to grant it.. The parties dispute whether our precedents create a hard rule that no site-specific EIS is ever required until the permitting stage, or a flexible test requiring a site-specific analysis as soon as practicable. If the latter, they dispute whether a site-specific EIS was practicable, and thus required, before issuance of the July 20 lease.

The parties' claims are primarily a dispute over the interpretation of NEPA and the CEQ regulations, which provide that assessment of a given environmental impact must occur as soon as that impact is "reasonably foreseeable," and must take place before an "irretrievable commitment of resources" occurs, We do not pursue this interpretation with a clean slate, however, as we have already applied these provisions to the leasing context in several past cases.

This court first addressed the tiering of impacts analysis in the oil and gas leasing context in *Park County Resource Council, Inc. v. U.S. Department of Agriculture,* 817 F.2d 609 (10th Cir. 1987), *overruled in part on other grounds by Village of Los Ranchos,* 956 F.2d 970. In that case, BLM had prepared an "extensive" EA before issuing leases, concluded that leasing would have no immediate environmental impacts, and issued a FONSI concluding that an EIS was unnecessary at that stage. Reviewing the decision to issue a FONSI rather than an EIS, we noted that no exploratory drilling had occurred in the entire plan area at the ime the lease was issued, and there was no evidence that full field development was likely to occur. Moreover, the leased parcel consisted of over 10,000 acres (more than six times the size of the BRU Parcel). Thus, as a common sense matter, a pre-leasing EIS would have "result[ed] in a gross misallocation of resources" and "diminish[ed] [the] utility" of the assessment process, and we affirmed the FONSI. We concluded that preparation of both plan-level and site-specific environmental impacts analysis was permissibly deferred until after leasing:. . .

After leasing and prior to issuance of an APD, the agency had drafted an EIS, and NEPA was thus satisfied,. IPANM argues that under *Park County,* BLM may routinely wait until the APD stage to conduct site-specific analysis, even without issuing a FONSI.

We next had occasion to consider tiering in the oil and gas context in *Pennaco Energy.* [Pennaco Energy, Inc. v. U.S. Department of Interior, 377 F.3d 1147, 161 O.&G.R. 417 (10th Cir. 2004).] In that case, BLM issued leases for coal bed methane ("CBM") extraction on public lands in Wyoming. A plan-level EIS for the area failed to address the possibility of CBM development, and a later EIS was prepared only after the leasing stage, and thus "did not consider whether leases should have been issued in the first place." Because the issuance of leases gave lessees a right to surface use, the failure to analyze CBM development impacts before the leasing stage foreclosed NEPA analysis from affecting the agency's decision. Accordingly, we held that in the circumstances of that case, an EIS assessing the specific effects of coal bed methane was required before the leasing stage. [42] As in *Park County*, the operative inquiry was simply whether all foreseeable impacts of leasing had been taken into account before leasing could proceed. Unlike in *Park County*, in *Pennaco Energy* the answer was "no."

Taken together, these cases establish that there is no bright line rule that site-specific analysis may wait until the APD stage. Instead, the inquiry is necessarily contextual. Looking to the standards set out by regulation and by statute, assessment of all "reasonably foreseeable" impacts must occur at the earliest practicable point, and must take place before an "irretrievable commitment of resources" is made. Each of these inquiries is tied to the existing environmental circumstances, not to the formalities of agency procedures. Thus, applying them necessarily requires a fact-specific inquiry. Both the Ninth Circuit and the District of Columbia Circuit have reached the same conclusion. *See N. Alaska Envtl. Ctr. v. Kempthorne*, 457 F.3d 969, 973, 977-78 (9th Cir. 2006) (concluding that an agency's failure to conduct site-specific analysis at the leasing stage may be challenged, but that a "particular challenge" lacked merit when environmental impacts were unidentifiable until exploration narrowed the range of likely drilling sites); *Sierra Club v. Peterson*, 230 U.S. App. D.C. 352, 717 F.2d 1409, 1415 (D.C. Cir. 1983) (concluding that an agency may wait to evaluate environmental impacts until after the leasing stage if it lacks information necessary to evaluate them, "provided that it reserves both the authority to preclude all activities pending submission of site-specific proposals and the authority to prevent proposed activities if the environmental consequences are unacceptable").

Applying these standards to the July 20 lease, we first ask whether the lease constitutes an irretrievable commitment of resources. . . . we conclude that issuing an oil and gas lease without an NSO stipulation constitutes such a commitment. The same regulation we cited in *Pennaco Energy* remains in effect and provides that HEYCO cannot be prohibited from surface use of the leased parcel once its lease is final. *See* 43 C.F.R. § 3101.1-2 ("A lessee shall have the right to use so much of the leased lands as is necessary to explore for, drill for, mine, extract, remove and dispose of all the leased resource in a leasehold subject to: Stipulations attached to the lease . . . [and other] reasonable measures . . . ."). Because BLM could not prevent the impacts resulting from surface use after a lease issued, it was required to analyze any foreseeable impacts of such use before committing the resources.

Accordingly, the next question is whether any environmental impacts were reasonably foreseeable at the leasing stage. Considerable exploration has already occurred on parcels adjacent to the BRU Parcel, and a natural gas supply is known to exist beneath these parcels. Based on the production levels of existing nearby wells, the record reveals that HEYCO has concrete plans to build approximately 30 wells on the BRU Parcel and those it already leases, and it has obtained the necessary permits for a gas pipeline connecting these wells to a larger pipeline in Texas. We agree with the district court that the impacts of this planned gas field were reasonably foreseeable before the July 20 lease was issued. Thus, NEPA required an analysis of the site-specific impacts of the July 20 lease prior to its issuance, [45] and BLM acted arbitrarily and capriciously by failing to conduct one.

V

[The court rejects New Mexico's claim that BLM violated FLPMA by failing to consult with it during the preparation of the RMP.

VI

For the foregoing reasons, we **VACATE** as moot that portion of the district court's order disposing of NMWA's ESA challenge. We **AFFIRM** the district court's determination that BLM complied with FLPMA, **AFFIRM** its finding that NEPA requires BLM to conduct further site-specific analysis before leasing lands in the plan area, and **REVERSE** its conclusion that BLM complied with NEPA in its plan-level analysis.

NOTES

1. Under the 1987 Reform Act BLM was given the authority to issue oil and gas leases on Forest Service land, but the Forest Service was given a veto over any lease issuance in National Forests. 30 U.S.C.A. § 226(h). The Forest Service also was authorized to manage the lessee operations. *Id.* at § 226(f). The Forest Service has issued regulations relating to NEPA compliance at four separate stages of the oil and gas leasing process. 36 C.F.R. §§ 219.10(b), (c), 219.12(a); 228.102. Stage I occurs when management plans are developed for each of the national forests as mandated under the National Forest Management Act (16 U.S.C. §§ 1601-14). Stage II occurs when Forest Service lands are made available for leasing. Stage III occurs when specific lands are considered for leasing while Stage IV occurs at the APD decision-making stage. The Forest Service procedures are criticized in Jan Laitos, *Paralysis by Analysis in the Forest Service Oil and Gas Leasing Program*, 26 Land & Water L.Rev. 105 (1991).

2. The difficulties encountered by the federal land management agencies in dealing with environmental and land use mandates are described in Mansfield, *Through the Forest of the Onshore Oil and Gas Leasing Controversy: Toward a Paradigm of Meaningful NEPA Compliance*, XXIV Land & Water L. Rev. 85 (1989).

3. In Pennaco Energy, Inc. v. U.S. Department of Interior, 377 F.3d 1147, 161 O.&G.R. 417 (10th Cir. 2004) cited by the court in the principal case, the Tenth Circuit in addition to its NEPA analysis dealt with the issue of whether the reviewing court had subject matter jurisdiction over the appeal from the Interior Board of Land Appeals decision which may or may not have been final agency action. In Potash Association of New Mexico v. United States Department of the

Interior, 367 Fed.Appx. 960 (10th Cir. 2010), the court found that the IBLA decision relating to whether the Department could lease tracts of land for oil and gas development that were within certain designated areas where potash mining was the preferred resource use was not the final agency decision so that the court lacked subject matter jurisdiction to review the IBLA decision.

4. Due to the multiple use mandates placed on many of the federal agencies who manage the various types of federal lands, the role of NEPA is quite important in the several stages of planning that must be accomplished before a drilling rig will ever be placed on the surface of federal lands. This oftentimes leads to litigation challenging decisions at all stages of planning and development. See e.g., Bob Marshall Alliance v. Hodel, 852 F.2d 1223, 103 O.&G.R. 525 (9th Cir. 1988), *cert. denied*, 489 U.S. 1066 (1989); Conner v. Burford, 836 F.2d 1521 (9th Cir. 1988). It is not uncommon for federal land managers to impose no surface occupancy or NSO restrictions in federal oil and gas leases which may minimize the environmental impact of the oil and gas exploration and production activities. NSO provisions are often included as stipulations in the oil and gas lease and may cover specific types of surfaces, e.g., steep slopes or specific types of biota which may be a critical habitat for endangered species. For examples of NSO and other surface use stipulations see 1 *Law of Federal Oil and Gas Leases*, ch. 15, appx. A-B.

NOTES ON POST-LEASING
OBSTACLES FOR FEDERAL LESSEES

In addition to agency compliance with federal leasing and environmental statutes, a federal oil and gas lessee faces a number of unique obstacles and hazards that may delay, make more costly or even preclude her ability to explore and develop the leasehold after obtaining the lease.

(1) *Royalty Computation and Enforcement*. The Federal Oil and Gas Royalty Management Act of 1982, 30 U.S.C.A. § 1701 et seq., contains a number of provisions designed to strengthen the monitoring and enforcement of proper royalty payments. The Act requires the Secretary of the Interior to establish a comprehensive inspection, collection, accounting and auditing system for oil and gas royalties and to audit and reconcile "to the extent practicable" all current and past lease accounts. 30 U.S.C.A. § 1711(a)(c). It also permits the Secretary to conduct investigations and hearings and to authorize lease site inspections. Id. §§ 1717, 1718. Penalties are established for various criminal and civil violations. Id. §§ 1719, 1720. With the consent of the Secretary, states and Indian tribes may carry out several inspection, auditing and investigation activities. *Id.* §§ 1732, 1735. The Act has been interpreted to authorize the Secretary to require information about gas processing facilities on private lands which process federal gas and deduct the costs from the amount owed. Norfolk Energy, Inc. v. Hodel, 898 F.2d 1435, 110 O.&G.R. 273 (9th Cir. 1990). In 1996, Congress further

amended the royalty payment system through the adoption of the Federal Oil and Gas Royalty Simplication and Fairness Act, Pub.L.No. 104-185, 110 Stat. 1700, codified at 30 U.S.C. § 1701 et seq.. The 1996 Act, however, only applies to production occurring after September 16, 1996 and is not applicable to royalties owed on Indian oil and gas leases.

The federal government can determine the value of oil and gas production for purposes of computing its own royalty. 43 C.F.R. Parts 202, 206 . The courts grant substantial deference to these determinations. Marathon Oil Co. v. United States, 807 F.2d 759, 90 O.&G.R. 6 (9th Cir. 1986), *cert. denied,* 480 U.S. 940 (1987); Hoover & Bracken Energies, Inc. v. United States Department of the Interior, 723 F.2d 1488, 1489, 79 O.&G.R. 282 (10th Cir. 1983), *cert. denied,* 469 U.S. 821 (1984). The valuation standards contained in the regulations delegate "considerable discretion" to the Department to determine the value of the oil and gas produced. 30 C.F.R. § 206.101. Marathon Oil Co. v. United States, supra. In the case of Indian leases, the Secretary's trust responsibility may require him to value royalties at levels that exceed the prices that the lessee could legally obtain. Jicarilla Apache Tribe v. Supron Energy Corp., 782 F.2d 855, 80 O.&G.R. 352 (10th Cir. 1986), *modified,* 793 F.2d 1171, 88 O.&G.R. 519 (10th Cir.1986), *cert. denied,* 479 U.S. 970 (1986). A BLM decision to include within the definition of gross proceeds from the sale of gas for which the federal lessee is to pay royalty the amount of state severance tax reimbursements made to the lessee/producers from the first purchasers of the gas was upheld in Enron Oil & Gas Co. v. Lujan, 978 F.2d 212, 124 O.&G.R. 643 (5th Cir. 1992). See also Mesa Operating Limited Partnership v. U.S. Department of the Interior, 931 F.2d 318, (5th Cir. 1991), *cert. denied,* 502 U.S. 1058 (1992). Since 1988, the regulations have provided a distinction between valuing royalty depending on whether the oil or gas has been sold in arms'-length or non-arms'-length contracts. 30 C.F.R. § 206.101(oil); 30 C.F.R. § 206.151 (gas). There are substantial restrictions on what post-production costs may be used to calculate royalty using the netback methodology. 1 *Law of Federal Oil and Gas Leases* § 13.04[4-7].

In some areas, however, the courts have not been particularly deferential to Interior Department decisions and/or promulgation of rules. When the Interior Department attempted to assess royalties for take or pay or settlement payments made to federal lessees the courts have invalidated such decisions. Diamond Shamrock Exploration Co. v. Hodel, 853 F.2d 1159, 103 O.&G.R. 38 (5th Cir. 1988). In Independent Petroleum Association of America v. Armstrong, 91 F.Supp.2d 117, 144 O.&G.R. 46 (D.D.C. 2000), the court invalidated a regulation of the Minerals Management Service dealing with the transportation allowance that may be computed in determining royalties on natural gas production. For a discussion of the scope of judicial review of federal agency regulations in the context of public land decisions see generally Kramer & Martin, *The Law of Pooling & Unitization* § 24.05[2][c].

(2) *Access.* The Mineral Leasing Act, 30 U.S.C.A. § 181 et seq. provided no

right of access to minerals subject to its provisions. Neither the courts nor the Department of Interior have recognized an implied right of access across federal lands to leased federal oil and gas. The longstanding policy of the Bureau of Land Management has been:

> The Bureau does not guarantee access to mineral lease areas, either through the construction of BLM roads or the acquisition of rights-of-way across private or non-BLM lands that may control access to BLM mineral lease areas. In effect, BLM mineral leases are issued on a caveat emptor basis, and the Bureau makes no claims that guaranteed access exists.

2 *Law of Federal Oil & Gas Leases* § 22.01 (2000). See also, Martz, Love & Kaiser, *Access to Mineral Interests by Right, Permit, Condemnation or Purchase*, 28 Rocky Mtn.Min.L. Inst. 1075 (1983).

The Secretary of Interior has discretionary authority to grant access across BLM lands for federal oil and gas lessees. 43 U.S.C.A. § 1761. But access can be denied for a number of reasons, including that the proposed right-of-way would not be in the public interest. 43 C.F.R. § 2802.4. Special problems exist for access across Wilderness Study Areas. 43 U.S.C.A. § 1782(c). See e.g., State of Utah v. Andrus, 486 F.Supp. 995 (D. Utah 1979). Access across different public lands, such as national parks and national wildlife refuges may be governed by different statutory provisions. 2 *Law of Federal Oil and Gas Leases* § 22.02[4] (2006). Pipeline rights-of-way are not governed by the usual Interior regulations, 43 C.F.R. § 2802.2-1(c) (2006), but are governed by a separate set of regulations promulgated under the aegis of the Mineral Leasing Act. 43 C.F.R. Part 2880 (2006). The presence of Indian lands adds additional complexities to the problem. See Note, *Tribal Sovereignty and Congressional Dominion: Rights-of Way For Gas Pipelines on Indian Reservations*, 38 Stan. L.Rev. 196 (1985).

If access is needed across private lands, may a federal oil and gas lessee condemn a right-of-way under state private condemnation statutes? See Coquina Oil Corp. v. Harry Kourlis Ranch, 643 P.2d 519, 72 O.&G.R. 21 (Colo.1982) (lessee has no standing); Coronado Oil Co. v. Grieves, 603 P.2d 406, 74 O.&G.R. 545 (Wyo.1979) (summary judgment against lessee reversed and remanded; condemnation for "mining" purposes includes oil and gas exploration); Comment, *Eminent Domain and the Federal Oil and Gas Lessee-Lessee's Standing to Condemn a Right-of-way*, 1984 Utah L.Rev. 391; 2 *Law of Federal Oil & Gas Leases* § 22.06[4]. The recently enacted Energy Policy Act of 2005 requires the Secretary of the Interior to review current policies to determine their effect on privately owned surface estates. Pub.L.No. 109-58, § 1835, 119 Stat. 594 (Aug. 8, 2005).

(3) *Pooling and Unitization.* Under the Mineral Leasing Act, the Secretary of the Interior has broad discretion in approving lessees' unitization and communitization plans. 30 U.S.C.A. § 226(m). Communitization is the federal equivalent of pooling. Kramer & Martin, *The Law of Pooling and Unitization* §16.01. Communitization is permitted "[w]hen separate tracts cannot be

independently developed and operated in conformity with an established well-spacing or development program . . .," provided that the communitization is "determined by the Secretary of the Interior to be in the public interest . . . " 30 U.S.C.A. § 226(m).

Unitization, as we have seen earlier in Chapter 7, involves agreements by lessees and royalty owners to jointly operate and develop all or part of an entire reservoir. The Mineral Leasing Act permits formation of such unit plans "[f]or the purpose of more properly conserving the natural resources of any oil or gas pool, field, or like area, or any part thereof . . .," provided that the plans are "determined and certified by the Secretary of the Interior to be necessary or advisable in the public interest." 30 U.S.C.A. § 226(m).

The predominant type of federal unit is the exploratory unit. It usually involves the joint exploration and development of an area that is potentially productive of oil and gas. Kramer & Martin, *The Law of Pooling & Unitization* §16.02. The Rocky Mountain Mineral Law Foundation has held three special institutes devoted to federal onshore pooling and unitization. The materials provided in those institutes provide a wealth of information and reference sources for issues relating to the communitization and unitization of federal oil and gas leases. See also, Coffield *Selected Problems with Federal Exploratory Units*, 31 Rocky Mtn.Min.L.Inst. 13-1 (1986). These unit plans are subject to extensive government supervision before and after they are executed. There is a model form unit agreement which is widely used for exploratory units. 43 C.F.R. § 3186.1. Federal agency practice normally limits the size of exploratory units to 25,000 acres where there is only one exploratory well drilled to a depth of 5000 feet or deeper. The model unit agreement requires the unit operator to complete a test well within 6 months of the approval of the unit by the federal official. The agency normally retains a veto power over well location and it is also quite normal to have a multiple well completion requirement contained in a large-size unit. Federal exploratory units use the concept of "participating areas" which are defined as:

> That part of a unit area which is considered reasonably proven to be productive of unitized substances in paying quantities or which is necessary for unit operations and to which production is allocated in the manner prescribed in the unit agreement. 43 C.F.R. § 3180.0-5 (2006).

Most private unit agreements allocate costs and benefits throughout the entire unit based on an agreed-to formula. Under the federal exploratory unit, only the owners of interests in the participating area immediately share in the costs and benefits of production. The model unit agreement uses a surface acreage allocation formula to pay all royalties owed in the participating areas.

A federal lessee can avail himself of the state compulsory pooling laws, but only with the consent of the federal government. In Ohmart v. Dennis, 188 Neb. 260, 196 N.W.2d 181, 42 O.&G.R. 621 (1972), the United States sought to force pool a private interest under the Nebraska statutes. An earlier attempt to force pool by the federal lessee alone was dismissed because the court had concluded that the United States had not consented to have its interest subject to the state agency and therefore the lessee lacked standing. Once the federal government consented to the pooling of its interest under state law, the lessee could take advantage of the compulsory pooling process.

A federal lessee should not rely on production from a pooled unit well which is off of the federal leasehold estate to extend the federal lease if the state pooling or spacing order has not been approved by the federal government. For example, is a federal lease extended by production from a well included in an Oklahoma spacing unit, where under Oklahoma law the creation of the spacing unit automatically pools the royalty owners' interests? In Kirkpatrick Oil & Gas Co. v. United States, 675 F.2d 1122, 73 O.&G.R. 351 (10th Cir. 1982) the court said production from the well would not extend the federal lease because Kirkpatrick Oil had not received the consent of the federal agency prior to the expiration of the primary term. See also, Bruce Anderson, 91 Interior Dec. 203 (1984) (lease expired even though the lessee had contributed $ 115,000 to the costs of drilling a well.) Current regulations soften the harsh results by only requiring the "authorized officer" to approve the communitization agreement and making the agreements "effective from the date of the agreement or from the date of the onset of production . . . whichever is earlier." 43 C.F.R. § 3105.2-3.

What is the impact on private lands owners if a well is commenced on Indian-leased land within a state spacing unit? Would the activities on the Indian leases operate to the benefit of the private lessees so as to extend the lease into the secondary term? The Oklahoma Supreme Court answered both questions no in Kardokus v. Walsh, 797 P.2d 322, 113 O.&G.R. 117 (Okla. 1990). The Indian lease was on a section that was part of a Corporation Commission spacing unit. The Interior Department, however, had not consented to the spacing order, or a later pooling order affecting the working interest owners until after the expiration of the private lease's primary term. Since the federal lease could not be pooled, either as to royalty or working interest owners without the consent of the Department of the Interior, the activities of the federal lessee would not inure to the benefit of the private lessee even though consent was eventually granted. See also Samedan Oil Corp. v. Cotton Petroleum Corp., 466 F.Supp. 521, 64 O.&G.R. 519 (W.D. Okla. 1978).

B. UNIQUE PROBLEMS WITH FEDERAL NATURAL RESOURCE DEVELOPMENT

As the following three cases indicate, unique problems may arise in dealing with the relationship of private and federal ownership interests, including the transfer by deed or lease of federally-owned oil and gas resources. The first case, Amoco Production Co. v. Southern Ute Indian Tribe, 526 U.S. 865, 142 O.&G.R. 437 (1999) raises the issue of whether coalbed methane gas was reserved by the

United States when it issued a "patent" that reserved the "coal." This issue, as it applies to fee or private deeds is discussed in Chapter 5, Section 2. The second case, BP America Production v. Burton, 127 S.Ct. 38, 163 O.&G.R. 807 (2006), deals with a statute of limitations issue regarding the alleged under-payment of royalty. The third case, Dunn-McCampell Royalty Interest, Inc. v. National Park Service, 630 F.3d 431 (5th Cir. 2011) raises interesting questions regarding federal surface ownership acquired with the consent of a State and how subsequent surface regulation may impact privately-owned mineral interests that underlie the federally-owned surface estate.

AMOCO PRODUCTION CO. v. SOUTHERN UTE INDIAN TRIBE

Supreme Court of the United States
526 U.S. 865, 142 O.&G.R. 437 (1999)

JUSTICE KENNEDY delivered the opinion of the Court.

Land patents issued pursuant to the Coal Lands Acts of 1909 and 1910 conveyed to the patentee the land and everything in it, except the "coal," which was reserved to the United States. Coal Lands Act of 1909 (1909 Act), 35 Stat. 844, 30 U.S.C. § 81; Coal Lands Act of 1910 (1910 Act), ch. 318, 36 Stat. 583, 30 U.S.C. §§ 83-85. The United States Court of Appeals for the Tenth Circuit determined that the reservation of "coal" includes gas found within the coal formation, commonly referred to as coalbed methane gas (CBM gas). See 151 F.3d 1251, 1256 (1998) (en banc). We granted certiorari, 525 U.S. (1999), and now reverse.

I

During the second half of the nineteenth century, Congress sought to encourage the settlement of the West by providing land in fee simple absolute to homesteaders who entered and cultivated tracts of a designated size for a period of years. See, e.g., 1862 Homestead Act, 12 Stat. 392; 1877 Desert Land Act, ch. 107, 19 Stat. 377, as amended, 43 U.S.C. §§ 321-323. Public lands classified as valuable for coal were exempted from entry under the general land-grant statutes and instead were made available for purchase under the 1864 Coal Lands Act, ch. 205, § 1, 13 Stat. 343, and the 1873 Coal Lands Act, ch. 279, §1, 17 Stat. 607, which set a maximum limit of 160 acres on individual entry and minimum prices of $10 to $20 an acre. Lands purchased under these early Coal Lands Acts -- like lands patented under the Homestead Acts -- were conveyed to the entryman in fee simple absolute, with no reservation of any part of the coal or mineral estate to the United States. The coal mined from the lands purchased under the Coal Lands Acts and from other reserves fueled the Industrial Revolution.

At the turn of the twentieth century, however, a coal famine struck the West. . . . At the same time, evidence of widespread fraud in the administration of federal coal lands came to light. Lacking the resources to make an independent assessment of the coal content of each individual land tract, the Department of the Interior in classifying public lands had relied for the most part on the affidavits of entrymen. Watt v. Western Nuclear, Inc., 462 U.S. 36, 48, 76 L. Ed. 2d 400, 103 S. Ct. 2218, and n. 9 (1983). Railroads and other coal interests had exploited the system to avoid paying for coal lands and to evade acreage restrictions by convincing individuals to falsify affidavits, acquire lands for homesteading, and then turn the land over to them.

In 1906, President Theodore Roosevelt responded to the perceived crisis by withdrawing 64 million acres of public land thought to contain coal from disposition under the public land laws. Western Nuclear, 462 U.S. at 48-49. As a result, even homesteaders who had entered and worked the land in good faith lost the opportunity to make it their own unless they could prove to the land office that the land was not valuable for coal.

President Roosevelt's order outraged homesteaders and western interests, and Congress struggled for the next three years to construct a compromise that would reconcile the competing interests of protecting settlers and managing federal coal lands for the public good. President Roosevelt and others urged Congress to begin issuing limited patents that would sever the surface and mineral estates and allow for separate disposal of each. Although various bills were introduced in Congress that would have severed the estates -- some of which would have reserved "natural gas" as well as "coal" to the United States -- none was enacted. . . .

Finally, Congress passed the Coal Lands Act of 1909, which authorized the Federal Government, for the first time, to issue limited land patents. In contrast to the broad reservations of mineral rights proposed in the failed bills, however, the 1909 Act provided for only a narrow reservation. The Act authorized issuance of patents to individuals who had already made good-faith agricultural entries onto tracts later identified as coal lands, but the issuance was to be subject to "a reservation to the United States of all coal in said lands, and the right to prospect for, mine, and remove the same." 30 U.S.C. § 81. The Act also permitted the patentee to "mine coal for use on the land for domestic purposes prior to the disposal by the United States of the coal deposit." Ibid. A similar Act in 1910 opened the remaining coal lands to new entry under the homestead laws, subject to the same reservation of coal to the United States. 30 U.S.C. §§ 83-85.

Among the lands patented to settlers under the 1909 and 1910 Acts were former reservation lands of the Southern Ute Indian Tribe, which the Tribe had ceded to the United States in 1880 in return for certain allotted lands provided for their settlement. Act of June 15, 1880, ch. 223, 21 Stat. 199. In 1938, the United States restored to the Tribe, in trust, title to the ceded reservation lands still owned by the United States, including the reserved coal in lands patented under the 1909 and 1910 Acts. As a result, the Tribe now has equitable title to the coal in lands within its reservation settled by homesteaders under the 1909 and 1910 Acts.

We are advised that over 20 million acres of land were patented under the 1909 and 1910 Acts and that the lands -- including those lands in which the Tribe owns the coal -- contain large quantities of CBM gas. Brief for Montana et al. as Amici Curiae 2. At the time the Acts were passed, CBM gas had long been considered a dangerous waste product of coal mining. By the 1970's, however, it was apparent that CBM gas could be a significant energy resource, see Duel & Kimm, Coalbed Gas: A Source of Natural Gas, Oil & Gas J., June 16, 1975, p. 47, and, in the shadow of the Arab oil embargo, the Federal Government began to encourage the immediate production of CBM gas through grants, see 42 U.S.C. §§ 5901-5915 (1994 ed. and Supp. III), and substantial tax credits, see 26 U.S.C. § 29 (1994 ed. and Supp. III).

Commercial development of CBM gas was hampered, however, by uncertainty over its ownership. "In order to expedite the development of this energy source," the Solicitor of the Department of the Interior issued a 1981 opinion concluding that the reservation of coal to the United States in the 1909 and 1910 Acts did not encompass CBM gas. See Ownership of and Right to Extract Coalbed Gas in Federal Coal Deposits, 88 Interior Dec. 538, 539. In reliance on the Solicitor's 1981 opinion, oil and gas companies entered into leases to produce CBM gas with individual landowners holding title under 1909 and 1910 Act patents to some 200,000 acres in which the Tribe owns the coal.

In 1991, the Tribe brought suit in Federal District Court against petitioners, the royalty owners and producers under the oil and gas leases covering that land, and the federal agencies and officials responsible for the administration of lands held in trust for the Tribe. The Tribe sought, inter alia, a declaration that Congress' reservation of coal in the 1909 and 1910 Acts extended to CBM gas, so that the Tribe -- not the successors in interest of the land patentees -- owned the CBM gas.

The District Court granted summary judgment for the defendants, holding that the plain meaning of "coal" is the "solid rock substance" used as fuel, which does not include CBM gas. 874 F. Supp. 1142, 1154 (Colo. 1995). On appeal, a panel of the Court of Appeals reversed. 119 F.3d 816, 819 (CA10 1997). The court then granted rehearing en banc on the question whether the term "coal" in the 1909 and 1910 Acts "unambiguously excludes or includes CBM." 151 F.3d 1251 at 1256. Over a dissenting opinion by Judge Tacha, joined by two other judges, the en banc court agreed with the panel. Ibid. The court held that the term "coal" was ambiguous. Ibid. It invoked the interpretive canon that ambiguities in land grants should be resolved in favor of the sovereign and concluded that the coal reservation encompassed CBM gas. Ibid.

The United States did not petition for, or participate in, the rehearing en banc. Instead, it filed a supplemental brief explaining that the Solicitor of the Interior was reconsidering the 1981 Solicitor's opinion in light of the panel's decision. Brief for Federal Respondents 14, n. 8. On the day the Government's response to petitioners' certiorari petition was due, see id. at 47, n. 37, the Solicitor of the Interior withdrew the 1981 opinion in a one-line order, see Addendum to Brief for Federal Respondents in Opposition 1a. The United States now supports the Tribe's position that CBM gas is coal reserved by the 1909 and 1910 Acts.

II

We begin our discussion as the parties did, with a brief overview of the chemistry and composition of coal. Coal is a heterogeneous, noncrystalline sedimentary rock composed primarily of carbonaceous materials. . . . It is formed over millions of years from decaying plant material that settles on the bottom of swamps and is converted by microbiological processes into peat. Over time, the resulting peat beds are buried by sedimentary deposits. As the beds sink deeper and deeper into the earth's crust, the peat is transformed by chemical reactions which increase the carbon content of the fossilized plant material. The process in which peat transforms into coal is referred to as coalification.

The coalification process generates methane and other gases. Because coal is porous, some of that gas is retained in the coal. CBM gas exists in the coal in three basic states: as free gas; as gas dissolved in the water in coal; and as gas "adsorped" on the solid surface of the coal, that is, held to the surface by weak forces called van der Waals forces. These are the same three states or conditions in which gas is stored in other rock formations. Because of the large surface area of coal pores, however, a much higher proportion of the gas is adsorped on the surface of coal than is adsorped in other rock. When pressure on the coalbed is decreased, the gas in the coal formation escapes. As a result, CBM gas is released from coal as the coal is mined and brought to the surface.

III

While the modern science of coal provides a useful backdrop for our discussion and is consistent with our ultimate disposition, it does not answer the question presented to us. The question is not whether, given what scientists know today, it makes sense to regard CBM gas as a constituent of coal but whether Congress so regarded it in 1909 and 1910. In interpreting statutory mineral reservations like the one at issue here, we have emphasized that Congress "was dealing with a practical subject in a practical way" and that it intended the terms of the reservation to be understood in "their ordinary and popular sense." Burke v. Southern Pacific R. Co., 234 U.S. 669, 679, 58 L. Ed. 1527, 34 S. Ct. 907 (1914) (rejecting "scientific test" for determining whether a reservation of "mineral lands" included "petroleum lands"); see also Perrin v. United States 444 U.S. 37, 42, 62 L. Ed. 2d 199, 100 S. Ct. 311 (1979) ("Unless otherwise defined, words will be interpreted as taking their ordinary, contemporary, common meaning" at the time Congress enacted the statute). We are persuaded that the common conception of coal at the time Congress passed the 1909 and 1910 Acts was the solid rock substance that was the country's primary energy resource.

A

At the time the Acts were passed, most dictionaries defined coal as the solid fuel resource. For example, one contemporary dictionary defined coal as a "solid and more or less distinctly stratified mineral, varying in color from dark-brown to black, brittle, combustible, and used as fuel, not fusible without decomposition and very insoluble." 2 Century Dictionary and Cyclopedia 1067 (1906). See also American Dictionary of the English Language 244 (N. Webster 1889) (defining "coal" as a "black, or brownish black, solid, combustible substance, consisting, like charcoal, mainly of carbon, but more compact"); 2 New English Dictionary on Historical Principles 549 (J. Murray ed. 1893) (defining coal as a "mineral, solid, hard, opaque, black, or blackish, found in seams or strata in the earth, and largely used as fuel"); Webster's New International Dictionary of the English Language 424 (W. Harris & F. Allen eds. 1916) (defining coal as a "black, or brownish black, solid, combustible mineral substance").

In contrast, dictionaries of the day defined CBM gas -- then called "marsh gas," "methane," or "fire-damp" -- as a distinct substance, a gas "contained in" or "given off by" coal, but not as coal itself. See, e. g., 3 Century Dictionary and Cyclopedia 2229 (1906) (defining "fire-damp" as "the gas contained in coal, often given off by it in large quantities, and exploding, on ignition, when mixed with atmospheric air"; noting that "fire-damp is a source of great danger to life in coal-mines").

As these dictionary definitions suggest, the common understanding of coal in 1909 and 1910 would not have encompassed CBM gas, both because it is a gas rather than a solid mineral and because it was understood as a distinct substance that escaped from coal as the coal was mined, rather than as a part of the coal itself.

B

As a practical matter, moreover, it is clear that, by reserving coal in the 1909 and 1910 Act patents, Congress intended to reserve only the solid rock fuel that was mined, shipped throughout the country, and then burned to power the Nation's railroads, ships, and factories. Cf. Leo Sheep Co. v. United States, 440 U.S. 668, 682, 59 L. Ed. 2d 677, 99 S. Ct. 1403 (1979) (public land statutes should be interpreted in light of "the condition of the country when the acts were passed" (internal quotation marks omitted)). In contrast to natural gas, which was not yet an important source of fuel at the turn of the century, coal was the primary energy for the Industrial Revolution. . . . As the history recounted in Part I, supra, establishes, Congress passed the 1909 and 1910 Acts to address concerns over the short supply, mismanagement, and fraudulent acquisition of this solid rock fuel resource. Rejecting broader proposals, Congress chose a narrow reservation of the resource that would address the exigencies of the crisis at hand without unduly burdening the rights of homesteaders or impeding the settlement of the West.

It is evident that Congress viewed CBM gas not as part of the solid fuel resource it was attempting to conserve and manage but as a dangerous waste product, which escaped from coal as the coal was mined. Congress was well aware by 1909 that the natural gas found in coal formations was released during coal mining and posed a serious threat to mine safety. Explosions in coal mines sparked by CBM gas occurred with distressing frequency in the late nineteenth and early twentieth centuries. Congress was also well aware that the CBM gas needed to be vented to the greatest extent possible. Almost 20 years prior to the passage of the 1909 and 1910 Acts, Congress had enacted the first federal coal-mine-safety law which, among other provisions, prescribed specific ventilation standards for coal mines of a certain depth "so as to dilute and render harmless . . . the noxious or poisonous gases." 1891 Territorial Mine Inspection Act, § 6, 26 Stat. 1105.

That CBM gas was considered a dangerous waste product which escaped from coal, rather than part of the valuable coal fuel itself, is also confirmed by the fact that coal companies venting the gas to prevent its accumulation in the mines made no attempt to capture or preserve it. The more gas that escaped from the coal once it was brought to the surface, the better it was for the mining companies because it decreased the risk of a dangerous gas buildup during transport and storage.

. . .

There is some evidence of limited and sporadic exploitation of CBM gas as a fuel prior to the passage of the 1909 and 1910 Acts. . . . It seems unlikely, though, that Congress considered this limited drilling for CBM gas. To the extent Congress had an awareness of it, there is every reason to think it viewed the extraction of CBM gas as drilling for natural gas, not mining coal.

That distinction is significant because the question before us is not whether Congress would have thought that CBM gas had some fuel value, but whether Congress considered it part of the coal fuel. When it enacted the 1909 and 1910 Acts, Congress did not reserve all minerals or energy resources in the lands. It reserved only coal, and then only in lands that were specifically identified as valuable for coal. It chose not to reserve oil, natural gas, or any other known or potential energy resources.

The limited nature of the 1909 and 1910 Act reservations is confirmed by subsequent congressional enactments. When Congress wanted to reserve gas rights that might yield valuable fuel, it did so in explicit terms. In 1912, for example, Congress enacted a statute that reserved "oil and gas" in Utah lands. Act of Aug. 24, 1912, 37 Stat. 496. In addition, both the 1912 Act and a later Act passed in 1914 continued the tradition begun in the 1909 and 1910 Acts of reserving only those minerals enumerated in the statute. . . . It was not until 1916 that Congress passed a public lands act containing a general reservation of valuable minerals in the lands. See Stock-Raising Homestead Act, ch. 9, 39 Stat. 862, as amended, 43 U.S.C. § 299 (reserving "all the coal and other minerals in the lands" in all lands patented under the Act). See also Western Nuclear, 462 U.S. at 49 ("Unlike the preceding statutes containing mineral reservations, the [1916 Stock-Raising Homestead Act] was not limited to lands classified as mineral in character, and it did not reserve only specifically identified minerals").

C

Respondents contend that Congress did not reserve the solid coal but convey the CBM gas because the resulting split estate would be impractical and would make mining the coal difficult because the miners would have to capture and preserve the CBM gas that escaped during mining. See, e.g., Brief for Respondent Southern Ute Indian Tribe 46; see also id. at 25-26 (emphasizing that the reservation includes the right to "mine" the coal, "indicating that Congress reserved all rights needed to develop the underlying coal" including the right to vent CBM gas during mining). We doubt Congress would have given much consideration to these problems, however, because -- as noted above -- it does not appear to have given consideration to the possibility that CBM gas would one day be a profitable energy source developed on a large scale.

It may be true, nonetheless, that the right to mine the coal implies the right to release gas incident to coal mining where it is necessary and reasonable to do so. The right to dissipate the CBM gas where reasonable and necessary to mine the coal does not, however, imply the ownership of the gas in the first instance. Rather, it simply reflects the established common-law right of the owner of one mineral estate to use, and even damage, a neighboring estate as necessary and reasonable to the extraction of his own minerals. See, e.g., Williams v. Gibson, 84 Ala. 228, 4 So. 350 (1888); Rocky Mountain Mineral Foundation, 6 American Law of Mining § 200.04 (2d ed. 1997). Given that split estates were already common at the time the 1909 and 1910 Acts were passed, see, e.g., Chartiers Block Coal Co. v. Mellon, 152 Pa. 286, 25 A. 597 (1893), and that the common law has proved adequate to the task of resolving the resulting conflicts between estates, there is no reason to think that the prospect of a split estate would have deterred Congress from reserving only the coal.

Were a case to arise in which there are two commercially valuable estates and one is to be damaged in the course of extracting the other, a dispute might result, but it could be resolved in the ordinary course of negotiation or adjudication. That is not the issue before us, however. The question is one of ownership, not of damage or injury.

In all events, even were we to construe the coal reservation to encompass CBM gas, a split estate would result. The United States concedes (and the Tribe does not dispute) that once the gas originating in the coal formation migrates to surrounding rock formations it belongs to the natural gas, rather than the coal, estate. See Brief for Federal Respondents 35; Brief for Respondent Southern Ute Indian Tribe 3, n. 4. Natural gas from other sources may also exist in the lands at issue. Including the CBM gas in the coal reservation would, therefore, create a split gas estate that would be at least as difficult to administer as a split coal/CBM gas estate. If CBM gas were reserved with the coal estate, those developing the natural gas resources in the land would have to allocate the gas between the natural gas and coal estates based on some assessment of how much had migrated outside the coal itself. There is no reason to think Congress would have been more concerned about the creation of a split coal/CBM gas estate than the creation of a split gas estate.

Because we conclude that the most natural interpretation of "coal" as used in the 1909 and 1910 Acts does not encompass CBM gas, we need not consider the applicability of the canon that ambiguities in land grants are construed in favor of the sovereign or the competing canons relied on by petitioners.

The judgment of the Court of Appeals is reversed.

It is so ordered.

JUSTICE BREYER took no part in the consideration or decision of this case.

JUSTICE GINSBURG, dissenting.

I would affirm the judgment below substantially for the reasons stated by the Court of Appeals and the federal respondents. As the Court recognizes, in 1909 and 1910 coalbed methane gas (CBM) was a liability. See ante, at 4, 9-10. Congress did not contemplate that the surface owner would be responsible for it. More likely, Congress would have assumed that the coal owner had dominion over, and attendant responsibility for, CBM. I do not find it clear that Congress understood dominion would shift if and when the liability became an asset. I would therefore apply the canon that ambiguities in land grants are construed in favor of the sovereign. See Watt v. Western Nuclear, Inc., 462 U.S. 36, 59, 76 L. Ed. 2d 400, 103 S. Ct. 2218 (1983) (noting "established rule that land grants are construed favorably to the Government, that nothing passes except what is conveyed in clear language, and that if there are doubts they are resolved for the Government, not against it" (internal quotation marks omitted)).

NOTES

1. The state law decisions on whether coalbed methane gas is included in a grant or reservation of "coal" or "gas" are not consistent. The two earliest decisions seemed to conclude that the owner of the coal and not the owner of the oil and gas owned the CBM. *See* United States Steel Corp. v. Hoge, 503 Pa. 140, 468 A.2d 1380, 79 O.&G.R. 96 (1983); Vines v. McKenzie Methane Corp., 619 So.2d 1305, 1308-09, 122 O.&G.R. 34 (Ala. 1993). But the Alabama Supreme Court in NCNB Texas National Bank, N.A. v. West, 631 So.2d 212, 127 O.&G.R. 209 (Ala. 1993), *on later appeal*, 646 So.2d 1356 (Ala. 1994) found that where the coal had been conveyed and the gas reserved, the coal owner owns

the CBM located within the coal seams but the gas owner owns the CBM that has migrated into other strata.. Wyoming, in a series of cases, has determined that extrinsic evidence is admissible to determine the ownership of CBM, but has always concluded that the coal owner does not own the CBM. *See* Mullinnix, LLC v. HKB Royalty Trust, 2006 WY 14, 126 P.3d 909, 162 O.&G.R. 14 (page 568 *supra*); Hickman v. Groves, 2003 WY 76, 71 P.2d 256, 160 O.&G.R. 281.

2. Should special rules or canons of construction apply where a governmental entity is the grantor of a patent or deed? In *Southern Ute* the Supreme Court did not apply the long-recognized canon of construction to construe statutes and patents in favor of the government. Andrus v. Charlestone Stone Products Co., 436 U.S. 604 (1978). This canon traces its history to Charles River Bridge Co. v. Warren Bridge, 36 U.S. 420, 548 (1837). There is also a canon of construction that says where statutes and patents are plain on their face they should not be interpreted to deprive private owners of their interests. Leo Sheep Co. v. United States, 440 U.S. 668 (1979). See Bruce M. Kramer, *Amoco Production Co. v. Southern Ute Indian Tribe: Restatement or Revolution?* 45 Rocky Mtn.Min.L.Inst. 7-1 (1999). In Carbon County v. Union Reserve Oil Co., 271 Mont. 459, 898 P.2d 680, 135 O.&G.R. 260 (Mont. 1995), the Montana Supreme Court eschewed applying any special canons even though a County was the grantor but nonetheless held that the County had reserved the CBM when it granted "coal and coal rights" to the grantee.

BP AMERICA PRODUCTION CO. v. BURTON

Supreme Court of the United States
127 S.Ct. 638, 163 O.&G.R. 807 (2006)

JUSTICE ALITO delivered the opinion of the Court.

This case presents the question whether administrative payment orders issued by the Department of the Interior's Minerals Management Service (MMS) for the purpose of assessing royalty underpayments on oil and gas leases fall within 28 U.S.C. § 2415(a), which sets out a 6-year statute of limitations for Government contract actions. We hold that this provision does not apply to these administrative payment orders, and we therefore affirm.

I

A

The Mineral Leasing Act of 1920 (MLA) authorizes the Secretary of the Interior to lease public-domain lands to private parties for the production of oil and gas. 41 Stat. 437, as amended, 30 U.S.C. § 181 et seq. MLA lessees are obligated to pay a royalty of at least "12.5 percent in amount or value of the production removed or sold from the lease." § 226(b)(1)(A).

In 1982, Congress enacted the Federal Oil and Gas Royalty Management Act (FOGRMA), 96 Stat. 2447, as amended, 30 U.S.C. § 1701 et seq., to address the concern that the "system of accounting with respect to royalties and other payments due and owing on oil and gas produced from such lease sites [was] archaic and inadequate." § 1701(a)(2). FOGRMA ordered the Secretary of the Interior to "audit and reconcile, to the extent practicable, all current and past lease accounts for leases of oil or gas and take appropriate actions to make additional collections or refunds as warranted." § 1711(c)(1). The Secretary, in turn, has assigned these duties to the MMS. 30 CFR § 201.100 (2006).

Under FOGRMA, lessees are responsible in the first instance for the accurate calculation and payment of royalties. 30 U.S.C. § 1712(a). MMS, in turn, is authorized to audit those payments to determine whether a royalty has been overpaid or underpaid. §§ 1711(a) and (c); 30 CFR §§ 206.150(c), 206.170(d). In the event that an audit suggests an underpayment, it is MMS' practice to send the lessee a letter inquiring about the perceived deficiency. If, after reviewing the lessee's response, MMS concludes that the lessee owes additional royalties, MMS issues an order requiring payment of the amount due. Failure to comply with such an order carries a stiff penalty: "Any person who -- (1) knowingly or willfully fails to make any royalty payment by the date as specified by [an] order . . . shall be liable for a penalty of up to $ 10,000 per violation for each day such violation continues." 30 U.S.C. § 1719(c). The Attorney General may enforce these orders in federal court. § 1722(a).

An MMS payment order may be appealed, first to the Director of MMS and then to the Interior Board of Land Appeals or to an Assistant Secretary. 30 CFR §§ 290.105, 290.108. While filing an appeal does not generally stay the payment order, § 218.50(c), MMS will usually suspend the order's effect after the lessee complies with applicable bonding or financial solvency requirements, § 243.8.

Congress supplemented this scheme by enacting the Federal Oil and Gas Royalty Simplification and Fairness Act of 1996 (FOGRSFA), 110 Stat. 1700, as amended, 30 U.S.C. § 1701 et seq. FOGRSFA adopted a prospective 7-year statute of limitations for any "judicial proceeding or demand" for royalties arising under a federal oil or gas lease. § 1724(b)(1). The parties agree that this provision applies both to judicial actions ("judicial proceedings") and to MMS' administrative payment orders ("demands") arising on or after September 1, 1996. Ibid. This provision does not, however, apply to judicial proceedings or demands arising from leases of Indian land or underpayments of royalties on pre-September 1, 1996, production. FOGRSFA §§ 9, 11, 110 Stat. 1717, notes following 30 U.S.C. § 1701.

There is no dispute that a lawsuit in court to recover royalties owed to the Government on pre-September 1, 1996, production is covered by 28 U.S.C. § 2415(a), which sets out a general 6-year statute of limitations for Government contract actions. That section, which was enacted in 1966, provides in relevant part:

> Subject to the provisions of section 2416 of this title, and except
> as otherwise provided by Congress, every action for money

damages brought by the United States or an officer or agency thereof which is founded upon any contract express or implied in law or fact, shall be barred unless the complaint is filed within six years after the right of action accrues or within one year after final decisions have been rendered in applicable administrative proceedings required by contract or by law, whichever is later.

Whether this general 6-year statute of limitations also governs MMS administrative payment orders concerning pre-September 1, 1996, production is the question that we must decide in this case.

B

Petitioner BP America Production Co. holds gas leases from the Federal Government for lands in New Mexico's San Juan Basin. BP's predecessor, Amoco Production Co., first entered into these leases nearly 50 years ago, and these leases require the payment of the minimum 12.5 percent royalty prescribed by 30 U.S.C. § 226(b)(1)(A). For years, Amoco calculated the royalty as a percentage of the value of the gas as of the moment it was produced at the well. In 1996, MMS sent lessees a letter directing that royalties should be calculated based not on the value of the gas at the well, but on the value of the gas after it was treated to meet the quality requirements for introduction into the Nation's mainline pipelines. Consistent with this guidance, MMS in 1997 ordered Amoco to pay additional royalties for the period from January 1989 through December 1996 in order to cover the difference between the value of the treated gas and its lesser value at the well.

Amoco appealed the order, disputing MMS' interpretation of its royalty obligations and arguing that the payment order was in any event barred in part by the 6-year statute of limitations in 28 U.S.C. § 2415(a). The Assistant Secretary of the Interior denied the appeal and ruled that the statute of limitations was inapplicable.

Amoco, together with petitioner Atlantic Richfield Co., sought review in the United States District Court for the District of Columbia, which agreed with the Assistant Secretary that § 2415(a) did not govern the administrative order. *Amoco Production Co. v. Baca*, 300 F. Supp. 2d 1, 21 (2003). The Court of Appeals for the District of Columbia Circuit affirmed, *Amoco Production Co. v. Watson*, 366 U.S. App. D.C. 215, 410 F.3d 722, 733 (2005), and we granted certiorari, 547 U.S.--- , 126 S. Ct. 1768, 164 L. Ed. 2d 515 (2006) , in order to resolve the conflict between that decision and the contrary holding of the United States Court of Appeals for the Tenth Circuit in *OXY USA, Inc. v. Babbitt*, 268 F.3d 1001, 1005 (2001) (en banc). We now affirm.

II

A

We start, of course, with the statutory text. *Central Bank of Denver, N. A. v. First Interstate Bank of Denver, N. A.*, 511 U.S. 164, 173, 114 S. Ct. 1439,

128 L. Ed. 2d 119 (1994). Unless otherwise defined, statutory terms are generally interpreted in accordance with their ordinary meaning. Perrin v. United States, 444 U.S. 37, 42, 100 S. Ct. 311, 62 L. Ed. 2d 199 (1979). Read in this way, the text of § 2415(a) is quite clear.

The statute of limitations imposed by § 2415(a) applies when the Government commences any "action for money damages" by filing a "complaint" to enforce a contract, and the statute runs from the point when "the right of action accrues." The key terms in this provision -- "action" and "complaint" -- are ordinarily used in connection with judicial, not administrative, proceedings. In 1966, when § 2415(a) was enacted, a commonly used legal dictionary defined the term "right of action" as "the right to bring suit; a legal right to maintain an action," with "suit" meaning "any proceeding . . . in a court of justice." Black's Law Dictionary 1488, 1603 (4th ed. 1951) (hereinafter Black's). Likewise, "complaint" was defined as "the first or initiatory pleading on the part of the plaintiff in a civil action." Id., at 352, 47 S. Ct. 389, 71 L. Ed. 676. . . . The phrase "action for money damages" reinforces this reading because the term "damages" is generally used to mean "pecuniary compensation or indemnity, which may be recovered in the courts." Black's 466 (emphasis added).

Nothing in the language of § 2415(a) suggests that Congress intended these terms to apply more broadly to administrative proceedings. On the contrary, § 2415(a) distinguishes between judicial and administrative proceedings. Section 2415(a) provides that an "action" must commence "within one year after final decisions have been rendered in applicable administrative proceedings." Thus, Congress knew how to identify administrative proceedings and manifestly had two separate concepts in mind when it enacted § 2415(a).

B

In an effort to show that the term "action" is commonly used to refer to administrative, as well as judicial, proceedings, petitioners have cited numerous statutes and regulations that, petitioners claim, document this usage. These examples, however, actually undermine petitioners' argument, since none of them uses the term "action" standing alone to refer to administrative proceedings. Rather, each example includes a modifier of some sort, referring to an "administrative action," a "civil or administrative action," or "administrative enforcement actions." This pattern of usage buttresses the point that the term "action," standing alone, ordinarily refers to a judicial proceeding.

Petitioners contend that their broader interpretation of the statutory term "action" is supported by the reference to "every action for money damages" founded upon "any contract." 28 U.S.C. § 2415(a) (emphasis added). But the broad terms "every" and "any" do not assist petitioners, as they do not broaden the ordinary meaning of the key term "action."

. . .

For these reasons, we are not persuaded by petitioners' argument that the term "action" in § 2415(a) applies to the administrative proceedings that follow the issuance of an MMS payment order.

C

We similarly reject petitioners' suggestion that an MMS letter or payment order constitutes a "complaint" within the meaning of § 2415(a). Petitioners point to examples of statutes and regulations that employ the term "complaint" in the administrative context. See, e.g., 15 U.S.C. § 45(b) (requiring the Federal Trade Commission to serve a "complaint" on a party suspected of engaging in an unfair method of competition); 29 CFR § 102.15 (2006) (a "complaint" initiates unfair labor practice proceedings before the National Labor Relations Board). But the occasional use of the term to describe certain administrative filings does not alter its primary meaning, which concerns the initiation of "a civil action." Black's 356. Moreover, even if the distinction between administrative and judicial proceedings is put aside, an MMS payment order lacks the essential attributes of a complaint. While a complaint is a filing that commences a proceeding that may in the end result in a legally binding order providing relief, an MMS payment order in and of itself imposes a legal obligation on the party to which it is issued. As noted, the failure to comply with such an order can result in fines of up to $ 10,000 a day. An MMS payment order, therefore, plays an entirely different role from that of a "complaint."

D

To the extent that any doubts remain regarding the meaning of § 2415(a), they are erased by the rule that statutes of limitations are construed narrowly against the government. E. I. DuPont de Nemours & Co. v. Davis, 264 U.S. 456, 44 S. Ct. 364, 68 L. Ed. 788 (1924). This canon is rooted in the traditional rule quod nullum tempus occurrit regi -- time does not run against the King. Guaranty Trust Co. v. United States, 304 U.S. 126, 132, 58 S. Ct. 785, 82 L. Ed. 1224 (1938). A corollary of this rule is that when the sovereign elects to subject itself to a statute of limitations, the sovereign is given the benefit of the doubt if the scope of the statute is ambiguous. . . .

E

We come now to petitioners' argument that interpreting § 2415(a) as applying only to judicial actions would render subsection (i) of the same statute superfluous. Subsection (i) provides as follows:

> The provisions of this section shall not prevent the United States
> or an officer or agency thereof from collecting any claim of the
> United States by means of administrative offset, in accordance
> with section 3716 of title 31. 28 U.S.C. § 2415(i).

An administrative offset is a mechanism by which the Government withholds payment of a debt that it owes another party in order to recoup a payment that this party owes the Government. 31 U.S.C. § 3701(a)(1). Thus, under subsection (i), the Government may recover a debt via an administrative offset even if the Government would be time barred under subsection (a) from pursuing the debt in court.

Petitioners argue that, if § 2415(a) applies only to judicial proceedings and not to administrative proceedings, there is no need for § 2415(i)'s rule protecting a particular administrative mechanism (i.e., an administrative offset) from the statute of limitations set out in subsection (a). Invoking the canon against reading a statute in a way that makes part of the statute redundant, see, e.g., TRW Inc. v. Andrews, 534 U.S. 19, 31, 122 S. Ct. 441, 151 L. Ed. 2d 339 (2001), petitioners contend that subsection (i) shows that subsection (a) was meant to apply to administrative, as well as judicial, proceedings. We disagree.

As the Court of Appeals noted, subsection (i) was not enacted at the same time as subsection (a) but rather was added 16 years later by the Debt Collection Act of 1982. 96 Stat. 1749 . This enactment followed a dispute between the Office of the Comptroller of the Currency (OCC) and the Department of Justice's Office of Legal Counsel (OLC) over whether an administrative offset could be used to recoup a debt where a judicial recoupment action was already time barred.

In 1978, in response to a question from the United States Civil Service Commission, OLC opined that an administrative offset could not be used to recoup a debt as to which a judicial action was already time barred. OLC reached this conclusion not because it believed that § 2415(a) reached administrative proceedings generally, but rather because of the particular purpose of an administrative offset. "Where [a] debt has not been reduced to judgment," OLC stated, "an administrative offset is merely a pre-judgment attachment device." Memorandum from John M. Harmon, Assistant Attorney General, OLC, to Alan K. Campbell, Chairman, U.S. Civil Service Commission Re: Effect of Statute of Limitations on Administrative Collection of United States Claims 3 (Sept. 29, 1978), Joint Lodging. OLC opined that a prejudgment attachment device such as this exists only to preserve funds to satisfy any judgment the creditor subsequently obtains. Id., at 4 (citing cases). OLC therefore concluded that, where a lawsuit is already foreclosed by § 2415(a), an administrative offset that is the functional equivalent of a pretrial attachment is also unavailable. Id., at 3.

The OCC disagreed. See In the Matter of Collection of Debts -- Statute of Limitations on Administrative Setoff, 58 Comp. Gen. 501, 504-505 (1979). In its view, the question was answered by

> [t]he general rule . . . that statutes of limitations applicable to
> suits for debts or money demands bar or run only against the
> remedy (the right to bring suit) to which they apply and do not

discharge the debt or extinguish, or even impair, the right or obligation, either in law or in fact, and the creditor may avail himself of every other lawful means of realizing on the debt or obligation. See Mascot Oil Co. v. United States, 42 F.2d 309, 70 Ct. Cl. 246 (Ct. Cl. 1930), affirmed 282 U.S. 434, 51 S. Ct. 196, 75 L. Ed. 444, 71 Ct. Cl. 783, 1931-1 C.B. 190; and 33 Comp. Gen. 66 (1953). See also Ready-Mix Concrete Co. v. United States, 130 F. Supp. 390, 131 Ct. Cl. 204 (Ct. Cl. 1955). Ibid.

That Congress had time-barred the judicial remedy, OCC reasoned, imposed no limit on the administrative remedy.

The OLC-OCC dispute reveals that, even under the interpretation of subsection (a) -- the one we are adopting -- that considers it applicable only to court proceedings, subsection (i) is not mere surplusage. It clarifies that administrative offsets are not covered by subsection (a) even if they are viewed as an adjunct of a court action.

To accept petitioners' argument, on the other hand, we would have to hold either that § 2415(a) applied to administrative actions when it was enacted in 1966 or that it was extended to reach administrative actions when subsection (i) was added in 1982. The clear meaning of the text of § 2415(a), which has not been amended, refutes the first of these propositions, and accepting the latter would require us to conclude that in 1982 Congress elected to enlarge § 2415 to cover administrative proceedings by inserting text expressly excluding a single administrative vehicle from the statute's reach. It is entirely unrealistic to suggest that Congress would proceed by such an oblique and cryptic route.

III

Petitioners contend that interpreting § 2415(a) as applying only to judicial actions results in a statutory scheme with peculiarities that Congress could not have intended. For example, petitioners note that while they are required by statute to preserve their records regarding royalty obligations for only seven years, 30 U.S.C. § 1724(f) , the interpretation of § 2415(a) adopted by the Court of Appeals permits MMS to issue payment orders that reach back much farther.

We are mindful of the fact that a statute should be read where possible as effecting a "'symmetrical and coherent regulatory scheme,'" FDA v. Brown & Williamson Tobacco Corp., 529 U.S. 120, 133, 120 S. Ct. 1291, 146 L. Ed. 2d 121 (2000), but here petitioners' alternative interpretation of § 2415(a) would itself result in disharmony. For instance, under FOGRSFA, MMS payment orders regarding oil and gas leases are now prospectively subject to a 7-year statute of limitations except with respect to obligations arising out of leases of Indian land. Consequently, if we agreed with petitioners that § 2415(a) applies generally to administrative proceedings, payment orders relating to oil and gas royalties owed under leases of Indian land would be subject to a shorter (i.e., 6-year) statute of limitations than similar payment orders relating to leases of other public-domain lands (which would be governed by FOGRSFA's new 7-

year statute). Particularly in light of Congress' exhortation that the Secretary of the Interior "aggressively carry out his trust responsibility in the administration of Indian oil and gas," 30 U.S.C. § 1701(a)(4), it seems unlikely that Congress intended to impose a shorter statute of limitations for payment orders regarding Indian lands.

Petitioners contend, finally, that interpreting § 2415(a) as applying only to judicial actions would frustrate the statute's purposes of providing repose, ensuring that actions are brought while evidence is fresh, lightening recordkeeping burdens, and pressuring federal agencies to assert federal rights promptly. These are certainly cogent policy arguments, but they must be viewed in perspective.

For one thing, petitioners overstate the scope of the problem, since Congress of course can enact and has enacted specific statutes of limitations to govern specific administrative actions. See, e.g., 42 U.S.C. § 5205(a)(1) (statute of limitations for an administrative action to recover payments made to state governments for disaster or emergency assistance). Indeed, in 1996, FOGRSFA imposed just such a limitation prospectively on all non-Indian land, oil, and gas lease claims.

Second, and more fundamentally, the consequences of interpreting § 2415(a) as limited to court actions must be considered in light of the traditional rule exempting proceedings brought by the sovereign from any time bar. There are always policy arguments against affording the sovereign this special treatment, and therefore in a case like this, where the issue is how far Congress meant to go when it enacted a statute of limitations applicable to the Government, arguing that an expansive interpretation would serve the general purposes of statutes of limitations is somewhat beside the point. The relevant inquiry, instead, is simply how far Congress meant to go when it enacted the statute of limitations in question. Here prior to the enactment of § 2415(a) in 1966, contract actions brought by the Government were not subject to any statute of limitations. See Guaranty Trust Co., 304 U.S., at 132, 58 S. Ct. 785, 72 L. Ed. 1224. Absent congressional action changing this rule, it remains the law, and the text of § 2415(a) betrays no intent to change this rule as it applies to administrative proceedings.

In the final analysis, while we appreciate petitioners' arguments, they are insufficient to overcome the plain meaning of the statutory text. We therefore hold that the 6-year statute of limitations in § 2415(a) applies only to court actions and not to the administrative proceedings involved in this case.

. . .

For these reasons, the judgment of the Court of Appeals for the District of Columbia Circuit is affirmed.

It is so ordered.

THE CHIEF JUSTICE and JUSTICE BREYER took no part in the consideration or decision of this case.

NOTES

1. The apparent harshness of the result from the perspective of federal oil and gas lessees is lessened by its legislative overruling in the Federal Oil and Gas Royalty Simplication and Fairness Act of 1996, Pub.L.No. 104-185, 100 Stat. 1700 which clearly sets forth a seven-year statute of limitations for any action, administrative or judicial, relating to claims by the Federal Government for royalty.

2. Did the court rely on canons of statutory construction? Did they favor the Government's position or the royalty payor?

3. Since 1988, the United States, has adopted a variant of the "first marketable product" or "marketable condition" rule for determining how federal royalty is to be calculated for natural gas production. *See e.g.,* 30 C.F.R. §303.152 (2006). The federal regulatory scheme differentiates between arms'-length and non-arms'-length contracts. *Id.*

C. THE RELATIONSHIP BETWEEN STATE AND FEDERAL AUTHORITY

We have already studied in Chapter 7 supra the extensive state regulation of oil and gas development. Conflicts between the exercise of federal and state regulatory powers can and do occur. Traditional issues of federalism and preemption are raised in several different contexts for the federal oil and gas lessee. See generally, G. Coggins & R. Glicksman, *Public Natural Resources Law* (2004 with supplements); Kramer & Martin, *The Law of Pooling and Unitization,* ch. 16, 24.

Congress receives its authority to regulate the public lands from the Property Clause of the U.S. Constitution. It provides:

> The Congress shall have power to dispose of and make all needful rules and regulations respecting the territory or other property belonging to the United States; and nothing in this Constitution shall be so construed as to prejudice any claims of the United States or of any particular state. U.S.Const. art. IV, § 3, cl. 2.

In Kleppe v. New Mexico, 426 U.S. 529 (1976) the Supreme Court interpreted the Property Clause to give plenary or sovereign power over the public lands. But the Supreme Court also made clear that state and local governments were free to regulate federal lands and federal permittees until such time as Congress exercises those plenary powers. *Id.* at 543-44. For the oil and gas lessee the issues arise in the context of the Mineral Leasing Act of 1920 and its impact on state conservation regulation. The following two cases suggest that all may not be settled in determining the scope and extent of federal preemption of state oil and gas conservation laws.

VENTURA COUNTY v. GULF OIL CORP.

United States Court of Appeals, Ninth Circuit, 1979.
601 F.2d 1080, 64 O.&G.R. 19, affirmed without opinion, 445 U.S. 947, 65 O.&G.R. 169 (1980).

Before HUFSTEDLER and WRIGHT, CIRCUIT JUDGES, and CALLISTER, DISTRICT JUDGE.

HUFSTEDLER, CIRCUIT JUDGE: The question on appeal is whether the County of Ventura ("Ventura") can require the federal Government's lessee, Gulf Oil Corporation ("Gulf"), to obtain a permit from Ventura in compliance with Ventura's zoning ordinances governing oil exploration and extraction activities before Gulf can exercise its rights under the lease and drilling permits acquired from the Government. The district court denied Ventura's motion for a preliminary injunction, and dismissed Ventura's second amended complaint. Ventura appeals. We uphold the district court because the local ordinances impermissibly conflict with congressional regulation of Gulf's activities on government land.

On January 1, 1974, the Department of the Interior, Bureau of Land Management, pursuant to the Mineral Lands Leasing Act of 1920 (30 U.S.C. §§ 181 *et seq.*), leased 120 acres located within the Los Padres National Forest in Ventura for purposes of oil exploration and development. A subsequent assignment of this lease to Gulf was approved by the Department of the Interior, effective April 1, 1974. On February 25, 1976, the United States Department of the Interior, Geological Survey, issued a permit approving Gulf's proposal to drill an oil well pursuant to its lease. On March 8, 1976, and April 15, 1976, the United States Department of Agriculture, Forest Service, also granted its permission, and on March 8, 1976, the California Resources Agency, Division of Oil and Gas, approved the proposed exploration. After drilling operations were commenced on April 28, 1976, Gulf pursued activities related to oil exploration and extraction on the leased property, and it intends to continue development of both its present and other drill sites.

Throughout this period the leased property has been zoned Open Space ("O-S") by Ventura. Under its zoning ordinance, oil exploration and extraction activities are prohibited on O-S property unless an Open Space Use Permit is obtained from the Ventura County Planning Commission in accordance with Articles 25 and 43 of the Ventura County Ordinance Code. The O-S Use Permits are granted for such time and upon such conditions as the Planning Commission considers in the public interest. The permits contain 11 mandatory conditions and additional conditions are committed to the Planning Board's discretion.

On May 5, 1976, Ventura advised Gulf that it must obtain an O-S Use Permit if it wished to continue its drilling operations. Gulf refused to comply, and on May 20, 1976, Ventura brought suit . . .

Although Ventura and amicus argue extensively that congressional enactments under the Property Clause generally possess no preemptive capability, we believe that Kleppe v. New Mexico (1976) 426 U.S. 529, 96 S.Ct. 2285, 49 L.Ed.2d 34, is dispositive. . . .

Ventura next contends that even if Congress had the power to enact overriding legislation, there is no evidence of either a congressional intent to preempt local regulation or a conflict between local and federal law that can be resolved only by exclusion of local jurisdiction. We need not consider the extent to which local regulation of any aspect of oil exploration and extraction upon federal lands is precluded by federal legislation; the local ordinances impermissibly conflict with the Mineral Lands Leasing Act of 1920 and on this basis alone they cannot be applied to Gulf.

The extensive regulation of oil exploration and drilling under the Mineral Leasing Act is evident from the present record. The basic lease assigned to Gulf in 1974 contains approximately 45 paragraphs including requirements of diligence and protection of the environment as well as reservation of a one eighth royalty interest in the United States. Because the lands lie within a National Forest, the lease requires Gulf's acceptance of additional Department of Agriculture conditions designed to combat the environmental hazards normally incident to mining operations. Specific drilling permits were also required from the Department of the Interior, Geological Survey, and the Department of Agriculture, Forest Service. The Geological Survey, which has formalized its procedures in accordance with the National Environmental Policy Act of 1969 (42 U.S.C. §§ 4321 *et seq.*), approved the proposed drilling on February 25, 1976, subject to 10 conditions which assure continued and detailed supervision of Gulf's activities. And on March 8, and April 15, 1976, the Forest Service issued a drilling permit subject to conditions focusing upon protection of the National Forest. Finally, Gulf is subject to the extensive regulations governing oil and gas leasing (43 C.F.R., Part 3100) and both subsurface and surface operations (30 C.F.R., Part 221) promulgated by the Secretary of the Interior under his authority "to prescribe necessary and proper rules and regulations to do any and all things necessary to carry out and accomplish the purposes" of the act. (30 U.S.C. § 189.) And since the lease concerns lands within a National Forest, Secretary of Agriculture regulations governing oil and gas development are also applicable. (36 C.F.R., Part 252.)

Despite this extensive federal scheme reflecting concern for the local environment as well as development of the nation's resources, Ventura demands a right of final approval. Ventura seeks to prohibit further activity by Gulf until it secures an Open Space Use Permit which may be issued on whatever conditions Ventura determines appropriate, or which may never be issued at all. The federal Government has authorized a specific use of federal lands, and Ventura cannot prohibit that use, either temporarily or permanently, in an attempt to substitute its judgment for that of Congress.

The present conflict is no less direct than that in *Kleppe v. New Mexico, supra.* Like *Kleppe,* our case involves a power struggle between local and federal governments concerning appropriate use of the public lands. That the New Mexico authorities wished to engage in activity that Congress prohibited, while the Ventura authorities wish to regulate conduct which Congress has authorized is a distinction without a legal difference.

Relying upon Huron Portland Cement Co. v. Detroit (1960) 362 U.S. 440, 80 S.Ct. 813, 4 L.Ed.2d 852, Ventura argues that this conclusion is unjustified because, unlike *Kleppe,* the present conflict has not ripened. . . . Ventura prefers the Court's observation: "To hold otherwise would be to ignore the teaching of this Court's decisions which enjoin seeking out conflicts between state and federal regulation where none clearly exists." (*Id.* at 446, 80 S.Ct. at 817-818.) It argues that since Gulf has never applied for a use permit, an adjudication of preemption is in conflict with Huron. We disagree. This argument ignores the very relief that Ventura sought when it asked to enjoin Gulf's operations. Ventura cannot bring an action seeking immediate recognition and implementation of its veto power over the Government's lessee, and successfully contend that an adverse ruling would require "seeking out conflict" unnecessarily. The issue is whether Ventura has the power of ultimate control over the Government's lessee, and this issue persists whether or not a use permit would eventually be granted.

Federal Power Commission v. Oregon (1955) 349 U.S. 435, 75 S.Ct. 832, 99 L.Ed. 1215, presented a similar question . . .

Ventura attempts to distinguish *Federal Power Commission v. Oregon* on the basis of reservations of local jurisdiction contained in sections 30 and 32 of the Mineral Lands Leasing Act (30 U.S.C. §§ 187, 189).[113] It contends that

---

[113]   Before its 1978 amendment, section 30 provided in pertinent part: "Each lease shall contain provisions for the purpose of insuring the exercise of reasonable diligence, skill, and care in the operation of said property; a provision that such rules for the safety and welfare of the miners and for the prevention of undue waste as may be prescribed by said Secretary shall be observed, including a restriction of the workday to not exceeding eight hours in any one day for underground workers except in cases of emergency; provisions prohibiting the employment of any boy under the age of sixteen or the employment of any girl or woman, without regard to age, in any mine below the surface; provisions securing the workmen complete freedom of purchase; provisions requiring the payment of wages at least twice a month in lawful money of the United States, and providing proper rules and regulations to insure the fair and just weighing or measurement of the coal mined by each miner, and such other provisions as he may deem necessary to insure the sale of the production of such leased lands to the United States and to the public at reasonable prices, for the protection of the interests of the United States, for the prevention of monopoly, and for the safeguarding of the public welfare. None of such provisions shall be in conflict with the laws of the states in which the leased property is situated." Section 32 provides: "The Secretary of the Interior is authorized to prescribe necessary and proper rules and regulations and to do any and all things necessary to carry out and accomplish the purposes of this chapter, also to fix and determine the boundary lines of any structure, or oil or gas field, for the purposes of this chapter. Nothing in this chapter shall be construed or held to affect the rights of the States or

although preemption was perhaps appropriate in light of the narrow reservations of local jurisdiction in the Federal Power Act, a similar finding in the present case is unwarranted given the broad savings provisions contained in the Mineral Lands Leasing Act.

The proviso in § 187 provides that "[n]one of Such provisions shall be in conflict with the laws of the states in which the leased property is situated." (30 U.S.C. § 187.) But, as Gulf points out, by the use of the language "such provisions," the proviso relates only to the provisions of the preceding sentence. These provisions relate to employment practices, prevention of undue waste and monopoly, and diligence requirements. There is no mention of land use planning controls. Moreover, the proviso assures only that the Secretary of the Interior shall observe state standards in drafting the lease's terms. It is not a recognition of concurrent state jurisdiction.

Nor is the savings clause in § 189 of any avail. After delegating to the Secretary of the Interior broad authority to prescribe rules and regulations necessary to effect the purposes of the act, the section continues:

> Nothing in this chapter shall be construed or held to affect the rights of the States or other local authority to exercise any rights which they may have, including the right to levy and collect taxes upon improvements, output of mines, or other rights, property, or assets of any lessee of the United States. (30 U.S.C. § 189.)

The proviso preserves to the states only "any rights which they may have." While this is an express recognition of the right of the states to tax activities of the Government's lessee pursuant to its lease . . . and has been relied upon in part to uphold forced pooling and well spacing of federal mineral lessee operations (Texas Oil & Gas Corp. v. Phillips Petroleum Co. (W.D.Okl.1967) 277 F.Supp. 366, 371, aff'd (10th Cir.1969) 406 F.2d 1303, 1304), the proviso cannot give authority to the state which it does not already possess. Although state law may apply where it presents "no significant threat to any identifiable federal policy or interest" (*Texas Oil & Gas Corp. v. Phillips Petroleum Co., supra,* at 371), the states and their subdivisions have no right to apply local regulations impermissibly conflicting with achievement of a congressionally approved use of federal lands and the proviso of §189 does not alter this principle.

Finally, we are reassured in the correctness of our decision by policy considerations implicitly reflected in the structure and operation of the Mineral Lands Leasing Act of 1920 and the National Environmental Policy Act of 1969 (42 U.S.C. §§4321 et seq.). As Ventura recognized in filing its second amended complaint, the National Environmental Protection Act ("NEPA") and the guidelines, regulations, and Executive Orders issued in pursuance of that act,

---

other local authority to exercise any rights which they may have, including the right to levy and collect taxes upon improvements, output of mines, or other rights, property, or assets of any lessee of the United States."

mandate extensive federal consideration and federal-local cooperation concerning the local, environmental impact of federal action under the Mineral Lands Leasing Act. If federal officials fail to comply with these requirements, Ventura has a remedy against those officials.

Our decision does not mean that local interests will be unheard or unprotected. In rejecting a local veto power while simultaneously guarding local concerns under NEPA, local interests can be represented, the integrity of the federal leases and drilling permits reconciling national energy needs and local environmental interests can be protected, and the ultimate lessee will be responsible to a single master rather than conflicting authority.

Although we recognize that federal incursions upon the historic police power of the states are not to be found without good cause . . ., we must affirm because "under the circumstances of this particular case, (the local ordinances) stand as an obstacle to the accomplishment and execution of the full purposes and objectives of Congress." . . . "(W)here those state laws conflict . . . with other legislation passed pursuant to the Property Clause, the law is clear: The state laws must recede." (*Kleppe v. New Mexico, supra,* 426 U.S. at 543, 96 S.Ct. at 2293-2294.)

AFFIRMED.

NOTES

1. Can a state conservation agency condition a well permit issued to a federal oil and gas lessee to deny the lessee his preferred access to the site in order to prevent surface waste and protect the environment? Does *Ventura County* allow a federal oil and gas lessee to avoid all state regulation? In Gulf Oil Corp. v. Wyoming Oil and Gas Conservation Commission, 693 P.2d 227, 84 O.&G.R. 579 (Wyo. 1985), the Wyoming Supreme Court relied on 30 U.S.C.A. §§ 187 and 189 (cited in Ventura County) to find no federal preemption of either the well permit decision or the access decision. Gulf had argued that the Mineral Leasing Act, Forest Service regulations and the National Environmental Policy Act all evinced an intent of Congress to preempt state regulation of federal oil and gas drilling and exploratory activities. The Wyoming Supreme Court disagreed finding an important state interest in protecting its environment and natural resources which would not be preempted without a clearer statement of federal intent to preempt.

2. Can a state conservation agency apply its well spacing and compulsory pooling statutes to a federal lessee? The Tenth Circuit concluded the regulations of the Oklahoma Corporation Commission applied to a federal lessee. Texas Oil & Gas Corp. v. Phillips Petroleum Co., 406 F.2d 1303, 32 O.&G.R. 477 (10th Cir. 1969), *cert. denied,* 396 U.S. 829 (1969).

3. In addition to the Mineral Leasing Act, several state supreme courts tackled the preemption issue as it affected the Mining Law of 1872 applying to hard rock minerals with mixed results. *See e.g.,* Brubaker v. Board of County

Commissioners, El Paso County, 652 P.2d 1050, 75 O.&G.R. 35 (Colo. 1982) (county zoning regulations preempted); State ex rel. Andrus v. Click, 97 Idaho 791, 554 P.2d 969 (1976)(state dredge permit law not preempted); State ex rel. Cox v. Hibbard, 31 Or.App. 269, 570 P.2d 1190 (1977)(State laws not preempted).

———

CALIFORNIA COASTAL COMMISSION v. GRANITE ROCK COMPANY

Supreme Court of the United States, 1987.
480 U.S. 572, 107 S.Ct. 1419, 94 L.Ed.2d 577.

JUSTICE O'CONNOR delivered the opinion of the Court.

This case presents the question whether Forest Service regulations, federal land use statutes and regulations, or the Coastal Zone Management Act (CZMA), 16 U.S.C. § 1451 et seq. (1982 ed. and Supp. III), pre-empt the California Coastal Commission's imposition of a permit requirement on operation of an unpatented mining claim in a national forest.

I

Granite Rock Company is a privately owned firm that mines chemical and pharmaceutical grade white limestone. Under the Mining Act of 1872, 17 Stat. 91, codified, as amended, at 30 U.S.C. § 22 et seq., a private citizen may enter federal lands to explore for mineral deposits. If a person locates a valuable mineral deposit on federal land, and perfects the claim by properly staking it and complying with other statutory requirements, the claimant "shall have the exclusive right of possession and enjoyment of all the surface included within the lines of their locations," 30 U.S.C. § 26, although the United States retains title to the land. The holder of a perfected mining claim may secure a patent to the land by complying with the requirements of the Mining Act and regulations promulgated thereunder, see 43 CFR § 3861.1 et seq. (1986), and, upon issuance of the patent, legal title to the land passes to the patent-holder. Granite Rock holds unpatented mining claims on federally owned lands on and around Mount Pico Blanco in the Big Sur region of Los Padres National Forest.

From 1959 to 1980, Granite Rock removed small samples of limestone from this area for mineral analysis. In 1980, in accordance with federal regulations, see 36 CFR § 228.1 et seq. (1986), Granite Rock submitted to the Forest Service a 5-year plan of operations for the removal of substantial amounts of limestone. The plan discussed the location and appearance of the mining operation, including the size and shape of excavations, the location of all access roads and the storage of any overburden. The Forest Service prepared an Environmental Assessment of the plan. The Assessment recommended modifications of the plan, and the responsible Forest Service Acting District Ranger approved the plan with the recommended modifications in 1981. Shortly after Forest Service approval of the modified plan of operations, Granite Rock began to mine.

Under the California Coastal Act (CCA), Cal.Pub.Res.Code Ann. § 30000 et seq. (West 1986), any person undertaking any development, including mining, in the State's coastal zone must secure a permit from the California Coastal Commission. §§ 30106, 30600. According to the CCA, the Coastal Commission exercises the State's police power and constitutes the State's coastal zone management program for purposes of the federal CZMA, described infra. In 1983 the Coastal Commission instructed Granite Rock to apply for a coastal development permit for any mining undertaken after the date of the Commission's letter.

Granite Rock immediately filed an action in the United States District Court for the Northern District of California seeking to enjoin officials of the Coastal Commission from compelling Granite Rock to comply with the Coastal Commission permit requirement and for declaratory relief under 28 U.S.C. § 2201 (1982 ed., Supp. III). Granite Rock alleged that the Coastal Commission permit requirement was pre-empted by Forest Service regulations, by the Mining Act of 1872, and by the CZMA. Both sides agreed that there were no material facts in dispute. The District Court denied Granite Rock's motion for summary judgment and dismissed the action. 590 F.Supp. 1361 (1984). The Court of Appeals for the Ninth Circuit reversed. 768 F.2d 1077 (1985). The Court of Appeals held that the Coastal Commission permit requirement was pre-empted by the Mining Act of 1872 and Forest Service regulations. The Court of Appeals acknowledged that the statute and regulations do not "go so far as to occupy the field of establishing environmental standards," specifically noting that Forest Service regulations "recognize that a state may enact environmental regulations in addition to those established by federal agencies," and that the Forest Service "will apply [the state standards] in exercising its permit authority." 768 F.2d at 1083. However, the Court of Appeals held that "an independent state permit system to enforce state environmental standards would undermine the Forest Service's own permit authority and thus is pre-empted." Ibid.

. . .

III

Granite Rock does not argue that the Coastal Commission has placed any particular conditions on the issuance of a permit that conflict with federal statutes or regulations. Indeed, the record does not disclose what conditions the Coastal Commission will place on the issuance of a permit. Rather, Granite Rock argues, as it must given the posture of the case, that there is no possible set of conditions the Coastal Commission could place on its permit that would not conflict with federal law–that any state permit requirement is per se pre-empted. The only issue in this case is this purely facial challenge to the Coastal Commission permit requirement.

. . . We agree with Granite Rock that the Property Clause gives Congress plenary power to legislate the use of the federal land on which Granite Rock holds its unpatented mining claim. The question in this case, however, is whether

Congress has enacted legislation respecting this federal land that would preempt any requirement that Granite Rock obtain a California Coastal Commission permit. To answer this question we follow the pre-emption analysis by which the Court has been guided on numerous occasions:

> [S]tate law can be pre-empted in either of two general ways. If Congress evidences an intent to occupy a given field, any state law falling within that field is pre-empted. . . .. If Congress has not entirely displaced state regulation over the matter in question, state law is still pre-empted to the extent it actually conflicts with federal law, that is, when it is impossible to comply with both state and federal law, . . . or where the state law stands as an obstacle to the accomplishment of the full purposes and objectives of Congress, . . .

A

Granite Rock and the Solicitor General as amicus have made basically three arguments in support of a finding that any possible state permit requirement would be pre-empted. First, Granite Rock alleges that the Federal Government's environmental regulation of unpatented mining claims in national forests demonstrates an intent to pre-empt any state regulation. Second, Granite Rock and the Solicitor General assert that indications that state land use planning over unpatented mining claims in national forests is pre-empted should lead to the conclusion that the Coastal Commission permit requirement is pre-empted. Finally, Granite Rock and the Solicitor General assert that the CZMA, by excluding federal lands from its definition of the coastal zone, declared a legislative intent that federal lands be excluded from all state coastal zone regulation. We conclude that these federal statutes and regulations do not, either independently or in combination, justify a facial challenge to the Coastal Commission permit requirement.

Granite Rock concedes that the Mining Act of 1872, as originally passed, expressed no legislative intent on the as yet rarely contemplated subject of environmental regulation. In 1955, however, Congress passed the Multiple Use Mining Act, 69 Stat. 367, 30 U.S.C. § 601 et seq., which provided that the Federal Government would retain and manage the surface resources of subsequently located unpatented mining claims. 30 U.S.C. § 612(b). Congress has delegated to the Secretary of Agriculture the authority to make "rules and regulations" to "regulate [the] occupancy and use" of national forests. 16 U.S.C. § 551. Through this delegation of authority, the Department of Agriculture's Forest Service has promulgated regulations so that "use of the surface of National Forest System lands" by those such as Granite Rock, who have unpatented mining claims authorized by the Mining Act of 1872, "shall be conducted so as to minimize adverse environmental impacts on National Forest System surface resources." 36 CFR §228.1, §228.3(d) (1986). It was pursuant to these regulations that the Forest Service approved the Plan of Operations submitted by Granite Rock. If, as Granite Rock claims, it is the federal intent that Granite Rock conduct its mining unhindered by any state environmental regulation, one would expect to find the expression of this intent in these Forest Service regulations. . . .

Upon examination, however, the Forest Service regulations that Granite Rock alleges pre-empt any state permit requirement not only are devoid of any expression of intent to pre-empt state law, but rather appear to assume that those submitting plans of operations will comply with state laws. The regulations explicitly require all operators within the National Forests to comply with state air quality standards, 36 CFR § 228.8(a) (1986), state water quality standards, §228.8(b), and state standards for the disposal and treatment of solid wastes, §228.8(c). The regulations also provide that, pending final approval of the plan of operations, the Forest Service officer with authority to approve plans of operation "will approve such operations as may be necessary for timely compliance with the requirements of Federal and State laws. . . ." § 228.5(b) (emphasis added). Finally, the final subsection of § 228.8, "[r]equirements for environmental protection," provides:

> (h) Certification or other approval issued by State agencies or other Federal agencies of compliance with laws and regulations relating to mining operations will be accepted as compliance with similar or parallel requirements of these regulations.

It is impossible to divine from these regulations, which expressly contemplate coincident compliance with state law as well as with federal law, an intention to pre-empt all state regulation of unpatented mining claims in national forests. Neither Granite Rock nor the Solicitor General contends that these Forest Service regulations are inconsistent with their authorizing statutes.

Given these Forest Service regulations, it is unsurprising that the Forest Service team that prepared the Environmental Assessment of Granite Rock's plan of operation, as well as the Forest Service officer that approved the plan of operation, expected compliance with state as well as federal law. The Los Padres National Forest Environmental Assessment of the Granite Rock plan stated that "Granite Rock is responsible for obtaining any necessary permits which may be required by the California Coastal Commission." The Decision Notice and Finding of No Significant Impact issued by the Acting District Ranger accepted Granite Rock's plan of operation with modifications, stating:

> The claimant, in exercising his rights granted by the Mining Law of 1872, shall comply with the regulations of the Departments of Agriculture and Interior. The claimant is further responsible for obtaining any necessary permits required by State and/or county laws, regulations and/or ordinance.

B

The second argument proposed by Granite Rock is that federal land management statutes demonstrate a legislative intent to limit States to a purely advisory role in federal land management decisions, and that the Coastal Commission permit requirement is therefore pre-empted as an impermissible state land use regulation.

In 1976 two pieces of legislation were passed that called for the development of federal land use management plans affecting unpatented mining claims in national forests. Under the Federal Land Policy and Management Act (FLPMA), 90 Stat. 2744, 43 U.S.C. § 1701 et seq. (1982 ed. and Supp. III), the Department of Interior's Bureau of Land Management is responsible for managing the mineral resources on federal forest lands; under the National Forest Management Act (NFMA), 90 Stat. 2949, 16 U.S.C. §§ 1600-1614 (1982 ed. and Supp. III), the Forest Service under the Secretary of Agriculture is responsible for the management of the surface impacts of mining on federal forest lands. Granite Rock, as well as the Solicitor General, point to aspects of these statutes indicating a legislative intent to limit States to an advisory role in federal land management decisions. For example, the NFMA directs the Secretary of Agriculture to "develop, maintain, and, as appropriate, revise land and resource management plans for units of the National Forest System, coordinated with the land and resource management planning processes of State and local governments and other Federal agencies," 16 U.S.C. § 1604(a). The FLPMA directs that land use plans developed by the Secretary of the Interior "shall be consistent with State and local plans to the maximum extent [the Secretary] finds consistent with Federal law," and calls for the Secretary, "to the extent he finds practical," to keep apprised of state land use plans, and to "assist in resolving, to the extent practical, inconsistencies between Federal and non-Federal Government plans." 43 U.S.C. § 1712(c)(9).

For purposes of this discussion and without deciding this issue, we may assume that the combination of the NFMA and the FLPMA preempt the extension of state land use plans onto unpatented mining claims in national forest lands. The Coastal Commission asserts that it will use permit conditions to impose environmental regulation. See Cal.Pub.Res.Code Ann. § 30233 (West) (1986) (quality of coastal waters); § 30253(2) (erosion); § 30253(3) (air pollution); § 30240(b) (impact on environmentally sensitive habitat areas).

While the CCA gives land use as well as environmental regulatory authority to the Coastal Commission, the state statute also gives the Coastal Commission the ability to limit the requirements it will place on the permit. The CCA declares that the Coastal Commission will "provide maximum state involvement in federal activities allowable under federal law or regulations. . . ." Cal.Pub.Res.Code Ann. § 30004 (West). Since the state statute does not detail exactly what state standards will and will not apply in connection with various federal activities, the statute must be understood to allow the Coastal Commission to limit the regulations it will impose in those circumstances. In the present case, the Coastal Commission has consistently maintained that it does not seek to prohibit mining of the unpatented claim on national forest land. See 768 F.2d at 1080 ("The Coastal Commission also argues that the Mining Act does not preempt state environmental regulation of federal land unless the regulation prohibits mining altogether. . . ."); 590 F.Supp. at 1373 ("The [Coastal Commission] seeks not to prohibit or 'veto,' but to regulate [Granite Rock's] mining activity in accordance with the detailed requirements of the CCA. . . . There is no reason to find that the [Coastal Commission] will apply the CCA's regulations so as to deprive [Granite Rock] of its rights under the Mining Act. . . ."); Defendants' Memorandum of Points and Authorities in Opposition to Plaintiff's Motion for Summary Judgment 41-42, California Coastal Commission v. Granite Rock Co., No. C-835137 (N.D.Cal.1983) ("Despite Granite Rock's characterization of Coastal Act regulation as a 'veto' or ban of mining, Granite Rock has not applied for any coastal permit, and the State . . . has not indicated that it would in fact ban such activity. . . . [T]he question presented is merely whether the state can regulate uses rather than prohibit them. Put another way, the state is not seeking to determine basic uses of federal land: rather it is seeking to regulate a given mining use so that it is carried out in a more environmentally sensitive and resource-protective fashion.")

The line between environmental regulation and land use planning will not always be bright; for example, one may hypothesize a state environmental regulation so severe that a particular land use would become commercially impracticable. However, the core activity described by each phrase is undoubtedly different. Land use planning in essence chooses particular uses for the land; environmental regulation, at its core, does not mandate particular uses of the land but requires only that, however the land is used, damage to the environment is kept within prescribed limits. Congress has indicated its understanding of land use planning and environmental regulation as distinct activities. As noted above, 43 U.S.C. § 1712(c)(9) requires that the Secretary of Interior's land use plans be consistent with state plans only "to the extent he finds practical." The immediately preceding subsection, however, requires that the Secretary's land use plans "provide for compliance with applicable pollution control laws, including State and Federal air, water, noise, or other pollution standards or implementation plans." §1712(c)(8). Congress has also illustrated its understanding of land use planning and environmental regulation as distinct activities by delegating the authority to regulate these activities to different agencies. The stated purpose of Part 228, subpart A of the Forest Service regulations, 36 CFR § 228.1, is to "set forth rules and procedures" through which mining on unpatented claims in national forests "shall be conducted so as to minimize adverse environmental impacts on National Forest System surface resources." The next sentence of the subsection, however, declares that "[i]t is not the purpose of these regulations to provide for the management of mineral resources; the responsibility for managing such resources is in the Secretary of the Interior." Congress clearly envisioned that although environmental regulation and land use planning may hypothetically overlap in some instances, these two types of activity would in most cases be capable of differentiation. Considering the legislative understanding of environmental regulation and land use planning as distinct activities, it would be anomalous to maintain that Congress intended any state environmental regulation of unpatented mining claims in national forests to be per se pre-empted as an impermissible exercise of state land use planning. Congress' treatment of environmental regulation and land use planning as generally distinguishable calls for this Court to treat them as distinct, until an actual overlap between the two is demonstrated in a particular case.

Granite Rock suggests that the Coastal Commission's true purpose in enforcing a permit requirement is to prohibit Granite Rock's mining entirely. By choosing to seek injunctive and declaratory relief against the permit requirement before discovering what conditions the Coastal Commission would have placed on the permit, Granite Rock has lost the possibility of making this argument in this litigation. Granite Rock's case must stand or fall on the question whether any possible set of conditions attached to the Coastal Commission's permit requirement would be pre-empted. As noted in the previous section, the Forest Service regulations do not indicate a federal intent to pre-empt all state environmental regulation of unpatented mining claims in national forests. Whether or not state land use planning over unpatented mining claims in national forests is pre-empted, the Coastal Commission insists that its permit requirement is an exercise of environmental regulation rather than land use planning. In the present posture of this litigation, the Coastal Commission's identification of a possible set of permit conditions not pre-empted by federal law is sufficient to rebuff Granite Rock's facial challenge to the permit requirement. This analysis is not altered by the fact that the Coastal Commission chooses to impose its environmental regulation by means of a permit requirement. If the Federal Government occupied the field of environmental regulation of unpatented mining claims in national forests–concededly not the case–then state environmental regulation of Granite Rock's mining activity would be pre-empted, whether or not the regulation was implemented through a permit requirement. Conversely, if reasonable state environmental regulation is not pre-empted, then the use of a permit requirement to impose the state regulation does not create a conflict with federal law where none previously existed. The permit requirement itself is not talismanic.

C

Granite Rock's final argument involves the CZMA, 16 U.S.C. § 1451 et seq., through which financial assistance is provided to States for the development of coastal zone management programs. Section 304(a) of the CZMA, 16 U.S.C. § 1453(1), defines the coastal zone of a State, and specifically excludes from the coastal zone "lands the use of which is by law subject solely to the discretion of or which is held in trust by the Federal Government, its officers or agents." The Department of Commerce, which administers the CZMA, has interpreted § 1453(1) to exclude all federally-owned land from the CZMA definition of a state's coastal zone. 15 CFR § 923.33(a) (1986).

Granite Rock argues that the exclusion of "lands the use of which is by law subject solely to the discretion of or which is held in trust by the Federal

Government, its officers or agents" excludes all federally-owned land from the CZMA definition of a State's coastal zone, and demonstrates a congressional intent to pre-empt any possible Coastal Commission permit requirement as applied to the mining of Granite Rock's unpatented claim in the national forest land.

. . .

Because Congress specifically disclaimed any intention to pre-empt pre-existing state authority in the CZMA, we conclude that even if all federal lands are excluded from the CZMA definition of "coastal zone," the CZMA does not automatically pre-empt all state regulation of activities on federal lands.

IV

Granite Rock's challenge to the California Coastal Commission's permit requirement was broad and absolute; our rejection of that challenge is correspondingly narrow. Granite Rock argued that any state permit requirement, whatever its conditions, was per se pre-empted by federal law. To defeat Granite Rock's facial challenge, the Coastal Commission needed merely to identify a possible set of permit conditions not in conflict with federal law. The Coastal Commission alleges that it will use its permit requirement to impose reasonable environmental regulation. Rather than evidencing an intent to pre-empt such state regulation, the Forest Service regulations appear to assume compliance with state laws. Federal land use statutes and regulations, while arguably expressing an intent to pre-empt state land use planning, distinguish environmental regulation from land use planning. Finally, the language and legislative history of the CZMA expressly disclaim an intent to pre-empt state regulation.

Following an examination of the "almost impenetrable maze of arguably relevant legislation," Justice Powell concludes that "[i]n view of the Property Clause . . ., as well as common sense, federal authority must control. . . ." As noted above, the Property Clause gives Congress plenary power over the federal land at issue; however, even within the sphere of the Property Clause, state law is preempted only when it conflicts with the operation or objectives of federal law, or when Congress "evidences an intent to occupy a given field," Silkwood v. Kerr-McGee Corp., 464 U.S., at 248, 104 S.Ct., at 621. The suggestion that traditional pre-emption analysis is inapt in this context can be justified, if at all, only by the assertion that the state regulation in this case would be "duplicative." The description of the regulation as duplicative, of course, is based on the conclusions of the dissent that land use regulation and environmental regulation are indistinguishable, and that any state permit requirement, by virtue of being a permit requirement rather than some other form of regulation, would duplicate federal permit requirements. Because we disagree with these assertions we apply the traditional pre-emption analysis which requires an actual conflict between state and federal law, or a congressional expression of intent to preempt, before we will conclude that state regulation is preempted.

Contrary to the assertion of Justice Powell that the Court today gives States power to impose regulations that "conflict with the views of the Forest Service," we hold only that the barren record of this facial challenge has not demonstrated any conflict. We do not, of course, approve any future application of the Coastal Commission permit requirement that in fact conflicts with federal law. Neither do we take the course of condemning the permit requirement on the basis of as yet unidentifiable conflicts with the federal scheme.

The judgment of the Court of Appeals is reversed and the case is remanded for further proceedings consistent with this opinion.

It is so ordered.

JUSTICE POWELL, with whom JUSTICE STEVENS joins, concurring in part and dissenting in part.

JUSTICE SCALIA, with whom JUSTICE WHITE joins, dissenting.

NOTES

1. What happened to the *Ventura County* decision? Only Justice Scalia in his dissenting opinion referred to *Ventura County* but did not use it to support a conclusion that the Coastal Commission's regulatory program was preempted by the applicable federal laws and regulations.

2. Was it critical to the majority's analysis that the Coastal Commission specifically eschewed a veto power over federal permittees? Did Ventura County claim that the federal permittee could not drill for oil and gas without a county permit, which could be withheld if it would violate the county zoning ordinance?

3. While eschewing a veto power could the Coastal Commission impose "reasonable" environmentally based regulations which would make mining on the federal lands financially infeasible? What forum are federal permittees now required to use if a state or local government imposes difficult conditions? Will they be required to exhaust their state administrative and judicial remedies before they can vindicate their federal rights?

4. In the first significant decision dealing the preemption issue relating to mining activities on public lands, the Tenth Circuit invalidated a county ordinance regulating mining activities on federal unpatented mining claims. South Dakota Mining Ass'n v. Lawrence County, 155 F.3d 1005 (10th Cir. 1998). The federal statutory provisions were the same in the *Lawrence County* case as in *Granite Rock*. Should the court have applied the regulatory/prohibitory analysis of *Granite Rock* ? What if the county ordinance totally prohibited mining activities without a variance mechanism? *See generally*, Kramer & Martin, *The Law of Pooling and Unitization* § 24.04[1]. There has been little judicial activity on the federalism issue in the past 10 years. Is there an explanation for the lack of controversy?

5. How clear a line can be drawn between land use and environmental issues? In the context of federal land managers and federal statutes aren't environmental issues required to be considered? *See generally*, G. Coggins, *Public Natural Resources Law* §4.03[1][d] (2004); Freyfogle, *Granite Rock:*

Institutional Competence And The State Role in Federal Land Use Planning, 59 U. Colo. L.Rev. 475 (1988); Leshy, *Granite Rock and The States' Influence Over Federal Land Use*, 18 Envtl. L. 99 (1987).

6. The Mineral Leasing Act provides in part:

When separate tracts cannot be independently developed and operated in conformity with an established well-spacing or development program, any lease, or a portion thereof, may be pooled with other lands, whether or not owned by the United States, under a communitization or drilling agreement providing for an apportionment of production or royalties among the separate tracts of land comprising the drilling or spacing unit when determined by the Secretary of the Interior to be in the public interest and operations or production pursuant to such an agreement shall be deemed to be operations or production as to each such lease committed thereto.

30 U.S.C.A. § 226(m).

Does this section preempt the application of state conservation laws in the absence of specific consent of the Interior Department after *Granite Rock*? In Kirkpatrick Oil & Gas Co. v. United States, 675 F.2d 1122, 73 O.&G.R. 351 (10th Cir.1982), the Tenth Circuit concluded that without the Interior Secretary's consent to the inclusion of any federal lease into a pooling agreement state powers were preempted. Does that conclusion survive Granite Rock? Does the statute evince an intent to preempt state compulsory pooling laws unless the Secretary consents to their application? *In accord*, Texas Oil & Gas Corp. v. Phillips Petroleum Co., 406 F.2d 1303, 32 O.&G.R. 477 (10th Cir. 1969), *cert. denied*, 396 U.S. 829 (1969); *but cf.*, Currey v. Corporation Commission, 617 P.2d 177, 68 O.&G.R. 274 (Okl. 1979), *cert. denied*, 452 U.S. 938 (1981), *reh. denied*, 453 U.S. 927 (1981) (state order requiring well plugging on Indian leases not preempted). *See generally*, Kramer & Martin, *The Law of Pooling and Unitization* § 24.04.

7. A state statute regulates "real estate brokers." It defines a broker as one who negotiates, offers or advertises the sale, rental or leasing of real estate. A business in the state, operating without a license, provides assistance to those who want to participate in the non-competitive federal leasing program. Can the state licensing agency prevent the business from operating without a license? Is the state agency preempted by the Mineral Leasing Act and the non-competitive bidding regulations from licensing this type of business activity? In Arizona State Real Estate Department v. American Standard Gas & Oil Leasing Service, Inc., 119 Ariz. 183, 580 P.2d 15 (Ariz.App.1978), the court found that the business engaged in activities relating to the leasing of real estate and that the Mineral Leasing Act did not preempt state licensing activities.

DUNN-MCCAMPELL ROYALTY INTEREST INC. v. NATIONAL PARK SERVICE

United States Court of Appeals, 2011
630 F.3d 431

Before JOLLY and DENNIS, CIRCUIT JUDGES and BOYLE, DISTRICT JUDGE

JOLLY, CIRCUIT JUDGE. Before 1963, there was no Padre Island National Park off the coast of the State of Texas. It took a lot of maneuvering between the State of Texas and the United States to create the national park out of these coastal island lands, much belonging to the State of Texas, some belonging to private parties. The Texas Consent Statute, the deeds of conveyance, the federal Enabling Act of 1962, and the Oil and Gas Management Plan of 2001, as well as the Energy Policy Act of 2005, are all involved in this appeal.

Now, almost fifty years later, this appeal presents a conflict between the National Park Service (the "Service") and owners of certain mineral estates in the Padre Island National Seashore (the "Seashore"), with respect to those mineral owners' rights of ingress and egress over the Seashore's surface; such rights, if recognized, would allow the owners to exploit the subsurface minerals contained on the Island. The Service must manage the Seashore to preserve the environment for recreational use while respecting the legal rights of the mineral estate owners to extract oil and natural gas. In 2001, the Service attempted to strike this balance through its Oil and Gas Management Plan (the "Plan"). In this federal action, three related companies (collectively, "Dunn-McCampbell") seek declaratory relief under the Administrative Procedure Act ("APA"), , arguing that the Plan exceeds the Service's regulatory power over the Seashore because it denies Dunn-McCampbell its rights of ingress and egress as provided by the special provisions of state and federal law that established the Seashore. The district court agreed and entered a declaratory judgment in Dunn-McCampbell's favor. The Service now appeals. Although we assume that the Service's normally broad regulatory authority over park lands is limited by the agreements between Texas and the Service that were made when the Seashore was established, we hold that these limitations do not provide the relief Dunn-McCampbell seeks today. We reverse, vacate, and remand.

I.

Padre Island is a narrow barrier island that stretches from Corpus Christi, Texas, nearly to the Mexican border. Long barren and inaccessible, the island began to draw interest from real estate developers after causeways were completed at either end. Developers and the federal government were not the only ones interested in the island. Oil companies had discovered the island's oil and gas resources, and by the time the Seashore was created, there was extensive mineral exploitation on the Island.

Congress authorized the Seashore's creation in 1962. The Enabling Act provides that the Service is to administer the Seashore consistent with the law widely known as the National Park Service Organic Act ("Organic Act"), except as otherwise provided in the Enabling Act. Congress authorized the Service to acquire private property and interests in such property by purchase, condemnation, or otherwise, but provided that it could obtain state lands from Texas only with the state's "concurrence."

Thereafter, on April 4, 1963, Texas's Legislature passed the "Consent Statute," authorizing the federal government to acquire public and private lands within the State "subject to the limitations contained in this Act." TEX. REV. CIV. STAT. art. 6077t § 3. Texas reserved its "entire mineral estate [with] the right of occupation and use of so much of the surface of the land or waters as may be required for all purposes reasonably incident to the mining, development, or removal of the minerals ...." Texas also concurred in the Service's acquisition of private land, "provided that the acquisition of lands in such area shall not deprive the grantor or successor in title the right of ingress and egress for the purpose of exploring for, developing, processing, storing and transporting minerals from beneath said lands and waters with the right of housing employees for such purposes."

The Texas legislature directed the School Land Board to execute a deed incorporating the conditions set forth in the Consent Statute. *Id.* at § 3. The deed by which the State conveyed the State's land expressly provided that such conveyance of State lands was "subject to certain limitations, exceptions, and reservations set forth in the" Consent Statute, which, as we have just noted, addressed the acquisition of private lands as well. The Service, by virtue of this deed, acquired Texas's lands, and the Service separately acquired private lands. The Service acquired only surface, not mineral, estates.

In 1979, the Service implemented nationwide regulations concerning exploitation of mineral rights not owned by the Service within all national parks and seashores. Those regulations are not at issue here, although the Service argued below that the current suit should be barred under res judicata principles, an argument that the district court rejected, and that is not appealed.

The regulations at issue stem from the Service's 2001 Oil and Gas Management Plan (the "Plan"). The Plan designates certain areas of the Seashore as Sensitive Resource Areas (SRAs) that contain "particularly rare and/or vulnerable resources." These areas cover 52.7 percent of the Seashore and carry with them various restrictions. The Plan notes that it "effectively close[s] surface use ... [for] drilling operations" in 7.6 percent of the Seashore. On the other hand, it projects that "all oil and gas would be accessible," although there would likely be "increased costs for operators to design operations to avoid or reduce impacts to SRAs." Further, the Plan notes that these increased costs might discourage resource exploitation.

Dunn-McCampbell brought suit in the Southern District of Texas under the APA, seeking a declaratory judgment that the Plan unlawfully violates the Enabling Act by closing certain areas of the Seashore to oil and gas activities and otherwise impairing Dunn-McCampbell's rights of ingress and egress.

The district court held that the Consent Statute is assimilated into federal law, and thus is binding on the Service; that the Consent Statute protects Dunn-McCampbell's rights of ingress and egress; and that the designation of the Sensitive Resource Areas, and accompanying regulations, deprive Dunn-McCampbell of that right. The district court entered a declaratory judgment, declaring the Plan invalid insofar as it "close[s] certain areas of the [Seashore] or otherwise deprive[s] Dunn-McCampbell's rights of ingress and egress for the purpose of developing their oil and gas interests." The district court, however, also held the Enabling Act does not otherwise protect Dunn-McCampbell, regardless of any provisions in the Energy Policy Act of 2005.

II.

The overarching question presented by the Service's appeal is whether the trial court erred in granting summary judgment to Dunn-McCampbell on the grounds that the Plan, under the APA, 5 U.S.C. § 1, *et seq,* "transgressed the bounds fixed by Congress."

III.

Dunn-McCampbell presents three arguments to support its position that the Plan violates the APA. First, it argues that the Plan is inconsistent with the Congressional grant of power to the Service to promulgate regulations, insofar as the Plan violates Dunn-McCampbell's rights of ingress and egress that are protected in the Consent Statute, and hence in the Enabling Act. Second, Dunn-McCampbell argues that because the Service does not own the mineral estate which lies beneath the park, and because the mineral estate is thus *outside* the Seashore's boundaries, under the terms of the Enabling Act, it has special rights of ingress and egress. Third, and finally, Dunn-McCampbell argues that Congress, speaking through a Sense of Congress provision contained in the Energy Policy Act of 2005, has expressly provided that the Service, under the terms of the Enabling Act, must respect Dunn-McCampbell's rights of ingress and egress.

IV.

It is of course basic that the Constitution affords Congress the power to make laws that apply on federal lands, and it has been accepted that such "power over the public land ... is without limitations." *Kleppe v. New Mexico,* 426 U.S. 529, 540, 96 S.Ct. 2285, 49 L.Ed.2d 34 (1976); *United States v. San Francisco,* 310 U.S. 16, 29, 60 S.Ct. 749, 84 L.Ed. 1050 (1940). Unlike Congress, however, agency power is not so broad; the Service can act only within its statutory authority. We therefore must apply the Enabling Act, which created the Seashore, authorized the Service to acquire land for the Seashore, and set out the Service's regulatory authority. It provides, in pertinent part:

Except as otherwise provided in sections 459d to 459d-7 of this title, the property acquired by the Secretary under such sections shall be administered by the Secretary, subject to the provisions of sections 1 and 2 to 4 of this title, as amended and supplemented, and in accordance with other laws of general application relating to the areas administered and supervised by the Secretary through the National Park Service; except that authority otherwise available to the Secretary for the conservation and management of natural resources may be utilized to the extent he finds such authority will further the purposes of sections 459d to 459d-7 of this title.

The Enabling Act thus provides that the Organic Act, subject to certain exceptions, shall apply to the Seashore. Two of these exceptions are crucial to our analysis, and we discuss each below.

A.

The first relevant exception to the Organic Act contained in the Enabling Act provides that "[a]ny property, or interest therein, owned by the State of Texas or political subdivision thereof may be acquired only with the concurrence of such owner." The Service contends that the Consent Statute does not limit its authority to regulate the Seashore. For the purposes of this case, however, we will assume that the terms of the Consent Statute bind the Service. We further assume that, although the Enabling Act requires the concurrence only of the State of Texas, the Consent Statute's special protections extend to certain private mineral interests referenced in the Statute.

We thus turn to determine whether, after applying these assumptions, Dunn-McCampbell's rights of ingress and egress are, under the Consent Statute, excepted from the Service's regulations. In making this determination, as set out more fully below, we will first examine the plain language of the Consent Statute; only then, if it is necessary, will we consider the legislative history.

Section 3 of the Consent Statute provides, in relevant part, that

> [i]n all conveyances of said park property under Sections 3 and 6 hereof to the United States of America, the Secretary of the Interior shall permit a reservation by the grantor of all oil, gas, and other minerals in such land or waters with the right of occupation and use of so much of the surface of the land or waters as may be required for the purposes of reasonable development of oil, gas, and other minerals ....

Tex. Rev. Civ. Stat. art. 6077t § 3. Section 6 similarly provides that the Service "shall not deprive the grantor or successor in title of the right of ingress and egress for the purpose of exploring for, developing, processing, storing and transporting minerals from beneath said lands and waters with the right of housing employees for such purposes." Tex. Rev. Civ. Stat. art. 6077t § 6.

Dunn-McCampbell suggests that Sections 3 and 6 of the Consent Statute are ambiguous, and the intent of the statute is effectively to recognize Dunn-McCampbell's rights of ingress and egress. The Service argues that the Consent Statute, in express terms, protects only those private mineral owners who conveyed surface land to the Service, or their successors in title; consequently, Dunn-McCampbell is not protected, because Dunn-McCampbell concedes that neither it nor any of its predecessors ever transferred surface land to the Service; that is to say, the mineral estate owned by Dunn-McCampbell had been severed from the surface estate before the surface estate was conveyed to the Service. The key question, then, is whether, the term, "grantor or successor in title," can be construed-or ignored-so as to allow Dunn-McCampbell's rights of ingress and egress to come within coverage of the Consent Statute.

The plain language of the Consent Statute affords Dunn-McCampbell no protection. Dunn-McCampbell had no ownership interest in the surface estate above its land, did not convey or "grant" any land to the Service, and is not a successor of any party that did convey land to the Service; in short, Dunn-McCampbell, unambiguously, is simply not a grantor or a successor in title. Consequently, we must interpret the Consent Statute to exclude Dunn-McCampbell's interest, unless to do so is absurd or causes an absurd result.

Dunn-McCampbell argues that it is absurd to assume that Texas's Consent Statute only extends its protections to the unsevered private mineral estates. Dunn-McCampbell contends that, under this interpretation, the Consent Statute fails to address the majority of the mineral estates within the Seashore because, when the Seashore was created, "[o]wnership of the mineral interests in Padre Island ha[d] been separated largely, if not completely, from the ownership of the surface interests."

I   t is certainly true that we should "avoid any interpretation that would lead to absurd or unreasonable outcome[,]" but we cannot agree that applying the plain language of the Consent Statute leads to an absurd result. Legislative history might well support an argument that some members of the Texas Legislature and some members of Congress intended specially to protect the rights of ingress and egress held by all mineral estate owners, including those of the character of Dunn-McCampbell. We, however, do not turn to legislative history in this case, as the statute is not ambiguous, and applying the literal language does not create an absurd result.

It cannot be labeled absurd that the State of Texas referred only to grantors and successors in title, when severed mineral interest holders were not parties or potential parties to the transactions establishing the park; such interests were not involved in obtaining parklands; and negotiations with the federal government here did not immediately threaten those private interests at the time.

Thus, in sum, we have assumed that Section 459d-1(a) of the Enabling Act, implicitly reflecting the terms of the Consent Statute, provides an exception to the Service's Organic Act authority to regulate mineral interests in national parks. We have concluded, however, that because Dunn-McCampbell is not a grantor or successor in title as provided in Sections 3 and 6, the plain language of the Consent Statute affords Dunn-McCampbell no protection; and we further have concluded that applying the plain language of the Consent Statute is not an absurdity. We have thus rejected Dunn-McCampbell's argument that we should resort to legislative or congressional history for interpretative guidance.

Consequently, we hold that Dunn-McCampbell is not protected by the terms of the Consent Statute.

B.

Having determined that Dunn-McCampbell is not protected under the Texas Consent Statute, we must decide whether Dunn-McCampbell is, as it argues, protected by a second relevant exception to the Organic Act contained in the Enabling Act, which provides that:

> Any acquisition hereunder shall exclude and shall not diminish any right of occupation or use of the surface under grants, leases, or easements existing on April 11, 1961, which are reasonably necessary for the exploration, development, production, storing, processing, or transporting of oil and gas minerals *that are removed from outside the boundaries* of the national seashore and the Secretary may grant additional rights of occupation or use of the surface for the purposes aforesaid upon the terms and under such regulations as may be prescribed by him.

16 U.S.C. § 459d-3(b) (emphasis added).

The Service urges that this exception does not apply because Dunn-McCampbell's mineral estate is *within* the Seashore's boundaries. Dunn-McCampbell counters that because the Service owns only the surface estate, its subsurface mineral estate is outside the property that the Service owns and thus outside the park boundaries of the Seashore, and that this means that the minerals removed from these estates "are removed from outside the park boundaries." Dunn-McCampbell thus argues that its easements under Texas law are protected by the provisions of Section 459d-3(b).

At this point, it is worthwhile for us to refer to the facts underlying this argument. Under the terms of the Texas Consent Statute, the Service was permitted to acquire only the surface estate; it was specifically not permitted to acquire, and did not acquire, the subsurface mineral estates. Therefore, the question raised, for purposes of Dunn-McCampbell's argument, is whether the land beneath the surface estate is within the Seashore's boundaries. The Service, acknowledging that it does not own the mineral estates, contends that precedent from other circuits makes clear that privately owned property can exist within the boundaries of a national park; thus, whether the mineral estates are privately owned does not determine the boundaries of this national park. Dunn-McCampbell disagrees, arguing that because the Service did not acquire the mineral estates, these estates are necessarily outside the park boundaries. Dunn-McCampbell further argues that because the authority relied on by the Service concerns only horizontal, and not vertical boundaries, this authority is irrelevant to the question raised by the mineral estates located beneath the surface estate. Dunn-McCampbell's argument is, in effect, that when the Service acquired the surface estate, the conveyance included no subsurface soil or space, and that all subsurface land, including the mineral estates underlying the Seashore, is therefore "outside the boundaries of the national seashore," within the meaning of the Enabling Act. Dunn-McCampbell offers no authority that would directly support this novel proposition.

Although it is true that Dunn-McCampbell and others own mineral estates beneath the Seashore's surface, the conveyance of mineral rights ownership does not convey the entirety of the subsurface. As the Texas Supreme Court has stated, "[t]he minerals owner is entitled, not to the molecules actually residing below the surface, but to a fair chance to recover the oil and gas ...." *Coastal Oil & Gas Corp. v. Garza Energy Trust,* 268 S.W.3d 1, 15 (Tex.2008). In other words, if there are no minerals beneath the surface of the Seashore, Dunn-McCampbell owns the legal fiction of an estate that is nothing.

Here, there was a conveyance of land to the Service. "[L]and includes the surface of the earth and everything over and under it, including minerals in place ..." *Averyt v. Grande Inc.,* 717 S.W.2d 891, 894 (Tex.1986). In this case, the minerals were not "in place" since they had been severed or were reserved. Although "[t]here is a difference ... between the estate granted and the land described [in that] [l]and is the physical earth in its natural state, while an estate in land is a legal unit of ownership in the physical land[,]" it unbearably strains credulity to suggest that a surface estate, conveyed in a deed describing the land in horizontal terms, only touches a millimeter of the surface, and excludes all other land below the surface. If, as here, the surface estate alone is conveyed, and a mineral reservation is made, the conveyance "vests in the grantee such rights to the use thereof as are usually exercised by owners in fee subject *only* to the right of the grantor to remove the minerals reserved.". As we have noted, the mineral estate owner does not own the "molecules actually residing below the surface." It thus stands to reason that the Service, not Dunn-McCampbell, owns all non-mineral "molecules" of the land, i.e., the mass that undergirds the surface of the National Seashore.

With respect to Dunn-McCampbell's argument that its privately owned property cannot be within the boundaries of a public park, it undoubtedly is true that Dunn-McCampbell privately owns the mineral estate beneath the publicly owned surface of the Park. The ownership of property, however, does not establish a park's boundaries, as has been made clear by at least three circuits, which have held that land that is not owned by the Service can still exist within the boundaries of a national park. Although Dunn-McCampbell is correct that these cases deal with horizontal boundaries, the reasoning of these cases is applicable to defining park boundaries in whatever abstraction of space the question may be presented.

To summarize, Dunn-McCampbell does not own the land below the Seashore's surface, and, even if it did, the subsurface land would still be within the park's boundaries. Texas law establishes that the holder of a mineral estate has the right to exploit minerals, but does not own the subsurface mass. We have also relied on precedent from other circuits to hold that land that is not owned by the National Park Service can nonetheless be within the boundaries of a National Park. We thus hold Dunn-McCampbell's mineral estate is within the Seashore's boundaries, and that Section 459d-3(b) therefore does not provide Dunn-McCampbell with any special right of ingress and egress over the Seashore's surface.

V.

We will now wrap up. We have assumed that the Service, pursuant to 16 U.S.C. § 459d-1(a), is bound by the terms of Texas's concurrence when it deeded its land to the Service. The terms of the concurrence, which are set out in the Consent Statute, provide, in relevant part, that owners of private land who convey land to the Service may preserve their mineral rights and the rights of ingress and egress to exploit their mineral estates, for themselves and for their successors in title. We have held, however, that these provisions do not apply to Dunn-McCampbell, as it is neither a grantor nor a successor in title as referred to in the Consent Statute.

We have also assumed that 16 U.S.C. § 459d-3(b) requires the Service to recognize the rights of ingress and egress possessed at the time of Texas's conveyance by those who remove minerals from outside the Seashore's boundaries. We have held, however, that the mineral estate owned by Dunn-McCampbell, although not owned by the Service, and beneath the surface of the Seashore, is within the Seashore's boundaries.

We thus conclude that because Dunn-McCampbell does not fall under any of the special protections provided in the Enabling Act, the trial court erred in granting Dunn-McCampbell summary judgment, and accordingly we thus reverse and vacate the district court's declaratory judgment and remand for entry of judgment in favor of the Service.

NOTES

1. There are many circumstances where the surface and mineral estate have been severed and one of the owners is the United States. Many of the national forests in the Eastern United States have split estates because they were acquired in the early part of the 20th Century from private owners. The issue of whether and to what extent the private mineral estate owner may be restricted in its use of the federally-owned surface estate is not that clear. *Compare* Duncan Energy Co. v. U.S. Forest Service, 50 F.3d 584, 132 O.&G.R. 144 (8th Cir. 1995), on later appeal, 109 F.3d 497, 136 O.&G.R. 93 (8th Cir. 1997) *with* Minard Run Oil Co. v. U.S. Forest Service, 2009 U.S.Dist. LEXIS 116520 (W.D.Pa., Dec. 15, 2009).

2. Does the Fifth Circuit eviscerate the absolute ownership rule in Texas when it analyzes whether or not the severed mineral interests are inside or outside of the boundaries of the National Seashore?

3. While the United States may have the power of eminent domain to purchase any privately-owned mineral or surface estate, that is not universally the case. As in the principal case, the United States may be legislatively prohibited from condemning privately-owned mineral estates.

D. OFFSHORE OIL AND GAS DEVELOPMENT

The Outer Continental Shelf (OCS) has been a substantial source of hydrocarbons. See generally, 2 *Law of Federal Oil & Gas Leases* ch. 25 (2006) Approximately 24,000 wells have been drilled and over 8000 wells have produced oil or gas. There are over 39 million acres under lease in the Gulf of Mexico and over 4000 offshore production platforms. OCS productions accounts for an increasing share of total production of domestic oil and gas. As defined by the Submerged Lands Act (43 U.S.C.A. § 1301, the OCS consists of submerged lands located seaward and outside of lands beneath navigable waters. The Outer Continental Shelf Lands Act (OCSLA), 43 U.S.C.A. § 1331 et seq. is the statute that governs oil and gas exploration on the OCS. Under the OCSLA the government retains broad supervisory powers over the activities of federal lessees. As the Supreme Court has put it:

> Under the amended OCSLA, the purchase of a lease entitles the purchaser only to priority over other interested parties in submitting for federal approval a plan for exploration, production, or development. Actual submission and approval or disapproval of such plans occurs separately and later. . . . Lease purchasers acquire the right to conduct only limited "preliminary" activities on the OCS-geophysical and other surveys that do not involve seabed penetrations greater than 300 feet and that do not result in any significant environmental impacts.

Secretary of the Interior v. California, 464 U.S. 312, 337-9, 79 O.&G.R. 448 (1984).

The OCSLA and the accompanying regulations (30 C.F.R. Part 250) are extraordinarily complex and lengthy. *See* D. Fant, *An Analysis and Evaluation of Rules and Policies Governing OCS Operations* (ABA Monograph Series 1990) where the author in 220 pages attempts to discuss the rules relating to OCS oil and gas operations. A five-year plan or leasing schedule must be prepared by the Secretary of the Interior. 43 U.S.C.A. § 1344(a). In addition to the five-year plan relating to the size, timing and location of OCS leasing, Congress has through the appropriations process prohibited oil and gas leasing in specified areas. For example, the 1991 Interior Department Appropriations bill banned leasing in areas off of the costs of California, Florida, Washington, Oregon, the New England states and the Mid-Atlantic states. See Yates, *The 101st Congress and Energy Issues: A Legislative Update*, 127 No. 1 Pub. Util. Fort. 35 (1/1/91). In addition, Congress has prohibited drilling in certain offshore areas through the enactment of legislation, sometimes with unintended consequences. See Mobil Exploration and Producing Southeast, Inc. v. United States, *infra*, and its discussion of the Outer Banks Protection Act, 104 Stat. 955 (1990). Recent efforts by the Bush Administration to expand drilling in the OCS areas may or may not be successful in the face of continued state and congressional opposition to offshore drilling in many areas outside of the Gulf of Mexico.

Each lease sale must go through several different steps before completion.

The first step is a call for information whereby the Secretary invites comments regarding areas that may be favorable for oil and gas development. 30 C.F.R. §§ 256.23-256.25 . This formal step is usually preceded by governmental studies and reports about various OCS areas. After the call for information, the Mineral Management Service (MMS) announces an area identification. *Id.*, §§ 256.26-256.28. The third step is the preparation of a draft environmental impact statement. The fourth step, following a 45 day notice and comment period, the publication of the final environmental impact statement. MMS may then prepare a Secretarial Issue Document to assist the Secretary in determining whether to have a sale and the parameters of the sale area. If the Secretary agrees that a sale should take place a proposed notice of sale is issued in the Federal Register. 30 C.F.R. §§ 256.29-256.31. The proposed notice is held open for a comment period and is then followed by a final notice of sale which will list the date on which bids will be open. 30 C.F.R. §§ 256.32 .

The MMS receives the bids submitted by the prospective lessees. The OCSLA was amended in 1976 to give greater flexibility to the Secretary by authorizing a wider range of minimally acceptable bids. Prior to 1976 the predominant bid had been for a royalty of 1/6 with a bonus bid being the determining factor. After 1976 between 20 and 60 percent of the lands made available for leasing can have various combinations of royalty and bonus percentages, along with other revenue sharing methods such as a fixed share of net profits. 43 U.S.C.A. § 1337(a)(5)(B).

MMS utilizes a standard lease form which may be individually modified to deal with special problems. Lease Form MMS 2005. The primary terms may vary from five to ten years depending on the predicted difficulty and expense of exploring for and producing the hydrocarbons. The lease provides for a rental payment that must be made during the primary term and prior to discovery of oil or gas. Various issues have been raised regarding the determination of royalty payments under OCS leases. At a minimum the royalty must reflect the fair market value of the gas. 30 C.F.R. § 206.150 et seq..

OCSLA permits the government to disapprove an exploration plan or development and production plan, to suspend any lease activities or operations and even cancel the lease (with compensation) if certain criteria are met, including criteria related to threatened or probably serious harm to "life (including fish and other aquatic life), to property, to any mineral (in areas leased or not leased), to the national security or defense, or to the marine, coastal, or human environment." 43 U.S.C.A. §§ 1334(a)(2)(A), 1340(c)(1). 1351(h)(1)(C) & (D), 1334(a)(1)(B); 30 C.F.R. §§ 250.34-1(e), 250.12(a)(1)(ii); 2 *Law of Federal Oil & Gas Leases* § 25.06[4] (2000).

OCSLA also presents some federalism problems. The governor of an affected state may submit recommendations to the Secretary of the Interior "regarding the size, timing, or location of a proposed lease sale or with respect to a proposed development and production plan." 43 U.S.C.A. § 1345. The

Secretary must accept these recommendations if he determines they provide for "a reasonable balance between the national interest and the well-being of the citizens of the affected State." The Secretary has the discretion to ignore the recommendations although his decision cannot be arbitrary or capricious. Commonwealth of Massachusetts v. Clark, 594 F.Supp. 1373 (D.Mass.1984).

States may also influence development of OCS oil and gas if they adopt coastal zone management plans under the Coastal Zone Management Act, 16 U.S.C.A. §§ 1451 et seq. A lessee must certify, upon the submission of his exploration, development or production plan, that the plan complies with, and is consistent with the state's coastal zone program. The Secretary of the Interior may not approve the lessee's plan unless the state concurs in the lessee's certification, or the state is presumed to concur because of its failure to act, or the Secretary of Commerce finds that the lessee's plan is consistent with the objectives of CZMA, "or is otherwise necessary in the interest of national security." 16 U.S.C.A. §§ 1456(c)(3)(B). See Secretary of the Interior v. California, 464 U.S. 312, 340 (1984).

Must the federal government's sale of leases be consistent with a state's coastal zone program? In a 5-4 decision the Supreme Court focusing on the legislative history and the segmented nature of OCS lease development found that the actual leasing did not "directly affect" the coastal zone. Secretary of the Interior v. California, 464 U.S. 312, 79 O.&G.R. 448 (1984). See also, Miller, *Offshore Federalism: Evolving Federal-State Relations in Offshore Oil and Gas Development*, 11 Ecology L.Q. 401 (1984).

NEPA also applies to offshore oil and gas leasing. An EIS may be prepared at several stages in the OCS leasing and development process. An EIS is prepared when the Secretary of the Interior prepares his five-year nationwide OCS leasing program. 2 *Law of Federal Oil & Gas Leases* § 25.05[1]; Jones, *The Development of Outer Continental Shelf Energy Resources*, 21 Pub.Land & Resources L.Dig. 36, 99 (1984). As noted above, another EIS is then done after identification of particular areas proposed for leasing. After a lease is issued an EIS may be required prior to approval of an exploration plan, a development plan, a production plan or a revision of those plans. 30 C.F.R. § 250.32(j). The plans normally require the applicant to develop substantial environmental assessments of the proposed activities. Id. at §250.34 (b) (8) (v).

As noted above, Congress often intrudes itself into the decision-making process relating to both off- and on-shore oil and gas development on federal lands. As the next case indicates, it does so at its own peril.

———

MOBIL OIL EXPLORATION & PRODUCING SOUTHEAST, INC. v. U.S.

Supreme Court of the United States, 2000,
530 U.S. 98, 150 O.&G.R. 98

JUSTICE BREYER delivered the opinion of the Court.

Two oil companies, petitioners here, seek restitution of $156 million they paid the Government in return for lease contracts giving them rights to explore for and develop oil off the North Carolina coast. The rights were not absolute, but were conditioned on the companies' obtaining a set of further governmental permissions. The companies claim that the Government repudiated the contracts when it denied them certain elements of the permission-seeking opportunities that the contracts had promised. We agree that the Government broke its promise; it repudiated the contracts; and it must give the companies their money back.

I

A

A description at the outset of the few basic contract law principles applicable to this case will help the reader understand the significance of the complex factual circumstances that follow. "When the United States enters into contract relations, its rights and duties therein are governed generally by the law applicable to contracts between private individuals." . . .

As applied to this case, these principles amount to the following: If the Government said it would break, or did break, an important contractual promise, thereby "substantially impair[ing] the value of the contract[s]" to the companies, ibid., then (unless the companies waived their rights to restitution) the Government must give the companies their money back. And it must do so whether the contracts would, or would not, ultimately have proved financially beneficial to the companies.

B

In 1981, in return for up-front "bonus" payments to the United States of about $158 million (plus annual rental payments), the companies received 10-year renewable lease contracts with the United States. In these contracts, the United States promised the companies, among other things, that they could explore for oil off the North Carolina coast and develop any oil that they found (subject to further royalty payments) provided that the companies received exploration and development permissions in accordance with various statutes and regulations to which the lease contracts were made "subject."

The statutes and regulations, the terms of which in effect were incorporated into the contracts, made clear that obtaining the necessary permissions might not be an easy matter. In particular, the Outer Continental Shelf Lands Act (OCSLA), 67 Stat. 462, as amended, 43 U.S.C. § 1331 et seq. (1994 ed. and Supp. III), and the Coastal Zone Management Act of 1972

(CZMA), 16 U.S.C. § 1451 et seq., specify that leaseholding companies wishing to explore and drill must successfully complete the following four procedures.

First, a company must prepare and obtain Department of the Interior approval for a Plan of Exploration. 43 U.S.C. § 1340(c). Interior must approve a submitted Exploration Plan unless it finds, after "consider[ing] available relevant environmental information," § 1346(d), that the proposed exploration "would probably cause serious harm or damage to life (including fish and other aquatic life), to property, to any mineral . . ., to the national security or defense, or to the marine, coastal, or human environment." § 1334(a)(2)(A)(i). Where approval is warranted, Interior must act quickly–within " thirty days" of the company's submission of a proposed Plan. § 1340(c)(1).

Second, the company must obtain an exploratory well drilling permit. To do so, it must certify (under CZMA) that its Exploration Plan is consistent with the coastal zone management program of each affected State. If a State objects, the certification fails, unless the Secretary of Commerce overrides the State's objection. If Commerce rules against the State, then Interior may grant the permit.

Third, where waste discharge into ocean waters is at issue, the company must obtain a National Pollutant Discharge Elimination System permit from the Environmental Protection Agency. 33 U.S.C. §§ 1311(a), 1342(a). It can obtain this permit only if affected States agree that its Exploration Plan is consistent with the state coastal zone management programs or (as just explained) the Secretary of Commerce overrides the state objections.

Fourth, if exploration is successful, the company must prepare, and obtain Interior approval for, a Development and Production Plan–a Plan that describes the proposed drilling and related environmental safeguards. Again, Interior's approval is conditioned upon certification that the Plan is consistent with state coastal zone management plans–a certification to which States can object, subject to Commerce Department override.

C

The events at issue here concern the first two steps of the process just described–Interior's consideration of a submitted Exploration Plan and the companies' submission of the CZMA "consistency certification" necessary to obtain an exploratory well drilling permit. The relevant circumstances are the following:

1. In 1981, the companies and the Government entered into the lease contracts. The companies paid the Government $158 million in up-front cash "bonus" payments.

2. In 1989, the companies, Interior, and North Carolina entered into a memorandum of understanding. In that memorandum, the companies promised that they would submit an initial draft Exploration Plan to North Carolina

before they submitted their final Exploration Plan to Interior. Interior promised that it would prepare an environmental report on the initial draft. It also agreed to suspend the companies' annual lease payments (about $250,000 per year) while the companies prepared the initial draft and while any state objections to the companies' CZMA consistency certifications were being worked out, with the life of each lease being extended accordingly.

3. In September 1989, the companies submitted their initial draft Exploration Plan to North Carolina. Ten months later, Interior issued the promised ("informal" pre-submission) environmental report, after a review which all parties concede was "extensive and intensive." Interior concluded that the proposed exploration would not "significantly affec[t]" the marine environment or "the quality of the human environment."

4. On August 20, 1990, the companies submitted both their final Exploration Plan and their CZMA "consistency certification" to Interior.

5. Just two days earlier, on August 18, 1990, a new law, the Outer Banks Protection Act (OBPA), § 6003, 104 Stat. 555, had come into effect. That law prohibited the Secretary of the Interior from approving any Exploration Plan or Development and Production Plan or to award any drilling permit until (a) a new OBPA-created Environmental Sciences Review Panel had reported to the Secretary, (b) the Secretary had certified to Congress that he had sufficient information to make these OCSLA-required approval decisions, and (c) Congress had been in session an additional 45 days, but (d) in no event could he issue an approval or permit for the next 13 months (until October 1991). OBPA also required the Secretary, in his certification, to explain and justify in detail any differences between his own certified conclusions and the new Panel's recommendations.

6. About five weeks later, and in light of the new statute, Interior wrote a letter to the Governor of North Carolina with a copy to petitioner Mobil. It said that the final submitted Exploration Plan "is deemed to be approvable in all respects." It added:

> [W]e are required to approve an Exploration Plan unless it is inconsistent with applicable law or because it would result in serious harm to the environment. Because we have found that Mobil's Plan fully complies with the law and will have only negligible effect on the environment, we are not authorized to disapprove the Plan or require its modification.

But, it noted, the new law, the "Outer Banks Protection Act (OBPA) of 1990 . . . prohibits the approval of any Exploration Plan at this time." It concluded, "because we are currently prohibited from approving it, the Plan will remain on file until the requirements of the OBPA are met." In the meantime a "suspension has been granted to all leases offshore the State of North Carolina."

About 18 months later, the Secretary of the Interior, after receiving the new Panel's report, certified to Congress that he had enough information to consider the companies' Exploration Plan. He added, however, that he would not consider the Plan until he received certain further studies that the new Panel had recommended.

7. In November 1990, North Carolina objected to the companies' CZMA consistency certification on the ground that Mobil had not provided sufficient information about possible environmental impact. A month later, the companies asked the Secretary of Commerce to override North Carolina's objection.

8. In 1994, the Secretary of Commerce rejected the companies' override request, relying in large part on the fact that the new Panel had found a lack of adequate information in respect to certain environmental issues.

9. In 1996, Congress repealed OBPA. § 109, 110 Stat. 1321-177.

D

In October 1992, after all but the two last-mentioned events had taken place, petitioners joined a breach-of-contract lawsuit brought in the Court of Federal Claims.

We granted certiorari to review the Federal Circuit's decision.

II

The record makes clear (1) that OCSLA required Interior to approve "within thirty days" a submitted Exploration Plan that satisfies OCSLA's requirements, (2) that Interior told Mobil the companies' submitted Plan met those requirements, (3) that Interior told Mobil it would not approve the companies' submitted Plan for at least 13 months, and likely longer, and (4) that Interior did not approve (or disapprove) the Plan, ever. The Government does not deny that the contracts, made "pursuant to" and "subject to" OCSLA, incorporated OCSLA provisions as promises. The Government further concedes, as it must, that relevant contract law entitles a contracting party to restitution if the other party "substantially" breached a contract or communicated its intent to do so. Yet the Government denies that it must refund the companies' money. This is because, in the Government's view, it did not breach the contracts or communicate its intent to do so; any breach was not "substantial"; and the companies waived their rights to restitution regardless. We shall consider each of these arguments in turn.

A

The Government's "no breach" arguments depend upon the contract provisions that "subject" the contracts to various statutes and regulations. Those provisions state that the contracts are "subject to" (1) OCSLA, (2) "Sections 302 and 303 of the Department of Energy Organization Act," (3) "all regulations issued pursuant to such statutes and in existence upon the effective date of" the contracts, (4) "all regulations issued pursuant to such statutes in the future which provide for the prevention of waste and the conservation" of Outer

Continental Shelf resources, and (5) "all other applicable statutes and regulations." The Government says that these provisions incorporate into the contracts, not only the OCSLA provisions we have mentioned, but also certain other statutory provisions and regulations that, in the Government's view, granted Interior the legal authority to refuse to approve the submitted Exploration Plan, while suspending the leases instead.

. . . Second, the Government refers to 30 CFR § 250.110(b)(4) (1999), formerly codified at 30 CFR § 250.10(b)(4) (1997), a regulation stating that "[t]he Regional Supervisor may . . . direct . . . a suspension of any operation or activity . . . [when the] suspension is necessary for the implementation of the requirements of the National Environmental Policy Act or to conduct an environmental analysis." The Government says that this regulation permitted the Secretary of the Interior to suspend the companies' leases because that suspension was "necessary . . . to conduct an environmental analysis," namely, the analysis demanded by the new statute, OBPA.

The "environmental analysis" referred to, however, is an analysis the need for which was created by OBPA, a later enacted statute. The lease contracts say that they are subject to then-existing regulations and to certain future regulations, those issued pursuant to OCSLA and §§ 302 and 303 of the Department of Energy Organization Act. This explicit reference to future regulations makes it clear that the catchall provision that references "all other applicable . . . regulations," *supra*, at 2433, must include only statutes and regulations already existing at the time of the contract, see 35 Fed.Cl., at 322-323, a conclusion not questioned here by the Government. Hence, these provisions mean that the contracts are not subject to future regulations promulgated under other statutes, such as new statutes like OBPA. Without some such contractual provision limiting the Government's power to impose new and different requirements, the companies would have spent $158 million to buy next to nothing. In any event, the Court of Claims so interpreted the lease; the Federal Circuit did not disagree with that interpretation; nor does the Government here dispute it.

Instead, the Government points out that the regulation in question–the regulation authorizing a governmental suspension in order to conduct "an environmental analysis"–was not itself a future regulation. Rather, a similar regulation existed at the time the parties signed the contracts, 30 CFR §250.12(a)(iv) (1981), and, in any event, it was promulgated under OCSLA, a statute exempted from the contracts' temporal restriction. But that fact, while true, is not sufficient to produce the incorporation of future statutory requirements, which is what the Government needs to prevail. If the pre- existing regulation's words, "an environmental analysis," were to apply to analyses mandated by future statutes, then they would make the companies subject to the same unknown future requirements that the contracts' specific temporal restrictions were intended to avoid. Consequently, whatever the regulation's words might mean in other contexts, we believe the contracts before us must be interpreted as excluding the words "environmental analysis" insofar as those words would incorporate the requirements of future statutes and future regulations excluded by the contracts' provisions. Hence, they would not incorporate into the contracts requirements imposed by a new statute such as OBPA.

Third, the Government refers to OCSLA, 43 U.S.C. § 1334(a)(1), which, after granting Interior rulemaking authority, says that Interior's "regulations . . . shall include . . . provisions . . . for the suspension . . . of any operation . . . pursuant to any lease . . . if there is a threat of serious, irreparable, or immediate harm or damage to life . . ., to property, to any mineral deposits . . ., or to the marine, coastal, or human environment." (Emphasis added.)

The Government points to the OBPA Conference Report, which says that any OBPA-caused delay is "related to . . . environmental protection" and to the need "for the collection and analysis of crucial oceanographic, ecological, and socioeconomic data," to "prevent a public harm." At oral argument, the Government noted that the OBPA mentions "tourism" in North Carolina as a "major industry . . . which is subject to potentially significant/ disruption by offshore oil or gas development." From this, the Government infers that the pre-existing OCSLA provision authorized the suspension in light of a "threat of . . . serious harm" to a "human environment."

The fatal flaw in this argument, however, arises out of the Interior Department's own statement–a statement made when citing OBPA to explain its approval delay. Interior then said that the Exploration Plan " fully complies" with current legal requirements. And the OCSLA statutory provision quoted above was the most pertinent of those current requirements. The Government did not deny the accuracy of Interior's statement, either in its brief filed here or its brief filed in the Court of Appeals. Insofar as the Government means to suggest that the new statute, OBPA, changed the relevant OCSLA standard (or that OBPA language and history somehow constitute findings Interior must incorporate by reference), it must mean that OBPA in effect created a new requirement. For the reasons set out supra, however, any such new requirement would not be incorporated into the contracts.

We conclude, for these reasons, that the Government violated the contracts. Indeed, as Interior pointed out in its letter to North Carolina, the new statute, OBPA, required Interior to impose the contract-violating delay. It therefore made clear to Interior and to the companies that the United States had to violate the contracts' terms and would continue to do so.

Moreover, OBPA changed pre-existing contract-incorporated requirements in several ways. It delayed approval, not only of an Exploration Plan but also of Development and Production Plans; and it delayed the issuance of drilling permits as well. It created a new type of Interior Department environmental review that had not previously existed, conducted by the newly created Environmental Sciences Review Panel; and, by insisting that the

Secretary explain in detail any differences between the Secretary's findings and those of the Panel, it created a kind of presumption in favor of the new Panel's findings.

We do not say that the changes made by the statute were unjustified. We say only that they were changes of a kind that the contracts did not foresee. They were changes in those approval procedures and standards that the contracts had incorporated through cross-reference. The Government has not convinced us that Interior's actions were authorized by any other contractually cross- referenced provision. Hence, in communicating to the companies its intent to follow OBPA, the United States was communicating its intent to violate the contracts.

B

The Government next argues that any violation of the contracts' terms was not significant; hence there was no "substantial" or "material" breach that could have amounted to a "repudiation." In particular, it says that OCSLA's 30-day approval period "does not function as the 'essence' of these agreements." The Court of Claims concluded, however, that timely and fair consideration of a submitted Exploration Plan was a "necessary reciprocal obligation," indeed, that any "contrary interpretation would render the bargain illusory." 35 Fed.Cl., at 327. We agree.

We recognize that the lease contracts gave the companies more than rights to obtain approvals. They also gave the companies rights to explore for, and to develop, oil. But the need to obtain Government approvals so qualified the likely future enjoyment of the exploration and development rights that the contract, in practice, amounted primarily to an opportunity to try to obtain exploration and development rights in accordance with the procedures and under the standards specified in the cross-referenced statutes and regulations. Under these circumstances, if the companies did not at least buy a promise that the Government would not deviate significantly from those procedures and standards, then what did they buy?

The Government's modification of the contract-incorporated processes was not technical or insubstantial. . . .

The upshot is that, under the contracts, the incorporated procedures and standards amounted to a gateway to the companies' enjoyment of all other rights. To significantly narrow that gateway violated material conditions in the contracts. The breach was "substantia[l]," depriving the companies of the benefit of their bargain. And the Government's communication of its intent to commit that breach amounted to a repudiation of the contracts.

C

The Government argues that the companies waived their rights to restitution. [The court rejects the government's claim that the oil companies continued to perform under the terms of the contract amounting to a waiver that the contract had been wrongfully terminated. Eds.].

D

Finally, the Government argues that repudiation could not have hurt the companies. Since the companies could not have met the CZMA consistency requirements, they could not have explored (or ultimately drilled) for oil in any event. Hence, OBPA caused them no damage. As the Government puts it, the companies have already received "such damages as were actually caused by the [Exploration Plan approval] delay," namely, none. This argument, however, misses the basic legal point. The oil companies do not seek damages for breach of contract. They seek restitution of their initial payments. Because the Government repudiated the lease contracts, the law entitles the companies to that restitution whether the contracts would, or would not, ultimately have produced a financial gain or led them to obtain a definite right to explore. If a lottery operator fails to deliver a purchased ticket, the purchaser can get his money back–whether or not he eventually would have won the lottery. And if one party to a contract, whether oil company or ordinary citizen, advances the other party money, principles of restitution normally require the latter, upon repudiation, to refund that money.

III

Contract law expresses no view about the wisdom of OBPA. We have examined only that statute's consistency with the promises that the earlier contracts contained. We find that the oil companies gave the United States $158 million in return for a contractual promise to follow the terms of pre-existing statutes and regulations. The new statute prevented the Government from keeping that promise. The breach "substantially impair[ed] the value of the contract[s]." And therefore the Government must give the companies their money back.

For these reasons, the judgment of the Federal Circuit is reversed. We remand the cases for further proceedings consistent with this opinion.

It is so ordered.

NOTE

Under the Coastal Zone Management Act and the Outer Continental Shelf Lands Act was it likely that the oil companies would have received a permit to drill off of the North Carolina coast? Would that affect the determination of whether there was a breach of contract or merely the amount of damages? In a portion of the opinion that the editors have excised the court noted that after holding the leases for 9 years without submitting any exploration plans, the companies filed their plans two days prior to the effective date of the Outer Banks Protection Act. Is that a coincidence or does it reflect the companies' own decision as to the likelihood that drilling would have been approved?

———

E. INDIAN OIL AND GAS LEASES

There are special problems relating to the leasing of oil and gas that is owned by Indian Tribes or individual members of Indian Tribes. Leasing is complicated by the history of restrictions placed on the alienation and devolution of land held either by tribes or individuals. Further complications arise from the relationship of the Indian tribes and the Federal Government. This issues are fully explored in 2 *Law of Federal Oil and Gas Leases* ch. 26 (2006). The following case is designed to illustrate some of those problems as they relate to the pooling, or in the federal language, communitization, of Indian leases with other privately owned leases that are subject to state drilling and/or spacing regulation.

WOODS PETROLEUM CORP. v. U.S. DEPARTMENT OF THE INTERIOR

United States Court of Appeals, Tenth Circuit (en banc), 1994.
47 F.3d 1032, 129 O.&G.R. 72.

EBEL, CIRCUIT JUDGE. We granted the request for rehearing en banc in this case to clarify the Secretary of the Interior's authority under 25 U.S.C. § 396 to disapprove a proposed oil and gas communization agreement that includes Indian-owned mineral interests. Woods Petroleum Corporation and other oil companies ("Woods Petroleum") commenced this action to challenge the Secretary's order rejecting a proposed agreement to communize Indian and non-Indian mineral interests for oil and gas production in the Anadarko region of Oklahoma. The district court upheld the Secretary's order. A panel of this court reversed and remanded with instructions to reverse the Secretary's order and to approve the proposed communization agreement. The United States Department of the Interior, Tomlinson Properties, Inc., and certain of the Indian lessors petitioned for rehearing en banc on the grounds that our panel decision clashed with earlier Tenth Circuit authority.

Today, we make explicit that the Secretary acts arbitrarily and abuses his discretion under § 396d when he (1) rejects a proposed communization agreement for the sole purpose of causing the expiration of a valid Indian mineral lease and allowing the Indian lessors to enter into a new, more lucrative, lease, and then (2) approves essentially the identical communization agreement, with the new lessee of the Indian lands simply substituted for the old lessee, and permits the Indian lessors to collect royalties retroactively to the date of first production under the original unit plan. Accordingly, we adhere to our panel decision's reversal of the district court order and we remand with instructions to approve the Woods Petroleum communization agreement.

I. Background

Our panel opinion provides a full recitation of the factual background to this action, Woods Petroleum, 18 F.3d at 855-57, but we review the critical events and rulings to frame our analysis. In February 1977, the Indians named in this action leased their undivided interest in 117.5 net mineral acres in Custer County, Oklahoma to National Cooperative Refinery Association ("National"). The three leases granted National an exclusive right to drill for and extract all oil and gas underlying the land for a primary term of five years and "as much longer thereafter as oil and/or gas is produced in paying quantities." The Concho Agency Superintendent of the Bureau of Indian Affairs ("BIA"), an agency in the United States Department of Interior ("Interior"), approved the leases. Next, the BIA approved National's assignment of its interest in the leases to Woods Petroleum.

On May 18, 1979, the Oklahoma Corporation Commission ("Commission") established a 640-acre drilling and spacing unit that included this Indian land. On December 1, 1981, all working interest owners in the unit area executed a communization agreement, naming Woods Petroleum as the unit operator ("Woods Petroleum communization agreement"). Communization permits the development of several contiguous leaseholds as a single unit, so that "operations conducted anywhere within the unit area are deemed to occur on each lease within the communitized area and production anywhere within the unit is deemed to be produced from each tract within the unit." Kenai Oil & Gas, Inc. v. Dep't of the Interior, 671 F.2d 383, 384 (10th Cir. 1982).

Pursuant to the communization agreement–and six weeks prior to the expiration date of the leases–Woods Petroleum commenced drilling the authorized well on a non-Indian tract within the unit on January 5, 1982. On February 17, 1982, also prior to the expiration of the Indian leases, Woods Petroleum submitted the communization agreement to Interior for its approval. An approved communization agreement, in conjunction with the "commence drilling" clause, would extend Woods Petroleum's leases beyond the primary term for as long as there is production in paying quantities.

On April 12, 1982, the Anadarko Area Director of the BIA approved the proposed Woods Petroleum communization agreement over the Indian lessors' objections. The Indian lessors had unsuccessfully argued that communization was not in their economic best interest because the Secretary's disapproval of the agreement would trigger the expiration of the Indians' existing leases and facilitate the execution of more profitable leases.

On September 6, 1983, the Indians filed an administrative appeal to the Assistant Secretary of Interior. Notably, the Indians did not object to any particular provision in the communization agreement. Instead, they simply argued that their mineral interests could easily be released "on a competitive bid basis and with a stroke of a pen a Communization Agreement can be signed by the Area Director joining validly leased Indian Allotments to the spacing Unit."

To evaluate the Indians' appeal, the Deputy Assistant Secretary requested the BIA Area Director to prepare a "best interest assessment." The BIA's "best interest assessment" advised the Assistant Secretary to affirm the Area Director's approval of the Woods Petroleum communization agreement for many reasons. First, the BIA opined that prevailing market conditions could stifle the separate development of the Indian interests under new leases and that the Oklahoma Corporation Commission may not approve independent production, transportation and sale of product. Second, the BIA noted that the Indians would forfeit nearly $400,000 in unit royalties currently held in escrow under the state-approved spacing unit in which production had already occurred. And finally, the BIA presciently anticipated that rejection of the communization agreement could invite protracted litigation.

After receiving the BIA's recommendation, the Assistant Secretary solicited comments from both the Indians and Woods Petroleum. Again, the Indians expressed their desire to be able to release their mineral interests for a bonus exceeding $400,000 and then to "be immediately joined with the 640 acre spacing by a Communization Agreement." Notwithstanding the BIA's recommendation, the Assistant Secretary reversed the BIA Area Director's approval of the Woods Petroleum communization agreement on May 15, 1986. The Assistant Secretary did not challenge the appropriateness of the communization agreement as establishing a proper geological and economic unit. Instead, the Assistant Secretary focused on the underlying Indian leases and concluded that it would be in the Indians' best interest to allow the Woods Petroleum leases to expire so that the Indians could enter into new leases with Defendant- Appellee Tomlinson Properties, Inc. ("Tomlinson"), who was willing to pay the Indians a $400,000 bonus. The Assistant Secretary did not address the BIA's specific rationales in support of communization and thus did not analyze the factors enumerated in the BIA guidelines. Relying exclusively on Tomlinson's bonus offer, the Assistant Secretary disapproved the Woods Petroleum communization agreement and issued an order declaring that the Woods Petroleum leases had expired for failure to drill, produce, or communitize during the primary term.

[After seeking judicial review, a panel of the Tenth Circuit reversed the Secretary's decision denying approval to the communitization agreement. 18 F.3d at 660. Ed's]. We granted the petition for en banc consideration to clarify the Secretary's authority to approve and reject communitization agreements. In so doing, we adhere to the legal principles articulated in our panel decision, Cotton Petroleum [Cotton Petroleum v. U.S. Department of the Interior, 870 F.2d 1515, 105 O.&G.R. 296 (10th Cir. 1989] Kenai, and Cheyenne-Arapaho [Cheyenne-Arapaho Tribes of Oklahoma v. United States, 966 F.2d 583, 119 O.&G.R. 312 (10th Cir. 1992), *cert. denied*, 113 S.Ct. 1642, 1643 (1993)].

II. Discussion

Under the Administrative Procedure Act, we may set aside an agency decision that is "arbitrary, capricious, an abuse of discretion, or otherwise not in accordance with the law." 5 U.S.C. § 706(2)(a). An agency decision may be arbitrary and capricious if it fails to consider important relevant factors. Alternatively, the decision is arbitrary and capricious if there is no "rational connection between the facts found and the choice made."

Congress has empowered the Secretary of the Interior with an important supervisory role over oil and gas communization agreements involving Indian mineral interests. See Indian Mineral Leasing Act of 1938, 25 U.S.C. § 396a et seq. Pursuant to 25 U.S.C. § 396d, the Secretary has the discretion to approve or prescribe a "reasonable cooperative unit."

> All operations under any oil, gas, or other mineral lease issued pursuant to the terms of sections 396a to 396g of this title or any other Act affecting restricted Indian lands shall be subject to the rules and regulations promulgated by the Secretary of the Interior. In the discretion of said Secretary, any lease for oil or gas issued under the provisions of sections 396a to 396g of this title shall be made subject to the terms of any reasonable cooperative unit or other plan approved or prescribed by said Secretary prior or subsequent to the issuance of any such lease which involves the development of oil or gas from land covered by such lease.

Id.; see also 25 C.F.R. §§ 211.21(b) (governing tribal lands) & 212.24(c) (governing allotted lands). Thus, inclusion of Indian mineral interests in a cooperative plan, including a state-ordered spacing unit, requires the Secretary's approval.

To guide the Secretary's evaluation of a proposed communization agreement involving Indian mineral interests, the BIA promulgated guidelines in 1982 that identify three critical factors: (1) whether the long term economic effects of the proposed agreement are in the Indian lessor's best interest; (2) whether the engineering and technical aspects of the agreement adequately protect the Indian lessors; and (3) whether the lessee has complied with the terms of the lease in all respects prior to its expiration date. Cotton Petroleum, 870 F.2d at 1518 (quoting the BIA Guidelines). The first two of these factors require the Secretary to focus on the communization agreement, and the third factor simply asks if the lessee is in default under the existing lease.

In evaluating the Secretary's actions, we must keep in mind that the Secretary and his delegates act as the Indians' fiduciary and thus must represent the Indians' best interests. The power to manage and regulate Indian mineral interests carries with it the duty to act as a trustee for the benefit of the Indian landowners. Yet, as with any trustee-beneficiary relationship, the Secretary's fiduciary duty to the Indians under § 396d is not boundless and cannot be exercised in a manner that exceeds or flouts the authorizing statute and regulations. When the Secretary deviates from firmly established procedures, or exceeds the limits of his fiduciary duty, we have found an abuse of discretion and have reversed the Secretary.

For example, we have consistently admonished the Secretary to analyze all relevant factors and have reversed rulings that either disregarded certain factors or treated one factor as determinative. In Kenai, the lessee alleged that the Secretary's rejection of a communization agreement constituted an abuse of

Cheyenne-Arapaho and Kenai had not been placed in a state-ordered spacing unit at the time the lessees presented the proposed communization agreement to the Secretary for his approval. Here, as in Cotton Petroleum the Oklahoma Corporation Commission had already approved a 640-acre drilling and spacing unit, signifying that the Indian mineral interests were situated within a common source of supply.

Because the Secretary must approve any state-ordered spacing unit that includes Indian mineral interests, 25 C.F.R. § 212.24(c), the Corporation Commission's spacing order does not protect the Indian mineral owners until the Secretary approves the plan. Absent a communization agreement, Indian owners of tracts adjacent to a producing well may have no right to share in the revenues generated by that well, even if oil and gas has been drained from their property. Thus, both here and in Cotton Petroleum—where the only producing wells were off Indian land—the Secretary's rejection of the communization agreement risked forfeiture of the Indian mineral owners' right to share unit revenues generated from the producing wells on non-Indian tracts within the unit and jeopardized the independent development of the Indian tracts. The Secretary sought to avert these adverse ramifications in the instant case and in Cotton Petroleum by promptly approving new communization agreements and awarding the Indian mineral owners royalties retroactive to the date of first production in the unit. In so doing, however, the Secretary merely demonstrated the arbitrariness of his action in disapproving the communization agreement.

III. Conclusion

For the foregoing reasons, we REVERSE the district court's order and REMAND with instructions to reverse the Secretary's order, to reinstate the BIA Area Director's approval of the Woods Petroleum communization agreement, to declare that the Woods Petroleum leases have not expired, and to declare void both the Tomlinson communization agreement and the Tomlinson leases. Finally, we direct the court to conduct an accounting of all funds involved, including bonuses, with distribution and/or return to follow as if the Area Director's approval of the Woods Petroleum communization agreement had been timely adopted and AFFIRMED.

HENRY, CIRCUIT JUDGE, dissenting, with whom SEYMOUR, CHIEF CIRCUIT JUDGE, joins:

Acknowledging the Secretary of the Interior's fiduciary obligations, the majority overturns a decision that resulted in the negotiation of more lucrative leases on behalf of Indian mineral owners. I find the majority's reasoning inconsistent with both the paramount obligation of trust that our government owes to its indigenous peoples and with the great deference that we generally afford to agency decisions. By imposing requirements on the Secretary that we have not required of other agencies, the majority undermines the generally established standard of review for administrative decisions. The majority opinion also does not resolve the conflict in our prior cases regarding the Secretary's evaluation of communization agreements. Therefore, I must respectfully dissent.

In my judgment, in disapproving the Woods communization agreement, the Assistant Secretary made a reasoned decision to which we should defer. I base this conclusion on an examination of the Secretary's fiduciary obligations to Indian mineral owners, our prior decisions regarding the evaluation of communization agreements governing tribal and allotted lands, and the record of the Interior Department proceedings, which reflects extensive efforts by the Assistant Secretary to obtain and review relevant facts and law.

A. The Secretary's Fiduciary Obligations

In its dealings with Indian peoples, the federal government "has charged itself with moral obligations of the highest responsibility and trust." As a result, the relationship between the Indians and the federal government "is marked by peculiar and cardinal distinctions which exist nowhere else" and "resembles that of a ward to his guardian."

In addition to this general trust relationship, there are "other, context-specific trust relationships of varying depth and responsibility" that arise in particular areas of federal government regulation involving Indian people. One scholar has observed that "the more specific the obligation, the higher the duty of care."

Accordingly, in deciding whether to approve or disapprove communization agreements governing allotted lands such as those at issue here, the Secretary must act as a fiduciary for Indian mineral owners, guided by the same exacting standards that we outlined in Jicarilla [Jicarilla Apache Tribe v. Supron Energy Corp., 728 F.2d 1555, 80 O.&G.R. 352 (10th Cir. 1984) Ed's]. In exercising this duty to the Indian owners, "[n]ot honesty alone, but the punctilio of an honor the most sensitive, is then the standard of behavior." That fiduciary duty of utter loyalty, combined with our narrow standard of review, should guide our analysis of the Assistant Secretary's decision to disapprove the Woods communization agreement.

B. Kenai, Cheyenne-Arapaho, and Cotton Petroleum

We have discussed the Secretary's obligation in evaluating communization agreements affecting leases of tribal and allotted lands in several prior cases: Cheyenne-Arapaho Tribes v. United States, 966 F.2d 583 (10th Cir. 1992), cert. denied, 113 S.Ct. 1642, 123 L.Ed.2d 265 and 113 S.Ct. 1643, 123 L.Ed.2d 265 (1993); Cotton Petroleum Corp. v. Department of the Interior, 870 F.2d 1515 (10th Cir. 1989); and Kenai, *supra*, 671 F.2d 383. The majority reads those decisions as establishing a uniform body of law that provides that, in assessing a communization agreement, the Secretary must consider all the relevant factors and may not treat one factor as determinative. I agree that we have consistently directed the Secretary to consider all the relevant factors. However, I believe that our cases are in conflict as to whether particular factors may be determinative. An examination of our prior decisions illustrates this conflict.

NOTES

1. Did the majority opinion resolve the differences between Kenai/Cotton Petroleum and Cheyenne-Arapaho? Are there inconsistencies in the results that follow from applying the statutory or regulatory guidelines? Does the majority create a priority or hierarchical approach? Does it minimize the trust relationship that exists between the Federal Government and the Indian tribes? See generally, 1 Kramer & Martin, *The Law of Pooling and Unitization* § 16.06 for a discussion of these cases.

2. Must the Secretary approve all communitization plans that are technologically sound even if they would cause an injury to the interests of the Indian tribe or individual? In *Cheyenne/Arapaho*, the lessee argued that the "quintessential component of communitization of leases is whether the proposed plan is reasonable and appropriate for the purpose of proper development and conservation of natural mineral resources." 966 F.2d at 590. The court there went on to say that the "[lessee's] argument completely ignores the fiduciary relationship between the Secretary and the Tribe." Does the Secretary act arbitrarily and capriciously when he ignores the fiduciary obligation to the Indian tribes? Does the Secretary act arbitrarily and capriciously when he ignores the reasonableness and appropriateness of the technical grounds for communitizing?

3. There is an elaborate administrative procedure for dealing with issues relating to Indian oil and gas leases. Must one exhaust those administrative remedies before seeking judicial relief? Coosewoon v. Meridian Oil Co., 25 F.3d 920, 129 O.&G.R. 490 (10th Cir. 1994) required an Indian lessor to utilize those administrative procedures prior to bringing a civil action for terminating the lease due to non-payment of royalties.

4. There are two principal statutes relating to the leasing of tribal oil and gas rights. The 1938 Omnibus Leasing Act, 25 U.S.C. §§ 396a-g, principally applies to the leasing of unallotted tribal lands held in trust inside or outside of reservations. The second statute is the Indian Mineral Development Act of 1982, 25 U.S.C. §§ 2101-08. See generally, 2 *Law of Federal Oil and Gas Leases*, ch. 26 (2006).

———

SECTION 2. STATE LANDS

In number of the major producing states substantial mineral acreage is owned by the state or a political subdivision of the state. It is not feasible in this casebook to discuss the leasing procedures adopted for publicly owned lands. In general, state leases are similar to the leases of federal lands, or of privately owned mineral estates, although in some instances, by state statute or regulation, some special lease provisions are mandatory. In addition, lawyers need to be aware that even within a single state there may be more than one agency involved in the leasing of state lands. Careful attention to the details of applicable statutes and regulations is required. It must suffice here merely to list some of the secondary authority on leasing procedures and problems in the various states.

A detailed analysis of the oil and gas leasing laws of the Rocky Mountain states (Arizona, Colorado, Idaho, Montana, Nebraska, New Mexico, North Dakota, South Dakota, Utah and Wyoming) is reported in 7 Rocky Mountin Mineral Law Institute 353-476 (1962). California state lease problems are discussed in Krueger, *State Tidelands Leasing in California*, 5 U.C.L.A. L.Rev. 427 (1958). See also, Madden, *Leasing of Public Lands in Louisiana for Oil, Gas and Mineral Development*, Seventh Annual Tulane Tidelands Institute 47 (1963) and O'Keefe, *Exploration and Leasing of State Land for Oil and Gas*, 7 S.D.L.Rev. 142 (1962). The Florida, Maryland and Montana state land leasing statutes are discussed in Anderson, *Recent State Legislation Affecting Oil and Gas Law*, 41 Inst. on Oil & Gas L. & Taxn 2-1, 2-11-15 (1990).

Texas has a rather unique system for the leasing of some of its minerals. Under the Texas Relinquishment Act. Tex. Nat. Res. Code Ann. § 52.171 et seq. (Vernon 1978 & 2000 Supp.) some seven million acres of state owned oil and gas are leased by the surface owner. The surface owner shares in the lease benefits along with the state. This private/public partnership has spawned substantial litigation and criticism since it was instituted in 1919. See generally, Walker, *The Texas Relinquishment Act*, 1 Inst. on Oil & Gas L. & Taxn 245 (1949); Whitworth, *Leasing and Operating State-Owned Lands for Oil and Gas Development*, 16 Tex. Tech L.Rev. 673 (1985). In addition to Relinquishment Act lands, the General Land Office leases other state lands through a more typical governmental leasing procedure. A model state lease is used in Texas which contains the usual private lease conditions. Typically, however, the lease will not contain a pooling clause although the General Land Office will approve a pooling of interests if it determines that it is in the best interest of the state.

———

APPENDICES

The forms reproduced hereinafter are typical of those in current use or reflected in the cases. The student should remember that the reports are full of cases in which difficulty has been created by blind use of form books by draftsmen without an understanding of the significance of the language used. He or she should carefully analyze any form used as a model, clause by clause. Because of ambiguities inherent therein or the possibility of mistake by the parties, certain of the forms which follow are not recommended for use. In analyzing these forms, in the light of the material studied, the student should search out such defects.

Appendix 1

OIL AND GAS LEASE: AN "UNLESS" FORM

Producers 88—Rocky Mountain

OIL AND GAS LEASE

This Agreement, made and entered into this _____ day of _____, 20__ by and between _____, of _____ herein called lessor (whether one or more) and _____, hereinafter called lessee:

Witnesseth: That the lessor, for and in consideration of $_____ cash in hand paid, receipt of which is hereby acknowledged, and of the covenants and agreements hereinafter contained on the part of the lessee to be paid, kept and performed, has granted, demised, leased and let and by these presents does grant, demise, lease and let exclusively unto said lessee, with the exclusive right of mining, exploring by geophysical and other methods and operating for and producing therefrom oil and all gas of whatsoever nature or kind, and laying pipe lines, telephone and telegraph lines, housing and boarding employees, building tanks, power stations, gasoline plants, ponds, roadways, and structures thereon to produce, save, market and take care of said products and the exclusive surface and sub-surface rights and privileges related in any manner to any and all such operations and any and all other rights and privileges necessary, incident to, or convenient for the economical operation alone or conjointly with neighboring land for such purposes, all that certain tract or tracts of land situated in the County of _____, State of _____, described as follows, to wit:

_____ _____

of section _____ Township _____ Range _____ and containing _____ acres, more or less.

It is agreed that this lease shall remain in force for a term of ten years from date and as long thereafter as oil, or gas of whatsoever nature or kind, or either of them is produced from said land or drilling operations are continued as hereinafter provided. If, at the expiration of the primary term of this lease, oil or gas is not being produced on or from said land, but lessee is then engaged in drilling or reworking operations thereon, then this lease shall continue in force so long thereafter as drilling or reworking operations are being continuously prosecuted on said land or on a drilling or development or operating unit which includes all or a part of said land; and drilling or reworking operations shall be considered to be continuously prosecuted if not more than sixty days shall elapse between the completion or abandonment of one well and the beginning of operations for the drilling or reworking of another well. If oil or gas shall be discovered and/or produced from any such well or wells drilled, being drilled or reworked at or after the expiration of the primary term of this lease, this lease shall continue in force so long thereafter as oil or gas is produced from the leased premises or from any such unit which includes all or a part of said lands.

In consideration of the premises the said lessee covenants and agrees:

1st. To deliver to the credit of lessor, free of cost in the pipeline to which lessee may connect his wells, the equal one-eighth part of all oil produced and saved from the leased premises, or at the lessee's option, may pay to the lessor for such one-eighth royalty, the market price for oil of like grade and gravity prevailing on the day such oil is run into the pipe line or into storage tanks.

2nd. To pay lessor for gas of whatsoever nature or kind produced and sold, or used off the premises, or used in the manufacture of any products therefrom, one-eighth, at the market price at the well for the gas sold, used off the premises, or in the manufacture of products therefrom, said payments to be made monthly.

3rd. Lessor shall have fuel gas free of cost from any well producing fuel gas, for all stoves and all inside lights in the principal dwelling house on said land by making his own connections with the wells at his own risk and expense.

If no well be commenced on said land on or before one year from the date hereof, this lease shall terminate as to both parties, unless the lessee on or before that date shall pay or tender to the lessor or to the lessor's credit in the _____ Bank at _____, or its successors, which shall continue as the depository for rental regardless of changes in the ownership of said land, the sum of _____ Dollars, ($_____) which shall operate as a rental and cover the privilege of deferring the commencement of a well for twelve months from said date. In like manner and upon like payments or tenders the commencement of a well may be further deferred for like periods of the same number of months successively. All payments or tenders may be made by check or draft of lessee or any assignee thereof, mailed or delivered on or before the rental paying date. It is understood and agreed that the consideration first recited herein, the down payment, covers not only the privilege granted to the date when said first rental is payable as aforesaid, but also the lessee's right of extending that period as aforesaid, and any and all other rights conferred. Should the depository bank hereafter close without a successor, lessee or his assigns may deposit rental or royalties in any National bank located in the same county with first named bank, due notice of such deposit to be mailed to lessor at last known address. If the first well drilled on said land is dry and a second well is not commenced thereon within twelve months from the expiration of the last rental period for which rental has been paid, this lease shall terminate as to both parties unless the lessee on or before the expiration of said twelve months shall resume the payment of rentals in the same amount and manner as above provided and following such resumption of rental payments the lease shall continue in force as though there had been no interruption in such payments by said drilling.

Lessee may at any time release this lease as to part or all of the lands above described, after which all payments and liabilities thereafter to accrue, as to the lands released, shall cease and determine. In the event of a partial release, the annual delay rental above mentioned shall be reduced proportionately.

No part of the surface of the leased premises shall, without the written consent of the lessee, be let, granted, or licensed by the lessor to any other party for the erection, construction, location or maintenance of structures, tanks, pits, reservoirs, equipment or machinery to be used for the purpose of exploring, developing, or operating adjacent lands for oil or gas.

Lessee shall have the right to use, free of cost, gas, oil and water produced on said land for its operation thereon, except water from ditches, ponds, reservoirs, or wells of lessor.

When requested by the lessor, lessee shall bury its pipe lines on cultivated portions below plow depth.

No well shall be drilled nearer than 200 feet to the house or barn now on said premises, without the written consent of the lessor.

Lessee shall pay for damages caused by his operation to growing crops on said lands.

Lessee shall have the right at any time to remove all machinery and fixtures placed on said premises, including the right to draw and remove casing.

If the estate of either party hereto is assigned, and the privilege of assigning in whole or in part is expressly allowed, the covenants hereof shall extend to their heirs, executors, administrators, successors or assigns, but no change in the ownership of the land or assignment of rentals or royalties shall be binding on the lessee until after the lessee has been furnished with certified copies of muniments of title deraigning title from lessor; and it is hereby agreed in the event this lease shall be assigned as to a part or parts of the above described lands and the assignee or assignees of such part or parts shall fail or make default in the payment of the proportionate part of the rents due from him or them, such default shall not operate to defeat or affect this lease insofar as it covers a part or parts of said lands as to which the said lessee or any assignee thereof shall make due payment of said rental.

Lessor hereby warrants and agrees to defend the title to the lands herein described, and agrees that the lessee shall have the right at any time to pay for lessor, any mortgage, taxes or other liens on the above described lands in the event of default of payment by lessor, and be subrogated to the rights of the holder thereof, and lessor hereby agrees that any such payments made by the lessee for the lessor may be deducted from any amounts of money which may become due the lessor under the terms of this lease.

If said lessor owns a less interest in the above described land than the entire and undivided fee simple estate therein, then the royalties and rentals herein provided shall be paid the lessor only in proportion which his interest bears to the whole and undivided fee. Any interest in the production from the lands herein described to which the interest of lessor may be subject shall be deducted from the royalty herein reserved.

All of lessee's obligations and covenants hereunder, whether express or implied, shall be suspended at the time or from time to time as compliance with any thereof is prevented or hindered by or is in conflict with Federal, State, County, or municipal laws, rules, regulations or Executive Orders asserted as official by or under public authority claiming jurisdiction, or Act of God, adverse field, weather, or market conditions, inability to obtain materials in the open market or transportation thereof, war, strikes, lockouts, riots, or other conditions or circumstances not wholly controlled by lessee, and this lease shall not be terminated in whole or in part, nor lessee held liable in damages for failure to comply with any such obligations or covenants if compliance therewith is prevented or hindered by or is in conflict with any of the foregoing eventualities. The time during which lessee shall be prevented from conducting drilling or reworking operations during the primary term of this lease, under the contingencies above stated, shall be added to the primary term of the lease; provided, however, that delay rentals as herein provided shall not be suspended by reason of the suspension of operations and if this lease is extended beyond the primary term above stated by reason of such suspension, lessee shall pay an annual delay rental on the anniversary dates hereof in the manner and in the amount above provided. Lessor agrees that lessee or its assigns may include said land or any part thereof in any unit plan of development or operations which is approved by the Secretary of the Interior or to which lessee may voluntarily subscribe, and lessor agrees to execute any such unit plan in order to make it effective as to the interests covered by this lease. In such event, royalty will be paid to lessor at the rate set forth above, as to the land covered hereby and included in such unit, based upon the production allocated pursuant to the unit plan to said land; and the drilling or completion or continued operation of a well on any portion of the area included within such a plan shall be construed and considered as the drilling or completion or continued operation of a well under the terms of this lease as to all of the land covered by the lease.

Should any person, firm or corporation having an interest in the above described land not leased to lessee, or should any one or more of the parties named above as lessors not execute this lease, it shall nevertheless be binding upon the party or parties executing the same.

The undersigned lessors for themselves and their heirs, successors, and assigns, hereby expressly release and waive all rights under and by virtue of the homestead exemption laws of said state, insofar as the same may in any way affect the purposes for which this lease is made as recited herein.

In Witness Whereof, the undersigned execute this instrument as of the day and year first above written.

[Blanks for signatures of witnesses and parties and for acknowledgments are omitted here and in the examples following.]

Appendix 2

OIL AND GAS LEASE: AN "OR" FORM

Oil Age Form 86–C—Revised

OIL AND GAS LEASE

This Agreement, made and entered into this _____ day of _____, 20__, by and between _____, party of the first part, herein styled "Lessor," and _____, party of the second part, herein styled "Lessee."

Witnesseth: That for and in consideration of _____ Dollars lawful money of the United States of America, to the Lessor paid, and of other valuable considerations, the receipt of all of which is hereby acknowledged, and in consideration of the covenants and agreements hereinafter contained by the Lessee to be kept and performed, the Lessor has granted, leased, let and demised, and by these presents does grant, lease, let and demise unto the Lessee, its grantees, successors and assigns, the land and premises hereinafter described, with the sole and exclusive right to the Lessee to drill for, produce, extract, take and remove oil, gas, asphaltum and other hydrocarbons (and water without cost for its operations) from, and to store the same upon, said land during the term hereinafter provided, with the right of entry thereon at all times for said purposes, and to construct, use, maintain, erect, repair and replace thereon and to remove therefrom all pipe lines, telephone and telegraph lines, tanks, machinery, buildings and other structures which the Lessee may desire in carrying on its business and operations on said land, or adjoining or neighboring premises operated by Lessee, with the further right to the Lessee or any of its subsidiaries to erect, maintain, operate and remove a plant with all necessary appurtenances, for the extraction of gasoline from gas produced from said land and/or other premises in the vicinity of said land, including all rights necessary or convenient thereto, together with rights-of-way for passage over, upon and across, and ingress and egress to and from, said land, for any or all of the above mentioned purposes. The possession by the Lessee of said land shall be sole and exclusive, excepting only that the Lessor reserves the right to occupy said land or to lease the same for agricultural, horticultural, or grazing uses, which uses shall be carried on subject to, and with no interference with, the rights or operations of the Lessee hereunder. The land which is the subject of this lease is situated in the County of _____, State of California, and is described as follows, to-wit:

_____

_____

_____

and contains _____ acres, more or less.

To Have and to Hold the same for a term of _____ years from and after the date hereof and so long thereafter as oil or gas, or casinghead gas, or other hydrocarbon substances, or either or any of them, is produced therefrom.

In consideration of the premises it is hereby mutually agreed as follows:

1. Lessee shall pay Lessor as royalty on oil the equal _____ part of the proceeds of all oil produced, saved and sold from the leased premises, after making the customary deductions for temperature, water and b.s. at the posted available market price in the district in which the premises are located for oil of like gravity the day the oil is run into purchaser's pipe line or storage tank, and settlement shall be made by Lessee on or before the 25th day of each month for accrued royalties for the preceding calendar month. At Lessor's option exercised not oftener than once in any one calendar year upon _____ (_____) days' previous written notice, Lessee shall deliver into Lessor's tanks on the leased premises, or at mouth of well to pipe line designated by Lessor free of cost, Lessor's royalty oil, provided that Lessee may at any time purchase and take Lessor's royalty oil at said posted available market price. No royalty shall be due to the Lessor for or on account of oil lost through evaporation, leakage or otherwise prior to the marketing of the same or delivery to Lessor if royalty oil is being taken in kind.

2. For all gas produced, saved and sold from said land by Lessee, the Lessee shall pay as royalty the _____ part of the net proceeds from the sale of such gas, but nothing herein contained shall be deemed to obligate the Lessee to produce, save, sell or otherwise dispose of gas from said land. For the purpose of having gasoline extracted from gas produced from said land, the Lessee may transport, or cause to be transported, to a gasoline extraction plant located either on said land or on other lands, all or any portion of such gas where it may be commingled with gas from other properties. Lessee shall meter such gas so transported and such meter readings, together with the results of content tests by recognized methods made at approximately regular intervals, at least once every month, shall furnish the basis for computation of the amounts of gasoline and residue gas to be credited to this lease. Gas used or consumed, or lost in the operations of any such plant, shall be free of charge, and Lessee shall not be held accountable to the Lessor for the same or for any royalty thereon. Lessee shall not be required to pay royalty for or on account of any gas used for repressuring any oil-bearing formation which is being produced from a well or wells on the leased premises, even though such repressuring is done by injecting such gas into wells not situated on the leased premises. The Lessor shall be entitled to gas free

of charge from any gas wells on the leased premises for all stoves and inside lights in the principal dwelling houses on said land by making his own connections at a point designated by Lessee, the taking and use of said gas to be at the Lessor's sole risk and expense at all times.

3. Any casinghead gasoline extracted from gas produced from said land shall, at the option of the Lessee, be returned to the oil produced therefrom and shall be treated as a part thereof; otherwise the Lessee shall pay to the Lessor as royalty for such extracted gasoline the equal _____ part of the net proceeds of the sale thereof after deducting transportation and extraction costs, or of the Lessee's portion thereof if extracted on a royalty basis. If there shall be no available market and/or no public or open market price for the gasoline at the place of extraction, then the Lessee shall be entitled to sell and/or dispose of all the gasoline for the best price and on the best terms obtainable, but in no case shall settlement of royalty be at a less price than that obtained by the Lessee for its portion of the gasoline.

4. The Lessee shall not be required to account to the Lessor for, or pay royalty on, oil, gas or water produced by the Lessee from said land and used by it in its operations hereunder, but it may use such oil, gas and water free of charge.

5. Commencing with the _____ of the term hereof, if the Lessee has not theretofore commenced drilling operations on said land or terminated this lease as herein provided, the Lessee shall pay or tender to the Lessor _____ in advance, as rental, the sum of _____ Dollars per acre per _____ for so much of said land as may then still be held under this lease, until drilling operations are commenced or this lease terminated as herein provided.

6. The Lessee agrees to commence drilling operations on said land within _____ from the date hereof (unless the Lessee has sooner commenced the drilling of an offset well on said land as herein provided) and to prosecute the same with reasonable diligence until oil or gas is found in paying quantities, or to a depth at which further drilling would, in the judgment of the Lessee, be unprofitable; or it may at any time within said period terminate this lease and surrender said land as hereinafter provided. No implied covenant shall be read into this lease requiring the Lessee to drill or to continue drilling on said land, or fixing the measure of diligence therefor. The Lessee may elect not to commence or prosecute the drilling of a well on said land as above provided, and thereupon this lease shall terminate.

7. If the Lessee shall elect to drill on said land, as aforesaid, and oil or gas shall not be obtained in paying quantities in first well drilled, the Lessee shall, within _____ (_____) months after the completion or abandonment of the

first well, commence on said land drilling operations for a second well, and shall prosecute the same with reasonable diligence until oil or gas is found in paying quantities, or until the well is drilled to a depth at which further drilling would, in the judgment of the Lessee, be unprofitable; and the Lessee shall in like manner continue its operations until oil or gas in paying quantities is found, but subject always to the terms and conditions hereof and with the rights and privileges to the Lessee herein given.

8. If oil or gas is found in paying quantities in any well so drilled by the Lessee on said land, the Lessee, subject to the provisions hereof and to the suspension privileges hereinafter set forth, shall continue to drill additional wells on said land as rapidly as one string of tools working with reasonable diligence can complete the same, until there shall have been completed on said land as many wells as shall equal the total acreage then held under this lease divided by _____; whereupon the Lessee shall hold all of the land free of further drilling obligations; provided, that the Lessee may defer the commencement of drilling operations for the second or any subsequent well for a period not to exceed _____ months from the date of completion of the well last preceding it. Except as herein otherwise provided, it is agreed that the Lessee shall drill such wells and operate each completed oil well with reasonable diligence and in accordance with good oil field practice so long as such wells shall produce oil in paying quantities while this lease is in force as to the portion of said land on which such well or wells are situated; but in conformity with any reasonable conservation or curtailment program affecting the drilling of wells or the production of all oil and/or gas from said land, which the Lessee may either voluntarily or by order of any authorized governmental agency subscribe to or be subject to. Drilling and producing operations hereunder may also be suspended while the price offered generally to producers in the same vicinity for oil of the quality produced from said land is _____ (_____) cents or less per barrel at the well, or when there is no available market for the same at the well.

9. If the Lessee shall complete a well or wells on said land which shall fail to produce oil in paying quantities but which produces gas in paying quantities, the Lessee shall either sell so much of said gas as it may be able to find a market for, and pay the Lessor the royalty provided herein on the volume of gas so sold, or Lessee may, if it so elects, suspend the operation of such gas well or wells from time to time and during the period of such suspension pay or tender to the Lessor as rental _____ in advance, a sum equal to _____ per acre for so much of the acreage then held under this lease, such rental to continue until producing operations are resumed and royalties are paid to the Lessor for gas sold as above provided. It is further understood and agreed that if the Lessee shall complete a well which shall fail to produce oil in paying quantities, but which produces gas in paying quantities, it shall not be obliged to conduct any further

drilling operations on said land (except the drilling of offset wells as hereinafter provided) unless and until, in its judgment, the drilling of such additional wells under the provision of this lease is warranted in view of existing or anticipated market requirements.

10. If it should hereafter appear that the Lessor at the time of making this lease owns a less interest in the leased land than the fee simple estate or the entire interest in the oil and gas under said land, then the rentals and royalties accruing hereunder shall be paid to the Lessor in the proportion which his interest bears to the entire fee simple estate or to the entire estate in said oil and gas.

11. There is hereby expressly reserved to the Lessor, and as well to the Lessee, the right and privilege to convey, transfer or assign in whole or in part its interest in this lease or in the leased premises or in the oil and/or gas therein or produced therefrom, but if the Lessor shall sell or transfer any part or parts of the leased premises or any interest in the oil and/or gas under any part or parts thereof the Lessee's drilling obligations shall not thereby be altered, increased or enlarged, but the Lessee may continue to operate the leased premises and pay and settle rents and royalties as an entirety.

12. In the event a well is drilled on adjoining property within _____ (_____) feet of the exterior limits of any land at the time embraced in this lease and oil or gas is produced therefrom in paying quantities and the drilling requirements as specified in paragraph 8 hereof are not fully complied with, and the owner of such well shall operate the same and market the oil or gas produced therefrom, then the Lessee agrees to offset such well by the commencement of drilling operations within ninety days after it is ascertained that the production of oil or gas from such well is in paying quantities and that the operator thereof is then producing and marketing oil or gas therefrom. For the purpose of satisfying obligations hereunder such offset well or wells shall be considered as other wells required to be drilled hereunder.

13. The obligations of the Lessee hereunder shall be suspended while the Lessee is prevented from complying therewith, in whole or in part, by strikes, lockouts, actions of the elements, accidents, rules and regulations of any Federal, State, Municipal or other governmental agency, or other matters or conditions beyond the control of the Lessee, whether similar to the matters or conditions herein specifically enumerated or not.

14. The Lessee shall pay all taxes on its improvements and all taxes on its oil stored on the leased premises on the first Monday of March in each year, and _____ of the taxes levied and assessed against the petroleum mineral rights. Lessor agrees to pay all taxes levied and assessed against the land as such

and _____ of the taxes levied and assessed against the petroleum mineral rights. In the event the State, United States or any municipality levies a license, severance, production or other tax on the oil produced hereunder, or on the Lessee's right to operate, then and in that event the Lessee shall pay _____ of said tax and Lessor shall pay _____ of said tax.

15. The Lessee agrees not to drill any well on said land within _____ (_____) feet of the now existing building thereon without the written consent of the Lessor. The Lessee agrees to pay all damages directly occasioned by its operations to crops on said land.

16. The Lessor may at all reasonable times examine said land, the work done and in progress thereon, and the production therefrom, and may inspect the books kept by the Lessee in relation to the production from said land, to ascertain the production and the amount saved and sold therefrom. The Lessee agrees, on written request, to furnish to the Lessor copies of logs of all wells drilled by the Lessee on said land.

17. All the labor to be performed and material to be furnished in the operations of the Lessee hereunder shall be at the cost and expense of the Lessee, and the Lessor shall not be chargeable with, or liable for, any part thereof; and the Lessee shall protect said land against liens of every character arising from its operations thereon.

18. Upon the written request of the Lessor, the Lessee agrees to lay all pipe lines which it constructs through cultivated fields, below plow depth, and upon similar request agrees to fence all sump holes or other excavations to safeguard livestock on said land.

19. The Lessee shall have the right at any time to remove from said land all machinery, rigs, piping, casing, pumping stations and other property and improvements belonging to or furnished by the Lessee, provided that such removal shall be completed within a reasonable time after the termination of this lease. Lessee agrees after termination of this lease to fill all sump holes and other excavations made by it.

20. If royalty oil is payable in cash, Lessee may deduct therefrom a proportionate part of the cost of treating unmerchantable oil produced from said premises to render same merchantable. In the event such oil is not treated on the leased premises, Lessor's cash royalty shall also bear a corresponding proportionate part of the cost of transporting the oil to the treating plant. Nothing herein contained shall be construed as obligating Lessee to treat oil produced from the herein described premises. If Lessor shall elect to receive royalty oil in

kind, such royalty oil shall be of the same quality as that removed from the leased premises for Lessee's own account, and if Lessee's own oil shall be treated before such removal, Lessor's oil will be treated therewith before delivery to Lessor and Lessor in such event will pay a proportionate part of the cost of treatment.

21. Upon the violation of any of the terms or conditions of this lease by the Lessee and the failure to begin to remedy the same within _____ after written notice from the Lessor so to do, then, at the option of the Lessor, this lease shall forthwith cease and terminate, and all rights of the Lessee in and to said land be at an end, save and excepting _____ (_____) acres surrounding each well producing or being drilled and in respect to which Lessee shall not be in default, and saving and excepting rights-of-way necessary for Lessee's operations; provided, however, that the Lessee may, at any time after such default, and upon payment of the sum of _____ (_____) Dollars to the Lessor as and for fixed and liquidated damages, quitclaim to the Lessor all of the right, title and interest of Lessee in and to the leased lands in respect to which it has made default, and thereupon all rights and obligations of the parties hereto one to the other shall thereupon cease and terminate as to the premises quitclaimed.

22. All royalties and rents payable in money hereunder may be paid to the Lessor by mailing or delivering a check therefor to _____ Bank at _____ its successors and assigns, herein designated by the Lessor as depositary, the Lessor hereby granting to said depositary full power and authority on behalf of the Lessor, his heirs, executors, administrators, successors and assigns, to collect and receipt for all sums of money due and payable from the Lessee to the Lessor hereunder. No change in the ownership of the land or minerals covered by this lease, and no assignment of rents or royalties shall be binding on the Lessee until it has been furnished with satisfactory written evidence thereof.

23. Lessor hereby warrants and agrees to defend title to the land herein described, and agrees that the Lessee, at its option, may pay and discharge any taxes, mortgages, or other liens existing, levied or assessed on or against the above described land; and, in the event it exercises such option, it shall be subrogated to the rights of any holder or holders thereof and may reimburse itself by applying to the discharge of any such mortgage, tax, or other lien, any royalty or rentals accruing hereunder.

24. If and when any oil produced from the demised premises shall for any reason be unmarketable at the well at the price mentioned in paragraph 8 hereof, the Lessor agrees in such case to take and receive his royalty in kind, and should he fail or refuse so to do, then the Lessee may sell the same at the best price obtainable, but not less than the price which the Lessee may be receiving for its own oil of the same quality.

25. The words "drilling operations" as used herein shall be held to mean any work or actual operations undertaken or commenced in good faith for the purpose of carrying out any of the rights, privileges or duties of the Lessee under this lease, followed diligently and in due course by the construction of a derrick and other necessary structures for the drilling of an oil or gas well, and by the actual operation of drilling in the ground.

26. On the expiration or sooner termination of this lease, Lessee shall quietly and peaceably surrender possession of the premises to Lessor and deliver to him a good and sufficient quitclaim deed, and so far as practicable cover all sump holes and excavations made by Lessee. Before removing the casing from any abandoned well Lessee shall notify Lessor of the intention so to do, and if Lessor within _____ (_____) days thereafter shall inform Lessee in writing of Lessor's desire to convert such well into a water well, and for that purpose to retain and purchase casing therein, Lessee will leave therein such amount of casing as Lessor may require for said purpose, provided such procedure is lawful and will not violate any rule or order of any official, commission or authority then having jurisdiction in such matters, and provided further that Lessor pay to Lessee _____ (_____) per cent of the original cost of the casing on the ground.

27. Lessee may at any time quitclaim this lease in its entirety or as to part of the acreage covered thereby, with the privilege of retaining _____ (_____) acres surrounding each producing or drilling well, and thereupon Lessee shall be released from all further obligations and duties as to the area so quitclaimed, and all rentals and drilling requirements shall be reduced pro rata. All lands quitclaimed shall remain subject to the easements and rights-of-way hereinabove provided for. Except as so provided, full right to the land so quitclaimed shall revest in Lessor, free and clear of all claims of Lessee, except that Lessor, his successors or assigns, shall not drill any well on the land quitclaimed within _____ (_____) feet of any producing or drilling well retained by Lessee.

28. If this lease shall be assigned as to a particular part or as to particular parts of the leased premises, such division or severance of the lease shall constitute and create separate and distinct holdings under the lease of and according to the several portions of the leased premises as thus divided, and the holder or owner of each such portion of the leased premises shall be required to comply with and perform the Lessee's obligations under this lease for, and only to the extent of, his portion of the leased area, provided that nothing herein shall be construed to enlarge or multiply the drilling or rental obligations, and provided further that the commencement of the drilling operations and the prosecution thereof, as provided in paragraph 6 hereof, either by the Lessee or any assignee hereunder, shall protect the lease as a whole.

29. This lease and all its terms, conditions and stipulations shall extend to and be binding upon the heirs, executors, administrators, grantees, successors and assigns of the parties hereto.

30. Any notice from the Lessor to the Lessee must be given by sending the same by registered mail addressed to the _____, and any notice from the Lessee to the Lessor must be given by sending the same by registered mail, addressed to the Lessor at _____.

In Witness Whereof, the parties hereto have caused this agreement to be duly executed as of the date first hereinabove written.

Appendix 3

A MINERAL DEED FORM

The State of Texas,   )
)   Know all Men by These Presents:
County of _____   )

That _____ hereinafter called Grantor, of _____ County, Texas, for and in consideration of the sum of _____ Dollars ($_____) cash in hand paid by _____ hereinafter called Grantee, the receipt of which is hereby acknowledged, have granted, sold, conveyed, assigned and delivered, and by these presents do grant, sell, convey, assign and deliver unto the said Grantee, an undivided _____ interest in and to all of the oil, gas and other minerals in and under, and that may be produced from the following described land situated in _____ County, Texas, to-wit:

_____

_____

_____

Together with the right of ingress and egress at all times for the purpose of mining, drilling and exploring said land for oil, gas and other minerals, and removing the same therefrom.

Said land being now under an oil and gas lease executed in favor of _____, it is understood and agreed that this sale is made subject to the terms of said lease and/or any other valid lease covering same, but covers and includes _____ of all of the oil royalty and gas rental or royalty due and to be paid under the terms of said lease, in so far as it covers the above described land.

It is understood and agreed that _____ of the money rentals, which may be paid, on the above described land, to extend the term within which a well may be begun under the terms of said lease, is to be paid to the said Grantee; and, in event that the above described lease for any reason becomes canceled or forfeited, then and in that event, Grantee shall own _____ of all oil, gas and other minerals in and under said lands, together with a like _____ (_____) interest in all bonuses paid, and all royalties and rentals provided for in future oil, gas and mineral leases covering the above described lands.

To Have and to Hold the above described property, together with all and singular the rights and appurtenances thereto in anywise belonging unto the said Grantee herein, and Grantee's successors, heirs, and assigns forever; and Grantor does hereby bind _____ self, _____ successors, heirs, executors and administrators, to warrant and forever defend all and singular the said property unto the said Grantee herein, and Grantee's successors, heirs and assigns, against every person whomsoever lawfully claiming or to claim the same, or any part thereof _____.

Witness _____ hand__ this the _____ day of _____, 20__.

Appendix 4

A DEFEASIBLE TERM ROYALTY DEED FORM

ROYALTY DEED

(Non–Participating)

The State of Texas,	)	
	)	Know all Men by These Presents:
County of _____	)	

That _____ hereinafter called Grantor (whether one or more) for and in consideration of the sum of _____ Dollars, cash in hand paid by _____, hereinafter called Grantee, the receipt of which is hereby acknowledged, have granted, sold, conveyed, assigned and delivered, and by these presents do grant, sell, convey, assign, set over and deliver unto the said grantee an undivided _____ interest in and to all of the oil royalty, gas royalty, and royalty in casinghead gas, gasoline, and royalty in other minerals in and under, and that may be produced and mined from the following described land situated in the County of _____, State of Texas, to-wit:

together with the right of ingress and egress at all times for the purpose of mining, drilling and exploring said lands for oil, gas and other minerals, and removing the same therefrom. This grant shall run, and the rights, titles and privileges hereby granted shall extend to grantee herein, and to grantee's heirs, administrators, executors and assigns for a period of _____ years from date hereof, and as long thereafter as oil, gas or other minerals, or either of them, is produced or mined from the lands described herein, in paying or commercial quantities. If, at the expiration of said term, oil, gas or other minerals, or either of them, is not being produced or mined from said land or any portion thereof in paying or commercial quantities, this contract shall be null and void, and the grantee's rights hereunder shall terminate.

Said lands or portions thereof, being now under oil and gas lease executed in favor of _____; it is understood and agreed that this sale is made subject to the terms of said lease, but covers and includes the same interest as first hereinabove named, of all the oil royalty and gas royalty and casinghead gas and gasoline royalty, and royalty from other minerals or products, due and to be paid under the terms of said lease, only insofar as it or they cover the above described land.

And it is further understood and agreed that notwithstanding the Grantee does not by these presents acquire any right to participate in the making of future oil and gas mining leases on the portion of said lands not at this date under lease, nor of participating in the making of future leases, should any existing or future lease for any reason become cancelled or forfeited, nor of participating in the bonus or bonuses which Grantor herein shall receive for any future lease, nor of participating in any rental to be paid for the privilege of deferring the commencement of a well under any lease, now or hereafter;

Nevertheless, during the term of this grant, neither the Grantor nor the heirs, administrators, executors and assigns of the Grantor shall make or enter into any lease or contract for the development of said land or any portion of same for oil, gas or other minerals, unless each and every such lease, contract, leases or contracts shall provide for at least a royalty on oil of the usual one-eighth to be delivered free of cost in the pipeline and a royalty on one-eighth of the natural gas of value of same when sold or used off the premises, or one-eighth of the net proceeds of such gas, and one-eighth of the net amount of gasoline manufactured from natural or casinghead gas; and in the event Grantor, or the heirs, administrators, executors and assigns of the Grantor, or as in the status of the fee owners of the land and minerals, or as the fee owner of any portion of said land, shall operate and develop the minerals therein, Grantee herein shall own and be entitled to receive as a free royalty hereunder, an undivided one-eighth of the percent interest first hereinabove named, of all the oil produced and saved from the premises delivered to Grantee's credit free of cost in the pipeline, and the same percent interest and portion of the value or proceeds of the sales of natural gas when and while the same is used or sold off the premises, and the same percent interest of the net amount of gasoline or other products manufactured from gas or casinghead gas produced from wells situated on the premises, during the term hereof.

To Have and to Hold the above described property and rights, together with all and singular the rights and appurtenances thereto in any wise belonging, unto the said Grantee, and the Grantee's heirs, administrators, executors and assigns forever; and the Grantor does hereby bind _____ heirs, administrators, executors, and assigns to warrant and forever defend all and singular, the said property and rights unto the said Grantee, and Grantee's heirs, administrators, executors and assigns, against every person whomsoever lawfully claiming or to claim the same or any part thereof.

Witness the following signatures, this the _____ day of _____, 20__.

Appendix 5

A FORM FOR ASSIGNING A LEASE WITH RETENTION OF OVERRIDING ROYALTIES

State of _____   )
)   Know All Men by These Presents:
County of _____   )

That _____, herein called "Assignor", for and in consideration of the sum of Ten ($10.00) Dollars and other good and valuable considerations cash in hand paid by _____, the receipt of which is hereby acknowledged and confessed, has granted, sold, transferred and assigned, and subject to the reservations and exceptions hereinafter set out, does hereby grant, sell, transfer and assign, unto _____, herein called "Assignee", that certain Oil, Gas and Mineral Lease dated _____, from _____, as Lessors, to _____, as Lessee, recorded in Volume _____, Page _____ of the Deed Records of _____ County, _____, said lease covering _____ acres of land, more or less, more fully described as

_____ _____ _____ _____

Reference to the foregoing lease and the record thereof is here made for all purposes.

There is excepted from this assignment and reserved unto Assignor, _____, his heirs and assigns, out of production from the above described land under said lease or under any renewal or extension thereof, the following overriding royalties to be paid and delivered free and clear of all expenses of production, drilling and operating the premises, to wit:

(a) On oil, 1/8 of 8/8 of that produced and saved from the above described land under said lease, the same to be delivered at the option of Assignor free of cost at the well, or to his credit into the pipe lines to which the wells may be connected.

(b) On gas, casinghead gas or other gaseous substances sold or used off the premises, the market value at the well of 1/8 of 8/8 of the gas so sold or used provided that where gas, including casinghead gas or other gaseous substances, is produced from the premises and used or processed in the derivation or manufacture of gasoline, carbon black, hydrocarbons, or other products or by-products, liquid, solid or gaseous, the royalty herein reserved shall be 1/8 of 8/8 of the market value of such products or by-products (including residue gas) at the point of manufacture.

(c) On all other minerals (except sulphur) mined and marketed 1/8 of 8/8 thereof, either in kind or value at the well or point of production at the election of the Lessee.

(d) On sulphur, One ($1.00) Dollar per long ton produced from the premises.

The royalty on oil and gas shall be computed after deducting any used by Assignee in normal development and drilling operations on the leased premises.

In the event the above described lease covers less than the full and undivided fee simple estate in the lands described therein, or in the event the title of Assignor thereunder should fail in whole or in part, then as to production from the tract under which title should have so failed, the overriding royalties herein reserved shall be proportionately reduced.

As a part of the consideration for this assignment, Assignee agrees it will comply with all of the expressed and implied covenants contained in said lease, and that it will not surrender or abandon the same without tendering to Assignor, his heirs or assigns, a reassignment thereof prior to such surrender or abandonment and at least thirty (30) days prior to the next ensuing rental paying date thereunder.

Assignor shall upon request be furnished with complete information with respect to all drilling operations conducted by Assignee on the leased premises, including copies of all logs, drilling reports and electrical surveys made thereon.

To Have and to Hold the above described lease and leasehold estate, together with all and singular the rights and appurtenances thereto in any wise belonging or appertaining unto the said _____, Assignee, its successors and assigns, and the said _____, Assignor, does for himself, his heirs and assigns, agree to warrant and forever defend said leases and leasehold estates unto the said _____, Assignee, its successors and assigns, against any person or persons whomsoever lawfully claiming or to claim the same or any part thereof, by, through or under Assignor but no further and not otherwise.

Appendix 6

A BOTTOM HOLE LETTER

(Date)

Mr. John Doe
P.O. Box _____
Kansas City, Kansas

Dear Mr. Doe:

You propose to drill a test well in search of oil or gas at a location described as _____, to a depth sufficient to adequately test the _____ Formation.

If you drill said well, and within ninety (90) days of the date of this letter, determine that said well can not be completed as a well commercially productive of oil or gas from the _____ Formation, (expected to be encountered at a depth of approximately _____ feet) and in so doing adequately test the _____ Sand, we will pay you the sum of _____, subject, however, to the conditions hereinafter stated.

You are not bound to drill said well nor are you bound to deepen said well for a test of the _____ Sand, but if you do deepen said well to a depth sufficient to test the _____ Sand, or if you obtain commercial production from the _____ Sand at a lesser depth than the top of the _____ Formation, the above mentioned sum shall be due and payable.

You shall furnish us copies of daily drilling reports and full information relative to the well as drilling progresses, including parts of cores taken and samples of formations encountered. Our agents and employees shall have access to the derrick floor at all times and shall be timely notified so that we may have representatives present to witness tests of showings encountered, production tests, and measurement for final depth. You shall furnish us a copy of the driller's log, sworn to by someone having actual knowledge of the facts, copies of any and all electrical or other surveys made on said well, and copies of any and all core analyses made. All well information, logs, reports and correspondence shall be sent to the attention of _____ at our office at _____.

All showings of oil or gas which reasonably justify testing shall be adequately tested by you, and you will use your best efforts in accordance with good oil field practice to complete said well as a commercial producer.

We shall have no control over the well, which shall be drilled, plugged and abandoned, or completed as a producer, at your sole cost, risk, and expense; and the drilling and any subsequent operations in connection therewith shall not be considered as a joint undertaking, mining partnership or otherwise, and we shall not be liable for any part of the cost. You agree to protect, indemnify and save us harmless from all claims, demands, and causes of action arising directly or indirectly out of or in connection with such operations.

This agreement must be accepted by you and a signed copy returned to us within ten (10) days of this date, otherwise it shall not be a binding contract between us.

This agreement is not assignable by you without our written consent first obtained.

Please indicated your acceptance in the space provided below and return one copy of this letter to us for our file.

Very truly yours,

_____ Oil Company

By _____,

Vice President.

Agreed to and Accepted this _____ day of _____, 20__.

_____
John Doe

Appendix 7

A Dry Hole Letter

(Date)

Mr. John Doe
P.O. Box _____
Kansas City, Kansas

Dear Mr. Doe:

Subject to the conditions hereinafter stated, and when, if, and after you have drilled, plugged, and abandoned as a dry hole, within the time and in the manner herein provided, the well hereinafter provided for, we will pay you the sum of _____ Dollars ($_____).

You agree to commence on or before _____, 20__, a test well for the production of oil or gas at a location in the _____ _____, and to drill said well with due diligence and reasonable dispatch to completion at a depth sufficient adequately to test thoroughly the _____ Sand (expected to be encountered at a depth of approximately _____ feet), unless oil or gas is encountered and the well is completed at a lesser depth. _____ expressly reserves the right to be the sole judge as to whether the well has been drilled to a depth sufficient thoroughly to test the _____ Sand.

If you commence said well on or before _____, 20__, drill the same diligently to the depth above specified, plug and abandon the well as a dry hole, and furnish us a certified copy of the driller's log and any electrical log, if one be run in the well, we agree to pay you the total sum of _____ Dollars ($_____).

We shall be furnished copies of daily drilling reports and full information relative to the well as drilling progresses, including parts of cores taken and samples of formations encountered. Our agents and employees shall have access to the derrick floor at all times; and our representative shall be timely notified so that he may be present to witness tests of showings encountered, production test, and measurement for final depth. All well information, logs, reports, and correspondence shall be sent to the attention of _____ at our office at _____ .

You will use your best efforts in accordance with good oil field practice to complete the well as a commercial producer, and all showings of oil or gas which reasonably justify testing shall be adequately tested by you.

We shall have no control over the well, which shall be drilled, plugged and abandoned, or completed, at your sole cost, risk, and expense; and the drilling and any subsequent operations in connection therewith shall not be considered as a joint undertaking, mining or otherwise, and we shall not be liable for any part of the cost. You agree to protect, indemnify and save us harmless from all claims, demands and causes of action arising directly out of or in connection with such operations.

Unless and until this agreement has been accepted by you and two accepted copies returned to us by _____ , it shall not be a binding contract between us. This agreement is not assignable without our written consent.

If this dry hole letter is acceptable to you, please indicate your acceptance in the space provided below and return two copies for our file.

Very truly yours,

_____ Oil Company

By _____ ,

Vice President.

Agreed to and Accepted this _____ day of _____ , 20__ .

_____

John Doe

Appendix 8

A FARM–OUT LETTER

(Date)

Mr. John Doe
P.O. Box _____
Kansas City, Kansas

Dear Mr. Doe:

You agree to commence on or before _____, 20__, the actual drilling of a test well at a lawful location in the _____, said well to be drilled to a total depth of _____ feet or to a depth sufficient to test the sand appearing at _____ feet in the _____ Well located in said Section _____, unless oil or gas in paying quantities is found at a lesser depth or unless at a lesser depth heaving shale, salt, or other impenetrable substances are encountered through which you are unable to drill after diligent effort is made to do so.

We will be furnished copies of daily drilling reports and full information relative to the well as drilling progresses, including parts of cores taken and samples of formations encountered. Our agents and employees shall have access to the derrick floor at all times, and our representative shall be timely notified so that he may be present to witness tests of showings encountered production tests, measurement for final depth, electric logs, and any other testing operations performed by you or at your request. We shall be furnished with one true copy of the driller's log sworn to by someone having actual knowledge of the facts and one copy of the Schlumberger to be made from the bottom of the surface casing to the total depth of the well. All well information, logs, reports, and correspondence shall be sent to the attention of _____ at our office at _____.

You will use your best efforts in accordance with good oil field practice to complete the well as a commercial producer. This shall include the taking of side wall cores and making of drill stem tests of showings of oil and gas, or either of them, when necessary or desirable to adequately appraise or test saturated formations of good showings of oil or gas when encountered. All showings of oil or gas, which in the opinion of a reasonably prudent operator under the same or similar circumstances justify testing, shall be adequately tested.

We shall have no control over the well, which shall be drilled, plugged and abandoned, or completed at your sole cost, risk, and expense; and the drilling and any subsequent operations in connection therewith shall not be considered as a joint undertaking, mining or otherwise, and we shall not be liable for any part of the cost; and you agree to protect, indemnify, and save us harmless from all claims, demands, and causes of action arising directly or indirectly out of or in connection with such operations.

If and when you have drilled such well under the aforesaid terms and conditions, we agree to assign to you, subject to the reservations hereinafter mentioned, all of our right, title, and interest in and to the following described oil, gas and mineral leases, to wit:

_____ _____

If this letter agreement is in accordance with your understanding of our trade, please indicate your acceptance by your signature in the space provided below and return two copies of this letter agreement for our file.

Very truly yours,

_____ Oil Company

By _____,

Vice President.

Agreed to and Accepted this _____ day of _____, 20__.

_____

John Doe

Appendix 9

OPERATING AGREEMENT

This Agreement made and entered into this _____ day of _____, 20__, by and between _____, a _____ corporation, herein designated as "Operator," and the undersigned party or parties other than "Operator"; Witnesseth That:

Whereas said parties are the owners of oil and gas interests or oil and gas leases covering lands described in "Exhibit A", attached hereto and made a part hereof, as particularly set forth therein; and

Whereas it is desired to enter into an operating agreement with respect to such interests;

Now, Therefore, it is agreed as follows:

1.

Each of the parties hereto other than Operator shall, within _____ (_____) days from date hereof, submit to Operator abstracts certified to recent date and all title papers relating to his or its respective interest(s) to be subjected hereto for the purpose of causing such abstracts and title papers to be examined by Operator's attorneys on behalf of all the parties hereto. Abstracts certified to recent date and title papers relating to the interest(s) to be subjected hereto by Operator shall, within a similar period, be submitted to _____ for examination by his or its attorneys. All such examinations shall be made without charge to the other party or parties. Such examinations shall include an examination of the form and conditions of the leases to be subjected to this agreement, it being understood that such leases shall be on a customary form, shall contain no express drilling commitment unless hereinafter otherwise stated, and shall have primary terms expiring not sooner than _____ (_____) year(s) from the date hereof. The opinion of said attorneys as to such titles shall be binding and conclusive on the parties.

After submission of abstracts and title papers as aforesaid, the examining party shall be allowed a period of _____ (_____) days in which to examine title and submit title requirements. The party or parties whose interests are affected by such requirements shall have _____ (_____) days after submission of title requirements within which to meet such requirements. If any title, upon examination, be rejected, then unless the defect or defects in such title be waived by the parties, this agreement shall terminate. Upon approval or acceptance of titles, the following provisions of this agreement shall become operative.

2.

The following land, to-wit:

_____ _____

shall be developed and operated for oil and gas purposes by Operator, subject to the provisions herein contained.

3.

The respective interests of the parties hereto which are subjected to this agreement are set forth in "Exhibit A," attached hereto and made a part hereof. If any interest listed in "Exhibit A" hereof is shown by such exhibit to be an unleased interest in oil and gas rights, then such unleased interest shall be treated for all of the purposes of this agreement as if it were an oil and gas lease covering such unleased interest on a form providing for the usual and customary one-eighth (1/8) royalty and containing the usual and customary "lesser interest clause." This agreement shall in no way affect the right of the owner of any such unleased interest to receive an amount or share of production equivalent to the royalty which would be payable or due if such unleased interest were subject to an oil and gas lease as provided in the preceding sentence.

4.

All costs, expenses and liabilities accruing or resulting from the operation of the premises pursuant to this agreement, shall be determined, shared and borne by the parties hereto in the following respective proportions:

_____ _____

5.

Operator shall have full control of the premises subjected hereto and, subject to the provisions hereof, shall conduct and manage the development and operation of said premises for the production of oil and gas therefrom. Operator shall pay and discharge all costs and expenses incurred pursuant hereto, and shall charge each of the parties hereto with his or its respective proportionate share upon the cost and expense basis provided in the Accounting Procedure attached hereto, marked "Exhibit B" and made a part hereof. Each party hereto other than Operator, will promptly pay Operator such costs as are hereunder chargeable to him or it. All production of oil and gas from said land, subject to the payment of applicable royalties thereon, and all materials and equipment acquired pursuant hereto, shall be owned by the parties hereto in the respective proportions set out in the preceding section hereof.

6.

If any of the oil and gas leases held by the parties on the acreage covered hereby be subject to any overriding royalty, production payment or other charge in addition to the usual one-eighth (1/8) royalty, the party contributing any such lease shall bear, assume and discharge any such overriding royalty, production payment or other charge out of the interest attributable to him or it hereunder.

7.

Each party holding an oil and gas lease subjected to this agreement shall, before the due date, pay all delay rentals which may become due under the lease or leases contributed by him or it, and each party paying such rentals shall, within ten (10) days after such rentals have been paid, but at least ten (10) days prior to the rental date, notify Operator of such rental payment. Operator shall furnish similar information as to its own lease(s) to any party hereto who requests such information. The burden of paying such rentals shall fall entirely upon the party required to make payment thereof hereunder. In event of failure to make proper payment of any delay rental through mistake or oversight where such rental is required to continue the lease in force (it being understood that any such failure shall not be regarded as a title failure within the meaning of any other section of this agreement), there shall be no money liability on the part of the party failing to pay such rental, but such party shall make a bona fide effort to secure a new lease covering the same interest and in event of failure to secure a new lease within a reasonable time the interests of the parties hereto shall be revised so that the party failing to pay any such rental will not be credited with the ownership of any lease on which rental was required but was not paid.

8.

Without the consent of parties hereto whose interests aggregate at least _____% of the total interests subject to this agreement:

(a) No well shall be drilled on the premises except any well expressly provided for by this agreement; and

(b) No expenditure shall be made by Operator in developing and operating the premises or for capital investment in excess of $_____ except in connection with a well the drilling of which has been previously authorized pursuant to this agreement.

9.

In the event the parties hereto cannot mutually agree upon the drilling of a particular well on the premises subject hereto, then the party or parties desiring to drill such well shall give the other party or parties written notice thereof, specifying the location, proposed depth, and estimated cost. The other party or parties shall have thirty (30) days after receipt of such notice within which to notify the party or parties desiring that said well be drilled whether or not he, it or they elect to participate in the cost of drilling said well. The failure to give such notice within said period of thirty (30) days shall be construed as an election by said parties not to participate in the cost of drilling said well. Any well drilled pursuant to this paragraph shall be drilled by and at the cost, risk and expense of the party or parties electing to drill. Any such well shall conform to the then-existing well-spacing program. If any party shall elect not to participate in the drilling of said well, then, within thirty (30) days after the expiration of said period of thirty (30) days, the party or parties desiring to drill shall commence the actual drilling of said well at said location, and thereafter complete said well with due diligence, in order to be entitled to the benefit of this paragraph. If any such well be completed as a producer, it shall be taken over and operated by Operator, and the parties hereto shall have the same rights with respect to the production from such well as are hereinafter set forth in Section 14 hereof, except that the proportionate share or shares of the non-drilling party or parties in the oil and gas produced from such well shall be sold and the non-drilling party or parties shall direct the purchaser thereof to pay to the drilling party or parties (and the drilling party or parties shall be entitled to receive) all the proceeds from the sale thereof, after deducting all royalty interests, overriding royalty interests and production payments, if any, until such drilling party or parties shall have been reimbursed in an amount equal to the total accrued expense of operating such well plus _____% of the cost of drilling, testing, completing and equipping such well. Any amount realized from the sale or disposition of equipment acquired in connection with the drilling, completing, equipping and operating of any well drilled pursuant to this Section shall be credited against the total unreturned cost of drilling, completing, equipping and operating said well. Until the drilling party or parties shall have been so reimbursed, the cost of operating any such well shall be borne wholly by the party or parties who drilled it and thereafter, all expenses of operating such well shall be borne by the parties hereto in the same proportions and manner as in the case of a well drilled pursuant to mutual agreement of all parties.

The provisions of this section shall have no application whatever to the initial test well which may be drilled on the premises subject hereto but shall apply only in event such initial test well has been previously drilled.

10.

No well which is producing or has once produced shall be abandoned without the mutual consent of the parties hereto; provided, however, if the parties are unable to agree as to the abandonment of any well, then the party or parties not desiring to abandon the well shall tender to each of the parties desiring to abandon the proportionate share of each party desiring to abandon of the reasonable value of the material and equipment in and on said well, such value to be determined, so far as possible, in accordance with the attached "Exhibit B"—Accounting Procedure. Upon receipt of said sum, each party desiring to abandon such well shall, without express or implied warranty of title, assign to the party or parties tendering said sum his or its interest in said well and the equipment therein, together with all of his or its rights in all working-interest production therefrom which may be produced from the formation or formations from which such well is producing. If there is more than one non-abandoning party, such assignment shall run in favor of the non-abandoning parties in proportion to their respective interests.

11.

The number of employees, the selection of such employees, the hours of labor, and the compensation for services to be paid any and all such employees, shall be determined by Operator. Such employees shall be the employees of Operator.

12.

Operator shall carry such Workmen's Compensation and Employers' Liability insurance as may be required by the laws of the state in which the above described land is located, provided that Operator shall be a self-insurer as to either or both of such risks if permissible under the laws of such state. No other insurance shall be carried by Operator for the benefit of the parties hereto except by mutual consent of the parties.

13.

Operator shall have a lien on the interest or interests of each party hereto other than Operator subjected to this agreement, the oil and gas produced therefrom, the proceeds thereof, and the material and equipment attributable thereto, to secure Operator in the payment of any sum due to Operator hereunder from any such party. The lien herein provided for shall not extend to any royalty rights attributable to any interest subjected hereto.

14.

Each of the parties hereto shall take in kind or separately dispose of his or its proportionate share of the oil and gas produced from the premises exclusive of production which may be used in development and producing operations on said premises and in preparing and treating oil for marketing purposes and production unavoidably lost, and shall pay or cause to be paid all applicable royalties thereon. Any extra expenditure incurred by the taking in kind or separate disposition by any party hereto of his or its proportionate share of the production shall be borne by such party. Each party hereto shall be entitled to receive directly payment for his or its proportionate share of the proceeds from the sale of all oil or gas produced, saved and sold from said premises, and on all purchases or sales, each party shall execute any division order or contract of sale pertaining to his or its interest. In event any party hereto shall fail to make the arrangements necessary to take in kind or separately dispose of his or its proportionate share of the oil or gas produced from said premises, Operator shall have the right, subject to revocation at will by the party owning same, to purchase such oil and gas or sell the same to others for the time being at not less than the market price prevailing in the area and not less than the price which Operator receives for its own portion of such oil or gas, any such purchase or sale to be subject always to the right of the owner of such oil or gas to exercise, at any time, his or its right to take in kind or separately dispose of his or its share of such oil or gas not previously delivered to a purchaser pursuant hereto.

15.

Surplus material and equipment from the premises, which in the judgment of Operator is not necessary for the development and operation thereof, may be sold by Operator to any of the parties to this agreement or to others for the benefit of all parties hereto, or may be divided in kind between such parties. Proper charges and credits shall be made by Operator as provided in the Accounting Procedure, marked "Exhibit B," attached hereto.

16.

Each of the parties hereto shall have access to the premises at all reasonable times to inspect and observe any operations thereon, and shall have access at reasonable times to information pertaining to the development or operation thereof including Operator's books and records relating thereto, and Operator, upon request, shall furnish each of the other parties hereto with copies of all drilling reports, well logs, tank tables, daily gauge and run tickets and reports of stock on hand at the first of each month, and shall make available samples of any cores or cuttings taken from any wells drilled hereunder.

17.

All wells drilled on the premises shall be drilled on a competitive contract basis at the usual rates prevailing in the area. Operator, if it so desires, may employ its own tools and equipment in the drilling of wells, but in such event, the charge therefor shall not exceed the prevailing rate in the field, and such work shall be performed by Operator under the same terms and conditions as shall be customary and usual in the field in the contracts of independent contractors who are doing work of a similar nature.

18.

The leases covered by this agreement shall not be surrendered in whole or in part unless the parties mutually consent thereto. Should any party at any time desire to surrender the leases subject hereto and the other party or parties should not agree or consent to such surrender, the party desiring to so surrender shall assign without express or implied warranty of title all of his or its interest in such leases and the wells, material and equipment located thereon to the party or parties not desiring to surrender and thereupon such assigning party shall be relieved from all obligations thereafter accruing (but not theretofore accrued) hereunder. From and after the making of such assignment the assigning party shall have no further interest in the leases assigned and the equipment thereon but shall be entitled to be paid for his or its interest in any material on said leases so assigned at its reasonable value determined, so far as possible, as provided in the attached "Exhibit B." If such assignment shall run in favor of more than one party hereto, the interest covered thereby shall be shared by such parties in the proportions that the interest of each party assignee bears to the total interest of all parties assignee.

19.

In the event any party desires to sell all or any part of his or its interest subject to this agreement, the other party or parties hereto shall have a preferential right to purchase the same. In such event, the selling party shall promptly communicate to the other party or parties hereto the offer received by him or it from a prospective purchaser ready, willing and able to purchase the same, together with the name and address of such prospective purchaser, and said party or parties shall thereupon have an option for a period of ten (10) days after the receipt of said notice to purchase such undivided interest for the benefit of such remaining parties hereto as may agree to purchase the same. Any interest so acquired by more than one party hereto, shall be shared by the parties purchasing the same in the proportions that the interest of each party so acquiring bears to the total interest of all parties so acquiring. The limitations of this paragraph shall not apply where any party hereto desires to mortgage his or its interest or to dispose of his or its interest by merger, reorganization, consolidation or sale of all his or its assets, or a sale of his or its interest hereunder to a subsidiary or parent company or subsidiary of a parent company or to any company in which any one party hereto owns a majority of the stock.

In event of a sale by Operator of the interests owned by it which are subject hereto, the holders of a majority interest in the premises subject hereto shall be entitled to select a new operator but unless such selection is made the transferee of the present Operator shall act as operator hereunder.

20.

The liability of the parties hereunder shall be several and not joint or collective. Each party shall be responsible only for his or its obligations, as herein set out, and shall be liable only for his or its proportionate share of the cost of developing and operating the premises subject hereto.

21.

Operator shall not be liable for any loss of property or of time caused by strikes, riots, fires, tornadoes, floods or for any other cause beyond the control of Operator through the exercise of reasonable diligence. All of the provisions of this agreement are hereby expressly made subject to all applicable Federal or State laws, orders, rules and regulations, and in the event this contract or any provision hereof is found to be inconsistent with or contrary to any such law, order, rule or regulation, the latter shall be deemed to control and this contract shall be regarded as modified accordingly and as so modified shall continue in full force and effect.

22.

This agreement shall remain in full force and effect for the life of the oil and gas leases which are subjected hereto and any extensions or renewals thereof whether by production or otherwise.

23.

In the event of a failure or partial failure of title to any interest subjected hereto, any and all losses occasioned by reason thereof shall be borne and shared by the parties hereto proportionately to their respective interests as set forth in Section 4 hereof.

24.

The term "oil and gas" as herein used shall include casinghead gas and any other mineral covered by any oil and gas lease subjected hereto.

25.

All notices that are required or authorized to be given hereunder, except as otherwise specifically provided herein, shall be given in writing by United States mail or Western Union telegram, postage or charges prepaid, and addressed to the party to whom such notice is given as follows:

_____ _____

The originating notice to be given under any provision hereof shall be deemed given only when received by the party to whom such notice is directed, and the time for such party to give any response thereto shall run from the date the originating notice is received. The second or any subsequent responsive notice shall be deemed given when deposited in the United States post office or with Western Union Telegraph Company, with postage or charges prepaid.

26.

Other conditions, if any, are:

_____

_____

This Agreement shall be binding upon the parties hereto, their successors and assigns, and the terms hereof shall constitute a covenant running with the lands and leasehold estates covered hereby.

In Witness Whereof, the parties hereto have signed this agreement the day and year first above written.

[Exhibits A and B omitted.]

INDEX

References are to Pages
